This work is dedicated to my past graduate students; my wife, Lois; and to the memory of my teachers, master entomologists Leland Chandler and Neil Walker.

***Larry P. Pedigo***

The sixth edition is dedicated to my wife and best friend, Kay; thanks for allowing me to indulge my passion for all things entomological; and to my parents, Robert and Phyllis, who always encouraged my boyhood tendency to capture a menagerie of small creatures in jars.

***Marlin E. Rice***

# BRIEF CONTENTS

# ENTOMOLOGY
## AND
# PEST MANAGEMENT

**SIXTH EDITION**

Larry P. Pedigo
Iowa State University

Marlin E. Rice
Iowa State University

**PHI Learning Private Limited**
Delhi-110092
2014

**This Indian Reprint—₹ 695.00**
(Original U.S. Edition—₹ 7374.00)

**ENTOMOLOGY AND PEST MANAGEMENT, 6th ed.**
by Larry P. Pedigo and Marlin E. Rice

Original edition, entitled *Entomology and Pest Management, 6th ed.*, by Larry P. Pedigo and Marlin E. Rice, published by Pearson Education, Inc., publishing as Pearson Prentice Hall.



**ISBN-978-81-203-3886-9**



Indian edition published by PHI Learning Private Limited.

Published by Asoke K. Ghosh, PHI Learning Private Limited, Rimjhim House, 111, Patparganj Industrial Estate, Delhi-110092 and Printed by Mohan Makhijani at Rekha Printers Private Limited, New Delhi-110020.

# CONTENTS

# LIST OF DIAGNOSTIC BOXES

# PREFACE

Writing the original version of *Entomology* and *Pest Management* was an enjoyable but daunting task. Deciding on how to combine basic entomology with applied aspects of the science was particularly difficult. That difficulty remains, even after six editions.

Based on recommendations gathered from comprehensive reviews, we decided to continue approximately the same mix of basic and applied topics in this sixth edition. Yet, we wanted to place increased emphasis on advances in the technology of management. In an attempt to accomplish this goal, we developed a new chapter entitled *Biopesticides for Management*. This chapter (Chapter 12) embodies new writing plus a reorganization of previously treated subjects.

Chapter 12, *Biopesticides for Management*, covers a rapidly growing list of pesticides registered by the Environmental Protection Agency. These materials include microbial pesticides, biochemical pesticides, and plant-incorporated protectants. They have the great benefit of being environmentally friendly and safe to handle. This chapter will be of particular interest to students and others involved in organic-food production and horticultural crops.

However, the addition of the biopesticides chapter does not diminish the importance of conventional insecticides, and this edition has ample information for students interested in use of these materials. The chapter on traditional insecticides (Chapter 11) has been completely updated to reflect newly registered compounds and provides improved explanations of established ones.

To bring *Entomology and Pest Management* in line with current thinking in insect systematics, the sixth edition also addresses a new classification scheme. This change in Chapter 3 focuses on the orthopteroid and hemipteroid groups. A new key to the insect orders also reflects this change in classification.

Other changes in the sixth edition include a new emphasis on horticultural crops in the case-histories chapter (Chapter 18), with a detailed discussion of insect management in California almonds. Moreover, new insect diagnostic boxes, color photographs, and enhanced black-and-white photographs improve student comprehension in this and other areas.

Lastly, in addition to updating information in almost every chapter, *Favorite Web Sites* at the end of each chapter have been verified and updated, making the Internet a valuable companion in student learning. Also, the popular section *World Wide Web Sites of Entomological Resources* (Appendix 3) has been reviewed and completely updated, offering even greater direction in locating specific entomological Internet sites and the wealth of information they provide for students.

## TARGET AUDIENCE

*Entomology and Pest Management* can be used as an introduction to applied entomology for undergraduates or beginning graduate students. For undergraduates with only an elementary biology background, early chapters provide a basis for understanding the remaining content on insect ecology, surveillance, and management. Students with at least one course in entomology may wish to omit early chapters and focus on the strategy and tactics of management found in later chapters. Omitting Chapters 1 through 4 for graduate courses will not result in a loss of continuity.

### Content and Organization

The book consists of eighteen chapters, three appendices, and a glossary. Concepts and principles are emphasized and supported by factual detail and specific examples. Beginning chapters (1 through 3) concentrate on general entomology for the novice. Chapters 4 and 5 synthesize the elements of insect biology and ecology required for understanding insect pest management. Chapter 6 covers techniques and principles of sampling for problem assessment. Chapter 7 builds on this knowledge by outlining types of reactions of crops to insect densities. It also features the concept of adding environmental costs in the decision-making process for management. The ideas and history of insect pest technology are reviewed, and the concept and philosophy of modern pest management is introduced in Chapter 8.

With this basic information presented in the first eight chapters, the student is introduced, chapter by chapter, to the individual tactics used as elements in pest management programs. The order of tactics presented is based on their relative importance in existing pest management programs. Consequently, natural enemies and ecological management of the environment, primarily preventive tactics, are mentioned first, followed by conventional insecticides and biopesticides; the premier elements in curative tactics. The remaining tactics discussed in Chapters 13 through 15 are more specialized but, nevertheless, convey some of the newest and most innovative ideas.

Chapter 16 discusses the ways pest management and pest technology are practiced. This chapter draws the analogy between human medicine and pest management, and it emphasizes the idea of prevention and therapy in combining several management tactics. Area-wide pest technology and recent successes with the cotton boll weevil in the United States are discussed.

Chapter 17 is unique among entomology texts because it integrates the problems of resistance, resurgence, replacement, and recent phenomena, such as enhanced microbial degradation of insecticides, into a single concept—ecological backlash. The chapter suggests to students that applying the tactics discussed does not always result in sustainable pest management, and it recommends ways of reducing or avoiding such problems.

The book ends (Chapter 18) by presenting examples of successful insect pest management programs in the context of diverse commodities.

### Special Features

**Basic and applied entomology.** The primary purpose of the book is to promote an understanding of major elements of general entomology and relate them to modern principles of insect pest management. Both theory and practice are emphasized in a conceptual approach to the topics, and numerous examples are presented to facilitate learning.

**Ecological approach.** Pest management topics are discussed as aspects of applied ecology, and solutions to pest problems are presented with regard to environmental quality, profitability, and durability.

**Insect diagnostic boxes.** Sixty-eight insect diagnostic boxes are presented throughout the text. Each box contains detailed information on distribution, importance, appearance, and life cycle of a species or species group. Insects chosen are from examples mentioned in the text. Grouping specific data in boxes provides background information about a species through examples and case histories without detracting from the main discussion. Students not familiar with the species can consult the boxes to better understand and appreciate the examples in the text. Information in the boxes is referenced in the index. Additional information about major pests are given in Chapter 3 as insect families of major economic importance are included in presenting information on insect classification.

**Boldface type.** This feature allows the student to recognize new terms and important concepts quickly and serves as a basis for topic review.

**Appendices.** Three appendices facilitate learning and serve as reference material. Appendix 1 presents a key to the orders of insects, allowing identification of both adult and immature insects. Appendix 2 contains a list of insect common names, scientific names, and classifications. Appendix 3 is a comprehensive list of World Wide Web sites of entomological resources that can be used for customized computer searches.

**Glossary.** An expanded glossary for quick reference appears at the end of the book.

**Favorite Web Sites.** The Web Sites accessible through the Internet are presented as URL addresses along with a short description of the site's content. Readers can receive updates on a topic by consulting these sites and navigating links to other related sites for additional information.

## ACKNOWLEDGMENTS

We owe many thanks to several of our colleagues here at Iowa State University for their valuable input in preparing the sixth edition of *Entomology and Pest Management*. Of particular note is the valuable review and advice of Joel Coats and his graduate student, Gretchen Palauch, in developing the new chapter on biopesticides and refining the chapter on traditional insecticides. We thank Les Lewis, Research Leader, USDA/ARS, Corn Insect and Crop Genetics Research Unit, for his excellent, review and suggestions on the discussion of microbial pesticides. Special thanks also go to Carol Pilcher, Iowa State University Coordinator, Pest Management and the Environment, for her help with the section on insect pest management in almonds as a model management system. Additionally, we thank John VanDyk, Adjunct Assistant Professor and Systems Analyst, Department of Entomology, Iowa State University, for preparing and allowing us to publish his list of World Wide Web addresses in this edition. We also thank Scott Hutchins, Dow AgroSciences, and Bob Peterson, Montana State University, for their involvement in the preparation of insect diagnostic boxes; and Laura Karr, Dow AgroSciences, for her work on the glossary.

Thanks to the following reviewers for their valuable feedback: John J. Brown, Washington State University; Lynn A. DuPuis, Alfred State College; Henry Fadamiro, Auburn University; and Allan S. Felsot, Washington State University. Lastly, we thank Prentice Hall and our editor William Lawrensen for providing us the opportunity to update this work.

Larry P. Pedigo
Marlin E. Rice

CHAPTER 1

# INTRODUCTION

"YOU STUDY WHAT?" "Insects? You mean bugs?" "Yuck. What for?" These are common utterances of people in casual conversations at parties and other gatherings. Although we are usually taken aback and a little disgusted by some of the remarks and guffaws, the question is certainly legitimate. The standard answer, of course, is that insects represent one of the most important forms of life on this planet. They have influenced human existence since its very beginning and continue to control many of our daily activities. Therefore, we need to know about them so we can deal with them, usually in the context of a threat. When we begin to study insects, we find that they are fascinating creatures and that learning about insect life is engrossing.

## INSECT ABUNDANCE

The fact is, today's human population is adrift in a sea of insects. If we look at numbers alone, the estimated ratio of insects to humans is 200 million to 1, and insects average about 40 million per acre of land. Being much larger than insects, we might be tempted to argue that humans are ecologically more successful, making up in mass for our lower numbers. However, analysts estimate that the United States is home to some 400 pounds of insect biomass per acre, compared with our 14 pounds of flesh and bone. Another amazing statistic is that in the Brazilian Amazon, ants alone outweigh the total biomass of all vertebrates by four to one. Based solely on numbers and biomass, insects are the most successful animals on earth!

## INSECT DIVERSITY

Along with humans, insects live in almost every habitable place on the earth, except the ocean depths. According to the distinguished entomologists Eisner and Wilson (1977), insects all but own the land. They are the chief consumers of plants; they are the major predators of plant eaters; they play a major role in decay of organic matter; and they serve as food for other kinds of animals.

Knowing these ecological facts, we might expect these organisms to be diverse and adaptable—and they are. Today, more than 900,000 kinds (species) have been described, and many believe that five to seven times as many are yet to be discovered.

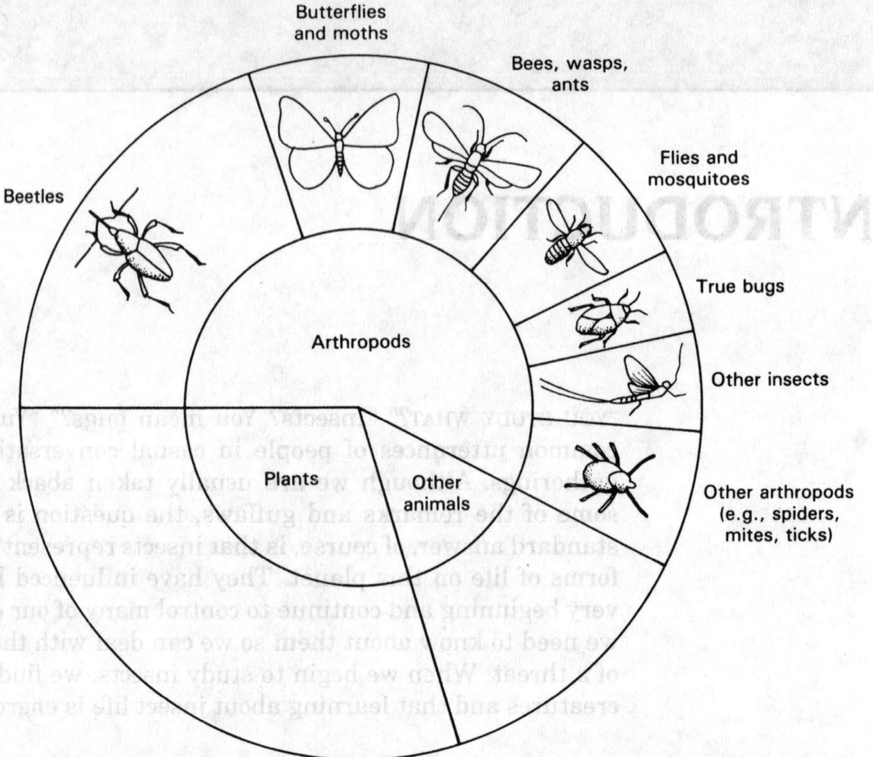


**Figure 1.1** Chart showing size of various arthropod groups and other life forms. Pie slices are proportional to the number of species in the group. (Redrawn from Daly et al., 1978, *Introduction to Insect Biology and Diversity.* McGraw-Hill Book Company)

These statistics make other animal groups seem small by comparison (Fig. 1.1). For instance, species in the class Mammalia, of which humans are members, count less than 1 percent of those in the class Hexapoda. Furthermore, insects boast more species than any other kind of organism, making up about three-fourths of the total number known. Indeed, it has been estimated that every fifth animal is a beetle, only one group in the class Hexapoda. Without a doubt, this extraordinary diversity is yet another mark of insects' ecological success.

## WHAT AN INSECT IS

Insects are grouped with other animals sharing similar characteristics in the phylum Arthropoda. Arthropods, as they are known, are characterized as having a body divided by grooves to form **segments** and a well-developed covering, the **integument,** which makes up the outer shell, or **exoskeleton.** Some of the arthropod body segments possess one or more pairs of jointed appendages from which the phylum name was derived (from *arthro,* joint; *poda,* foot). In addition, arthropods have a heart at the top and a nerve cord at the bottom of their body cavity. Examples of arthropods other than insects include spiders, shrimp, and centipedes.

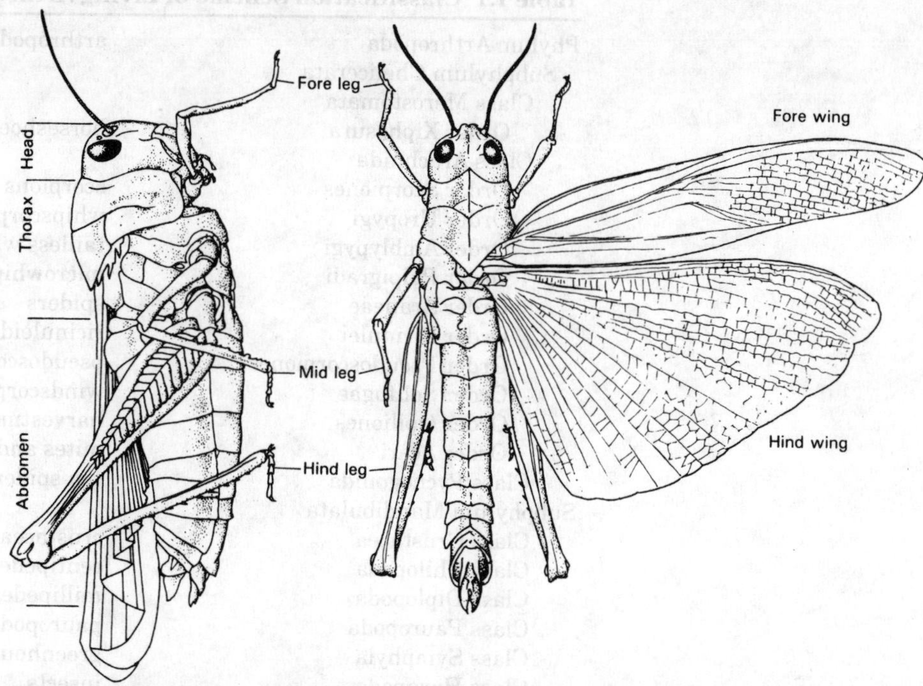


**Figure 1.2** Side (*left*) and top aspects of a grasshopper, showing important external body features of insects. (Redrawn from T. Nolan, 1970, Commonwealth Scientific and Industrial Research Organization, *The Insects of Australia*. Melbourne University Press)

More specifically, insects belong to the class Hexapoda (formerly Insecta), forming the most important category of the Arthropoda. The class Hexapoda has specific characteristics that set its members apart from other arthropod classes. These characteristics include (Fig. 1.2):

1. A body divided into three distinct regions: head, thorax, and abdomen;
2. A middle region, the thorax, bearing three pairs of legs and, most often, two pairs of wings; and
3. A system for breathing composed of air tubes.

## OTHER ARTHROPODS

Other arthropod classes have fewer total numbers and species than insects. Although a detailed survey of all classes is beyond the scope of this book, a brief overview of the four most important classes may help clarify the uniqueness of insects. These other important arthropods include those in the classes Crustacea, Arachnida, Diplopoda, and Chilopoda. A complete classification of living arthropods is outlined in Table 1.1.

**Class Crustacea.** This class includes many common animals such as crayfish, lobsters (Fig. 1.3), pillbugs, crabs, and shrimp (Fig. 1.4). Crustaceans as a group are mostly aquatic, and they play major ecological roles in marine habitats that are devoid of insects. A few species of terrestrial crustaceans, such as pillbugs and sowbugs, are commonly found in basements and other humid places of the home. When disturbed, they curl up into a compact ball (Fig. 1.5).

**Table 1.1 Classification Scheme of Living Arthropods.**

| | |
|---|---|
| Phylum Arthropoda | arthropods |
| Subphylum Chelicerata | |
| Class Merostomata | |
| Order Xiphosura | horseshoe crabs |
| Class Arachnida | |
| Order Scorpiones | scorpions |
| Order Uropygi | whipscorpions |
| Order Amblypygi | tailless whipscorpions |
| Order Palpigradi | microwhipscorpions |
| Order Araneae | spiders |
| Order Ricinulei | ricinuleids |
| Order Pseudoscorpiones | pseudoscorpions |
| Order Solifugae | windscorpions |
| Order Opiliones | harvestmen |
| Order Acari | mites and ticks |
| Class Pycnogonida | sea spiders |
| Subphylum Mandibulata | |
| Class Crustacea | crustaceans |
| Class Chilopoda | centipedes |
| Class Diplopoda | millipedes |
| Class Pauropoda | pauropods |
| Class Symphyla | greenhouse centipedes |
| Class Hexapoda | insects |
| (orders listed in Chapter 3) | |

SOURCE: Following Kaestnfer, 1968.

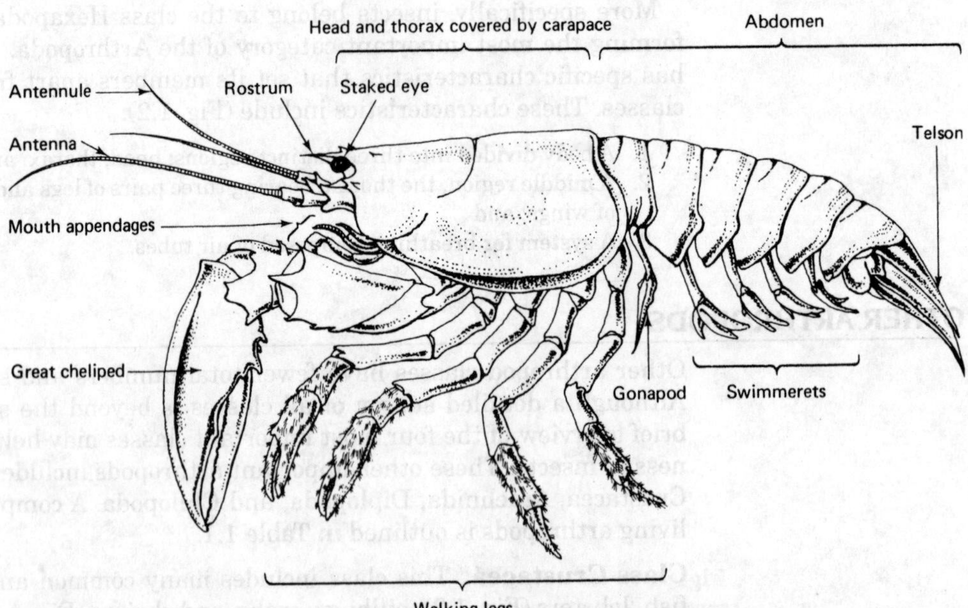


**Figure 1.3** Side aspect of the American lobster, *Homaris americana*, showing the various modifications in appendages and other body structures. (Reprinted with permission of Macmillan Publishing Company from *Insects in Perspective* by Michael D. Atkins. © 1978 by Michael D. Atkins)

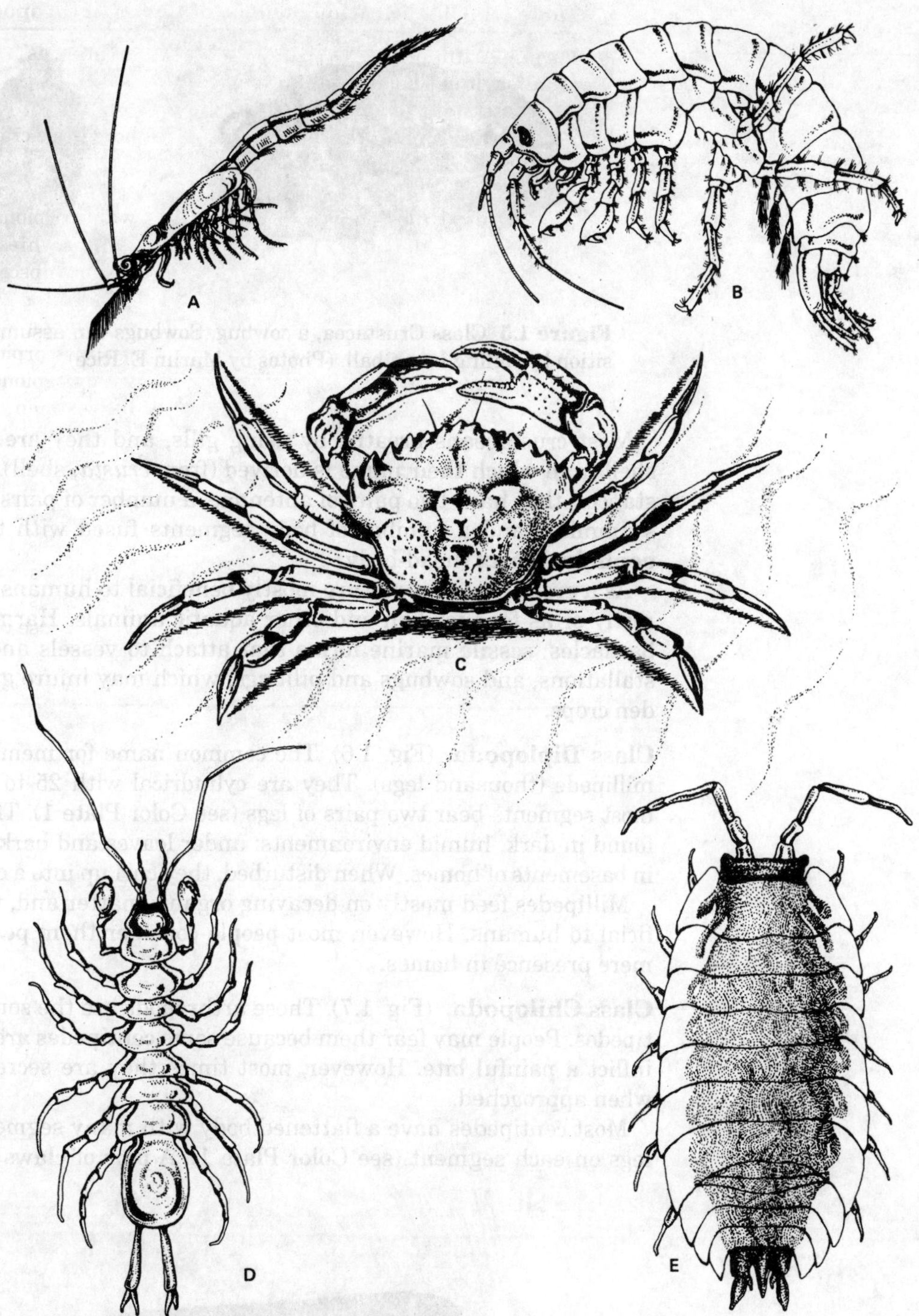


**Figure 1.4** Representatives of important groups in the class Crustacea. A. Freshwater mysid shrimp. B. Sand flea (amphipod). C. Shore crab (decapod). D. Freshwater isopod. E. Sowbug or pillbug (terrestrial isopod). (Reprinted with permission of Macmillan Publishing Company from *Insects in Perspective* by Michael D. Atkins. © 1978 by Michael D. Atkins)

**Figure 1.5** Class Crustacea, a sowbug. Sowbugs can assume a defensive position by rolling into a ball. (Photos by Marlin E. Rice)

Most crustaceans breathe by using gills, and they are covered by a hard shell from which their name is derived (from *crusta*, shell). In a number of instances, they have two pairs of antennae, a number of pairs of legs modified for swimming, and a number of body segments fused with the head to form a **cephalothorax** (head-body).

As a group, crustaceans are mostly beneficial to humans, serving as food directly or as food for fish and other aquatic animals. Harmful species include barnacles, sessile marine forms that attach to vessels and destroy shore installations, and sowbugs and pillbugs, which may injure greenhouse and garden crops.

**Class Diplopoda.** (Fig. 1.6). The common name for members of this class is millipede (thousand legs). They are cylindrical with 25 to 100 segments, and most segments bear two pairs of legs (see Color Plate 1). These arthropods are found in dark, humid environments: under leaves and bark, in woodlands, and in basements of homes. When disturbed, they curl up into a characteristic spiral.

Millipedes feed mostly on decaying organic matter and, therefore, are beneficial to humans. However, most people consider them pests because of their mere presence in homes.

**Class Chilopoda.** (Fig. 1.7). These arthropods are the sometimes-feared centipedes. People may fear them because some centipedes are venomous and can inflict a painful bite. However, most times they are secretive and run away when approached.

Most centipedes have a flattened body with many segments and one pair of legs on each segment (see Color Plate 1). A pair of claws behind the head is

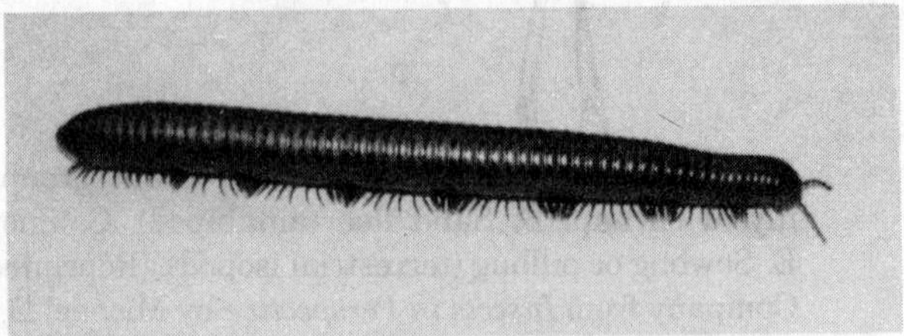

**Figure 1.6** Class Diplopoda, a millipede. (Photo by Marlin E. Rice)

**Figure 1.7** Class Chilopoda, a centipede. (Photo by Marlin E. Rice)

used to inject venom and paralyze insects and other invertebrates that serve as food for the centipede.

**Class Arachnida.** (Fig. 1.8). Next to the Hexapoda, this is the most diverse class of terrestrial arthropods. Members of this class possess a cephalothorax, as do crustaceans, but they lack antennae. Spiders breathe through structures that act like bellows, called **book lungs,** but other arachnids breathe through the skin or with air tubes. Most arachnids live on land.

Based on their relationship to humans, the most significant of the Arachnida are the spiders, mites, ticks, and scorpions. Other orders that contribute to the diversity of the class Arachnida include the Opiliones (harvestmen), Pseudoscorpiones (pseudoscorpions), and Uropygi (whipscorpions).

**Spiders** belong to the order Araneae and are represented by thousands of species. They are distinguished by their unsegmented abdomen, which is attached to the cephalothorax by a slender stalk, or pedicel (see Color Plate 1). Spiders feed mainly on insects by using mouthparts that crush their prey and allow these predators to suck out the body fluids. Silk-spinning organs are located on the underside of the abdomen, permitting spiders to build webs. Spider webs may be orb-shaped, funnel-shaped, triangle-shaped, or irregular. Insects and other prey become snared in the web and are killed outright or are paralyzed by venom from the spider's bite. As a group, spiders benefit humans by serving as natural enemies of insect pests. Some, however, are medically hazardous pests because of their dangerous bite. In North America, these include the black widow (*Latrodectus mactans*) and the brown recluse (*Loxosceles reclusa*).

From the standpoint of harm to humans, the most important order of the Arachnida is the Acari, comprising the mites and ticks. Here can be found pests of humans, other animals, and plants. Their saclike body and unsegmented abdomen broadly joined to the cephalothorax make the Acari distinctive. Mouthparts of mites and ticks pierce tissues and suck out the contents.

**Mites** are usually very small arthropods (1 to 3 mm long) that feed on plants, animals, and organic debris. Some mites are important predators of small insects and, particularly, other plant-feeding mites. One plant-feeding

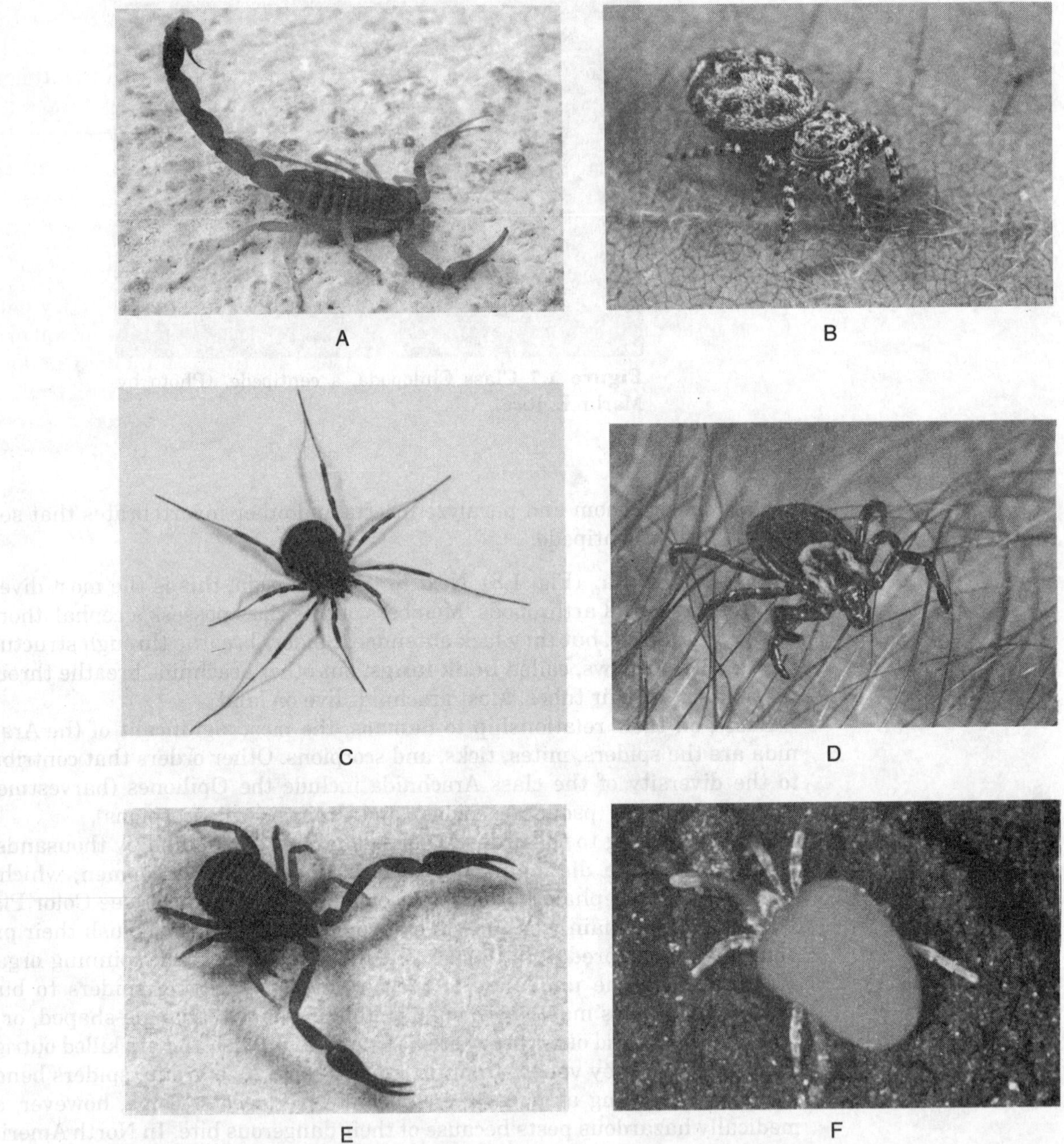

**Figure 1.8** Representatives of important groups in the class Arachnida. A. Scorpion. B. Spider. C. Harvestman (daddy longlegs). D. Tick. E. Pseudoscorpion. F. Mite. (Photos by Marlin E. Rice)

mite is the twospotted spider mite (*Tetranychus urticae*), which injures many crops in dry climates or during droughts in wetter regions (see Color Plate 1). Chiggers (larvae of *Trombicula alfreddugesi* and others) infest humans, causing intense itching when they inject enzymes to dissolve skin tissue on which they feed. Mange mites (*Sarcoptes scabiei*) are examples of mites that feed on many animals, including hogs, horses, dogs, and humans. These mites feed in the skin and produce burrows in which eggs are laid and young develop. In humans, these mites cause scabies, a skin condition often noted in the elderly.

**Ticks** are acarines larger than mites (10 to 20 mm long), with leathery integument or skin (see Color Plate 1). Ticks feed only on animals (humans and other mammals, birds, and reptiles), sucking blood from, and sometimes transmitting disease-causing organisms to, their hosts. Examples of important tick pests in North America include the lone star tick (*Amblyomma americanum*) and the American dog tick (*Dermacentor variabilis*). Both of these species attack humans, dogs, and livestock and are transmitters of the causal agent of Rocky Mountain spotted fever.

**Scorpions** are arachnids belonging to the order Scorpiones. They can be found in tropical and temperate regions but are most abundant in warm, dry climates. These arthropods are distinguished by their segmented abdomen, broadly joined to a cephalothorax, and long front appendages with strong pincers at the tip (see Color Plate 1). Scorpions have a well-known structure at the end of the abdomen, the sting, used in immobilizing and killing prey, often an insect, before feeding. When humans are stung by most species of scorpions, the effect of the poison, a neurotoxin, is painful but not usually dangerous. However, the sting of some species found in Mexico, for example, the Durango scorpion (*Centruroides suffusus*), can be fatal.

## WHAT ENTOMOLOGY IS

Entomology is the science of insects. It is an organized study to obtain knowledge of all phases of insect life and to understand insects' roles in nature. As in any science, entomology assumes that elements of the universe interact with one another in a predictable way and that insects, being one of these elements, can be understood relative to these interactions. From the knowledge obtained through *science,* insect events can be predicted and, if desired, modified with *technology*. Therefore, we can learn about insect feeding behavior and developmental rate from scientific study and apply this knowledge to alter these processes for plant and animal protection with technology.

Entomology is a biological science. It is the study of life, with its particular focus on one life form, the insect. More specifically, entomology is treated as a division of zoology, the study of animal life. In this context, we can think of entomology's position as one in a hierarchy of disciplines that become increasingly specific in their subject matter.

SCIENCE—study of the universe
BIOLOGY—study of life
ZOOLOGY—study of animals
ENTOMOLOGY—study of insects

Following this hierarchy, entomology is usually taught as a separate course(s) in universities, based in the departments of biology, zoology, or, in many instances, separate departments of entomology.

Although, in the strictest sense, entomology is the study of insects, other arthropods usually are treated by the science. In particular, these include the terrestrial arthropods, such as mites and ticks, that share habitats and interact with insects and cause similar pest problems.

**Entomologists.** Entomologists are people who specialize in entomology. They have background and skills in biology and particular interests in and knowledge of zoology (Fig. 1.9).

The first entomologists were mostly general biologists who made significant contributions to entomology and laid the groundwork for its development as a separate discipline. Today, many people make a living working with insects, and some simply study them for fun. According to British entomologist Sir Richard Southwood (1977), insects are of interest to people in one of three main ways: as a branch of zoology, as the scientific basis for pest control, or as a pastime. The first two ways are of concern to professional entomologists who teach, conduct research, and deal with pest problems. The last way is mainly of concern to amateur entomologists but also to many professionals who enjoy observing, photographing, and collecting insects because of their beauty and fascinating behavior. Consequently, both professionals and amateurs contribute

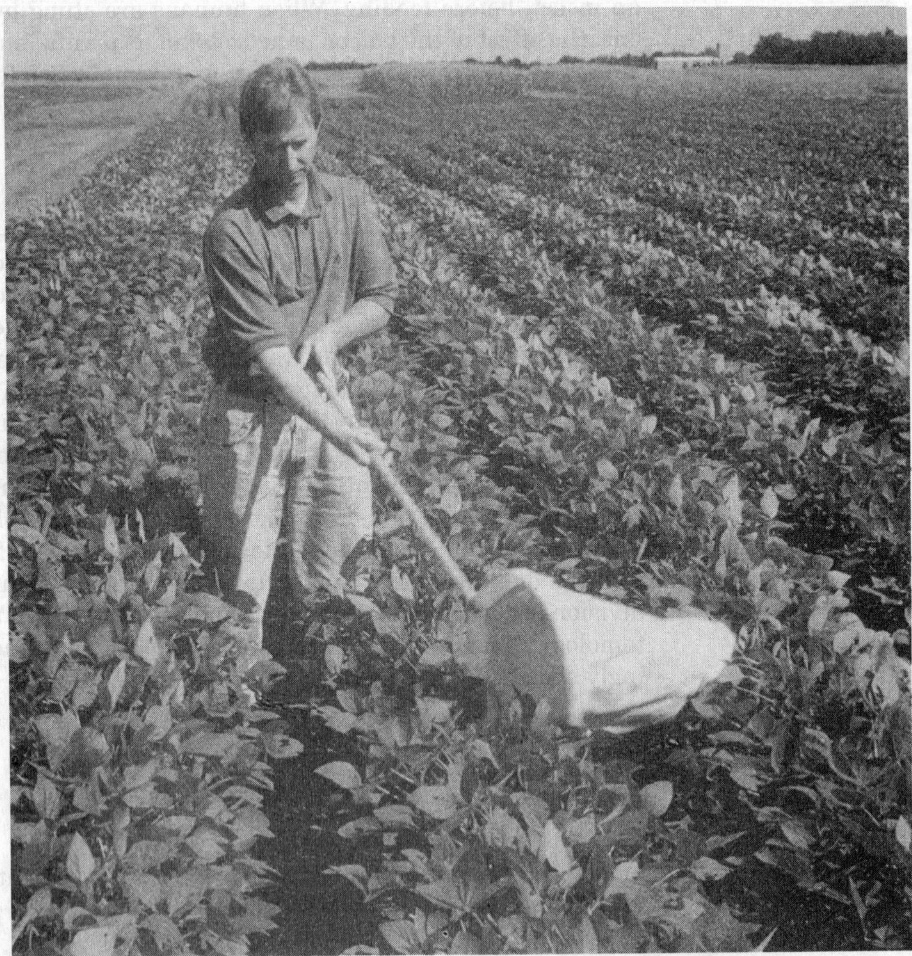

**Figure 1.9** Entomologist sampling insects with a sweep net. In addition to their special knowledge of insects, entomologists have basic backgrounds in zoology and other biological sciences. (Photo by Marlin E. Rice)

to the overall body of knowledge we call entomology, and all can be considered entomologists.

**Producers of entomological information and services.** Professional entomologists supply the greatest share of entomological information and services. They conduct their activities either through public or private agencies.

Public sources of information and services include universities, institutes, and governmental agencies. In the United States, most land-grant universities (those established by the Morrill Act of 1862) have a department of entomology that acquires information through research and disseminates it by teaching students, publishing scientific and popular articles, and using mass media. An important aspect of this educational system is the Cooperative Extension Service, a state and federally supported organization attached to the university, with a mission of providing up-to-date information and advice to farmers and the general public (Fig. 1.10). In addition to the universities in the United States and other countries, research organizations such as the Boyce Thompson Institute for Plant Research (Ithaca, N.Y.), the Rothamsted Experiment Station (Harpenden, Hertfordshire, United Kingdom), and the Institute of Animal Resource Ecology (Vancouver, British Columbia, Canada) supply information on insects that adds significantly to our current body of knowledge.

**Figure 1.10** A university entomologist discussing insect problems in soybeans with farmers. Entomologists funded through the Cooperative Extension Service at land-grant universities in the United States are important sources of insect-control information. (Photo by Marlin E. Rice)

Major governmental agencies supplying entomological information in the United States include the Agricultural Research Service, the Forest Service, and the Animal and Plant Health Inspection Service of the U.S. Department of Agriculture. Equivalents of these federal organizations exist in many other parts of the world; the Canadian Department of Agriculture and the Commonwealth Scientific and Industrial Research Organization (C.S.I.R.O.) in Australia are examples.

Not to be overlooked are entomological information and services available from private, mainly profit-oriented companies. Insect control services for households and businesses are obtained from structural pest control firms, which rid buildings of termites and other insect pests. Some of these companies also provide local mosquito control services and deal with pests of turf and ornamentals. Insecticide manufacturing firms also conduct entomological research and provide a good deal of information about insect control to their customers and the general public. Other private firms that consult in the management of crop pests contract to assess insect and other pest conditions on growers' property and give expert advice for dealing with pests.

The acceptance of the Internet as a method of communication has greatly changed the mode of delivering information about insects. The field of entomology was an early adopter of Internet-based methods using the delivery vehicles Gopher and the World Wide Web. Such methods have greatly increased the amount of information presented and the speed at which it is distributed to the public. We can anticipate this computer-based delivery to expand even more and to influence how we deal with insects in our daily lives.

**Users of entomological information and services.** Directly or indirectly, all of us use information about insects and benefit from the services of entomologists. On a personal basis, the general public uses entomological information and services to save dwellings, ward off mosquitoes, keep ticks off pets, keep flies in check, help grow vegetable gardens, and deal with insects in other ways.

Other direct beneficiaries are the producers of raw agricultural products. These individuals depend on a wide background of knowledge and techniques to avoid or reduce economic losses from insects. For some crops, particularly those of high value like apples, managing insect pests can mean the difference between profit or loss for the season.

Likewise, food processing firms greatly benefit from entomological information. In this industry, contamination from insects in processed foods can necessitate discarding huge quantities of an otherwise marketable product. The processor must absorb the loss. Knowing about insect behavior and sanitation procedures can minimize or prevent such losses.

Because insects affect human health and the health of livestock, both physicians and veterinarians are primary users of entomological information and techniques. Problems ranging from head louse epidemics in schools to transmission of disease organisms by mosquitoes are addressed by physicians who depend on their own and others' knowledge of insects. These professionals may consult medical entomologists before making a diagnosis and recommending therapy. Veterinarians presented with problems such as hog mange or cattle grubs rely on entomological research and often seek advice from livestock entomologists in suggesting cures.

Recently, even police have used entomological information to investigate homicides. To solve these crimes, estimating the time of death becomes crucial to determine circumstances and suspects. For example, knowledge of the behavior and developmental rates of carrion fly maggots inhabiting corpses allows estimates of egg-laying time and, therefore, the time of death of the victim. Such information, developed through **forensic entomology,** is expected to contribute significantly to advancements in criminology.

These are only a few examples of the users and uses of entomological information and services. Certainly, the quality of human life has been and will continue to be enhanced through our knowledge of insects.

## RELATIONSHIPS BETWEEN INSECTS AND PEOPLE

It is usually with a great deal of hesitation that human beings share their habitats with insects. Because we have no choice in the matter, we have adjusted our existence because of insects, and many customs and lifestyles reflect the intimacy of our relationship with them.

### Brief History of Relationships

**Insects in antiquity.** Insects have lived unhampered by human contrivances for most of their existence on earth. According to Speight, Hunter, and Watt (1999), the earliest known insect fossils date back to the late Devonian period, some 370 million years ago. Furthermore, most of the major groups (orders) we have today are distinguishable in the fossil record as far back as 250 million years (Fig. 1.11).

**Success of insects.** The class Hexapoda is one of the dominant life forms on earth, but what reasons can we advance for their great success? What characteristics account for their tremendous diversity and numbers? Most specialists on this topic believe that insect success is the result of several biological features acting in combination.

Probably first and foremost of the biological features is arthropod *body architecture*, which emphasizes an integument that is light and strong, forming a shell to protect inner tissues and the attachment of muscles. Moreover, this shell, which usually includes an outermost wax layer, helps to prevent water loss from evaporation, a critical problem for small animals living on the land. Arthropod body architecture also includes jointed appendages that, in insects, have been profusely adapted into legs for locomotion, mouthparts for feeding, structures of reproduction, and other uses.

Insects are also animals of relatively *small size*. Most vary from about 1/16 inch (about 2 mm) to 1 inch (about 3 cm). Some may be smaller, however, and a few, such as the 4-inch (about 15 cm) goliath beetle of Africa and the 13-inch (about 35 cm) walkingstick of Malaysia, definitely fall outside this normal range. Nevertheless, the small size of most insects facilitates dispersal, allows them to escape from birds and other predators, and enables them to use food present only in small amounts.

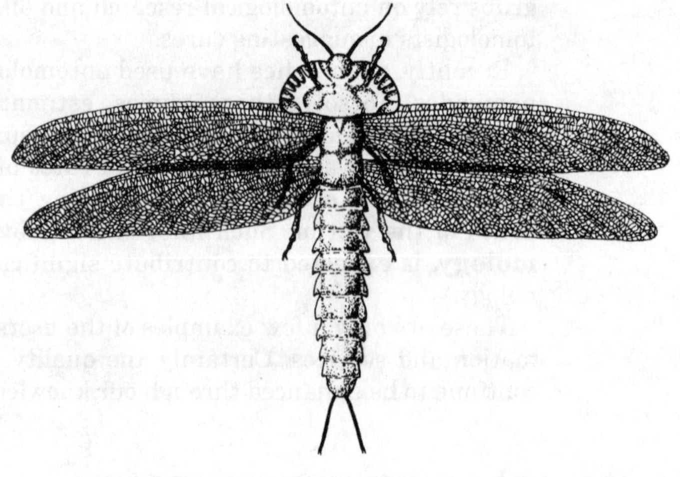

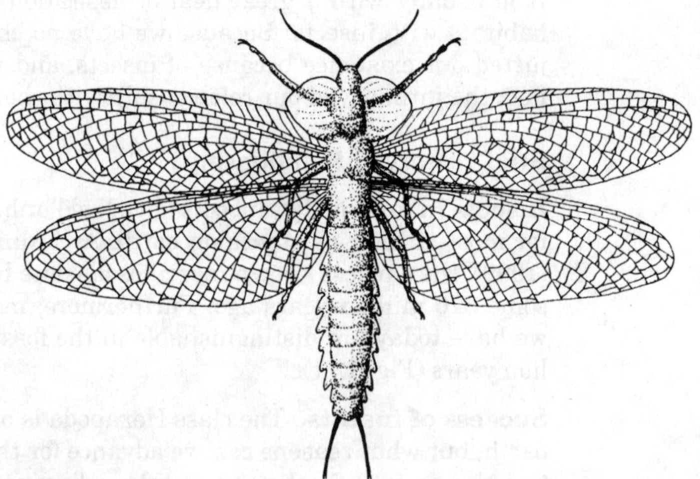

**Figure 1.11** Reconstructions of Carboniferous fossil insects. (Redrawn from Jeannel, 1960, *Introduction to Entomology,* Hutchinson and Company)

Differing from all other arthropods and invertebrates, most insects can fly. This *ability to fly* (Fig. 1.12) is one of the most important reasons for success of the class as a whole. Flying aids insects in escaping predators and, perhaps more important, enables widespread dispersal of species. This dispersal promotes colonization of new habitats.

Finally, the great *reproductive capacity* (Fig. 1.13) of insects and features of their growth and development have enhanced their ability to persist even in unfavorable environments. The ability to lay large numbers of eggs, combined with a relatively short generation time, produces a great amount of genetic variability that can be tested against the environment. The result is rapid adaptation of populations to changing environmental conditions. These major features, collectively unique to insects, contribute to their great ecological success.

**Figure 1.12** The ability to fly is one of the most important reasons for the success of the class Hexapoda. (Photo by Marlin E. Rice)

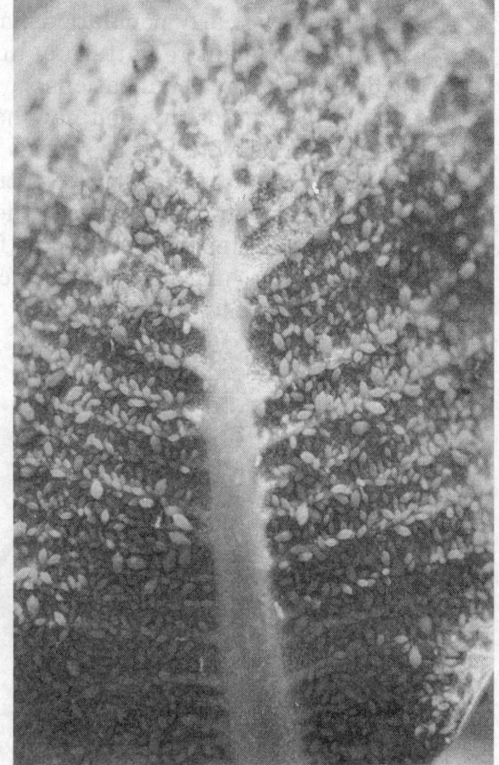

**Figure 1.13** Aphids (Hemiptera: Aphididae) on a milkweed leaf. The great reproductive capacity in these and other insects has enhanced their ability to persist even in unfavorable environments. (Photo by Marlin E. Rice)

**Prehistoric times.** At the dawn of human existence, insects were already entrenched in every conceivable land habitat. They fed on plants, all sorts of animals (including other insects), and all forms of nonliving organic debris. Therefore, it is easy to imagine that they were already feeding as parasites on our earliest ancestors.

Human hairlessness seems to point to a tropical environment where protohumans probably originated and where they fed on nuts and fruits, vegetable shoots, and probably insects. On the basis of abundance in the tropical habitat, insects may have been one of the first animals that humans ate on a regular basis!

Because prehistoric humans probably coexisted with monkeys and arboreal apes, we can assume that they shared the same parasites. Additionally, some scientists believe that human associations with bed bugs (species closely related to bat parasites) arose from humans sharing caves with bats during the ice ages. Therefore, we can suppose that early humans were infested with the array of parasites known from modern primates, including mites, fleas, ticks, flies, and worms. Humans were also subject to diseases caused by many microorganisms, including the so-called **arboviruses,** or arthropod-borne viruses transmitted by mosquitoes (Fig. 1.14) and other bloodsucking arthropods. Further, plasmodia, transmitted by mosquitoes to humans and causing epidemics of **malaria** (Fig. 1.15), must have had horrendous effects on those early populations, as they still do today. Malaria undoubtedly has altered the entire pattern of human history and probably contributed to the decline of some civilizations.

As early humans invaded new habitats, they were exposed to greater and greater contact with insect species previously not encountered. This was the case when humans began to venture into uninhabited regions of Africa in search of game and encountered the tsetse fly. This fly was (and still is) capable of transmitting a trypanosome from antelopes and other ungulates to humans, causing African sleeping sickness, a debilitating disease that can be lethal within a few weeks.

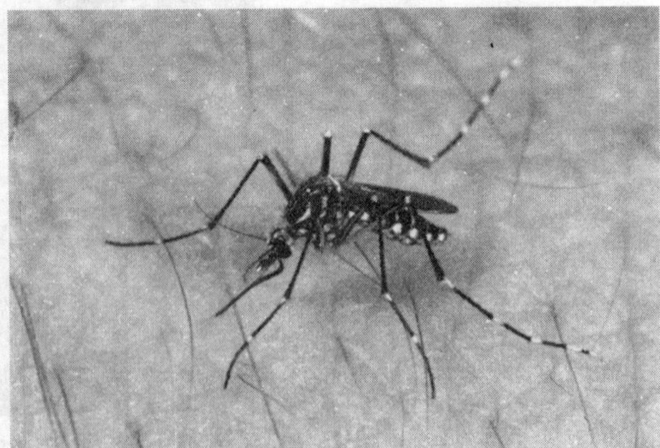

**Figure 1.14** A mosquito, *Aedes aegypti* (Diptera: Culicidae), on human skin. Ticks, mites, fleas, flies, and worms are important pests of humans. (Photo by Marlin E. Rice)

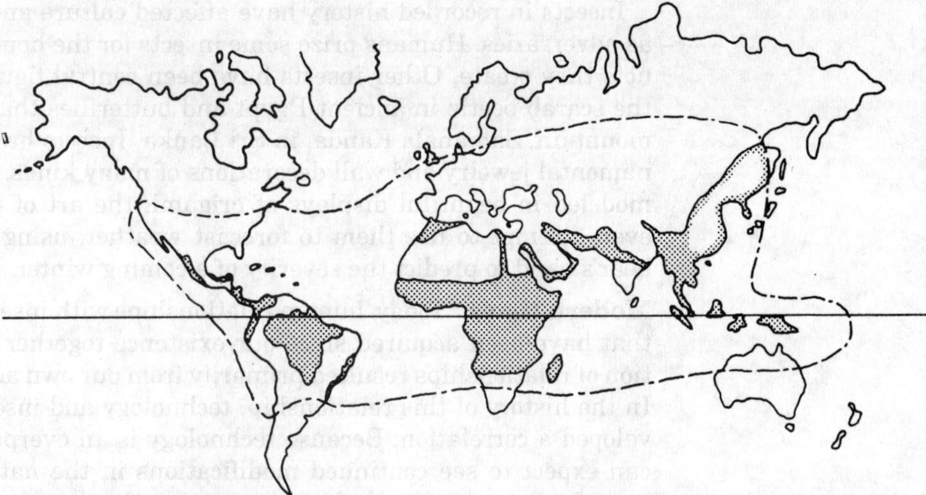

**Figure 1.15** Map of the world showing the present distribution of significant numbers of cases of malaria (*shaded*) and estimated extent of its distribution in the past (*dashed line*). (Reprinted with permission of Macmillan Publishing Company from *Insects in Perspective* by Michael D. Atkins. © 1978 by Michael D. Atkins)

Therefore, early mankind had pervasive encounters with insects, some good and some bad. Certainly, it would be difficult to overstate the impact of arthropod-borne disease not only in terms of human suffering but also in terms of retarding the development of civilization. In total, the relationships that developed left an indelible mark on human development.

**The rise of agriculture and civilization.** As early agriculture developed, humans remodeled the landscape significantly by encouraging some animals and plants to multiply and others to be displaced. The result was local areas with reduced biological diversity and, as might be expected, a greater confrontation of humans with insects. Not only were humans to contend with insects feeding on their bodies and transmitting diseases, they also had to concern themselves with insects competing for a desired resource: their crops. Subsequently, when agriculture developed and greater areas of land were utilized for farming, pressure from insect populations (and other pests like weeds) increased disproportionately, making pest control a major preoccupation.

In addition to the competition with insects fostered by agricultural technology came an added intensity of parasite problems. For example, in western Africa, slash-and-burn methods of cultivation created more intensive epidemics of malaria. This phenomenon occurred when clearings made for cultivation provided additional breeding habitats (more standing water) for the mosquito, *Anopheles gambiae,* a transmitter of malarial plasmodia. This mosquito displaced other mosquito species that fed on animals and became a menace to every human who ventured into the new forest clearings.

Problems with parasitic insects also became more prevalent with urbanization, the congregation of people to form cities. This urbanization was the direct result of agriculture, when one person began to supply more food than the family could consume. Urbanization created greater opportunities for insects to be transferred from one person to the next, as with lice and fleas, and from one household to the next, as with cockroaches and bed bugs.

Insects in recorded history have affected culture and lifestyle, but not only as adversaries. Humans prize some insects for the honey, silk, and other products they create. Other insects have been central figures in religion, such as the scarab beetle in ancient Egypt and butterflies that migrate to the sacred mountain, Samanala Kanda, in Sri Lanka. Insects have also been used in ornamental jewelry and wall decorations of many kinds. Also, insects have been modeled in beautiful displays of origami, the art of paper folding. Humans even attempt to use them to forecast weather, using the width of a woollybear's band to predict the severity of a coming winter.

**Modern times.** Today, human relationships with insects are the sum of those that have been acquired since our existence together began. This accumulation of relationships resulted primarily from our own activities to tame nature. In the history of this relationship, technology and insect encounters have developed a correlation. Because technology is an overpowering force today, we can expect to see continued modifications in the natural landscape. Consequently, it seems likely that new associations, and perhaps confrontations, with insects will also develop.

## The Ledger

In analyzing the history of human and insect relationships, it might be tempting to believe that insects are our enemies incarnate and that their elimination would result in a better world. However, insects have redeeming values both for human existence and the overall ecology of our planet. An analysis, therefore, should picture human/insect associations as profits and losses, as would a business enterprise. To do this, we might prepare a ledger (Fig. 1.16) showing black ink, the profits, and red ink, the losses, and then form a balance sheet to arrive at our conclusions about insects.

**Black ink: The benefits.** Insects can benefit humans by providing a product desired for human consumption, a **primary resource,** or by interacting with elements of our environment to yield a benefit, an **intermediate resource.** Probably the most valued primary resources insects provide today are **silk, honey, wax**, and their **bodies** for human consumption and experimentation.

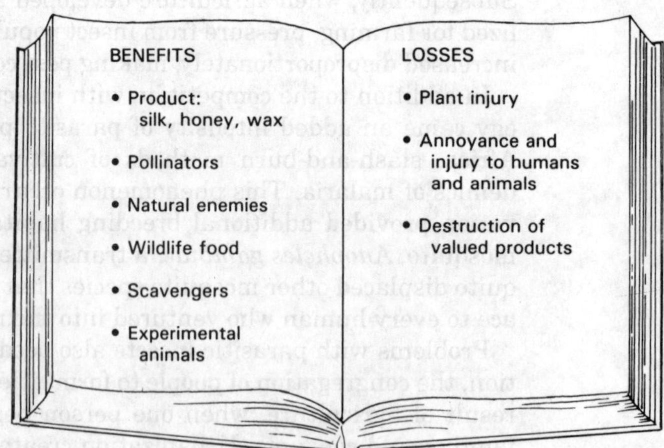


**Figure 1.16** The ledger of insect relationships with humans, indicating benefits and losses.

Among the intermediate resources are insect activities as **pollinators, natural enemies of pests, food for wildlife,** and **scavengers.**

Silk, the queen of fabrics, is woven from thread of fine strands secreted from the salivary glands of silkworms, caterpillars of the moth *Bombyx mori*. The propagation of silkworms for production of this luxurious fabric is termed *sericulture* and is believed to have been developed in China as far back as 2500 B.C. With more than 45,000 metric tons of silk produced on an annual basis in Asia and Europe, an estimated 2,000 billion silkworms must be reared on more than 90 million tons of mulberry leaves each year! These statistics and the price of silk on the world market support the argument that the silkworm is one of our most significant insect species and is truly a valued resource.

Honey and wax, of course, are produced by the honey bee, *Apis mellifera*. This species is probably even more important than the silkworm. Wild honey bee hives are believed to have been raided by humans as long ago as 7000 B.C., and management of the species to obtain honey and beeswax, called beekeeping (or apiculture), has developed into a major agricultural industry in much of the world. Honey is produced by worker bees that feed on minute quantities of nectar from flower blossoms. The nectar mixes with a bee's saliva, and this mixture, after reacting with certain enzymes, is regurgitated into individual hexagonal wax cells in the hive. These cells remain open until much of the water in the mixture has evaporated, and the cells with cured honey are sealed with a thin capping of wax by the workers. This stored honey serves as food for young and developing bees. An important by-product of extracting honey from a hive is beeswax, which is melted down and sold for use in furniture polish, candles, car wax, cosmetics, and other products.

The remaining products of insects, although small by comparison with silk and honey, have nonetheless been important in certain instances. They include raw material from lac scales (*Lacifer lacca*) used in making shellac and the coloring of other scale insects for making red and purple dyes.

As mentioned earlier, whole insects, in addition to their products, have been eaten by humans since the beginning of human life. Today, insects continue to serve as food directly, and edible insects are a source of protein and fat in otherwise deficient diets of people in many parts of the world (Fig. 1.17). Among

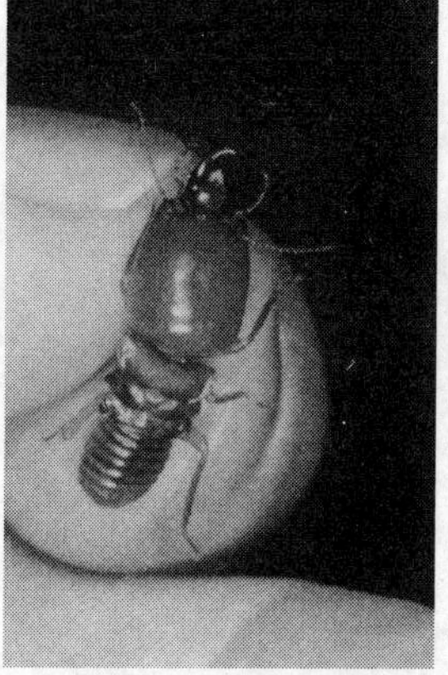

**Figure 1.17** A young girl in Zimbabwe uses a grass stem to extract soldier termites from their mound. The termites will be fried and eaten as a supplemental source of protein. (Photos by Marlin E. Rice)

the edible insects, termites and grasshoppers are probably the most widely consumed, with caloric value sometimes exceeding 500 calories per 100 insect grams. Europeans and their descendants are the only major group of people not consuming insects in any appreciable amount. This may be surprising when considering that the Greeks and Romans, who laid the foundations of modern European culture, ate a wide variety of insects and reared some as specialty foods.

---

RECIPE FOR SAUTEED ALFALFA WEEVIL LARVAE

1/4 cup butter
4 garlic cloves, crushed
1 cup cleaned alfalfa weevil larvae

Rinse larvae and pat dry. Melt butter in fry pan. Sauté garlic in butter for 5 minutes. Add insects, continue to sauté for 10 to 15 minutes. (Recipe used by the Iowa State University Entomology Club for serving at the Annual Insect Horror Film Festival)

---

A final example of a primary resource is the role insects play today as experimental animals (Fig. 1.18). Many scientists use these six-legged guinea pigs to test theories and solve biological problems. For laboratory work, insects are small, require little space, and many, such as large cockroaches, demand little care and maintenance. Indeed, insects used as experimental animals have been indispensable in such fields as genetics, toxicology, and neurobiology; they have allowed advances that would have been prohibitively costly or otherwise impossible. The science of genetics, in particular, has relied on insects, and probably no other species has been more important than the fruit, or pomace, fly, *Drosophila melanogaster*, in unraveling the complexities of inheritance. In applied ecology, scientists have relied on the presence and absence of certain insect species as indicators of pollution. This is possible with insects like immature mayflies, which are particularly sensitive to chemical changes in water.

**Figure 1.18** An American cockroach, *Periplaneta americana*. Such large cockroaches are often used as experimental subjects for laboratory research. (Photo by Marlin E. Rice)

**Figure 1.19** A honey bee, *Apis mellifera*, feeding on a flower. Insect feeding on flower nectar is an important means of transferring pollen from flower to flower. (Photo by Dennis Schotzko)

Of all the ways insects serve as intermediate resources, probably none is more important than their plant pollination activity (Fig. 1.19). Butterflies, bees, flies, and other groups of insects inadvertently pollinate plants when they feed on pollen and nectar of flowers. Moving from flower to flower to collect nectar, insects like bumble bees come into contact with pollen grains that adhere to their legs and body hair. With each visitation, some grains are left on the sticky female stigmata, thereby cross-fertilizing the plant. In the process, other pollen grains are picked up and subsequently transferred to other plants. Such relationships play an important role in the success of both insects and many plants. Although some of our important crops such as wheat, corn, and other cereals are wind-pollinated, most fruits and vegetable crops are pollinated by insects. Berries, pears, apples, citrus, melons, peas, beans, and tomatoes are but a few of our important crops that would not exist in their present form, if at all, without insects. In fact, species like alkali bees, *Nomia melanderi,* and leafcutter bees, *Megachile rotundata,* are managed to improve alfalfa pollination and, consequently, seed production in the western United States and Canada. These wild bees, particularly leafcutter bees, are cultured by alfalfa seed producers and provided through mobile field-nesting sites. Larger and more successful producers may produce bee populations of more than 30 million females.

Equally important to human welfare is the activity of some insects that kill and eat other kinds of insects. These insects, known as natural enemies, function as predators or parasites in reducing the numbers of harmful insects. For example, lady beetles feed voraciously on many species of aphids, sometimes averting the need for any pest management action. Parasites such as tiny *Trichogramma* wasps lay eggs inside the eggs of many moth pests (Fig. 1.20), producing larvae that consume pest-egg contents and sometimes prevent pest populations from causing economic damage. Although the exact magnitude of natural enemy effects on all insect pests and potential pests cannot be measured, it is believed momentous: Probably 30 to 40 percent of all insect species are natural enemies of other insects.

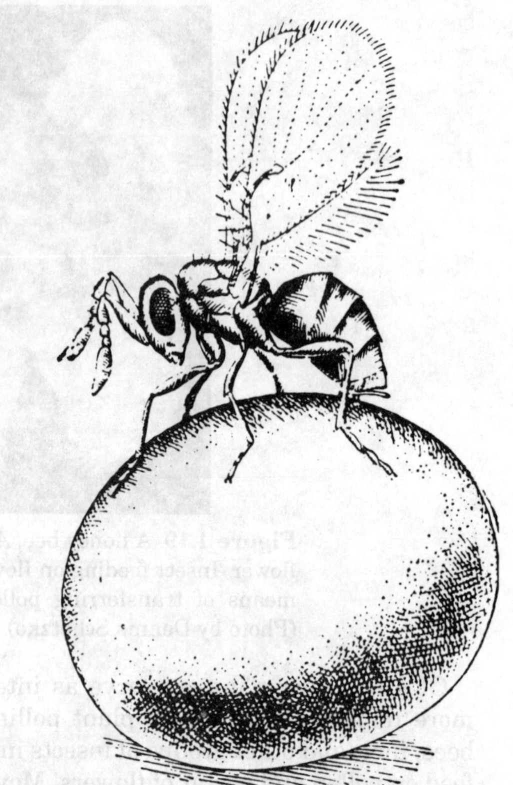

**Figure 1.20** A *Trichogramma* wasp (Hymenoptera: Trichogrammatidae) laying an egg inside a moth egg. (Courtesy USDA)

Yet another activity of insects that benefits humans is scavenging. When animals and plants die, the organic matter that makes up their bodies possesses energy in the form of chemical bonds. That energy is released via a step-by-step breakdown of body tissue constituents. Insects, in feeding on dead plant and animal tissue, often carry out the first stage of decomposition by predisposing matter for enhanced decay and ultimate breakdown by microorganisms. Some prominent examples of insect decomposers are termites that break down wood, springtails that assist in the decomposition of dead leaves, and carrion beetles and many fly maggots that feed on dead animals.

The role of insects as food for wildlife and, in some instances, for domesticated animals is not to be overlooked. Insects also are consumed by an array of fish, amphibians, reptiles, birds, and small mammals, in addition to other insects. Indeed, insects are the major food source for many small-bodied vertebrates, and effective colonization of certain areas by these animals depends on the insect food source. For example, mayflies are thought to be the single most important dietary component of trout, and the skilled angler would not expect good fishing without these insects. In other instances, insects are reared for sale as fish bait and pet food, and some, including crickets and locusts, are collected during mass outbreaks for swine and poultry feed in some countries.

Taken in total, the products and services provided by insects to humankind are nearly immeasurable. Only recently are we beginning to understand just what the benefits amount to. In a study by Losey and Vaughan (2006), the services provided by insects, which include food for wildlife, pest destruction, crop pollination, and scavenging, are estimated at $57 billion in the United States alone. The authors believe this estimate is conservative, and it does not include many other uses where insects are a primary resource.

**Red ink: The losses.** Although insects are usually thought of as harmful to humans, probably fewer than 1 percent of all insect species fall into the pest category. From this group, probably 3,500 species require regular attention, and in the United States, as few as 600 species can be considered significant.

Nevertheless, these species we call pests have imposed and continue to impose burdens on human populations by causing:

1. Injury to crop plants, forests, and ornamentals;
2. Annoyance, injury, and death to humans and domesticated animals; and
3. Destruction or value depreciation of stored products and possessions.

Injury to crop plants, forests, and ornamentals is one of the most conspicuous ways insects cause economic losses to humans. Because approximately one-half of all known insects are plant feeders, our greatest number of insect problems involve plants.

Insects injure plants by feeding on them and laying eggs in plant tissues. By far the more dangerous of these injuries is feeding. Insects consume plant tissues with various types of chewing mouthparts and remove plant juices with piercing-sucking mouthparts. In the process of feeding, some insects—many aphids and leafhoppers, for example—also transmit hundreds of kinds of plant pathogens, including bacteria, fungi, viruses, and mycoplasmas that subsequently cause losses from diseases. These insects are known as **vectors.** Other insects such as the potato leafhopper, *Empoasca fabae,* inject toxins that influence the plant's physiology, resulting in yield and quality changes.

Losses from insect feeding may be **direct** or **indirect** (Fig. 1.21). Direct losses occur when insects feed on potentially harvestable produce, eliminating it or causing devaluation. Significant examples of this injury include that of the boll weevil, *Anthonomus grandis,* which destroys cotton squares and bolls;

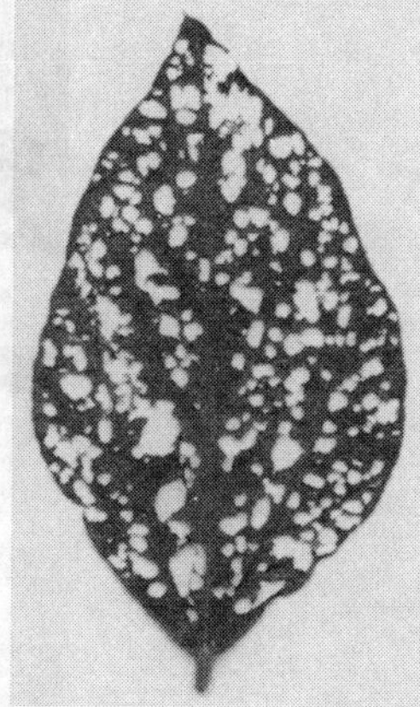

**Figure 1.21** Direct injury, represented by codling moth larva, *Cydia pomonella,* feeding on apple (*left*), and indirect injury, represented by bean leaf beetle adults, *Ceratoma trifurcata,* feeding on soybean (*right*). (Photos by Marlin E. Rice)

the codling moth, *Cydia pomonella,* which feeds inside apples; and bark beetles, *Scolytus* species and others, which bore into bark and cambium of trees. Direct losses may occur merely from the presence of insects on produce, causing value depreciation, or **dockage.** Oystershell scales, *Lepidosaphes ulmi,* on the rind of grapefruit are one example. Indirect losses arise when insects feed on roots, stems, and leaves, thereby causing a decline in quantity or quality of potentially harvestable produce, usually seeds, fruits, or tubers. However, because of the recuperative potential of most plants, large populations of indirect feeders may be required to cause a significant loss. Examples of these pests include the Colorado potato beetle, *Leptinotarsa decemlineata,* consuming leaf tissue of potatoes; northern corn rootworms, *Diabrotica barberi,* eating roots of corn; and Hessian fly, *Mayetiola destructor,* boring in wheat stems.

Annoyance and injury to humans and domesticated animals are equally important ways that insects affect humans. However, probably fewer insect species are involved in causing these kinds of problems than are involved with plants. Insects of medical importance may affect humans directly or indirectly. Directly, a wide variety of species are parasitic on humans, primarily feeding on blood. Common examples are mosquitoes, lice, fleas, and biting flies. Others cause annoyance and discomfort simply by their presence. Some insects and other arthropods inject venoms when they bite or sting, often causing a local reaction to the wound, but sometimes causing death. Although rare, death from envenomization occurs most frequently from bee stings to persons hypersensitive to the venom.

Insects as medical pests are much more important in an indirect way, as vectors of human pathogens. Here, some insects, ticks, and mites harbor nematodes (Fig. 1.22), rickettsiae, viruses, protozoa, and bacteria. In some instances, part of the microorganism's life cycle, often including multiplication, occurs in the insect body. Among the most important diseases transmitted by insect bites are

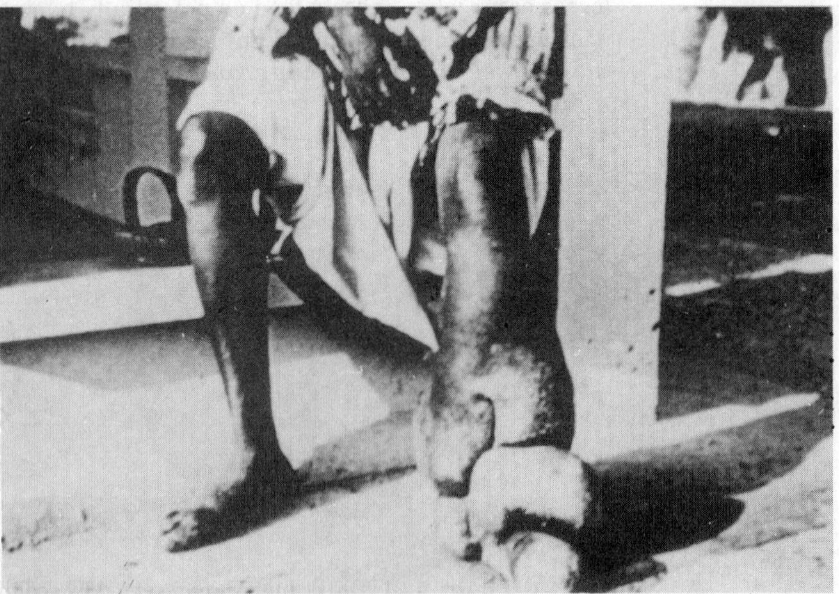

**Figure 1.22** Leg elephantiasis caused by infection with filarial nematodes. These nematodes are transmitted by biting mosquitoes, and enlargement of the legs is caused by blockage of lymph drainage. Arthropod-borne diseases remain some of the most important medical problems in many developing nations.

**malaria, yellow fever, filariasis, dengue,** and several types of **encephalitis.** The causes are malaria from a protozoan, filariasis from filarial worms, and dengue and encephalitides (plural of encephalitis) from viruses called **arboviruses,** an abbreviation of the term *arthropod-borne viruses*. Of these diseases, malaria is the most significant. Of the total world population (6 billion people), millions are exposed to malarial infections in about 90 countries. It has been estimated that between 300 and 500 million clinical cases of malaria occur each year, with over 90 percent of these in tropical Africa. What's more, malaria causes 1.4 to 2.6 million deaths worldwide, mostly in Africa, and the problem has become worse with increasing resistance of the plasmodia to antimalarial drugs.

Additionally, a dreaded African arbovirus, the **West Nile virus,** was detected in the United States in 1999. This virus, transmitted by mosquitoes, causes severe neurological illnesses such as encephalitis and meningitis. Infection may result in death, particularly in elderly persons.

Another important insect-transmitted disease causing concern in the United States is **Lyme disease.** Lyme disease was named after the site where an epidemic was reported in 1976, Old Lyme and Lyme, Connecticut. The disease is caused by a spirochete bacterium that is transmitted by so-called deer ticks, *Ixodes ricinus* and *Ixodes scapularis,* in North America. Lyme disease affects tissues of skin, joint, nervous system, and heart in all age groups, but it is not life threatening. It is the most prevalent arthropod-transmitted disease in the United States.

In comparison with other human diseases, arthropod-transmitted diseases are not of major importance in North America and Europe. However, these diseases remain some of the greatest medical problems in many developing nations.

Insects affect domesticated animals in the same way they affect humans. In most instances today, insect parasites of livestock and poultry cause monetary losses by weakening animals, rather than killing them (Fig. 1.23). Animals in a

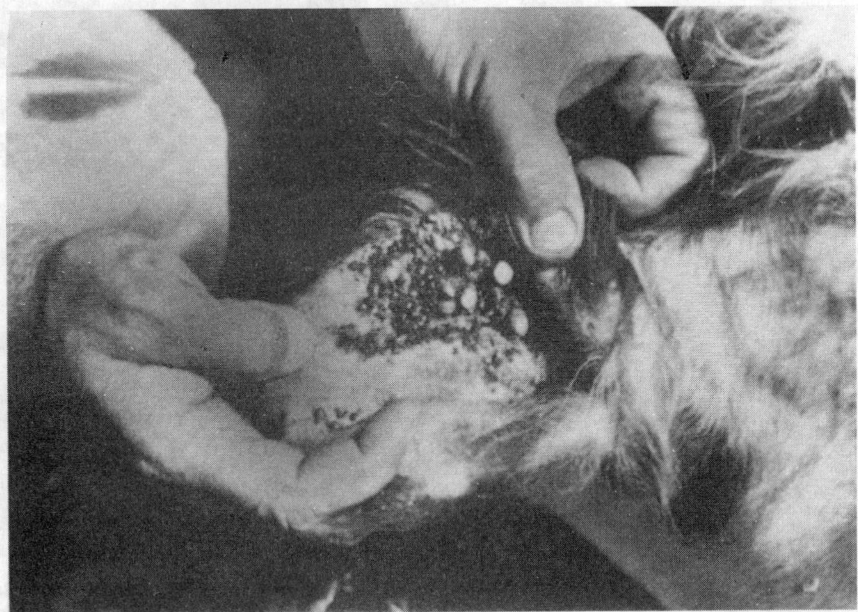

**Figure 1.23** Gulf Coast tick, *Amblyomma maculatum*, on ear of cow. Such arthropod ectoparasites cause significant livestock and poultry losses by weakening and sometimes killing the host. (Courtesy W. Rowley, Iowa State University)

weakened or agitated state eat less and gain less weight. Some of the most important pests of livestock are the true flies, including the stable fly, *Stomoxys calcitrans;* horse fly, *Tabanus* species; horn fly, *Haematobia irritans;* and face fly, *Musca autumnalis*. Another fly, the screwworm, *Cochliomyia hominivorax,* found in the southern United States, Mexico, and Central and South America, is a serious livestock pest that is now eradicated in the United States and Mexico. Because of the spectacular success of a sterile-fly release program, this fly problem was eliminated from Florida, Texas, and other western states, where millions of dollars in losses have been averted. Chewing lice are probably the most important insect pests of poultry, with heavy infestations causing skin irritation that results in reduced weight gain and reduced egg production.

Destruction or value depreciation of stored products and possessions is sometimes understated. Losses can amount to millions of dollars annually, even in countries such as the United States, where storage facilities are good and control technology is readily available. Indeed, in some instances, losses to food in storage may exceed those inflicted by insects before crops are harvested.

Most of the important insect pests of stored grains and processed foods in North America are beetles (for example, granary weevil, *Sitophilus granaries* (Fig. 1.24), and red flour beetle, *Tribolium castaneum*) and moths (for example, Angoumois grain moth, *Sitotroga cerealella,* and Indianmeal moth, *Plodia interpunctella*).

Other insects affect household possessions. Clothes moths and carpet beetles eat anything containing animal fibers, including woolens, upholstery, and carpet. Other insects such as cockroaches spot and taint food and living areas. Termites are another frequent household problem in many localities, consuming wood on foundations, in floors, and in walls that often require expensive repairs.

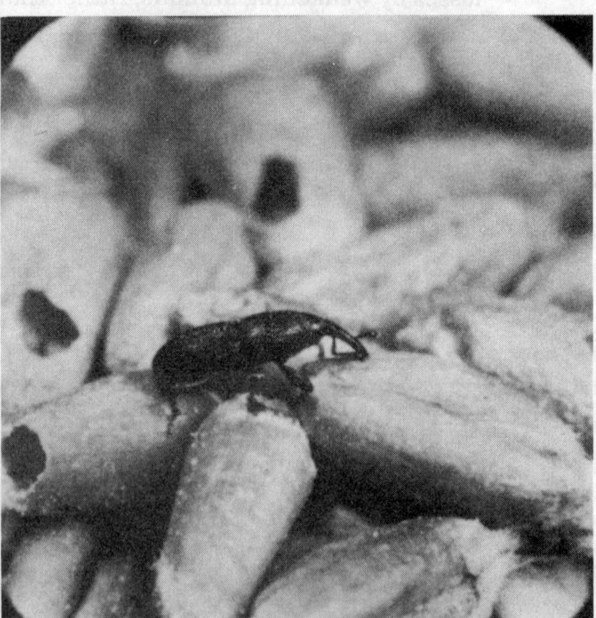

**Figure 1.24** Granary weevil, *Sitophilus granarius,* infesting stored wheat. Beetles are some of the most important pests of stored grain in North America. (Courtesy USDA)

**The balance sheet.** When we weigh the good and bad, would the world really be better for humans without insects? This question could arouse a debate in which the outcome may be less important than our having argued the question. The fact is, insects are part of our everyday lives and are here to stay. Perhaps, having considered the question, our perception may change, if ever so slightly, toward a greater understanding of nature. More understanding can result in a greater appreciation and enjoyment of the things we like and greater tolerance of the things we dislike.

Regrettably, many people believe all insects are bad. We might say that insects suffer from poor public relations. This view was expressed by Mertins (1986) in a review of insects and other arthropods that have been used as subject matter in moviemaking. In films, arthropods usually represent threatening, sinister figures or images of danger or death. They are also shown as distasteful or shocking images for the squeamish or images of silliness for ridicule by "rational" persons. In a similar vein, Mertins found that entomologists are usually shown as "detached from reality, as eccentric buffoons, often as psychotics, or, at best, ineffectual dupes." He found no instances where popular movies present the scientific study of insects in a positive light. Such movies may not only tend to reflect public opinion but also may help to form and direct it. Unfortunately, commercial films in countries such as the United States have portrayed insects mostly in an inaccurate, unflattering way (Fig. 1.25), and this same medium serves to educate masses of impressionable young people.

**Figure 1.25** In the film *Mothra* (1962), Japan is subjected to the catastrophic backwash from a gigantic moth. Most commercial films have portrayed insects in an inaccurate, unflattering way. (From *A Pictorial History of Science Fiction Films* by Jeff Rovin, 1975, published by arrangement with Lyle Stuart)

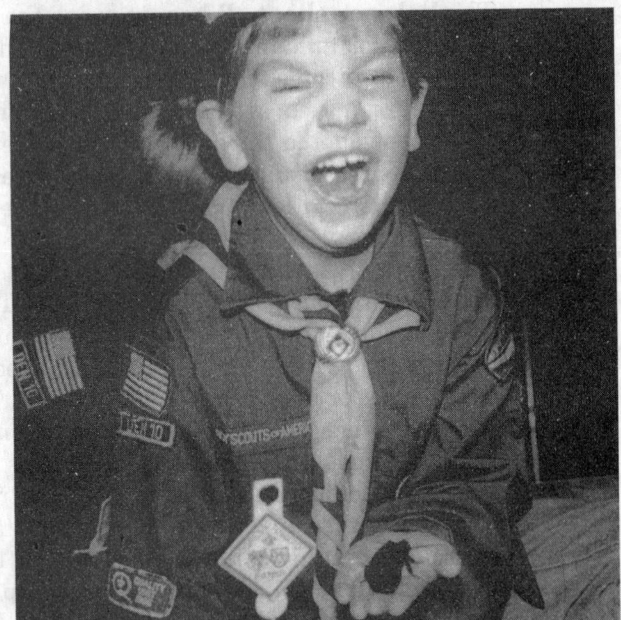

**Figure 1.26** Most children are not inherently fearful of insects and other arthropods. They are taught fear and disgust by adults. (Photo by Marlin E. Rice)

The movie theater is not the only place that the young are taught an aversion to insects. Parents often scold curious youngsters to throw away a caterpillar tucked in the hand or encourage them to step on a beetle running across the sidewalk. Unfounded admonitions such as "It will bite you, get rid of it" and "Kill it before it gets away" serve to create attitudes of both fear and confrontation that are passed along from generation to generation (Fig. 1.26).

The ultimate development of fear of insects can be seen in certain psychological disturbances found in a small number of people. An irrational and lasting fear of real insects, resulting in unconventional behavior, is known as **entomophobia.** A somewhat different fear is expressed by persons suffering from **delusory parasitosis;** sufferers have hallucinations of "bugs" on their bodies. In either instance, professional help is usually required to overcome the affliction.

A strong confrontational attitude and technological ability to effectively kill insects has given rise to "the only-good-bug-is-a-dead-bug or if-it-moves-kill-it" syndrome. Such attitudes have created the fetish of an insect-free environment that exists not only around the home, where owners spray chemicals in the yard to eliminate mosquitoes and fleas, but also in agriculture and forest production. Although some improvements in attitude have been made, this fetish continues to be the source of certain water quality and other environmental problems.

Although maligned in general, a few insects have a relatively good image. These include the producers of useful materials, such as bees and silkworms, and famous natural enemies of pests, particularly lady beetles (Fig. 1.27). Anthropomorphic images of industry ("busy as a bee") and cooperation ("the farmer's friend") understandably are ascribed to these insects.

**Figure 1.27** A sevenspotted lady beetle, *Coccinella septempunctata,* feeding on corn leaf aphids, *Rhopalosiphum maidis*. A good image is usually ascribed to predatory insects that feed on pest species. (Photo by Marlin E. Rice)

Other insects have positive images as objects of beauty. Colorful butterflies, for example, are found in collections among other beautiful objects. Today, so-called "butterfly gardens," plantings that attract a variety of butterflies, are becoming increasingly popular. But why do we find butterflies beautiful, even symbolic of virtue and the human spirit (*psyche,* from Greek mythology, a young girl with the wings of a butterfly)? Is it their color? Other insects are also beautiful and yet are considered repugnant by some people. Is it the seemingly warm appearance of butterflies, a delusion created by their furlike covering of scales?

Clearly, our culture and environment could be improved with a general change in attitude about insects. If we viewed other insect species as we do butterflies, we could not believe the world would be better off without insects. With such an outlook, we might accept that the benefits of insects far outweigh the losses caused by them, and our fetish of creating insect-free environments would disappear, along with significant quantities of environmentally harmful chemicals.

The key to this change in attitude is knowledge and understanding about insects and the important part insects play in the ecology of our planet. From past mistakes, it would seem that the most advantageous attitude should be to live in harmony as much as possible with all elements of nature, including insects. As T. Eisner has put it, "Bugs are not going to inherit the earth. They own it now, so we might as well make peace with the landlord." Just how we can promote an attitude of harmony is the basis for insect pest management, the sensible approach to insect problems.

## INSECT PEST MANAGEMENT

Humans have attempted a great many approaches to alleviate insect pest problems through the years. Recently, no approach has been more popular than **pest management,** also called **integrated pest management** (Fig. 1.28). Pest management is a general approach to deal with all kinds of pests: insects, mites, and other arthropods; plant-parasitic nematodes, microbial and viral plant pathogens; weeds; and vertebrates. The primary objective of the approach is to

**Figure 1.28** Pest management is a general approach to dealing with all kinds of pests, and many sources of information are available on this topic. (Photo by Donald Lewis)

reduce losses from pests in ways that are effective, economically sound, and ecologically compatible.

Pest management sometimes is equated with plant protection, an approach that focuses more on the object of value than on the pest. However, the scope of pest management is broader because in addition to crop plants, it addresses pest problems with livestock, urban dwellings, and landscape. It also covers certain aspects of human health, including mosquitoes and other vectors of human disease.

Pest management is characterized by the way several techniques are employed simultaneously to solve specific pest problems. Therefore, a single pest management program may involve the use of pesticides, host-plant resistance, tillage, sanitation, and biological control. To use these tools effectively, the approach relies on an intimate understanding of ecology, pest biology, and the appropriate integration of information. An important goal of pest management programs is to arrive at long-term solutions to problems, rather than simply to derive short-term protection.

**Insect pest management,** the primary topic of this book, is a division of integrated pest management that emphasizes insects and other arthropods. It seeks to reduce the status of pests by following principles of ecology and using the latest advancements in technology. Much of the foundation of insect pest management rests on determining whether an insect is truly a pest and, if so, just how serious a problem it causes—in other words, determining its status.

### The Concept of Pest

Pest species are those that interfere with human activities. According to Australian entomologist P. W. Geier and his colleagues (1983), the quality of being a pest is **anthropocentric** (considering humans as the central fact or final aim of a system) and circumstantial. Termites (Box 1.1) feeding on dead wood

## BOX 1.1 EASTERN SUBTERRANEAN TERMITE

**SPECIES:** *Reticulitermes flavipes* (Kollar) (Isoptera: Rhinotermitidae); also western subterranean termite, *Reticulitermes hesperus* Banks, and Formosan subterranean termite, *Coptotermes formosanus* Shiraki

**DISTRIBUTION:** Several species of subterranean termites occur in the United States. They occur in all states except Alaska, but they are more common in the southern and middle-latitude states.

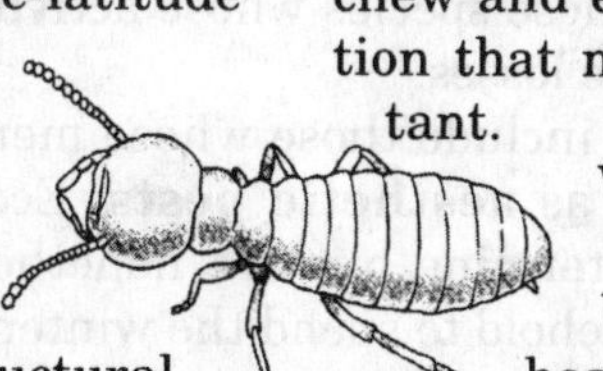

**IMPORTANCE:** Subterranean termites can be extremely damaging to wood and other material containing cellulose. Their damage often ruins the structural integrity of houses and other buildings. Damage to wood is mostly confined to the softer, springwood growth. Tunnels tend to follow the wood grain.

**APPEARANCE:** Subterranean termites are social insects that live in colonies in the soil. Colonies contain three forms or castes: reproductives, workers, and soldiers. Individuals of each caste have several stages, including the egg, nymph or soldier, and the adult (with three stages). Reproductive females and males can be winged or wingless. Primary reproductives, also called alates or swarmers, vary in color, by species, from pale yellowish-brown to dark black. Wings may be pale or smoky brown to gray, with a few distinct veins. Swarmer termites are about 1/4 to 3/8 inch (6 to 10 mm) long. Secondary and tertiary reproductives are white to cream-colored and may have short wing buds. Workers are wingless, white to creamy white, and 1/4 to 3/8 inch long. Soldiers defend the colony from invaders, such as ants, and resemble workers, except that they have large, well-developed brownish heads with strong mandibles. Another type of soldier, called the nasute, has a long, tubelike projection on the front of the head, which exudes a sticky substance to entangle their enemies.

**LIFE CYCLE:** Workers make up the largest proportion of individuals within the colony. They do most of the work—excavating tunnels, feeding the other castes, grooming the queen, and cleaning the nest. Because termites obtain their nutrition from the cellulose, the workers must chew and eat wood, which causes the destruction that makes termites economically important.

When the termite colony matures in 2 to 4 years, swarmers are produced. Usually after rainfall, factors such as heat, light, and moisture trigger the swarmer emergence from the colony. Both male and female swarmers fly into the sky where they pair up and fall back to the ground. The pair quickly seeks protective cover. Here, they make a small nest before mating. The queen then lays a few eggs. The king remains with the queen because periodic mating is required for continued egg production. The colony initially grows slowly, but in subsequent years the queen grows larger and lays more eggs.

Nymphs hatch from the eggs within several weeks and are cared for by the queen and king. The nymphs molt into pseudergate workers, and then into presoldiers or wingless nymphs. The colony stabilizes in the size when the queen reaches maximum egg production.

Subterranean termites derive their nutrition from wood and other material containing cellulose, including paper and burlap. Subterranean termites cannot digest cellulose, but must depend on large numbers of one-called animals (protists) living in the termite hindgut. The protists break down the cellulose to simple acetic acid, which termites can then digest. Worker termites and older nymphs consume wood and share the nourishment with all other castes in the colony.

in a forest serve an important ecological function, degradation in the process of returning nutrients to the soil. Clearly, termites are not pests in this context; they are beneficial to humankind. The same species performing the same ecological function in the environment of a human home is a pest. An understanding of the often significant ecological roles played by "pest" species in both unmanaged and agricultural environments frequently gives insights into how to deal with them. Moreover, this understanding can aid in developing more tolerant attitudes about insect presence.

Probably our greatest share of insect pest problems are encountered in agricultural and forest production systems. Here, problems arise because of large numbers of insects, not simply because of the presence of a species. Most insect pests of agriculture (including agronomy, animal science, horticulture, and forestry) are those species whose activities, enhanced by population numbers, cause economic losses.

Other pests include those whose mere presence is objectionable. We might refer to these as **aesthetic pests.** Economic losses based on aesthetics are difficult to determine but are nonetheless real. In instances where insects enter the household to spend the winter—boxelder bugs (*Leptocoris trivittatus*) (Fig. 1.29), for example—some persons may be motivated to spend money on insecticidal control and, therefore, suffer an economic loss. Others may not be as bothered by insect presence. Presence of insects or insect parts in food, however, is another matter and is regulated by law. Insect presence in food processing and storage requires costly sanitation measures and often the disposal of tainted products. Such losses can be measured objectively and are significant.

Another group of insects can be termed **medical pests.** Losses caused by medical pests also are very significant but, like some aesthetic pests, are extremely difficult to measure. Loss of work or work efficiency may be measurable in economic terms, but how can economic loss be assigned to discomfort, pain, and even loss of life? Such problems make management of medical pests and some aesthetic pests quite difficult.

**Figure 1.29** Boxelder bug, *Leptocoris trivittatus*. Such insects are sometimes considered aesthetic pests when they enter homes to overwinter. They do no harm there, but their mere presence is objectionable to many people. (Photo by Marlin E. Rice)

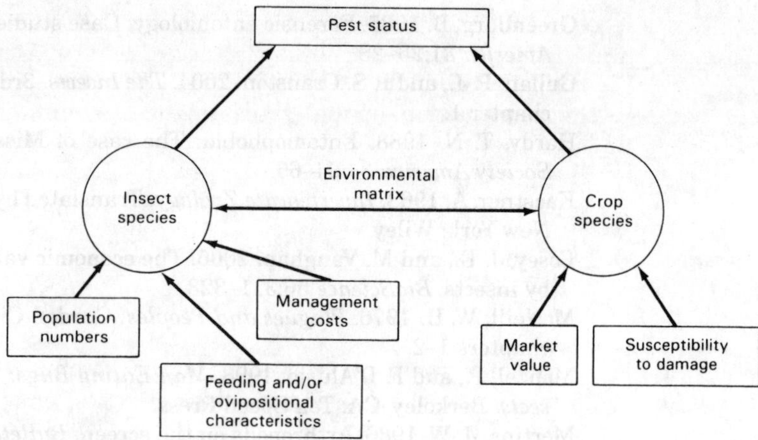


**Figure 1.30** Diagrammatic representation of the major factors contributing to pest status.

## The Concept of Pest Status

**Pest status** is the ranking of a pest relative to the economics of dealing with the species. This aspect is variable for a pest and depends on many factors. With agricultural pests, status depends greatly on the crop involved and the environment in which the pest-crop interaction occurs (Fig. 1.30).

Among the major factors contributing to pest status, market value is one of the most variable. With all other factors held constant, a pest can assume higher or lower levels of importance because of changing economics. Susceptibility of the crop to pest injury is another important, but often little understood, factor. Variable weather factors (particularly moisture) and cultural procedures, such as fertilization, can profoundly affect crop vigor. Under optimal conditions, pest status can be lowered when plants compensate for damage.

Ultimately, the environment, including the human social environment, is the basis of change in the factors governing pest status. Environmental factors mediate the importance of pests and determine the measure and constraints of insect pest management programs.

In the following chapters, we will discuss the approach of pest management to the solution of insect problems. To do this, we first need to learn about insect biology and life processes and then proceed to understand modern protection technology and the optimal application of this knowledge.

## Further Reading


Anderson, J. F., and L. A. Magnarelli. 1994. Lyme disease: A tick-associated disease originally described in Europe, but named after a town in Connecticut. *American Entomologist* 40:217–227.

Atkins, M. D. 1978. *Insects in Perspective*. New York: Macmillan, chapters 18–22.

Eisner, T., and E. Wilson. 1977. General introduction: The conquerors of the land, pp. 2–15. In T. Eisner and E. Wilson, eds., *The Insects. Readings from Scientific American*. San Francisco: Freeman.

Geier, P. W., L. R. Clark, and D. T. Briese. 1983. Principles for the control of arthropod pests, I. Elements and functions involved in pest control. *Protection Ecology* 5:1–96.

Greenburg, B. 1985. Forensic entomology: Case studies. *Bulletin Entomological Society America* 31:25–28.

Gullan, P. J., and P. S. Cranston. 2004. *The Insects.* 3rd ed. Oxford: Blackwell Publishing, chapter 1.

Hardy, T. N. 1988. Entomophobia: The case of Miss Muffet. *Bulletin Entomological Society America* 34:64–69.

Kaestner, A. 1968. *Invertebrate Zoology*. Translated by H. W. Levi and L. R. Levi, vol. 2. New York: Wiley.

Losey, J. E., and M. Vaughan. 2006. The economic value of ecological services provided by insects. *BioScience* 56:311–323.

McNeill, W. H. 1976. *Plagues and Peoples*. Garden City, N.Y.: Anchor Press/Doubleday, chapters 1–2.

Menzel, P., and F. D'Alusio. 1998. *Man Eating Bugs: The Art and Science of Eating Insects.* Berkeley, CA: Ten Speed Press.

Mertins, J. W. 1986. Arthropods on the screen. *Bulletin Entomological Society America* 32:85–90.

Park, Y. L., and J. Bradshaw. 2003. Insect origami. *American Entomologist* 49:210–215.

Paskewitz, S. M., and M. J. Gorman. 1999. Mosquito immunity and malaria parasites. *American Entomologist* 45:80–94.

Pedigo, L. P. 1985. Integrated pest management, pp. 22–31. In *McGraw-Hill Yearbook of Science and Technology*. New York: McGraw-Hill.

Skeleton, T. E. 1976. Insects and human welfare. *American Biology Teacher* 38:208–210.

Speight, M. R., M. D. Hunter, and A. D. Watt. 1999. *Ecology of Insects*. Oxford, U.K.: Blackwell Science, chapter 1.

Southwood, T. R. E. 1977. Entomology and mankind. *American Scientist* 65:30–39.

World Health Organization. 1995. Vector control for malaria and other mosquito-borne diseases. *World Health Organization Publication No. 857*.

VanDyk, J. K. 2000. Impact of the Internet on extension entomology. *Annual Review of Entomology* 45:795–802.

Zenger, J. T., and T. J. Walker. 2000. Impact of the Internet on entomology teaching and research. *Annual Review of Entomology* 45:747–767.

## Favorite Web Sites

http://www.ent.iastate.edu/List/
Entomology Index of Internet Resouces. A directory and search engine of insect-related resources on the Internet.

http://www.ag.ohio-state.edu/~ohioline/hyg-fact/2000/2160.html
Hosted by entomologists, farmers, and chiefs who promote edible insects. They refer to insects as "microlivestock." Gives recipes and nutritional information.

http://www.ucmp.berkeley.edu/arthropoda/arthropoda.html
Has color photos showing the diversity in the phylum Arthropoda.

http://www.ent.iastate.edu/imagegal/
Many close-up color photos of common pest and beneficial insects.

http://bugguide.net/node/view/15740
An online resource devoted to North American insects, spiders, and their kin, offering identification, images, and information.

CHAPTER 2

# INSECT STRUCTURES AND LIFE PROCESSES

BY ANY MEASURE, the success of organisms is due to the particular biological features they possess. These features have allowed the animals to overcome an array of inhospitable environments through effective solutions to the problems of body maintenance, reproduction, and dispersal. In large part, these important adaptations are derived from an amazing body comprising specialized structures that function in harmony with one another and with their external environment.

Consequently, our understanding of insects in nature and our explanation of them as pests depends on a knowledge of their biological features. As we shall see, pest management relies heavily on predicting pest behavior, and basic to predicting behavior is understanding insect structures and their functions.

## THE INSECT BODY

### General Organization

**Tagmosis and the body wall.** Insects have segmented bodies with certain segments fusing to form three usually well-defined regions: **head, thorax,** and **abdomen** (Fig. 2.1). The grouping of segments into functional regions is known as **tagmosis.**

The head shows few signs of segmentation in the adult stage. The thorax consists of three distinctly modified segments, each bearing a pair of segmented legs. The least modified of the body regions is the abdomen, which may have as many as eleven segments.

These insect structures are formed from the body wall, which also supplies a support system known as the **exoskeleton.** The exoskeleton provides a rigid foundation for the body and serves as a point for the attachment of muscles, something like the internal skeleton of vertebrates. The flexible exoskeleton is similar to a spring, but it does not stretch, except at certain times during growth. In addition to its support function, the exoskeleton serves as a covering to protect internal organs and helps prevent desiccation.

Certain parts of the body wall are hardened, or **sclerotized,** in the form of plates called **sclerites.** Between these plates, the body wall is membranous, or

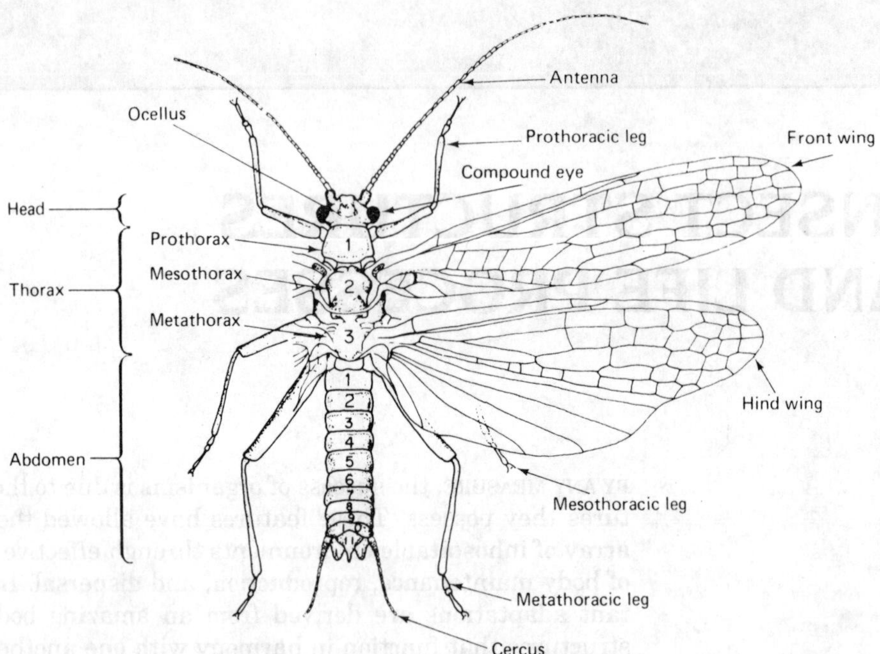


**Figure 2.1** Generalized insect body showing tagmosis and major anatomical features. (Reprinted with permission of Macmillan Publishing Company from *Insects in Perspective* by Michael D. Atkins. © 1978 by Michael D. Atkins)

soft, which allows movement and expansion of the body during eating and egg development. Other types of membranous connections found in insect bodies and appendages include telescoping ring segments, simple appendage joints, and ball-and-socket articulated joints.

In addition to sclerites and membranes, the body wall is characterized by other external and internal features. Externally, these features include wrinkles, spurs, scales, spines, and hairs (Fig. 2.2). Internally, inpushings of the body wall called **apodemes** are prominent in the head and thorax. These strengthen the exoskeleton and provide areas for muscle attachment.

**Detail of the body wall.** The body wall, or **integument,** of insects is composed of a single layer of cells called the **epidermis** that is bound on the inside by the **basement membrane** and on the outside by the **cuticle** (Fig. 2.3). Among the epidermal cells are scattered wax-secreting cells and other cells that secrete molting fluid involved in the growth process. The cuticle covers the entire outer body surface and also lines the insect's air tubes, salivary glands, and parts of the digestive tract. It is produced by the epidermis.

The cuticle is made up of three primary layers: a thin outer **epicuticle,** a thicker **exocuticle,** and a still thicker **endocuticle.** The epicuticle itself is divided into four layers: an innermost **homogeneous** layer that is overlain in succession by layers of **cuticulin** (a lipoprotein), **wax** of exceptionally long hydrocarbon chains, and, in most instances, **cement,** a tanned protein substance that serves as a varnishlike covering. The cuticulin layer is critical to the growth process, for it is permeable to the chemicals and nutrients needed for growth and impermeable to enzymes that break down parts of the old cuticle

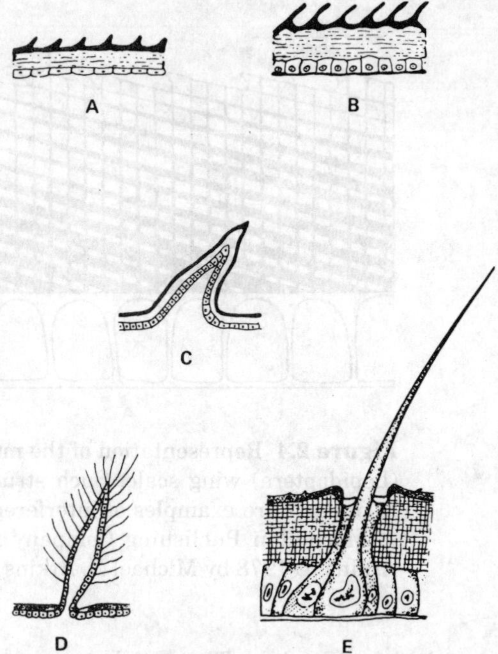


**Figure 2.2** External features of the body wall. A. and B. Noncellular surface configurations include ridges and spurs. C., D., and E. Cellular structures include setae and multicellular spines. (Reprinted with permission of Macmillan Publishing Company from *Insects in Perspective* by Michael D. Atkins. © 1978 by Michael D. Atkins)

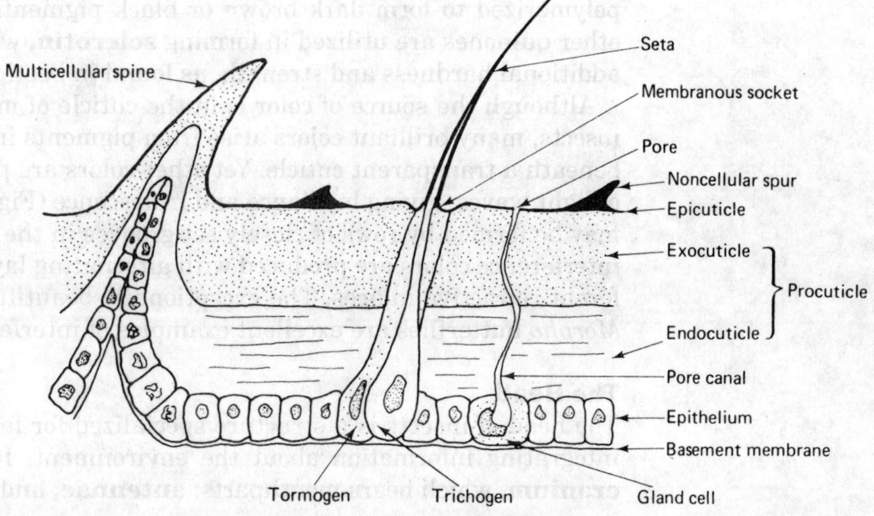


**Figure 2.3** Insect integument and associated structures. (Reprinted with permission of Macmillan Publishing Company from *Insects in Perspective* by Michael D. Atkins. © 1978 by Michael D. Atkins)

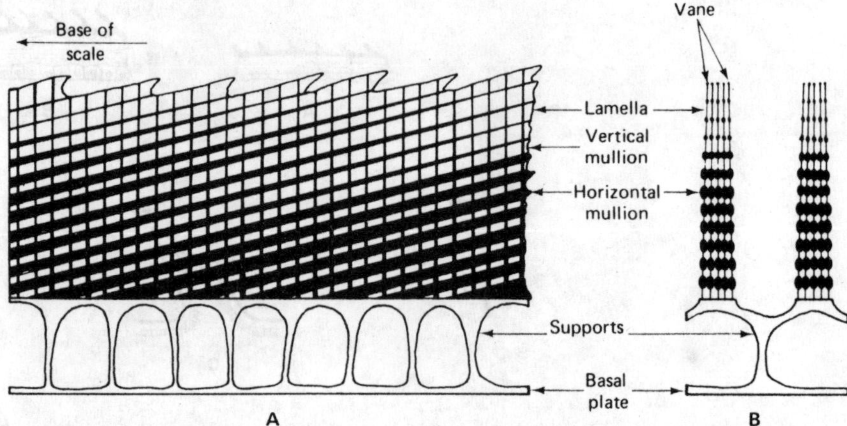


**Figure 2.4** Representation of the multilayered structure of a *Morpho* butterfly (Lepidoptera) wing scale. Such structured scales produce beautiful iridescent colors and are examples of interference coloration. (Reprinted with permission of Macmillan Publishing Company from *Insects in Perspective* by Michael D. Atkins. © 1978 by Michael D. Atkins)

before it is shed. The cuticle has numerous vertical channels for sensory nerve fibers and ducts for wax glands. Helical **pore canals** that transport materials to the body surface are also found in the cuticle of many insects.

The exocuticle gives the cuticle its characteristic strength and resilience. It is formed of **chitin,** a polymerized compound (nitrogenous polysaccharide linked to a protein) found not just in insects but also in other arthropods. The molecules of polymerized compounds are linked in long chains; they are common in nature as a base for materials like wood, hair, and horn.

Other constituents of the cuticle include **quinones.** Some quinones, derived mainly from the blood, are incorporated directly into the cuticle, and some are polymerized to form dark brown or black pigments known as **melanin.** Yet other quinones are utilized in forming **sclerotin,** which gives the exoskeleton additional hardness and strength, as found in beetles.

Although the source of color is in the cuticle of most black and dark brown insects, many brilliant colors arise from pigments in the epidermis or fat cells beneath a transparent cuticle. Yet other colors are produced by the diffraction of light waves, giving brilliance and iridescence (Fig. 2.4). Diffraction gratings may be formed by rows of closely set grooves in the surface of the cuticle, and interference colors are produced with alternating layers of the cuticle with different refractive indices. The exceptionally beautiful iridescent blue wings of *Morpho* butterflies are excellent examples of interference colors.

### The Head

The head of insects is a structure specialized for feeding and for sensing and integrating information about the environment. It is made up mostly of a **cranium,** which bears mouthparts, **antennae,** and **eyes** (Fig. 2.5).

**Cranium.** The insect cranium is a hardened capsule with an opening leading to the mouth and thorax. It is attached to the thorax by way of a short neck, or **cervix.** The orientation of the cranium varies among different insects,

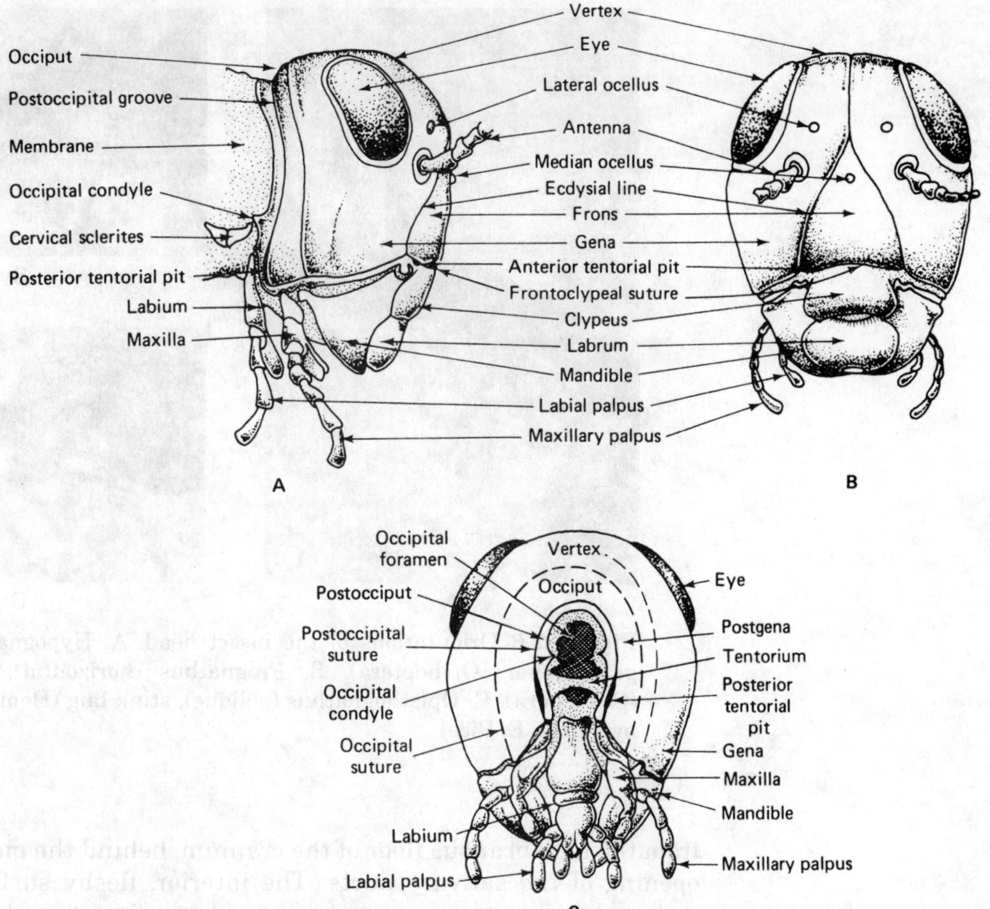


**Figure 2.5** Detail of a typical insect head. A. Side aspect. B. Frontal aspect. C. Rear aspect. (Reprinted with permission of Macmillan Publishing Company from *Insects in Perspective* by Michael D. Atkins. © 1978 by Michael D. Atkins)

positioned vertically (mouthparts pointed down, hypognathous), horizontally (mouthparts pointed forward, prognathous), or obliquely (mouthparts pointed backward, opisthognathous or opistorhynchous of some authors) (Fig. 2.6). Insects like ground beetles and grasshoppers with chewing mouthparts have heavy crania, adapted for muscles involved in capturing prey and biting off leaf tissue. Insects like leafhoppers and mosquitoes, which pierce and suck liquid food, often have much more delicate crania. Internally, the cranium is strengthened by a set of invaginations of the body wall fused to form the **tentorium.**

**Mouthparts.** Mouthparts are some of the most distinctive features of insects, and their structure tells a great deal about the feeding habits of a species. Mouthparts (Fig. 2.7) are greatly varied, but in their simplest form, for example, in grasshoppers and beetles, they include a **labrum** (upper lip), a pair of chewing **mandibles** (jaws), a pair of **maxillae** (second jaws), and a **labium** (lower lip). These structures surround the mouth and form the pre-oral cavity. In addition, a central tonguelike **hypopharynx** drops

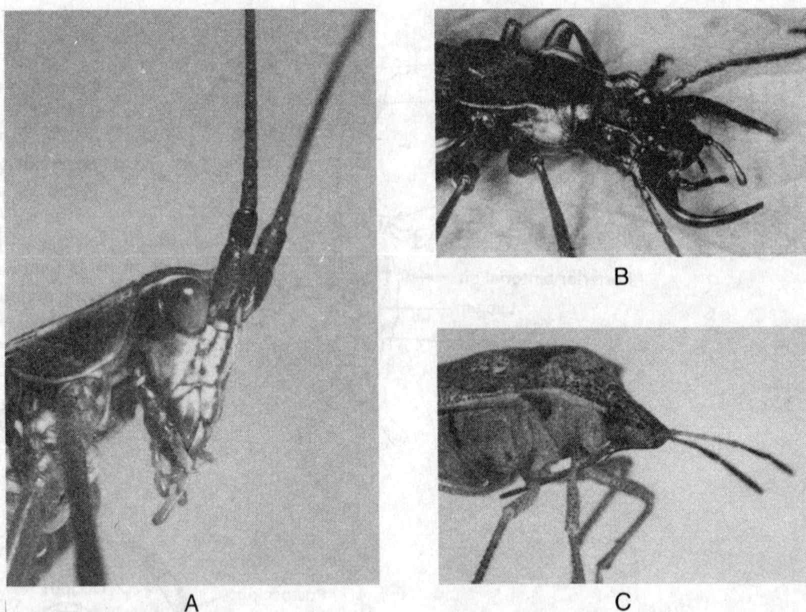


**Figure 2.6** Orientations of the insect head. A. Hypognathus (vertical), grasshopper (Orthoptera). B. Prognathus (horizontal), ground beetle (Coleoptera). C. Opisthognathus (oblique), stink bug (Hemiptera). (Photos by Marlin E. Rice)

from the membranous floor of the cranium, behind the mouth, and bears the opening of the salivary ducts. The interior, fleshy surface of the labrum, endowed with numerous sensory structures, is referred to as the **labrum-epipharynx.**

This arrangement of mouthparts, called chewing mouthparts (Fig. 2.7), enables the insect to bite off pieces of food with the cutting area of the mandible. Both mandibles and maxillae then orient and manipulate the food, and, to a degree, the grinding area of the mandible masticates it. During this activity, saliva is applied, after which the food bolus is guided to the mouth opening with the aid of the maxillae, labium, and hypopharynx.

Several other types of mouthparts occur in insects that differ from the basic chewing type. In addition to chewing, insects have mouthpart types that include piercing-sucking, rasping-sucking, siphoning, sponging, cutting-sponging, and chewing-lapping.

Piercing-sucking mouthparts (Fig. 2.8) are very common among insects. They pierce the epidermis of plants or the skin of animals and suck up sap or blood. In plant feeders like cicadas, spittlebugs (Box 2.1), and aphids (Hemiptera), the piercing-sucking needle is formed from four hairlike stylets fitted closely together. The outer stylets are derived from mandibles, and the inner stylets are derived from maxillae. The maxillae are double-grooved on the inner face and, when held together, form two channels; one is for the passage of saliva into the plant to facilitate food flow and digestion, and the other is for the uptake of plant juices. The labium forms a protective sheath for the stylets. Other plant feeders like leafhoppers, aphids, and scales have the same

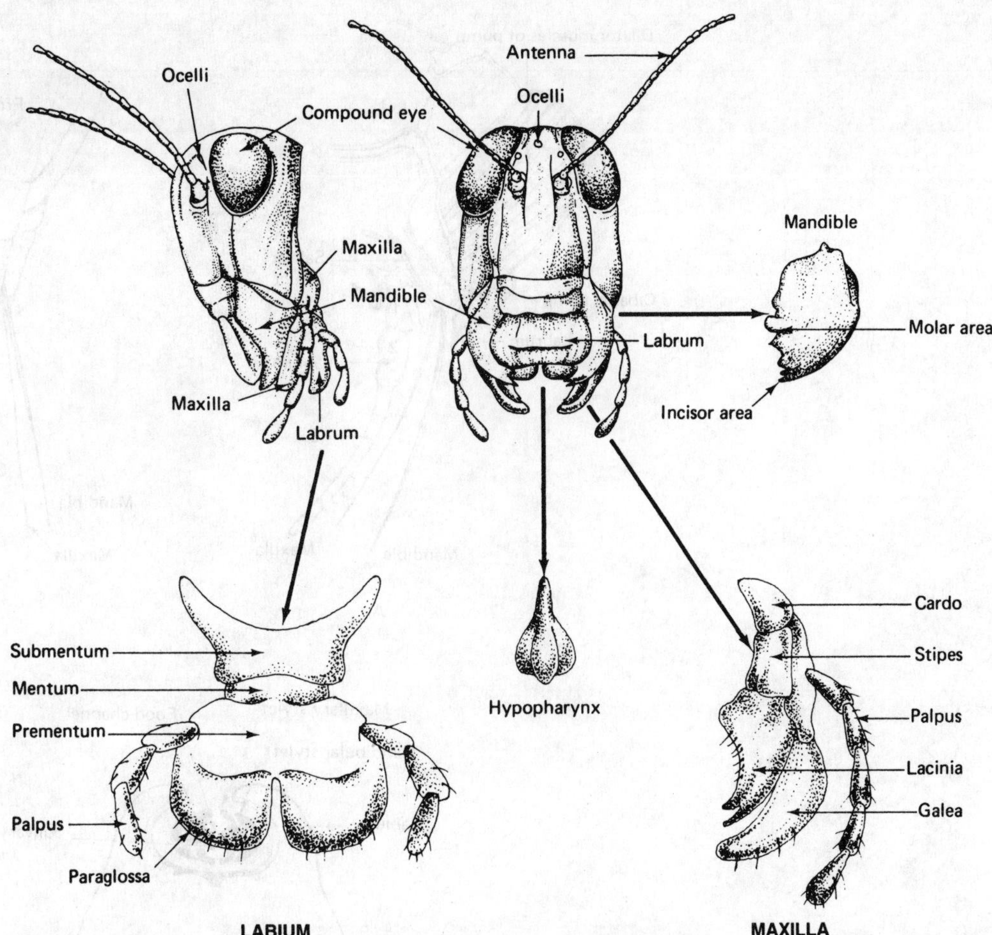


**Figure 2.7** Generalized chewing mouthparts of a grasshopper (Orthoptera). (Reprinted with permission of Macmillan Publishing Company from *Insects in Perspective* by Michael D. Atkins. © 1978 by Michael D. Atkins)

basic mouthpart structure, and many inject viruses and other disease-causing organisms into the plant along with their saliva.

In blood-feeding insects like mosquitoes (Diptera: Culicidae), an elongate labium also forms a protective sheath for six, rather than four, stylets (Fig. 2.9). Here, the mandible- and maxilla-formed stylets are accompanied by an additional pair, modified from the hypopharynx and labrum-epipharynx, which forms a food channel. The back of the food channel is closed by the hypopharynx with its salivary duct. This duct carries saliva with enzymes and anticoagulants that reduce blood clotting in the host and improve the flow of blood into the mosquito. In feeding, the maxillary and mandibular stylets work together as a needle to penetrate the host's skin.

Rasping-sucking mouthparts (Fig. 2.10) are a form of the piercing-sucking type and are found in thrips (Thysanoptera). Thrips' mouthparts have a cone-shaped beak formed from the clypeus (a plate above the labrum in chewing insects), labrum, parts of the maxillae, and the labium. This beak contains the

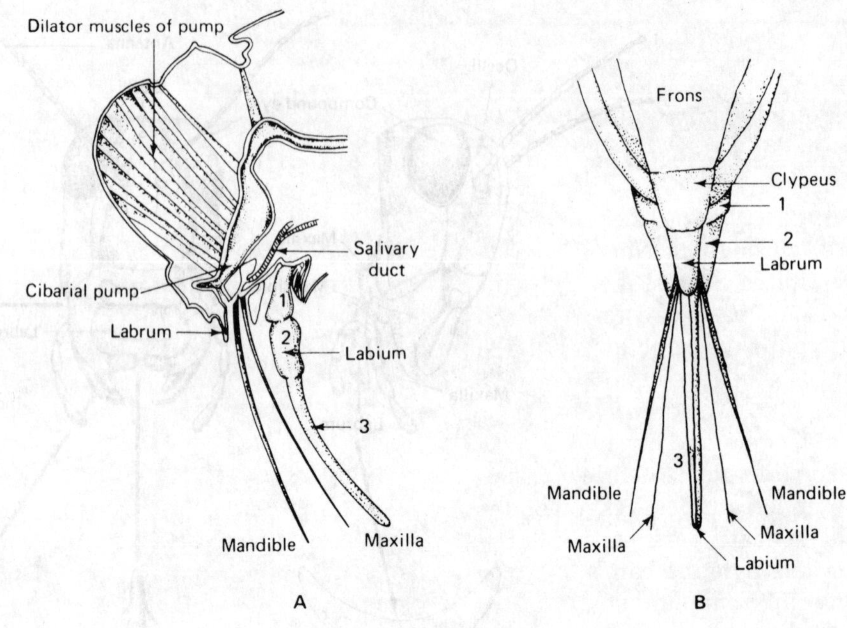


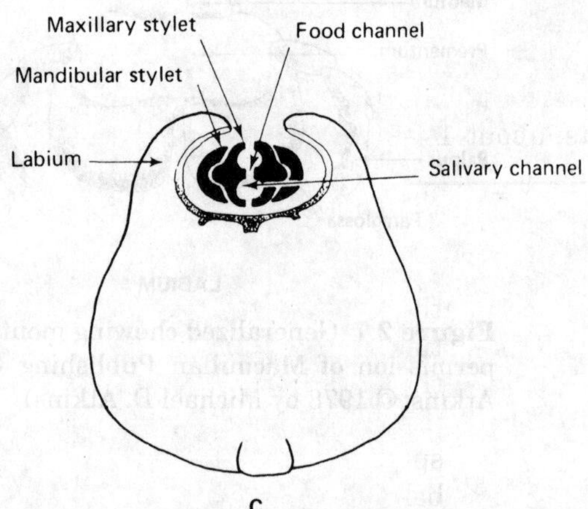


**Figure 2.8** Piercing-sucking mouthparts of a cicada (Hemiptera). A. Lateral cross sectional aspect. B. Frontal aspect. C. Section through mouthparts showing relative position of structures and channels. (Reprinted with permission of Macmillan Publishing Company from *Insects in Perspective* by Michael D. Atkins. © 1978 by Michael D. Atkins)

maxillae, hypopharynx, and the left mandible; together these structures form a stylet. Thrips use the beak to rasp host tissues and take up liquid food through the stylet.

Siphoning mouthparts (Fig. 2.11) are a specialized type for the uptake of flower nectar and other liquids by butterflies and moths (Lepidoptera). Here, a long **proboscis** composed of maxillary elements forms a tube through which food passes. The tube is held in a coiled-spring fashion when not in use.

## BOX 2.1 MEADOW SPITTLEBUG

**SPECIES:** *Philaenus spumarius* (L.) (Hemiptera: Cercopidae)

**DISTRIBUTION:** The meadow spittlebug is distributed throughout North America and Europe. In the United States, it is common from the Atlantic Coast eastward to states that border the Mississippi River, and along the Pacific Coast.

**IMPORTANCE:** Meadow spittlebugs are general feeders and have been recorded on over 400 plant species. They can be important pests of red clover, alfalfa, wheat, oats, and strawberries. Both adults and nymphs can stunt plant growth by sucking sap from plant stems. It has been estimated that an average of one nymph per stem can reduce dry hay yields by 30 pounds per acre.

**APPEARANCE:** Adults are thick-bodied, wedge-shaped insects, about 1/4 inch (6 to 7 mm) long. They are usually mottled brown and gray, varying from light to dark shades of these colors. When disturbed, they jump quickly, The nymphs are yellow to orange or yellow to light green. They are wingless and have red eyes. The nymphs can be found behind leaf sheaths or on the leaves and stems in masses of white froth or spittle. The spittle is produced by the nymphs as they feed on the plant and is a defensive mechanism against predators and desiccation.

**LIFE CYCLE:** The insect overwinters in the egg stage, hatching in early spring in the central United States. After hatch, the nymphs quickly begin sucking sap from the plant, producing the pittle that envelops them. Several nymphs may be found in the same spittle mass. Nymphs mature to adults by June, and adults on a variety of plants until late August. Adults frequently congregate in red clover and alfalfa fields where they lay their eggs through late summer. There is one generation each year.

Sponging mouthparts (Fig. 2.12) occur in house flies (*Musca domestica*) and their close relatives. With this type, the mandibles and maxillae are nonfunctional, and the remaining parts form a proboscis with a fan-shaped sponge at the tip. Liquid food is "mopped up" by the capillary action of this sponge, and, if not liquid, salivary secretions through the mouthparts make it so. Cutting-sponging mouthparts (Fig. 2.13) of biting flies, for example, in horse flies (Tabanidae), are similar to the sponging type but have well-developed mandibles, forming sharp blades, and maxillae, forming probing stylets. Together, these structures cut and tear the skin of mammals, causing blood to flow. Subsequently, the blood is taken up by the sponge and conducted to the fly's gut.

Another mouthpart type used in eating liquid food is the chewing-lapping type (Fig. 2.14) in bees and wasps (Hymenoptera). With this type, the mandibles and labrum are similar to the chewing type and are used in molding wax, manipulating nest materials, and grasping prey. The maxillae and labium form a channeled proboscis through which saliva is discharged and nectar is drawn up.

**Antennae.** A pair of variously shaped appendages, called **antennae,** are found on the head of insects, below or between the eyes. These are movable and contain sensory structures that allow insects to detect odors, vibrations, and other environmental stimuli. Antennal shape is strikingly different among insect species and often between sexes of the same species. For this reason, antennae frequently are used in insect identification and sex determi-

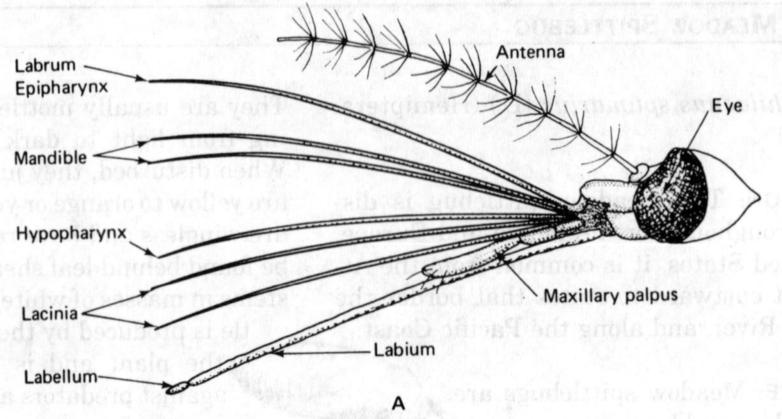


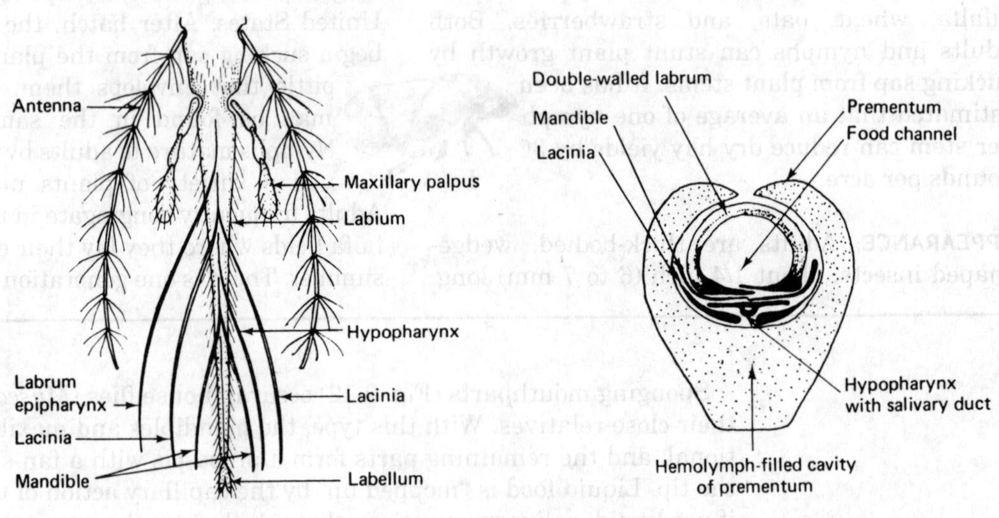


**Figure 2.9** Piercing-sucking mouthparts of a mosquito (Diptera). A. Side aspect. B. Frontal aspect. C. Section through mouthparts showing relative position of structures and food channel. (Reprinted with permission of Macmillan Publishing Company from *Insects in Perspective* by Michael D. Atkins. © 1978 by Michael D. Atkins)

nation. Anatomically, the antenna is divided into three parts, consisting of two basal segments, the **scape** and the **pedicel,** and a tip usually of several subsegments, the **flagellum.** Major antennal types (Fig. 2.15) include **filiform** (threadlike, for example, ground beetle, cockroach), **moniliform** (beadlike, for example, bark beetle), **serrate** (sawlike, for example, cowpea weevil, click beetle), **clavate** (club-shaped, for example, lady beetle), **capitate** (head-shaped, for example, sap beetle), **lamellate** (platelike, for example, June beetle), **pectinate** (comb-shaped, fire-colored beetles), and **plumose** (featherlike, for example, male mosquito, some moths, see Color Plate 1).

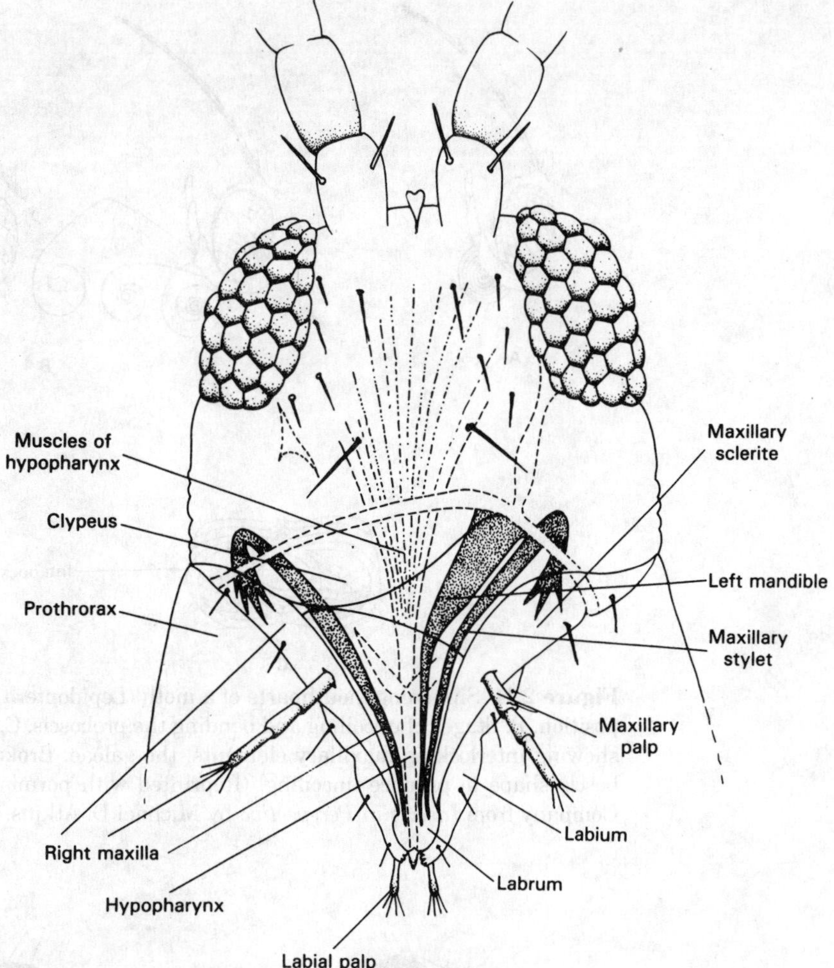


**Figure 2.10** Rasping-sucking mouthparts of a thrips (Thysanoptera). (Redrawn from Metcalf et al., 1962, *Destructive and Useful Insects,* McGraw-Hill Book Company)

**Eyes**. Located on each side of the head of most adult insects are prominent **compound eyes,** which consist of many hexagonal elements. These elements, known as **facets** (Fig. 2.16), number from only a few to as many as 28,000 in dragonflies (Odonata). In addition to compound eyes, most adult insects have several simple eyes, or **ocelli.** The ocelli are located between compound eyes on the front of the head. Most often, they are small and have a single lens. Many immature insects, for example, caterpillars (Lepidoptera), have only simple eyes, and other insects may have no eyes at all, as in some springtails (Collembola).

## The Thorax

The **thorax** is a rigid "box" of three distinct segments. From the head back, these segments are the **prothorax, mesothorax,** and **metathorax.** Each thoracic segment bears a pair of jointed legs, and in most adult insects the mesothorax and metathorax each has a pair of wings.

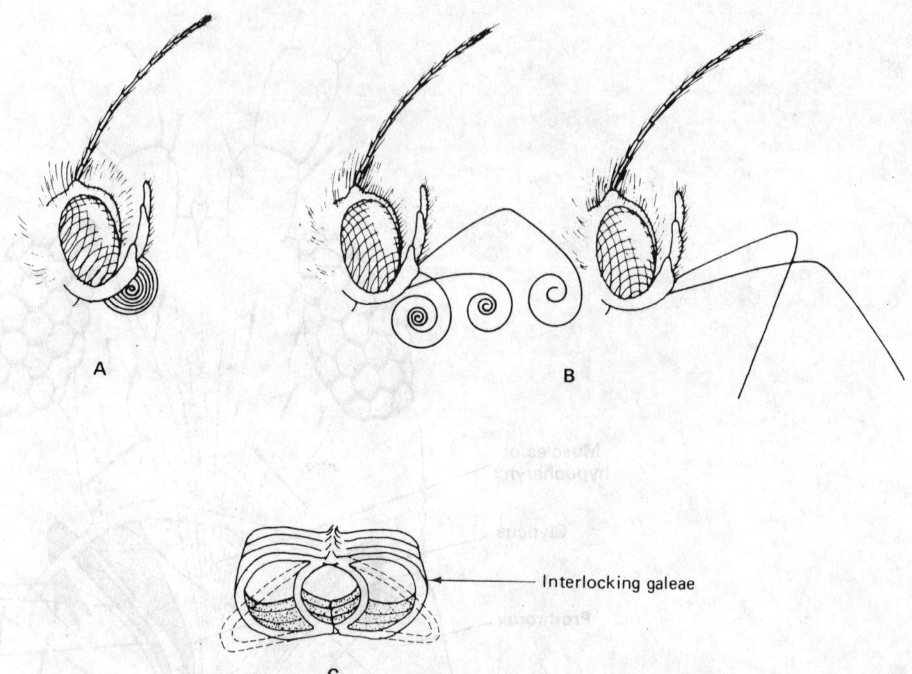


**Figure 2.11** Siphoning mouthparts of a moth (Lepidoptera). A. The proboscis in its coiled position. B. Stages of uncoiling and bending the proboscis. C. Section through the proboscis showing interlocking maxillary elements, the galeae. Broken line shows changes in proboscis shape to produce uncoiling. (Reprinted with permission of Macmillan Publishing Company from *Insects in Perspective* by Michael D. Atkins. © 1978 by Michael D. Atkins)

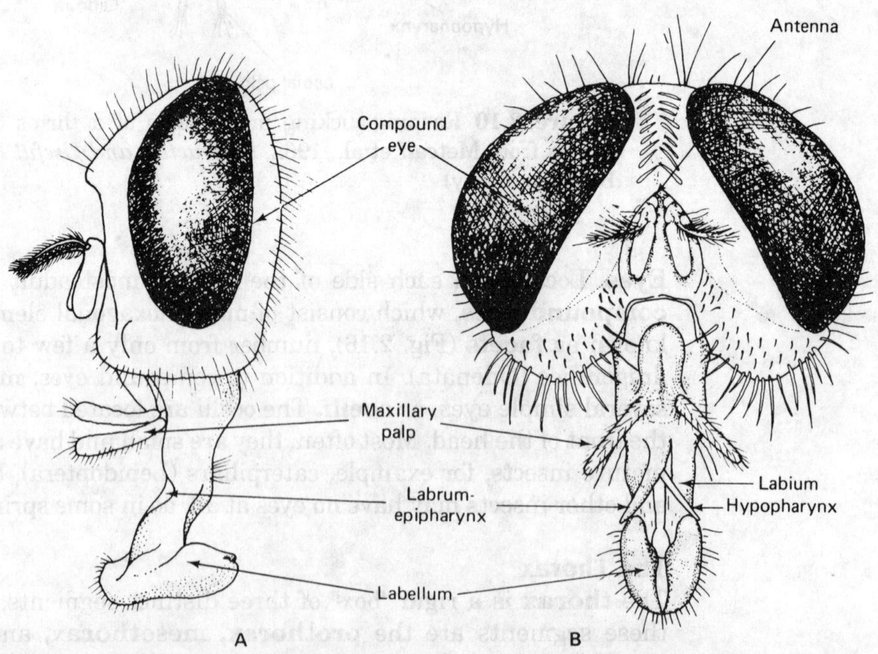


**Figure 2.12** Sponging mouthparts of a house fly (Diptera). A. Side aspect. B. Frontal aspect. (Reprinted with permission of Macmillan Publishing Company from *Insects in Perspective* by Michael D. Atkins. © 1978 by Michael D. Atkins)

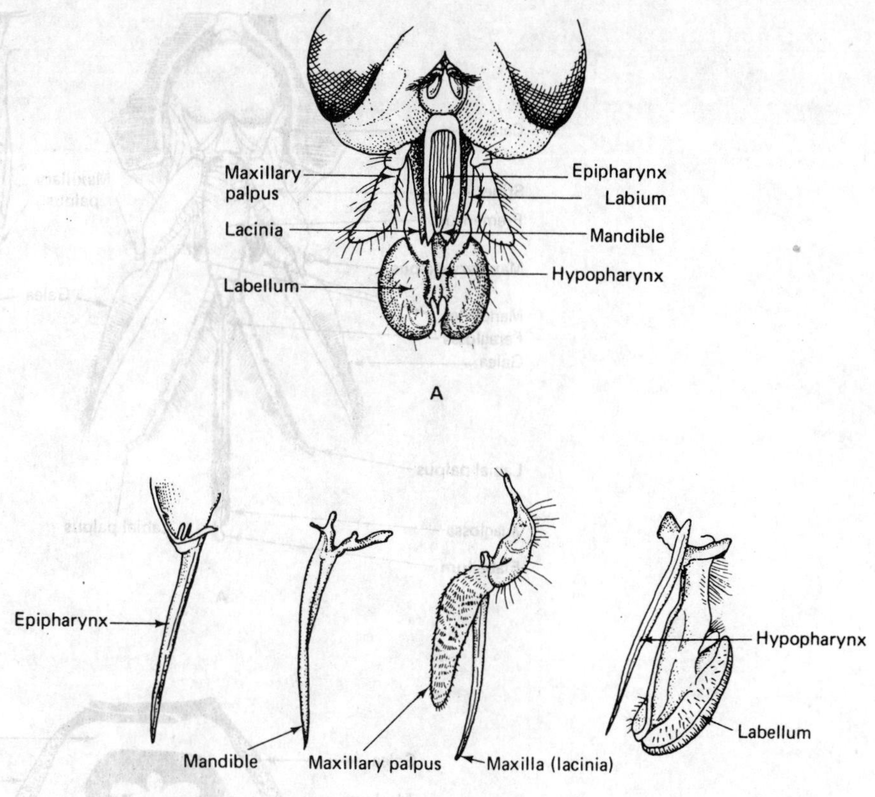


**Figure 2.13** Cutting-sponging mouthparts of a horse fly (Diptera). A. Frontal aspect. B. Side aspect with structures separated from front (*left*) to back. (Reprinted with permission of Macmillan Publishing Company from *Insects in Perspective* by Michael D. Atkins. © 1978 by Michael D. Atkins)

Each thoracic segment is composed of hardened plates that give it rigidity. The upper plate is called the **notum,** and the lower plate is called the **sternum** (Fig. 2.17). Side plates called **pleura** (sing. **pleuron**) also are present and are most well developed on the winged segments.

**Legs.** Insect legs are articulated appendages comprising five segments. From the body outward, these segments are the **coxa, trochanter, femur, tibia,** and **tarsus** (Fig. 2.18). The coxa fits into a cuplike depression in the body, allowing multidirectional movement. The trochanter connects the coxa with the femur, usually the largest leg segment. The tibia is often a long, slender structure with downward-pointing spines that aid in climbing. The tarsus, usually made up of several subsegments called **tarsomeres,** terminates in a **pretarsus.** The pretarsus usually includes a pair of claws. A pad called the **arolium** may be found between the claws for adhesion. If a pad is present at the base of each claw, the pads are called **pulvilli.**

**Wings.** Along with birds and bats, insects are the only other animals with wings. Insect wings show much diversity, varying in shape, texture, and coloration (Fig. 2.19). Most insects have two pairs, except flies and a few other

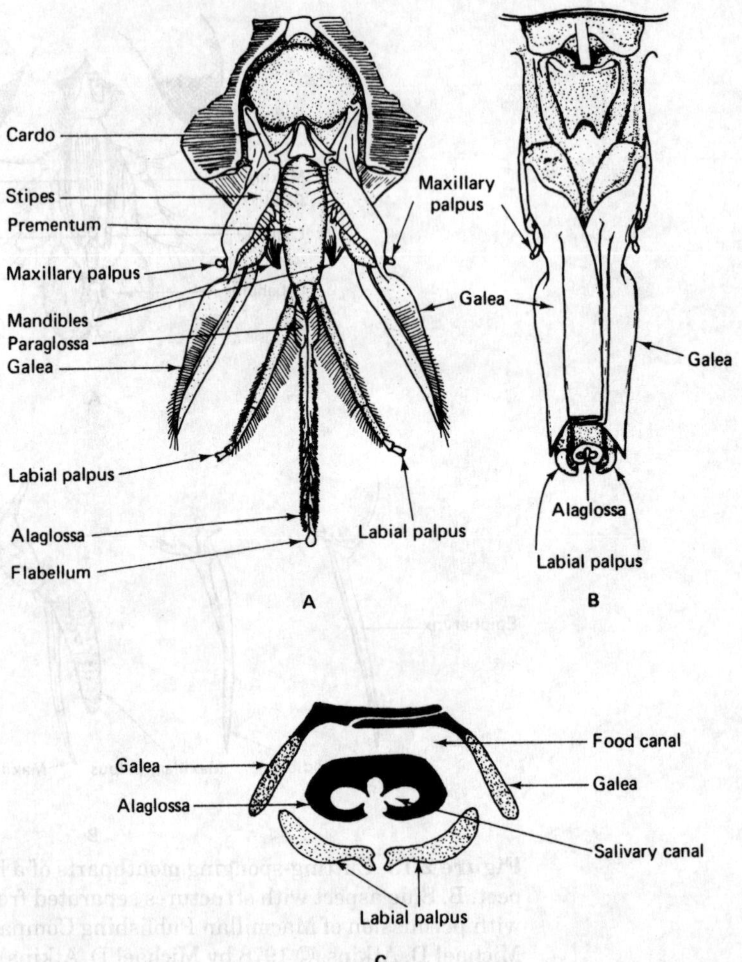


**Figure 2.14** Chewing-lapping mouthparts of a honey bee (Hymenoptera). A. Rear aspect with structures spread apart. B. Frontal aspect with mouthparts folded together. C. Section through mouthparts showing relative position of structures and canals. (Reprinted with permission of Macmillan Publishing Company from *Insects in Perspective* by Michael D. Atkins. © 1978 by Michael D. Atkins)

groups, which have only one pair. In flies, the thorax has a knobbed guidance organ called a **haltere.** In beetles (Coleoptera), the front wings form a tough protective cover, the **elytra.** Yet other insects, such as thrips, have fringed wings, and the wings of butterflies, moths, and mosquitoes are covered with scales.

However, the most common wing is a membranous one with a system of veins running throughout. These veins give strength and rigidity to the wing and are used widely in the classification and identification of insects. A system of naming veins, devised for a hypothetical wing, serves as a sort of road map in analyzing and identifying certain insect groups.

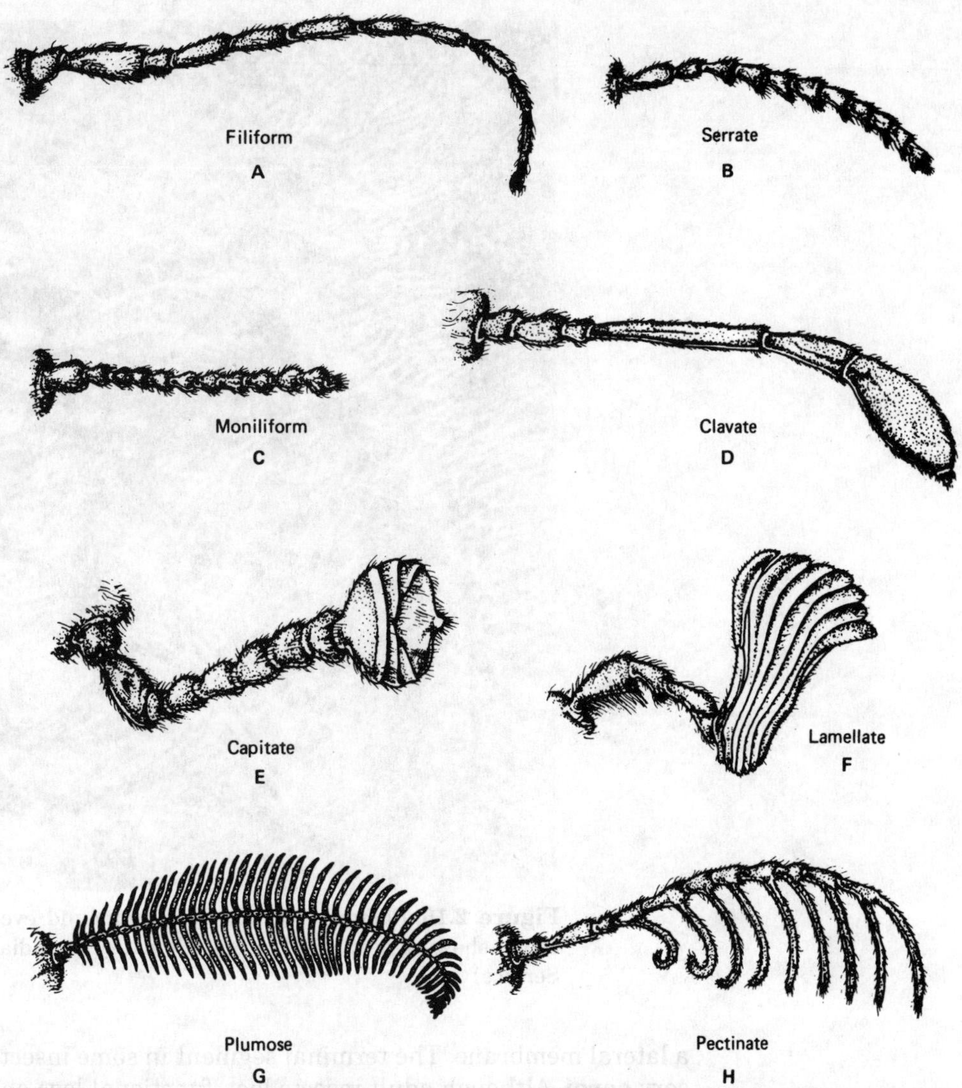


**Figure 2.15** Major antennal types found in insects. (Reprinted with permission of Macmillan Publishing Company from *Insects in Perspective* by Michael D. Atkins. © 1978 by Michael D. Atkins)

Some insects have no wings. Aphids have nonwinged forms only at certain times of the year. Except for mayflies (Ephemeroptera), all immature insects lack functional wings.

### The Abdomen

The insect abdomen usually consists of six to ten segments and terminates in the **paraproct,** where the anus opens. This paraproct may form a lobelike **epiproct** above the anus and a pair of lateral paraprocts around it (Fig. 2.20).

As with the thorax, each abdominal segment has a **tergum** (usually called a notum in the thorax) and a **sternum,** and these are connected on each side by

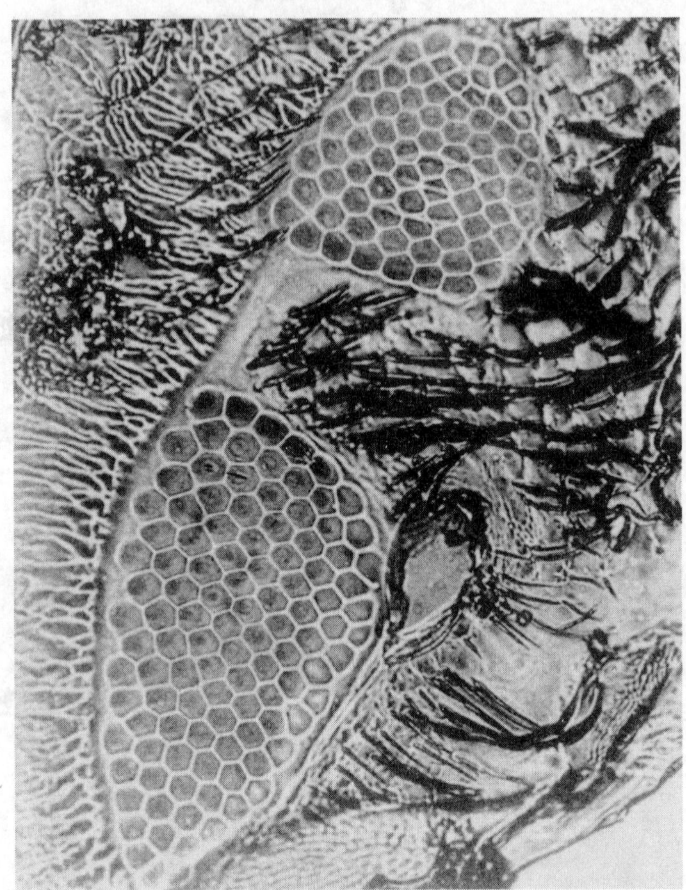

**Figure 2.16** Photomicrograph of a compound eye of a bark beetle showing hexagonal facets. (Courtesy Canadian Forestry Service)

a lateral membrane. The terminal segment in some insects bears a pair of sensory **cerci.** Although adult insects lack functional legs on the abdomen, many immatures such as caterpillars have fleshy prolegs that aid in locomotion. Also, some aquatic immatures such as mayflies have respiratory gills associated with the abdominal segments.

Paired reproductive structures, arising from the primitive eighth and ninth abdominal segments, form an egg-laying device, the **ovipositor,** in most females (Fig. 2.21). Most males have a copulatory organ, the **aedeagus,** developed from a pair of appendages of the primitive ninth segment and sometimes parts of the tenth (Fig. 2.22). Because of their tremendous diversity and distinctiveness, copulatory organs are often used to identify insect species.

## MAINTENANCE AND LOCOMOTION

As with other forms of life, the insect body is a cosmos of interdependent and interacting systems. Basic to the organism is the insect cell. Collections of cells make up tissues, which make up organs, and organs in turn form organ systems. For energy, insect cells require food, water, oxygen, and minerals, which

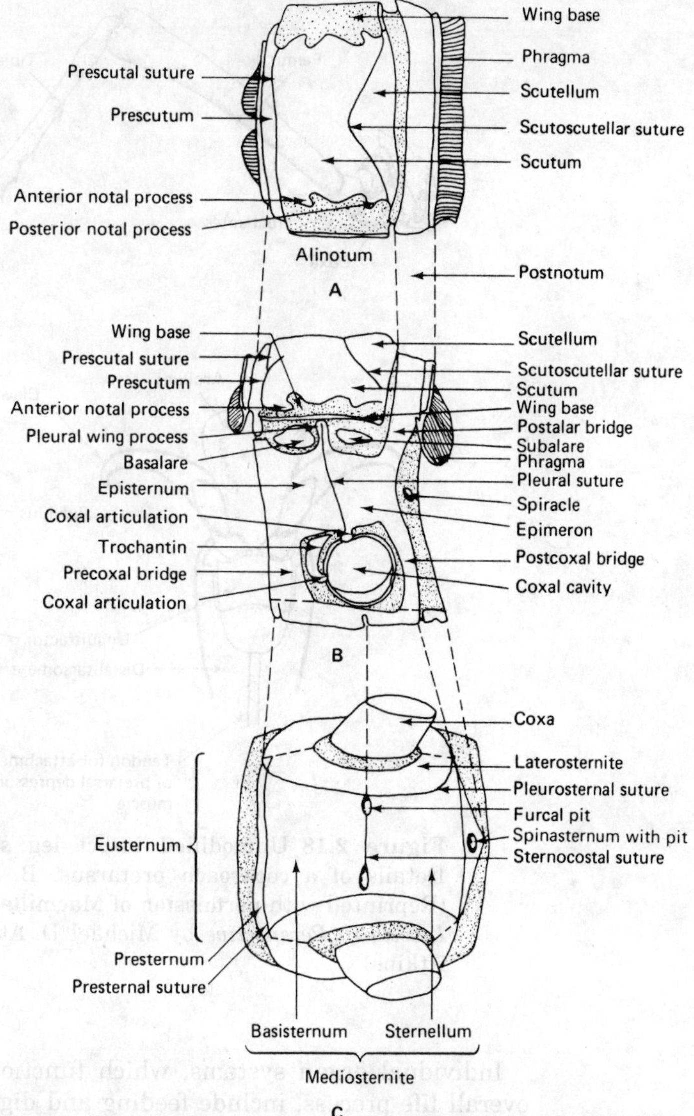


**Figure 2.17** Structural details of a generalized mesothoracic segment. A. Top (dorsal) aspect. B. Side (lateral) aspect. C. Bottom (ventral) aspect. Major plates or sclerites are separated by membranous areas. Sclerites are subdivided into areas by lines and sutures, which frequently are used as identifying characteristics. (Reprinted with permission of Macmillan Publishing Company from *Insects in Perspective* by Michael D. Atkins. © 1978 by Michael D. Atkins)

are obtained from eating, drinking, and respiratory activities. Organic molecules and oxygen are then disseminated within the insect body, where metabolism occurs. The by-products of metabolism are excreted and wastes disposed of, and appropriate salt and water balances are maintained.

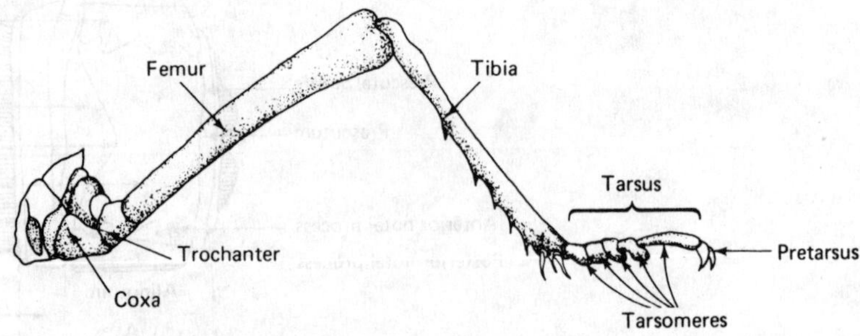


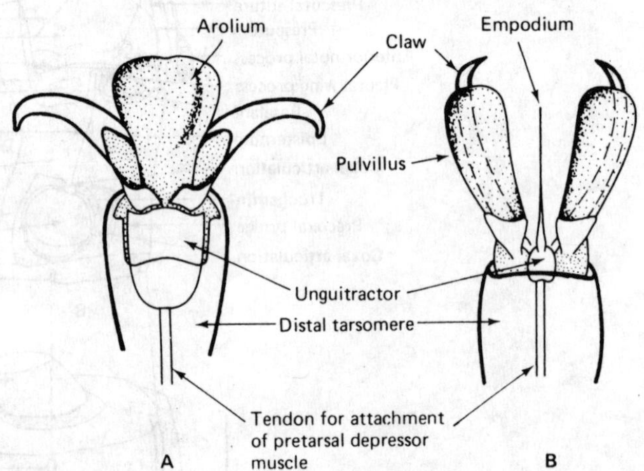


**Figure 2.18** Unmodified insect leg showing segments (top). A. Details of a cockroach pretarsus. B. Details of a fly pretarsus. (Reprinted with permission of Macmillan Publishing Company from *Insects in Perspective* by Michael D. Atkins. © 1978 by Michael D. Atkins)

Individual organ systems, which function simultaneously to achieve the overall life process, include feeding and digestion, respiration, blood circulation, waste excretion, musculature, and nerves and sensory receptors. Although we discuss the details of these systems separately, we should not lose the perspective that they operate interactively and concurrently.

## Feeding and Digestion

**Feeding.** As a class, insects feed on and digest a wide variety of foods. Part of their success as a group derives from their diverse menu. According to the type of food they eat, insects can be described as **phytophagous, zoophagous,** or **saprophagous.**

Phytophagous insects feed on plants; as mentioned earlier, many insect species fall into this category. They may feed on any part of the plant by removing tissues of leaves, stems, roots, and reproductive structures. Insects such as butterflies that feed on nectar or other plant products also are considered phytophagous, along with those that puncture the plant to feed on sap, such as leafhoppers (Hemiptera).

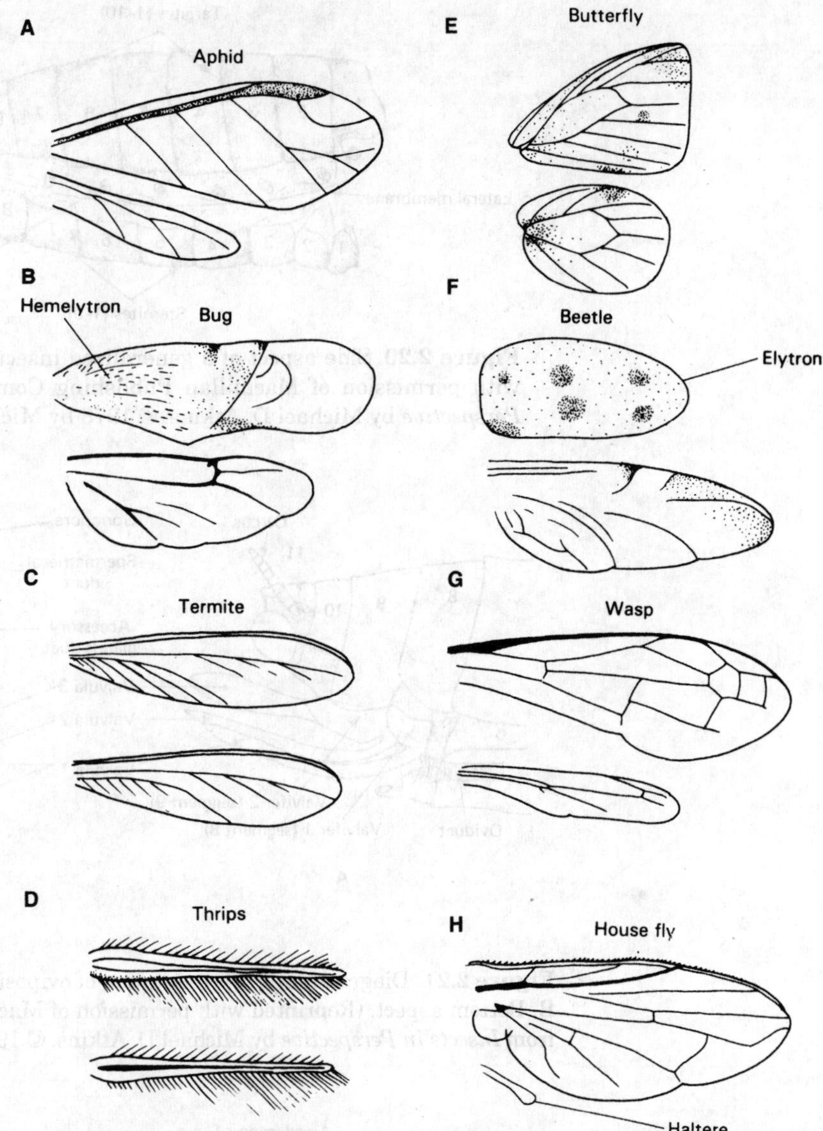


**Figure 2.19** Major wing types found in insects showing front and hind wing. A. Aphid (Hemiptera). B. Bug (Hemiptera). C. Termite (Isoptera). D. Thrips (Thysanoptera). E. Butterfly (Lepidoptera). F. Beetle (Coleoptera). G. Wasp (Hymenoptera). H. House fly (Diptera), showing haltere. (Redrawn from Fenemore, 1982, *Plant Pests and Their Control,* Butterworths Publishing Company)

Zoophagous insects feed on other animals, including almost all vertebrates and invertebrates. Those feeding on vertebrates are mostly parasites like fleas (Siphonaptera) or mosquitoes (Diptera). A few, such as predaceous diving beetles (Coleoptera), kill and eat small fish. Most zoophagous insects, however, kill and eat or parasitize other invertebrates, including insects.

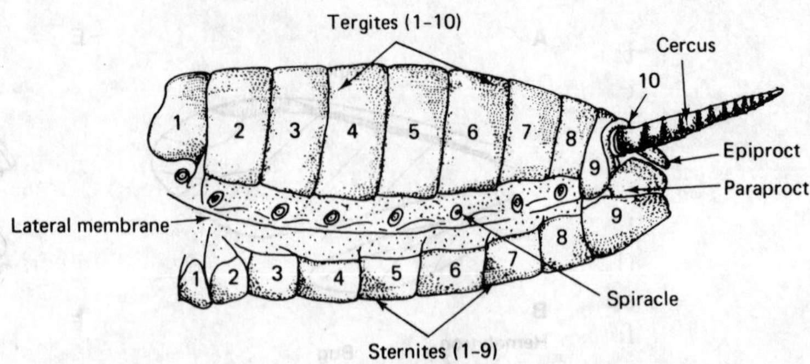


**Figure 2.20** Side aspect of a generalized insect abdomen. (Reprinted with permission of Macmillan Publishing Company from *Insects in Perspective* by Michael D. Atkins. © 1978 by Michael D. Atkins)

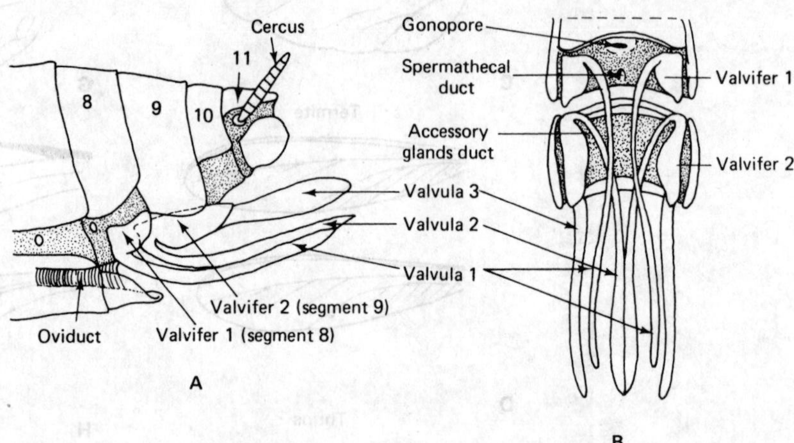


**Figure 2.21** Diagrammatic representation of ovipositor structure. A. Side aspect. B. Bottom aspect. (Reprinted with permission of Macmillan Publishing Company from *Insects in Perspective* by Michael D. Atkins. © 1978 by Michael D. Atkins)

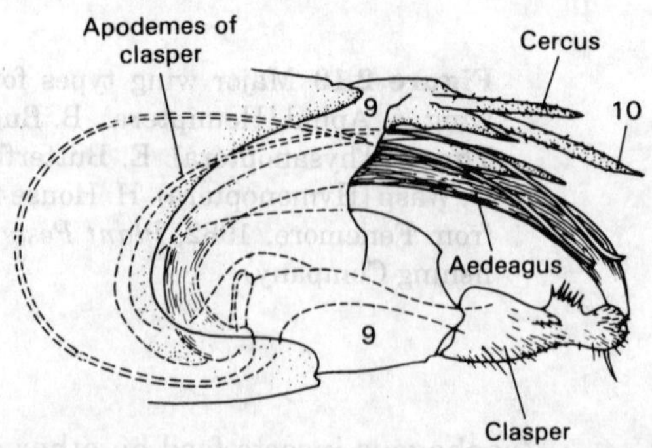


**Figure 2.22** Tip of abdomen of a male caddisfly (Trichoptera) showing side aspect of the aedeagus. Numbers identify abdominal segments. (Courtesy Illinois Natural History Survey)

Saprophagous insects feed on nonliving organic matter. They play an important role in cycling nutrients in the environment. Saprophagous insects include general scavengers (cockroaches, Orthoptera), dung feeders (dung beetles, Coleoptera), dead plant feeders (termites, Isoptera), carrion feeders (carrion beetles, Coleoptera), and humus feeders (springtails, Collembola).

Although these categories are a reasonable representation of insect food habits, some insects may cross categories. For instance, some stink bugs (Hemiptera) that normally eat other insects may switch to feeding on plant tissue if prey is not found. Such insects are considered **omnivorous** in their feeding habits.

**The digestive system.** The insect gut, or **alimentary canal,** is the basic structure of the digestive system; essentially, it is a tube that extends from the mouth to the anus. The gut can be differentiated into three parts: foregut (**stomodaeum**), midgut (**mesenteron**), and hindgut (**proctodaeum**) (Fig. 2.23). These divisions are usually separated by valves, the **cardiac valve** in the front and the **pyloric valve** in the rear. Both the foregut and the hindgut are formed from inpushings of the body wall during embryonic development (**ectoderm**), but the midgut arises from a separate embryonic cell layer, the **endoderm.**

The foregut is commonly described as having divisions, some poorly defined. Beginning at the mouth and moving toward the back, these divisions include the **pharynx** (throat), the **esophagus** (gullet), an expanded **crop,** and a valvelike **proventriculus.** In some insects such as cockroaches, the proventriculus may be muscular or gizzardlike and possess teeth for masticating and straining food.

The midgut has no divisions but often has associated with it **gastric caeca,** blind sacs that open into its forward end. Internally, the midgut often has a semipermeable, chitinous **peritrophic membrane** (Fig. 2.24) that forms a protective layer between food material and delicate epithelial cells. The peritrophic membrane is absent in many fluid-feeding insects.

The hindgut varies greatly among insects but is normally divided into a tubular **intestine** and a short, expanded **rectum** connected to the anus.

Other structures associated with the gut include paired salivary glands, with ducts that open into the preoral cavity by way of the hypopharynx, and a number of slender **Malpighian tubules** that join the intestine just behind the pyloric valve. These tubules make up the main excretory organ.

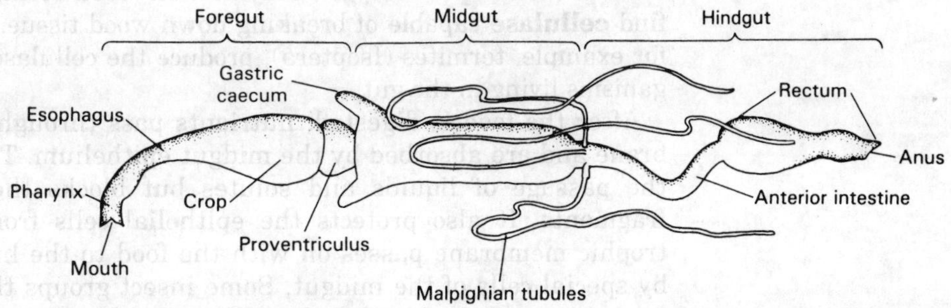


**Figure 2.23** Diagrammatic representation of the gut or alimentary canal of insects. (Reprinted by permission of John Wiley & Sons, Inc., from *A Textbook of Entomology* by H. H. Ross, C. A. Ross, and J. R. Ross. © 1987)

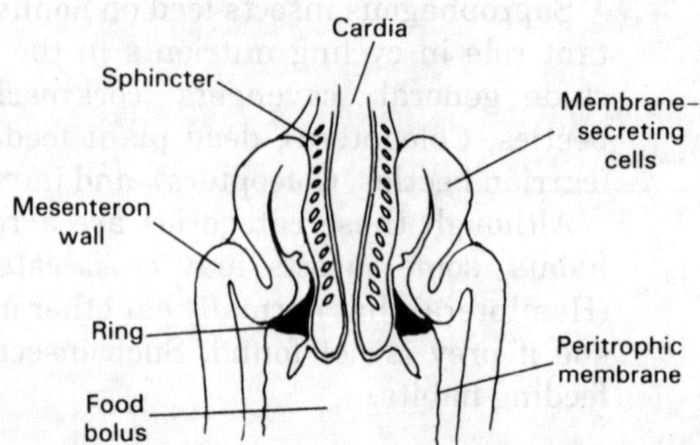


**Figure 2.24** Portion of the midgut of an earwig (Dermaptera), showing the peritrophic membrane. (Redrawn from Wigglesworth, 1953, *The Principles of Insect Physiology,* Methuen and Company)

**Digestion.** Limited digestion of food begins when it comes in contact with enzymes present in the saliva. Some of this digestion may occur externally before food is ingested and when saliva is injected into host tissues (for example, leafhoppers, Hemiptera) or exuded over the food (for example, house flies, Diptera). In extreme cases, much of the digestion occurs externally, as with some diving beetles (Coleoptera), which regurgitate digestive juices from the gut and inject them with the mandibles into prey.

Usually, slightly digested food is passed along to the crop, where it can be stored or moved to the proventriculus. In insects with a well-developed proventriculus, food is ground or strained, thereby increasing surface area for action by digestive enzymes. When food reaches the midgut, it comes into contact with the major digestive enzymes. Much enzyme activity occurs within and in the area of the gastric caeca.

Insects possess most of the major enzymes found in other animals, but all may not be found in any one species. Therefore, we find an array of proteases, lipases, and carbohydrases, each breaking down a basic food group. Insects that feed on a variety of food groups have such enzymes as **amylase, maltase, lipase, invertase,** and **exo-** and **endopeptidases.** Blood-feeding species have primarily **proteolytic** enzymes. In wood-boring insects we may even find **cellulase** capable of breaking down wood tissue. Many of these insects, for example, termites (Isoptera), produce the cellulase by symbiotic microorganisms living in the gut.

After the food is digested, nutrients pass through the peritrophic membrane and are absorbed by the midgut epithelium. This membrane permits the passage of liquids and solutes but blocks the movement of larger fragments. It also protects the epithelial cells from abrasion. The peritrophic membrane passes on with the food to the hindgut and is replaced by special cells of the midgut. Some insect groups that feed only on liquid food, for example, blood-sucking flies and butterflies, lack the peritrophic membrane.

In the hindgut, undigested wastes are moved along and eliminated through the anus. Here, the rectum plays a particularly important role by resorbing water from fecal pellets and thereby helping to maintain the balance of body water and salts. Some aquatic insects such as dragonfly naiads (Odonata) also have **tracheal gills** in the rectum that aid in respiration (Fig. 2.25).

**Nutrition.** Insects have similar nutritional requirements to other animals. **Carbohydrates** are necessary for energy and are usually supplied as glucose or sucrose. **Amino acids** are necessary for protein synthesis and the development of new tissue. Fats and oils are not generally considered essential in insect nutrition. The **vitamins** A, C, D, E, and those of the B complex are used in various metabolic functions, but all are not required in all insects. Insects do not synthesize sterols; thus, they require supplements of **cholesterol** in their diet. Water also is required along with a large number of minerals, including phosphorus, potassium, iron, zinc, calcium, copper, and cobalt.

## Excretion

When nutrients are taken up and metabolism occurs, a number of waste products are produced that are toxic to insect cells and tissues. These are eliminated from the insect body through a process called excretion. Although there are others, the main waste materials usually considered in excretion are nitrogenous wastes, which accumulate from the degradation of proteins and nucleic acids.

The system that serves in the excretion of nitrogenous wastes also plays an important role in the maintenance and regulation of salts and water balance in body fluids. This regulation is crucial in allowing the insect to perform its life processes and survive in dry environments.

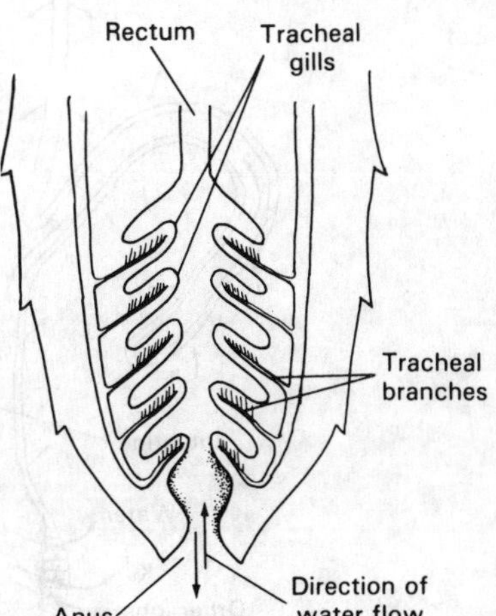


**Figure 2.25** Diagrammatic representation of the tracheal gills in an immature dragonfly (Odonata). (Redrawn from Wigglesworth, 1953, *The Principles of Insect Physiology,* Methuen and Company)

**The excretory system.** The most important organs of the excretory system are the Malpighian tubules and the rectum (Fig. 2.26). The Malpighian tubules vary in number from 2 in some scale insects (Hemiptera) to 200 or more in honey bees (Hymenoptera). In most instances these tubules are closed at their tips and float freely in the body cavity but sometimes are embedded in the outer part of the midgut wall.

**Excretion.** Excretion occurs when blood flows over the Malpighian tubules and substances diffuse or are actively transported into the tubules. Most of the nitrogen taken up by the tubules is in the form of uric acid salts, which are formed in the fat body and passed into the blood before removal. Other substances picked up by the tubules include amino acids, various ions, and water. A potassium-ion gradient in the tubule causes the urine to flow.

The urine then is discharged from the tubules into the hindgut, where it passes to the rectum. Here, most water resorption occurs, producing relatively dry feces in many insects. The lining of the rectum also serves to remove needed inorganic ions and amino acids from the urine and return them to the blood in controlled amounts, thereby maintaining an appropriate ionic balance.

**Other excretory modes.** In some insects, uric acid is not eliminated from the body but is inactivated and stored in special **urate cells** in the **fat body** (a group of loosely organized cells in the body cavity that also function in intermediate metabolism), for example, in immatures of parasitic wasps (Hymenoptera). In other insects, the uric acid is deposited in the cuticle as white crystals, producing patterns of white color.

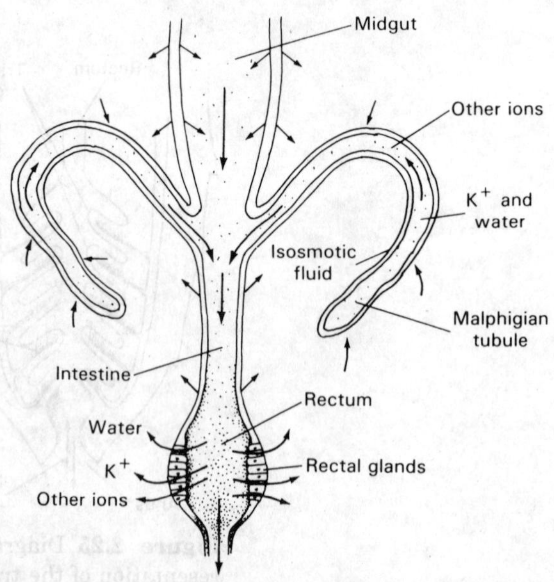


**Figure 2.26** Diagrammatic representation of water and salt circulation in an insect excretory system. (Redrawn from Ross et al., 1982, *A Textbook of Entomology,* Wiley & Sons)

**Circulation of Blood**

In many animals, including vertebrates, blood circulates by movement through a series of vessels (veins, arteries, and capillaries), which is termed a closed system. Insects, on the other hand, have an open system of circulation. Their blood flows through the open body cavity, the **hemocoel,** thereby supplying organs and tissues.

**The circulatory system.** The **dorsal vessel** is the main organ of circulation in insects. This vessel lies at the top of the hemocoel and extends from the back of the abdomen to the head. It comprises two parts, a **heart** at the rear and an **aorta** in the front (Fig. 2.27). The heart is a chambered organ, drawing blood in through openings in each chamber, called **ostia,** and contracting to pump it forward to the aorta. The aorta is usually a simple tube that carries blood forward and dumps it into the head capsule.

The heart of most insects is bordered below by bands of wing-shaped muscles known as **alary muscles.** These connect the heart to lateral portions of the terga. These muscles can form an almost complete partition between the heart and the body cavity. This partition is known as the **dorsal diaphragm** and the chamber occupied by the heart as the **dorsal sinus.**

In some insects, accessory pulsating organs exist in the thorax and lower body cavity. **Thoracic pulsating organs,** found in hawk moths (Lepidoptera), occur in the mesothorax and draw blood through the wings. In many

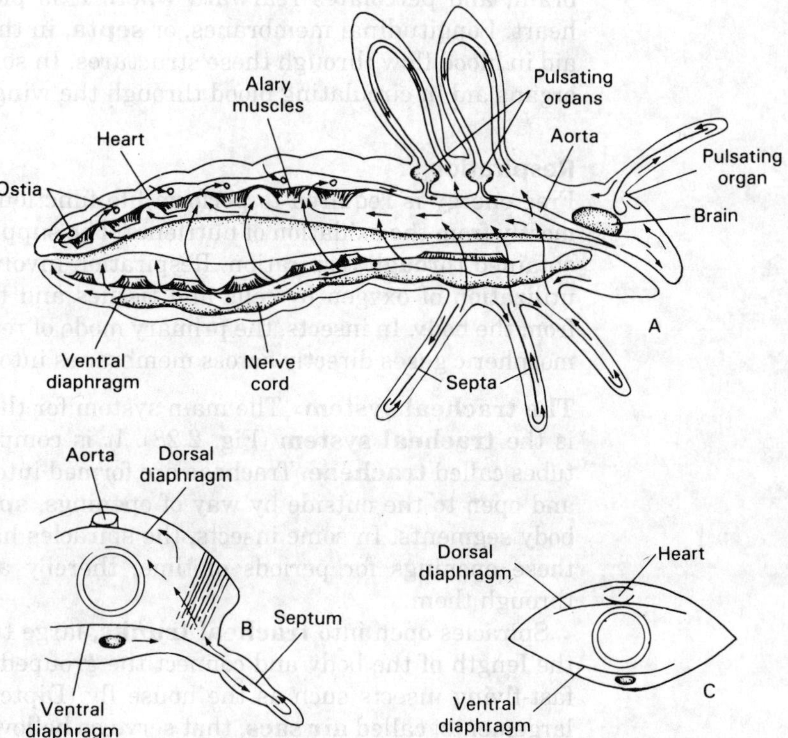


**Figure 2.27** Diagrammatic representation of the circulatory system and blood circulation in insects. (Redrawn from Ross et al., 1982, *A Textbook of Entomology,* Wiley & Sons)

insects there is also a **ventral diaphragm,** similar in form to the dorsal diaphragm, which guides blood flow back and to the sides.

**Blood.** The blood of insects, called **hemolymph,** is a watery fluid that may be colored yellow or green. It comprises liquid plasma, and about 10 percent of its volume is made up of blood cells, or **hemocytes.**

Unlike vertebrates, most insect blood cells contain no hemoglobin, the transporter of oxygen. Therefore, insect blood does not play a role in respiration, except in certain groups such as bloodworms (immatures of *Chironomus* species).

The main function of blood in insects is to transport nutrients, wastes, and hormones. The blood absorbs nutrients from the digestive system and carries them through the body cavity, feeding tissues and organs. Wastes from metabolism at cell sites also are absorbed by the blood and carried to excretory organs, where they are removed. Hormones from glands are transported in the blood to sites of activity. Moreover, blood has a role in the immune system of insects, which have specialized cells that dispose of harmful microorganisms; it also functions in the healing of wounds. Additionally, in intermediate metabolism it stores and converts substances (for example, trehalose stored and converted to glucose) and serves as a pressure source (hydraulic function) to expand body parts during hatching and growth.

**Circulation.** Insect blood taken in by the heart circulates forward because of the peristaltic heart movements. It is carried to the head, flows over the brain, and percolates rearward where it is picked up and recycled by the heart. Longitudinal membranes, or **septa,** in the appendages of some insects aid in blood flow through these structures. In some insects, thoracic pulsating organs aid in circulating blood through the wing veins.

### Respiration

Free energy is required to perform life functions in insects, and it is derived mostly from the oxidation of nutrients. The supply of oxygen for this purpose is obtained through respiration. Respiration involves the intake, transport, and utilization of oxygen by cells and tissues and the removal of carbon dioxide from the body. In insects, the primary mode of respiration is the diffusion of atmospheric gases directly across membranes into the cells.

**The tracheal system.** The main system for the transport of gases in insects is the **tracheal system** (Fig. 2.28). It is composed of a series of branching tubes called **tracheae.** Tracheae are formed into groups in each body segment and open to the outside by way of openings, **spiracles,** on each side of most body segments. In some insects, the spiracles have operable valves that close these openings for periods of time, thereby avoiding excessive water loss through them.

Spiracles open into **tracheal trunks,** large tracheae on each side that run the length of the body and connect the grouped tracheae of each segment. In fast-flying insects such as the house fly (Diptera), tracheal trunks have enlargements, called **air sacs,** that serve as bellows to increase ventilation. Tracheal branches arise from the trunks in each segment and become increasingly fine and branched. Ultimately, the fine tips of the tracheae divide into minute (less than 1 micron in diameter) **tracheoles** (Fig. 2.29). These fine capillary

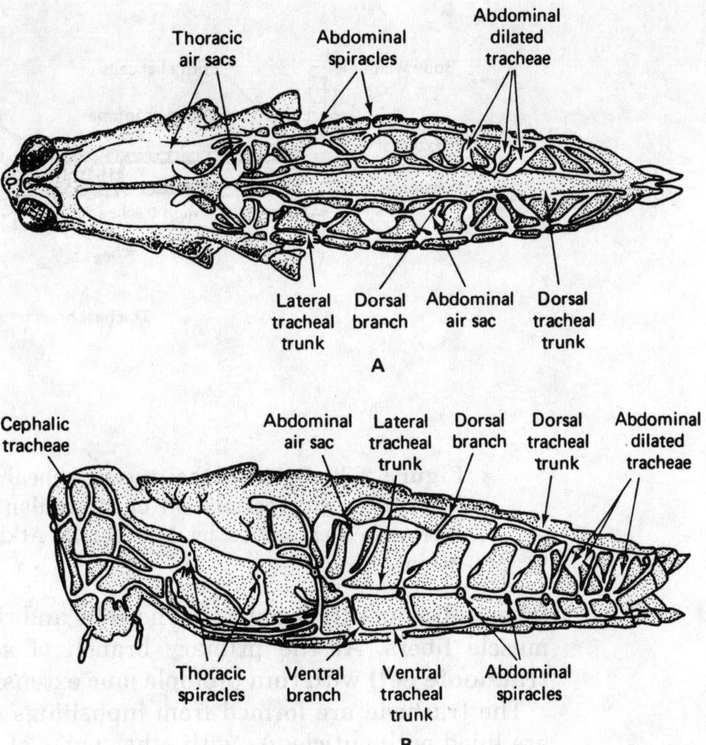


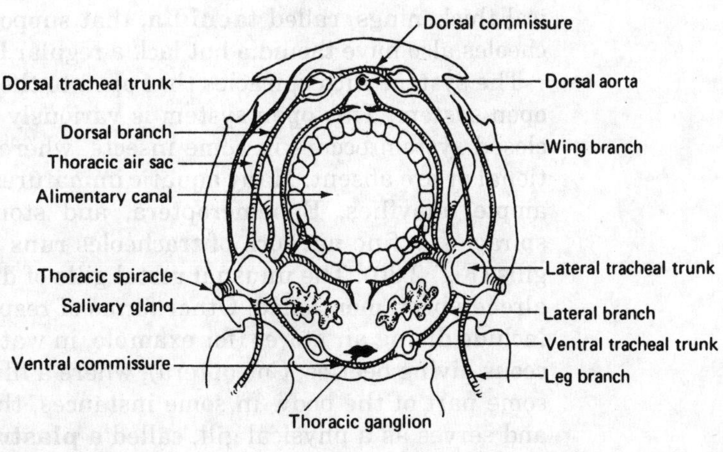


**Figure 2.28** Generalized tracheal system as seen in a grasshopper (Orthoptera). A. Top aspect. B. Side aspect. C. Section through body. (Reprinted with permission of Macmillan Publishing Company from *Insects in Perspective* by Michael D. Atkins. © 1978 by Michael D. Atkins)

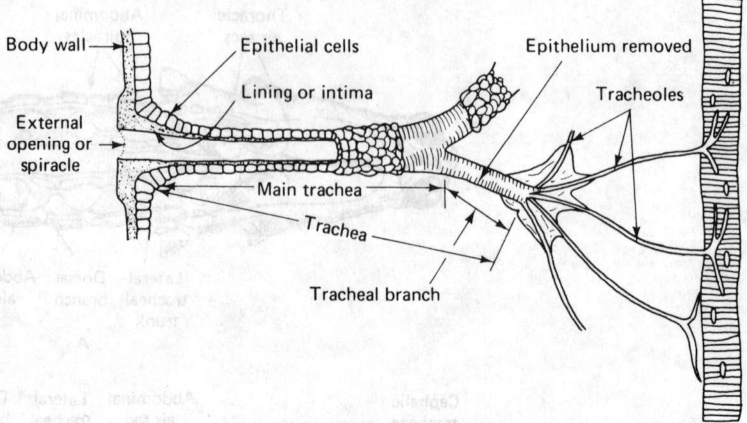


**Figure 2.29** Detail of the insect tracheal system showing tracheoles. (Reprinted with permission of Macmillan Publishing Company from *Insects in Perspective* by Michael D. Atkins. © 1978 by Michael D. Atkins)

tubes subsequently branch around cells and tissues and penetrate deeply into muscle fibers. At the primary branch of several tracheoles is a weblike **tracheole cell** with thin protoplasmic extensions.

The tracheae are formed from inpushings of the body wall and, therefore, are lined with cuticle. As with other parts of the integument, this lining, the **intima,** is secreted by the epithelial cells but has different characteristics of permeability than the body wall. The inner surface of the intima contains helical thickenings, called **taenidia,** that support and shape the tracheae. Tracheoles also have taenidia but lack a regular layer of epithelium.

The system with spiracles that open to the outside of the body is called an open system. This open system is variously modified in different insects. A closed system occurs in some insects, where the spiracles become nonfunctional or are absent. Many aquatic immatures have the closed system (for example, mayflies, Ephemeroptera, and stoneflies, Plecoptera); instead of spiracles, a fine network of tracheoles runs under the skin or into external gills (Fig. 2.30). The unusual rectal gills of dragonfly (Odonata) naiads have already been mentioned. Other forms of respiration found in aquatic insects include diving air stores (for example, in water scavenger beetles and predaceous diving beetles, Coleoptera), where a film or bubble of air is attached to some part of the body. In some instances, the air bubble does not dissipate and serves as a physical gill, called a **plastron,** across which gases diffuse. Other aquatic insects possess air tubes used like a snorkle to siphon air from the surface (for example, immature mosquitoes and water scorpions). Examples of open respiratory systems in aquatic insects are shown in Fig. 2.31.

**Respiratory process.** It is believed that respiration occurs by the diffusion of oxygen and carbon dioxide through the tracheal system, aided by mechanical ventilation of abdominal tracheae and air sacs. Diffusion of oxygen into the system is triggered by a drop in oxygen pressure at the tip of the tracheoles. Similarly, carbon dioxide diffuses out through the tracheal system, and because insect tissues are more permeable to it, significant amounts of carbon dioxide also escape through the body wall.

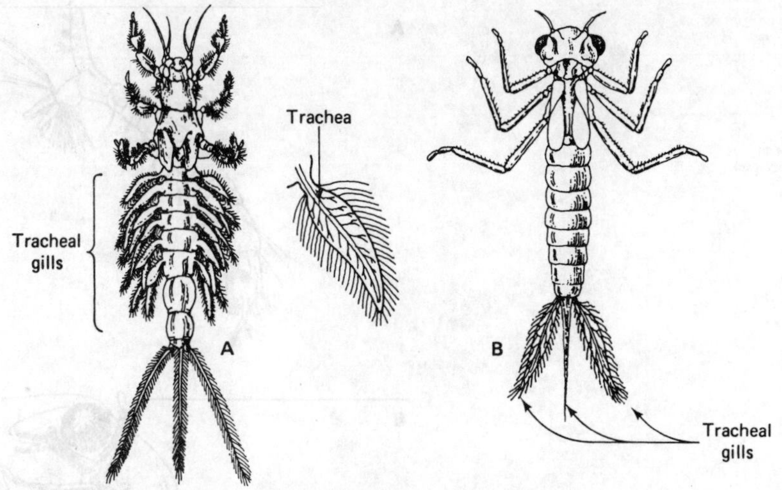

Trachea

Tracheal gills

A

B

Tracheal gills

**Figure 2.30** Examples of closed respiratory systems found in aquatic insects. A. Immature mayfly (Ephemeroptera), showing side (lateral) gills. B. Immature dragonfly (Odonata), showing terminal gills. (Reprinted with permission of Macmillan Publishing Company from *Insects in Perspective* by Michael D. Atkins. © 1978 by Michael D. Atkins)

## Musculature and Locomotion

Movement of the insect body and its appendages is produced by an intricate system of musculature. The coordinated movement of muscles and appendages is critical to important activities such as feeding, walking, jumping, and flying.

**Muscle system**. The musculature of insects occurs as conspicuous layers and bands distributed differently in various parts of the body. For convenience, muscles can be grouped into three categories: **visceral, segmental,** and **appendage** muscles.

Visceral muscles occur in circular, longitudinal, or oblique bands around the digestive tract. Here, they produce peristaltic movements of the gut that move food and waste along its length. Special muscles also are associated with valves of the spiracles and are responsible for their operation. Others cause peristalsis of the heart.

Segmental muscles occur as a series of bands that connect body segments (Fig. 2.32). In the abdomen, tergites are connected to one another by longitudinal bands, as are the sternites below. The tergite of an abdominal segment is connected to the sternite of the same segment by oblique muscles on each side of the body. In the thorax, the predominant muscles are large cordlike bands that move the legs and wings. Other smaller muscles form complicated patterns of attachment.

Movable appendages have muscles either in them or attached to them. Usually, appendages like the legs, which are divided into segments, have attachments for muscles housed in the body as well as for muscles housed directly in the appendage (Fig. 2.33). The body muscles generally move the whole appendage, and the segmental muscles move parts of the appendage. Muscles of such nonsegmented appendages as mandibles have operating muscles housed only in the body proper.

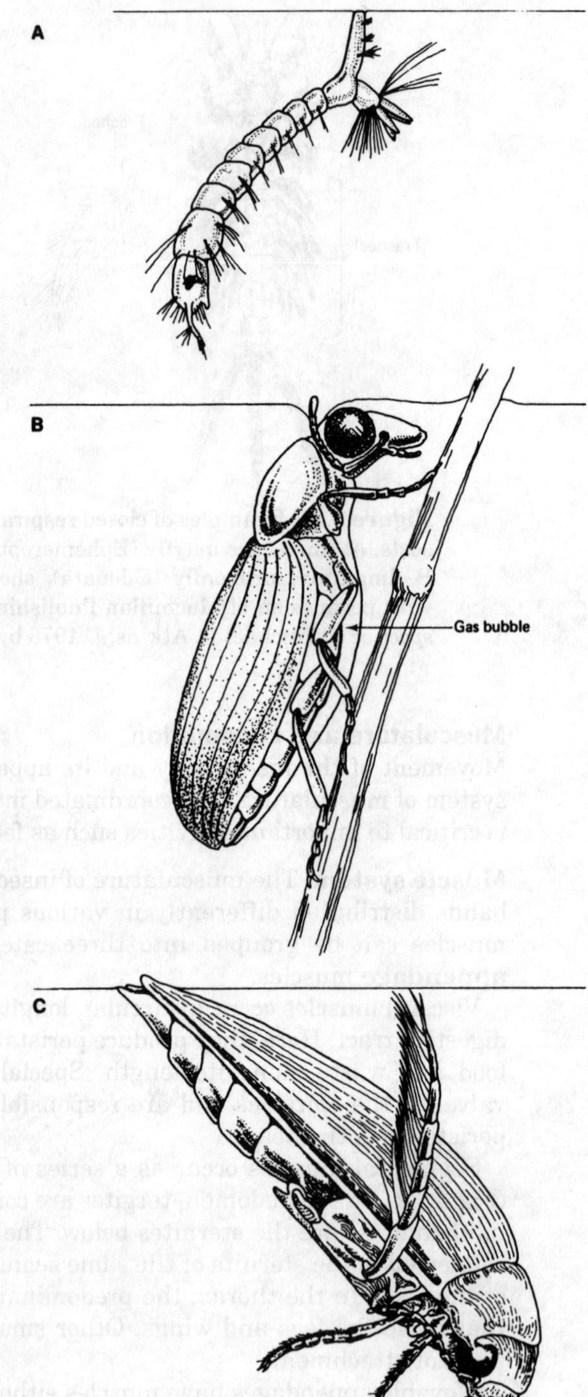


**Figure 2.31** Examples of open respiratory systems in aquatic insects. A. Mosquito larva (Diptera) siphoning air. B. Water scavenger beetle (Coleoptera) replenishing gas bubble by breaking water-surface film with antennae. C. Predaceous diving beetle replenishing air store located beneath front wing covers. (Reprinted with permission of Macmillan Publishing Company from *Insects in Perspective* by Michael D. Atkins. © 1978 by Michael D. Atkins)

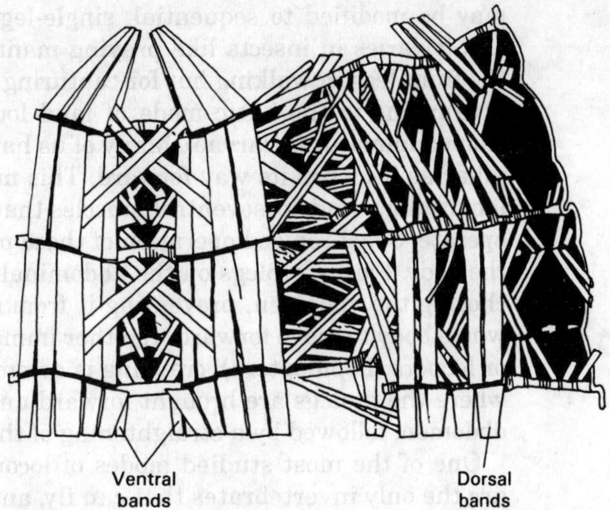


**Figure 2.32** Segmental musculature in mesothorax and metathorax of a caterpillar (Lepidoptera). (Redrawn from Snodgrass, 1935, *Principles of Insect Morphology,* McGraw-Hill Book Company)

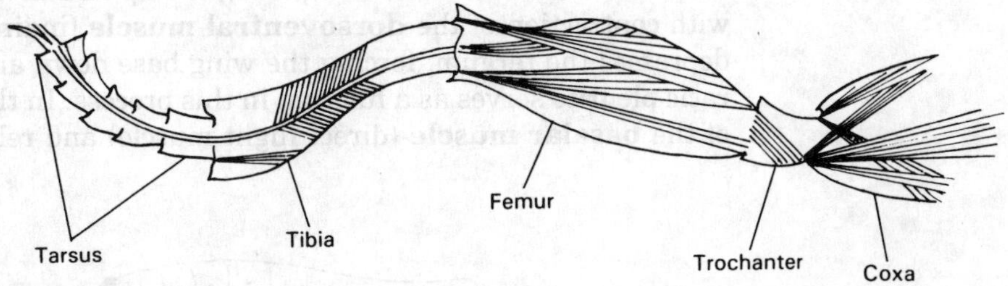


**Figure 2.33** Diagrammatical representation of an insect leg and muscle attachments. (Redrawn from Ross et al., 1982, *A Textbook of Entomology,* Wiley & Sons)

**Muscle function.** All muscles in insects are composed of **striated fibers.** Each of the fibers is made up of many **myofibrils** situated in a matrix of multi nucleated cytoplasm, **sarcoplasm.** Along with the myofibrils are numerous **mitochondria.** These organelles contain enzymes for the oxidation process that releases free energy for muscle reaction.

When stimulated by a nerve cell impulse, insect muscles contract similarly to those of other animals. The period of contraction is followed by a period of recovery. During recovery, the muscle assumes its original elongated shape.

**Locomotion.** The basic mode of locomotion for insects on land is walking. Adult insects, with three pairs of legs, take each step on a tripod; that is, two legs are on the ground on one side and one leg on the other side. The other three legs are raised as the grounded legs push backward. The alternation of pushing tripods results in a slightly zigzag path of forward motion. This gait may be modified to sequential, single-leg movements in very slow walking, and it varies in insects like praying mantids (Orthoptera), which do not use the front legs in walking but for capturing and manipulating prey.

Crawling is the basic mode of land locomotion commonly seen in worm-shaped immatures (larvae). Many of us have seen the rippling body of a caterpillar as it inches its way forward. This movement is accomplished when the longitudinal and dorsoventral muscles that oppose each other in each segment operate. Synchronized operation of these produces a wave of contraction down the body. Fleshy prolegs on the abdominal segments assist this motion by anchoring the abdomen, preventing it from slipping backward and causing the whole body to move forward. In other immatures, such as so-called inchworms or loopers (Lepidoptera), crawling is accomplished in a sort of looping motion, where the prolegs are brought forward and anchored, causing a "loop" in the abdomen, followed by a straightening of the whole body (Fig. 2.34).

One of the most studied modes of locomotion in insects is flying. Insects are the only invertebrates that can fly, and only adults can accomplish flight (with the exception of mayflies, Ephemeroptera, where weak flights may occur just before full sexual maturity). The flight of insects involves the large cordlike muscles of the thorax that attach to tergites and sternites, **indirect flight muscles,** and others that attach to the wing bases, **direct flight muscles.**

In some insects (for example, dragonflies, Odonata, and cockroaches, Orthoptera) and in beetles (Coleoptera), the up-and-down wing-beat cycle begins with contraction of the **dorsoventral muscle** (indirect flight muscle), which depresses the tergum, forcing the wing base down and the wing up. The thoracic pleurite serves as a fulcrum in this process. In the next cycle, contraction of the **basalar muscle** (direct flight muscle) and relaxation of the dorsoven-

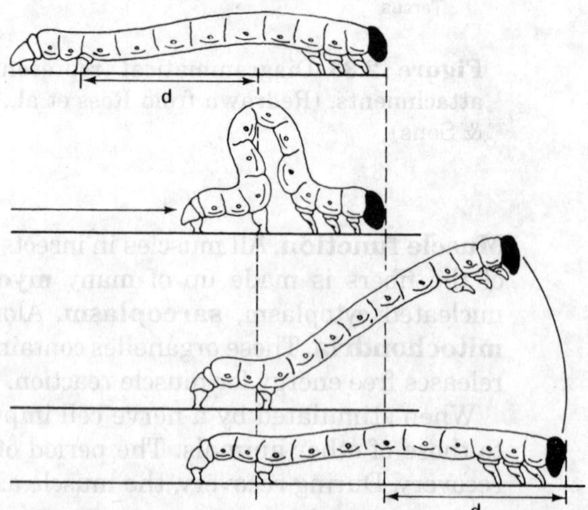


**Figure 2.34** Crawling locomotion used by inchworms or loopers (Lepidoptera), with *d* being distance traveled in each step. (Reprinted with permission of Macmillan Publishing Company from *Insects in Perspective* by Michael D. Atkins. © 1978 by Michael D. Atkins)

tral muscle forces the wing down, and the tergum is raised (Fig. 2.35). In more advanced fliers such as honey bees, up-and-down movement is produced only by indirect muscles (dorsoventral and **dorsal longitudinal**) that cause flexing of the thoracic segment to produce wing beat (Fig. 2.36). In addition to these up-and-down movements, wings display a synchronized rotation that gives more surface on the downstroke than the upstroke and pushes air backward to cause forward movement. This rotation is accomplished by muscles

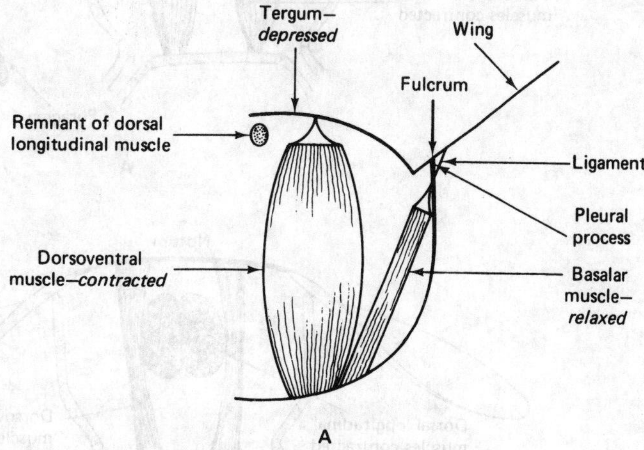


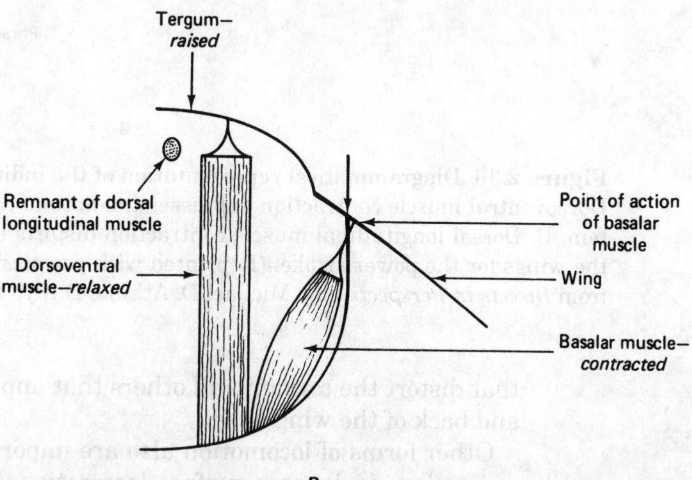


**Figure 2.35** Diagrammatical representation of the direct flight muscle system in insects. A. Contraction of the dorsoventral muscle depresses tergum, which pushes wing bases below the pleural wing process (fulcrum), thereby causing wing upstroke. B. Relaxation of the dorsoventral muscles and contraction of the basal muscle pulls wing down against the fulcrum for the power stroke. (Reprinted with permission of Macmillan Publishing Company from *Insects in Perspective* by Michael D. Atkins. © 1978 by Michael D. Atkins)

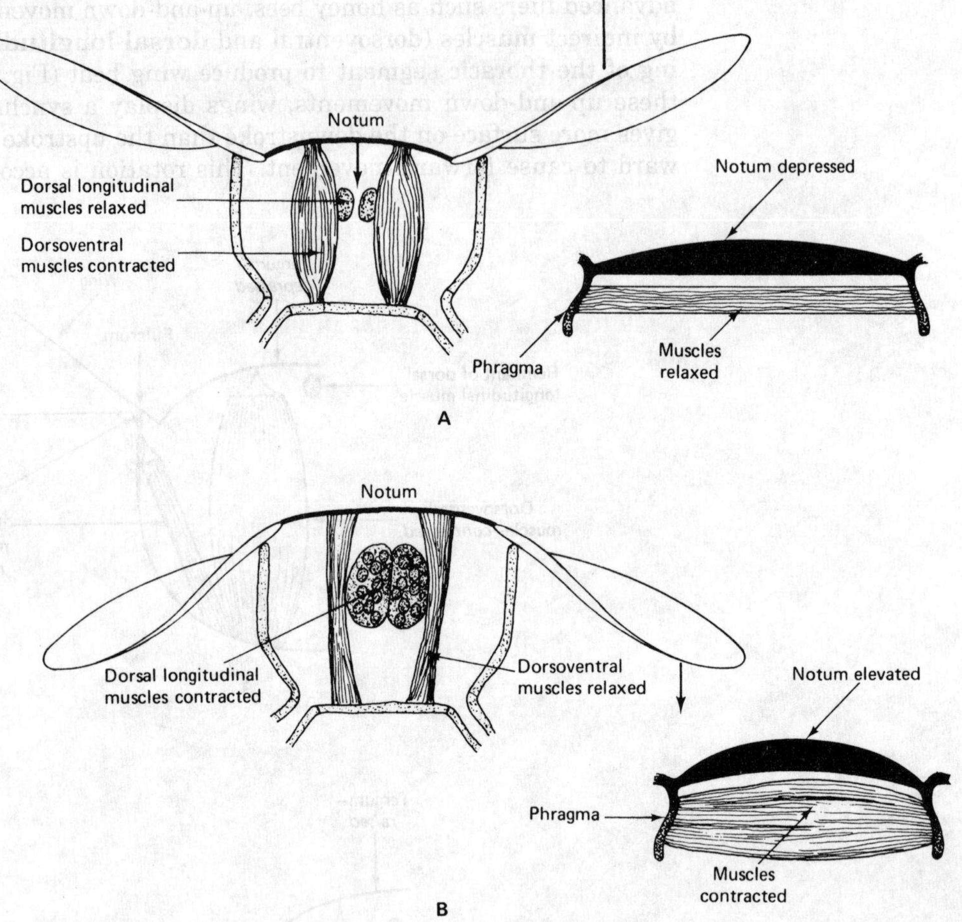


**Figure 2.36** Diagrammatical representation of the indirect flight muscle system in insects. A. Dorsoventral muscle contraction depresses the notum, raising the wings as in the indirect system. B. Dorsal longitudinal muscle contraction distorts the notum in such a way as to depress the wings for the power stroke. (Reprinted with permission of Macmillan Publishing Company from *Insects in Perspective* by Michael D. Atkins. © 1978 by Michael D. Atkins)

that distort the tergum and others that apply pressure alternately to the front and back of the wing base.

Other forms of locomotion also are important to insects, including jumping, swimming, and water surface locomotion. These somewhat more specialized activities will not be discussed here.

## SENSING THE ENVIRONMENT AND INTEGRATING ACTIVITIES

For insects to exist, they must sense the realities of their environment and govern their activities accordingly. They must locate food, find mates, avoid enemies, make nests, and perform internal functions. They can accomplish these tasks only by perceiving conditions and integrating this information to perform the appropriate behavior. Perception is achieved by a number of different sense organs, and behavior results from integration of information and stimulation by the nervous system.

## Sense Organs

The basic sense organs of insects include those involved in sight, smell, taste, touch, and hearing. Entomologists usually group these organs by category, the most basic of which are **photoreceptors, chemoreceptors,** and **mechanoreceptors.**

**Photoreceptors.** Photoreception is the sensing of light, and when images are produced, it is called sight. To a certain extent, all animal cells are light-sensitive, but certain ones in insects are specialized to sense light presence, day length, light intensity, color, and other aspects.

Probably the most complex photoreceptors in insects are those involved in forming images, the eyes. Compound eyes and simple eyes have already been mentioned as prominent organs on the head of many insects. Compound eyes are not found in some wingless insects, and they are also lacking in immatures of higher-order insects (division Endopterygota). Only simple eyes are present in many of these species, and most insects with compound eyes also have simple eyes. Some insects have no eyes and sense light through the cuticle (**dermal photoreception**).

The compound eye is made up of a number of individual sensory units called **ommatidia** (Fig. 2.37). Light is gathered in the ommatidium by the lens or **cornea** that focuses light through a **crystalline cone** to a light-receptor

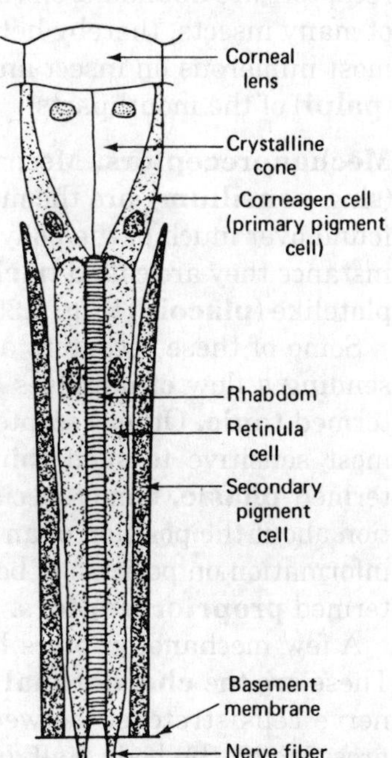


**Figure 2.37** Diagrammatical representation of a compound eye unit, the ommatidium. (Reprinted with permission of Macmillan Publishing Company from *Insects in Perspectives* by Michael D. Atkins. © 1978 by Michael D. Atkins)

apparatus. This apparatus comprises six to eight **retinula cells** that combine to produce a central light sensor, the **rhabdom.** Both the crystalline cone and the light-sensing apparatus are surrounded by pigment cells that isolate, to various degrees, one ommatidium from another. The retinula cells turn light into electrical energy that is in turn carried by nerve fibers to the brain. Images produced by the collection of ommatidia are believed to result in an overall mosaic of the object, with each ommatidium supplying only a small piece of vision.

Simple eyes, or ocelli, are believed to perceive images, color, and movement, as do compound eyes, although they probably produce a less complete mosaic. However, some insects with only ocelli "scan" the environment by moving their head back and forth. The structure of ocelli varies among insects, with some similar to individual ommatidia (as in caterpillars, Lepidoptera) and others having a single cornea that overlays several retinula cells and rhabdoms.

**Chemoreceptors.** The senses of taste (**gustation**) and smell (**olfaction**) are based on the detection of certain molecules by receptor organs that subsequently produce a nerve impulse. The distinction between these anthropomorphic senses is trivial, separated almost solely on the basis of distance from a source. In other words, food is tasted only when in the mouth; when it is not in the mouth (farther away), it is smelled. In either instance, the mechanism of chemoreception is the same.

Chemoreceptors usually occur in the form of short pegs or hairs on various body parts (Fig. 2.38). Taste receptors sense molecules from liquids and are often hairlike and have fine nerve endings exposed to the environment at the hair's tip. Smell receptors may be more peglike and have a greater number of nerve endings at the surface than taste receptors. As might be expected, taste receptors are abundant on the mouthparts, but they are also dense on the tarsi of many insects, thereby helping the insect to detect food. Smell receptors are most numerous on insect antennae but may also be abundant on appendages (**palpi**) of the mouthparts.

**Mechanoreceptors.** Mechanoreceptors, most often in the form of **sensilla** (sing. **sensillum**), are the most numerous sensory structures on insects, being found over much of the body surface. These sensilla may be hairlike, in which instance they are called **trichoid,** or they may be domelike (**campaniform**) or platelike (**placoid**) (Fig. 2.39).

Some of these receptors are sensitive to touch and respond to pressure by sending a flow of impulses to the nervous system. These sensilla have been termed **tonic.** Other receptors respond mostly to the rate of bending and are most sensitive to such things as air or water vibrations. These are often termed **phasic.** Some specialized sensilla also function in relaying information about the position of an insect in its environment, whereas others provide information on position of body parts relative to one another. These have been termed **proprioreceptors.**

A few mechanoreceptors have no external structure associated with them. These are the **chordotonal sensilla** that are composed of a bundle of bipolar nerve cells stretched between two surfaces of the integument. These detect pressure on the body wall and movements of the insect. Specialized groups of chordotonal sensillae make up **Johnston's organ,** a structure on the second antennal segment of adult insects that responds to movement of the antennae and may be involved in hearing. The main hearing organ of insects, however,

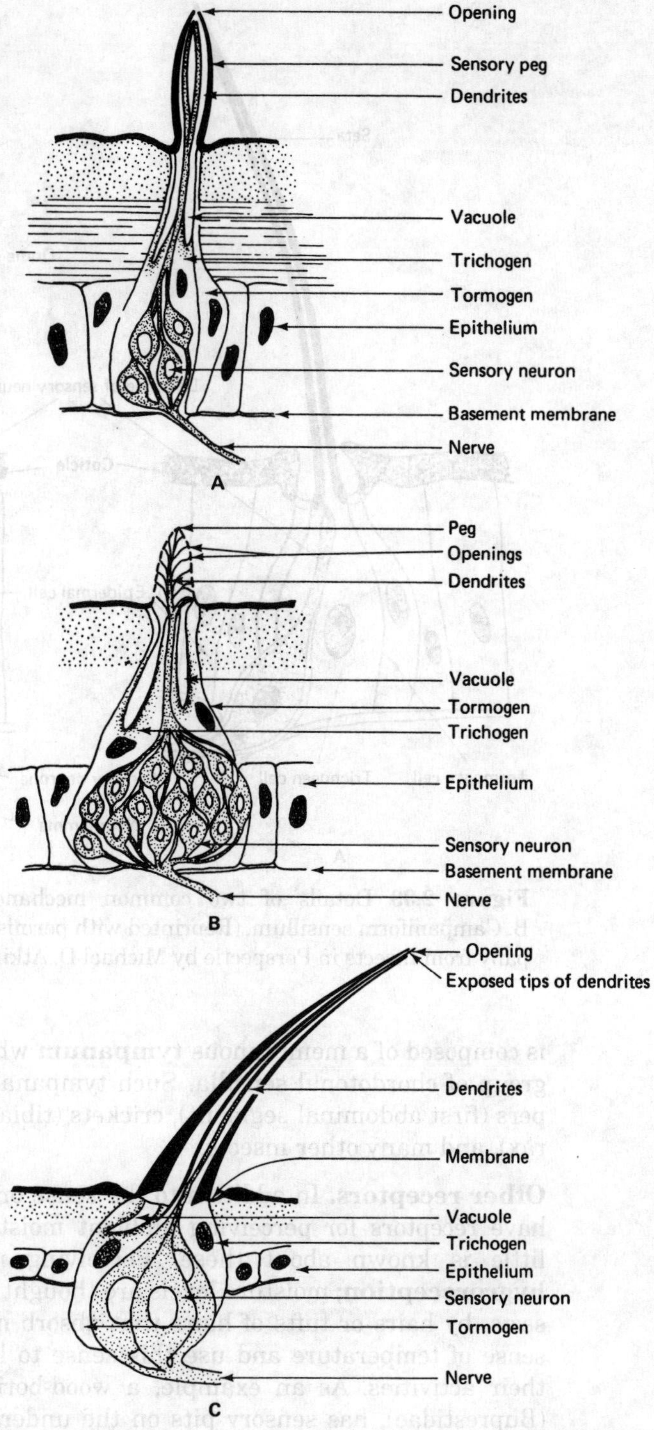


**Figure 2.38** Typical insect chemoreceptors. A. Thick-walled peg that serves in general chemical perception. B. Thin-walled peg that often serves in smell. C. Setal-type receptor frequently involved in taste. (Reprinted with permission of Macmillan Publishing Company from *Insects in Perspective* by Michael D. Atkins. © 1978 by Michael D. Atkins)

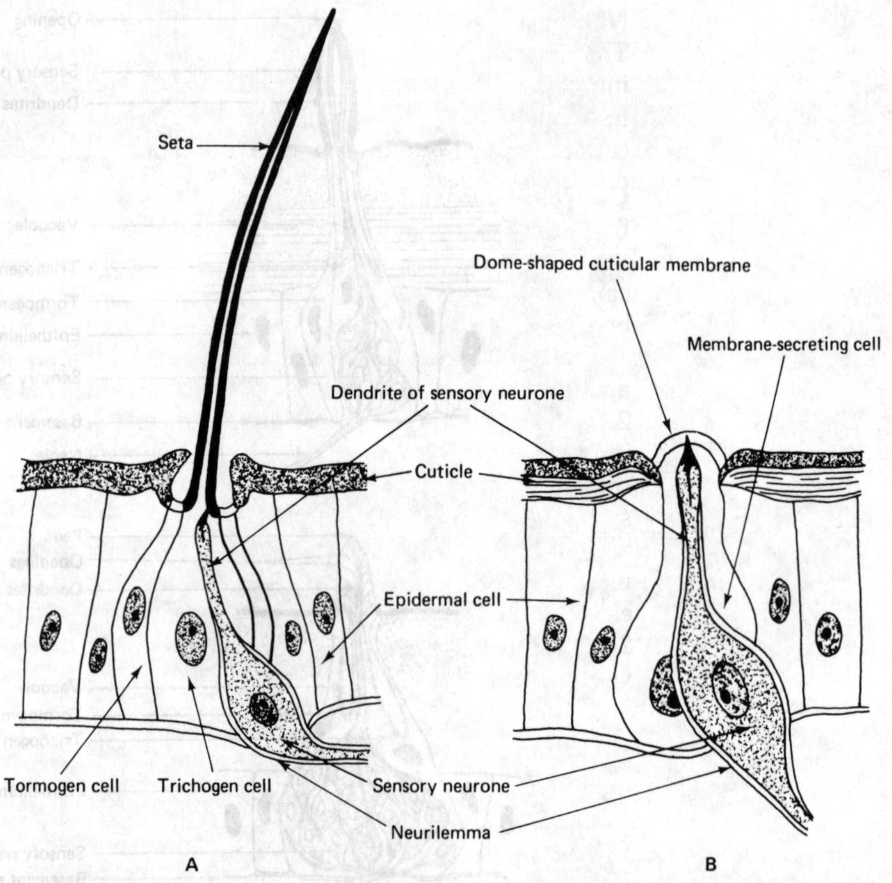


**Figure 2.39** Details of two common mechanoreceptors. A Hairlike sensillum. B. Campaniform sensillum. (Reprinted with permission of Macmillan Publishing Company from Insects in Perspectie by Michael D. Atkins. © 1978 by Michael D. Atkins)

is composed of a membranous **tympanum** whose vibrations are detected by a group of chordotonal sensilla. Such tympanal organs are found in grasshoppers (first abdominal segment), crickets (tibiae), moths (abdomen or metathorax), and many other insects.

**Other receptors.** In addition to the sense organs just described, insects also have receptors for perceiving ambient moisture and temperature, although little is known about these. Perceiving moisture in the air is called **hygroreception;** moisture levels are thought to be sensed, at least in some insects, by hairs or tufts of hairs that absorb moisture. Insects have a general sense of temperature and use this sense to locate suitable environments for their activities. As an example, a wood-boring beetle, *Melanophila* species (Buprestidae), has sensory pits on the underside of the mesothorax that are sensitive to the heat of recently fire-damaged trees, which the beetle prefers to attack. **Geomagnetic receptors** that sense magnetic fields also occur in insects such as honey bees, which use them in orientation and other activities. However, the actual sense organs used in geomagnetic reception have not been identified by electrophysiology.

## Nervous System

The nervous system of insects functions to generate and transport electrical impulses, to integrate information received, and to stimulate muscles for movement. This system can be divided into a central nervous system and a visceral nervous system.

**Central nervous system.** Basically, the central nervous system is formed from a brain, located in the head, and a ventral nerve cord running from the brain through the abdomen along the base of the body cavity (Fig. 2.40). The central nervous system supervises and coordinates activities of the insect body.

The brain is located in the upper part of the cranium above the esophagus and, therefore, is often referred to as the **supraesophageal ganglion** (Fig. 2.41). The brain is structured as paired parts consisting of the **protocerebrum,** innervating the compound eyes and ocelli; the **deutocerebrum,** innervating the antennae; and the **tritocerebrum,** which connects to the visceral nervous system.

The brain connects to a large nerve **ganglion** (nerve cell bundle) located under the esophagus by way of a paired nerve cord that branches around the esophagus. This special ganglion, called the **subesophageal ganglion** (Fig. 2.41), innervates the mouthparts and the salivary duct and attaches to the ventral nerve cord.

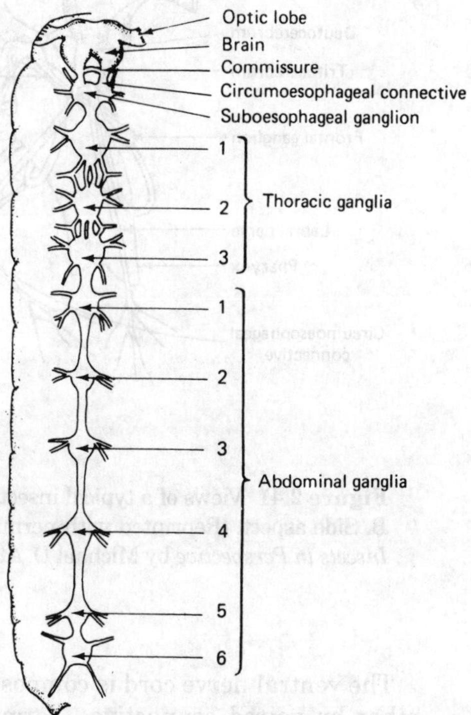


**Figure 2.40** Organization of the central nervous system in generalized form. (Reprinted with permission of Macmillan Publishing Company from *Insects in Perspective* by Michael D. Atkins. © 1978 by Michael D. Atkins)

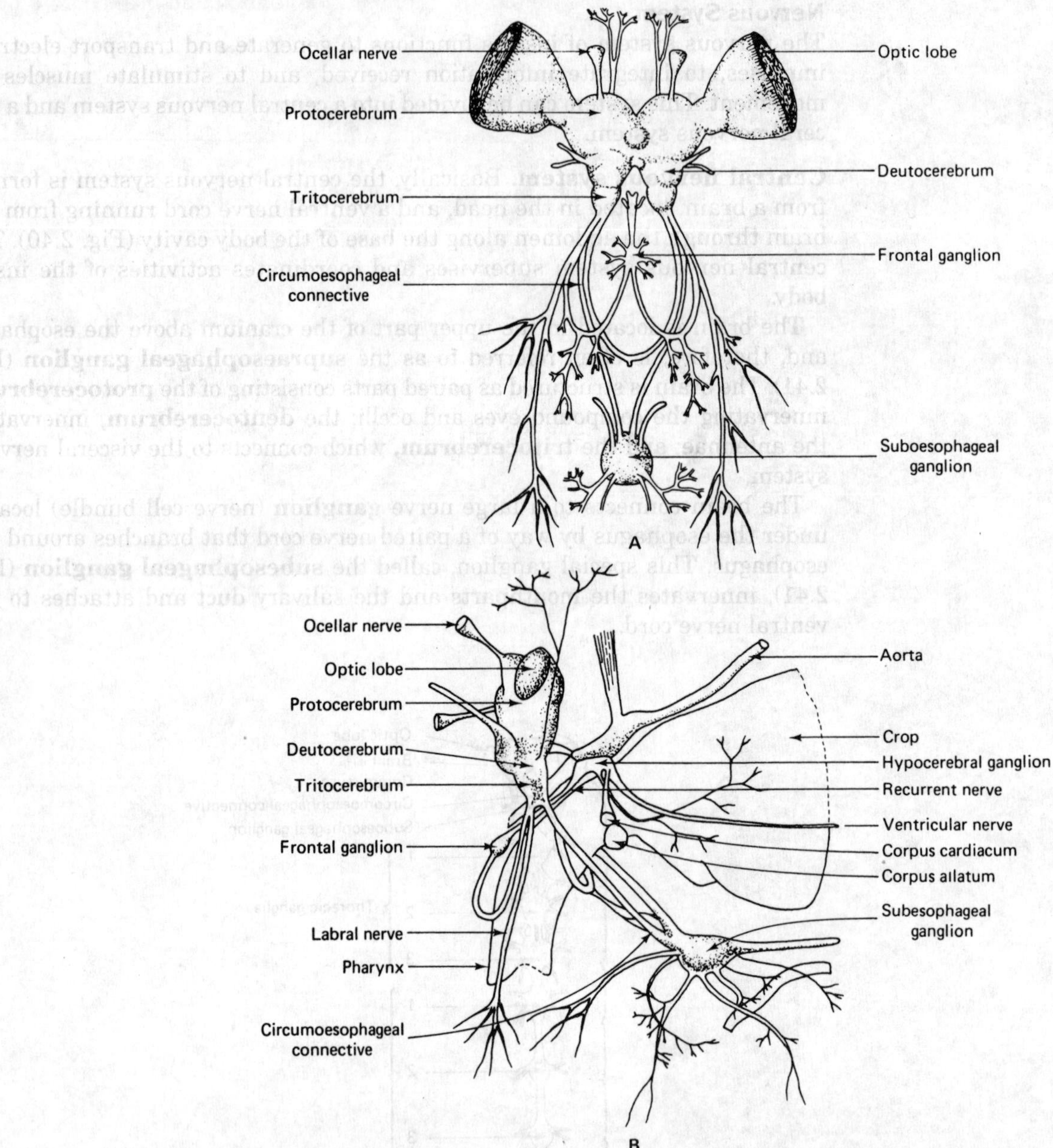


**Figure 2.41** Views of a typical insect brain and associated structures. A. Top aspect. B. Side aspect. (Reprinted with permission of Macmillan Publishing Company from *Insects in Perspective* by Michael D. Atkins. © 1978 by Michael D. Atkins)

The ventral nerve cord is composed of a series of ganglia attached to one another by paired connectives, forming a kind of chain (Fig. 2.40). Typically, there is a ganglion in each thoracic segment and, in many species, one in each abdominal segment. In some groups of insects, the thoracic ganglia fuse to form a larger unit, and abdominal ganglia become reduced in number or are lost almost completely. Nerves arise from each ganglion and branch and innervate both sensory structures and muscles.

**Visceral nervous system**. The main component of the visceral nervous system is the so-called **stomodeal nervous system.** The stomodeal system controls activities of the anterior gut and the dorsal vessel. This system consists of a **frontal ganglion** connected to the brain and then to other small ganglia. These ganglia give rise to paired nerves that innervate mainly the digestive tract and neuroendocrine glands, the **corpora cardiaca** and the **corpora allata.** Both glands are involved in insect growth, which will be discussed in chapter 4. Other parts of the visceral nervous system include the ventral sympathetic system, which innervates the spiracles of each segment in some insects, and a caudal sympathetic system, which is involved in activities of internal sexual organs.

### Nerve-Impulse Transmission and Integration

**Nerve-impulse transmission.** The basic unit of the nervous system that functions in nerve-impulse transmission is the nerve cell, or **neuron.** A neuron is composed of a **cell body,** one or more receptor fibrils, and an **axon** that branches at the tip. Three types of neurons are classified by function: **sensory neurons, motor neurons,** and **interneurons** (Fig. 2.42). Sensory neurons have receptor fibers arising directly from the cell body that are connected to sense organs. The axon of sensory neurons carries impulses to ganglia in the central nervous system. The cell body and receptor fibrils of motor neurons are located within the central nervous system and have an axon that branches into muscle tissue. The axon of motor neurons carries nerve impulses away from the central nervous system. Interneurons are situated totally within the central nervous system and connect sensory and motor neurons.

Nerve impulses traveling along axons are electrical. These impulses arise from the flow of positive sodium ions through the cell membrane, causing

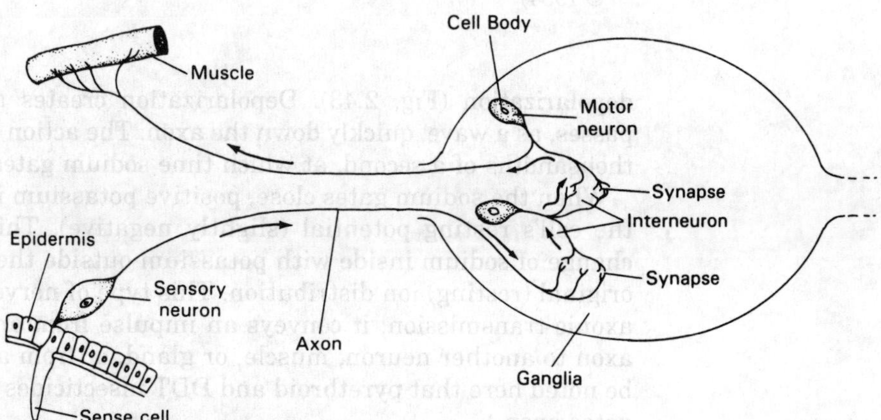


**Figure 2.42** Arrangement of nerve cells for impulse transmission in reflex reactions. Arrows indicate direction of nerve impulse. (Redrawn from Wigglesworth, 1953, *The Principles of Insect Physiology,* Methuen and Company)

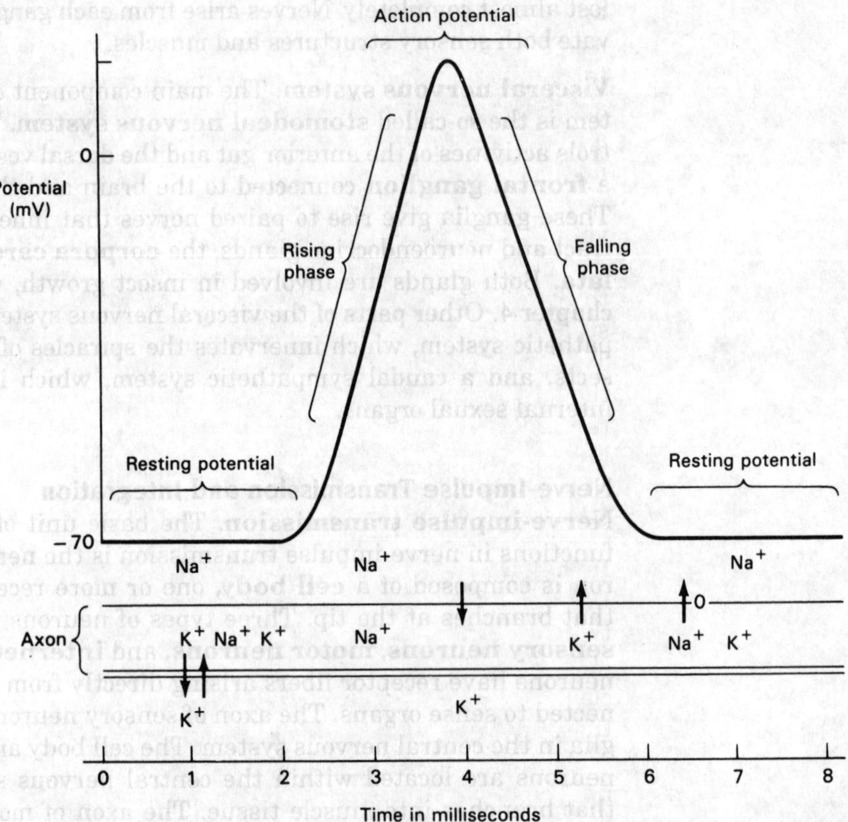


**Figure 2.43** A nerve axon showing electrical dynamics during axonic transmission. At rest, the axon has a high potassium level and a low sodium level; the reverse is found outside. The action potential occurs (center) when the first sodium ions rush in (depolarization), followed by potassium ions rushing out (repolari zation). After the action potential has occurred, sodium ions inside the axon are exchanged with potassium ions outside. (Reprinted with permission of Benjamin/Cummings Publishing Company from *Insect Biology* by Howard E. Evans. © 1984)

depolarization (Fig. 2.43). Depolarization creates an action potential that passes, as a wave, quickly down the axon. The action potential lasts only a few thousandths of a second, at which time sodium gates of the membrane close.

When the sodium gates close, positive potassium ions flow out and restore the cell's resting potential (slightly negative). This is followed by an exchange of sodium inside with potassium outside the cell and a return of the original (resting) ion distribution. This type of nerve transmission is termed axonic transmission; it conveys an impulse from an arrival point along the axon to another neuron, muscle, or gland, or from a receptor cell. (It should be noted here that pyrethroid and DDT insecticides work by keeping sodium gates open.)

Another type of impulse transmission is synaptic transmission, which is mainly chemical. A **synapse** is the junction between neurons and other cells. Among other places, a synapse occurs at the junction between an interneuron and a sensory neuron, and another occurs between the same interneuron and a motor neuron. A completed reaction from a sensory neuron through an interneuron and directly to a motor neuron, causing muscle contraction, is called a **reflex** reaction. An example of this reaction occurs in the human body when a finger is jerked back from a hot surface with no thought involved. Here, the nerve impulse travels a short route from the finger to the spinal cord and directly to an arm muscle; integration of information by the brain is not involved.

When an impulse moving along the axon reaches a synapse, it dies out. As this happens, a chemical is secreted that crosses the synapse and induces and stimulates an impulse in an adjacent neuron or stimulates a muscle or gland. The most well known of these chemical transmitters is **acetylcholine,** although there are many others. Following transmission, the synapse is returned to the resting state by enzymes such as **acetylcholinesterase,** which break down the chemical transmitter acetylcholine. As we will see in chapter 11, many modern insecticides (carbamates and organophosphates) act by upsetting chemical reactions at nerve synapses.

**Integration.** Many life-and-death activities of insects involve reflex reactions. Large nerve fibers allow rapid passage of the impulse, producing a galvanic twitch of legs for jumping or flipping the body to escape an enemy.

Usually, smaller nerves carry impulses that may cause a slow wave of muscle contraction, resulting in more delicate and precise movements, including those involved in feeding, walking, and nest building.

The brain is a coordinating center. It receives information from antennae, compound and simple eyes, and labrum and other sensory structures and sends impulses for appropriate actions. Unlike the brain of vertebrates, the insect brain inhibits rather than initiates locomotion. If the brain is removed, the insect will walk steadily forward but cannot climb or go backwards.

The subesophageal ganglion is the coordinating center for feeding. It receives information and controls the movements of the mandibles, maxillae, and labium.

Generally, each thoracic and abdominal ganglion organizes and coordinates the behavior of its own segment. A striking example of this segmental behavior can be seen in praying mantids (Orthoptera) when the male continues to copulate as the female eats first the male's head and then his thorax. The insect can be seen behaving as a set of individually functioning segments, each coordinated by a nerve ganglion. The brain acts as a center for receiving stimuli from important sensory structures of the head and for governing the release of segmental and intersegmental activities.

## INSECT REPRODUCTION

The power of insects to reproduce is one of their most important biological features, accounting for much of their success. Most insects are **dioecious;** that is, there is a male and a female that mate to produce a **zygote** (fertilized egg). The reproductive organs of females and males are somewhat similar, usually occurring near the rear of the abdominal cavity.

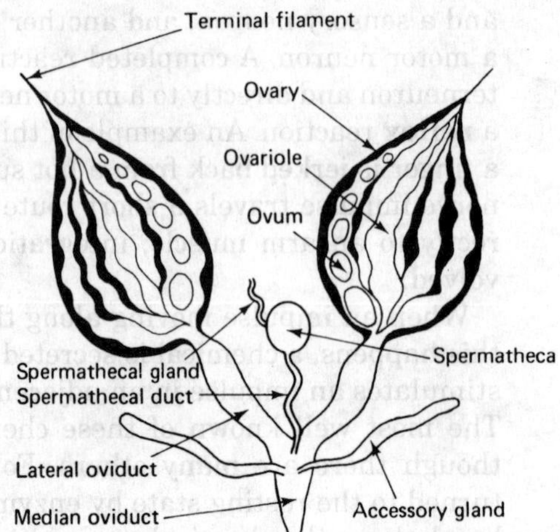


**Figure 2.44** Diagrammatical representation of a typical female reproductive system. (Reprinted with permission of Macmillan Publishing Company from *Insects in Perspective* by Michael D. Atkins. © 1978 by Michael D. Atkins)

### The Female System

The main organs of the female reproductive system are the paired **ovaries** (Fig. 2.44). Each ovary is usually made up of a bundle of several tubular **ovarioles** in which eggs are formed. Each ovariole is attached with a thread called the **terminal filament,** and germ cells form in the ovariole just below this filament. The germ cells develop as they move along and arrive as fully formed eggs at the ovariole base, or **pedicel.**

Eggs pass through the pedicels (collectively, the calyx) into paired **lateral oviducts** and then into a common **oviduct.** From the oviduct, they move to the **vagina,** where they are fertilized and held for laying. The organ involved in fertilization is the **spermatheca,** which receives and stores sperm after copulation. A **spermathecal gland** attached to the spermatheca supplies nutrients for maintaining the sperm before it is dispensed, and paired **accessory glands** secrete adhesive and a protective covering for the egg after it has been fertilized. Many variations of this basic system occur among insect groups.

### The Male System

The primary organs of the male reproductive system are paired **testes,** found in approximately the same position as the female ovaries (Fig. 2.45). Each testis is made up of a number of sperm tubes enclosed in a sheath. Sperm are produced in the sperm tubes and move through narrow **vasa efferentia,** which empty into a common duct, the **vas deferens.** Sperm continue through the vas deferens and are held in a storage structure, the **seminal vesicle,** where they combine with secretions of paired accessory glands to form semen. In some insects, the sperm are enclosed in a capsule, the **spermatophore.** During copulation, **semen** from the seminal vesicle moves through the **ejaculatory duct** and out the **penis.** The aedeagus, mentioned earlier, is associated with the penis and aids in the copulatory process.

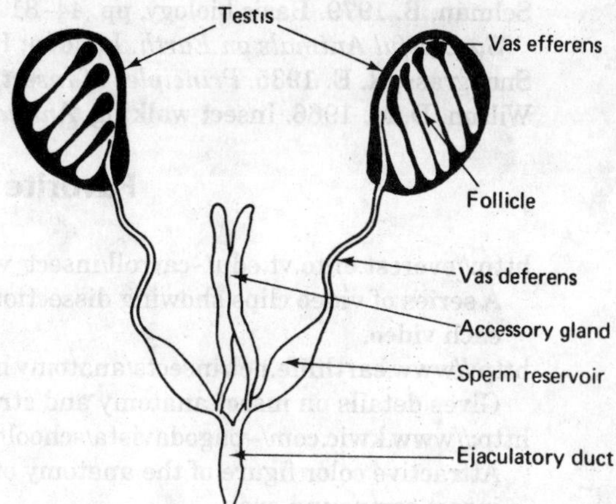


**Figure 2.45** Diagrammatical representation of a typical male reproductive system. (Reprinted with permission of Macmillan Publishing Company from *Insects in Perspective* by Michael D. Atkins. © 1978 by Michael D. Atkins)

## Further Reading


Atkins, M. D. 1978. *Insects in Perspective*. New York: Macmillan, chapters 5–10.

Barrington, E. J. W. 1979. *Invertebrate Structure and Function*, 2nd ed. New York: Wiley.

Blum, M. S., ed. 1985. *Fundamentals of Insect Physiology*. New York: Wiley.

Chapman, R. F. 1998. *The Insects, Structure and Function*, 4th ed. Cambridge, UK: Cambridge University Press.

Dethier, V. G. 1963. *The Physiology of Insect Senses*. New York: Wiley.

Elzinga, R. J. 2004. *Fundamentals of Entomology,* 6th ed. Upper Saddle River, N.J.: Prentice Hall, chapters 2–3.

Gilmour, D. 1970. General anatomy and physiology. In Commonwealth Scientific and Industrial Research Organization (C.S.I.R.O.), *The Insects of Australia*. Carlton, Victoria, Australia: Melbourne University Press, chapter 2.

Happ, G. M. 1984. Major life systems. In H. E. Evans, ed., *Insect Biology*. Reading, Mass.: Addison-Wesley, chapter 3.

Horn, D. J. 1976. *Biology of Insects*. Philadelphia: Saunders, chapters 4–7.

Matsuda, R. 1970. Morphology and evolution of the insect thorax. *Memoirs Entomological Society Canada* no. 76.

Matsuda, R. 1976. *Morphology and Evolution of the Insect Abdomen*. New York: Pergamon Press.

Pringle, J. W. S. 1965. Locomotion: Flight, pp. 283–329. In M. Rockstein, ed., *The Physiology of Hexapoda*, vol. 2. New York: Academic Press.

Richards, O. W., and R. G. Davies. 1977. *Imms' General Textbook of Entomology*, vol. 1, 10th ed. London: Chapman and Hall.

Romoser, W. S., and J. G. Stoffolano, Jr. 1998. *The Science of Entomology*, 4th ed. Boston: McGraw-Hill, chapters 2–8.

Ross, H., C. A. Ross, and J. R. P. Ross. 1982. *A Textbook of Entomology*, 4th ed. New York: Wiley, chapters 3–6.

Schaefer, C. W., and R. A. B. Leschen. 1993. Functional morphology of insect feeding. *Thomas Say Publications in Entomology Proceedings*. Lanham, Md.: Entomological Society of America.

Selman, B. 1979. Basic biology, pp. 44–81. In *Insects, An Illustrated Survey of the Most Successful Animals on Earth*. London: Hamlyn Pub. Group.
Snodgrass, R. E. 1935. *Principles of Insect Morphology*. New York: McGraw-Hill.
Wilson, D. M. 1966. Insect walking. *Annual Review of Entomology* 11:103–122.

## Favorite Web Sites

http://everest.ento.vt.edu/~carroll/insect_video_dissection.html
A series of video clips showing dissection of a cockroach. Descriptions accompany each video.

http://www.earthlife.net/insects/anatomy.html#Pete
Gives details on insect anatomy and structure function.

http://www.kwic.com/~pagodavista/schoolhouse/species/insects/bugparts.htm
Attractive color figure of the anatomy of the jewel beetle, as well as details of an insect compound eye.

http://www.rothamsted.ac.uk/pie/DeBug/Anatomy.html
Allows a virtual tour inside of a honey bee, giving names of internal structures.

CHAPTER 3

# INSECT CLASSIFICATION

THE NAMING AND ORDERING of objects into groups is probably the most fundamental step in the development of scientific principles. For instance, in chemistry the discovery and naming of elements and organizing them into the periodic table became the basis for the development of the science.

In biology, the naming of organisms is referred to as **nomenclature,** and ordering them into a hierarchy of categories is known as **classification.** A related science, **taxonomy,** involves the theoretical basis for classification and the study of classification schemes. Specialists working in these areas usually are referred to as *systematists*; their overall activity, **systematics,** is the study of the diversity and classification of organisms.

Because of the tremendous size of the class Hexapoda, the naming and classification of all insects would seem to be a difficult, if not impossible, achievement. However, hundreds of insect systematists around the world work daily on this task, and great strides have been made in understanding insect diversity. Indeed, some experts estimate that nearly 7,000 species new to science are discovered, named, and classified each year! However, there is still much systematics work to be done before insects will be as well known as birds, mammals, and other animals.

## OBJECTIVES OF CLASSIFICATION

Like the periodic table of chemistry, classification allows us to order what we know about insects and to compare and contrast characteristics. From these comparisons, we formulate predictions about relationships, including those with both evolutionary and ecological meaning. For instance, members of the same **species** are expected to behave similarly in their food habits, tolerances to environmental extremes, developmental patterns, and other ways. A group of similar species, put together in a higher category called a **genus,** also could be predicted to share somewhat similar ecologies and to have evolved from the same ancestor. Moving to higher and higher groupings in classification, we expect more and more diversity within the grouping.

A major application of classifications is in *identification* of insect specimens. Identifications of major groups such as insect **orders** can usually be made at a glance; however, finer identifications often require the use of **keys.** Most keys comprise a sequence of paired statements and questions that allow the user to eliminate alternative options and eventually associate the unknown specimen with a name (Table 3.1). Many keys exist for orders and families of insects. Some of the most useful are those written by C. A. Triplehorn and N. F. Johnson (*Borror and DeLong's Introduction to the Study of Insects*, 7th ed., 2005) for the insects of North America, and by C. T. Brues, A. L. Melander, and F. M. Carpenter (*Classification of Insects*, 1954) for the insects of the world.

Correct identification is the first step and probably the most important one in dealing with a pest. It allows us to retrieve the information required for insect pest management. Without identification, we have no basis for predicting injury and advising action.

**Table 3.1 An example of an identification key as represented by one for the classes of Arthropoda. (To use this key, start with the couplet and follow couplet routing according to the characteristics of the specimen until the class is identified)**

*Key to Classes of Arthropoda*

This key includes all living classes of Arthropoda except Pentastomida (or Linguatulida) and Tardigrada. They are small groups of small to minute, seldom-seen animals that lack antennae and are not likely to be taken for Arthropoda. In fact, some authorities do not consider them such.

| | | |
|---|---|---|
| 1. | Without antennae (in immature stages or in some adults of certain insects in which antennae are much reduced or lacking; such forms will not key out well in this part of the key) | 2 |
| | With antennae | 4 |
| 2. | Usually with 7 pairs of appendages, 5 of them legs; abdomen rudimentary (sea spiders) | PYCNOGONIDA |
| | With 6 (rarely fewer) pairs of appendages, 4 (rarely 5) of them legs; abdomen usually well developed as a body region but sometimes fused with cephalothorax | 3 |
| 3. | Abdomen with booklike gills on underside; large animals up to 50 cm long, with hard expanded shell and long spinelike tail (horseshoe crabs) ...... class MEROSTOMATA, order or subclass | XIPHOSURA |
| | Abdomen without booklike gills; smaller forms rarely over 7 cm long, body not as above (spiders, scorpions, mites, etc.) | ARACHNIDA |
| 4. | With 2 pairs of antennae (1 pair may be rudimentary in sowbugs); head merged with thorax to form cephalothorax; breathing by gills (crabs, lobsters, shrimps, sowbugs, etc.) | CRUSTACEA |
| | With 1 pair of antennae; head separate from thorax; breathing by tracheae | 5 |
| 5. | With 3 pairs of legs at some stage in life cycle; body in 3 divisions—head, thorax, and abdomen; true legs only on thorax, but abdomen sometimes with appendages resembling but differing from those on thorax; wings often present (all insects) | HEXAPODA |
| | With 9 or more pairs of legs; most body segments behind head with legs; wings lacking; body more or less wormlike (myriapodan classes) | 6 |
| 6. | Most segments of body with double pairs of legs; slow-moving animal (millipedes) | DIPLOPODA |
| | Legs not in double pairs; most segments of body with 1 pair | 7 |
| 7. | Body more or less flattened; not minute, with 15 or more pairs of legs; rapidly moving animals (centipedes) | CHILOPODA |
| | Body usually cylindrical; small to minute forms not over 8 mm long, with 9 to 12 pairs of legs | 8 |
| 8. | Antenna branched; 9 pairs of legs; minute animals 1 to 1.5 mm long, found in leaf litter, etc. (pauropods) | PAUROPODA |
| | Antenna not branched; 10 to 12 pairs of legs; white cylindrical, centipedelike animals 1 to 8 mm long (symphylans) | SYMPHYLA |

SOURCE: Steyskal et al. 1986, courtesy USDA.

## ELEMENTS OF CLASSIFICATION

The classification of organisms is based on a hierarchy of categories, with the most inclusive occurring at the top and the least inclusive at the bottom. The major categories used in animal classification are **phylum, class, order, family, genus,** and **species.** But for added distinction in large, diverse groups, many other categories fall between these major ones. For example, a *subclass* category is commonly present below the class category and a *superfamily* category above the family category.

An example of the major categories for the European corn borer shows the following classification:

Phylum—Arthropoda
Class—Hexapoda
Order—Lepidoptera
Family—Crambidae
Genus—*Ostrinia*
Species—*Ostrinia nubilalis*

Note that each category consists of only one word except the species category. The scientific name of a species is a **binomen,** or **binomial name;** it is composed of two names, a genus name, which is always capitalized, (but can be abbreviated after first mention) and a specific name, also called a *specific epithet*. Unlike all the higher categories, the specific name cannot stand alone; it must be used with the genus name. In zoology, it is conventional, but not required, for the species name to have the name of the person who first described the species as a suffix. Therefore, we might see the species name written as *Ostrinia nubilalis* (Hübner). The parentheses around the author name indicates that when Hübner originally described the species, he placed it in another genus. An author name without parentheses means that the species remains in the genus originally used by the describing author. Author names are used often in technical systematic literature; they are omitted in this book.

The system of binomial nomenclature we use today for classification was advanced by Swedish naturalist Carolus Linnaeus, who first used it consistently in 1758. Strict rules and conventions apply for name assignment, and these are stated for zoology in the *International Code of Zoological Nomenclature*. Although they may be based in any language, scientific names are Latinized and usually refer to some characteristic of the animal or group named. The binomial name of a species is always printed in italics or, if handwritten, is underlined to indicate italics. The names of genera and higher categories begin with a capital letter, but the specific name of the species and subspecific names always begin with a lowercase letter, as in the subspecies *Diabrotica undecimpunctata undecimpunctata* Mannerheim. In this example, the third name indicates a subspecies, a group of individuals of the same species that have differences in body form or color and geographical distribution.

A frequently used name for the species, the common name, is not covered by the formal rules of nomenclature. The name "European corn borer" is a common name, somewhat similar to a nickname. Such names are often used in insect pest management because they are more easily pronounced and remembered compared with the Latin name. However, because no formal rules govern these names, we may find different names or versions used in different localities for the same species, and confusion may arise. Because there is only one scientific name, it is always the safest name to use to avoid problems of semantics. In the United States, a list of common names approved by the Entomological Society of America is presented in *Common Names of Insects and Related Organisms*. These names are merely recommended and are not necessarily followed by the international scientific community.

The species category plays the central role in systems of classification. It is the only category that is real; in other words, it is the only natural group. The species is a complex unit that is sometimes difficult to define, but a widely used definition applied to the species concept is as follows: a group of interbreeding individuals that are similar in body structure and that produce fertile offspring; moreover, these groups are reproductively isolated from other such groups. This is the biological species definition.

All other categories of the hierarchy are based on the species category but are abstractions. Even though they are derived from suspected pathways of insect evolution, these other categories are human contrivances used for convenience. Because this is so, we find many differences of opinion as to where to "lump" and where to "split" categories; therefore, many different schemes of insect classification result.

Consequently, there are no "correct" and "incorrect" systems. Rather, we have "successful" and "unsuccessful" systems based on their usefulness. In this book, we mostly use the widely adopted scheme of C. A. Triplehorn and N. F. Johnson (2005).

## GENERAL CLASSIFICATION OF INSECTS

Clearly, no one person can know all the kinds of insects in our environment. Not even trained entomologists instantly recognize every insect specimen presented to them. However, a good working knowledge of the major insect groups is possible with a modicum of reading and laboratory study. The usual level of insect identification for communication with specialists is the order category. There are 31 orders in the scheme we will use, and discussions will focus on identifying characteristics and biological properties of each order. Seven of the 31 orders are the most diverse and account for the greatest share of our pest problems. Major families will be named and discussed in these orders. An overall scheme of insect ordinal classification follows, with major orders indicated by an asterisk (*). In general, this list proceeds from what many entomologists consider to be the most primitive insects to the most highly evolved.

Class Hexapoda (formerly Insecta)

Subclass Apterygota—wingless insects

1. Protura—proturans
2. Collembola—springtails
3. Diplura—diplurans
4. Thysanura—bristletails
5. Microcoryphia—jumping bristletails

Subclass Pterygota—winged and some wingless insects

Division Exopterygota—simple body change during growth

6. Ephemeroptera—mayflies
7. Odonata—dragonflies and damselflies
*8. Orthoptera—grasshoppers and crickets
9. Phasmatodea—walkingsticks
10. Grylloblattaria—rock crawlers
11. Mantophasmatodea—gladiators
12. Dermaptera—earwigs
13. Plecoptera—stoneflies
14. Embiidina—webspinners
15. Zoraptera—zorapterans
16. Isoptera—termites
17. Mantodea—mantids
*18. Blattodea—cockroaches
*19. Hemiptera—bugs, aphids, scale insects, hoppers, cicadas, psyllids, and whiteflies
*20. Thysanoptera—thrips
21. Psocoptera—psocids
22. Phthiraptera—chewing lice and sucking lice

Division Endopterygota—complex body change during growth

*23. Coleoptera—beetles
*24. Neuroptera—alderflies, antlions, dobsonflies, fishflies, lacewings, snakeflies, and owlflies
*25. Hymenoptera—ants, bees, wasps, and sawflies
26. Trichoptera—caddisflies
*27. Lepidoptera—butterflies and moths
*28. Siphonaptera—fleas
29. Mecoptera—scorpionflies
30. Strepsiptera—twisted-winged parasites
*31. Diptera—flies and mosquitoes

A summary of characteristics of pest and beneficial insect orders is given in Table 3.2, and a key for identification of specimens to order is provided in Appendix 1.

## Subclass Apterygota

The Apterygota are an assemblage of the most primitive insects, consisting of the orders Protura, Collembola, Diplura, Thysanura, and Microcoryphia. Insects in all these orders are wingless. Moreover, internal structures that strengthen the thorax for flight in winged insects are absent in these orders. All members of this subclass grow with very little change in form during development, a pattern called *no metamorphosis*.

Entomologists differ considerably on the classification of this group. Many systematists feel that the orders Protura, Collembola, and Diplura are sufficiently different from the rest of the insects to be excluded from the true insects and that they should be placed in three separate classes. Here, they will be included as insects in the broad sense. In particular, insects in these orders have mouthparts housed within a cavity, and only the tips protrude when the insects

**Table 3.2 Usual Characteristics of Major Pest and Beneficial Insect Orders.**

| *Order* | *Common Name* | *Example* | *Front Wings* | *Hind Wings* | *Antennae* | *Mouthparts* | *Importance* |
|---|---|---|---|---|---|---|---|
| **Orthoptera** | Grasshoppers, cockroaches, crickets, katydids | | Elongate, thickened, parchmentlike, many veins | Wider than front wings, folded, fanlike | Threadlike, many segmented, short to long, Fig. 2.15 A | Chewing, Fig. 2.7 | Pests of crops; leaf and fruit feeders |
| **Phasmatodea** | Walkingsticks | | Much reduced or entirely absent | Similar to front wings | Threadlike, many segments, long | Chewing | Plant feeders |
| **Isoptera** | Termites | | Long, narrow, weak venation, Fig. 2.19 C; absent in workers | Similar to front wings, Fig. 2.19 C | Beadlike or hairlike, moderately long, Fig. 2.15 A, C | Chewing | Significant destroyers of wooden structures |
| **Blattodea** | Cockroaches | | Generally present with many veins, sometimes much reduced | Similar to front wings | Threadlike, many segments, long | Chewing | Household presence, contaminate food |
| **Hemiptera** | True bugs aphids, scales, leafhoppers, cicadas | | Thickened at base, membraneous at tip, Fig. 2.19 B or when present, same texture throughout; at rest, held rooflike over body, Fig. 2.19 A | Membranous, shorter and sometimes wider than front wings, Fig. 2.19 A, B | Threadlike, many segments, short to long, Fig. 2.15 A | Piercing-sucking, arising underside of head | Plant feeders; suck sap; parasites of humans and animals; insect predators |
| **Thysanoptera** | Thrips | | Slender, fringed with hairs, Fig. 2.19 D | Same as front wings, Fig. 2.19 D | Short, 6 to 10 segments | Rasping-sucking type, Fig. 2.10 | Pests of field greenhouse, vegetable, fruit crops; rasp flowers, leaves, buds |

Table 3.2 *(Continued)*

| *Order* | *Common Name* | *Example* | *Front Wings* | *Hind Wings* | *Antennae* | *Mouthparts* | *Importance* |
|---|---|---|---|---|---|---|---|
| **Phthiraptera** | Chewing lice, sucking lice | | None | None | Short, few segments | Modified chewing type or piercing-sucking type, can retract in head | Poultry, livestock, pet and parasites; human transmit diseases |
| **Coleoptera** | Beetles | | Thick, hard, cover of hind wings, pair meets in straight line down back, Fig. 2.19 F | Membranous, fold beneath front wings at rest | Usually with 11 segments, often clubbed, sawlike, or platelike, Fig. 2.15 B, D, F | Chewing type | Largest order; pests of all parts of most plants, stored grain, and natural fibers; significant predators of other insects |
| **Neuroptera** | Lacewings, ant lions | | Long, slender, net veined | Similar to front wings, slightly smaller | Short to long, many segments, some clubbed | Chewing but modified into sucking in immatures of some | Immatures important predators of other insects |
| **Hymenoptera** | Sawflies, ants, wasps, bees | | Membranous with few veins, Fig. 2.19 G | Similar to front but smaller, capable of attachment, Fig. 2.19 G | Often threadlike, few to many segments, Fig. 2.15 A | Chewing or chewing-lapping (honey bee), Fig. 2.14 | Sawflies important plant pests, particularly woody species; significant as insect predators, pollinators, and honey producers |

**Table 3.2** ***(Continued)***

| *Order* | *Common Name* | *Example* | *Front Wings* | *Hind Wings* | *Antennae* | *Mouthparts* | *Importance* |
|---|---|---|---|---|---|---|---|
| **Lepidoptera** | Butterflies, moths | | Slender to broad, covered with scales, few cross-veins, Fig. 2.19 E | Similar to front wings but shorter, broader, and more rounded, Fig. 2.19 E | Long, many segments, some clubbed or featherlike, Fig. 2.15 D, G | Siphoning type, Fig. 2.11; caterpillars with chewing type | Caterpillars very significant pests of all crops; adults important pollinators |
| **Siphonaptera** | Fleas | | Absent | Absent | Short, three-segmented, in grooves on side of head | Sucking-piercing type | Blood-feeding parasites of humans, pets, and livestock; transmitters of disease |
| **Diptera** | Flies, gnats, midges | | Membranous, few veins and cross-veins, Fig. 2.19 H | Not functional, modified into knobbed structures, Fig. 2.19 H | Often threadlike, few to many segments | Chewing but some with modifications for sponging (Fig. 2.12), cutting-sponging (Fig. 2.13), and piercing-sucking (Fig. 2.9) | Important parasites and disease transmitters of humans and livestock; pests of crop seeds and growing plants |

are feeding. Such mouthparts are called **entognathous.** All other insects have exposed, or **ectognathous,** mouthparts. Whether or not they are considered insects, they are still commonly dealt with in the science of entomology.

### ORDER PROTURA—proturans

Proturans (Fig. 3.1) are some of the most unusual insects. They are small (0.6 to 2.0 mm long), and whitish with styletlike mandibles and no eyes or **antennae.** However, the front legs are held out in front of the head, possess many sensillae, and function somewhat like antennae. Another unusual feature of proturans is that the young begin with nine segments and add three more during development, a phenomenon known as **anamorphosis.** The final segment of the adult abdomen is the **telson;** hence the sometimes-used name telsontails.

Proturans live in soil and decomposing plant material, where they feed on organic debris and fungal spores. They are not known to be pests.

### ORDER COLLEMBOLA—springtails

The Collembola (Fig. 3.2) possess a central extensible, tubelike structure attached to the underside of the first abdominal segment called the **collophore;** hence their ordinal name. The collophore, or "glue bar," is eversible and maintains secretions at the tip. Entomologists once believed that the structure was used as an aid in climbing on smooth surfaces, but more recent evidence suggests it may be used in water uptake or preening activity.

Collembolans are small, ranging from 0.2 to 10 mm in length, and may be gray, off-white, or more brightly colored green or purple. Body shape is either long and slender or rounded (globular). Mouthparts are usually the chewing type, but in some species, the maxillae and mandibles are modified into long, sharp stylets. Each eye occurs as a loose aggregation of single ommatidia, but eyes are absent in many true soil-dwelling and cave-dwelling forms.

The common name springtail comes from the **furcula** arising from the underside of the abdomen, near the tip. Most of the time, the furcula is "cocked" by a clasp, the **tenaculum.** When a springtail is disturbed, the furcula is released, propelling the insect for a distance several times its body length. The furcula is absent in species that inhabit deeper levels of soil.

Springtails are found most frequently in moist environments, including soil, decaying leaves and wood on forest floors, at the edges of ponds, on snow, and among fleshy fungi. Some species of the family Sminthuridae (globular forms) are found on green plants, where they are occasional pests. The most infamous in this regard is the Lucerneflea, *Sminthurus viridis*, which has been a notable

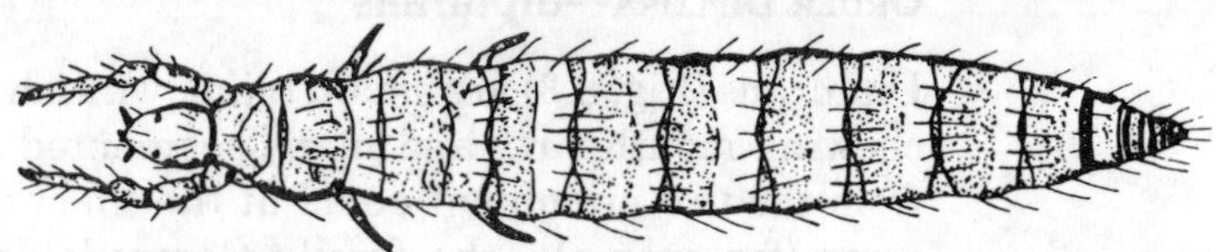

**Figure 3.1** Order Protura, proturans. Insects in this order lack eyes and antennae. (Reprinted with permission of Macmillan Publishing Company from *Insects in Perspective* by Michael D. Atkins. © 1978 by Michael D. Atkins)

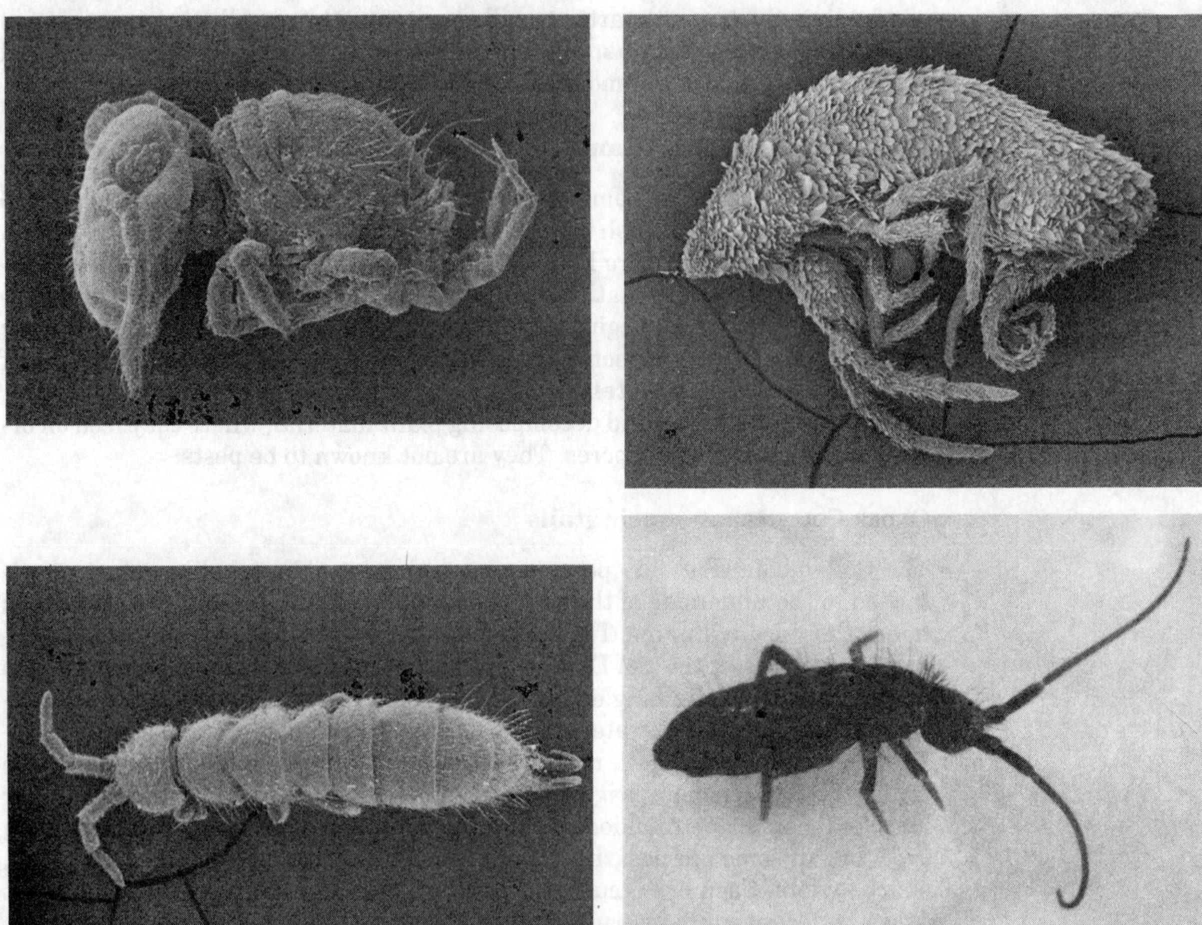

**Figure 3.2** Order Collembola, springtails. Scanning electron micrographs (SEMs) of Sminthuridae (*upper left*), Entomobryidae (*upper right*), Isotomidae (*lower left*), and living specimen of Isotomidae. Insects in this order have a ventral tube (collophore), and most have a ventral furcula (shown as the curved structures on the right end of the micrographs). (SEMs by Royce Bitzer; photo by Marlin E. Rice)

pest of alfalfa in Australia, and the garden springtail, *Bourletiella hortensis*, a pest of garden vegetables. Other species, for example, *Hypogastrura armata*, are also pests of commercially grown mushrooms.

## Order Diplura—diplurans

Diplurans (Fig. 3.3) are small, blind, whitish insects, less than 7 mm long, with chewing mouthparts and many-segmented antennae. Their name is derived from the two prominent **cerci** at the tip of the abdomen that occur as styli in some (for example, the family Campodeidae) and forceps in others (family Japygidae).

Because of their small size, low abundance, and secretive habits, diplurans are rarely seen. They exist most often in soil and soil surface debris. Very little is known about the ecology of these insects, and no species is known to be a pest.

**Figure 3.3** Order Diplura, family Japygidae, diplurans. Insects in this order are blind and have two cerci at the tip of the abdomen. (Photo by Dennis Schotzko)

## ORDER THYSANURA—bristletails

Thysanura (Fig. 3.4) are a group of medium-sized insects, 7 to 19 mm long, that have a somewhat flattened body and ectognathous chewing mouthparts. The most distinctive features of these insects are the two long cerci and a median caudal filament that occur at the end of the abdomen, giving the appearance of three "tails." Small compound eyes and scales, which cover the body and often give a silver sheen, occur in many species. Eyes or scales may be lacking in others.

Some bristletails are found in woodlands under decaying bark, but others may live in caves, mammal burrows, or in ant and termite nests. The most well known in North America occur in households and other human dwellings. Here, the two most frequently encountered species are silverfish, *Lepisma saccharina*, and firebrats, *Thermobia domestica*; both are members of the family

**Figure 3.4** Order Thysanura, bristletails, as represented by the common household pest, the firebrat, *Thermobia domestica*. The most common features in this order are the two long cerci and the median filament at the tip of the abdomen. (Photo by Dennis Schotzko)

Lepismatidae. Temperature requirements may vary from cool, damp situations favored by silverfish to warm, dry areas like furnace rooms, inhabited by firebrats.

Although they are minor household pests, silverfish and firebrats do cause damage to possessions by feeding on starchy substances. Where books are stored, they feed on starch in bindings and on labels. In other instances they have been known to feed on starched clothing, curtains, and starch paste in wallpaper.

### ORDER MICROCORYPHIA—jumping bristletails

Jumping bristletails (Fig. 3.5) are very much like the bristletails in size and appearance. They differ from bristletails in having a cylindrical body and several abdominal styli arising beneath the abdomen, in addition to the three "tails." All species have large, compound eyes and chewing mouthparts.

The name of this order is derived from the ability of these insects to jump (sometimes distances of 25 to 30 cm) when disturbed. They occur most frequently in wooded habitats under leaves, rocks, and bark of decaying logs. Jumping bristletails are believed to feed mainly on algae, but they also may eat lichens, mosses, and other materials. No species of Microcoryphia is considered a pest.

## Subclass Pterygota

The most important characteristic of the subclass Pterygota is the possession of wings in the adult stage. Although some adult Pterygota lack wings, for example, fleas, evolutionists believe they evolved this condition secondarily and that their ancestors probably had wings.

Pterygota are variously divided by systematists into large groups of orders based on the presence or absence of wing-flexing mechanisms, internal or external growth of wings, or complexity of the developmental process. Orders of insects whose wings do not flex back over the body are usually placed in the infraclass **Paleoptera;** orders with folding wings are placed in the infraclass **Neoptera.** Neoptera comprise by far the largest infraclass. In other classification schemes, orders with individuals that have a simple form of body change

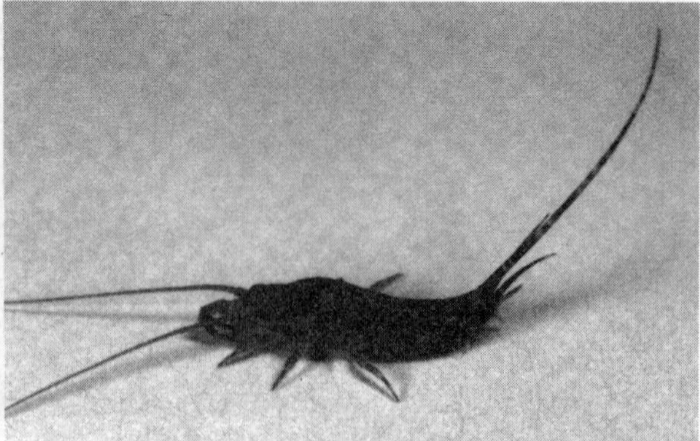

**Figure 3.5** Order Microcoryphia, jumping bristletails. Although these insects look somewhat like bristletails, their body is more cylindrical, and their compound eyes are large and contiguous. (Photo by Marlin E. Rice)

during growth and external wing development are placed in the division **Exopterygota;** those with complex changes in body form and internal wing development are placed in the division **Endopterygota.**

### ORDER EPHEMEROPTERA—mayflies

Mayflies (Fig. 3.6) are characterized by triangular, net-veined, membranous wings and two or three filamentous appendages at the tip of the abdomen. These insects are soft-bodied, 10 to 40 mm long, with short, bristlelike antennae and mouthparts that are reduced and nonfunctional in the adult. Frequently, adults are found near lakes, ponds, and streams and are attracted in large numbers to lights.

Mayflies lay eggs on the surface of water or on objects in the water. The young remain and develop in the water, feeding on algae and detritus. An immature looks like the adult in overall body shape but has well-developed chewing mouthparts, no functional wings, leaflike gills on the sides of the abdomen, and usually three "tails."

Adult mayflies emerge in huge swarms over bodies of water and mate; females lay eggs within a few minutes to a few hours after mating. The adults do not feed and die within a day or two, therefore the name Ephemeroptera (*ephemera;* for a day, short-lived). Upon death, mayfly bodies may pile up to a depth of over a meter along shorelines, and sometimes their presence makes roadways and bridges slick and dangerous for traffic.

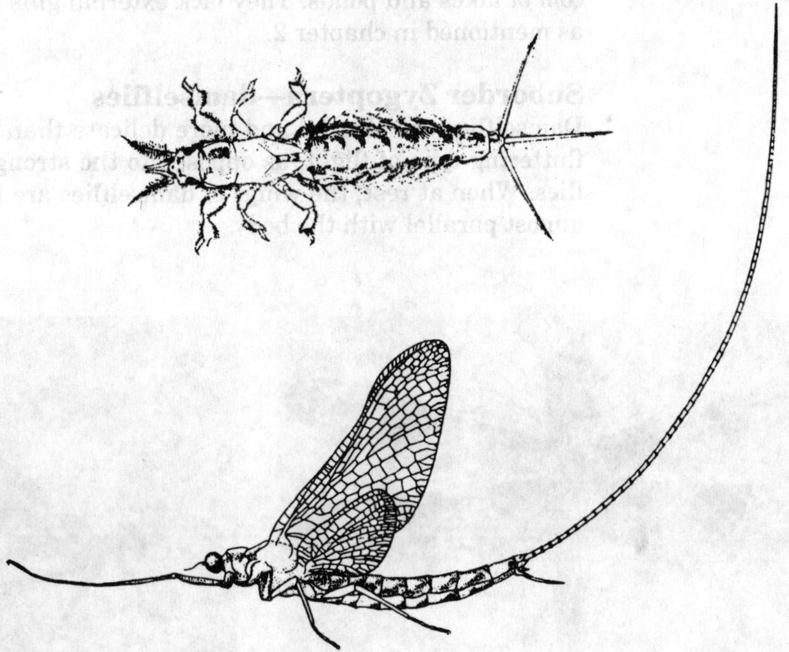

**Figure 3.6** Order Ephemeroptera, mayflies. Immature (naiad) (*top*) and adult. The most common features of this order are leaflike abdominal gills and three "tails" on naiads. Adults have triangular net-veined wings and two or three filamentous abdominal appendages. (After Steyskal et al., 1986, courtesy USDA)

The major importance of mayflies is food for freshwater fish and other aquatic animals. Indeed, many fishermen recognize the importance of mayflies as fish food, designing fishing "flies" modeled after mayflies. Because of different habitat preferences among species, mayflies also are useful as indicators of aquatic conditions and concentrations of pollutants upstream.

### ORDER ODONATA—dragonflies and damselflies (see Color Plate 1)

Members of the Odonata (Fig. 3.7) are large insects that are usually conspicuous because of their sometimes brightly colored bodies and banded wings. Dragonflies and damselflies have four net-veined, membranous wings and large heads, with large, protruding eyes and long, slender bodies. Antennae are short and inconspicuous, and mouthparts are the well-developed chewing type.

Adults feed on insects caught in flight, including mosquitoes and midges. Females lay eggs in water. Immatures are aquatic and also prey on other insects and small animals in ponds, lakes, and backwaters of streams. These immatures do not swim but walk along the bottom among vegetation and debris in search of prey. Development to the adult stage requires from 1 to 4 years.

## Suborder Anisoptera—dragonflies

Adult dragonflies are heavy-bodied and have strong, well-controlled flight. Males often establish territories that they patrol in search of food and defend against other dragonfly males. The rustle of wings is often heard when two males clash.

Dragonfly immatures are robust-bodied and usually found in mud at the bottom of lakes and ponds. They lack external gills but have internal rectal gills, as mentioned in chapter 2.

## Suborder Zygoptera—damselflies

Damselflies are slender and more delicate than dragonflies. Adults display a fluttering type of flight, as opposed to the strong directional flight of dragonflies. When at rest, the wings of damselflies are held together above the back, almost parallel with the body.

**Figure 3.7** Order Odonata, dragonfly (*top*) and damselfly. The most common characteristics of this order are four equal-sized, net-veined wings; large, protruding eyes; and well-developed chewing mouthparts. Dragonflies (suborder Anisoptera) are more heavy-bodied than damselflies (suborder Zygoptera) and hold their wings out when at rest. Damselflies hold their wings together above their backs. (Photos by Marlin E. Rice)

Immature damselflies are also more slender than immature dragonflies. They possess three leaflike gills at the tip of the abdomen and are more often found near submerged vegetation than on the very bottom of lakes and ponds.

### *Order Orthoptera (see Color Plates 1, 5, 7)

This is the first of the major orders that contain important pest species. Although some are wingless, most have two pairs of wings, with parchmentlike front wings, the **tegmina,** and membranous, fanlike hind wings. These wings are folded over the back when not in use. Mouthparts are the typical chewing type, and a pair of short cerci is present at the top of the abdomen. There is little change in body form during growth and development (gradual metamorphosis). In temperate climates, Orthoptera often have only one generation each year, and the winter is usually passed in the egg stage.

Sound production is very common among the Orthoptera, particularly in grasshoppers and crickets. Sounds are made by rubbing their body parts together. This behavior, called **stridulation,** usually is involved in mate finding.

**Family Acrididae—grasshoppers or locusts.** Grasshoppers (Fig. 3.8) are some of the best-known insects. They are characterized by threadlike antennae that are shorter than the body and by hind legs modified for jumping. The ovipositor of females is short, and the tarsi are three-segmented.

Grasshopper eggs are usually laid in soil and frequently in grass sod. In temperate regions, winters are spent in the egg stage, and hatching occurs in the spring. Grasshoppers feed on a wide array of green plants and are some of the most serious agricultural pests in the world. Outbreaks occur most often in the western United States and in drier parts of the world or following protracted droughts in wetter areas.

**Family Tettigoniidae—longhorn grasshoppers and katydids.** Longhorn grasshoppers (Fig. 3.9) have antennae longer than the body and some, such as the well-known Mormon cricket (*Anabrus simplex*), are wingless. All have four-segmented tarsi, and most females possess a saberlike ovipositor. Many longhorn

**Figure 3.8** Order Orthoptera, family Acrididae, represented by a bandwinged grasshopper, *Arphia* sp. (Photo by Marlin E. Rice)

**Figure 3.9** Order Orthoptera, family Tettigoniidae, represented by a katydid. Sometimes called longhorned grasshoppers, members of this family have longer antennae than body and legs and four-segmented tarsi. (Photo by Marlin E. Rice)

grasshoppers are green, as are most katydids and false katydids, and several make elaborately patterned sounds. Eggs are laid singly or in groups on leaves or twigs of trees and shrubs, within plant tissues, or in the soil.

Very few species of the Tettigoniidae are known as pests. Occasionally, katydids injure orange trees, but the most outstanding pest in this family is the Mormon cricket. This pest occurs in the Great Basin region of the Rocky Mountains in North America, where outbreaks have caused severe injury to rangeland and cultivated crops. The most famous outbreak occurred in 1848 when this pest threatened the complete destruction of crops in the Mormon settlement. However, flocks of seagulls moved in to consume these insects, and disaster was averted. A monument to commemorate this event was erected in Utah. A recent outbreak occurred in eighteen Utah counties in 2001.

**Family Gryllidae—crickets.** Crickets (Fig. 3.10) are known by almost everyone, as are their chirping sounds. These insects are usually recognized by their long antennae, a lance-shaped ovipositor, and three-segmented tarsi.

Field crickets, *Gryllus* species, are some of the most widely distributed gryllids in North America. Most are black or brown. They feed on a wide array of foods, ranging from plants to other insects, and sometimes they are cannibalistic. They may damage field crops or invade homes, where they may chew holes in clothing.

Other pests in the Gryllidae include the tree crickets (Fig. 3.11), *Oecanthus* species, which insert eggs in bark and stems of trees and shrubs, sometimes causing young twigs to die.

### Order Phasmatodea

**Family Phasmatidae—walkingsticks.** Walkingsticks (Fig. 3.12) are some of the most unusual-looking American insects; they are usually wingless and bear a striking resemblance in shape and color to the tree twig or grass stem on which they are found. They move slowly through the vegetation, consuming leaf material. Females simply drop their eggs to the ground. Usually only one generation is produced per year.

**Figure 3.10** Order Orthoptera, family Gryllidae, a cricket. (Photo by Marlin E. Rice)

**Figure 3.11** Order Orthoptera, family Gryllidae, a tree cricket. (Photo by Dennis Schotzko)

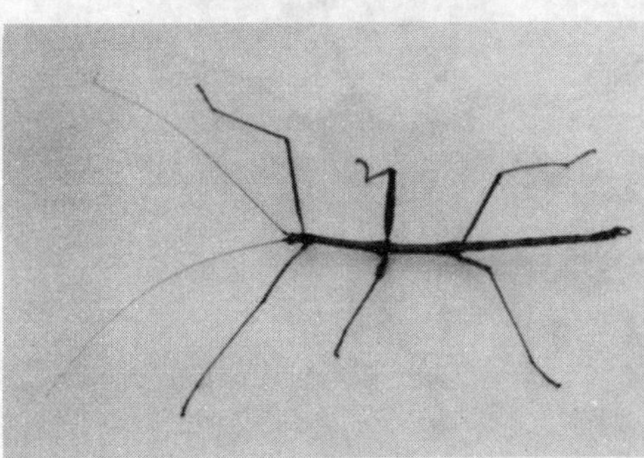

**Figure 3.12** Order Phasmatodea, family Phasmatidae, walkingsticks. These insects resemble tree twigs or grass stems in color and shape. (Photo by Marlin E. Rice)

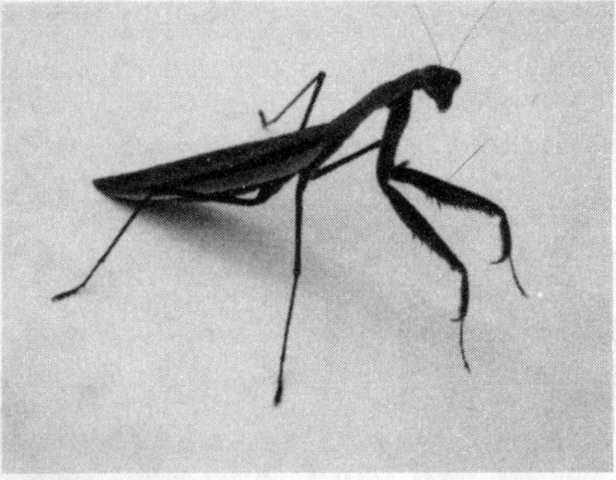

**Figure 3.13** Order Mantodea, family Mantidae, as represented by the Carolina mantid, *Stagmomantis carolina*. Members of this family have front legs for grasping prey and an elongated prothorax. (Photo by Marlin E. Rice)

The longest insect in the United States is a walkingstick, *Megaphasma denticrus*, reaching lengths of 15 to 18 cm. Tropical walkingsticks may exceed 30 cm in length.

### ORDER MANTODEA

**Family Mantidae—mantids (see Color Plate 2).** The mantids, or praying mantids, (Fig. 3.13) also have a striking appearance because of their raptorial front legs and elongated prothoraces. The common name, praying mantis (soothsayer, holy or wise man, prophet), comes from the habit of holding the front legs up in a "praying" manner while waiting for prey. They grasp other insects (including other mantids) with the spined front legs and hold them with these while they eat. Being wholly predaceous, mantids are the only group of the Orthoptera beneficial to humans outside of the laboratory and are sometimes collected as "pets."

Eggs are laid in a grayish egg case, which is glued to tree branches, fence posts, weeds, or other objects. One generation occurs each year.

### ORDER BLATTODEA

**Families Blattidae and Blattellidae—cockroaches.** Many people find cockroaches (Fig. 3.14) the most repulsive of all insects. This feeling arises because some species live in close association with people, inhabiting cupboards, sink cabinets, and other places where food is prepared or stored. However, not all cockroaches are found in human dwellings, and most species exist out of doors in tropical or subtropical climates. Cockroaches prefer warm, dark, humid habitats, where they breed year-round. Their common name is derived from the Spanish *cucaracha*.

Cockroaches have an oval shape and a somewhat flattened body and are usually brown or brownish-black. The head is oriented downward and is barely

**Figure 3.14** Order Blattodea, family Blattellidae, represented by a female German cockroach with ootheca (egg case) protruding from ovipositor. Insects in this order have an oval, flattened shape and head concealed under the pronotum. (Photo by Marlin E. Rice)

visible from above because an extended **pronotum** covers most of it. Very often, two pairs of fully developed wings are present, but occasionally they are reduced or lacking. Cockroaches feed on many kinds of plant and animal products, including human food, sewage, garbage, binding and sizing of books, and wallpaper paste. Most species lay eggs in egg pods, or **oothecae,** and reproduction occurs year-round. Most urban species cannot survive out-of-doors in temperate climates.

Cockroaches are economically important because of the loss of revenue in managing them. Many home dwellers attempt pest control themselves using relatively safe household pesticides. Elimination of cockroaches in commercial establishments is usually achieved by contracting the services of structural pest control firms. Most pest species in the United States are members of the families Blattidae and Blattellidae.

### ORDER MANTOPHASMATODEA—gladiators

Gladiators were described as recently as 2002, representing the first new insect order since 1914. Species representing this order were found in the Brandberg Mountains of Namibia, on the west coast of southern Africa. Gladiators have characteristics similar to walkingsticks, mantids, and grasshoppers. The insects differ from walkingsticks, in that the first body segment is the largest. They differ from mantids in using both fore and mid legs to catch prey, and they cannot jump as do grasshoppers. Gladiators are about 1.6 inches long (4 centimeters), and are carnivorous and nocturnal. They live in clumps of grass growing in rock crevices. Fossils of this insect had been seen previously in 40-million-year-old golden amber, and the impression was that it was extinct. Studies are continuing on species of the order and their geographical distribution.

### ORDER DERMAPTERA—earwigs

Earwigs (Fig. 3.15) are medium-sized insects that superficially resemble certain types of beetles. They are elongate insects with a pair of pincerlike cerci at the tip of the abdomen. These insects may or may not have wings. Those with

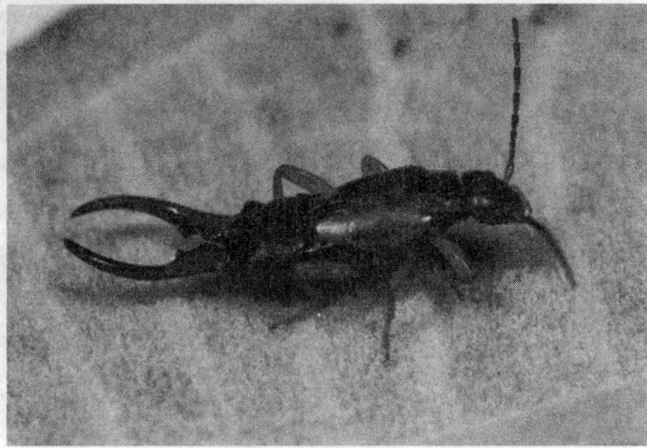

**Figure 3.15** Order Dermaptera, earwigs, represented by the European earwig, *Forficula auricularia*. Insects in this order have short front wings that form leathery coverings for the hind wings and a pair of pincerlike cerci. (Photo by Marlin E. Rice)

wings have front wings that form short, leathery coverings for the large membranous hind wings. Earwigs have chewing mouthparts and three-segmented tarsi and show little change in body form during growth (gradual metamorphosis). Some species have glands with openings on the abdomen that emit a foul-smelling liquid believed to serve as protection from enemies.

Earwigs are active at night and hide during the day. They feed mostly on decaying organic matter, but a few occasionally feed on plants, and others are predaceous. Eggs are laid in a cell in soil, and females watch over the eggs and young. In temperate climates, there is only one generation per year.

The name earwig is believed to have come from the erroneous belief that these insects crawl into people's ears. They do not bite but will pinch with the cerci, sometimes painfully so, when handled. The greatest concern with earwigs is as a nuisance when they enter households.

### ORDER ISOPTERA—termites

Termites (Fig. 3.16) are the most dreaded insect pests of human dwellings and other structures because of their ability to destroy wood and wood products. Termites are often called "white ants" because of their color and social habit. Although about the same size, their bodies differ from those of ants by having an abdomen broadly joined to the thorax versus one joined by a narrow waist. The front and hind wings of termites are the same shape and size, but in ants the front wings are the larger. Moreover, termites spend their lives in wood, soil, or earthen tunnels they build, but ants move openly above ground. Termite mouthparts are the chewing type, and there is little body change during growth (gradual metamorphosis). Antennal segments are beadlike.

Termites are social insects, living in communal nests and having a division of labor among individuals. The termite society is based on castes, with members of each caste differing from those of other castes in body structure and behavior. In most common species, these castes include primary reproductives (kings and queens), supplementary reproductives, workers, and soldiers.

**Figure 3.16** Order Isoptera, termites. Winged reproductives and workers. These insects superficially resemble ants but their abdomen is broadly joined to the thorax. (Photo by Dennis Schotzko)

New termite colonies are formed, usually in the spring or fall, by winged primary reproductives that swarm from the parent colony. After a short flight, the wings break off, and males and females form pairs. Each pair excavates a cell in soil or wood, and they mate. The original king and queen remain mated for life, and both are involved in rearing the first termite brood. Workers and soldiers are produced in the first brood, after which the queen's primary activity is egg laying.

Workers are immatures, undifferentiated individuals that are wingless and lack compound eyes. They do the work of the colony: gathering food; feeding reproductives, soldiers, and the very young; constructing nests, associated tunnels, and galleries; and completing other tasks.

Soldiers have enlarged heads and/or mandibles, and they may or may not have compound eyes. When present in a colony, soldiers attack intruders or protect the structural integrity of the nest. In some species soldiers are called **nasuti;** these individuals use a glandular snout to spew a sticky material over intruders. In colonies where they occur, supplementary reproductives replace the reproduction of the primary king and queen when the latter decline and die or are lost.

Most termites are saprophagous, feeding on dead wood and other nonliving plant materials. The cellulose of wood is a major termite food component. As mentioned in chapter 2, cellulose is digested with the aid of symbiotic microorganisms that supply the needed enzymes for this purpose. Because they can break down woody tissues, termites have extraordinary ecological importance in the humification of soil.

Two groups of termites in the United States have a major economic impact; drywood termites and subterranean termites.

**Family Kalotermitidae—drywood termites.** Drywood termites are found in the southwestern United States and attack sound, dry wood of buildings, stored lumber, and furniture. They do not require external moisture or contact with the ground, as do subterranean termites. Another group in this family, powderpost termites, occurs in the southeastern United States and also attacks dry wood, turning it into sawdust.

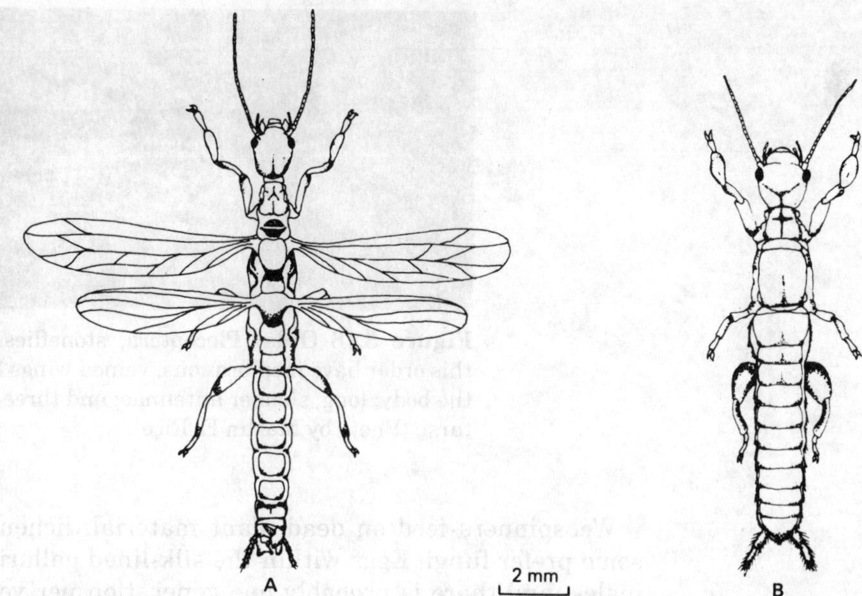


**Figure 3.17** Order Embioptera, webspinners. A. Winged male. B. Wingless female. Members of this order have threadlike antennae, three-segmented tarsi, and short cerci. (Reprinted with permission of Macmillan Publishing Company from *Insects in Perspective* by Michael D. Atkins. © 1978 by Michael D. Atkins)

**Family Rhinotermitidae—subterranean termites.** Subterranean termites cause damage throughout most of the United States and account for approximately 95 percent of monetary losses from termites. Subterranean termite colonies almost always have contact with the soil. They move above ground, usually by building earthen tubes that lead to wood not touching the soil surface. These tubes allow the insects to get food and obtain moisture for survival.

Damage from subterranean termites is prevented by proper construction—keeping wood away from soil. The soil of the building site can be treated with insecticides as an added preventive. Drywood termite damage is usually prevented by insecticide fumigation of buildings. This fumigation involves enclosing the entire structure in a plastic envelope.

## ORDER EMBIOPTERA—webspinners

Webspinners (Fig. 3.17) are small insects, 4 to 7 mm long, that are rarely encountered. Males may be winged or wingless, and females are always wingless. Mouthparts are the chewing type, antennae are threadlike, and tarsi are three-segmented. A pair of short cerci is found at the tip of the abdomen. An unusual characteristic of webspinners is the presence of silk glands and spinning hairs on the first segment of each front tarsus. Body form changes little from young to adult stages (gradual metamorphosis).

These insects are most common in tropical and subtropical areas. They are gregarious, and some build communal silken tunnels under stones, tree bark, logs, and in other environments. Small numbers of eggs are laid in these tunnels.

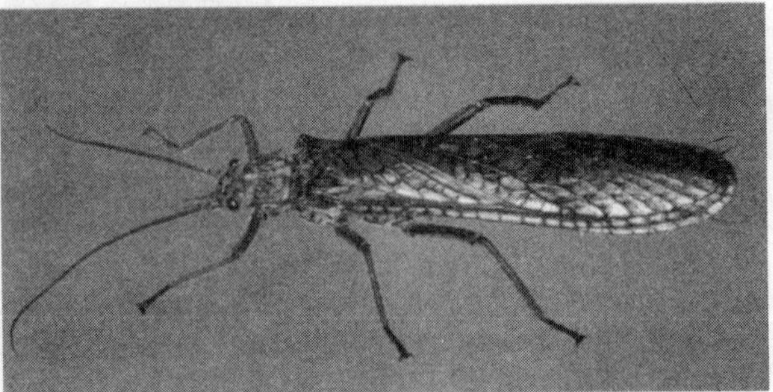

**Figure 3.18** Order Plecoptera, stoneflies. Adults in this order have membranous, veined wings longer than the body; long, slender antennae; and three-segmented tarsi. (Photo by Marlin E. Rice)

Webspinners feed on dead plant material, lichens, moss, and bark, and some prefer fungi. Eggs within the silk-lined gallaries are tended by the females, and there is probably one generation per year. No webspinners are known as pests.

### ORDER PLECOPTERA—stoneflies

Like mayflies, stoneflies (Fig. 3.18) are associated with aquatic environments. They are drab gray, green, brown, or black, and adults have two pairs of membranous, multiveined wings that exceed the length of the body. The wings fold flat over the abdomen at rest. Antennae are long and slender, and tarsi are three-segmented. A prominent pair of cerci is at the tip of the abdomen, and mouthparts, although reduced in the adults of some species, are the chewing type.

Body form is similar in both young and adults. Stonefly young resemble those of mayflies but have two tails rather than three and have branched gills on the thorax and leg bases rather than along the abdomen.

The name stonefly comes from the habits of the immatures, which crawl under stones on the bottoms of streams and along lake shores. The young of some species eat plants, but other species are predaceous or even omnivorous. As their name implies, winter stoneflies are active as adults and reproduce during the winter. Adults are weak fliers and, in some species, do not feed. The smaller stonefly species produce one generation per year, but those with larger individuals may require more than one year to complete development.

Stoneflies are largely beneficial insects, serving as food for fish and other aquatic animals. None is known to be a pest.

### ORDER ZORAPTERA—zorapterans

Zorapterans (Fig. 3.19) are very rare, minute insects (fewer than 3 mm long) that superficially resemble termites. Adults of either sex may be winged or wingless. When present, both pairs of wings are membranous with few veins. Wingless forms lack eyes. Antennae are beadlike, and a pair of short cerci that each terminate in a bristle are found at the tip of the abdomen. The tarsi are two-segmented, and mouthparts are the chewing type. Immatures resemble adults (gradual metamorphosis).

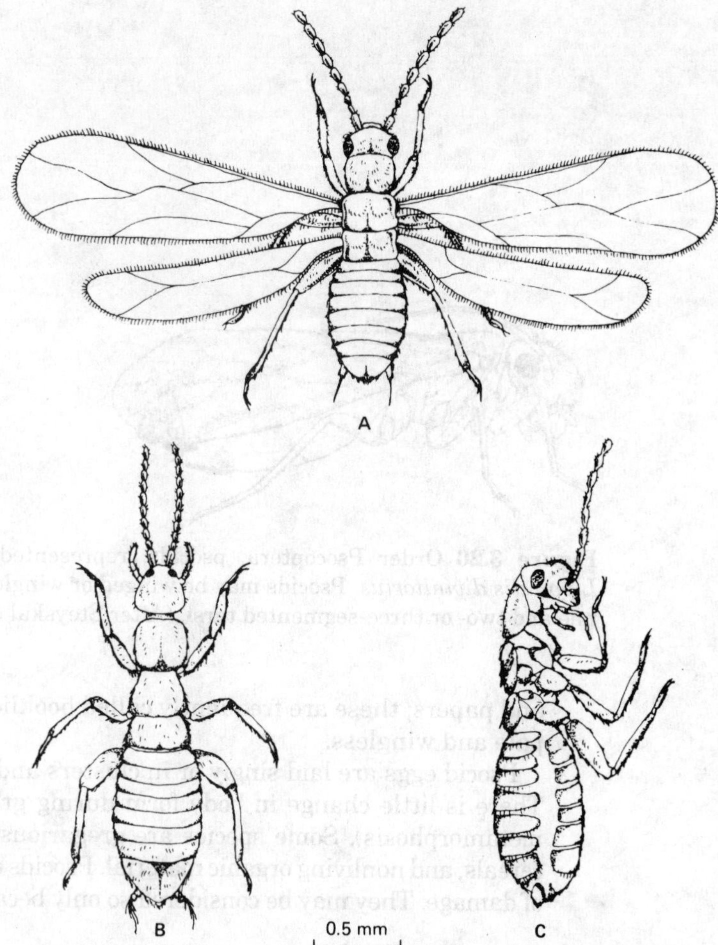


**Figure 3.19** Order Zoraptera, zorapterans, represented by *Zoroptypus hubbardi.* A. Winged female. B. Wingless female. C. Dewinged female. (Reprinted with permission of Macmillan Publishing Company from *Insects in Perspective* by Michael D. Atkins. © 1978 by Michael D. Atkins)

Aggregations of zorapterans may be found in rotten logs, under bark, and in piles of sawdust. Their food seems to be fungal spores and dead arthropod tissue. Zorapterans are not known as pests.

## ORDER PSOCOPTERA—psocids and booklice

Psocids (Fig. 3.20) are small, delicate insects, less than 6 mm long. Adults usually have two pairs of membranous, sparsely veined wings that are held rooflike over the body when not in use and antennae that are long and threadlike. These insects have chewing mouthparts and two- or three-segmented tarsi.

Most psocids occur on tree bark, in the foliage of trees and shrubs, and under stones and dead leaves. These species are often called barklice. However, the psocids most often seen by the average person occur in buildings among books

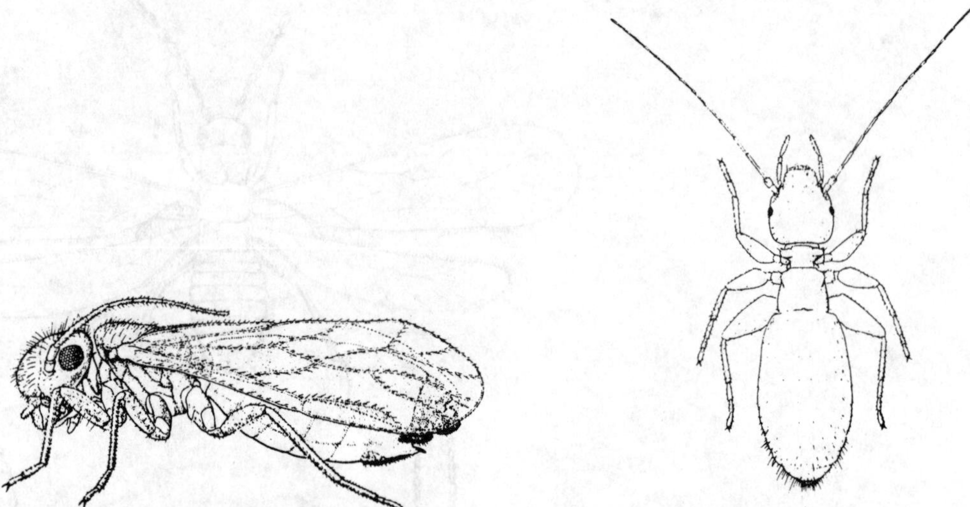

**Figure 3.20** Order Psocoptera, psocids, represented by *Ectopsocus pumilis* (*left*) and *Liposcelis divinitorius*. Psocids may be winged or wingless and have long, threadlike antennae and two- or three-segmented tarsi. (After Steyskal et al., 1986, courtesy USDA)

and papers; these are frequently called booklice. The indoor species are mostly pale and wingless.

Psocid eggs are laid singly or in clusters and may be covered by a silken web. There is little change in body form during growth to the adult stage (gradual metamorphosis). Some species are gregarious, and most feed on pollen, fungi, cereals, and nonliving organic material. Psocids cannot be considered pests because of damage. They may be considered so only because of their unwanted presence.

### ORDER PHTHIRAPTERA—chewing lice and sucking lice

Chewing lice (Fig. 3.21) are parasites of both birds and mammals, but because many species are found on birds, they are often called bird lice. These small, flattened parasites are wingless and have a large head that is wider than the thorax. They have reduced eyes and short antennae. As the common name implies, they have chewing mouthparts. The young greatly resemble adults (gradual metamorphosis).

The tarsi of chewing lice have either one or two claws at their tips. Although a few exceptions exist, those with one claw usually parasitize mammals, and those with two claws parasitize birds. These claws form clasping organs adapted for clinging to hair and feathers.

Chewing lice are ectoparasites; they live and feed on the outside of their hosts. Food of these insects consists mainly of feathers, hair, skin scales, blood clots, and other surface materials of the host. They are usually quite specific in choosing hosts, and the entire life is spent on the host. Eggs, commonly called **nits,** most often are glued near the bases on feathers or hairs, and 3 to 7 weeks may be required for development of adults.

The principal injury caused by these insects is irritation of the host. In domestic animals, such irritation can reduce the rate of weight gain to cause a subsequent loss in profit. However, deaths of young chickens are known as a

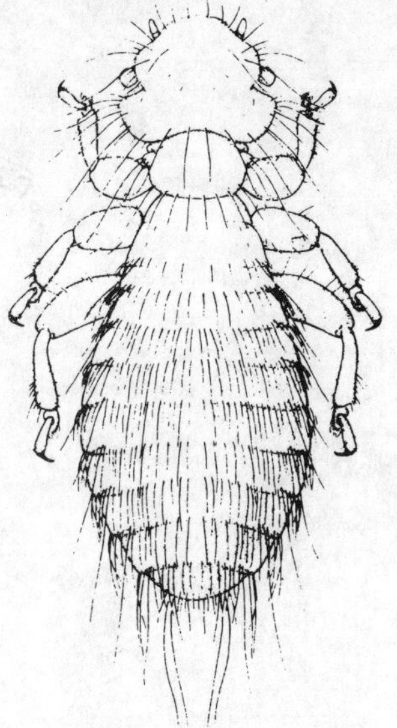

**Figure 3.21** Order Phthiraptera, chewing lice, represented by *Meromenopon meropis.* Chewing lice are wingless ectoparasites with a head wider than the thorax. (After Steyskal et al., 1986, courtesy USDA)

result of louse infestations. Chewing lice do not transmit disease organisms and do not parasitize humans. Two of the more important pests of poultry are the chicken head louse, *Cuclotogaster heterographus*, and the large turkey louse, *Chelopistes meleagridis*. Both are members of the family Philopteridae.

Infestations by these parasites are usually managed by dipping animals or dusting them with suitable insecticides.

Sucking lice (Fig. 3.22) are small, wingless ectoparasites that feed on the blood of humans and other mammals are distinguished by their narrow heads (narrower than the thorax), mouthparts formed for piercing-sucking, and flattened bodies. Eyes are reduced or absent, and legs bear a single claw that folds back against the tibia. This claw and tibia arrangement forms an efficient mechanism for grasping hair. There is little change in body form during growth (gradual metamorphosis), and the entire life is spent on the host.

These parasites are pests because of the irritation they cause, but some can also transmit disease organisms. The most important lice that infest humans are two species: head and body lice and crab lice.

**Family Pediculidae—head and body lice.** Head and body lice belong to the same species, *Pediculus humanus*, but are divided into subtaxa, according to the region of the body they infect. The head louse, *Pediculus humanus capitis*, is found only on the head and is transmitted from person to person by the sharing of caps, hair brushes, combs, and other such items. Body lice, or "cooties," *Pediculus humanus humanus*, are found on the body and are transmitted by shared clothing and bedding or migration from one pile of clothes to another.

Head lice attach their eggs to hair, and body lice lay them along seams of clothing. These lice feed at frequent intervals by piercing the host's skin to suck a blood meal. The irritation from their activities causes itching. The body louse

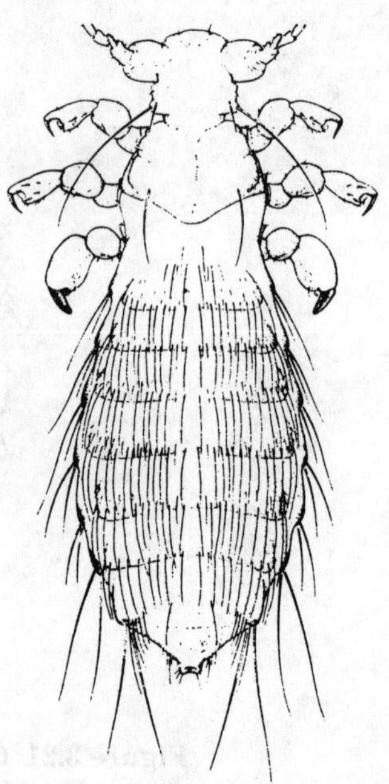

**Figure 3.22** Order Phthiraptera, sucking lice, as represented by *Polyplax stephensi.* Sucking lice are wingless ectoparasites, but have a head narrower than the thorax and mouthparts formed for piercing and sucking. (After Steyskal et al., 1986, courtesy USDA)

transmits microorganisms that cause epidemic typhus in humans. Infection occurs when louse feces or the crushed body of the louse is scratched into the skin but not by the louse's bite. Until the development of DDT, typhus epidemics associated with war often killed more people than combat. Other diseases associated with body louse transmission are relapsing fever and trench fever, the latter important in World War I.

Today, body lice are not as common as in the past because of more frequent bathing and laundering of clothes. However, in many cities of the United States, head louse epidemics continue to be frequent among young school children.

**Family Pthiridae—crab lice.** Crab lice, *Pthirus pubis*, are shorter and more rounded than head and body lice. These parasites occur mainly in human pubic hair but may be present on other parts of the body, including eyelashes and facial hair. Eggs are glued to body hair. This species is irritating but does not transmit disease organisms.

**Family Haematopinidae—wrinkled sucking lice.** Several sucking lice are important pests of domestic animals. Some of the more significant are the hog louse, *Haematopinus suis*, on hogs and the shortnosed cattle louse, *Haematopinus eurysternus*, on cattle. Most lice on domestic animals are managed with appropriate insecticidal sprays and dips.

### ORDER THYSANOPTERA—thrips

Thrips (the singular is also thrips) are very small insects, usually 2 to 3 mm long (Fig. 3.23). The most striking features of thrips are the two pairs of wings

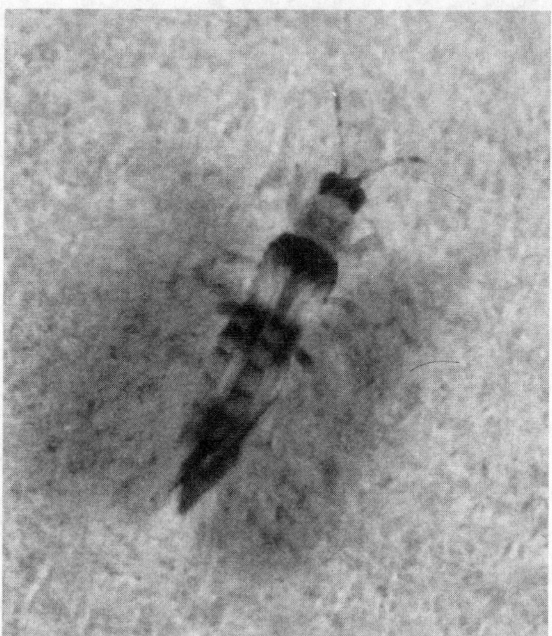

**Figure 3.23** Order Thysanoptera, thrips. Major features found in this order include wings fringed with long hairs; conelike, rasping-sucking mouthparts; and two-segmented tarsi without claws. (Photo by Marlin E. Rice)

that are fringed with long hairs and the conelike rasping-sucking mouthparts. The tarsi are also quite unusual; each is one- or two-segmented, lacks claws at the tip, and has a bladderlike organ that protrudes when the tarsus touches an object. The body form of immatures is generally the same as that of adults, and a quiescent period occurs in full-grown immatures just before they reach adulthood; still, their development is often considered a gradual metamorphosis. Thrips have relatively large compound eyes and six- to nine-segmented antennae.

Most thrips reproduce sexually; however, some need not mate to produce young. Thrips may or may not have an ovipositor; those that do lay eggs in plant tissues, and those that do not lay eggs in crevices or under loose bark. Several generations are usually produced each year.

Thrips often occur in great numbers in flowers and feed on a wide array of cultivated plants. They can be pests on onions, pears, cotton, soybeans, citrus, and greenhouse plants. Some are predaceous. Plant injury is produced when these insects rasp the surface of leaves, stems, buds, and flowers, often causing necrotic strips (silvery appearance) or the dieback of stems. In some instances, injury is most severe in the seedling stage, when feeding is concentrated on a small surface. Some plant disease organisms are vectored by thrips, including the virus that causes tobacco ringspot disease in soybeans.

Some of the important pest species are onion thrips, *Thrips tabaci*, attacking onions, melons, and other garden crops; flower thrips, *Frankliniella tritici*,

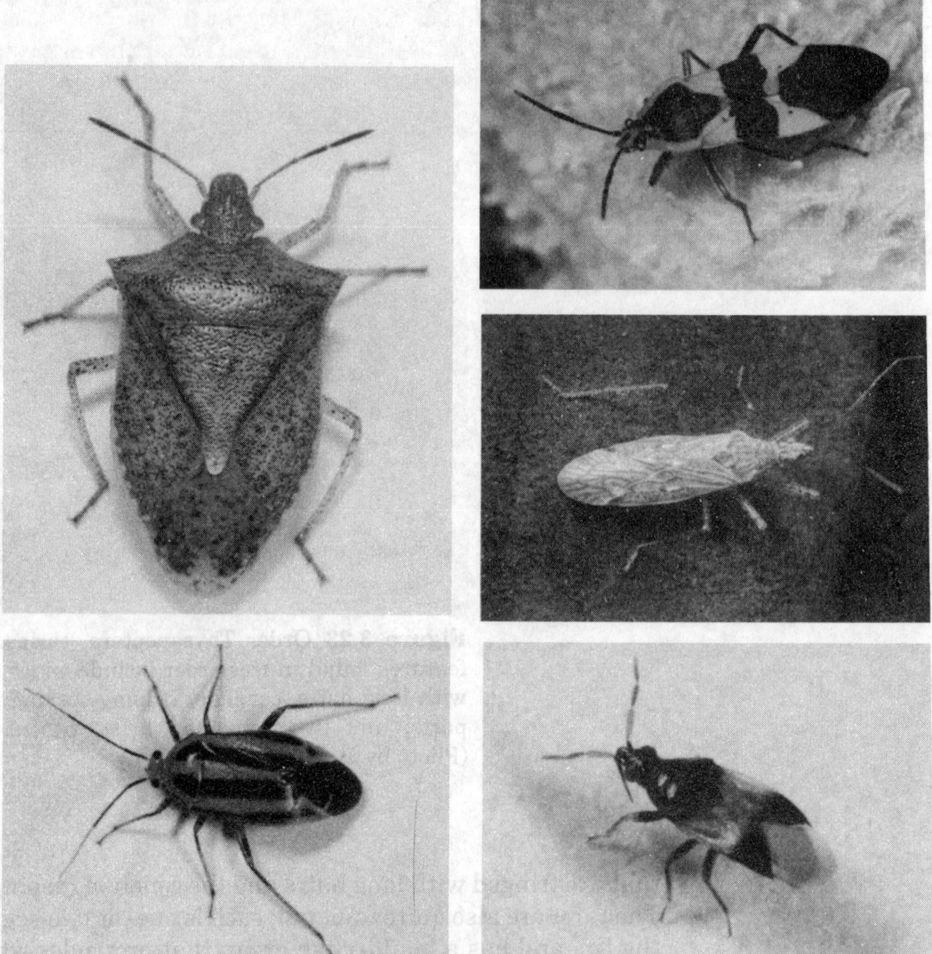

**Figure 3.24** Suborder Heteroptera showing representative families (not to scale). Top row (*left to right*): Family Pentatomidae (stink bug) and family Lygaeidae (milkweed bug). Bottom row: Family Nabidae (damsel bug), family Anthocoridae (insidious flower bug), and family Miridae (plant bug). The major features of the Heteroptera include an abdomen broadly joined to the thorax, and front wings (hemelytra) that are thickened at the base and membranous at the tip. (Photos by Marlin E. Rice)

attacking flowers, wheat, strawberries, and other crops; and citrus thrips, *Scirtothrips citri*, feeding on buds, new growth, and fruit of citrus trees.

## *Order Hemiptera—true bugs, aphids, hoppers, and scales

### Suborder Heteroptera

Although many people refer to all insects as "bugs," the only true bugs are heteropterans (Fig. 3.24). These are some of the most common insects, occurring in both terrestrial and aquatic habitats. Although many of our most important plant pests are bugs, several are also important natural enemies that help to destroy insect pests. Yet other bugs are ectoparasites of humans and other warm-blooded animals.

Bugs have a somewhat flattened body, and they have an abdomen that is broadly joined to the thorax. Most have two pairs of wings; the front wings, called **hemelytra,** are the most distinctive characteristic of the order. Hemelytra are thickened at the bases and membranous at the tips, and when at rest, they overlay the membranous hind wings. Between the bases of the wings, there is usually a triangular plate, called a **scutellum,** that may cover much of the body in some species.

The mouthparts of heteropterans are the piercing-sucking type, with which bugs pierce tissues of plants or animals and extract sap or blood. Many bugs have scent glands from which they secrete disagreeable odors, presumably for protection. Heteropteran antennae are moderately long, and compound eyes are usually well developed.

Eggs are laid most often in plant tissues, on the surface of plants, or in crevices. Growth and development is gradual, with immatures resembling adults but lacking fully developed wings and, sometimes, adult coloration.

The Heteroptera comprises fifty-four families, but only a few of the prominent terrestrial families are mentioned here.

**Family Pentatomidae—stink bugs.** Stink bugs are relatively large, triangular bugs, with five-segmented antennae and a large scutellum. The common name is derived from the disagreeable odor they produce. Most overwinter as inactive adults in the northern United States, but they may be active year-round in the South. Eggs are usually barrel-shaped and are laid in clusters.

Stink bugs are largely plant feeders, sucking sap from leaves, stems, and reproductive parts. An imported species, the southern green stink bug, *Nezara viridula*, is a particularly important pest in the southern United States, where it injures soybeans, peas, beans, cotton, tomato, pecan, citrus, and other crops. On the other hand, the spined soldier bug, *Podisus maculiventris*, is beneficial to humans by feeding on many plant pests and particularly on caterpillars. Other species, such as the green stink bug, *Acrosternum hilare*, are both phytophagous and predaceous.

**Family Coreidae—leaffooted bugs.** The leaffooted bugs are so named for the hind tibia, which is expanded and leaflike in some species. These insects have four-segmented beaks and antennae. Most are plant feeders, and the squash bug, *Anasa tristis*, which injures cucurbits, is probably the most important pest.

**Family Lygaeidae—seed bugs.** Lygaeidae is a large family of small bugs much like the Coreidae, but they have only a few veins in the membrane of the hemelytra. The most important pest of this group is the chinch bug, *Blissus leucopterus*, which has been a devastating pest of wheat, corn, and other cereals. Most recently it has become an important pest of turf grasses. Another group of lygaeids, the Geocorinae (*Geocoris* species and others), is important as natural enemies of insect pests.

**Family Nabidae—damsel bugs (see Color Plate 2).** Damsel bugs are slender and have a number of small cells formed by veins around the membranous margins of the hemelytra. Damsel bugs, *Nabis americoferus* and other species, are predaceous on many different insects, particularly aphids and caterpillars.

**Family Anthocoridae—pirate bugs.** Pirate bugs are very small bugs (2 to 5 mm long) that often have black-and-white markings and a three-segmented beak. Many of the species are predaceous on other small insects and insect eggs. One of the most beneficial of these natural enemies is the insidious flower bug, *Orius insidiosus*, an important predator of the corn earworm, *Helicoverpa zea*, and soybean aphid, *Aphis glycines*.

**Family Miridae—plant bugs.** The Miridae family contains more species than any other in the order. These insects are recognized by the presence of a **cuneus,** a special part of the hemelytron set off by a crease, and one or two closed cells in the membranous part of the wing. Although the name implies that these are all plant feeders, a few species are predaceous. One of the most important pests in this group is the tarnished plant bug, *Lygus lineolaris*, which causes damage to alfalfa, vegetables, and flowers in the eastern United States (see Color Plate 6). A related species, *Lygus hesperus*, is an important pest of alfalfa and cotton in the western United States.

### Suborder Auchenorrhyncha

Insects in this suborder (Fig. 3.25) are a very large, important group of insects that are so diverse that no single common name represents the whole suborder.

Auchenorrhynchan mouthparts are piercing-sucking like those in the Heteroptera, but unlike the Heteroptera, wherein the beak arises from the front of the head, the auchenorrhynchan beak joins to the back of the head. Auchenorrhynchans usually have two pairs of wings; the front pair is membranous (or slightly thickened) and uniformly thick throughout, whereas the back pair is often thinner and membranous. When at rest the wings are held rooflike over the body. Winged or wingless individuals may occur in one sex or the other in certain species; in other species, both conditions may be found in the same sex.

Antennae vary from short and bristlelike to long and threadlike. Compound eyes are usually present and well developed, but ocelli may be absent in some groups.

Most auchenorrhynchans develop in a simple manner from immatures that resemble adults (gradual metamorphosis). However, whiteflies have a quiescent stage just before adulthood, as found in more complex forms of development. All auchenorrhynchans are plant feeders, and many are important agricultural pests.

There are fifteen auchenorrhynchan families, of which only a few representatives are mentioned here.

**Family Cicadidae—cicadas (see Color Plate 2).** Cicadas are widely known by many people who recognize their calling sound coming from trees in summer. They are sometimes mistakenly called "locusts," but this name is appropriate only for some of the shorthorned grasshoppers.

The cicadas are the largest of the suborder, ranging from 25 to 50 mm in length. They can be recognized by their large size and two pairs of membranous wings. The most common types are the large dogday cicadas (several species) and the smaller periodical cicadas (*Magicicada* species). Periodical cicadas require 13 or 17 years to develop, depending on the particular brood. Mass emergence of these insects is a particularly noticeable event because some broods are numerous and loud.

Cicadas injure woody plants by inserting eggs into twigs. This injury may cause the terminal parts of twigs to die.

**Family Cicadellidae—leafhoppers.** Leafhoppers usually are less than 13 mm long and resemble some of the other "hoppers" of the suborder. Some are

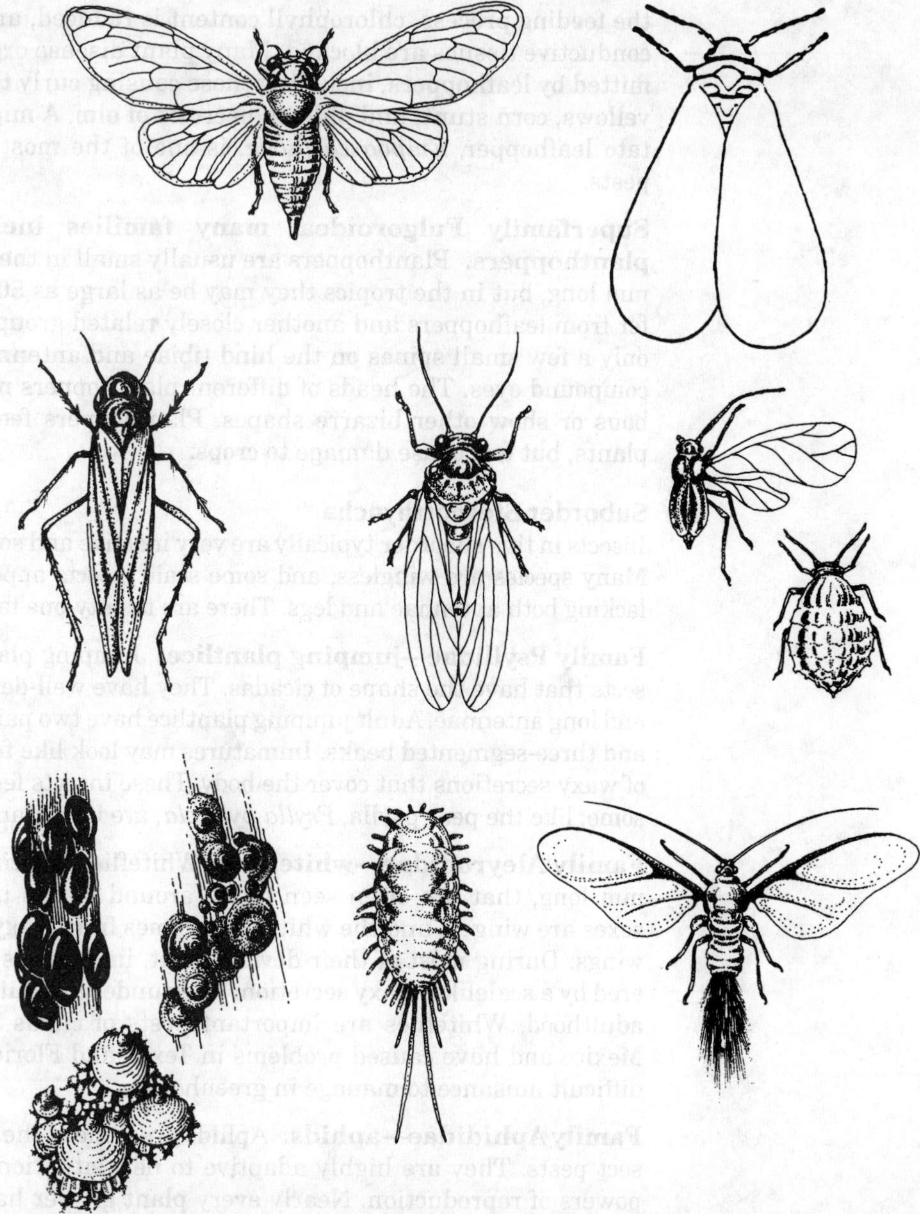

**Figure 3.25** Order Hemiptera showing representative families. Top row (*left to right*): Family Cicadidae (cicada) and family Aleyrodidae (whitefly). Middle row: Family Cicadellidae (leafhopper), family Psyllidae (jumping plantlouse), and family Aphididae (aphid). Bottom row: Superfamily Coccoidea, (*left and center*) female scale insects and (*right*) a winged male scale insect. (Reprinted with permission of Macmillan Publishing Company from *Insects in Perspective* by Michael D. Atkins. © 1978 by Michael D. Atkins)

brightly colored and patterned, but all are distinguished by the one or more rows of small spines on the hind tibiae.

Leafhoppers feed on all types of plants, from woody to herbaceous species. The feeding site is usually leaves, where the insect sucks plant juices. During the feeding process, chlorophyll content is reduced, and in certain instances, conductive tissues are blocked. Many plant disease organisms also are transmitted by leafhoppers, including those causing curly top in sugar beets, aster yellows, corn stunt, and phloem necrosis of elm. A migratory species, the potato leafhopper, *Empoasca fabae*, is one of the most ubiquitous leafhopper pests.

**Superfamily Fulgoroidea, many families including Fulgoridae—planthoppers.** Planthoppers are usually small in the United States, 10 to 13 mm long, but in the tropics they may be as large as 50 mm. Planthoppers differ from leafhoppers and another closely related group, spittlebugs, in having only a few small spines on the hind tibiae and antennae that arise below the compound eyes. The heads of different planthoppers may look pointed or bulbous or show other bizarre shapes. Planthoppers feed on a vast number of plants, but few cause damage to crops.

### Suborder Sternorrhyncha

Insects in this suborder typically are very inactive and some are nearly sedentary. Many species are wingless, and some scale insects appear unlike insects at all, lacking both antennae and legs. There are twenty-one families in the suborder.

**Family Psyllidae—jumping plantlice.** Jumping plantlice are very small insects that have the shape of cicadas. They have well-developed legs for jumping and long antennae. Adult jumping plantlice have two pairs of membranous wings and three-segmented beaks. Immatures may look like fuzzy, white balls because of waxy secretions that cover the body. These insects feed on all kinds of plants; some, like the pear psylla, *Psylla pyricola*, are important pests of fruit trees.

**Family Aleyrodidae—whiteflies.** Whiteflies are tiny, white insects, 2 to 3 mm long, that are often seen flying around plants in the greenhouse. Both sexes are winged, and the white color comes from waxy powder that covers the wings. During most of their development, immatures are stationary and covered by a scalelike, waxy secretion. They undergo a quiescent period just before adulthood. Whiteflies are important pests of citrus in the West Indies and Mexico and have caused problems in Texas and Florida (Box 3.1). They are a difficult nuisance to manage in greenhouses.

**Family Aphididae—aphids.** Aphids are some of the most important of all insect pests. They are highly adaptive to new situations and have tremendous powers of reproduction. Nearly every plant grower has problems with aphids at some time.

Aphids are small, delicate insects, usually less than 5 mm long. They can be recognized by their characteristic pearlike shape and the pair of tubelike **cornicles** that protrude from the back of the abdomen. The cornicles secrete a waxlike substance believed to function in protection. As do some other homopterans, aphids secrete **honeydew,** a sugary liquid exuded through the anus. When excreted, honeydew falls onto leaves, twigs, fruit, or other surfaces below, attracting ants that sometimes care for and defend the aphids from predators. Honeydew on plants also may produce an unsightly appearance when sooty mold grows on it.

The seasonal changes of aphid generations may be very complex, with the occurrence of various forms and various reproductive patterns.

## BOX 3.1 Sweetpotato Whitefly

**SPECIES:** *Bemisia tabaci* Gennadius (Hemiptera: Aleyrodidae)

**DISTRIBUTION:** The sweetpotato whitefly is found in the tropics and subtropics worldwide. Its origin is unknown, but the pest was discovered in Florida in 1894. It has been found from Florida to California in the United States.

**IMPORTANCE:** Two sweetpotato whitefly biotypes exist, the so-called cotton strain and the poinsettia strain. The poinsettia strain has caused much concern in recent years, with widespread outbreaks occurring in several crops. The poinsettia strain was first detected in Florida greenhouses in 1986 on poinsettias, then on field crops and in California greenhouses in 1987. The species spread to Arizona between 1988 and 1990.

Sweetpotato whitefly nymphs suck plant sap and excrete a covering of sticky honeydew. Additionally, they transmit viruses, including lettuce infectious yellows and squash silver leaf. Host plants of this pest may number more than 400 species. The insect can cause severe damage to melons, alfalfa, cotton, early-planted lettuce, cauliflower, broccoli, and eggplant. Moderate damage can occur in citrus, grapes, lettuce, sugar beets, tomatoes, and green beans. In California, annual losses have been estimated at more than $100 million during severe outbreaks.

**APPEARANCE:** Adults are four-winged insects, 1.5 to 2.0 mm long, that resemble tiny moths. This appearance comes from the white dust or waxy powder that covers the wings. Some infestations are so heavy that, when the insects are disturbed, a white cloud appears to rise from plants. Immatures appear as transparent oval forms, later developing two yellow spots about one-third of the way along the body. Eggs are oval-shaped and arranged in a semicircle on the underside of leaves. Pupae are opaque yellow. The sweetpotato whitefly adult resembles that of the greenhouse whitefly but can be distinguished by its yellow body and white wings, as opposed to the white body and white wings of the greenhouse whitefly.

**LIFE CYCLE:** Reproduction in this pest is sexual, and females lay between 30 and 300 eggs. Development is somewhat unusual in that there is an active first instar, followed by sessile forms, which look like scale insects in later stages. There is an outwardly quiescent stage, which is usually called a pupa. Adults live between 8 and 12 days, and the entire life cycle requires 14 to 21 days, depending on temperature. In California the sweetpotato whitefly overwinters on desert weeds, colonizes cucurbits in January and February, and migrates to cotton in April and May.

---

Aphids well may be the most destructive pests in the Hemiptera. In addition to direct injury, they are the most important vectors of plant viruses, most of which cause mosaic symptoms. A single species may transmit only a single virus, or it may, as in the case of the green peach aphid, *Myzus persicae*, transmit several different viruses. A recently introduced species, the soybean aphid (Box 3.2) (*Aphis glycines*), causes plant stunting in soybeans and also transmits soybean mosaic virus.

**Superfamily Coccoidea—scale insects.** Scale insects are a group of small, greatly diverse Hemiptera that may be divided into sixteen separate families. Adult female scale insects are wingless and usually lack legs. They remain stationary through much of their lives and are covered with waxy or resinous secretions, thickened integuments, or waxy scales. Males have legs and usually one pair of wings on the mesothorax. The males resemble small gnats but lack

## BOX 3.2 Soybean Aphid (See Color Plate 6)

**SPECIES:** *Aphis glycines* Matsumura (Hemiptera: Aphididae)

**DISTRIBUTION:** The soybean aphid is found in China, Japan, far-eastern Russia, Korea, Thailand, Borneo, Malaya, Philippines, Indonesia, and Australia. In North America, the species has an eastern distribution, being found in at least twenty states and three Canadian provinces. Highest densities have been located in southeast Minnesota, northeast Iowa, southeast and northwest Wisconsin, and northwest Illinois and Indiana.

**IMPORTANCE:** Winged soybean aphids colonize soybeans in early season, producing wingless females that feed especially on young and developing leaves. Aphids feed by sucking plant sap, which can cause leaf curling and plant stunting. As the plants grow, aphid populations expand to the middle of the plant and feed on the underside of leaves. Losses of up to 52 percent have been quantified from this injury with early-season experimental infestations. The aphid has the ability to transmit viruses wherever it occurs. These viruses include soybean mosaic virus. This virus can cause significant yield loss, particularly when plants are also infected with other viruses such as bean pod mottle virus (transmitted by the bean leaf beetle).

**APPEARANCE:** Soybean aphids are small insects, adults being about 1/16 inch long. They are pale yellow and have dark-tipped tubes called cornicles (siphunculi) on the back of the abdomen. A projection on the lower tip of the abdomen, called the cauda . . . , is pale and contrasts with the color of the cornicles. These aphids feed though piercing-sucking mouthparts and have both wingless and winged forms.

**LIFE CYCLE:** The seasonal cycle of soybean aphids is complex, showing alternation of hosts and reproductive mode. The primary host is buckthorn, *Rhamnus* spp. Eggs are produced on buckthorn in fall and overwinter there. The eggs hatch in spring, giving rise to wingless females. These wingless females are parthenogenetic (reproduce without mating) and produce winged females that migrate to soybeans. Those founding females produce wingless females that also reproduce without mating and give rise to active young (viviparous reproduction) on soybean plants in late May and June.

The species is reported to pass through about fifteen generations in soybean during the growing season, producing both winged and wingless forms. Two seasonal peaks may occur, one in July and the other late August to September. The greatest damage usually is from the first peak, about the time of flowering. Near soybean maturity, winged aphid males are produced, as well as winged parthenogenetic females, both of which migrate back to buckthorn. The winged parthenogenetic females then produce wingless sexual females that mate with the winged males. These mated females subsequently lay eggs, beginning a new seasonal cycle.

---

mouthparts (they do not feed as adults) and have a styletlike protrusion at the end of their abdomens.

Scale insects have various modes of development, most of which are very complex. Newly emerged immatures, called **crawlers,** have legs and antennae and are active. As development proceeds, females lose their legs and antennae, and their bodies are covered by secretions. Females complete development under these coverings and produce young there. Males develop much like females except for a quiescent period just before adulthood.

These insects are often divided into three informal groups, including mealybugs, soft scales, and armored scales. Scale insects attack a wide variety of

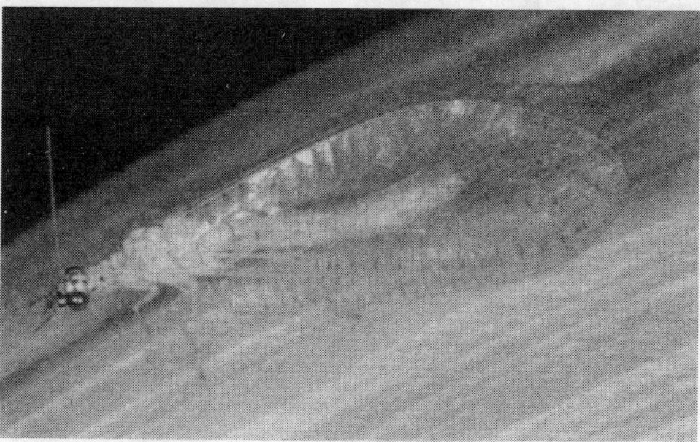

**Figure 3.26** Order Neuroptera, nerve-winged insects, represented by the families Sialidae (alderfly, *left*) and Chrysopidae (lacewing). The most distinguishing feature of this order is the membranous wings with numerous longitudinal and cross veins. These veins give the appearance of branching nerves. (Photos: alder fly, Dennis Schotzko; green lacewing, Marlin E. Rice)

plants; some of the most important pests injure orchard crops, particularly citrus, and ornamentals.

### ORDER NEUROPTERA—nerve-winged insects

Nerve-winged insects (Fig. 3.26) are medium to large insects, with two pairs of membranous wings held rooflike over the abdomen when not in use. The most distinctive feature of adults is the wing venation made up of numerous longitudinal and cross-veins that give a netlike appearance (as a group of branching nerves).

The mouthparts of neuropterans are the chewing type, and antennae are usually long and many-segmented. The tarsi have five segments, and there are no cerci at the tip of the abdomen.

Neuropterans are the first insects discussed whose members undergo a dramatic change from young to adult. This type of change is called complete metamorphosis, which is discussed in more detail in chapter 4.

Nearly all neuropterans are predaceous in both the immature and adult stages. They are found in terrestrial and aquatic habitats. Some of the more important terrestrial groups include the lacewings (family Chrysopidae, see Color Plate 8), which consume pests such as aphids, scale insects, and thrips; antlions (family Myrmeleontidae, see Color Plate 2), whose larvae sometimes build pits for capturing ants; and mantisflies (family Mantispidae), which as adults superficially resemble praying mantids but as immatures are mainly parasitic in the egg sacs of spiders.

The major aquatic neuropterans (suborder Megaloptera) are the dobsonflies, fishflies, and alderflies. The immatures of these groups live in lakes and streams and have gills along the sides of the abdomen. They feed on other aquatic insects and themselves serve as food for fish. Indeed, large larvae of dobsonflies, called **hellgrammites,** are a favorite fish bait.

### *ORDER COLEOPTERA—beetles

Beetles (Fig. 3.27) are by far the largest order of the Hexapoda—with more than one-quarter million described species! In North America, beetles range in length from less than 1 mm to approximately 75 mm.

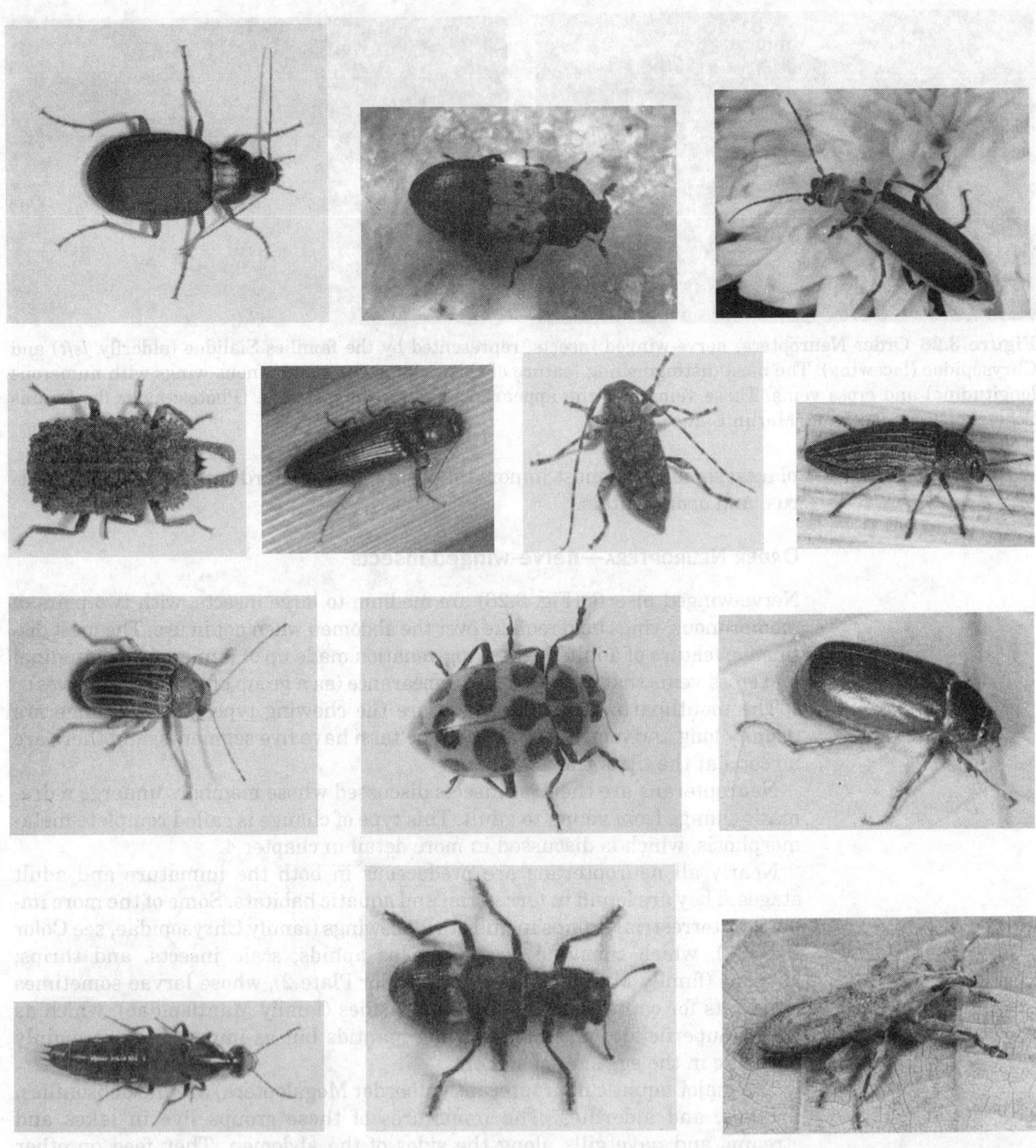

**Figure 3.27** Order Coleoptera, beetles (not to scale). Top row (*left to right*): Family Carabidae (ground beetle), family Dermestidae (dermestid), and family Meloidae (blister beetle). Second row: Family Tenebrionidae (darkling beetle), family Elateridae (click beetle), family Cerambycidae (longhorned beetle), and family Buprestidae (metallic wood-boring beetle). Third row: Family Chrysomelidae (leaf beetle), family Coccinellidae (lady beetle), and family Scarabaeidae (scarab). Bottom row: Family Staphylinidae (rove beetle), family Cleridae (checkered beetle), and family Curculionidae (weevil). The most conspicuous feature of this order is the front wings (elytra), which form leathery or hard coverings over the hind wings. (Photos by Marlin E. Rice)

Most beetles possess two pairs of wings; the front pair is specialized and the most unifying characteristic of the order. These wings are thickened to form leathery or hard coverings called **elytra.** When the insect is not flying, these elytra are folded to form a protective cover over the back. The hind wings are membranous and used for flight. They fold under the elytra when not in use.

Most beetles are hard-bodied, with unusually tough, thick exoskeletons. Both immatures (Fig. 3.28) and adults have typical chewing mouthparts, although in some the head is formed into a long snout, with mouthparts located at the tip. The growth of immatures follows the pattern of complete metamorphosis.

Beetles are found in almost any type of habitat: on plants, on the soil surface, in soil, in water, inside seeds and fruits, and in ant nests, to name a few. Most are phytophagous or predaceous, but a few are scavengers, fungus consumers, or parasites.

Beetle species are divided by authorities into more than one hundred families and four suborders. The two most common suborders are the Adephaga and the Polyphaga. The Adephaga is the smaller group, characterized by hind coxae that divide the first abdominal segment. The Polyphaga is by far the larger group and has hind coxae not dividing the first abdominal segment. Only one family, the Carabidae, is mentioned to represent the Adephaga. The remaining families discussed are a small sampling of the Polyphaga.

**Family Carabidae—ground beetles (see Color Plate 2).** The family Carabidae is a large family of beetles that provide a benefit. There are 2,402 species in the United States and Canada. Most species feed on other insects in both adult and immature stages. Many species are dark and shiny, but a few are bright green, blue, or other colors. These insects hide under rocks and surface

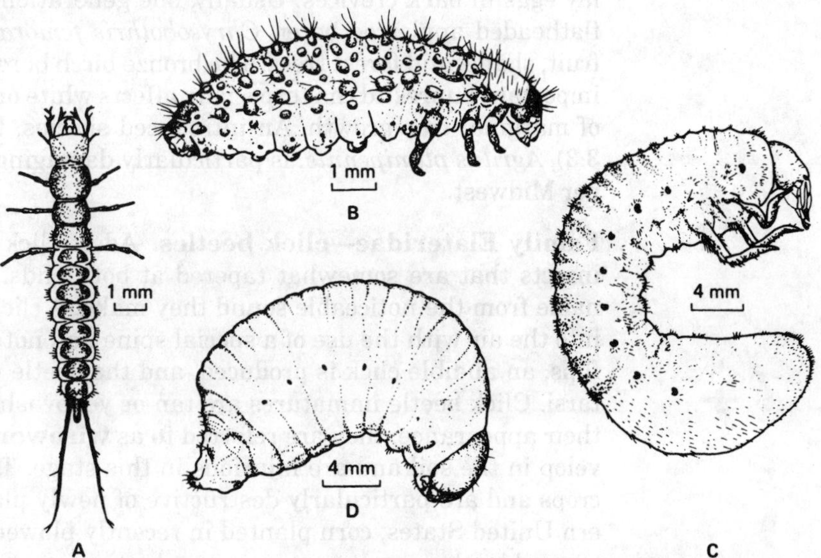


**Figure 3.28** Major forms of beetle larvae. A. Carabaeiform. B. Campodeiform. C. Scarabaeiform. D. Apodus. (Reprinted with permission of Macmillan Publishing Company from *Insects in Perspective* by Michael D. Atkins. © 1978 by Michael D. Atkins)

debris during the day, and most search for prey on the ground at night. They run rapidly when disturbed.

**Family Staphylinidae—rove beetles (see Color Plate 2).** Rove beetles are usually small, elongate insects with short elytra exposing several abdominal segments when viewed from above. Rove beetles run rapidly when disturbed and hold the abdomen up like a scorpion. Most species are predaceous and can be found under stones and other objects, near water, and in nests of birds, mammals, or other insects. Some are parasitic. The Staphylinidae is the largest beetle family in North America north of Mexico, with 4,153 described species. This family represents 16.5 percent of all beetle species in the United States and Canada.

**Family Scarabaeidae—scarab beetles.** The Scarabaeidae family consists of a large number of diverse species. Generally they are oval and bulky or somewhat elongate beetles with five-segmented tarsi and antennae that form plate-like lamellae or compact clubs at the tips. As a group, scarabs feed on plants, dung, decomposing plants and animals, and other material. Plant-feeding pests injure foliage, roots, fruits, and flowers. Common pests like white grubs, *Phyllophaga* species and others, cause serious damage to lawns and golf greens. The Japanese beetle, *Popillia japonica*, an introduced species that causes widespread damage to horticultural and agronomic crops, is the most important American scarab pest.

**Family Buprestidae—metallic wood borers.** Insects in the Buprestidae family are mostly metallic green, blue, or black as adults and fly readily when approached. Buprestid grubs live in shallow burrows beneath tree bark and have a broad, flat thorax; therefore, they are also known as flatheaded wood borers. Adults are usually attracted to unhealthy and dying trees where they lay eggs in bark crevices. Usually, one generation is produced each year. The flatheaded appletree borer, *Chrysobothris femorata*, is an important pest of fruit, shade, and forest trees. The bronze birch borer, *Agrilus anxius*, is another important buprestid that especially affects white or paper birch trees in regions of marginal tree growth. An introduced species, the emerald ash borer (Box 3.3), *Agrilus planipennis*, is particularly damaging to species of ash in the upper Midwest.

**Family Elateridae—click beetles.** Adult click beetles are often brownish insects that are somewhat tapered at both ends. They derive their common name from the noticeable sound they make. A click beetle on its back will flip into the air with the use of a special spine and notch on its underside. When it flips, an audible click is produced, and the beetle usually lands upright on its tarsi. Click beetle immatures are tan or yellowish and cylindrical. Because of their appearance, they are referred to as **wireworms.** Wireworms usually develop in the soil and are injurious in this stage. They attack many cultivated crops and are particularly destructive of newly planted seed. In the midwestern United States, corn planted in recently plowed sod may be completely destroyed by these pests, requiring replanting.

**Family Dermestidae—dermestids.** Dermestids are small, plump, oval beetles. Many are covered with minute scales that may be brown, red, or white; some are uniform brown or black and hairy. Immature dermestids are bristly

## BOX 3.3 Emerald Ash Borer

**SPECIES:** *Agrilus planipennis* Fairmaire (Coleoptera: Buprestidae)

**DISTRIBUTION:** The emerald ash borer is native to Asia, occurring in China, Korea, Japan, Mongolia, far-eastern Russia, and Taiwan. This introduced pest was discovered feeding on ash (*Fraxinus* spp.) trees in southeastern Michigan and identified from this discovery in July 2002.

**IMPORTANCE:** The Chinese have reported damaging populations of this borer primarily from ash forests, but additional Asian tree species have been listed as hosts. In Michigan, the borer has been observed only on ash trees and has been responsible for killing green ash, white ash, black ash, and several horticultural ash varieties. Larvae feed on phloem and the outer sapwood of trees, forming S-shaped feeding galleries that wind back and forth and become progressively wider as larvae grow. Galleries are packed with fine frass. Individual galleries usually extend over an area that is 20 to 30 cm in length, though the length of the affected area can range from 10 to 50 cm. Infested branches in the canopy die when they are girdled by the serpentine tunnels excavated by feeding larvae. Infested trees may lose 30 to 50 percent of the canopy in 1 year, and the trees are often killed after 2 to 3 years of infestation.

**APPEARANCE:** Adults are slender and elongate and 7.5 to 13.5 mm long. The adult body is brassy or golden green overall, with darker, metallic, emerald green wing covers, or elytra. The top of the abdomen under the elytra is metallic coppery red (seen only when the wings are spread). The prothorax, to which the first pair of legs is attached, is slightly wider than the head but the same width as the base of the elytra. The back edges of the covering on the prothorax are wavy, and the top is sculptured with tiny, transverse wavy ridges. The surfaces of the elytra are granularly roughened. Tips of the elytra are rounded, with small teeth along the edge. Larvae reach a length of 26 to 32 mm, are cream-colored, and have a body flattened top-to-bottom. The larval head is brown and mostly retracted into the prothorax, with only the mouthparts visible. The ten-segmented larval abdomen has a pair of brown, pincerlike appendages on the last segment.

**LIFE CYCLE:** The emerald ash borer seems to have a 1 year life cycle in southern Michigan, but 2 years may be required in colder regions. Adult emergence begins in mid to late May, peaks in early to mid June, and continues into late June. The adults are active during the day, particularly when conditions are warm and sunny. Information from China indicates that male adults live an average of 13 days, and females live about 21 to 22 days. Females can mate multiple times, and egg laying begins 7 to 9 days after the initial mating. Females lay a total of sixty-five to ninety eggs. Eggs are deposited individually on the bark surface or in bark crevices on the trunk or branches. In southeastern Michigan, the egg-laying period probably extends into mid to late July.

Eggs hatch in 7 to 10 days. After the eggs hatch, first-stage larvae chew through the bark and into the sapwood of the tree. The insect overwinters as a full-grown larva in a shallow chamber excavated in the sapwood. Pupation begins in late April or early May. Newly emerged adults may remain in the pupal chamber for 1 to 2 weeks before exiting, head-first, through D-shaped holes 3 to 4 mm in diameter.

---

or hairy grubs and are important pests. Although most adults feed on plant pollen, grubs are scavengers of dried organic materials, primarily of animal origin, including leather, furs, skins, wool, silk, and stored food. One important pest, the larder beetle, *Dermestes lardarius*, is particularly a problem in kitchens. Here, it infests nearly any stored food. Grubs of some carpet beetles,

*Anthrenus* species, feed on and ruin household possessions made of animal fibers, including carpet, clothing, upholstery, and curtains. They may even ruin specimens in insect collections.

**Family Coccinellidae—lady beetles.** Probably no insect is more accepted by the general public than a lady beetle. Lady beetles are often colorful little insects well known for their benefits to the farmer and home gardener. Most adults are oval-shaped and convex. They are often tan, black, or red, and spotted with contrasting colors of yellow, white, black, or red. Immatures are elongate, have a tapered abdomen, and are covered with spines. They are often dark with bright orange or red markings.

With few exceptions, both adults and immatures are beneficial, feeding voraciously on any arthropod they can handle, but particularly on aphids, scale insects, and spider mites. Several species, including the famous vedalia beetle, *Rodolia cardinalis*, which saved the California citrus industry from disaster, have been imported and released as a pest management tactic. The native convergent lady beetle, *Hippodamia convergens*, was one of the most common coccinellid species in North America, but has recently been replaced by an introduced species, the multicolored Asian lady beetle (Box 3.4), *Harmonia axyridis*.

An exception to the beneficial nature of lady beetles is the Mexican bean beetle, *Epilachna varivestis*. This coccinellid is an important defoliator of snap beans, lima beans, and soybeans in some regions.

**Family Meloidae—blister beetles.** Blister beetles are medium-sized, cylindrical insects with leathery elytra. The thorax is narrower than the elytra, and they have a complex mode of development referred to as hypermetamorphosis (discussed in chapter 4). Blister beetles are often black, gray, or tan with black stripes. Their common name comes from the effect of their body fluids on human skin. They contain **cantharidin,** a substance that sometimes causes blisters. As a **keratolytic,** cantharidin preparations are prescribed by physicians for the removal of benign epithelial growth (wart removal).

Some immature blister beetles are predators of grasshopper eggs, and others eat eggs of solitary bees along with associated food stores. Adults are pests that often feed gregariously, causing patchy areas of defoliation in plantings. All types of vegetable and agronomic crops may be injured. A recent concern comes from the presence of blister beetle bodies in baled hay. Horses feeding on this hay can die from the ingestion of cantharidin from only a few dead beetles. Such occurrences have resulted in substantial economic losses of thoroughbred race horses.

**Family Cerambycidae—longhorned beetles (see Color Plate 2).** Longhorned beetles have usually large, cylindrical bodies, which are sometimes strikingly marked or patterned. The antennae are long, at least the length of the combined head and thorax and often longer than the entire body. Like the metallic wood borers, cerambycid grubs bore into wood, but they differ in that their bodies are straight and cylindrical, with swollen thoraces. Many of these insects attack unhealthy trees, but some, like the locust borer, *Megacyllene robiniae*, inhabit healthy ones.

**Family Chrysomelidae—leaf beetles.** Leaf beetles are some of the most important agricultural pests in North America. They are rather small insects, usually less than 12 mm long, that are closely related to the cerambycids but are smaller and have much shorter antennae. Many common species are yellow

## BOX 3.4 MULTICOLORED ASIAN LADY BEETLE (SEE COLOR PLATE 8)

**SPECIES:** *Harmonia axyridis* (Pallas) (Coleoptera: Coccinellidae)

**DISTRIBUTION:** This lady beetle is native to Asia but occurs in many areas of the United States, including the Midwest, East, South, and Northwest.

**IMPORTANCE:** This insect was imported and released as early as 1916 in attempt to naturally control certain insect pests. But the first populations were not found in this country until 1988 in Louisiana near New Orleans. Over the years, entomologists released the insect at a number of locations. In addition, accidental entries have occurred via imported nursery items at ports in Delaware and South Carolina. This variably colored and spotted lady beetle is an effective, natural control for harmful plant pests such as aphids, scales, and other soft-bodied arthropods. They have been particularly effective as biological control agents in pecan orchards, allowing for decreases in insecticide use. However, the lady beetle's tendency to overwinter in homes and other buildings, sometimes in large numbers, often makes them a nuisance. If agitated or squashed, the beetles may exhibit a defensive reaction known as "reflex bleeding," in which a yellow fluid with an unpleasant odor is released from leg joints. This reaction usually prevents birds and other predators from eating them, but in the home, the fluid may stain walls and fabrics.

**APPEARANCE:** Adult multicolored lady beetles are large by lady beetles standards, being about 0.2 to 0.3 inch long. They look like other lady beetles, having a domed and oval shape. They are colored yellow, orange, or red and may or may not have spots. The most common U.S. form is mustard to red with sixteen or more black spots. On the white pronotum (top covering near the head), many individuals have several spots that fuse into a regularly or sometimes irregularly shaped "M" pattern.

**LIFE CYCLE:** As other lady beetles, this species displays complete metamorphosis. Adults begin laying eggs on host plants in early spring. Eggs hatch in about 3 to 5 days, and larvae begin searching on plants for aphids and other soft-bodied arthropods on which they feed. Adults and larvae typically feed upon the same prey. Larvae molt four times, becoming larger after each molt, and enter an immobile pupal stage after the last molt. After several days, adult beetles emerge from pupal cases. Development time from egg to adult requires about 15 to 25 days, depending on temperature and food availability. Later in the fall, near the time of killing frosts, the adults seek shelter to spend the winter.

---

or tan with black markings. In many species, such as the well-known potato pest, the Colorado potato beetle, *Leptinotarsa decemlineata*, both immatures and adults feed on plant foliage and flowers. In others, such as the northern corn rootworm, *Diabrotica barberi*, and the bean leaf beetle, *Cerotoma trifurcata*, immatures feed in the soil on roots and associated structures. In a few species, the immatures feed by mining leaves.

**Family Curculionidae—weevils.** Weevils are an economically important group of Coleoptera that, along with other families in the superfamily Curculionoidea, have long snouts with chewing mandibles at the tips. This is the largest insect family. The antennae of curculionids are attached to the snout about halfway along its length. These weevils use the snout to feed internally on plant tissues and notch out egg-laying sites. Immature weevils are light-colored and without legs, and most feed within plant tissues. Adults drill holes and feed in seeds, fruits, and other reproductive parts; some feed on leaves. Weevils rank near the top as pests of certain commodities. Some of the most notable include the boll weevil, *Anthonomus grandis*, on cotton, the alfalfa weevil, *Hypera postica*, on alfalfa, and the rice weevil, *Sitophilus oryzae*, in stored grain.

**Family Scolytidae—bark beetles.** Scolytids are small cylindrical beetles less than 9 mm long. The antennae are elbowed and clubbed at the ends, and the end of the abdomen appears squared off. Immatures are small, white, legless grubs found in tunnels beneath tree bark or burrowing into the heartwood. These insects spend most of their lives in the tunnels, and adults usually leave for only a short time to find new trees to colonize.

Members of this family are often divided into two groups: bark beetles, which feed internally on tree tissues, and ambrosia beetles, which occupy sapwood and feed on ambrosia (fungus) that they cultivate in their burrows. In North America, these beetles cause greater destruction of timber than any other insect. The most important pests occur in the genera *Dendroctonus, Ips*, and *Scolytus*. Elm bark beetles, *Scolytus multistriatus* and *Hylurgopinus rufipes*, are principally important in spreading the fungus *Ceratocystis ulmi*. This organism is the cause of Dutch elm disease, a fatal disease of certain elms in the United States.

## ORDER STREPSIPTERA—twisted-winged parasites

The Strepsiptera (Fig. 3.29) are minute insects, 3 mm long and smaller, that usually are internal parasites of certain Diptera, Hemiptera, Hymenoptera, Orthoptera, and Thysanura.

Male and female strepsipterans differ greatly in their appearance. Males have two pairs of wings, but the front wings are reduced to a pair of paddlelike vestiges. The hind wings are broad and triangular. Males have large, protruding eyes and antennae with lobes extending from some of the segments. They are free-living.

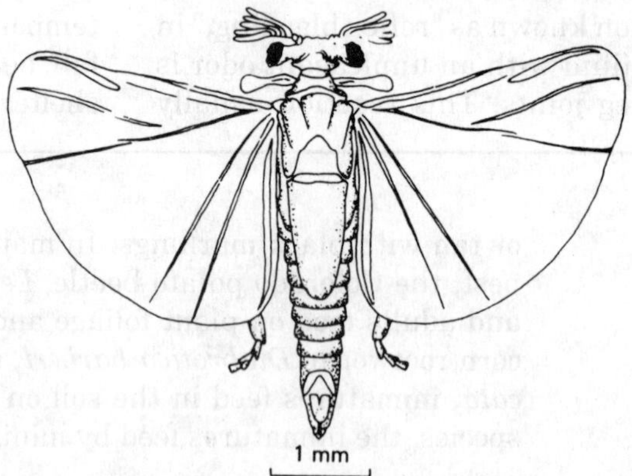


**Figure 3.29** Order Strepsiptera, twisted-winged parasites. The most notable feature of this order is the paddlelike front wings and larger triangular hind wings of males. Females (*not shown*) are wingless and often have no legs. (Reprinted with permission of Macmillan Publishing Company from *Insects in Perspective* by Michael D. Atkins. © 1978 by Michael D. Atkins)

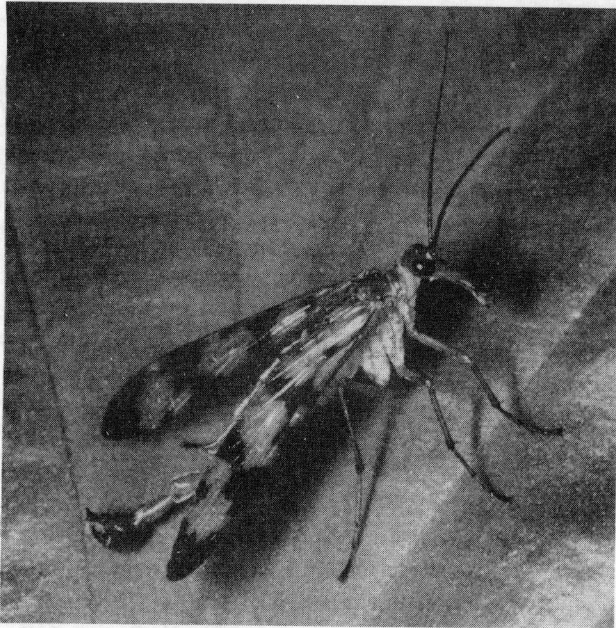

**Figure 3.30** Order Mecoptera, family Panorpidae, scorpionflies. The most notable features of this order are the elongated head that forms a beak and the pincerlike organs at the tip of the male abdomen. (Photo by Marlin E. Rice)

Females are wingless and often without legs; many lack eyes and antennae and have heads fused with thoraces. Most remain within hosts.

All strepsipterans have complete metamorphosis. Because their development resembles that of some beetles, a few authorities group strepsipterans as a family of the Coleoptera. They are of little importance in insect pest management.

### ORDER MECOPTERA—scorpionflies

The Mecoptera (Fig. 3.30) are medium-sized insects, 18 to 25 mm long, and most have an elongated head that forms a beak with chewing mouthparts at the tip. Usually mecopterans have two pairs of multiveined membranous wings and antennae that are long and slender. The name scorpionfly comes from the pincerlike copulatory organ at the tip of the male abdomen that resembles the stinging organ of a scorpion.

Immature scorpionflies look a little like caterpillars but have more abdominal legs. Development is by complete metamorphosis. Both adults and immatures live in moist places with rank vegetation. Most are scavengers, feeding on dead insects and other organic debris, but adult hangingflies (family Bittacidae) are predators. No scorpionflies are pests.

### ORDER TRICHOPTERA—caddisflies

Caddisflies (Fig. 3.31) are mothlike insects with two pairs of membranous, usually hairy wings. In the resting position the wings are held rooflike over the abdomen. Antennae are long and threadlike, and mouthparts are the chewing type; however, mandibles are greatly reduced, and palpi are well developed. Adults are found near water, where they lay eggs.

**Figure 3.31** Order Trichoptera, caddisflies. Adults of this order have two pairs of membranous wings that are usually hairy. (Photo by Marlin E. Rice)

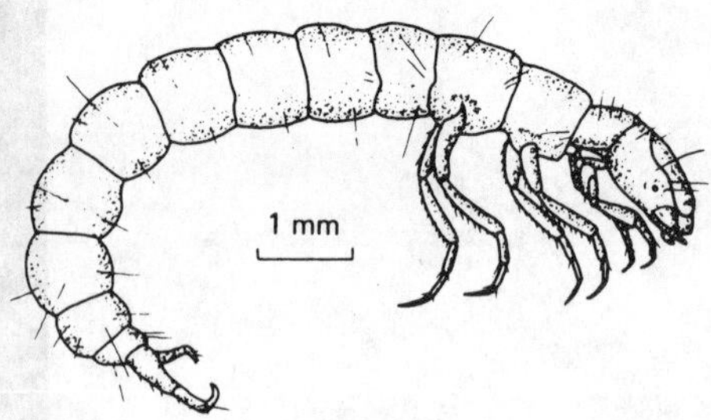


**Figure 3.32** Order Trichoptera, a caddisfly larva. Some larvae construct nets for capturing food, and others build portable cases (*not shown*) in which they live. (Reprinted with permission of Macmillan Publishing Company from *Insects in Perspective* by Michael D. Atkins. © 1978 by Michael D. Atkins)

Immatures (Fig. 3.32) develop in water and show complete metamorphosis. They look somewhat like caterpillars but have only one pair of abdominal legs and usually a series of abdominal gills. Mouthparts are the well-developed chewing type, and food is animal and vegetable debris found in the water. A few species are predaceous, feeding on other aquatic insects.

Some immature caddisflies build portable cases, and others construct nets for capturing food. The case-building immatures build their cases with silk from modified salivary glands. In some species, characteristic materials like sand pebbles, shells, or debris are incorporated into the case, but in others only silk is used. Case builders live inside these protective coverings, coming partly out to feed and retracting into the cases at other times. Net spinners live in fast-flowing water or along lakeshores, where they build tunnellike nets of silk. The nets are tied to rocks or other objects and held open by water flowing through them. The immatures hide near these nets and feed on materials caught in them. Predaceous caddisfly immatures do not build cases or nets.

As a group, caddisflies are considered beneficial in that they are food for fish. Although adults may be a nuisance when attracted to lights, none is considered a pest.

### *Order Lepidoptera—butterflies and moths

Order Lepidoptera is a very large order that includes some of the most beautiful species and some of the most important pests in the Hexapoda. Butterflies and moths are distinguished by their large, membranous wings covered with minute, overlapping scales. These scales give wings rigidity and color. Only occasionally are wings absent or nearly without scales.

Adult mouthparts are usually of the siphoning type and are coiled under the head when not in use. These mouthparts are adapted for feeding on flower nectar and plant juices, and flower pollination often accompanies the feeding process. Some lepidopterans have spines at the tip of the siphoning tube that allow the insects to break the skin of fruit and suck the juices. In a few instances, mouthparts may be reduced, and the adult does not feed. In another rare instance (Micropterigidae), the mouthparts are developed for chewing.

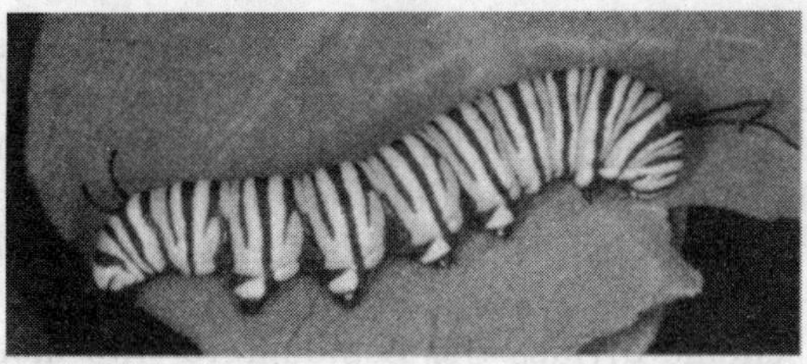

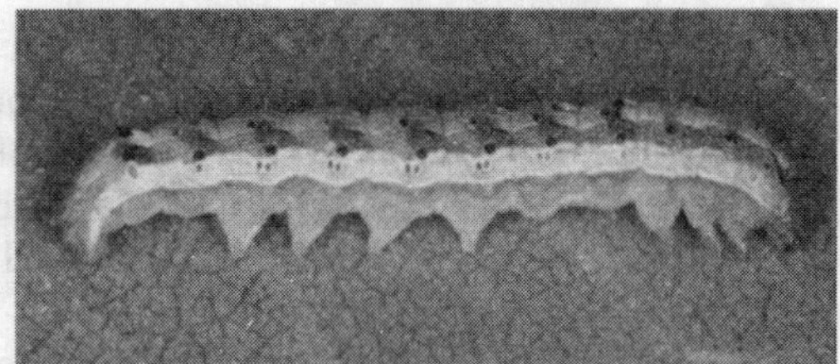

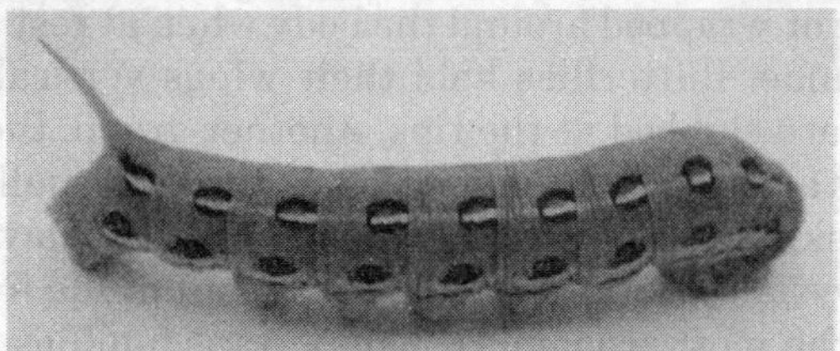

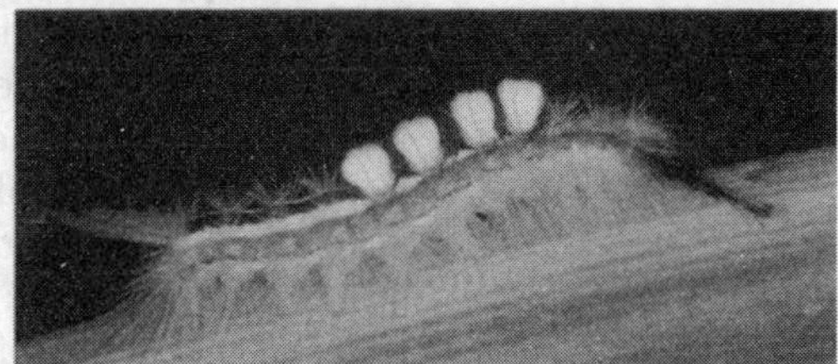

**Figure 3.33** Examples of caterpillars, order Lepidoptera. Top row (*left to right*): Family Danaidae (monarch butterfly) and family Noctuidae (corn earworm). Bottom row: Family Sphingidae (whitelined sphinx) and family Lymantriidae (whitemarked tussock moth). (Photos by Marlin E. Rice)

Immature butterflies and moths (Fig. 3.33) are usually called caterpillars, and development in the order is through complete metamorphosis. Caterpillars have a wormlike appearance and three pairs of jointed legs on the thorax; usually two to five pairs of fleshy prolegs are present on the abdomen. Hooks or claws, known as **crochets,** at the tips of the prolegs aid in locomotion. Caterpillars have short antennae, simple eyes, and chewing mouthparts.

Almost all immatures have well-developed silk glands formed from modified salivary glands. Silk may be used in tying plant parts together to form a sheltered environment for feeding, to make nests for aggregations of caterpillars, or to build cocoons. Cocoons are shelters built by full-grown moth caterpillars for transformation to the adult stage. Most butterfly caterpillars do not build cocoons. Instead, the transformation stage, often called a **chrysalis,** is naked and held in place on a twig or other object by a silk pad and a spiny **cremaster** at the tip of the abdomen. Some chrysalids also employ silken threads around their midsections that hold them upright.

With the exception of clothes moths (Fig. 3.34) (family Tineidae) and a few predators and scavengers, caterpillars feed on plants. Almost all plant species

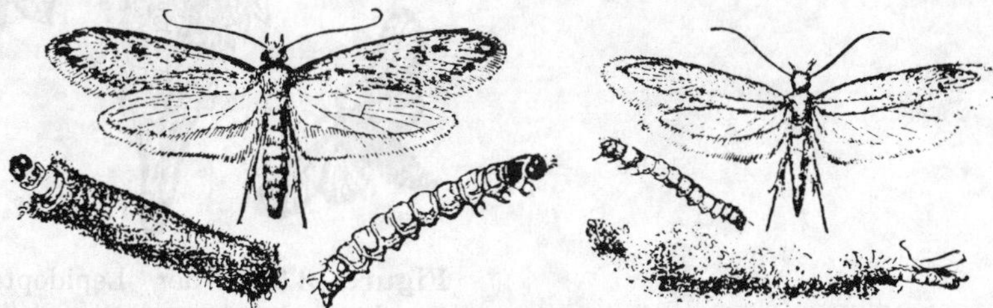

**Figure 3.34** Order Lepidoptera, family Tineidae, clothes moths. A casemaking clothes moth, *Tinea pellionella* (*left*), and a webbing clothes moth, *Tineola bisselliella.* (Courtesy USDA)

have at least one caterpillar pest feeding on them, making this order of great concern to insect pest management.

Many schemes of classification exist for the Lepidoptera. Some divide the order into five suborders. A more informal method divides the order into the Microlepidoptera and the Macrolepidoptera according to average body size of the included species.

Another informal scheme divides the Lepidoptera into moths, butterflies, and skippers. Usually moths are distinguished from butterflies by having wings folded rooflike or wrapped around the body when at rest and filamentous or featherlike antennae. Butterflies hold their wings vertically when at rest and their antennae are clubbed at the tips. Another group, the skippers, have antennae with hooks at the tips. Behaviorally, moths are usually active at night and butterflies and skippers are usually active during the day.

The suborder Ditrysia is by far the most important of the five formal suborders. Only a few of the most economically important families in this suborder are mentioned.

**Family Tineidae—clothes moths and others.** Tineids (Fig. 3.34) are small moths (wingspan about 10 mm), usually with large maxillary palps. The most important family member is the webbing clothes moth, *Tineola bisselliella*, whose caterpillars feed on wool, furs, feathers, and other organic products, causing considerable damage to fabrics. These caterpillars produce a weblike mass over fabric as they feed, whereas larvae of another clothes moth, the casemaking clothes moth, *Tinea pellionella*, spin a portable case in which they reside during development.

**Family Psychidae—bagworm moths.** Bagworm moths (Fig. 3.35) are so named because of the bag they construct and inhabit. Males have wings with very few scales. Females are wingless and lay eggs that overwinter in silken bags. Immatures leave the mother's bag in the spring and build their own bags that become larger and larger as caterpillars grow. The bags are characterized

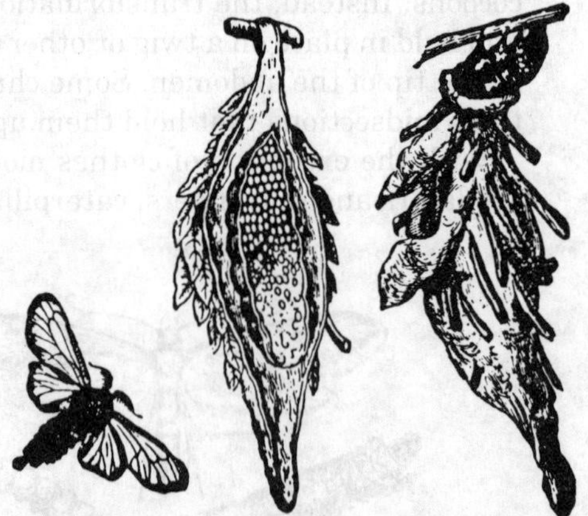

**Figure 3.35** Order Lepidoptera, family Psychidae, bagworm, represented by *Thyridopteryx ephemeraeformis*. Adult (*left*), case containing eggs (*middle*), and larva in case. (Courtesy USDA)

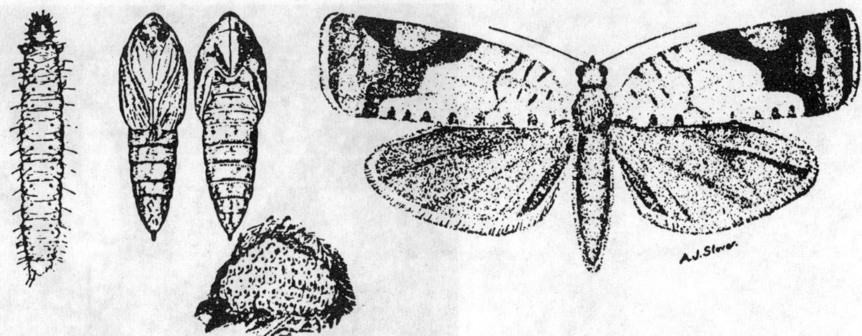

**Figure 3.36** Order Lepidoptera, family Tortricidae, tortricid moths, represented by the fruittree leafroller, *Archips argyrospilus.* Larva (*left*), pupae (*middle, top*), egg mass (*middle, bottom*), and adult. (Courtesy Oregon Agricultural Experiment Station)

by bits of leaves, twigs, and other plant material incorporated into them. Immatures carry their bags with them as they move and extend their heads and legs out of their bags to feed. Adult males leave their bags, but females remain in theirs and lay eggs there. Mating occurs through the lower opening of the bag. These insects are particularly important pests of woody ornamentals.

**Family Tortricidae—tortricid moths.** The tortricids (Fig. 3.36) are small moths. Tortricid caterpillars sometimes roll leaves and feed, inside out, on the rolled leaf. Others feed on fruits, buds, or stems. Adults are characterized by having broad front wings squared off at the tips. Many pest species in this family are destructive of fruit tree and forest crops. One of the most important of the fruit tree groups is the codling moth, *Cydia pomonella*, whose caterpillars enter the blossom end of apples and pears and feed internally on the fruit. Two of the most destructive forest pests are the spruce budworm, *Choristoneura fumiferana*, in the East, and the western spruce budworm, *Choristoneura occidentalis*, in the West. These species feed on buds and new foliage, severely defoliating spruces and firs.

**Families Pyralidae and Crambidae—snout and grass moths.** Insects in these two families are similar morphologically, often biologically, and were, until recently, included in the Pyralidae. The name "snout moth" comes from the well-developed labial palps found in some members of this group. Most pyralids are small, fragile moths that have a scaled **proboscis.** One significant crambid pest is the European corn borer, *Ostrinia nubilalis*, a species introduced into North America from Hungary or Italy in shipments of broom corn about 1908 or 1909. It ranks as one of the most destructive pests of corn, but it also damages peppers and other garden crops. In corn, early-season larval tunneling disrupts normal plant physiological processes, and late-season tunneling causes structural weakening of the plant and dropped ears.

**Family Papilionidae—swallowtail butterflies.** Members of this group (Fig. 3.37) are medium to large butterflies. The swallowtails are particularly well known as very colorful insects with long, taillike extensions on the hind wings. Some of the largest butterflies in the world belong to this group, having wingspans as large as 255 mm! One of the most common in North America is the black and yellow tiger swallowtail, *Papilio glaucus*.

**Figure 3.37** Order Lepidoptera, family Papilionidae, swallowtails, represented by the tiger swallowtail, *Papilio glaucus.* (Photo by Dennis Schotzko)

**Figure 3.38** Order Lepidoptera, family Nymphalidae, brushfooted butterflies, represented by the painted lady butterfly. (Photo by Dennis Schotzko)

**Family Nymphalidae—brushfooted butterflies.** Members of this large family of many common butterflies, along with some other families, have reduced front legs without claws. Only the middle and hind legs are used in walking. The painted lady butterfly (Fig. 3.38), *Vanessa cardui*, a migratory butterfly found throughout the world, is one of the most common representatives of this family in North America. Occasionally it is a pest.

**Family Lasiocampidae—tent caterpillars and lappet moths.** Tent caterpillars (Fig. 3.39) and lappet moths are medium-sized moths that have heavy bodies and feathery antennae. The caterpillars also are covered with long hairs and feed on the foliage of a number of deciduous trees, including fruit trees. Tent caterpillars, for example, the eastern tent caterpillar, *Malacosoma americanum,* obtain

**Figure 3.39** Order Lepidoptera, family Lasiocampidae. Shown are the webs of the eastern tent caterpillar, *Malacosoma americanum,* on crabapple. (Photo by Marlin E. Rice)

**Figure 3.40** Order Lepidoptera, family Sphingidae, sphinx moths. Shown is the whitelined sphinx, *Hyles lineata.* (Photo by Dennis Schotzko)

their name from building silken nests in the crotches of cherry, apple, peach, plum, and other trees. Caterpillars move back and forth from the tents to feeding areas during development, using the tents for shelter. Caterpillars of a related species, the forest tent caterpillar, *Malacosoma disstria,* do not make tents.

**Family Sphingidae—sphinx moths.** The front wings of sphingids are narrow and pointed at the tips; with a spindle-shaped body, the family is readily recognizable (Fig. 3.40). These moths often have large, stout bodies and hover in front of flowers, like hummingbirds, as they feed. The caterpillars are usually called

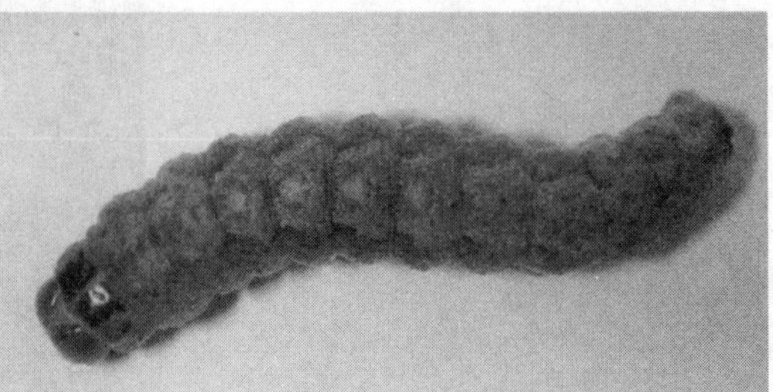

**Figure 3.41** Order Lepidoptera, family Noctuidae, noctuid moths, represented by western bean cutworm, *Striacosta albicosta.* Adult (*top*) and larva. (Photos by Marlin E. Rice)

hornworms because of the large curved horn arising on the upper surface and at the end of the abdomen in most species. One of the common hornworms in this family is the tomato hornworm, *Manduca quinquemaculata,* a large green caterpillar that is one of the most important insect pests of tomatoes and potatoes.

**Family Noctuidae—noctuid moths.** The Noctuidae family (Fig. 3.41) is the largest family of the Lepidoptera. Probably most of the moths seen around street and yard lights in summer are noctuids. Most adults are somewhat triangle-shaped when the wings are at rest, and they are often dull gray, brown, or black (with the notable exception of underwing moths, which have colorfully striped hind wings). Most noctuid caterpillars are foliage feeders, but some bore into stems and fruits. Many of our most devastating agricultural pests are found in this family, including several species of armyworms, cutworms, and loopers.

Perhaps the most destructive pest in this group, or in the order for that matter, is the corn earworm, *Helicoverpa zea.* Caterpillars of this noctuid are most destructive when they feed on the fruits of plants and on flower buds. However, they also feed on foliage and young stems. Important crops attacked include corn, cotton, and tomatoes.

Other important noctuid pests worthy of mention are armyworms and cutworms. Like the corn earworm, species in these groups attack a wide variety of crop plants.

### *Order Diptera—flies

Many kinds of insects have the word *fly* as part of their common names. Some of the ones we have discussed include dragonflies, mayflies, and butterflies. However, none of these is a "true" fly, and entomologists conventionally designate this fact by writing the word *fly* in combination with a descriptor to form one word. True flies occur only in the order Diptera, and in this instance, *fly* is conventionally written as a separate word from the descriptor, as in "house fly" and "horse fly."

Diptera are very diverse and successful insects. They are significant pests that directly and indirectly affect the health of humans, domestic animals, and

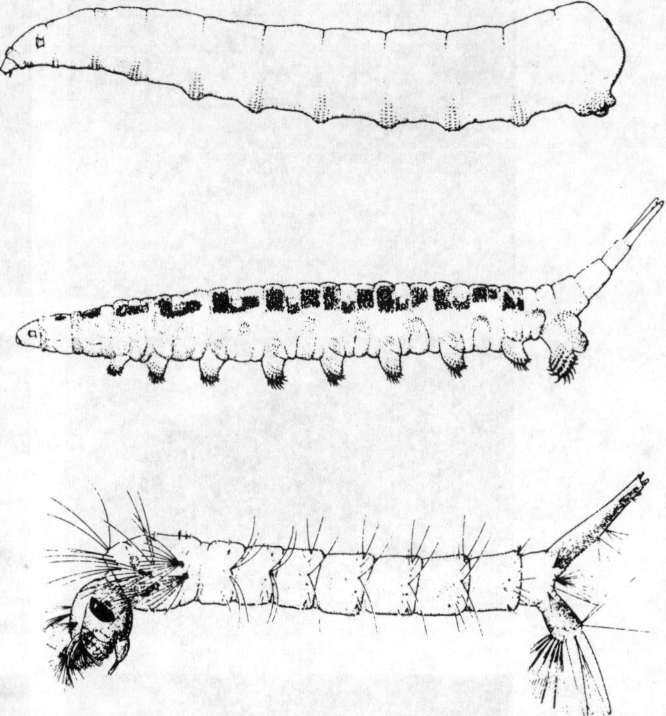

**Figure 3.42** Examples of Diptera larvae. Muscidae (house fly, *top*), Ephydridae (shore fly, *middle*), and Culicidae (northern house mosquito). (After Steyskal et al., 1986, courtesy USDA)

plants. Others are important natural enemies of insects, and some are considered beneficial as plant pollinators. They occur in every conceivable terrestrial and freshwater habitat; they even occur in puddles as well as in brackish and alkaline water and in pools of crude oil.

True flies are easily distinguished from the rest of the Hexapoda on the basis of their wings. They have only one pair of membranous wings arising from the **mesothorax,** and the hind pair forms balancing organs called **halteres.** Wings are absent in some Diptera, but halteres are usually present. Only male scale insects have one pair of wings and metathoracic balancing structures, but these insects have a stylus at the end of the abdomen.

The mouthparts of the Diptera are some of the most diverse in the Hexapoda. As previously mentioned, mosquitoes have piercing-sucking mouthparts, house flies and related insects have sponging mouthparts, and biting flies have cutting-sponging mouthparts. Mouthparts are not functional in a few adult Diptera. Development is by complete metamorphosis, and the immatures take on a variety of shapes and sizes (Fig. 3.42).

The Diptera are often divided into two suborders: Nematocera and Brachycera. Adult nematocerans, like mosquitoes and gnats, have antennae with six or more segments, and immatures have a well-developed head and mandibles. In the Brachycera (for example, horse flies and house flies), the antennae are usually three-segmented. Only a few important families (Fig. 3.43) representative of the Diptera are discussed here.

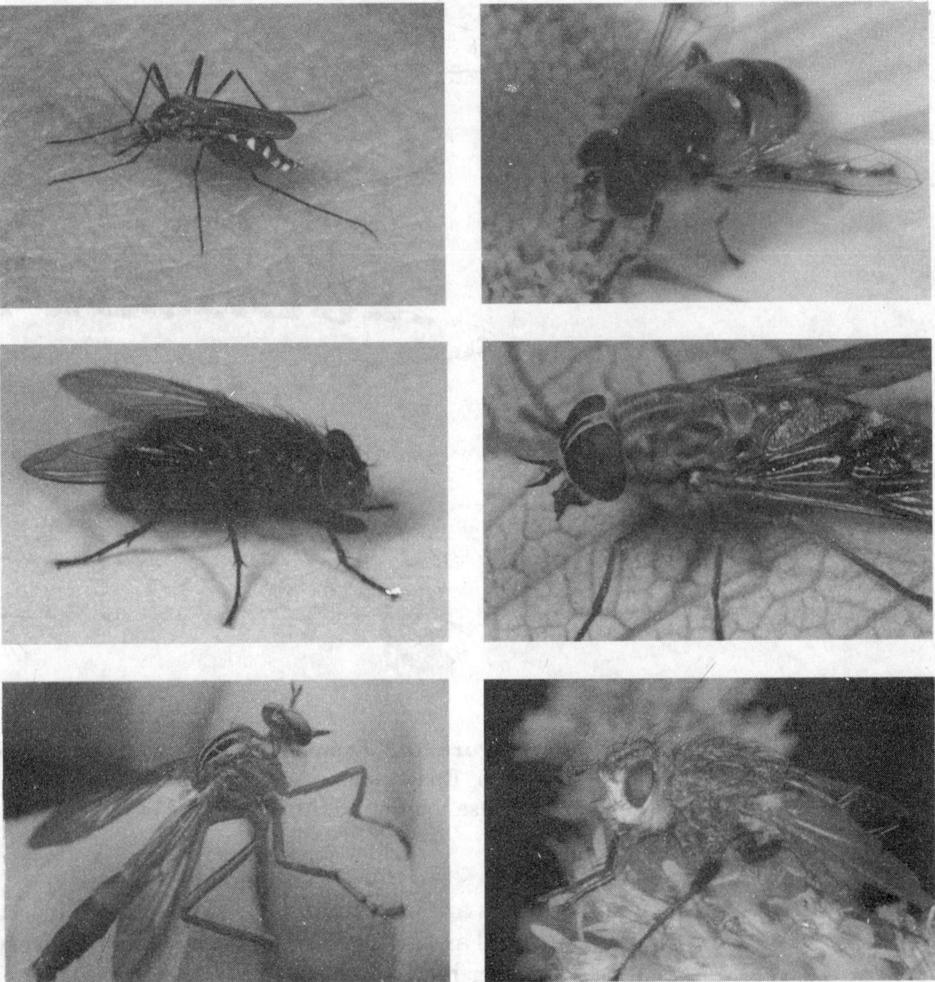

**Figure 3.43** Order Diptera showing representative families. Top row (*left to right*): Family Culicidae (mosquito) and family Syrphidae (flower fly). Middle row (*left to right*): Muscidae (blue bottle fly) and family Tabanidae (horse fly). Bottom row (*left to right*): Family Asilidae (robber fly) and family Tachinidae (parasitic tachinid). Adult dipterans are distinguished by their one pair of membranous wings and presence of balancing organs (halteres) in place of the hind wings. (Photos: tachinid, Dennis Schotzko; all others, Marlin E. Rice)

**Family Culicidae—mosquitoes.** Mosquitoes have been some of the most significant pests in the history of humankind. Their importance as medical pests continues today, particularly through vectoring organisms that cause malaria and other diseases such as Bancroftian filariasis.

Adult mosquitoes are distinguished by their long proboscis and scales along the wing veins. Adult females feed on blood and are responsible for transmitting disease organisms (see Color Plate 6). Adult males (and occasionally females) feed on nectar and other plant exudates. The feeding immatures (larvae or wrigglers) are found in water, where most consume algae and organic debris.

Members of the genus *Anopheles* are known as the malaria mosquitoes and lay their eggs singly on the water surface. Those in the genus *Culex* lay eggs in clusters as rafts on the water surface. Members of another important genus, *Aedes* (floodwater mosquitoes), lay eggs in depressions, along high water lines, and in almost any container of rainwater. These eggs lie dormant until flooded with water, after which they hatch, and wrigglers develop in the flooded area. As a vector of the yellow fever pathogen, *Aedes aegypti* is one of the most important floodwater mosquitoes.

**Family Cecidomyiidae—gall midges.** Gall midges are small, fragile nematocerans with long antennae and reduced wing venation. Almost all gall midges feed on plants as immatures, and many cause abnormal swellings in plant tissues called galls. A given species forms galls only on certain plant parts, and the gall is usually very distinctive. One of the most important pests in this family is the Hessian fly, *Mayetiola destructor*, a species introduced to North America from Europe that attacks wheat. Winter wheat is injured in the spring when maggots feed between the leaf sheath and the stem, weakening plants and sometimes causing them to die.

**Family Tabanidae—horse flies and deer flies.** Horse flies and deer flies are representatives of the suborder Brachycera that are medium to large flies with cutting-sponging mouthparts. Like mosquitoes, adult females feed on blood, and adult males consume pollen and nectar of flowers. Tabanid larvae feed on small invertebrates in water or in moist soil.

The two most common genera are *Tabanus,* the horse flies, and *Chrysops,* the deer flies. Horse flies are known to transmit pathogens causing anthrax and tularemia, but most harm is done through their vicious bite and extraction of blood. Deer flies are usually smaller than horse flies and will buzz around the back of people's heads and attempt to get into their hair. The bites of both types of flies are painful.

**Family Syrphidae—flower flies.** Adult flower flies are often mistaken for bees because of their bright yellow bands and their habit of hovering about over flowers. Flower fly larvae are of special importance because many feed on aphids and other pests of cultivated crops. Others feed in polluted water and have long breathing tubes; these are called rattailed maggots.

**Family Tephritidae—fruit flies.** Tephritid fruit flies (not to be confused with flies in the family Drosophilidae) are small- to medium-sized flies that usually have banded or patterned wings (Fig. 3.44). Adults occur on vegetation, and females often lay eggs in fruit. Developing maggots feed inside the fruit, often causing serious losses to apples, cherries, and citrus. An important pest of apples in the northern United States is the apple maggot, *Rhagoletis pomonella.* Another species, the Mediterranean fruit fly, *Ceratitis capitata,* is a devastating pest of fruit in tropical and subtropical parts of the world. It has been introduced into Florida and California on several occasions, and eradication programs seemingly eliminated it each time in Florida. Persistent sightings in California have required continued surveillance and eradication activities.

**Family Muscidae—muscid flies.** Muscid flies are a large group of common flies that are very important as pests of humans and domestic animals. The family includes the house fly, *Musca domestica,* whose maggots feed on all kinds of decaying matter. The fly is known to carry organisms that cause

**Figure 3.44** Order Diptera, family Tephritidae, fruit flies, represented by an adult Mexican fruit fly, *Anastrepha ludens*. (Photo by Dennis Schotzko)

typhoid fever, yaws, anthrax, dysentery, and forms of conjunctivitis. Another pest similar to the house fly is the face fly, *Musca autumnalis,* a species that clusters on the faces of cattle. Neither the house fly nor the face fly bites.

Some other species in this family are biting flies, including stable flies, *Stomoxys calcitrans* (see Color Plate 5), and horn flies, *Haematobia irritans,* both important pests of cattle.

**Family Tachinidae—tachinid flies.** The Tachinidae family is a very large group of Diptera. Many tachinids look similar to house flies and other muscids, but others are larger and covered with strong bristles. They are parasitic on other insects, particularly immature Coleoptera, Lepidoptera, and Hymenoptera. In most instances, females lay their eggs on the body of the host, and after eggs hatch, the maggots tunnel inward through the body wall. These maggots feed internally, eventually killing their host, and when fully grown, they emerge to develop into adults. In some species, eggs are laid on plant foliage, and larvae attach to appropriate hosts as these hosts pass by. There are also many other life-cycle variations.

### ORDER SIPHONAPTERA—fleas

Fleas (Fig. 3.45) are very small, wingless insects that, in the adult stage, feed on the blood of mammals and birds. Most adult fleas are characterized by a body that is flattened from side to side and covered with rows of bristles that point backwards. Flea antennae are short and lie in grooves on the sides of the head. The mouthparts, modified for piercing and sucking blood, are composed of three stylets. Eyes may be present or absent.

The legs of adult fleas are long with long coxae, which allow a flea to jump distances many times its own body length. Such a capability permits fleas to effectively move from host to host.

Fleas develop by complete metamorphosis, and the eggs are usually laid on the ground or in the nest of the host. When laid on the host, they fall off into the nest or onto the ground. Egg hatch produces white, legless immatures. These are worm-shaped, with a well-developed head, no eyes, chewing mouth-

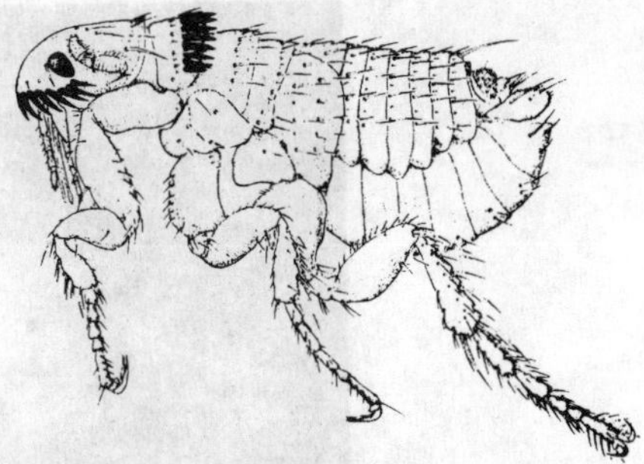

**Figure 3.45** Order Siphonaptera, fleas, represented by a cat flea, *Ctenocephalides felis*. Adult fleas are characterized by a body flattened from side to side and covered with rows of bristles that point backwards. (After Steyskal et al., 1986, courtesy USDA)

parts, and small hooks at the tip of the abdomen. These immatures feed mostly on organic debris and feces of adult fleas. When fully grown, they spin cocoons in which transformation to the adult occurs.

Although some have common names referring to hosts, most fleas are not very specific feeders. For instance, two common species, the cat flea, *Ctenocephalides felis,* and the dog flea, *Ctenocephalides canis,* both are parasitic on cats, dogs, humans, and other hosts. Similarly, the human flea, *Pulex irritans,* attacks humans, dogs, cats, rats, horses, and other animals.

Fleas are probably most important as vectors of disease-causing organisms in humans. The most important disease resulting from transmission is **plague**, or black death. Plague transmitted by fleas is called bubonic plague, caused by the bacterium *Yersinia pestis*. Through the ages, it has resulted in tremendous loss of life. The reservoir for the disease is rats, and fleas such as the oriental rat flea, *Xenopsylla cheopis*, move off of rats dying of the disease and onto humans. Here, they transmit the bacterium, mostly through regurgitation at the time of biting.

### *ORDER HYMENOPTERA—ants, bees, and wasps

The Hymenoptera is another of the major orders and is considered by many people to be the most beneficial. This view is based on activities of the Hymenoptera in plant pollination, in producing honey and wax, and in killing pest insects. However, certain sawflies, (Fig. 3.46) ants, and other forms are pests and can be quite destructive.

A great diversity of form occurs in the Hymenoptera, but most species have two pairs of membranous wings, with the hind pair smaller than the front. The front and hind wings are held together by rows of tiny hooks on the hind wings called **hamuli.** Wing venation is often reduced or almost lacking in some groups. Mouthparts are basically of the chewing type but are modified for sucking or lapping in bees. Most hymenopteran females have a well-developed ovipositor, and in some groups, it has been modified to form an effective stinging organ used for protection.

Development in the Hymenoptera occurs by complete metamorphosis, and the immatures (larvae) have well-developed head capsules and chewing mouthparts. On attaining full growth, the immatures sometimes build silken or parchmentlike cocoons.

**Figure 3.46** Example of a sawfly larva. Family Cimbicidae, elm sawfly (*Cimbex americana*). (Photo by Marlin E. Rice)

The order Hymenoptera is usually divided into two suborders, the Symphyta and the Apocrita. The Symphyta include sawflies and horntails, which have abdomens broadly joined to their thoraces. The ovipositor is usually well developed and sawlike in form. Larvae of the Symphyta look much like the caterpillars of moths but have more than five pairs of abdominal prolegs and no crochets. Most Symphyta feed externally on plants.

The Apocrita are characterized by a "thin waist," with the abdomen attached to the thorax by a stalklike **petiole.** This suborder includes ants, wasps, bees, and many forms of parasitic wasps. Most larvae of this suborder are legless and maggotlike or grublike. Food habits of this large suborder are very diverse. As larvae, most species are parasitic on other insects, but others are predators, feed on plant materials, or eat other products. Adults often consume nectar, pollen, sap, and other plant materials. Females of some parasitic species may ingest body fluids of hosts they sting when laying eggs.

**Family Tenthredinidae—tenthredinid sawflies.** Tenthredinid sawflies (Fig. 3.47) compose a very large family of medium-sized (less than 21 mm long) and sometimes brightly colored species. Adults are wasplike, and many larvae curl over the edge of a leaf when feeding. Most species feed on trees and shrubs, and some can be quite destructive. The cherry fruit sawfly, *Hoplocampa cookei,* is an example of an important pest in this family, attacking fruit crops of cherries and plums in the western United States.

**Family Cephidae—stem sawflies.** Stem sawflies (Fig. 3.48) are elongate insects, about 9 mm long, which are flattened from side to side. Cephid immatures burrow in the stems of grasses, shrubs, and other plants, in many instances causing them to break over. One of the most important pests in this family is the wheat stem sawfly, *Cephus cinctus,* which is found in the Great Plains area of the United States.

**Family Braconidae—braconids.** Braconids (Fig. 3.49) are a family of small wasps, usually less than 15 mm long, that parasitize and kill many different kinds of insects. Depending on the species, one to more than a hundred braconid

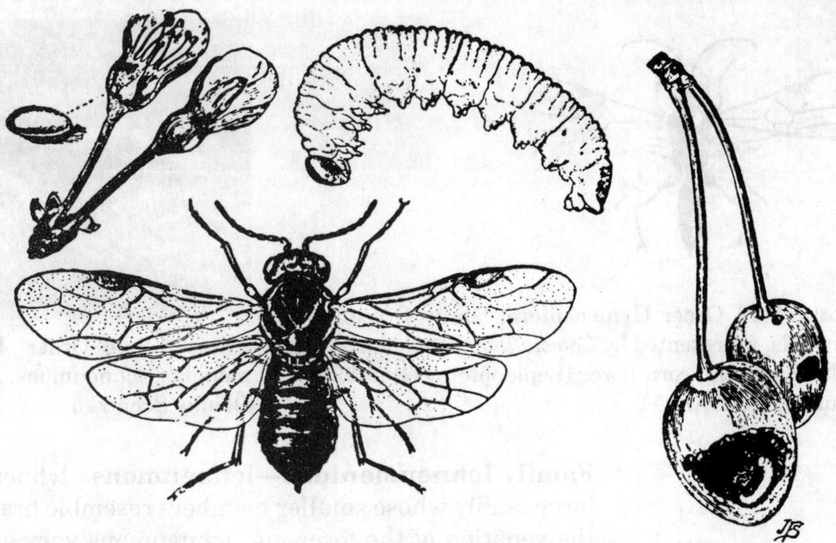

**Figure 3.47** Order Hymenoptera, family Tenthredinidae, common sawflies, represented by the cherry fruit sawfly, *Hoplocampa cookei*. Egg and egg position on blossom (*upper left*), larva (*upper right*), adult (*lower middle*), and infested cherries. (Courtesy USDA)

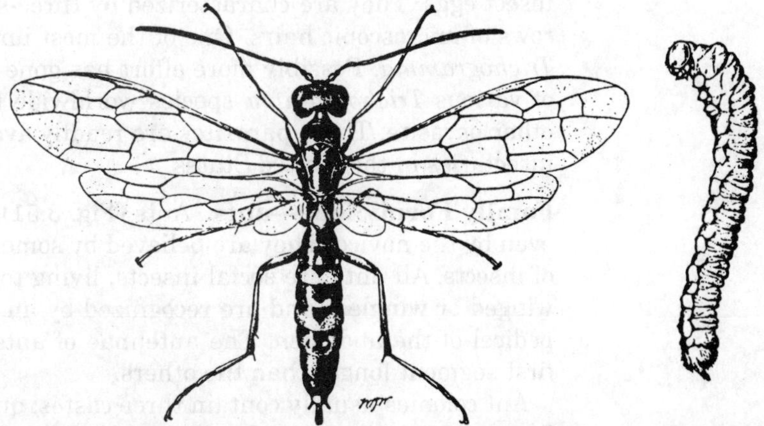

**Figure 3.48** Order Hymenoptera, family Cephidae, stem sawflies, represented by the wheat stem sawfly, *Cephus cinctus*. Adult (*left*) and larva. (Courtesy USDA)

larvae can develop in a single host. Some braconid larvae leave the host when full grown and build silken cocoons, where they transform into adults. Others complete adult development inside the host. Braconids are important in the natural regulation of many insect populations and have been reared and released in programs of pest management. One of the most important genera in this family for pest management is *Cotesia*.

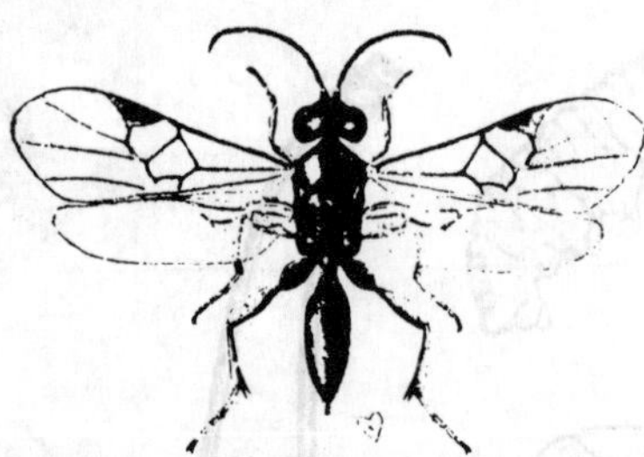

**Figure 3.49** Order Hymenoptera, family Braconidae, braconids, represented by *Cotesia thompsoni,* a parasite of the European corn borer (Lepidoptera: Grambidae). (Courtesy USDA)

**Figure 3.50** Order Hymenoptera, family Ichneumonidae, ichneumons. A female ichneumon. (Photo by Dennis Schotzko)

**Family Ichneumonidae—ichneumons.** Ichneumons (Fig. 3.50) are a very large family whose smaller members resemble braconids. They mainly differ in the venation of the forewing. Ichneumons vary greatly in size, with some attaining lengths of 40 mm or more. The ovipositors of many ichneumons are long, sometimes longer than their bodies. All ichneumons are parasitic, and most insect groups have species with at least one ichneumon parasite. Certain ichneumons are also parasites of spiders.

**Family Trichogrammatidae—minute egg parasites.** Members of the Trichogrammatidae family are tiny wasps, 0.3 to 1.0 mm long, that parasitize insect eggs. They are characterized by three-segmented tarsi and wings with rows of microscopic hairs. One of the most important genera in this family is *Trichogramma.* Possibly more effort has gone into the production and release of various *Trichogramma* species worldwide for pest management than any other parasite. These parasites are readily available from several commercial insectaries in the United States.

**Family Formicidae—ants.** Ants (Fig. 3.51) are widespread and recognized even by the novice. They are believed by some to be the most successful group of insects. All ants are social insects, living together in colonies. Ants may be winged or wingless and are recognized by an upright lobe that occurs on the pedicel of the abdomen. The antennae of ants are usually elbowed, with the first segment longer than the others.

Ant colonies usually contain three castes: queens, males, and workers. Mating occurs in swarms of flying queens and males. After mating, males die and females establish new colonies. Usually, the young queen's wings break off, and she makes a cavity in the soil under a stone or bark. After a month or more in the sealed cavity, the queen lays eggs and feeds developing larvae her saliva until they mature. The first adults produced are all small workers (sterile females), which, on becoming adults, take over the duties of the colony. Worker activities include nest construction, food gathering, and care of the young. As the colony expands, more and more chambers are added to the nest. After several years, winged males and females (queens) are produced to leave the nest and establish new colonies of their own. Queens are usually long-lived; some may live 25 years or more. Colonies may exist for as long as a human generation. In addition to soil, ants may nest in plant parts (for example, stems and nuts), or in damp, decaying wood, which is sometimes excavated for galleries.

**Figure 3.51** Order Hymenoptera, family Formicidae, ants represented by the carpenter ant, *Camponotus castaneus.* (Photo by Marlin E. Rice)

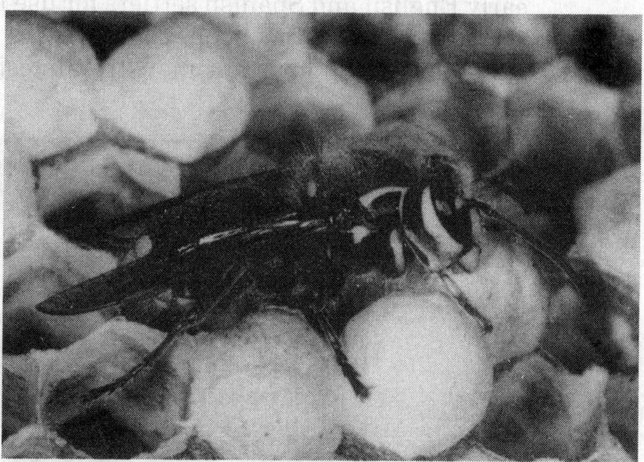

**Figure 3.52** Order Hymenoptera, family Vespidae, vespids, represented by the baldfaced hornet *Dolichovespula maculata*. (Photo by Marlin E. Rice)

Several species of ants are household pests but mainly exist as a nuisance. Others, such as the Texas leafcutting ant, *Atta texana,* may cause serious damage to trees, shrubs, and cultivated crops by stripping their foliage. The foliage is cut into pieces and carried inside the nest to certain galleries where it is used as a medium for growing fungus. This cultivated fungus then is eaten by the ants.

**Family Vespidae—vespid wasps.** Vespid wasps (Fig. 3.52) are a small but common group of the Hymenoptera that includes paper wasps, yellowjackets, and hornets. Many of these insects are familiar black or reddish brown wasps with yellow or off-white markings. Like ants, these social insects have queens, males, and workers.

Unlike ants, however, new colonies are established annually. Queens and workers have well-developed stinging organs and can inflict painful stings. These wasps construct nests of papery material that they produce by chewing wood fibers with saliva. Overwintered queens begin a colony in the spring by starting a new nest or using one of a previous year. The first brood produced is composed of workers (sterile females), which take over nest duties when they mature. The immatures are fed with insects (often pests) and other invertebrates. Paper wasp nests are commonly attached to the eaves of houses, and many people believe that wasp activities outside the home are a nuisance.

**Family Apidae—bumble bees, honey bees, and others.** Common bees are some of the most beneficial insects. Most are social and important as plant pollinators.

Bumble bees (genera *Bombus* and *Psithyrus*) are relatively large, hairy bees, 20 mm or longer, that are usually colored black and yellow. They build nests in ground depressions, deserted rodent nests, hollow logs, and other places. Like vespids, colonies in temperate regions are established on an annual basis by overwintered, fertilized queens. Workers produced in the first brood take over nest duties during the production of subsequent broods. The proboscis of bumble bees is very long, making them effective pollinators of plants with long corollas, such as red clover.

The honey bee, *Apis mellifera,* is a species introduced into North America by early English and Spanish settlers for use in beekeeping. Beekeeping, the culturing of bees for their honey and wax, is one of the oldest agricultural pursuits and is an important industry almost everywhere (Fig. 3.53). However,

**Figure 3.53** Order Hymenoptera, family Apidae, represented by the honey bee, *Apis mellifera,* on a hive. (Photo by Marlin E. Rice)

the pollination service provided by honey bees far outweighs the value of their material products.

Unlike bumble bees, honey bees maintain colonies on a perennial basis. A colony consists of three castes: a single queen, males (called **drones**), and workers (sterile females). Nests in nature are built in logs, hollow trees, caves, and other such places. Inside the nests, workers construct and maintain vertical **combs** of wax, with hexagonal cells for the storage of honey, pollen, and immature bees. The primary function of the queen in this colony is to lay eggs, and she sometimes produces 1,500 to 2,000 daily. Fertilization of queens occurs during mating flights, and a queen may live for several years.

New queens are reared in special cells, called queen cells, and are fed exclusively on **royal jelly,** a glandular secretion of workers. Queens are produced from fertilized eggs. Drones are reared in drone cells and develop from unfertilized eggs. After the mating season, drones are driven out of the hive to die, thereby helping to conserve the store of honey for the winter. Workers develop from fertilized eggs in other cells of the comb; they are fed royal jelly initially but complete their development on pollen and nectar. New workers feed and care for the queen and brood, secrete wax, build combs, clean the hive, convert nectar to honey, act as guards, and perform other duties. After about 3 weeks, workers become field bees, mainly gathering nectar and pollen and bringing water to the hive. New colonies are formed when new queens are produced, and the old queen moves out with a group of workers to swarm and build a new nest.

## MITE AND TICK CLASSIFICATION

Although mites and ticks are not insects, entomologists often are called upon to deal with them because of their agricultural importance. Therefore, at least an elementary knowledge of their classification can be useful.

### ORDER ACARI—mites and ticks

Mites and ticks belong to the arthropod class **Arachnida** and to the order **Acari.** The Acari are a large group, comprising more than 30,000 species. Mites and ticks usually have an oval shape, with two general body regions. There may be little or no differentiation between the two body regions. After eggs hatch, the young, called larvae, have only three pairs of legs but add another pair after they shed the first skin (molt). After first-stage larvae molt, they are called nymphs.

As a group, the Acari occur in most of the habitats occupied by insects. Although they have no wings and do not fly, they may be found in the atmosphere, being dispersed passively by wind. Members of the order are both aquatic and terrestrial and are some of the most abundant arthropods in soil and on soil surfaces. Soil forms help in decomposition of plant and animal debris. Many species are external parasites of vertebrates and invertebrates, including insects. Others, such as the mite *Amblyseius fallacis,* are predators and can be important in managing plant-feeding pests. Yet, in agriculture, the plant-feeding behavior of many species, as well as parasitism of livestock, attracts the most attention.

In classification, the Acari are sometimes divided into suborders, of which six are commonly recognized. These include the suborders Holothyrina, Mesostigmata, Ixodida, Prostigmata, Astigmata, and Oribatida. Only those directly important to agriculture will be mentioned here.

## Suborder Mesostigmata

The Mesostigmata is a large suborder that includes predators, parasites, and decomposers. Parasitic mites are important pests in this group and include the chicken mite (Fig. 3.54), *Dermanyssus gallinae*. Chicken mites are free living, hiding during the day and attacking poultry at night. This species is widely distributed but is of greatest importance in warm and dry regions. The mites feed on the blood of their hosts, sometimes killing young chickens and setting hens.

## Suborder Ixodida

The Ixodida are known as ticks and are represented by two families in North America. The Ixodidae are the hard ticks characterized by a tough leathery covering, the scutum (Fig. 3.55). Their mouthparts protrude forward. The Argasi-

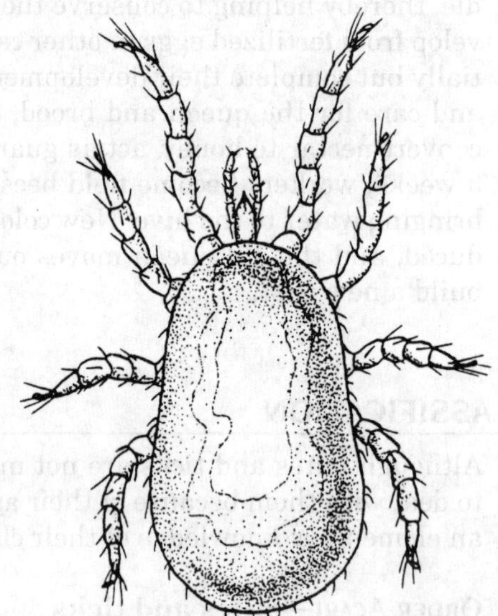

**Figure 3.54** A chicken mite, *Dermanyssus gallinae*, of the suborder Mesostigmata. (Courtesy USDA)

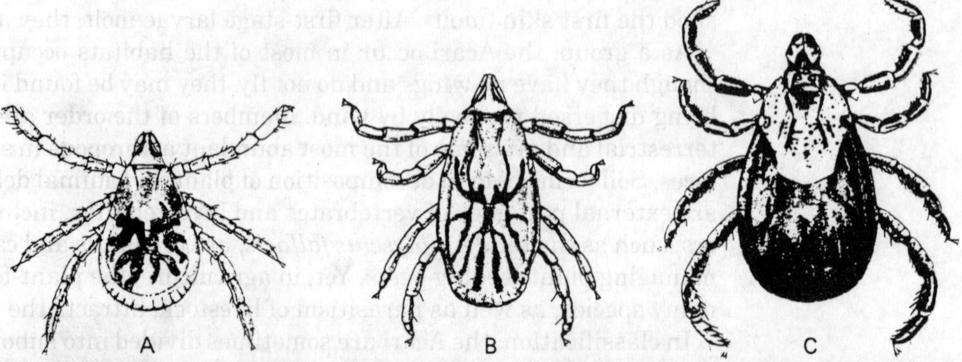


**Figure 3.55** The American dog tick, *Dermacentor variabilis*, of the suborder Ixodida. These ticks have a tough leathery covering. A. Larva. B. Nymph. C. Adult. (Courtesy USDA)

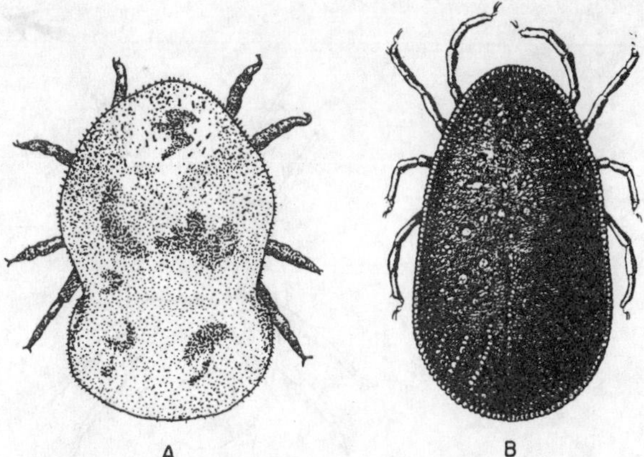


**Figure 3.56** A. An ear tick, *Otobius megnini.* B. A fowl tick, *Argas persicus.* Both species (suborder Ixodida) are soft-bodied and have ventrally located mouthparts. (Courtesy USDA)

dae are the soft ticks (Fig. 3.56), which are soft-bodied, and have ventrally located mouthparts.

Ticks are large compared to mites, and feed on blood of mammals, birds, and reptiles. Ticks are important vectors of disease to humans and domestic animals. Some of the most important tick-borne diseases are spotted fever, Lyme disease, tularemia, Texas cattle fever, Colorado tick fever, and Rocky Mountain spotted fever.

Ticks usually lay eggs in vegetative habitats, and larvae and nymphs must find hosts on which to feed. The life cycle includes a series of feeding bouts, where immatures feed and drop off the host, which is followed by host finding, feeding, and dropping-off behaviors that occur several more times.

### Suborder Prostigmata

This is a very large suborder whose members vary widely in habitat and ecology. Of considerable interest to agriculture are the plant-feeding spider mites of the family Tetranychidae. These mites, feeding on a wide array of foliage and fruit, cause serious problems in greenhouses, orchards, and field crops. Many species overwinter in the egg stage, with immatures active in early spring. Eggs are laid on plants during the growing season and several generations occur each year. Spider mites usually form mats of webbing on plant leaves that help shelter and protect the mite colony, hence their name. Some of the most widespread spider mite problems include those caused by the two-spotted spider mite (Fig. 3.57), *Tetranychus urticae,* and the European red mite (Fig. 3.58), *Panomychus ulmi.* Chiggers, whose immatures feed on human skin and cause itching and irritation, also occur in this suborder.

### Suborder Astigmata

Most of the mites in this suborder are of little importance to agriculture. However, a few species are parasitic and are significant pests of humans and

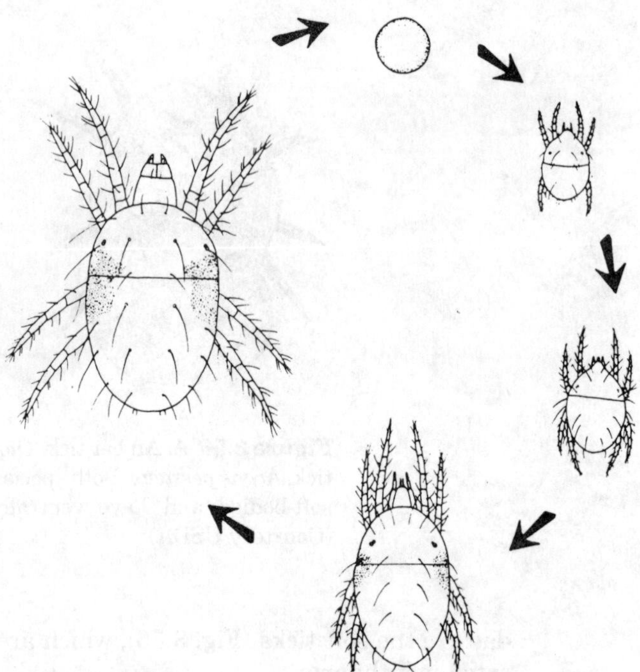

**Figure 3.57** The two-spotted spider mite, *Tetranychus urticae,* showing the life cycle. Stages clockwise from top are egg, larva, protonymph, deutonymph, and adult. (Courtesy Illinois Agricultural Experiment Station)

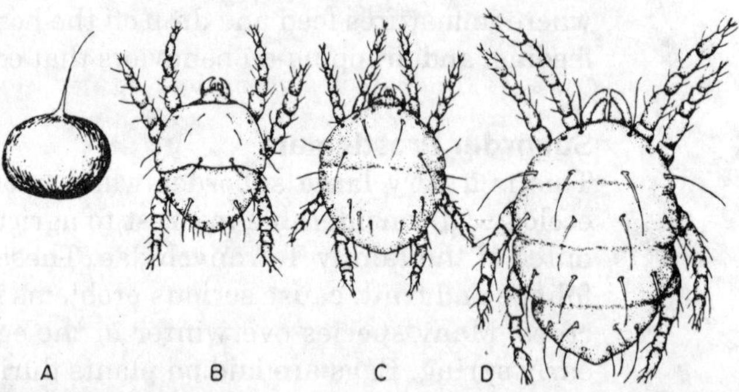


**Figure 3.58** The European red mite, *Panomychus ulmi,* suborder Prostigmata. A. Egg. B. Six-legged larva. C. Nymph. D. Adult. (Courtesy USDA)

domestic animals. These are grouped into two families, the Sarcoptidae and the Psoroptidae. The Sarcoptidae, called itch or mange mites (Fig. 3.59), feed in the skin and produce distinct burrows in which eggs are laid. Human itch and hog mange are forms of dermatitis caused by these mites. The Psoroptidae have been called scab mites (Fig. 3.60) and attack sheep, goats, cattle, and other

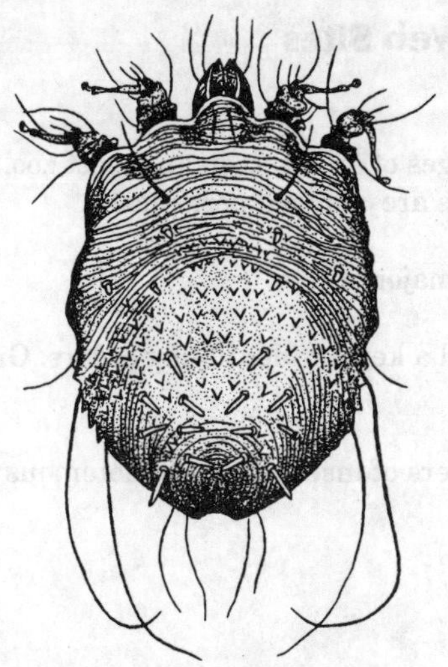

**Figure 3.59** A female sarcoptic mange mite, suborder Astigmata. (Courtesy USDA)

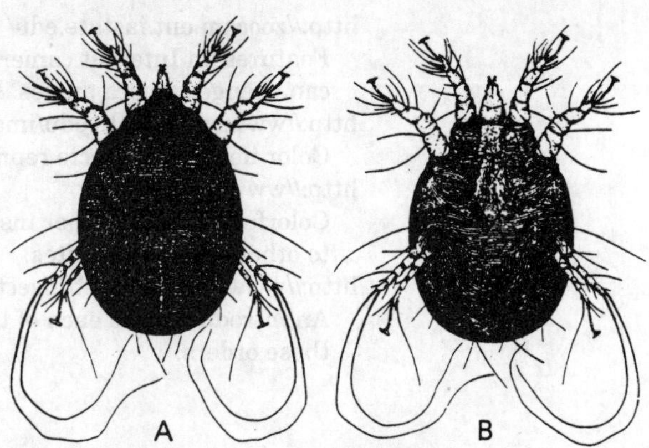


**Figure 3.60** A female psoroptic mite, the sheep scab mite, *Psoroptes ovis,* suborder Astigmata. A. Dorsal view. B. Ventral view. (Courtesy USDA)

animals. Psoroptic mites live at the base of hairs, but they do not burrow in the skin. These mites feed by piercing the skin with their mouthparts, which produces an exudate that piles up, hardens, and forms a scab. All stages of the mites can be found amongst the scabs.

## Further Reading

Atkins, M. D. 1978. *Insects in Perspective.* New York: Macmillan.

Brues, C. T., A. L. Melander, and F. M. Carpenter. 1954. Classification of insects. *Bulletin of the Museum of Comparative Zoology,* vol. 108. Cambridge, Ma.: Harvard University.

Davidson, R. H., and W. F. Lyon. 1987. *Insect Pests of Farm, Garden, and Orchard,* 8th ed. New York: Wiley.

Klass, K. D., O. Zompro, N. P. Kristensen, and J. Adis. 2002. Mantophasmatodea: A new insect order with extant members in the Afrotropics. *Science* 296:1456–1459.

Little, V. A. 1972. *General and Applied Entomology,* 3rd ed. New York: Harper & Row.

Marske, K. A., and M. A. Ivie. 2003. Beetle fauna of the United States and Canada. *Coleopterists Bulletin* 57:495–503.

Pfadt, R. E., ed. 1985. *Fundamentals of Applied Entomology,* 4th ed. New York: Macmillan.

Ross, H. H., C. A. Ross, and J. R. P. Ross. 1982. *A Textbook of Entomology,* 4th ed. New York: Wiley.

Steyskal, G. C., W. L. Murphy, and E. M. Hoover, eds. 1986. *Insects and Mites: Techniques for Collection and Preservation.* United States Department of Agriculture, Agricultural Research Service, Miscellaneous Publication 1443.

Stoetzel, M. B., ed. 1989. *Common Names of Insects and Related Organisms.* College Park, Md.: Entomological Society of America.

Triplehorn, C. A., and N. F. Johnson. 2005. *Borror and DeLong's Introduction to the Study of Insects,* 7th ed. Thomson Brooks/Cole. Belmont, CA.

## Favorite Web Sites

http://zoocam.ent.iastate.edu/
Features an Internet camera for live images of an insect in an insect zoo. Viewers can change camera angles. Insect subjects are changed periodically.

http://www.ent.iastate.edu/imagegallery/
Color images of insects representing the major orders.

http://www.insects.org/
Colorful photos of major insect orders and a key for identifying orders. Gives links to other entomology sites.

http://www.earthlife.net/insects/orders.html
An introduction to each of thirty-two orders of insects with dichotomous keys to these orders.

CHAPTER 4

# THE INSECT LIFE CYCLE

INSECTS PERFORM AN AMAZING FEAT: survival and persistence in the face of seemingly insurmountable odds. The bitterness of winter, the harshness of drought, and the punishment of wind and rain would seem to argue against the success of these small, cold-blooded animals. But they do survive, and at certain times of the year, they become a major force to be reckoned with in our daily lives.

The story of insect survival and persistence really begins with innate patterns of living—biological templates for species characterized by reproduction, growth, and development of individuals in the population. We refer to these patterns or templates as life cycles.

More formally, a **life cycle** is the chain or sequence of biological events that occurs during the lifetime of an individual insect. The cycle is usually considered to begin with deposition of the egg and to end with egg laying by the adult female. A life cycle lies in the boundaries of a single generation.

For convenience, we will consider the life cycle as having three major divisions and discuss it from this aspect. These divisions include: (1) reproduction and embryonic development, (2) postembryonic growth and development, and (3) maturity.

Related to the life cycle is the insect seasonal cycle (Fig. 4.1). The seasonal cycle is the sequence of life cycles of a species that occurs over a 1-year period. By addressing the topic of seasonal cycles, we can more fully understand some of the ways insects can survive adverse times of the year.

An understanding of both life cycles and seasonal cycles for a pest species is required for proper management. Prediction of injurious stages, times when insects are vulnerable, and probable habitats all depend on this knowledge. Without it, development of effective, long-term solutions to insect pest problems would be virtually impossible.

## REPRODUCTION AND EMBRYONIC DEVELOPMENT

### Types of Reproduction

By far the most frequently observed form of reproduction in insects is sexual reproduction. In most species, both females and males are prevalent, and mating occurs with the eggs fertilized in the female oviduct as they pass the duct of the sperm storage sac (**spermatheca**) (Fig. 4.2). Infrequently, fertilization may occur while eggs are still in the ovarioles or in the hemocoel.

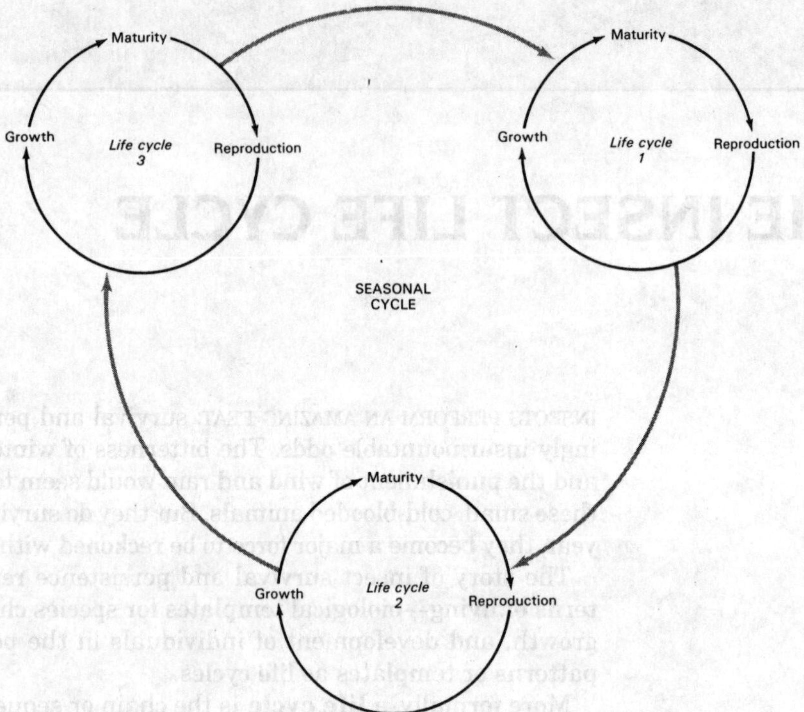


**Figure 4.1** Diagram showing the relationship of a life cycle to the insect seasonal cycle. Most insect species undergo one or more life cycles during a seasonal cycle.

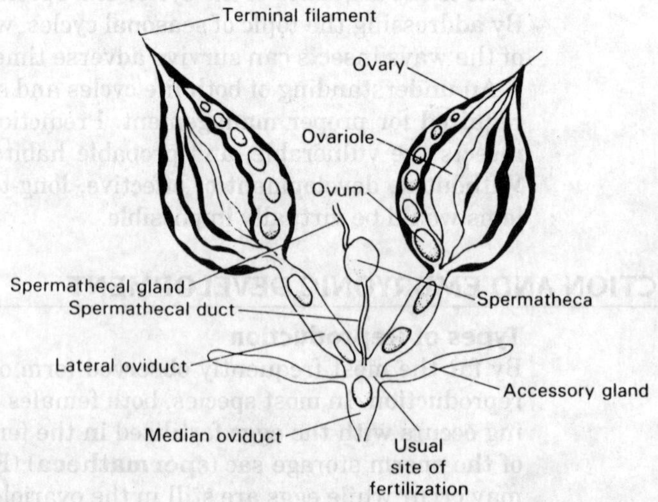


**Figure 4.2** Typical female insect reproductive system showing usual site of fertilization. (Redrawn from Atkins, 1978, *Insects in Perspective*, Macmillan Publishing Company)

Reproduction in the absence of male gametes is not as common. This form, asexual reproduction, or **parthenogenesis,** is a derived and specialized characteristic in some insects. Males are not found in species with obligate parthenogenesis, and unfertilized eggs are females. Examples of this type of reproduction include the pear sawfly, *Caliroa cerasi*, several black flies (Diptera: Simuliidae), a number of weevils (Coleoptera: Curculionidae), and many aphids (Hemiptera: Aphididae).

A combination of sexual and asexual modes, **haplodiploidy,** occurs in a few species of four insect orders: Coleoptera, Hymenoptera, Thysanoptera, and Hemiptera. The honeybee (Box 4.1) is one of the best known examples of this mode, which is sometimes called facultative parthenogenesis. Here, unfertilized eggs (haploids) are males who become drones, and fertilized eggs (diploids) are females, most of whom become workers. The queen bee, or reproductive female, has the ability to control the sex of her offspring, and this characteristic was an important factor in the development of social life in bees, wasps, and ants.

## BOX 4.1 HONEY BEE

**SPECIES:** *Apis mellifera* Linnaeus (Hymenoptera: Apidae)

**DISTRIBUTION:** The honey bee is an introduced species to the United States, with most colonies residing in human-made hives. Escaped swarms, however, survive in the wild and usually live in hollow trees. The species is completely domesticated (as opposed to the African honey bee, *Apis mellifera adansoni*) and has a cosmopolitan distribution.

**IMPORTANCE:** Honey bees play a vital and indispensable role in agriculture and horticulture as pollinators. Although wild flowers are often sufficiently pollinated by wild bees, large-scale agricultural plantings often require that domesticated bees be placed in the proximity of the crop for adequate pollination and subsequent grain/fruit production. In addition, raising bees (apiculture) for honey and beeswax is a multimillion dollar industry in the United States. Honey, as a commodity, is a popular food source and is often used as a sugar substitute. Beeswax is another important by-product of apiculture and has been used in shaving creams (to prevent drying), cold creams, cosmetics, crayons, candles, and other similar items.

**APPEARANCE:** Adult queens are 16 to 20 mm long; the workers (sterile females) are 11 to 13 mm long; and the drones (males) are 15 to 16 mm long. The bodies of workers are generally dark brown with a reddish brown abdomen and black apical segments (banded). Adults are lightly covered with hair.

**LIFE CYCLE:** Workers are the most abundant caste, providing the bulk of colony maintenance. Upon eclosion, workers care for the queen within the hive by feeding her and taking her eggs to cells where they develop as larvae. The workers care for developing larvae by providing food during development. Toward the end of their lives, workers leave the hive and begin to forage for nectar and pollen. In this activity, adult workers pick up nectar from several flowers and transfer it back to the hive, where it is converted to honey for larvae and the queen. Workers also defend the hive by attacking and stinging hostile invaders. Once they expend their venom, however, they soon die. The single queen has sole responsibility for egg production in the hive and is fertilized by drones. New colonies are formed when a queen, accompanied by a portion of the workers, swarms from the old hive to establish a new location.

Another unusual form of reproduction—occurring in one beetle family (Coleoptera: Micromalthidae), several genera of gall and fungus gnats (Diptera: Cecidomyiidae), and a common flower fly (Diptera: Syrphidae)—is **paedogenesis,** which refers to reproduction by the juvenile form. In paedogenic flower flies, the ovaries become functional in females with immature (larval) body form, and eggs develop parthenogenetically. Eventually, the mother larva becomes filled with developing daughter larvae, and these eventually emerge from the mother. Under certain environmental conditions, winged males and females are produced and, in turn, reproduce sexually.

### Fertilization

As mentioned, fertilization of eggs usually occurs in the female oviduct. Here, the sperm enters the egg through an opening in the egg covering (**chorion**) called the **micropyle** (Fig. 4.3). One to several sperm enter, with one reaching the nucleus to form a zygote. Subsequently, cleavage of the egg nucleus begins.

### Development of the Embryo

Unlike vertebrates and some other animals, most insects (except Collembola) do not undergo complete cell division during cleavage. Instead, the nucleus divides into a number of daughter nuclei (**energids**). These gather a mass of cytoplasm, further divide, and migrate to the periphery of the egg (Fig. 4.4). On reaching the periphery, a membrane forms around each nucleus and cytoplasm mass to form a cell. These cells create a wall around the interior of the egg one cell-layer thick, forming a stage called the **blastula.** A crowding of cells then begins in a ventral area of the egg, forming a thickened area known as the **germ band,** or, if small and platelike, the germ disk. It is from the germ band or germ disk that the new embryo develops (Fig. 4.5).

Development in most species proceeds when the germ band forms an infolding that eventually produces a three-layered embryo (**gastrula**) (Fig. 4.6). The inner layer (**mesoderm**) gives rise to muscles, body fat, heart, blood cells, and reproductive organs. The outer cell layer (**ectoderm**) produces the body wall, fore- and hindgut, nervous system, respiratory system, and many glands. Rudiments at either or both ends of the body give rise to a third embryonic tissue (**endoderm**) that forms the midgut.

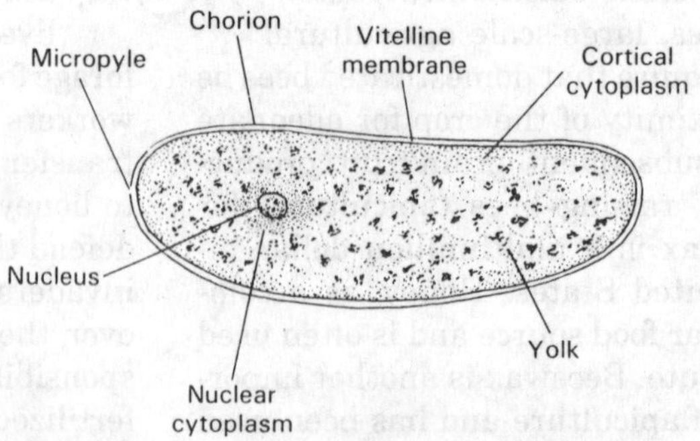


**Figure 4.3** Section of a typical insect egg showing important components. (Redrawn from Snodgrass, 1935, *Principles of Insect Morphology*, McGraw-Hill Publishing Company)

Segmentation (transverse creases) of the body begins at an early stage of embryonic development, and segments fuse later to form the head (Fig. 4.7). Appendages develop soon after segmentation, some of which become legs, mouthparts, and posterior abdominal structures.

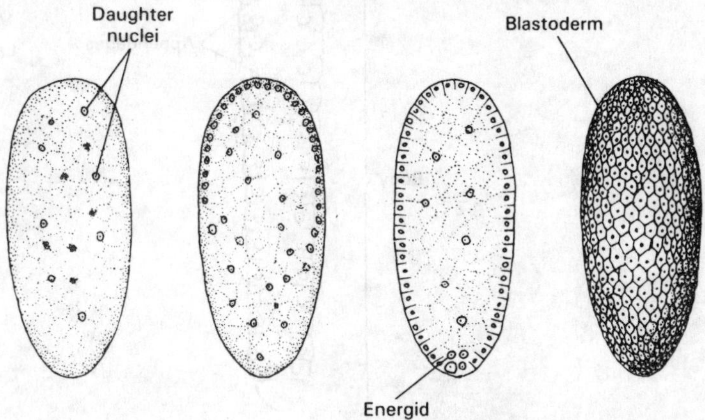


**Figure 4.4** Diagrams of insect eggs showing division of a nucleus into daughter nuclei (*far left*), migration of energids to egg periphery (*middle*), and final formation of multicellular blastoderm. (Redrawn from Snodgrass, 1935, *Principles of Insect Morphology*, McGraw-Hill Publishing Company)

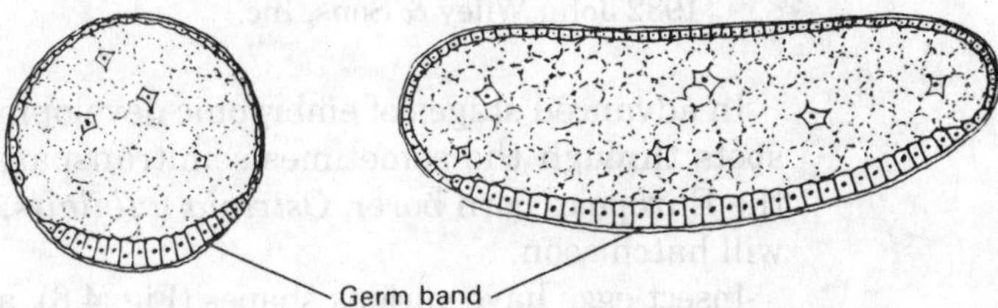


**Figure 4.5** Formation of the germ band in the blastoderm. The germ band later gives rise to the insect embryo. Transverse section (*left*) and longitudinal section. (Redrawn from Snodgrass, 1935, *Principles of Insect Morphology*, McGraw-Hill Publishing Company)

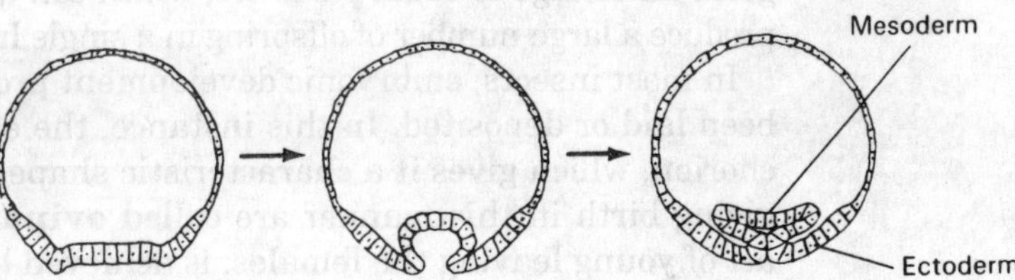


**Figure 4.6** Cross section of a blastoderm (*left*), followed by successive development toward the two-layered embryo (gastrula). In this instance, mesoderm is formed by invagination; however, other processes of mesoderm formation are also found in insects. (Redrawn from Snodgrass, 1935, *Principles of Insect Morphology*, McGraw-Hill Publishing Company)

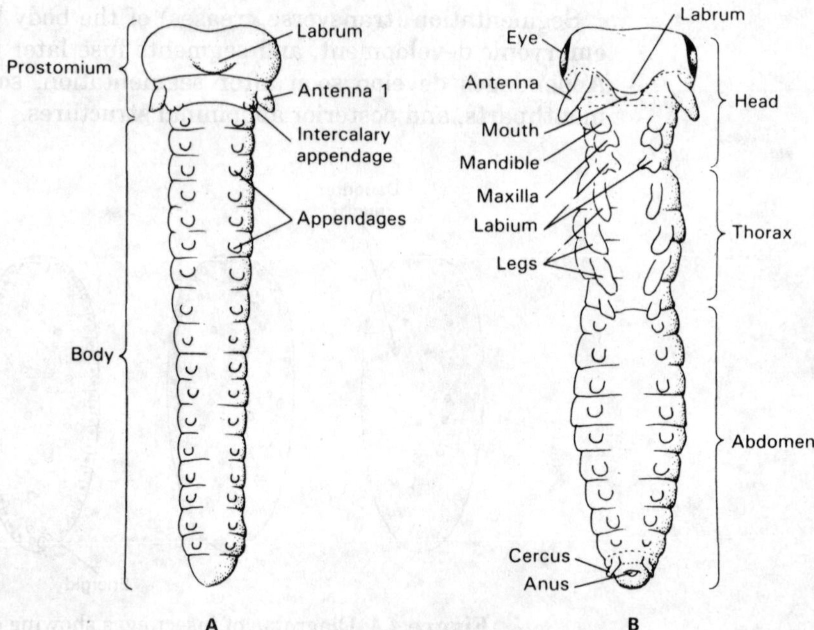


**Figure 4.7** Segmentation of the embryo and early stages of appendage development. A. Very young embryo with rudimentary appendages on all but the last segment. B. Older embryo showing development of body regions, as well as appendages. (Reprinted by permission of John Wiley & Sons, Inc., from *A Textbook of Entomology* by H. H. Ross, C. A. Ross, and J. R. Ross. © 1982 John Wiley & Sons, Inc.)

In advanced stages of embryonic development, an observer can often see eye spots through the sometimes-semitransparent chorion. In some species, like the European corn borer, *Ostrinia nubilalis*, this is an indication that the egg will hatch soon.

Insect eggs have various shapes (Fig. 4.8), and in most insect species, only one embryo forms in a single egg. In some parasitic Hymenoptera, however, more than one embryo may be formed through a process of asexual division. The numbers of embryos produced in this process of **polyembryony** usually vary from two to a few. However, in some species, 100 to 3,000 embryos are produced from one egg. In polyembryony, each embryo develops into an active larva. Such a phenomenon offers a great advantage to small parasites, which can quickly insert only one egg and yet produce a large number of offspring in a single host.

In most insects, embryonic development proceeds within the egg after it has been laid or deposited. In this instance, the egg is surrounded by a protective chorion, which gives it a characteristic shape and surface appearance. Insects giving birth in this manner are called **oviparous.** In other words, birth, the act of young leaving the females, is achieved by egg laying.

In a few species of many insect orders, eggs with a chorion are held within the female until embryonic development is nearly complete. Hatching usually occurs when these eggs are still in the female egg-laying organs. These insects—for example, many flesh flies (Diptera: Sarcophagidae)—are referred to as **ovoviviparous.** Such an adaptation is particularly advantageous to insects whose food supply deteriorates quickly, before the eggs, if laid, would hatch.

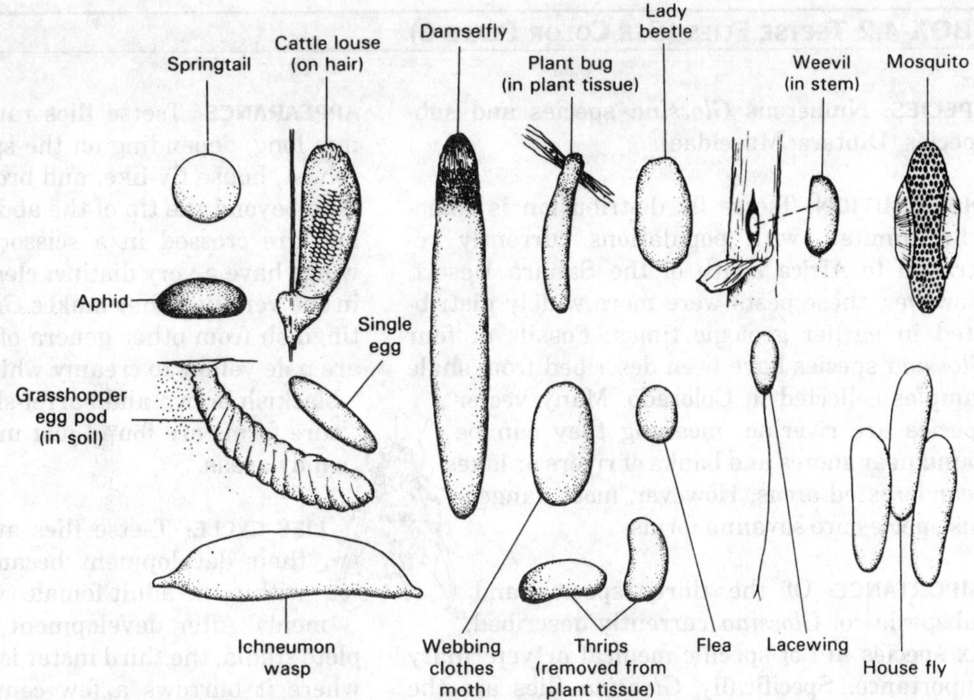


**Figure 4.8** Various examples of insect eggs. (Redrawn after Essig and others from Ross et al., 1982, *A Textbook of Entomology*, John Wiley & Sons)

A very specialized form of birth occurs in **viviparous** species of insects. Members of the fly genus, *Glossina*, which includes the tsetse fly (Box 4.2), exhibit viviparity. Here, one larva develops at a time in a specialized "uterus." Each larva grows by receiving nutriments from special uterine ("milk") glands. Tsetse fly larvae are fully developed when born and begin adult development soon after birth. Most aphid species are also viviparous, but only at certain times of the year.

## POSTEMBRYONIC GROWTH AND DEVELOPMENT

### Egg Hatching

When embryonic development is complete in oviparous insects, eggs are predisposed to hatch. The hatching process may begin immediately on completion of embryonic development or may be delayed for a time, thereby taking advantage of more favorable environmental conditions.

At this point in the life cycle, the embryo within the egg must break out of its enclosure. The hatching process often begins when the embryo in the egg swallows fluids or air. This action gives the embryo more bulk and turgidity.

Subsequently, the embryo must rupture the egg covering to escape. Ruptures may be caused when the insect produces rhythmic muscular activity and presses or strikes against the covering with its head. The eggshell rupture may occur irregularly along the egg surface, as in grasshoppers (Orthoptera: Acrididae), or

## BOX 4.2 TSETSE FLIES (SEE COLOR PLATE 6)

**SPECIES:** Numerous *Glossina* species and subspecies (Diptera: Muscidae)

**DISTRIBUTION:** Tsetse fly distribution is somewhat limited, with populations currently restricted to Africa south of the Sahara Desert. However, these pests were more widely distributed in earlier geologic times. Fossils of four *Glossina* species have been described from shale samples collected in Colorado. Many vector species are riverine, meaning they can be found near shores and banks of rivers or lakes near forested areas. However, most dangerous species are savanna forms.

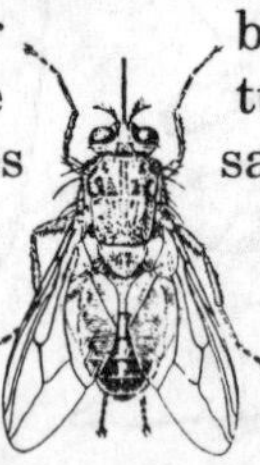

**IMPORTANCE:** Of the thirty species and subspecies of *Glossina* currently described, six species are of specific medical or veterinary importance. Specifically, *Glossina* flies are the primary vectors of pathogens that cause sleeping sickness in Africa (also known as African trypanosomiasis). The disease, natively known as Nagana, also affects domestic cattle. These pathogens are flagellate protozoa, which are transmitted with the bite of a tsetse fly. Sleeping sickness is a disabling and usually fatal disease. It is a primary factor limiting the development of Africa south of the Sahara Desert and is responsible for about 7,000 human deaths annually.

**APPEARANCE:** Tsetse flies range from 7.5 to 14 mm long, depending on the species. The body is robust, house fly-like, and brown. The wings extend beyond the tip of the abdomen when at rest and are crossed in a scissorslike fashion. The wings have a very distinct cleaver or hatchet cell in the venation that makes *Glossina* easy to distinguish from other genera of Muscidae. Larvae are pale yellow to creamy white, but puparia are blackish brown and barrel-shaped. Both immature forms are found just under the surface in sandy areas.

**LIFE CYCLE:** Tsetse flies are very atypical in their development because larvae develop within the adult female (viviparous development). After development through two complete stadia, the third instar is deposited on sand, where it burrows a few centimeters and completes its development. Larvae are susceptible to predation or parasitization in this stage but form puparia within a few hours and are relatively safe from natural enemies. Fourth instars develop within the puparia prior to true pupation, which lasts 30 days or more. Only one larva is carried by the female at a time, with 8 to 25 days needed for development. Adult flies emerge and continue the cycle. Breeding is continuous in the warm latitudes of tropical Africa.

---

along preformed lines of weakness. The green stink bug (Box 4.3) (Hemiptera: Pentatomidae) represents the latter type; its eggs have easily ruptured "caps" that the insects open like a lid. In some insect groups, such as dragonflies (Odonata), a T- or Y-shaped central "tooth" (egg burster) forces the rupture, and in others, spines or teeth on the head or body are used to "saw" through the eggshell. Many butterflies and moths (Lepidoptera) simply chew their way out of the eggshell (Fig. 4.9).

Upon causing the rupture, the insect works its way out of its enclosure. Insects that possess a special embryonic cuticle or skin shed it as they leave the egg or shortly thereafter. The act of the larva leaving the egg is called **eclosion.**

### Growth of Immatures

After eclosion, the new free-living insect may be similar to its parents or have an entirely different form. In either instance, the immature stage is usually characterized by feeding behavior. The immature is a virtual "feeding machine" at this point in the life cycle. Frequently, injury by pest species is most

## BOX 4.3 GREEN STINK BUG (SEE COLOR PLATE 4)

**SPECIES:** *Acrosternum hilare* (Say) (Hemiptera: Pentatomidae)

**DISTRIBUTION:** The green stink bug is found in both North and Central America, but is more injurious in the southern United States. Individuals of this species can be found in field borders before fruit is set and move into fields as fruits and pods become available.

**IMPORTANCE:** The green stink bug attacks many wild and cultivated plants, including alfalfa, lima bean, peach, cotton, and soybean. With cotton, the piercing-sucking mouthparts are inserted into the bolls, and sap is removed from immature seeds. The punctures cause shedding of small bolls, whereas larger ones have stained lint. In soybeans, nymphs and adults pierce pods and feed directly on the seeds. This feeding results in smaller seeds and shriveled seed coats. Such injury results in a downgrading of seed quality at sale, reducing the market value of the entire harvest.

**APPEARANCE:** Adults are shield-shaped, typical pentatomids, with a large triangular scutellum. This species has a distinctive long spine between the thoracic legs, which distinguishes it from a similar species, the southern green stink bug, *Nezara viridula* (Linnaeus). Nymphs resemble adults, except they are smaller and wingless. In addition, nymphs are dark with white lines on the abdomen (southern green stink bugs have white dots on the abdomen). Stink bug eggs are characteristically barrel-shaped and are typically found in compact clusters of regular rows. It is not uncommon to find parasitized stink bug eggs. When parasitized, eggs are black and become inviable.

**LIFE CYCLE:** Adults overwinter in sheltered areas such as fence rows where plant refuse is abundant. In spring, they become active, and females oviposit 300 to 500 eggs, in clusters of about 30, on foliage during a 30-day period. Hatching occurs about 7 days after oviposition, and five stadia follow, showing gradual metamorphosis. The adult stage is reached in approximately 6 weeks. Multiple generations occur annually, with overlap typical. Later generations, however, frequently are the most damaging to crops because of the presence of plant fruiting structures.

**Figure 4.9** Hatched and unhatched eggs of western bean cutworm, *Striacosta albicosta*. (Photo by Marlin E. Rice)

**Figure 4.10** Stink bug (*left*) and plant bug molting their exuvium. (Photos by Dennis Schotzko)

severe at this time and, in many instances, occurs exclusively during this stage (for example, caterpillar pests).

This feeding behavior results in substantial growth in size of the insect. Growth, as measured by weight gain, usually progresses in a stepwise fashion, with each step delineated by the shedding of the immature's old skin **(exuvium).** This shedding process, called *molting,* (Fig. 4.10) results from the inability of the insect's skin or cuticle to expand. Therefore, the insect grows to a size where its outer covering becomes limiting. It sheds that covering, develops a new larger one, grows to fill up the new skin, sheds it, and so on, until growth is complete.

The molting process usually begins with the cessation of feeding and the clearing of gut contents. The process is initiated and controlled by hormones that circulate in the blood. The major hormones involved are **brain hormone** (AH, activation hormone, or PTTH, prothoracicotropic hormone, as it is called by some authors) and **ecdysone** (Fig. 4.11). Brain hormone is produced by neurosecretory cells in the brain and dispensed into the blood cavity through brain accessory structures. Once in the blood, brain hormone circulates to an activity site in the insect's prothorax. A small gland in the prothorax, the **prothoracic gland,** is stimulated to secrete ecdysone (molting hormone, also called prothoracic gland hormone), which initiates the growth and molting activities of cells.

Upon stimulation by ecdysone, cells in the epidermis (the cellular part of the exoskeleton) divide and become closely packed. As a result, the cuticle (noncellular part of the exoskeleton) becomes separated from the epidermis, a process called **apolysis.** As the cuticle separates from the epidermis, **molting fluid** is produced. Molting fluid contains proteinase and chitinase enzymes that digest up to 90 percent of the old cuticle. This digested material is absorbed and metabolized. While the old cuticle is being digested, a new cuticle is started. It grows by forming different layers, with the outer layer resistant to enzymes of the molting fluid, thus protecting new layers from the digestive process.

Once the new cuticle is complete and digested material is absorbed, ecdysis occurs. **Ecdysis** is the process of shedding the remainder of the old cuticle, which is split along weakened lines by movements of the insect inside. The

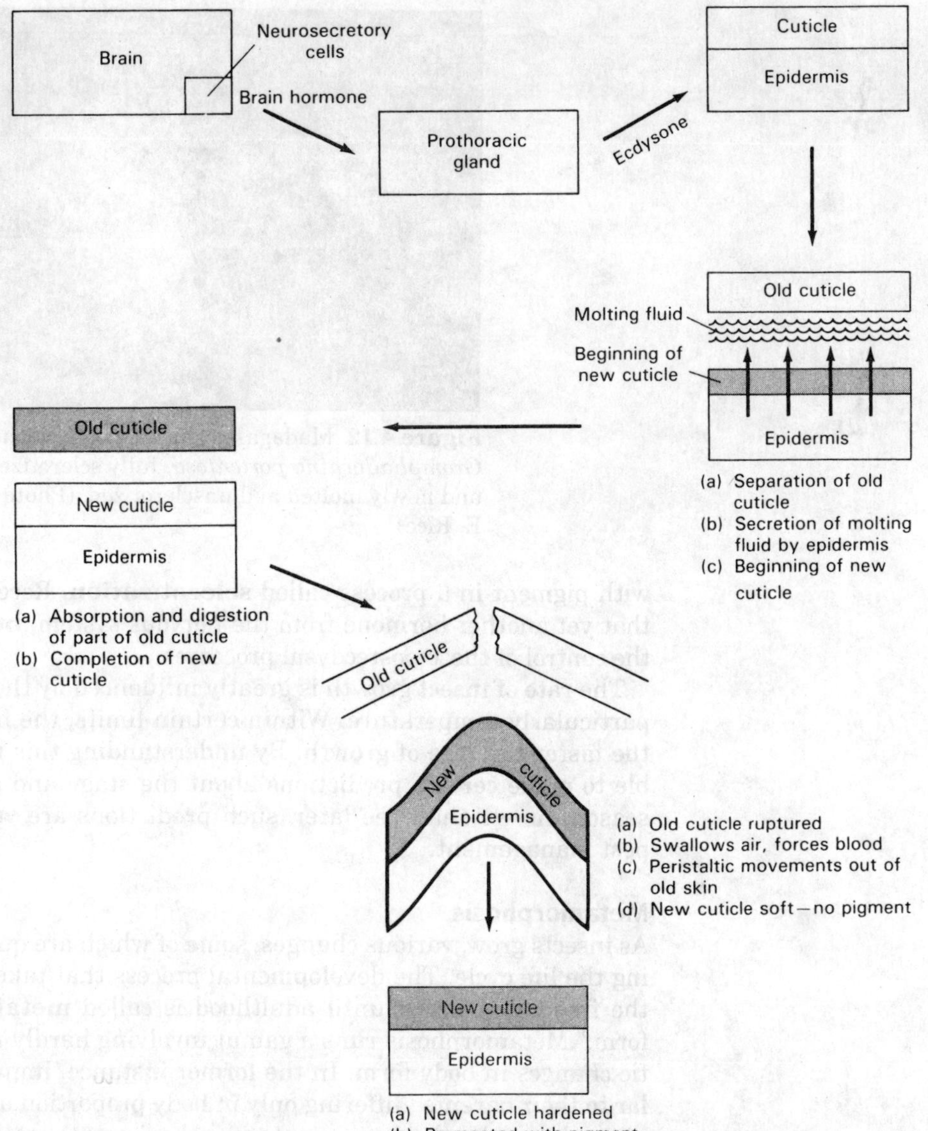


**Figure 4.11** Schematic diagram of molting in insects and principal hormones involved.

initiation of these movements in many insects is believed to be caused by another hormone called **eclosion hormone.** Eclosion hormone is secreted by specific brain cells at a certain time of day. In many instances, the insect swallows air or water and thereby increases blood pressure. Muscle action pumps blood to a certain part of the body, usually the thorax, to expand it. This expansion causes the old cuticle to rupture. The insect then exits through this rupture, usually head and thorax first, followed by the abdomen and appendages.

Immediately after emergence, the new cuticle is soft and pliable (Fig. 4.12). The insect expands this new covering before it hardens. This process may again involve swallowing air or water and using blood pressure to expand the body, one part at a time. Once expanded, the new cuticle hardens and is permeated

**Figure 4.12** Madagascar hissing cockroaches, *Gromphadorhina portentosa,* fully sclerotized *(left),* and newly molted and unsclerotized. (Photo by Marlin E. Rice)

with pigment in a process called **sclerotization.** Recently, it has been shown that yet another hormone from the nervous system, **bursicon,** plays a role in the control of these postecdysal processes.

The rate of insect growth is greatly influenced by the physical environment, particularly temperature. Within certain limits, the higher the temperature, the faster the rate of growth. By understanding this relationship, it is possible to make certain predictions about the stage and size of insects during a season. As we shall see later, such predictions are very important in insect pest management.

## Metamorphosis

As insects grow, various changes, some of which are quite dramatic, occur during the life cycle. The developmental process that takes place from eclosion of the free-living insect until adulthood is called **metamorphosis,** "change in form." Metamorphosis runs a gamut involving hardly noticeable to quite drastic changes in body form. In the former instance, immature insects look similar to their parents, differing only in body proportion and lacking reproductive appendages. In the latter instance, the immatures are very different from their parents, often changing from a wormlike, crawling form to an ovoid, inactive form and, finally, to a winged, flying form.

Metamorphosis represents orderly, genetically programmed changes in insect form during completion of the life cycle. Insects with radical changes in body form, such as butterflies, undergo internal processes that cause rearrangements within cells, tissues, and organs. These processes can be ordered into two phases, **histolysis** and **histogenesis.** Histolysis is the breakdown of body tissues, which often begins when the insect is at the end of its primary growth stage; it continues into the transformation stage. During histolysis, stored nutrients (carbohydrates, fats, glycogen) supply energy for numerous biochemical reactions, and enzymes with the aid of special blood cells (**phagocytes**) convert the insect fat body and muscle tissue into a nutritive mix, which is transported through the blood to growing tissues. Histogenesis is the formation of new body tissues; it occurs at the

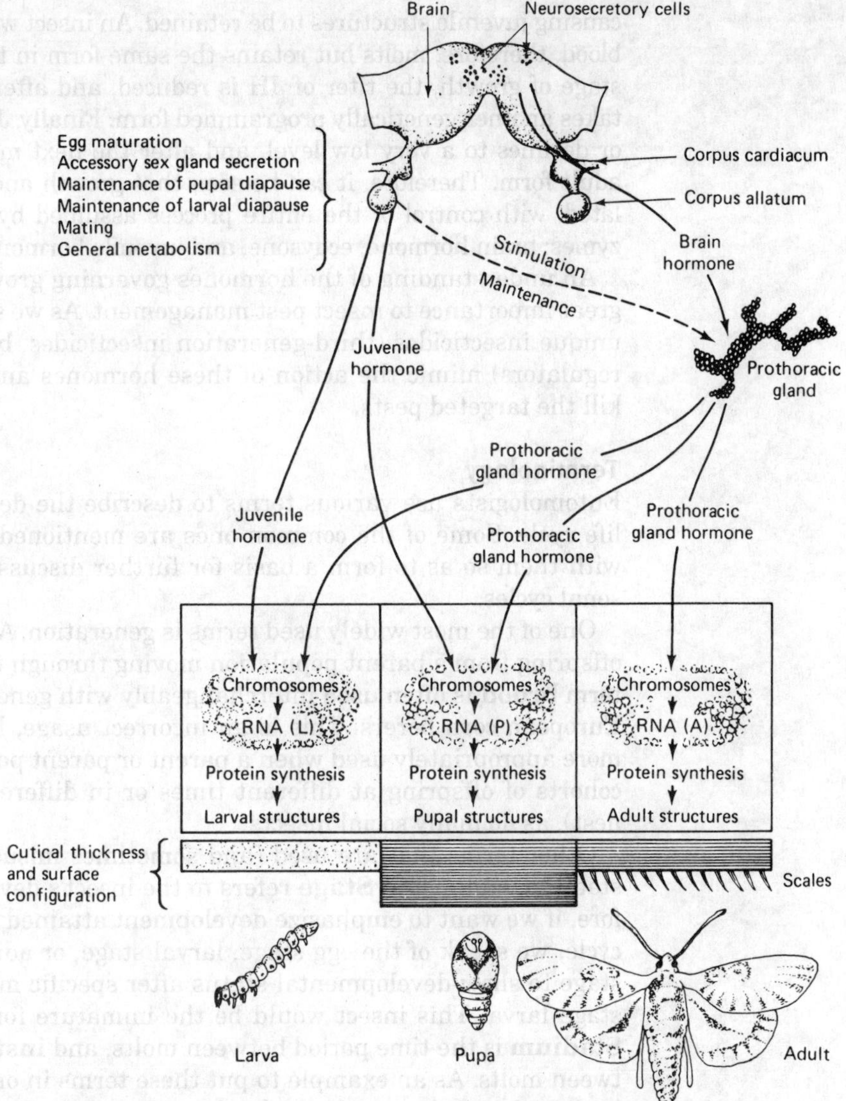


**Figure 4.13** Diagrammatic representation of the principal hormones involved in insect growth and metamorphosis and their action. (Reprinted with permission of Macmillan Publishing Company from *Insects in Perspective* by Michael D. Atkins. © 1978 by Michael D. Atkins)

same time as histolysis. In particular, complete exchanges of immature muscle tissues for adult muscle tissues occur during this time, as well as exclusive adult organs such as wings and reproductive structures. The result, in many instances, is an adult that is quite different in appearance from the immature form. Specific metamorphic patterns are discussed later in this chapter.

Metamorphosis is primarily under the control of another insect hormone called **juvenile hormone** (JH). JH is produced by glands accessory to the brain, the **corpora allata** (Fig. 4.13). Production is believed to be mediated by brain hormone. Secretion of JH in the blood functions to suppress adult characteristics, causing juvenile structures to be retained. An insect with a high titer of JH in the blood, therefore, molts but retains the same form in the next stage. At a critical stage of growth, the titer of JH is reduced, and after the next molt, the insect takes another genetically programmed form. Finally, JH production is suspended or declines to a very low level, and after the next molt, the insect assumes an adult form. Therefore, it can be seen that growth and development are interrelated, with control of the entire process assumed by three very important enzymes: brain hormone, ecdysone, and juvenile hormone.

An understanding of the hormones governing growth and development is of great importance to insect pest management. As we shall see, some of our most unique insecticides (third-generation insecticides, biorationals, insect-growth regulators) mimic the action of these hormones and, when applied properly, kill the targeted pests.

### Terminology

Entomologists use various terms to describe the developing insect during its life cycle. Some of the common ones are mentioned here to acquaint readers with them so as to form a basis for further discussion of life cycles and seasonal cycles.

One of the most widely used terms is generation. A **generation** is a cohort of offspring from a parent population moving through the life cycle together. The term **brood** is often used interchangeably with generation, as in second-brood European corn borers. This is an incorrect usage, however, because brood is more appropriately used when a parent or parent population produces several cohorts of offspring at different times or in different places (more than one nest), as do many social insects.

Other terms that are used (and sometimes misused) frequently are stage, stadium, and instar. **Stage** refers to the insect's developmental status. Therefore, if we want to emphasize development attained at a given point in the life cycle, we speak of the egg stage, larval stage, or adult stage. We also can use stage to show developmental status after specific molts, for instance, a third-stage larva. This insect would be the immature form after the second molt. **Stadium** is the time period between molts, and **instar** is the actual insect between molts. As an example to put these terms in order, we might say the following: the sixth-stage larva displays the greatest rate of consumption during the life cycle, and its stadium is 3 days. We have counted ten sixth-instars in the sample, and they are green with white stripes. Never use the term *sixth-instar larvae*, for example, because elements of the term are redundant.

## MATURITY

### Emergence of the Adult

Adults emerge from the last-immature exuviae in a way similar to immatures exiting their old skins. After splitting the old cuticle, the new adult pulls itself out and, if winged, expands its newly formed wings by pumping blood through them. As with many newly molted immatures, the new adult at first is soft and may lack much of its final pigmentation. At this time, it is very vulnerable to natural enemies and other environmental adversities. Within a short time, however, the cuticle hardens and becomes pigmented, and the wings are fully expanded and functional.

The degree of sexual maturation at the time of adult emergence varies between sexes and among species. Most male insects emerge with a store of sperm, and some are fully capable of mating within a short time. However, in some species the female may require a longer time to become sexually mature. During female reproductive maturation, eggs form, the reproductive tract completes development, and other glands associated with reproduction develop. In certain short-lived species—for example, mayflies (Ephemeroptera)—both sexes are mature within a day or so, and mating and egg laying occur shortly thereafter.

### Mating Behavior

Mating behavior consists of all the events involved in the transfer of sperm from the male to the female. The two most important events in this behavior are finding mates and copulation (Fig. 4.14).

Mate finding is an important behavior in all but a very few insects. For example, some Collembola transfer sperm indirectly; the male simply places a sperm packet, called a **spermatophore,** on a substrate and it is picked up subsequently by a female, presumably by chance. In most instances, however, locating a mate is necessary, and this act may involve complex processes, among which are certain cues to locate and recognize an appropriate mate. Such cues may be visual, olfactory, auditory, tactile, or combinations of these.

Olfactory cues are probably the most commonly used type in mate finding. Many insect species secrete volatile chemical attractants that are highly specific and can be used by one sex to attract its opposite over considerable distances. Such volatiles are called **sex pheromones.** Male-attracting sex pheromones are the common type found in many insect orders, particularly Coleoptera and Lepidoptera. In many moths, females begin "calling" (everting the pheromone-releasing glands and releasing the volatile) about 1 to 2 hours before dawn. Males detect differences in concentration of the volatile and tend to follow the gradient of increasing concentration until they locate the female. When the male gets close to a female, he hovers and lands, then moves about until he collides with her. In other insects, like the boll weevil, *Anthonomus grandis*, the male produces a sex pheromone, and the female is attracted.

Copulation may follow immediately after mate finding or, in some species, after specific forms of courtship behavior. Depending on the species, a female may copulate several times during her lifetime or only once. Following the act of copulation (joining of male and female genital structures), **insemination** (introducing of sperm into the female reproductive tract) occurs.

As we will see in later chapters, an understanding of mating behavior has important implications for insect pest management. By taking advantage of the compelling nature of sex pheromones, entomologists have developed more effective insect sampling programs, and they have attempted to utilize pheromones and synthetic mimics to disrupt the mating process.

### Oviposition

Once mating has occurred and eggs are matured and fertilized, egg deposition is the ultimate event in the life cycle of an insect. The act of laying eggs by the adult female is called **oviposition.**

Insects differ greatly in the ways they lay their eggs, or oviposit. Eggs may be laid singly (for example, lacewings), in clumps (stink bugs), or in a contiguous

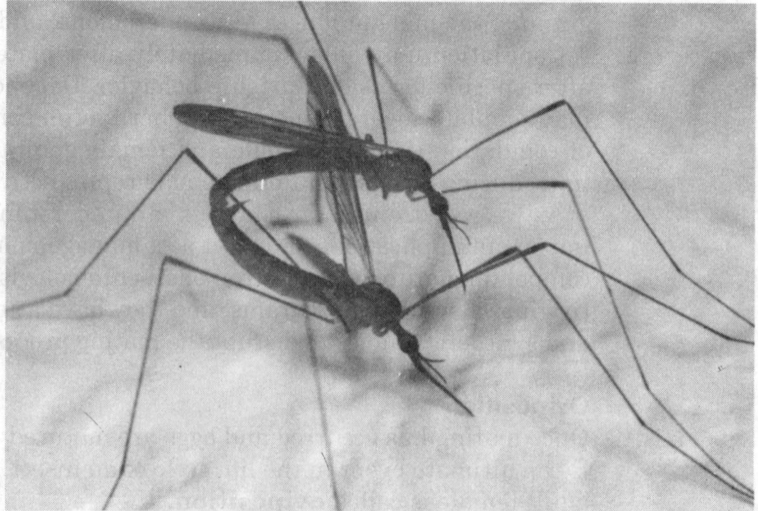

**Figure 4.14** Typical copulatory position of grasshoppers *(top, with female on bottom)*, brown stink bugs, and crane flies. (Photos by Marlin E. Rice)

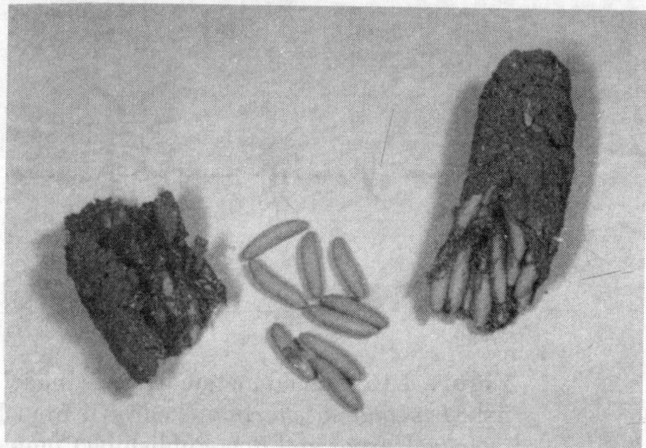

**Figure 4.15** Egg pod of grasshopper laid in soil. (Photo by Marlin E. Rice)

mass (European corn borer). Some insects, such as cockroaches, produce an **ootheca,** which is a protective pod that contains several eggs "glued" together by a secretion.

Insect eggs simply may be dropped on the ground, as with walkingsticks (Phasmatodea: Phasmatidae), or they may be placed on or in host tissues where newly eclosed immatures have ready access to food. In the latter instance, these eggs may be glued to plant leaves or stems, as with moths and butterflies, or inserted inside stems and leaf petioles, as with many leafhoppers, including the potato leafhopper, *Empoasca fabae*. Another important placement pattern for many insects is oviposition in soil; for example, grasshoppers lay masses of eggs in special chambers in grass sod (Fig. 4.15). Oviposition in plant and animal tissues and soil provides a more or less protected environment for egg development.

Female insects lay their eggs with a specialized structure called an **ovipositor.** Ovipositor structure varies from a simple extensile tube through which eggs pass (for example, in Lepidoptera and Diptera) to a well-developed "saw" found in sawflies (Hymenoptera: Cephidae). This saw-toothed ovipositor cuts into plant tissues for egg insertion. Highly developed ovipositors with dual functions are found in other Hymenoptera. They are used to "sting" a host, usually another insect, and place eggs inside (for example, parasitic wasps) or sting another insect (or other animal) for defensive or paralyzing purposes (for example, bees, ants, and wasps). In the latter instance, a venom is injected when the "stinger" is inserted. With hunting wasps, the stung prey (for example, a caterpillar, grasshopper, or cricket) becomes stupefied and is carried back to the wasp nest; subsequently, eggs are laid on it. After hatching, newly emerged immatures begin feeding on the stunned but still living prey. When bees, wasps, and ants sting people, pain, discomfort, or even death may result from the venom.

## GENERAL MODELS OF THE LIFE CYCLE

A study of insects shows that each life cycle is species specific; in other words, each species has its own peculiar mode of living, and it is difficult to generalize from one species to the next. But we do see patterns in metamorphosis

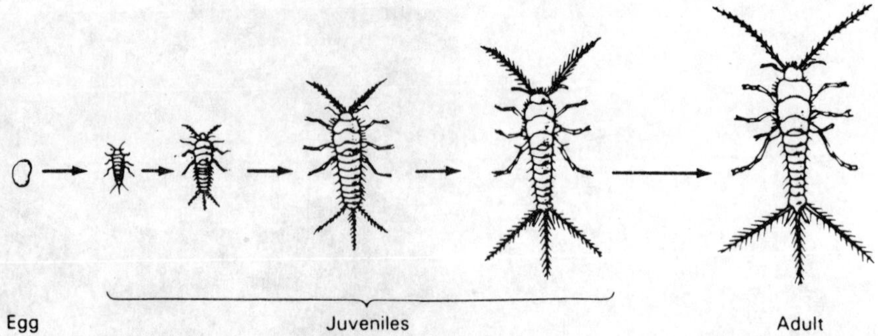


**Figure 4.16** The no-metamorphosis model of development represented by a silverfish, *Lepisma saccharina*. (Redrawn from Little, 1972, *General and Applied Entomology*, Harper and Row Publishing Company)

among insect species that relate to their activities and modes of living in the environment, and we can use these patterns as general models of the life cycle.

By emphasizing degree of complexity and metamorphic pattern, we can recognize four insect life cycle models: (1) no metamorphosis, (2) gradual metamorphosis, (3) incomplete metamorphosis, and (4) complete metamorphosis. These categories are given for convenience in discussing life cycles, not for physiological classification of metamorphosis. For such classification, most authorities would group models 2 and 3 into a single category, gradual metamorphosis.

### The No-Metamorphosis Model—Ametabolous Development

The no-metamorphosis model (Fig. 4.16) is found in primitively wingless insects (subclass Apterygota), for example, springtails, silverfish, firebrats, and other noninsect arthropods. In these groups, the life cycle progresses from egg to **juvenile** to adult. The transition from the first juvenile to the adult is gradual, with several molts taking place in the juvenile stage. During the life cycle, the juvenile appears very similar to the adult, differing mainly in body size and proportion and lacking functional genitalia. In these insects, all stages can be found in the same habitat, and juveniles feed similarly to adults. Unlike other insects, molting continues in the adult stage, and because females lose the lining of the spermatheca with each molt, they are inseminated several times during the life cycle.

### The Gradual-Metamorphosis Model—Paurometabolous Development

The life cycle of insects with gradual metamorphosis (Fig. 4.17) has three major stages: egg, **nymph,** and adult. Common examples of this model are found in grasshoppers, bugs, cockroaches, and crickets. Here the nymph resembles the adult but lacks fully formed wings and external genitalia. Nymphs may be colored quite differently from adults (for example, some stink bugs), and they show the beginnings of wings, called **wing pads,** in later developmental stages. In most instances, eggs, nymphs, and adults with this life cycle are found in the same habitat, and nymphs and adults feed similarly.

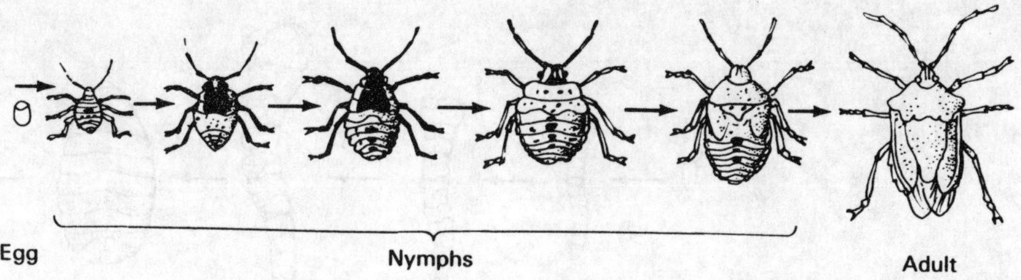


**Figure 4.17** The gradual-metamorphosis model of development represented by a stink bug (Hemiptera: Pentatomidae). (Redrawn from Little, 1972, *General and Applied Entomology,* Harper and Row Publishing Company)

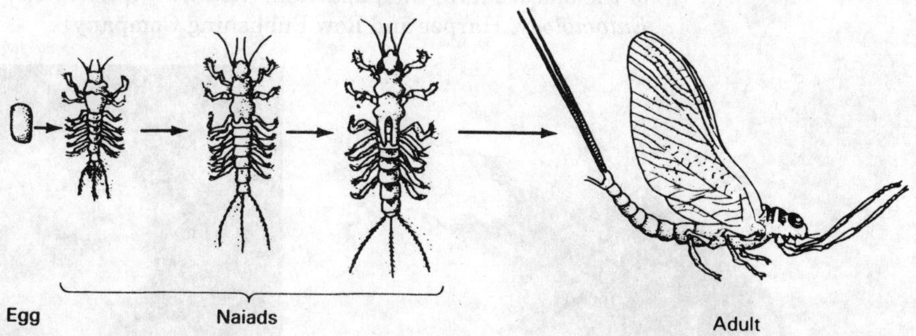


**Figure 4.18** The incomplete-metamorphosis model of development represented by a mayfly (Ephemeroptera). (Redrawn from Little, 1972, *General and Applied Entomology,* Harper and Row Publishing Company)

### The Incomplete-Metamorphosis Model—Hemimetabolous Development

The incomplete-metamorphosis model (Fig. 4.18) is one in which immatures may or may not resemble adults, have external wing pads, and may have special "tracheal gills" that allow aquatic respiration. There are many variations in this type of life cycle, but in general, eggs are found in or near the water and immatures, called **naiads,** feed and develop in water. Adults of these insects may be found flying above or near the water or, with some dragonflies, quite some distance from it.

Unlike the gradual-metamorphosis model, immatures and adults of species with incomplete metamorphosis spend most of their time in very different habitats and feed quite differently. Although they do not represent a cohesive group in an evolutionary sense, mayflies (Ephemeroptera), dragonflies and damselflies (Odonata), and stoneflies (Plecoptera) all follow this life cycle model.

### The Complete-Metamorphosis Model—Holometabolous Development

The complete-metamorphosis life cycle (Fig. 4.19) has four distinct life stages in the cycle: egg, **larva, pupa,** and adult. The presence of the larval and pupal stages are the most notable aspects of this life cycle model. With few exceptions, larvae (Fig. 4.20) are very different from adults. They have a different form, lack compound eyes, have reduced antennae, and lack external evidence of wing formation. In most, wing development begins in the early larval stages

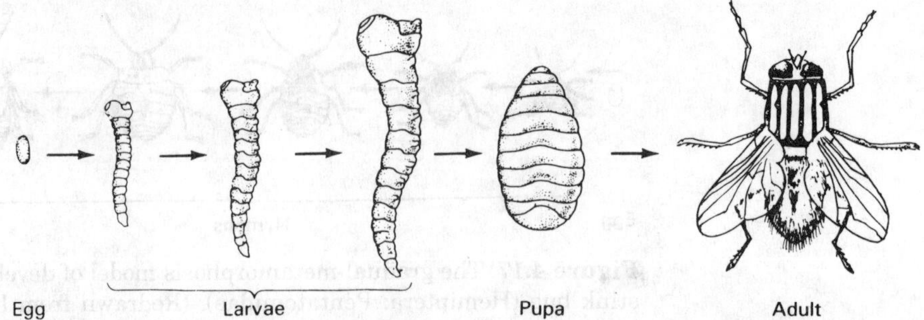


**Figure 4.19** The complete-metamorphosis model of development represented by a house fly, *Musca domestica*. The pupa of this insect is found within the hardened skin of the oldest larva, the puparium. (Redrawn from Little, 1972, *General and Applied Entomology,* Harper and Row Publishing Company)

A B C D E F

**Figure 4.20** Various forms of insect larvae. A. Antlion (Neuroptera). B. Scarab beetle (Coleoptera). C. Ground beetle (Coleoptera). D. Weevil (Coleoptera). E. Fly (Diptera). F. Looper (Lepidoptera). (Photos by Marlin E. Rice)

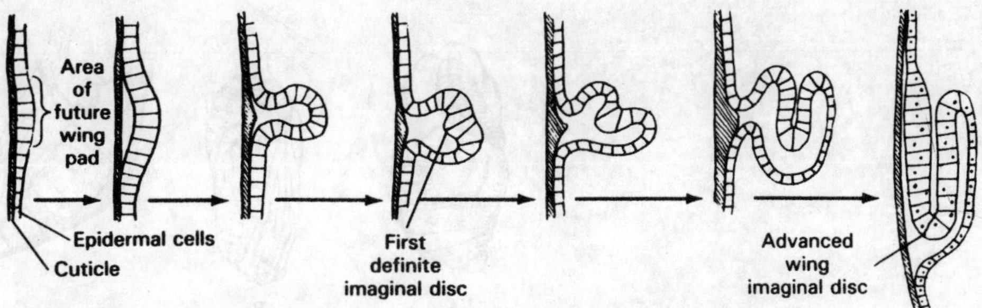


**Figure 4.21** Stages of growth of imaginal disks. (Redrawn from Ross et al., 1982, *A Textbook of Entomology*, John Wiley & Sons)

but occurs internally from rudiments known as **imaginal disks** (Fig. 4.21). The pupal stage is represented by an insect usually in a quiescent state (a few insect species have active pupae) (Fig. 4.22). Most quiescent pupae are found in hidden, protected habitats. Some, like the moths, are present in **cocoons,** a covering of silk or silk-bound debris constructed by the last larval instars (**prepupae**). Pupae of the more advanced Diptera are found within the last larval skin, which hardens to form a protective **puparium.** Wing pads appear on the pupa, and the active physiological processes of histolysis and histogenesis take place internally during this stage.

The pupal stage allows for the transformation of specialized larval structures that are not continued in the adult stage. With this arrangement, larvae and adults often have very different ecological roles and behaviors, with larvae specializing in food gathering and adults developing advanced means of reproduction and dispersal. In addition to larvae consuming much more food than most adults, they also may consume different food, thereby eliminating much of the competition between stages. For instance, moth larvae (caterpillars) feed on green plant tissue for growth, but adult moths feed on nectar for maintenance. When both larvae and adults are specialized for food gathering, their food source may differ. For example, mosquito larvae are found in water and feed on microorganisms there, but adult females are specialized to find and extract blood from birds and mammals.

In the majority of life cycles with complete metamorphosis, larvae in all stages behave similarly; they live in the same habitat and consume the same food. However, in some insects such as blister beetles (Coleoptera: Meloidae) or many parasitic Hymenoptera, body form, feeding behavior, and sometimes habitat change as larvae age. This phenomenon is termed **hypermetamorphosis.** For example, in the black blister beetle, (Box 4.4) *Epicauta pennsylvanica*, the first-stage larva, the **triungulin,** (Fig. 4.23) actively seeks out a grasshopper egg, punctures it, and feeds on its contents. The second, third, fourth, and fifth instars are inactive and continue to feed on the egg contents. When fully grown, the fifth instar leaves the food mass, burrows farther into the soil, and builds an earthen cell. There, it molts to become a sixth instar, a hardened oval form (**coarctate** form) with exceptionally small legs that does not feed and overwinters. This form is particularly resistant to drought, and, if the following summer is abnormally dry, it will not progress to the seventh stage for 1 or 2 years. When appropriate conditions occur, a seventh instar appears, and this

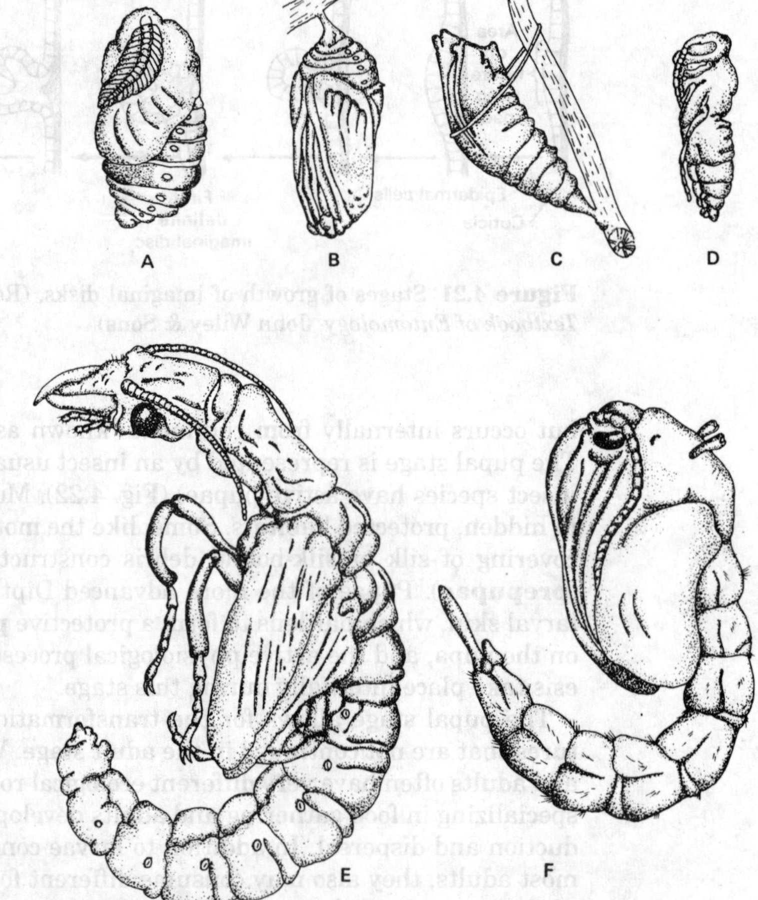


**Figure 4.22** Various forms of pupae. A., B., C. Lepidoptera. D. Hymenoptera. E. Neuroptera (dobsonfly). F. Diptera (mosquito). E. and F. are active pupae. (Reprinted with permission of Macmillan Publishing Company from *Insects in Perspective* by Michael D. Atkins. © 1978 by Michael D. Atkins)

larva molts to become the pupa. Adult blister beetles then may feed on and become important pests of potatoes, tomatoes, squash, some legumes, and other crops.

Other insects exhibiting hypermetamorphosis include mantidflies, flat bark beetles, wedge-shaped beetles, and twisted-wing parasites.

## INSECT SEASONAL CYCLES

Thus far, we have focused attention on insects within a single generation and emphasized the major events taking place in a life cycle. A logical extension of our focus would be to examine the cycles of insects from an annual perspective—in other words, look at the cycles as they pass through different seasons

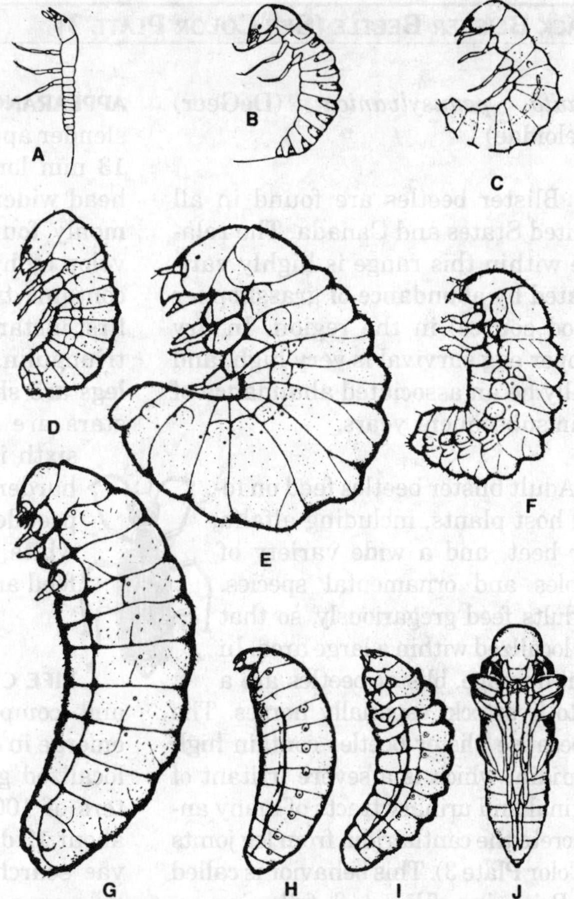


**Figure 4.23** Hypermetamorphosis as seen in a black blister beetle, *Epicauta pennsylvanica*. A. Unfed first instar. B. Fully fed first instar. C., D., E. Second, third, and fourth instars. F. Newly molted fifth instar. G. Gorged fifth instar. H. Sixth instar. I. Seventh instar. J. Pupa. (Reprinted by permission of John Wiley & Sons, Inc., *from A Textbook of Entomology* by H. H. Ross, C. A. Ross, and J. R. Ross. © 1982 John Wiley & Sons, Inc.)

of the year. This progression of one or more life cycles occurring during a 1-year period (for example, winter to winter) is termed the insect **sea-sonal cycle.**

Seasonal cycles are very important to the survival of insects because most environments of which insects are a part undergo annual cycles. Therefore, to achieve success, insects have behavioral and physiological ways to time their activities to take advantage of environmental resources and avoid unfavorable extremes. Knowledge of patterns in seasonal cycles and the timing of important biological events (for example, dormancy, migration, development, and reproduction) relative to environmental cycles is crucial when attempting to reduce pest damage.

Seasonal cycles can be grouped according to the number of generations that occur in a year; this number is referred to as a population's **voltinity.** Voltinity types are univoltine, multivoltine, and delayed voltine.

## BOX 4.4 Black Blister Beetle (See Color Plate 3)

**SPECIES:** *Epicauta pennsylvanica* (DeGeer) (Coleoptera: Meloidae)

**DISTRIBUTION:** Blister beetles are found in all parts of the United States and Canada. The relative abundance within this range is highly variable and regulated by abundance of grasshopper eggs (larval food source) in the region. In dry years, grasshopper egg survival is very high, and there will usually be an associated abundance of blister beetles in subsequent years.

**IMPORTANCE:** Adult blister beetles feed on foliage of several host plants, including alfalfa, soybean, sugar beet, and a wide variety of garden vegetables and ornamental species. Behaviorally, adults feed gregariously, so that damage may be localized within a large area. In addition to plant damage, blister beetles are a serious threat to livestock, especially horses. The danger arises because blister beetles contain high levels of cantharidin, which is a severe irritant of the gastrointestinal and urinary tracts of many animals. Adults secrete the cantharidin from leg joints when injured (Color Plate 3). This behavior is called reflex bleeding. Poisoning of livestock following accidental consumption of alfalfa hay containing dead blister beetles is occasionally reported. Ironically, larvae are considered to be beneficial because they feed almost exclusively on grasshopper eggs.

**APPEARANCE:** Adults have a distinct elongate and slender appearance. Black blister beetles are 7 to 13 mm long, relatively soft-bodied, and have a head wider than the pronotum. Adults are commonly found on goldenrod flowers. Larvae develop by hypermetamorphosis (each larval stage contains two or more distinct body types). The first instar is an active, long-legged form called a triungulin. The second instar is similar, but its legs are shorter. The third, fourth, and fifth instars are robust and somewhat grublike. The sixth instar lacks functional appendages, is harder and darker, and is frequently called a pseudopupa. The seventh instar is small, white, and active (still legless) but does not feed and develops into the pupa.

**LIFE CYCLE:** Pseudopupae overwinter in soil and complete development in spring. Adults emerge in June or July and feed on plant hosts in localized groups. Females oviposit eggs in clusters of 100 to 200 in holes prepared in soil. In about 12 days the eggs hatch, and triungulin larvae search for grasshopper egg masses, where they gnaw into the egg pods and feed on the eggs throughout larval development. Typically, there is one generation per year, with the survival of larval populations dependent on the availability of grasshopper egg pods.

### Univoltine Cycles

The **univoltine** cycle refers to a single generation each year (Fig. 4.24). This type is common among most grasshoppers, corn rootworms (*Diabrotica* species), and many other insects. In this type, the generations are easily distinguished in nature, with generation overlap occurring when adults of one generation are laying eggs of the next. Therefore, when observed in the field at any point in time (except during egg laying), most of the population would be in the same growth stage. Because of this simplicity, these populations are often easiest to study and understand.

### Multivoltine Cycles

**Multivoltine** cycles have more than one generation per year (Fig. 4.25). Numbers of generations per year may range from two to four or more, depending on developmental time requirements and environmental conditions. Examples of multivoltine insects include house flies (*Musca domestica*), thrips (Thysanoptera),

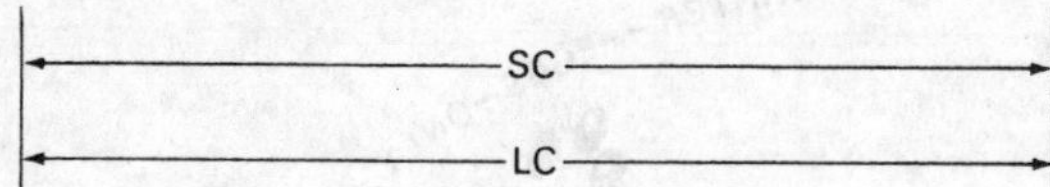


**Figure 4.24** Schematic drawing of the relationship between a seasonal cycle (SC) and a life cycle (LC) in univoltine populations.

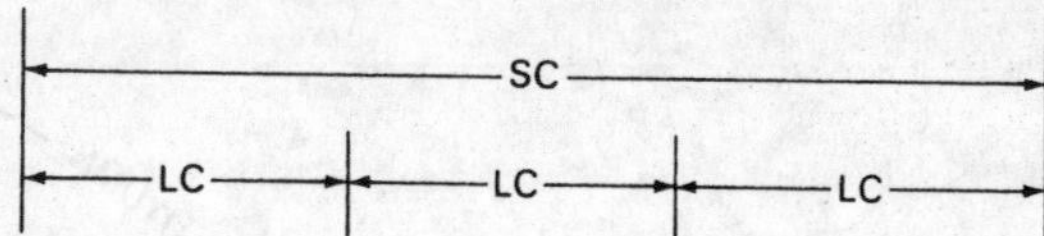


**Figure 4.25** Schematic drawing of the relationship between a seasonal cycle (SC) and life cycles (LC) in multivoltine populations.

aphids, and the European corn borer. The European corn borer is somewhat unusual in having both multivoltine (for example, bivoltine in Iowa) and univoltine (for example, in Nova Scotia, Canada) types. As numbers of generations per year increase, overlap between these becomes greater with seasonal progression. This overlap may be great enough toward late season that all stages of the insect are present, and generations may become indistinguishable (Fig. 4.26). Such a phenomenon usually occurs with thrips and mites; this complexity makes the study of their populations very difficult.

Multivoltine cycles may be subdivided into types, depending on their pattern. The most common, simplest pattern occurs with species having repetitious generations. With these, each successive life cycle is essentially the same. Therefore, each generation of cockroach has similar characteristics: the same morphology, eating habits, and reproductive behavior. Periods of dormancy that may occur with some insects in certain life cycles are not considered a major alteration in the fundamental life cycle pattern. Consequently, a multivoltine species whose overwintering generation differs from others only by virtue of its dormancy is still considered to have repetitious generations.

Multivoltine species in which succeeding generations are different in mode of reproduction or morphology are said to have alternating generations. One of the best examples of alternating generations is aphids (Fig. 4.27). In a relatively simple aphid life cycle, winter is spent in the egg stage. Eggs hatch in early spring, and all new nymphs are wingless females. When mature, these so-called **stem mothers** reproduce parthenogenetically and viviparously, giving rise to other generations of parthenogenetic, viviparous females. By midsummer, generations of both winged (**alate**) and wingless (**apterous**) females are produced, with the winged forms dispersing to begin new populations

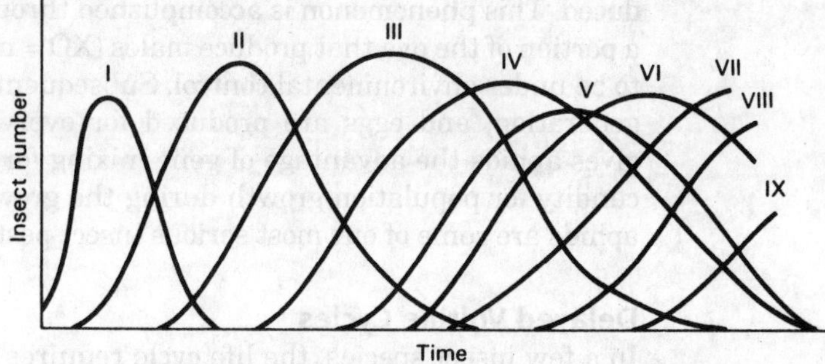


**Figure 4.26** A hypothetical example of population curves of a multivoltine insect population. Roman numerals indicate generations during a season. Note the increasing degree of overlap as the season progresses.

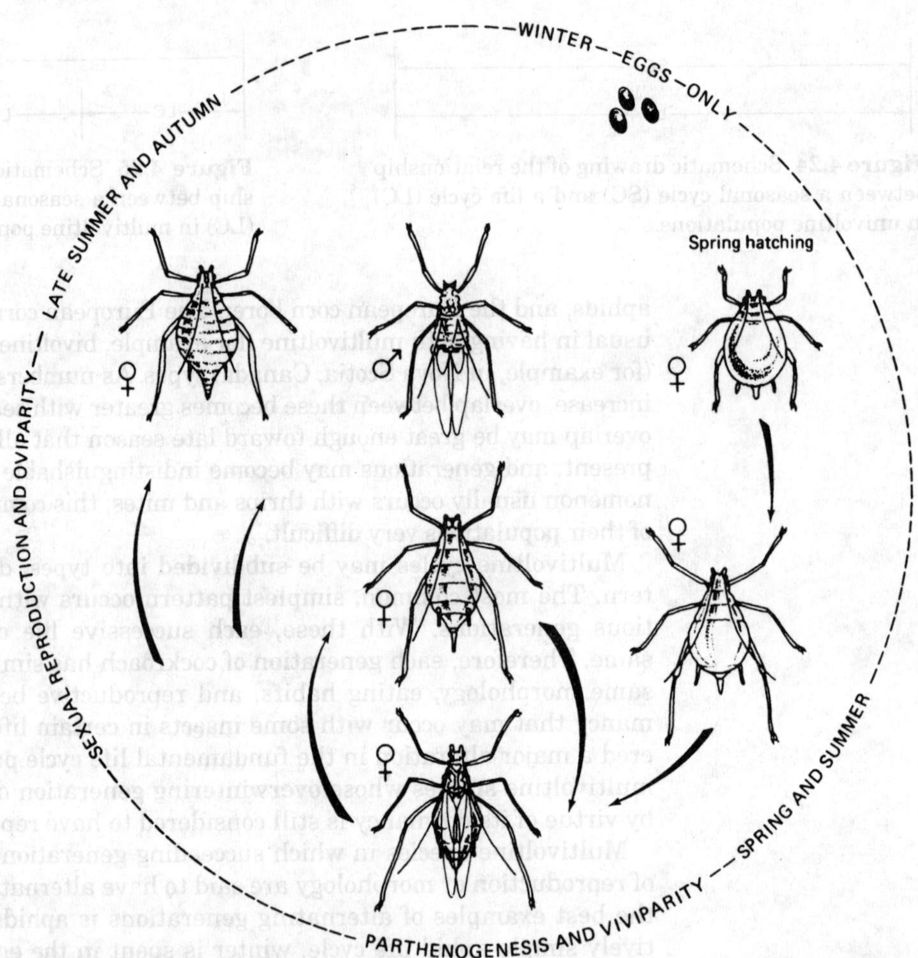


**Figure 4.27** Diagrammatic representation of a relatively simple seasonal cycle in aphids (Hemiptera: Aphididae). (Reprinted with permission of Macmillan Publishing Company from *Insects in Perspective* by Michael D. Atkins. © 1978 by Michael D. Atkins)

elsewhere. In late summer and fall, winged males and wingless females are produced. This phenomenon is accomplished through a loss of the X chromosome in a portion of the ova that produce males (XO = male, XX = female) and is believed to be under environmental control. Subsequently, mating occurs in this "sexual" generation, and eggs are produced for overwintering. Such an arrangement gives aphids the advantage of gene mixing for hybrid vigor and tremendous fecundity for population growth during the growing season. It is one reason why aphids are some of our most serious insect pests.

### Delayed Voltine Cycles

In a few insect species, the life cycle requires more than 1 year for completion (Fig. 4.28). In this instance, the insect must pass through more than one seasonal cycle to complete a single generation. Examples of delayed voltinism can be found in wireworms (Coleoptera: Elateridae) and many June beetles (Coleoptera: Scarabaeidae), with larvae requiring 2 to 3 years to mature. An

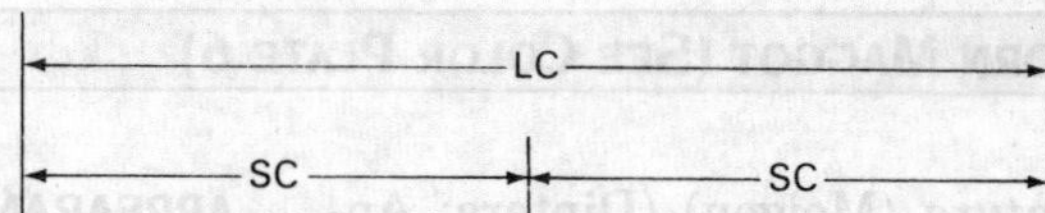


**Figure 4.28** Schematic drawing of the relationship between seasonal cycles (SC) and a life cycle (LC) in delayed voltine populations.

extreme instance is the 17-year cicada (Hemiptera: Cicadidae), which requires 17 years, longer than any other insect, to develop. In most instances of delayed voltinism, generations overlap so that adults are found every year in nature, and the activities of combined generations are often viewed in describing their seasonal cycles.

### Seasonal Adaptations

During the seasonal cycle, insects adapt to many types of adverse environmental conditions. These conditions include irregular and unpredictable extremes in temperature, moisture, food, crowding, and other factors. They are usually localized and temporary. Insect response to these stresses tends to be immediate and may involve becoming **quiescent** (torpid) or moving out of the area. Other types of adverse conditions occur regularly with the seasons and therefore are predictable; for example, freezing temperatures in temperate climates, rains in the tropics, and droughts in deserts. Most insects avoid these predictable adversities in their environment by undergoing physiological and behavioral alterations that condition them for the approaching season. These alterations help to determine the particular pattern of an insect's seasonal cycle. The major forms of this conditioning may include dormancy, changes in body form and color, and migration.

**Dormancy.** Dormancy is particularly prevalent among insects as a means for escaping regular seasonal adversities. Dormancy can be defined as a seasonally recurring period in the life cycle when growth, development, and reproduction are suppressed. It may occur during summer, fall, winter, or spring; during these periods, dormancy is termed aestivation, autumnal dormancy, hibernation, and vernal dormancy, respectively. Some insects may be dormant through more than one season. For those insects, dormancy terms are combined; for example, aestivo-hibernation refers to larvae of the seedcorn maggot, (Box 4.5) *Delia platura*, which aestivate most of summer and go directly into winter hibernation.

In temperate climates, hibernation is one of the most common characteristics of the insect seasonal cycle, and cold hardiness usually accompanies it. Cold hardiness is achieved in insects through the physiological processes of supercooling and freezing tolerance. **Supercooling** is a resistance to freezing by lowering the temperature at which freezing of body fluids begins, whereas **freezing tolerance** is survival despite freezing of body fluid.

Dormancy differs from quiescence in that during dormancy, growth and reproduction may be suppressed even during periods when conditions are temporarily favorable for these functions. During quiescence, growth and reproduction resume immediately upon the return of favorable conditions.

## BOX 4.5 SEEDCORN MAGGOT (SEE COLOR PLATE 6)

**SPECIES:** *Delia platura* (Meigen) (Diptera: Anthomyiidae)

**DISTRIBUTION:** The seedcorn maggot is widely distributed in the temperate regions of the world and was first reported in North America in 1855. Surveys have shown that this species is found on all continents except Antarctica.

**IMPORTANCE:** The larvae (maggots) develop in the soil and feed on decaying organic matter. In the process of feeding, larvae readily feed on newly planted seeds, either prior to emergence or soon after. Feeding before plant emergence may damage the seed embryo and cause inviability, which translates into reduced plant density. Feeding after the seed has germinated may result in different plant responses. If the feeding is minor, the plant may be delayed in development, or it may be exposed to pathogenic soil organisms. If the feeding is moderate, the growing tip may be injured or destroyed, and the plant loses apical dominance, which results in a "Y"-plant (two main stems). Under severe feeding, the seedling may simply die from tissue loss. The host range is vast, including bean, pea, corn, cabbage, cauliflower, spinach, turnip, radish, and onion. Actually, any seed is susceptible if it is located near larvae when they are feeding.

**APPEARANCE:** Adults are small, gray flies with black legs. They are about 5 mm long and have fine hairs on the undersurface of the scutellum. Larvae are pale and yellowish white. They are 6 to 7 mm long, tapering from posterior to anterior, legless, and tough-skinned. Pupae are dark brown, about 6 mm long, and oval.

**LIFE CYCLE:** Pupae overwinter in northern regions of the United States. Adults emerge in spring and are common on flowing heads of grasses. After mating, female flies oviposit eggs singly or in small clusters just beneath the surface of the soil near larval food. Preferred sites are found in moist soil containing decaying organic matter, such as newly cultivated fields. Over approximately 6 weeks, females produce an average of 270 eggs each. Eggs develop and hatch within 9 days. Larvae develop rapidly when food is abundant and temperatures are appropriate. Three larval stages develop in 7 to 21 days, after which pupation occurs in the soil. Puparia develop initially and last about 2 days, followed by the true pupal stage, which lasts from 7 to 26 days. The number of generations per year varies with region, depending on local temperatures. In some areas, populations aestivate in mid to late summer before going into winter hibernation.

**Diapause.** Dormancy, as well as seasonal migration and morphological change, are controlled by a physiological condition called diapause. **Diapause** is usually characterized by low metabolism, little or no development, increased resistance to environmental extremes, and altered (often greatly reduced) behavioral activity. The mechanism found to underlie and orchestrate diapause is hormone activity, with some of the following involved: (1) larval and pupal diapause, lack of brain hormone, and consequently, ecdysone; (2) adult diapause, lack of brain hormone and juvenile hormone; and (3) egg diapause, release of a neurosecretory hormone from the subesophageal ganglia of the parent.

Diapause is induced by specific environmental cues called **token stimuli.** They are given this term because these environmental factors in themselves are neither favorable nor unfavorable but signal forthcoming changes in the environment. The most important token stimulus for insects is day length, or **photoperiod.** Temperature, moisture, and biotic factors may also act as token stimuli, either alone or in combination. The onset of diapause occurs well before most environmental conditions change, whereas dormancy is a concurrent response or reaction to changing conditions. Once diapause is induced, it continues, even though favorable environmental conditions prevail.

Usually diapause occurs at a specific, genetically fixed stage of the insect's life cycle. All life stages have been known to diapause, depending on the insect species. For example, most grasshoppers and corn rootworms diapause as eggs; European corn borers and many bark beetles (Coleoptera: Scolytidae) diapause as larvae; cecropia moths, *Hyalophora cecropia*, diapause as pupae; and boll weevils, *Anthonomus grandis*, diapause as adults. The stage sensitive to token stimuli may be the diapause stage (some crickets), certain periods in the parental generation (silkworms), or even periods in the grandparent generation (some aphids).

Once diapause is initiated, it continues for several weeks in most insects but may last much longer in some. Commonly, diapause is maintained by specific patterns in day length, temperature, or both, and these patterns regulate its intensity. Once ended, the insect enters a postdiapause transitional period of dormancy (postdiapause quiescence) that is not sensitive to token stimuli. Subsequently, the return of favorable conditions signals the end of the diapause syndrome, and development and reproduction are resumed.

The diapause phenomenon mediates the progression of insect generations and, in doing so, enhances survival in unfavorable environments. In univoltine species, every generation has a diapause stage, and interruptions in development and reproduction occur in every life cycle. In multivoltine species, the phenomenon occurs only in a single stage of one of the generations, although any generation is capable of diapause given the appropriate token stimulus. Clearly, an understanding of these and other aspects of the diapause syndrome, especially genetic regulation, is necessary before we can efficiently rear insects for experimentation and predict the seasonal cycle with the precision required for insect pest management.

## Further Reading

Atkins, M. D. 1978. *Insects in Perspective*. New York: Macmillan, chapter 9.

Denlinger, D. L. 2007. Regulation of diapause. *Annual Review of Entomology* 47:93–122.

Elzinga, R. J. 2004. *Fundamentals of Entomology,* 6th ed. Upper Saddle River, N.J.: Prentice Hall, chapter 4.

Gillot, C. 1980. *Entomology*. New York: Plenum Press, chapters 19–21.

Happ, G. M. 1984. Development and reproduction. In H. E. Evans, ed., *Insect Biology*. Reading, Mass.: Addison-Wesley, chapter 4.

Little, V. A. 1972. *General and Applied Entomology*, 3rd ed. New York: Harper & Row, chapter 3.

Matthews, R. W., and J. R. Matthews. 1978. *Insect Behavior*. New York: Wiley.

Ross, H. H., C. A. Ross, and J. R. P. Ross. 1982. *A Textbook of Entomology*, 4th ed. New York: Wiley, chapter 7.

Tauber, M. J., C. A. Tauber, and S. Masaki. 1986. *Seasonal Adaptations of Insects*. New York: Oxford University Press.

## Favorite Web Sites

http://www.hortnet.co.nz/publications/hortfacts/Icindex.htm
Life cycles of many pest insects of New Zealand crops. Black-and-white drawings of pests accompany life cycle descriptions. Many pests are found worldwide.

http://classes.entom.wsu.edu/348/Life_cycle.htm
Explanation of insect life cycles, with photos of various stages of development.

http://www.kendall-bioresearch.co.uk/life.htm
Presents a brief guide to the main types of growth and development among insects, with color figures.

CHAPTER 5

# INSECT ECOLOGY

THERE IS LITTLE QUESTION that insects are one of the most prevalent features of our environment. No one can deny their presence; no one can avoid dealing with them. Insects' lives are inextricably intertwined with our own.

Although some of us are fascinated and intrigued by them, others see insects only as a hindrance to human activities and desires. In either instance, it is usually insect numbers, not simply the presence of a solitary individual, that attracts our attention. We are either appalled or in awe at seeing insects virtually explode in numbers as the growing season develops. We question why they can be so abundant in one area and not in another or why they are numerous one year and not the next.

To understand insect numbers is to understand their inherited traits and the particular environment of their life cycles. The interaction between inherited traits and environment results in given numbers of individuals, and because these aspects are constantly changing, insect numbers are also dynamic.

A major objective of insect ecology is to explain the dynamics of insect numbers in time and space. For this explanation, insect ecology relies on an understanding of the physiology and behavior of insects as affected by their environment. The focal point of study can be the individual, collections of individuals of the same species, or interactions between or among species.

In addition to the dynamics of insect numbers, environment also affects timing of biological events in insects (**phenology**) and the diversity of species in an area. These additional topics addressed by insect ecology are no less important than the question of abundance.

By providing explanations of how environment (including the agricultural environment) affects abundance, timing, and diversity of insects, ecology forms the very basis of pest management. Because pest problems usually develop from too many insects occuring at inappropriate times, dealing with them necessitates a knowledge of ecological processes. With this knowledge, the occurrence and severity of potential problems can be predicted and appropriate management activities formulated.

In this chapter, we consider the ecological role of pests in agricultural environments and survey the major determinants of insect abundance. Then we learn how environmental factors affect timing of seasonal events and function to regulate insect numbers.

## THE ECOLOGICAL ROLE OF INSECT PESTS

### The Idea of Populations

Insect populations are groups of individuals set in a frame limited in time and space. Often the boundaries of time and space for a population are somewhat vague and are fixed for convenience by the ecologist. Therefore, it is not uncommon (or inappropriate) to speak of, for example, the black cutworm population in Boone County, Iowa, in 2008. The aspects of time and space are of utmost importance in studying the population and need to be defined for any consideration of population dynamics.

Every insect population has unique characteristics; to understand population dynamics it is necessary to define and quantify these. The characteristics of a population are group attributes not possessed by single individuals in the group. These population attributes include: density, dispersion, natality, mortality, age distribution, and growth form.

Density and dispersion are complementary attributes. **Density** is the number of individuals per some unit of measure (for example, number of redlegged grasshoppers (Box 5.1) per square meter), whereas **dispersion** is the spatial arrangement of those numbers (Fig. 5.1). Most insect populations are said to have clumped or aggregated dispersions, although in relatively uniform habitats (for example, grain fields with even terrain) dispersion may seem random. Randomness means that an insect is as likely to be found in one place as another.

Natality and mortality are rate processes. **Natality** is birth rate, often measured as the total number of eggs or eggs per female laid per unit time

### BOX 5.1 REDLEGGED GRASSHOPPER (SEE COLOR PLATE 5)

**SPECIES:** *Melanoplus femurrubrum* (DeGeer) (Orthoptera: Acrididae)

**DISTRIBUTION:** The redlegged grasshopper is distributed from central Mexico through all the United States and into southern Canada. Individuals are found in roadside ditches, along field margins, and in meadows. The southern redlegged grasshopper, *Melanoplus femurrubrum propinquus* Scudder, a subspecies, is the most common grasshopper in the southeastern United States.

**IMPORTANCE:** Under climatic conditions where rainfall is less than 30 inches per year, redlegged grasshoppers are very destructive pests of clover, alfalfa, and soybean. These grasshoppers defoliate legumes and feed on pods of soybean, exposing the seeds to pathogens and reducing yield and quality. This pest also feeds on small grains and corn, and when populations are high, it attacks almost any plant.

**APPEARANCE:** Adults are generally less than 25 mm long and are brownish red with the hind tibiae pinkish red. Nymphs are similar to adults, differing in size and lack of wings. Twenty-five to thirty eggs, enclosed in a pod, are deposited in the soil.

**LIFE CYCLE:** Eggs overwinter, and hatching occurs in early spring after the ground has warmed and received adequate rainfall. Nymphs begin feeding near egg-laying sites and eventually move to other areas where food is more plentiful. After molting five to six times (40 to 60 days), the adults emerge and mate, and females begin ovipositing. Adults continue to feed until they are killed by cold temperatures in fall. There is typically one generation per year in the temperate regions of the United States.

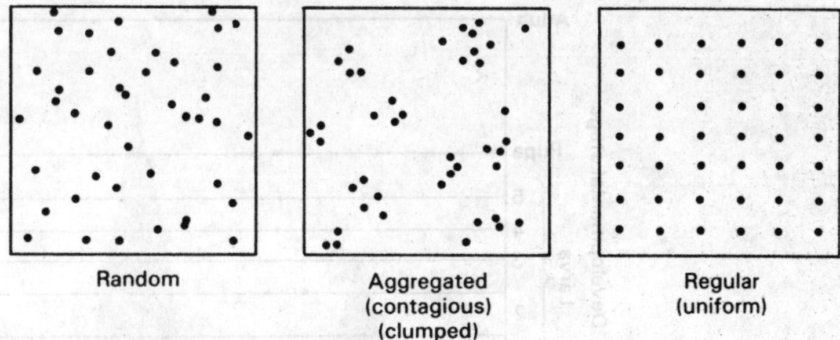


**Figure 5.1** Diagrams showing types of spatial dispersions among insects and other organisms.

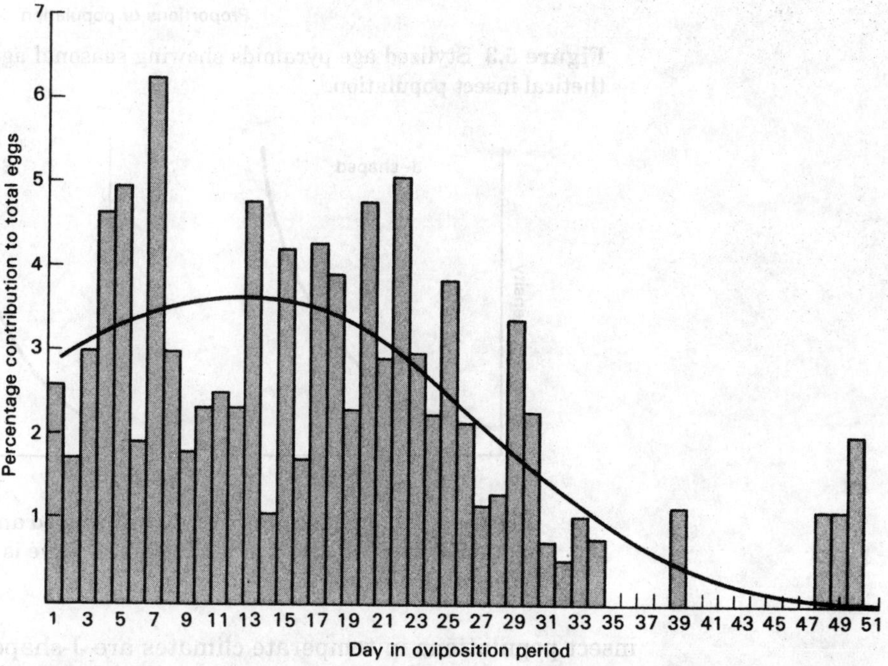


**Figure 5.2** Pattern of egg laying (natality) found in the green cloverworm, *Hypena scabra.* The curve in the graph is a fit to the data points, represented by bars. Such natality patterns are typical of many insect populations. (Redrawn from Wellik and Pedigo, 1978, *Environmental Entomology* 7:171–177)

(Fig. 5.2). **Mortality** is the death rate, or numbers dying per unit of time. These processes act in concert, natality adding numbers and mortality subtracting them, in determining population density and dispersion.

**Age distribution** is the particular proportions of individuals in different age groups at a given time. For instance, in early season the age distribution of an insect population may be adults, 75 percent; eggs, 20 percent; and first-stage larvae, 5 percent. This distribution would change to mostly eggs and larvae as time progressed, and so on (Fig. 5.3).

**Growth form** of a population refers to the particular shape of the density curves during a season or over a longer time period. Growth curves of many

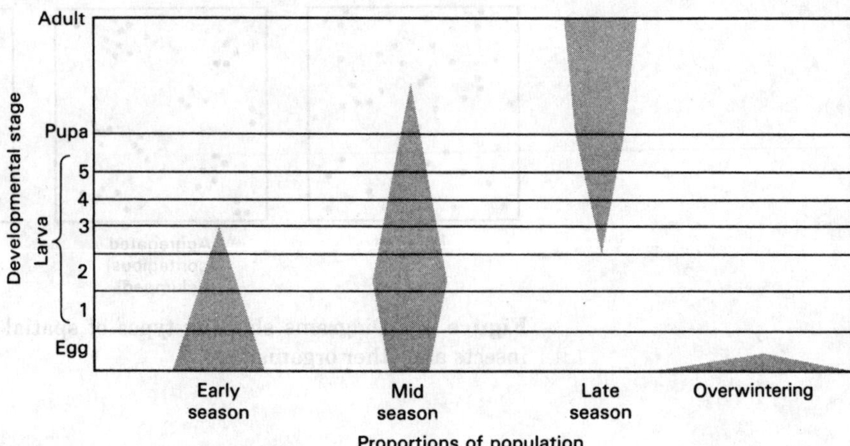


**Figure 5.3** Stylized age pyramids showing seasonal age distribution of a hypothetical insect population.

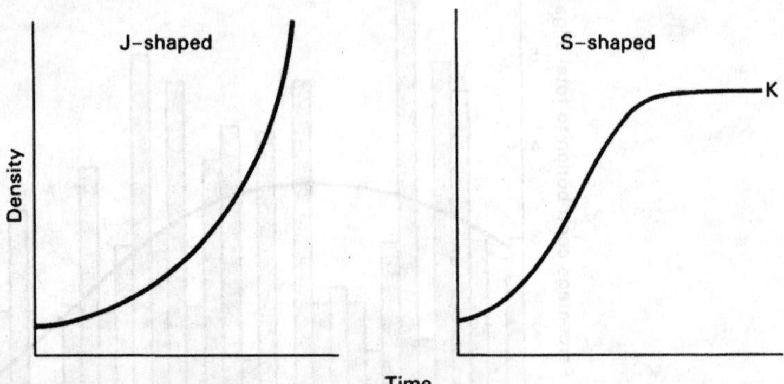


**Figure 5.4** Simplified growth curves, J-shaped and S-shaped, found in animal populations. K in the S-shaped curve is the upper asymptote, or limit of growth.

insect populations in temperate climates are J-shaped (Fig. 5.4); the population grows exponentially, then falls off abruptly with the advent of winter. The particular shape of growth curves determines in part the amount of overlap between numbers of different life stages within a generation and between generations.

## Ecosystems and Agroecosystems

When studying insect populations in a locale, one very quickly discovers that no population exists as an isolated entity. Individuals not only interact among themselves for mating, feeding, and other purposes, but also interact with individuals of other animal and plant populations. At the very least, individuals in a population feed upon and, in turn, are fed upon by individuals of other populations. This interacting "web" of populations in an area is called a **community** (Fig. 5.5).

The interaction of populations in a community is strongly influenced by the physical environment. These assemblages of elements, communities, and physical environments are usually referred to as **ecosystems** (Fig. 5.6). Common examples of natural ecosystems are ponds, lakes, forests, and prairies. Ecosystems

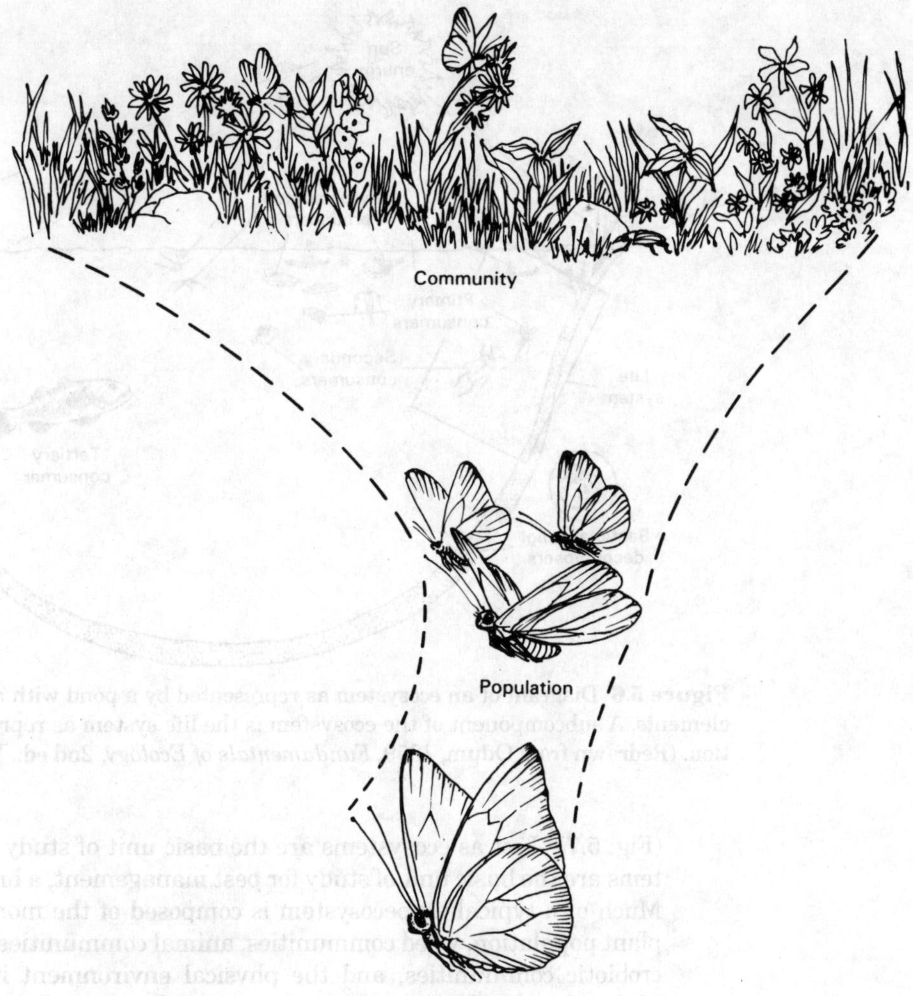


**Figure 5.5** Diagram of the relationship between the individual, the population, and the community.

are usually thought of as the ultimate unit for study in ecology because they are composed of both organisms and the nonliving environment.

But because of the great complexity of an entire ecosystem, it may be more convenient or desirable to consider subdivisions of it. Such subdivisions have been referred to as **life systems.** A life system is a subject species and its **effective environment.** The effective environment is composed of those elements in an ecosystem that have a direct influence on reproduction, survival, and movements of the subject species. Therefore, we can study the life system of an insect within an ecosystem and later combine our knowledge of several life systems to better understand that ecosystem.

Just as ponds and forests are systems of interacting elements, so are crop areas. Consequently, we can think of the interacting biotic and abiotic elements of an orchard or grain field as an ecosystem. Any ecosystem largely created and maintained to satisfy a human want or need is called an **agroecosystem**

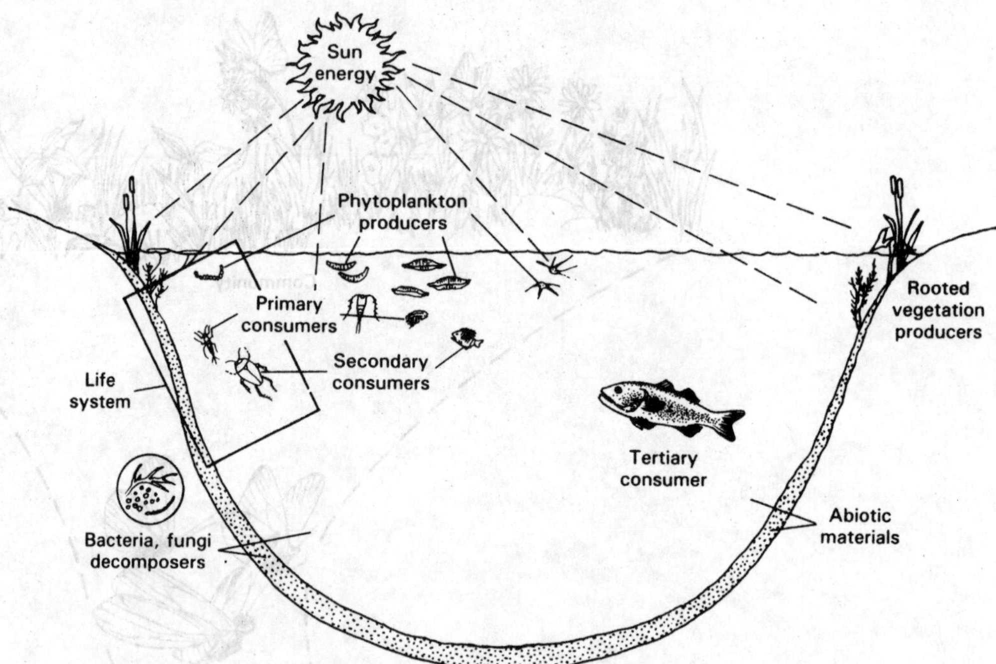


**Figure 5.6** Diagram of an ecosystem as represented by a pond with all its living and nonliving elements. A subcomponent of the ecosystem is the life system as represented by the boxed portion. (Redrawn from Odum, 1959, *Fundamentals of Ecology,* 2nd ed., W. B. Saunders)

(Fig. 5.7). Just as ecosystems are the basic unit of study for ecology, agroecosystems are the basic unit of study for pest management, a branch of applied ecology. Much of a typical agroecosystem is composed of the more or less uniform crop-plant population, weed communities, animal communities (including insects), microbiotic communities, and the physical environment in which they interact. Much of the land mass in major agricultural areas is occupied by agroecosystems.

Although the theory of ecosystems extends to agroecosystems, several features make plant-based agroecosystems unique:

1. Agroecosystems often lack temporal continuity. Their existence may be of limited duration and may undergo immense, abrupt changes in microclimate because of cutting, plowing, disking, burning, chemical application, and other cultural practices.
2. Agroecosystems are dominated by plants selected by humans, many consisting of imported genetic material. Other crop plants not from imported material have been in protective cultivation so long that they hardly resemble the parent stock from which they were derived.
3. Most agroecosystems have little species diversity, and the crop species has little intraspecific diversity; in other words, the crop tends to be genetically uniform. Usually a single species dominates an agroecosystem, and the elimination of weed species further simplifies the environment.
4. With crop plants of similar type and age in the system, the vegetative structure is uniform, and a given phenological event (for example, flowering or podding) occurs in almost all the plants at the same time.
5. Nutrients usually are added to agroecosystems, which results in crop plants with uniformly succulent, nutrient-rich tissues.
6. Agroecosystems often have frequently occurring insect, weed, and disease outbreaks.

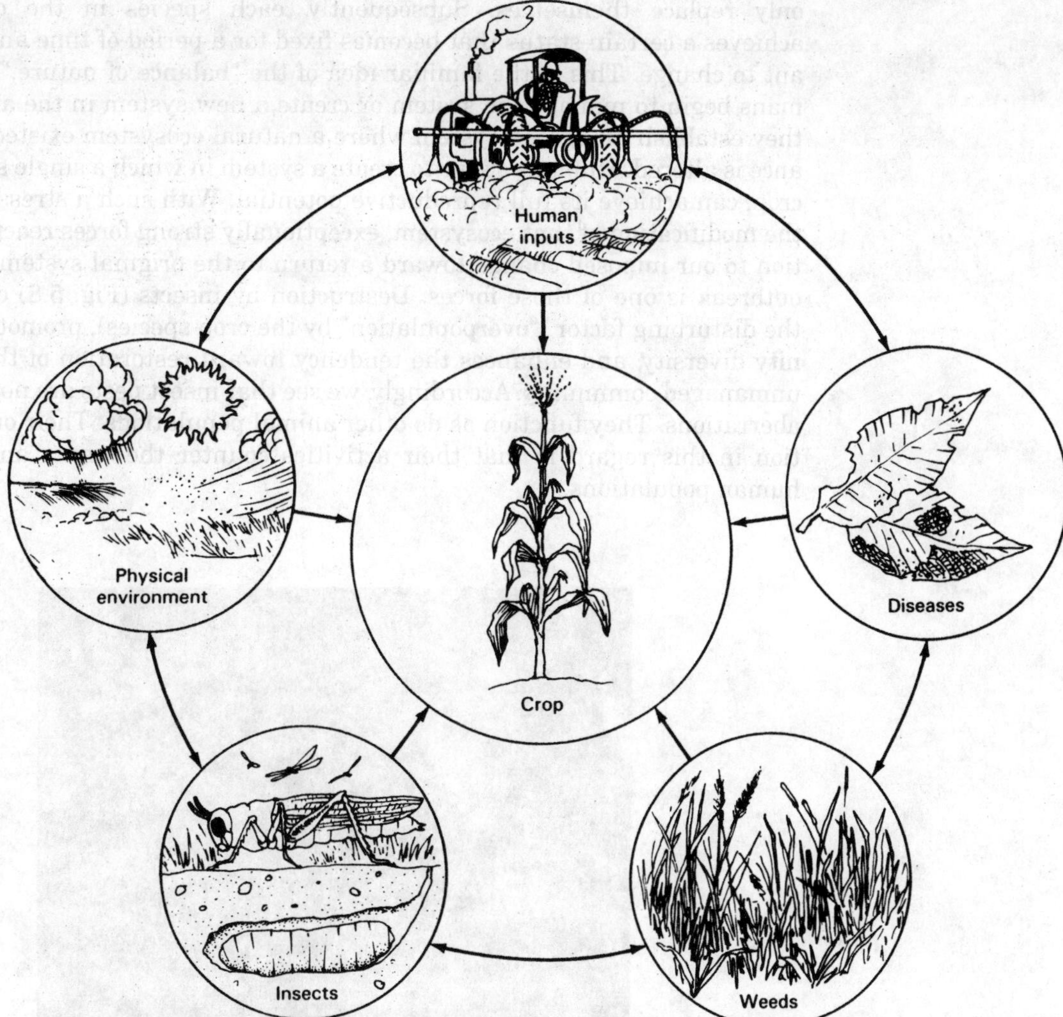


**Figure 5.7** Diagram showing the major interacting elements of an agroecosystem.

The last feature, pest outbreaks, largely results from the preceding five features. With insects, only a small fraction of the entire community finds the crop environment optimal, but those species that do can reproduce and survive very well, creating populations that can cause devastating losses.

### The Ecological Role of Insect Outbreaks

We often ask, why do we have insects that cause damage to our crops? What is the real role of damaging insects in the great plan of nature? Of course, an answer to these questions is teleological. However, a consideration of the ecological mechanisms operating in natural ecosystems may give some insights.

Some ecologists believe that in unmanaged ecosystems, a state of balance exists or will be reached; that is, species interact with each other and with their physical environment in such a way that, on average, individuals can only replace themselves. Subsequently, each species in the community achieves a certain status that becomes fixed for a period of time and is resistant to change. This is the familiar idea of the "balance of nature." When humans begin to manage the system or create a new system in the area—when they establish an agroecosystem where a natural ecosystem existed—the balance is altered. Humans strive to create a system in which a single species, the crop, can achieve its full reproductive potential. With such a stress caused by the modification of that ecosystem, exceptionally strong forces react in opposition to our imposed change toward a return to the original system. An insect outbreak is one of those forces. Destruction by insects (Fig. 5.8) counteracts the disturbing factor ("overpopulation" by the crop species), promotes community diversity, and enhances the tendency toward restoration of the original, unmanaged community. Accordingly, we see that insect pests are not ecological aberrations. They function as do other animal populations. Their only distinction in this regard is that their activities counter the wants and needs of human populations.

**Figure 5.8** Severe insect injury as represented by armyworm, *Pseudaletia unipuncta,* on corn. (Photo by Marlin E. Rice)

## DYNAMICS OF INSECT LIFE SYSTEMS

### Determinants of Insect Abundance

The determinants of insect abundance are found within a species' life system (Fig. 5.9). They comprise the inherited properties of individuals in the species and attributes of the effective environment. These factors operate either to reduce or to promote insect numbers. Together, they explain differences in abundance among habitats and numerical change over time in the same habitat.

Inherited properties involve characteristics transferred genetically from individuals of one generation to those of the next. The inherited properties of individuals in a population combine to give the population its unique characteristics and mainly affect their ability to reproduce and survive.

Reproductive ability is often expressed as potential natality. Potential natality is the reproductive rate of individuals in an optimal environment. Potential natality varies greatly among insects. For example, house flies, *Musca domestica,* have a potential natality of approximately 500 eggs per female, whereas in sheep keds, *Melophagus ovinus,* potential natality is only about 15 eggs per female.

Survival rate also varies widely among insect species and depends much on feeding habits and protection of young. Insects that feed on a wide variety of food sources and protect their young, to a degree, have greater survival potential. In the latter instance, viviparous and ovoviviparous birth types generally have less mortality of newborns than oviparous birth types. This is because the offspring are in a protected environment, inside the females, much longer than are those of oviparous species.

Reproductive rate and survival rate are usually related in that species with high potential natalities tend to have relatively low survival rates and vice versa. Insect pests with high reproductive rates and low survival rates are called ***r* strategists,** named after the statistical parameter *r,* the symbol for the growth-rate coefficient (Fig. 5.10). Such pests may overwhelm the environment with new individuals, and although their losses are great, they can succeed

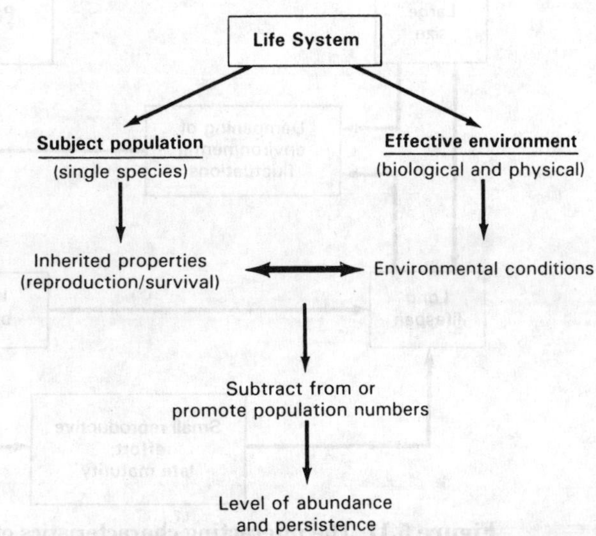


**Figure 5.9** Diagram of the determinants of insect abundance in a life system.

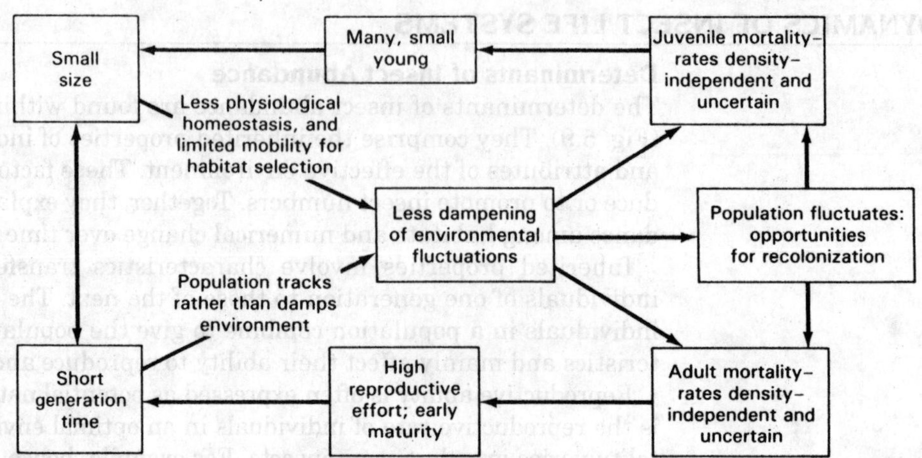


**Figure 5.10** The interacting characteristics of an *r* strategist. (Reprinted with permission of Blackwell Scientific Publications, Ltd., from *Population Ecology,* 2nd ed., by M. Begon and M. Mortimer. © 1986)

because of sheer numbers. These pests, for example, aphids (Hemiptera: Aphididae), increase astronomically in a favorable environment and can rebound quickly after environmental catastrophes. In contrast, ***K* strategists,** named after the symbol for the asymptote or flattened portion of a population growth curve, reproduce slowly. However, they compete effectively for environmental resources, and therefore their survival rate is high (Fig. 5.11). Pests like the

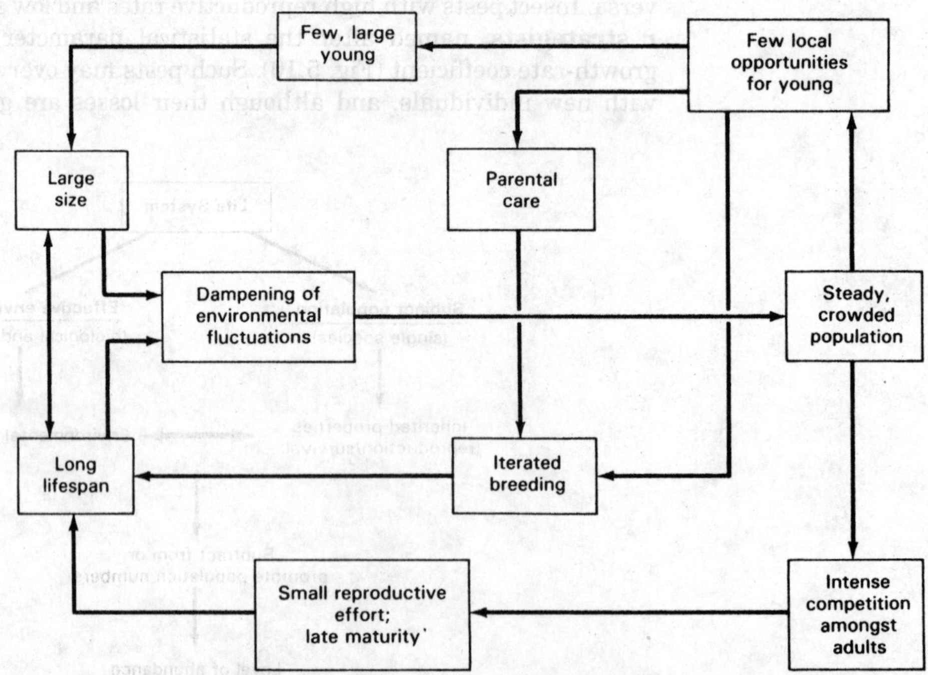


**Figure 5.11** The interacting characteristics of a *K* strategist. (Reprinted with permission of Blackwell Scientific Publications, Ltd., from *Population Ecology,* 2nd ed., by M. Begon and M. Mortimer. © 1986)

codling moth, *Cydia pomonella,* are sometimes considered $K$ strategists. Such species are often hard to deal with because of their resiliency to induced and natural hazards in their environment.

The effective environment is a force of equal importance to inherited properties in determining population numbers. Factors such as weather, food quality and quantity, and living space may either support or inhibit population growth. Conversely, natural enemies, a prominent part of the effective environment, are strictly a subtractive force.

Overall, the actual numbers of a population are sometimes thought of as a sort of algebraic sum of the innate ability of the population to reproduce and survive and the degree of restraint imposed by the environment. R. N. Chapman, an early ecologist, referred to this innate ability as **biotic potential,** and the restraint as **environmental resistance.** He reasoned that populations grow at a rate determined by the dynamic power of a species pitted against the subtractive forces of the environment. His idea can be shown as:

Actual Abundance = Biotic Potential − Environmental Resistance

## Population Change

Change in numbers from one time to the next and from one place to the next is one of the most prevalent properties of any population. A widely used expression with the primary factors of this change is:

$$N_t = N_0 e^{(b-d)t} - E_t + I_t$$

where

$N_t$ = number at the end of a short time period
$N_0$ = number at the beginning of the time period
$e$ = base of natural logarithms = 2.7183
$b$ = birth rate
$d$ = death rate
$t$ = time period
$E$ = emigration = movement out of an area
$I$ = immigration = movement into an area

The expression is a general model of change in any population and shows the mathematical relationship among the primary factors of this change: births, deaths, and movements.

In attempting to explain population numbers, we must account quantitatively for these primary factors. To attempt prediction of change, however, we must understand how environment modifies these primary factors.

The effective environment of a population can be thought of as having an array of secondary factors that determine the magnitude, duration, and frequency of the primary factors. Some secondary factors are weather, natural enemies, breeding habitat, and overwintering space. These affect the primary factors by altering the supply of necessary resources or directly affecting the existence of individuals (Fig. 5.12).

## Birth Rate

Birth rate is a major process for adding new individuals to a population. It is often expressed as numbers born per female or per 1,000 females in the population during a specific time period.

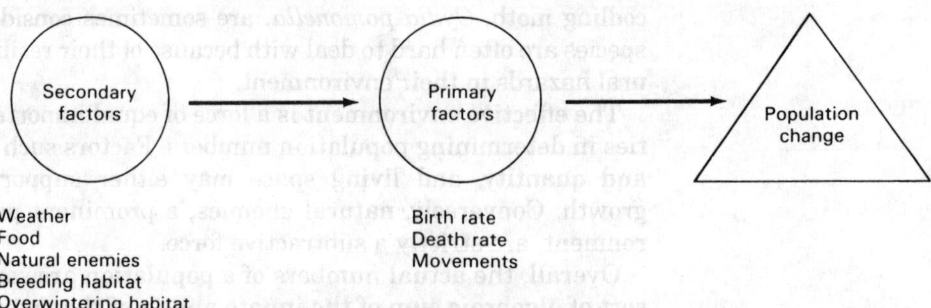


**Figure 5.12** Diagram of secondary and primary factors affecting population change.

The major factors determining birth rate are fecundity, fertility, and sex ratio. **Fecundity** is the rate at which females produce ova, whereas **fertility** is the rate at which they produce new individuals, for example, fertilized eggs. With fecundity, the focus is strictly on female capability to produce reproductive units. Fertility focuses on the mating and fertilization process, or the production of zygotes. **Sex ratios** in most populations are 1:1, male to female. Exceptions to this ratio are most common in parasitic Hymenoptera and, of course, parthenogenetic species. Overall, changes in birth rate may be caused by changes in the average production of eggs per female, changes in mating success, changes in the proportion of female individuals in the population, or a combination of these.

Both fecundity and fertility may vary greatly for an insect population. Secondary factors like temperature, moisture, and food may strongly influence the number of eggs produced by a female. For instance, in controlled experiments with the European corn borer, *Ostrinia nubilalis,* 708 fertile eggs were produced per female at a temperature of 21°C, but only 533 were produced at 32°C. Likewise, female egg production can be greatly influenced by nutrition during the immature stage. For example, egg production by the Colorado potato beetle, *Leptinotarsa decemlineata,* depends on the variety of potato on which the larvae feed. In the spruce budworm (Box 5.2), *Choristoneura fumiferana,* egg production increases to about 170 eggs per female when larvae are fed succulent new foliage versus about 100 eggs per female when larvae consume late-season foliage. Also, habitat and the physical environment are especially important during mating activities. It is well established that conditions essential to reproduction have a much narrower range than for other life functions, for example, movement and feeding.

Variability in birth rate of a population is often overlooked as a cause of population change. Although usually difficult to study, attempts to document and explain this variability are important because the factors responsible may be the essence of a population outbreak.

### Death Rate

Probably more has been written about death rates and causes of death than any other aspect of insect ecology. Numbers of insects dying over a period of time are often represented in **survivorship curves** (Fig. 5.13) or in **life tables** (Table 5.1), a tabular form for accounting for deaths. In many insect populations, high death rates (mortalities) are the rule, with sometimes fewer than 1 percent of the individuals of a generation reaching adulthood.

## BOX 5.2 Spruce Budworm

**SPECIES:** *Choristoneura fumiferana* (Clemens) (Lepidoptera: Tortricidae)

**DISTRIBUTION:** The spruce budworm is found throughout the coniferous forests of southeastern Canada and the northeastern United States. This species was once thought to inhabit all the coniferous areas in North America, but recent studies show that several different, though related, species occur in western Canada and the western United States.

**IMPORTANCE:** Spruce budworm larvae primarily are pests of balsam fir but also feed on larch, pine, hemlock, and white, red, and blue spruce. Where populations of the caterpillars are large, all new needles produced by the tree may be consumed. If the outbreak populations remain for 3 to 5 years, top-killing and tree mortality occur. Overwintering larvae emerge in early spring, attack new buds, and bore into older needles. Half-grown larvae form distinctive nests of twigs tied together with silk.

**APPEARANCE:** Adult moths, which have a gray mottled appearance, have forewings with yellow or reddish brown blotches and hind wings that are dark brown with a white fringe. Mature larvae are 25 to 30 mm long and have a brown head capsule and two rows of white spots along the back. These spots contain many short spinules. Eggs are pale green and oviposited in overlapping masses on the undersides of needles. Pupae are concealed in loose silken cocoons.

**LIFE CYCLE:** Second instars overwinter in silken hibernaculae on the twigs and branches of trees and emerge in early spring to feed on or in needles and new buds. As the larvae grow, they form twig and silk nests, and pupation occurs in late June. After approximately 10 days, adults emerge and are active from late June to early August. During this time, female moths oviposit shinglelike masses containing about 60 eggs. The eggs hatch in about 8 to 10 days, and the larvae feed until they become second instars. Subsequently, they prepare for overwintering by spinning the silken hibernaculae.

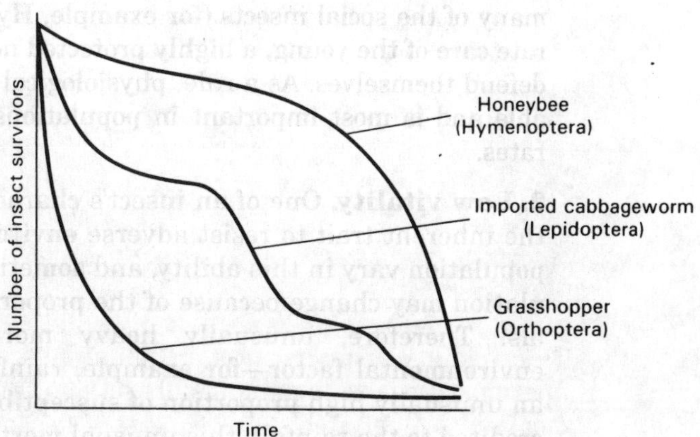


**Figure 5.13** Graph of stylized survivorship curves found in insect populations. (Based on data from Ito, 1961, *Bulletin of Natural and Agricultural Science, Series C,* 13:57–88)

Mortality factors may be biotic or abiotic and may operate in a single life stage or over many stages. For example, intense rain causes mortality of all stages of aphids, but the parasitoid *Rogas* (a small wasp) parasitizes and kills only middle-aged green cloverworm (*Hypena scabra*) larvae.

**Table 5.1 Life Table for the Spruce Budworm at Green River, New Brunswick, Canada.**[a]

| Stage | Number Living Beginning of Stage | Number Dying by End of Stage | Cause of Death | Percent Reduction During Stage |
|---|---|---|---|---|
| Egg | 200 | 10.0 | Parasites | |
| | | 20.0 | Other | 15 |
| Early larva | 170 | 136.0 | Dispersal | 80 |
| Late larva | 34 | 13.6 | Parasites | |
| | | 6.8 | Disease | |
| | | 10.2 | Other | 90 |
| Pupa | 3.4 | 0.3 | Parasites | |
| | | 0.5 | Other | 25 |
| Adult | 2.5 | 0.5 | Miscellaneous | 20 |

[a]Numbers are insects per 10 ft$^2$ of conifer branch surface area.
SOURCE: Modified from Morris, 1957, The interpretation of mortality data in studies of population dynamics, *Canadian Entomologist* 89:49–69.

Like birth rates, death rates can vary greatly from one time and place to the next. Documenting change in birth rate and causes of death has been a major preoccupation among entomologists in attempting to explain insect numbers. Indeed, in many instances, relaxation of an important mortality factor may be the primary cause of a population outbreak.

L. R. Clark and his colleagues (1967) conveniently grouped the causes of mortality of insects into seven major categories.

**1. Aging.** Aging and death from old age is sometimes referred to as "physiological death." In most insect populations, only a small percentage of individuals live to achieve their longevity potentials—a "ripe old age." Exceptions are many of the social insects (for example, Hymenoptera), where there is elaborate care of the young, a highly protected nest, and adults that can effectively defend themselves. As a rule, physiological death is probably not highly variable and is most important in populations with relatively low reproductive rates.

**2. Low vitality.** One of an insect's characteristics is the ability to survive, the inherent trait to resist adverse environmental factors. Individuals of a population vary in this ability, and sometimes the average vitality of a population may change because of the proportions of certain types of individuals. Therefore, unusually heavy mortality may be caused by an environmental factor—for example, rainfall—acting on a population with an unusually high proportion of susceptible individuals. Rather than being credited to the rainfall, this unusual mortality may be credited to low vitality, a genetic trait.

**3. Accidents.** Accidents encompass abnormal events in the insect life cycle resulting in death. Such events range from the physiological (for example, inability to completely shed the old larval skin or failure to expand wings after molting) to the purely ecological (for example, leaf-feeding insects being eaten along with leaves by a browsing deer). The proportion of individuals dying from accidents may be fairly constant in a population.

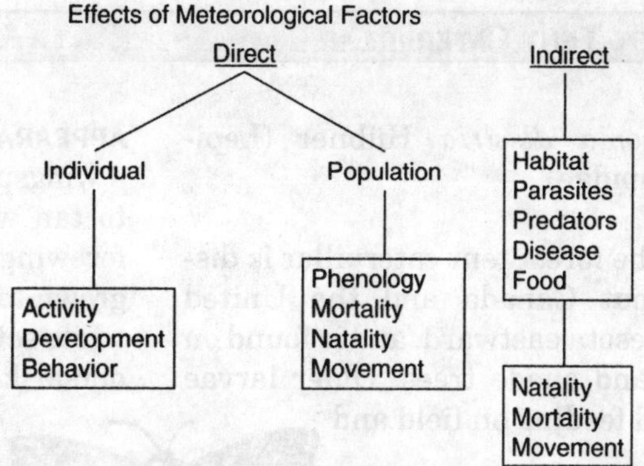


**Figure 5.14** Diagram showing direct and indirect effects of meteorological factors on insects.

**4. Physicochemical conditions.** Physicochemical conditions involve the physical and chemical conditions of air, water, and substrates in or on which the insect population lives. A particular feature with which we are concerned for most pest insects is weather. Components of weather, mainly temperature and moisture, can either promote population growth or cause population decline. Directly, severe weather increases death rates and also may retard or arrest physiological and behavioral processes so as to reduce birth rates (Fig. 5.14). As a rule of thumb, cool, wet extremes are the most deleterious to insects because they promote disease, slow growth rates, and other adversities. But caution must be exercised when applying this rule because some species—for example, mosquitoes (Diptera: Culicidae); the seedcorn maggot, *Delia platura;* and the forest tent caterpillar (Box 5.3), *Malacosoma disstria*—survive best under such conditions.

In temperate climates, variable winter temperatures may be a major reason for changes in death rates of overwintering species. Indeed, outbreaks following a "mild" winter are often believed to be the result of lower death rates. Again, however, caution is the rule when correlating these events because some "mild" winters may cause more mortality than "harsh" winters. This condition can occur because during the "mild" winter abnormally warm days may cause some species to become active early. If this early activity is followed by days of normally cold temperatures, unusually high mortality may result because the insects are no longer protected by their previous state of dormancy.

The long-term influence of weather on insect mortality ultimately results in a particular geographical distribution of a species. However, this distribution may change during the course of a year, with some species occupying regions during the growing season that will not allow survival during the winter. Well-known examples of pest species that cannot survive winters of the upper midwestern United States but invade this area during the growing season include the potato leafhopper (*Empoasca fabae*), corn earworm (*Helicoverpa zea*), and black cutworm (*Agrotis ipsilon*). Many more pests probably exhibit this behavior than are known to exist today.

Because of its omnipotence, weather also affects death rate indirectly. For instance, rainfall can affect plant phenology and nutrient quality for insects. Moreover, asynchrony between a flowering or podding event of a plant and

## BOX 5.3 Forest Tent Caterpillar

**SPECIES:** *Malacosoma disstria* Hübner (Lepidoptera: Lasiocampidae)

**DISTRIBUTION:** The forest tent caterpillar is distributed throughout Canada and the United States, from Minnesota eastward, and is found on deciduous forest and shade trees. Older larvae may even be found feeding on field and vegetable crops.

**IMPORTANCE:** The young larvae (caterpillars) attack expanding buds, and older larvae consume the leaves of many deciduous trees. When outbreaks occur, larvae defoliate trees and migrate in huge "swarms" to new areas, often invading buildings and moving into field and vegetable crops. Outbreaks frequently continue for more than 3 years, resulting in widespread defoliation of deciduous trees in forests and urban areas.

**APPEARANCE:** Adults are large, hairy moths with a wingspan of 25 to 35 mm. The wings are orange to tan with two parallel brown lines across the forewing. Larvae, about 25 mm long when full grown, are light blue and sparsely covered with white setae. Each abdominal segment has a yellow dorsal keyhole-shaped marking and pale yellow horizontal stripes on the sides of the body. The 100 to 350 eggs are cemented together in masses on twigs and branches.

**LIFE CYCLE:** Eggs overwinter and hatch in mid-May. Larvae feed gregariously and form silken mats on branches and trunks in which they are protected after feeding. About 6 weeks after hatching, pupation occurs in cocoons, and adults emerge in 10 to 12 days. Swarms of adults can be seen in mid-July, and eggs are laid through August. There is only one generation per year throughout the pest's range.

occurrence of a particular stage of the insect can also result in unusually heavy mortality. Asynchrony occasionally occurs with the apple blossom weevil, *Anthonomus pomorum*, which lays its eggs in flower buds and suffers high mortalities when blossoms open early in relation to egg hatch. In this instance, new larvae fall out of the blossom before they have time to tie the petals together.

Another indirect effect of weather is on natural enemies. Here, high humidities are known to enhance the growth and virulence of disease-causing organisms such as fungi, and these in turn form widespread epidemics in insect populations. Conversely, abnormally low mortalities from parasitism may result in insect host populations (for example, some Diptera) because of unusually low summer temperatures; in other words, low temperature has a more deleterious effect on the parasite species than on the host species.

**5. Natural enemies.** Organisms that prey upon insects (Fig. 5.15) or parasitize them are called natural enemies. Natural enemies can be grouped, for convenience, as predators, parasites, and pathogenic microorganisms. These are discussed in greater detail in chapter 9, but a brief mention here will allow their orientation among the other mortality factors.

Natural enemies are a very important component in the population dynamics of insect species. They have a strictly subtractive influence on total insect numbers, and sometimes their impact is related to the density of their prey or host; their action is **density-dependent** (Fig. 5.16). Therefore, in some life systems they become an all-important part of ensuring the

**Figure 5.15** Spined soldier bugs, *Podisus maculiventris,* attacking an alfalfa caterpillar, *Colias eurytheme.* Such insect predators are important natural enemies. (Photo by Marlin E. Rice)

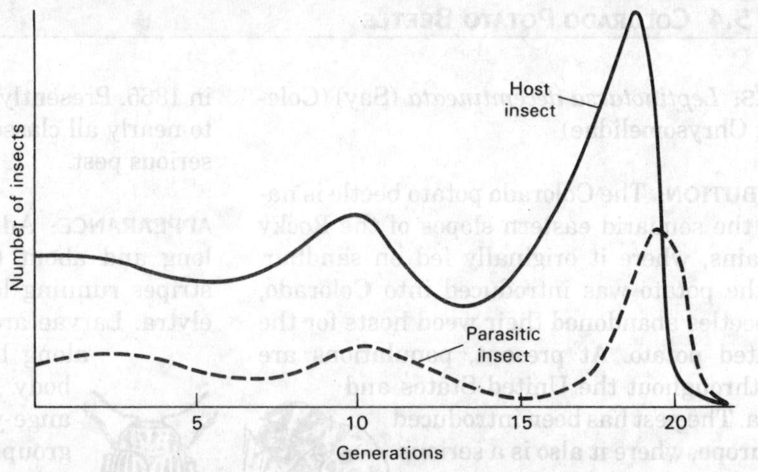


**Figure 5.16** Hypothetical relationship between numbers of a parasite and those of its host, showing density-dependent action.

stability of insect numbers. In these instances, when numbers become large, death rate from natural enemies is great, and when numbers are small, death rate is low.

Because all pest species have natural enemies, understanding mortality caused by these enemies is critical to managing the pest. Indeed, the primary principle of all pest management systems is to conserve natural enemies in the system. We will return to this principle several times throughout the book.

**6. Food shortage.** Food for insects largely depends on prevailing conditions in the ecosystem. In agroecosystems, the major food source, at least for plant-feeding insects, is the crop species, and to a major extent, supply of this food is determined by the grower. Except when weather seriously limits crop production, lack of food is not a major cause of insect pest mortality in agroecosystems. This is not true, however, for many insects that prey on or parasitize other insects.

These natural enemies of insects must search through the habitat for prey or hosts of select species and often select age. Particularly when physicochemical conditions limit the numbers of prey or hosts or the natural enemy decimates its own food source, starvation can be widespread. The latter phenomenon (growth of the natural enemy population when its host is abundant, subsequent decline of the host, followed by a decline of the natural enemy) is the very basis for classical biological control. Here competition for food and starvation (or reduced reproduction) in natural enemy populations are the mechanisms by which regulation of some host populations is achieved.

**7. Lack of shelter.** Lack of shelter influences death rate indirectly. Protective habitats allow insects to avoid exposure to weather extremes and natural enemies; the absence of such may produce considerable mortality in the population. For example, Colorado potato beetles (Box 5.4) succumb to winter temperatures below −12°C. In sandy soils, beetles burrow to depths of 36 to 91 cm, thereby escaping the lethal temperatures (Fig. 5.17). In some soils, an impervious layer

## BOX 5.4 Colorado Potato Beetle

**SPECIES:** *Leptinotarsa decemlineata* (Say) (Coleoptera: Chrysomelidae)

**DISTRIBUTION:** The Colorado potato beetle is native to the semiarid eastern slopes of the Rocky Mountains, where it originally fed on sandbur. After the potato was introduced into Colorado, some beetles abandoned their weed hosts for the cultivated potato. At present, populations are found throughout the United States and Canada. The pest has been introduced into Europe, where it also is a serious pest of potato.

**IMPORTANCE:** This species is the most destructive potato pest in North America. In the past, the beetles have destroyed entire potato crops in several states. Injury is caused when adults and larvae feed on foliage and stems of potato plants, resulting in poor yields and/or death of the plants. Adults can also vector plant diseases such as spindle tuber, bacterial wilt, and root rot. The need to control these beetles resulted in the first use of Paris green, an insect stomach poison, in 1865. Presently, however, the beetle is resistant to nearly all classes of insecticides and remains a serious pest.

**APPEARANCE:** Adults are approximately 9.5 mm long and about 6.4 mm wide, with ten black stripes running lengthwise on yellow to orange elytra. Larvae are red with black spots in rows along both sides of the humpbacked body. Pupae are yellow. Eggs are orange-yellow and are oviposited in groups of about twelve on the undersides of leaves.

**LIFE CYCLE:** Adults overwinter in the soil, becoming active about the time that potato plants emerge. Adults will feed and mate, and the females oviposit up to 300 eggs over a 5-week period. The eggs hatch in 4 to 9 days, and larvae begin feeding, molting four times in 2 to 3 weeks. Subsequently, larvae burrow into the soil and pupate, and adults emerge in about 3 days. There are typically one to three generations per year, depending on climate.

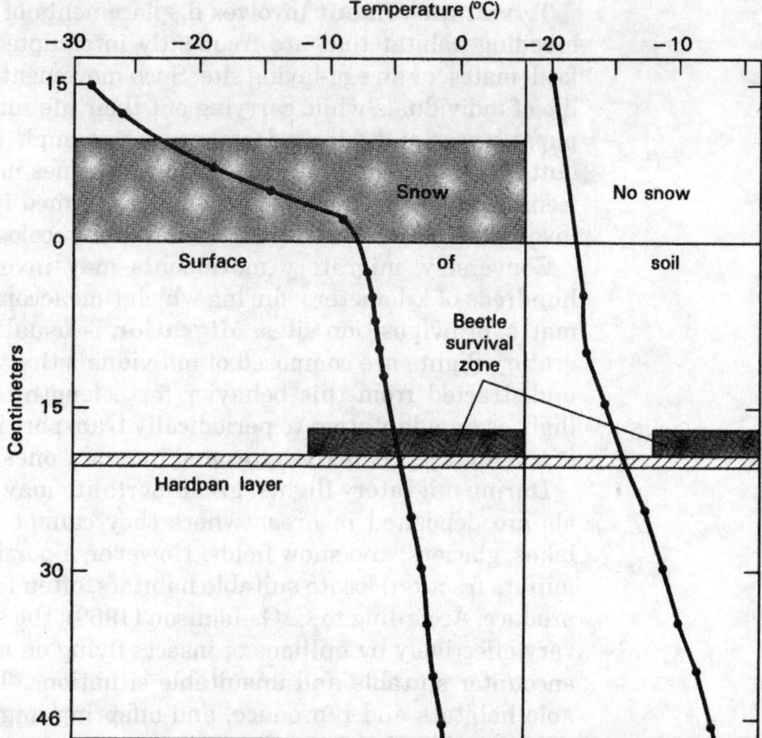


**Figure 5.17** Graph showing soil temperatures relative to snow cover and level of a "hardpan." Overwintering Colorado potato beetles, *Leptinotarsa decemlineata,* cannot burrow through the hardpan, and survival in these soils is possible only in years and locations with adequate snow cover. (Redrawn from Mail and Salt, 1933, *Journal of Economic Entomology* 26:1068–1075)

of clay, a hardpan, occurs at 20 cm, preventing beetles from reaching required shelter. Unless adequate snow cover insulates the soil and moderates temperatures, widespread beetle mortality occurs. The effect of overwintering sites on insect mortality is discussed in detail by Leather et al. (1993).

## Movements

Knowledge of movements into an area (**immigration**) and out of an area (**emigration**) is vital to understanding population dynamics. Indeed, movements, rather than births or deaths, may be the main cause of rapid population change within a season; that is, predictable movement can be a major population characteristic. Information obtained from studying insect movements may allow such activities as pinpointing sources of plant-disease vectors or predicting the seasonal arrival of an important pest that does not overwinter locally.

Movement of all kinds is a rule during the life cycle of most insects. It may involve persistent crawling, walking, or hopping along terrestrial surfaces; swimming on or in the water; or flying or being carried passively (as with wingless insects) in the air. Most movements can be categorized as trivial (nonmigratory) or migratory.

**Trivial movement** involves displacements of insects within or close to the breeding habitat that are frequently interrupted when the insects encounter food, mates, or an egg-laying site. Such movements usually occur throughout the life of individuals while carrying out their life functions, except during egg and pupal stages and during dormancy. An example of trivial movement is when a butterfly flits from flower to flower, sometimes moving among several habitats, feeding on nectar. Such movements are termed trivial because of the distances involved, not because they are trivial to the ecology and survival of the insect.

Conversely, migratory movements may involve great distances, perhaps hundreds of kilometers, during which time locomotion is not inhibited by food, mates, or oviposition sites. **Migration** is usually accomplished by flight. Migratory flights are composed of individuals that are predisposed to fly and are undistracted from this behavior for a lengthy period. This peculiar type of flight is an adaptation to periodically transport insects beyond the boundaries of their old reproductive sites and into new ones.

During migratory flights, great mortality may occur because many individuals are deposited in areas where they cannot survive, including open seas, lakes, glaciers, and snow fields. However, a portion of the migrants (perhaps a minute fraction) locate suitable habitats, often fortuitously, where they can reproduce. According to C. G. Johnson (1969), the surface of the earth is scanned very effectively by millions of insects flying on air currents, who continuously encounter suitable and unsuitable situations. They stop temporarily in suitable habitats and reproduce, and offspring migrate in directions determined either by the wind or by themselves.

The height of migration flights above the surface can vary from only a few meters (for example, some butterflies) to hundreds of meters (some leafhoppers, aphids, and moths). Migrating butterflies, flower flies, and dragonflies flying at lower levels are readily seen by the casual observer and appear to maintain a steady course. These insects are said to migrate within their **boundary layer.** The insect boundary layer (not to be confused with meteorologists' use of the term) is the layer of air that extends from ground level upward through increasing wind speeds to a height where wind speed and insect flight speed are equal. Within this level, insects determine their own flight track. Insects migrating at higher altitudes, above their boundary layer, may not be seen. The flight track of these insects is determined by the flight orientation of the insect and wind direction; in other words, they are largely transported by the wind.

One of the most famous insect migrations in North America is that of the monarch butterfly, (Box 5.5) *Danaus plexippus*. This migration occurs each spring when butterflies fly northward from overwintering sites in Mexico, California, and other southern locations. They reach destinations in the northern United States and Canada in May and June, repopulating more than 1 million square miles in 2 months. Reproduction occurs, and three to five generations develop by September. The final generation then migrates back, often in large aggregations, to overwintering sites (Fig. 5.18). It is this generation that begins the northerly migration each spring. By marking, releasing, and recapturing butterflies in the southerly migration, researchers have determined that between 18 and 109 days are required to make the fall return flight, which may cover as many as 3,465 kilometers (2,153 miles). The difference in time is probably determined by the distance that individuals travel to their destination and weather along the migration route. During the

## BOX 5.5 MONARCH BUTTERFLY (SEE COLOR PLATES 4, 5)

**SPECIES:** *Danaus plexippus* (Linnaeus) (Lepidoptera: Danaidae)

**DISTRIBUTION:** The monarch butterfly is distributed throughout nearly all North America, except for Alaska and the Pacific Northwest coastal region. This species can also be found throughout South America, Australia, and the Hawaiian Islands.

**IMPORTANCE:** The monarch butterfly is one of the most recognizable insects in the United States. This species is famous for its annual migrations both north and south along predictable routes. In the midwestern and eastern United States, the butterflies group and fly south to central Mexico, where they overwinter in dense concentrations in the Sierra Madre Mountains. Western monarchs fly to California, where they spend the winter in areas near Pacific Grove. The following spring, the individuals fly north, reproducing along the return route. Remarkably, the adults that overwintered in the south do not return to their original location. Instead, their offspring return and reproduce, and the cycle is repeated. Currently, conservation of monarch overwintering grounds in both Mexico and California has received great attention. This species is also known for its ability to ingest toxic chemicals (cardioglycosides) produced by its milkweed host plants and then store these toxins for its own protection. Both larvae and adults are toxic to potential predators and display this toxicity through bright warning coloration. The viceroy butterfly, *Basilarchia archippus* (Cramer), which is palatable to most predators, mimics monarch coloration and thereby avoids some predation.

**APPEARANCE:** Adult monarchs are large, with a wingspan of 76 to 100 mm. They have rust-orange wings with black veins and a black marginal band, offset by many pairs of white spots. A pair of black spots appears on the hind wings of the male, visibly differentiating it from the female. Larvae, about 51 mm long, are white with alternating bands of black and yellow. Two filamentous appendages are projected from the front and the rear of the body. The chrysalides (pupal cases), about 28 mm long, are pale green and display gold spots. Eggs are pale green, ribbed, and about 1.2 mm long by 0.9 mm wide.

**LIFE CYCLE:** Females fly north in early spring, each ovipositing 400 to 500 eggs on milkweed plants. The developing larvae feed on milkweed leaves and, when full grown, larvae prepare chrysalides, where metamorphosis occurs. There are typically three generations in the northern part of the monarch's range and up to five generations in the southern part of its range.

---

southern migration, butterflies aggregate (roost) in trees at night and during poor weather.

Another famous example of migration is that of the desert locust, *Schistocerca gregaria.* This insect occurs across north and central Africa to the Middle East, Arabia, and India, causing plagues that have been reported since biblical times. Vast swarms may occur across hundreds of kilometers and may be composed of as many as 10 billion individuals, with a weight of 15,000 tons. Such a swarm is estimated to have the same daily consumption as 1.5 million people. When they alight, the locusts consume all above-ground vegetation. The insect has both solitary and migratory phases, with phase determined by degree of crowding of nymphs and their mothers. The migratory swarm appears as a mass flying purposively in a fixed direction. However, orientation of individuals in the swarm is random, and members at the

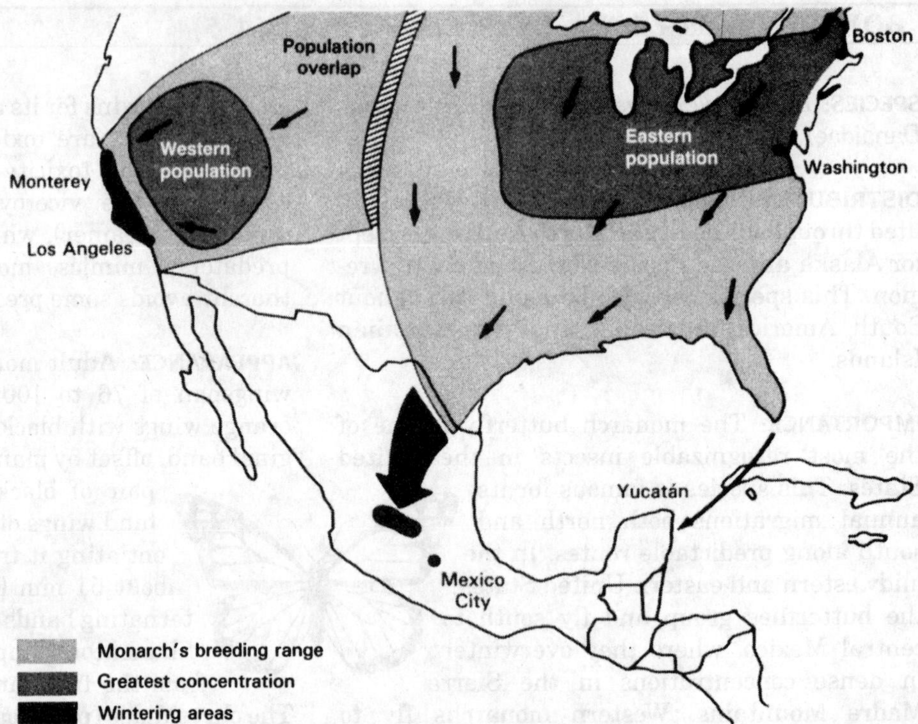


**Figure 5.18** Map of North America showing summer breeding ranges and return migration of the monarch butterfly, *Danaus plexippus*. (Reprinted by permission of John Wiley & Sons, Inc., from *Insect Behavior* by R. W. Matthews and J. R. Matthews. © 1978 John Wiley & Sons, Inc.)

swarm's edge move toward the center of the swarm. This flight behavior is adaptive because the whole swarm is displaced downwind toward zones of wind convergence.

One such area is the Intertropical Convergence Zone, where winds of both sides of the equator meet. Borne on these winds, locusts accumulate in the convergence zone, and swarm displacement ends there. The environment in the convergence zone is favorable for the locusts because rising air here causes rainfall (Fig. 5.19). The rainfall in turn allows the growth of vegetation, and locusts in these areas find environments appropriate for successful reproduction. The Intertropical Convergence Zone moves back and forth across the equator once each year, and because of individual flying behavior and meteorological conditions, locust swarms follow (Fig. 5.20) and take advantage of the resulting favorable environments.

The desert locust (Box 5.6) example shows that crowding and subsequent shortage of food can be an important stimulus for insects to migrate. Other important stimuli may include photoperiod (unusually long day length) and weather factors.

In particular, weather factors may be important in determining the exodus flight. Convection caused by warming of the earth's surface provides thermal updrafts that transport some migrants to altitudes required for horizontal transport by the wind; the opportunity for night-flying insects to take advantage of such updrafts is not as great, and these insects often must gain altitude by their own flight activity. Rapid changes in atmospheric pressure may also

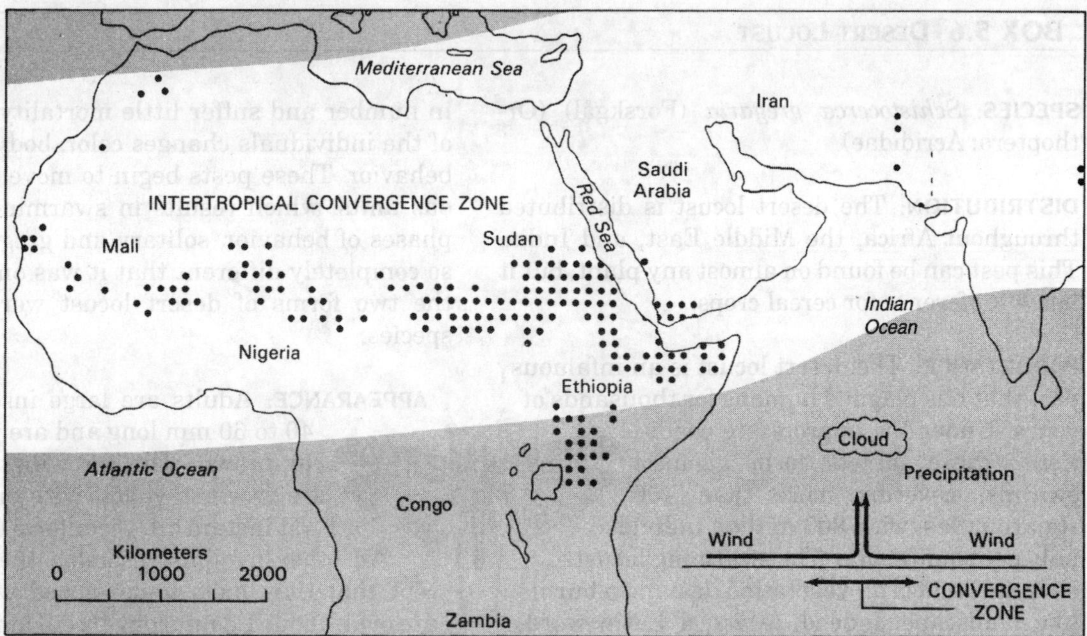


**Figure 5.19** Distribution of desert locust (*Schistocerca gregaria*) swarms (black dots) relative to the Intertropical Convergence Zone. Winds from the north and south converge in this zone, causing rising air currents and rain (*lower right*). Rain results in vegetation growth required for reproduction of locusts. (Redrawn and adapted from Rainey, 1969, *Quarterly Journal of the Royal Meteorological Society* 95:42–434)

**Figure 5.20** Ground view of migrating desert locusts, *Schistocerca gregaria*. (Courtesy FAO and Anti-Locust Research Centre)

## BOX 5.6 Desert Locust

**SPECIES:** *Schistocerca gregaria* (Forskgål) (Orthoptera: Acrididae)

**DISTRIBUTION:** The desert locust is distributed throughout Africa, the Middle East, and India. This pest can be found on almost any plant, but it has a preference for cereal crops.

**IMPORTANCE:** The desert locust is an infamous pest that has plagued humans for thousands of years. Under the appropriate conditions, these insects form gigantic swarms, covering more than 100 square miles, with 300 million individuals per square mile. The swarming locusts strip the land of all vegetation, leaving a burnt-like landscape. Indeed, *locust*, a Latin word, translates as "burnt place." Centuries ago, after the locusts devoured all the vegetation, famine and pestilence followed. Numerous references to this pest can be found in the Bible and other ancient writings. Under normal environmental conditions, the desert locust is a serious yet sporadic pest throughout its range. Outbreaks and swarming occur only when flooding provides ample food and is followed by a period of dry conditions. The locusts will increase in number and suffer little mortality. Crowding of the individuals changes color, body form, and behavior. These pests begin to move as gregarious units, which results in swarming. The two phases of behavior, solitary and gregarious, are so completely different that it was once thought the two forms of desert locust were different species.

**APPEARANCE:** Adults are large insects, about 40 to 60 mm long and are pale yellow or brown. The forewings (tegmina) are greenish yellow with brown spots. First instars are vermiform (wormlike). All other nymphs resemble the adult, except that they have undeveloped wings. Eggs are oval, about 1.3 mm long by 0.8 mm wide, and are contained in pods.

**LIFE CYCLE:** Female locusts lay four to five pods in the soil, with each pod containing 70 to 100 eggs. The wormlike first instars crawl to the surface and quickly molt into the more recognizable locust form. Nymphs undergo three more molts before they reach adult stage. Depending on environmental conditions, the desert locust either enters a solitary or gregarious phase.

---

stimulate migratory flight and allow insects to take advantage of certain wind movements; for example, many aphids and flies are known to increase their takeoff with decreasing barometric pressure.

Once aloft, migrants seemingly respond to weather forces like wind speed and direction, temperature, barometric pressure, and dew point. These factors change with altitude, and migrant insects may actively find an altitude where conditions are favorable. Finding the proper altitude, insects may continue to fly or assume a flight attitude and be carried for hundreds of kilometers with the wind.

Transport by the wind continues for hours or days, until the migrant becomes grounded. Deposition may be gradual or abrupt. Gradual deposition, grounding of the migrants over a period of time and over a considerable area, often is caused by factors intrinsic to the population. These include changes in neurophysiological state, inefficient flight in cold air, exhaustion, or flight inhibition triggered by light intensity. Sudden deposition of migrants probably occurs as a result of atmospheric conditions, which may include downdrafts in cold fronts or cooling of the insect to the point that it cannot support itself in the air. Of all possible weather conditions, cold fronts seem to be the most important in deposition (Fig. 5.21).

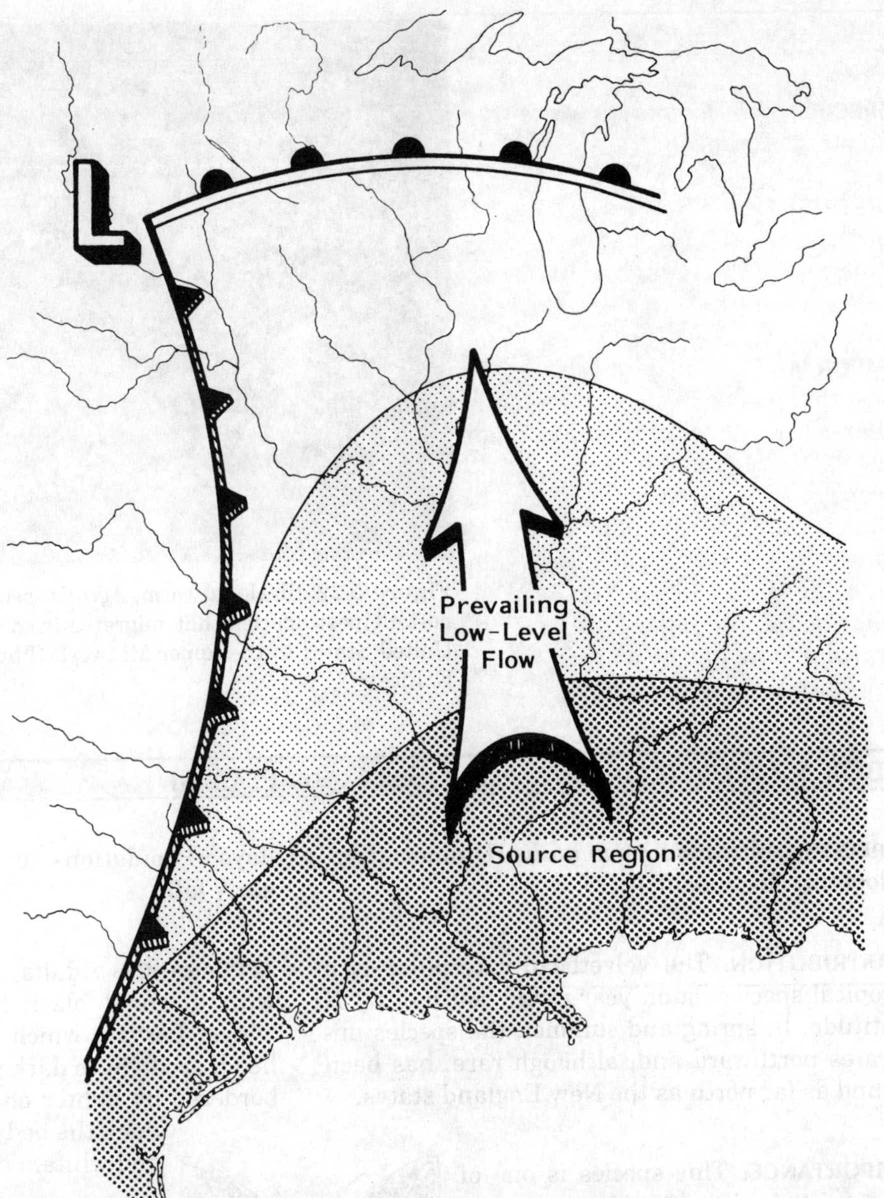


**Figure 5.21** A typical weather system for transporting migrating insects from the southern United States into the upper Midwest. The letter L represents a low-pressure center. A trailing cold front is indicated by the spiked line and the upper warm front by the domed line. On the map, insects are advancing (lightly stippled area) from the source region (heavily stippled area). (Courtesy R. Smelser, Lowa State University)

We thus see that migration is an important adaptive characteristic of insects, one that can strongly influence their population dynamics. As the study of insect migration continues, it seems that many more insect species than heretofore believed use wind flow to relocate and colonize favorable habitats. Some insect pests whose migration patterns are well known are the potato leafhopper, black cutworm (Fig. 5.22), velvetbean caterpillar (Box 5.7)

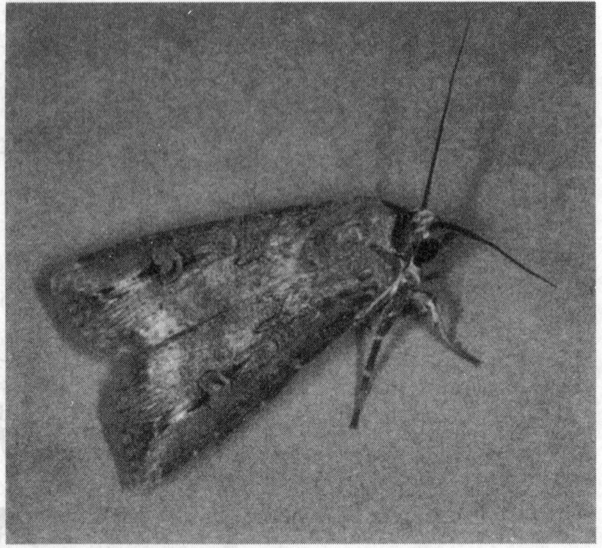

**Figure 5.22** Black cutworm, *Agrotis ipsilon,* is an important pest species that migrates from the southern United States to the upper Midwest. (Photo by Marlin E. Rice)

## BOX 5.7 VELVETBEAN CATERPILLAR (SEE COLOR PLATE 6)

**SPECIES:** *Anticarsia gemmatalis* (Hübner) (Lepidoptera: Noctuidae)

**DISTRIBUTION:** The velvetbean caterpillar is a tropical species found year-round south of 28°N latitude. In spring and summer, the species migrates northward and, although rare, has been found as far north as the New England states.

**IMPORTANCE:** This species is one of the most significant soybean pests in the western hemisphere. In the Gulf Coast region of the United States, the larvae are key defoliators of soybean, velvetbean, peanut, alfalfa, and cowpea. Larvae attack newly produced leaves first, then move to and feed on older leaves. All tissues on the leaves are eaten, except the veins. Older larvae often attack stems of host plants. Where outbreaks of the pest occur, entire fields can be defoliated in 5 to 7 days. Populations are the most damaging in early fall.

**APPEARANCE:** Adults are light brown to black, with diagonal black lines running across the wings. Larvae, which have a yellow or orange head, are light to dark green and have dark lines bordered by lighter ones, which run lengthwise along the body. The eggs are very small and white.

**LIFE CYCLE:** Moths migrate north and appear at the Gulf states in June or July. Females oviposit, and the eggs hatch in about 4 days. Larvae feed for approximately 2 days, then pupate just below the soil surface. Adults emerge about 10 days later and the cycle is repeated. There are typically three to four generations per year in the southern United States.

## Box 5.8 Fall Armyworm (See Color Plate 4)

**SPECIES:** *Spodoptera frugiperda* (Smith) (Lepidoptera: Noctuidae)

**DISTRIBUTION:** The fall armyworm is indigenous to Central America, South America, and the southeastern United States. However, this species migrates north each year and can disperse as far north as Canada.

**IMPORTANCE:** The caterpillars attack the leaves and tender stems of many cultivated crops, including cotton, tobacco, peanut, alfalfa, and garden vegetables. Larvae also feed on ears of corn, causing injury similar to that of corn earworms. Fall armyworms are most damaging in growing seasons preceded by a cold, wet spring. Late in the season, populations of the pest build up, and larvae move in large numbers seeking food. When outbreaks occur, entire areas may be defoliated.

**APPEARANCE:** Adult moths are gray with mottled black and white blotches on the wings. Each forewing is marked with a distinctive white tip. Larvae vary from brown to green to black, and the caterpillars have stripes running lengthwise along the body. A white inverted Y-shaped marking on the front of the larval head distinguishes the species. Larvae reach 40 mm long when full grown.

**LIFE CYCLE:** Adults migrate northward in the spring and summer, and females oviposit about 1,000 eggs in dome-shaped masses of 150 eggs on the host plant. In 2 to 10 days, the eggs hatch, and the larvae feed gregariously. After six molts (20 days), larvae pupate under about one inch of soil. Adults emerge in about 14 days. Female moths fly many miles before ovipositing. Typically, there are one or two generations in the north and as many as six generations in the south.

---

(*Anticarsia gemmatalis*), fall armyworm (Box 5.8) (*Spodoptera frugiperda*), and spruce budworm. It is particularly important that we recognize this potential when predicting pest outbreaks because rapid population increase may well be caused by migration alone.

## EFFECTS OF ENVIRONMENT ON INSECT DEVELOPMENT

Like other organisms, a given insect species is capable of survival only within certain environmental limits, and when possible, individuals actively seek out preferred temperatures, humidities, and light intensities. Within this favorable range, these environmental factors usually influence rate responses of activities such as feeding, dispersal, egg laying, and development. In recent years, an understanding of insect developmental rates has played a particularly important role in insect pest management. Timing of management activities is crucial to implementation of pest management tactics. Knowing or predicting that an insect population is in the egg stage or larval stage, for example, may be one of the cues to begin insect sampling, initiate spraying, destroy crop residues, or plant a crop.

Of the environmental factors, temperature probably has the greatest affect on insect developmental rates. Primarily, this is because insects are **poikilothermic,** or cold-blooded. Within certain limits the higher the temperature, the faster the development of the insect. Developmental rate increases with temperature because chemical reactions occur more frequently

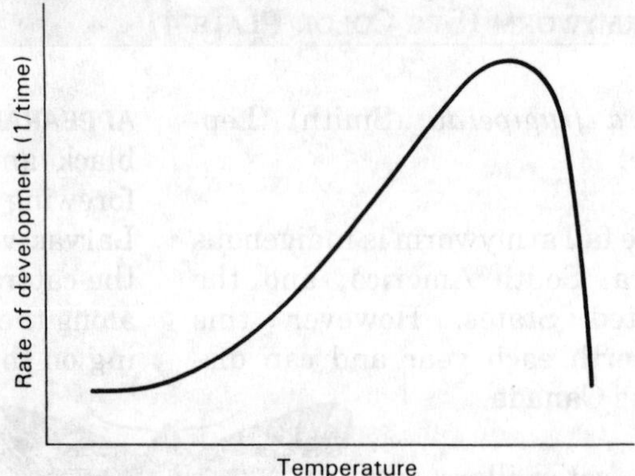

**Figure 5.23** Graph of the relationship between rate of development (1/time) in insects and temperature.

and proceed more rapidly at higher temperatures. As temperature increases, diffusion rates for substrates and enzymes also increase, resulting in greater formation of enzyme-substrate complexes. In addition, higher temperatures provide more thermal energy for the requirements of biochemical reactions.

## Predicting Biological Events: The Degree-Day Method

Because temperature is crucial to biochemical reactions, it can be used to predict rates of insect development (Fig. 5.23). The idea of using temperature and time to describe development of cold-blooded animals is more than 250 years old, and several methods have been advanced to utilize the relationship for prediction. Of these, the most prevalent method of measuring and estimating this so-called physiological time is the **degree-day** method.

Degree days represent the accumulation of heat units above some temperature for a 24-hour period. Below this minimum, no development takes place, but above it, heat units are accumulated toward development. For degree-day accumulation, if the **developmental minimum,** or **threshold,** of an insect is 15°C and the average temperature for the day is 27°C, then 12 degree days, (symbolized as DD or °D) would have accumulated on that day.

To predict the stage of development from degree days, the thermal constant for an event must have been established. This **thermal constant** is the number of degree days required for an event to occur. By accumulating degree days each day and relating these to the thermal constant for an event, one can estimate whether the event occurred on the day in question. The most common events for which insect thermal constants are established are hatching, each larval or nymphal molt, pupation, and adult emergence.

Both thermal constants and developmental thresholds vary among different species of insects. For instance, the thermal constant for egg-to-adult development of the painted lady butterfly (Box 5.9), *Vanessa cardui,* is 440 Celsius degree days, but for the seedcorn maggot, it is 376 Celsius degree days. Developmental thresholds for these insects are 12°C and 4°C, respectively. Because of significant differences, threshold values must be determined for each species of interest.

## BOX 5.9 Painted Lady Butterfly (See Color Plate 5)

**SPECIES:** *Vanessa cardui* (Linnaeus) (Lepidoptera: Nymphalidae)

**DISTRIBUTION:** The painted lady butterfly is found worldwide, except for South America and Australia. In North America, the species cannot survive cold temperatures; instead, it migrates northward each spring from the southwestern United States and Mexico and returns in the fall.

**IMPORTANCE:** Although the painted lady cannot overwinter in temperate areas, the species is among the most common butterflies of the world. In early spring, individuals begin migrating northward, and by late spring the species succeeds in recolonizing the areas where it could not successfully overwinter. The caterpillars attack more than 100 species of plants, including Canada thistle, soybean, sunflower, garden vegetables, and ornamentals. Occasionally, when large migrations occur, the species can achieve pest status, causing widespread defoliation of host plants. However, the painted lady, additionally called the thistle caterpillar, is also considered a beneficial species because of its preference for Canada thistle, a common weed.

**APPEARANCE:** Adult butterflies are brown with red and orange mottling and have black and white spots. The hindwings have four eyespots in a row. Adults are approximately 25 mm long, with wingspans from 51 to 57 mm. Larvae are about 32 mm long when full grown and are brown to black with a yellow stripe along each side of the body, which is covered with small, spiny hairs. The eggs are pale green and have several longitudinal ridges. Pupae (chrysalides) are about 22 mm long and are gold-colored.

**LIFE CYCLE:** Females oviposit eggs singly on host plants, and in about 7 days the eggs hatch. After 2 to 6 weeks of feeding, the larvae pupate. In 7 to 17 days, adults emerge and the cycle is repeated. There are typically two generations per year in the northern part of the species range, but generations are continuous in the tropical and desert regions.

---

The following formula has been devised to calculate degree days (DD) for a specific date:

$$DD = \text{Average Daily Temperature} - \text{Developmental Threshold}$$

The main difference among workers in making degree-day calculations is in computing average daily temperature. Although four basic procedures have been developed to estimate average daily temperature (rectangle method, sine-wave method, cosine curve, and triangle method), the most widely used is the rectangle, or simple-average, method. Although somewhat crude, this method probably offers enough precision for most practical programs of pest management. With simple averaging, the degree-day formula becomes:

$$DD = \frac{\text{Maximum Temp.} + \text{Minimum Temp.}}{2} = \text{Developmental Threshold}$$

By simply adding degree days, a straight-line relationship between average daily temperature and development is assumed. In reality, however, the relationship is a straight line with curved ends. Because much of the relationship is straight line, we can set limits on the upper curve and ignore minor deviations caused by the lower curve. In so doing, we gain simplicity but introduce error.

Several conventions or rules are used in making degree-day calculations, mostly for the purpose of reducing this error. These rules can be listed as follows:

1. If the maximum temperature did not exceed the developmental threshold, no degree days are accumulated. For example,

Developmental Threshold = 10°C
Maximum Temperature = 9°C
Minimum Temperature = 4°C
DD = 0

2. If the maximum temperature was above the developmental threshold, but the minimum temperature is below it, the minimum is set equal to the developmental threshold in calculating the average. For example,

Developmental Threshold = 10°C
Maximum Temperature = 20°C
Minimum Temperature = 8°C
Transformed Minimum = 10°C
DD = 5

3. If the maximum temperature exceeded that of the developmental optimum (the temperature at which developmental rate is highest), it is set equal to the optimum. For example,

Developmental Threshold = 10°C
Developmental Optimum = 27°C
Maximum Temperature = 29°C
Transformed Maximum = 27°C
Minimum Temperature = 21°C
DD = 14

Rule number 1 is always applied in calculating degree days, and rules number 2 and 3 are applied when so instructed by the particular degree-day program.

Degree-day programs (for example, see Table 5.2) for insect species are usually developed by research entomologists who record the rate of development of insects placed in laboratory growth chambers, each set at a different temperature. Time required for development of the egg stage, each larval or nymphal stage, and pupal stage are recorded for each temperature. The reciprocal of time (1/days) for an event to occur is plotted for each temperature, and a straight line is fitted to these points using a statistical procedure called linear regression analysis. The point at which this line crosses the horizontal axis is the developmental threshold (Fig. 5.24). As can be seen, this point is a statistical estimation of development and not a true biological point; in other words, some development can occur below the estimated developmental threshold, but usually it is trivial for the purpose of estimation. Once a developmental threshold is determined, the experimental data can be used further to establish thermal constants for important events in the insect life cycle.

In addition to manual calculations for degree-day accumulations, several programs have been written for computer calculation. One such program, DEGDAY, is functional on personal computers and derives degree-day accumulations by rectangle, sine-wave, and triangle methods (see Higley et al. 1986).

**Table 5.2 Accumulated Degree-Days (Developmental Threshold of 50°F) from Initial Capture of Moths in the Spring for First Occurrence of Life Stages and General Activity of European Corn Borer.**

| Accumulated Degree-Days[a] | First Occurrence of Stage or Event | Days to First Occurrence[b] | Mean Daily Temperature | General Activity |
|---|---|---|---|---|
| 0 | First spring moth | | | Mating and egg laying |
| First generation | | | | |
| 212 | Egg hatch (first instar) | 16.3 | 63 | Pin hole leaf feeding |
| 318 | Second instar | 6.6 | 66 | Shot hole leaf feeding |
| 435 | Third instar | 6.5 | 68 | Midrib and stalk boring[c] |
| 567 | Fourth instar | 6.6 | 70 | Stalk boring |
| 792 | Fifth instar | 10.2 | 72 | Stalk boring |
| 1,002 | Pupa | 8.8 | 74 | Changing to adult |
| 1,192 | Adult moths | 7.6 | 75 | Mating and egg laying |
| Second generation | | | | |
| 1,404 | Egg hatch (first instar)[d] | 8.2 | 76 | Pollen and leaf axil feeding |
| 1,510 | Second instar | 4.1 | 76 | Leaf axil feeding |
| 1,627 | Third instar | 4.3 | 77 | Sheath, collar, and midrib boring |
| 1,759 | Fourth instar | 5.1 | 76 | Stalk boring[c] |
| 1,984 | Fifth instar | 9.0 | 75 | Stalk boring |

[a]Based on populations from Lowa, North Dakota, Missouri, Delaware, and Pennsylvania that were reared in the laboratory on stalk sections from whorl-stage corn by using a minimum developmental threshold of 50°F.

[b]Average number of days of development in most two-generation regions in order to reach the first occurrence of the stage or event since initiation of the previous stage that is listed, based on the mean daily temperature for the time of year when the previous life stage normally occurs.

[c]First-generation larvae bore into stalks earlier than second-generation larvae because the younger stalks are more tender than those of older, more mature plants.

[d]Peak egg hatch occurs 10 days or approximately 200 to 250 degree-days later than first hatch.

SOURCE: From Mason et al., 1996, *European Corn Borer Ecology and Management*. North Central Regional Extension Publication No. 327.

### Degree-Day Programs in Insect Pest Management

As mentioned previously, degree-day programs are a basic component of many insect pest management programs. In most programs, degree-day accumulations are made from the beginning of a growing season and are continuously compared with thermal constants that indicate time of potential crop injury. Thereafter, samples are taken, density estimates are compared with economic threshold levels, and decisions are made as to whether suppression is necessary.

Although the degree-day method has been useful in the management of many insect pests, it is not always applicable. For such pests as corn rootworms, *Diabrotica* species, developmental time can be predicted best by date, without considering temperature. In such instances, some factor other than temperature may be more important, or temperatures in the microhabitat are relatively uniform, making them unimportant in prediction. Additionally, research information on degree-day requirements is not yet available for many insect species.

Other limitations of the degree-day method relate to accuracy. Accuracy of degree-day accumulations depends on the temperature measurements used in

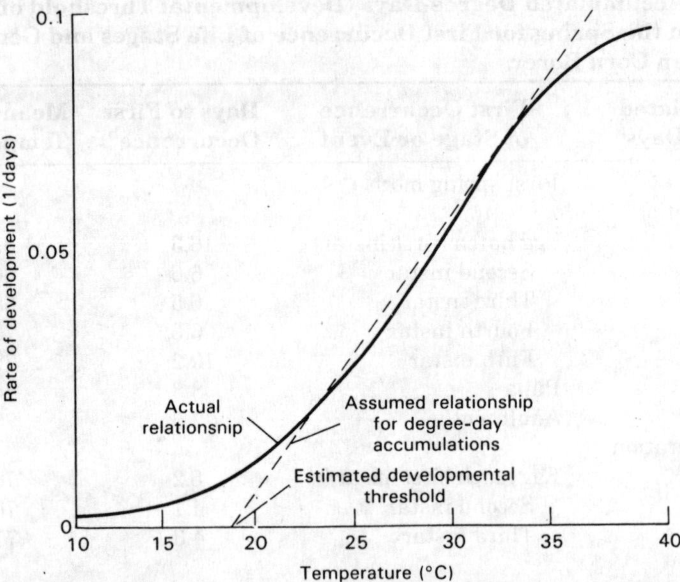


**Figure 5.24** Graph showing the actual relationship of insect development and temperature (solid line) and the assumed relationship used with degree-day accumulations (broken line). Although some error is introduced in its use, the straight-line relationship is practical for predicting insect development. Slope and intercept of the straight line are calculated from data of the curved line.

the calculations. Consequently, degree days should be calculated with temperatures that represent environments where the species is present. Also, temperatures at one site give only a rough estimate of insect development at another site several miles away.

## REGULATION OF INSECT POPULATIONS

Anyone who has seen hundreds of aphids crowded on a single plant leaf or witnessed the swarms of pomace flies hovering over ripened fruit will not doubt the reproductive capabilities of insects. Indeed, it would seem that population growth might cause numbers to overrun their food supply and thus cause their complete destruction from starvation. Likewise, at certain times of the year or at certain locations, the numbers of a species may seem so small that extinction would be imminent.

Neither of these catastrophes usually occurs to a species, at least in the duration of a human lifetime. Rather, population numbers fluctuate within certain bounds for long periods of time, neither increasing (overrunning resources) nor wasting away to extinction. The natural phenomenon of numbers being arrested short of extinction has been called **regulation.** Therefore, it might be said that all existing populations are regulated in one way or another. This regulation may occur at levels not bothersome to humans or at levels that are inconvenient to them, as with pests.

How regulation occurs or, indeed, if it occurs at all has been the subject of much biological debate. Some popular theories to explain how animal regulation occurs or, in some instances, simply why numbers change have been advanced by W. R. Thompson, A. J. Nicholson, H. G. Andrewartha and L. C. Birch, D. Chitty, V. C. Wynne-Edwards, D. Pimentel, A. Milne, and T. R. E. Southwood.

In particular, Milne's (1957) theory provides a useful way of looking at natural regulation in insect populations, and data from many studies seem to support his tenets. As Milne's theory goes, there are basically three types of natural factors acting to regulate population numbers: (1) perfectly density-dependent factors, (2) imperfectly density-dependent factors, and (3) density-independent factors (Fig. 5.25).

According to Milne, a **perfectly density-dependent factor** is one that never fails to control the increase in population numbers. Such a factor acts strongly to subtract numbers from a population when density is high and increasing, and the factor reduces its pressure when density is low and decreasing. The only natural factor influenced solely by numbers in the subject population is intraspecific competition.

**Intraspecific competition** is a phenomenon whereby individuals in a population vie for a resource in limited supply, such as food. As population growth occurs, more and more of the resource is eliminated, causing greater and greater intraspecific competition. As individuals compete, some are eliminated, in this case from starvation, and density is reduced. With falling numbers, competition is relaxed and the rate of decrease drops.

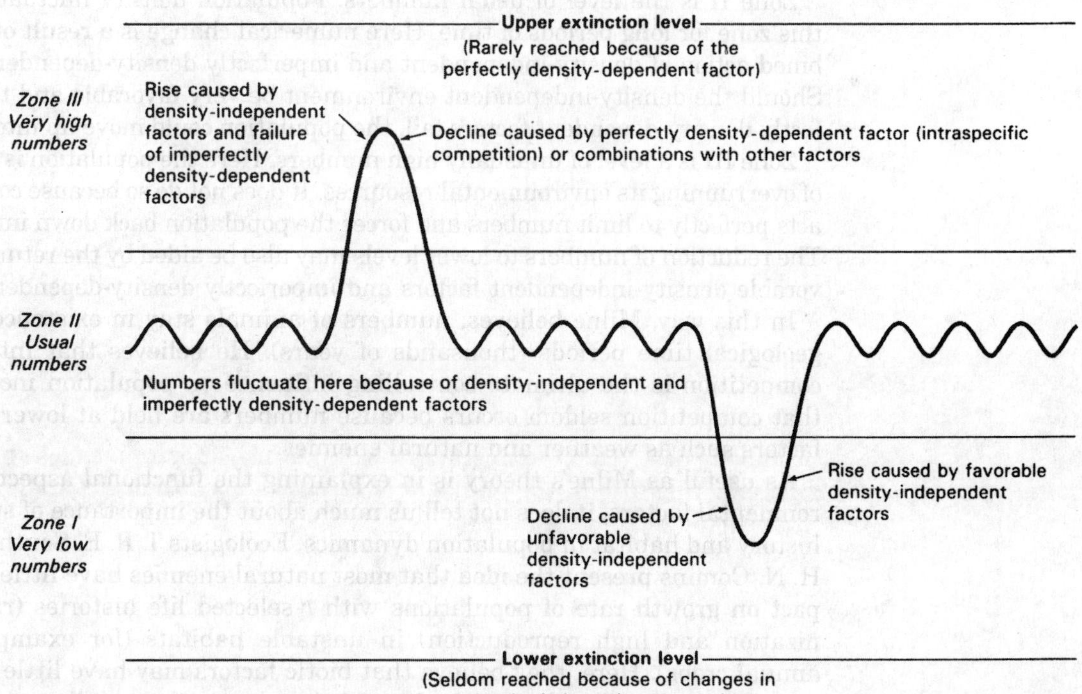


**Figure 5.25** Diagram of insect population regulation based on the theory of Milne. (Redrawn from Milne, 1957, *Canadian Entomologist* 89:193–213)

An **imperfectly density-dependent factor** functions similarly to a perfectly density-dependent factor, except that the imperfect factor may sometimes fail to limit numerical increase. Agents in this category include predators, parasites, and pathogenic microorganisms that cause insect death. Because the lives of these organisms are influenced by many environmental factors, one or more may limit their influence on a prey or host population. For example, a population of parasitic wasps may be very effective in limiting the size of a caterpillar population, and the action of the parasite acts in a density-dependent manner; it causes a greater proportion of deaths when the host population is large than when that population is small. Occasionally, however, wet, cool weather has a severe effect on the parasite and not on its host. Therefore, the density-dependent action of the parasite fails, and the host population increases.

**Density-independent** factors do not vary in their impact on a population according to its density. The primary example of this kind of factor is weather. Weather factors, including rainfall, temperature, and humidity, may have a strong effect on insect survival, but the size of the population has no significant influence on the degree of impact from these factors.

Milne visualizes annual numbers as fluctuating within three zones. Zone I is a level of very low numbers, and populations seldom fall here from Zone II. When they do, it is because of an unusual occurrence (duration and/or intensity) of unfavorable density-independent factors. Extinction from being in this zone usually does not occur in a human lifetime because density-independent factors, acting in random fashion, are not sufficiently deleterious long enough for this catastrophe to occur. On the return of favorable density-independent factors, the population moves back into Zone II.

Zone II is the level of usual numbers. Population density fluctuates within this zone for long periods of time. Here numerical change is a result of the combined action of density-independent and imperfectly density-dependent factors. Should the density-independent environment be very favorable and the imperfectly density-dependent factors fail, the population could move up into Zone III.

Zone III is a level of unusually high numbers. Here the population is in danger of overrunning its environmental resources. It does not do so because competition acts perfectly to limit numbers and forces the population back down into Zone II. The reduction of numbers to lower levels may also be aided by the return of unfavorable density-independent factors and imperfectly density-dependent factors.

In this way, Milne believes, numbers of animals stay in existence through geological time periods (thousands of years). He believes that intraspecific competition is the ultimate controlling influence on population increase but that competition seldom occurs because numbers are held at lower levels by factors such as weather and natural enemies.

As useful as Milne's theory is in explaining the functional aspects of environmental factors, it does not tell us much about the importance of species life history and habitat in population dynamics. Ecologists T. R. E. Southwood and H. N. Comins present the idea that most natural enemies have little or no impact on growth rate of populations with *r*-selected life histories (rapid colonization and high reproduction) in unstable habitats (for example, many annual crops). Here, they believe that biotic factors may have little influence in preventing outbreak growth. In such instances, abnormally rapid growth rates are believed to be brought down by food shortages, disease, and emigration. However, the ecologists feel that populations in stable habitats (for example,

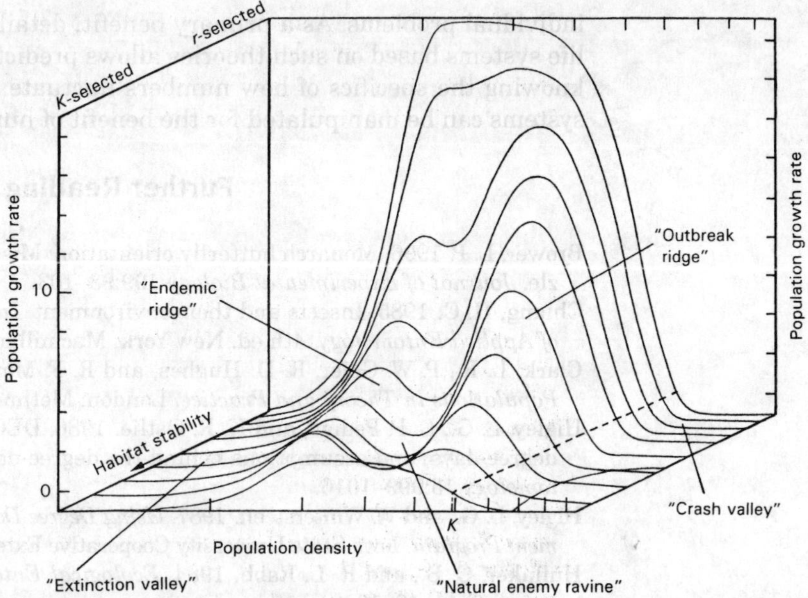


**Figure 5.26** Diagram showing the effect of natural enemies on growth rates of animals with *K*- to *r*-selected life histories in stable to unstable habitats. A synoptic model of population change by T. R. E. Southwood and H. N. Comins. (Redrawn from Southwood and Comins, 1976)

some perennial crops) that tend toward *K*-selected life histories (poor colonizers and low reproduction) are more influenced by natural enemies. These extremes can be seen graphically, with curves responding to three axes, population density, population growth, and a combined axis, representing *K/r* selection and habitat stability (Fig. 5.26). The shapes of the curves are controlled largely by the power of natural enemies to counteract growth. Therefore, a major feature of the graph shows a natural enemy "ravine," or depression in growth rates, that occurs at moderate population densities in most curves. Note that the growth-rate curve of strongest *r*-selection in an unstable habitat is very large and has no natural enemy ravine. Outbreak growth is intense in these populations, indicating a sort of "boom or bust" situation, and natural enemies can seldom prevent outbreaks from occurring. However, an ever-deepening natural enemy ravine and lower intensity of outbreak growth occurs with increasing habitat stability and tendency toward a *K*-selected life history. In the most stable habitat, the curve shows no ravine or outbreak ridge at all. Here natural enemies influence growth rate over a wide range of densities.

This theory helps us understand how insect species may differ in their response to density-dependent factors of the environment, depending on their inherited life history. It also implies that these factors may operate differently within a species, depending on the habitat occupied. In practical terms, the theory might suggest that the utility of at least some biological control agents will depend on a pest's life history and the particular habitat occupied by the pest at a given time.

As can be seen, these population theories give a general explanation of the mechanism of population regulation. As such, they play an important role in pest management by orienting researchers and practitioners in the study of individual problems. As a primary benefit, detailed analyses of specific insect life systems based on such theories allows prediction of outbreaks. Ultimately, knowing the specifics of how numbers fluctuate gives clues to how insect life systems can be manipulated for the benefit of humankind.

## Further Reading


Brower, L. P. 1996. Monarch butterfly orientation: Missing pieces of a magnificent puzzle. *Journal of Experimental Biology* 199:93–103.

Chiang, H. C. 1985. Insects and their environment. In R. E. Pfadt, ed., *Fundamentals of Applied Entomology,* 4th ed. New York: Macmillan, chapter 5.

Clark, L. R., P. W. Geier, R. D. Hughes, and R. F. Morris. 1967. *The Ecology of Insect Populations in Theory and Practice.* London: Methuen, chapters 1–4.

Higley, L. G., L. P. Pedigo, and K. R. Ostlie. 1986. DEGDAY: A program for calculating degree-days, and assumptions behind the degree-day approach. *Environmental Entomology* 15:999–1016.

Higley, L. G., and W. Wintersteen. 1987. *Using Degree Days in an Integrated Pest Management Program.* Iowa State University Cooperative Extension Service Pamphlet Pm-1296.

Huffaker, C. B., and R. L. Rabb. 1984. *Ecological Entomology.* New York: Wiley, chapters 1, 2, 11, 12, 13, and 17.

Johnson, C. G. 1969. *Migration and Dispersal of Insects by Flight.* London: Methuen, chapters 1, 2, 10, 22, and 24.

Kennedy, G. G., and N. P. Storer. 2000. Life systems of polyphagous arthropod pests in temporally unstable cropping systems. *Annual Review of Entomology* 45:467–493.

Kogan, M., and J. D. Lattin. 1999. Agricultural systems as ecosystems. In J. R. Ruberson, ed., *Handbook of Pest Management.* New York: Marcel Dekker, chapter 1.

Leather, S. R., K. F. A. Walters, and J. S. Bale. 1993. *The Ecology of Insect Overwintering.* Cambridge, Great Britain: Cambridge University Press.

Mattson, W. J., and R. A. Haack. 1987. The role of drought in outbreaks of plant-eating insects. *BioScience* 37:110–118.

Milne, A. 1957. The natural control of insect populations. *Canadian Entomologist* 89:193–213.

Price, P. W. 1997. *Insect Ecology,* 3rd ed. New York: Wiley, chapters 13 and 17.

Price, P. W., and G. P. Waldbauer. 1994. Ecological aspects of pest management, pp. 35–64. In R. L. Metcalf and W. H. Luckmann, eds., *Introduction to Insect Pest Management,* 3rd ed. New York: Wiley.

Schowalter, T. D. 2006. *Insect Ecology: An Ecosystem Approach.* San Diego: Academic Press, chapters 1–11.

Southwood, T. R. E., and H. N. Comins. 1976. A synoptic population model. *Journal Animal Ecology* 45:949–965.

Turnbull, A. L. 1969. The ecological role of pest populations. *Proceedings Tall Timbers Conference on Ecological Animal Control by Habitat Management* 1:219–232.


## Favorite Web Sites

http://ipmworld.umn.edu/chapters/ecology.htm

Gives background on population ecology and summarizes population regulation theories. Text is supported by color graphs.

http://MonarchWatch.org

Discusses monarch migration and many aspects of the species' biology. Collecting, tagging, and participating in the monarch migration project is described.

http://www.gypsymoth.ento.vt.edu/~sharov/PopEcol/popecol.html

Presents an introduction to population ecology, with emphasis on analytical methods.

CHAPTER 6

# SURVEILLANCE AND SAMPLING

THE IMPORTANCE OF UNDERSTANDING insect activity to make decisions in insect pest management cannot be overstated. Indeed, this knowledge is indispensable to the paramount principle of pest management: to take no action against a pest unless that pest is known to be present and is an actual or potential threat.

Such insect activities as crop invasion, long-range migration, local movement, feeding, and reproduction are detected and documented through pest surveillance. Pest surveillance is the watch kept on a pest for the purpose of decision making. Depending on the kind of pest (native, newly introduced, potential invader), surveillance programs attempt to determine if a pest species is present, to estimate the numbers in a population and their distribution, and to assess how these factors change over time. More succinctly stated, the major objectives of surveillance are detection of species presence and determination of population density, dispersion, and dynamics.

Aside from private individuals conducting programs, pest surveillance is a major activity for several governmental and commercial agencies. The U.S. Department of Agriculture conducts programs of the widest scope in the United States, particularly through the Animal and Plant Health Inspection Service (APHIS). An administrative subdivision of APHIS, Plant Protection and Quarantine Programs (PPQ), is charged with detection of new pests entering the United States, determination of the spread of introduced pests into new areas, and determination of levels of infestation of established agricultural pests (Box 6.1). Some of the major surveillance programs of PPQ are those for the gypsy moth (Box 6.2) (*Lymantria dispar*), Asian longhorned beetle (*Anoplophora glabripennis*), imported fire ant (*Solenopsis saevissima*), Japanese beetle (*Popillia japonica*), Mediterranean fruit fly (*Ceratitis capitata*), Mexican fruit fly (*Anastrepha ludens*), pink bollworm (*Pectinophora gossypiella*), khapra beetle (*Trogoderma granarium*), and emerald ash borer (Box 3.3) (*Agrilus planipennis*). PPQ also has responsibility in coordinating its surveillance activities with those of other agencies, including the Agricultural Research Service, the United States Forest Service, and state and private establishments.

Many states employ a state entomologist whose primary responsibility is surveillance of established and potential pest species and inspection of certain commodities moving into and out of the state. Most states also have a land-grant

## BOX 6.1 Gypsy Moth

**SPECIES:** *Lymantria dispar* (Linnaeus) (Lepidoptera: Lymantriidae)

**DISTRIBUTION:** The gypsy moth was introduced into North America (Massachusetts) from Europe in 1869 by a scientist interested in experimenting with silk production. The moths were accidentally released and soon spread to the New England states. Presently, the species covers most of the northeastern United States and southeastern Canada. In 1976, gypsy moths were discovered as far west as San Jose, California.

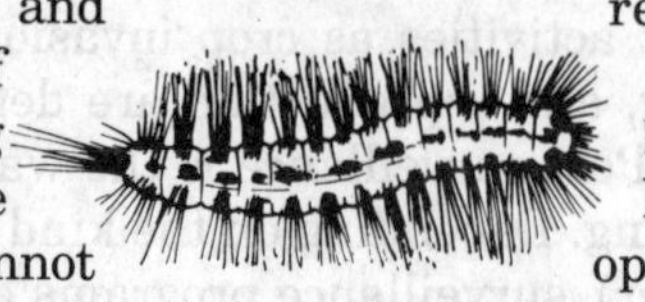

**IMPORTANCE:** Gypsy moth larvae are serious defoliators of trees in forests and urban areas. More than 500 species of trees are attacked by this pest, causing widespread defoliations often severe enough to kill trees. Female moths cannot fly; thus humans have been instrumental in their long-distance movements. Shipment of nursery stock and transport by automobile have aided the spread of the species. Young larvae that produce silken threads are also subject to transport because they may be carried long distances by the wind.

**APPEARANCE:** Adult females are white with yellow hairs on the abdomen and have black wavy bands across the forewings. Adult males are smaller than females and are grayish brown with black wavy bands across the forewings. Larvae are very hairy and about 60 mm long when full-grown. They have distinctive yellow markings on the head and six pairs of double red spots, followed by five pairs of double blue spots on the back. The eggs, about 1 mm in diameter, are oviposited in hair-covered masses of 100 to 1,000 on tree trunks and limbs. Pupae are reddish brown and lightly covered with red hairs.

**LIFE CYCLE:** The eggs overwinter, and hatching begins in early May. Developing larvae devour leaves of their hosts for about 40 days and become full-grown by mid-July. Larvae then spin loose, silken cocoons, and adults emerge 10 to 14 days later. The adults mate, and the females oviposit on host trees during their 6- to 10-day existence. This species produces only one generation per year.

---

university where entomologists of the Cooperative Extension Service and those in research conduct regular surveillance programs. Additionally, a few states have county-based programs and mosquito abatement districts that conduct surveillance of specific pests.

Private firms can be another important source of pest surveillance information. Most of these are agricultural consulting firms that contract with producers to monitor pest levels and recommend where, when, and how to deal with problems. This type of consulting is rapidly growing, as indicated by the number of new firms established in the last decade.

The most successful surveillance programs carried out by the agencies mentioned are achieved through regular survey activities. An insect pest survey is a detailed collection of insect population information at a particular time in a given area. The survey program may be carried on for an entire growing season or at certain critical periods in the insect life cycle. The area involved in a survey may be as small as a field, a pond, or a woodlot, or it may be as extensive as a state or region of the country.

Surveys may be classified as qualitative or quantitative. **Qualitative** surveys are the least complex and are generally aimed at pest detection. Usually, qualitative surveys yield lists of pest species discovered, along with a subjective

## BOX 6.2 IMPORTED FIRE ANTS (SEE COLOR PLATE 4)

**SPECIES:** Black imported fire ant, *Solenopsis richteri* Forel, red imported fire ant, *Solenopsis invicta* Buren (Hymenoptera: Formicidae)

**DISTRIBUTION:** The black imported fire ant was introduced into the United States (Mobile, Alabama) from South America in 1918. In the 1940s, the red imported fire ant was imported into the same region. Since the initial introductions, imported fire ants have spread into the southeastern United States as far north as North Carolina and into the southwestern United States as far west as New Mexico. Presently, both species continue to expand their range.

**IMPORTANCE:** Both species are a nuisance to humans, livestock, and wildlife because they viciously bite and sting when they are disturbed or threatened. Fire ants get their name from the effects of the characteristic sting, which is painful, concentrated, and firelike. The ants first bite into the skin and then proceed to sting the victim. During the stinging, a necrotizing poison is injected into the skin, resulting in pus formation in the wound. Wounds often result in scarring of the skin. Occasionally, victims are stung extensively by many fire ants, causing nausea and dizziness. Humans allergic to fire ant venom have died in situations in which medical treatment was not available. The species also attack, kill, and eat young poultry and their unhatched eggs. Fire ants attack germinating seedlings such as soybean and corn, and the distinctive hard-crusted mounds damage farm machinery in the field. Ironically, though, fire ants are considered beneficial insects because they prey on bollworms and boll weevils in cotton fields.

**APPEARANCE:** Adults are polymorphic, consisting of majors (6 mm long) and minors (3 mm long). Black imported fire ants are light brown to dark brown, whereas red imported fire ants are light reddish brown to dark brown. The abdomen of both species has a two-segmented petiole, and the antennae are slightly clubbed at the end. Larvae are white and grublike. Fire ant mounds are very hard and reach from 38 to 92 cm high.

**LIFE CYCLE:** Both species are active year-round. Newly mated queens dig brood chambers and oviposit from 75 to 100 eggs, which hatch in 8 to 10 days. Larvae are fed for 6 to 12 days, then pupate. After 9 to 16 days, adult workers emerge. Queens rear the first brood, but all subsequent broods are reared by workers. There are many broods per year.

---

reference to density, for example, abundant, common, or rare. Such surveys are typical at international ports of entry, state boundaries, and similar areas where agricultural commodities are inspected. Qualitative surveys are also employed with newly introduced pests; in such instances attempts are made to understand the extent of the infestation. Often, qualitative surveys precede quantitative ones.

**Quantitative** surveys are the most common type employed in insect pest management. The quantitative survey attempts to define numerically the abundance of an insect population in time and space. Such information is used to predict future population trends and to assess damage potentials.

To collect information in the quantitative survey, a count of insects or a measure of their presence is required. Because of the great number and/or secretive nature of many insects, it is not feasible or even desirable to take a census by counting every individual in the population. Usually, the more efficient method is to estimate population density by sampling.

## SAMPLING UNITS AND SAMPLES

The nature of insect sampling is greatly characterized by the sampling units selected. A **sampling unit** is a proportion of the habitable space from which insect counts are taken. Therefore, the insect population can be envisioned as being composed of a finite number of distinct sampling units. Among other criteria, the size of the habitable proportion is determined by the sampler, but the units themselves must be distinct and not overlap. Together, all sampling units contain the population.

The sampling-unit concept is most easily explained when total counts are taken from a unit-area of land surface. For example, direct counts of all the caterpillars in 1 $m^2$ of alfalfa could be considered a sampling unit. If the caterpillar population occupies 100 $m^2$, the habitable space is composed of 100 sampling units.

It is more difficult to visualize sampling units when total counts from unit areas are not used. For instance, a sampling unit based on collecting looper caterpillars in cotton with an insect sweep-net may consist of taking 20 sweeps down the row. Yet, the sweep-net makes contact with only a portion (for example, the upper third) of each plant that is swept. However, the whole canopy, or the part inhabited by loopers, can be divided into a finite number of nonoverlapping 20-sweep units. Even though not all loopers are captured when a unit of habitable space is swept, the count taken is relative to other counts taken in a similar manner and therefore is useful for comparisons of population densities.

Another example of sampling units related to relative counts involves the use of insect light traps. Here, the light trap has a certain effective range of attracting and capturing cutworm adults. The total area occupied by flying adults encompasses a finite number of spaces for trap placement, so that one trap does not influence the catch of another. Consequently, the sampling unit becomes the trap area of capture and duration of operation. The resulting estimate is used to compare numbers in time and space with other traps of the same configuration and operating specifications.

Because it is usually impractical to count all insects in all the sampling units, a group of such units is delineated, which is then used to characterize the whole population. This group of sampling units is referred to as a **sample,** and it is from the sample that an estimate is made. Both sampling-unit size and number taken for a sample are dictated by the sampling design. Information on design and procedures to establish sampling-unit size and number will be discussed later in this chapter.

## SAMPLING UNIVERSE

A frequently used term in insect sampling is the *sampling universe.* In statistics, the term often refers to the whole population from which samples are taken. In other words, universe and population are synonymous. However, in insect sampling, the **sampling universe** has come to represent the habitat in which the population occurs. Consequently, sampling units and samples are taken within the sampling universe. For example, we determine that essentially all of a population of bean leaf beetle eggs is found in the soil within 8 cm on either side of a soybean row and no deeper than 4 cm. The sampling universe can then be designated as a band 16 cm wide × 4 cm deep over the row, and soil sampling units can be drawn from this location. Precise determination of sampling-universe dimensions ensures that no sampling units are taken in noninhabited areas. In this manner, variability among egg counts and, subsequently, sampling costs are reduced.

## SAMPLING TECHNIQUES AND SAMPLING PROGRAMS

To achieve our sampling objectives, it is necessary to count insects in sampling units from a sampling universe and make an estimate of population density. To achieve this, both a sampling technique and a sampling program are required. Although these terms are often used interchangeably, it is useful for us to distinguish between them.

A **sampling technique** is the method used to collect information from a single sampling unit. Therefore, the focus of a sampling technique is on equipment and/or the way insects are counted. Examples of sampling techniques include taking direct counts of all face flies on the face of a Hereford, or taking 20 sweeps in a pendulum fashion in alfalfa to obtain a sampling unit of adult lady beetles.

In contrast to the sampling technique, a **sampling program** is the procedure that employs the sampling technique to obtain a sample and make a density estimate. Sampling programs direct how a sample is to be taken, including: (1) insect stage to sample, (2) sampling-unit number, (3) spatial pattern to obtain sampling units, and (4) timing of samples.

In general, sampling programs comprise two types, extensive programs and intensive programs. **Extensive programs** are conducted over broad areas to determine such information as species distributions or the status of injurious insect stages. Usually in extensive programs only a single insect stage is sampled and only one or a few samples are taken per season. Quite often, only moderate levels of precision are required for extensive programs, with the primary emphasis on low cost. Conversely, **intensive programs** usually are conducted as part of research in population ecology and dynamics. Here sampling is done frequently, often more than once per week in a small area (a few fields). Usually all or most stages in the life cycle of the insect are sampled, and a high degree of precision is sought.

## COMMON SAMPLING TECHNIQUES IN INSECT PEST MANAGEMENT

A vast array of different techniques are used for sampling insects. Each has advantages and disadvantages for the insect species being studied, and more than one technique may be appropriate for development of a sampling program. Only some of the most common techniques are discussed in this chapter.

### *In Situ* Counts

*In situ* (Latin for "in place") counts, also known as direct counts and direct observation, often require no special equipment but rely on a good eye. With this method, large and mostly conspicuous insects are viewed in the habitat, and counts are recorded (Fig. 6.1). Usually the habitat viewed with this method is the plant canopy, a specific plant part, or a certain region on an animal (for example, the face of a cow). If numbers of insects are relatively low and the plant is small and isolated from adjacent plants (as with seedlings), all insects on the plant may be counted. Therefore, this procedure is frequently used with early-season pests

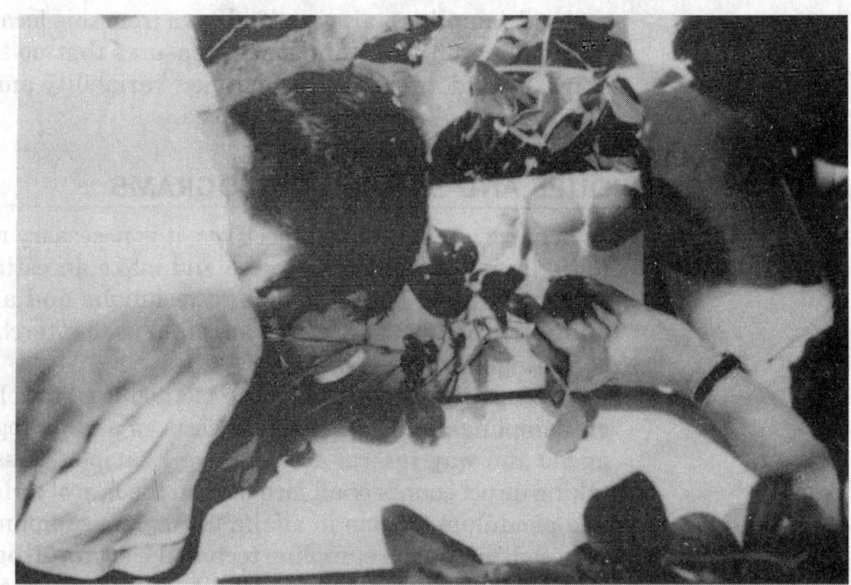

**Figure 6.1** *In situ* counts of insects on plants using a magnifying lens. White poster board placed on either side of plants provides a good background for observations.

**Figure 6.2** An A-frame trap with detachable plastic bag (at right). The frame is placed over a plant canopy and closed. The plants are then cut off at the soil surface, and the contents are shaken down into the bag. Sample bags are returned to the laboratory for insect sorting and counting. (Photo by E. Bechinski)

of agronomic crops, or in later season with crops where herbage is removed and regrowth is inspected, as in the case with alfalfa grown for hay. When the plant canopy is continuous and individual plant sampling is impractical, all the plants in an area—for example, 1 $m^2$ or 30 cm of row—may be inspected. In this case, however, a piece of equipment such as a square metal frame tossed over the area is used (to count grasshoppers in 1 $m^2$ of pasture, for instance). In some instances, plants in an area are removed (cut off at ground level) and bagged, and the insects are counted later (Fig. 6.2). Specialized equipment such as an A-frame or clam trap is sometimes used for this purpose.

When plants are large, such as mature trees, shrubs, and late-season herbaceous vegetation, only certain numbers of leaves, stems, flowers, buds, or pods may be counted. Orchard and forest pests are frequently sampled by counting numbers per shoot or branch. Similarly, agronomic crop pests may be counted as numbers per so many leaves, stems, or reproductive structures. For example, rootworm beetles may be counted in the "ear zone" of a corn plant and this information related to total numbers on the plant (Fig. 6.3). When only part of the plant is inspected, it is important to know if insect numbers may be strati-

**Figure 6.3** Counting the number of corn rootworm adults, *Diabrotica* species, in the ear zone of a corn plant. Subsequently, this information may be related to the total number of insects per plant. (Photo by L. Higley)

fied in certain parts of the canopy, and to account for any stratification when making estimates.

For any sampling technique, it is also crucial to assess the state of the plant and density of the sampling units present. This assessment is relatively simple if the whole plant is the sampling unit: count the number of plants per unit row or area. An estimate of the numbers of leaves or shoots available for insect habitation becomes much more complex and time consuming, and estimation errors become greater. Still, these estimates should be made because plant growth causes changes in the amount of available living space. Such changes do not allow the direct comparison of insect numbers from one time to the next. For instance, if numbers of aphids per leaf of a potato plant are falling during a period of sampling dates, is this due to decreasing size of the aphid population, more leaf area on the potato plant, or both? With only the aphid per leaf information, we have a measure of population intensity, perhaps all that is needed if only a plant damage assessment is desired. If the actual trend in the aphid population is needed for predictive purposes, then a measure of insect population density is necessary. The population density estimate could be determined by multiplying the average number of aphids per leaf by the average number of habitable leaves per square meter.

The *in situ* count method is the most widely used with plants; for obvious reasons, use of the technique for damaging stages of poultry and livestock pests has been limited. Movements by animals being inspected and feathers or hair make direct counting difficult, at best. The method probably can be used only with large parasites, for example, ticks (Acari), or those that are exposed or very active, for example, sheep keds, *Melophagus ovinus*. Other parasites that can be counted directly on livestock include face flies (Box 6.3) on

## BOX 6.3 FACE FLY

**SPECIES:** *Musca autumnalis* DeGeer (Diptera: Muscidae)

**DISTRIBUTION:** The face fly is native to Europe, Africa, and Asia and was subsequently introduced into Nova Scotia, Canada, from Europe about 1952. The species is currently distributed throughout Canada and the United States.

**IMPORTANCE:** The face fly is primarily a cattle pest but also attacks other domesticated animals. Adult flies feed on secretions produced from the eyes, nose, and lips of cattle and also on blood from open wounds or sores. The cattle become irritated and graze less, decreasing their weight gain and milk production. The species also transmits pinkeye, certain eyeworms, such as *Thelazia,* hemorrhagic filariasis, and perhaps brucellosis. During outbreaks, there are often more than 100 face flies on each animal.

**APPEARANCE:** Adults are slightly larger than the house fly. Females have a black abdomen, whereas males have an orange one. Larvae (maggots) are yellow and from 8 to 19 mm long when full-grown. Eggs are white, and each egg has a black respiratory stalk. Puparia are oval and grayish white.

**LIFE CYCLE:** Adults overwinter in masses in buildings where they are also known as cluster flies. In early spring, females oviposit about 25 eggs in fresh dung. Females oviposit about 50 eggs in total. Usually after only 1 day, eggs hatch, and larvae feed in the dung, undergoing two molts in 2 to 4 days. Pupation occurs in puparia in the soil, and adults emerge in about 10 days. There are several generations per year.

**Figure 6.4** Face flies, *Musca autumnalis,* on the face of a Hereford. *In situ* counts are sometimes used as a sampling technique for this pest. (Courtesy Iowa State University Extension Service)

the faces of cattle (Fig. 6.4) and horn flies on the shoulders, backs, bellies, or horns of cattle or other livestock. Because of the many difficulties involved, *in situ* counting is most valuable for detection and qualitative surveys of these parasites on their hosts and for quantitative estimates in the parasites' free-living stages.

### Knockdown

Knockdown is closely related to *in situ* counting, but in this instance, the insects are removed from the habitat by jarring, chemicals, or heating, and then counted. Jarring is probably the most common method of knockdown from plants. It has been used particularly when sampling insects on the lower branches of trees and shrubs. In this method, a cloth, tray, or other receptacle is placed on the ground, a branch is pulled down over the receptacle, and the branch is struck a prescribed number of times with a stick (Fig. 6.5). Insects knocked off fall into the receptacle and are counted. Versions of this jarring approach also are used with some agronomic crops where the plant canopy can be bent over a ground cloth and shaken vigorously for a given time. In soybeans, for instance, this is the standard technique for surveying defoliating insects throughout most of the season (Fig. 6.6). In alfalfa, alfalfa weevil (*Hypera postica*) samples are taken by picking stems and beating them against the side of a container to dislodge larvae for counting. The jarring approach is particularly useful with insects like adult weevils (Coleoptera: Curculionidae) and many beetles that "play dead" when disturbed.

Certain chemicals may be used in conjunction with the jarring technique to facilitate knockdown. This approach has been found most useful with plant parts placed in a container with the chemical. Here, the container is often shaken, after which insects drop off and are counted. Examples of this include

**Figure 6.5** Using a stick to beat an apple tree branch, a surveyor jars loose insects, which fall to the white drop cloth on the ground. Counts are taken of those on the drop cloth. (Photo by L. Higley)

**Figure 6.6** Soybean plants are shaken over a white ground cloth to dislodge insects. Counts are recorded from insects on the cloth. (Photo by Marlin E. Rice)

aphids (Hemiptera: Aphididae) exposed to vapors of methyl isobutl ketone and thrips (Thysanoptera) exposed to turpentine vapors. When whole plants or several plants in an area are sampled, they may be enclosed in a polyethylene envelope and treated with a quick-knockdown insecticide such as pyrethrum. After exposure, the plants may be shaken to facilitate knocking dead or stunned insects off the plant and onto a ground cloth or other receptacle (Fig. 6.7).

Heating is most often used when plants, plant parts, or plant debris are removed and taken to a facility for insect counting. Here, plant samples may be placed in a special device, a Berlese funnel, that heats the sample, causing it to dry out (Figs. 6.8, 6.9). Finding the plant habitat increasingly inhospitable, insects move out of the plant material, fall through a funnel, and are retrieved from a receptacle for counting. This approach has the advantage of saving time and effort usually devoted to sorting plant material to locate insects. It has the disadvantage that some insects may die during the extraction process and not be counted. A closed modification of the Berlese funnel has been commonly used to extract alfalfa weevil larvae from alfalfa samples.

**Figure 6.7** Polyethylene enclosures are sometimes used with insecticides to kill or knock down insects for sampling purposes. After insecticidal action, the lid is removed, and dead insects are shaken into the receptacle at the base of the plants. Counts are made from insects in the receptacle.

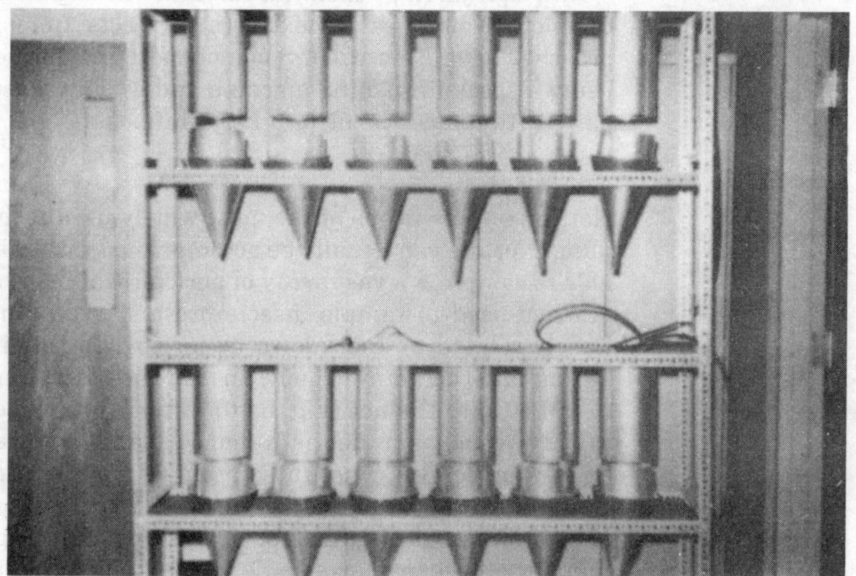

**Figure 6.8** A bank of modified Berlese funnels. Soil cores containing arthropods are placed in a screened tray at the opening of the funnel. A light bulb (or other heating element) in the shroud above the sample tray heats and dries the soil core, driving out the arthropods. These extracted arthropods then fall through the funnel into a vial of alcohol attached to the tip of the funnel.

**Figure 6.9** Soil-core sampler used in obtaining cores for the Berlese funnel. The sampler is constructed in two pieces, which are held together with a clamp. After obtaining a core from the soil, the clamp is removed and the corer is taken apart. This provides a soil section that is not compressed, allowing arthropods easy escape in the Berlese funnel.

Although jarring, chemical knockdown, and heating are some of the most common techniques for extracting insects from plant material, other approaches have been used. Some of these include removing mites from leaves using a special brushing machine and washing insects from plants and plant parts with soap, alcohol, or other solutions.

### Netting

Netting insects is one of the most widely used techniques in population sampling. Netting can usually be accomplished rather inexpensively and is adaptable to sampling a vast array of agricultural pests and their natural enemies.

When used to sample insects present in a plant canopy, the technique is closely related to knockdown by jarring. In this instance, a muslin "net" is swung into the plant canopy, jarring the plants and causing insects on them to fall off into the net (Fig. 6.10). After a prescribed number of such swings, called sweeps, plant debris is removed, and the insects are counted (Fig. 6.11). Such a procedure constitutes a single sample. Compared with many other techniques, the sweep-net sample is inexpensive, and adequate insect numbers often can be obtained for analysis of small populations by increasing the number of sweeps per sample. A frequent disadvantage of the technique is the great sampling variability among different surveyors from differences in swing force, net angle, and other factors. The technique also has been criticized for producing variable efficiency with different species, habitats, time of day, and weather, but most techniques can be criticized on these points.

Insect samples can also be taken from plants by a more sophisticated net, the vacuum net. This technique, often referred to by a trade name, D-Vac, consists

**Figure 6.10** Sweep-net sampling in soybeans. The net is swung through the plant canopy, and the insects that are jarred loose fall into the net. (Photos by Marlin E. Rice)

of a removable net attached inside a plastic cone, which in turn is connected to an engine-powered fan via a flexible hose (Fig. 6.12). With the engine running, a strong vacuum is created in the cone; by placing the cone over or into the plant canopy, insects are sucked into the net. A sample may result from holding the net at one place for a prescribed period (for example, 15 seconds), placing the net in several locations, or using it to "sweep" through vegetation. Both backpack and handheld models of this net are available. A recent version of the vacuum net, the Allen-vac, uses a modified gardener's blower/vac for efficiency and economy. The vacuum net has been used most widely with light-bodied insects like leafhoppers. Equipment efficiency declines rapidly as insect body size and clinging ability increases. Because of limited efficiency, cost, and the cumbersome nature of the equipment, vacuum nets are used only in special circumstances, often in research.

Other netting of insects is performed with "true" mesh-bag nets that capture insects in flight. The most common aerial net is a lightweight, handheld net that is swept through the air to capture an insect. This is really more of a collector's tool than a sampling technique; nonetheless, it is useful in surveillance for detection. Aerial nets used for sampling more commonly include the rotary net and the tow net. The rotary net, used especially with forest pests, consists of one or two nets at the ends of beams that are rotated by a motor-driven shaft. The insects collected in the net(s) are counted after a given period of time (Fig. 6.13). By accumulating the volume of air encountered by the net(s) per hour, aerial density can be calculated. The tow net is a less widely used

**Figure 6.11** Contents of a sweep-net bag can be inverted and inspected after taking several sweeps through a plant canopy. Both insects and plant debris can be seen. Counts may be made directly from this sample, or the sample may be bagged and returned to the laboratory for processing later. (Photo by L. Higley)

technique that usually involves the attachment of nets to moving aircraft, automobiles, or ships. Such nets are most often used for detection.

### Trapping

Trapping involves some of the most important sampling techniques for insect survey. All trapping procedures have two basic requirements: the insects must move, and the trap must hold captured insects. Most traps are set, left unattended for a period of time, and then visited to pick up the "catch." They are often operated continuously, integrating time, insect density, and insect activity to produce the numbers collected.

Traps can be either attractive or passive in their mode of collection. Attractive traps rely on a physical or chemical stimulus to lure insects into them; passive traps collect insects incidentally. It is difficult to categorize most trapping equipment as attractive or passive because, obviously, a passive trap can be made into an attractive trap by adding an attractant. As commonly used,

**Figure 6.12** Sampling insects with a vacuum insect net. Here, a backpack model has been mounted on a cart for ease of handling. With the device, insects are sucked into a cone that supports an internal net. After a sample is taken, the net is removed for insect counting.

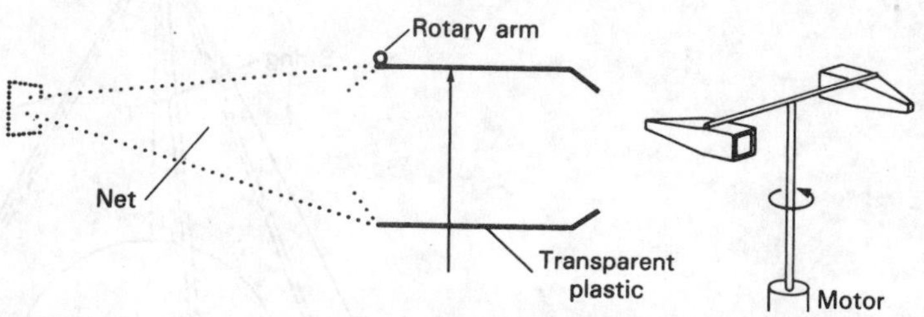


**Figure 6.13** A rotary aerial net. (Redrawn from Southwood, 1978, *Ecological Methods,* Chapman and Hall Publishing Company)

visual traps and bait traps are the attractive types, and pitfall, window, Malaise, some sticky, water pan, and suction traps often lack attractants.

The light trap (Fig. 6.14) is the most widely used visual trap, having been particularly important in surveillance programs for moths (Lepidoptera) and mosquitoes (Diptera: Culicidae). Because many moths (particularly species of Noctuidae) and other insects are attracted by short wavelengths of the light spectrum, so-called blacklight lamps emitting ultraviolet light are widely used in trap design. A commonly used trap for collection of mosquitoes, the New Jersey trap, combines a blacklight and dry ice ($CO_2$) for attraction and a fan for sucking the insects into the trap when they fly near. Once inside the trap, insects may be taken live for counting or other uses or killed directly in the receptacle. Dichlorvos-impregnated vinyl strips (Vapona®) are used frequently for the latter purpose. Other visual traps rely on insect perception of shapes for attraction. These include the Manitoba trap, utilizing a black or red sphere suspended under a transparent cone for collection of horse flies (Fig. 6.15, Box 6.4), and the

**Figure 6.14** A standard blacklight trap with collecting funnel and screened receptacle (*right*) showing part of the insects collected. (Photos by Marlin E. Rice)

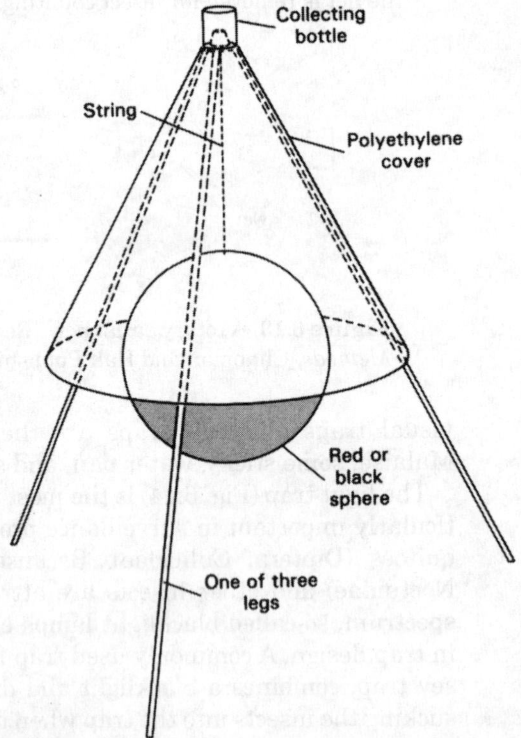


**Figure 6.15** A Manitoba trap used for collecting horse flies (Diptera: Tabanidae). Flies are visually attracted to the suspended sphere, fly up into the polyethylene canopy, and are captured in the collecting bottle. (Courtesy USDA)

## BOX 6.4 Horse Flies and Deer Flies (See Color Plates 2, 4)

**SPECIES:** Many species, especially in the genera *Tabanus* and *Chrysops* (Diptera: Tabanidae)

**DISTRIBUTION:** Horse flies and deer flies are found throughout the world, with different species dominating specific regions.

**IMPORTANCE:** Both horse flies and deer flies are significant livestock pests in all areas of the world. Females inflict deep wounds in animals, causing substantial blood loss that the flies sponge up with their specialized mouthparts. Under heavy infestations, an average of 100 ml of blood may be lost per animal per day. Males feed only on nectar and honeydew. Fly attacks result in both reduced weight gain and reduced milk production. The flies also vector many pathogens, such as bacteria, viruses, helminths, and protozoans. Members of the genus *Chrysops,* commonly called deer flies, often annoy and attack humans.

**APPEARANCE:** Many species differ greatly in appearance, but typically, horse flies are larger and less colorful than deer flies. Adult horse flies are from 10 to 25 mm long and are dark brown to black, with large bright-green or black eyes. Adult deer flies are yellow to black, with stripes on the abdomen and wings mottled with dark patches. Larvae (maggots) are tapered at both ends and are white to tan. Many species have black bands around each segment of the body, and some species are 50 mm long when full-grown. Eggs are cylindrical, from 1 to 2.5 mm long and are found in layers of up to 1,000 on objects near water.

**LIFE CYCLE:** Typical tabanids overwinter as larvae in mud at the bottom or edge of streams, ponds, and lakes. In spring, larvae begin feeding on vegetable matter or on small animals, depending on the species. When fully grown, larvae move to dry soil to pupate. After 5 to 35 days (depending on the species) adults emerge and mate, and females begin taking blood meals. Females then oviposit on objects near water, and in 5 to 7 days, eggs hatch. Newly hatched larvae burrow into the mud and begin feeding. Generation times vary greatly among species. Some species produce two generations per year, whereas others may produce only one every 2 to 3 years.

---

Missouri cutworm trap (Fig. 6.16), which depends in part on a vertical screen for attraction. The latter trap also combines wheat seedlings and wheat bran bait for the attraction of black cutworm (*Agrotis ipsilon*) larvae in corn.

Bait traps rely on an insect's olfaction, or sense of smell, for attraction. A common attractant is food; for example, a mixture of yeast and molasses in a cone trap to sample seedcorn maggot (*Delia platura*) adults (Fig. 6.17). Scents from food that attract insects or cause other behavioral responses are referred to as **kairomones;** study of these is an active field of entomological research. Other intensively researched attractants are **sex pheromones.** Traps containing pheromones or synthetic mimics of them, along with sticky surfaces for holding trapped insects, are sold commercially and have been useful in forest and orchard pest management programs (Fig. 6.18). For example, traps baited with the synthetic sex pheromone **codlemone** are used to indicate potential damage by the codling moth in apple, pear, plum, and walnut orchards. Pheromone-baited traps are also used widely in detection programs for exotic pests like the Mediterranean fruit fly, oriental fruit fly *Dacus dorsalis*, and melon fly *Dacus cucurbitae* in California and Florida.

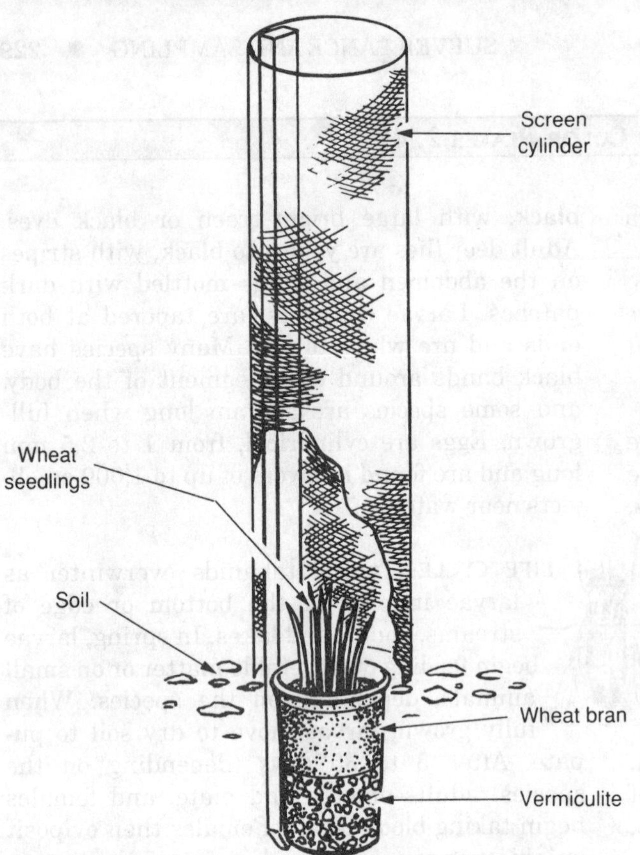


**Figure 6.16** A Missouri cutworm trap for sampling black cutworms, *Agrotis ipsilon.* The vertical screen is a visual attractant, and larvae crawl underneath and into the container. Wheat seedlings and bran in the container also aid in attraction. (Courtesy W. Showers and USDA)

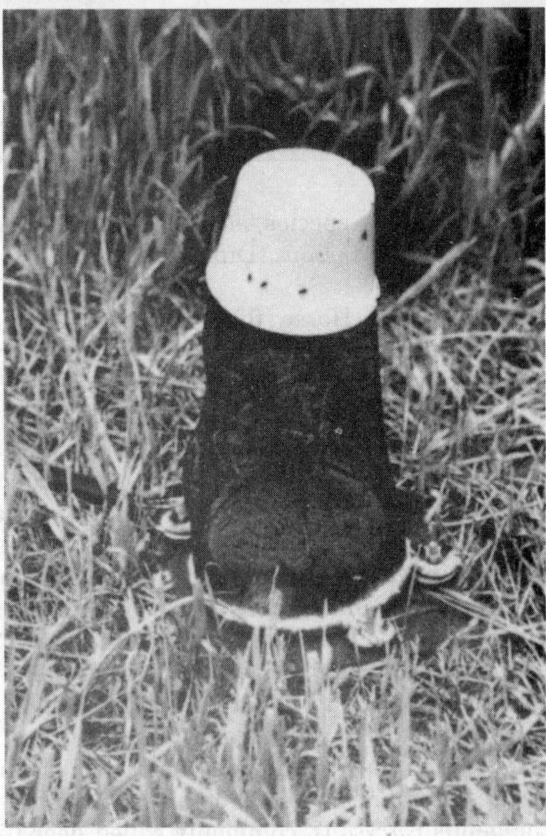

**Figure 6.17** A screened-cone trap for sampling adult seedcorn maggots, *Delia platura.* The trap is baited with a container of yeast and molasses. Flies attracted to the bait fly up and are captured in sticky material applied to the inner surface of the plastic container (top of trap). (Photo by J. Funderburk)

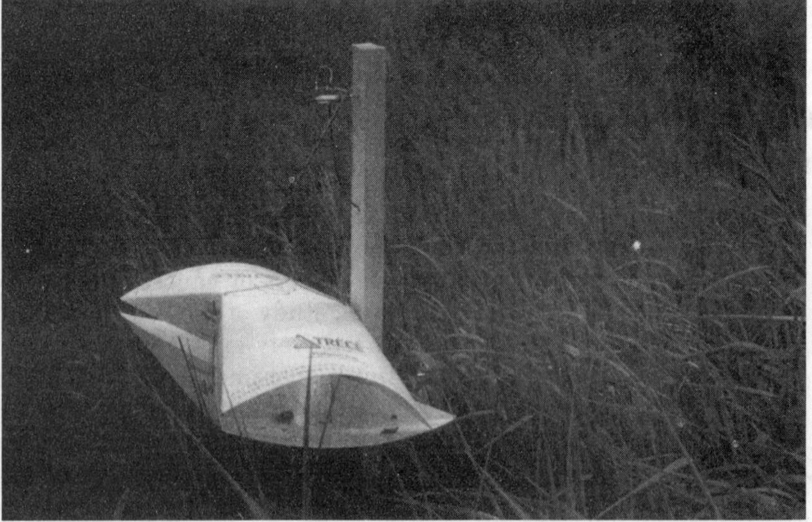

**Figure 6.18** A typical trap baited with synthetic pheromone for monitoring crop pests. Insects are attracted to pheromone in the trap and become caught on sticky material applied to the inside surface. (Photo by Marlin E. Rice)

**Figure 6.19** A Malaise trap. Insects fly into the screened baffles and are directed upward. They are captured in the collecting jar at the top. (Photo by Marlin E. Rice)

Most traps commonly used without attractants collect flying insects. Chief among these are the Malaise trap and the suction trap. The **Malaise trap** is basically an open-fronted tent made of cotton or nylon mesh that intercepts flying insects (Fig. 6.19). The roof of the tent slopes upward to a peak where a receptacle is located. Insects tend to move upward and into the receptacle, where they are held for counting. The suction trap is more "active" in collecting, relying on suction to capture flying insects. Most suction traps are basically a wire-gauze funnel leading to a collecting jar, with a motor-driven fan situated below the funnel to create the suction (Figs. 6.20, 6.21). These traps have been particularly useful for sampling alate (winged) aphids and leafhoppers (Hemiptera: Cicadellidae).

Window, water-pan, and sticky traps are examples of interception traps. The window trap (Fig. 6.22) consists of a large sheet of glass that sits in a collecting trough supported by wooden legs. Insects fly into the glass, are knocked down into the trough containing soapy water, and are counted. The water-pan trap (Fig. 6.23) consists of an open pan, often a crankcase-oil drain pan, mounted on a wooden post and filled with soapy water. An omnidirectional baffle is usually set in the pan for the interception of insects. In some instances, as with aphids, the baffle and interior of the pan are painted (yellow for aphids) to add physical attraction to the trap. Several types of sticky traps exist and also may be painted to attract certain insects. Such traps may be cylindrical (omnidirectional) or flat (unidirectional) (Fig. 6.24) and are usually mounted on a wooden stake at various heights above a plant canopy. Adhesives applied to the traps' surfaces include such materials as Tanglefoot®, Boltac®, and Tack Trap®. Pheromone traps are really a kind of baited sticky trap.

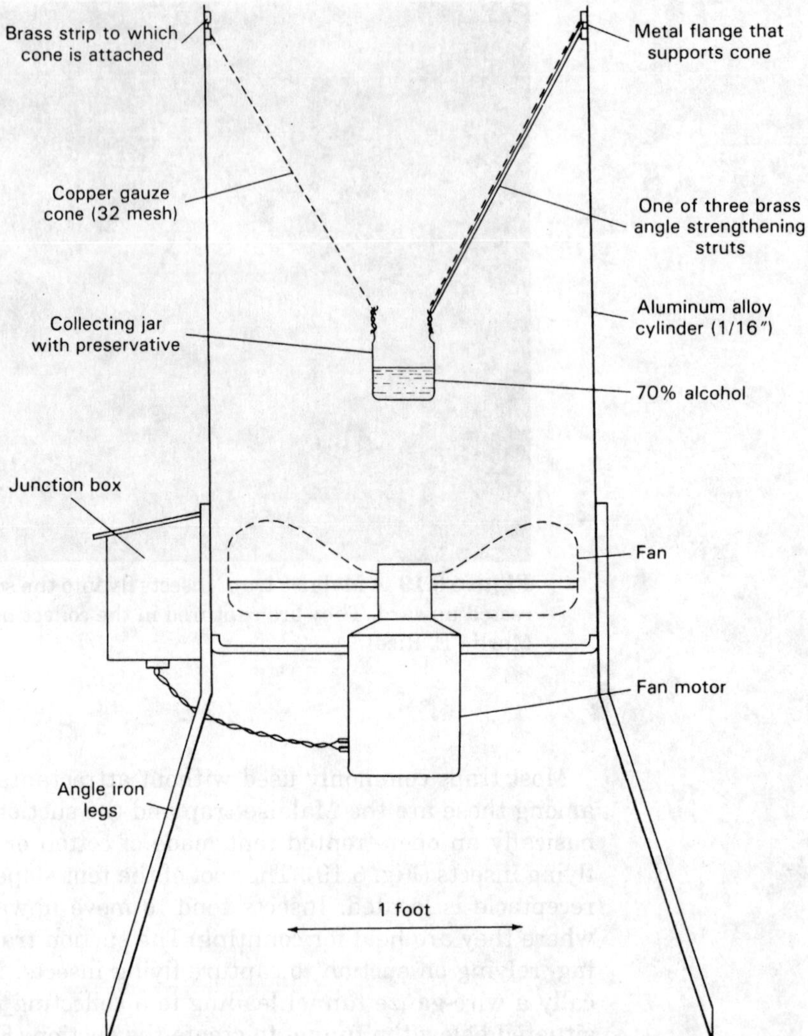


**Figure 6.20** A suction trap for sampling flying aphids (Hemiptera: Aphididae) and leafhoppers (Hemiptera: Cicadellidae). (Redrawn from Southwood, 1978, *Ecological Methods,* Chapman and Hall Publishing Company)

Whereas the traps previously mentioned sample mostly flying and, therefore, mature insects, **pitfall traps** capture ground-moving insects. The pitfall trap basically consists of a collecting jar or bottle buried in the ground with a funnel at the soil surface that empties into the bottle (Fig. 6.25). The landing to the funnel opening is smoothed so that walking and crawling invertebrates are not impeded and fall into the funnel. Consequently, a combination of both mature and immature insects is collected. If collected insects are to be killed for counting, a preservative, usually ethanol, is placed in the receptacle. A problem with pitfall trapping is that runoff from rainfall may cause overflow of the receptacle's content and loss of collected specimens. This problem can be overcome somewhat with a more elaborate trap that includes rain deflectors.

**Figure 6.21** A suction trap, partly constructed from a PVC pipe, extends approximately 25 feet upward to catch flying insects such as aphids. (Photo by Marlin E. Rice)

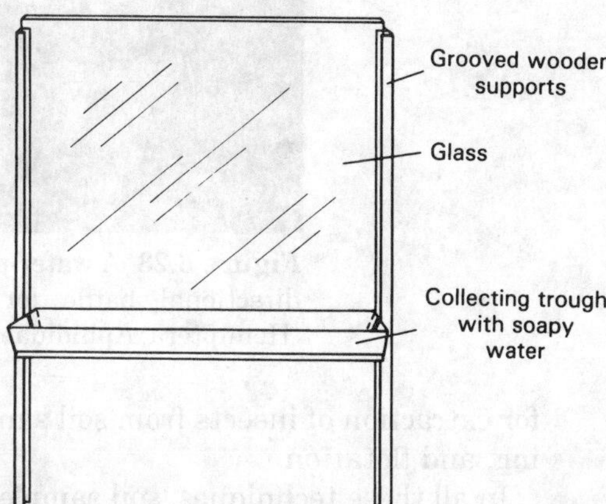


**Figure 6.22** A window trap for sampling flying insects. Insects run into the window and fall into soapy water in the collecting trough. Soap is used as a wetting agent, and insects drown. (Redrawn from Southwood, 1978, *Ecological Methods,* Chapman and Hall Publishing Company)

## Extraction from Soil

Soil is one of the most difficult environments for sampling insects because it is dense, and insects usually cannot be seen or easily extracted from it. Therefore, soil sampling techniques tend to be more complex, and sampling more expensive, than in other environments. Yet sampling here is very important because more than 90 percent of insect species spend at least one stage in the soil or at the soil surface. Probably the most often used techniques

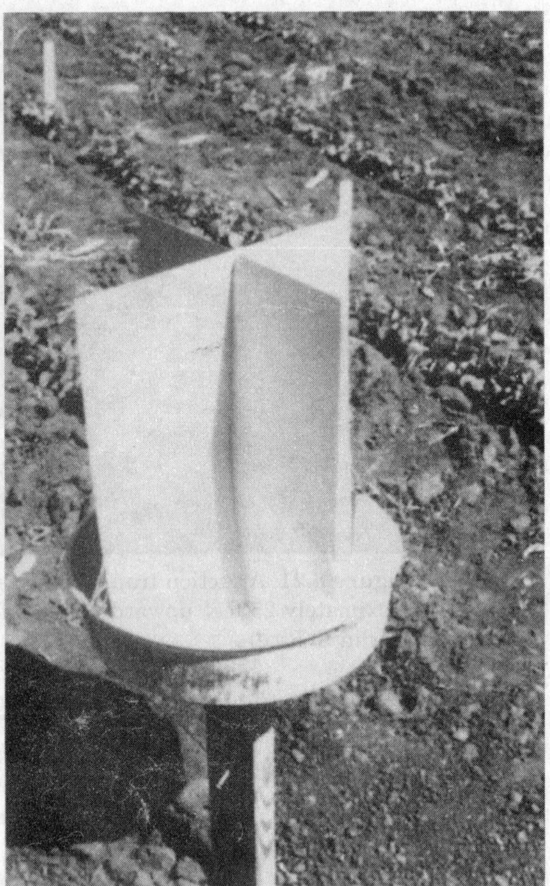

**Figure 6.23** A water-pan trap with yellow, omnidirectional baffle for sampling flying aphids (Hemiptera: Aphididae). (Photo by R. Hammond)

for extraction of insects from soil samples are Berlese funnels, sieving, washing, and flotation.

In all these techniques, soil samples must first be taken, then delivered to a facility for extraction. Soil samples may be taken with conventional soil corers (Fig. 6.26), golf-hole borers, bulb cutters, shovels, trowels, metal frames, or other equipment. Even commercial trenching machines have been used to collect soil for sampling rootworm eggs. A major consideration for selecting a soil sampling technique is whether the extracting equipment requires live insects, as does the Berlese funnel. In this instance, compacting the soil core when removing it from the corer may kill insects or close passages required for escape and collection. Here, sampling efficiency may be improved by using a split corer (Fig. 6.9).

The Berlese funnel described earlier is one of the most common devices for extracting insects from soil cores. This technique, originally developed by Italian entomologist A. Berlese in the early 1900s, utilized a water jacket surrounding a funnel to heat and dry the soil sample contained inside. A modification by A. Tullgren, a Swede, replaced this heat source with an incandescent lamp suspended above the sample. Other modifications for efficiency of extracting certain

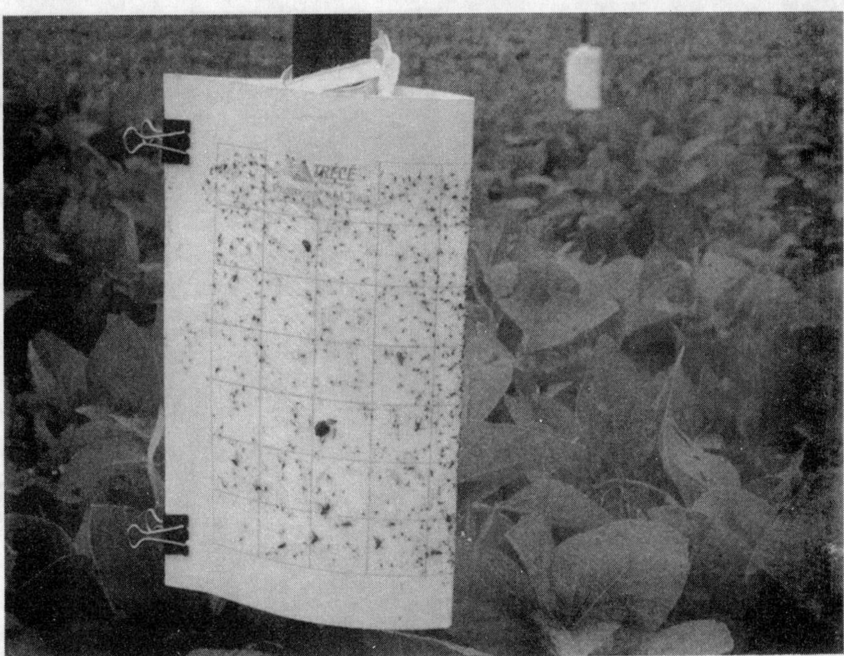

**Figure 6.24** A unidirectional sticky trap showing insects adhering to the sticky surface. The trap can be removed from the field and taken to the laboratory for insect counts. (Photo by Marlin E. Rice)

**Figure 6.25** A typical pitfall trap located in a plant row. Here, a metal can is buried in soil and supplied with a jar of alcohol or auto antifreeze. A funnel is inserted into the opening of the metal can, and the funnel extends into the collecting jar.

**Figure 6.26** Taking a soil core to sample wireworm (Elateridae) larvae. Once the soil is removed from the soil corer and placed in a bag, it is returned to a facility where it is washed through sieves to recover the larvae.

insect groups are available, but all such funnels operate on the same principles: making the soil environment so adverse that the insects inside are forced to move out, or providing a direct stimulus to invoke withdrawal. With this technique, care must be taken to prevent undue mortality before insects leave the core. Cores are usually inverted in the funnel so insects can leave by the same passages used in entering the soil, and the core is heated gradually so they are not killed from desiccation. Berlese funnels have been used most frequently to sample small, abundant arthropods in the soil—for example, mites (Acari), springtails (Collembola), and beetles (Coleoptera)—with efficiency varying greatly among species and physical characteristics of the soil. Obviously, this technique is useless for immobile insect stages in the soil like eggs, most pupae, and dormant stages.

Sieving techniques are mechanical and, unlike the Berlese funnel, do not depend on insect movement for extraction. Both dry- and wet-sieving techniques have been used to sample insects. Dry sieving employs one or more sieves of varying coarseness, beginning with a coarse mesh (for example, number 20) and ending with a fine mesh (for example, number 80 or 100). First sieves are

**Figure 6.27** Wet sieving a soil sample (in malt cup) with rootworm (*Diabrotica* species) eggs (*left*). Sieved material is placed in a separatory funnel with a magnesium sulfate solution, and eggs float to the surface (*right*). Heavy debris is drawn off the bottom by opening the funnel stopcock. Water is added to the supernatant, and eggs sink to the bottom of the funnel. Subsequently, this sediment with eggs is drawn off into a petri dish for counting purposes. (Courtesy Jon Tollefson)

usually employed to screen out larger soil particles and other debris and to allow insects to pass through. Subsequent screens are then used to collect specific groups of insects, depending on body size. Standardized soil sieves may be purchased and used to sample insects or may be fabricated by constructing a wooden frame and attaching common hardware cloth. Seedcorn maggot pupae can be sampled with "homemade" hardware cloth sieves. Wet sieving follows the same principles as dry sieving, except that water is used to facilitate the movement of particles, including insects, through the sieves (Fig. 6.27). With this technique, the soil sample and sieves may be placed in a container of water, with the water being drained after a given time. Another method is to use a forced stream of water to wash particles through the sieves.

The flotation technique is another widely used extraction procedure that may be used alone or in conjunction with sieving and other techniques. Flotation operates on the principle that particles of lower specific gravity than their environmental medium float on the surface of that medium (like ice cubes float in a glass of tea). In certain instances, for example, extraction of rootworm (*Diabrotica* species) larvae, water is a sufficient flotation medium (Fig. 6.27). In other situations, however, the efficiency of flotation is increased by adding a salt such as magnesium sulfate to increase the medium's specific gravity. Compromises usually are made in the specific gravity chosen to give reasonable extraction of insects without an inordinate amount of extraneous organic debris. Usually, different salts and concentrations are tried before deciding on an appropriate one.

**Figure 6.28** Washing corn roots (*left*) so they can be rated. A rating scale of zero to three, or one to six (shown), is often used (*right*). Such ratings can be used as an index of crop injury and insect population size. (Courtesy Jon Tollefson)

## Indirect Techniques

In all sampling techniques discussed previously, data for estimating population size were expressed as numbers of insects. Estimates can also be made from data on insect effects or products. Such estimates are often called **population indices.**

Measuring the effects of insects is probably the most common method of indexing an insect population. Root rating schemes indicating the feeding of corn rootworms, percentage defoliation of soybeans, numbers of plants cut by cutworms, and number of plants with tillers or "deadheart" caused by borers are all examples of index data. Such data, of course, are also paramount in determining the economic impact of insects on the crop. The sampling technique involved with most of these measures is direct observation of results of insect injury and, sometimes, skilled ranking of the observations. A few techniques like corn rootworm injury ratings require special procedures: in this case, pulling up corn plants, washing roots, and then rating them according to amount of feeding (Fig. 6.28).

Data on insect products include measures of larval and pupal skins, frass (droppings including insect excrement), and nests. With some of the larger insects having conspicuous skins, as with cicada (Hemiptera: Cicadidae) exuviae on tree branches, directly counting these may be an efficient method of sampling. Measuring the amount of frass may also be an efficient means of indexing an insect population. The measurement of frass drop in a collecting tray has been used to index population size of several moth (Lepidoptera) and sawfly (Hymenoptera) pests in forests. Nests of colonial insects, like webworms (Lepidoptera: Crambidae) and tent caterpillars (Lepidoptera: Lasiocampidae), also may be easily counted, with these counts serving to index the population in a locality.

## Auxiliary Survey Equipment

In addition to the equipment used to collect insect samples, surveyors or scouts need several other items. Although it is difficult to develop a complete list of these for all sampling programs, the following outline of equipment may be helpful.

1. Clothing—field clothes, comfortable boots, rain suit, rubber boots
2. Containers—knapsack, small plastic bags, plastic vials with alcohol, ice cream cartons, ice chest with ice (for storing samples)
3. Measuring equipment—hand counter, hand lens, tape measure, thermometer
4. Recording materials—clipboard, columned paper or scouting forms
5. Tools—trowel, utility knife
6. Other—mosquito repellent

Of course, not all these materials can be carried by the surveyor on the survey route. However, they should be kept easily accessible, perhaps at the edge of the field or in the car.

## THE SAMPLING PROGRAM

R. F. Morris (1955), a noted Canadian entomologist, once expressed the opinion that the sampling program is a mixture of art, science, and drudgery and that the art and science are in the sampling design. Design involves selecting a sampling technique and detailing all the steps required to obtain an estimate. Usually, several designs are devised for an insect species or group, and these are tested to determine the most appropriate ones for a particular pest management program. The time and effort expended in obtaining good sampling designs are more than repaid later when the pest management program is implemented.

### Kinds of Estimates

Selection of an appropriate sampling technique depends initially on the kind of estimate desired. Generally, estimates can be divided into two broad categories—absolute and relative.

**Absolute estimates.** Absolute estimates measure actual numbers in the insect population. Such estimates are expressed in numbers per ground surface area; for example, number per acre, number per hectare, number per square meter. These are often the most difficult estimates to make and frequently the most costly. Absolute estimates are of utmost importance in research in insect population dynamics, but because of their cost, they are not used widely in pest management practice. Some examples of techniques used to make absolute estimates include: (1) suction traps where the volume of air passing through the trap is known, (2) Berlese funnels where efficiency of extraction is known, (3) vegetation removal and laboratory processing of the sample, and (4) fumigation cage sampling. In all instances when a habitat unit is sampled, numbers of habitat units per ground surface area must be known to make the final absolute estimate. For example, after finding the number of cabbage loopers per head of cabbage, the number of cabbage plants per acre is used to arrive at the number of loopers per acre.

If, in the previous example, the first type of estimate meets our needs by providing the number of loopers per cabbage head, we would be using another type of estimate—the **population intensity estimate.** Population intensity is the number of insects per habitat unit, and it is an absolute-type estimate. The population intensity estimate is used frequently in pest management programs because it relates closely to crop injury. Such estimates have been particularly useful in establishing economic injury levels, and subsequently, these are expressed in population-intensity terms, such as number of aphids per leaf, number of borers per stem, or number of horn flies per side of steer or cow. *In situ* counts and other techniques used for true absolute estimates also are used for population intensity estimates.

Another estimate related to the absolute estimate is the *basic population estimate.* This estimate is intermediate between the absolute estimate and the population intensity estimate. It combines the habitat unit with a unit of measure. In forest entomology, number of insects per 10 square feet of branch surface has been used. A common measure in row crops has been number of insects per row foot.

**Relative estimates.** Relative estimates differ greatly from absolute and related estimates because they do not translate directly into numbers per ground surface area. Instead, relative estimates are based on the kind of sampling technique used. For example, the sweep-net technique is used to make relative estimates, which are expressed as number of insects per sweep or insects per some fixed number of sweeps. These estimates do not directly answer the question of how many insects are present per acre, per plant, or per foot of row. Rather, they are used to compare population sizes in time and space according to a specific sampling technique. Thus, we can compare potato leafhopper, *Empoasca fabae,* populations per sweep of alfalfa in Baker County in 2006 and 2007, and we can compare the same between Baker and Jones counties in 2007, but we cannot determine directly the number of leafhoppers per acre for the comparison. Why not? Because the estimate is relative; it depends on the technique used to gather the data. Relative estimates usually can be converted to absolute estimates, but considerable research and analysis is required first. Many techniques have been used to make relative estimates, including several kinds of netting, all types of trapping, some *in situ* methods, and knockdown approaches. Relative estimates have the great advantage of being inexpensive and are often the most appropriate for insect pest management.

Population indices, as discussed earlier, are usually employed similarly to relative estimates for population comparisons. They have added value when used as an indicator of existing plant damage and as a basis for management decisions. They are expressed in numerous ways, depending on the insect effect measured, such as cast skins, webbing, root feeding, or leaf feeding. Percentage defoliation of a plant canopy is one of the most popular population indices in insect pest management.

## Converting Relative Estimates to Absolute Estimates

As some ecologists have pointed out, there is no hard and fast distinction among methods to achieve absolute and relative estimates. Most absolute procedures seldom yield all insects in a potential sampling unit, and relative estimates can be transformed to estimate the absolute population. These transformations can be accomplished in one of two ways, statistically or experimentally.

A statistical transformation can be created by making paired relative and absolute estimates from a series of populations (dense and sparse), followed by statistical analyses that regress the relative estimates on the absolute. This procedure has been used frequently with sweep-net, vacuum net, and shake cloth methods.

The statistical approach is exemplified in a study of two grasshopper species, *Melanoplus femurrubrum* and *M. differentialis,* in soybean to transform

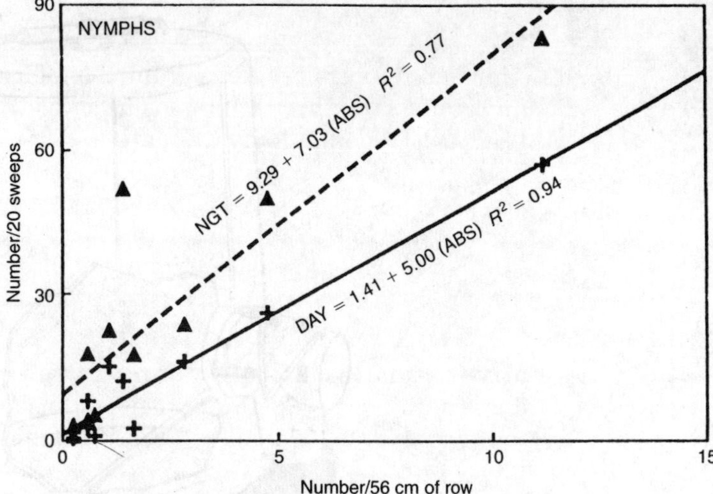


**Figure 6.29** Sweep-net regression lines for grasshopper nymphs and adults using mean catch of redlegged and differential grasshoppers. ABS - absolute estimate from nighttime plant-removal technique; DAY - estimate from day sweep-net; NGT - estimate from night sweep-net. Equations can be used to give absolute estimates from relative (sweep-net) estimates. (From Browde et al., 1992, *Journal of Economic Entomology* 85:2270. With permission)

sweep-net estimates into estimates of the absolute population (Fig. 6.29). Assessments of absolute density were made at night by using a plant-removal method (similar to Fig. 6.30), and these were compared to both day and night sweep-net samples using regression analysis. The models make it possible to predict absolute population densities from either day or night sweep samples.

Although the statistical approach focuses on studies of natural populations, the experimental approach utilizes artificially established populations with a known density. This approach was used in a study of grasshoppers in burned and unburned tallgrass prairie. Here, a night trap was used to capture marked grasshoppers placed inside. Marked individuals constituted the experimental population, and it was determined that about 75 percent of the marked individuals were captured. Such a procedure allows calibration of the method to give estimates of absolute density. Experimental populations represent one of the few methods available to substantiate the efficiency of methods designed to give absolute estimates.

## Descriptive Statistics

Whether population estimates are expressed in absolute or relative terms, the proper use of statistics in summarizing data and designing sampling programs is extremely important. Statistics, of course, is the discipline of collecting, analyzing, and interpreting numerical data. Descriptive statistics used in insect sampling involve analysis of data to obtain a comprehensive, quantitative summary of insect population characteristics. Probably the most widely used statistics in describing insect population density are the mean, the standard deviation, and the standard error.

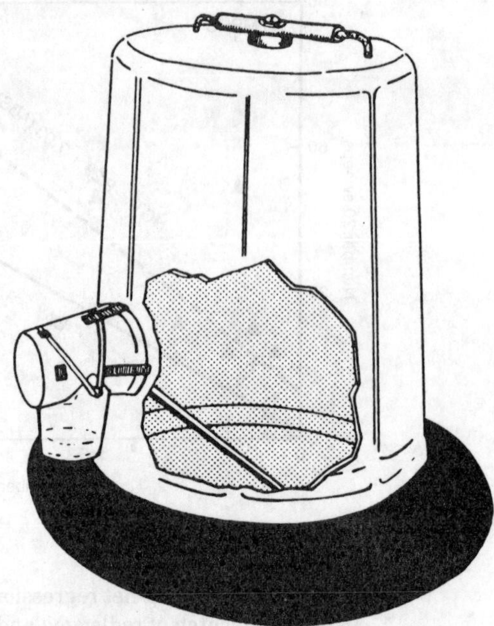

**Figure 6.30** Night trap for obtaining absolute estimates of grasshopper populations in tall-grass prairie. The trap is constructed from a 68-liter plastic trash can (covering 0.152 square meters). Cut-away shows a crawling ramp leading to a transparent-plastic receptacle with soapy water. (From Evans et al., 1983, *Environmental Entomology* 12:1449. With permission)

The **mean** is simply the arithmetic average of the sample numbers. For a set of $N$ sampling units, where the individual sample numbers are represented by $X_1, X_2, X_3, X_4, \ldots, X_N$, the sample mean $\bar{x}$ is calculated as:

$$\bar{x} = \frac{X_1 + X_2 + X_3 + X_4 + \cdots + X_N}{N}$$

For example, if six sampling units are taken and insect counts are 10, 13, 8, 12, 9, and 11, the mean is 10.5. If several of these means are obtained from a number of different sample sets and we calculate a grand mean, as above, from these sample means, we have *m,* our best estimate of the population size. By computing such values, we then can compare insect populations at different times and in different places.

Although the mean gives us one of the most important characteristics of an insect population—central tendency—it does not tell us anything about variation in the samples—numerical differences among different samples. This characteristic, referred to as distribution of the data, is most often expressed as the range or as the standard deviation. The **range** is the simpler of the two expressions, being the difference between the smallest and the largest sample

**Table 6.1 Computation of the Standard Deviation Directly from Raw Data.**

| Sample Number | Sample Count $X$ | $X^2$ |
|---|---|---|
| 1 | 27 | 729 |
| 2 | 15 | 225 |
| 3 | 23 | 529 |
| 4 | 18 | 324 |
| 5 | 9 | 81 |
| 6 | 12 | 144 |
| 7 | 10 | 100 |
| 8 | 18 | 324 |
| 9 | 30 | 900 |
| 10 | 16 | 256 |
| | SUM $X$ = 178 | SUM $X^2$ = 3,612 |

$$\bar{x} = 178/10 = 17.8$$

$$s = \sqrt{\frac{1}{N-1}\left(\text{SUM } X^2 - \frac{(\text{SUM } X)^2}{N}\right)}$$

$$= \sqrt{(1/9)(3{,}612 - \tfrac{(178)^2}{10})}$$

$$= 6.99$$

number. Although the range gives some idea as to the spread of data about the mean, it depends only on extremes, some of which may be accidental and/or very rare.

The **standard deviation** of a set of sample numbers eliminates the problem of sole dependence on extremes by averaging deviations of sample counts from the mean value. The standard deviation $s$ of a set of insect samples can be readily calculated on handheld calculators or laptop computers, many of which are programmed for this statistic. If manual calculation is necessary, the raw data formula for $s$ is most practical. The example in Table 6.1 gives the formula and how it is used.

The value 6.99 is one standard deviation and gives us an idea of the magnitude of variation in our sample. If our data were distributed normally (described by a bell-shaped curve), then it would be expected that 68 percent of our sample numbers would fall within one standard deviation on either side of the mean; in other words, 17.8 minus and plus $s$ gives a range from 10.8 to 24.8, and the majority of samples should fall within this range.

Although the standard deviation gives us a measure of variation in a set of samples, it is not our best estimate of the standard deviation of the insect population. For this estimate, we must consider the number of sampling units taken. This estimate is called the standard error of the mean or simply **standard error.** It is calculated as:

$$s_x = \frac{s}{\sqrt{N}}$$

where

$s_x$ = standard error, sometimes designated SE
$s$ = standard deviation of the sample numbers
$N$ = number of sampling units taken

As can be seen, an important characteristic of the standard error is that its magnitude decreases as the number of sampling units increases. This statistic is particularly important in insect sampling programs because it can be used as a measure of precision, and we are striving for the greatest precision practicable in our estimates. Because the lower the standard error, the more precise the sampling program, it is possible to increase samping-unit number and thereby increase precision. Greater detail on increasing precision will be discussed later.

Although the statistics discussed tell us something about population density and the variation in our samples, they do not describe the dispersion of insects in a population. Dispersion (not dispersal) is the pattern or arrangement of insects in space, or how they are spread out in an area. An understanding of dispersion is important because it gives us information about population dynamics and may influence the way we sample the population in an area.

As discussed earlier, the most common ways insects are dispersed in an environment are "random" and "clumped." A random dispersion means that an insect has as good a chance of being present in one place as in another, and individuals do not affect each other's presence in a place. Random does not mean that the population has a uniform dispersion over an area. Random dispersions in insects probably occur most often in relatively uniform environments, such as a field of soybeans on flat terrain. Clumped (contagious) dispersions probably are the most common type in insect populations. Clumping of individuals in a population means that if one insect of the species is found, chances are good that others are in the same vicinity. Clumping may be caused by either behavioral (for example, mating and feeding) or environmental (for example, uneven habitat) factors or both. Statistically, random and clumped dispersions can be described by mathematical models. The most common models used in entomology are the **Poisson model** to describe a random dispersion and the **negative binomial model** to describe a clumped dispersion. Methods of analyzing dispersion pattern are beyond the scope of this discussion but may be obtained from textbooks that discuss ecological methods (for example, Andrewartha 1961, Pedigo and Buntin 1994, Southwood 1978).

### Criteria of Estimates

Estimates obtained from newly designed sampling programs must be evaluated for their usefulness before final recommendations are made. Usually, several preliminary sampling plans are devised and used, with the data from these trials employed in the evaluation.

The major criteria for these evaluations are fidelity, precision, and cost. **Fidelity** is the accuracy with which the estimates follow actual numbers in the insect population. Fidelity is usually determined by comparing a program of unknown fidelity with one known to be accurate; numbers known through releases in an area or habitat removal sampling are examples.

**Precision** measures the degree of error in making estimates and is usually expressed as percentage standard error of the mean. This value, known as relative variation (RV), is calculated as:

$$RV = (SE/\bar{x})\ 100$$

where

$SE$ = standard error of the mean
$\bar{x}$ = mean.

With RV as a measure of precision, a good criterion for practical pest management programs is to obtain values near 25. For population research, greater precision is desirable and values near 10 are sought. Because *SE* decreases with increasing numbers of samples, precision can be improved by taking more sampling units.

The remaining, often overriding, criterion is *cost.* Population estimates made with great fidelity and precision are useless if they are too expensive. Therefore, great attention is given in accounting for costs of tentative sampling programs, and final design recommendations are those that yield the greatest fidelity and precision for the lowest cost.

## Program Dimensions

After determining the kinds of estimates needed for a pest management program and sampling criteria, several programs are usually designed and tried. The design of a program involves at least four basic dimensions: (1) insect stage to sample, (2) number of sampling units to take, (3) time to sample, and (4) pattern of sampling.

**Insect stage.** In pest management, the insect stage sampled depends on the particular insect involved and on the management strategy for that insect. In making management decisions, the stage most often sampled is the damaging stage. This is especially true when little lead time is needed to suppress the population. When substantial lead time is needed, a stage prior to the damaging stage may be selected so that early predictions can be made. For example, sampling the nonfeeding adult stage can help predict the subsequent damaging larval population.

**Number of sampling units.** The number of sampling units taken in a program is usually the result of a compromise between precision on the one hand and cost on the other. In many instances, the most affordable number is the prime determinant. When the greatest affordable number of sampling units yields too little precision, the program should be modified or abandoned in favor of other designs. Considerable research may be required to find a program where the number of sampling units needed is economically feasible. As a rule of thumb, many pest management sampling programs call for taking at least five sampling units in an average-size cropping area (about 35 acres). When greater precision is desired, the sampling unit number should be increased. A useful formula to calculate number of sampling units required to achieve a given degree of precision is:

$$N = [(t \times s)/(D \times \bar{x})]^2$$

where

$N$= number of sampling units required
$\bar{x}$ = mean density
$D$= required precision (RV) expressed as a decimal (for example, 0.25)
$s$ = standard deviation
$t$ = student's $t$ value = a value obtained from statistical tables (usually the value of $t$ for 0.05 probability and sample number used to calculate the mean)

The mean and standard deviation values in the formula are obtained by using data from preliminary sampling.

**Time to sample.** The timing of sampling depends on the occurrence of the insect stage being sampled. For management decisions, it is most important for the sampling to coincide with the time of peak numbers of the stage. Such timing may be predicted using degree-day or other phenological models or by trial sampling. To trial sample, a conservative date of appearance is chosen, and samples are taken regularly (once or twice per week) until a peak is obvious from the cumulative data—in other words, when successive dates of increase are followed by successive dates of decrease. The greatest problem with the latter approach is that unacceptable damage may occur before a peak is detected. This problem does not occur if sampling is conducted early and frequently and management decisions are based on threshold values before a potential peak occurs. It has been suggested that outdoor cages containing a control population be set up for observations and that timing of sampling be gauged from these observations. Time of day for sampling may also be a problem with some insects. If the sampling technique depends on a certain insect activity (for example, flight) and that activity occurs mostly at a certain time of day (for example, dusk to dawn), then samples should be taken during that activity period.

**Pattern of sampling.** The spatial sampling pattern at a location can vary widely, depending on the uniformity of the habitat and population dispersion. In the strictest sense, unbiased estimates of population density should be taken "at random"; that is, every sampling spot in the area has an equal chance of being selected for sampling. For this sampling, two random numbers (measurements of distance) along two coordinates in the area determine the sample site. For example, coordinate 1 might be row 12, and coordinate 2 might be 10 meters down the row (Fig. 6.31).

Often, such sampling is too laborious (expensive) for pest management surveillance, and subjective selections (less expensive) of sampling sites are usually made. Usually, pest management scouting calls for walking a prescribed route through a growing area and taking samples somewhat regularly along this route. Such sampling is known as systematic sampling. Samples from such a pattern are usually treated as if they were random in later analyses. This approach is more efficient when sampling many growing areas in a single day and is satisfactory unless cumulative errors arise because of the sampling pattern. A problem might arise, for example, if the pattern results in frequent sampling sites of very low or very high density. In early stages of sampling program development, it is worthwhile to test for cumulative errors by comparing strict random patterns with systematic patterns. As a rule of thumb, spread sampling units out over an area, making sure ridges and low areas are included (Fig. 6.32). The edges of a growing area (first 5 meters or so) are usually avoided in making estimates unless the pest species normally invades and/or is more important at the edge, for example, armyworms (Lepidoptera: Noctuidae), grasshoppers (Orthoptera: Acrididae), and stalkborers (Lepidoptera: Noctuidae).

## Pest Management Scouts and Scouting Records

Samples for making pest management decisions in agriculture may be taken by entomologists or other specialists, and reports are made to grower or grower agencies. Growers may do their own sampling or supplement the sampling of specialists. Persons who sample growing areas regularly for the purpose

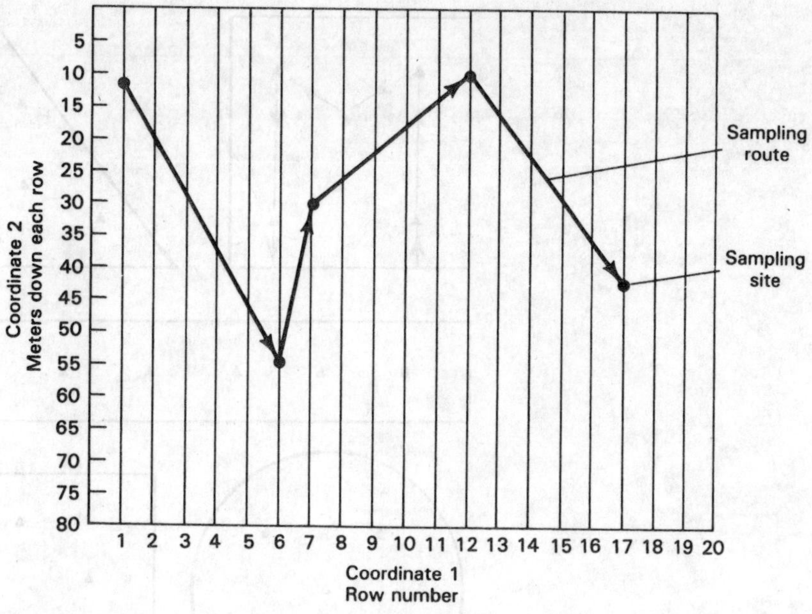


Random sampling program

| Sample | Coordinate 1 | Coordinate 2 |
|---|---|---|
| 1 | 12 | 10 |
| 2 | 7 | 30 |
| 3 | 1 | 12 |
| 4 | 6 | 55 |
| 5 | 17 | 42 |

**Figure 6.31** An example of a random sample pattern using two coordinates. The coordinates are selected from a table of random numbers or generated by computer.

of making pest management decisions are called **pest management scouts,** and their activity is called **scouting.** Therefore, we often hear of such descriptions as a "rootworm scouting program" or such activities as "scouting an apple orchard." Many pest management scouts are employed by private consulting firms that contract with growers to sample plantings for pests and provide timely advice.

To provide this advice, scouting records must be kept and data must be analyzed to demonstrate pest population trends and make predictions. These records not only include the insect species present and its numbers but may also include information such as field location, insect growth stage, current weather conditions, degree of damage to date, crop sampled, variety grown, crop condition and growth stage, and other pests present (Fig. 6.33). An historical account of these factors helps in understanding both insect population dynamics and potential impact on crop yield. The records also help in interpreting the quality of the sampling estimates themselves because many of the factors recorded affect sampling efficiency.

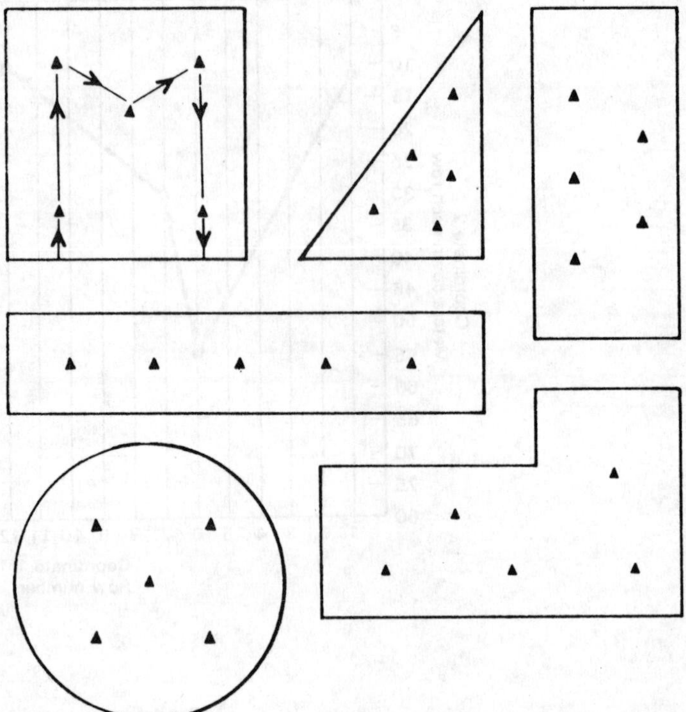

**Figure 6.32** Some common patterns of systematic sampling. Each darkened triangle represents a sampling location. (After *Corn Pest Management for the Midwest,* North Central Publication 98)

## Alfalfa Weevil Sampling: An Example

Alfalfa weevils, particularly larvae, are early-season defoliators of alfalfa in the eastern half of the United States, causing significant damage to the first-cut crop of the season (Box 6.5). Management of this insect involves scouting and, when appropriate, either early cutting or application of insecticides. In fact, scouting is the crucial element in this pest management program.

The typical scouting program in northern regions (areas where most weevil eggs are laid in the spring) involves keeping a record of accumulated degree days (48°F, minimum threshold temperature) from January 1. When 300 degree days have accumulated, the sampling program begins, with samples taken approximately weekly. To scout a field, 30 stems are picked at equal intervals across a field. A U-shaped pattern is walked to pick the stems, and each stem is shaken vigorously over a pail to dislodge larvae (Fig. 6.34). After removing larvae, the length of every third stem is measured. After all samples have been taken, larvae are transferred to a shallow pan and counted. The total larval count is divided by 30 to obtain an average, and plant height is also averaged. Using average larvae per stem and average plant height, tables are consulted to obtain pest management recommendations (Table 6.2). Information in the table will advise the grower to spray, to harvest the crop, or to wait and resample the field in a specified number of days (Table 6.3).

This example of sampling design has been widely recommended for use in alfalfa weevil management. Similar programs are the backbone of insect pest management practice.

## CORN INSECT SURVEY (Example)

Grower ____________ County ____________ Date ____________

Township ____________ Section ____________ Field No. ____________ Acres ____________

Crop stage ____________ Extended leaf height ____________

**PLANT POPULATION:** (count number of plants in 20 feet at 5 locations)

| Set 1 | | Set 2 | | Set 3 | | Set 4 | | Set 5 | Total | | Row width |
|---|---|---|---|---|---|---|---|---|---|---|---|
| ____ | + | ____ | + | ____ | + | ____ | + | ____ | ____ | X A | PLANTS PER ACRE ____ |

Row width "A" factors: 30" = 174.24; 36" = 145.2; 38" = 137.561; 40" = 130.681

**GENERAL FORM**____(Black cutworms, corn borer; for miscellaneous insects encountered, use this section with a sample size of 20 plants.)

| | Set 1 | | Set 2 | | Set 3 | | Set 4 | | Set 5 | | Total | |
|---|---|---|---|---|---|---|---|---|---|---|---|---|
| Damaged plants/set | ____ | + | ____ | + | ____ | + | ____ | + | ____ | = | ____ | ÷ C = % infestation ____ |
| No. of insects | ____ | + | ____ | + | ____ | + | ____ | + | ____ | = | ____ | ÷ D = No. of insects ____ |
| Size | | | | | | | | | | | | Size of insects ____ |

**BLACK CUTWORM:** Use General Form above with 50 plants per set, count worms around all cut plants per set, size worms in inches, C and D = 2.5.

Threshold: When 3% or more of the plants are being cut and worms are ½" - 1" in length, treatment may be necessary.

**EUROPEAN CORN BORER:** Egg masses/25 plants ____ X 4 Egg masses per 100 plants ____

Also use General Form above with 20 plants per set; count the number of insects on 2 infested plants per set; size is in instar (IV); C = 1; D = 10.

1st Brood Tassel height ____ ÷ Plant height ____ = Tassel ratio ____ TR avg. ____

Threshold 1st Brood: If 50% or more of the plants have whorl feeding and live worms are present, treatment may be justified. **2nd Brood:** treatment may be justified if there are 1 or more egg masses or larvae per plant.

**CORN LEAF APHIDS:** Each set consists of 10 plants. Note beneficial insects found.

| | Set 1 | | Set 2 | | Set 3 | | Set 4 | | Set 5 | | Total | |
|---|---|---|---|---|---|---|---|---|---|---|---|---|
| None present-few | ____ | + | ____ | + | ____ | + | ____ | + | ____ | = | ____ | × 2 = % low ____ |
| ▶50/Plant | ____ | + | ____ | + | ____ | + | ____ | + | ____ | = | ____ | × 2 = % moderate-high ____ |
| tassel covered | ____ | + | ____ | + | ____ | + | ____ | + | ____ | = | ____ | × 2 = % tassel covered ____ |

Threshold: If 50% or more of the plants have 50 or more aphids per plant, treatment may be justified; also if 3% of the plants have the tassels covered and corn is under moisture stress.

**CORN ROOTWORM BEETLE POPULATIONS:** Each set consists of 5 plants for stalks (____), and 5 different plants for ear zone ____

| | Set 1 | | Set 2 | | Set 3 | | Set 4 | | Set 5 | | Total | | Ear zone | Stalk |
|---|---|---|---|---|---|---|---|---|---|---|---|---|---|---|
| NCR | ____ ( ) | + | ____ ( ) | + | ____ ( ) | + | ____ ( ) | + | ____ ( ) | = | ____ | × .04 = NCR's/ | ____ | ( ____ ) |
| WCR | ____ ( ) | + | ____ ( ) | + | ____ ( ) | + | ____ ( ) | + | ____ ( ) | = | ____ | × .04 = WCR's/ | ____ | ( ____ ) |

All ear zones + stalks combined = Beetles/plant ____

Threshold: If corn rootworms beetle populations exceed 5 per plant and silks have not begun to turn brown, pollination may be affected.

NOTES:

**Figure 6.33** Portion of a typical scouting form for insect pest management, using corn as an example. (After *Corn Pest Management for the Midwest,* North Central Publication 98)

## BOX 6.5 ALFALFA WEEVIL (SEE COLOR PLATE 3)

**SPECIES:** *Hypera postica* (Gyllenhal) (Coleoptera: Curculionidae)

**DISTRIBUTION:** The alfalfa weevil was originally imported into North America from southern Europe around 1900. In 1904, the first alfalfa weevils were discovered in Utah. The species was found in Maryland in 1951 as a result of a second introduction. Currently the pest is found throughout the United States and regions of Canada and Mexico. A similar species, the Egyptian alfalfa weevil, *Hypera brunneipennis* (Boheman), occurs in the western United States.

**IMPORTANCE:** The alfalfa weevil is one of the most serious alfalfa pests in North America. The species primarily damages first-growth alfalfa, but the second growth can also be damaged. The most significant injury is caused by larvae that damage the top growth of the plant by entering and feeding on the developing buds and by feeding on the leaves. Their dried, damaged leaves make injured plants appear frosted. Adult weevils feed on older, lower leaves of alfalfa. Growth is stunted in infested alfalfa, resulting in both reduced quality and yield. Also, injured alfalfa is unsuitable for seed production.

**APPEARANCE:** Adults are about 5 mm long and are light brown, with a dark brown stripe running down the center of the back. As adults age, their color varies from brown to gray. Larvae are pale green with a black head and are 10 mm long when full-grown. They have a distinctive white stripe running down the back of the body. The eggs, laid in clusters of about ten, are initially yellow but turn black before hatching.

**LIFE CYCLE:** Adults overwinter in plant debris around the edges of fields; however, some individuals overwinter under leaf litter in wooded areas. In early spring, adults become active and begin feeding on alfalfa leaves. Females oviposit 400 to 1,000 eggs from April to May. Initially, egg clusters are deposited in dead alfalfa stems and field litter, but later they are laid in holes chewed into the live stems by the females. Eggs hatch in 7 to 21 days, depending on temperature. After three molts (25 days), larvae spin a cocoon that is either attached to alfalfa leaves or attached to plant debris on the ground. In 7 to 14 days, adults emerge and feed on alfalfa for a short period of time. Then the adults aestivate for the duration of the summer, becoming active in the fall only to prepare for winter hibernation. Typically, there is one generation per year, but some experts believe two generations occur in specific regions.

### Sequential Sampling

A potentially important type of sampling program in insect pest management is the sequential sampling or sequential decision program. Such programs are based on insect dispersion patterns and economic decision levels. Sequential sampling was first developed during World War II for the inspection of war goods; it allows the placement of a given population into one of two or more categories. For example, these categories efficiently tell the pest management scout if the population is an outbreak type requiring management or an endemic type requiring no action. Any good sampling program can do this if the economic decision level is known, but sequential sampling usually allows the decision to be made with fewer samples. Therefore, sequential sampling saves money in the pest management program.

With sequential sampling, the total number of samples to be taken in a field is variable and unknown when the sampling is begun. Using a table (or less

**Figure 6.34** After picking 30 alfalfa stems, each is shaken vigorously over a pail to dislodge alfalfa weevil larvae. Larvae in the pail are then counted.

often, a graph), one can start sampling and continue until the population can be classified. The number of sampling units required is relatively few if the insect population is either very large or very small. More sampling units are required when the population has a density near the dividing line between endemic and outbreak levels. This means we can be confident of classifying a population with only a few samples when it is at its high or low extremes, but we have to take more samples to be sure of our classification when it is at an intermediate level. Sequential sampling tells us how many sampling units to take before we can be confident about making a pest management decision (Fig. 6.35).

An example of a sequential sampling table is shown using the green cloverworm (*Hypena scabra*) program for soybeans. This insect defoliator is sampled by shaking 60 cm of soybean row (30 cm on each side of the scout) over a shake cloth and counting larvae in each sample. Sample numbers are accumulated, and a table is consulted with the cumulative number (Fig. 6.36). If the cumulative number is less than the lower critical value, sampling is discontinued, and no action is recommended. If the cumulative number is greater than the higher critical value, sampling also is discontinued, and the population is treated with an insecticide. If the cumulative number is between the upper and lower critical values, another sample is taken. Sampling continues in this manner until a decision is made or ten samples are taken in the field. If a decision cannot be made after ten samples are taken, the scout stops sampling and returns to the field in 2 or 3 days to classify the population.

Sequential sampling plans have been developed for many insect pests and development continues for many others. Although intermediate population levels may still require considerable sampling, average savings of 50 percent can be expected of successful sequential sampling programs.

**Table 6.2 Alfalfa Weevil Pest Management Recommendations—Number of Larvae Collected from a Thirty-Stem Sample.**[a]

| | Alfalfa Height (inches) | | | | | | | | | | | | | | | | |
|---|---|---|---|---|---|---|---|---|---|---|---|---|---|---|---|---|---|
| **Total Degree Days (dd)** | **2** | **3** | **4** | **5** | **6** | **7** | **8** | **9** | **10** | **11** | **12** | **13** | **14** | **15** | **16** | **17** | **18 or more** |
| 290–310 | | | | | | | | | | | | | | | | | |
| Spray | | 25 | 37 | 52 | 67 | 75 | 83 | 94 | 105 | 105 | 105 | | | | | | |
| Resample in 50 dd | | 0–24 | 0–36 | 0–51 | 0–66 | 0–74 | 0–82 | 0–93 | 0–104 | 0–104 | 0–104 | | | | | | |
| 340–360 | | | | | | | | | | | | | | | | | |
| Spray | | | | | 82 | 82 | 82 | 82 | 82 | 82 | 82 | 82 | 82 | 82 | 82 | | |
| Resample in 50 dd | | | | | 14–81 | 14–81 | 14–81 | 14–81 | 14–81 | 14–81 | 17–81 | 17–81 | 17–81 | 17–81 | 17–81 | | |
| Resample in 100 dd | | | | | 0–13 | 0–13 | 0–13 | 0–13 | 0–13 | 0–13 | 0–16 | 0–16 | 0–16 | 0–16 | 0–16 | | |
| 390–510 | | | | | | | | | | | | | | | | | |
| Spray | | | | | | | | | | 52 | 52 | 58 | 64 | 68 | 72 | 76 | 80 |
| Resample in 50 dd | | | | | | | | | | 8–51 | 8–51 | 8–57 | 14–63 | 14–67 | 14–71 | 18–75 | 18–79 |
| Resample in 100 dd[a] | | | | | | | | | | 0–7 | 0–7 | 0–7 | 0–13 | 0–13 | 0–15 | 0–17 | 0–17 |
| 540 to harvest (see Table 6.3) | | | | | | | | | | | | | | | | | |
| 100 after harvest | | | | | | | | | | | | | | | | | |
| Spray | 23 | 33 | 43 | 48 | 53 | 58 | 63 | | | | | | | | | | |
| Resample in 50 dd | 17–22 | 17–32 | 17–42 | 20–47 | 23–52 | 23–57 | 23–62 | | | | | | | | | | |
| Resample in 100 dd[b] | 0–16 | 0–16 | 0–16 | 0–19 | 0–22 | 0–22 | 0–22 | | | | | | | | | | |
| 150 or more after harvest (see Table 6.3) | | | | | | | | | | | | | | | | | |

[a]If this field was sprayed more than 7 days ago, you can wait 200 degree days to resample.
[b]If last preharvest sample had less than 30 larvae, the weevil season is over and you can quit sampling.

SOURCE: After Iowa State University Cooperative Extension Service Publication IC-428, 1986.

**Table 6.3 Alfalfa Weevil Pest Management Recommendations.**

| | *Change in Number of Larvae Since Last Sample* | | |
|---|---|---|---|
| **Total Degree Days (dd)** | **Decreased 10 or More** | **Within 10** | **Increased 10 or More** |
| 540 to harvest | | | |
| Spray or harvest | 73 | 63 | 53 |
| Resample in 50 dd | 23–72 | 18–62 | 13–52 |
| Resample in 100 dd[a] | 0–22 | 0–17 | 0–12 |
| 150 or more after harvest | | | |
| Spray | 78 | 58 | 48 |
| Resample in 50 dd | 28–77 | 18–57 | 0–47 |
| Quit sampling | 0–27 | 0–17 | |

[a]If this field was sprayed more than 7 days ago, you can wait 200 degree days to resample.

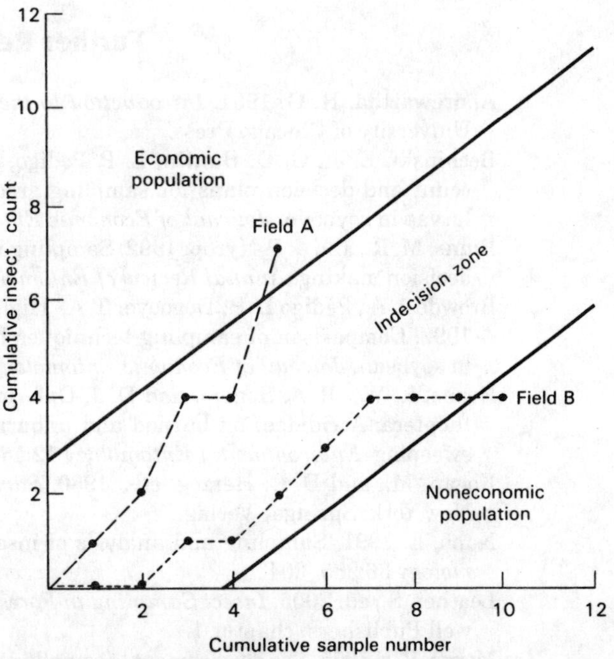


**Figure 6.35** Graph of a sequential sampling program for a hypothetical pest. In the graph, samples were taken as long as the cumulative numbers were within the indecision zone. As soon as a number fell outside this zone, sampling was discontinued, and appropriate action was recommended. In field A, the pest population was classified as "economic," in which case insecticide application might be recommended. In field B, the population was classified "noneconomic," and no action would be advised.

**Sequential decision plans for GCW larval management in soybeans**

| $n^b$ | Cumulative total no. of GCW larvae[a] | | | | | | | | |
|---|---|---|---|---|---|---|---|---|---|
| | Soybean stage V7–R1[c] | | | Soybean stage R2–R3[d] | | | Soybean stage R4–R5[e] | | |
| | < | | > | < | | > | < | | > |
| | Noneconomic population | Continue sampling | Treatment required | Noneconomic population | Continue sampling | Treatment required | Noneconomic population | Continue sampling | Treatment required |
| 1 | — | | — | — | | — | — | | — |
| 2 | — | | — | — | | — | — | | — |
| 3 | 37 | | 71 | 67 | | 113 | 52 | | 92 |
| 4 | 53 | | 91 | 94 | | 146 | 73 | | 119 |
| 5 | 68 | | 112 | 121 | | 179 | 94 | | 146 |
| 6 | 84 | | 132 | 148 | | 212 | 116 | | 172 |
| 7 | 100 | | 152 | 175 | | 245 | 138 | | 198 |
| 8 | 117 | | 171 | 203 | | 277 | 160 | | 224 |
| 9 | 133 | | 191 | 231 | | 309 | 182 | | 250 |
| 10 | 150 | | 210 | 259 | | 341 | 204 | | 276 |

[a]Cumulative number of first- to sixth-stage GCW larvae.
[b]$n$ = Number of 60-cm plant shake samples.
[c]V7–R1 = Late vegetative to early flower bloom.
[d]R2–R3 = Late flower bloom to early pod development.
[e]R4–R5 = Late pod development to early bean development.

**Figure 6.36** A table for use in sequential sampling of the green cloverworm (GCW), *Hypena scabra*.

## Further Reading


Andrewartha, H. G. 1961. *Introduction to the Study of Animal Populations.* Chicago: University of Chicago Press.

Bechinski, E. J., G. D. Buntin, L. P. Pedigo, and H. G. Thorvilson. 1983. Sequential count and decision plans for sampling green cloverworm (Lepidoptera: Noctuidae) larvae in soybean. *Journal of Economic Entomology* 76:806–812.

Binns, M. R., and J. P. Nyrop. 1992. Sampling insect populations for the purpose of IPM decision making. *Annual Review of Entomology* 37:427–453.

Browde J. A., Pedigo L. P., Degooyer T. A., Higley L. G., Wintersteen W. K. & Zeiss M. R. 1992. Comparison of sampling techniques for grasshoppers (Orthoptera: Acrididae) in soybean. *Journal of Economic Entomology* 85:2270–2274.

Evans, E. W., R. A. Rogers, and D. J. Opfermann. 1983. Sampling grasshoppers (Orthoptera: Acrididae) on burned and unburned tallgrass prairie: night trapping vs. sweeping. *Environmental Entomology* 12:1449–1454.

Kogan, M., and D. C. Herzog, eds. 1980. *Sampling Methods in Soybean Entomology.* New York: Springer-Verlag.

Kuno, E. 1991. Sampling and analysis of insect populations. *Annual Review of Entomology* 36:285–304.

Leather, S., ed. 2005. *Insect Sampling in Forest Ecosystems.* Williston, Vermont: Blackwell Publishers, chapter 1.

Morris, R. F. 1955. The development of sampling techniques for forest insect defoliators, with particular reference to the spruce budworm. *Canadian Journal of Zoology* 33:225–294.

Morris, R. F. 1960. Sampling insect populations. *Annual Review of Entomology* 5:243–264.

Pedigo, L. P., and G. D. Buntin, eds. 1994. *Handbook of Sampling Methods for Arthropods in Agriculture*. Boca Raton, Fla: CRC Press.

Ruesink, W. G., and M. Kogan. 1994. The quantitative basis of pest management: Sampling and measuring, pp. 355–391. In R. L. Metcalf and W. H. Luckmann, eds., *Introduction to Insect Pest Management,* 3rd ed. New York: Wiley.

Southwood, T. R. E. 1978. *Ecological Methods,* 2nd ed. New York: Chapman & Hall.


## Favorite Web Sites

http://www.ento.vt.edu/~sharov/PopEcol/lec2/sampling.html
Discusses the many aspects of insect sampling for population estimates and includes elements of geostatistics.

CHAPTER 7

# ECONOMIC DECISION LEVELS FOR PEST POPULATIONS

WITHOUT A DOUBT, economic decision levels are the keystone of insect pest management programs. Such levels are indispensable because they indicate the course of action to be taken in any given pest situation. D. L. Chant, in 1966, termed the study of economic decision levels *bioeconomics* and argued that sensible pesticide use is possible only with an understanding of the insect population level that causes economic damage. Indeed, without such knowledge we risk making absurd economic blunders, even spending more to suppress an insect than the value of the commodity the pest could destroy. Conversely, understanding and properly using economic decision levels in dealing with pests can increase producer profits and conserve environmental quality.

Economic decision levels usually are expressed as the number of insects per area, plant or animal unit, or sampling procedure. Less commonly, such levels are given as degree of plant damage or combinations of numbers and damage. The levels are unique in that they have both biological and economic attributes, and they are most often used for management decisions for private concerns—for example, on the farm.

## CONCEPTS OF ECONOMIC LEVELS

In 1959, V. M. Stern and colleagues formally proposed the concepts and terminology of bioeconomics that we use today. Specifically, they developed the ideas of economic damage, economic-injury level, and economic threshold, collectively called the **economic-injury level (EIL) concept.** Although this concept was originally proposed in 1959, some of the ideas expressed had been discussed years earlier (1934) by W. D. Pierce in a particularly farsighted article. Pierce raised questions that became one incentive for developing economic-injury levels (EIL) "Is all insect attack to be computed as assessable damage? If not, at what point does it become assessable? Is control work warranted when damage is below that point?"

Although fundamental to the concept, such questions may not have been the ultimate impetus for EIL development. The concept actually emerged as an encouragement for more rational use of insecticides. In discussing the EIL concept, Stern and colleagues emphasized the concerns of many persons regarding excessive and other inappropriate uses of insecticides. They highlighted problems of insecticide resistance, residues, and effects on nontarget organisms. These ideas were a critical part of the concept of **integrated control,** a new approach at the time, recommended as a replacement for the overly simplistic strategy of "identify and spray."

### Economic Damage and the Damage Boundary

**Economic damage** was originally defined as the amount of injury which will justify the cost of artificial control measures. To understand this term, we distinguish between injury and damage. **Injury** is the effect of pest activities on host physiology that is usually deleterious. **Damage** is a measurable loss of host utility, most often including yield quantity, quality, or aesthetics. Therefore, injury is centered on the pest and its activities, and damage is centered on the crop and its response to injury (Fig. 7.1).

As the concept applies to pest management, economic damage begins to occur when money required for suppressing insect injury is equal to the potential monetary loss from a pest population. The term **gain threshold** has been used to express this beginning point of economic damage. The gain threshold can be expressed as follows:

$$\text{Gain threshold} = \frac{\text{Management costs (\$/acre)}}{\text{Market value (\$/bushel)}} = \text{bushels/acre}$$

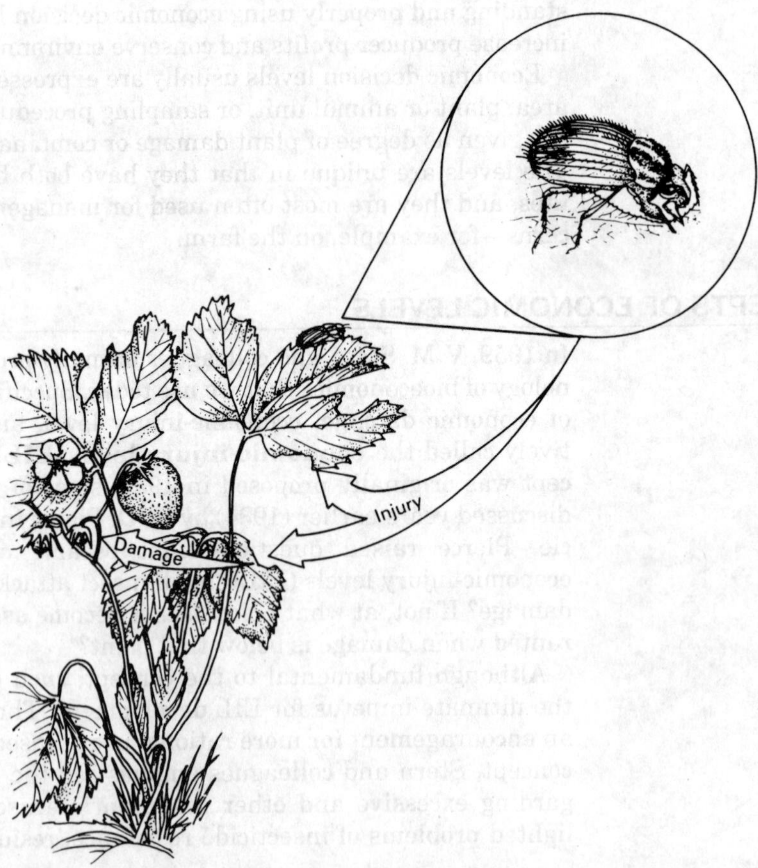


**Figure 7.1** Diagram relating insect leaf injury to damage of the harvestable produce.

Here, gain threshold is expressed as a unit of measure of the marketable product per a specified land area. For example, if management costs for application of an insecticide are \$10 per acre and harvested corn is marketed for \$2 per bushel, the gain threshold would be five bushels per acre:

$$\text{Gain threshold} = \frac{\$10/\text{acre}}{\$2/\text{bushel}} = 5 \text{ bushels/acre}$$

In other words, an insecticide application would need to save at least five bushels per acre for the activity to be profitable.

The gain threshold, therefore, is a very important measure because it is a worksheet standard: our margin for determining benefits of management and establishing decision indices.

Although not recognized by Stern and his colleagues, another important damage level to consider is the **damage boundary,** defined as the lowest level of injury where damage can be measured. The level is reached before economic damage occurs and is a necessary complement to the idea of economic damage. Together, these principles allow Pierce's original questions to be answered. Specifically, no injury level below the damage boundary merits control, and injury estimated to result in economic damage does warrant action. These concepts are illustrated in Fig. 7.2.

## Economic-Injury Level

The **economic-injury level** (EIL) is defined as the lowest number of insects that will cause economic damage, or the minimum number of insects that would reduce yield equal to the gain threshold. The relationship of the EIL to the damage boundary is shown in Fig. 7.3. Although expressed as numbers of

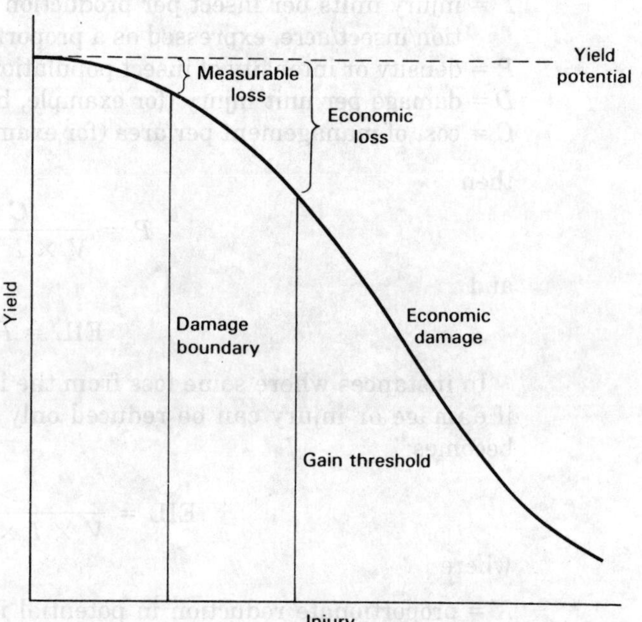


**Figure 7.2** Graph showing relationships between the damage boundary and the gain threshold.

insects per unit area, the EIL, as its name implies, is really a level of injury. Because injury is usually difficult to measure in a field situation, numbers of insects are used as an index of that injury. For example, it is usually easier to count insect numbers than it is to estimate the area of foliage removed by a pest population or the amount of juices (photosynthates) sucked from plants.

In some instances, particularly when several pest species causing similar injury are present, **insect equivalents** may be considered instead of insect numbers. An insect equivalent is the amount of injury that could be produced by one pest through its complete life cycle. By understanding feeding activity of the various species and comparing these according to a standard, EILs can be developed and decisions made for managing the whole complex. See Pedigo et al. (1986) for more details on insect equivalents.

To understand the development of an EIL using conventional insect numbers, consider the previous example of a 5-bushel per acre gain threshold for pest management in corn. If 1 insect per plant causes 1 bushel per acre loss, then the EIL for the pest is 5 insects per plant. In this example, 5 insects per plant potentially could consume enough plant tissue to reduce yields by 5 bushels per acre. Therefore, such an insect population is considered economic, and management activities are justified. Insect populations below this level and whose potential growth will not allow them to reach this level are considered subeconomic; no management is advised.

If management action (insect suppression) can be taken quickly and loss can be averted completely, then the EIL can be expressed as follows:

$$V \times I \times P \times D = C$$

where

$V$ = market value per unit of produce (for example, \$/bushel)
$I$ = injury units per insect per production unit (for example, percent defoliation/insect/acre, expressed as a proportion)
$P$ = density or intensity of insect population (for example, insects/acre)
$D$ = damage per unit injury (for example, bushels lost/acre/percent defoliation)
$C$ = cost of management per area (for example, \$/acre)

then

$$P = \frac{C}{V \times I \times D}$$

and

$$\text{EIL} = P$$

In instances where some loss from the insect is unavoidable, for example, if damage or injury can be reduced only 80 percent, then the relationship becomes:

$$\text{EIL} = \frac{>C}{V \times I \times D \times K}$$

where

$K$ = proportionate reduction in potential injury or damage (for example, 0.8 for 80 percent)

With some insect pests, particularly piercing-sucking insects, the separation of the $I$ and $D$ variables presents a problem. This is because the $I$ variable for

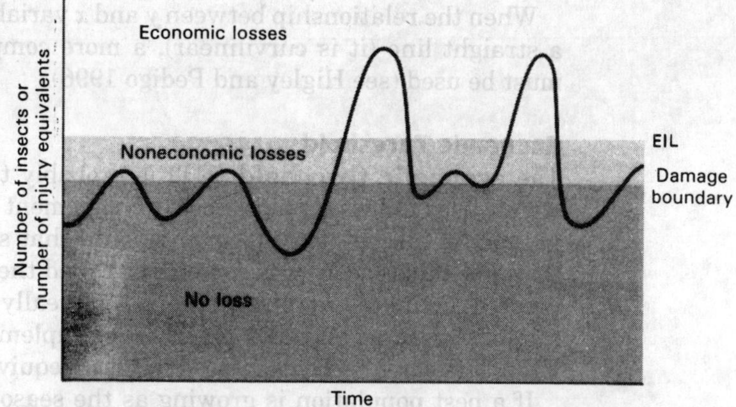


**Figure 7.3** Graph showing relationships of a hypothetical insect population with the damage boundary and the economic-injury level.

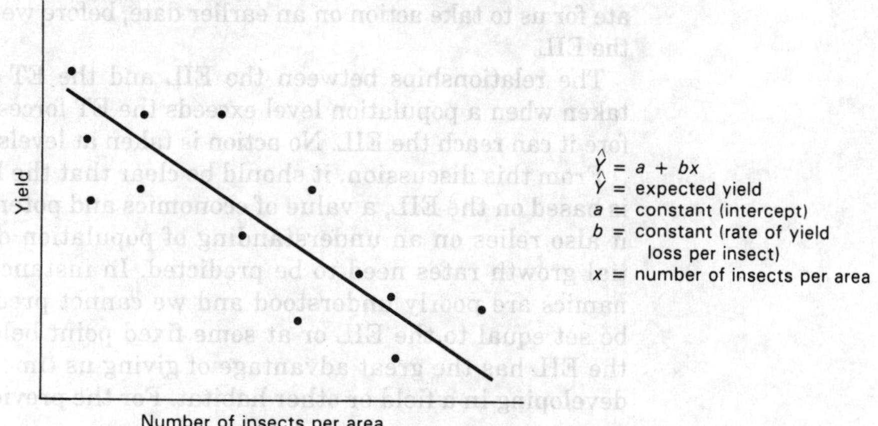


**Figure 7.4** A linear regression used in obtaining *b*, the loss per insect.

plants would represent photosynthate (sap) removed per insect and the *D* variable would represent yield loss per unit of photosynthate removed. Because these variables would be difficult to measure, a coefficient *b* representing loss per insect is substituted. In other words, $b = I \times D$ in the EIL equation. These *b* coefficients are obtained from statistical regression analyses of data by using experimental populations and measuring yield losses. The *b* coefficient can be obtained from the following expression (see Fig. 7.4):

$$Y = a + bx$$

where

$Y$ = yield/area
$a$ = a constant (the $y$ intercept)
$b$ = yield loss/insect
$x$ = number of insects/area

and therefore

$$\text{EIL} = \frac{C}{V \times b} \quad \text{or} \quad \text{EIL} = \frac{C}{V \times b \times K}$$

When the relationship between $y$ and $x$ variables cannot be approximated by a straight line (it is curvilinear), a more complex form of the EIL equation must be used (see Higley and Pedigo 1996).

### Economic Threshold

The **economic threshold (ET)** is probably the best-known term and most widely used index in making pest management decisions. The ET indicates the number of insects (density or intensity) that should trigger management action. For this reason, it is sometimes called the *action threshold.* Although expressed in insect numbers, the ET is really a time parameter, with pest numbers used as an index for when to implement management. Just as with EILs, ETs can also be expressed in insect equivalents.

If a pest population is growing as the season progresses, growth rates are predicted, and the ET is set below the EIL (Fig. 7.5). By setting the ET at a lower value, we are predicting that once the population reaches the ET, chances are good that it will grow to exceed the EIL. Therefore, it is appropriate for us to take action on an earlier date, before we accrue losses in reaching the EIL.

The relationships between the EIL and the ET demonstrate that action taken when a population level exceeds the ET forces down the population before it can reach the EIL. No action is taken at levels below the ET.

From this discussion, it should be clear that the ET is a complex value. It is based on the EIL, a value of economics and potential for injury. However, it also relies on an understanding of population dynamics because potential growth rates need to be predicted. In instances where population dynamics are poorly understood and we cannot predict growth, the ET may be set equal to the EIL or at some fixed point below it. An ET lower than the EIL has the great advantage of giving us time to respond to a problem developing in a field or other habitat. For the previous example with an EIL

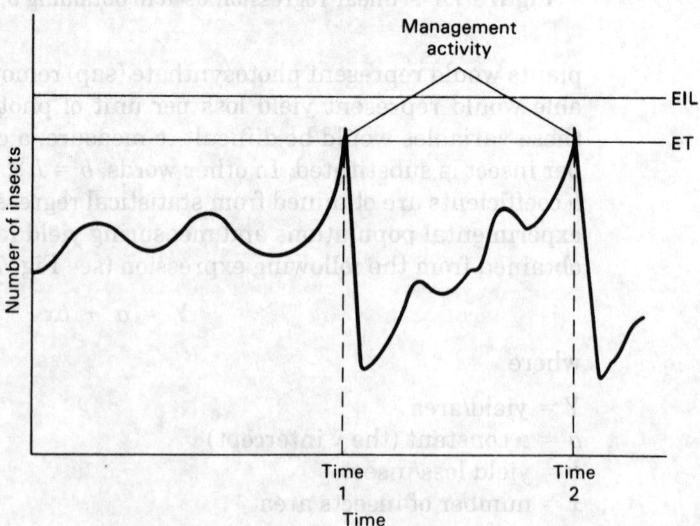


**Figure 7.5** Graph showing the relationship between the economic threshold (ET) and the economic-injury level (EIL) in taking action against an insect population.

of 5 insects per plant, we might set the ET at 4 insects per plant to obtain added reaction time and avoid some of the early losses. This type of an ET has been referred to as a **fixed ET** because the percentage of the EIL is fixed. This does not mean that the fixed ET is unchanging; indeed, it shifts with changes in the value of the EIL.

If the growth rate of the pest population is known from past studies, a more objective ET, or **descriptive ET,** can be calculated. In contrast to the fixed ET, the descriptive ET takes into account possible changes in pest population growth rate. Such an ET was developed by F. DuToit for Russian wheat aphids, *Diuraphis noxia,* on winter wheat as:

$$\text{ET} = \text{EIL} \times C^{-x}$$

where

$C$ = factor of increase per unit time (for example, 0.5/week, increasing 50 percent per week)

$x$ = time period in question expressed as weeks (for example, 4 weeks)

Although this type of ET is more objective than the fixed ET, pest population growth rates are seldom predictable. Therefore, the fixed ET is more often used.

## CALCULATION OF ECONOMIC DECISION LEVELS

In developing economic indices for pest management decisions, the principal level to estimate is the EIL because the EIL includes the basic damage potential of a given insect population. It can be used as the ET, or an ET can be determined from knowledge of the EIL and population dynamics. In either situation, the EIL must be known first.

The calculation of the EIL for an insect is a continuing process because new values are required with changes in the input variables. Consequently, when market value, management costs, and plant susceptibility change, recalculation is necessary. Additionally, several EILs are usually required for any one season because of crop development and consequent changes in susceptibility.

The basic steps required to calculate the EIL are as follows:

1. Estimate the loss per insect.
2. Determine the gain threshold as described previously.
3. Determine the loss that can be avoided by applying the management tactic (this is a proportion, for example, 0.8 for 80 percent mortality or 1.00 if all loss is avoided).
4. Calculate the EIL as:

$$\text{EIL} = \frac{\text{Gain threshold}}{\text{Loss per insect} \times \text{Amount of loss avoided}}$$

Of these steps, the first is by far the most difficult. Crude estimates of losses are usually obtained from field observation and experimentation with various-sized insect populations on a crop at specific times. Subsequently, yields are measured, and losses caused by the insects are determined.

An example of the steps required in EIL calculation can be shown with the potato leafhopper, *Empoasca fabae,* on soybeans. This insect is very common in the crop and may cause losses when another preferred crop, alfalfa, is clipped. Lacking suitable habitat and food after the alfalfa harvest, leafhoppers migrate into soybeans, where they suck plant juices and can cause a leaf necrosis (hopperburn), especially in young plants.

To estimate the damage per insect, small soybean plots were caged and infested with various leafhopper numbers at three plant growth stages. Yields were measured at season's end, and statistical procedures (regression analyses) determined the loss per insect at each plant growth stage. These values, based on the loss caused by one insect per plant, were as follows: early vegetative stage = 1.55 bushels/acre, beginning flowering stage = 0.17 bushel/acre, and beginning pod fill = 0.08 bushel/acre. Loss per insect decreased throughout the season because of plant growth and greater leaf area, which decreased insect intensity.

To calculate the gain threshold, we estimate the cost of applying 1 pound (actual ingredient) of malathion insecticide by aircraft at \$9.50 per acre. Further, we predict that we can sell the harvested grain for \$4.15 per bushel and estimate that our insecticide application heads off all loss. First, we calculate the gain threshold as:

$$\text{Gain threshold} = \frac{\$9.50\text{/acre}}{\$4.15\text{/bushel}} = 2.29 \text{ bushels/acre}$$

Next, EILs can be calculated for each stage as follows:

$$\text{Seedling stage: EIL} = \frac{2.29 \text{ bushels/acre}}{1.55 \text{ bushels/acre/insect} \times 1.00}$$
$$= 1.48 \text{ (about 2) leafhoppers/plant}$$

$$\text{Beginning flowering stage: EIL} = \frac{2.29 \text{ bushels/acre}}{0.17 \text{ bushel/acre/insect} \times 1.00}$$
$$= 13.47 \text{ (about 14) leafhoppers/plant}$$

$$\text{Beginning pod–fill stage: EIL} = \frac{2.29 \text{ bushels/acre}}{0.08 \text{ bushel/acre/insect} \times 1.00}$$
$$= 28.63 \text{ (about 29) leafhoppers/plant}$$

Finally, we can use these EILs as our "action levels" (set the ET equal to the EIL), or we may choose to set ETs at levels conservatively below the EILs, say at 75 percent of the EILs.

## DYNAMICS OF ECONOMIC-INJURY LEVELS

Economic levels are very dynamic. They vary with changes in costs, values, and production environments. An insect pest feeding on a crop at a given time can be expected to have a different EIL when feeding on the crop at another time during the same season or in another season (Fig. 7.6). Variations in the levels may be insignificant at times or vary severalfold at other times. The major forces behind change in economic decision levels, as shown in the previous equations, are: (1) crop value, (2) management costs, (3) degree of injury per insect, and (4) crop susceptibility to injury.

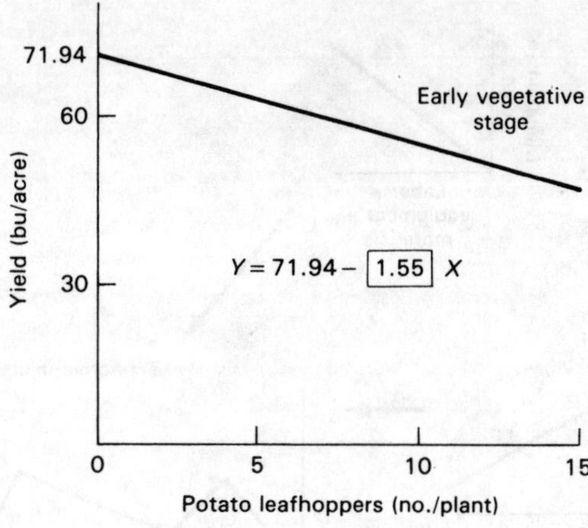


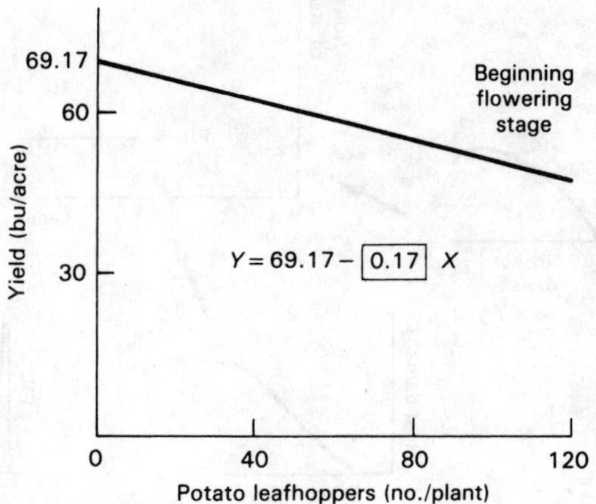


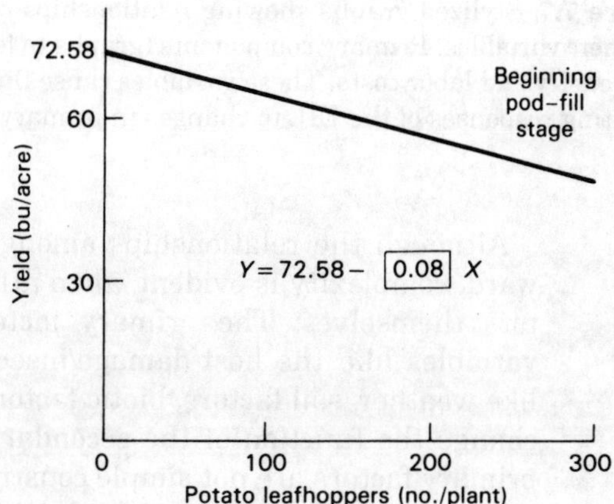


**Figure 7.6** Statistical regression analysis of soybean yield on number of potato leafhoppers, *Empoasca fabae,* present per plant at three plant stages.

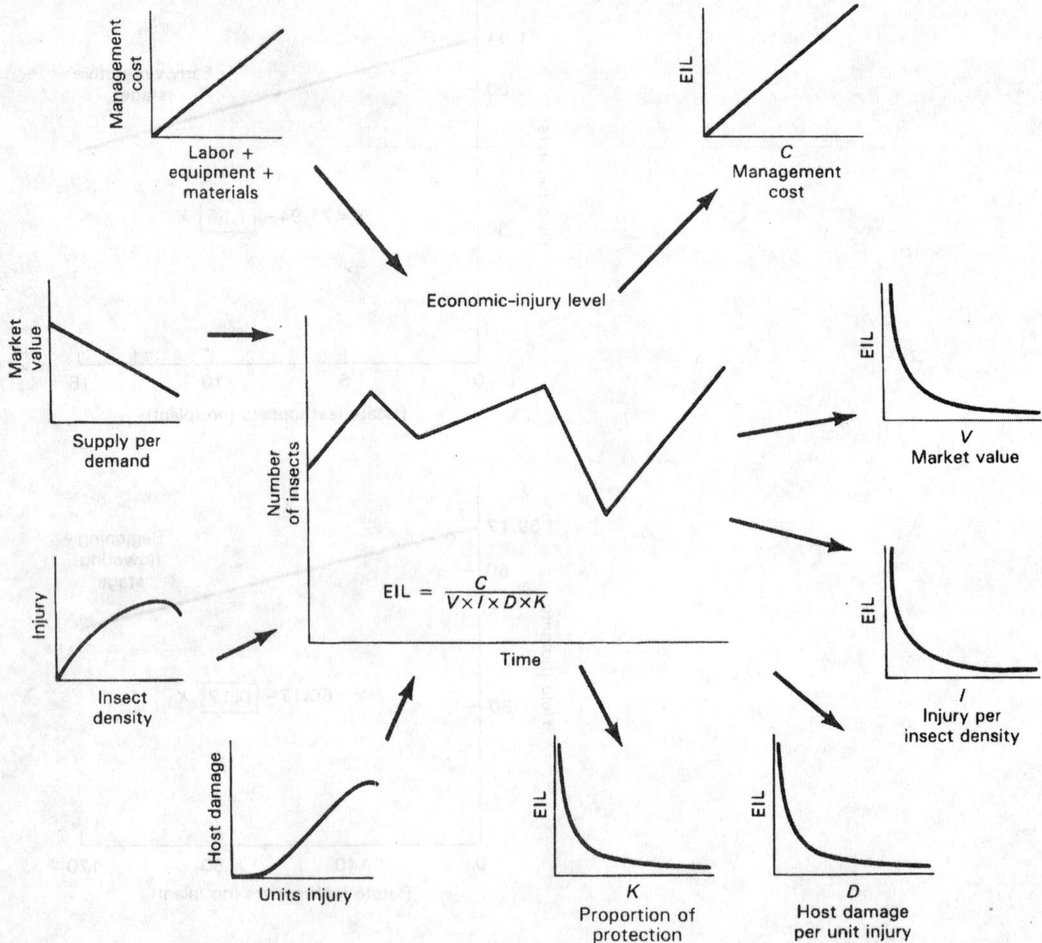


**Figure 7.7** Stylized graphs showing relationships of the economic-injury level (EIL) components and their variables. Primary components (graphs to left) are affected by secondary variables like insect density and labor costs. These variables cause fluctuations in the EIL with time (center graph). Resulting responses of the EIL to changes in primary components are shown in graphs at right.

Although the relationships among these factors are relatively straightforward, complexity is evident when attempting to predict variability in the factors themselves. The primary factors are affected by complex secondary variables like the host-damage/insect-injury relationship. Tertiary variables like weather, soil factors, biotic factors, and the human social environment all change the function of the secondary variables (Fig. 7.7). Consequently, the primary factors are not simple constraints; rather, they are complex processes that operate through time. In dealing with these components, entomologists have tended simply to account for the economic aspects and conduct research on the biological components.

### Market Value (*V*)

Of the primary factors, crop value (*V*) is one of the most variable, and it alone accounts for much of the change in EILs. Crop market values are notorious for fluctuations and unpredictability. For instance, grain prices may fluctuate 70 to 80 percent in a single year, with a profound influence on EILs of insect pests. The relationship between EIL and market value is inverse; as market value increases, EIL decreases, and vice versa.

As a general rule, estimates for EIL calculation are based on current or past records of crop value. But these values should be forecasted for the anticipated date of sale to reflect the expected increases or decreases. If, for example, the value of a crop is forecasted to increase from one year to the next, then the EIL for the forecast year should be reduced appropriately.

The quality of a commodity may be of overriding importance in determining market values in many instances. In situations where several specific grades (and prices) exist, the value of the desired grade should be used in the EIL calculations. Similarly, when quality is associated with appearance (cosmetics), the level of control should reflect the desired appearance. A target price, based on the expected quality of production, should be used whenever feasible.

When no clear system of marketing exists for a product, some estimate about the utility of the output to the producer must be made. Forage and pasture crops, for example, often have no established markets but are necessary inputs in animal production. Because most forage and pasture crops are produced for on-farm use, their value depends on the relative contribution that they make to the growth and production of the animal. One way to estimate this value is to determine the substitution price for other, more marketable feeds. To be valid, however, these must be equal in nutritional quality to the on-farm feeds.

As with each of the other primary factors of the EIL, variability in market values cannot be controlled by the producer. In many instances, however, an effective marketing plan can be used to help estimate this factor.

### Management Costs (*C*)

The cost of suppressing a pest population must be estimated before profitability of an action can be assessed. As management cost (*C*) increases, net benefit of control decreases. Consequently, EILs must be raised to accommodate the higher gain thresholds.

Most years, management costs tend to be more stable than crop market values. These costs include labor (scouting and application of insecticides), materials (insecticides), and equipment (hydraulic sprayer and sweep-nets). Management costs usually change gradually, depending on inflation rates, and therefore are usually predictable.

Because the expenses associated with equipment and labor may equal or exceed the cost of the control material, care should be taken to formulate the best estimate for these application expenses. If, for example, airplane application at $8.50 per acre must be substituted for the usual ground application at $5.50 per acre, the additional cost should be reflected with a higher EIL. The variability in cost associated with a particular management measure probably is not significant within a season but could be between seasons. Therefore, most of the uncertainty with this variable in the EIL is caused by differences in costs between management options rather than between seasons. For this reason, management costs for calculating EILs are probably the simplest of the primary factors to estimate.

### Degree of Injury per Insect (*I*)

Injury is a dual-sided phenomenon, governed both by insect and host populations. The insect aspect concerns a particular act or behavior of individuals that causes impairment of a host's ability to survive, grow, and reproduce. The host, as the recipient of the behavior, plays a major role in determining the kind and degree of injury.

Most insect injury is caused by insect feeding on host tissues or fluids, although other major causes include injecting toxins and vectoring pathogenic microorganisms. Chewing and sucking are the most common feeding behaviors, producing injuries to plants such as leaf skeletonizing, leaf mining, stem boring, and fruit scarring.

Some of the most detailed studies of insect injury have been conducted with plants. Plant pests can be placed in at least six different categories: stand reducers, leaf-mass consumers, assimilate sappers, turgor reducers, fruit feeders, and architecture modifiers. These categories include pests that kill plants outright or impair physiological processes.

**Stand reducers.** Insects that reduce stand (for example, cutworms (Box 7.1), Lepidoptera: Noctuidae) produce an immediate loss in plant biomass and decreased photosynthesis of the crop (Fig. 7.8). Depending on the severity of plant loss, however, compensation in the stand occurs when uninjured plants adjacent to dead plants grow and yield at increased rates. For instance, under some conditions soybean stands reduced to four or five plants per foot of row may yield as well as stands with ten plants per foot of row. The total effect of stand reduction is governed by number, timing, and dispersion of plants lost.

**Leaf-mass consumers.** Leaf consumption by insects is believed to directly affect absolute photosynthesis of the remaining plant canopy. Effect of the injury on plant physiology can be accounted for by measuring leaf mass consumed per unit of land area, timing of leaf consumption, and vertical distribution or location of the defoliation.

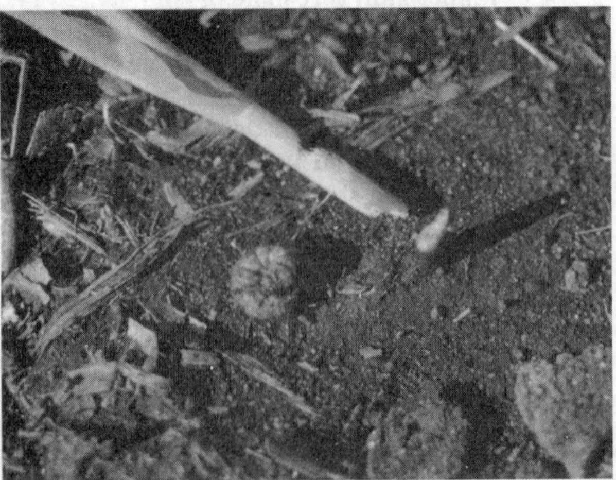

**Figure 7.8** Black cutworm, *Agrotis ipsilon*, injury to field corn. Such insects can reduce plant stand. (Photo by Marlin E. Rice)

## BOX 7.1 Black Cutworm (See Color Plate 3)

**SPECIES:** *Agrotis ipsilon* (Hufnagel) (Lepidoptera: Noctuidae)

**DISTRIBUTION:** The black cutworm is widely distributed throughout the world, including the United States and southern Canada. Northern regions of the United States must be recolonized each year by migrant moth populations from southern regions where overwintering is possible. Outbreaks within infested regions often are associated with high moisture conditions. Hence, river bottoms or low areas in fields are often very susceptible.

**IMPORTANCE:** The species is chiefly a pest of seedling plants, causing damage to many species. Most notable plants attacked are corn, cotton, tobacco, and various vegetables. In most instances, damage is seen as a "cutting" of young seedlings, often causing plant death. This type of damage may be very destructive because several plants can be cut by a single larva. In fact, relatively small populations of cutworms are capable of destroying entire stands of cotton or corn. Even when seedlings are large enough to avoid cutting, foliar feeding reduces plant vigor and yield. One practical limitation to controlling black cutworm injury is that larvae live under the soil surface and feed at night. This makes assessment difficult, especially because much damage can be sustained in a short period.

**APPEARANCE:** Adult moths have a wingspan of about 38 to 50 mm. Forewings are long, narrow, and usually darker than hind wings. In addition, the forewings are marked with a characteristic black "dagger." Larvae vary from light gray to black. Full-grown larvae are 38 to 50 mm long. Eggs are ribbed and about 0.45 mm high.

**LIFE CYCLE:** Pupae overwinter in the southern United States and Mexico. In spring, adults emerge, and females oviposit on plant foliage or on the soil surface in the proximity of hosts. In northern latitudes adults fly north and are carried by southerly winds. Females oviposit in the soil near host plants. Larvae hatch in 3 to 6 days and move into soil where they remain during the day. At night, larvae move to the surface and feed on young plants. They develop through six stadia (occasionally seven). Depending on temperature, larvae begin to pupate in 25 to 35 days. Pupation occurs in the soil and lasts about 12 to 15 days. Several generations per year are possible, but the spring generation (usually the first), which places eggs near the young plants, is the most damaging.

---

**Assimilate sappers.** These comprise mostly piercing-sucking and rasping insects that remove plant carbohydrates and nutrients after carbon is taken up and before the plant can convert it to tissue. As mentioned earlier, measuring assimilate removal per pest is a problem, as is the effect of injection of toxic substances during the feeding process, such as the effects of the tarnished plant bug, *Lygus lineolaris,* on peaches (Fig. 7.9).

**Turgor reducers.** These pests, represented by soil insects and stem feeders, influence plant water and nutrient balances at root and stem sites. Insects like corn rootworms, *Diabrotica* species, prune corn roots, thus reducing rooting depth and density. Others, like the threecornered alfalfa hopper, *Spissistilus festinus,* girdle soybean stems, thus destroying conductive tissues (Fig. 7.10). Severe instances of reduced water uptake result in decreased plant turgor, followed by reduced development of leaves, stems, and fruits and reduced photosynthesis.

**Fruit feeders.** When insects injure fruit (Fig. 7.11), they cause direct damage to the harvestable produce. Such injury can affect quality (appearance and/or

**Figure 7.9** Soybean stems girdled by threecornered alfalfa hoppers, *Spissistilus festinus* (see Color Plate 5). The girdling destroys conductive tissues, causing the plants to break and lodge near the soil surface. (Photo by Marlin E. Rice)

**Figure 7.10** Tarnished plant bug, *Lygus lineolaris*, injury (cat-facing) to peaches. These pests are assimilate sappers and inject toxins when they feed. (Courtesy Iowa State University Extension Service)

**Figure 7.11** European corn borer, *Ostrinia nubilalis*, damage to bell pepper. This pest is an example of a fruit feeder. (Photo by Marlin E. Rice)

makeup), yield, or both, depending on utility of the produce. Injury to fruit may seem simple and straightforward to measure, but quantification is complicated by the fact that yield losses usually are not proportional to percentage loss of reproductive sites. In particular, plant compensation needs to be understood to quantify relationships of losses to this type of injury.

**Figure 7.12** Southern corn billbug, *Sphenophorus callosus,* and injury to corn. This is an example of plant architecture modification. (Photos by Marlin E. Rice)

**Architecture modifiers.** Injury from these types of insects change plant morphology to reduce yield. For example, injury caused by the threecornered alfalfa hopper, in addition to destroying conductive tissues, also causes plants to lodge (Fig. 7.9). Lodged plants may die outright or continue to live and grow in a "gooseneck" fashion. This change can reduce physiological yield of the plant as well as harvestable yield. Other examples of architecture modifiers include the seedcorn maggot, *Delia platura,* which consumes the plumules of soybean seedlings and causes "Y" plants, and the tunneling of the stalk borer, *Papaipema nebris,* in young corn plants, destroying the growing tip and causing tillering and low yielding or barren plants. In these instances, pests not only destroy a quantity of tissue but also reduce the quality. Such injury results in potentially drastic changes in pattern and, usually, in rate of plant growth (Fig. 7.12). Because of complexity added by morphological changes in plant structure, this type of injury is difficult to understand and quantify on an injury-per-insect basis.

**Injury measurements in EIL calculations.** With the majority of insect pests, direct quantification of insect feeding has not been attempted. Most studies have emphasized yield losses, comparing these with observed insect numbers. With data from such studies along with regression analysis, estimates of loss per insect can be made. Before the production system can be modeled, however, quantification of tissue destroyed per insect must be known. Only by quantifying tissue destroyed and understanding physiological responses of costs to this destruction can we begin to explain yield losses caused by a diversity of pests.

To date, the most detailed studies to quantify tissue consumption by insects have been done for plant defoliators. The simplest approach has been to visually estimate percent defoliation of the plant. Then, by estimating the leaf area present and number of insects (of a single species), consumption per insect can be calculated. The major problem with this approach is the error involved in estimating defoliation. Such errors can be great, with a tendency toward overestimation. More accurate studies have been conducted by presenting premeasured leaves to insects and remeasuring leaf area after defoliation. Recent developments of efficient photoplanimeters have made leaf area measurement practical for defoliation studies in the laboratory and the field.

For the purpose of calculation, injury per insect usually has been assumed to have a straight-line relationship with insect numbers. In other words, injury is expected to double with twice as many insects, triple with three times the insects, and so on. However, crowding that occurs at high densities has been shown to reduce injury per insect with some insects because of interference between individuals and/or relative shortage of food. Although this phenomenon is a concern, for many species the density/injury relationship is straight line at intensities up to and including those in the vicinity of the EIL. Therefore, unless there are other indications to the contrary, insects should be considered in an additive manner in estimating injury for EILs. Such an approach yields conservative estimates in decision making.

### Crop Susceptibility to Injury (*D*)

The relationship between injury and crop yield or utility is the most fundamental factor of the EIL. This relationship provides the biological foundation on which economic and practical constraints can be superimposed.

Virtually all theoretical and practical attention to the injury/crop response has been limited to plants. Indeed, our frequent inability to describe these responses with veterinary and medical pests is one reason EILs are usually unavailable or inappropriate for such pests. Therefore, discussion will be limited to injury and plant responses.

In his book, P. G. Fenemore (1982) outlined four major factors involved in the injury/plant-response relationship. They include: (1) time of injury with respect to plant growth, (2) type of injury, (3) intensity of injury, and (4) environmental influences on the plant's ability to withstand injury.

**Time of injury.** The time in a plant's growth cycle when injury occurs has an obvious influence on response to that injury. Usually, seedlings are the most susceptible; older plants are better able to tolerate or compensate for injury. Similarly, while yield-producing organs are forming, plants again become very susceptible, but when they mature, injury usually has much less influence (unless fruit is injured directly). The timing of pest injury is most often accommodated in EILs by recognizing a separate yield loss-to-injury

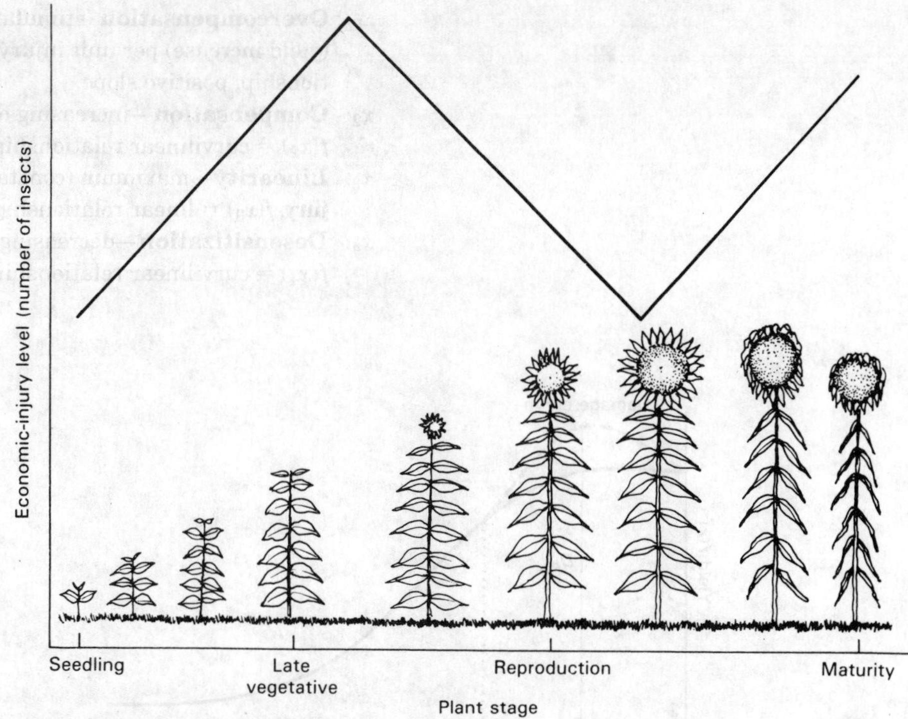


**Figure 7.13** Graph showing typical changes in the economic-injury level with changes in plant stage. Many plants are particularly vulnerable during reproductive stages.

relationship (and therefore, separate EIL) for each different stage of crop development (Fig. 7.13).

**Plant part injured.** The part of the plant injured also influences the plant's response to injury. Usually, a distinction is made between injury to yield-forming organs (**direct injury**), and injury to nonyield-forming organs (**indirect injury**). In general, EILs are calculated for only one injury type because with only a few exceptions, insects do not produce both direct and indirect injury simultaneously.

**Injury types.** The common types of injury were discussed earlier. It is important to note that the nature of the injury is fundamental to how a plant responds to different intensities of a particular form.

**Intensity of injury.** The relationship between the intensity or amount of injury and plant yield is the most important factor in the injury/crop-response relationship. In seeking to explain this relationship, it is of value to consider a generalized response curve of insect injury and plant yield (curve A in Fig. 7.14). Not all plants undergo an injury response that includes every portion of the curve, but all potential responses can be described by some part of it.

Specific areas of this curve can be distinguished according to their response rate to injury. In Figure 7.14, each area of the generalized response curve is described by a different $x$. These areas and their descriptions are as follows:

$x_1$: **Tolerance**—no damage per unit injury, yield with injury equals yield without injury, $f(x_1)$ = a constant (zero) slope ($f$ = function of).

$x_{2a}$: **Overcompensation** (stimulation)—negative damage (yield increase) per unit injury, $f(x_{2a})$ = curvilinear relationship, positive slope.

$x_2$: **Compensation**—increasing damage per unit injury, $f(x_2)$ = curvilinear relationship, negative slope.

$x_3$: **Linearity**—maximum (constant) damage per unit injury, $f(x_3)$ = linear relationship, negative slope.

$x_4$: **Desensitization**—decreasing damage per unit injury, $f(x_4)$ = curvilinear relationship, negative slope.

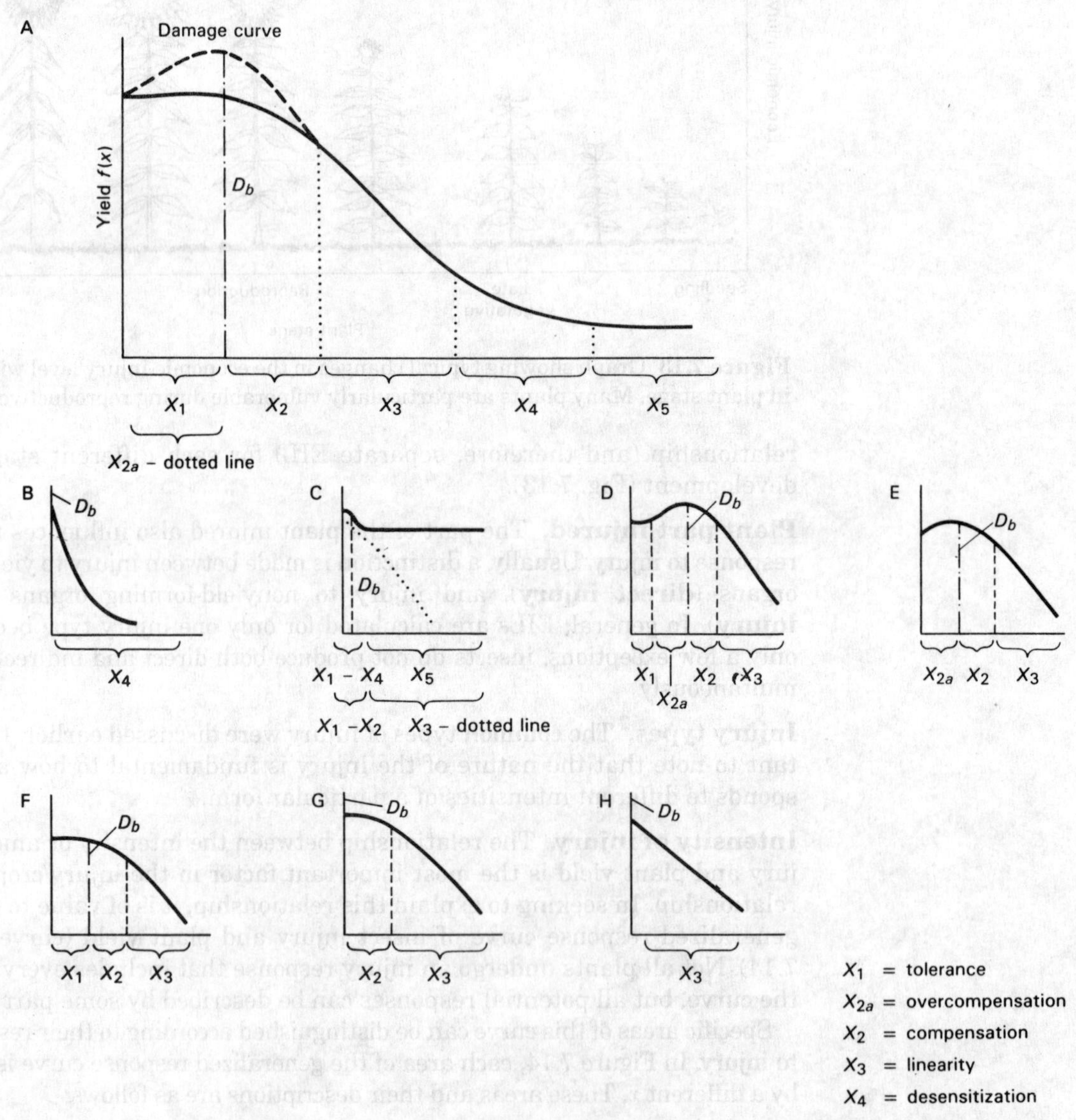


**Figure 7.14** General (A) and specific (B–H) forms of the damage curve of plants. (See text for details.)

$x_5$: **Inherent Impunity**—no damage per unit injury, yield with injury less than yield with no injury, $f(x_5)$ = a constant (zero) slope.

These responses apply equally to individual plants and to plant stands; however, the responses displayed by a plant and its plant stand at any one time may differ. Usually, individual plants display less of the early portion of the damage curve, but plant stands have a greater ability for tolerance, compensation, and overcompensation. Because EILs usually are developed for plant stands and not single plants, the stand response to injury is of primary importance in calculation.

The last two responses, desensitization and inherent impunity, are not common. When they do occur, the injury levels that cause them are much higher than those found with EILs and are unimportant in calculations. However, they could be important in quantifying total crop losses. An example of desensitization (curve B in Fig. 7.14) occurs with aphids (Hemiptera: Aphididae) when they vector plant pathogens. Inherent impunity (curve C in Fig. 7.14) is exemplified by the citrus rust mite, *Phyllocoptruta oleivora,* on citrus. This pest may produce only a slight yield reduction (fresh weight), followed by little further reduction regardless of mite density. This response is shown by the solid line in the figure. But if yield is redefined to include appearance, the damage curve is so radically altered that inherent impunity may not appear at all (dotted line in the figure). Thus, the precise definition of yield, better termed *utility,* is important in determining the shape of the damage curve.

Tolerance and overcompensation probably occur more frequently than desensitization or inherent impunity, but these responses may be very small and masked by environmental effects. These occur mostly at low injury levels relative to economic injury and are undetected. Probably all plants display some degree of tolerance to indirect injury. Overcompensation seems not as widespread but, when present, produces a response curve as shown in curves D and E in Figure 7.14.

More commonly, responses such as those shown in curves F, G, or H in Figure 7.14, which include combinations of tolerance, compensation, and linearity, are obtained for calculating EILs. Most plants respond to some level of indirect injury with compensation. Compensation implies that the plant or plant stand can prevent the injury from having its maximum influence on yield. In contrast, linearity represents a direct relationship between injury and yield loss. Overcompensation and compensation are primarily limited to indirect injury. An entirely linear response, on the other hand, is expected with direct injury but may also be found with indirect injury.

**Environmental effects.** Factors in the environment often play an important role in determining how plants respond to injury. Within a given season, environmental factors may influence how long a plant remains susceptible to a specific type of injury. Similarly, between seasons, a plant's response to the same level of injury may be drastically changed. For example, as much as a twofold difference in soybean yield-loss per green cloverworm, *Hypena scabra,* larva has been found between growing seasons with different weather conditions (Fig. 7.15). Such extreme variability emphasizes the need to calculate EILs over a range of environmental conditions. A better approach is to develop EILs for specific sets of weather and environmental factors.

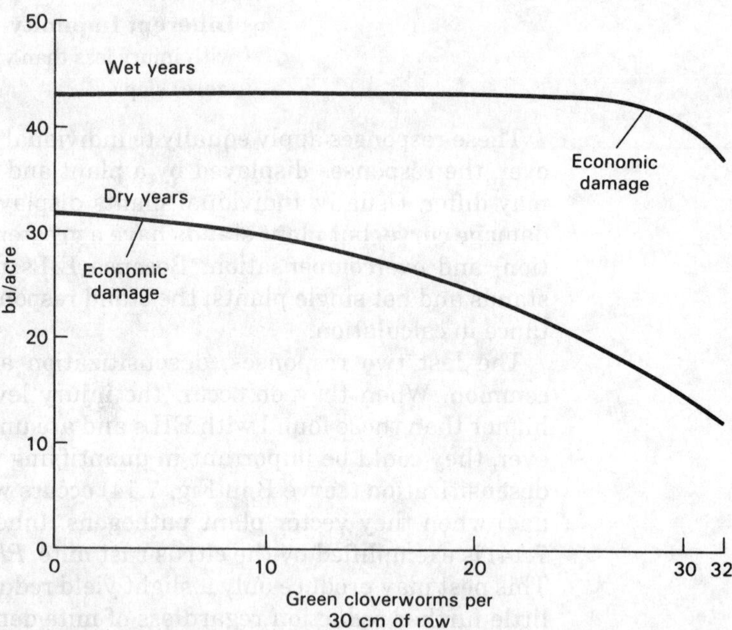


**Figure 7.15** Stylized soybean-yield response curves to defoliation of the green cloverworm, *Hypena scabra,* in years of normal rainfall (wet) and those with below normal rainfall (dry). Note the differences in number of insects required to cause economic damage between the two types of years.

## Amount of Damage Avoided (*K*)

The variable *K* from the EIL equation previously has been defined as the proportional reduction in pest attack. Although not formally stated as such, most practitioners simply equate *K* with the expected proportion of the population killed by a management tactic, usually a pesticide. Such a narrow view of *K* belies its true meaning and makes it seem not applicable to protective tactics that do not kill, including barriers, repellents, and antifeedants.

A better definition of *K*, which expresses its true meaning, expands *K* and, indeed, the EIL concept, to consider an array of tactics, rather than merely pesticides. Therefore, *K* should be considered the proportion of total damage averted by the timely application of a management tactic. In addition to the examples already given, this definition would apply to and allow a more accurate quantification of *K* for such insect biological controls as *Bacillus thuringiensis,* which may protect plants by causing cessation of feeding long before pests die.

The attainment of *K* at a level of 1.00, avoiding all damage, is usually the goal of most agriculturalists. Lower values usually are not a matter of choice; rather, they reflect less-than-desired effectiveness of the tactic employed, achieving only 60 percent mortality, for example, when 95 percent or more is desired. Because most producers are risk averse, pest recommendations are based on avoiding damage by keeping injury well below the damage boundary; in short, action is taken before significant damage is done, and the pests remaining after the activity cause no detectable loss. Strategies that do not achieve this goal are often considered ineffective and, unless technology is lacking, usually are not adopted.

### Experimental Techniques to Determine Plant Damage Response

Techniques to determine the type and magnitude of plant response to insect injury vary greatly. The most common approaches can be grouped into four categories: (1) observation of natural insect populations, (2) modification of natural populations, (3) establishment of artificial populations, and (4) injury simulation.

**Observation.** Observation of natural insect populations is the simplest, but perhaps least precise, approach. The method usually involves making intensity estimates on plants from several fields and correlating these with resulting yield losses in those fields. This approach has been widely used in all types of crops and most often with perennial or severe pests (pests that are present and cause losses most years).

**Modification of natural populations.** This approach is also used with perennial and severe pests; however, in this instance experiments are designed to produce populations of desired sizes in designated plots. Usually, selective insecticides or selective insecticidal rates are used to create different population densities on plants for analysis (Fig. 7.16). This procedure is more precise than simple observation and usually allows insect-crop loss relationships to be determined more effectively. However, the effect of the insecticide on plant physiology must be known and accounted for with this approach.

If insect numbers are low, the problem in experimentation is to increase numbers rather than reduce them to desired levels. In this instance, increases to obtain various insect levels in plots may be attempted by using baits, attractant crops, or attractant isolines (Fig. 7.17) of the same crop. **Isolines** are cultivars that differ in only a single gene and could be expected to yield similarly to the cultivar in question.

**Creating artificial populations.** This approach can be used when precise control of numbers is sought or numbers in the natural population are low.

**Figure 7.16** Applying insecticides with a small-plot sprayer. Insecticides may be used to create different-sized insect populations in establishing damage response curves. (Photo by L. Higley)

**Figure 7.17** Stunting of a glabrous isoline of "Clark" soybeans in response to potato leafhopper (Box 7.2), *Empoasca fabae,* injury (*left* and *right*), as compared with the pubescent "Clark" variety (*center*). Such isolines can sometimes aid in establishing damage response curves. (Photo by Marlin E. Rice)

**Figure 7.18** Field cages in alfalfa plots. Insects are introduced into the cages at various levels to establish damage response curves. (Photo by S. Hutchins)

Most often, this approach involves rearing or collecting the insect, then artificially infesting small plots. The usual procedure is to place feeding larvae or other stage in plots and cover them with cages (Fig. 7.18). In some instances, as with the soybean looper, *Pseudoplusia includens,* the adult stage may be released in cages to produce damaging larval populations. Desired damage levels from the artificial populations then can be achieved by properly timed insecticide applications.

## BOX 7.2 POTATO LEAFHOPPER (SEE COLOR PLATE 5)

**SPECIES:** *Empoasca fabae* (Harris) (Hemiptera: Cicadellidae)

**DISTRIBUTION:** The potato leafhopper occurs throughout the eastern half of the United States and southern Canada in summer. To inhabit this range, however, individuals must migrate annually from an overwintering region in the southern United States. They accomplish this by moving vertically into winds conducive for northern migration. This adaptive movement terminates when precipitation causes "dropout" of individuals.

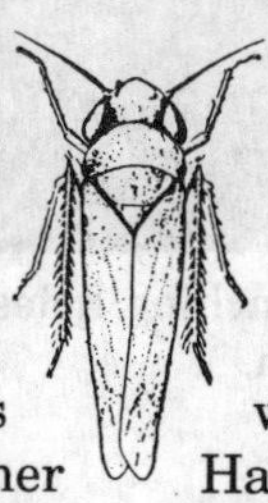

**IMPORTANCE:** This species is one of the most important alfalfa pests in the eastern and northcentral United States. In addition, this pest is capable of feeding on numerous other host species, including soybean and potato. Feeding with piercing-sucking mouthparts results in phloem blockage within the plant and produces an inverted V-shaped yellowing of the leaf (chlorosis). This symptom is referred to as "hopperburn." Injury to plants results in reduced photosynthesis, which translates into shorter, less productive plants. In the case of alfalfa, leaf proteins also are severely reduced. Before hopperburn was associated with potato leafhopper feeding, the chlorosis was believed to be the symptom of a leaf disease.

**APPEARANCE:** Adults are green, wedge-shaped, and about 3 mm long, with a row of six round white spots along the pronotum. Nymphs, which develop by gradual metamorphosis, are smaller than adults and are wingless. Eggs are about 1 mm long.

**LIFE CYCLE:** Upon arrival and mating in the northern United States and Canada, females oviposit 60 to 100 eggs over a 30-day period within stems and larger veins of host leaves. Hatching occurs in 6 to 9 days, and nymphs emerge to develop through five stadia. Soon after the adults appear, mating occurs and oviposition soon follows. Approximately 21 days are required for development from egg to adult in warmer summer temperatures. There are several overlapping generations per year, with the exact number depending on time of arrival and temperatures.

**Injury simulation.** Simulating injury by an insect is another method used to assess losses by insects. Also known as *surrogate injury,* this approach seeks to mimic insect feeding behavior. It has been applied most frequently with defoliators, where workers have picked leaves or punched holes (Fig. 7.19) in them in a real time and rate sequence. The primary advantage of this method is that degree of injury can be precisely controlled and the biology of the crop response followed intensively. However, much preliminary information regarding fidelity of the surrogate injury to actual insect feeding must be gathered before the technique is employed.

When we consider the diverse circumstances involved with insect injury, it is clear that no single method of estimating crop response is appropriate in all instances. The choice of an approach depends on the insect species and crop investigated and the resources available for the investigation.

## ENVIRONMENTAL EILS

In recent years, the use of EILs has had an important influence on environmental quality, particularly where these levels were the basis for insecticide use. Regular use of conventional EILs can result in reduced pesticide residues by decreasing the frequency of application. Indeed, it has been estimated that

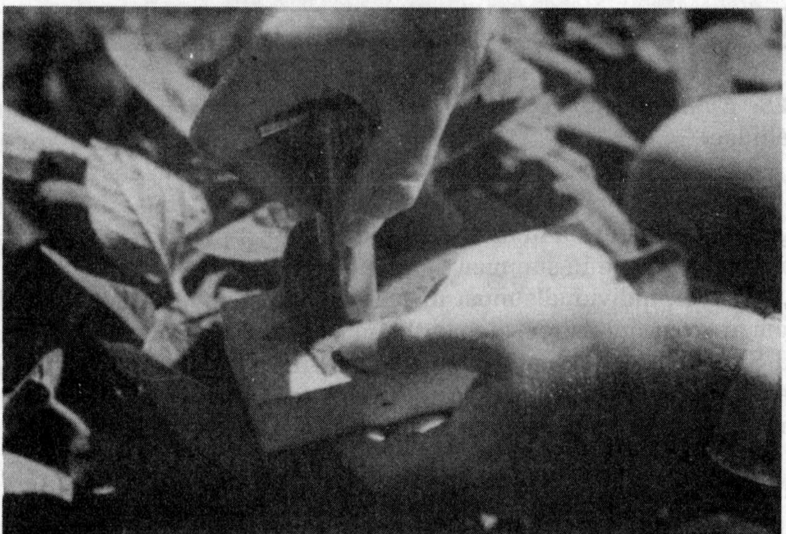

**Figure 7.19** Punching holes in leaves with a cork borer to simulate insect defoliation.

pest monitoring, establishment of proper decision levels, and reduced dosage can lower pesticide use by as much as 30 to 50 percent. Therefore, the expanded use of EILs can be considered crucial in conserving environmental quality. However, aside from development and increased use of the conventional EIL, the idea that a special EIL could be crafted to directly address environmental concerns has been put forth. This special type of EIL has been termed an *environmental EIL*.

An environmental EIL evaluates management procedures not only on their direct costs and benefits to users but also on their environmental effects. Moreover, the EIL equation integrates many management elements, each of which may have a role in making pest management more environmentally compatible. To develop an environmental EIL, each of the EIL variables should be considered for possible manipulation.

**Assigning Realistic Management Costs (*C*)**

Although *C* accounts for the direct costs of management actions, neither *C* nor other components of the EIL address the indirect environmental costs associated with environmental risks from management activities. Entomologists L. Higley and W. Wintersteen (1992) suggested using an economic technique called contingent valuation, which attempts to assign monetary values to nonmarket goods such as environmental quality. To assign realistic costs to the use of pesticides, they analyzed levels of risk for thirty-two field-crop insecticides to environmental elements (surface water, ground water, aquatic organisms, birds, mammals, beneficial insects) and to human health (toxicity levels). Furthermore, they used survey information on farmer opinion regarding the importance of the risks and how much they would be willing to pay in higher pesticide costs or yield losses to avoid different levels of risk. Finally, this information was used to calculate environmental costs for each pesticide.

Using such costing information, environmental EILs can be calculated as follows:

$$\text{EIL} = \frac{C + EC}{V \times I \times D \times K}$$

where *EC* is environmental cost and the remaining variables are as previously defined in the conventional EIL equation.

As can be seen, adding environmental costs to the equation raises the value of the EIL. With a higher EIL (more pests required to cause an economic loss), pesticide application would be less frequent, thereby lowering the amount of pesticide residues and conserving environmental quality to a degree.

A practical example of how *C* can be adjusted to develop environmental EILs is shown in Table 7.1. In the table, conventional EILs are calculated for different insecticides as well as environmental EILs. Here, the environmental EILs contain not only direct pesticide costs but also environmental costs that reflect the level of environmental risk of a single use. By including environmental costs in *C*, note that the EIL is increased. In this example, EILs for the bean leaf beetle, *Cerotoma trifurcata,* more than double for many insecticides. Therefore, greater yield losses from the bean leaf beetle would be tolerated when using EILs with environmental costs as compared to conventional EILs. In using the environmental EILs, the soybean producer would sacrifice some yield to reduce environmental risk.

### Reducing Damage per Pest (*D*)

Recall that the *D* variable is the amount of damage resulting from a given level of insect injury. If plant cultivars or animal breeds can be found that are less sensitive to insect injury than previously grown, the level of *D* is reduced and the EIL value is increased, resulting in fewer pesticide applications. Therefore, use of tolerance and/or compensatory characteristics can contribute to environmental EIL development. An added advantage in using these characteristics is the avoidance of insect resistance to the management tactic. Insect resistance is discussed in more detail in Chapter 17.

### Developing an Environmentally Responsible *K* Value (*K*)

Earlier, we noted that the goal of most pest managers is to achieve a *K* of 1.00—in other words, to avoid all damage. Although this may be a worthy goal from a profit perspective, attempts to achieve it may foster overuse of pesticides and result in pest overkill. Pest overkill occurs most often when the goal of killing insects replaces the more appropriate goal of avoiding significant damage.

Pest overkill can occur even when producers follow labeled rates of pesticides. Label rates form the basis for most recommendations in insect pest management and usually are established for an array of environmental conditions and application methods. Unfortunately, because of product liability fears and other factors, even the minimum rate recommended may be too high. The goal in developing an environmental EIL is to determine the lowest pesticide rate to achieve a *K* value that is virtually equal to 1.00. Achieving this value may mean reducing rates to attain 60 to 70 percent mortality, while producing a crop yield not significantly different from that with 100 percent pest mortality. In addition to reducing pesticide use, rate reductions could make pesticide applications more compatible with natural and biological controls.

**Table 7.1 Various Pesticides, *C*, and EILs Previously Applied to Second-Generation Bean Leaf Beetle on Soybeans.**

| Pesticide | Trade Name and Formulation | Minimum Application Rate[a] | *C, $/ha* Pesticide[a] | Environmental[b] | Total | *EIL*[c] Conventional | Environmental | % Increase |
|---|---|---|---|---|---|---|---|---|
| Acephate | Orthene 75S | 0.56 | 26.51 | 16.26 | 42.77 | 8.7 | 14.0 | 61 |
| Carbaryl | Sevin XLR+ | 0.56 | 19.12 | 20.78 | 39.90 | 6.3 | 13.1 | 108 |
| Chlorpyrifos | Lorsban 4E | 0.56 | 23.54 | 25.15 | 48.69 | 7.8 | 16.0 | 105 |
| Dimethoate | Cygon 400 | 0.56 | 21.54 | 25.23 | 46.77 | 7.0 | 15.4 | 120 |
| Methomyl | Lannate 1.8L | 0.25 | 24.61 | 27.53 | 52.14 | 8.1 | 17.2 | 112 |
| Permethrin | Ambush 2EC | 0.06 | 18.55 | 20.39 | 38.94 | 6.0 | 12.8 | 113 |
| Permethrin | Pounce 3.2EC | 0.06 | 17.98 | 20.39 | 38.37 | 6.0 | 12.5 | 108 |
| Thiocarb | Larvin 3.2F | 0.42 | 25.27 | 19.82 | 45.09 | 8.4 | 14.8 | 76 |

[a]Insecticide application rate is lowest rate listed on label in kg (AI)/ha, and pesticide costs include an application rate of $12.35/ha.
[b]Environmental costs from Higley and Wintersteen (1992).
[c]Standard EILs for $0.23/kg ($6.00/bu) soybeans in beetles per 30 cm of row.

### Manipulating Other EIL Variables

The EIL variables least amenable to purposeful change are *V* and *I*. In a free-market system, crop market values vary with supply and demand and thus cannot be easily manipulated. If market values were controlled, higher values would result in more pesticide use rather than less, and few producers would tolerate lower market values simply to attain higher EILs for less pesticide use. One suggested manipulation to improve environmental quality is to set a higher market value for pesticide-free produce; however, it remains to be seen if buyers would be willing to pay a sufficiently high premium for pesticide-free goods to justify the potentially large economic losses caused by unmanaged pests.

Likewise, injury per pest would be difficult to manipulate because doing so would likely require changes in pest-population genetics, which may be environmentally undesirable. Changing injury per pest is an approach considered for some medical pests, such as by identifying genotypes in vector populations that cannot transmit pathogens. However, because entire pest populations must be changed, such approaches are rare and unlikely to have much broad application.

## USING ECONOMIC LEVELS

As mentioned at the beginning of this chapter, economic decision levels are an important, if not indispensable, part of most insect pest management programs. Of the levels discussed, the ET is the practical operational level recommended to producers for making management decisions. J. D. Mumford and G. A. Norton (1984) describe the ET as a usable, if not ideal, decision rule. It has been adopted in many situations.

To date, the most widespread use of the ET has been where **curative** (therapeutic) management tactics (mainly insecticides) are used. Here, insect populations are sampled on a regular basis, and when needed, suppressive action is taken. Usually less value is placed on the ET when the pest always causes economic damage, and a preventive management strategy is used.

### Implementation Categories

The place and ultimate use of ETs in pest management programs become clearer when their state of development is categorized. Accordingly, most of the decision rules currently used can be placed in one of four categories: (1) no thresholds, (2) nominal thresholds, (3) simple thresholds, and (4) comprehensive thresholds (Fig. 7.20).

**No thresholds.** Decisions in this category were common in applied entomology prior to the late 1950s, when "identify and spray" was the common practice. Although using thresholds is considered an advance for most insect problems today, there are certain situations when they are not appropriate:

1. Pest sampling cannot be done economically.
2. Practical response to cure a problem cannot be implemented in a timely way.
3. Once detected, the problem cannot be cured.
4. ET is immeasurably low (some quality losses, disease transmission, rapid growth potential).
5. Populations are intense, with a general level of density always above the EIL.

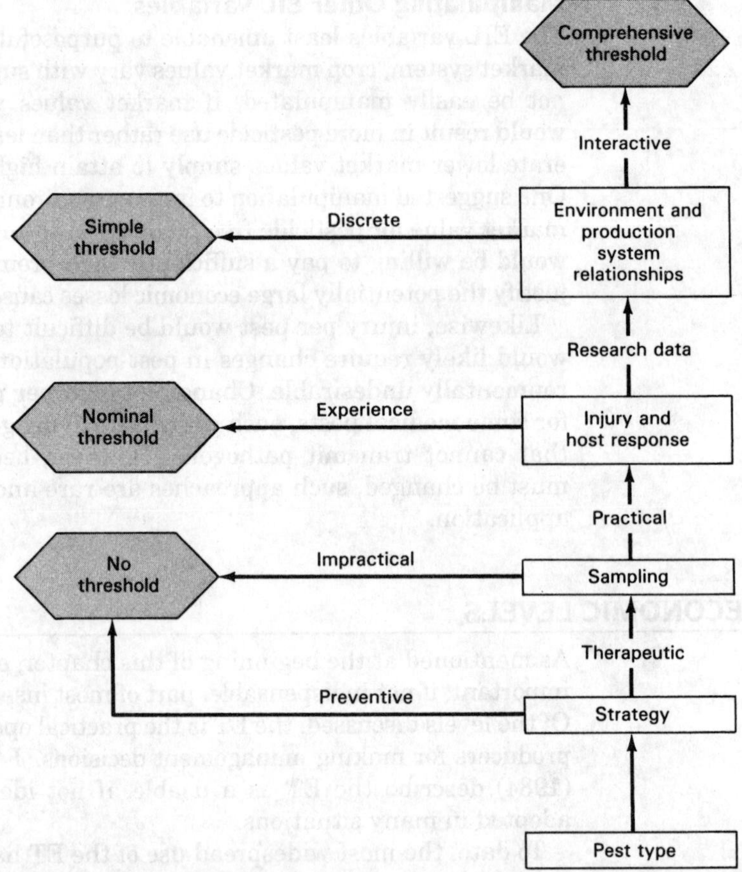


**Figure 7.20** Diagram of the implementation categories of economic thresholds.

Management implementation in these situations usually must depend on prevention rather than cure. When insecticides are used for prevention, scheduled treatments according to calendar date or cultural operation are the rule. More advanced programs may predict treatment dates using degree-day models and computer-based delivery systems.

**Nominal thresholds.** This category represents decision rules that are declared on the basis of a manager's experience. Historically, these were the first thresholds used and are still the most frequent type implemented. It is common for nominal thresholds to be developed by experienced entomologists of a state's Cooperative Extension Service and to be published in insecticide recommendation bulletins. Such levels, which may be used over broad regions, tend to be static, unchanging with changes in the variables mentioned previously. Although sometimes criticized for not being based on rigorous research, nominal thresholds are an advance over using none at all because such decision rules tend to be conservative (that is, they err on the side of taking action when it is not needed), and their use has resulted in less frequent insecticide applications.

**Simple thresholds.** This is the type of threshold described at the beginning of the chapter. These levels are based on calculated EILs, which are developed from average responses of hosts to injury caused by an insect. The five major inputs of market value, management costs, tissue destroyed (or damage done) per insect, yield (or quality) reduction per tissue destroyed, and amount of damage avoided are used to make calculations. Such levels may be implemented by using a simple formula to make calculations, consulting tables that consider the input variables, or using software packages with programmable pocket calculators or personal computers. Although these thresholds may be our best current practice, they usually fail to consider possible interactions of several pests and changes in the cropping environment that influence decisions.

**Comprehensive thresholds.** Decision levels in this category are still under development. Only recently have entomological studies been completed that address topics like effects of insect and weed interactions on plant stress and that include weather factors in calculating EILs. At the core of establishing truly comprehensive thresholds for crops is an intimate understanding of the host plant and its reaction to combined stressors, both biotic and physical. Such an understanding should include quantification of such processes as dry-matter partitioning throughout the growing season and not only a simple measurement of yield, as has been done in the past. Only in this way can the effects of a pest community be understood and thresholds developed in the context of the entire production system. Opportunities for implementing such thresholds probably will lie with advances in computer-based information delivery systems and the acquisition of on-farm computers.

### Limitations of the EIL Concept

Although the economic levels discussed have been used on a practical basis for more than 35 years, they are not without drawbacks. Their limitations relate to the types of pests or injury, management tactics selected, research requirements, and desirability of multiple inputs in making decisions.

Decision levels for management of some types of pests cannot be determined with EILs. Many vectors, medical pests, veterinary pests, and pathogens often do not have a quantitative relationship (or only a weak one at best) between damage and injury; consequently, they are not amenable to calculated EILs. Moreover, because we cannot put a "market value" on human health and life, it is virtually impossible to put an economic limit on the control of most medical pests.

Economic considerations also limit our ability to use EILs for aesthetic pests. It is very difficult to place a monetary value on the reduction in "aesthetic value" associated with a given type of pest injury. Usually any assigned values are subjective, which greatly limits their usefulness in calculating EILs.

A similar problem exists with forest pests. Pests of a few woody species like fruit trees and Christmas trees may be amenable to EILs; however, forest pests are not easily described by them. Almost all the components of EILs are difficult to estimate for forest insects; accurate market values are a problem because projections often must be made many years in advance. Management costs may vary greatly and frequently must include more environmental and social costs than in other pest management programs. In addition, the injury/crop-response relationship may be difficult to determine because the growth of the crop spans many years.

Some pests have a quantitative relationship to yield, but they are still difficult to manage with EILs. Pathogenic microorganisms on plants, although not entomological, are a good example of this quandary. Here, the yield reduction produced by many pathogens is usually quantitatively related to pathogen number. Unfortunately, sampling and quantifying the number or amount of these pathogens tends to be impractical. Thus, the question of whether pests can be economically sampled is important in determining practicality of EILs. Furthermore, management tactics of these pathogens are more often preventive, not therapeutic; therefore, determining whether a pathogen population is at the EIL after infection may not be of significant value if the only management options available must be applied before infection.

In fact, this last example highlights an important limitation of EILs. As we have noted, the EIL concept originally was developed with the objective of reducing insecticide use. Consequently, both the EIL and the ET can be used most appropriately when a single, curative management action can be made. They also are useful with some of the new tactics for killing insects like microbial insecticides and insect growth regulators; however, these new tactics are used similarly to conventional insecticides. Except from an assessment standpoint, EILs are of limited use with preventive measures, such as host-plant resistance and most forms of ecological management. These tactics are implemented as a matter of course, according to historical problems with pests.

Finally, the significant amount of research required and relative unsuitability when several diverse pests (those causing different kinds of injury) occur may present serious obstacles in attempting to employ the EIL concept. However, if injuries from different pests produce the same host response and can be placed on a common basis, or if effects of different injuries are additive, the concept may find application for pest complexes.

## CONCLUSIONS

The EIL concept has been one of the most important innovations in the development of insect pest technology and is the only truly unifying principle of pest management. Although not without drawbacks, EIL decision rules are the most widely accepted and practical tools for their purpose. However, these decision rules are yet to be established for many species and need refinement in others.

Refinement of the EIL concept for agriculture is possible in the form of environmental EILs and comprehensive thresholds, which account for multiple pests as well as other environmental stresses. Future development of environmental EILs will depend on our capability to quantify and include environmental costs in calculations, develop tolerant plant and animal varieties, and discover environmentally responsible management tactics. Future development of comprehensive thresholds needs to be founded on the total production system of a given agricultural enterprise. In large part, advancements here will depend on our ability to understand how diverse pests stress plants and animals and on quantification and detection of these stresses.

## Further Reading

Andow, D. A., and K. Kiritani. 1983. The economic injury level and the control threshold. *Japan Pesticide Information* 43:3–9.

Bardner, R. R., and K. E. Fletcher. 1974. Insect infestations and their effects on the growth and yield of field crops: A review. *Bulletin on Entomological Research* 64:141–160.

Boote, K. J. 1981. Concepts for modeling crop response to pest damage. *ASA Paper* 81–4007. St. Joseph, Mo.: American Society of Agricultural Engineering.

Capinera, J. L., G. W. Wheatley, D. C. Thompson, and J. Jenkins. 1983. Computer assisted crop loss assessment: A micro-computer model for estimation of sugarbeet insect effects to facilitate decision-making in pest management. *Protection Ecology* 5:319–326.

Chiang, H. C. 1979. A general model of the economic threshold level of pest populations. *United Nations F.A.O. Plant Protection Bulletin* 27:71–73.

Dutoit, F. 1986. Economic thresholds for *Diuraphis noxia* (Hemiptera: Aphididae) on winter wheat in the eastern orange free state. *Phytophylactica* 18:107–109.

Fenemore, P. G. 1982. *Plant Pests and Their Control.* Wellington, New Zealand: Butterworths, pp. 125–144.

Higley, L. G., and L. P. Pedigo. 1996. *Economic Thresholds for Integrated Pest Management.* Lincoln, Neb.: University of Nebraska Press, chapter 2.

Higley, L. G., and W. K. Wintersteen. 1992. A new approach to environmental risk assessment of pesticides as a basis for incorporating environmental costs into economic injury levels. *American Entomologist* 38:34–39.

Mumford, J. D., and G. A. Norton. 1984. Economics of decision making in pest management. *Annual Review of Entomology* 29:157–174.

Norton, G. A., and J. D. Mumford, eds. 1993. *Decision Tools for Pest Management.* Wallingford, U.K.: CAB International, pp. 56–57.

Ostlie, K. R., and L. P. Pedigo. 1985. Soybean response to simulated green cloverworm (Lepidoptera: Noctuidae) defoliation: Progress toward determining comprehensive economic injury levels. *Journal of Economic Entomology* 78:437–444.

Pedigo, L. P., and L. G. Higley. 1992. The economic-injury level concept and environmental quality: A new perspective. *American Entomologist* 38:12–21.

Pedigo, L. P., S. H. Hutchins, and L. G. Higley. 1986. Economic injury levels in theory and practice. *Annual Review of Entomology* 31:341–368.

Pierce, W. D. 1934. At what point does insect attack become damage? *Entomological News* 45:1–4.

Poston, F. L., L. P. Pedigo, and S. M. Welch. 1983. Economic injury levels: Reality and practicality. *Bulletin of the Entomological Society of America* 29:49–53.

Stern, V. M., R. F. Smith, R. Van Den Bosch, and K. S. Hagen. 1959. The integrated control concept. *Hilgardia* 29:81–101.

## Favorite Web Sites

http://extension.usu.edu/files/gardpubs/ipm03.pdf
Site of the Utah State Extension Service. Discusses the economic-injury level concept. Graphics and calculation methods are presented.

http://txipmnet.tamu.edu/index.html
Site of the Texas IPM Program. Gives recommendations and decision guidelines for various crops.

# CHAPTER 8

# PEST MANAGEMENT THEORY

FURNISHED WITH THE KNOWLEDGE of basic insect biology, ecology, and decision making, we appropriately turn our attention to the overall concept of managing insect populations. An understanding of this topic requires a look at the basic nature of pest control.

**Pest control** is the application of technology, in the context of biological knowledge, to achieve a satisfactory reduction of pest numbers or effects. As such, it has been described as a two-strand approach. The technological aspect includes such tools as insecticides and the equipment used to apply them. Biological knowledge allows us to know where, when, and how to apply the technology.

A simple example of pest control can be made with a common household pest, the German cockroach (Box 8.1), *Blattella germanica*. Biological knowledge necessary for pest control includes these facts: (1) the species is tropical or subtropical in origin and survives only indoors in temperate locations; (2) it prefers warm, dark, humid environments; (3) it feeds widely on all sorts of plant and animal products, particularly spilled food and cooking debris in the kitchen; and (4) eggs of the species require 3 or 4 weeks for incubation. The technology selected for pest control could appropriately be the application of premixed insecticide sold in a container, complete with a hand-pump sprayer. Satisfactory pest control in this instance usually means eliminating the pest from the household. According to biological knowledge, sprays would be applied to dark corners, cracks and crevices, cabinet shelves, underside of sinks, and other hiding places in the kitchen and bathrooms. Better yet, we could place roach traps in various locations, then treat only those places where cockroaches are found. Because the spray kills only nymphs and adults, it would need to be applied two or three times at 2-week intervals to kill new nymphs as the eggs hatch. In addition, food spills and debris should be thoroughly removed to limit the pest's food supply and help prevent further introduction and establishment.

Admittedly, this is a very elementary example, but it illustrates the dual nature of pest control. If either biological knowledge or technology is deficient, the control program probably will not succeed.

## BOX 8.1 GERMAN COCKROACH (SEE COLOR PLATE 4)

**SPECIES:** *Blattella germanica* (Linnaeus) (Blattodea: Blattellidae)

**DISTRIBUTION:** Although native to Africa, the German cockroach is found throughout the temperate areas of the world, usually in human dwellings.

**IMPORTANCE:** The German cockroach is one of the most common, prolific household cockroaches. Both adults and nymphs are a major annoyance to humans because individuals eat almost anything, excrete on food, and leave a characteristic bad odor. The species is active only at night and prefers warm areas of buildings. The pest also is suspected of being able to vector disease organisms.

**APPEARANCE:** Adults, about 12 mm long when fully grown, are yellowish brown with two parallel stripes on the pronotum. The antennae are long and the wings cover the entire abdomen. Nymphs are similar to adults, except they are smaller and lack wings. Oothecae (egg capsules) are light brown and contain 25 to 30 eggs.

**LIFE CYCLE:** Females carry oothecae for about 21 days, until just before hatching occurs. Nymphs develop for about 3 months before finally becoming adults. Each female can produce up to five oothecae. There are typically two to three generations per year.

## HISTORICAL HIGHLIGHTS OF PEST TECHNOLOGY

### Pre-Insecticide Era

Agricultural pest problems and actions to alleviate them are nearly as old as the beginnings of crop cultivation. The earliest record of pest technology seems to be the use of sulfur (an insecticide still used) by ancient Sumerians about 2500 B.C. In prebiblical times, both the Egyptians and Chinese used insecticides formulated from herbs and oils, particularly for protection of seeds and stored grain.

A very important advancement, occurring around 300 B.C., was the recognition of phenology as a science. This led to the principle of timely planting of crops to avoid pest losses. During this time, the Chinese discovered the use of natural enemies to control insects; for example, they placed ants on citrus to reduce pest infestations (Fig. 8.1).

Few developments in pest control were recorded during the Middle Ages (from the end of the Roman Empire until about 1000 A.D.). However, by 1101 the Chinese had discovered the use of soap to control pests, and by the late 1600s tobacco infusions and insecticides from other herbs, as well as arsenic, were commonly used.

With the expansion of agriculture in the 1700s, Réaumur published on the importance of temperature summation in determining insect phenology. This idea was to be used almost two centuries later to predict insect events and determine timing of pest control activities. Also in the 1700s, the concept of plant resistance to insects was advanced for the Hessian fly, *Mayetiola destructor*, in the United States, and botanical insecticides (derived from plants) were rediscovered.

During the mid-1800s, imperial expansion, accompanied by developing technology, resulted in countries sending citizens and materials far beyond their own borders. Along with this movement, insect species were introduced into new regions. Examples of notable introductions include San Jose scale, *Quadraspidiotus perniciosus*, introduced into the United States from China, and the Colorado potato beetle, *Leptinotarsa decemlineata*, introduced into Europe from the United States. In many countries, significant losses from introduced pests gave impetus to developing inspection and quarantine procedures that would be instrumental in slowing the spread of insect pests around the world.

The late 1800s and early 1900s saw the rapid development of insecticide application equipment. A French company, Vermorel, claims to have invented the first commercial sprayer in 1880, and the first power sprayer was developed in Germany in 1904. These commercial sprayers were used in Europe primarily for pesticide applications on fruit and vine crops. In the United States, the use of insecticides increased rapidly on cotton during this period to combat mounting problems with the boll weevil, *Anthonomus grandis*; dusts were the primary insecticide formulation applied (Fig. 8.2).

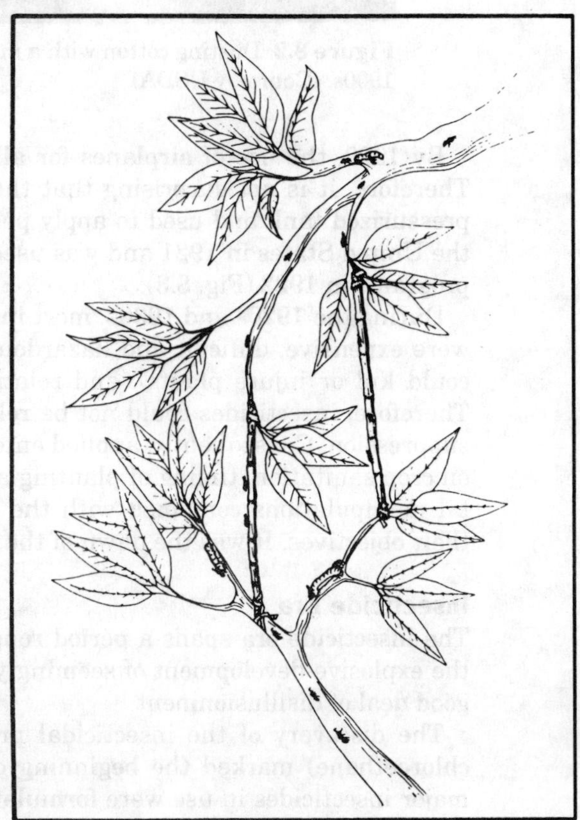

**Figure 8.1** The earliest known example of using natural enemies is believed to be in China, where ants were placed on citrus to suppress caterpillars and beetles. Bamboo bridges were tied between branches to encourage ant movement from tree to tree. (Reprinted with permission of Plenum Press from *Introduction to Integrated Pest Management* by M. L. Flint and R. van den Bosch. © 1981)

**Figure 8.2** Dusting cotton with a mule-drawn power duster in the early 1900s. (Courtesy USDA)

By 1920, the use of airplanes for all sorts of activities was commonplace. Therefore, it is not surprising that they were fitted with a spray boom and pressurized tank and used to apply pesticides. This development occurred in the United States in 1921 and was used in emergency boll weevil pest control programs in 1922 (Fig. 8.3).

During the 1920s and 1930s, most insecticides were not very efficient. They were expensive, difficult and hazardous to apply, somewhat phytotoxic (they could kill or injure plants), and relatively ineffective by today's standards. Therefore, insecticides could not be relied on, by themselves, to achieve pest suppression. Consequently, applied entomologists developed approaches based on crop sanitation, timing of planting and harvesting, and other environmental manipulations combined with the low-key use of insecticides to achieve their objectives. It was the dawn of the insecticide era.

### Insecticide Era

The insecticide era spans a period roughly from 1939 to 1962. It began with the explosive development of seemingly miraculous insecticides. It ended in a good deal of disillusionment.

The discovery of the insecticidal properties of DDT (*d*ichloro*d*iphenyl*t*richloroethane) marked the beginning of this era. Prior to its discovery, the major insecticides in use were formulated from petroleum and coal tar distillates, plants, or inorganic compounds, particularly arsenic. The trend in applied chemistry in the late 1930s was toward synthesis of new compounds, and pest control workers were seeking to synthesize moth-proofing agents. Dr. Paul Müller, a Swiss chemist with Geigy Corporation, examined DDT while seeking to develop a new, purely synthetic insecticide. This obscure compound had been synthesized in 1873 by a German graduate student, Othmar Zeidler, who did not realize its insecticidal properties. After testing, Müller found the compound extraordinarily effective against many insects.

**Figure 8.3** Dusting for boll weevils, *Anthonomus grandis,* with an airplane in the early 1920s. (Courtesy USDA)

During World War II, DDT was revealed to the Western Allies and was used primarily for mosquito, flea, and louse suppression (Fig. 8.4). Its application saved thousands of lives that would have been lost to malaria and typhus. Indeed, DDT had such an impact on human health that in 1948 Müller received the Nobel Prize in medicine for his discovery. After the war, DDT had phenomenal successes against agricultural pests, and many other effective insecticides were developed from the chlorination of hydrocarbons.

While the Allies were working with DDT and related compounds, the Germans accelerated their efforts to find suitable replacements for nicotine and other scarce insecticides. Dr. Gerhard Schrader with Farbenfabriken Bayer AG, having worked actively with organophosphates as possible nerve gases, discovered the insecticidal properties of these compounds.

At the close of World War II, insecticide development proceeded at a very rapid pace. Many firms, particularly petroleum companies, became involved in the development of agricultural chemicals. The trend in pest control was toward the testing of new and even more effective insecticides. As this trend continued, agricultural entomologists turned their attention away from multifaceted programs involving sanitation, planting dates, and the like, in favor of a more narrow, purely insecticidal approach to solving pest problems.

The insecticidal approach usually entailed identifying the pest and finding the most "effective" chemical to apply, with little concern for other factors important in the agroecosystem. Spraying was often performed according to a

**Figure 8.4** Dusting a child with DDT for typhus control during World War II. (Courtesy World Health Organization)

calendar schedule and without knowledge of pest phenology, density, or damage potential. The low cost and high performance of these chemicals fostered the abandonment of earlier combined approaches for suppression. The insecticidal approach relied only on ability to kill insects, and killing, rather than crop protection, became a major preoccupation in pest technology.

Such a practice worked well in the short run, but in time insecticide resistance and other problems began to occur with surprising regularity. In addition, residues of DDT and related compounds were discovered in milk and other foods as well as in food chains of predator birds and other wildlife (see Chapter 11).

The realization of unintentional pesticide movement in, and its effects on, ecological systems was brought to the fore in 1962 with the publication of *Silent Spring* by Rachel Carson. In her book (Fig. 8.5), Carson made strong accusations against pesticides, causing a public outcry in the United States for environmentally safe approaches to pest control (Anelli et al. 2006). This reaction resulted in appointment of the Presidential Science Advisory Committee on Environmental Quality, and a report recommending improved pest control practices was issued in 1965 (Krupke et al. 2007).

## Emergence of Pest Management

The concept of pest management grew out of the discontent with the purely insecticidal approach to pest control of the 1950s. Although the term was coined

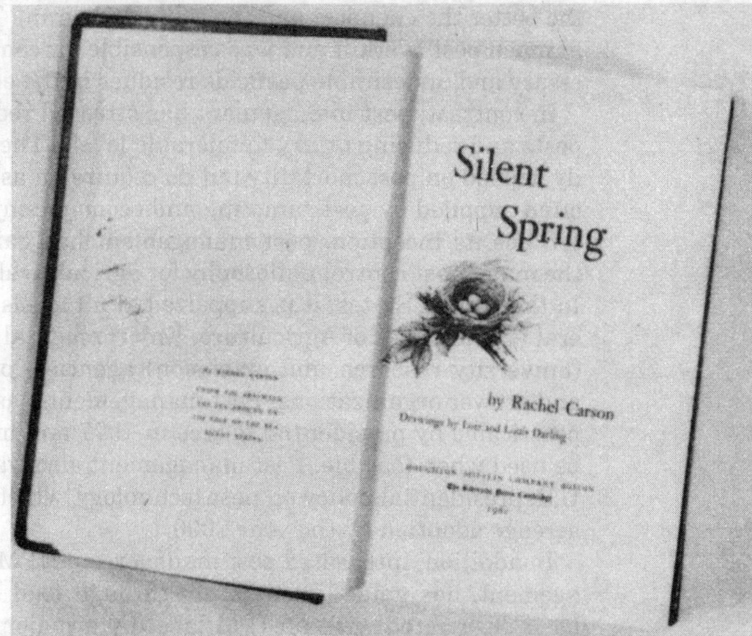


**Figure 8.5** The publication of *Silent Spring* in 1962 had a significant influence on the direction of pest control in the United States.

in the early 1960s, pest management had its roots in earlier theories of biological control. Early successes and lasting solutions with devastating pests encouraged the development of biological control, which emphasized the significant impact of effective natural enemies on pest populations.

With many pests, however, natural enemies were undependable for satisfactory containment. Subsequently, the concept of **integrated control** was developed, emphasizing the selective use of insecticides so that natural enemies were conserved in the agroecosystem. This integration of control techniques was expanded in later years to include other techniques (for example, resistant hosts and sanitation), but natural enemy conservation was always an axiom of integrated control.

Against this background, Australian entomologists L. R. Clark and P. W. Geier clearly outlined the principles of pest management in 1961. They suggested the terms *protective population management* or *pest management* for their ideas. **Pest management** differed from earlier approaches in its holistic viewpoint, its synthesis of ideas, and the inclusion of basic population theory in its design.

Pest management was also different philosophically from previous pest technologies. Before the development and acceptance of pest management, pest technology focused on *control* rather than *management*. Control refers to having power over something; the true meaning of control is best exemplified in the use of conventional pesticides, which constitutes chemical control. Conversely, management refers to a judicious use of means to accomplish a desired end. The main objective of pest control is to reduce pest impact, which usually means killing pests. Ultimately, killing may become a major control objective in itself, rather than crop protection, and a compulsion for 100 percent mortality may result. With this thinking, the greater the mortality—in other words, "percent control"—

the better the chemical and the control program. This earlier mind-set resulted in much pest overkill and was responsible for considerable quantities of unnecessary and undesirable pesticide residues in the environment.

In contrast, pest management has stressed reducing or modifying *impact* of pests and reducing *injury* to tolerable levels. These objectives do not necessarily depend on pest mortality but do require an assessment of pest status, most often supplied by pest sampling and economic-injury levels.

Since its inception, pest management has gained momentum and become the major pest control philosophy for agricultural pests in developed countries. In the United States, it is supported at all levels of agriculture, including federal (Department of Agriculture, Environmental Protection Agency) and state (university research and extension) agencies, private crop consulting firms, and grower organizations. Pest management was ordered to be used wherever practicable by presidential decree in 1977 and mandated by California law to be used when feasible. Pest management also was the cornerstone of previous U.S. presidential policy on pest technology, which called for a 75 percent farm-acreage adoption by the year 2000.

In addition, integrated pest management (IPM), which includes insect management, has gained momentum through food labeling. Food products have been designated with an IPM label by a major food chain in New York and Pennsylvania to educate consumers on how their food is produced. Some of the commodities so labeled include apples, mushrooms, fresh-market sweet corn, and tomatoes. Consumers' favorable reaction to canned and frozen products carrying IPM labels has encouraged expansion of the labeling program in these states and will conceivably spread to other parts of the United States. Some specialists believe that the extent of consumers' acceptance of IPM labeling could significantly impact IPM research and implementation.

## THE CONCEPT OF PEST MANAGEMENT

### Definition and Characteristics of Pest Management

Probably as many definitions of pest management have been suggested as there are authors on the topic. Most descriptions mention three elements: multiple tactics (for example, natural enemies, resistant varieties, and insecticides) used in a compatible manner, pest populations maintained below levels that cause economic damage, and conservation of environmental quality.

Additional characteristics that help us conceptualize pest management were outlined by Geier in 1966. He contended that pest management programs should be: (1) selective for the pest, (2) comprehensive for the production system, (3) compatible with ecological principles, and (4) tolerant of potentially harmful species but within economically acceptable limits.

For our purposes, we will consider these foregoing elements and define pest management as *a comprehensive pest technology that uses combined means to reduce the status of pests to tolerable levels while maintaining a quality environment*. This definition carries with it several implications that set pest management apart from other pest technologies. First, it seeks to deal with pests in the context of the whole production system rather than as a separate collection of individual problems. It proposes that several techniques be employed to alleviate problems rather than relying on a single tactic (for example, insecticides alone).

The objectives of pest management are also clear from our definition. The main objective of the approach is to reduce pest status. Although reducing status may be achieved by killing pests, killing certainly is not the objective—preventing economic loss is. Indeed, pest status may also be reduced by avoiding or repelling pests or by reducing their reproductive rates.

Another implication of our definition is that pest populations or their effects should be reduced to tolerable levels. In this instance, tolerance means that humans should accept the presence of pest species, although at levels that are not economically important. This aspect admits that complete elimination of pests may not be feasible or even desirable. Particularly, this acceptance of pest presence sets pest management apart from pest control.

Finally, the capstone objective of pest management is the maintenance of a quality environment. This objective clearly refers to conservation, and it includes quality of both cropping and nonagricultural environments (air, water, soil, wildlife, and plant life). With past approaches, environmental quality was largely ignored, except for the attention paid to crop vigor and soil elements supporting the crop. The principles of pest management emphasize that cropping systems behave similarly to unmanaged or natural systems; that is, they have a diversity of interacting elements like natural insect and weed enemies that influence ecology of the crop environment. By maintaining the quality of this environment, lasting solutions can be achieved. Most approaches that conserve the quality of crop environments also aid in conserving surrounding, nonagricultural environments.

## Pest Management Strategies and Tactics

Elementary to the pest management concept is the development of a strategy. A **pest management strategy** is the overall plan to eliminate or alleviate a real or perceived pest problem. The particular strategy developed depends on the particular life system of the pest and the crop involved.

In addressing problems using pest management, we aim to reduce pest status. Because pest status is determined by both the insect and the crop, our management program may emphasize modification of either or both of these. Therefore, several types of strategies might be developed, based on economics and pest characteristics:

1. Do nothing.
2. Reduce pest population numbers.
3. Reduce crop susceptibility to pest injury.
4. Combine reduced population numbers with reduced crop susceptibility.

After an appropriate strategy has been developed, the methods of implementing the strategy must be chosen. These methods usually are called **pest management tactics.** Optimally, several tactics are used to implement a management strategy.

### Do-Nothing Strategy

The "do-nothing" strategy at first may seem absurd. Obviously, we know when a pest is doing us harm—but do we? It is possible that insect injury only seems as if it is causing a loss, when in reality the crop species tolerates the injury without economic damage. The phenomenon of mistaking trivial insect injury for economically significant injury occurs all too often and is usually the result

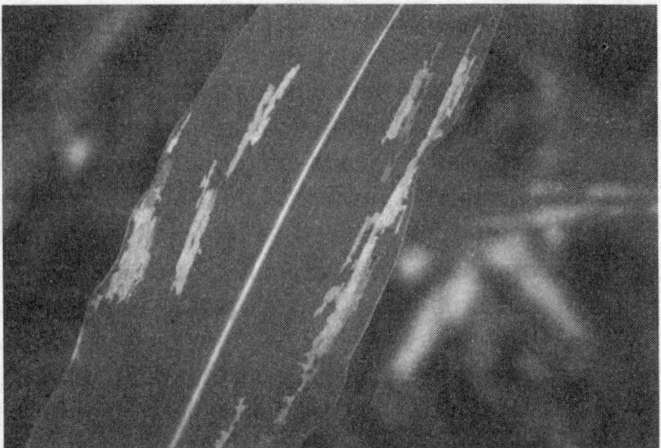

**Figure 8.6** Western corn rootworm, *Diabrotica virgifera,* injury (leaf scraping) to corn. Indirect injury often results in no measurable loss in the valued commodity. (Photo by Marlin E. Rice)

of not assessing population density in relation to the economic threshold. When pest densities are below the economic threshold, "do nothing" definitely is the strategy to follow; otherwise, more money is spent on management than is gained in utility. In other words, a net loss occurs from management. The most frequent need for the "do-nothing" strategy usually arises when insects cause indirect injury (for example, see Fig. 8.6). This strategy may also be the ultimate one following a successful pest management program. In this instance, only surveillance of the resulting pest population is required.

Obviously, tactics are not implemented in the "do-nothing" strategy. However, this does not mean that little activity is involved or that pest suppression is not occurring. Indeed, considerable sampling is required to assure that taking no action is appropriate, and significant pest suppression may have occurred as a result of natural environmental factors.

### Reduce-Numbers Strategy

Reducing insect numbers to alleviate or prevent problems is probably the most widely used strategy in pest management. This strategy is usually employed in a therapeutic manner when densities reach the economic threshold or in a preventive manner based on a history of problems.

In seeking to reduce numbers, one of two objectives may be desirable. Where the pest's long-term average density, the **general-equilibrium position** (GEP), is low compared with the economic threshold (problems are not particularly severe), the best strategy would be only to dampen population peaks (Fig. 8.7). This action would not change the GEP appreciably yet would prevent economic damage from occurring during outbreaks.

On the other hand, severe pest problems call for more drastic population reductions. With these pests, the GEP lies very close to, or above, the economic threshold. What is required for these populations is a general lowering of the GEP so that the highest population peaks never reach the economic threshold. The most appropriate method of accomplishing this is to either: (1) reduce the environmental **carrying capacity** (the maximum number a given environ-

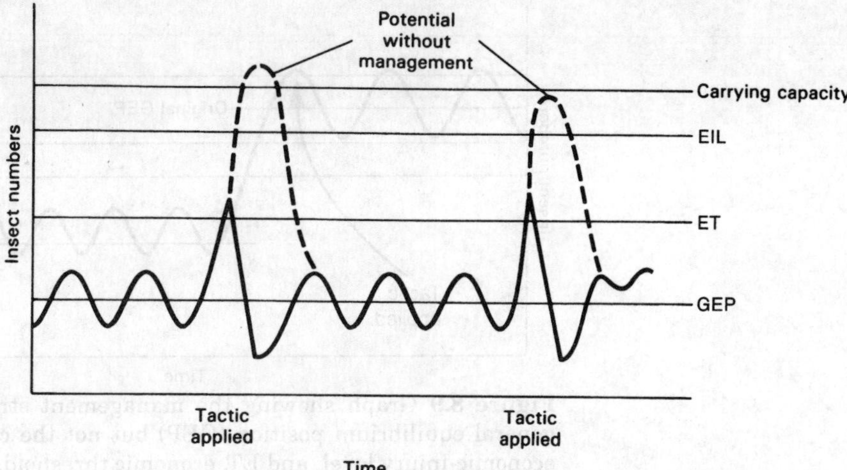


**Figure 8.7** Graph showing the management strategy of dampening peaks. EIL, economic-injury level; ET, economic threshold; and GEP, general equilibrium position.

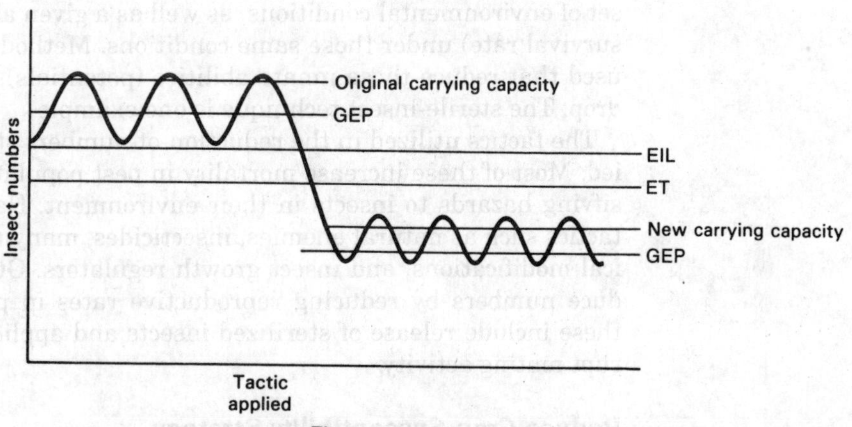


**Figure 8.8** Graph showing the management strategy of lowering the carrying capacity of a pest. EIL, economic-injury level; ET, economic threshold; and GEP, general equilibrium position.

ment will support for a sustained period) (Fig. 8.8) or (2) reduce the inherited reproductive and/or survival potentials of populations (Fig. 8.9).

To reduce carrying capacity, we are dealing exclusively with the environment, trying to reduce the favorableness of the habitat. For instance, if numbers of cattle allowed to graze in an area of pasture are reduced, we reduce the carrying capacity of the habitat for populations of cattle tick, *Boophilus annulatus*. The same is true when crops are rotated from a pest's host species to a nonhost species (for example, a legume rotated with a grass) or a field is tilled to eliminate overwintering habitat. In all of these situations, we attempt to reduce the GEP of the pest population over time by reducing the inherent favorableness of the pest's environment.

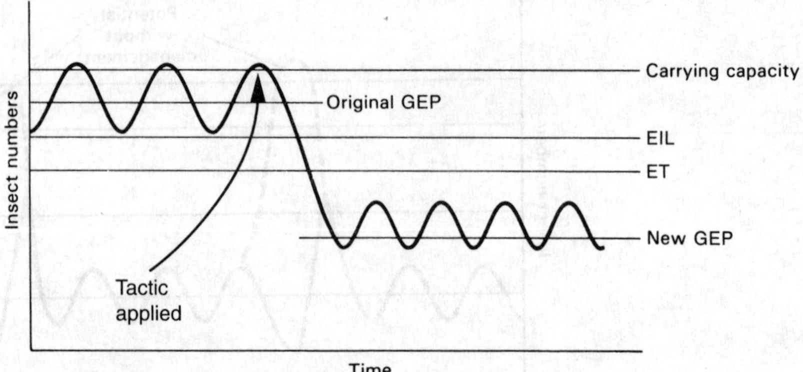


**Figure 8.9** Graph showing the management strategy of lowering the general equilibrium position (GEP) but not the carrying capacity. EIL, economic-injury level, and ET, economic threshold.

With the second approach, we do not deal with the environment but with the characteristics of the pest itself. Depending on genetic makeup, each pest population has a specific ability to reproduce (potential birth rate) under a given set of environmental conditions, as well as a given ability to survive (potential survival rate) under those same conditions. Methods have been proposed and used that reduce those innate abilities (potentials), thus causing the GEP to drop. The sterile-insect technique is one example.

The tactics utilized in the reduction-of-numbers strategy are many and varied. Most of these increase mortality in pest populations by creating or intensifying hazards to insects in their environment. Hazards are increased with tactics such as natural enemies, insecticides, many resistant cultivars, ecological modifications, and insect growth regulators. Other tactics attempt to reduce numbers by reducing reproductive rates in pest populations. Some of these include release of sterilized insects and application of chemicals to disrupt mating activity.

### Reduce-Crop-Susceptibility Strategy

Reducing crop susceptibility to insect injury is often one of the most effective and environmentally desirable strategies available. For this strategy, the insect population is not modified at all. Rather, we rely on changes made in the host plant or animal that render it less susceptible to an otherwise damaging pest population.

The tactics involved in the reduce-crop-susceptibility strategy usually involve elements of host-plant and livestock resistance or ecological management (crop environment manipulations). In the first instance, a certain form of resistance, referred to as **tolerance,** occurs in some plant cultivars or animal breeds that have a greater degree of impunity to insect injury than others do. When such cultivars or breeds are utilized, insect population numbers are not reduced; however, losses are minimized because the tolerant crop is less affected by a given level of injury. Some of the tactics in ecological management for reducing crop susceptibility are improving plant vitality through fertilization and changing planting dates to upset the synchrony between a pest and a susceptible plant stage.

**Combined Strategies**

The remaining, and perhaps most obvious, strategy is one that combines objectives of all previously mentioned strategies to produce a pest management program with several tactics. In many ways, this is the most desirable strategy when feasible. The diversification resulting from such programs tends to produce a greater degree of consistency in pest control than can be achieved with a single strategy, particularly a single tactic. Past experience has shown that pest control programs using a single tactic are subject to failure when, either gradually or abruptly, that tactic fails. On the other hand, multifaceted programs do not rely on a single tactic, and if one tactic fails, others are present to help moderate losses. As we will see, the use of multiple strategies and tactics is a basic principle in developing insect pest management programs.

**Kinds of Pests and Likely Strategies**

The appropriate strategy and subsequent complexity of a pest management program is primarily determined by the status of an insect pest in the production system. Four pest types may be designated according to status, and identifying a pest species as one of these is useful in developing an appropriate strategy. The pest types include noneconomic populations (called subeconomic pests by some authors), and occasional, perennial, and severe pests.

**Subeconomic pests.** The subeconomic pest may seem to be a contradiction in terms. However, these insects are pests in a true sense, even if they cause insignificant losses. The GEP with this pest type is far below the economic-injury level, and the highest pest-population fluctuations do not reach that level (Fig. 8.10). As mentioned earlier, attempting to reduce injury from such pests would cost more than the losses they inflict.

Many insects fall within the subeconomic category. An example is the alfalfa caterpillar (Box 8.2), *Colias eurytheme*, in central Iowa. This insect is a defoliator of alfalfa and causes direct losses of forage. But densities are low and commodity values modest, so that the most appropriate action is to do nothing about the pest. As discussed previously, the pest management activity involved with this kind of pest is to monitor activity levels for changes that may

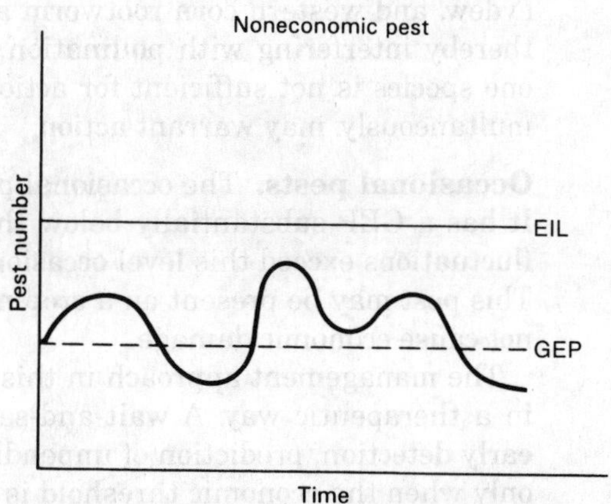


**Figure 8.10** Graph representing a noneconomic or subeconomic pest. EIL, economic-injury level, and GEP, general equilibrium position.

## BOX 8.2 Alfalfa Caterpillar (See Color Plate 3)

**SPECIES:** *Colias eurytheme* Boisduval (Lepidoptera: Pieridae)

**DISTRIBUTION:** The alfalfa caterpillar is native to North America and is found throughout the United States and Southern Canada. However, the pest is most abundant and significant in the southwestern United States.

**IMPORTANCE:** The species is an important alfalfa pest, especially in the irrigated areas of the southwestern United States. Larvae defoliate host plants and can also attack stems. The pest also feeds on other legumes.

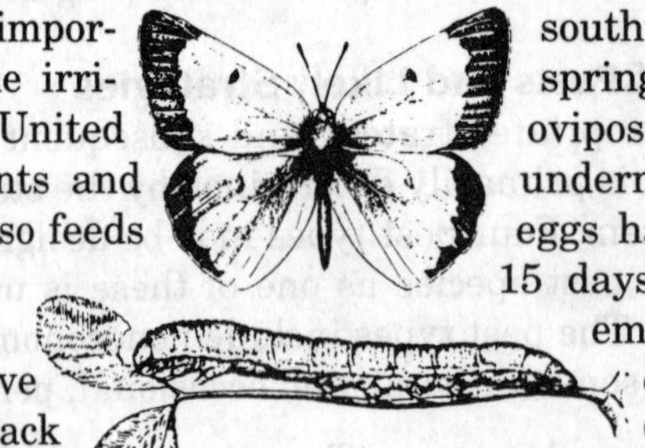

**APPEARANCE:** Adults typically have yellow to orange wings with a black margin on the edges. Albinism is common among females, which have white wings instead of orange. Wingspans are approximately 50 mm, and the antennae are slightly clubbed. Larvae are about 38 mm long when full-grown and are green with a pink and white stripe running lengthwise along the side of the body. Eggs are spindle shaped, with numerous ridges.

**LIFE CYCLE:** In the northern parts of their distribution, pupae overwinter in plant debris; however, all stages overwinter in the southwestern United States. In early spring, adults emerge, and females oviposit from 200 to 500 eggs singly underneath leaves. In about 3 days, eggs hatch, and larvae feed for 12 to 15 days. Pupation occurs, and adults emerge in 5 to 7 days. In the northern United States and Southern Canada, there are typically two generations per year. In the southwestern United States, there may be as many as seven generations per year.

---

alter the pest's status. Changes in the pest biology, crop biology, or economics can cause status to change, thereby requiring other strategies.

Another consideration with subeconomic pests is their contribution to overall injury. Should a complex of several subeconomic insect pests occur, management may be necessary when the total injury threatens economic damage. Such a complex sometimes occurs in corn when corn leaf aphids (Box 8.3), *Rhopalosiphum maidis*, feeding on tassels and silks, covering them with honeydew, and western corn rootworm adults, *Diabrotica virgifera;* cliping silks, thereby interfering with pollination. In many instances, the activity of only one species is not sufficient for action, but injury from the two, occurring simultaneously, may warrant action.

**Occasional pests.** The occasional pest is a very common type of insect pest. It has a GEP substantially below the economic-injury level, but the highest fluctuations exceed this level occasionally and usually sporadically (Fig. 8.11). This pest may be present on a crop most years but, more often than not, does not cause economic damage.

The management approach in this instance is usually to deal with the pest in a therapeutic way. A wait-and-see attitude is assumed, with reliance on early detection, prediction of impending outbreaks, and employment of tactics only when the economic threshold is reached. The objective of this strategy is to dampen outbreak peaks, with no effort toward reducing the GEP. Therefore, pest management programs for pests in this category tend to be less complex than those of more serious pests and may rely on only two or three tactics.

## BOX 8.3 CORN LEAF APHID

**SPECIES:** *Rhopalosiphum maidis* (Fitch) (Hemiptera: Aphididae)

**DISTRIBUTION:** The corn leaf aphid is found throughout the world and is distributed in North America east of the Rocky Mountains. Individuals can be found on corn, sorghum, sugarcane, barley, and several species of wild grasses.

**IMPORTANCE:** The species is an occasional corn pest during the late growing season. Individuals suck juices from leaves of hosts, causing a mottled yellow appearance. More importantly, the aphids feed on tassels and silk, coating them with honeydew, which interferes with pollination. Severely infested tassels eventually drop from the plants. The pest also vectors several plant pathogens, including maize dwarf mosaic virus of corn, mosaic disease of sugarcane, and yellow dwarf virus of barley.

**APPEARANCE:** Individuals are soft-bodied, about 3 mm long, and greenish-blue, with black antennae and appendages. Both winged and wingless forms exist during the growing season.

**LIFE CYCLE:** Corn leaf aphids migrate to the northcentral United States each year from southern regions. Individuals in early spring are all wingless females. At certain times in the growing season, winged males and females are produced and then disperse and reproduce. Reproduction is parthenogenetic. Normally, up to nine generations per year occur in the northcentral United States, and as many as fifty generations per year have been recorded in Texas.

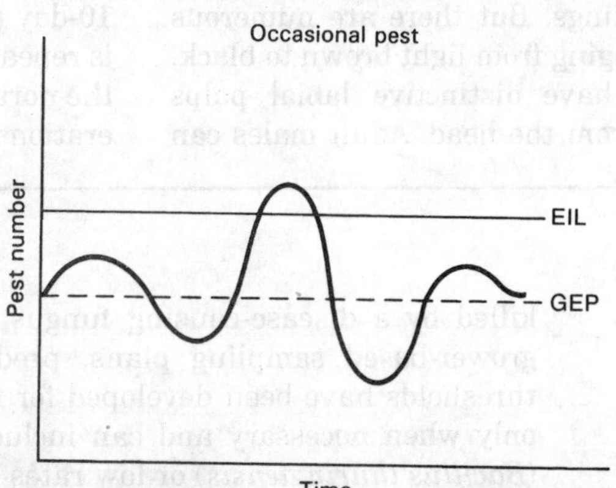


**Figure 8.11** Graph representing an occasional pest. EIL, economic-injury level, and GEP, general equilibrium position.

Pesticides often fit well in the strategy for occasional pests. When used judiciously, these can serve for long-term management because application is infrequent, and consequently, problems with undesirable side effects are minimized.

An example of this type of pest is the green cloverworm, (Box 8.4) *Hypena scabra,* a defoliator of soybeans in the midwestern United States. This insect is present each year in nearly all soybean fields, but unusually heavy moth flights from the South in late spring set the stage for midseason problems with larvae. The insect is rarely a problem in late season because it is attacked and

## BOX 8.4 Green Cloverworm (See Color Plate 4)

**SPECIES:** *Hypena scabra* (Fabricius) (Lepidoptera: Noctuidae)

**DISTRIBUTION:** The green cloverworm is distributed in North America from the Great Plains eastward and from southeastern Canada to the Gulf of Mexico.

**IMPORTANCE:** The green cloverworm is considered a serious yet sporadic defoliator of soybeans, alfalfa, and other legumes. One of the most widespread of soybean insect pests, green cloverworm populations occasionally cause economic damage in the midwestern and southeastern United States. Larvae defoliate the host plant by chewing the tissues between the veins of the leaf, leaving a ragged appearance.

**APPEARANCE:** Adult moths appear triangular when at rest and typically have dark brown wings with lighter markings. But there are numerous color variants, ranging from light brown to black. The moths also have distinctive labial palps pointed forward from the head. Adult males can be differentiated from females in that the eyes of male moths are much larger. The larvae are pale green with two thin white stripes placed horizontally along each side of the body. Three pairs of ventral prolegs (four pair including the anal pair) distinguish these larvae, which are approximately 25 mm long when fully grown. The eggs, which are 0.5 mm in diameter and have longitudinal ridges, are translucent green but turn light brown about 48 hours before hatching. Pupae are brown and lightly wrapped in a silken cocoon.

**LIFE CYCLE:** Adults overwinter south of 41° N latitude. Near the Gulf coast, feeding activity and reproduction occur year-round. Northern areas are colonized each spring by migrating moths transported by southerly winds. Females oviposit eggs singly on the undersides of leaves. The eggs hatch in about 3 days, and the larvae undergo six to seven molts in about 14 days. After a 7- to 10-day pupal stage, moths emerge, and the cycle is repeated. There are two generations per year in the northern United States and three to four generations per year in the southern United States.

killed by a disease-causing fungus, an important natural enemy. Efficient grower-based sampling plans, predictive indices, and definitive economic thresholds have been developed for the pest. Tactics of suppression are used only when necessary and can include insect-selective microbial insecticides (*Bacillus thuringiensis*) or low rates of conventional insecticides. In the green cloverworm pest management program, natural enemy conservation is achieved, environmental problems averted, and grower profits sustained.

**Perennial and severe pests.** Perennial and severe pests cause the most serious and difficult problems in crop production. Only a few insects belong in these categories, and these are often referred to as **key pests.** Problems created by these pests are usually caused by relatively high market value of the crop and/or very dense insect populations. In this category are insects and other arthropods that attack the harvested produce directly and those that almost always occur in great numbers. Some of the worst pests may cause only blemishes on produce, making it unacceptable to consumers and resulting in serious economic losses. An example of losses from blemishes occurs with the citrus rust mite, *Phyllocoptruta oleivora,* which causes a serious russeting of oranges.

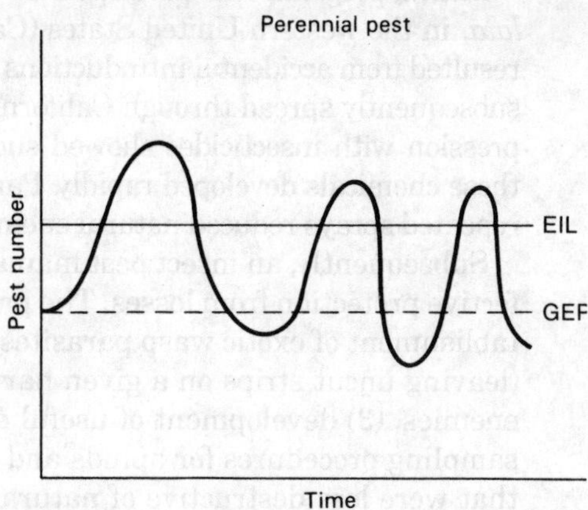


**Figure 8.12** Graph representing a perennial pest. EIL, economic-injury level, and GEP, general equilibrium position.

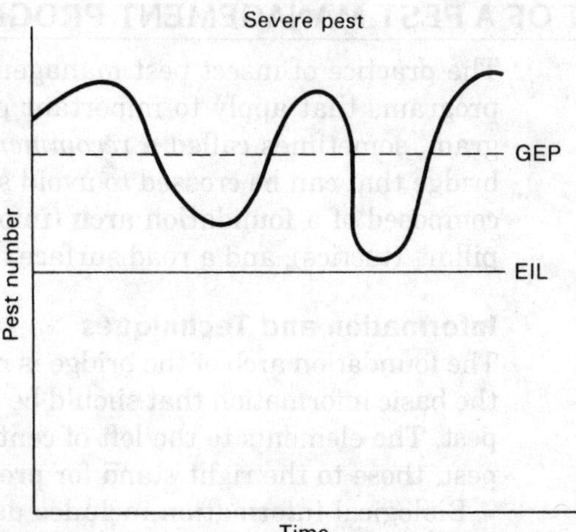


**Figure 8.13** Graph representing a severe pest. EIL, economic-injury level, and GEP, general equilibrium position. Note that this pest's GEP is above the EIL.

Pests in these categories are characterized by a very high average density in relation to the economic-injury level. With the **perennial pest,** the GEP is below the economic-injury level, but so close that economic damage occurs more years than not (Fig. 8.12). Only infrequently do population peaks of this pest type not reach the economic-injury level. **Severe pests** have a GEP that is actually above the economic-injury level, making them a constant problem (Fig. 8.13).

Unlike the strategy for occasional pests, the primary insect pest management strategy for perennial and severe pests is to reduce the GEP of the population. More complex programs are required to produce lasting solutions. These programs usually comprise combined strategies involving several tactics to lower the environmental carrying capacity and/or the GEP of the pest population.

Such a program was developed for the spotted alfalfa aphid, *Therioaphis maculata,* in the western United States (California Central Valley). This aphid problem resulted from accidental introductions from the Middle East in 1954, and the species subsequently spread through California and to other states. Early attempts at suppression with insecticides showed success, but resistance (loss of effectiveness) to these chemicals developed rapidly. Particularly in areas with resistant populations, repeated sprays reduced natural enemies, and aphid damage became acute.

Subsequently, an insect pest management program was developed, offering effective protection from losses. The program involved: (1) the introduction and establishment of exotic wasp parasites that attack the aphids, (2) strip harvesting (leaving uncut strips on a given harvest date) to conserve and enhance natural enemies, (3) development of useful economic thresholds and implementation of sampling procedures for aphids and natural enemies, and (4) use of insecticides that were less destructive of natural enemies. Later developments of aphid-resistant alfalfa varieties have almost eliminated the need for insecticide sprays in this program. The lasting value of such an integrated approach lies in its reliance on several compatible tactics rather than on a single one.

## DEVELOPMENT OF A PEST MANAGEMENT PROGRAM

The practice of insect pest management depends on a series of well-designed programs that apply to important pests in the production system. Each program, sometimes called a *recommendation algorithm,* can be visualized as a bridge that can be crossed to avoid significant losses to pests. Such a bridge is composed of a foundation arch (information and techniques), several vertical pillars (tactics), and a road surface (avoidance of losses) (Fig. 8.14).

### Information and Techniques

The foundation arch of the bridge is made up of several elements and represents the basic information that should be known and techniques needed to manage a pest. The elements to the left of center stand for biological information about a pest; those to the right stand for procedures needed to obtain this information.

Biological information includes data on habitat requirements showing how individuals feed, grow, develop, reproduce, and disperse. Pest management seeks particularly vulnerable points in the life cycle to exploit. The seasonal cycle is studied along with pest developmental rates to provide predictions of pest occurrence and relationship to biological events of the crop.

Population dynamics is another critical element in the biological information needed for insect pest management. It seeks to explain the forces (natural enemies, weather, food, and habitat) causing population change by quantifying these forces and presenting the numbers as mathematical models. Such models, consisting of component equations, often have been organized to represent an ecological system and have been programmed for computer simulation. Such simulations are believed useful for understanding population dynamics in various environments. Population dynamics information builds on the understanding of behavior, life cycle, and seasonal cycle, and it allows prediction of infestations and outbreaks. It also indicates environmental factors that might be exploited to reduce pest status.

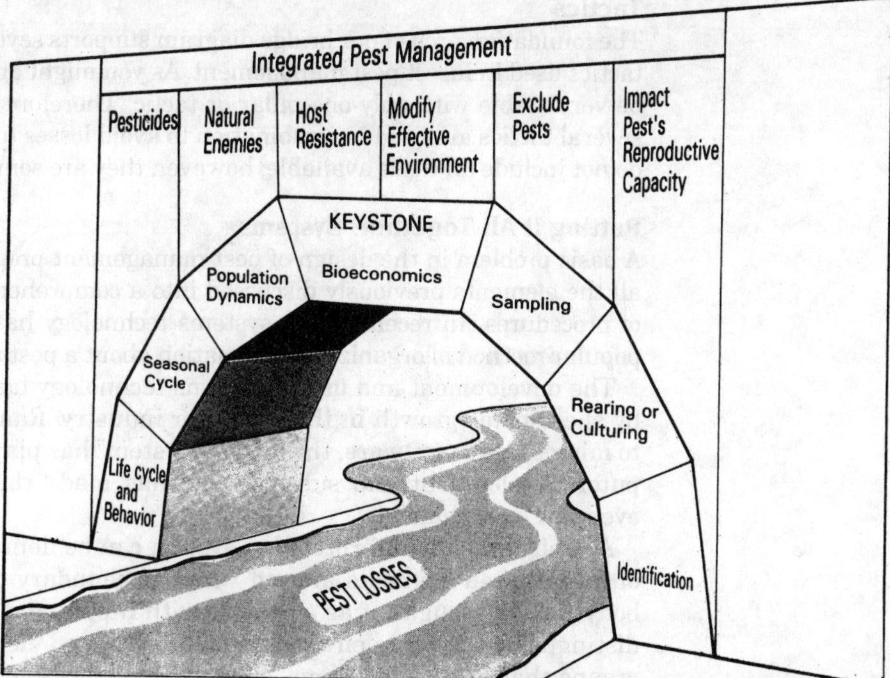


**Figure 8.14** Schematic representation of the major components of an insect pest management program.

The techniques in the foundation arch include species identification, rearing or culturing, and sampling. Identification involves assigning the correct name to a species and knowing its taxonomic classification. Because every species behaves differently, it is critical to make accurate identifications; they are a key to the published literature about a pest. Species identifications aid in research and development of pest management programs and serve as a reference for delivering such programs to practitioners. Not only is the ability to rear, culture, or colonize a pest useful for obtaining biological information about a species, but also laboratory and field experimentation with management tactics can proceed more rapidly than by having to depend completely on natural pest occurrence.

Sampling techniques and programs play a particularly important role in insect pest management. As mentioned earlier, such techniques allow the estimation of numbers for population dynamics studies and later serve to assess the population for management decisions. Without precise, economical sampling programs, it is doubtful that an effective pest management program can be designed.

The keystone of the pest management foundation arch is **bioeconomics.** As you will recall, this is the relationship of pest numbers to losses. Bioeconomics ties together biological information of the pest and host crop, sampling programs, and economics. Determination of bioeconomics depends greatly on a knowledge of pest-plant or pest-livestock relationships and the crop's response to pest damage. Coupling known crop responses with pest densities and accounting for market values and pest management costs allow the development of economic-injury levels and economic thresholds.

### Tactics
The foundation arch of our bridge diagram supports several pillars that represent tactics used in insect pest management. As you might agree, the bridge would not be very stable with only one pillar or tactic. Therefore, the diagram shows that several tactics are used in combination to avoid losses to pests. The tactics shown do not include all those available; however, they are some of the most useful ones.

### Putting It All Together: Systems
A basic problem in the design of pest management programs is how to assemble all the elements previously discussed into a comprehensive, understandable set of procedures. In recent years, systems technology has become an increasingly popular method of organizing information about a pest and delivering it to users.

The development and use of systems technology has been greatly enhanced by spectacular growth in the computer industry. Ranging from circuit boards to microchips to software, the idea of "system" has played a central role in computer development and subsequently has made the word a part of nearly everyone's vocabulary.

Although widely interpreted, a system can be defined as an identifiable set of interrelated elements around which a boundary can be drawn. Your own body can be thought of as a system with inputs and outputs, and it is easily distinguished from its environment. Interrelated elements in this system are organs that form subsystems, such as the heart and circulatory system, that contribute to the functioning of the overall system.

Likewise, a pest management system has interrelated elements like the processes of obtaining information about potential pest problems, decision making, and strategies. W. G. Ruesink (1982) presented the pest management system as a diagram with the interrelated elements and associated inputs and outputs (Fig. 8.15). The boundary of the pest management system is shown

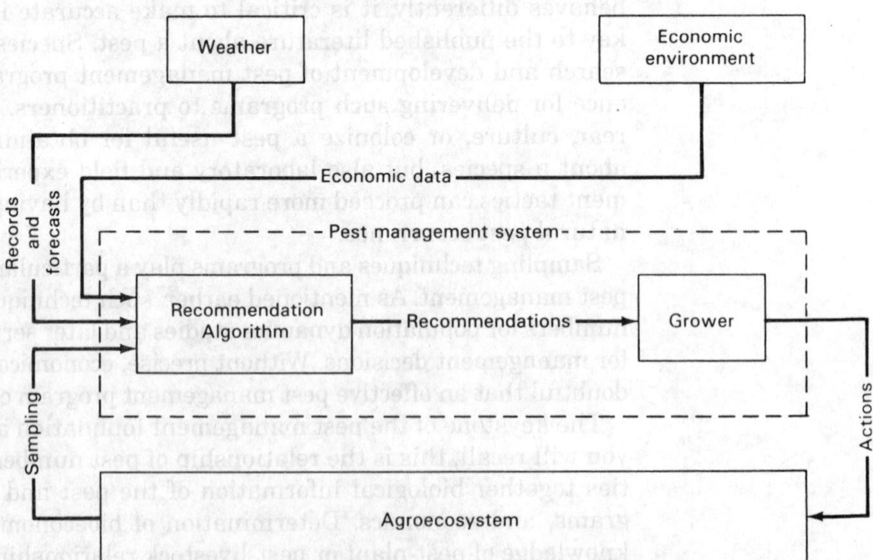


**Figure 8.15** Diagram of a pest management system. (Redrawn from Ruesink, 1982, in Metcalf and Luckmann, eds., *Introduction to Insect Pest Management*, John Wiley & Sons)

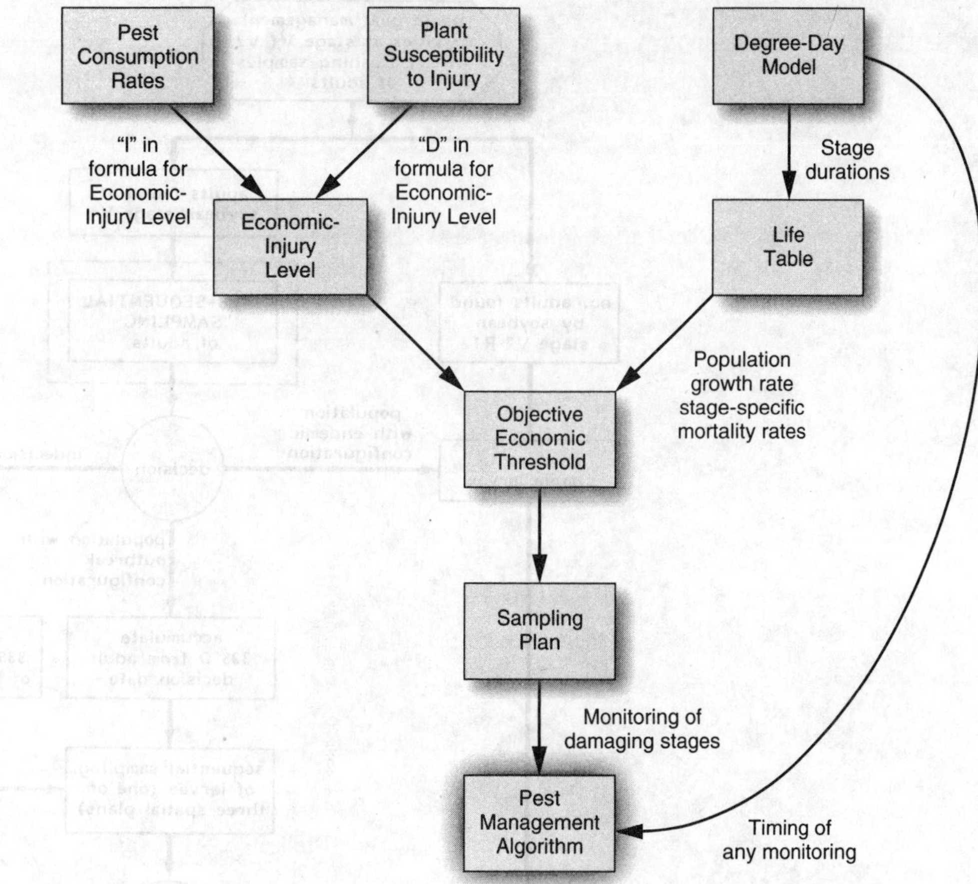


**Figure 8.16** Possible steps for developing a pest management algorithm. Labels beside arrows are data that allow progress to the next box.

with dashed lines, and information for its operation comes from weather agencies, business sources, and sampling of the agroecosystem. Outputs of the system are actions of the grower directed toward the agroecosystem.

At the heart of the pest management system is the so-called recommendation algorithm. An **algorithm** is a precisely defined sequence of rules stating how to produce specific outputs from specific inputs through a given number of steps. Therefore, a recommendation algorithm is the set of steps for managing a pest. Such an algorithm would exist for each insect pest or group of pests important to the agroecosystem. Each recommendation algorithm produces a specific pest management scheme for the inputs received. An example of steps in developing a recommendation algrorithm are shown in Figure 8.16.

An example of a recommendation algorithm is shown in Figure 8.17 for the green cloverworm in soybeans. Although all details are not presented, a number of pest management activities are assembled in an organized manner, requiring inputs on insect population (adults and larvae), weather (degree-day accumulations), plant growth (stage designations), and economics (included in the larval

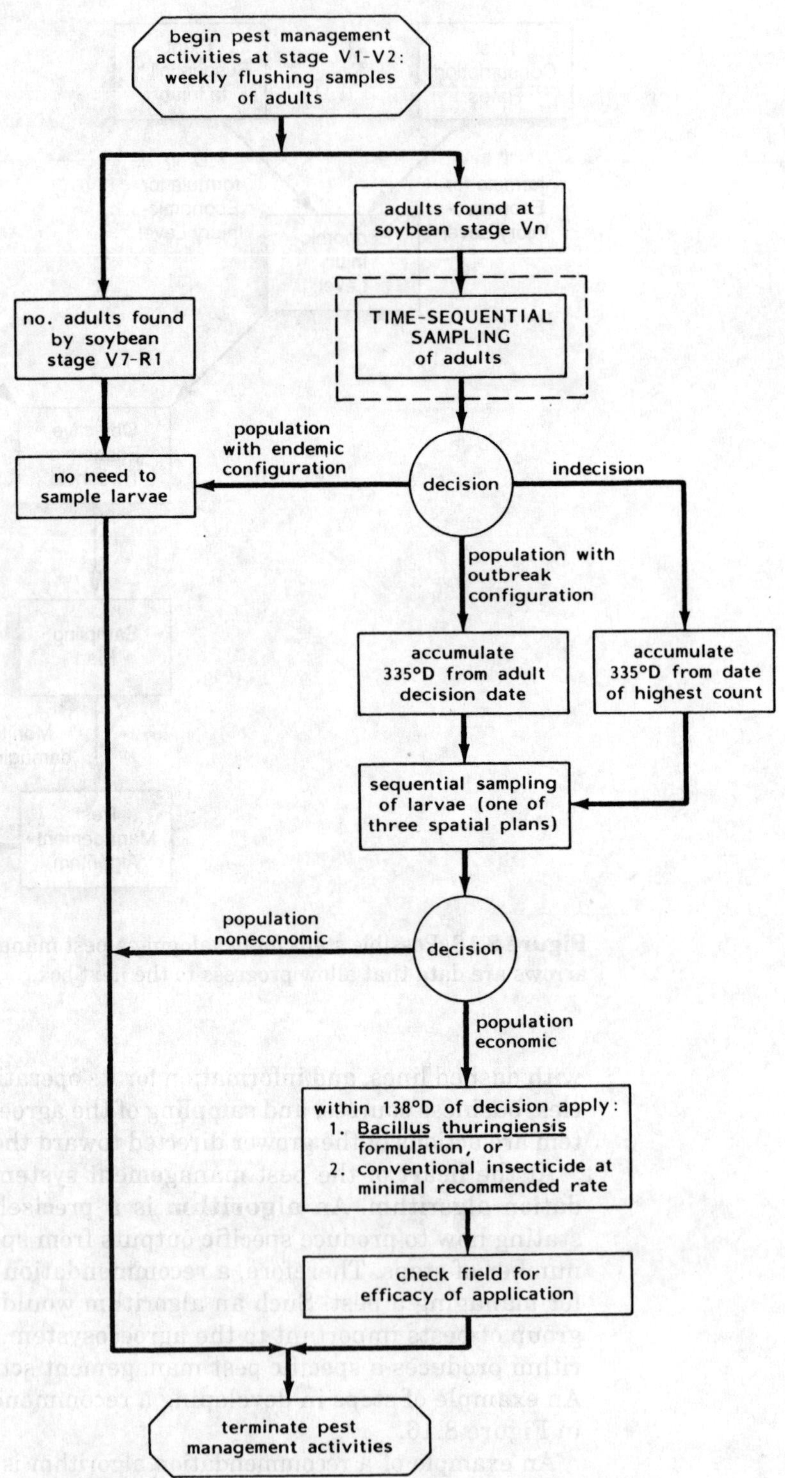


**Figure 8.17** An example of a recommendation algorithm as represented by that for the green cloverworm, *Hypena scabra,* on soybeans.

sequential sampling plans). After several steps in the sequence, decisions are made and recommendations (outputs) are advanced for a particular set of circumstances.

Such recommendation algorithms can conveniently be programmed for solution by computers. In the United States, there is a growing trend in this direction, with the delivery of pest management recommendations to growers via on-farm personal computers and computer networks sponsored by the State Cooperative Extension Service. Today, both on-line and stand-alone systems exist that deliver information about current pest conditions and present management information (see Bechinski 1994).

The advantage of using the systems approach is obvious. It provides a mechanism for integrating complex information from many sources to produce optimal pest management programs that are most economically and ecologically sound. In this way systems technology helps to fulfill the objectives set down by the principles of pest management.

## Further Reading

Allen, G. E., and J. E. Bath. 1980. The conceptual and institutional aspects of integrated pest management. *Bioscience* 30:658–664.

Anelli, C. M., C. H. Krupke, and R. P. Prasad. 2006. Professional entomology and the years since *Silent Spring,* Part 1. *American Entomologist* 52:224–233.

Bechinski, E. J. 1994. Designing and delivering in-the-field scouting programs, pp. 683–706. In L. P. Pedigo and G. D. Buntin, eds., *Handbook of Sampling Method for Arthropods in Agriculture*. Boca Raton, Fla.: CRC Press.

Cate, J. R., and M. K. Hinkle. 1993. *Integrated Pest Management: The Path of a Paradigm*. Washington, D.C.: National Audubon Society.

Chant, D. A. 1964. Strategy and tactics of insect control. *Canadian Entomologist* 96:182–201.

Funderburk, J. E., L. G. Higley, and G. D. Buntin. 1993. Concepts and directions in arthropod pest management. *Advances in Agronomy* 51:125–172.

Geier, P. W. 1966. Management of insect pests. *Annual Review of Entomology* 11:471–490.

Jones, D. P. 1973. Agricultural entomology, pp. 307–332. In R. F. Smith, T. E. Mittler, and C. N. Smith, eds., *History of Entomology*. Palo Alto, Calif.: Annual Reviews.

Kogan, M. 1998. Integrated pest management: Historical perspectives and contemporary development. *Annual Review of Entomology* 43:243–270.

Krupke, C. H., R. P. Prasad, and C. M. Anelli. 2007. Professional entomology and the years since *Silent Spring,* Part 2. *American Entomologist* 53:18–26.

Lewis, W. J., J. C. Vanlenteren, S. C. Phatak, and J. H. Tumlinson, III. 1997. A total system approach to sustainable pest management. *Proceedings National Academy of Science USA* 94:12243–12248.

Matthews, G. A. 1984. *Pest Management*. Essex, England: Longman, pp. 165–182.

Rabb, R. L., G. R. Defoliart, and G. G. Kennedy. 1984. An ecological approach to managing insect populations, pp. 697–728. In C. B. Huffaker and R. L. Rabb, eds., *Ecological Entomology*. New York: Wiley.

Royer, T. A., P. G. Mulder, and G. W. Cuperus. 1999. Renaming (redefining) integrated pest management: Fumble, pass, or play? *American Entomologist* 45:136–139.

Ruesink, W. G. 1982. Analysis and modeling in pest management, pp. 353–373. In R. L. Metcalf and W. H. Luckmann, eds., *Introduction to Insect Pest Management,* 2nd ed. New York: Wiley.

Stern, V. M., R. F. Smith, R. van den Bosch, and K. S. Hagen. 1959. The integrated control concept. *Hilgardia* 29:81–101.

## Favorite Web Sites

http://www.pmac.net/

A broad-based book on integrated pest management (IPM) issues published in 1996 and subsequently presented on the Web. Emphasis placed on the IPM continuum and biointensive approaches to management.

http://www.ipmworld.umn.edu/

The site contains a comprehensive textbook on IPM, discussing principles, tactics, and programs for specific crops. Click on "contributed chapter" to access specific topics.

http://www.ipm.iastate.edu/ipm/nipmn/

This is the site of the northcentral regional IPM program. Provides links to all states in the northcental region. Other regions have similar sites. Use a search engine to reach other regions with the key words "National IPM."

CHAPTER 9

# MANAGEMENT WITH NATURAL ENEMIES

UNNOTICED BY THE CASUAL OBSERVER, thousands of insect and mite species attack crops, forests, and livestock, but they occur in very low numbers. Most of these species cannot be considered pests at all or are subeconomic pests because their activity does not result in an economic loss. The many reasons for their low numbers include unfavorable weather, lack of material requisites (for example, food and nesting sites), and, very often, **natural enemies.**

As discussed previously, natural enemies are living organisms found in nature that kill insects outright, weaken them and thereby contribute to their premature death, or reduce their reproductive potential. A natural enemy usually reduces the subject insect population, the **host** or **prey,** by feeding on individuals, thereby promoting its own population at the expense of the population fed upon. Not only do these natural enemies help prevent some insects from attaining pest status, but they also play a role in reducing the damage potential of significant pests. Established pests, as damaging as they may be, would cause even more damage were it not for the presence of natural enemies.

Virtually all insect populations are affected to a greater or lesser extent by natural enemies. This includes insects from tropical to subarctic regions and those in nonmanaged habitats as well as in agroecosystems. For many species, natural enemies are the primary regulating force in the dynamics of their populations. For this reason alone, it is very important to know what natural enemies are affecting an insect-pest population and to obtain an estimate of their impact. Such information may be the basis for explaining pest population density and predicting outbreaks.

Perhaps even more significant to pest management programs is an understanding of natural enemies so that they can be manipulated to our advantage. Here we would use a strategy that seeks to use nature as a partial ally, not as a total adversary. Indeed, a basic tenet of pest management is to consider the natural enemy component first in developing a strategy.

The pest management tactic involving purposeful natural enemy manipulation to obtain a reduction in a pest's status is called **biological control, biocontrol,** or *biologically intensive management.* Biological control differs from **natural control** in that the latter may involve agents other than natural enemies, such as weather or food, and no purposeful manipulation is involved. Some authorities broaden biological control to include all methods that involve living organisms as part of the tactic, including host plant resistance, sterile insect releases, and genetic manipulation. For the purpose of discussion, we consider biological control in the conventional sense, involving only the manipulation of natural enemies.

## BRIEF HISTORY OF BIOLOGICAL CONTROL

Biological control is one of the oldest, most effective means of achieving insect control. As already mentioned, the earliest record dates back to fourth-century China (Kwantung Province), when ants were used to suppress pests in citrus. Many other early attempts were made throughout the world using insects to control other insects; birds, lizards, and toads to control insects; and vertebrates to control other vertebrates (mainly rats). These attempts all met with varying degrees of success.

It was not until 1888 that biological control became firmly established in the United States as a significant method. That year marked the phenomenal success of an introduced predatory insect, the vedalia beetle, *Rodolia cardinalis,* in eliminating the threat to citrus of the cottony cushion scale (Box 9.1) *Icerya purchasi* (Fig. 9.1). This scale insect was introduced and became established in the Los Angeles area sometime before 1876. By 1885, it was threatening to wipe out the citrus industry in southern California. C. V. Riley, previously the state entomologist for Missouri, examined specimens and determined that the pest had been introduced from Australia. After becoming chief entomologist for the U.S. Department of Agriculture, Riley arranged for an expedition to Australia to collect natural enemies of the cottony cushion scale for possible introduction. Subsequently, another American entomologist, Albert Koebele, traveled to Australia and found two enemies of the scale, a parasitic fly and the vedalia beetle. These were shipped to San Francisco, reared, and sent to Los Angeles. The natural enemies were released on scale-infested trees covered by canvas tents. The vedalia beetle population increased rapidly and was allowed to spread to adjacent trees. With the help of growers and state agencies, the vedalia beetle dispersed throughout the scale-infested areas. Complete economic suppression of the cottony cushion scale had occurred by 1889, and that suppression continues even today. Additionally, the fly parasite also became established and is today an important natural enemy of the scale in an area around San Francisco and some parts of southern California. The total cost of this management program was less than $2,000! For a detailed account of this story, see Sawyer (1996).

Many other successes followed, including the use of insects to control such weeds as the prickly pear cactus in Australia, Klamath weed in the western United States, and the musk thistle in Oklahoma. In a survey of biological control literature to 1980 (*Bulletin of the Entomological Society of America* 26:111–114), 602 attempts of classical biological control were found on a worldwide basis. The rate of complete success was 16 percent, and partial success (reduction but not elimination of the pest problem) was 58 percent. In this survey, the most complete successes were obtained with pests in the orders Hemiptera (now suborders Auchenorrhncha and Sternorrhyncha) (30 percent), Hemiptera (now suborder Heteroptera) (15 percent), Lepidoptera

## BOX 9.1 COTTONY CUSHION SCALE

**SPECIES:** *Icerya purchasi* Maskell (Hemiptera: Margarodidae)

**DISTRIBUTION:** The cottony cushion scale was introduced into the United States (California) about 1868. The species later was introduced into Florida but has never been a serious pest there.

**IMPORTANCE:** The cottony cushion scale is a citrus pest, but it also attacks shade trees, ornamentals, and other fruit trees. Soon after its introduction, the species became a serious pest, and by 1890, it had killed thousands of trees. With no effective control measures, the scale threatened the entire California citrus industry. The scale was found in New Zealand (where it was a serious pest) and in Australia (where it was not a pest). In Australia, a lady beetle, *Rodolia cardinalis* (Mulsant), commonly called the vedalia beetle, was observed to feed voraciously on the cottony cushion scale, effectively suppressing the populations. Vedalia beetles were subsequently captured and introduced into California in 1888. After 18 months, the vedalia beetle had successfully suppressed the cottony cushion scale population, and currently the species is considered only an occasional pest. The vedalia beetle has been successful against the cottony cushion scale in almost all areas of the world where it has been introduced, making it one of the primary examples of classical biological control.

**APPEARANCE:** Females are yellow to reddish brown and covered with white to yellow waxlike filaments. They are usually found carrying yellow egg masses. Males are brown and have two wings. Eggs are oval, pale red, and contained in cottonlike masses, with up to 1,000 eggs per mass. Nymphs are minute, about 0.7 mm long, and red with black legs.

**LIFE CYCLE:** Females oviposit egg masses on trees, and eggs hatch in 3 to 60 days, depending on temperature. Nymphs feed and darken as they mature. Waxy filments are excreted, covering the back and sides of the developing nymphs.

**Figure 9.1** A vedalia beetle, *Rodolia cardinalis,* eating a citrus scale. (Courtesy USDA/ARS)

(6 percent), and Coleoptera (4 percent). In 1990, it was estimated that the 722 biological control agents previously introduced in the United States had resulted in at least some level of suppression of 63 insect and mite pests (Greathead and Greathead 1992). The degree of success for the amount of money invested in biological control has been particularly impressive and has stimulated much interest in the approach for use in pest management programs.

### The Theory Behind Classical Biological Control

The theory of classical biological control is actually no different from that in ecology and population dynamics. As we have noted, many environmental factors regulate population density, thereby keeping insects within certain bounds of fluctuation. These include density-independent and perfectly and imperfectly density-dependent factors.

In the context of biological control, we think initially of the pest population as the one being regulated, and natural enemies as imperfectly density-dependent factors responsible for that regulation. The object of biological control is either to introduce natural enemies or to manipulate existing ones to cause the pest population to fluctuate (be regulated) at a density below the economic-injury level.

A goal of many of these biological control programs is to establish a self-sustaining system. For example, a natural enemy is introduced in an area with hopes that it will become established, cause the pest to fluctuate below the economic-injury level, and continue to hold the density down indefinitely without further manipulation. Of course, this strategy does not mean total elimination of the pest in an area because achieving that would also mean elimination of the food supply for the natural enemy.

The mechanism for the self-sustaining system, in theory, is based on food supply for and reproductive capability of the natural enemy. An increasing pest population provides plenty of food for the natural enemy, and the natural enemy population expands. As expansion occurs, an increasing proportion of the pest population is destroyed, reducing food for the natural enemy. This food shortage worsens until natural enemy reproduction rates decrease, causing a downward trend in enemy population numbers. As natural enemy numbers fall, less pressure is placed on the pest population, and pest numbers turn upward, followed by an increase in natural enemy numbers (Fig. 9.2).

In reality, the regulating mechanism is usually not so perfectly dependable nor smoothly operating; thus, regulation by natural enemies is imperfect at best. Therefore, this theoretical description is a rough but reasonable approximation of successful biological control programs, like the vedalia beetle and the cottony cushion scale.

In the foregoing discussion, only one aspect of natural enemy response to pest density was mentioned, namely, changes in natural enemy numbers. This aspect has been called the *numerical response.* Natural enemies also may respond to increasing pest numbers by destroying more pests per natural enemy. This aspect is referred to as the *functional response.* Numerical responses are usually of more interest because often they represent the responses most accountable for overall suppression of pest populations.

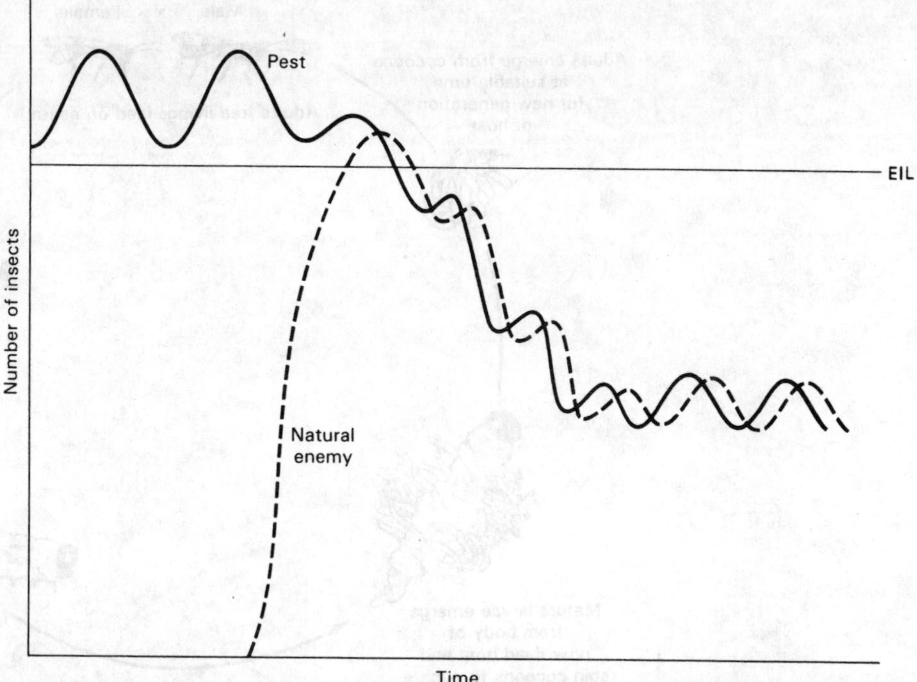


**Figure 9.2** Graphical representation of the theoretical relationship between a pest and the introduction of an effective natural enemy.

## AGENTS OF BIOLOGICAL CONTROL

Most, if not all, species of insects are preyed upon or serve as hosts for other life forms. These diverse natural enemies of insects include vertebrates, nematodes, microorganisms, invertebrates other than insects, and, perhaps most important, insects themselves. Indeed, other insects may be an insect population's worst enemy. As a group, natural enemies may function as parasites, predators, or pathogens.

### Parasites and Parasitoids

A **parasite** is an animal that lives on or within a larger animal, its host. The parasite feeds on its host, usually weakening it and sometimes killing it. A parasite requires only one or part of one host to reach maturity. Frequently, there are many parasites in a single host. Parasites with the greatest impact on insect populations are insects and nematodes. These parasites also attack mammals, as shown in Color Plate 7. Mites also parasitize insects but have a lesser impact on host populations.

**Parasitoids.** Insects that parasitize other insects and arthropods are most appropriately called **parasitoids.** A parasitoid is parasitic in its immature stages but is free-living as an adult. In all instances, parasitoids kill their hosts, but in some circumstances, the host may live much of its full life before dying. Parasitoids may attack any host stage, but the adult stage is the least frequently parasitized. These natural enemies have been used more frequently in biological control than any other kind of agent.

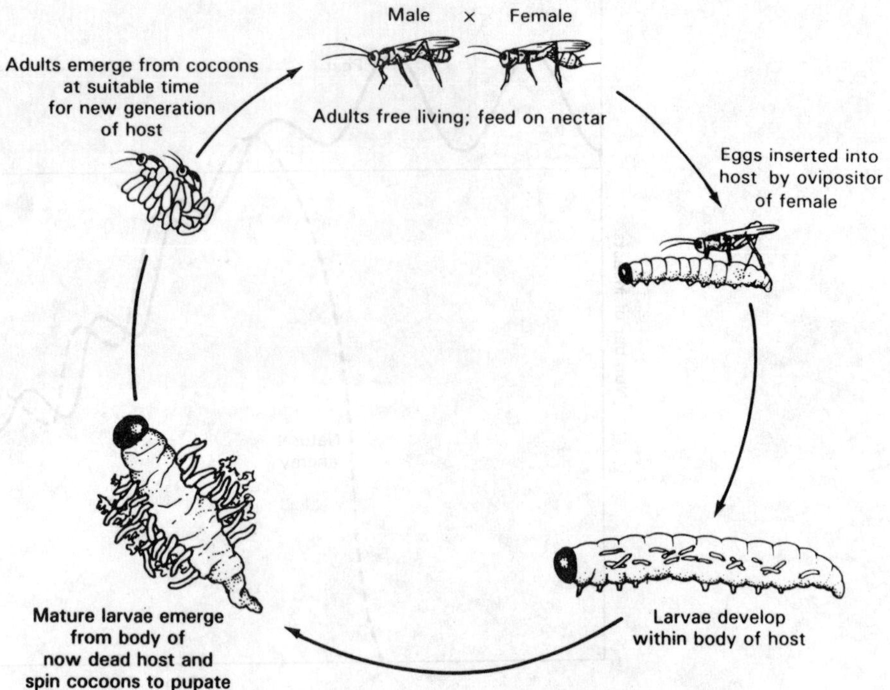
Male × Female
Adults emerge from cocoons at suitable time for new generation of host
Adults free living; feed on nectar
Eggs inserted into host by ovipositor of female
Larvae develop within body of host
Mature larvae emerge from body of now dead host and spin cocoons to pupate

**Figure 9.3** Typical life cycle of a parasitic wasp (Hymenoptera). (Redrawn from Finemore, 1984, *Plant Pests and Their Control,* Butterworths Publishing Company)

Six orders of insects, including eighty-six families, have been listed as parasitoids. These orders include the Coleoptera, Diptera, Hymenoptera, Lepidoptera, Neuroptera, and Strepsiptera. However, the most important of these by far are the Hymenoptera and Diptera, in that order (see Color Plate 7). The most significant groups of parasitoids in the Hymenoptera include small parasitic wasps in the Ichneumonidae, Braconidae, and Chalcidoidea (several families). In the Diptera, the Tachinidae stand out as the most significant parasitoids.

Parasitoids either penetrate the body wall and lay eggs inside (Fig. 9.3) the host or attach eggs to the outside of the host's body (Fig. 9.4). The latter phenomenon requires newly emerged larvae to burrow inside or somehow break through the host's exoskeleton to feed.

Parasitoids are often effective biological control agents because: (1) survival is usually good, (2) only one (or parts of one) host is required for complete development of a parasitoid, (3) populations can be sustained at low host levels, and (4) most parasitoids have a narrow host range, often resulting in a good numerical response to host density. The most frequent disadvantages of parasitoids in biological control are that: (1) host searching capacity may be strongly reduced by weather or other factors, (2) only the female searches, and (3) often the best searchers lay few eggs.

Synchronization is also a difficult problem with some parasitoids. To be effective, the parasitoid life cycle must coincide closely with that of its host before establishment and suppression can occur. Synchronization may be upset by certain environmental conditions, causing the parasitoid to fail to reduce host numbers significantly.

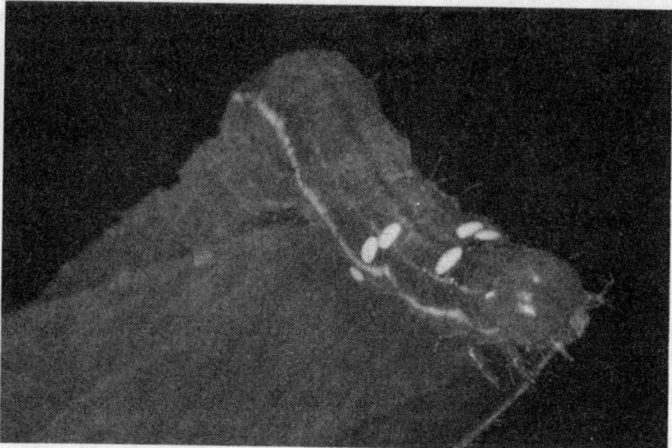

**Figure 9.4** A green cloverworm, *Hypena scabra,* parasitized with eggs of a fly (Tachinidae). The parasitoid eggs will hatch and feed inside the living caterpillar, eventually killing it. (Photo by Marlin E. Rice)

**Insect parasitic nematodes.** Nematodes are thin, unsegmented roundworms, some of which parasitize insects (Fig. 9.5). Many other species are parasitic on humans, other animals, and plants. In biological control, particular attention has been given to nematodes as parasites of bark beetles (Coleoptera: Scolytidae), grasshoppers (Orthoptera: Acrididae), and black flies (Diptera: Simuliidae). The most important families of the Nematoda affecting insects include the Mermithidae, Neotylenchidae, and Steinernematidae.

Nematode parasites of insects are now available commercially for several horticultural and agricultural applications. The primary genera used for these biological pesticides are *Steinernema* and *Heterorhabditis.* Formulations of these products include alginate gels, clays, flowable gels, and water-dispersible granules. The water-dispersible granule formulation extends the shelf life of the material, thus making the biological agents more practical and competitive with conventional insecticides. Nematode-based pesticides have been used for high- and medium-value crops such as mushrooms, berries, artichokes, citrus, mint, and turfgrass. Potential exists, however, for extending use in lower-value agricultural crops such as cotton and corn. Some current examples of nematode-based pesticides include Ecomask® (*Steinernema carpocapsae*) for borers, cutworms, root maggots, and others, and Nemasys® (*Steinernema feltiae*) for wax moth, mosquito larvae, and soil insects.

**Nonbeneficial parasites.** Although parasites and parasitoids of insects are usually thought of as beneficial, some are not. Like pests, natural enemies have enemies themselves. Parasitoids have parasites (most often, tiny wasps) that may kill them while the parasitoids are feeding on their hosts. This phenomenon has been termed **hyperparasitism.** Here, the original parasite of the pest is the *primary parasite,* and the hyperparasite is the *secondary parasite.* Primary parasites are also not beneficial to biological control when they parasitize insect predators or plant-feeding insects introduced to suppress weeds. In some situations, the potential of biological control programs can be threatened by the presence of these detrimental species, and they must be considered before major biological control programs are implemented.

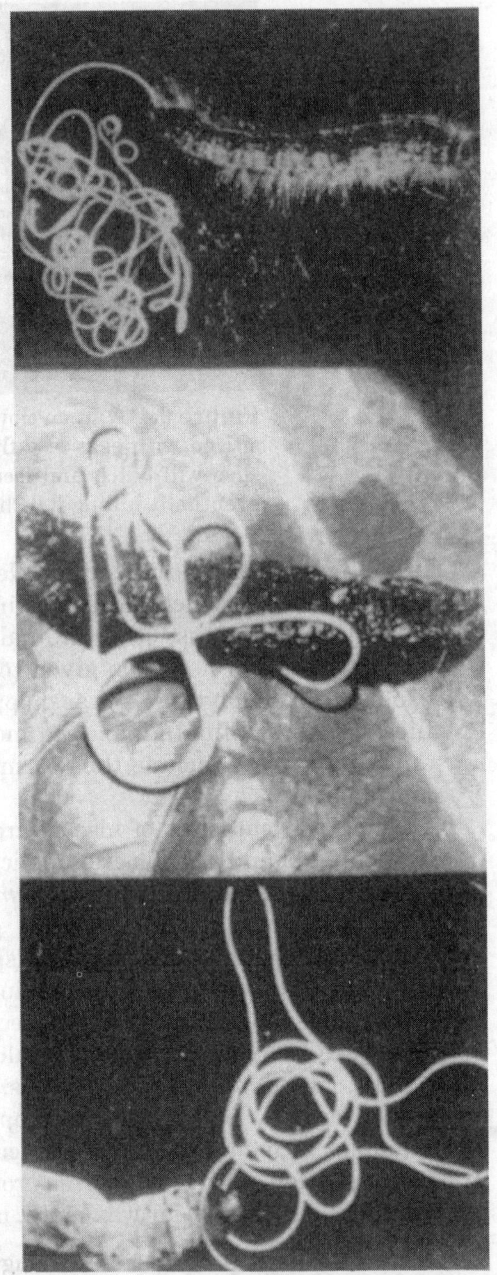

**Figure 9.5** Parasitic nematodes (Mermithidae) emerging from Lepidoptera: tent caterpillar (*top*), mahogany shoot borer (*middle*), and fall armyworm. (From G. C. Papavizas, ed., 1982, *Biological Control in Crop Production,* Beltsville Agricultural Research Center, Totowa, N.J.: Allanheld, Osmun & Co, 1982)

**Figure 9.6** Mosquito fish feeding on a mosquito larva. (After Copple and Mertins, 1977, *Biological Insect Suppression*, Springer-Verlag Publishers, with permission)

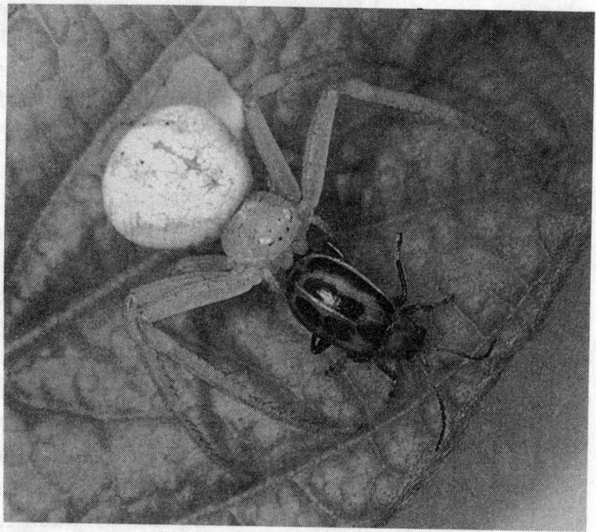

**Figure 9.7** A crab spider (Thomsidae) preying upon a bean leaf beetle, *Ceratoma trifurcata*. (Photo by Marlin E. Rice)

## Predators

**Predators** (see Color Plates 2, 8) are free-living organisms that feed on other animals, their prey, sometimes devouring them completely and usually rapidly. Predators may attack prey both as immatures and adults, and more than one prey individual is required for a predator to reach maturity. Major predators of insects include birds, fish (Fig. 9.6), amphibians (toads, frogs, salamanders), reptiles (lizards, snakes, turtles), mammals (bats, rodents, shrews, and others), and arthropods (insects, mites, spiders) (Fig. 9.7). Even some plants are predators of insects; for example, the Venus flytrap has clustered leaves with blades that close and trap insects for food.

Many of the most important predators in biological control programs have been insects and mites. The habit of predation is very common among insects, and some of our most successful instances of biological control have been with insect predators and insect pests.

**Figure 9.8** The convergent lady beetle, *Hippodamia convergens,* is an important polyphagous predator of many pest species, especially aphids. (Photo by Marlin E. Rice)

It is difficult to rank insect orders with regard to significance in predation because nearly every order has important species. However, in terms of diversity and significance of biological control, the Coleoptera, Neuroptera, Hymenoptera, Diptera, and Hemiptera (see Color Plate 7) are outstanding. Some particularly important species groups include lady beetles (Fig. 9.8, Box 9.2) (Coleoptera: Coccinellidae), ground beetles (Coleoptera: Carabidae), lacewings (Neuroptera: Chrysopidae), ants (Hymenoptera: Formicidae), flower flies (Diptera: Syrphidae), damsel bugs (Hemiptera: Nabidae), and assassin bugs (Hemiptera: Reduviidae).

Some predaceous insects like the vedalia beetle have a narrow prey range, feeding almost exclusively on a single species. These predators are **monophagous.** Other predators such as syrphid larvae have a narrow host range, feeding on only a few prey species. These predators are **oligophagous.** As a group, however, predators tend to feed on a wide range of prey and are considered **polyphagous.**

Being polyphagous has both advantages and disadvantages for biological control. It is an advantage that polyphagous predators can survive by shifting to alternate prey when densities of the pest species are low. It is disadvantageous, however, when the polyphagous habit results in lack of responsiveness to changes in pest density—in other words, where the predator does not suppress a growing pest population. It is difficult to predict the density response of a natural enemy to a given pest from food specificity alone. See Symondson et al. (2002) for a detailed discussion of this topic.

Other characteristics make predators effective biological control agents: (1) prey is killed rapidly, (2) often all individuals in the population (males, females, immatures, and adults) search for prey, and (3) synchronization in predator/prey life cycles is not a frequent problem.

Much debate has focused on whether predators or parasitoids (for example, see Fig. 9.9) are the more efficient biological control agents. Certainly, parasitoids have been used more frequently than predators in attempts at biological

## BOX 9.2 LADY BEETLES (SEE COLOR PLATE 8)

**SPECIES:** Numerous coccinellid species (Coleoptera: Coccinellidae)

**DISTRIBUTION:** Lady beetles are widely distributed throughout the world, with many species represented in North America. Throughout their range, they are found on a wide variety of trees, shrubs, weeds, grasses, and cultivated crops.

**IMPORTANCE:** Lady beetles represent one of the most beneficial and recognizable group of insects known to humankind. For most species, both adults and larvae are voracious predators on a wide range of prey, including aphids, scales, mites, and thrips. One of the most common, *Hippodamia convergens* Guerin, can consume its weight in aphids daily as a larva and as many as 50 aphids daily as an adult. Lady beetles have been imported to control outbreaks of aphids and scale insects. The most successful of these introductions was in 1888, when *Rodolia cardinalis* (Mulsant) was imported from Australia to control the cottony cushion scale, *Icerya purchasi* Maskell, a pest of California citrus groves. For 100 years this beetle, commonly called the vedalia beetle, has kept cottony cushion scale populations below economic levels. Two species of Coccinellidae, however, are plant-feeding and considered pests: the Mexican bean beetle, *Epilachna varivestis* Mulsant, and the squash beetle, *Epilachna borealis* (Fabricius).

**APPEARANCE:** Adults range from 3 to 8 mm long and are oval. Both males and females are brightly colored, with the most common species red or brown with black spots. A few species have black elytra with red spots. Eggs of most species are orange and deposited in clusters of ten to fifty. Larvae are elongate and usually patched with bright colors.

**LIFE CYCLE:** Adults overwinter, often aggregating in large masses on plants and then under rocks, bark, leaf debris, and the southern side of structures. In spring, the beetles become active and mate. Females oviposit as many as 1,000 eggs in a 1- to 3-month period. Incubation time in most instances is 2 to 6 days. Most lady beetles undergo three larval molts in about 20 days. Pupation lasts from 3 to 10 days, and adults can live from a few months to more than a year. Larvae and adults can be seen throughout the summer on the same plant, feeding on aphids and other small insects and their eggs.

**Figure 9.9** An armyworm, *Pseudaletia unipuncta,* with a cluster of newly emerged parasitoid larvae (Hymenoptera) on the side of its body. The parasitoids will form cocoons and the caterpillar will die from the internal injuries. (Photo by Marlin E. Rice)

control. Ratios of parasitoids to predators used in programs have been estimated at from 2:1 to 4:1. But the most successful type of agent for any one pest problem is almost impossible to predict. Success actually depends on the particular agroecosystem, the pest(s) involved, and the environment in which these interact. Consequently, the best design usually has been management programs that rely on the actions of both parasitoids and predators.

### Pathogenic Microorganisms

Just like people and other animals, insects contract diseases, and natural populations are often strongly influenced by disease epidemics. Insect diseases and their symptoms have been recognized as far back as 2700 B.C. in China with the honey bee, *Apis mellifera,* and the silkworm, *Bombyx mori.* Early scientific studies on insect diseases include investigations of white muscardine disease caused by the fungus *Beauveria bassiana* and pebrine disease caused by the microspordian *Nosema bombycis* on silkworms. The idea of using a microorganism (a fungus) to control a pest (the sugarbeet curculio, *Cleonus punctiventus*) dates back to the eighteenth century. Today, the science of insect pathology is an established discipline, contributing significantly to the biological control of insects.

The major microorganisms causing diseases in insects include bacteria, viruses, protozoans, fungi, and rickettsiae. They may cause diseases that kill insects outright, reduce their reproductive capabilities, or slow their growth and development (see Color Plate 7). These organisms cause disease epidemics in insect populations naturally, and understanding their action is an important part of predicting the population dynamics in many pests.

Most of these microorganisms have been used, or studied for use, as so-called **microbial insecticides.** Microbial insecticides are biological preparations that are often sprayed or delivered in ways similar to those of conventional chemical insecticides; the various formulations include baits, dusts, granules, and sprays. Such biological preparations have label directions like insecticides and must be registered with the United States Environmental Protection Agency before they can be sold. More than twelve insect pathogens have been registered since the first registration of *Bacillus popilliae* in 1948.

To date, the microorganisms most widely used as microbial insecticides are bacteria and, to a lesser extent, viruses and fungi. Neither protozoa nor rickettsiae are used extensively, having the disadvantage of being slow to kill insects, if they kill at all. Some bacteria and many viruses kill insects relatively quickly (days to a few weeks rather than several weeks or months). Although ultraviolet radiation can strongly reduce their persistence, they are fairly resilient to other weather extremes. Microbial insecticides are discussed in detail in Chapter 12.

## THE PRACTICE OF BIOLOGICAL CONTROL

Through the years, many diverse methods have been developed for practicing biological control, but most programs have a common thread. They all seek to reduce pest status by increasing pest mortality, frequently by increasing existing natural-enemy-to-prey ratios. Attempts at increasing ratios include placing natural enemies in the environment and modifying the environment to ecologically promote increases in natural enemy numbers. Another method used to raise ratios is modifying established crop production practices so that natural enemies otherwise destroyed are conserved.

To reflect these practices, many specialists group biological control practices into three categories: introduction, augmentation, and conservation. Although there is some overlap between these categories, we use them for the sake of discussion. It also should be noted that more than one practice may be used in a single biological control program.

### Introduction

**Introduction,** also known as importation, is often considered the "classical" practice in biological control because most early programs used this approach. The basis for this practice is to identify the natural enemies that regulate a pest in its original location and introduce these into the pest's new location; thus, natural enemies are reassociated with their prey and hosts. The hope is that the natural enemies, once introduced, will become established and permanently reduce the pest's general equilibrium position (average population level) below the economic-injury level.

One of the most successful examples of this approach has already been mentioned: the introduction of the vedalia beetle to control the cottony cushion scale. Other substantial successes include citrus blackfly, *Aleurocanthus woglumi,* controlled by *Eretmocerus serius,* a wasp parasitoid; woolly apple aphid, *Eriosoma lanigerum,* controlled by *Aphelinus mali,* a wasp parasitoid; purple scale, *Lepidosaphes beckii,* controlled by *Aphytis lepidosaphes,* a wasp parasitoid; and olive scale, *Parlatoria oleae,* controlled by *Aphytis maculicornis* and *Coccophagoides utilis,* both wasp parasitoids.

Spectacular successes from the practice of introduction have occurred in several types of agroecosystems but particularly in those of orchards and forests. Biological control of the alfalfa weevil in alfalfa is an example of the approach in an important forage crop (Fig. 9.10). Eleven parasitoids and one predator were introduced in the northeastern United States, most of them in 1959. By 1970, six parasitoids (all Hymenoptera) were widely established and caused

**Figure 9.10** A parasitic wasp, *Bathyplectes* sp., laying an egg inside an alfalfa weevil larva. The parasitoid larva will hatch, feed inside the weevil, and kill it. The parasitoid then emerges and spins a cocoon where it will transform into the adult. (Photo by Dennis Schotzko)

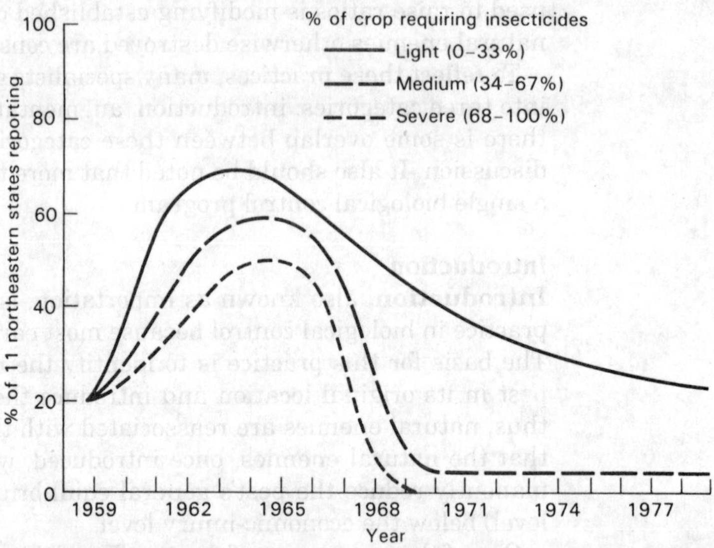


**Figure 9.11** Effect of biological control in reducing insecticide use with the alfalfa weevil, *Hypera postica,* in eleven northeastern states. (From G. C. Papavizas, ed., 1982, *Biological Control in Crop Production,* Beltsville Agricultural Research Center, Totowa, N.J.: Allanheld, Osmun & Co, 1982)

significant mortality in egg, larval, and adult stages. By the late 1970s, these parasitoids were responsible for weevil mortality rates of about 70 percent, with about 73 percent of the alfalfa acreage in the Northeast no longer requiring insecticide treatments (Fig. 9.11). Similar introductions by USDA/APHIS, particularly using *Microctonus aethiopoides,* have been made in the Midwest and other parts of the United States, with hopes of achieving similar results.

One of the most important successes of introduced biological agents occurred in central Africa with the cassava mealybug, *Phenacoccus manihoti* (Hemiptera: Pseudococcidae). This pest was introduced in the 1970s from South America and threatened the production of a major food staple for more than 200 million people. The cassava mealybug caused up to 80 percent losses of the crop by feeding on the growing tips of plants, resulting in severe stunting and limited tuber development. Following identification of this pest, exploration for natural enemies was conducted in Mexico and Central and South America. Several natural enemies were exported to Africa, tested, and released for suppression of the pest. One parasitoid species, *Epidinocarsis lopezi* (Hymenoptera: Encyrtidae), quickly became established. Releases throughout the African cassava belt were conducted in more than 100 areas between 1981 and 1990. Major reductions of cassava mealybug numbers to tolerable levels occurred in all regions, except those with poor growing conditions. The spread of this biological control agent exceeds that of any other agent introduced into Africa, and adaptability of the agent to different ecological conditions is without precedent there (Herren and Neuenschwander 1991).

This tremendous success is credited with preventing the malnutrition of millions of Africans and may well be the most important example of classic biological control ever. For this achievement, Dr. Hans Herren, an entomologist headquartered in Kenya, received the 1995 World Food Prize.

The pattern for introducing natural enemies in the preceding examples begins with determining the geographical origin of the pest species. If the pest has itself been introduced from another country (that is, it is **exotic**), the introduction of natural enemies from its original location may hold promise. For native pests, this approach usually is not practical. If the species is exotic, foreign exploration is conducted to collect natural enemies. International exploration is aided by such agencies as the European Biological Control Laboratory (USDA/ARS), headquartered in Montpellier, France, Commonwealth Institute of Biological Control, headquartered in Trinidad, West Indies, and the International Organization for Biological Control of Noxious Animals and Plants, headquartered in Zurich, Switzerland. Natural enemies are quarantined for study and processing. If approved for release, they are colonized in appropriate agroecosystems, and efforts are made to assist their dispersal. Finally, evaluations are conducted to determine natural enemy impact on pest populations.

Although there have been some outstanding and well-known successes, most introductions for classical biological control have not been successful. Statistics show that for arthropod pests, only about 50 percent of the introductions, or less, lead to establishment of the biological control agent, and in only about 10 percent of the cases do they actually contribute to control of the target pest.

However, there has been value in natural enemy introduction. When successful, introduction has been long-lasting and economical. However, certain potential limitations should be mentioned. First, introduction is primarily useful only for exotic pests. In the United States, exotic pests are very important, but they make up only about 39 percent of the 600 most important arthropod-pest species. A few examples show some success of an exotic natural enemy against a native pest. Although there is some debate on this issue, it seems unlikely that the majority of native pests can be controlled in this way because introduced enemies must compete with an established complex of native natural enemies, thereby reducing their chances of establishment.

Agroecosystem stability also may hamper success through introduction. The majority of successes with introduced natural enemies have been with pests in rather stable systems. These usually include crops of perennials, for example, orchards and forests. Less success has been shown in greatly disturbed systems such as corn, cotton, wheat, and other annual row crops. When clean tillage is practiced, the above-ground habitat must be recolonized each year from outside the field. Whereas rapid recolonization has been observed with many native natural enemies, establishment of exotic natural enemies under these circumstances is difficult. However, the status of agroecosystem stability is changing at the present time, with shifts away from conventional tillage toward no-till, chisel-plow, and other conservation tillage. Such changing practices may have important effects on the potential of classical biological control in annual row crops.

Another concern voiced by many ecologists is that the intentional introduction of alien species into complex biological communities may upset their ecological stability and dynamics. Specific questions involve the potential for host switching, dispersal of introduced species into nonagricultural habitats, irreversibility of introductions, potential efficacy and ecological impacts, possible adaptation to new hosts, and difficulty of predicting interaction outcomes (Louda et al. 2003). Future research into introductions for biological control will likely focus on these concerns.

### Augmentation

Augmentation is a biological control practice that includes any activity designed to increase numbers or effect of existing natural enemies. These objectives may be achieved by releasing additional numbers of a natural enemy into a system or modifying the environment in such a way as to promote greater numbers or effectiveness.

Releasing additional numbers of a natural enemy is similar to the introduction practices discussed previously, except that augmentive releases are expected to result only in temporary (usually one season or less) suppression. Therefore, the ecological result is most likely the dampening of pest population peaks, rather than significantly changing the general equilibrium position. Because of their temporary effect, these releases must be made periodically. Periodic releases may be considered as either inundative or inoculative.

**Inundative releases.** These releases depend on propagation of massive numbers of natural enemies and their widespread distribution. Subsequently, the release of the actual enemies suppresses the pest population, with little or no impact expected from progeny of the released individuals. Massive releases have been attempted in several programs involving natural enemies like *Trichogramma* species, a tiny wasp that parasitizes insect eggs. This natural enemy was commercially marketed in Europe by BASF France, with production adequate to treat 12,000 to 16,000 hectares of corn against the European corn borer. Improved application methods of the wasps include paper walnut-shaped packets, Trichocaps®, that protect the wasps and serve to disperse them in crops (Fig. 9.12). Moreover, an estimated 1 million acres are treated each year with this natural enemy in the People's Republic of China. In addi-

**Figure 9.12** Trichocaps® used in augmentation programs of biological control. These cardboard packets contain about 500 eggs of the Mediterranean flour moth parasitized by a small wasp, *Trichogramma brassicae.* They are held in cold storage until needed and applied to the crop by hand or airplane. Once developed, the wasps emerge from a hole in the capsule and parasitize pests such as the European corn borer.

## BOX 9.3 GREEN LACEWING (SEE COLOR PLATE 8)

**SPECIES:** *Chrysoperla carnea* (Stephens) (Neuroptera: Chrysopidae)

**DISTRIBUTION:** Green lacewings are found as larvae and adults throughout North America on the leaves of many species of weeds, grasses, trees, and shrubs. Adults are especially attracted to lights at night.

**IMPORTANCE:** Green lacewing larvae are important predators of a wide variety of small insects and their eggs. The larvae are very active and feed on almost any prey their size or smaller. Although many species of adult lacewings are predaceous, common green lacewing adults feed on honeydew. This species, which can be reared relatively easily, is generally tolerant to pesticides and has been used in biological control to manage injurious moth larvae and eggs.

**APPEARANCE:** Adults have green cylindrical bodies with light green veins on large, transparent wings. The antennae are of the long filiform type, and the eyes are golden colored. Pale white eggs are oviposited at the ends of long stalks that elevate them above plant surfaces. These stalks offer protection to the eggs from predators. The pale larvae, often called aphidlions, have pronounced, hollow sickle-shaped jaws that they use to capture and suck the contents of their prey. When larvae are preparing to pupate, they spin white parchmentlike cocoons from which adults emerge.

**LIFE CYCLE:** Adults overwinter in dry, dark areas, such as leaf litter, abandoned wasp nests, barns, and stables. The adult life-span is from 4 to 6 weeks. Females oviposit from 100 to 200 eggs, which hatch after a 3- to 6-day incubation. Lacewing larvae undergo three molts, which total approximately 14 days. Prepupal and pupal stages last about 8 days.

---

tion to this parasitoid, general predators like green lacewings (Box 9.3), *Chrysoperla carnea,* and lady beetles, *Hippodamia convergens,* are frequently used in augmentation programs. Several of these polyphagous natural enemies are available commercially (see Hunter 1994).

In many instances, however, inundative releases of insect predators and parasitoids have been unsuccessful. The lack of success is usually ascribed to insufficient coverage over a great enough area or augmentation in environments not supportive of the numbers released. Another reason is the movement of released natural enemies out of the targeted area.

Some of the most successful inundative releases have been with pathogenic microorganisms like *B. thuringiensis.* Microbial insecticides have been applied mostly against caterpillars (Lepidoptera), beetles (Coleoptera), mosquitoes, and black flies (Diptera); they suppress a pest population quickly, much the same as a conventional insecticide. After the initial impact of the "release," little subsequent suppression can be expected. If additional pest outbreaks occur during the season, more applications of the microbes must be made. A number of pesticides are registered as microbial insecticides (see Chapter 12).

**Inoculative releases.** These releases differ from inundative releases in that, once they occur, the natural enemy is expected to colonize and spread throughout an area naturally. An inoculation is often made only once in the growing season,

## BOX 9.4 MEXICAN BEAN BEETLE (SEE COLOR PLATE 4)

**SPECIES:** *Epilachna varivestis* Mulsant (Coleoptera: Coccinellidae)

**DISTRIBUTION:** The Mexican bean beetle is native to the semiarid southwestern United States, with a natural range in and around Mexico. The pest has successfully spread throughout much of the United States since 1920, when it was found in Birmingham, Alabama. Most of the expansion has occurred east of the Mississippi River; however, an infestation was found in California in 1946.

**IMPORTANCE:** The Mexican bean beetle is unusual as a pest because it is one of only two plant-eating species of lady beetles, which are, as a group, typically beneficial insects. Adults and larvae feed on the underside of leaves and are often economically damaging to beans, including bush, pole, lima, and soybean. The skeletonized appearance of leaves injured by the Mexican bean beetle is quite distinct. This appearance results from adults and larvae scraping off the leaf epidermis and soft leaf tissue while leaving the more rigid structured leaf components intact. Eventually, the netting may also die and fall out of the injured area. Larvae are subject to attack by several parasites, including a eulophid wasp, which has been successfully reared and mass-released as a pest management tactic.

**APPEARANCE:** Adults have sixteen black spots arranged in three rows on yellowish bronze elytra. Adults are rounded and convex and about 8 mm long. Eggs are yellow and found in clusters of forty to sixty on the underside of host leaves. Larvae are pale yellow to orange and have six longitudinal rows of branched spines on their back. Pupae are orange-yellow and remain attached to the underside of leaves by utilizing the last larval skin until adult emergence.

**LIFE CYCLE:** Adults overwinter in plant residue on the ground. They begin feeding on available foliage in spring, and females oviposit on the undersides of leaves. Eggs hatch in about 7 days, and new larvae begin to feed on leaf epidermis of the plant host. There are four stadia, each lasting about 6 days (depending on temperature). Pupation occurs and lasts about 7 days, with total generation time averaging 33 days. There may be one to four generations per year, depending on latitude and temperature.

and the progeny of the released individuals have the most significant impact on the pest population. Some of the most successful programs utilizing this approach with predators and parasitoids have been reported from field and greenhouse crops in the People's Republic of China and Russia. In the United States, the largest programs involving inoculative releases have included *Aphytis* species used on about 4,450 hectares against California red scale, *Aonidiella aurantii; Chrysoperla* species (lacewings) used on about 2,385 hectares against mealybugs (Hemiptera: Pseudococcidae); *Cryptolaemus montrouzieri* (a lady beetle) used on about 5,670 hectares against the black scale, *Saissetia oleae;* and the egg parasite, *Trichogramma* species, used on more than 100,000 hectares of crops against several lepidopterous pests. In past years, a small program in the eastern United States for soybeans used an exotic parasitoid, *Pediobius foveolatus,* imported from India, against a pest native to Central America and Mexico, the Mexican bean beetle (Box 9.4), *Epilachna varivestis.*

Other significant uses of inoculative releases include those against greenhouse (glasshouse) pests in Great Britain, Holland, and other European countries. In particular, planned releases of predatory mites, *Phytoseiulus persimilis,* against plant-feeding mites; and parasitoids, *Encarsia formosa* and *Aphidius matricariae,* against whiteflies (Hemiptera: Aleyrodidae) and aphids (Hemiptera: Aphididae) have gained credibility in recent years.

Inoculative releases have been useful in the past against the Japanese beetle in turf grass. Two bacteria are responsible for causing milky disease in this insect. To suppress beetle populations, commercial formulations of spore powder, containing *Bacillus popilliae,* are deposited on the turf surface and washed in artificially or by rainfall. When successful, the treatment causes an epidemic in the local grub (larval) population. Currently, there is one registered product containing this bacterium in the United States.

**Environmental manipulations.** Another method of augmenting numbers of natural enemies, besides releasing them, involves environmental manipulation. Here, the crop itself or surrounding crop areas are utilized as "field insectaries" to increase numbers or make existing numbers more efficient. Most practical have been attempts to make requisites (for example, alternate nutrients, nesting habitats, overwintering sites) more available, thereby enhancing natural enemy reproduction. Examples of this approach include providing nesting sites for predaceous wasps, *Polistes* species, and maintaining wild blackberries or French prune tree orchards adjacent to vineyards as food for alternate hosts of a parasitoid, *Anagrus* spp. In the latter instance, the wasp parasitoid increases populations in the wild blackberries and prune trees, then moves into vineyards, thereby suppressing grape leafhoppers, *Erythroneura elegantula,* a pest of imported grapes (Fig. 9.13).

Another important example of environmental manipulation is the planting of cover crops (for example, broadbean, *Vicia faba*) in apple orchards to reduce pest populations. In California, maintaining a cover crop, periodically mowing it, and leaving the mulch was found to lower infestations of aphids, leafhoppers (Hemiptera: Cicadellidae) (Fig. 9.14), and codling moth, *Cydia pomonella.* Lowered infestations were accompanied by increased numbers of surface-dwelling arthropod predators. The cover crop and related activities are believed to interfere with the host-finding ability of the pests, as well as to provide alternate food and habitat for natural enemies. Van Driesche and Bellows (1996) and Landis et al. (2000) give detailed information on the many forms of habitat modifications for enhancing natural enemy cropping systems.

Although somewhat futuristic, certain chemicals also may play a role in natural enemy augmentation. Presently, substances called **kairomones** are being researched that attract natural enemies and/or stimulate them to become more efficient. For instance, kairomones derived from the scales of corn earworm moths have been shown to stimulate searching behavior of adult *Trichogramma* parasitoids. When these kairomones were applied to field plots, wasp searching increased, and this behavior was followed by increased parasitization rates. It is possible that additional research with kairomones may provide practical substances that enhance the effect of natural enemies already present in agroecosystems.

## Conservation of Natural Enemies

Probably the most widely practiced form of biological control is conservation of natural enemies. The objective of conservation is to protect and maintain

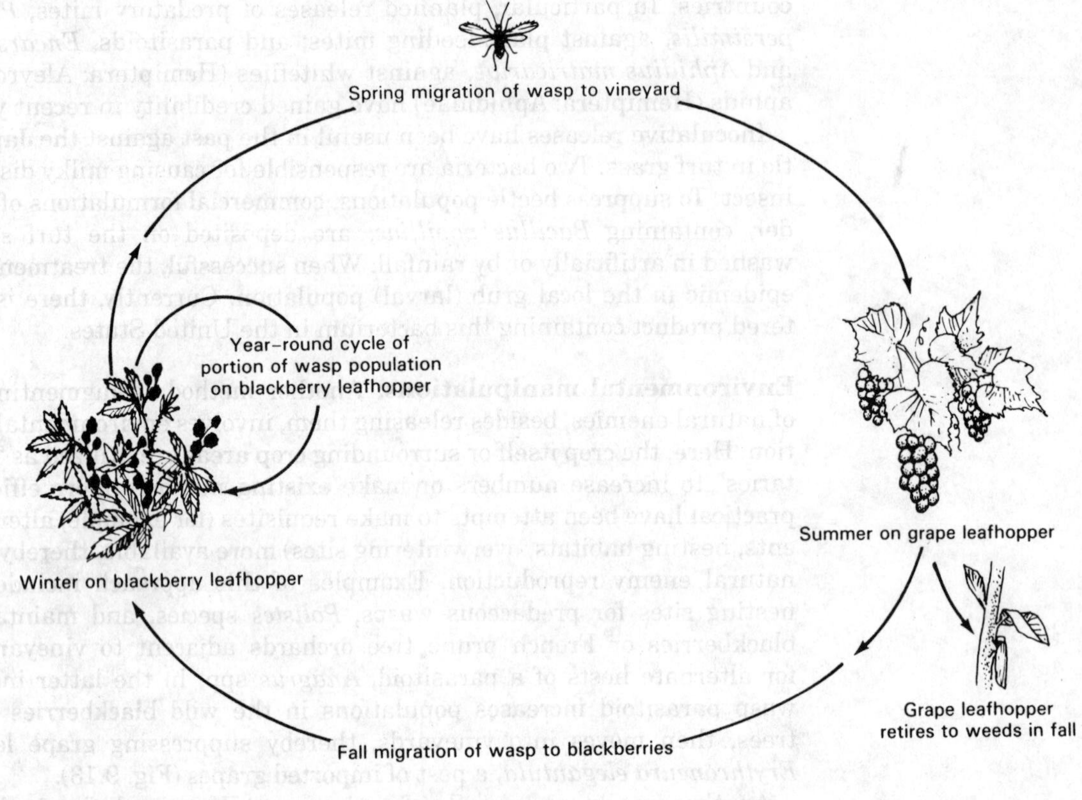


**Figure 9.13** Diagram of the life cycle of a parasitic wasp, *Anagrus epos,* which attacks the grape leafhopper, *Erythroneura elegantula.* The parasitoid overwinters in California blackberries, parasitizing eggs of the blackberry leafhopper, *Dikrella cruentata,* when the grape leafhopper is inactive. In spring, some wasps move into vineyards to parasitize the grape leafhopper. (Reprinted with permission of Plenum Press from *Introduction to Integrated Pest Management* by M. L. Flint and R. van den Bosch. © 1981)

these existing populations, particularly the insect predators and parasitoids, in an agroecosystem.

Basically, this approach requires knowledge about all aspects of the natural enemy community, including species present, population numbers, phenology, and impact on pest populations—in other words, the life systems of natural enemies in a given area. With this knowledge, crop production activities, including existing pest management practices, are modified to avoid natural enemy destruction.

Methods to conserve natural enemies may include less frequent mowing of field edges and clipping alfalfa in strips at different times to maintain habitat and alternate food sources for their populations. Perhaps more important as a conservation approach is the use of insecticides to avoid significant natural enemy mortality. This method usually includes using chemicals less toxic to natural enemies, reducing numbers of applications, and reducing dosage levels.

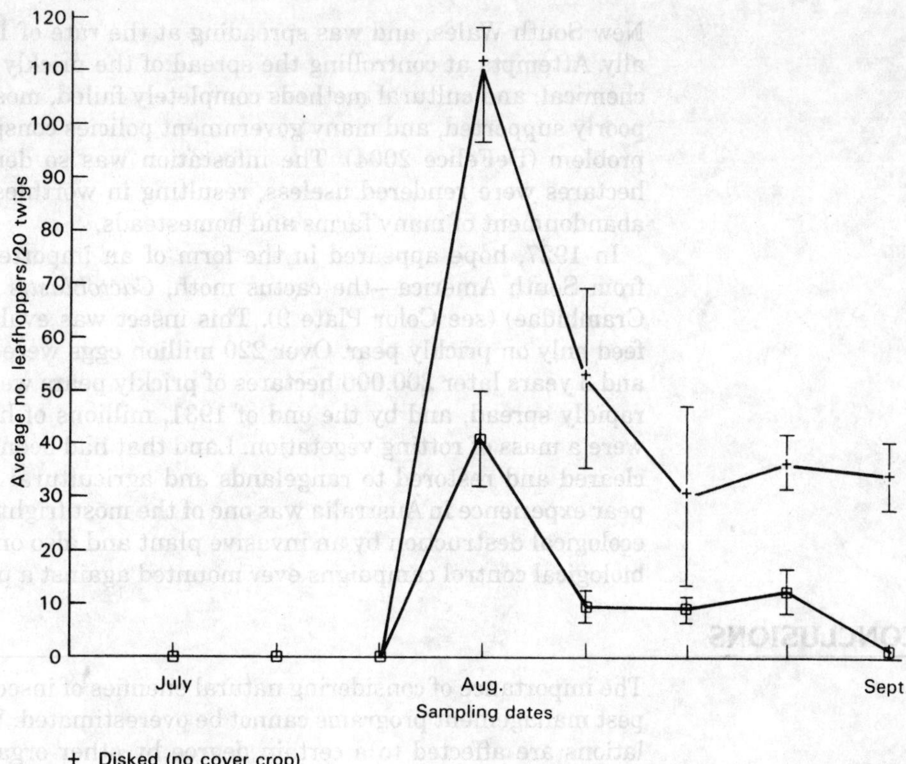


**Figure 9.14** Differences in leafhopper populations in California apple orchards with and without cover crops. (After Altieri and Schmidt, *California Agriculture,* January–February 1986)

## CASE STUDY

### Prickly Pear Cactus and Cactus Moths in Australia

Prickly pears, or prickly pear cactus (*Opuntia* spp.), are native to the Americas but have become serious invasive weeds in suitable habitats around the world. Around 1840, cuttings of prickly pears were brought to Queensland, Australia, for use as a hedge around fields and homesteads, as a botanical curiosity, and for production of cochineal—a dark reddish dye produced by scale insects that feed on the plant. Livestock and native birds quickly spread prickly pear seeds across overgrazed grasslands, where competition was reduced during droughts, whereas during heavy rainfall, broken pieces of prickly pears were carried into the interior of westward flowing rivers (DeFelice 2004). The climate and soil of eastern Australia was ideal for prickly pear, and the weed quickly spread. Attempts were made by farmers and ranchers in the 1880s to control the weed, but were without success. In 1893, it was declared a noxious weed in Queensland.

By 1913, prickly pear was estimated to cover 1.4 million hectares with dense infestations and another 4.9 million hectares with scattered infestations. By 1926, the prickly pear had infested 24 million hectares in Queensland and

New South Wales, and was spreading at the rate of 1 million hectares annually. Attempts at controlling the spread of the prickly pear using mechanical, chemical, and cultural methods completely failed, mostly because control was poorly supported, and many government policies conspired only to worsen the problem (DeFelice 2004). The infestation was so dense that the 12 million hectares were rendered useless, resulting in worthless grazing land and the abandonment of many farms and homesteads.

In 1927, hope appeared in the form of an imported plant-parasitic insect from South America—the cactus moth, *Cactoblastis cactorum* (Lepidoptera: Crambidae) (see Color Plate 9). This insect was evaluated and confirmed to feed only on prickly pear. Over 220 million eggs were reared and distributed, and 3 years later 200,000 hectares of prickly pears were destroyed. The insect rapidly spread, and by the end of 1931, millions of hectares of prickly pears were a mass of rotting vegetation. Land that had been useless for decades was cleared and restored to rangelands and agricultural production. The prickly pear experience in Australia was one of the most frightening cases in history of ecological destruction by an invasive plant and also one of the most successful biological control campaigns ever mounted against a pest (DeFelice 2004).

## CONCLUSIONS

The importance of considering natural enemies of insect pests when developing pest management programs cannot be overestimated. Virtually all insect populations are affected to a certain degree by other organisms in their environment, and some of our most efficient approaches to pest management include using these organisms to our advantage. This can mean simply modifying a crop culture technique to conserve the enemies already present or supplementing their numbers with releases. If we are dealing with an introduced pest, searching for a native natural enemy and introducing it also holds promise.

Although much has been accomplished through biological control, its potential has not been fully developed for most crops. Much remains to be learned about natural enemy ecology and methods of manipulation before this tactic can be fully integrated with other management tactics. With more detailed knowledge, we can look forward to pest management programs that have greater effectiveness, lower costs, and less disruption of nonagricultural environments.

### Further Reading

Bellows, T. S., and T. W. Fisher, eds. 1999. *Handbook of Biological Control.* New York: Academic Press.

Clausen, C. P. 1940. *Entomophagous Insects.* New York: McGraw-Hill.

Coppel, H. C., and J. W. Mertins. 1977. *Biological Insect Suppression.* New York: Springer-Verlag.

Day, W. H. 1982. Biological control of the alfalfa weevil in the northeastern United States, pp. 361–390. In G. C. Papavizas, ed., *Biological Control in Crop Production.* London: Allanheld, Osmun Publishers.

DeFelice, M. S. 2004. Prickly pear cactus, *Opuntia* spp.—A spine-tingling tale. *Weed Technology* 18:869–877.

Flexner, J. L., and D. L. Belnavis. 1998. Microbial insecticides, pp. 35–62. In J. E. Rechcigl and N. A. Rechcigl, eds., *Biological and Biotechnological Control of Insect Pests,* Boca Raton, Fla.: Lewis Publishers.

Frisbie, R. E., and J. W. Smith, Jr. 1989. Biologically intensive integrated pest management: The future, pp. 151–164. In J. J. Menn and A. L. Steinhauer, eds., *Progress and Perspectives for the 21st Century.* Lanham, Md.: Entomological Society of America.

Greathead, D. J., and A. H. Greathead. 1992. Biological control of insect pests by parasitoids and predators: The Biocat data base. *Biocontrol News and Information* 13(4):61N–68N.

Hagler, J. R. 2000. Biological control of insects, pp. 207–241. In J. E. Rechcigl and N. A. Rechcigl, eds., *Insect Pest Management.* Boca Raton, Fla.: Lewis Publishers.

Hall, F. R., and J. W. Barry, eds. 1995. *Biorational Pest Control Agents.* Washington, D.C.: American Chemical Society.

Herren, H. R., and P. Neuenschwander, 1991. Biological control of cassava pests in Africa. *Annual Review of Entomology* 36:257–283.

Hoy, M., and D. C. Herzog, eds. 1985. *Biological Control in Agricultural IPM Systems.* Orlando, Fla.: Academic Press.

Huffaker, C. B., and P. S. Messenger, eds. 1976. *Theory and Practice of Biological Control.* Orlando, Fla.: Academic Press.

Hunter, C. D. 1994. *Suppliers of Beneficial Organisms in North America.* Sacramento, Calif.: California Environmental Protection Agency.

Khetan, S. K. 2000. *Microbial Pest Control.* New York: Marcel Dekker.

Landis, D. A., S. D. Wratten, and G. M. Gurr. 2000. Habitat management to conserve natural enemies of arthropod pests in agriculture. *Annual Review of Entomology* 45:175–201.

Louda, S. M., R. W. Permerton, M. T. Johnson, and P. A. Follett. 2003. Nontarget effects—the Achilles heel of biological control. *Annual Review of Entomology* 48:365–396.

Maddox, J. V. 1994. Insect pathogens as biological control agents, pp. 199–244. In R. L. Metcalf and W. H. Luckmann, eds., *Introduction to Insect Pest Management,* 3rd ed. New York: Wiley.

Nordlund, D. A. 1984. Biological control with entomophagous insects. *Journal Georgia Entomological Society* 19 (2nd supp.):14–27.

Reuveni, R., ed. 1995. *Novel Approaches to Integrated Pest Management.* Boca Raton, Fla.: Lewis Publishers.

Ridgway, R. L., and S. B. Vinson, eds. 1977. *Biological Control by Augmentation of Natural Enemies.* New York: Plenum Press.

Roduner, M., G. Cuperus, P. Mulder, J. Stritzke, and M. Payton. 2003. Successful biological control of the musk thistle in Oklahoma using the musk thistle head weevil and the rosette weevil. *American Entomologist* 49:112–119.

Ruberson J. R., J. R. Nechols, and M. J. Tauber. 1999. Biological control of arthropod pests, pp. 417–448. In J. R. Ruberson, ed., *Handbook of Pest Management.* New York: Marcel Dekker.

Sawyer, R. C. 1996. *To Make a Spotless Orange: Biological Control in California.* Ames, Ia.: Iowa State University Press.

Sheck, A. L. 1991. Biotechnology and plant protection, pp. 199–211. In D. Pimentel, ed., *CRC Handbook of Pest Management in Agriculture,* vol. 2, 2nd ed. Boca Raton, Fla.: CRC Press.

Stehr, F. W. 1982. Parasitoids and predators in pest management, pp. 135–173. In R. L. Metcalf and W. H. Luckmann, eds. *Introduction to Insect Pest Management,* 2nd ed. New York: Wiley.

Symondson, W. O. C., K. D. Sunderland, and M. H. Greenstone. 2002. Can generalist predators be effective biocontrol agents? *Annual Review of Entomology* 47:561–594.

Tanada, Y., and H. K. Kaya. 1993. *Insect Pathology.* New York: Academic Press. U.S. Congress Office of Technology Assessment. 1995. *Biologically Based Technologies for Pest Control.* Washington, D.C.: U.S. Government Printing Office.

Vail, P. V., J. R. Coulson, W. C. Kauffman, and M. E. Dix. 2001. History of biological control programs in the United States Department of Agriculture. *American Entomologist* 47:24–50.

van den Bosch, R., P. S. Messenger, and A. P. Gutierrez. 1982. *An Introduction to Biological Control.* New York: Plenum Press.

Van Driesche, R. G., and T. S. Bellows. 1996. *Biological Control.* New York: Chapman & Hall, pp. 117–127.

## Favorite Web Sites

http://www.ent.iastate.edu/list/biological_control.html
Presents an index of many biological control sites.

http://entowww.tamu.edu/images/slideset/biocontrol/bio005.html
Gives role of biological control in integrated pest management and shows detailed color slides of insect predators, parasitoids, and pathogens.

CHAPTER 10

# ECOLOGICAL MANAGEMENT OF THE CROP ENVIRONMENT

IN ADDITION TO NATURAL ENEMIES, many other aspects of the crop environment greatly influence insect numbers and have the potential to aid in pest management. To understand these other factors, we must recount the ecological basis of insect problems.

Crop production in an area usually results in significant alteration in the species occupying the space, as well as various modifications to soil, water, and topography. Natural flora and fauna may be displaced from the area in favor of mostly exotic plants and animals that are the focus of agricultural production. These crop species are established by clearing the area of undesired species and replacing them with those that produce human food and fiber. The crop species are maintained by high-energy inputs that involve such practices as tillage, fertilization, cultivation, and irrigation.

These practices of crop growth and livestock production are designed to create a favorable environment for the desired species. This environment is usually ecologically simple and also provides a favorable habitat for some insects and other organisms (for example, unwanted plants and microorganisms). These unwanted species may have existed in the area previously or may have been attracted from other areas. With the supply of new requisites or additional amounts of existing requisites, reproductive rates and survival rates of unwanted populations increase, and pest activities result in reduced crop output.

Reduced output does not always occur immediately when a new crop is introduced in an area. At first, producers are preoccupied with the new crop and the cultural activities required to raise it. At this time, there is little concern for potential pests, and for a period, insects may be of little importance. This lag in problems often belies the future importance of injurious populations and may give a false sense of security. As species become adapted to the new environment, pest populations grow, losses increase, and attention finally must be given to alleviating these difficulties.

One of the oldest approaches to dealing with these emerging problems is to modify the established production schemes and techniques to make the crop environment less favorable for the pest or at least to minimize losses of the commodity in spite of injury.

## ECOLOGICAL MANAGEMENT

Historically, entomologists have called modifying the crop environment through changes in production techniques *cultural control*. **Cultural control** can be defined as the purposeful manipulation of the environment to reduce rates of pest increase and damage. Cultural control manipulations traditionally involve environmental factors that already exist, as opposed to adding new factors (for example, insecticides or natural enemies). In keeping with the idea of insect pest management and to broaden the notion of cultural control somewhat, we use the term **ecological management** here to describe the approach.

Ecological management is based on a thorough understanding of pest ecology as it relates to the crop in production. In general, we must understand a pest's ecological requisites, know how available these requisites are in the agroecosystem, and understand insect behavior in attaining requisites. Major pest requisites include food, appropriate space for feeding, mating, and egg laying, and shelter from weather extremes and enemies (Fig. 10.1). When pests are major problems, these requisites may be found totally within the crop area or supplemented from sources in nearby areas. For example, grasshoppers, *Melanoplus* species and others, find an abundant supply of nutritious food in annual field crops like corn and soybeans but must return to grass sod at the edge of fields for appropriate egg-laying sites. Likewise the European corn borer, *Ostrinia nubilalis,* finds satisfactory food and egg-laying sites in corn fields, but adults seek out waterways and other grassy areas outside the crop for the proper mating environment.

The idea behind ecological management is to find "weak links" in the insect seasonal cycle and exploit them. These weak links may be behavior patterns to complete development, like crawling over the ground surface to reach adequate feeding sites, as do the larvae of the stalk borer, *Papaipema nebris,* or locating proper shelter for overwintering, as do adult boll weevils, *Anthonomus grandis*.

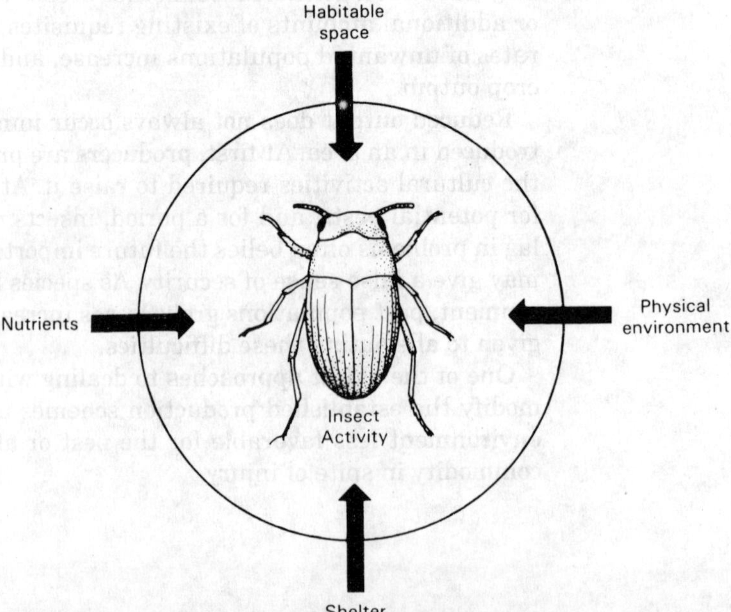


**Figure 10.1** Diagram of the major environmental factors to manipulate for reducing pest proliferation and damage.

Food is often the environmental factor involved in the weak link. Insects vary greatly in their food requirements; some are quite selective (for example, larvae of the northern corn rootworm, *Diabrotica barberi,* on corn), and others display seasonal shifts from one type of food to another (for example, chinch bugs, *Blissus leucopterus,* moving from maturing wheat to corn or armyworms (Box 10.1), *Pseudaletia unipuncta,* moving from grass in ditch bank into emerged corn). Such habits provide potential opportunities for exploiting the insect's seasonal cycle.

Not only is food vital to the insect, but it may also be one of the most practical factors in the production scheme that can be manipulated. Modifying the food source may produce profound effects on the pest species because it often provides both nutrients and habitable space for reproduction and other activities.

The physical environment, also critical in insect ecology, may be manipulated through ecological management. Plant culture activities like soil tillage and debris removal can expose insects to intolerable weather extremes and subsequent death. Manipulating temperatures and humidities above or below those tolerated by insects in stored grain are procedures that modify the physical environment to obtain ecological management.

### BOX 10.1 ARMYWORM (SEE COLOR PLATE 3)

**SPECIES:** *Pseudaletia unipuncta* (Haworth) (Lepidoptera: Noctuidae)

**DISTRIBUTION:** The armyworm occurs throughout most of the United States east of the Rocky Mountains. However, this pest also has been found in Arizona, California, and New Mexico.

**IMPORTANCE:** Larvae of armyworm moths readily feed on corn, oats, barley, and rye. In most years, damage to these crops is insignificant, but in outbreak years large populations can completely devour the original host and "march" to new hosts (hence the term "armyworm"). Outbreaks of this pest tend to occur following a cold, wet spring. Larvae of the first generation are commonly associated with outbreak occurrences. The degree of crop injury is frequently heavier in field borders. Young leaves are completely devoured, but older, less palatable leaves may be only skeletonized, leaving the vein structure.

**APPEARANCE:** Adult moths are brownish gray, with a distinct white spot in the center of the forewing. The hind wing is a lighter brown and has darker veins; wingspan is about 40 mm. Eggs are laid in masses or rows that are commonly folded in leaves or leaf sheaths. Larvae of the armyworm are green to brown and have a distinctive pattern of longitudinal stripes. A dark stripe can be seen along each side, and a broad stripe is found along the back. This dorsal stripe has a fine, light-colored broken line running down the center. The head capsule is pale brown with green and dark brown mottling.

**LIFE CYCLE:** Partially grown larvae overwinter in the soil or refuse at the surface and complete their growth the following spring. Larval development continues in early spring, and pupation soon follows. Adults emerge, mate, and oviposit within 14 to 21 days. Females lay large masses of eggs, often in the same general location, at night. Hatching occurs within about 8 days after deposition, and the neonate larvae begin feeding on leaf tissue. Fully grown larvae are about 40 mm long when they pupate and a second generation of moths emerges. There may be up to three generations in most regions of the United States, but it is uncommon for successive outbreaks to occur in the same general location. There is continuous breeding in extreme southern locations, where generation overlap is common. This pest is subject to population regulation by numerous larval and egg parasitoids, which often aid in keeping armyworm populations below economic levels.

Depending on the objective, most procedures of ecological management can be grouped into one of four categories: (1) reducing the average favorability of the ecosystem, (2) disrupting the continuity of requisite food sources, (3) diverting pest populations from the crop, and (4) reducing the impact of insect injury. One or more approaches may be used in ecological management to keep pest populations "off balance" and prevent them from achieving their full destructive potential.

## REDUCING AVERAGE FAVORABILITY OF THE ECOSYSTEM

Ecosystems include interacting biotic and abiotic elements both in the crop and in associated nonagricultural habitats. The level of requisites available to pest species in an agroecosystem determines, to a degree, the average density of the pest and severity of pest problems. Procedures in this category mainly aim at reducing pest density by reducing the average availability of food, shelter, and habitable space. Only the level of the requisite is reduced because complete elimination is unwarranted or impractical.

### Sanitation

Sanitation is one of the most elementary procedures in reducing favorability of the agroecosystem for pest species. Because many species breed and overwinter in all sorts of debris, removal of debris from habitats can reduce rates of reproduction and survival.

**Crop residue destruction and utilization.** Destroying or removing crop residues from crop sites is one of the most basic ways to eliminate pest overwintering sites and reduce the spread of infestations. Destruction may be achieved by plowing directly or shredding and chopping with special field implements before plowing. Burning residues in place or raking and scooping them into piles for burning or chemical treatment is another method used to eliminate infestations from residues. Perhaps an even better approach with some residues (for example, dropped fruit) is to feed them to livestock, thereby obtaining an economic benefit. In some operations, livestock can be pastured at appropriate times to consume residues and trample hibernating pests.

Probably the greatest limitations to residue cleanup are cost of destruction and usefulness of residues in soil conservation. If much labor is involved in destruction, the approach quickly becomes impractical for commercial enterprises. For instance, the collection and disposal of dropped fruit is an effective method for reducing infestations of the codling moth (*Cydia pomonella*), apple maggot (Fig. 10.2) (*Rhagoletis pomonella*), and plum curculio (*Conotrachelus nenuphar*) in small orchards. However, it is usually impractical for large commercial orchards. Additionally, the practice of leaving plant residues on the soil surface to reduce wind and water erosion rules out this measure where soil conservation is a top priority.

In agronomic crops, some of the most widespread uses of crop residue destruction for insect management have occurred with cotton, corn, and wheat. In particular, cotton production in the southern and southwestern United States emphasizes destruction of infested cotton plants after harvest and before the first killing frost. Area-wide sanitation programs in Texas cotton have been especially successful. Here, defoliants and mechanical strippers have allowed removal of infested plant parts from the field, and remaining debris is plowed under. Such activities have been particularly important in reducing infestations of the pink bollworm, *Pectinophora gossypiella,* and boll weevil (Box 10.2).

## BOX 10.2 Boll Weevil (See Color Plate 9)

**SPECIES:** *Anthonomus grandis grandis* Boheman (Coleoptera: Curculionidae)

**DISTRIBUTION:** The boll weevil is not native to the United States; rather, it has moved north from its natural range in Central America. It was first reported in the United States in 1894 in Brownsville, Texas. From there, it became established in southern Texas and began to spread throughout the Cotton Belt at a rate of 40 to 160 miles per year. This pest continued to spread and adapt to new areas and eventually occupied the entire Cotton Belt.

**IMPORTANCE:** The boll weevil is a major pest on much of the cotton grown in the United States. Lint and cottonseed losses between 1909 and 1949 averaged $203 million annually. In spite of ongoing eradication activities, cotton growers in the Mid-South and Texas continue to suffer substantial losses to this pest. Feeding by adults on early flower buds (referred to as "squares") causes the bracts to flare, with eventual square drop. Oviposition into squares also results in shedding. After bloom and during boll (fruit) development, females continue to feed and oviposit within the bolls. At this point, the boll usually does not shed but larval feeding destroys the lint within the boll. In 1904 the United States Bureau of Plant Industry was charged with demonstrating to farmers how to control this pest. The program was successful and eventually led to the establishment of the Agricultural Extension Service.

**APPEARANCE:** Adults are about 6 mm long, brown, and hard-shelled. The distinctive snout is half as long as the body. Larvae are legless, about 12 mm long when full-grown, and grublike, with a brown head capsule. Eggs are white, elliptical, and about 0.8 mm long. Pupae are white. A recent taxonomic study has demonstrated that there are three closely related forms of weevils attacking cotton: the boll weevil, the thurberia weevil, *Anthonomus grandis thurberiae* Pierce, and the Mexican boll weevil, *Anthonomus hunteri* Burke and Cate.

**LIFE CYCLE:** Adults overwinter in ground litter in or around cotton fields, fence rows, woods, or other protected places. The adult diapause period may extend from March to mid-July, with most emergence occurring in June. Adults may disperse several miles in search of host plants. After locating fruiting cotton, adults begin to feed on squares and, less frequently, on leafbuds and growing terminals. Females mate and deposit eggs singly into cavities previously made with their chewing mouthparts. They seal the oviposition site with a frass "nipple," which makes a damaged square or boll easily identifiable. Larvae feed within the square or boll for 7 to 12 days and then pupate. Adults emerge within about 6 days and disperse to new hosts.

---

With field corn, overwintering European corn borer and southwestern corn borer, *Diatraea grandiosella,* populations have been reduced in previous years by residue shredding (Fig. 10.3) and deep plowing techniques. However, production systems emphasizing greater soil conservation in the Corn Belt region have reduced the current implementation of this approach.

In wheat, the plowing under and disposal of stubble may still be a practical procedure for reducing overwintering populations of important pests like the Hessian fly, *Mayetiola destructor*.

Sanitation can also be a very important pest management procedure in truck farming. Efficient producers remove vegetables dropped during packing operations and/or left in fields during harvest. These vegetables are removed and destroyed or fed to livestock to prevent buildup of several Lepidoptera,

**Figure 10.2** An apple damaged by the apple maggot, *Rhagoletis pomonella.* Collection and disposal of dropped fruit containing such pests can reduce future infestations. (Courtesy Iowa State University Extension Service)

**Figure 10.3** Shredding corn stalks with a tractor-drawn shredder. This operation has been used to reduce European corn borer, *Ostrinia nubilalis,* and southwestern corn borer, *Diatraea grandiosella,* infestations in corn. (Courtesy D. Erbach and USDA)

squash bugs (*Anasa tristis*), some aphids (Hemiptera: Aphididae), and other pests Removing vines and infested tubers also is an important practice in the management of the potato tuberworm, *Phthorimaea operculella,* and the sweetpotato weevil, *Cylas formicarius elegantulus,* by potato growers.

As we have noted, the frequent removal of dropped fruit can reduce infestations of several insect pests and is recommended for orchards when practical. Also, trees in abandoned orchards or those severely infested by scale insects (Hemiptera: Coccoidea) should be removed and destroyed. In yet another instance, navel orangeworm (*Amyelois transitella*) problems in California orchards can be reduced significantly by removal and destruction of almonds remaining on the trees and ground after harvest.

Removal of plant debris may also be important in silviculture. It is recommended that slash be destroyed following logging operations, particularly in pine stands of the southern and western United States. Such sanitation procedures help prevent reproduction and buildup of bark beetle populations and reduce damage to trees growing near the logged area. The destruction (burning) of elm logs and branches infested with the smaller European elm bark beetle, *Scolytus multistriatus,* and the native elm bark beetle (Box 10.3), *Hylurgopinus*

## BOX 10.3 NATIVE ELM BARK BEETLE

**SPECIES:** *Hylurgopinus opaculus* (LeConte) (Coleoptera: Scolytidae)

**DISTRIBUTION:** The native elm bark beetle originated in the United States and occurs in most of the eastern states from Maine to Virginia, in Mississippi, Kansas to Minnesota, and in eastern Canada. Its range corresponds to that of the American elm tree within the United States. It is mostly monophagous on elm trees but has been reported to also feed on basswood trees.

**IMPORTANCE:** This pest, in addition to the smaller European elm bark beetle, *Scolytus multistriatus* (Marsham), has played an important role in the near extinction of American elm in the midwestern United States. The beetles cause minimal injury by feeding on foliage, but they transmit the fungal organism *Ceratocystis ulmi,* which is responsible for Dutch elm disease. The disease is present throughout most of the United States and has been reported in Canada. Because of the mobility of elm bark beetles, the disease was able to spread unchecked, eventually killing all but a few American elms in the Midwest.

**APPEARANCE:** Adults are dull brown and less than 3 mm long. The head is somewhat blunt, and body shape is typical of scolytid beetles. Eggs are small, globular, and pearly white. Larvae are about 3 to 4 mm long and creamy white, with a brown head capsule. They are legless, C-shaped grubs. Damaged trees can be identified by feeding galleries under the bark. Parent galleries consist of two branches diverging from the entry points. Adult galleries extend across the grain, whereas larval galleries usually run with the grain of the wood.

**LIFE CYCLE:** Larvae and adults overwinter. Adults appear in spring after emerging from bark that was fed upon the previous year. Only dying or injured elm trees are attacked by the beetles, which bore into bark and form galleries. Females deposit eggs in niches along the sides of the galleries. Young larvae begin feeding in the cambial region of trees and form larval galleries independent of adults. The width of the galleries increases with larger larvae and greater consumption. Pupation occurs at the end of larger galleries, followed by adult emergence. The life cycle requires 45 to 60 days for completion, and two to three generations can occur per year.

**Figure 10.4** Spreading liquified manure achieves fertilization and allows manure to dry out. The drying helps prevent fly colonization. (Courtesy D. Erbach and USDA)

*opaculus,* is another sanitation practice recommended to reduce the spread of Dutch elm disease. Pruning and destruction of insect-infested twigs and branches also have been used as management tactics for the European pine shoot moth, *Rhyacionia buoliana,* a pest of pine plantations in the Great Lakes region of the United States.

**Elimination of animal wastes.** It goes without saying that good housekeeping is an important aspect of fly and cockroach control in households. Likewise, disposal of animal wastes in and around livestock confinement areas reduces food and habitat for developing fly larvae and is a must if adequate management of house flies, *Musca domestica,* and stable flies, *Stomoxys calcitrans,* is to be achieved. Methods have included scattering pulverized or liquified manure with spreaders (Fig. 10.4) to achieve drying and prevent fly colonization. At the same time, the manure fertilizes. With other organic wastes, composting facilities and solid-waste treatment plants turn otherwise useless material into products of value as fertilizers and fuel, while eliminating potential insect habitats.

**Efficient storage and processing.** On a worldwide basis, as much as 10 percent of the stored cereal grain is estimated to be lost through insect infestation. To avoid losses, cleaning facilities and eliminating spillage is of utmost importance in storing grain and other commodities. Such sanitation, along with maintaining cool, dry atmospheres, is absolutely necessary in preventing infestations of stored-grain pests such as the Angoumois grain moth (Box 10.4), *Sitotroga cerealella.* Most food processing plants have sanitation programs designed to thoroughly clean bulk storage bins and ensure an insect-free environment each time before they are filled. Grain elevators reduce the incidence of infestations by removal of waste grain and chaff from grain augers, elevator equipment, and support facilities (Fig. 10.5). Often heavy-duty vacuum cleaners are used for this purpose, as well as for sanitation in food plants, bakeries, and warehouses. Regularly scheduled inspections and cleaning "field days" are the best approaches to this type of sanitation.

### Destruction or Modification of Alternate Hosts and Habitats

Many insect pests have requirements that cannot be met by the crop itself and need to move out or switch to other foods during certain parts of the year. If these alternate hosts and habitats can be destroyed or limited in extent, insect numbers can be reduced.

Many pests of annual crops leave the unprotected, barren crop area for field borders during the fall. Here, they may find an environment with dense vegetation and other cover in which to spend the winter. Insects sheltered in these areas and in a state of dormancy can survive the adversities of winter and are a source of infestations the next spring. Burning or plowing this habitat may be a practical means of reducing such sources.

### BOX 10.4 Angoumois Grain Moth

**SPECIES:** *Sitotroga cerealella* (Olivier) (Lepidoptera: Gelechiidae)

**DISTRIBUTION:** This pest is worldwide in distribution. It was first reported in the United States in 1728 and has since spread throughout North America. Larvae attack grains in storage and in the field.

**IMPORTANCE:** The Angoumois grain moth is the most common of the moths that infest whole grains. Indeed, this species is second in importance only to rice and granary weevils as a stored-grain pest. The pest is of greatest importance in southern North America and in portions of the eastern and central United States. This pest has caused millions of dollars in damage to grain stored in central locations like grain elevators. It can also be a severe household pest by infesting grain products during processing or after purchase.

**APPEARANCE:** Adults are light grayish brown or straw-colored moths with a satiny luster and a wingspan of about 16 mm. Hind wings are fringed, with long, dark hairs, and they have a point at the tip like a finger, which distinguishes this pest from clothes moths. Newly laid eggs are white but turn red as they near hatching. Larvae are white, with a brown head capsule. They feed within individual kernels of wheat, corn, or other grains and are about 5 mm long when fully grown.

**LIFE CYCLE:** Larvae overwinter in the northern United States (two generations per year), but in heated buildings and in the southern United States, breeding is continuous. Eggs are deposited (in spring if overwintering) on the stored grains or on developing grain heads while hosts are still in the field. Females oviposit an average of 40 eggs but may produce as many as 400 per individual under ideal conditions. Upon hatching, larvae chew into grain and consume the inside portions, usually rendering it inviable and useless. Prior to pupation, larvae prepare exit holes through the seed coats for the moths to exit. Total development may be completed in 5 weeks under warm conditions.

**Figure 10.5** Sanitation applied to grain-storage facilities includes removing residual grain (*top left*), cleaning grain bins with a vacuum cleaner (*top right*), removing chaff from grain-loading equipment (*bottom left*), and removing weeds around grain bins. (Courtesy University of Nebraska Extension Service)

One of the most frequently mentioned pests for this kind of management is the sorghum midge, *Contarinia sorghicola*. The most important source of infestations in sorghum and Sudan grass in the United States seem to be from Johnsongrass in or along borders of fields (Fig. 10.6). Significant reductions of pest numbers have been reported where weed burning and other forms of pest habitat destruction were used.

Many bugs, like squash bugs and stink bugs, and other garden pests also overwinter in trash and plant cover at the edge of plantings. Elimination of these hibernating habitats can significantly reduce infestations in cucurbits, garden beans, and other vegetables.

Another source of insect infestations is volunteer plants of the crop that remain in turnrows and in parts of a field after harvest. These plants can harbor large numbers of pests at times when pest presence otherwise would be impossible. The destruction of these volunteer plants is particularly important when crop rotation is practiced to eliminate pests. For example, volunteer corn in soybean fields should be removed to prevent corn rootworm adults, *Diabrotica* species, from egg laying and producing larvae that would colonize corn the following season. Other pests where destruction of volunteer plants is

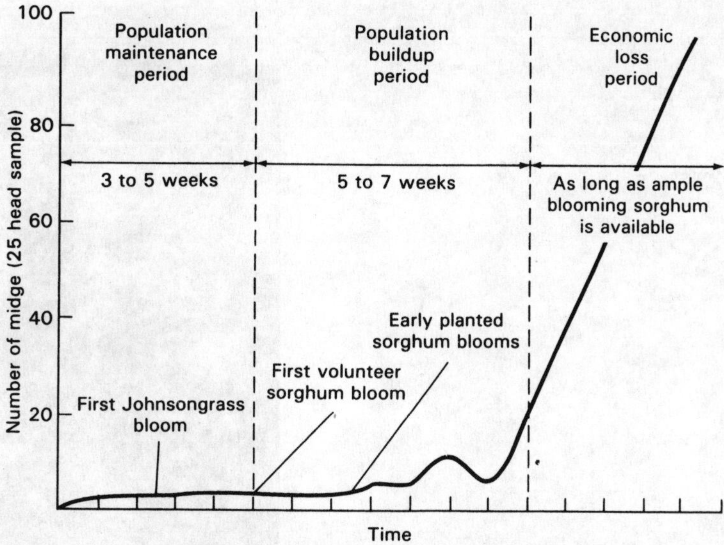


**Figure 10.6** Relationship between sorghum midge, *Contarinia sorghicola*, densities and the blooming of Johnsongrass and sorghum. The midge lays eggs in floral spikelets during flowering, and larval feeding prevents seed development. Sorghum is the preferred host, but spring- and early-summer generations of the midge are maintained on Johnsongrass. (Courtesy Texas Agricultural Experiment Station)

recommended for suppression include grasshoppers, Hessian fly, sweetpotato weevil (Fig. 10.7), potato tuberworm, potato aphid (*Macrosiphum euphorbiae*), and the wheat curl mite (*Eriophyes tulipae*).

Although destruction of alternate hosts and habitats may be practical for some insect pests, it may also eliminate important habitat for beneficial insects

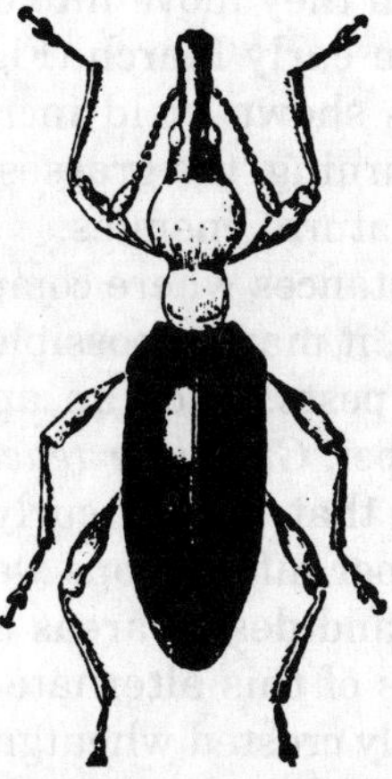

**Figure 10.7** The sweetpotato weevil, *Cylas formicarius elegantulus*, one of the pests for which destruction of volunteer crop plants is recommended. (Courtesy USDA)

**Figure 10.8** Burning grass at the edge of a corn field to kill stalk borer eggs. This approach can be used to reduce losses from this pest, which overwinters in field borders and invades fields in late spring. (Photo by Marlin E. Rice)

and wildlife. Therefore, before widespread destruction is implemented, surveys of beneficial organisms should be conducted, after which the benefits of this fauna should be determined and weighed against the benefits of reducing pest numbers. In some instances, a decision against habitat destruction may be the most appropriate one.

Burning may be a practical means of managing insects that require alternate habitats. An example involves the stalk borer, *Papiapema nebris,* which lays eggs on stems and leaves of grasses and broadleaf weeds surrounding cornfields. The eggs hatch in the spring, larvae bore into these plants to feed, and then they move into corn as they mature. Burning grasses and broadleaf weeds in early March (Fig. 10.8) can reduce infestations by 82 to 97 percent and has shown yield increases in adjacent rows by as much as 68 percent. After burning, the grasses regrow, conserving the burned area as a habitat for insect natural enemies.

In instances where complete elimination of cover in alternate habitats is ill-advised, it may be possible to replace the vegetation with plantings less favorable to pests. Such an approach succeeded several years ago with the beet leafhopper, *Circulifer tenellus,* in southern Idaho. The beet leafhopper vectors a virus that causes curly top disease in several crops. The source of beet leafhopper infestations was found to be from Russian thistle growing in large ranges and desert areas of the region (Fig. 10.9). In 1959, more than 47,000 hectares of this alternate habitat were seeded with perennial range grasses, primarily crested wheatgrass, to replace the thistle. This program was an outstanding success, providing a substantial reduction in beet leafhopper and, subsequently, a reduction of curly top incidence.

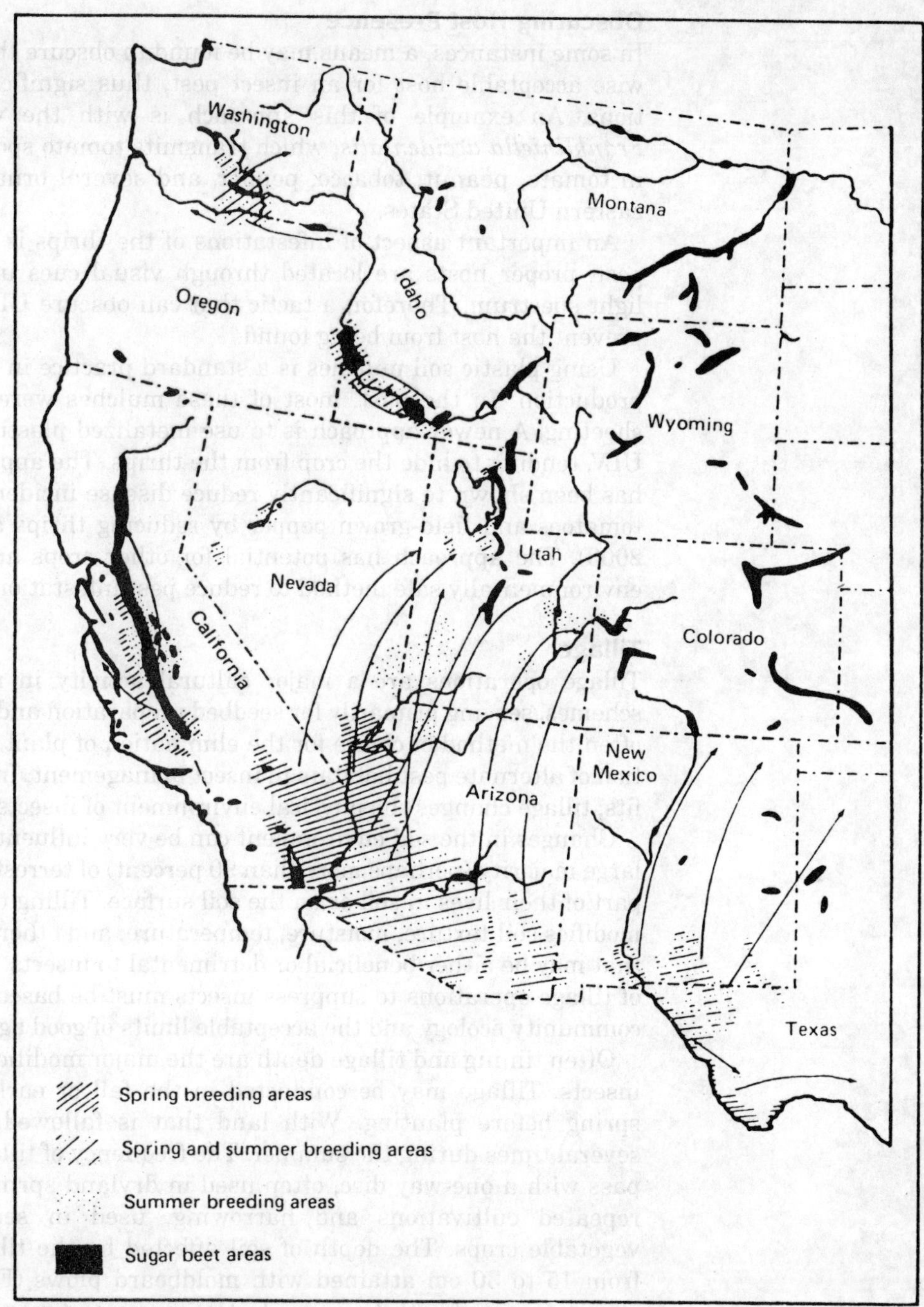


**Figure 10.9** Map showing migration routes of the beet leafhopper, *Circulifer tenellus*, from breeding locations into sugar beet cropping areas. (Courtesy USDA)

## Obscuring Host Presence

In some instances, a means may be found to obscure the presence of an otherwise acceptable host for an insect pest, thus significantly reducing infestations. An example of this approach is with the western flower thrips, *Frankliniella accidentalis*, which transmits tomato spotted wilt virus (TSWV) in tomato, peanut, tobacco, pepper, and several ornamentals in the southeastern United States.

An important aspect of infestations of the thrips is host finding. With this pest, proper hosts are located through visual cues in the ultraviolet (ULV) light spectrum. Therefor, a tactic that can obscure ULV light can potentially prevent the host from being found.

Using plastic soil mulches is a standard practice in southeastern vegetable production. In the past, most of these mulches were made of black plastic sheeting. A newer approach is to use metalized plastic sheeting that reflects ULV, tending to hide the crop from the thrips. The application of this sheeting has been shown to significantly reduce disease incidence caused by TSWV in tomatoes and field-grown pepper by reducing thrips abundance (Reitz et al. 2003). The approach has potential for other crops and offers a unique and environmentally safe method to reduce pest infestations.

## Tillage

Tillage operations are a major cultural activity in many plant production schemes, serving primarily for seedbed preparation and weed control. Also, it is often the method of choice for the elimination of plant refuse and the destruction of alternate pest habitats in insect management. In addition to these benefits, tillage changes the physical environment of insects inhabiting the soil.

Changes in the soil environment can be very influential on insects because a large majority (perhaps more than 90 percent) of terrestrial species spend some part of their lives in soil or on the soil surface. Tilling the soil for plant culture modifies soil texture, moisture, temperature, and other characteristics in ways that may be either beneficial or detrimental to insects. Therefore, modification of tillage operations to suppress insects must be based on a knowledge of soil community ecology and the acceptable limits of good agronomic practice.

Often timing and tillage depth are the major modifications made to manage insects. Tillage may be conducted in the fall or early winter and/or in the spring before planting. With land that is fallowed, it may be conducted several times during the summer. The frequency of tillage varies from a single pass with a one-way disc, often used in dryland spring wheat production, to repeated cultivations and harrowing, used in seedbed preparation for vegetable crops. The depth of soil affected by the tillage procedure may be from 15 to 30 cm attained with moldboard plows (Fig. 10.10) to a shallow stirring created with harrows. In California and many areas of the southern United States, subsoiling, or the disturbance of hard soil beneath normal plow depth, is another tillage method that has been used. Newer tillage systems employed in soil conservation include such procedures as fall chisel plow, till plant, and no till.

If tillage is deliberately used for reducing insect infestations, it should be timed with the insect's life cycle. The objective is to move insects when they are helpless (usually meaning immobile) from a favorable location to an unfavorable one. Therefore, timing is usually determined by the period of the life

**Figure 10.10** Moldboard plowing (*top*) and corn plants emerging in a no-till environment. Conventional methods of sanitation like deep plowing are not possible in many conservation tillage systems. (Courtesy D. Erbach and USDA)

cycle when either pupation or dormancy occurs, and depth is recommended by the location of this helpless stage.

As an example of timing and depth, ecological management of the European corn borer can be considered. This insect undergoes diapause as a full-grown larva in corn residue (stalks and earshanks) on or above the soil surface. Chopping and shredding (Fig. 10.3), followed by deep plowing with a moldboard plow in the spring, was historically used but is much less frequent today because of increased fuel costs and concerns about surface residue needs and subsequent soil erosion. With this approach, some mortality occurs from

**Figure 10.11** Wheat stem sawfly, *Cephus cinctus,* and injured wheat. Spring plowing has been recommended as an ecological management tactic to reduce infestations of this pest. (Courtesy USDA)

exposure to low winter temperatures above ground, mortality is added by the chopping and shredding process, and many of the remaining survivors are entombed by the plowing process.

In addition to reducing European corn borer populations, spring plowing is recommended to reduce infestations of wheat stem sawfly (Fig. 10.11), *Cephus cinctus,* in spring wheat. Moreover, it is used to aid in destruction of grasshopper eggs and to eliminate early weeds that are food for newly emerged nymphs.

In this instance, the grasshopper eggs are moved from soil to the surface, where they are subject to desiccation and predation by other insects and birds.

In vineyards, spring plowing is recommended to cover overwintering cocoons of the grape berry moth, *Endopiza viteana,* under grape trellises and thereby to reduce infestations. Yet another example of spring plowing for management has been reported for infestations of grass grub, *Costelytra zealandica,* in New Zealand. Plowing in September and October (spring in the southern hemisphere) disrupts pupal cells in the soil that cannot be reformed, causing insect mortality.

Fall plowing is often most appropriate for insects that are attracted to an undisturbed surface for egg laying and/or overwinter in soil. For insects like the corn earworm, *Heliocoverpa zea,* and several species of cutworms (Lepidoptera: Noctuidae) that undergo diapause in soil during winter, fall plowing moves them to the soil surface, where they are exposed to winter temperatures and vertebrate predators like birds and mice.

In many instances there is a strong relationship between the effectiveness of tillage on insect populations and winter temperatures. In studies of corn rootworms, *Diabrotica* species, egg survival in the northcentral United States was reduced by about 70 percent during severe winters in plots that had been paraplowed or moldboard plowed in autumn. There was no significant reduction in survival in no-till and chisel-plowed plots. During moderate winters, rootworm survival was not affected by any tillage method.

In addition to using tillage as an agent to increase winter mortality, it also can be used to help prevent pest colonization. An example of this approach is found in managing the black cutworm, *Agrotis ipsilon,* in corn. This pest migrates into the upper Midwest each spring and is attracted to lay eggs in weedy unplanted fields. Eggs hatch and larvae move from weeds to corn after the corn has emerged. To reduce the problem, it is recommended that weeds be tilled under at least 8 days before planting, thus reducing the favorability of the site for colonization.

Some tillage operations do not always benefit pest management and actually may encourage pest problems. For instance, tillage of fallowed land during August and September in Canada fosters successful oviposition by the pale western cutworm, *Agrotis orthogonia,* a pest of small grains. Avoiding tillage during these months promotes crusting of the soil surface and prevents females from penetrating soil to lay eggs. In another instance, parasitoids of the cereal leaf beetle (Box 10.5), *Oulema melanopus,* a pest of small grains, can be severely affected by tillage operations, which have little effect on the pest.

### Irrigation and Water Management

Irrigation is a primary plant culture activity in many regions, and little emphasis has been placed on its use in preventing insect problems. An exception to this statement is a technology called **chemigation** that uses the irrigation system to dispense insecticides and other pesticides over the area (Fig. 10.12). However, this is not an ecological management approach.

Water management is practiced as a form of ecological management in the muckland Everglades agricultural area of Florida. Here, fallow land is flooded prior to the production of winter vegetables. Of approximately 300,000 hectares, about 20 percent is flooded annually to suppress soil pests and to conserve and stabilize organic soils in the region. Primary targets of pest management in this operation are immature wireworms (Coleoptera: Elateridae),

## BOX 10.5 Cereal Leaf Beetle (See Color Plate 6)

**SPECIES:** *Oulema melanopus* (Linnaeus) (Coleoptera: Chrysomelidae)

**DISTRIBUTION:** The cereal leaf beetle is distributed throughout Europe, the Middle East, and North Africa. The species was first detected in the United States (Michigan) in 1962 and is currently found in the eastern United States and southern Canada.

**IMPORTANCE:** Both adults and larvae attack young leaves of grasses, including wheat, oats, barley, rye, and corn. Adults feed on grain shoots, whereas larvae consume tissues between leaf veins, giving the injured plants a silvery appearance. The pest is capable of vectoring the corn pathogen maize chlorotic mottle virus. Quarantines have been initiated to aid in preventing movement of the pest.

**APPEARANCE:** Adults are 4 mm long and have a metallic bluish black head and elytra with red legs and prothorax. Larvae are pale yellow with a black head. Often, larvae are covered with fecal excrement, giving them a dark appearance. Eggs are initially yellow but darken to black before hatching. Pupae are yellow but change to adult coloration before emergence.

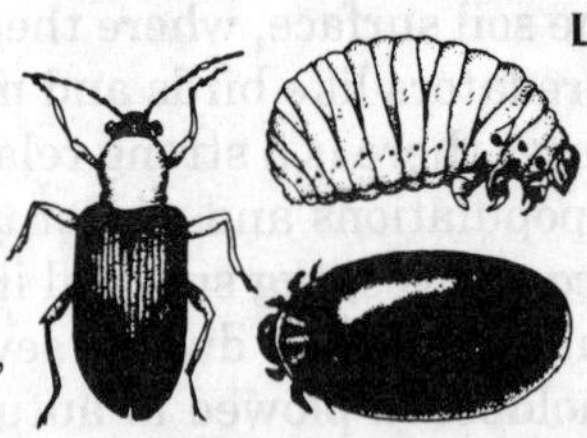

**LIFE CYCLE:** Adults overwinter in plant debris on the ground and in early spring they begin feeding on wild grasses. They eventually move to cultivated fields. Females oviposit on leaves, and hatching occurs in about 5 days. Larvae feed for about 10 days and then pupate under 50 mm of soil. After 20 to 25 days, adults emerge, feed for a short time, then aestivate until fall. The beetles become active again in fall and locate a suitable overwintering site. A typical life cycle takes 46 days, and there is only one generation per year.

**Figure 10.12** Overhead irrigation may be used to dispense insecticides (chemigation) and may be useful in suppressing some insect populations—even without chemicals. (Courtesy L. Hodgden and Dow Chemical Company)

which attack germinating plants and underground plant parts. To be effective against this pest, production areas are flooded for 6 weeks or more.

In another instance, sprinkler irrigation has been effective in suppressing certain foliage-feeding insects. The potato moth, *Phthorimaea operculella,* is effectively suppressed by frequent overhead irrigation of potatoes in New Zealand. The moist conditions appear to deter egg laying and cause mortality of newly eclosed larvae before they tunnel into the plants. Another pest managed by sprinkler irrigation is the diamondback moth (Box 10.6), *Plutella xylostella,* on watercress in Hawaii and head cabbage in Indiana. Here, intermittent applications of water are made during early evening hours, which is believed to disrupt moth mating and egg laying.

Of course, water management can also be used to grow more vigorous plants and thereby reduce losses. This aspect, however, is not a means of reducing pest numbers or decreasing the average favorability of the pest environment.

Another very significant use of water management in insect suppression has been with mosquitoes. In the United States, the control of water levels in water impoundments and other sources contributes significantly to reductions in both nuisance mosquitoes and those transmitting human pathogens. In addition, thousands of hectares of tidal marsh area have been diked and ditched to reduce mosquito breeding habitats—and thereby mosquito incidence. Modifications to some marsh areas is questionable, however, because of the tradeoff in eliminating habitat for certain fish and wildlife species.

## BOX 10.6 DIAMONDBACK MOTH

**SPECIES:** *Plutella xylostella* (L.) (Lepidoptera: Plutellidae)

**DISTRIBUTION:** Originally from Europe, this insect occurs throughout the United States where broccoli, cabbage, cauliflower, Brussel sprouts, collards, and related crops are grown.

**IMPORTANCE:** Larvae may eat small holes in leaves, or feed on the leaf surface while leaving a thin layer of tissue, creating a "windowpane" effect. Larvae will feed on the older or outer leaves of older plants but by also feed on the floral stalk and flower buds. Larval feeding on young plants may damage the developing flower buds and prevent proper development, thereby rendering the head unmarketable.

**APPEARANCE:** Adult moths are slender and 8 mm long, gray or brown in color, and move rapidly when disturbed. Males have a dark yellow marking, vaguely shaped like three connected diamonds on the back when the wings are folded. Larvae are light green and covered with tiny, erect, black hairs, with each arising from a white tubercle. There are two prolegs on the last segment that form a distinctive V shape. Full-grown larvae are about 8 mm long. When disturbed, they wiggle rapidly and attempt to escape by falling from the plant while hanging by a line of silk.

**LIFE CYCLE:** Females lay eggs either singly or in groups of two or three on the leaves or stalk near the terminal bud. The small, round eggs are yellow to white and hatch in 5 to 6 days. The four larval stages require 10 to 30 days to develop, depending on food supply and temperature. The pupa is enclosed in loosely woven gauzelike material attached to a leaf. The pupal stage is 7 to 10 days long during the summer. There are five to six generations per year in the southern United States. The insect overwinters as an adult.

## DISRUPTING CONTINUITY OF PEST REQUISITES

Insect pest problems originate with the establishment of agroecosystems and the creation of environments favorable to certain species. These problems continue because of the uninterrupted supply of requisites provided at appropriate times and places relative to the seasonal cycle of these species.

Within the limits of good cultural practice, the supply of requisites can sometimes be interrupted to reduce infestations of pests. In this instance, we are usually dealing with the crop species itself, manipulating its presence in time and space to eliminate insect requisites. Accomplishing that goal would reduce insect reproductive rates or cause pests to search elsewhere for their needs.

### Reduce Continuity in Space

With this approach, time is held constant. Our perspective is the cropping layout over a given area during a season for annual crops or over the life of a single planting for perennial ones. The focus is on crop planning and where to place crop plants relative to each other and in relation to other crops and unmanaged habitats.

**Crop spacing.** The location of plants within a crop has always been of great interest to agriculturists concerned with maximum production. Too few plants or too many plants result in less than optimal yields and have been the subject of detailed study. Although much is known about crop spacing in relation to weed pests, little is known about this aspect in relation to insect infestations.

Spacing can affect the relative growth rate of plants and the development of environments favorable to insect population growth. Plants placed close together create closed canopies that aid in insect movement, a situation that may be favorable for pests as well as natural enemies. Therefore, the detriments and benefits of spacing may vary from pest to pest.

In the instance of a spruce weevil, *Thylacites incanus,* open sunlit areas seem to be a requirement for survival. If trees are placed well apart, eggs in the open areas receive enough sunlight and sufficient temperatures to hatch. A management approach to reduce weevil infestations involves establishing plantings of young trees as close as possible to achieve early closing of the canopy. In this way, the continuity of open areas is eliminated.

Soybean production is another instance where a closed canopy is important in reducing some insect problems. Corn earworms prefer open canopies and therefore colonize late-planted and wide-row soybeans most readily. Early planting in narrow rows causes canopy closure earlier and helps to reduce infestations. Moreover, the high humidities of closed soybean canopies have enhanced epidemics of the pathogenic fungus *Nomuraea rileyi,* an important natural enemy of several defoliating caterpillars.

In yet another instance, shading has been important with the citrus rust mite, *Phyllocoptruta oleivora,* in Florida citrus. Groves planted in hammocks under oak and palm trees, which are shaded in the morning, have much lower mite populations than those grown in unshaded areas. Presently, however, priorities involving grove equipment maneuverability have encouraged hedging and clearing, producing less shaded environments and more mite problems.

**Figure 10.13** Corn in border rows injured by stalk borers, *Papaipema nebris,* that migrate in from grassy ditches adjacent to the field. (Photo by Marlin E. Rice)

**Crop location.** In selecting the location for plantings, it is prudent to consider other crops and environments near the proposed site. Many insects can move quickly from one field to the next and between botanically related crops to attain their requisites. Considerations should also be given to overwintering pests that may move out of woodlots and unmanaged habitats adjacent to the crop.

In planning crop location, field size and sources of insects that will colonize the proposed crop should also be considered. The nature of insect colonization of a field varies with such uncontrollable factors as wind and windbreaks and source location. Less mobile and weak flying species, overwintering or initially feeding outside the field proper, cause injury distributed in a definite edge pattern (the edge effect) (Fig. 10.13). Active insects that are strong fliers invade and quickly produce injury throughout the field. Pests producing edge effects tend to cause disproportionately greater damage in small fields versus large ones because there is a greater edge-to-area ratio in the smaller fields. An example is the stalk borer, which overwinters in field borders but rarely damages conventionally tilled corn more than ten rows from the field's edge. Small fields or cultivated strips near waterways or terraces suffer greater damage than do larger fields. For highly mobile pests, field size usually makes little difference.

As a rule of thumb, the wise choice is to locate dissimilar crops adjacent to one another to moderate pest movement. For instance, locating soybean adjacent to corn is a good choice. Because of botanical dissimilarity (a legume next to a grass), few insect species would find requisites in both, and neither crop usually supports pest infestations of the other. On the other hand, planting soybeans next to alfalfa (a legume next to a legume) is not a good idea, as it invites injury (particularly early season in edge rows) from potato leafhoppers, *Empoasca fabae*, which may move into soybean when alfalfa is clipped for hay.

Likewise, seed potatoes are not grown adjacent to main potato plantings to limit aphid infestations. Such a tactic is used to avoid potato viruses transmitted by aphids, which is important in seed certification programs.

Other situations where certain adjacent plantings are not recommended include small grains next to corn because of potential chinch bug problems in corn, cotton and soybeans next to wheat because of southern masked chafer (*Cyclocephala immaculata*) problems in cotton and soybeans, and late-planted potatoes next to early planted potatoes because of potato tuberworm (*Phthorimaea operculella*) problems in the late plantings.

### Upset Chronological Continuity

The general idea of this approach is to create a gap in time in the insect food source. Our perspective is a single location (space is held constant). Activities to achieve this goal are implemented at different times in the growing season, from one growing season to the next, or, for perennials, from one planting to the next.

**Crop rotation.** Probably the most important method of providing discontinuity in a pest's requisites is to rotate from one crop to another in alternate plantings. The practice of crop rotation actually developed from a desire to improve soil structure and fertility. Benefits in pest management have been mostly coincidental to this main purpose.

This method has been widely used with a number of different crops and pest species. Rotation schemes work best, however, when: (1) the pest has a narrow host range, (2) eggs are laid before the new crop is planted, and (3) the feeding stage is not very mobile. The latter factor is characteristic mostly of pests that feed underground on plant roots. Insects with these characteristics usually are purged from an area when a crop lacking food requisites is planted in place of a crop that has them. Such insects cannot move out of the area to obtain requisites and will die by the time the rotation has completed a full cycle. Rotation cycles that involve two or three crops are common.

One of the best examples of eliminating insect pests by crop rotation has been with corn rootworms (Box 10.7), *Diabrotica virgifera virgifera* (western) and *Diabrotica barberi* (northern), on corn in the midwestern United States. The development of corn production technology in the 1950s and 1960s supported the practice of corn monocropping, the continuous production of the same crop in one area. Monocropping, in turn, promoted the spread of corn rootworms in the Corn Belt.

Rootworm larvae emerge in the spring from eggs laid in corn fields in mid to late summer of the previous season. They are monophagous, feeding on roots of developing corn that was planted into their habitat. A 2-year rotation cycle of corn and soybeans has been used effectively to eliminate the problem, because larvae cannot survive in the soybean environment. However, certain strains of the northern corn rootworm have developed in southeastern South Dakota, Iowa, and southern Minnesota that can remain dormant in the egg stage for 2 years. With the development of these strains, a 2-year rotation has become ineffectual. The solution now lies with newer resistant cultivars like those developed for nematodes that encyst (*Heterodera* species) and many plant-pathogenic microorganisms: Rotations with longer cycles and involving more crops can better manage these new rootworm strains.

## BOX 10.7 CORN ROOTWORMS (SEE COLOR PLATE 3)

**SPECIES:** *Diabrotica* species (Coleoptera: Chrysomelidae

**DISTRIBUTION:** Three species of *Diabrotica,* the northern corn rootworm (*D. barberi* [Smith and Lawrence]), the western corn rootworm (*D. virgifera virgifera* LeConte), and the southern corn rootworm (*D. undecimpunctata howardi* Barber) inhabit different regions of the United States. The northern corn rootworm once solely inhabited the Corn Belt states. The western corn rootworm, once restricted to Colorado, Kansas, and Nebraska, has spread northward and eastward to inhabit most of the Corn Belt states. The southern corn rootworm is found mainly in the South and lower Midwest, but because of migration it can be found throughout the eastern United States and into Canada. There are many regions where one or more of the species inhabit the same location.

**IMPORTANCE:** The three species of beetles, referred to as rootworms because of larval feeding habits, are major corn pests. Adults damage corn during silking by clipping corn silks and reducing the chances for complete fertilization of the plants. This results in yield loss when corn ears are not completely filled with kernels. Larvae, or rootworms, cause the greatest degree of damage to corn plants. They damage corn by feeding on, and then tunneling in, the roots of growing plants. This injury reduces the amount of food available for plant growth and ear development, which consequently lowers yield (often up to 30 percent or more). In addition, the damaged root system provides less support for the plant during wind and rain storms. This injury often causes lodging of plants and difficulty in harvesting.

**APPEARANCE:** Adults of the northern corn rootworm are about 6 mm long and yellow to pale green (no striping or spots). Adults of the western corn rootworm are 6 to 8 mm long and yellow, with a black stripe on the outside of each elytron. Adults of the southern corn rootworm are yellow to pale green, with three rows of black spots on the elytra. Larvae of all species are about 6 to 7 mm long, slender, legless, and white, with a brown head capsule and terminal segment.

**LIFE CYCLE:** Northern and western corn rootworm species have a similar life cycle. Adults oviposit in the soil near corn plants during the fall. The eggs overwinter and hatch the following spring, and larvae begin feeding on the young roots of corn. They develop through three stadia and pupate in the soil. Upon emergence, adults begin feeding on corn silks and oviposit in the soil. Only one generation per year occurs for these two species. The life cycle of the southern corn rootworm is similar except that more than one generation is typical, and it overwinters as adults or eggs in the southern part of its range.

---

Rotations also have been successful for arthropods like white grubs (Coleoptera: Scarabaeidae) and other insects feeding on roots in grass sod. For these insects, legumes (such as alfalfa) are planted and maintained in the rotation cycle for at least 2 years.

Whitefringed beetles, *Graphognathus* species, can also be managed with rotations. These weevils do not fly, and females become fecund after feeding on peanuts, soybeans, and velvetbean. When corn and small grains are planted in a rotation cycle, the weevils feed on them but do not obtain proper nutrition, and females lay few eggs. These grass crops are not greatly injured by the beetles,

## BOX 10.8 WIREWORMS (SEE COLOR PLATES 4, 6)

**SPECIES:** Numerous species in several genera (Coleoptera: Elateridae)

**DISTRIBUTION:** The term *wireworm* describes the larvae of adult click beetles (family Elateridae). Because the various genera of plant-feeding elaterids are widespread, wireworms are also cosmopolitan and are found in almost any kind of soil.

**IMPORTANCE:** Wireworms are polyphagous and feed readily on numerous hosts while in larval form. They injure crops by devouring seeds in the soil, cutting off small underground stems and roots, and boring into larger stems, roots, and tubers. Normally, the crop is planted in soil already containing wireworms, and the pests simply begin feeding on the new food source. This is especially true in cases where prairie or sod is newly cultivated. No crop is known to be entirely immune to wireworm attack, and the estimated loss to farmers annually runs into the millions of dollars.

**APPEARANCE:** Adult click beetles are hard-shelled, 7 to 25 mm long (depending on species), brown to black to gray, with a tapered body at each end. The name *click beetle* refers to the clicking sound emitted by adults as they flip their body to right themselves. Larvae (wireworms) are slender, cylindrical, 7 to 35 mm long, hardened, and yellow to brown. Adults and larvae vary in details of appearance, depending on species.

**LIFE CYCLE:** Wireworms have varied life histories depending on species, but some aspects of development are similar. Adult females lay eggs singly in the soil, 2 to 15 cm deep during the spring or early summer. Hatching occurs in 14 to 30 days, and young larvae move within the soil to locate a food source. Heavy mortality occurs during the larval stage if environmental conditions are too harsh or food is too scarce. Larvae feed in the soil for 1 to 6 years, with much overlap of generations. Pupation occurs in the soil, and adults leave the soil and complete the life cycle by mating and ovipositing. Survival of larvae is increased by females seeking grassy areas (indicative of the extensive root mass) and laying eggs there.

---

and populations of the pest can be reduced with the rotation and effective weed control.

Some mites are also managed effectively by rotations. The winter grain mite, *Penthaleus* major, on wheat, oats, and barley can be suppressed by rotating to nongrass crops every third year.

In almost all instances, rotations involving botanically similar crops should be avoided. For example, corn following grass sod may suffer serious damage from wireworms. But this is not to say that rotations of dissimilar crops are always safe. For example, wireworm (Box 10.8) problems can occur in potatoes when this crop is grown after red or sweet clover.

**Crop fallowing.** For many years fallowing practices have been used in dry regions to allow increases of soil moisture and fertility for growing crops. For instance, in western Kansas winter wheat is often planted in fields that had been idled and kept weed-free the previous season. Without vegetation growing on it, the fallowed land is able to store a significant share of a season's precipitation. This stored moisture along with precipitation of the next season can support a biennial wheat crop. Fallowing may also be practiced as part of set-aside programs in the United States for the reduction of commodity surpluses and for weed control.

Although fallowing is not always practical, it can be effective against certain insect pests. Ecologically, fallowing creates a hiatus in the supply of requisites for pests and may be a way of purging an area of difficult pests. Among other pests, wireworms are said to have been reduced by summer fallow practices in parts of Canada. With these insects, two or three fallows may be required to reduce severe infestations. Where more than one planting of the same crop occurs in a single growing season, short fallow periods between plantings may be adequate to reduce infestations. In California and Florida, it is sometimes customary for early and late plantings of celery to overlap in time. Aphids infesting early plantings have been a problem in the later plantings when they migrate and transmit celery mosaic virus. By employing a celery-free period to eliminate the overlapping presence of a host, aphid infestations and the mosaic disease have been effectively suppressed.

A somewhat similar procedure to fallowing also occurs with livestock. Here, cattle and other animals are moved from one pasture to another, leaving the first pasture idle for a period of time. This process is called pasture spelling, and it has the same effect on less mobile livestock parasites as fallowing has on some plant pests. In the United States, the spelling or rotation of cattle from pastures was particularly important against the cattle tick, *Boophilus annulatus,* a vector of pathogens that cause bovine piroplasmosis. Elimination of cattle in fields for a few months caused ticks to starve and allowed the cleanup of infested pastures. Other parasites, such as the lone star tick, *Amblyomma americanum,* in the southwestern United States, have a wider host range and are not as easily managed by pasture spelling because they may survive on birds and other wildlife in the area.

**Disrupting crop and insect synchrony.** Part of the reason that insects are pests on given crops is because the insect seasonal cycle is synchronized with the seasonal cycle of these crops. If crop phenology (timing of biological events like emergence, flowering, fruiting, and seed maturity) can be changed to be asynchronous with such insect events as egg laying and larval development, insect numbers and/or injury can be reduced.

Changes in crop phenology may sometimes be accomplished by planting alternate varieties of the same crop, changing planting dates, or both. With some crops, like clover, clipping may modify reproductive phenology.

Modifying planting dates has been one of the most widely used methods to cause asynchrony between crops and pest insects. One of the classic examples of this approach is with the Hessian fly and the planting of winter wheat. Winter wheat is planted and emerges in the fall as do Hessian fly adults. These adults live only 3 or 4 days, and females lay eggs on the leaves of early-sown wheat. Eggs hatch in the fall, and larvae develop into pupae ("flaxseed" stage), which overwinter. Ecological management can be practiced by understanding the adult emergence period and delaying wheat planting until all adults have

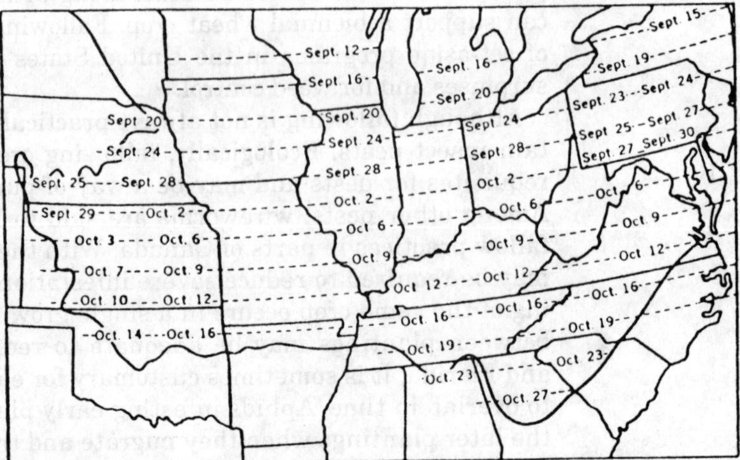


**Figure 10.14** Fly-free dates for planting wheat in the central United States to escape egg laying by the Hessian fly, *Mayetiola destructor.* (Courtesy USDA)

died. In so doing, the emerging wheat escapes infestation by the fall generation. Estimates of fly-free dates (Fig. 10.14) for planting are available for most northern states and are especially recommended when fly-resistant cultivars are unavailable.

In another example, modifying planting dates of kale (a type of cabbage) can reduce larval infestations of cabbage looper, *Trichoplusia ni,* on the crop. The pest is most abundant from late spring to early fall in the southern United States, and numbers on kale can be held to manageable levels by early planting of the spring crop and late planting of the fall crop. With this plan, the crop matures too early in the spring and too late in the fall to be affected by peak populations of the pest.

Another horticultural crop, the tomato, has also been shown to benefit from altering planting dates. By planting tomatoes early in North Carolina, for instance, ripening occurs before corn matures, and injury by the corn earworm is minimal. Late-planted tomatoes can be heavily infested by moths that move from the succeedingly less attractive corn into the fruiting tomatoes.

In a somewhat different situation, the formation of clover seed can be delayed by spring grazing or clipping. This has been done to avert damage by the clover seed midge, *Dasineura leguminicola,* in the northwestern United States, where clover is grown for seed. In this instance, the delayed phenology causes seeds to set after adults of the first fly generation have died. By the time second-generation flies are ready to oviposit, the seeds are too far advanced to serve as a host.

Timing of livestock production activities like dehorning and castration also may be modified to prevent problems from insects. Transmission of disease-producing organisms by flies is reduced by conducting these and other such procedures at cooler times of the year, when adults of the species are not active.

## DIVERTING PEST POPULATIONS AWAY FROM THE CROP

In many instances, it may not be practical to modify the crop itself, or modifications of existing environments may not prove very effective for pest management. Ineffectiveness may be prevalent when insects have a high dispersal rate.

As another approach to habitat management, it is sometimes possible to take advantage of insect dispersal abilities and/or their preferences for one host over another. With this procedure, we would attempt to divert the insect from a crop to be protected by presenting it with a more favorable substitute. Among the most common ways of diverting insects are trap cropping and strip harvesting.

### Trap Cropping

**Trap cropping** usually involves planting small areas of a crop or other species near the protected crop. The favorability of this alternate environment (the trap) compels the pest to move into the trap and stay away from the protected crop. Depending on the seasonal cycle, the insects are left to develop in the trap or are killed in place with an insecticide. Even if an insecticide is used, costs and environmental impact are reduced, because entire fields are not treated. In trap cropping, the trap may be a different species than the crop or the same species planted at a different time.

An instance showing feasibility of this technique has been developed with soybeans in the Mississippi delta region of the southern United States. Populations of the bean leaf beetle (Box 10.9), *Cerotoma trifurcata,* can be suppressed by planting 5 to 10 percent of a field with an early-maturing soybean variety about 10 days to 3 weeks earlier than the main planting. The main planting of the field is of a late-maturing variety. Early in the season, vegetative growth of the early planting attracts overwintered bean leaf beetles, which seek food and lay eggs of the first generation. Treatments of insecticides are made in the trap 7 to 10 days after emergence of first-generation adults to prevent infestations in the main planting. Later in the season, at soybean pod fill, southern green stink bugs, *Nezara viridula,* are also attracted to and concentrate in the trap. Subsequently, these may be treated with insecticide, if necessary. The net result of using this approach can be a reduction in both treatment costs and insecticide residues in the environment. An added benefit can be conservation of the insect natural enemies in the agroecosystem.

Planting strips of alfalfa in cotton fields in the western United States is another example of trap cropping. Several years ago this approach gave effective management for western lygus bugs, *Lygus hesperus,* in both experimental and commercial fields. With this technique, 5- to 10-meter strips of alfalfa are planted in every 100 to 120 meters of cotton (Fig. 10.15). When irrigated properly, the alfalfa effectively draws the bugs out of the cotton, and they remain in the alfalfa to feed. In this way, blossom abortion caused by bug feeding (Fig. 10.16) is significantly reduced in the cotton, and the alfalfa receives little injury. The technique was found most effective with low to moderate populations of plant bugs. Although a combination of economics, production problems, and federal regulations caused abandonment of the trap-cropping approach, this practice is being reevaluated for use against several pests in modern cotton production.

## BOX 10.9 Bean Leaf Beetle (See Color Plate 3)

**SPECIES:** *Cerotoma trifurcata* (Forster) (Coleoptera: Chrysomelidae)

**DISTRIBUTION:** The bean leaf beetle is widely distributed throughout the United States. Populations tend to be more abundant in the southeastern states but also occur regularly in the midwestern and eastern regions.

**IMPORTANCE:** Bean leaf beetles feed on several host species but are commonly found in soybeans, pea, cowpea, and several other bean species. Injury is predominately leaf consumption by adult beetles, but larvae also injure plants by feeding on roots and nodules. In many instances, adults do not cause economic damage by themselves, but their defoliation (rather symmetrical round holes) contributes to the feeding injury of other defoliating species. Adults also have been observed feeding on the mature bean pods. The consequence of this pod injury is seen as round necrotic areas on the pod directly over the seeds. Additionally, feeding on the pod peduncle (stem) causes pod loss. Overall, pod feeding can cause significant loss of seed quality and yield.

**APPEARANCE:** Adults are about 7 mm long and yellow, tan, or red. They usually have black spots on the elytra and have a black wing margin. This species is polymorphic, with exact color, number, and arrangement of spots varying among individuals. Orange, lemon-shaped eggs are commonly laid in clusters of 15 to 30. Larvae are white with a dark head and are distinctively segmented. Pupae are found in the soil within an earthen chamber.

**LIFE CYCLE:** Adults overwinter in the soil or under crop residue in and around the field or adjacent woodlots. Beetles become active in spring and begin feeding and laying eggs in the soil near stems of host plants. Hatching occurs within about 11 days, and larvae feed on plant nodules or root tissue through three stadia. Pupation occurs underground, and adults emerge and feed on available foliage. The complete life cycle requires 35 to 55 days, and one to three generations per year may occur, depending on temperature and latitude.

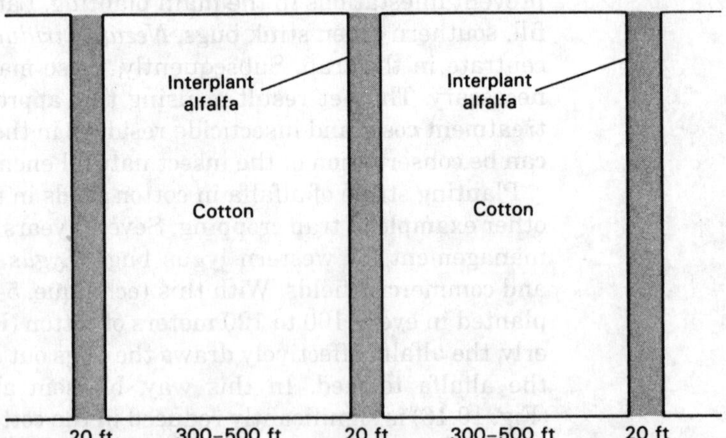


**Figure 10.15** Schematic drawing of restricted alfalfa plantings among cotton to trap populations of western lygus bugs, *Lygus hesperus.* (Reprinted with permission from V. Stern, 1981, in D. Pimentel, ed., *Handbook of Pest Management,* vol. 1, CRC Press Inc., Boca Raton, Fla.)

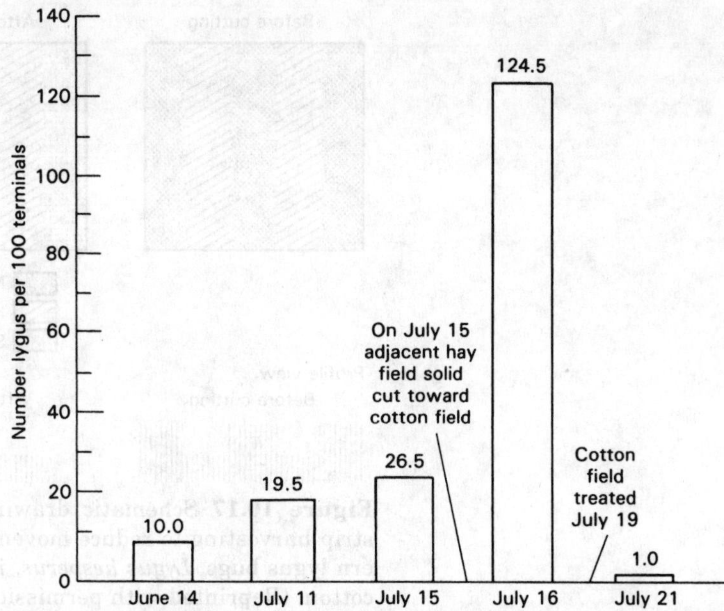


**Figure 10.16** Graph showing increases in densities of western lygus bugs, *Lygus hesperus,* in cotton following complete clipping of adjacent alfalfa. (Reprinted with permission from V. Stern, 1981, in D. Pimentel, ed., *Handbook of Pest Management,* vol. 1, CRC Press Inc., Boca Raton, Fla.)

### Strip Harvesting

**Strip harvesting** is similar to trap cropping except that a trap is created in a main crop by harvesting different areas at different times. With this approach, insects in the crop are not forced to search for requisites in adjacent crops.

Again, the main example of this approach is the western lygus bug, with cotton and alfalfa. In California, cotton and alfalfa may be produced adjacent to each other, and alfalfa fields have been the source of western lygus bugs. Movement of the bugs into cotton is particularly heavy after alfalfa fields have been clipped for hay. To reduce this movement, alfalfa can be harvested in alternate strips. When a strip is cut, strips on either side of it are half-grown (Fig. 10.17). Consequently, instead of leaving the field and moving to cotton, as they do in conventional operations, the bugs remain in alfalfa fields, moving over from harvested to unharvested plants. This approach is most applicable for forage crops, which are harvested several times each growing season.

### Intercropping

Yet another means of diverting pests away from a crop to areas more favorable is intercropping. Intercropping involves growing dissimilar crops in the same location, as is done in South American agriculture, with corn and beans grown in alternate rows.

An example of using intercropping in pest management is in forage cropping systems. Alfalfa-oat intercropping schemes have been shown to reduce potato leafhopper densities by 82 to 90 percent compared to alfalfa monocultures. Ad-

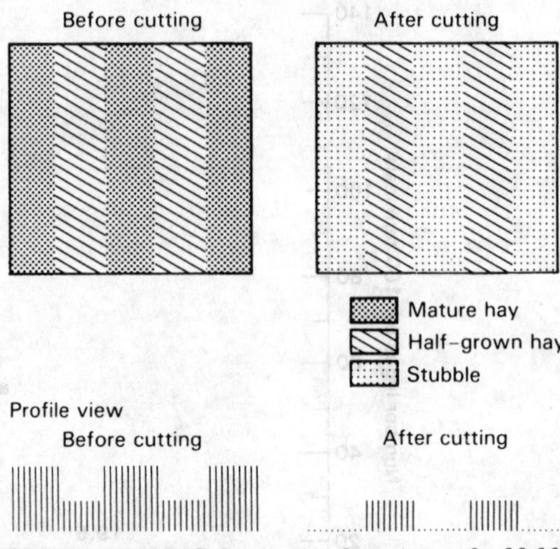


**Figure 10.17** Schematic drawing of alfalfa strip harvesting to reduce movement of western lygus bugs, *Lygus hesperus,* into adjacent cotton. (Reprinted with permission of Plenum Press from *Introduction to Integrated Pest Management* by M. L. Flint and R. van den Bosch. © 1981)

ditionally, alfalfa planted with forage grasses, such as smooth bromegrass and orchardgrass, significantly reduces densities of both potato leafhopper and alfalfa weevil compared to pure alfalfa stands. Although intercropping is useful as a pest management tactic and can eliminate some insecticide treatments, chemicals still may be required in some outbreak situations.

### Push-Pull Polycropping

Push-pull polycropping strategies in Africa use a combination of behavior-modifying stimuli to manipulate the distribution and abundance of pest or beneficial insects in pest management, with the goal of pest reduction on the protected host or resource. Pests are repelled or deterred away from the resource (push) by using stimuli that mask host apparency or are deterrent or repellant. Pests are simultaneously attracted (pull), using highly apparent and attractive stimuli, such as trap crops, where they are concentrated, facilitating their elimination (Cook et al. 2007).

The most successful push-pull strategy was developed for subsistence farmers in east Africa. Maize (*Zea mays*) and sorghum (*Sorghum bicolor*), two principal foods in east Africa, are attacked by stem borers (Lepidoptera), for example, *Busseola fuscus, Chilo partellus, Eldana saccharina*, and *Sesamia calamistis*, that cause 10 to 50 percent yield losses (Cook et al. 2007). Farmers combine the use of intercrops and trap crops, using plants that are appropriate for the farmers and exploit natural enemies. Stem borers are repelled from the maize and sorghum by nonhosts such as greenleaf desmodium (*Desmodium intortum*), silverleaf desmodium (*Desmodium uncinatumi*), and molasses grass (*Melinis minutiflora*), which are interplanted with the maize or sorghum (the push). Around the field edges are planted trap crops, mostly Napier grass

(*Pennisetum purpureum*) and Sudan grass (*Sorghum vulgare sudanese*), which attract and concentrate the pests (the pull). These grasses have a dual purpose, as they are also used as forage for livestock. Molasses grass, as an intercrop, reduces stem borer populations by producing stem borer repellent volatiles; it also increases parasitism by a parasitoid wasp. *Desmodium* also produces similar repellent volatiles, but also produces sesquiterpenes that suppress the parasitic African witchweed (*Striga hermonthica*), a major yield constraint of cropland in east Africa. The desmodium compounds stimulate germination of witchweed seeds and subsequent mortality of the seedlings. The push-pull strategy has contributed to increased grain yields and livestock production in east Africa, resulting in a significant impact on food security (Cook et al. 2007).

## REDUCING THE IMPACT OF INSECT INJURY

The purpose of this technique is to manage losses. We realize that insects are present in the crop and injury is being sustained. Instead of focusing on the insects themselves, we consider the crop and modify cultural techniques to minimize losses from the injury. Such an approach is often used in conjunction with other pest management tactics.

### Modify Host Tolerance

The modification of plants and animals to be more tolerant of insect injury is often achieved genetically. New forms of plants or animals are bred that yield well in spite of insect attack.

Host tolerance can also be modified in a cultivar or breed by nongenetic means. Most crops sustain a wide range in yield loss to a given degree of injury, depending on the vigor of crop growth. For instance, soybeans grown during periods of drought in the midwestern United States can suffer up to twice the yield loss from a given amount of defoliation as those in years with normal precipitation. Therefore, good production practices like proper irrigation, fertilization, and weed control can have a significant influence on the vigor of the crop and, consequently, on the amount of damage sustained from insect injury. By producing a vigorous crop, it is sometimes possible to overcome a degree of the damage that otherwise would be sustained from a pest population.

Although vigorous growth of plants and animals may reduce losses in many situations, there are documented instances where such growth attracts greater numbers. For example, leafhoppers (Hemiptera: Cicadellidae) have been found most abundant on rice fertilized with high rates of nitrogen, and mites (Acari) and aphids have been found on an array of well-fertilized crops. This does not mean that we should limit crop fertility, only that we should be cognizant of possible pest increases and be prepared to deal with them if they occur.

### Modify Harvest Schedules

With many crops, harvest time can vary within certain acceptable limits. When this is the case, harvest dates can sometimes be modified to avoid some types of insect losses. As a rule, crops injured by insects should be harvested at the earliest possible date.

In forage crops like alfalfa, damage from the potato leafhopper, *Empoasca fabae,* and the alfalfa weevil, *Hypera postica,* can be minimized by early cutting.

**Figure 10.18** A grain-drying system for corn. Early harvesting to reduce losses from dropped grain usually requires drying facilities. (Courtesy D. Erbach and USDA)

If significant numbers of the potato leafhopper are present, cutting is done at the very-early-bloom or late-bud stage. For the alfalfa weevil, decisions on early cutting versus insecticide spraying are based on numbers of larvae and plant height. If the crop is near harvest stage, cutting early may produce better economic returns than spraying weevils and delaying harvest.

Early harvest is also recommended for insect pests of other crops. One is the wheat stem sawfly, which tunnels in wheat stems, causing plant lodging and harvest loss. In South Dakota, some growers clip infested wheat before it lodges and windrow it. After drying, the windrowed plants are threshed with a special attachment on the combine, and harvest losses are reduced. The loss of yield from corn ears that drop early because of European corn borer tunneling is also reduced by early harvesting. However, early-harvested corn usually has unacceptably high moisture content and must be dried before it is stored (Fig. 10.18).

In forestry, logging operations are scheduled during specific times of the year to minimize damage from wood-boring insects. Timber cut from late summer to early winter is least likely to be damaged by wood borers. Losses from bark beetles, *Ips* species, in pine are least likely if logging operations are completed during the fall or winter months, thus avoiding the period of greatest beetle activity.

## CONCLUSIONS

Ecological management is one of the oldest, least expensive, and most ecologically compatible tactics for solving insect problems. To use this approach, weak links in the insect seasonal cycle are identified and exploited. Usually, food sources or physical factors of the crop environment are made unfavorable for insects through manipulating conventional production practices.

In most instances, the approach is a preventive tactic that anticipates problems before they occur and attempts to avoid or minimize their impact. It serves as a baseline procedure that is compatible and can be integrated with many other pest management tactics. As a rule, it is most effective with insects that have a narrow host range, a low rate of dispersal, and/or complex requirements in the seasonal cycle.

The major limitation of ecological management is that the most effective procedures to manage pests may not be practical in terms of crop production objectives. Although insect pests are an important part of crop production, other priorities like greater efficiency, higher yields, and soil conservation often dictate the confines in which ecological management can operate.

Additionally, new technologies are leading to changes in crop production and making older approaches to ecological management less viable. Particularly, the development of selective herbicides has permitted the refinement and acceptance of no-tillage or reduced-tillage practices. Such practices encourage soil conservation and save fuel, time, and labor. They also eliminate such options as sanitation and tillage for insect pest management.

Will such changes in crop culture cause significant changes in the status of insect pests? Probably so. For some pests, status may increase, but it may decrease with other pests. In either case, pest managers will need to accept these technologies and understand the ecology of new crop environments. Where pest infestations increase, a search for other ways to make crops less favorable for insects is needed.

## Further Reading


All, J. N. 1999. Cultural approaches to managing arthropod pests, pp. 395–415. In J. R. Ruberson, ed., *Handbook of Pest Management*. New York: Marcel Dekker.

Altieri, M. A. 1993. *Biodiversity and Pest Management in Agroecosystems*. New York: Food Products Press.

Cook, S. M., Z. R. Khan, and J. A. Pickett. 2007. The use of push-pull strategies in integrated pest management. *Annual Review of Entomology* 52:375–400.

DeGooyer, T. A., L. P. Pedigo, and M. E. Rice. 1999. Effect of alfalfa-grass intercrops on insect populations. *Environmental Entomology* 28:703–710.

Fenemore, P. G. 1982. *Plant Pests and Their Control*. Wellington, New Zealand: Butterworths, pp. 167–172.

Flint, M. L., and R. van Den Bosch. 1981. *Introduction to Integrated Pest Management*. New York: Plenum Press, pp. 46–48.

Funderburk, J. E., L. G. Higley, and G. D. Buntin. 1993. Concepts and directions in arthropod pest management. *Advances in Agronomy* 51:125–172.

Genung, W. G. 1974. Flooding in everglades soil pest management. *Proceedings Tall Timbers Conference on Ecological Animal Control by Habitat Management*, Gainesville, Fla. 6:165–172.

Hammond, R. B., R. A. Higgins, T. P. Mack, L. P. Pedigo, and E. J. Bechinski. 1991. Soybean pest management, pp. 341–472. In D. Pimentel, ed., *CRC Handbook of Pest Management in Agriculture*, vol. 3, 2nd ed. Boca Raton, Fla.: CRC Press.

Hanson, C. H., S. G. Turnipseed, N. T. Powell, J. M. Good, and D. L. Klingman. 1979. Cultural and preventive practices, pp. 91–101. In W. B. Ennis, Jr., ed., *Introduction to Crop Protection*. Madison, Wis.: American Society of Agronomy and Crop Science Society.

Muma, M. H. 1970. Preliminary studies on environmental manipulation to control injurious insects and mites in Florida citrus groves. *Proceedings Tall Timbers Conference on Ecological Animal Control by Habitat Management*, Tallahassee, Fla. 2:23–40.

National Academy of Sciences. 1969. *Insect Pest Management and Control. Principles of Plant and Animal Pest Control*, vol. 3. U.S. National Academy of Science Publication 1695, pp. 208–242.

Oseto, C. Y. 2000. Physical control of insects, pp. 25–102. In J. E. Rechcigl and N. A. Rechcigl, eds., *Insect Pest Management*. Boca Raton, Fla.: Lewis Publishers.

Rabb, R. L., G. K. Defoliart, and G. G. Kennedy. 1984. An ecological approach to managing insect populations, pp. 697–728. In C. B. Huffaker and R. L. Rabb, eds., *Ecological Entomology*. New York: Wiley.

Ralston, L. H., and C. E. McCoy. 1966. *Introduction to Applied Entomology*. New York: Ronald Press, pp. 102–107.

Reitz, S. R., E. L. Yearby, J. E. Funderburk, J. Stavisky, M. T. Momol, and S. M. Olson. 2003. Integrated management tactics for *Frankliniella* thrips (Thysanoptera: Thripidae) in field-grown pepper. *Journal of Economic Entomology* 96:1201–1214.

Sailer, R. I. 1991. Extent of biological and cultural control of insect pests of crops, pp. 3–12. In D. Pimentel, ed., *CRC Handbook of Pest Management in Agriculture*, vol. 1, 2nd ed. Boca Raton, Fla.: CRC Press.

Shelton, A. M., and F. R. Baenes-Perez. 2006. Concepts and applications of trap cropping in pest management. *Annual Review of Entomology* 51:285–308.

Stern, V. 1991. Environmental control of insects using trap crops, sanitation, prevention, and harvesting, pp. 157–182. In D. Pimentel, ed., *CRC Handbook of Pest Management in Agriculture*, vol. 1, 2nd ed. Boca Raton, Fla.: CRC Press.

Teetes, G. L. 1991. The environmental control of insects using planting times and plant spacing, pp. 169–216. In D. Pimentel, ed., *CRC Handbook of Pest Management in Agriculture*, vol. 1, 2nd ed. Boca Raton, Fla.: CRC Press.

Vincent, C., G. Hallman, B. Panneton, and F. Fleurat-Lessard. 2003. Management of agricultural insects with physical control methods. *Annual Review of Entomology* 48:261–281.

## Favorite Web Sites

http://ipmworld.umn.edu/chapters/ferro.htm

Site defines the concepts of cultural control and presents various approaches, with specific examples.

http://insects.tamu.edu/extension/bulletins/b1721.pdf

Site presents detailed, season-long, cultural tactics to suppress boll weevils in Texas cotton, an excellent example of ecological management.

CHAPTER 11

# CONVENTIONAL INSECTICIDES FOR MANAGEMENT

INSECTICIDES AND OTHER PESTICIDES are some of the most important chemicals used for the well-being of human populations. They are indispensable in maintaining high levels of health, nutrition, and quality surroundings. They rank with medicines in their impact on our existence.

In agricultural production, pesticides are a regular component of most systems, and their development has given rise to entirely new ways of growing crops. Indeed, we could not consider using such energy- and soil-conserving approaches as no-till cropping without them. Moreover, the quantity and quality of our food and fiber production could not be maintained without substantial pesticide inputs.

In the United States, more than 544,000,000 kilograms of pesticides are produced annually, with retail sales of more than $11 billion. Of these sales, it is estimated that agriculture accounts for 77 percent and industry and government about 14 percent. Home and garden use accounts for only about 9 percent (Fig. 11.1). A study of household, lawn, and garden use has indicated that 85 percent of American households use some kind of pesticide. This study reveals the pervasiveness of these chemicals in our society.

Insecticides are particular kinds of pesticides for killing insects and other invertebrates. They account for about 10 percent of the pesticide quantity used on major field and forage crops, but they represent more than 18 percent of that used for home and garden purposes. Furthermore, more than 220 insecticidally active ingredients with thousands of uses are registered with the United States Environmental Protection Agency (EPA).

Increasing insecticide use since World War II is the result of several advantages these chemicals have over other techniques for insect control. Of utmost importance is the effectiveness of insecticides. They can treat an insect problem while it is in progress, reducing numbers to insignificant levels. Insecticide action is rapid: It usually takes effect within hours and alleviates a problem within a few days. In particular, rapid action allows this tactic to cure a problem by dampening population peaks before major losses are sustained. Insecticides also are economical, at least in the short run, compared with many other pest management tactics. In terms of only the crop, cost/benefit ratios regularly are on the order of four to five dollars return for every dollar invested in insecticide applications. Finally, insecticides offer ease of application. Persons with minimal experience or training can apply them effectively.

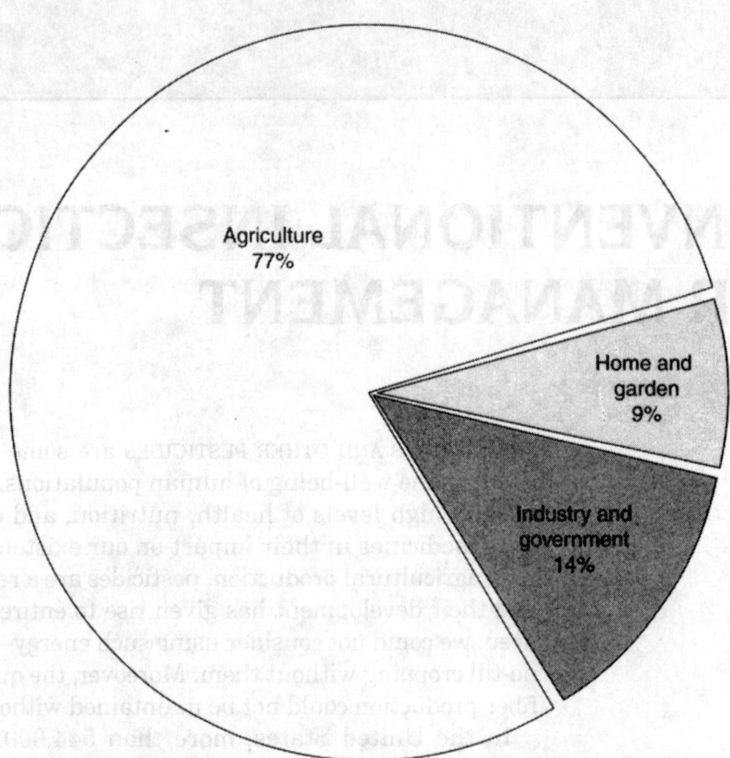


**Figure 11.1** Proportionate use of pesticides by groups in the United States. (Based on EPA data)

Unfortunately, insecticides have important disadvantages. Some problems occur within the cropping system itself from frequent use. Insecticide resistance, pest resurgence, and pest replacement can occur following repetitive applications of these compounds. Therefore, we may see decreasing effectiveness with their use. Moreover, insecticides may have a negative impact on nontarget species like honey bees, fish, and wildlife both within and outside agroecosystems. Destruction of these organisms is a cost often not included when calculating cost/benefit ratios. Furthermore, insecticides present risks to the user. Many insecticidal compounds are highly toxic to humans and can injure or kill when applied improperly or when accidents occur.

If we consider the alternatives, the benefits of insecticides probably outweigh the risks. But the goal of insecticide use to obtain maximal benefits with minimal risks has not yet been achieved in most situations. Therefore, a thorough understanding of these chemical tools and the consequences of their use is crucial if this approach is to be improved.

## INSECTICIDE NAMES AND FORMULAS

The word **insecticide** literally means killer of insects, derived from the Latin suffix, *cida,* or "killer." The term *pesticide* is broader, denoting a killer of pests in general. In addition to insecticides, other common pesticides include **acaricides** (mite and tick killers), **herbicides** (weed killers), **fungicides** (fungus killers), and **nematicides** (nematode killers). Other designations of pesticides have even more specific names. For example, some insecticides may be called aphicides when used for aphids and termiticides when used for termites. Legally, these are all classed as **economic poisons,** substances used for controlling, preventing, destroying, repelling, or mitigating any pest.

In most instances, insecticides are chemicals. Therefore, although fly swatters are used to kill insects, they are not called insecticides. In this chapter, we deal mostly with conventional or traditional insecticides, which are chemical compounds with wide use that kill insects quickly. Most insecticides are nerve poisons. Nonconventional insecticides include microbial insecticides and insect growth regulators with slower actions; death occurs by other means than those of conventional insecticides.

### Insecticide Nomenclature

Insecticide nomenclature is the formal process by which insecticides are named. Insecticides are designated by three names: the approved common name, the trade name, and the chemical name. In the United States, common names for insecticides are officially selected by the Entomological Society of America and approved by the American National Standards Institute and the International Organization for Standardization. The **trade name** (also the proprietary name or brand name) is given by the manufacturer or formulator of the insecticide. The chemical name provides a description of the insecticide's structure and is formed by following the conventions of the International Union of Pure and Applied Chemistry or those of the 9th Collective Index period of the Chemical Abstracts Service. Either of these conventions may be found defining an insecticide chemical name.

The following is an example of the names for a widely used insecticide:

Common name: carbaryl
Trade name: Sevin®
Chemical name: 1-naphthalenyl methylcarbamate

Note that the trade name has a **registered trademark** superscript, indicating that a patent exists or is pending and that the rights of the patent holder are protected by law. More than one manufacturer may hold patent rights to a compound; therefore, it is not uncommon to see more than one trade name for a single insecticide. In publications, the symbol is used the first time the insecticide is mentioned and not after that.

An understanding of the equivalence of these names is important because insecticide recommendations may be presented either as common names or as trade names. The state Cooperative Extension Service may publish either trade names or common names, usually depending on the name most frequently used for a compound. Lists of pesticides can be found in publications like the *Crop Protection Handbook* (Meister Pub. Co., Willoughby, Ohio), published annually, and equivalences of names can be determined.

### Chemical Formulas

Chemical formulas represent the composition of chemical compounds. They describe the components of a single molecule and are important in our characterizations of insecticides.

Both molecular formulas and structural formulas are used to describe insecticidal compounds. Molecular formulas use standard symbols of the elements and numbers to show the kind and quantity of atoms in a molecule. For example, $H_2O$ symbolizes water and indicates that two atoms of hydrogen are bonded by an atom of oxygen. On the other hand, the structural formula is more like a map, showing where atoms are located relative to each other.

| Molecular Formula | Structural Formula |
|---|---|
| $H_2O$ | H–O–H |

One of the most common structural components in insecticide chemistry is the benzene ring, also called the phenyl ring when other chemical groups are attached to it. This six-carbon ring with six hydrogens is presented as a hexagon with double bonds. For brevity, the benzene or phenyl ring is usually reduced to a simple hexagon with bonding symbols.

Benzene

Because molecules are actually three-dimensional structures, the structural formula does not give a complete representation; it is only two-dimensional. Therefore, *stereo,* or three-dimensional, formulas are sometimes given so that depth and true spatial form of a compound can be better perceived.

Dieldrin

Stereo Dieldrin

## SURVEY OF COMMON INSECTICIDES

Insecticides have been grouped or classified in several ways, according to application or chemical composition. If application is the main focus, they may be grouped according to the site of insect encounter (Fig. 11.2). In some of the

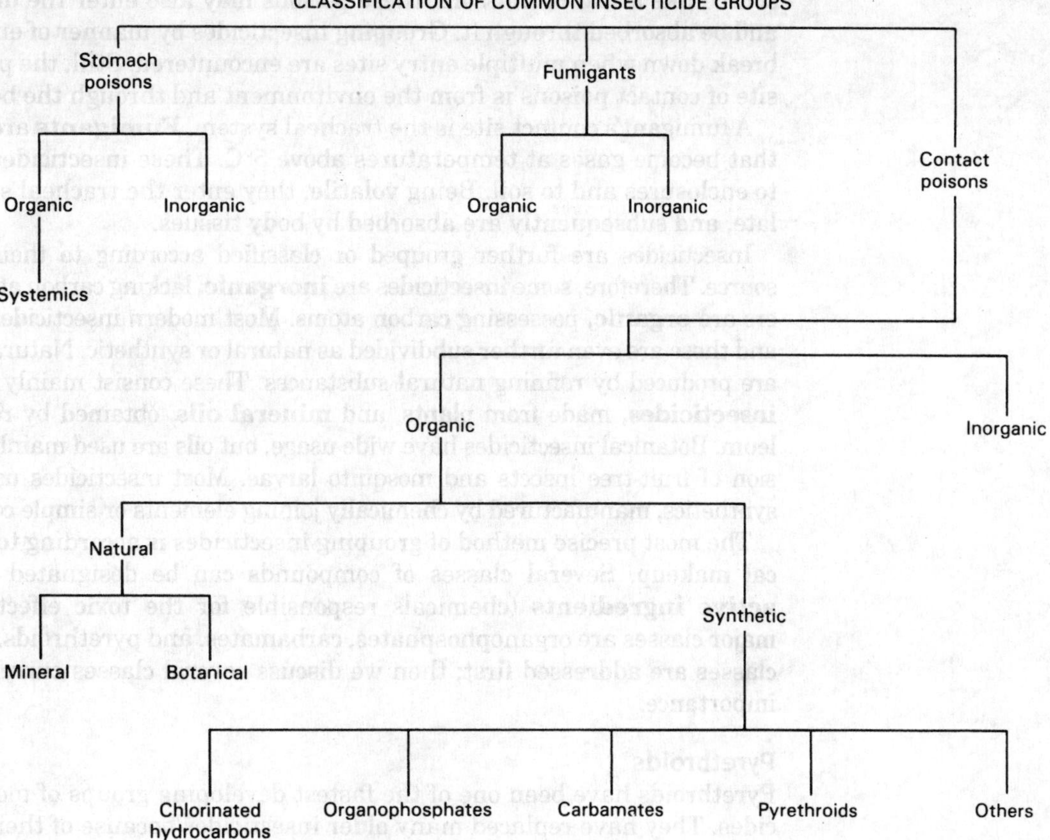


**Figure 11.2** Classification of common insecticide groups. Initial divisions are based on site of encounter.

established literature, insecticides are listed under the headings stomach poisons, contact poisons, and fumigants.

**Stomach poisons** enter the insect body through the gut and are fatal only after they are eaten. It is mainly the oldest insecticides that fall within this grouping, and few true stomach poisons are used today. One such stomach poison still used is boric acid ($H_3BO_4$), for use against cockroaches and other crawling insects in the household.

However, there are also modern insecticides whose main encounter with the insect is through the gut. These are so-called **systemic insecticides.** Systemics are taken up and translocated within plants and animals. Insects feeding on the protected host contact the insecticide through the gut, and susceptible individuals are killed. Systemics in plants mostly kill piercing-sucking insects; larger chewing insects usually are not affected. Piercing-sucking pests are killed more readily because they normally receive a greater insecticide dose than chewing insects feeding on the same plant. In livestock, systemics are often used against internal parasites like cattle grubs, *Hypoderma* species. Systemic insecticides also have properties of contact insecticides and are not usually thought of as true stomach poisons.

**Contact poisons** are the major group of modern insecticides. They usually enter the body when the insect walks or crawls over a treated surface. The insecticide is absorbed through the body wall. If the treated surface is a food source like a leaf or blossom, these poisons may also enter the digestive tract and be absorbed through it. Grouping insecticides by manner of entry begins to break down when multiple entry sites are encountered. Still, the primary entry site of contact poisons is from the environment and through the body wall.

A fumigant's contact site is the tracheal system. **Fumigants** are insecticides that become gases at temperatures above 5°C. These insecticides are applied to enclosures and to soil. Being volatile, they enter the tracheal system, circulate, and subsequently are absorbed by body tissues.

Insecticides are further grouped or classified according to their nature and source. Therefore, some insecticides are **inorganic,** lacking carbon atoms, and others are **organic,** possessing carbon atoms. Most modern insecticides are organic, and these are even further subdivided as natural or synthetic. Natural insecticides are produced by refining natural substances. These consist mainly of **botanical insecticides,** made from plants, and **mineral oils,** obtained by refining petroleum. Botanical insecticides have wide usage, but oils are used mainly for suppression of fruit-tree insects and mosquito larvae. Most insecticides used today are synthetics, manufactured by chemically joining elements or simple compounds.

The most precise method of grouping insecticides is according to their chemical makeup. Several classes of compounds can be designated according to **active ingredients** (chemicals responsible for the toxic effect). The three major classes are organophosphates, carbamates, and pyrethroids. These major classes are addressed first; then we discuss several classes currently of lesser importance.

### Pyrethroids

Pyrethroids have been one of the fastest developing groups of modern insecticides. They have replaced many older insecticides because of their great effectiveness and safety of application.

Pyrethroids are not new insecticides; the first, allethrin, was developed in 1949. Allethrin was synthesized to duplicate the insecticidal activity of a natural product, cinerin I, a component of the botanical insecticide **pyrethrum.** The term **pyrethroid** comes from "pyrethrum," and "-oid" means something that resembles something else. In other words, these insecticides resemble pyrethrum.

However, pyrethroids have certain advantages over their namesake. They are highly toxic to insects at very low rates. In addition to quick knockdown

Allethrin (Pynamin®)

2-methyl-4-oxo-3(2-propenyl)-2-cyclopenten-1-yl 2,2-dimethyl-3-(2-methyl-1-propenyl)cyclopropanecarboxylate

ability, there is less recovery of poisoned insects, compared with the natural product. Although pyrethrum is too expensive to be used in agriculture, modern pyrethroids compare favorably in price at low application rates with organophosphates and carbamates. Finally, the modern pyrethroids are broken down by ultraviolet wavelengths more slowly than is pyrethrum. Consequently, many pyrethroids are effective on plant foliage for 4 to 7 days.

The pyrethroids are often categorized into generations of development, of which there are four. Allethrin belongs to the first generation, and several compounds including resmethrin belong to the second generation. But the most widely used compounds today are from the third and fourth generations.

**Third-generation pyrethroids.** These include fenvalerate, introduced in 1972, and permethrin, introduced in 1973. These insecticides have been used in many crops, including cotton, corn, and soybeans, where they are effective against several above-ground insect pests. Rates of application in these crops are exceptionally low—for example, 0.11 kilogram or less of actual ingredient per hectare (0.1 pound or less per acre)—compared with many organophosphates and carbamates—for example, 1.1 to 2.3 kilograms of actual ingredient per hectare (1 to 2 pounds per acre).

**Fourth-generation pyrethroids.** These pyrethroids are even more potent insect poisons than are the third-generation chemicals. Application rates for these may be only one-tenth those of the third-generation pyrethroids for equal effectiveness. Much research is being conducted with fourth-generation compounds for greater usefulness and even more efficacy. Some of the insecticides in this category are cypermethrin, flucythrinate, fluvalinate, deltamethrin, bifenthrin, lambda-cyhalothrin, cyfluthrin, esfenvalerate, fenpropathrin, prallethrin, tefluthrin (Force®, not shown), tralomethrin, and imiprothrin (Pralle®, not shown).

Fenvalerate (Ectrin®)

$HC(CH_3)_2$ Cl C H C O O C C≡N O

cyano (3-phenoxyphenyl) methyl 4-chloro-α-(1-methylethyl) benzeneacetate

Permethrin (Ambush®, Pounce®)

$(CH_3)_2$ O C O $CH_2$ O $Cl_2C{=}CH$

(3-phenoxyphenyl)methyl(±)-*cis,trans*-3-(2,2-dichloroethenyl)-2,2-dimethylcyclopropanecarboxylate

Bifenthrin (Capture®, Talstar®)

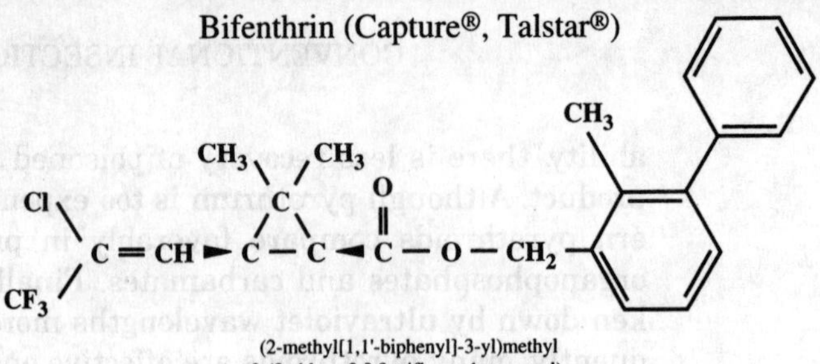

(2-methyl[1,1'-biphenyl]-3-yl)methyl
3-(2-chloro-3,3,3-trifluoro-1-propenyl).
2,2-dimethylcyclopropanecarboxylate.

Lambda-cyhalothrin (Karate®, Warrior®)

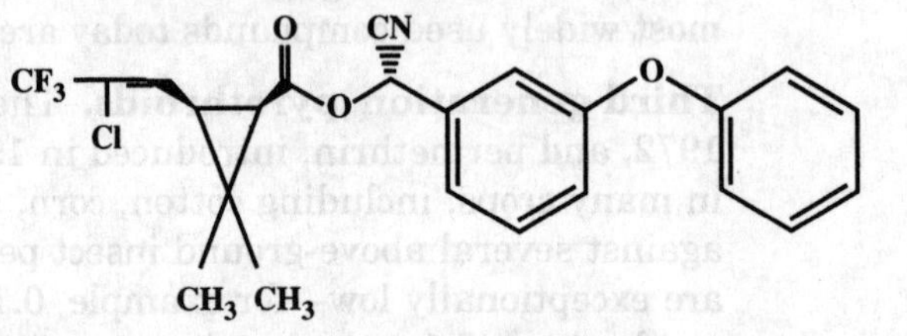

[1α(*S**),3α(*Z*)]-(±)-cyano(3-phenoxyphenyl)methyl
3-(2-chloro-3,3,3-trifluoro-1-propenyl)-2,2-
dimethlcyclopropanecarboxylate

Cyfluthrin (Baythroid®)

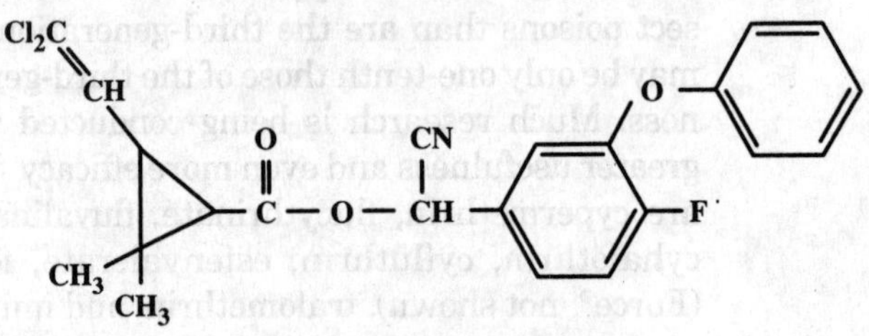

cyano(4-fluoro-3-phenoxyphenyl)methyl
3-(2,2-dichloroethenyl)-2,2-
dimethylcyclopropanecarboxylate

Fenpropathrin (Danitol®)

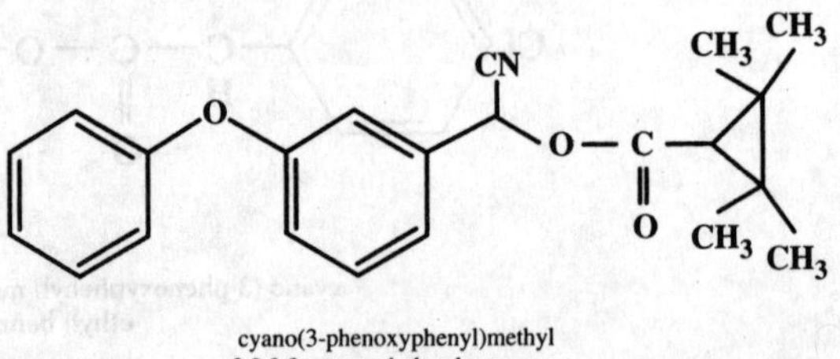

cyano(3-phenoxyphenyl)methyl
2,2,3,3-tetramethylcyclopropane-
carboxylate

Prallethrin (Etoc®)

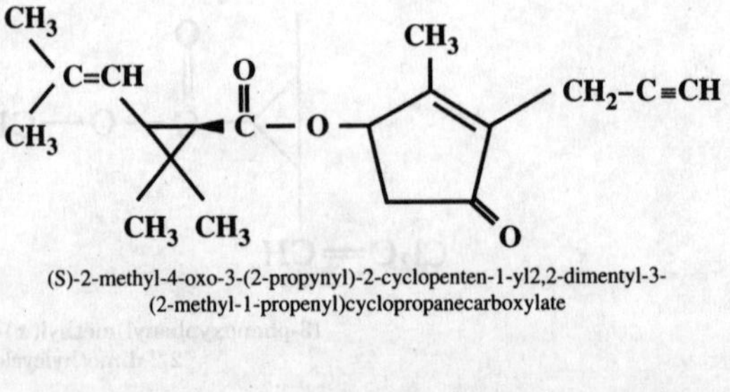

(S)-2-methyl-4-oxo-3-(2-propynyl)-2-cyclopenten-1-yl2,2-dimentyl-3-
(2-methyl-1-propenyl)cyclopropanecarboxylate

**Cypermethrin (Ammo®, Cymbush®, Cynoff®)**

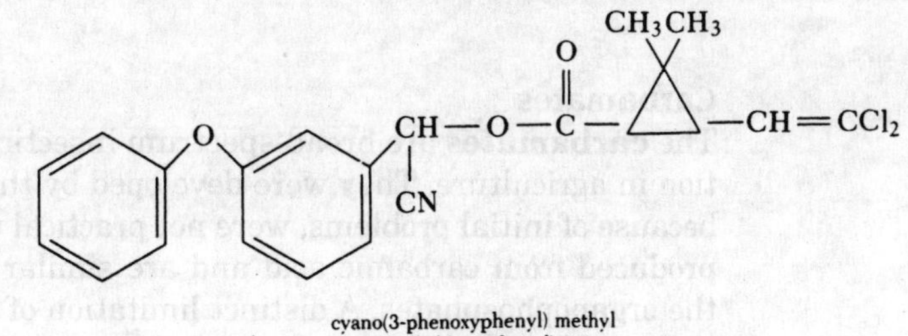

cyano(3-phenoxyphenyl) methyl 3-(2,2-dichloroethenyl)-2,2-dimethyl cyclopropanecarboxylate

**Flucythrinate (Cybolt®, Payoff®)**

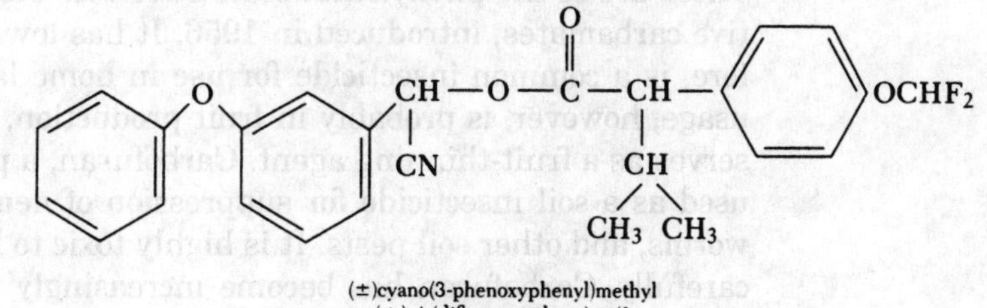

(±)cyano(3-phenoxyphenyl)methyl (+)-4-(difluoromethoxy)-α-(1 methylethyl)benzeneacetate

**Fluvalinate (Mavrik®)**

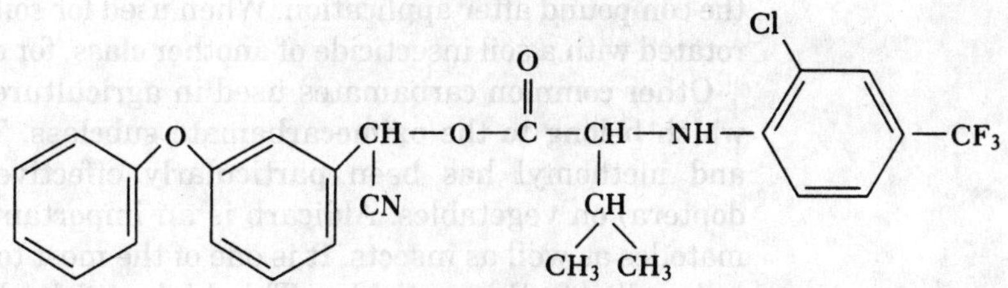

*N*-[2-chloro-4-(trifluoro methyl)phenyl]DL-valine, cyano (3-phenoxyphenyl)methyl ester

**Deltamethrin (Decis®)**

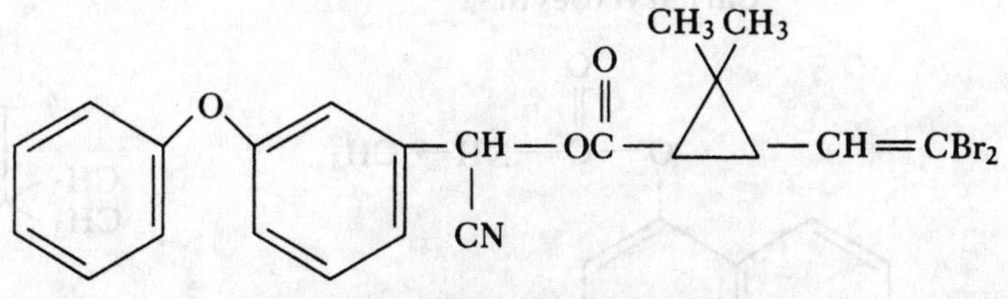

cyano(3-phenoxyphenyl) methyl 3-(2,2-dibromoethenyl)-2,2-dimethyl cyclopropanecarboxylate

Tralomethrin (Scout X-TRA®)

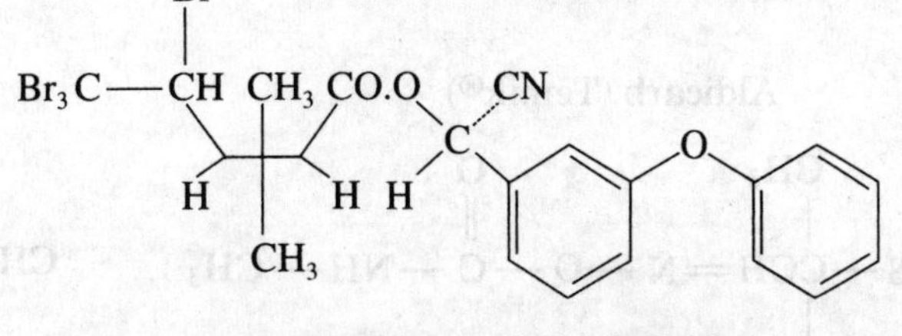

cyano(3-phenoxyphenyl)methyl 2.2-dimethyl-3-(1,2,2,2-tetrabromethyl)cyclopropanecarboxylate

## Carbamates

The **carbamates** are broad-spectrum insecticides that have had wide application in agriculture. They were developed by the Geigy Corporation in 1951 but, because of initial problems, were not practical until 1956. These insecticides are produced from carbamic acid and are similar in environmental persistence to the organophosphates. A distinct limitation of carbamates in pest management is their toxicity to Hymenoptera, including both pollinators and parasitoids.

Two carbamates widely used in agriculture are carbaryl and carbofuran. These are of the phenylcarbamate subclass. Carbaryl is the oldest of the effective carbamates, introduced in 1956. It has low toxicity to humans and, therefore, is a common insecticide for use in home lawns and gardens. Its greatest usage, however, is probably in fruit production, where it both kills insects and serves as a fruit-thinning agent. Carbofuran, a plant systemic, has been widely used as a soil insecticide for suppression of nematodes (Nematoda), corn rootworms, and other soil pests. It is highly toxic to humans and should be handled carefully. Carbofuran has become increasingly ineffective when used continuously in the same soil. This phenomenon, called *enhanced biodegradation,* is seemingly caused by increases in microorganisms capable of quickly degrading the compound after application. When used for soil insects, carbofuran should be rotated with a soil insecticide of another class, for example, an organophosphate.

Other common carbamates used in agriculture are aldicarb and methomyl, which belong to the oximecarbamate subclass. The first is a soil insecticide, and methomyl has been particularly effective against caterpillars (Lepidoptera) on vegetables. Aldicarb is an important systemic for controlling nematodes as well as insects. It is one of the most toxic and, therefore, dangerous to handle of all insecticides. This high toxicity has greatly limited aldicarb's usefulness. Under certain circumstances, aldicarb has been discovered in shallow groundwater, and its use must be monitored very carefully.

Carbaryl (Sevin®)

$O-C(=O)-NH-CH_3$

1-naphthalenyl methylcarbamate

Carbofuran (Furadan®)

$CH_3$, $CH_3$, O; $O-C(=O)-NH-CH_3$

2,3-dihydro-2,2-dimethyl-7-benzofuranyl methylcarbamate

Aldicarb (Temik®)

$CH_3-S-C(CH_3)_2CH{=}N-O-C(=O)-NH-CH_3$

2-methyl-2-(methylthio)propanal *O*-[(methylamino)carbonyl]oxime

Methomyl (Lannate®)

$CH_3-C(S-CH_3){=}N-O-C(=O)-NH-CH_3$

methyl *N*-[[(methylamino)carbonyl]oxy]ethanimidothioate

A remaining carbamate, propoxur, is used by professionals against cockroaches. It is particularly effective against species in restaurants and homes that have become resistant to organophosphate insecticides.

Propoxur (Baygon®)

$$\text{C}_6\text{H}_4(\text{OCH(CH}_3)_2)\text{—O—C(=O)—NH—CH}_3$$

2-(1-methylethoxy) phenyl methylcarbanate

### Organophosphates

The **organophosphate** insecticides were developed in Germany during World War II as a substitute for nicotine, an insecticide used against the Colorado potato beetle, *Leptinotarsa decemlineata.* The discovery of the insecticidal properties of this group was associated with other German studies on related chemicals, the so-called nerve gases (sarin, soman, and tabun). Since the passage of the Food Quality Protection Act (FQPA) in 1996, this class of conventional pesticides has been a primary focus of EPA reregistration activities.

Organophosphates (OP) are derived from phosphoric acid and are some of the most toxic insecticides. As opposed to the chlorinated hydrocarbons, a class of long-lasting insecticides, they are unstable in the presence of light and quickly break down into nontoxic compounds. Breakdown can occur within a few hours or days, as compared with months or years for many of the chlorinated hydrocarbons. For this reason and because they are effective, OPs were used to replace the chlorinated hydrocarbons in many programs. In fact, they are the most widely used group of insecticides today.

The amount of organophosphate insecticides used has declined 30 percent since 1980, from an estimated 131 million pounds in 1980 to 91 million pounds in 1999. Since 1980, however, organophosphate usage as a percent of total insecticide usage has increased, from 58 percent in 1980 to 72 percent in 1999. The increase in usage in 1999 was mainly due to the increased amount of malathion used as part of the USDA-sponsored Boll Weevil Eradication Program. Malathion's use in this program has increased substantially over the past few years as the program has expanded to include most of the major cotton-producing areas of the United States.

OPs are characterized as having different alcohols attached to their phosphorus atoms, and the various phosphorus acids produced are termed *esters.*

*o*-Phosphoric Acid

$$\text{HO—P(=O)(—OH)—O—H}$$

These esters have different combinations of oxygen, carbon, sulfur, and nitrogen, and OPs formed from them can be divided into three groups of derivatives: aliphatic, phenyl, and heterocyclic.

**Aliphatic derivatives.** The aliphatic derivatives are compounds with straight carbon chains, as opposed to those with ring structures. The oldest, and one of the most toxic, aliphatic derivatives is TEPP used for fly control in dairy barns because of its effectiveness and short residue (breakdown in 12 to 24 hours).

TEPP

$$(C_2H_5O)_2\overset{\overset{\displaystyle O}{\|}}{P}-O-\overset{\overset{\displaystyle O}{\|}}{P}(OC_2H_5)_2$$

tetraethyl pyrophosphate

The most widely used aliphatic is malathion, one of the safest and most effective of all OPs. Malathion has been used for all types of agricultural purposes and is effective against many insects. It is safe for home use, including most garden and household pests, and does not harm pets when used according to directions. It is commonly prescribed by physicians for head, body, and crab louse (Phthiraptera: Pediculidae) problems. Also, it is formulated as powders or dips for flea (Siphonaptera) control on pets and mange mites (Acari) on livestock. Ultra low volume (ULV) sprays, those applied as fine particles in concentrated form, also have been found effective and economical for management of such insects as grasshoppers on rangeland.

Malathion

$$\begin{array}{l} \qquad\qquad\qquad\qquad\quad\; O \\ \qquad\qquad\qquad\qquad\quad\; \| \\ \qquad\quad S \qquad\quad CH_2-C-OC_2H_5 \\ \qquad\quad \| \qquad\qquad\; | \\ (CH_3O)_2P-S-CH-C-OC_2H_5 \\ \qquad\qquad\qquad\qquad\quad\; \| \\ \qquad\qquad\qquad\qquad\quad\; O \end{array}$$

diethyl[(dimethoxyphosphinothioyl)thio]butanedioate

In addition, several aliphatics are plant systemics, including dimethoate, disulfoton, oxydemetonmethyl, and dicrotophos. Most of these insecticides are applied to the soil and are taken up by plants and translocated to above-ground plant parts. They are effectively used for many aphids, leafhoppers (Hemiptera: Cicadellidae), and other piercing-sucking insects. Dimethoate is also an effective contact poison against mites during outbreaks on soybeans and other crops.

Other common aliphatics include trichlorfon, methamidophos, and acephate. Trichlorfon has been used in crop insect management and around farms for fly control. It is one of the few modern insecticides that is insect-selective, killing

**Dimethoate (Cygon®)**

$$(CH_3O)_2\overset{S}{\overset{\|}{P}}-S-CH_2\overset{O}{\overset{\|}{C}}-NH-CH_3$$

*O,O*-dimethyl-*S*-[2-(methylamino)-2-oxoethyl] phosphorodithioate

**Oxydemeton-methyl (Metasystox-R®)**

$$(CH_3O)_2\overset{O}{\overset{\|}{P}}-S-CH_2CH_2-\overset{O}{\overset{\|}{S}}-C_2H_5$$

*S*-2-[(ethylsulfinyl)ethyl]*O,O*-dimethyl phosphorothioate

**Dicrotophos (Bidrin®)**

$$(CH_3O)_2\overset{O}{\overset{\|}{P}}-O-\overset{CH_3}{\overset{|}{C}}=CH\overset{O}{\overset{\|}{C}}-N(CH_3)_2$$

*O,O*-dimethyl-*O*-1-methylvinyl-*N,N*-dimethyl carbamoyl phosphate

**Disulfoton (Di-Syston®)**

$$(C_2H_5O)_2\overset{S}{\overset{\|}{P}}-S-CH_2CH_2-S-C_2H_5$$

*O,O*-diethyl-*S*-[2-(ethylthio)ethyl] phosphorodithioate

**Trichlorfon (Dylox®)**

$$(CH_3O)_2\overset{O}{\overset{\|}{P}}-\overset{OH}{\overset{|}{C}}HCCl_3$$

dimethyl (2,2,2-trichloro-1-hydroxyethyl) phosphonate

**Methamidophos (Monitor®)**

$$\begin{matrix} CH_3O \\ CH_3S \end{matrix} > \overset{O}{\overset{\|}{P}}-NH_2$$

*O,S*-dimethyl phosphoramidothioate

**Chlorethoxyfos (Fortress®)**

$$C_2H_5O-\underset{OC_2H_5}{\underset{|}{\overset{S}{\overset{\|}{P}}}}-O-\overset{Cl}{\overset{|}{C}}H-CCl_3$$

Phosphorothioate, *O,O*-diethyl *O*-(1,2,2,2-tetrachloroethyl)

**Acephate (Orthene®)**

$$\begin{matrix} CH_3O \\ CH_3S \end{matrix} > \overset{O}{\overset{\|}{P}}-NH-\overset{O}{\overset{\|}{C}}-CH_3$$

*O,S*-dimethyl acetylphosphoramidothioate

some insects but leaving others (many natural enemies) unharmed. Both methamidophos and acephate are more recent aliphatics used widely in agriculture and especially for management of vegetable pests.

Three other aliphatics that have been important as soil insecticides include phorate, terbufos, and ethoprop. Phorate is an older OP that is economical and effective against corn rootworms. Terbufos is another compound also widely used for protection of monocropped corn from corn rootworms. Both phorate and terbufos have plant-systemic properties. Ethoprop is used similarly but is not systemic. A more recent corn rootworm aliphatic is chlorethoxyfos (Fortress®), which is also used against wireworms, cutworms, seedcorn maggot, and white grubs.

Phorate (Thimet®)

$$C_2H_5O \quad S$$
$$(C_2H_5O)_2P(=S)—S—CH_2—S—C_2H_5$$

*O,O*-diethyl-*S*-2-(ethylthio)methyl phosphorodithioate

Terbufos (Counter®)

$$(C_2H_5O)_2—P(=S)—S—CH_2—S—C(CH_3)_3$$

*S*-[[(1,1-dimethylethyl)thio]methyl] *O,O*-diethyl phosphorodithioate

Ethoprop (Mocap®)

$$C_2H_5O—P(=O)(S—C_3H_7)(S—C_3H_7)$$

*O*-ethyl *S, S*-dipropyl phosphorodithioate

**Phenyl derivatives.** Phenyl OPs differ from aliphatic OPs in having a phenyl ring, which has one of the hydrogens displaced by a phosphorus moiety and others (one or more) displaced by $CH_3$, Cl, CN, $NO_2$, or S. Stability of phenyl OPs is greater, and residues usually last somewhat longer in the environment than the aliphatics.

One of the older, most widely used phenyls is parathion. The first form of parathion, ethyl parathion, was introduced in 1947, and early use was against aphids. It is very toxic to humans, and its use diminished and was canceled, later being replaced with the introduction in 1949 of the somewhat less toxic methyl parathion. Methyl parathion has a broad range of toxicity to many insect pests and had several agricultural applications. It was particularly important for difficult-to-kill insects and was sometimes mixed with other insecticides to increase overall effectiveness. Because of its toxicity, however, most uses were discontinued in 1999.

Ethyl Parathion (discontinued)

$$(C_2H_5O)_2P(=S)—O—C_6H_4—NO_2$$

*O,O*-diethyl *O*-*p*-nitrophenyl phosphorothioate

Methyl Parathion

$$(CH_3O)_2P(=S)—O—C_6H_4—NO_2 \quad NO_2$$

*O,O*-dimethyl *O*-*p*-nitrophenyl phosphorothioate

Much less toxic is stirofos, a household insecticide used for livestock parasites. Its toxicity is similar to that of malathion.

Stirofos (Tetrachlorvinphos)

$$(CH_3O)_2P(=O)—O—C(=CHCl)—C_6H_2Cl_3$$

*O,O*-dimethyl *O*-2-chloro 1-(2,4,5-trichlorophenyl) vinyl phosphate

Famphur (Warbex®)

$(CH_3O)_2P(=S)$—O—$C_6H_4$—$SO_2N(CH_3)_2$

*O*-[4-[(dimethylamino) sulfonyl]phenyl] *O,O*-dimethyl phosphorothioate

Fenthion (Baytex®)

$(CH_3O)_2P(=S)$—O—$C_6H_3(CH_3)$—S—$CH_3$

*O,O*-dimethyl *O*-[3-methyl-4-(methylthio)phenyl phosphorothioate

Some of the phenyl OPs also are used for animal systemics against cattle grubs. These include famphur and fenthion, which can be used on beef cattle and nonlactating dairy cattle. Usually, they are poured on the animal and are absorbed through the skin.

Other phenyl OPs used on field crops include profenofos and sulprofos. Yet others, for example, isofenphos, have been particularly useful against soil insects in both field and vegetable crops. However, isofenphos was canceled by the EPA in 1999.

Profenofos (Curacron®)

$(C_2H_5O)(C_3H_7S)P(=O)$—O—$C_6H_3(Cl)$—Br

*O*-(4-bromo-2-chlorophenyl)-*O*-ethyl-*S*-propyl phosphorothioate

Sulprofos (Bolstar®)

$(C_2H_5O)(C_3H_7S)P(=S)$—O—$C_6H_4$—$SCH_3$

*O*-ethyl-*S*-propyl *O*(4-methylthio) phenyl phosphorodithioate

Isofenphos (canceled)

$(C_2H_5O)((CH_3)_2CHNH)P(=S)$—O—$C_6H_4$—$COOCH(CH_3)_2$

1-methylethyl 2-[[ethoxy [(1-methylethyl) amino] phosphinothioyl] oxy]benzoate

**Heterocyclic derivatives.** Heterocyclic OPs, like phenyl OPs, have ring structures but differ in having one or more carbon atoms displaced by O, N, or S. Also, structural rings in this group may have three, five, or six atoms.

Compounds of this group are the most stable and long-lasting of the OPs. They have complex molecules that break down into many products. This characteristic makes residues difficult to measure in the laboratory, rendering these compounds somewhat more limited for use on food for human consumption.

Two heterocyclic OPs with wide use in agriculture include methidathion and phosmet. These have uses on field, forage, fruit, and nut crops against a variety of insects and mites. Phosmet is also used against two important weevils, the boll weevil on cotton and the plum curculio, *Conotrachelus nenuphar,* on fruit trees.

**Diazinon**

$(C_2H_5O)_2P(=S)—O—$ [pyrimidine ring with N, N; $—CH(CH_3)_2$; $CH_3$]

*O,O*-diethyl-*O*-(2-isopropyl-4-methyl-6-pyrimidinyl) phosphorothioate

**Azinphosmethyl (Guthion®)**

$(CH_3O)_2P(=S)—S—CH_2—N$ [benzotriazinone ring: O, N, N]

*O,O*-dimethyl-*S*(4-oxo-1,2,3-benzotriazin-3(4*H*)-ylmethyl) phosphorodithioate

**Chlorpyriphos (Dursban®, Lorsban®, Nufos®)**

$(C_2H_5O)_2P(=S)—O—$ [pyridine ring: N; Cl, Cl, Cl]

*O,O*-diethyl-*O*-(3,5,6-trichloro-2-pyridyl) phosphorothioate

**Methidathion (Supracide®)**

$(CH_3O)_2P(=S)—S—CH_2—N—N$ [thiadiazole ring: O, S, $OCH_3$]

*S*-[(5-methoxy-2-oxo-1,3,4-thiadiazol-3(2*H*)-yl)methyl] *O,O*-dimethyl phosphorodithioate

**Phosmet (Imidan®)**

$(CH_3O)_4P(=S)—S—CH_2—N$ [phthalimide ring: O, O]

*N*-(mercaptomethyl)-phthalimide *S*-(*O,O*-dimethylphosphorodithioate)

Previously, one of the most common heterocyclics was diazinon, an insecticide contained in many household and garden sprays. In the 1960s, it was used as a soil insecticide against corn rootworms but was replaced after resistance to the chemical developed. Also, most field crop and orchard uses of diazinon have been discontinued. Diazinon was also widely used for household, garden, and ornamental insect problems. However, under the Food Quality Protection Act (FQPA), allowable tolerances have been reduced as much as tenfold, placing diazinon in jeopardy of being canceled or severely limited in use, as has been done with azinphosmethyl (Guthion®).

In response to EPA assessments, the manufacturer agreed to terminate all indoor residential and indoor nonresidential uses of diazinon. Additionally, EPA and the manufacturer agreed to phase out and cancel outdoor residential lawn and garden uses. All told, these actions were aimed at reducing total diazinon use by 75 percent. Sale of the product for the home and garden ended in 2004.

Such EPA action has also been applied recently to other popular heterocyclics, specifically chlorpyriphos (Dursban®, Lorsban®). The Dursban over-the-counter product has been considered a relatively safe insecticide used in the structural pest control industry for cockroaches and termites and by homeowners for many home, yard, and garden pests. In applying risk levels allowed in the FQPA, household uses of chlorpyriphos deemed safe for children have been set at one one-thousandth of the "no-effect level" rather than the usual one one-hundredth. This stringent level of risk avoidance effectively ruled out continued use of Dursban in the home because consumer exposure could easily exceed this level. Although agricultural uses of chlorpyriphos (marketed as Lorsban® and other trade names) continue, allowable uses have been reduced. Such actions show the recent degree of scrutiny the EPA is giving the organophosphates and signal the tenuous state of their future.

### Neonicotinoids

Neonicotinoids are a relatively new class of synthetic insecticides. Just as pyrethroids resemble the natural product pyrethrum, neonicotinoids resemble (are analogs of) the natural product nicotine. The class has been prominently represented by the compound imidacloprid. Imidacloprid is a systemic and contact insecticide with primary activity on piercing-sucking insects such as aphids, leafhoppers, thrips, and whiteflies. It is also effective against termites, turf insects, soil insects, and some beetle species. Its mode of action is quite different from most other conventional insecticides, and therefore, it has potential for managing insects that have become insecticide-resistant. Additionally, this new insecticide has relatively low mammalian toxicity and generally good environmental characteristics.

Imidacloprid is marketed worldwide under several labels, including Gaucho®, Merit®, Admire®, Confidor®, Premier®, Premise®, and Provado®. Imidacloprid upsets the central nervous system by causing irreversible blockage of postsynaptic nicotinergic acetylcholine receptors.

Neonicotinoids are under intensive development. Newer products include nitenpyram (Bestguard®), thiamethoxam (Actara®, Platinum®, Cruiser®), dinotefuran (Starkle®), clothianidin (Poncho®), acetamiprid (Mospilan®), nithiazine (Quickstrike®), and thiacloprid (Calypso®). More compounds are likely to be forthcoming.

**Imidacloprid (Gaucho®, Merit®, Admire®)**

1-[(6-chloro-3-pyridinyl)methyl]-*N*-nitro-2-imidazolidinimine

**Nitenpyram (Bestguard®)**

(1*E*)-*N*-[(6-chloro-3-pyridinyl)methyl]-*N*-ethyl-*N*-methyl-2-nitro-1,1-ethenediamine

**Thiamethoaxam (Actara®, Platinum®, Cruiser®)**

3-[(2-chloro-5-thiazolyl)methyl]tetrahydro-5-methyl-*N*-nitro-4*H*-1,3,5-oxadiazin-4-imine

**Dinotefuran (Starkle®)**

*N*-methyl-*N*-nitro-*N*-[(tetrahydro-3-furanyl)methyl]guanidine

**Acetamiprid (Mospilan®)**

(1*E*)-*N*-[(6-chloro-3-pyridinyl)methyl]-*N*-cyano-*N*-methylethanimidamide

**Clothianidin (Poncho®)**

[*C*(*E*)]-*N*-[(2-chloro-5-thiazolyl)methyl]-*N*-methyl-*N*-nitroguanidine

**Nithiazine (Starbar® Quickstrike®)**

tetrahydro-2-(nitromethylene)-2*H*-1,3-thiazine

**Thiacloprid (Calypso®)**

(*Z*)-[3-[(6-chloro-3-pyridinyl)methyl]-2-thiazolidinylidene]cyanamide

### Phenylpyrazoles

Phenylpyrazoles are another new class of insecticides with a single material, fipronil. Fipronil acts as a potent blocker of the GABA (gamma-aminobutyric acid)-regulated chloride channel. It is effective against insects that are resistant or tolerant to pyrethroid, cyclodiene and organophosphorus and/or carbamate insecticides. Fipronil, registered as Regent® and others, is used on a variety of soil and foliar insects, including Colorado potato beetle and rice water weevil. Also, it is used as a seed treatment and developed as a bait for cockroaches, termites, and ants. Yet another use is treating soil around homes to eliminate termites. This product is marketed as Termidor®. Other phenylpyrazoles are under development.

Fipronil (Regent®)

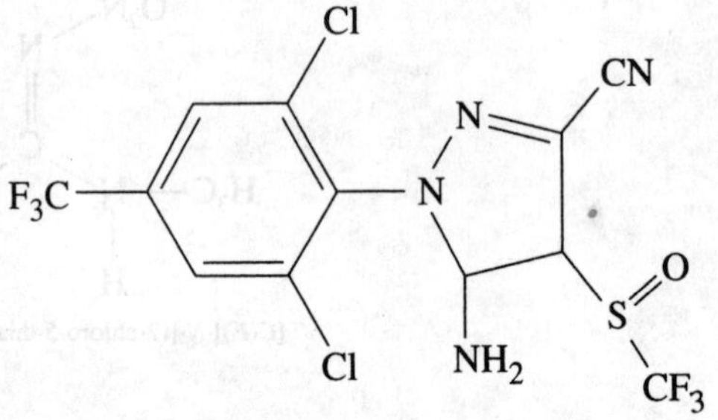

(+)-5-amino-1-[2,6-dichloro-4-(trifluoromethyl)phenyl]-4-[(trifluoromethyl)sulfinyl]-1*H*-pyrazole-3-carbonitrile

### Pyrroles

The pyrroles are a new group of insecticide/acaracides, having both contact and stomach modes of action. A single compound is found in this group, chlorfenapyr, labeled as Pylon®. The pesticide is used in cotton and is under development for whiteflies, thrips, leafminers, aphids, caterpillars, Colorado potato beetles, and mites in field crops, vegetables, and tree and vine crops. Chlorfenapyr kills by uncoupling oxidative phosphorylation in pests and has ovicidal action in some species.

Chlorfenapyr (Pylon®)

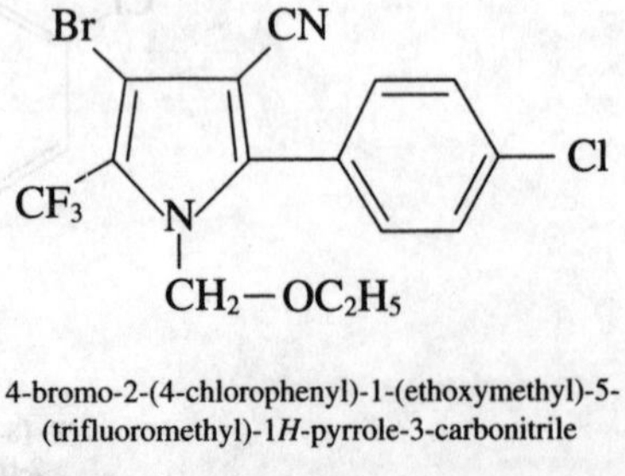

4-bromo-2-(4-chlorophenyl)-1-(ethoxymethyl)-5-(trifluoromethyl)-1*H*-pyrrole-3-carbonitrile

**Pyrazoles**

The pyrazoles are a new class of acaricides, with limited activity against some insect pests (aphids, whiteflies, thrips, and psyllids). They show contact and stomach modes of action. Development has focused on two compounds—tebufenpyrad (Pyranica®, Masai®) in field, vegetable, and fruit crops and fenpyroximate (Acaban®, Dynamite®, and others) for mites. The pesticides disrupt ATP formation in pests by inhibiting mitochondrial electron transport at the NADH-Co Q reductase site.

Tebufenpyrad (Pyranica®, Masai®)

4-chloro-*N*[[4-(1,1-dimethylethyl)phenyl]methyl]-3-ethyl-1-methyl-1*H*-pyrazole-5-carboxamide

Fenpyroximate (Acaban®, Dynamite®)

1,1-dimethylethyl 4-[[[(*E*)-[(1,3-dimethyl-5-phenoxy-1*H*-pyrazol-4-yl) methylene]amino]oxy]methyl]benzoate

**Pyridazinones**

The pyridazinones are represented by a single compound, pyridaben. This is a selective, contact insecticide labeled as Sanmite®, Nexter®, and others for mites, leafhoppers, psyllids, and whiteflies on fruit trees, vegetables, ornamentals, and field crops. Pyridaben has rapid knockdown and long residual properties. It acts by inhibiting mitochondrial electron transport.

Pyridaben (Nexter®, Sanmite®)

2-*tert*-butyl-5-(4-*tert*-butylbenzylthio)-4-chloropyridazin-3(2*H*)-one

### Pyridine Azomethines

This is a chemical class not previously used as an insecticide. It is represented by a single chemical, pymetrozine. The mode of action of pymetrozine has not been precisely determined biochemically, but it may involve effects on neuroregulation or nerve-muscle interaction. Physiologically, it seems to act as an antifeedant, preventing an insect from inserting its stylus into plant tissue. Pymetrozine is marketed as Fulfill® on vegetables and Endeaver® on nuts and ornamentals. It is practically nontoxic to terrestrial and aquatic vertebrates and honey bees, and only slightly toxic to aquatic invertebrates. It is a selective replacement for organophosphate insecticides.

Pymetrozine (Fulfill®, Endeaver®)

$H_3C$ N N H N O N C H N

4,5-dihydro-6-methyl-4-[(*E*)-(3-pyridinylmethylene)amino]-1,2,4-triazin-3(2*H*)-one

### Oxadiazines

The oxadiazines are a new class of insecticidal chemistry represented by a single compound, indoxacarb. Indoxacarb acts as a sodium channel blocker and was registered in the United States for use on apples, pears, and many other crops in May 2001. It is a reduced-risk pesticide with very low mammalian toxicity and has little effect on birds and aquatic organisms. Also, it has little

Indoxacarb (Steward®, Avaunt®)

O O N C C N N O $CH_3$ Cl O C O O $CH_3$ O C F F F

methyl (4*aS*)-7-chloro-2,5-dihydro-2-[[(methoxycarbonyl)[4-(trifluoromethoxy) phenyl]amino]carbonyl]indeno[1,2-*e*][1,3,4]oxadiazine-4*a*(3*H*)-carboxylate

effect on predatory insects. It is marketed as Steward® and Avaunt®. Indoxacarb is used against codling moth, white apple leafhopper, Pandemis leafroller, and Lacanobia fruitworm.

### Insect Growth Regulators

Insect growth regulators, also called IGRs, biorationals, and third-generation insecticides, are chemicals that cause death or sterility in insects by disrupting growth processes. These include methoprene (Altocid®), diflubenzuron (Dimilin®, Adept®, Micromite®), hexaflumuron (Sentricon®, Consult®), tebufenozide (Confirm®, Mimic®), buprofezin (Applaud®), and others. They are detailed in Chapter 14.

### Repellents

Repellents are chemicals that cause insects to orient their movements away from a source. Traditional repellents include dimethyl toluamide (DEET, Delphene®) for biting flies and mosquitoes. Other nontraditional repellents serve as antifeeding agents and include pymetrozine (Chess®, Fulfill®, Relay®). Specifics of these and other repellents are discussed in Chapter 14.

### Chlorinated Hydrocarbons

Chlorinated hydrocarbons are the oldest major insecticide class, having been the first widely used synthetic organic insecticides. All insecticides of this group contain chlorine, hydrogen, and carbon. Occasionally, these insecticides also contain oxygen and sulfur. Although very effective, the use of chlorinated hydrocarbons in the United States is negligible because of environmental and human safety concerns.

**DDT and relatives.** DDT is the most famous, perhaps infamous, of all insecticides. Its history has already been discussed, and today most of its uses have been canceled in the United States (official cancellation January 1, 1973). The primary reason for cancellation of DDT (and several other chlorinated hydrocarbons) is its characteristic stability and fat solubility.

DDT

$CCl_3$

Cl—CH—Cl

1,1,1-trichloro-2,2-bis(*p*-chlorophenyl)ethane

TDE (DDD)

$CHCl_2$

Cl—CH—Cl

1,1-dichloro-2,2-bis(*p*-chlorophenyl)ethane

Methoxychlor

$CCl_3$

$CH_3O$—CH—$OCH_3$

1,1,1-trichloro-2,2-bis(*p*-methoxyphenyl)ethane

Insecticides of the other major classes are relatively unstable; they are broken down in animals by enzymes and in the environment by microorganisms, heat, and/or ultraviolet light.

However, DDT is broken down slowly, sometimes over several years, and this stability provides active residues for animal uptake. Sublethal doses of the chemical are ingested by animals and, not being amenable to metabolism, are stored in body fat. In wide use for many years, DDT accumulated in animals that fed on residue-laden plants. Consequently, dairy cows fed DDT-treated hay produced milk with high residues in the milk fat (Fig. 11.3). Ultimately, humans ingested DDT through milk and other animal fats and stored these residues in their own tissues. Indeed, until recently Americans had an average of 10 parts per million of DDT in their body tissues. This concentration has fallen since the cancellation of most DDT uses.

Similar problems occurred with predators in many ecosystems. Invertebrates that consumed plants and detritus with DDT residues stored concentrations in body fat. These invertebrates were fed upon by rodents, reptiles, amphibians, fish, and other small insectivores, further concentrating the residues in their tissues. These primary predators, in turn, were eaten by yet other top predators in the food chain, such as hawks and owls, which received yet higher insecticide concentrations. This phenomenon, known as **biomagnification** (Fig. 11.4), was believed partly responsible for the decline of such predators as ospreys, falcons, eagles, seagulls, pelicans, and others. At least in some instances, the mode of decline was caused by thin egg shells and reproductive failure of the populations. Other chlorinated hydrocarbons that have caused biomagnification are DDE (a metabolite of DDT), TDE, dieldrin, aldrin, HCH and its isomers, mirex, heptachlor, and endrin.

The very characteristics of DDT that resulted in its cancellation also made it an effective insecticide. Because of stability, DDT's insecticidal activity

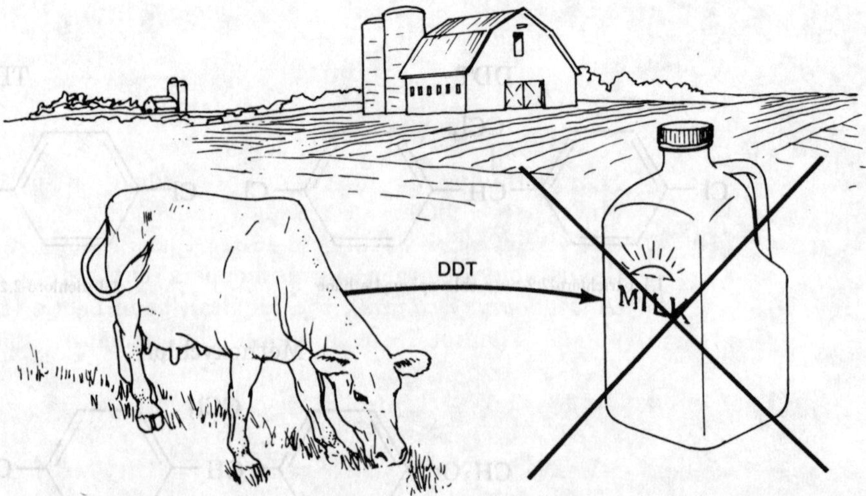


**Figure 11.3** Schematic drawing showing the path of DDT to dairy products. (Courtesy EPA)

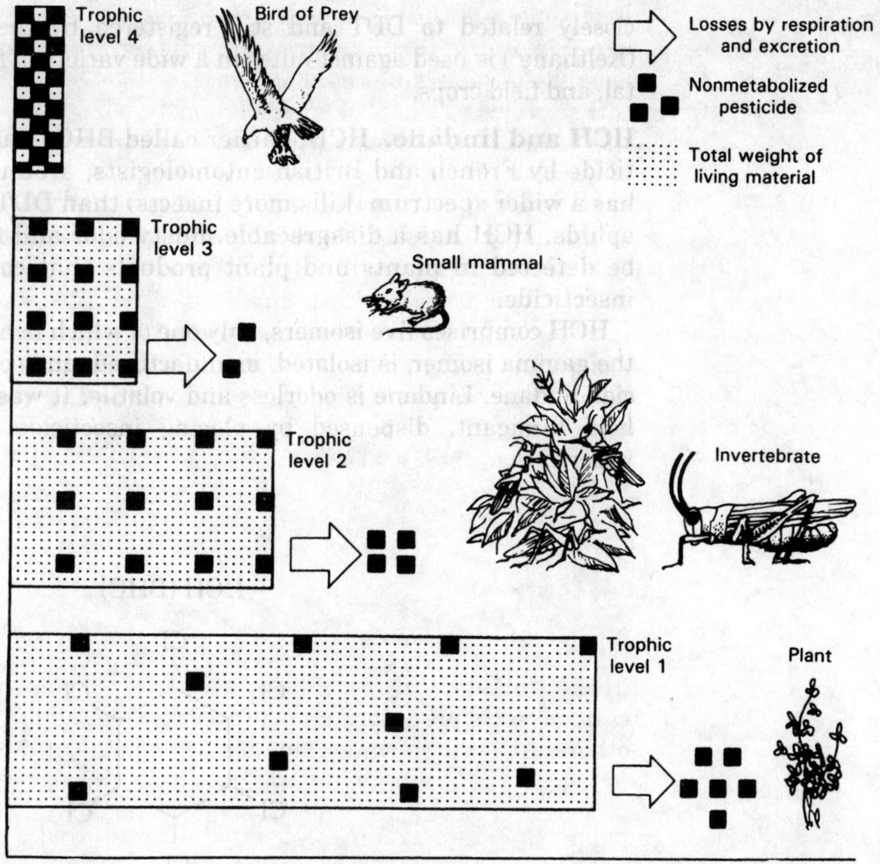


**Figure 11.4** Graph showing the accumulating levels of nonmetabolized pesticide residue as trophic levels are ascended. The size of the bars in the graph represents total biomass at a trophic level. Such is an example of biomagnification. (Redrawn from Flint and van den Bosch, 1981, *Introduction to Integrated Pest Management*, Plenum Press)

lasts considerably longer than many of the insecticides that have replaced it. Consequently, where one application had to be made previously, several may be needed now.

In addition to its importance in medicine (suppression of mosquitoes, lice, and fleas), DDT was used widely in agriculture and reached a peak of production in the United States in 1961. It was effective against many insects, with the notable exception of grasshoppers (Orthoptera: Acrididae), the boll weevil (*Anthonomus grandis*), the Mexican bean beetle (*Epilachna varivestis*), and aphids (Hemiptera: Aphididae). Unfortunately, the overwhelming effectiveness of DDT and its exceptionally low cost contributed to overuse and, subsequently, to its demise. It is still used in some parts of the world but at a fraction of former levels.

Insecticides closely related to DDT are TDE, ethylan, and methoxychlor. Methoxychlor was probably the most important of these other materials; it was used against flies after they had become resistant to DDT and as a spray to eliminate pests from empty stored-grain bins. Another organochlorine insecticide closely related to DDT and still registered by the EPA is dicofol. Dicofol (Kelthane®) is used against mites in a wide variety of fruit, vegetable, ornamental, and field crops.

**HCH and lindane.** HCH, earlier called BHC, was developed as an insecticide by French and British entomologists, around 1940. This insecticide has a wider spectrum (kills more insects) than DDT and is effective against aphids. HCH has a disagreeable musty odor and flavor, and its taste can be detected in plants and plant products that come in contact with the insecticide.

HCH comprises five isomers, only one of which is highly active. This isomer, the gamma isomer, is isolated, manufactured, and sold directly as the insecticide lindane. Lindane is odorless and volatile. It was used widely as a household fumigant, dispensed by placing insecticide pellets in electric wall vaporizers.

HCH (BHC)

Cl Cl Cl Cl Cl Cl

**1,2,3,4,5,6-hexachlorocyclohexane**

Presently, all registered products of HCH have been eliminated, as has lindane as a household fumigant. Lindane has been used mainly in sprays for commercial ornamentals and livestock, in dog shampoos, and as a human parasiticide.

**Cyclodienes.** The cyclodienes were developed after DDT and HCH, beginning about 1945. The major compounds in this group are aldrin, dieldrin, chlordane, heptachlor, endrin, mirex, endosulfan, and kepone. In particular, chlordane was widely used against termites and mirex for fire ant control. Endosulfan (Thiodan®) is still used in a variety of horticultural and field crops.

These insecticides are persistent chemicals, stable in soil and relatively so in sunlight. Therefore, many were used in great quantities against such soil insects as corn rootworms (*Diabrotica* species), wireworms (Coleoptera: Elateridae), and cutworms (Lepidoptera: Noctuidae). Most of the cyclodienes have higher levels of mammalian toxicity than DDT and are more dangerous to apply. Growing ineffectiveness from insecticide resistance and problems with residue uptake in harvested produce have caused the elimination of most cyclodienes in agriculture.

**Chlordane**

1,2,4,5,6,7,8,8-octachloro-3*a*,4,7,7*a*-tetrahydro-4,7-methanoindane

**Heptachlor**

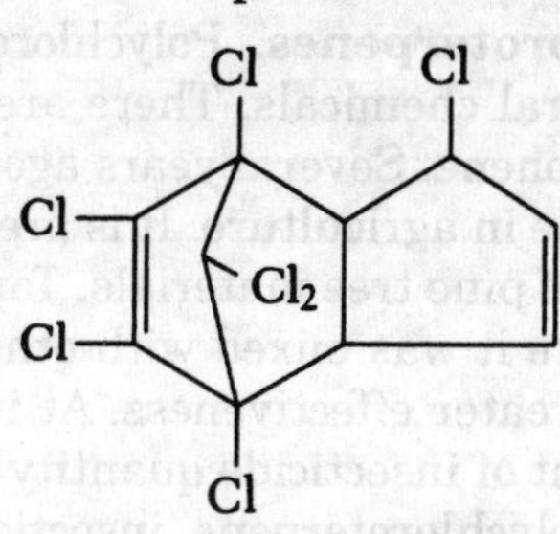


1,4,5,6,7,8,8-heptachloro-3*a*,4,7,7*a*-tetrahydro-4,7-methanoindane

**Aldrin**

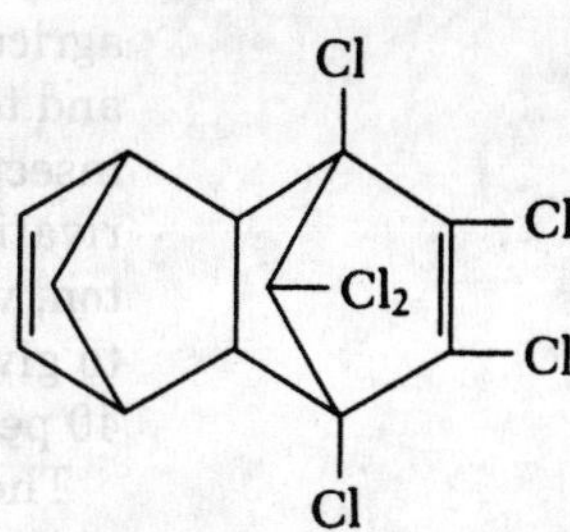


1,2,3,4,10-hexachloro-1,4,4*a*,5,8,8*a*-hexahydro 1,4-endo-*exo*-5,8-dimethanonaphthalene

**Endosulfan (Thiodan®)**

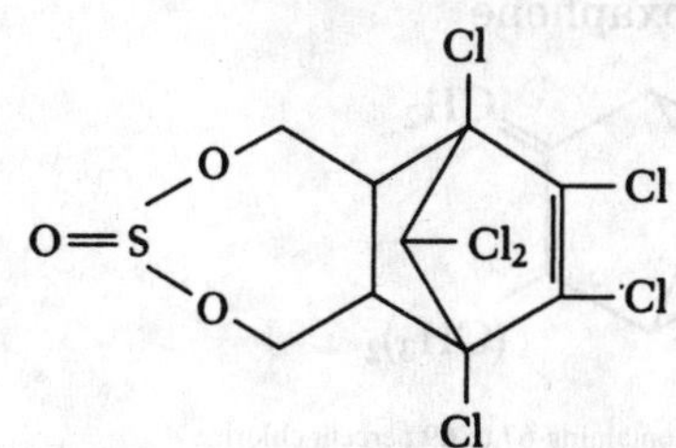


6,7,8,9,10,10-hexachloro-1,5,5*a*,6,9,9*a*-hexahydro-6,9-methano-2,4,3-benzodioxathiepin 3-oxide

**Mirex**

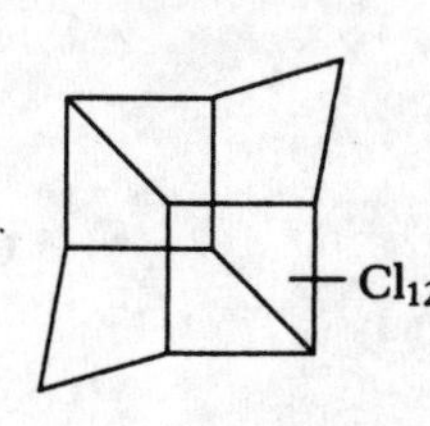


dodecachlorooctahydro-1,3,4-metheno-1*H* cyclobuta[*cd*] pentalene

**Chlordecone**

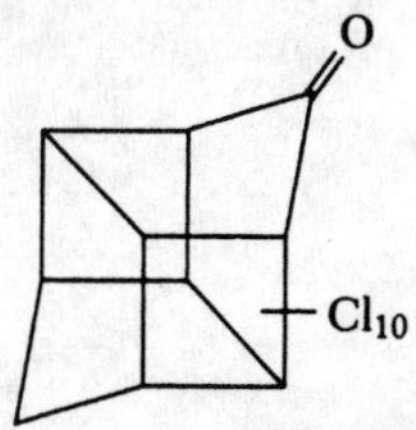


dodecachlorooctahydro-1,3,4-metheno-2*H* cyclobuta[*cd*] pentalene-2-one

**Dieldrin**

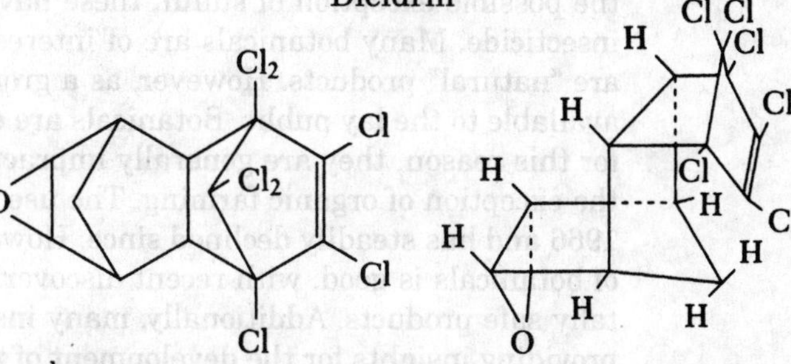

endo-exo

1,2,3,4,10,10-hexachloro-6,7-epoxy-1,4,4*a*,5,6,7,8,8*a*-octahydro-1,4-*endo-exo*-5,8-dimethanonaphthalene

**Endrin**

endo-endo

1,2,3,4,10,10-hexachloro-6,7-epoxy-1,4,4*a*,5,6,7,8,8*a*-octahydro-1,4-*endo-exo*-5,8-dimethanonaphthalene

**Polychloroterpenes.** Polychloroterpene insecticides are almost exclusively agricultural chemicals. There are only two materials in this group, strobane and toxaphene. Several years ago, toxaphene was the single most widely used insecticide in agriculture. It is prepared by the chlorination of camphene, a derivative of pine tree materials. Toxaphene was used widely, particularly in cotton, where it was mixed with other insecticides (especially methyl parathion) to give greater effectiveness. At its peak, toxaphene accounted for more than 40 percent of insecticide quantity in cotton.

The polychloroterpene insecticides are easily metabolized by birds and mammals, and storage in body fat is low. Even though not highly toxic to other animals, toxaphene is a potent fish poison. The use of toxaphene has been canceled, as have uses of strobane.

Toxaphene

$Cl_x$ $CH_2$ $(CH_3)_2$

chlorinated camphene containing 67 to 69 percent chlorine

## Botanicals

Botanical insecticides are derived directly from plants or plant products. With the possible exception of sulfur, these have been used longer than any type of insecticide. Many botanicals are of interest to organic gardeners because they are "natural" products. However, as a group they are no safer than synthetics available to the lay public. Botanicals are expensive to extract from plants, and for this reason, they are generally impractical in commercial agriculture, with the exception of organic farming. The use of these materials reached a peak in 1966 and has steadily declined since. However, the potential for the resurgence of botanicals is good, with recent discoveries of new effective and environmentally safe products. Additionally, many insecticidally active plant products are providing insights for the development of novel synthetic insecticides.

Botanical insecticides and other unlabeled natural products are often thought of as being safe to handle and use on food products. However, research has shown that plants contain many toxins, some of which can cause significant human or animal health effects, even death. Therefore, just as with synthetic poisons, great care should be exercised in choice, handling, and application of these insecticides.

**Pyrethrum.** Pyrethrum, which inspired the development of pyrethroids, is by far the most widely used botanical. This insecticide is extracted from flower petals of the *Chrysanthemum* species grown in Kenya, Ecuador, and other countries. Pyrethrum is made up of four compounds: pyrethrins I and II and cinerins I and II.

This insecticide is commonly contained in household aerosol sprays because it has a wide spectrum and rapid knockdown characteristics. Unless certain additives (synergists) are included, however, many insects are only stunned by the material and soon recover to full activity. Also, pyrethrum is available as a spray concentrate or dust for use on fruit trees, ornamentals, vegetables, and flowers.

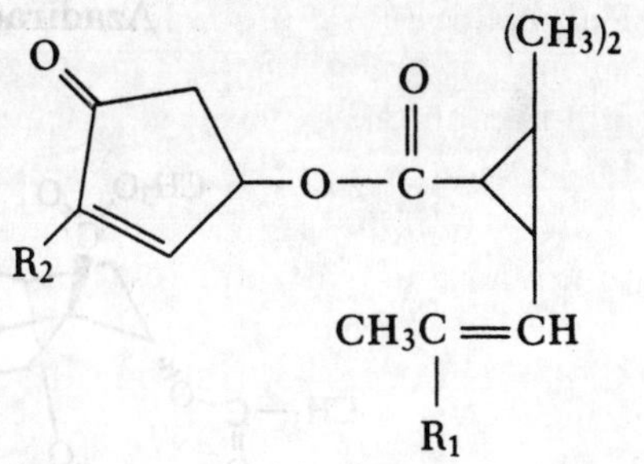

Pyrethrin I

$R_1 = -CH_3$
$R_2 = -CH_2CH{=}CHCH{=}CH_2$

Cinerin I

$R_1 = -CH_3$
$R_2 = -CH_2CH{=}CHCH_3$

Pyrethrin II

$R_1 = -C(=O)-OCH_3$
$R_2 = -CH_2CH{=}CHCH{=}CH_2$

Cinerin II

$R_1 = -C(=O)-OCH_3$
$R_2 = -CH_2CH{=}CHCH_3$

When it is used, fruits and other edibles can be harvested and eaten shortly after treatment. This insecticide breaks down quickly in the presence of sunlight.

**Azadiractins.** This class of chemicals is commanding much attention because of its safety and effectiveness against an array of insect pests. Azadiractins are extracted from seeds of the neem tree, *Azadirachta indica* (family Meliaceae), grown in many tropical and subtropical parts of the world. Often referred to as **neem insecticide,** azadiractins are believed to deter insect feeding and oviposition and interfere with growth, development, and reproduction. No adverse effects on mammals have been detected from neem oil; indeed, it has been shown to be therapeutic because of its anti-inflammatory and anti-ulcer activity. Evaluations with nontarget arthropods, fish, and livestock have shown excellent selectivity, and environmental residues are quickly broken down in sunlight. In addition to the azadiractins, novel insecticidal chemicals are still being isolated from the neem tree and its relative, the chinaberry tree, *Melia azedarach.* Neem insecticide is currently marketed under the trade names Azatin®, Align®, Neemix®, Neemactin®, and others.

**Nicotine.** Nicotine is obtained from extractions of tobacco leaves and was used as an insecticide as long ago as 1690. A favorite garden spray, Black Leaf 40®, had as its active ingredient 40 percent nicotine sulfate. Nicotine is an alkaloid, as are caffeine, quinine, morphine, LSD, cocaine, and strychnine.

Nicotine sulfate is very toxic to insects—and humans. In fact, it is the most dangerous of the botanicals to apply. This material was mostly sold as sprays because dusts are too dangerous to use. It was used mainly in small gardens against piercing-sucking insects and mites. Because of toxic residues, food cannot be harvested for at least 7 days after treatment. The most popular

Azadiractin (Azatin®, Neemactin®)

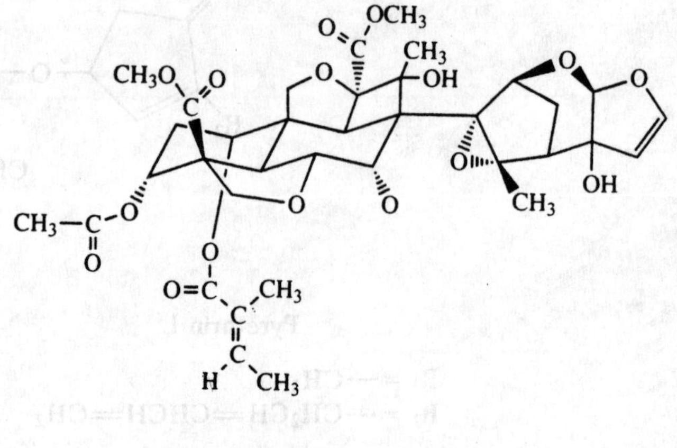

Nicotine Sulfate (NicoSoap®) (discontinued)

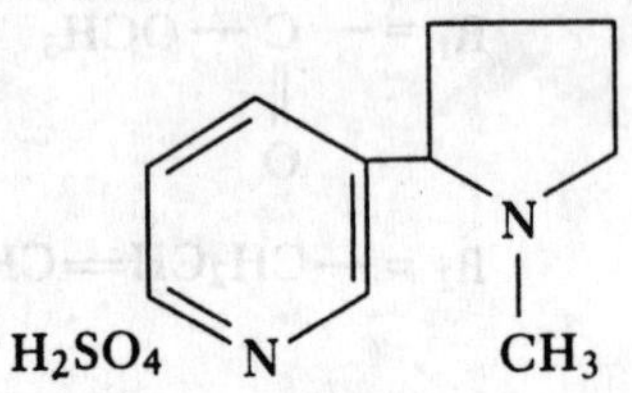

3-(1-methyl-2-pyrrolidyl) pyridine sulfate

nicotine product, Black Leaf 40, was discontinued by its producer as was a more current registration, Nico Soap®.

**D-limonene.** D-limonene is one of the most recent botanicals to reach the market. This chemical is extracted from citrus peel, constituting about 98 percent orange peel oil by weight. Limonene is used against external parasites of pets, including fleas, lice, mites, and ticks. Its toxicity to warm-blooded animals is negligible. D-limonene acts on the sensory nerves of the insect's peripheral nervous system, causing death.

D-limonene (Cide-Kick®, Kammo®)

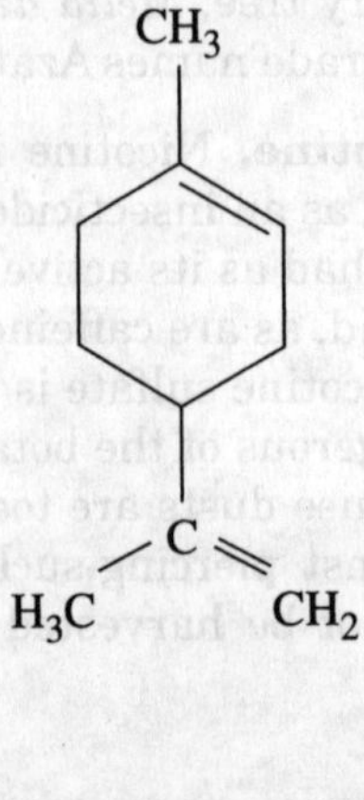

**Rotenone.** Rotenone is probably the second-most used botanical. It is extracted from the roots of the legumes *Derris* species, grown in Malaysia and the East Indies, and *Lonhocarpus* species, grown in South America. Rotenone has been applied as an insecticide since 1848 and as a fish poison by South Americans at least since 1649.

Rotenone (Prentox®, Derris®)

$OCH_3$ $CH_3O$ O O O O $C{=}CH_2$ $CH_3$

1,2,12,12*a*-tetrahydro-2-isopropenyl-8,9-dimethoxy-[1]benzopyrano-[3,4-*b*]furo[2,3-*b*][1]benzopyran-6(6*aH*)-one

Today, the insecticide is sold as a spray or dust for both chewing and piercing-sucking insects, mainly in garden and fruit crops. Another major use is as a fish poison for reclaiming lakes and ponds. It is applied to eliminate all fish, including rough species, before restocking with desirable game fish.

**Ryania.** Ryania is a botanical extracted from the stem and roots of *Ryania speciosa,* a shrub grown in Trinidad. Like nicotine, the active ingredient in this material is an alkaloid; however, it has low toxicity to humans. Ryania has been used most widely against caterpillars on fruit trees, particularly the codling moth, *Cydia pomonella,* on apples. Also, it is effective against such insects as the European corn borer, *Ostrinia nubilalis,* on corn, as well as many other insects in home gardens and fruit trees. A synthetic version of ryania, Rynaxpyr®, also has been released.

**Sabadilla.** This botanical is extracted from the seed of *Schoenocaulon officinale* and contains alkaloids as the active ingredients. Sabadilla has low toxicity to humans; however, it is known to cause eye irritation and sneezing in some individuals. This insecticide probably is the least used of the botanicals mentioned but is effective against most garden pests except aphids and mites. Sabadilla is the least toxic of the botanicals for handling and is used commonly by organic farmers.

### Fumigants

Fumigants are highly volatile pesticides, many of which contain one or more of the halogen gases: Cl, Br, or F. Effective fumigants have high penetrating ability and kill all stages of insects in enclosures, including eggs. Good fumigants are not corrosive and do not affect the quality of commodities that are treated. They desorb quickly from treated surfaces and leave no harmful residue. They are heavier than air.

Fumigants are used to eliminate all pests in grain elevators, greenhouses, homes, and warehouses, and also in packaged products like beans, grain, breakfast cereals, and dried fruits. They are also used against soil insects, nematodes, and pathogenic microorganisms. With soil applications, plastic sheets are sometimes laid down after application to reduce the rate of evaporation. The greatest hazards of using these insecticides are flammability of the gas and possible accidental poisoning of humans.

There are several fumigants registered for use in the United States. Only a few of the most common ones are mentioned here.

***p*-Dichlorobenzene and naphthalene.** *p*-Dichlorobenzene (PDB) and naphthalene are some of the most common fumigants used in the home. They are produced as solids that emit gas at a slow rate. In households they are used as moth balls or flakes for protection against clothes moths, *Tineola* species. They have also been used as soil fumigants.

**Inorganic phosphides and phosphine.** Several inorganic phosphides, including **aluminum phosphide** and magnesium phosphide, are applied to produce gaseous phosphine. This fumigant is used against insect pests in stored grains and to rid flour mills, railcars, ships and warehouses of insects. These insecticides are not used for soil fumigation in vegetable crops as are some other fumigants.

| | |
|---|---|
| Methyl bromide | $CH_3Br$ |
| Magnesium phosphide | $Mg_3P_2$ |
| Chloropicrin | $Cl_3CNO_2$ |
| Sulfuryl fluoride | $SO_2F_2$ |
| Metam-sodium | $CH_3NHC(=S)—S—Na$ |
| Ethylene oxide | $H_2C—CH_2$ (bridged by O) |
| Naphthalene (crystals) | |
| *p*-dicholorobenzene (PDB crystals) | Cl—$C_6H_4$—Cl |

Another related fumigant is phosphine gas. Registered initially in 1999, this fumigant is a mixture of phosphine and carbon dioxide, which is a propellant and flame inhibitor. It is marketed as Eco2fume®.

**Methyl bromide.** Methyl bromide is a highly volatile insecticide earlier used as a general fumigant. It is fairly stable and nonflammable and is very toxic to insects and some mites. Methyl bromide penetrates well and desorbs quickly. It was used in the fumigation of mills, granaries, and warehouses, but about 80 percent of its use was in soil fumigation for vegetables and production of other high-value crops. Also, it was frequently used to fumigate quarantined plants before they were shipped. Although it is a very useful insecticide, methyl bromide was determined to be an ozone-depleting substance. Accordingly, the EPA ordered incremental reductions of production and importation, when a phaseout took effect on January 1, 2005, except for allowable exemptions. These exemptions include the Quarantine and Preshipment (QPS) exemption, to eliminate quarantine pests, and the Critical Use Exemption (CUE), designed for agricultural users with no technically or economically feasible alternatives.

**Chloropicrin.** Chloropicrin is the active ingredient of "tear gas" used by police and is also frequently added to odorless fumigants as a safety precaution. When in a mixture, as with methyl bromide, the fumigant's presence repels users, and the chance of accidents is reduced. Chloropicrin is also a fumigant in its own right, effective against insects, fungi, nematodes, and weed seeds.

### Oils

Refined petroleum oils have long been an important source of insecticides used in management of mosquito larvae and fruit-tree pests. They are also used as carriers of other insecticides.

Because petroleum oil is **phytotoxic** (kills or injures plants), it must be highly refined before it can be applied to plants. The lower the viscosity (resistance to flowing) and distillation (boiling point) range, the less phytotoxic the oil. Light oils probably have less toxicity to plants because they volatilize more quickly than heavier oils and do not remain on the plant surface as long. Also, the amount of unsaturated hydrocarbons in an oil is another factor in its phytotoxicity. Unsaturated hydrocarbons are unstable and combine to form compounds toxic to plants. U.R. ratings (percent unsulfonated residue from laboratory tests) are a measure of these unsaturated hydrocarbons. Heavy oils (dormant oils) have U.R. ratings of 50 to 90 percent, and light oils (summer oils) have U.R. ratings of 90 to 96 percent.

Oils are usually emulsified with water for application. **Summer oils** are the most highly refined and can be applied to trees in full foliage. The summer oils are usually not as toxic to insects as dormant oils but can be useful against mites and scale insects (Hemiptera: Coccoidea) on citrus. **Dormant oils** are less refined and, to prevent phytotoxic effects, are applied when no foliage is present. These are applied to fruit trees and ornamentals during periods of mild weather in late winter before budswell occurs. They are applied mostly to kill partially grown scale insects and eggs of aphids and mites that are overwintering on branches and twigs.

The advantages of oil applications are many. They are inexpensive, usually result in good coverage, are simple to mix, and are safe to warm-blooded animals. Also, few insects have become resistant to these. Some disadvantages of use include phytotoxicity, instability in storage, and ineffectiveness against certain pests.

### Other Insecticides

Although the insecticide groups discussed so far make up the vast majority of those used, others are notable for specific purposes. These are mentioned briefly.

**Formamidines.** Formamidines are a group of insecticides that gained acceptance in the 1970s, mainly for use against organophosphate- and carbamate-resistant pests. They are quite effective against eggs and young caterpillars of several moth species and against most stages of mites and ticks.

Chlordimeform (Galecron®, Fundal®) (discontinued)

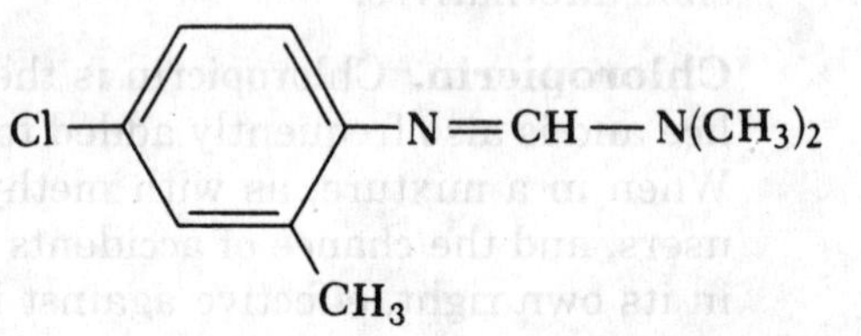

$Cl$, $CH_3$, $-N=CH-N(CH_3)_2$

*N′*-(4-chloro-*o*-tolyl)-*N*,*N*-dimethylformamidine

Amitraz (Agrotraz®, Ovasyn®)

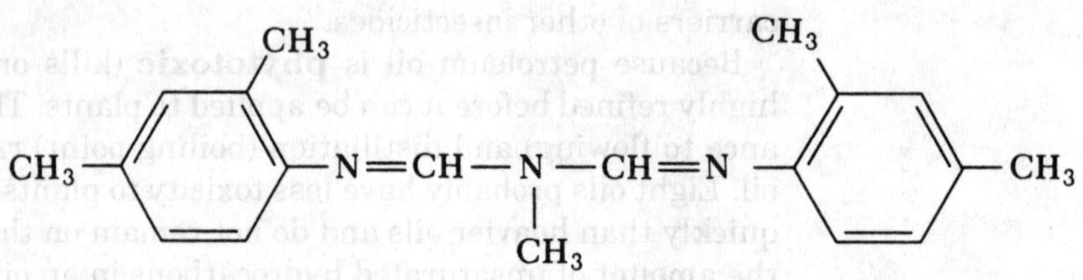

$CH_3$, $CH_3$, $-N=CH-N-CH=N-$, $CH_3$, $CH_3$, $CH_3$

*N′*-(2,4-dimethylphenyl)-*N*-[[2,4-dimethylphenyl) imino]methyl]-*N*-methylmethanimidamide

Dinitrocresol (DNOC) (discontinued)

$NO_2$, $NO_2$, $-O-Na$, $CH_3$

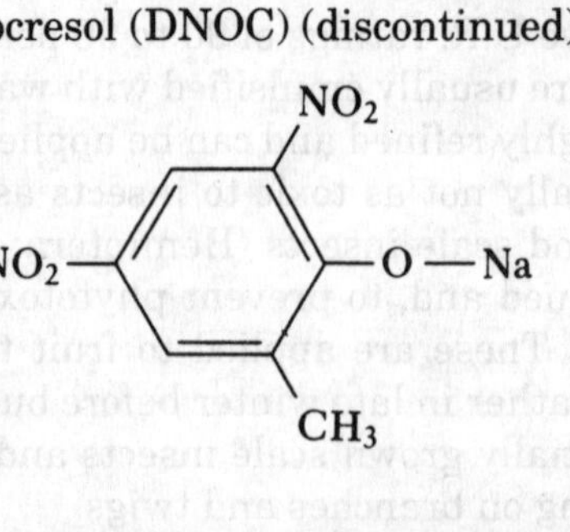

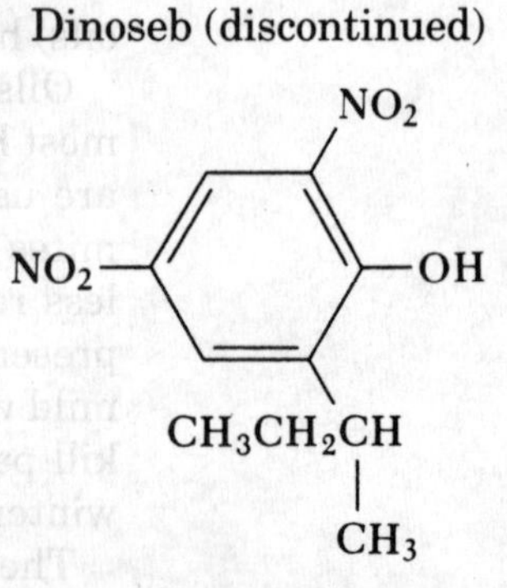

4,6-dinitro-*o*-cresol, sodium salt

Dinoseb (discontinued)

$NO_2$, $NO_2$, $OH$, $CH_3CH_2CH$, $CH_3$

2-*sec*-butyl-4,6-dinitrophenol

An early formamidine, chlordimeform, was one of the most widely used in this group. It was removed from the market in 1976, when laboratory studies indicated cancerous tumor formation in strains of mice that fed on it over their lifetime. It was brought back on the market in 1978 for use on nonfood crops, primarily cotton. However, the primary producers in the United States have discontinued this product. Formamidines still being marketed include amitraz (Agrotraz®, Ovasyn®, and others) and formetanate (Carzol®).

**Dinitrophenols.** Dinitrophenols have a broad range of toxicity for many kinds of organisms and have been developed as herbicides, fungicides, and insecticides. They are considerably toxic to humans. The oldest chemical of this group is DNOC (dinitrocresol); it has been used as an ovicide (egg killer), herbicide, fungicide, and blossom-thinning agent. DNOC was used mainly for killing all plants in an area. Another dinitrophenol, dinoseb, was used as a dormant spray against insects and mites on fruit. Because of its high toxicity, most uses have been canceled.

Tetradifon (Tedion®)

Cl O Cl—S—Cl O Cl

*p*-chlorophenyl 2,4,5-trichlorophenyl sulfone

Cyhexatin (Pennstyl®)

SnOH

tricyclohexylhydroxytin

**Organosulfurs and organotins.** The compounds of organosulfurs and organotins are often used as acaricides, effective against many species of plant-feeding mites. Organosulfurs have sulfur as their central atom but are more toxic to mites than sulfur alone. Of significance to pest management is the property of these compounds to suppress mites without causing ecological upsets in insect populations. Tetradifon is one of the older organosulfurs used.

The organotins are mainly used as selective acaricides. However, some serve a dual purpose as fungicides as well. Two of the most common organotins are cyhexatin and fenbutatin-oxide. Fenbutatin-oxide (Vendex®, Torque®, and others) is particularly effective against mites on citrus, deciduous fruits, and ornamentals, as well as in greenhouses.

**Inorganics.** Some of the oldest insecticides used in agriculture are inorganic compounds. Most were used in the early part of the century and have been replaced by the much more effective organic compounds.

Some of the inorganics that remain in use include sulfur, sodium fluosilicate, and cryolite. Sulfur is both a contact poison and stomach poison and is applied as a dust against mites and some fungi. Liquid lime-sulfur, made by boiling mixtures of sulfur and freshly slaked or hydrated lime, is used on fruit trees as a fungicide and pesticide against mites, aphids, and scale insects. Sodium fluosilicate is an insecticide used in cockroach and grasshopper baits, and cryolite is somewhat effective against a number of insects in truck crops. Both sodium fluosilicate and cryolite are stomach poisons, and as such, they usually do not have a strong effect on natural enemy populations. Newer inorganics include disodium octoborate (Tim-Bor®, Bora-Care®) for termites, and silica gels (Dri-Die®, Drianone®, Silikil Microcel®) for household pests.

**Insecticidal soaps.** Various kinds of soaps have been used by homeowners and gardeners for many years. Records of using soap sprays to control insects date back to the late 1800s and continue into the early 1900s. Spraying with soap declined with the advent of new effective insecticides in the mid-1940s. However, there has been a renewed interest in soaps with the recent growing desire for alternatives to synthetic insecticides.

The active ingredients in insecticidal soaps are fatty acids, which probably affect the insect nervous system and remove protective waxes on the surface of the insect cuticle. Commercial soap products, such as Safer® soap, usually comprise potassium salts of fatty acids. They are used against aphids, spider mites, mealybugs, and whiteflies on garden vegetables, shrubs, trees, and house plants. Soap sprays are effective only when the liquid contacts the insect and have little or no residual effect. Therefore, repeated applications at short intervals are usually necessary to achieve desired levels of pest suppression. Another insecticidal soap is M-Pede®.

**Other insecticides.** Some important insecticides not mentioned are the fermentation lactones of the spinosad and abamectin groups. These are discussed in Chapter 12.

## CHEMICALS USED WITH INSECTICIDES

When applying insecticides, often there are other chemicals added to give desired results. Some of these increase the toxicity of the insecticide directly and are called **synergists.** Others, in general called auxiliaries or **adjuvants,** serve to carry the insecticide or are added to improve adhesion, mixing, surface tension, or smell.

### Synergists

Synergists are chemicals that usually are not toxic to insects by themselves but, when added to another substance, make an insecticide mixture with enhanced toxicity. Synergists often are added to insecticides in ratios of 8:1 to 10:1 (synergist/insecticide).

Synergists were initially developed and used to increase the effectiveness of pyrethrum. Indeed, most aerosols with pyrethrum and some of the pyrethroids for household use today are enhanced with a synergist. In testing, synergists have also been found to increase the toxicity of some chlorinated hydrocarbons, organophosphates, carbamates, and other types of insecticides. In most instances, synergists function by inhibiting the **mixed-function oxidases (M-FOs),** enzymes that metabolize foreign substances.

Some of the most common synergists are piperonyl butoxide, sulfoxide, and MGK 264®. Many of these are used to prepare sprays for home and garden and stored-grain and dairy cattle facilities. Because of their cost, synergists are not frequently used for spraying broad cropping areas.

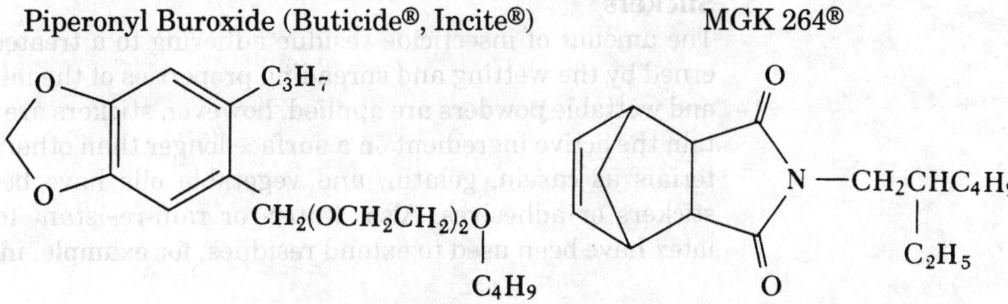


α-[2-(2-butoxyethoxy)ethoxy]-4,5-methylenedioxy-2-propyltoluene

N-(2-ethylhexyl)-5-norbornene-2,3-dicarboximide

### Solvents

Many of the organic compounds used for insecticides are insoluble in water. Before they can be turned into spray concentrates or aerosols, they must be dissolved. The solvent chosen depends on the planned use of the material. Such factors as solvency, phytotoxicity, animal toxicity, combustibility, odor, and cost are major considerations. Some examples of solvents used to dissolve insecticidal compounds include carbon tetrachloride, kerosene, and xylene.

### Diluents

Diluents are substances combined with concentrated insecticides as carriers and are necessary to obtain proper coverage of treated surfaces. Diluents can be either liquid or solid.

Liquid diluents of insecticides are usually water or refined oils. When water is used, it is necessary to add wetting and dispersing agents for proper suspension of the insecticide. Emulsifying agents are required where oil solutions are used with water.

Solid diluents are used to formulate insecticide dusts or granules. Coarse to finely ground particles of the diluent serve as a carrier of the insecticide.

Common solid diluents include organic flours (for example, soybean flour) and minerals (bentonite clay, talc, and volcanic ash).

### Surfactants

The term **surfactant** is a euphemism for several agents that aid or enhance the surface-modifying properties of a pesticide formulation. It is derived from the words *surface active agent.* Surfactants improve the emulsifying, wetting, and spreading properties of the mixture.

Usually, liquid insecticides, oils, and insecticides in water-insoluble solvents are formulated and applied as water emulsions. Emulsions are suspensions of microscopic droplets of one liquid in another. These can be formed by vigorous agitation but not very effectively. For effective suspensions, detergentlike materials are added to the insecticide formulation to enhance mixing. In most instances, when the insecticide and emulsifier are added to water, the oil carrier disperses immediately and uniformly, giving a milky appearance. A thoroughly emulsified liquid has efficient wetting and spreading properties, as opposed to a crude suspension.

### Stickers

The amount of insecticide residue adhering to a treated surface is partly governed by the wetting and spreading properties of the mixture. When emulsions and wettable powders are applied, however, stickers are sometimes added to retain the active ingredient on a surface longer than otherwise possible. Such materials as casein, gelatin, and vegetable oils have been used as insecticide stickers or adhesives. Also, water- or rain-resistant formulations containing latex have been used to extend residues, for example, in Sevin XLR®.

### Deodorants

Deodorants are materials added to insecticides to mask unpleasant odors. Many insecticides like pyrethrum and several organophosphates have strong odors that can be offensive. This is particularly unacceptable when formulated for home use. Therefore, various substances like cedar oil, pine oil, or flower scents are added to insecticide concentrates to disguise their odor.

## INSECTICIDE FORMULATIONS

The beginning stages of insecticide production involve the manufacture of **technical-grade material.** This material is a relatively pure form of the insecticidal compound that comprises the **active ingredient** (AI) in the final insecticide mixture. Before sale, however, the technical-grade material is mixed with auxiliaries to make the insecticide convenient to handle and easy to apply. Some of the auxiliaries may also be active against insects; however, many of them are inert. **Inert ingredients** have no direct effect on pests. The mixture of active and inert ingredients for killing insects is called an **insecticide formulation.** Some formulations are ready for use out of the container, but others must be diluted with water or oil.

There are many kinds of formulations available on the market, including liquids and solids, and a few are prepared for release of the active ingredient over a period of time. Only the most widely used formulations are discussed here.

### Liquid Formulations

Liquid formulations usually are sold in small cans and bottles, medium-sized pails, or large drums. If mixing is required, these formulations are the most convenient to use.

**Emulsifiable concentrates (EC *or* E).** It has been estimated that more than 75 percent of all pesticides are applied as sprays. The great majority of these sprays are applied as water-based emulsions using emulsifiable concentrates. With this kind of formulation, a detergentlike material is added to break up the insecticide into microscopic droplets, producing a milky liquid. ECs often contain 240 to 1,920 grams of AI per liter (2 to 8 pounds per gallon).

**Solutions (S).** Solutions are also liquid concentrates that may be used directly or require diluting. Those used directly from the container are low concentrates, usually containing fewer than 240 grams AI per liter (2 pounds per gallon). Most low concentrates are solutions in highly refined oils, and many have a convenient atomizing pump. Solutions are mainly used as household sprays, mothproofers, livestock sprays, and space sprays in barns. High concentrates usually contain 1,920 grams or more AI per liter (8 pounds per gallon). If dilution is required, oil is usually the diluent. A special kind of high-concentrate solution is the **ultra low volume (ULV) concentrate.** ULV formulations are applied without dilution with special aerial or ground equipment to produce an extremely fine spray. Volumes delivered by ULV are 0.6 liters to 4.7 liters per hectare (2 quarts or fewer per acre), as compared with more than twenty times that amount for conventional high-volume sprays. ULV application is particularly effective with certain insecticides because it reaches the insects in a concentrated dose. It is also economical because large areas can be treated without refilling the sprayer.

**Flowables (F *or* L).** In some instances, insecticidal compounds can be made only as a solid or semisolid. To give them some of the mixing advantages of ECs and solutions, they are wet-milled with a clay diluent and water. This leaves the technical material finely ground and wet, with a puddinglike consistency. This formulation can then be mixed with water for spraying. Flowables must be agitated constantly to prevent the insecticide from coming out of suspension and settling to the bottom of the spray tank.

**Aerosols (A).** Aerosols are the most frequently used formulation of insecticides around households. Insecticides used in aerosols are dissolved in volatile petroleum solvents. The solution is then pressurized in a can by a propellant gas like carbon dioxide or fluorocarbons. When sprayed, the solution is atomized and quickly evaporates, leaving microsized droplets suspended in air. Aerosols are sold in push-button or total-release containers. Although easy to use, aerosols have a low percentage of active ingredient and, therefore, are expensive. Most uses are in households and greenhouses and around barns.

**Liquefied gas (LG *or* F).** Several fumigants, when placed under pressure, turn into a liquid. These are stored in metal bottles under pressure and are released into structures like grain bins, into the soil by injection, or under tarps. Other insecticidal compounds remain liquid at normal atmospheric pressure but turn into a gas after they are applied. They are not stored under pressure and vaporize after they are placed in soil or in enclosures.

### Dry Formulations

Dry formulations are usually sold in paper bags or cans, which may be lined with plastic. Some are used directly from the container, but others require a diluent.

**Dusts (D).** Dusts are some of the oldest and simplest formulations of insecticides. They are prepared by milling the insecticidal compound into a fine powder. This powder is diluted with a dry diluent like organic flour or finely ground mineral. The percentage active ingredient in dusts usually ranges from 1 to 10 percent. Dusts are often easy to use in small areas because they can be shaken directly on a surface from the container or puffed into cracks and crevices with an applicator. However, dusts are the least effective, least economical insecticide formulations for outdoor use. This is because of wind-caused drift and poor rate of deposit on foliage and other surfaces. Sometimes only 10 to 40 percent of the insecticide reaches a crop when dust is applied by airplane, as opposed to 60 to 99 percent with an EC spray. Moreover, dusts are perhaps the most toxic formulation to honey bees and parasitic Hymenoptera. These characteristics make dusts rather poor pest management formulations for outdoor use.

**Granules (G).** Granular formulations are prepared by applying liquid insecticide to coarse particles of a porous material. These particles may be formed from corncobs, walnut shells, clay, or other materials. The insecticide is either absorbed into the granule or coats the outside, or it does both. The amount of active ingredient in granules ranges from 2 to 40 percent. Because of the size of the granular particle, this formulation is much safer—because it cannot be inhaled—to apply than dusts or even ECs. Granules are most often used against soil insects and may be placed directly with, or beside, the seed at planting time, banded over the top, or broadcast over an area. When dropped over plants, granules accumulate in leaf whorls. This type of application is useful against such insects as European corn borer larvae, which feed at the whorl before boring into the plant. As might be expected, proper coverage of plants may be a problem with this formulation.

**Wettable powders (WP *or* W).** Wettable powders look like dusts while in the container but are formulated to be mixed with water and sprayed on surfaces. A surfactant added to the dust allows wetting during the mixing processes. A particle suspension results when water is mixed. WPs are much more concentrated than dusts, containing 15 to 95 percent active ingredient. Like flowables, frequent agitation is required to keep the insecticide in suspension. WPs usually cause less phytotoxicity than ECs, but they are more abrasive to spray pumps and nozzles. WPs should never be used without dilution.

**Soluble powders (SP).** Unlike wettable powders, soluble powders dissolve in water, forming a true solution. Some are packaged in bags that look like plastic and dissolve when thrown into water. Some agitation is needed to get SPs into solution, but after they are dissolved, additional agitation is not needed. Usually, SPs are formulated with 50 percent or more active ingredient and always require dilution.

**Dry flowables (DF).** These are water-dispersible granules that are similar to wettable powders, but the insecticide is formulated on a granule instead of a powder. Dry flowables are easier to pour and mix than wettable powders and also require agitation during application. They are less of an inhalation hazard to the applicator than powders.

**Water-soluble packets (WSP).** These formulations are composed of wettable or soluble powder preweighed and delivered in water-soluble plastic bags. Bags are dropped into water-filled spray tanks, subsequently dissolving, and the insecticide contents are dispersed into the water. This formulation is very safe to the handler because no direct contact is made with the pesticide. Once mixed, this formulation is no safer than other mixtures.

**Poisonous baits (B).** This type of formulation combines an insect-edible or other attractive substance with the insecticide to improve effectiveness of treatment. Dried and pulverized fruit and other materials often are used to draw insects to a spot where they ingest or simply crawl across the insecticide. Baits can be used in buildings or outside for agricultural pests. Usually, active ingredient concentrations are very low in baits, on the order of 5 percent or less.

**Slow-release formulations (SR).** Along with the development of environmentally unstable insecticides have come problems of short-term effectiveness and the increased expense of several applications. Therefore, ways to extend the life of organophosphates and other chemicals have been the goal of considerable research. A degree of success in extending insecticide activity has been achieved through the development of slow-release insecticides. The first significant development of a slow-release formulation occurred in 1963 with the introduction of the Shell No-Pest Strip®. With this formulation, the volatile organophosphate, dichlorvos, was embedded into strips of polychlorovinyl resin. The polychlorovinyl resin slows the rate of volatilization of the insecticide, allowing it to kill most flying and some crawling insects in the vicinity. Other developments of the original "fly strip" idea now can be seen in ear tags and ear tape for fly control on livestock and flea collars for dogs and cats.

Another form of slow-release formulation has been in microencapsulation of insecticides. This formulation is accomplished by incorporating the insecticide in a permeable covering. This process forms microscopic spheres or microcapsules that permit the insecticide to escape at a reduced but effective rate. The life of an insecticide formulated in this manner is often two to four times that of an emulsifiable concentrate. A commonly used example of this formulation is Penncap-M®, a microencapsulated form of methyl parathion.

Yet another slow-release formulation is a paint-on type. Here, the insecticide is dissolved in a volatile petroleum solvent, along with certain plastics and lacquers. This formulation may be "painted" on directly or blended with latex or oil-base paints for application. When the solvents evaporate, the insecticide remains in the thin transparent film or the actual paint. Over a period of time the insecticide escapes or "blooms," causing the surface to be active against crawling insects.

## INSECTICIDE TOXICITY

Toxicity of a substance refers to its inherent poisonous potency under a given set of laboratory conditions. All insecticides are poisons, and the degree of toxicity varies greatly among them. To understand toxicity and how to use insecticides efficiently and safely, we must first understand their mode of action.

### Insecticide Modes of Action

Insecticide **mode of action** involves all the anatomical, physical, and biochemical responses to a chemical, as well as its fate in the organism. Of greatest focus in this discussion are the most fundamental physiological steps that result in the death of insects.

All insecticides block metabolic processes in insects, but different compounds do this in different ways. According to their mode of action, the major groups of the most frequently used insecticides are: (1) nerve poisons, (2) metabolic poisons, (3) alkylating poisons, (4) muscle poisons, and (5) physical toxicants.

**Nerve poisons.** By far most conventional insecticides act as nerve poisons. These affect the insect nervous system mostly as narcotics, axonic poisons, or synaptic poisons.

**Narcotics'** mode of action is mostly physical. Many fumigants (particularly those containing Cl, Br, and F) are narcotics, inducing unconsciousness in insects. These narcotics are fat-soluble and lodge in fatty tissues, including nerve sheaths and lipoproteins of the brain. An important characteristic of these insecticides is that their action is reversible; if a human being shows early symptoms of poisoning, removing him or her from the source of the fumigant allows recovery. This also means that poisoned insects can recover if the fumigated space is ventilated too quickly.

**Axonic poisons** act primarily by interrupting normal axonic transmission of the nervous system (Fig. 11.5). Remember from Chapter 2 that the axon of the nerve cell is an elongated extension of the cell body that transmits nerve impulses to other cells. These impulses are electrical and arise from the flow of positive sodium and potassium ions through the cell membrane, creating a wave-like action potential (the impulse). Subsequently, the action potential is followed by a resting potential. All the chlorinated hydrocarbons and pyrethroids are believed to disrupt normal transmission along the axon. Although many details are unknown, some of these, like the cyclodienes and pyrethroids, are believed to induce changes in axonic membrane permeability, causing repetitive discharges. Such discharges eventually result in convulsions, paralysis, and death.

**Synaptic poisons** act by interrupting normal synaptic transmission of the nervous system (Fig. 11.6). You may recall that a synapse is the junction between a neuron and another cell, including a muscle, gland, or sensory receptor cell or another neuron. In the central nervous system, the chemical acetylcholine forms and transmits an impulse across the synapse (gap) to the next cell. Subsequently, acetylcholine is broken down to form acetic acid and choline in the presence of an enzyme, acetylcholinesterase. Therefore, acetylcholinesterase clears the system before another nerve transmission occurs. Both organophosphates and carbamates inhibit or tie up acetylcholinesterase. Inhibition causes a buildup of acetylcholine and a malfunction of the transmitting system. Rapid nerve firing occurs, producing symptoms of restlessness, hyperexcitability, tremors and convulsions, paralysis, and death. One major difference in the

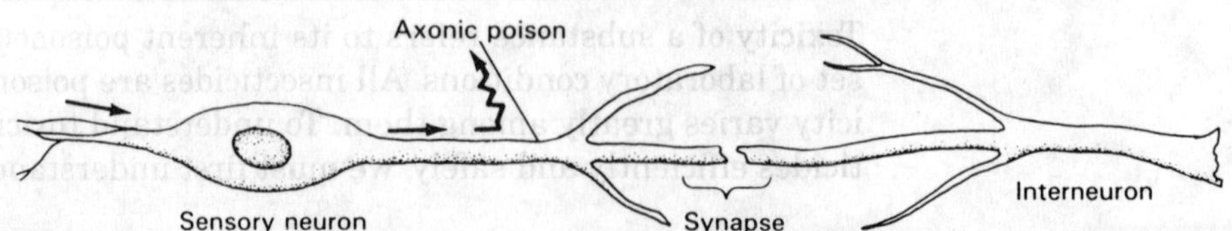


**Figure 11.5** Schematic representation of the interruption of axonic transmission of a nerve cell by cyclodiene and pyrethroid insecticides.

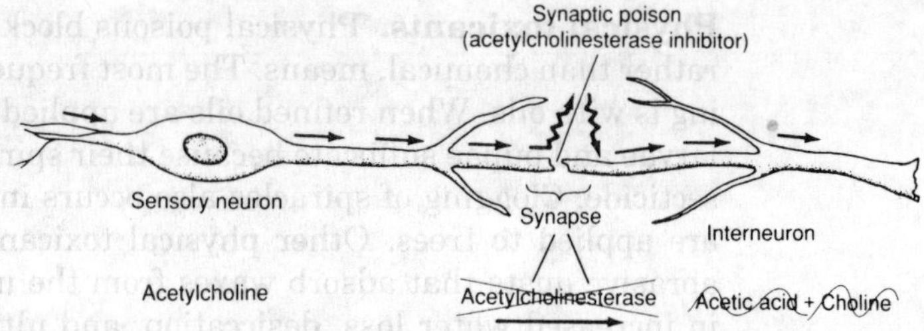


**Figure 11.6** Schematic representation of the interruption of normal synaptic transmission between nerve cells by organophosphate and carbamate insecticides. These insecticides inhibit acetylcholinesterase, which causes a buildup of acetylcholine and subsequent rapid nerve firing.

action of these two insecticidal groups is that inhibition of acetylcholinesterase by carbamates is reversible but not so with organophosphates.

Other synaptic poisons include nicotine, nicotine sulfate, neonicotinoids, spinosyns, and the formamidines. Nicotine, nicotine sulfate, and neonicotinoids poison by mimicking acetylcholine at the synapse; the receptors cannot distinguish between acetylcholine and nicotine. This phenomenon results in symptoms similar to the inhibition of acetylcholinesterase. Formamidines cause insect death by affecting enzymes involved with other chemical transmitters (including norepinephrine and octopamine) at certain nerve synapses.

Fipronil is another insecticide whose main mode of action is at the nerve synapse. It acts on GABA receptors, which function to increase chloride ion permeability of neurons, producing a "calming" effect after nerve firing. Fipronil prevents chloride ions from entering the neuron by blocking the GABA-regulated chloride channel.

**Metabolic poisons.** The action of these poisons is to disrupt a wide range of metabolic processes in insect cells. The site of the disruptions is the mitochondrion of the cell. The mitochondrion is central in providing energy for biochemical reactions through the production of ATP (adenosine triphosphate). A respiratory electron transport chain and respiratory enzymes are involved in the process, which is critical for the production of energy from the oxidation of carbohydrate and lipid molecules. Many of these poisons disrupt the process, blocking respiration and causing death of the insect. Those insecticides that act in this way include rotenone, pyrroles, pyrazoles, and pyridazinones.

**Alkylating poisons.** Insecticides in this group act by replacing active hydrogen in biologically significant compounds with an alkyl group (carbon chainlike structure). They attack loci on the nucleic acid molecule and deactivate critical enzymes, preventing the synthesis of nucleic acids. Some insecticides that have this mode of action are methyl bromide, ethylene dibromide, and some chemosterilants.

**Muscle poisons.** Muscle poisons have a direct influence on muscle tissue. A common example of their action is the botanical, ryania. In this insecticide the alkaloid, ryanodine, disrupts the excitable membrane of muscle. This disruption results in up to a tenfold increase in oxygen consumption, followed by flaccid paralysis and death. A similar mode of action is seen in another botanical, sabadilla.

**Physical toxicants.** Physical poisons block a metabolic process by physical, rather than chemical, means. The most frequent occurrence of physical poisoning is with oils. When refined oils are applied to the surface of water, mosquito larvae and pupae suffocate because their spiracles become clogged with the insecticide. Clogging of spiracles also occurs in scale insects when dormant oils are applied to trees. Other physical toxicants against insects include inert, abrasive dusts that adsorb waxes from the insect cuticle. Such action results in increased water loss, desiccation, and ultimately death of the insect from dehydration. For example, boric acid, in addition to being a stomach poison, acts as an abrasive dust in killing insects.

### Toxicity to Humans

Insecticides poison humans and insects similarly. A given product's **hazard** is the danger that injury will occur to people coming into contact with it. The degree of hazard depends on the toxicity of the active ingredient and the chance of exposure to toxic amounts of the product. According to the risks of insecticide use, two types of human poisoning can be distinguished, acute and chronic.

**Acute poisoning.** Acute poisoning causes illness or death from a single dose or exposure. This type of poisoning is of particular concern to persons involved directly in the manufacture and application of pesticides because they are at greatest risk. Acute poisonings also occur among nonprofessionals, however, usually as a result of accidents, ignorance, suicide, or crime. Accidental poisonings in the United States has decreased since 1984, with only 1.6 percent caused by pesticides. Regretfully, a substantial percentage of these poisonings occur in children under age six.

Antidotes are available for treatment of victims of organophosphate and carbamate insecticide poisoning. After first aid measures have been taken (explained on insecticide label), physicians may administer **atropine** and **2-PAM.** Atropine is often used for acute poisoning, and 2-PAM is given later (with organophosphate poisoning) to speed long-term recovery. Poison control centers are designated in each state to advise treatment for accidental poisoning.

**Chronic poisoning.** Chronic poisoning occurs from long-time exposure to low levels of a toxicant. This type of poisoning is often revealed only after several weeks of exposure and is of special concern to the general public. The main worry is that food will contain residues capable of causing sickness or death after repeated consumption over time.

Experiments with laboratory animals fed sublethal doses of pesticides over time have shown that several types of maladies are possible. Some of the most insidious effects are cancer (**carcinogenic** effects), genetic damage to future generations (**mutagenic** effects), and birth defects in offspring of exposed pregnant females (**teratogenic** effects). From animal studies, the EPA sets pesticide tolerances that are low enough that daily consumption of a particular food will not cause significant risk of such diseases.

The EPA believes that the most significant long-term source of pesticide exposure is through our diet. Most everyone agrees with this and the fact that there is a risk involved in eating food with any amount of pesticide residue in it. However, there is disagreement on what constitutes "acceptable risk." Many policymakers and consumer and environmental advocates believe that a "zero risk" policy should be enforced for all pesticides.

Tolerances for residues allowed by law are established from an acceptable daily intake (ADI). The ADI is the level of a residue to which daily exposure over a lifetime will not cause appreciable risk. This level typically has been set at 100 or more times less than the level where no observable effects of the poison can be detected (NOEL) in experimental animals. However, recent legislation allows this level to be decreased an additional tenfold to reduce risks to infants and children.

Previous, continuing studies of pesticide residues on food in the United States have shown that actual pesticide levels in a typical diet are far below the ADI and established tolerances. Therefore, it is unlikely that today's public is exposed to much risk from chronic poisoning associated with food consumption.

**Estimation of toxicity to humans.** The estimation of an active ingredient's inherent toxicity to humans is required before insecticides (and other pesticides) can be registered and sold. This estimate is made by the manufacturer using prescribed laboratory tests with white rats or rabbits. Toxicity in this instance is defined as the dose that will kill 50 percent of the test animals to which it is administered. This lethal dose is expressed as milligrams of toxicant per kilogram (mg/kg) of body weight and is referred to as the **$LD_{50}$** (Fig. 11.7).

Toxicity is established by feeding (oral), skin application (dermal), and inhalation (respiratory) tests. $LD_{50}$s (Figs. 11.8, 11.9) are determined from oral

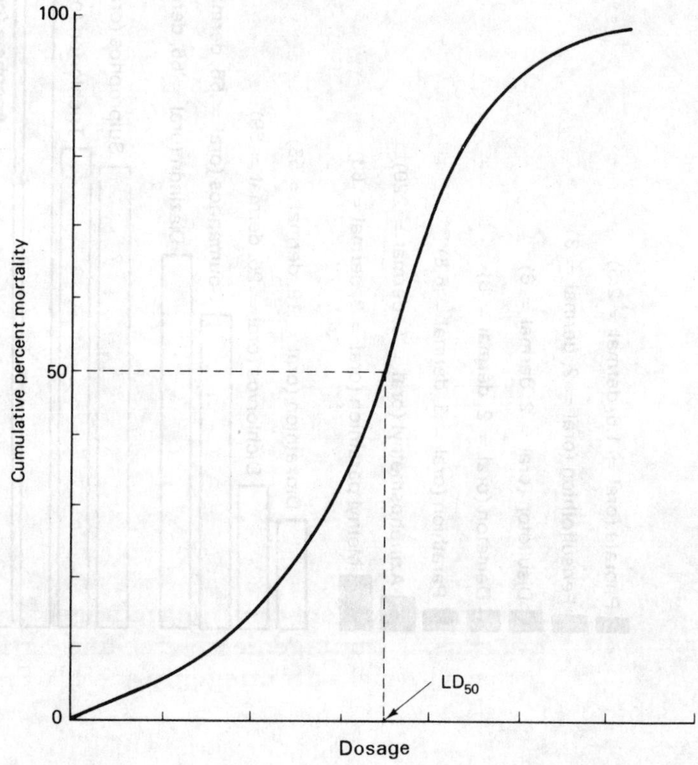


**Figure 11.7** Graph of the relationship between cumulative percent mortality and dosage of a pesticide. The $LD_{50}$ represents the dosage required to kill 50 percent of a population of laboratory animals, usually white rats or rabbits. (Reprinted with permission of Macmillan Publishing Company from *Insects in Perspective* by Michael D. Atkins. © 1978 by Michael D. Atkins)

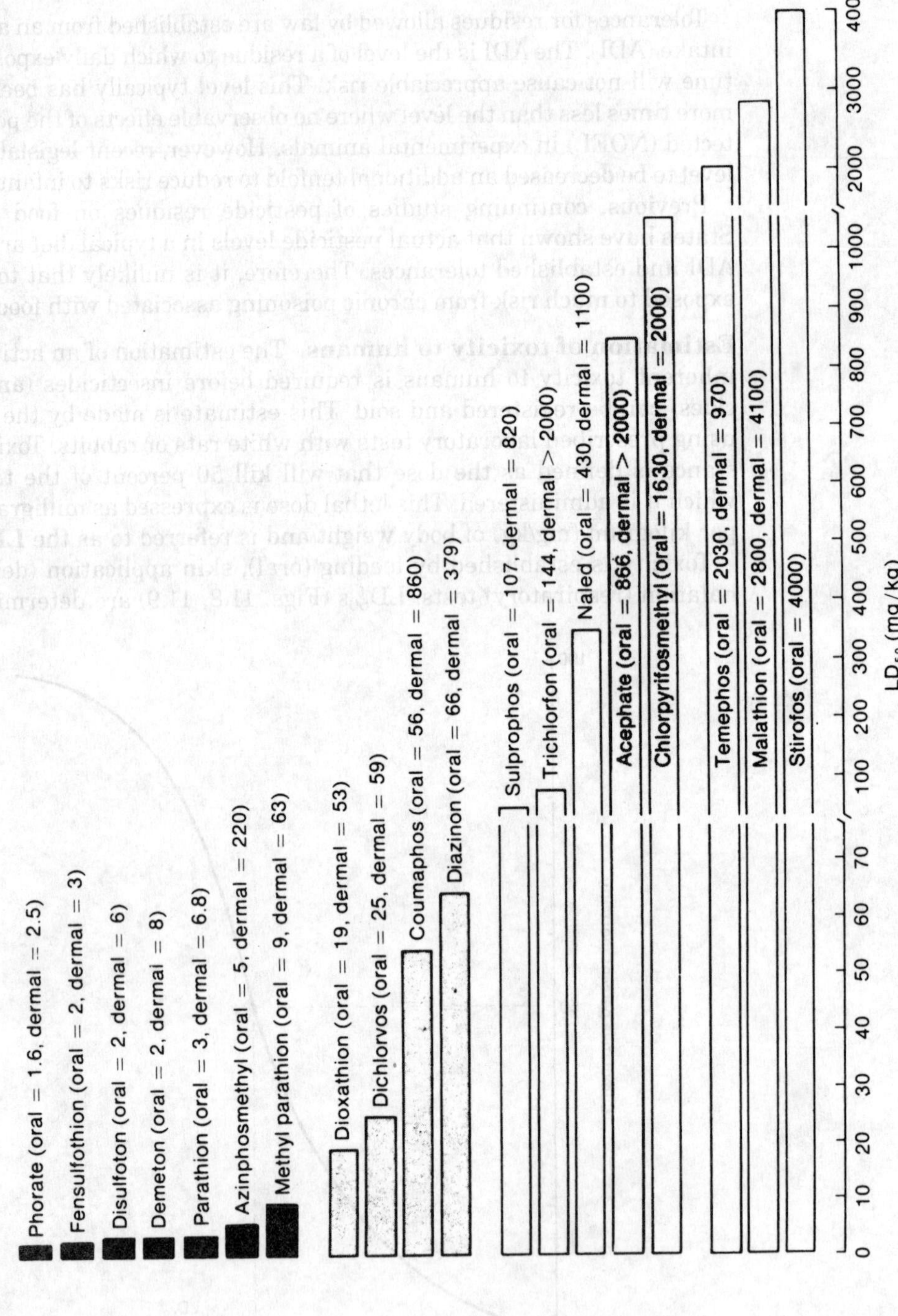


**Figure 11.8** $LD_{50}$s for common organophosphate insecticides. (From *Pesticides: Theory and Application* by George W. Ware. © 1978, 1983. W. H. Freeman and Company. Reprinted with permission.)

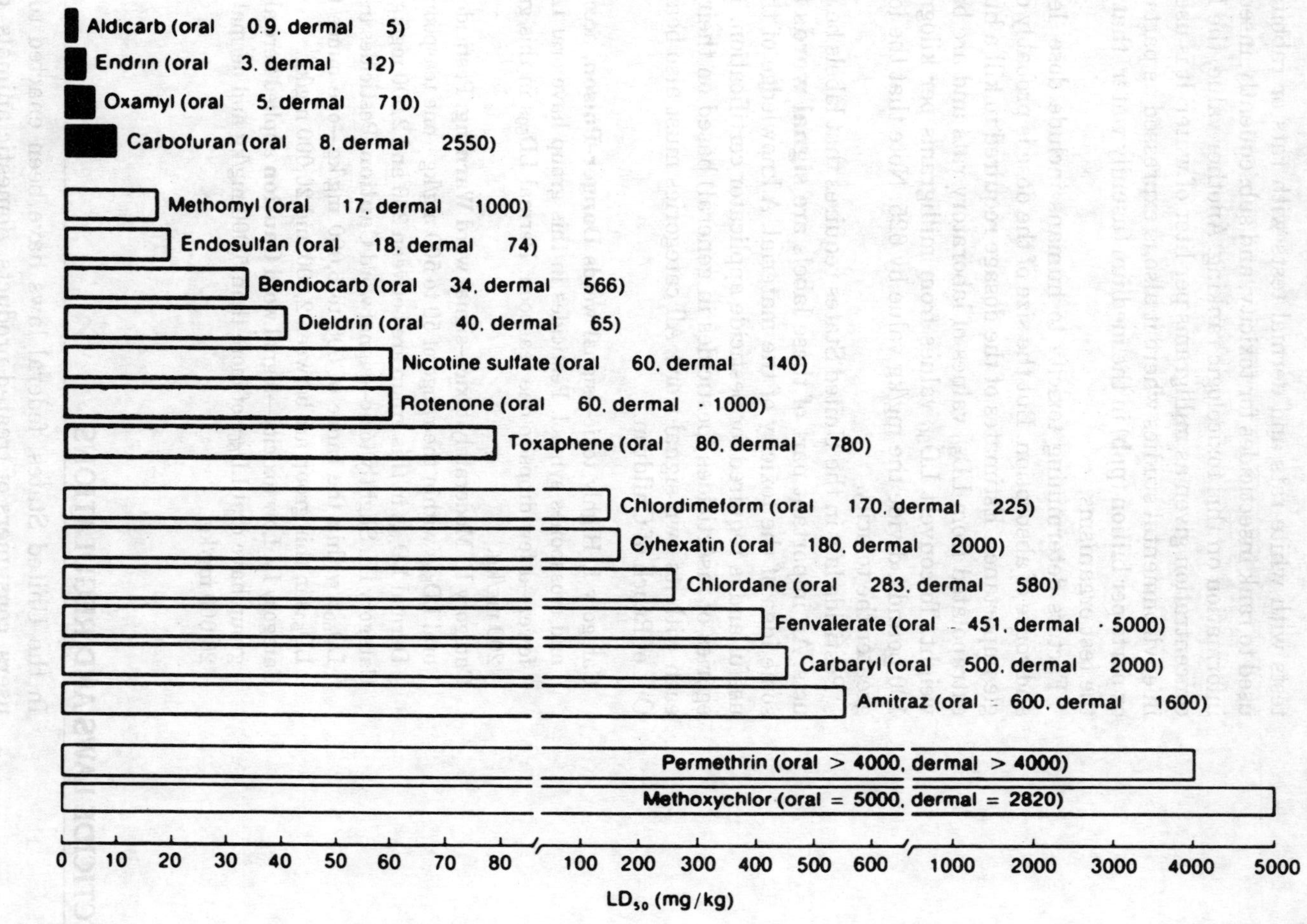


**Figure 11.9** $LD_{50}$s for various common insecticides. (From *Pesticides: Theory and Application* by George W. Ware. © 1978, 1983. W. H. Freeman and Company. Reprinted with permission)

tests with white rats and dermal tests with rats or rabbits. These tests are used to rank insecticides for toxicity; and subsequently, insecticide labels carry information on this toxicologic ranking. Another value, the **$LC_{50}$,** is the lethal concentration given as milligrams per liter of water. It is used most frequently in environmental studies where it also is expressed as parts per million (ppm) or parts per billion (ppb) in the medium (usually water) that kills 50 percent of the test organisms.

Factors determining toxicity to humans include dose, length of exposure, and route of absorption. But the size of the dose is probably of the greatest single importance. Estimates of the dosage required to kill a human being can be extrapolated from $LD_{50}$ values of laboratory rats and are based on a person's weight. To convert $LD_{50}$ values from milligrams per kilogram to ounces per 100 pounds, divide the mg/kg value by 625. Note that the lower the $LD_{50}$, the greater the toxicity.

Pesticide law in the United States requires that labels be placed on all products. An important part of these labels are **signal words** that give the user some idea of the toxicity of the material. A knowledge of these words is very useful and is required for pesticide applicator certification. There are four categories of insecticides (pesticides in general) based on their potential hazard, each with its own signal word. All categories must also bear the words "Keep Out of Reach of Children."

Category I. Highly toxic—signal words **Danger–Poison,** accompanied by a skull and crossbones symbol. Pesticides in this group have oral $LD_{50}$s of 50 mg/kg or fewer—a few drops to one teaspoon. Dermal $LD_{50}$s in this group are fewer than 200 mg/kg.

Category II. Moderately toxic—signal word **Warning.** Pesticides in this group have oral $LD_{50}$s within the range of 50 to 500 mg/kg—one teaspoon to two tablespoons. Dermal $LD_{50}$s in this group are between 200 and 2,000 mg/kg.

Category III. Slightly toxic—signal word **Caution.** Pesticides in this group have oral $LD_{50}$s within the range of 500 to 5,000 mg/kg—one ounce to one pint. Dermal $LD_{50}$s in this group are between 2,000 and 20,000 mg/kg.

Category IV. Low toxicity—signal word **Caution** applied here also. Pesticides in this group have oral $LD_{50}$s of more than 5,000 mg/kg and dermal $LD_{50}$s of more than 20,000 mg/kg.

## INSECTICIDE LAWS AND REGULATIONS

In the United States, federal laws have been enacted to protect pesticide users, consumers of treated products, domestic animals, pets, and the general environment. The first federal law to do this was the **Federal Insecticide Act** of 1910. This law, governing insecticides and fungicides, was designed mainly to protect farmers and other users from substandard or fraudulent products.

The next pesticide law was established in 1938 as an amendment to the **Pure Food Act** of 1906. The amendment forbade the presence of harmful residues of pesticides, primarily the old stomach poisons like lead arsenate and Paris green, on food. It also required the addition of coloring to pesticides to prevent consumers from mistaking them for food products like flour.

### Federal Insecticide, Fungicide, and Rodenticide Act

The Federal Insecticide Act of 1910 was superseded by the Federal Insecticide, Fungicide, and Rodenticide Act of 1947. This law, called FIFRA, extended coverage of the law to include rodenticides and herbicides and required that all pesticides for sale in interstate commerce be registered with the USDA. It did not cover products manufactured and sold within state borders. Primarily, the act of 1947 required adequate labeling and safety of the product, if label directions were followed.

In 1954, the Miller Amendment to the Food, Drug, and Cosmetic Act (1906, 1938) further strengthened the laws concerning pesticide residues on food. This amendment clearly set tolerances on all pesticides, which previously had been rather vague. In 1958, another amendment to the Food, Drug, and Cosmetic Act was passed that included the Delaney Clause. This clause disallowed any cancer-causing chemical (carcinogen) on food for human consumption.

A later amendment to FIFRA in 1959 extended coverage to include other chemicals in the category of economic poisons, including nematicides, plant regulators, defoliants, and desiccants. (Other poisons and repellents of vertebrates and invertebrates not previously included subsequently have been added in coverage.) FIFRA was amended again in 1964 to improve safety through specific label requirements.

As can be seen, both FIFRA and the Food, Drug, and Cosmetic Act operated together to achieve consumer protection of treated products, as well as that for the pesticide applicator. Until December 1970, these laws were administered jointly by the USDA and the Food and Drug Administration (FDA). With growing concerns of environmental quality, the EPA was established, and all responsibility for pesticide registration and establishing tolerances on food was transferred to this agency in 1970. The EPA remains in charge today, and the FDA is responsible for enforcing pesticide tolerances on food established by the EPA.

### FIFRA Amended

A significant change in FIFRA was enacted in 1972 with the **Federal Environmental Pesticide Control Act** (FEPCA), better known as **FIFRA Amended,** 1972 (Fig. 11.10). This is fundamentally our present law, comprising the following major provisions:

1. Use of any pesticide in a manner inconsistent with the label is prohibited.
2. All pesticides must be classified for general use or restricted use.
3. Restricted pesticides may be applied only by an applicator certified by the state in which that person works.
4. The states will certify pesticide applicators.
5. Farmers and other private applicators can be fined and/or imprisoned upon criminal conviction for a knowing violation of the law. Additional penalties apply to registrants, applicators, dealers, and other distributors.
6. All pesticide products must be registered by the EPA, including those shipped in interstate and intrastate commerce.
7. For registration, a manufacturer must prove that, when used as directed, a product will be effective; not injure humans, crops, livestock, or wildlife or damage the environment; and not produce illegal residues on food or feed.
8. Pesticide manufacturing plants must be registered and inspected by the EPA.

**An Act**

To amend the Federal Insecticide, Fungicide, and Rodenticide Act, and for other purposes.

*Be it enacted by the Senate and House of Representatives of the United States of America in Congress assembled*, That this Act may be cited as the "Federal Environmental Pesticide Control Act of 1972".

AMENDMENTS TO FEDERAL INSECTICIDE, FUNGICIDE, AND RODENTICIDE ACT

SEC. 2. The Federal Insecticide, Fungicide, and Rodenticide Act (7 U.S.C. 135 et seq.) is amended to read as follows:

"SECTION 1. SHORT TITLE AND TABLE OF CONTENTS.

"(a) SHORT TITLE.—This Act may be cited as the 'Federal Insecticide, Fungicide, and Rodenticide Act'.

"(b) TABLE OF CONTENTS.—

**Figure 11.10** The present pesticide laws in the United States are stipulated in the Federal Environmental Pesticide Control Act of 1972.

Several other amendments have been added to the law from 1975 through to the present, mostly to clarify the intent of the law. Two important amendments allow the efficacy provision to be waived and permit the use of pesticides at lower than labeled dosages. Also, certain other environmentally safe deviations from the label are allowed under the new provisions.

Other significant amendments were passed by Congress in 1988, which strengthened the EPA's regulatory authority. Some of the most important of these amendments were to improve the efficiency of the registration process, provide for fees in conducting registration and reregistration, and allow expedited registration for pesticides similar to those already registered. The 1988 amendments also authorized the EPA to establish labeling requirements for transportation, storage, and disposal of pesticides and pesticide containers. Yet later legislation created the Food Quality Protection Act (FQPA). The FQPA is discussed later in this chapter.

The EPA is responsible for interpreting and implementing FIFRA through duly processed regulations that carry the force of law. Regulations are developed in consultation with those persons most affected by them.

## Pesticide Label Regulations

The pesticide label is the printed information on, or attached to, the pesticide container; federal regulations stipulate exactly the information it must include. The label is a license to sell a product; to abide by the law, its directions must be followed. Label requirements and a key to label information is presented as follows (see location in Figs. 11.11, 11.12):

1. Trade name.
2. Manufacturer's name and address.
3. Net contents (total quantity).
4. EPA registration number.
5. EPA formulation manufacturer establishment number.
6. a. Ingredients statement.
   b. Pounds per gallon (for liquids).

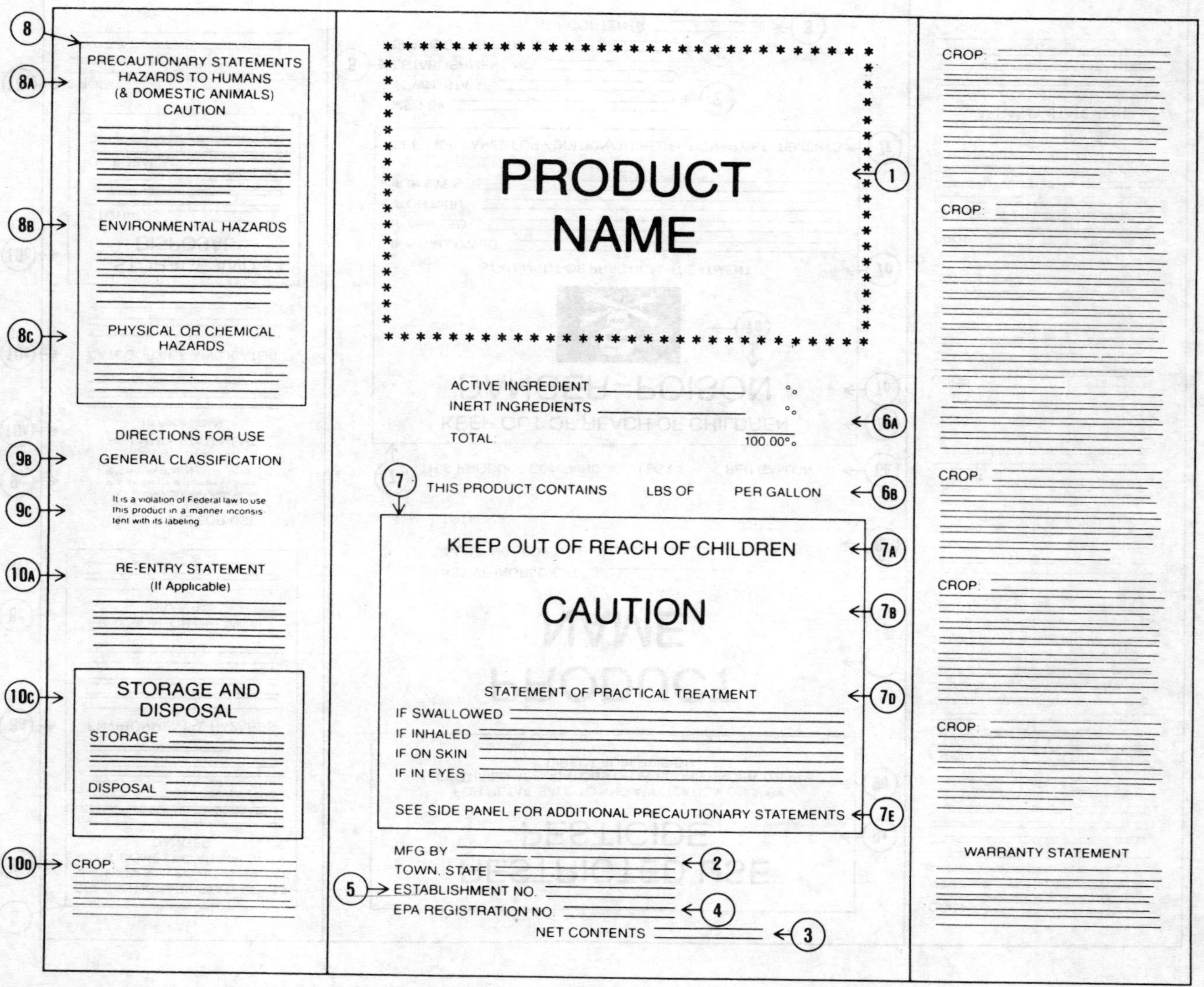


**Figure 11.11** Elements of a general-use pesticide label as required by law. See text for explanation of numbers.

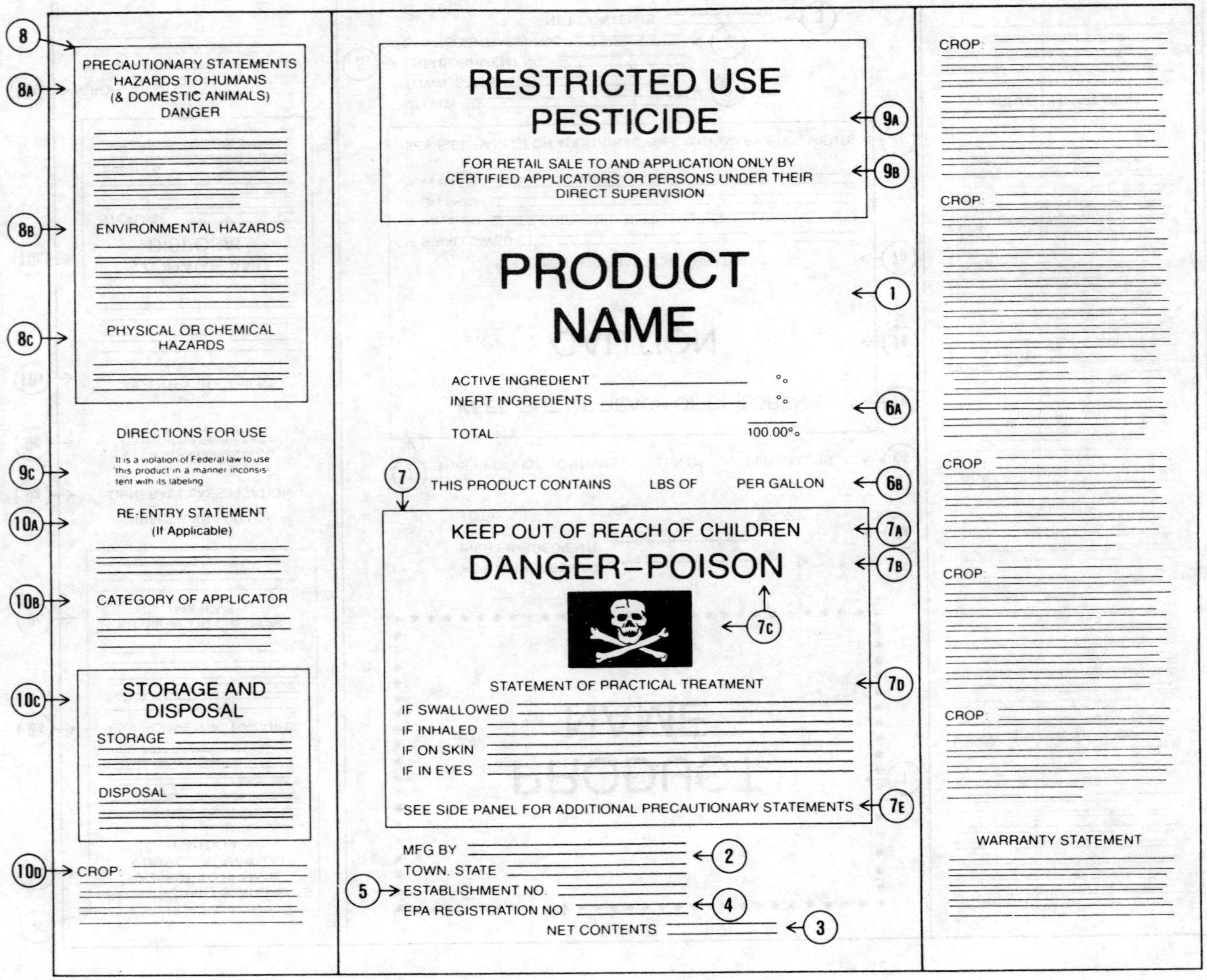

8 PRECAUTIONARY STATEMENTS
8A HAZARDS TO HUMANS
(& DOMESTIC ANIMALS)
DANGER
8B ENVIRONMENTAL HAZARDS
8C PHYSICAL OR CHEMICAL HAZARDS
DIRECTIONS FOR USE
9C It is a violation of Federal law to use this product in a manner inconsistent with its labeling
10A RE-ENTRY STATEMENT (If Applicable)
10B CATEGORY OF APPLICATOR
10C STORAGE AND DISPOSAL
STORAGE
DISPOSAL
10D CROP:

9A RESTRICTED USE PESTICIDE
9B FOR RETAIL SALE TO AND APPLICATION ONLY BY CERTIFIED APPLICATORS OR PERSONS UNDER THEIR DIRECT SUPERVISION
1 PRODUCT NAME
ACTIVE INGREDIENT %
INERT INGREDIENTS %
6A TOTAL 100.00%
7 THIS PRODUCT CONTAINS LBS OF PER GALLON 6B
7A KEEP OUT OF REACH OF CHILDREN
7B DANGER–POISON
7C
7D STATEMENT OF PRACTICAL TREATMENT
IF SWALLOWED
IF INHALED
IF ON SKIN
IF IN EYES
7E SEE SIDE PANEL FOR ADDITIONAL PRECAUTIONARY STATEMENTS
2 MFG BY
TOWN, STATE
5 ESTABLISHMENT NO.
4 EPA REGISTRATION NO.
3 NET CONTENTS

CROP:
CROP:
CROP:
CROP:
CROP:
WARRANTY STATEMENT

**Figure 11.12** Elements of a restricted-use pesticide label as required by law. See text for explanation of numbers.

7. Precautionary statements.
   a. Child hazard warning, **Keep Out of Reach of Children.**
   b. Signal word—**DANGER, WARNING,** or **CAUTION.**
   c. Skull and crossbones and word **POISON** in red.
   d. Statement of practical treatment in case of accident.
   e. Referral statement.
8. Side- or back-panel precautions statements.
   a. Hazards to humans and domestic animals.
   b. Environmental hazards.
   c. Physical or chemical hazards.
9. Box showing words **RESTRICTED-USE PESTICIDE.**
   a. Statement of pesticide classification.
   b. Sale and application restrictions.
   c. Statement on misuse.
10. Statement on how soon field can be reentered.
   a. Field reentry directions.
   b. Applicator category (limits of use to certain categories of commercial applicators).
   c. Instructions for storage and disposal.
   d. Directions for use followed by warranty statement.

In addition to these label specifications, those for another regulation, the **Worker Protection Standard** (WPS), may also be present in the Directions for Use section. The WPS is a federal regulation designed to protect agricultural workers and pesticide handlers from pesticide injury. Employers who must comply with this regulation include: (1) managers or owners of farms, forests, nurseries, or greenhouses; (2) labor contractors for farms, forests, nurseries, or greenhouses; and (3) custom pesticide applicators or crop consultants. Agricultural employees to be protected by the WPS fall under two categories, workers and pesticide handlers. Workers are those who perform tasks related to the production of agricultural crops, such as harvesting, weeding, and watering. Pesticide handlers are those who perform activities such as mixing and applying pesticides, loading open pesticide containers, servicing application equipment, or entering a treated area during a restricted-entry interval.

Not all pesticides are covered by WPS. If covered, the label will give specific instructions regarding restricted-entry intervals, personal protective equipment, and other safety provisions that must be followed.

### Applicator Certification

According to the provisions of FIFRA, all individuals applying restricted-use pesticides must be certified. Restricted-use pesticides are so designated by law because of their inherent toxicity or potential hazard to the environment. The basic divisions of applicators approved to handle these pesticides are **private** and **commercial.** The general distinction between these divisions is whether or not the individual is applying the pesticide for his or her personal use (private) or is in some way compensated for applying it (commercial). Training programs on pesticide application in preparation of certification are administered by the Cooperative Extension Service of most states. Testing for certification is often conducted by a state's department of agriculture.

The EPA has established ten categories of commercial applicator certification: (1) agricultural pest control (plant and animal); (2) forest pest control; (3) ornamental and turf pest control; (4) seed treatment; (5) aquatic pest control;

(6) right-of-way pest control; (7) industrial, institutional, structural, and health-related pest control; (8) public health pest control; (9) regulatory pest control; and (10) demonstration and research pest control. Some states have additional categories.

Examination for commercial-applicator certification emphasizes competency in: (1) label and labeling, (2) safety, (3) environment, (4) pests, (5) pesticides, (6) equipment, (7) application techniques, and (8) laws and regulations. Manuals and other study materials for exam preparation are available from the state Cooperative Extension Service. Certified applicators must be recertified at least every 5 years.

### Regulating Pesticides

The EPA regulates approximately 25,000 pesticide products, which involve fewer than 750 active ingredients. The regulated community includes all users and commercial suppliers. The latter includes about 30 major pesticide producers, 100 companies that market active ingredients, 3,300 product formulators, and 29,000 distributors.

**Regulation of new pesticides.** Under FIFRA, the EPA is charged with registering new pesticides. The EPA requires manufacturers to show that the new material, when used according to label directions, will not present unreasonable risks to human health or the environment. The law requires the EPA to take into account economic, social, and environmental costs and benefits in making decisions. Often, registration decisions made each year are for new formulations containing active ingredients that are already registered with the EPA or new uses of existing products.

The EPA bases registration decisions for new pesticides on its evaluation of test data provided by the applicants. Required studies include testing to show whether a pesticide has the potential to cause adverse effects in humans, fish, wildlife, and endangered species. Potential human risks include acute reactions like toxic poisoning and skin and eye irritation. Also, long-term risks like cancer, birth defects, or reproductive system disorders are assessed. Data on how a pesticide behaves in the environment (**environmental fate**) are also required so that threats to ground or surface water can be detected. In addition, if a food crop application is involved, the applicant must also petition the EPA for a tolerance and submit appropriate data so that a safe and realistic level can be established.

New active ingredients typically attain full registration within 8 to 9 years and may require over $25 million of capital investment. Total development can be $100 million, not including cost of the production facility.

The registration process begins with an experimental use permit (EUP) to allow premarket field testing of the pesticide. Subsequently, manufacturers typically submit an application for registration. This application must include appropriate health and safety data from experimental use, after which other data may be requested by the EPA. Registration is granted after study shows the new pesticide meets federal regulations.

In a policy change, the EPA instituted a Reduced-Risk Pesticides Initiative. The purpose of the initiative is to encourage the registration of pesticide products that reduce risks to human health and environmental quality. Under the initiative, manufacturers and formulators of new pesticides may seek to register a product containing novel active ingredients if the pesticide has a strong potential for improved human and environmental safety. If the new pesticide is approved for registration under this program, it will receive priority in the registration process and receive special considerations in registration.

The EPA also evaluates several other types of special registration submissions. Under section 18, the EPA may temporarily authorize state or federal agencies to combat emergencies with pesticides not permitted by usual registrations. Under another provision of the law, section 24(c), "special local need" registrations that take effect immediately can be granted directly by a state, but the EPA can override these registrations within 90 days.

Another type of special registration is one for biological pesticides or biopesticides. So-called **biopesticides** are pesticides derived from animals, plants, microorganisms, and certain minerals, for example, canola oil and baking soda. Today, there are over 240 biopesticide active ingredients and 1,100 products. These pesticides and their registrations are reviewed in greater detail in Chapter 12.

**Regulating existing pesticides.** In addition to regulating new pesticides, the EPA monitors and evaluates human and environmental risks from already registered pesticides. To ensure that these registered materials meet current scientific and regulatory standards, the law requires their "reregistration."

The reregistration program conducted by the EPA has been underway for many years. In 1988, FIFRA was amended to direct the EPA to complete the entire reregistration process by 1997. With the establishment of the Food Quality Protection Act of 1996, FIFRA was amended again to continue the reregistration process indefinitely by reviewing all pesticides every 15 years.

For reregistration, an originally registered pesticide must meet the same criteria that apply to new pesticides. Reregistration is accomplished in five phases. In Phase I, the EPA publishes lists of active ingredients subject to reregistration. Under Phase II, registrants notify the EPA on intent to seek reregistration, identify needed data, and pay necessary fees. As Phase III proceeds, the EPA provides guidelines for registrants, and registrants disclose data generated, identify potential adverse effects of compounds, and agree to fill data gaps. As Phase IV proceeds, the EPA identifies data gaps and issues a data call-in, requiring registrants to fill data gaps. Registrants then conduct required studies and submit additional data. In the final phase, Phase V, the EPA reviews all data for the active ingredients and determines eligibility for reregistration. The final result can be reregistration of the commercial product or denial of reregistration.

If a registered pesticide shows evidence of posing unacceptable personal or environmental risks, the EPA can also conduct a Special Review. A Special Review is a formal review process for obtaining risk/benefit information on a pesticide. Special Reviews can be initiated at any time against any pesticide the EPA believes meets or exceeds specified risk criteria. Once a pesticide is placed on the Special Review list, the manufacturer must produce data that shows its product does not cause adverse effects or risks. Subsequently, a Special Review team, made up of individuals from the federal and state governments, university personnel, and others, reviews the data and considers input from various user groups. The outcome of the Special Review may be to: (1) retain current registration, sometimes with label changes; (2) retain current registration but with restricted-use classification and/or geographical restrictions; or (3) issue a Notice of Intent to Cancel registration.

A **Notice of Intent to Cancel** is given to the manufacturer when a product poses unreasonable adverse effects. Those uses or registrations posing the risks are canceled and removed from the market. The EPA may also issue a **Suspension Order** or **Emergency Suspension Order** to quickly remove a pesticide from the market when it represents an imminent hazard or an imminent hazard posing an emergency, respectively. An example of an insecticide that received emergency suspension is ethylene dibromide.

Suspension orders are interim steps taken to quickly remove products from the market when they are found hazardous. Full cancellation proceedings that weigh the risks and benefits of product use are required to permanently eliminate a pesticide. At any time during these proceedings, a registrant may reach an agreement with the EPA to modify the terms and conditions of a registration, thereby reducing human and environmental risks. The modifications may include canceling uses, changing use patterns, changing application methods rates, and imposing personal protective measures.

## FOOD QUALITY PROTECTION ACT

The Food Quality Protection Act (FQPA) was signed into law on August 3, 1996. This act amends both FIFRA and the Federal Food, Drug, and Cosmetic Act (FFDCA). The purpose of FQPA is to improve food safety in the United States by addressing important pesticide issues, particularly the zero-tolerance provision created by the Delaney Clause and undue risk to children and infants by pesticide residues on food.

Recall that the Delaney Clause set a zero-risk cancer standard for pesticide residues on processed foods. In other words, residues of any pesticide known to cause cancerous tumors in laboratory animals at any rate was disallowed on processed foods. However, the law did not hold for nonprocessed foods. Therefore, a pesticide not allowed on processed foods potentially could be used on raw foods. This was termed the "Delaney Paradox." Another problem with the Delaney Clause was that detection of pesticides had improved greatly since its inception, making the standard very difficult to meet.

The pesticide issue dealing with children and infants emerged after a 1993 National Academy of Science study was published. The study emphasized that children are not small versions of adults and that some pesticides may have a greater impact on them than on adults. The report concluded that food tolerances for pesticides should be lowered to account for child sensitivity to the compounds. Charges from other sources contended that pesticide food tolerances were too high because some pesticides mimic or block human hormones and cause health problems. These additional public concerns gave support to the legislation and were considered in developing the food safety legislation.

Provisions of FQPA impact present laws by: (1) replacing the Delaney Clause with the standard of "reasonable certainty that no harm will result from aggregate exposure to pesticide residue"; (2) providing uniformity of state regulations on pesticide tolerances; (3) imposing civil penalties for marketing pesticide-adulterated food; (4) providing for a review of pesticide registrations at least every 15 years; (5) increasing data collection to ensure the health of children and infants; (6) requiring new screening procedures to determine estrogenic (hormone) effects of pesticides on women; and (7) promoting integrated pest management practices.

Under FQPA, the EPA was required to review the safety of all existing tolerances (maximum pesticide residue limits on food) that were in effect as of August 1996. Tolerance reassessment is being accomplished through the pesticide reregistration program by reviewing all existing uses of a pesticide when a new use is proposed and by revoking tolerances for pesticide uses that have been canceled. Of the 9,721 existing tolerances EPA was to reassess, over 99 percent were completed by August 2006. The EPA continues to make progress toward the completion of this mandate.

Perhaps the most important impact of this act will be to reduce an estimated two-thirds of existing tolerances, with the intention of protecting pregnant women, infants, and children. Another effect of the law likely will be the elimination of certain pesticides, particularly the organophosphates and carbamates.

## USING INSECTICIDES FOR PEST MANAGEMENT

Insecticides are some of the most potent, dependable substances that can be employed to manage insect pests. Indeed, many of the world's most serious insect problems can be contained only by these important compounds. Because of their broad-spectrum nature, however, they can be hazardous to humans and can cause undesirable side effects, both in agricultural and nonagricultural ecosystems.

Most often, the cause of undesirable side effects is the way insecticides are used. Side effects cannot be avoided totally, but they can be minimized by proper use, which includes applying insecticides as effectively as possible and having utmost regard for human and environmental safety.

### Effective Use

Using insecticides effectively begins with a proper strategy. With agricultural pests, insecticide use is a curative tactic that should be applied only after pest status has been assessed. Assessment involves accurate identification of the pest species present and estimates of its population level. Moreover, information on other potential pests and natural enemies in the agroecosystem should be a part of the overall assessment. If assessment shows that one or more pests will likely exceed the economic-injury level and no other tactic is practical, an appropriate insecticide should be chosen.

**Choosing an insecticide.** The choice of the most appropriate insecticide for a given pest situation depends on several factors including effectiveness, cost, formulations available, and equipment required. When choosing an insecticide, a place to start is with recommendations published by the state Cooperative Extension Service. Such publications are frequently organized according to crop and are oriented toward local needs and conditions. They are usually updated frequently so that users can take advantage of the latest knowledge and developments from research.

Often, several alternatives are presented in extension recommendations. The most appropriate from among these may depend on the particular equipment required for application. If ULV sprays are suggested and only low-pressure, high-volume sprayers are available, the ULV compound is obviously eliminated from the choice. In other instances, alternatives can be eliminated on the basis of time remaining until harvest. Some insecticides will be eliminated because unacceptable residues would be present on the crop at the expected harvest time. Also, if several pests are present that require suppression, some insecticides will not be effective against all. This requires cross-referencing of the recommendations to see if possible choices are effective against the entire complex of pests. Subsequent elimination of impractical alternatives should also be made after considering such factors as probable effects on natural enemies and wildlife.

The elimination process is continued until only one or a few acceptable choices are left. Final product acceptance can then be determined according to cost, availability, and convenience.

**Choosing a dosage.** Most insecticide recommendations list a dosage range for application, as does the insecticide label. As a general rule in insect pest management, least is best. To avoid insecticide resistance and other population counterresponses in agricultural applications, it is often best to reduce populations to insignificant levels rather than trying to eliminate them completely. This can be achieved most easily by keeping dosage levels as low as possible. Remember that EPA regulations allow pesticides to be used at rates lower than those stated on the label. Reducing dosage also cuts costs.

However, exceptions to the low-dosage rule do apply. For instance, where exotic pests have been newly introduced and occupy a limited area, higher doses for complete elimination may be necessary. Additionally, certain mites and a few insects have been shown to be reproductively stimulated by sublethal doses of pesticides. When this phenomenon, **hormoligosis,** is expected, highest legal doses may be necessary. In yet other instances, highest dosages may be prescribed when attempting to avoid certain types of insecticide resistance.

**Timing of applications.** No factor is more important in efficacy and environmental safety than proper timing of insecticide applications. The timing of applications is determined from characteristics and status of the target pest(s) and by conditions of the environment.

Insect life cycles and seasonal cycles greatly determine the general time when applications are made. Susceptible stages are usually targets of insecticide programs. For instance, sprays timed to kill small grasshopper nymphs are usually more effective than those applied against large nymphs and adults. In another instance, insecticide applications against second-generation European corn borer on corn must be timed to expose the peak occurrence of small larvae before they bore into the stalk.

Another important aspect of timing is to expose the population to an insecticide before significant injury has occurred. It may be consoling to kill insects that have injured a crop; however, unless significant damage can be avoided, it is economically unwise to do so. Usually, early surveillance and following recommended economic thresholds provide this proper timing information, which indicates estimates within a time period, perhaps 3 or 4 days, when application should take place.

The weather and concern with other production activities probably will determine when insecticides can be applied in this time period. To reduce drift and to concentrate most material in the target area, applications should be made under calm conditions. Windy days should be avoided and, when possible, applications should be made early in the morning or late in the evening before wind speed picks up. By avoiding periods from about 10 A.M. to 4 P.M., especially during hot weather, both insecticide drift and evaporation can be reduced.

Of course, if rain is occurring or imminent, additional limitations are placed on the application time. In addition, attention to production activities like cultivating, calving, and irrigating may take priority and further limit timing.

**Coverage and confinement of applications.** Thorough coverage of the target area is another very important factor for insecticide efficacy. Good coverage requires that the insecticide mixture be adequate to reach all pest feeding sites (Fig. 11.13). In some instances, as with some foliage feeders, insects

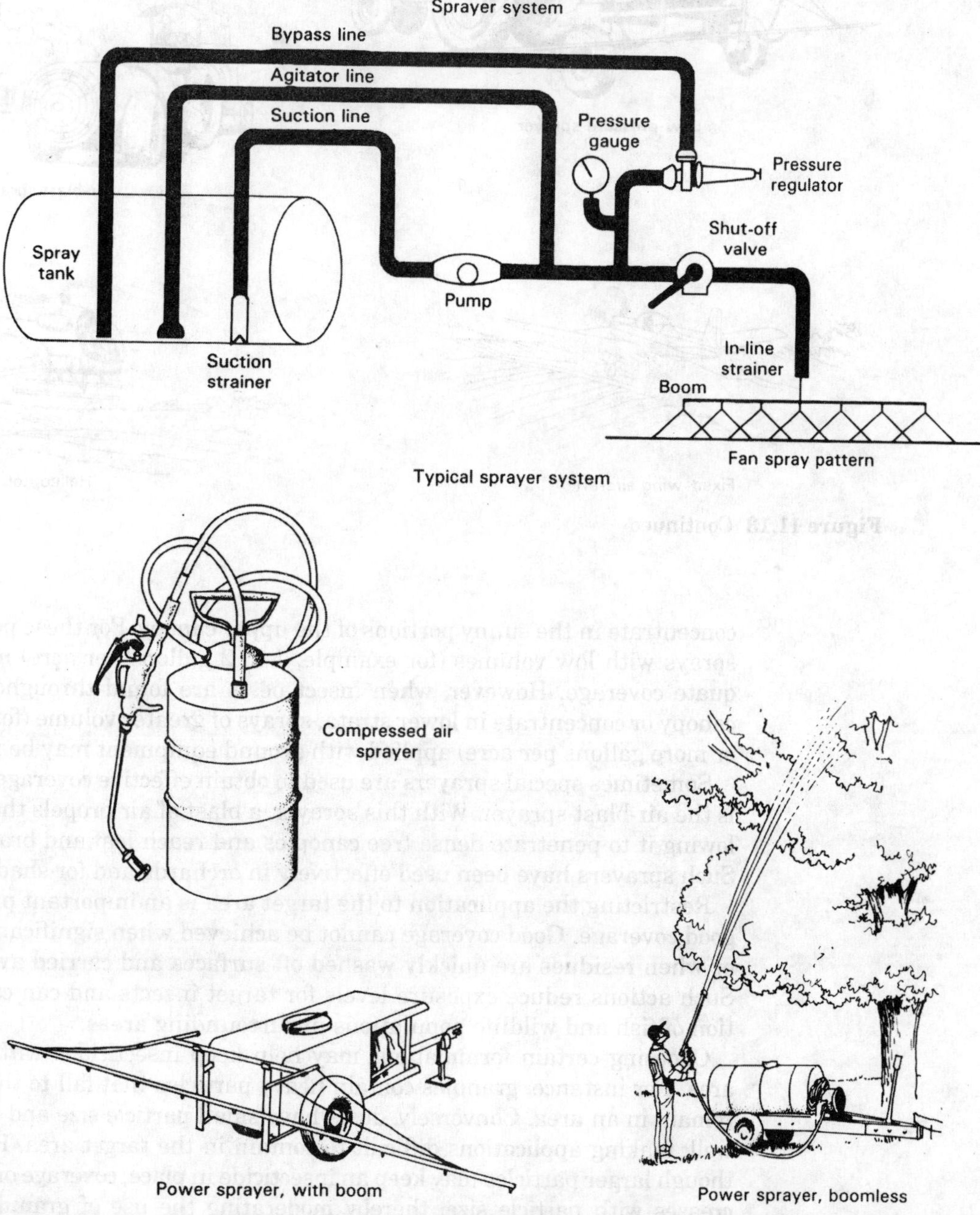


**Figure 11.13** A typical spraying system and common types of equipment for applying liquid formulations of pesticides.

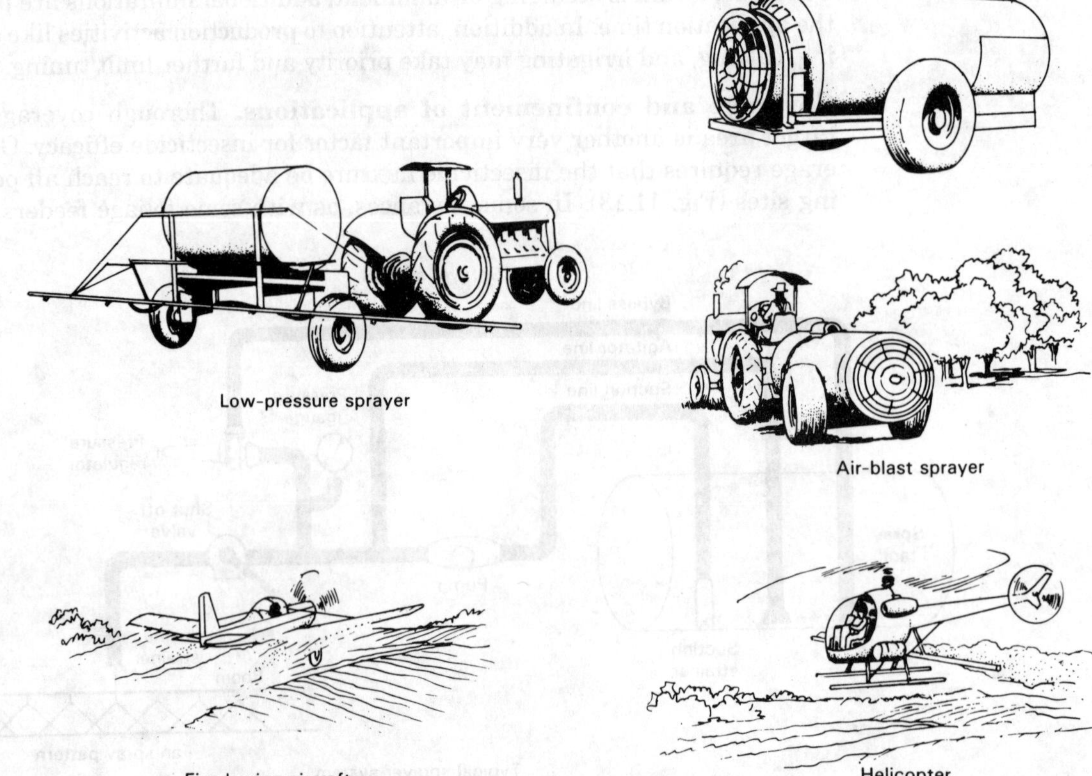


**Figure 11.13** Continued

concentrate in the sunny portions of the upper canopy. For these pests, airplane sprays with low volumes (for example, 1 to 2 gallons per acre) may give adequate coverage. However, when insect pests are found throughout the plant canopy or concentrate in lower strata, sprays of greater volume (for example, 20 or more gallons per acre) applied with ground equipment may be required.

Sometimes special sprayers are used to obtain effective coverage. An example is the air-blast sprayer. With this sprayer, a blast of air propels the mixture, allowing it to penetrate dense tree canopies and reach leaf and branch surfaces. Such sprayers have been used effectively in orchards and for shade trees.

Restricting the application to the target area is an important prerequisite to good coverage. Good coverage cannot be achieved when significant drift occurs or when residues are quickly washed off surfaces and carried away in runoff. Such actions reduce exposure levels for target insects and can cause destruction of fish and wildlife populations in surrounding areas.

Choosing certain formulations may help keep insecticides within the target area. For instance, granules contain heavy particles that fall to the surface and remain in an area. Conversely, dusts have small particle size and do not adhere well, making applications difficult to contain in the target area. However, even though larger particles may keep an insecticide in place, coverage on a surface decreases with particle size, thereby moderating the use of granules and large-droplet sprays (Fig. 11.14). Adding sticking agents or selecting formulations with

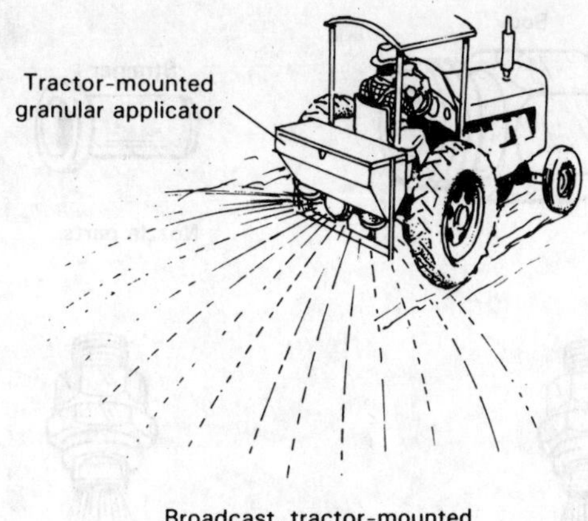

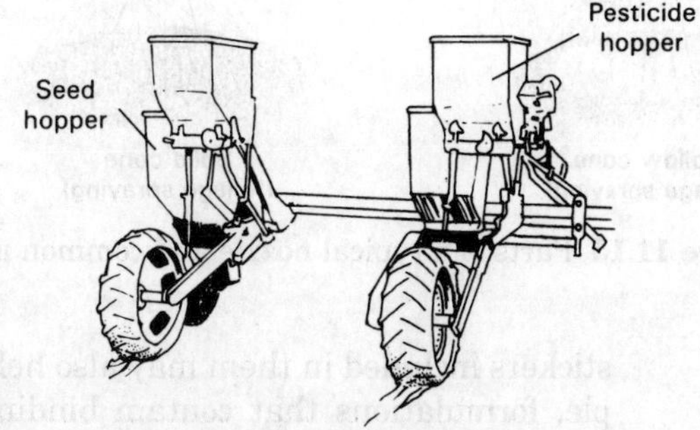


**Figure 11.14** Common types of equipment for applying pesticide granules.

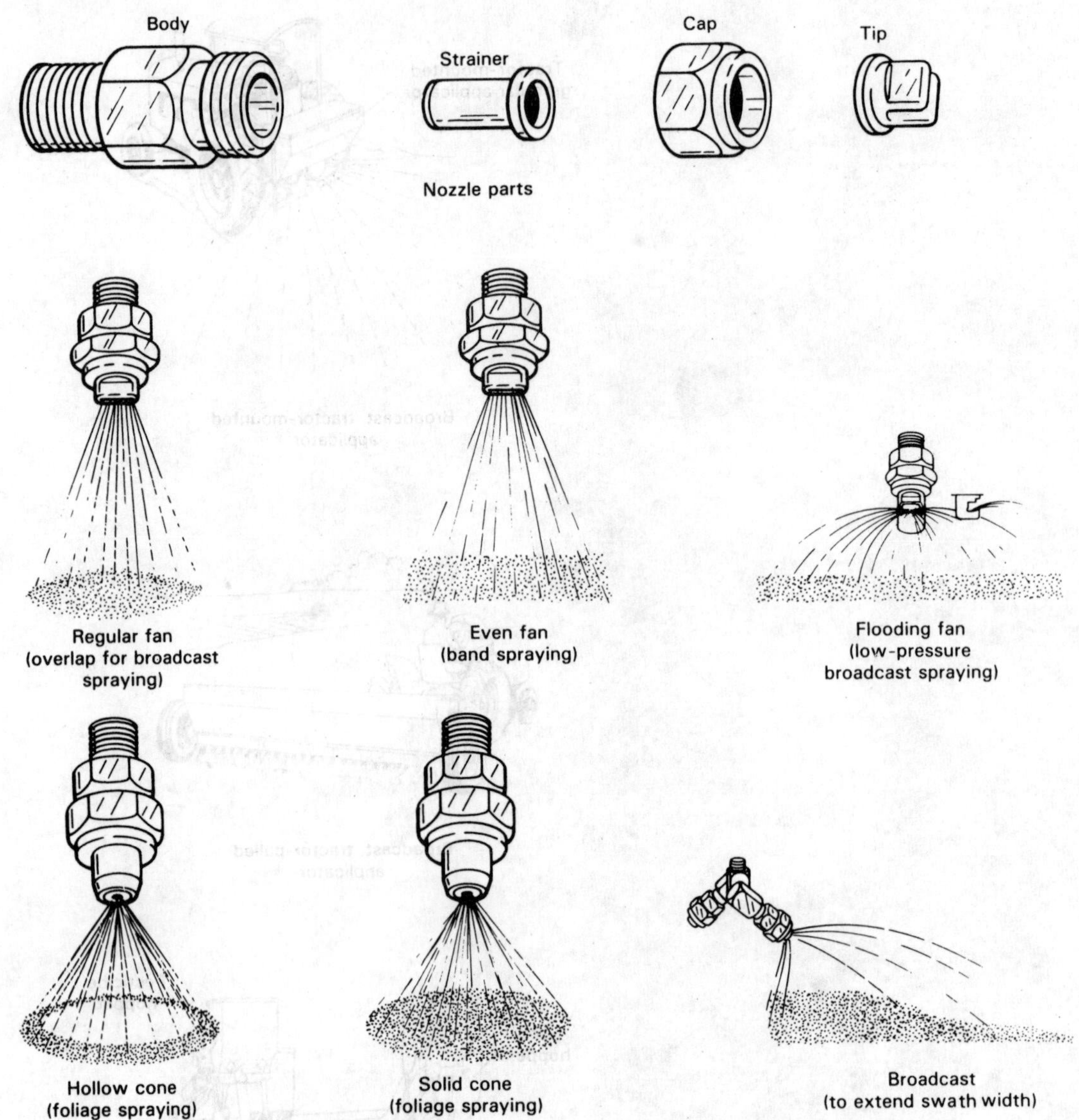


**Figure 11.15** Parts of a typical nozzle and common nozzle types.

stickers included in them may also help keep an insecticide on target. For example, formulations that contain binding and sticking agents can extend active residues several days by keeping the active ingredient, carbaryl, on plant leaves.

Choice of application equipment can also affect containment of insecticides (Fig. 11.15). In comparisons of airplane and ground sprays in cotton, high-clearance sprayers with three nozzles per row deposited an average 82 percent of the spray on target, versus only 54.5 percent for aerial sprays. But certain factors like wet fields, canopy closure, and problems of soil compaction may make ground applications impractical.

## Using Insecticides Safely

Personal safety with insecticides begins with a thorough reading of the label. The label gives explicit precautions and steps to be taken for mixing and applying an insecticide. It also gives information on disposal of containers and explains what to do if an accident occurs.

Any time insecticides or other pesticides are handled, protective clothing should be worn (Fig. 11.16). This clothing includes a long-sleeved shirt and long-legged trousers or coveralls. Trousers should be worn outside of boots. Unlined neoprene gloves should be worn in most instances, with shirt sleeves worn over the gloves. Other clothing for safe use includes a wide-brimmed hat, neoprene boots, and goggles or a face shield.

Clothing worn while handling pesticides should be laundered frequently and by itself. When it is soiled by an insecticide concentrate, the clothing should be safely destroyed.

Lungs and other parts of the respiratory system are much more absorbent of pesticides than skin. An approved respiratory device must be worn if the pesticide label so directs. Several types of respirators can be worn, and specific information as to type is also given in label directions.

Taking time to use pesticides safely is one of the most important investments that any user can make. Religiously following label directions and having respect for personal and environmental hazards is the only acceptable policy in using pesticides. It is also the law.

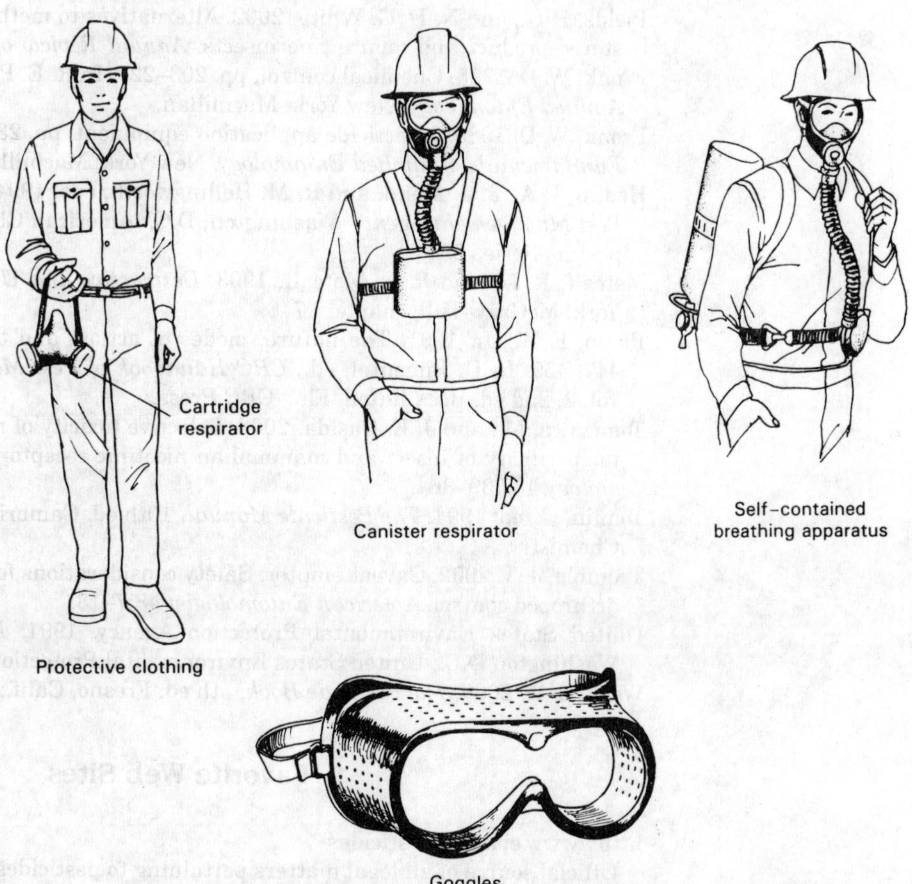


**Figure 11.16** Types of clothing and protective equipment used in applying pesticides. (Courtesy EPA)

## CONCLUSIONS

It is difficult to imagine a technology that would produce the amount of food and fiber and maintain the level of public health that we have today without pesticides. But their use presents a dilemma. Speaking on pesticides, M. J. Dover (1985) comments, "As their hazards become more apparent, so does the need to use them. Although designed to kill, they are often life-savers. Although increasingly costly, they bring economic benefits. And while they have opened up many possibilities for improving agriculture and public health, they have closed others, making us extremely dependent on them for our continued survival."

The challenge ahead for pest management is to reduce our dependency on hazardous pesticides and create more benign compounds.

### Further Reading


Coats, J. R. 1994. Risks from natural versus synthetic insecticides. *Annual Review of Entomology* 39:489–515.

Dover, M. J. 1985. *A Better Mousetrap—Improving Pest Management for Agriculture.* Washington, D.C.: World Resources Institute, Study No. 4.

Fields, P. G., and N. D. G. White. 2002. Alternatives to methyl bromide treatments for stored-product and quarantine insects. *Annual Review of Entomology* 47:331–359.

Fronk, W. D. 1985. Chemical control, pp. 203–229. In R. E. Pfadt, ed., *Fundamentals of Applied Entomology.* New York: Macmillan.

Fronk, W. D. 1985. Insecticide application equipment, pp. 230–246. In R. E. Pfadt, ed., *Fundamentals of Applied Entomology.* New York: Macmillan.

Hedlin, P. A., J. J. Menn, and R. M. Hollingworth, eds. 1994. *Natural and Engineered Pest Management Agents.* Washington, D.C.: American Chemical Society, ACS Symposium Series 551.

Metcalf, R. L., and R. A. Metcalf. 1993. *Destructive and Useful Insects,* 5th ed. New York: McGraw-Hill, chapters 7–8.

Plapp, F. W., Jr. 1991. The nature, modes of action, and toxicity of insecticides, pp. 447–459. In D. Pimentel, ed., *CRC Handbook of Pest Management in Agriculture,* vol. 2, 2nd ed. Boca Raton, Fla.: CRC Press.

Tomizawa, M., and J. E. Casida. 2003. Selective toxicity of neonicotinoids attributable to specificity of insect and mammalian nicotinic receptors. *Annual Review of Entomology* 48:339–364.

Tomlin, C., ed. 1994. *The Pesticide Manual,* 10th ed. Cambridge, U.K.: Royal Society of Chemistry.

Trumble, J. T. 2002. Caveat emptor: Safety considerations for natural products used in arthropod control. *American Entomologist* 48:7–13.

United States Environmental Protection Agency. 1991. *EPA's Pesticide Programs.* Washington D.C.: United States Environmental Protection Agency.

Ware, G. W. 2000. *The Pesticide Book,* 5th ed. Fresno, Calif.: Thomson Pub.


### Favorite Web Sites

http://www.epa.gov/pesticides

Official source of all legal matters pertaining to pesticides. The Office of Pesticide Programs/EPA posts the latest on pesticide registrations, special reviews, labeling, and other broad aspects of pesticide use.

http://ace.ace.orst.edu/info/extoxnet/ghindex.html
Referred to as the Extension Toxicology Network. Searches can be made for particular chemicals or allow browsing from a list of topics, including pesticide profiles, toxicology briefs and issues, fact sheets, and frequently asked questions.

http://ipmworld.umn.edu/chapters/bloomq.htm
Site focuses on the chemistry and mode of action of insecticides. Shows many structures and diagrams of action sites.

http://ipmworld.umn.edu/chapters/larson.htm
Site discusses recently developed insecticides with novel modes of action.

http://www.hclrss.demon.co.uk/index.html
This compendium is believed to be the only place where all of the ISO-approved standard names of chemical pesticides are listed. It also includes approved names from national and international bodies for pesticides that do not have ISO names.

CHAPTER 12

# BIOPESTICIDES FOR MANAGEMENT

THE VAST MAJORITY OF PESTICIDES used in insect pest management are the conventional materials reviewed in Chapter 11. However, with the desire to foster use and development of materials with greater safety and increased environmental compatibility, the U.S. Environmental Protection Agency (EPA) created a new category for registration of these pesticides and named them *biopesticides*. **Biopesticides** are certain types of pesticides derived from such natural materials as animals, plants, microorganisms, and certain minerals.

Biopesticides fall under the auspices of the Biopesticides and Pollution Prevention Division (BPPD) of the EPA's Office of Pesticide Programs (OPP) established in 1995. These products are distinguished from conventional chemical pesticides by all or many of the following characteristics: (1) unique mode of action, (2) narrow pest range, (3) low use volume, and (4) natural occurrence. The more characteristics a product possesses, the greater its chances of being placed in the biopesticide grouping. Today, there are over 241 active ingredients with insecticidal or acaricidal activity in 1,155 biopesticidal products.

Biopesticides are inherently less harmful and have fewer environmental effects than conventional pesticides. Therefore, the EPA generally requires much less data to register them than conventional pesticides. Indeed, new biopesticides require 1 or 2 years for registration compared to 5 to 7 years for conventional pesticides. Still, registrants must submit data on composition, toxicity, degradation, and other characteristics to show no adverse effects on human health or the environment.

Unlike many conventional pesticides, biopesticides may affect only a target pest and closely related species. For instance, some insect growth regulators directly affect only caterpillars feeding on a plant and not the beneficial insects and mammals living in the same environment. Moreover, biopesticides are often effective in small quantities and usually decompose quickly, not building up in the environment and avoiding the pollution found with some conventional pesticides. These safer, reduced-risk products are registered with the EPA under three major classes: (1) microbial pesticides, (2) biochemical pesticides, and (3) plant-incorporated protectants.

## MICROBIAL PESTICIDES

Most of the microbial pesticides registered with the EPA are active against insect pests, and these are more appropriately termed **microbial insecticides**. Microbial insecticides are biological preparations that are often sprayed or delivered in ways similar to those of conventional chemical insecticides. They are formulated as dusts, liquid drenches, liquid concentrates, wettable powders, or granules. The various formulations give a product specific properties to make it most effective. Such biological preparations have label directions like conventional insecticides.

As discussed in Chapter 9, insect populations can be strongly affected by diseases caused by microorganisms, including bacteria, viruses, fungi, protozoa, and rickettsiae. All, except the rickettsiae, can be found as active ingredients in microbial insecticides. To date, the microorganisms most widely used as microbial insecticides are bacteria. See Table 12.1 for selected examples of common microbial insecticides.

### Bacteria

Of the bacteria, the spore-forming species have been the most important in insect suppression. Particularly, the genus *Bacillus* has been used; some of the most notable examples are *B. popilliae* and *B. lentimorbus*. These bacteria

**Table 12.1 Selected Registered Microbial Insecticides**

| Organism | Trade Name | Manufacturer | Target Pest |
|---|---|---|---|
| | *Bacteria* | | |
| *Bacillus popilliae* | None | None | Japanese beetle |
| *B. thuringiensis* var. *kurstaki* | Dipel | Valent | |
| | Biobit | Valent | Several moths |
| | Thuricide | Valent | |
| | Condor | Certis | |
| *B. thuringiensis* var. *kurstaki* plus beta-exotoxin | Javelin | Certis | Armyworm and other moths |
| *B. thuringiensis* var. *aizawi* | Agree | Certis | Wax moth |
| *B. thuringiensis* var. *tenebrionis* | Novodor | Valent | Colorado potato beetle |
| *B. thuringiensis* var. *israelensis* | VectoBac | Valent | Mosquitoes, black flies |
| | *Fungi* | | |
| *Beauveria bassiana* | Naturalis | Troy Biosciences | Whiteflies scales, aphids |
| | Mycotrol | Ricon-Vitova | |
| | *Viruses* | | |
| *Spodoptera exigua* NPV | Spod-X | Certis | Beet armyworm |
| *Helicoverpa* NPV | Gemstar | Certis | Corn earworm, tobacco budworm |
| *Cydia pomonella* GV | ViroSoft | Biotepp | Codling moth |
| | *Microsporidian* | | |
| *Nosema locustae* | Nolo Bait | Ricon-Vitova Insectaries | Grasshopper nymphs Morman crickets |

cause milky disease in the Japanese beetle, *Popillia japonica* (Fig. 12.1, Box 12.1). Another is *B. thuringiensis,* causing diseases in many species of moths (Lepidoptera), mosquitoes (Diptera: Culicidae), and a few beetles (Coleoptera).

*Bacillus popilliae* and *B. lentimorbus* spores are ingested by Japanese beetle and other scarab larvae. The spores germinate and penetrate their gut and cause their blood to appear milky. Death follows soon afterward. Both these bacteria are *obligate* (they must have the host to develop) and therefore are

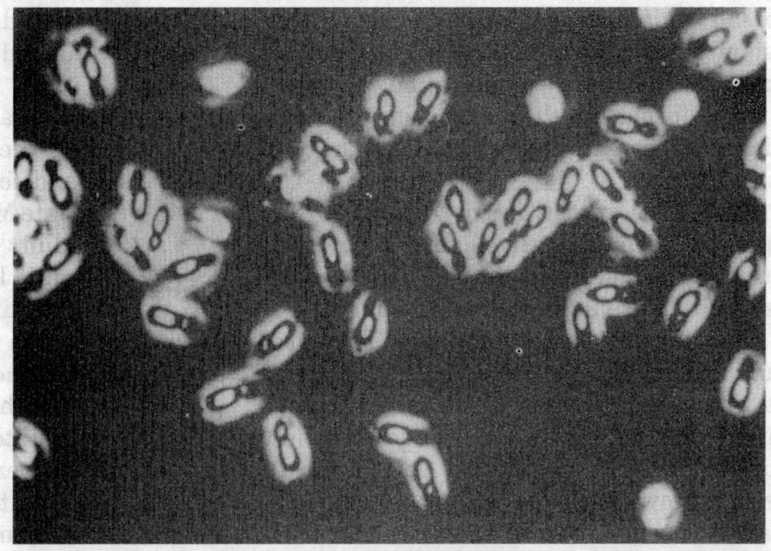

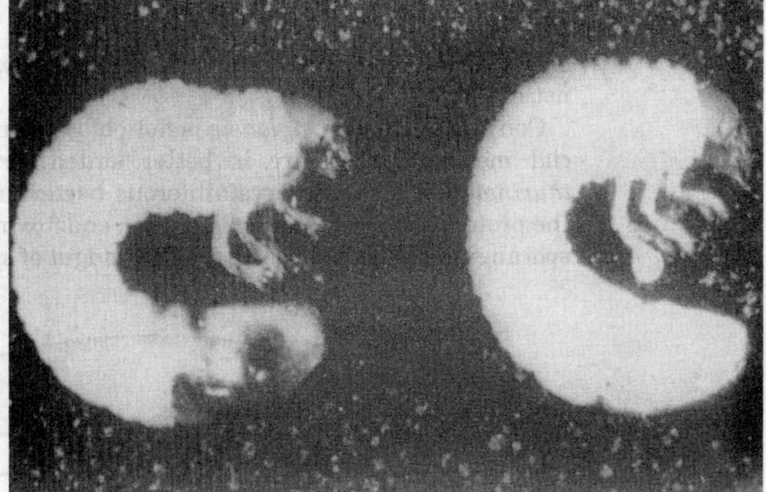

**Figure 12.1** Top: *Bacillus popilliae* sporangia under a phase-contrast microscope, each with a large spore and a smaller parasporal body. Bottom: a healthy *(left)* and *Bacillus popilliae*–infected Japanese beetle *(Popillia japonica)* larva. The metathoracic legs were clipped to produce a droplet of blood. Note that the blood color of the diseased larva on the right is white. (Top, courtesy USDA; bottom, after Copple and Mertins, 1977, *Biological Insect Suppression,* Springer-Verlag Publishers, with permission)

## BOX 12.1 Japanese Beetle (See Color Plate 4)

**SPECIES:** *Popillia japonica* Newman (Coleoptera: Scarabaeidae)

**DISTRIBUTION:** The Japanese beetle was introduced into the United States (New Jersey) from Japan about 1916, probably in perennial-plant nursery stock. Since its introduction, the species has spread to almost all states east of the Mississippi River and also portions of Ontario and Quebec in Canada. Two infestations have occurred in California, but the populations were destroyed in costly eradication programs.

**IMPORTANCE:** The species is only a minor pest in Japan, its native habitat, but in North America it is a serious pest. Larvae live in the soil and feed on the roots of many species of plants. They are especially damaging to pastures, lawns, golf courses, and parks. Grasses are stunted and yellowed under low infestations, but under high infestations (100 to 500 larvae per square yard) grasses turn brown and eventually die. Adults attack leaves, flowers, and fruits of more than 275 species of plants, such as fruit trees, shade trees, soybean, corn, flowers, and vegetables.

**APPEARANCE:** Adults are 8 to 10 mm long and are metallic green, with six white tufts of hair along the bronze elytra. Larvae, about 19 mm long when fully grown, are C-shaped white grubs with a brown head and a swollen tail. Eggs, about 1 mm, are white and deposited in 5 to 15 cm of soil.

**LIFE CYCLE:** Larvae overwinter in soil. In early spring, they move near the surface of the soil and feed on roots. Pupation occurs from late May to early June. In late June, adults begin emerging and are active until late September. Beetles mate during this time, and the females oviposit about fifty eggs in the soil. In about 14 days, the eggs hatch, and the larvae feed until the onset of cold weather. At this time larvae tunnel to about 20 cm in the soil to overwinter. There is usually one generation per year, but during excessively cold and wet conditions, 2 years may be needed to complete development.

---

difficult and expensive to propagate. Currently, *B. popilliae* is registered but not marketed.

Conversely, *B. thuringiensis* is not obligate and can be propagated on artificial media; it therefore is better suited for commercialization. *Bacillus thuringiensis* also is a crystalliferous bacterium (Fig. 12.2). After ingestion, the proteinaceous crystal (called delta endotoxin), associated with a spore in a sporangium, quickly dissolves in the midgut of susceptible species, causing gut

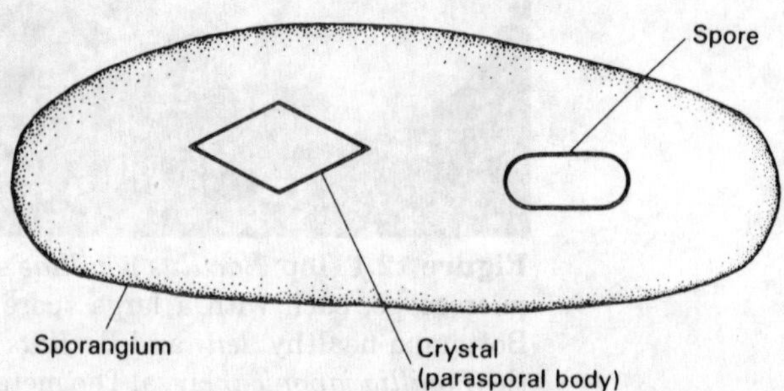


**Figure 12.2** Schematic of *Bacillus thuringiensis* sporangium with proteinaceous crystal and associated spore.

paralysis. The host larva then may be killed outright by the toxic crystal, or the crystal may enhance the penetration of the gut by the spores, which consequently cause a lethal septicemia.

*Bacillus thuringiensis,* usually referred to as Bt, is the most widely used biological control agent in North America. An example of its widespread use was witnessed in Canada in 1986, when more than 1.4 million hectares of forest area were treated with commercial Bt for spruce budworm management. Currently, there are more than 20 active ingredients registered for Bt products, representing several varieties and strains. Moreover, there are over 100 commercial products of Bt microbial insecticide.

The several Bt varieties (subspecies) discovered include *kurstaki* and *aizawai,* used against lepidopterous pests in crops, and *israelensis*, used against mosquitoes. Also, microbial formulations of the variety *tenebrionis* have been registered for use against the Colorado potato beetle and the elm leaf beetle and *morrisoni* for various Lepidoptera. In addition, another form of Bt containing another toxin (beta-exotoxin) has been developed that expands its use to other pests not managed with the Bt endotoxin. Presently, Bt has labeled uses in canola, cruciferous crops, cotton, corn, forage, forests and ornamentals, potatoes, stored grain, tree fruits, and tobacco, among others. In medical entomology it is used against mosquitoes and black flies.

Another *Bacillus* insecticide contains *Bacillus sphaericus.* Commercial products with this bacterium are applied to larval habitats (stagnant or turbid water) of *Culex* and other mosquitoes. Although most Bts work primarily through the toxic action of their endotoxins, *B. sphaericus* acts primarily through initiation of an infection, causing death of mosquitoes. Thus, it has longer activity than most Bts because it allows spread of disease in the pest population.

Two additional bacterial species have demonstrated good potential to produce microbial-based insect management agents, avermectin and spinosad. Avermectin, derived as a natural control agent from the species *Streptomyces avermitilis,* is registered for control of mites and leafminers and has been used as a key element for insect management in vegetable and orchard crops. Unlike existing microbial-based products, the toxicology and performance characteristics of this product are very similar to conventional synthetic insecticides (neurotoxins). A second discovery is spinosad, a natural product of the bacterium *Saccharopolyspora spinosa.* Spinosad is effective against armyworms, cotton bollworm, loopers, and tobacco budworm in cotton (Tracer®); pests in fruits and vegetables (SpinTor®, Success®); and insects in ornamentals and turf (Conserve®). This natural product has mammalian and nontarget characteristics similar to Bt-based products and performs similarly to conventional synthetic insecticides. Both of these biologically based products offer advantages to insect pest management because of their pest selectivity and new modes of action.

## Viruses

Although bacteria have been demonstrated as effective against many pest species, insect pathologists believe that viruses also hold much potential for insect management. To date, more than 1,200 insect/virus associations have been recorded, with the majority of these in the Lepidoptera, Hymenoptera, and Diptera.

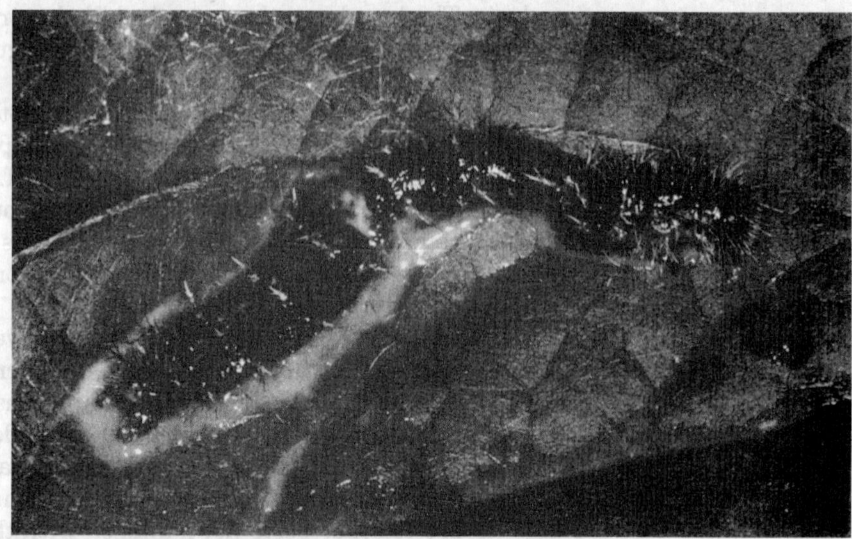

**Figure 12.3** Painted lady caterpillar, *Vanessa cardui,* probably killed by a nucleopolyhedro virus (NPV) and liquefying several days after death. (Photo by Marlin E. Rice)

The best-known viruses of insects are the baculoviruses, including the nucleopolyhedro viruses (NPV) (Fig. 12.3) and the granulosis viruses (GV). Although these have received the most attention for commercialization with Lepidoptera, the cytoplasmic polyhedrosis viruses (CPV) also show promise as microbial insecticides.

NPV viruses have posted more recorded successes in insect pest management than any other group (see Roberts et al. 1991). NPV viruses develop in the host cell nucleus where one or several virus rods occur singly or in groups encased in an envelope (Fig. 12.4). The envelopes are occluded (encased) in many-sided crystals called polyhedra. After ingesting the polyhedra, larvae show no outward symptoms for 4 days to 3 weeks. At this time, the larval skin darkens, and larvae climb to the highest point on their host plant, where they die. Dead, blackened larvae may be found hanging from the tops of plants. Subsequently, the integuments of these dead larvae rupture, and millions of polyhedra are released into the environment. Such diseases collectively have been called caterpillar wilt.

Today, there are seven viruses registered with the EPA for use against insects. The NPV microbials are labeled for use against celery looper, gypsy moth, Douglas fir tussock moth, corn earworm, and beet armyworm. Two GV microbials are registered for management of codling moth and Indianmeal moth.

## Fungi

Naturally occurring fungi are very important in insect population regulation, with more than 750 species known to infect insects. Some of the more widespread fungi causing disease epidemics in insects include those from the genera *Beauveria, Nomuraea* (Fig. 12.5), *Metarhizium, Entomophthora,*

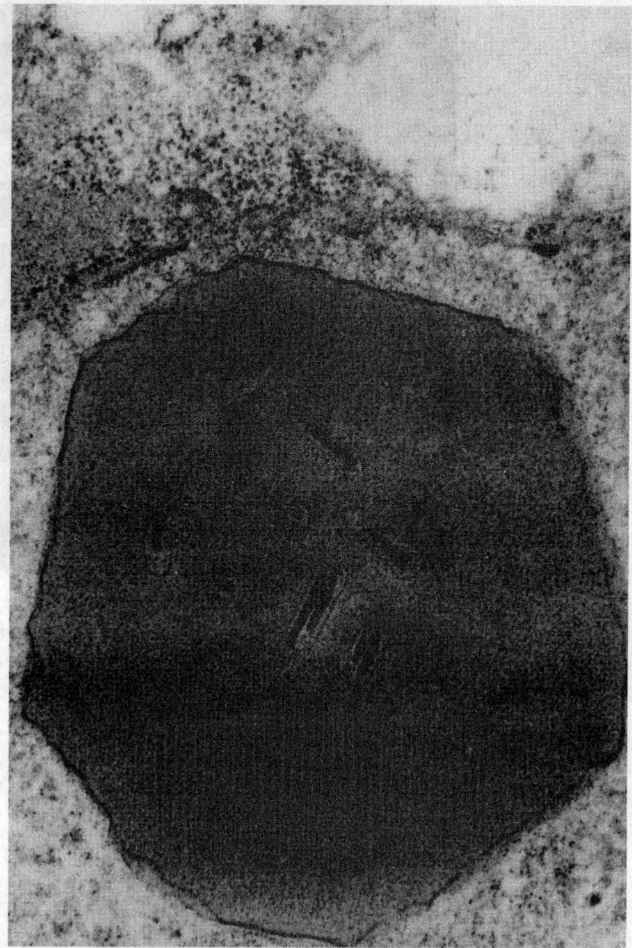

**Figure 12.4** Electron micrograph of nucleopolyhedro viruses showing packets of virus rods in one polyhedron (*bottom*) and single rods in the other. (Courtesy Les Lewis and USDA)

and *Zoophthora*. Unlike most of the other insect pathogens, fungi have the ability to attack insects through the cuticle. Usually a spore attaches to the cuticle, germinates, and penetrates the body wall. Once in the body cavity, the fungus spreads, colonizing the hemocoel, and sometimes producing toxins. When toxins are produced, death can be quite rapid; otherwise, death is delayed until body nutrients are depleted or internal organs destroyed (Fig. 12.6). Fungi have been used with much success in various parts of the world, most notably the People's Republic of China and Brazil. In the United States, the most widely commercialized fungus is *Beauveria bassiana*, marketed for whiteflies, thrips, aphids, mealybugs, and beetles as Mycotrol®, BotaniGard®, and Naturalis®. Another fungus, *Metarhizium anisopliae*, is registered for thrips and beetle larvae in ornamentals (Taenure®) and grubs and ticks (Tick–Ex®). Additionally, the fungus *Paecilomyces fumosoroseus* is registered for whiteflies, aphids, and thrips on ornamentals (PFR-97®).

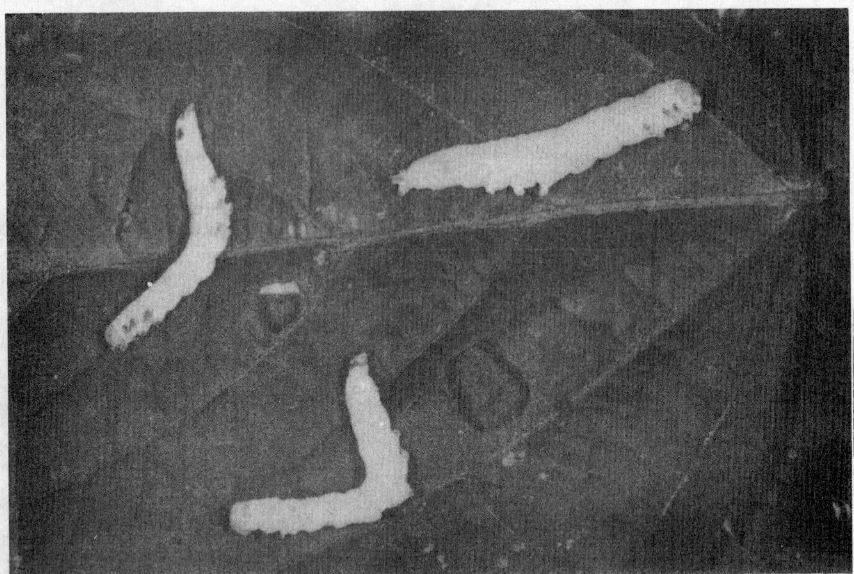

**Figure 12.5** Mummified bodies of green cloverworms, *Hypena scabra,* killed by the pathogenic fungus, *Nomuraea rileyi.* Conidiophores of the fungus form a white coating over the surface of the dead caterpillar. (Photo by Marlin E. Rice)

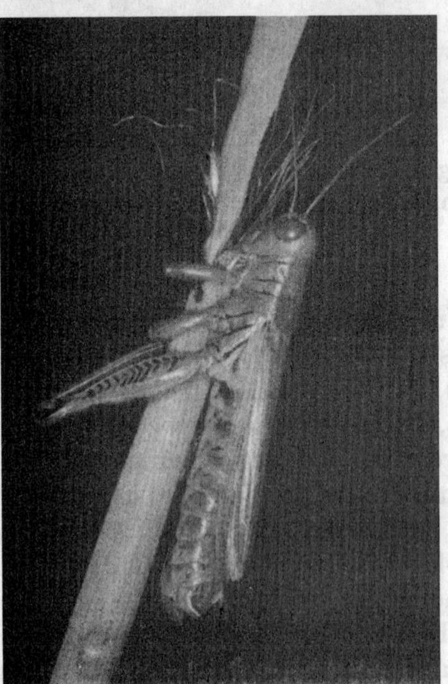

**Figure 12.6** Adult differential grasshoppers, *Melanoplus differentialis,* killed by the fungal pathogen, *Entomophaga* sp. *(left),* assume a rigid "death grip" on a plant, while grasshoppers killed by bacterial pathogens break apart and rot upon death (right). (Photos by Marlin E. Rice)

### Protozoa

Protozoa are single-celled animals. Some protozoa parasitize and kill insects when they are ingested. Therefore, protozoa are formulated as baits for use as microbial insecticides. Only one protozoan has been registered as a biopesticide. It is *Nosema locustae.* This naturally occurring protozoan infects and kills grasshoppers and Mormon crickets when these pests ingest bait containing the pathogen. When the bait is ingested, protozoa spores become active, the protozoa grow and replicate in the insect's digestive system, and the insect soon dies. The bait is most effective if used when the insects are still in their immature stages. Products containing the bait are marketed as Semaspore Bait®, Nolo-Bait®, and Hopper Stopper®.

### Biotechnology and the Future of Microbial Insecticides

The future of microbial insecticides appears to be assured. Increased problems of resistance and environmental contamination with conventional insecticides creates an impelling need for safer alternatives. Although no panacea, these materials may be used in the same way as insecticides, and this would make them very adaptable to many established pest management programs.

Also, advances in biotechnology should allow production costs of microbial insecticides to decrease and efficiency to increase. Then these agents would become even more appealing. Among the expectations of biotechnology, we might include: (1) modification of obligate microbial pathogens that allow them to reproduce on artificial media, (2) propagation of host-insect tissue from single cells that can be used as media for obligate pathogens, (3) expanding killing spectrum, and (4) improving environmental stability.

The latter two points were addressed for Bt by Ecogen, Inc., and Mycogen Corp. Ecogen has been able to engineer two products that combine more than one toxin in the Bt bacterium. One of these products is effective against both the gypsy moth and spruce budworm in forests and the other against both the Colorado potato beetle and European corn borer on potatoes. Mycogen Corp. (now owned by Dow AgroSciences) produced a product to improve Bt stability by engineering a rhizobacterium, *Pseudomona fluorescens,* to contain Bt toxin and then kill the cells (CellCap® technology). The dead cells of *P. fluorescens* encapsulate the toxin and make it much more resilient to environmental factors. In addition to these advancements in Bt use, much activity centers on producing transgenic plants that contain Bt toxin. Developments in this area are discussed in Chapter 13.

In addition to genetic engineering of Bt, similar research is directed toward improvements of virus efficacy. So far, one of the most promising developments is a recombinant virus from the alfalfa looper, *Autographa californica,* which was engineered to cause expression of an insect-specific toxin found in scorpions. When a lepidopteran is infected by the recombinant virus, its cells produce the scorpion toxin, which causes rapid cessation of feeding and death within 48 hours. This killing rate compares favorably to conventional insecticides and overcomes a major disadvantage of the unmodified virus, which has a killing rate of 7 to 10 days. Research with this virus and other exciting prospects using genetic engineering show much promise for the future development of practical viral insecticides.

## BIOCHEMICAL PESTICIDES

The biochemical category of biopesticides is composed of a large and diverse group of compounds, which often have unique uses in insect management programs. Biochemical pesticides are characterized primarily as materials that: (1) usually occur in nature and (2) do not have a direct and acute mode of action, that is, do not kill the pest outright. However, there are exceptions that allow materials to be categorized as biochemical pesticides that are not strictly "natural" in occurrence. For instance, a synthetic active ingredient can be placed in the category if it is structurally similar and functionally identical to a naturally occurring substance, for example, a compound synthesized to be similar to an insect sex pheromone. Other criteria are that a material will: (1) have little or no effect on nontarget organisms, (2) degrade rapidly in the environment, (3) require low application rates, and (4) have reasonable effectiveness. A special committee, the Biochemical Classification Committee (Office of Pesticide Programs, EPA), oversees the review and registration process to determine whether a material should be classified as a biochemical or a conventional pesticide.

Biochemical pesticides with insect activity should more properly be referred to as biochemical insecticides. These materials can be conveniently grouped and discussed according to their mode of action. Consequently, the groups are: (1) insect growth regulators, (2) attractants and repellents (excluding pheromones), (3) suffocating agents, (4) desiccants, (5) coatings, (6) pheromones, and (7) systemic acquired-response inducers. A review of these groupings follows.

### Insect Growth Regulators

Insect growth regulators, or IGRs, are compounds that are used to disrupt the normal growth and development of immature insects. Recall from Chapter 4 that a significant hormone involved in the molting process is *ecdysone,* secreted by a small gland in the insect's prothorax. Ecdysone stimulates separation of cells in the epidermis from the cuticle, which is followed by formation of a new cuticle and molting. Also, recall that development (progression of life stages) is controlled by another important hormone, *juvenile hormone,* secreted by accessory glands of the brain. Together, these important hormones control when an insect is to molt and what it is to become after molting, for example, another larval stage, a pupa, or an adult. When used as an insecticide, IGRs upset the molting process (chemical analogs of ecdysone), eventually causing death during molting, or disrupt metamorphosis (chemical analogs of juvenile hormone), eventually causing death or sterility of the new adult.

Not all IGRs are registered as biochemical pesticides. In particular, the chitin synthesis inhibitors (CSIs) are not. This failure to register these materials as biochemical pesticides is because of their broad-spectrum activity. Although CSIs have little effect on vertebrates in ecosystems, they can have deleterious effects on beneficial arthropods and aquatic organisms such as crustaceans. Conversely, those IGRs that mimic juvenile hormones (JHAs) are more species specific and environmentally benign. Therefore, they are the members of the IGR category that are registered biochemical pesticides. Greater detail of this process and development of practical IGRs is given in Chapter 14, but a brief listing of IGRs in the biochemical pesticide grouping is appropriate here.

There are four IGR active ingredients and over seventy-five products registered that have JHA activity. These include the following:

1. Methoprene—for suppression of beetles, flies, mosquitoes, ants, and others on food and nonfood crops, ornamentals, and livestock.
2. *S*-Methoprene (the S enantiomer of methoprene)—for similar pests and application sites as methoprene.
3. *S*-Hydroprene (the S enantiomer of hydroprene)—for cockroaches in food-handling establishments.
4. *S*-Kinoprene (the S enantiomer of kinoprene)—for whiteflies, gnats, aphids, mealybugs, and scales on ornamentals.

Another IGR registered as a biochemical pesticide that does not have JHA activity is 1*H*-Purine-2,6-diol. Marketed as Ecologix®, this product is termed a *nutritional metabolism disrupter*. It prevents the formation of uric acid in cockroaches, which interferes with development and causes sterility.

Not mentioned in this list is azadirachtin. This active ingredient of neem oil interferes with insect molting, thus killing insects. However, unlike many of the CSIs, it does not have many adverse effects on nontarget organisms and is therefore registered as a biochemical pesticide. It is not always listed as an IGR because of multiple modes of action; namely, it has antifeedant properties and also is said to disrupt mating and egg laying. Azadirachtin was discussed in detail as a botanical insecticide in Chapter 11.

### Attractants and Repellents

This is a very important grouping of the biochemical pesticides, including materials that attract pests to sites where they are killed, or repel them from a source. The category excludes insect pheromones and encompasses all other forms of semiochemicals (chemicals involved in communication between organisms). The category includes over eight active ingredients and over twenty-five products considered insect attractants and more than six active ingredients and over twenty-five products serving as repellents. Many of these materials are processed from plants.

Several plant oils are used to repel and/or kill certain insects. Plant oils used as pesticides are complex mixtures of compounds. Some of those registered as biochemical pesticides include the following:

1. Cedarwood—repels clothes moths.
2. Citronella—repels mosquitoes, flies, and fleas.
3. Eucalyptus—repels mites, fleas, mosquitoes, and other insects.
4. Jojoba—repels and kills whiteflies on many crops.
5. Lavandin—repels clothes moths.
6. Linalool—repels mosquitoes, fleas, mites, ticks, and spiders.

Other biochemical pesticides, which are isolated from flowers and other plant parts, have strong and distinctive odors. Frequently, these serve to attract certain insects to sites where they are killed. Some of these attractants include the following:

1. Cinnamaldehyde—attracts adult corn rootworms in food crops and ornamentals.
2. Citronellol—attracts mites in food crops and ornamentals.
3. Eugenol and methyl eugenol—attracts Japanese beetles in food crops and ornamentals.
4. Geraniol—attracts Japanese beetles in fruits, vegetables, and ornamentals.

5. Indole—attracts adult corn rootworms in fruits, vegetables, and corn for feed and food.
6. Ionone, alpha—attracts adult rose chafers (beetles) in plants.
7. 1-Octen-3-ol (from clover, alfalfa, other plants)—attracts mosquitoes and dragonflies to electronic insect killers.
8. 2-Phenylethyl-propionate (from peanuts)—attracts Japanese beetles and kills certain insects, mites, and ticks in food and feed crops, ornamentals, and various indoor sites.
9. 1,2,4-Trimethoxy-benzene (squash flowers)—attracts corn rootworms and cucumber beetles in fruit, vegetable, and feed crops.

For other attractants and repellents not registered as biochemical pesticides, see Chapter 14.

## Suffocating Agents

Certain materials registered as biochemical pesticides have unique modes of action in killing insects. Most often these are types of oils that affect various stages of soft-bodied insects, for example, aphids, mites, thrips, whiteflies, mealybugs, and psyllids. The primary mode of action for many of these oils is to block the respiratory system, thus causing suffocation. Additionally, they can break down the insect or mite cuticle and may penetrate body tissues, causing them to degrade. Oils derived from all sources may also alter the behavior of insects and mites and cause them to avoid laying eggs or disrupt their feeding. Oils are also widely used to control the egg stage of various mites and insects by preventing the normal exchange of gases through the egg surface or interfering with egg development.

One type of suffocating oil registered by the EPA as a biopesticide is mineral oil. This mineral oil is a highly refined petroleum product marketed as Summit Horticultural Spray Oil®, which can be used year-round against many light-bodied insects on fruit trees, shade trees, and vegetables. Other sprayable oils and details on their use were given in Chapter 11.

Soybean oil is another type of suffocating oil registered as a biochemical pesticide. This oil is extracted from soybean seeds. It is used against light-bodied insects on fruit trees, nut trees, evergreens, and woody shrubs. In addition, it is effective in killing gypsy moth eggs, found as masses, in shade tree and forest environments. A soybean oil product registered for these purposes is Golden Pest Spray Oil®.

## Desiccants

Insecticides that have a desiccating effect on insects fall into this category. A number of soft-bodied pests, including mites, aphids, certain caterpillars, and glassy-winged sharpshooters on food and nonfood crops, can be killed by these materials. Desiccants act by disrupting the waxy outer layer of the cuticle and causing water loss. This eventually results in death of the pest.

Two desiccants labeled as biopesticides are sucrose octanoate esters and sorbitol octanoate. Sucrose octanoate esters occur naturally in plants, and both materials decompose to become harmless substances in the environment. Some products with these active ingredients are Avachem® sorbitol octanoate and Avachem® sucrose octanoate.

### Coatings

Agents in this category act as pesticides by forming a nontoxic physical barrier between an insect pest and a leaf surface. Two such biopesticides acting in this way are kaolin and jojoba oil.

The active ingredient in kaolin or kaolin clay is an edible mineral used as an anticaking agent in processed food and has little effect on the environment. Kaolin is a natural clay that does not react with other materials and is insoluble in water. Most products (for example, Surround WP®) are sprayed on plants as a liquid, which evaporates, leaving a protective powdery film on plant parts. Tiny particles of film agitate and repel insects, leaving coated surfaces unsuitable for feeding and egg laying. Kaolin is labeled for use on a wide range of fruit and vegetable crops, including beans, beets, potatoes, eggplant, citrus fruits, apples, apricots, and berries.

Jojoba oil is a vegetable oil obtained from the jojoba bean, *Simmondsia chinensis* (Simmondsiaceae). When applied to crops, jojoba oil products (for example, Detur®) can control whiteflies. The oil has been used for decades in cosmetics, with no reported adverse effects. Although jojoba oil may affect insects directly, it also forms a physical barrier between the leaf surface and an insect pest, thus offering a degree of plant protection.

### Pheromones

Recall from Chapter 4 that volatile chemical attractants are often involved in insect mate finding. Such volatiles are called pheromones, and these are common in many insects, particularly the Lepidoptera and Coleoptera. Furthermore, male-attracting sex pheromones are those most often found.

Synthetic versions of natural pheromones have been manufactured and can be used effectively in insect pest management programs. These synthetics are being used to: (1) sample insect populations for surveillance; (2) confuse males, thus reducing fertility; and (3) attract and kill males, also reducing fertility. Pheromones labeled as biopesticides relate to points 2 and 3, aiming to disrupt pest reproduction and cause population density to fall.

The greatest number of pheromone biopesticides is directed at moth pests. There are about 25 active ingredients and over 200 products available for management of these insects. Some of the most common uses are among the following:

1. Codling moth (Box 12.2) on fruit, nuts, and ornamental trees and shrubs.
2. Omnivorous leaf roller on fruit, grapes, kiwi, and nuts.
3. Oriental fruit moth on fruit and nuts.
4. Peach twig borer on fruit, nuts, and agricultural crops.
5. Pink bollworm on cotton.

These are in addition to over fifteen other moth species or species complexes in which pheromone products are used as a tactic in the management program. Other than moth pests, pheromone products have been developed for the Japanese beetle (nuranone), house fly (tricosene), and California red scale (two active ingredients). More specific information on sex pheromone development and use in insect pest management is given in Chapter 14.

### Systemic Acquired-Response Inducers

This class of agents has resulted from research with induced natural plant resistance and the discovery of proteins that cause plants to tolerate certain degrees

## BOX 12.2 Codling Moth

**SPECIES:** *Cydia pomonella* (Linnaeus) (Lepidoptera: Olethreutidae)

**DISTRIBUTION:** The codling moth is found throughout the world except in Japan and western Australia. This pest was introduced into North America from southeastern Europe and now occurs wherever apples are grown. The species is also found on pear, English walnut, and quince.

**IMPORTANCE:** The codling moth is one of the most destructive apple pests in the world. Damage is caused by larvae that tunnel into the fruit and feed to the core, sometimes on the seeds. Larvae also initially chew into insecticide-treated apples, leaving blemish marks called "stings." Feeding injury also provides infection courts for diseases that further damage the fruit. Injury caused by larvae lowers the quality of the fruit and therefore lowers commercial value. Because of tunneling, larvae are very difficult to suppress. As a result, infestations affect 20 to 95 percent of the apples in a typical orchard.

**APPEARANCE:** Adult moths have a wingspan of about 18 mm. The wings are grayish brown and crossed with bands of light gray; they display a dark brown spot at the tip of each forewing. Larvae, which are from 12 to 20 mm long when full grown, are pinkish white with a brown head. Pupae are approximately 10 mm long and vary from yellow to brown. The minute white eggs are flattened and shaped like pancakes.

**LIFE CYCLE:** Full-grown larvae overwinter in silken cocoons concealed on or under the bark of apple trees. Pupation occurs in early spring, and adult moths emerge when apple trees are in bloom. The adults, living just 14 to 21 days, mate, and females oviposit from thirty to forty eggs singly on leaves, twigs, and developing fruits. After 5 to 15 days, the eggs hatch, and the larvae bore into fruits. Larvae feed for 3 to 5 weeks, then burrow out of the apple and spin silken cocoons under loose bark on the tree or under plant debris on the ground. In warm climates, adult moths emerge from the cocoon in 14 to 30 days, and the life cycle is repeated. The number of generations per year ranges from one to three, depending on climatic conditions.

---

of pest injury. These proteins are produced in nature by certain bacterial plant pathogens. They act by eliciting a complex natural defense mechanism in plants, analogous to a broad-spectrum immune response in animals. One such protein discovered is the Harpin protein.

In nature, Harpin is produced by *Erwinia amylovora,* a bacterium that causes the disease fire blight in apples and pears. A weakened strain of *Escherichia coli* was modified to produce Harpin on a commercial scale. Commercially produced Harpin protein is identical to the protein that occurs in nature, *Escherichia coli* K-12 is a nonpathogenic, nutritionally deficient bacterium that is unable to grow in the environment. Harpin is concentrated from the growth medium of the genetically modified *E. coli*, and the bacterial cells are killed and removed to produce the marketed product.

Harpin does not act directly on the insect pest, nor does it alter the DNA of treated plants. Instead, it activates a natural defense mechanism in the host plant, referred to as systemic acquired resistance (SAR). This active ingredient is currently the only broad-spectrum, proteinaceous elicitor of SAR commercially available. Harpin is effective against certain viral diseases for which there are no other controls or resistant plant varieties. It also protects against soilborne pathogens and pests, such as certain nematodes and fungal diseases, which have few effective controls except for methyl bromide, a discontinued pesticide.

In addition to its ability to protect plants against diseases, Harpin protein reduces infestations of selected insects and enhances plant growth, general vigor, and yield of many crops, including vegetables, traditional agronomic crops, and ornamentals. Harpin protein is marketed as Messenger® and its counterpart Harpin alpha beta protein (fragments of the Harpin protein) as Proact® (and others).

## PLANT-INCORPORATED PROTECTANTS

**Plant-incorporated protectants (PIPs)** are pesticidal substances that plants produce from genetic material added to the plant through biotechnology. For insect management, the biotechnology is used to insert a new segment of DNA (a gene) into a plant, thus causing the plant to produce a toxin, which kills certain insects feeding on it. The insect-resistant plants transformed with biotechnology are referred to as **transgenic plants.** In all instances, these transgenic plants have the gene responsible for producing a toxin from the insect pathogen *Bacillus thuringiensis,* or Bt. The toxin, sometimes called a protoxin, is the crystalline delta endotoxin of Bt, and it is the cry gene of the Bt that encodes this toxin in the plant.

There are thirteen insecticidal proteins of Bt strains or subspecies that kill insects and whose genetic material is inserted into plant genomes to produce over twenty-five products. Therefore, a given transgenic plant will kill only certain insect pests. Transgenic plants with a wider spectrum are created by **gene stacking,** producing plants resistant to widely different pests (for example, Lepidoptera and Coleoptera), or **gene pyramiding,** producing plants resistant to several closely related pests (for example, several Lepidoptera species). Each protein in stacking or pyramiding must be registered with the EPA.

The crystalline proteins registered for insertion in plants include the following:

1. Bt Cry1F—European corn borer, lesser corn stalk borer, southwestern corn borer, black cutworm, western bean cutworm, fall armyworm, corn earworm (partial control) on corn.
2. Bt Cry2Ab—cotton bollworm (same as corn earworm) on cotton.
3. Bt Cry34Ab1 and Cry35Ab1—corn rootworms (*Diabrotica* spp.) on corn.
4. Bt Cry3Bb—corn rootworms (*Diabrotica* spp.) on corn.
5. Bt Cry3Bb1—corn rootworms (*Diabrotica* spp.) on corn.
6. Bt Cry1Ab—corn rootworms (*Diabrotica* spp.) on corn.
7. Bt Cry3A—Colorado potato beetle on potatoes.
8. Bt KCry1Ab—European corn borer, corn earworm, fall armyworm on sweet and field corn.
9. Bt mCry 1F—black cutworm, corn earworm, western bean cutworm, fall armyworm, southwestern corn borer, European corn borer on corn.
10. Bt subsp. *kurstaki* (B.t.k.) Cry1Ac—cotton bollworm (same as corn earworm) on cotton.
11. Bt subsp. *kurstaki* strain HD1 (from Cry1Ab)—corn rootworms (*Diabrotica* spp.) on corn.
12. Bt var. *kurstaki* Cry1Ac (synpro)—synthetic protoxin—tobacco budworm, beet armyworm, cotton bollworm, soybean looper, pink bollworm on cotton.
13. Bt var. *aizawai* Cry1F (synpro)—synthetic protoxin—tobacco budworm, beet armyworm, cotton bollworm, soybean looper (Box 12.3) on cotton.

### BOX 12.3 SOYBEAN LOOPER (SEE COLOR PLATE 5)

**SPECIES:** *Pseudoplusia includens* (Walker) (Lepidoptera: Noctuidae)

**DISTRIBUTION:** The soybean looper migrates into the southeastern United States each year from Mexico, Central America, and Caribbean islands.

**IMPORTANCE:** The species is a common and destructive soybean pest in the Gulf coast area of the United States. Larvae defoliate soybeans and can cause serious yield losses when outbreaks occur. Usually, populations do not reach high levels until August and September. Damage can be especially serious in areas adjacent to cotton fields. Larvae also attack numerous vegetable crops.

**APPEARANCE:** Adults are brown with a wingspan of about 38 mm. The forewings display a distinctive silver marking. The hind wings are brown. Larvae are green with white stripes running lengthwise along the body, which is about 32 mm long when full-grown. They have black front legs and two pairs of abdominal prolegs (excluding the anal pair). Eggs are greenish white. Pupae are enclosed in silken cocoons.

**LIFE CYCLE:** Migrating females oviposit up to 600 eggs singly on leaves. After about 3 days, eggs hatch and the larvae feed for approximately 18 days (seven molts). Pupation occurs on the underside of leaves, and in about 8 days, adults emerge. There can be up to four generations per year.

In some instances, more than one of these cry-gene proteins are combined with others through traditional plant breeding to produce a cultivar with a wider spectrum of activity. A detailed discussion of transgenic plant development and use is given in Chapter 13.

## USING BIOCHEMICAL PESTICIDES IN INSECT MANAGEMENT

When applied correctly, biopesticides can often provide effective suppression of insect pest populations. They have the great benefit of offering safety in handling and allowing applications right up to the time of crop harvest. Additionally, they have a narrow spectrum of activity and usually do not strongly affect beneficial invertebrate or vertebrate animals in the agroecosystem or outside it. Moreover, some biopesticides (microbials and certain biochemicals) can be applied similarly and with the same equipment used to apply conventional pesticides. Furthermore, many biopesticides (microbials, natural biochemicals, pheromones) are allowed under the National Organic Program, USDA standards for organic crop production.

However, some disadvantages are possible with their use. Many biopesticides are more expensive than conventional synthetic poisons, and because of their narrow spectrum, supplemental applications of other products may be required to deal with the entire insect-pest complex. Additionally, various biopesticides have short residual activity, requiring more frequent than usual applications, and their activity may be specific to a given insect stage, as with most IGRs.

The use of biopesticides in effective insect pest management programs requires similar inputs as those using conventional pesticides. However, accuracy and detail of information that serve as a basis of the program are more critical; that is, there may be less room for error in achieving satisfactory results than with conventional pesticides. Therefore, management programs should be thought out well ahead of time, and detailed planning, including fallback alternatives, is absolutely necessary. Early detection and accurate identification of pest species present, as well as their life stages and predicted phenology, are also vital in selecting appropriate materials and applying them in a timely manner. Also, for many biopesticides, coverage of the protected commodity is more crucial than with conventional pesticides, requiring higher mixture volumes and efficient spray practices. Summing up, the insect pest management strategy with biopesticides is often complex and information intensive, requiring greater need for monitoring and reevaluation than for many conventional programs.

## Further Reading

Isman, M. B. 2006. Botanical insecticides, deterrents, and repellents in modern agriculture and an increasingly regulated world. *Annual Review of Entomology* 51:45–66.

O'Callaghan, M., T. R. Glare, E. P. J. Burgess, and L. A. Malone. 2005. Effects of plants genetically modified for insect resistance on nontarget organisms. *Annual Review of Entomology* 50:271–292.

Tanada, Y. 1959. Microbial control of insect pests. *Annual Review of Entomology* 4:277–302.

U.S. Congress, Office of Technology Assessment. 1995. *Biologically Based Technologies for Pest Control*. OTA-ENV-636. Washington, D.C.: U.S. Government Printing Office.

Zehnder, G., G. M. Gurr, S. Kühne, M. R. Wade, S. D. Wratten, and E. Wyss. 2007. Arthropod pest management in organic crops. *Annual Review of Entomology* 52:57–80.

## Favorite Web Sites

http://cls.casa.colostate.edu/TransgenicCrops/current.html
Gives statistics on adaptation of transgenic crops and detail on Bt transgenic development.

http://edis.ifas.ufl.edu/IN081
Very good discussion of microbial insecticides, pest uses, and a listing of current products by pathogen category.

http://www.epa.gov/oppbppd1/biopesticides/product_lists/bppd_products_by_AI.pdf
Presents a complete list of EPA registered biopesticides according to active ingredients.

http://www.epa.gov/pesticides/biopesticides/ingredients/
EPA listing of biopesticides by active ingredients, with links to Fact Sheets (detailed information) on these, if available.

CHAPTER 13

# MANAGING INSECTS WITH RESISTANT PLANTS

ONE OF THE MOST PROMISING ways to reduce dependence on pesticides in agriculture is to plant insect-resistant crops. Planting resistant cultivars when available is one of our most effective, economical, and environmentally safe management tactics.

The concept of using host resistance to our advantage comes from the knowledge that most plants and animals are resistant to most potential insect attackers. Certain physiological, morphological, and/or behavioral characteristics inherited by organisms form a core of defense against species that would otherwise attack them.

These defenses are the result of natural selection. In nature, crosses that produce highly susceptible plants and animals are not repeated because resultant progeny do not survive to reproduce. However, although resistant to most attacks, even well-adapted survivors are susceptible to a few forms with the ability to overcome their defenses. Therefore, attackers sustain their populations at the expense, though not usually the complete elimination, of the host. These successful attackers on wild hosts are equivalent to what we would call pests on domesticated hosts.

Just as wild plants and animals have degrees of susceptibility to enemies, so do domestic species. From the gene pool of a crop species, certain crosses produce **phenotypes** (visible expressions of the crosses) that vary from complete susceptibility to high levels of resistance. Susceptibility is the underlying cause of a pest problem, and the degree of it in a species forms the basis of developing useful, pest-resistant types.

By definition, pest resistance is any inherited characteristic of a host that lessens the effects of attack. From an evolutionary standpoint, such characteristics are preadaptive traits that have allowed the organism to overcome the pressures of insects and disease organisms and have thereby increased the organism's chances of survival and reproduction. The technology of plant and animal breeding is aimed, in part, at discovering these preadaptive traits and using them to develop pest-resistant cultivars and breeds.

Breeding for pest resistance is only one of many objectives of total breeding programs. Indeed, usually yield and quality are the two most important aims, with pest resistance an important complementary objective. Breeding for pest resistance may focus on any type of pest, but particular emphasis is placed on pathogenic microorganisms, nematodes, and arthropods.

Breeding programs emphasizing resistance to insects and other arthropods are some of the most important in agriculture. Presently, breeding for insect resistance in plants is far more advanced than that for domesticated animals. Seemingly, this is because of the lower costs, shorter time, and greater opportunity for hybridization involved in plant breeding. Because of its significance in plant pest management and lesser importance in animal production, only plant resistance is discussed here.

## BRIEF HISTORY

Some of the earliest observations of plant resistance to insects were recorded in the late eighteenth and early nineteenth centuries. As early as 1792, "Underhill" variety wheat resistant to the introduced Hessian fly, *Mayetiola destructor,* was reported in the United States by J. N. Havens. This is generally considered the earliest documented report of an insect-resistant plant variety. Somewhat later, in 1831, "Winter Majetin" apples were reported resistant to the woolly apple aphid (Box 13.1), *Eriosoma lanigerum.*

The first dramatic example of the value of plant resistance against insects occurred in the late 1800s. An insect species, the grape phylloxera (Fig. 13.1), *Daktulosphaira vitifoliae,* was inadvertently introduced into French vineyards and spread across Europe. The effect was catastrophic, and by 1880, the pest threatened to wipe out the French wine industry. Earlier, it had been found that American grapes were resistant to the phylloxera, and this knowledge led to the grafting of susceptible European grapevine scions to resistant

### BOX 13.1 Woolly Apple Aphid

**SPECIES:** *Eriosoma lanigerum* (Hausmann) (Hemiptera: Aphididae)

**DISTRIBUTION:** The woolly apple aphid is distributed throughout the world and is found on apple, elm, mountain ash, pear, and hawthorn.

**IMPORTANCE:** Woolly apple aphids feed on sap from large root knots, underground portions of trunks, and wounds on trunks and branches. Primary injury, however, is caused by root feeding, which causes stunting of growth. Under severe infestations, trees may die. An efficient wasp parasitoid, *Aph-elinus mali* (Haldeman), has successfully suppressed populations of the woolly apple aphid in practically all the areas in which the wasp has been released, including the Pacific Northwest of the United States. Here, chemical controls are now generally unnecessary.

**APPEARANCE:** Adults and nymphs are red to purple and covered with bluish white, cottonlike wax filaments. Winged and wingless forms appear during the year.

**LIFE CYCLE:** Eggs overwinter on elm bark; however, a number of wingless individuals also overwinter on apple trees. In spring, eggs hatch, and wingless females begin parthenogenetically reproducing at a rapid rate on elm trees. Winged individuals develop in early summer and disperse to apple and other plant hosts. Reproduction continues throughout the summer, and in fall winged individuals appear again, mate, and oviposit overwintering eggs.

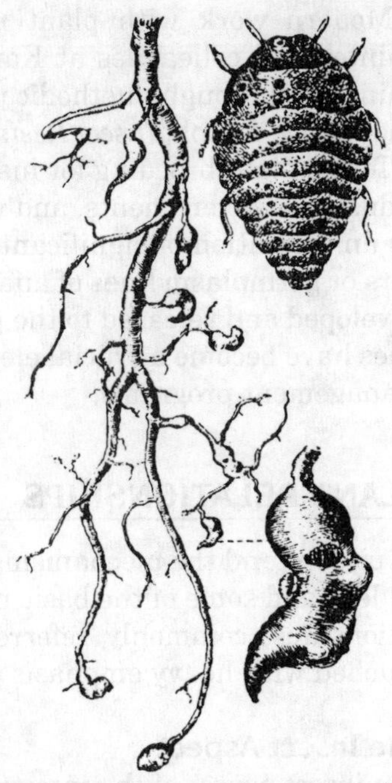

**Figure 13.1** Grape phylloxera, *Daktulosphaira vitifoliae,* female (*upper right*) and galls on terminal roots of grape caused by the pest. Gall at lower right is enlarged to show size relative to the insect. (Courtesy USDA)

American rootstocks. This produced vines resistant to grape phylloxera and yet allowed quality European wine grapes to be grown. Satisfactory insect control occurred by 1890, thereby saving the French wine industry. An American entomologist, C. V. Riley, was awarded a gold medal by the French government for making this recommendation.

Resistant grapevines have been the major means of preventing grape phylloxera injury for 100 years. However, grape phylloxera problems began to reoccur in Sonoma and Napa County, California, in the 1980s. It seems that a new biotype of the phylloxera (an asexual, totally root-inhabiting mutant) developed there that could overcome plant resistance. The previously resistant root stocks are being replaced by new resistant varieties, which growers hope will be effective for the next 100 years. An important lesson in this story is that insect pests have the ability to overcome nearly any single management tactic used against them.

Although the grape phylloxera example is often cited to illustrate the value of plant resistance, there is even more to the story. Ironically, plants introduced into France from North America carried the pathogenic fungus *Plasmopara viticola* (the causal agent of downy mildew), to which European grapes were highly susceptible. The American root stock saved French vineyards from grape phylloxera but exposed them to an even more dangerous risk. The devastating epidemic of downy mildew that followed threatened not only French wine production but also wine production throughout Europe. Ultimately the development of Bordeaux mixture (lime plus copper sulfate), an early fungicide active against the pathogen, saved European vineyards. Thus, the grape phylloxera/downy mildew episode offers yet another important lesson—the need to consider the whole pest complex and the implications of any management strategy.

Modern work with plant resistance to insects was pioneered by R. H. Painter and colleagues at Kansas State University, beginning in the 1920s. Painter's thorough, methodical studies paved the way for later achievements, and his 1951 book, *Insect Resistance in Crop Plants,* was the first on this topic.

Today, plant breeding for insect resistance is conducted by universities, seed industries, governments, and numerous national and international institutes. As an indication of significant progress, more than 400 insect-resistant cultivars or germplasm lines of alfalfa, cotton, corn, sorghum, and wheat have been developed and released to the public in past decades. Moreover, resistant varieties have become a crucial element in the success of many ongoing insect pest management programs.

## INSECT AND HOST-PLANT RELATIONSHIPS

To understand the mechanisms of plant resistance to insects, it is important to understand some of the basic relationships between these organisms. Such relationships, commonly referred to as insect/plant interactions, are usually couched with heavy emphasis on insects and plant response to attack.

### The Insect Aspect

The insect aspect of the insect/plant interaction is often described as a series of steps, in time and space, that lead to suitability of a plant for the insect. The major steps usually recognized by authorities include: (1) finding the general habitat; (2) finding the host plant; (3) accepting the plant as a proper host; and (4) sufficiency of the plant for survival and successful reproduction of the insect population.

**Finding the general habitat.** Insects locate the general area of the host by means usually unrelated to the plant. Physical stimuli such as light, wind, gravity, and perhaps temperature and humidity may help orient dispersing insects to the overall location of the host (for example, see Fig. 13.2). This step is most important when a species does not reside in an area year-round, as with many migrating forms.

**Finding the host plant.** Once in the general area the insect next must find a proper host. Most insects rely on vision and/or smell to locate a host plant. Remote factors in locating the plant include color, size, and shape. Much of the information on color in host finding is limited mainly to aphids (Hemiptera: Aphididae) and whiteflies (Hemiptera: Aleyrodidae), which are attracted to yellow-green surfaces. Color usually cannot be used in plant resistance because changing it affects fundamental physiological processes. In a few instances, however, red cultivars of cotton, cabbage, and oats have been shown less attractive to insects, and yet they have retained good agronomic characteristics. In addition to color, some insects, like fruit flies, *Ragoletis* species, are known to associate shape and size of trees in locating hosts.

Once insects are in contact with the plant, short-range stimuli arrest further movement. These stimuli are both physical, exciting tactile receptors, and chemical, exciting chemoreceptors on tarsi, antennae, and mouthparts.

**Accepting the plant as a proper host.** Subsequent to host finding, insects may take test bites, as do some caterpillars, to confirm host recognition. Continuous feeding is seemingly governed by the stimulation from various chemicals. In a monophagous insect, the silkworm (*Bombyx mori*), a series of substances

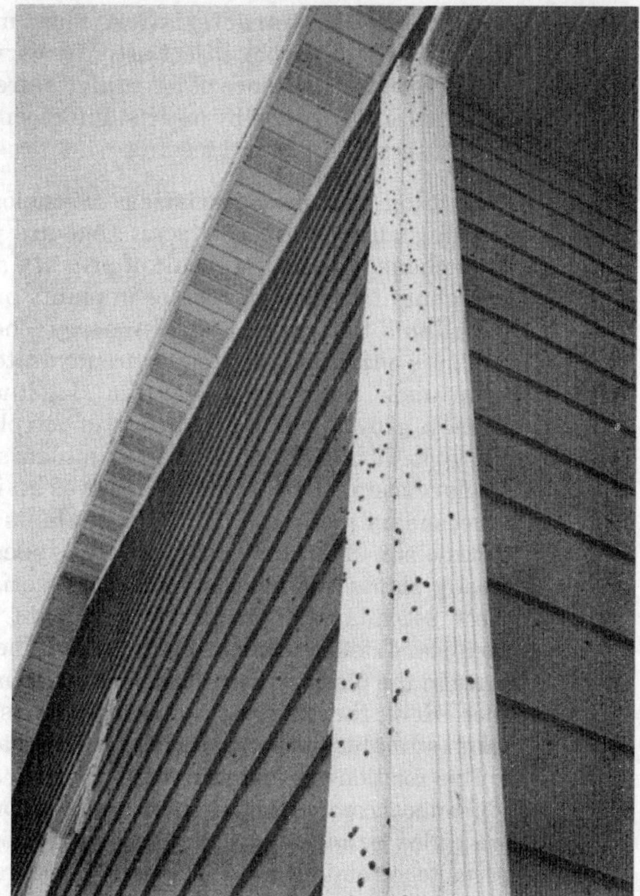

**Figure 13.2** Multicolored Asian lady beetles, *Harmonia axyridis,* are attracted to white vertical surfaces. Many insects use cues such as color and shape to locate general habitats, suitable hosts, or overwintering sites. (Photo by Marlin E. Rice)

are perceived in mulberry leaves that seem to mediate biting, swallowing, and continued feeding. Feeding to satiation then follows in the presence of appropriate chemicals.

Major physical factors involved in acceptance of a host may include leaf and stem toughness, leaf surface waxes, and pubescence (density and types of hairs). These factors may be important in relation to feeding and/or oviposition.

**Sufficiency of the plant for requisites.** Sufficiency of the plant as a host is finally determined during feeding. If nutrients are adequate and no toxicity occurs, the insect completes development within a normal time period and becomes an adult. Also, sufficiency is indicated in normal adult longevity and fecundity (the production of male and female gametes).

### The Plant Aspect

As the supplier of physical and chemical stimuli, the plant itself becomes an important participant in the insect and host-plant relationship. Both morphological and physiological characteristics of a plant elicit given insect responses.

**Morphological characteristics.** Plant morphological features may produce physical stimuli or bar insect activity. Variations in foliage size, shape, color, and presence or absence of glandular secretions may determine degree of acceptance or utilization by insects. Pubescence and tissue toughness sometimes limit insect mobility and feeding.

**Physiological characteristics.** Physiological characteristics influencing insects usually involve chemicals that are the products of plant metabolism. Such chemicals are the result of primary and secondary metabolic processes.

Primary metabolic processes in plants produce substances to catalyze reactions, build tissues, and supply energy. The plant requires inorganic ions and produces enzymes, hormones, carbohydrates, lipids, proteins, and phosphorus compounds for energy transfer. Together, these **primary metabolites** promote growth and reproduction of the plant. For insects, some of these primary metabolites are feeding stimulants, nutrients, and toxicants. Other primary metabolites are inert as far as an insect is concerned.

Secondary metabolic processes in plants seem to be coincidental to primary metabolism. The chemicals produced, **secondary metabolites,** vary widely among plants and are believed nonessential in primary metabolism. Some of these secondary metabolites are thought to have arisen as mechanisms for chemical defense against plant eating. They may be stored in any convenient place in the plant structure and often are exuded from outer layers of plant tissues. Here, they may be sensed by insects and function as **token stimuli.** A token stimulus elicits a response initially and afterward has no effect.

This relationship between plant chemical stimuli and insect response is a form of chemical communication between these organisms (Fig. 13.3). Such chemicals are called **semiochemicals.** Among semiochemicals are **pheromones,** which promote communication between members of the same species, and **allelochemics,** which promote communication between members of different species. Metabolites in plants that stimulate responses are kinds of allelochemics.

Allelochemics can be subdivided further into allomones and kairomones. **Allomones** are mostly defensive chemicals, producing negative responses in insects and reducing chances of contact and utilization. They include repellents, oviposition and feeding deterrents, and toxicants. Conversely, **kairomones** are advantageous to an insect, promoting host finding, oviposition, and feeding. They include attractants, arrestants, excitants, and stimulants.

## Host-Plant Selection

Host-plant selection by insects usually involves both primary and secondary metabolites. Some theorists emphasize secondary metabolites in this process; however, many contend that both play an important role. According to M. Kogan (1994), host-plant odor or taste for insects comes from nutrients and odd compounds that are intertwined as complex sensorial inputs. These inputs are interpreted by the insect's central nervous system to determine whether a given plant is a host. The genetically programmed "correct" signal supports a lasting host association.

In breeding programs for plant resistance, it is important to understand the nature of insect/plant relationships and their expression in host selection. Such understanding gives insights as to the underlying causes of the susceptibility/immunity gradient and enhances efficient development of resistant cultivars.

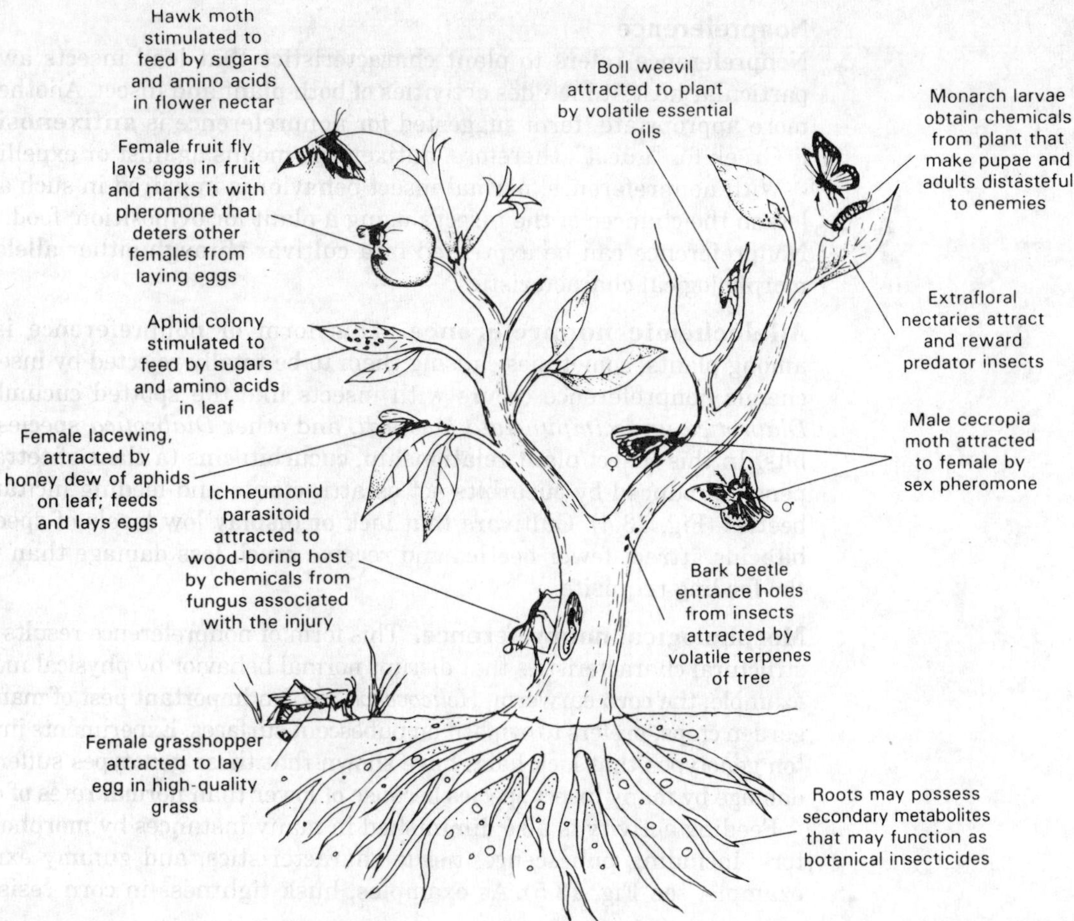
Hawk moth stimulated to feed by sugars and amino acids in flower nectar
Boll weevil attracted to plant by volatile essential oils
Monarch larvae obtain chemicals from plant that make pupae and adults distasteful to enemies
Female fruit fly lays eggs in fruit and marks it with pheromone that deters other females from laying eggs
Extrafloral nectaries attract and reward predator insects
Aphid colony stimulated to feed by sugars and amino acids in leaf
Male cecropia moth attracted to female by sex pheromone
Female lacewing, attracted by honey dew of aphids on leaf; feeds and lays eggs
Ichneumonid parasitoid attracted to wood-boring host by chemicals from fungus associated with the injury
Bark beetle entrance holes from insects attracted by volatile terpenes of tree
Female grasshopper attracted to lay egg in high-quality grass
Roots may possess secondary metabolites that may function as botanical insecticides

**Figure 13.3** Schematic drawing of the roles of semiochemicals associated with insects. (Redrawn from Hagen, Dadd, and Reese, 1984, in Huffaker and Rabb, eds., *Ecological Entomology,* John Wiley and Sons)

## MECHANISMS OF RESISTANCE

Resistant cultivars function in many different ways to reduce the effects of insect attack. Various steps are required in host selection by insects. For normal insect growth and development to occur, certain requisites, available in proper amounts and at specific times, are necessary. Resistant cultivars, by one means or another, do not supply these requisites and thereby interrupt the normal host-selection process. In some instances, the mechanism of resistance involves new allomones or increased levels of existing ones; in others, it may be based on reduced levels of kairomones. Also, physical factors may be involved.

Most authorities consider true plant resistance as being primarily under genetic control. In other words, the mechanisms of resistance are derived from preadapted inherited characters. Therefore, the expression of these characters always occurs, although they can be mediated by environmental conditions.

The most widely accepted classification of genetic resistance modes in plants is that of R. H. Painter. These modes or mechanisms include nonpreference, antibiosis, and tolerance.

### Nonpreference

Nonpreference refers to plant characteristics that lead insects away from a particular host; it includes activities of both plant and insect. Another, perhaps more appropriate, term suggested for nonpreference is **antixenosis.** *Xenosis* is Greek for "guest"; therefore, antixenosis means against or expelling guests.

With nonpreference, normal insect behavior is impaired in such a way as to lessen the chances of the insect's using a plant for oviposition, food, or shelter. Nonpreference can be expressed in a cultivar through either allelochemic or morphological characteristics.

**Allelochemic nonpreference.** This form of nonpreference is common among plants, sometimes causing them to be totally rejected by insects. Allelochemic nonpreference occurs with insects like the spotted cucumber beetle, *Diabrotica undecimpunctata howardi,* and other *Diabrotica* species on cucurbits. In this insect/plant relationship, cucurbitacins (a class of tetracyclic terpenes) produced by cucurbits act as attractants and feeding incitants for the beetles (Fig. 13.4). Cultivars that lack or display low levels of specific cucurbitacins attract fewer beetles and receive much less damage than those with the feeding requisite.

**Morphological nonpreference.** This form of nonpreference results from plant structural characteristics that disrupt normal behavior by physical means. As an example, the corn earworm, *Helicoverpa zea,* an important pest of many field and garden crops, prefers to oviposit on pubescent surfaces. Experiments involving cotton genotypes that lack hairs have shown that these genotypes suffer much less damage by many insect species because of lower than normal rates of oviposition.

Feeding activity is also diminished in many instances by morphological factors, including pubescence, tissue characteristics, and gummy exudates (for example, see Fig. 13.5). As examples, husk tightness in corn resists damage

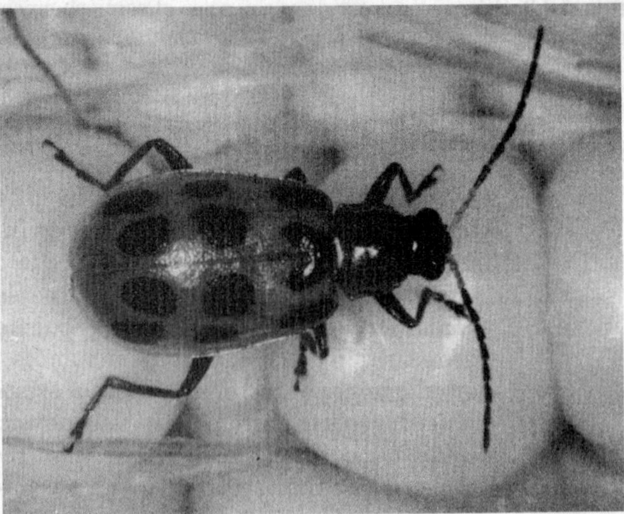

**Figure 13.4** Spotted cucumber beetle, also referred to as the southern corn rootworm, *Diabrotica undecimpunctata howardi* (see also Color Plate 5). Beetles are attracted to cucurbits and incited to feed by cucurbitacins. (Photo Marlin E. Rice)

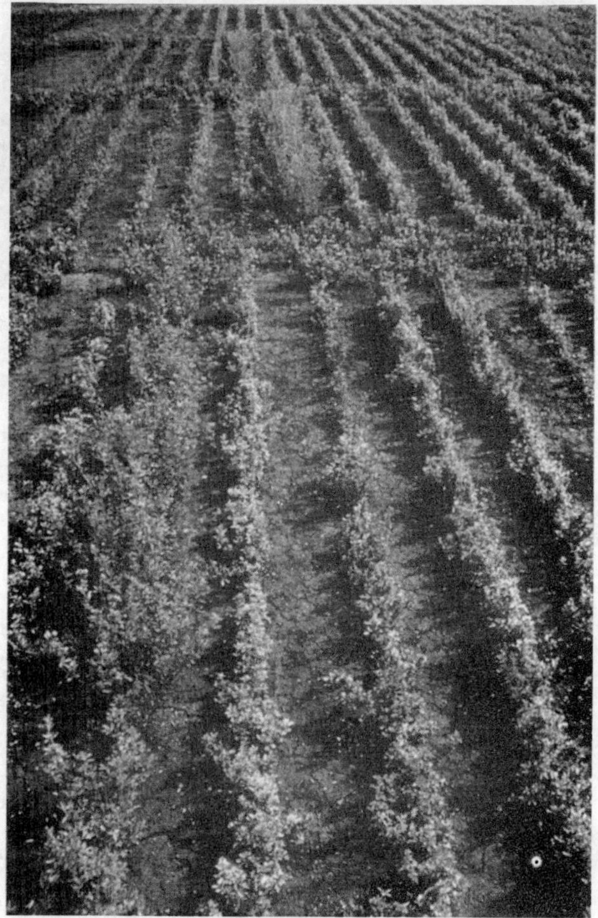

**Figure 13.5** Alfalfa susceptible (short rows) and resistant (tall rows) to potato leafhopper, *Empoasca fabae.* (Photo by Marlin E. Rice)

from corn earworms, stem density of pith and node tissues in wheat resists damage by the wheat stem sawfly (*Cephus cinctus*), and hard, woody stems with closely packed vascular bundles in cucurbits resist feeding of the squash vine borer, *Melittia cucurbitae* (Fig. 13.6).

**Use of nonpreference.** In a practical sense, the use of some nonpreference characteristics may be limited by given cultural environments. Many cultivars may show nonpreference if alternate hosts are in the vicinity, but in the absence of alternate hosts, the nonpreference may break down. The insect species does not particularly prefer the cultivar, but if nothing else is available, it will accept it. Because of the widespread practice of monocropping and the breakdown of resistance, allelochemic nonpreference is not a primary goal in plant breeding programs.

On the other hand, forms of morphological nonpreference that impair feeding behavior are very important and may be a first line of defense against many pests. This is partly because morphological nonpreference provides long-lasting effectiveness, compared with most chemically based resistance; that is, insect populations have a difficult time overcoming this form of resistance.

**Figure 13.6** Squash vine borer, *Melittia cucurbitae*, adult. Hard, woody stems of cucurbits resist feeding by larvae of this insect. (Photo by Marlin E. Rice)

### Antibiosis

By far, **antibiosis** is the most widely sought after objective of plant breeders. This mechanism usually impairs an insect's metabolic processes and often involves consumption of plant metabolites. As with nonpreference, both insect and plant factors are involved in the antibiosis mechanism.

Allelochemics frequently are associated with antibiosis. Some of the best-documented allelochemics include the cyclic hydroxamic acids in corn (DIMBOA, 2,4-dihydroxy-7-methoxy-1,4-benzoxazin-3-one) gossypol and related compounds in cotton, steroidal glycosides in potato, and saponins in alfalfa.

Quantity and quality of primary metabolites may also be important in conferring antibiosis. Particularly significant in this regard are imbalances of sugars and amino acids that result in nutritional deficiencies for insects feeding on the plant. For example, pea cultivars with low amino acid levels and increased sugar content show resistance to the pea aphid, *Acyrthosiphon pisum*, and rice cultivars deficient in asparagine (an amino acid) cause reduced fecundity in the brown planthopper, *Nilaparvata lugens*.

Symptoms of insects affected by antibiosis include:

1. Death of young immatures.
2. Reduced growth rate.
3. Increased mortality in pupal stage.
4. Small adults with reduced fecundity.
5. Shortened adult life span.
6. Morphological malformations.
7. Restlessness and other abnormal behavior.

### Tolerance

Unlike nonpreference and antibiosis, only a plant response is involved in tolerance. The plant has the ability to give satisfactory yields in spite of injury levels that would debilitate nonresistant plants. This is the least dramatic resistance mechanism, and some plant scientists do not consider it a form of resistance.

Many factors are involved in plant tolerance, yet the overall mechanisms are poorly understood. Known components of this form of resistance include general vigor, compensatory growth in individual plants and/or the plant population, wound healing, mechanical support in tissues and organs, and changes in photosynthate partitioning.

An important advantage of tolerance is that it places no selective pressure on insect populations, as do nonpreference and antibiosis mechanisms. Without selection pressure, variants do not develop that can overcome the resistance. Its disadvantage is that insect populations may be allowed to sustain epidemics in an area, causing problems in other crops. Also, producers are wary of recommendations that allow large populations of seemingly injurious species to build up. Another, perhaps more important, disadvantage is that tolerance is more strongly affected by environmental extremes than are other forms of resistance.

An example of tolerance is found in several corn genotypes that have the ability to repair and replace roots fed upon by the western corn rootworm, *Diabrotica virgifera*. Such tolerance allows the plants adequate water and nutrient uptake and anchorage despite heavy feeding. Surprisingly, tolerant genotypes developed greater root volume with rootworm feeding than without.

In addition to corn, tolerant cultivars have been observed in many crops including alfalfa, barley, cassava, cotton, rice, sorghum, and wheat.

Tolerant plants also may be developed through biotechnology. Recently, plant geneticists have been able to speed up plant growth by inserting the gene of a fast-growing plant into another plant that causes the receiver's cells to divide more quickly at the tips of roots and shoots. This has been done experimentally with a gene from the flowering weed *Arabidopsis,* which was inserted into the tobacco plant. The focus of this work is to make crop plants grow faster and be more competitive against weeds, thus reducing herbicide use. However, fast growth could also lead to producing tolerance to insect leaf feeding and other forms of injury. Therefore, we might look forward to much more use of tolerance as a resistance mechanism than has occurred in the past.

### Ecological Resistance

Ecological resistance, sometimes called apparent resistance or pseudoresistance, usually is not considered true resistance. This is because expression of ecological resistance relies more heavily on environmental conditions than on genetics.

The characteristics of this resistance are temporary, and the cultivars involved are potentially susceptible. Ecological resistance is important in insect pest management, but its use must be carefully synchronized with prevailing environmental conditions for effectiveness. The three types of ecological resistance recognized by most authorities include host evasion, induced resistance, and host escape.

**Host evasion.** With host evasion, the plant passes through a susceptible stage quickly or at a time such that its exposure to potentially injurious insects is reduced. Often, host evasion is accomplished by planting early maturing varieties. A good example is the planting of fast-fruiting, short-season cotton varieties in Texas to provide a long, host-free period for populations of the boll weevil, *Anthonomus grandis,* and pink bollworm, *Pectinophora gossypiella.* The same varieties also give considerable evasion from populations of corn earworm and tobacco budworm, *Helicoverpa virescens.*

Sometimes evasion with early-maturing varieties is confused with true resistance. To test early varieties for true resistance, they can be planted later than usual and inspected for late-season injury.

**Induced resistance.** **Induced resistance** is a form of temporary resistance derived from plant condition or the environment. Factors like fertilization or changes in soil moisture levels may make plants more tolerant of insects than under other circumstances. For instance, nitrogen and potassium levels are known to affect aphid populations on plants. High nitrogen levels usually allow increases in survival, but the opposite may occur for high levels of potassium. Providing a proper balance of these nutrients in fertilizers has been suggested as a means of inducing resistance to aphids.

Recently, some attention has been given to the role of **phytoalexins** in inducing plant resistance to insects. Phytoalexins are phenolic compounds produced by plants when they become diseased or are attacked by insects. These compounds enable plants, once fed upon, to resist further damage by the pests. The mechanism involved results from an accumulation of allomones triggered by the injury or some other environmental factor. For example, phytoalexin production in laboratory soybeans has been induced by inoculation with a fungus, *Phytophthora megasperma* variety *sojae*. Subsequently, phytoalexin levels in cotyledons increased and functioned as feeding deterrents against larvae of the Mexican bean beetle, *Epilachna varivestis*.

The potential of induced resistance is shown in research with spider mites that infest grapevines in the San Joaquin Valley of California. Leaves are injured both by the Pacific spider mite, *Tetranychus pacificus,* and the Willamette mite, *Eotetranychus willametti*. The Pacific spider mite is by far the most injurious. Researchers found that when vines were infested with Willamette mites early in the season, Pacific mite densities were much lower. Experiments confirmed that "vaccinating" vines with the less injurious species early could effectively reduce effects of the more injurious species, at least in some situations. Research is continuing on this "inoculation" approach.

Continuing research with induced natural plant resistance has produced a novel chemical that has recently been registered as a biochemical pesticide by the EPA. As mentioned in Chapter 12, the chemical common name is harpin, and the product is being marketed by EDEN Bioscience under the trade name Messenger®. Harpin is a naturally occurring protein that binds to specific plant receptors. This binding creates a signal that switches on natural defense and growth systems in the plant, thus the name Messenger®. By activating these natural defenses, plants become protected against certain bacteria, viruses, and fungi and can repel, suppress, or tolerate some insects, mites, and nematodes. Messenger® has shown no toxicity to humans and no adverse effects on nontarget animals and plants tested. Messenger® is registered for use on field crops, trees, turf, and ornamentals. This chemical shows great potential as a desirable tool for pest management and suggests an exciting future in pesticide development.

**Host escape.** This category explains the lack of infestation of susceptible plants in a population of otherwise infested plants. The principle of host escape recognizes that the presence of an uninfested plant may not mean that it is resistant and emphasizes that escapes occur in most plant populations, even with heavy insect infestations. The reason for escapes is rarely understood.

## GENETIC NATURE OF RESISTANCE

Plant breeding activities aimed at developing resistance rely heavily on a knowledge of genetic background for the resistance. Such knowledge provides a quantitative basis for designs to recombine genes and select for proper characters. It also allows the identification of stable resistance factors that are least likely to be overcome by a pest population.

### Epidemiological Types of Resistance

This classification of resistant types has been advanced by plant pathologists to express effectiveness and stability against a population of pests. Effectiveness and stability of resistant varieties are determined by the plant genes that confer resistance and the insect genes that allow the resistance to be overcome.

**The gene-for-gene relationship.** Many pest populations include individuals with **virulent genes,** which allow a pest species to overcome resistance and once more attack a plant. One or more virulent genes may be present that allow an individual pest to overcome the effects of one or more plant genes responsible for resistance. This principle has been called the **gene-for-gene relationship.**

In the gene-for-gene relationship, plant cultivars are resistant because they have a resistant allele at a gene locus that corresponds to an avirulent (susceptible) allele at an equivalent locus in the insect. Even though the resistant cultivar is effective against most insects in the population, an occasional insect may have a virulent allele instead of the normally avirulent allele. For example, the resistance gene in the host plant may code for a protein toxic to the insect, and a corresponding virulent gene in the insect may code for an enzyme that detoxifies the toxic protein of the plant. This circumstance allows virulent individuals to attack the otherwise resistant plant, and over a period of time, the virulent genotype can replace the avirulent genotype. Eventually, the effectiveness of the resistant cultivar would decrease.

Different populations of an insect species that vary in their virulence to a cultivar are referred to as **biotypes.** Some species of insects like the Hessian fly (Box 13.2), *Mayetiola destructor,* are known to have several biotypes. The term *biotype* is frequently used for certain insect populations that overcome plant resistance.

To date, most insect biotypes have developed among aphid pests (Fig. 13.7). Prominent examples include the spotted alfalfa aphid, *Therioaphis maculata,* on alfalfa; greenbug, *Schizaphis graminum,* on wheat and sorghum; corn leaf aphid, *Rhopalosiphum maidis,* on sorghum and corn; and pea aphid, *Acyrthosiphon pisum,* on peas and alfalfa. Other hemipterans with biotypes include grape phylloxera on grapes and brown planthopper on rice.

However, one of the best understood and most famous examples of biotypes is with a dipteran, the Hessian fly. To date, at least sixteen biotypes of this pest have been discovered.

**Vertical and horizontal types of resistance.** Judging from the range of effectiveness of a resistant plant variety, J. E. van der Plank (1963) recognized two types of resistance, vertical and horizontal. In entomological terms, **vertical resistance** refers to cultivars with resistance limited to one or a few pest genotypes. **Horizontal resistance** describes cultivars that express resistance against a broad range of genotypes (Fig. 13.8).

## BOX 13.2 HESSIAN FLY

**SPECIES:** *Mayetiola destructor* (Say) (Diptera: Cecidomyiidae)

**DISTRIBUTION:** The Hessian fly was introduced into the United States from Europe, supposedly in straw bedding material used by Hessian soldiers during the Revolutionary War. The species is currently distributed throughout the wheat-producing areas of the world.

**IMPORTANCE:** The Hessian fly is one of the most destructive wheat pests. Larvae (maggots) feed between the leaf sheath and stalk, causing stunted and weakened plants. Injury results from salivary gland secretions from larvae that enter the plants, interfering with normal metabolism. The first generation in the spring stunts wheat and interferes with normal seed production, whereas the second generation, in the fall, weakens the stems, killing many plants before spring arrives. During outbreaks, twenty to thirty larvae may infest a single plant. In 1915, Hessian flies caused more than $100 million damage to wheat in the United States. Currently, with the advent of fly-free planting dates, resistant wheat varieties, and new cultural practices, the pest causes about $16 million in damage to wheat each year. The species also attacks barley and rye but to a lesser extent.

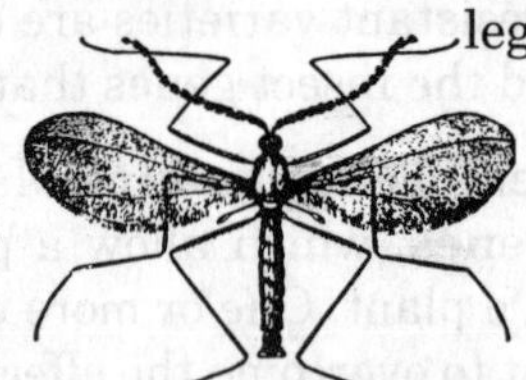

**APPEARANCE:** Adults are black, about 2.5 mm long, and resemble mosquitoes. Gravid females have red abdomens. Larvae are red when newly eclosed, but eventually turn white. A green stripe (visible gut contents) runs down the back of the legless larvae. Eggs are 0.5 mm long and red. Puparia are called "flaxseeds," which they resemble. They are dark brown and 3 to 5 mm long.

**LIFE CYCLE:** Larvae overwinter on plant debris, volunteer wheat, and winter wheat. In early spring, larvae pupate, and after 14 to 21 days, adults emerge. The time of adult emergence varies with latitude. Adults do not feed; they mate and live only 2 to 3 days. Females oviposit from 30 to 485 eggs in grooves of young wheat leaves. After 3 to 15 days, eggs hatch, and larvae feed at the leaf sheath base for 4 to 6 weeks. Larvae then become dormant in puparia during a portion of summer. In fall, adults emerge, and females oviposit. The eggs hatch and larvae feed for a short time, then overwinter as puparia (flaxseeds). There are typically two generations per year; however, some climatic conditions may be conducive to three or more generations.

---

Some authorities have argued against vertical resistance in breeding programs because of the potential development of biotypes. But it has been successful in many instances, as with the Hessian fly on wheat, and is easier to incorporate into new varieties than horizontal resistance. To manage insects by using vertical resistance, it has been suggested that resistant plant genes need to be identified and incorporated into varieties that are held, then released when biotypes appear.

Improvement in crops with horizontal resistance is a building process based on stepwise accumulation of genes with favorable additive effects. At present, the only known way to accumulate these favorable genes is by selective breeding over several generations, involving genetic recombinations (recurrent selection).

Horizontal resistance has low heritability and is difficult for plant breeders to incorporate. Nevertheless, some success with insects has been possible, for example, cultivars of corn with high resistance to both generations of the European corn borer, *Ostrinia nubilalis.* Furthermore, horizontal resistance

**Figure 13.7** Corn leaf aphids, *Rhopalosiphum maidis,* on a corn leaf. To date, most insect biotypes have developed among aphid species attacking crops. (Photo by Marlin E. Rice)

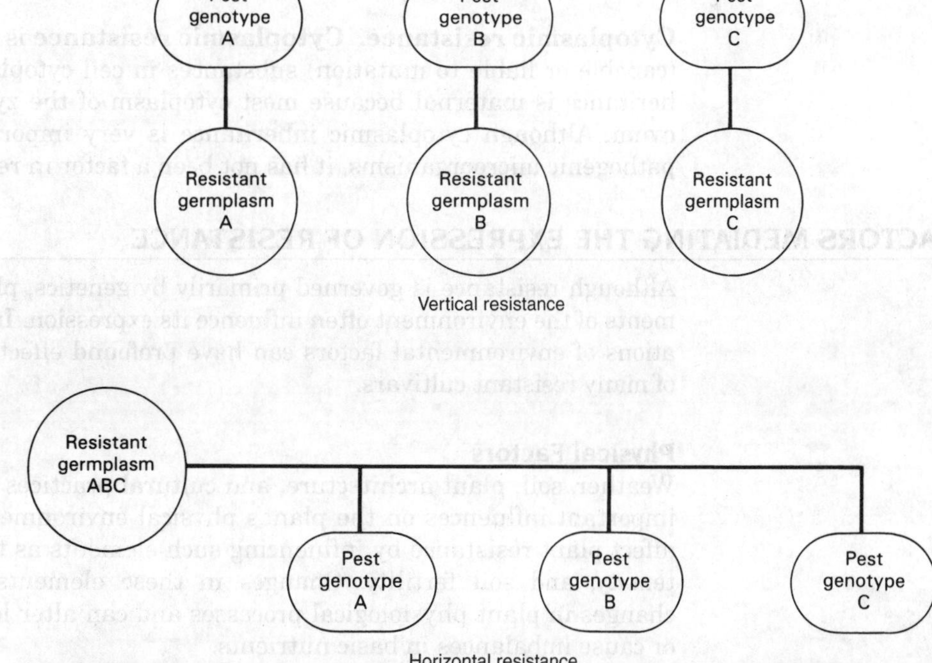


**Figure 13.8** Schematic representation of vertical and horizontal resistance.

may be the most desirable type of resistance to use in pest management because of its stability.

### Resistance Classes Based on Mode of Inheritance

Classes of plant resistance can also be distinguished according to the mode by which the resistance is inherited. In this regard, P. R. Day (1972) recognized three major resistance categories: oligogenic, polygenic, and cytoplasmic.

**Oligogenic resistance. Oligogenic resistance** is also called "major-gene resistance" and is conferred by one or only a few genes. This type usually produces vertical resistance against insects and may be inherited through dominant or recessive genes. Single-gene dominant resistance has been incorporated into varieties of such crops as apple, cotton, raspberry, rye, rice, and sweet clover. Single-gene recessive resistance can be found in corn lines resistant to the western corn rootworm and wheat resistant to the greenbug. Wheats resistant to the Hessian fly also should probably be considered oligogenic; in this instance, however, resistance is conferred by a series of dominant or partially dominant genes, as well as by several recessive genes. Resistance is often considered oligogenic in this instance because a well-understood gene-for-gene relationship exists between resistance genes in wheat and the corresponding virulence genes in the Hessian fly.

**Polygenic resistance. Polygenic resistance** is conferred by many genes, each contributing to the resistance effect. For this reason, it is also called "minor-gene resistance." Resistance inherited through the polygenic mode is usually very complex and may be associated with such quantitative traits as plant vigor and yield. Horizontal resistance is usually polygenic. An example of polygenic resistance, as already mentioned, occurs in corn varieties resistant to the European corn borer.

**Cytoplasmic resistance. Cytoplasmic resistance** is conferred by mutable (capable or liable to mutation) substances in cell cytoplasm. Cytoplasmic inheritance is maternal because most cytoplasm of the zygote comes from the ovum. Although cytoplasmic inheritance is very important in resistance to pathogenic microorganisms, it has not been a factor in resistance to insects.

## FACTORS MEDIATING THE EXPRESSION OF RESISTANCE

Although resistance is governed primarily by genetics, physical and biotic elements of the environment often influence its expression. Indeed, abnormal deviations of environmental factors can have profound effects on the performance of many resistant cultivars.

### Physical Factors

Weather, soil, plant architecture, and cultural practices are some of the most important influences on the plant's physical environment. These factors can affect plant resistance by influencing such elements as temperature, light intensity, and soil fertility. Changes in these elements cause fundamental changes in plant physiological processes and can alter levels of allelochemics or cause imbalances in basic nutrients.

**Temperature.** Abnormally high or low temperatures for a period of time may cause loss of resistance. As an example, exceptionally low temperatures have caused the loss of resistance of some alfalfa genotypes to spotted alfalfa aphid and pea aphid and of some sorghum genotypes to greenbug. Loss of resistance to the Hessian fly has been found in some wheats at temperatures above 18°C.

**Light intensity.** Shade-induced loss of resistance has been found in several instances. These include wheats resistant to the wheat stem sawfly, *Cephus cinctus;* sugar beets resistant to the green peach aphid, *Myzus persicae;* and potatoes resistant to the Colorado potato beetle, *Leptinotarsa decemlineata* (Fig. 13.9). With potatoes, shading was found to be associated with reduced levels of steroidal glycosides in leaves. These substances are known to retard feeding and development of the beetle.

**Soil fertility.** Changes in soil-nutrient levels may also mediate the expression of resistance in some plants, but little is known about the mechanisms involved. In an example of this phenomenon, clones of alfalfa resistant to the

**Figure 13.9** Colorado potato beetle, *Leptinotarsa decemlineata,* showing striped adult (*top*) and larvae (*bottom*). Shading has been found to be associated with reduced levels of steroidal glycosides in potato leaves. These compounds are known to retard development of the beetle. (Photos by Dennis Schotzko)

spotted alfalfa aphid were found to have reduced resistance if deficient levels of calcium or potassium or excess levels of magnesium or nitrogen were present. In the same example, resistance was increased by deficiency of phosphorus.

Another example is with the soybean aphid, *Aphis glycines,* on soybeans. It has been shown that these aphids produced 39 percent more nymphs on potassium-deficient plants than on healthy plants. A possible reason for this phenomenon is that potassium-deficient soybeans contain a greater amount of amino acids, which are required for optimal growth of the aphids. So, healthy plants are more resistant to the aphid than those without proper soil nutrients.

### Biological Factors

Just as physical factors can influence expressions of resistance, so can biological ones. The most significant biological factors are the selection of biotypes and changes in resistance with plant age.

**Biotypes.** Increases in the proportion of resistance-breaking biotypes already has been discussed. When resistant cultivars are grown widely, selection pressure is imposed by these hosts on the insect population. When capable, the insect population responds with genotypes having virulence to overcome the resistance. As time passes, the virulent genotypes with superior fitness increase in number, displacing the avirulent types. The result is a situation of growing ineffectiveness of the resistant cultivar. Presently, there are more than seventy documented insect biotypes.

Biotypes occur in all kinds of pests. In plant pathology biotypes are called "races," or in instances of plant-parasitic nematodes, "pathotypes."

G. E. Russell (1978) distinguishes among four main types of genotypic variants in relation to resistant cultivars. These are:

1. True resistance-breaking biotypes that can attack previously resistant varieties. There is a definite gene-for-gene relationship between the pest insect and the resistant plant. Most Hessian fly biotypes are examples of true resistance-breaking biotypes.
2. Unusually vigorous variants that have high reproductive potentials on all plant genotypes, with no definite gene-for-gene relationship. An example of this occurs with biotypes of the cabbage aphid, *Brevicoryne brassicae,* on rape.
3. Geographical variants that may or may not be true resistance-breaking biotypes. These have been observed among biotypes of the Hessian fly and seem to be produced by factors other than widespread planting of resistant varieties.
4. Mistaken identities believed to be biotypes that really are different species for which resistance was never developed. Examples here are often difficult-to-identify forms like some nematodes.

It is important to distinguish among these forms of "biotypes" in developing a proper counterresponse to the problem.

The time required for breakdown in resistance may be only a few years, as for example, brown planthopper resistance in rice. However, it has not been uncommon for a resistant cultivar to remain effective for 8 to 10 years. During this period, time is available to search for new resistant genes and incorporate them into new varieties, anticipating the next appearance of another biotype. As an example of excellent stability, some cotton cultivars in Africa with vertical leafhopper resistance have remained effective for more than 50 years.

Other than aphids, the development of biotypes has occurred infrequently in most insect groups. The frequent occurrence of aphid biotypes is partly the result of aphid reproduction, which has a parthenogenetic mode during parts of the seasonal cycle. Consequently, only one individual mutant, capable of feeding on a resistant variety, can produce a new biotype. Even including aphids, however, resistance-breaking insect biotypes have occurred much less frequently than have those among fungal and bacterial plant pathogens.

**Plant age.** Physiological responses in plants vary with age, and these can lead to changes in the expression of cultivar resistance. For instance, resistance in corn to the European corn borer results from the presence of the cyclic hydroxamic acid **DIMBOA**. DIMBOA levels are highest early in the growing season (midwhorl growth stage), thus offering maximum resistance to first-generation corn borers. DIMBOA levels decline as the season progresses (more rapidly in susceptible than in resistant cultivars), and many commercial cultivars have very little resistance to second-generation corn borers.

The expression of resistance corresponding to plant age has also been reported in alfalfa resistant to spotted alfalfa aphid and some tobacco resistant to green peach aphid.

## TRADITIONAL DEVELOPMENT OF INSECT-RESISTANT VARIETIES

Crop resistance programs are based on the involvement of specialists from several disciplines. With plant resistance to insects, the programs require input from plant geneticists, plant breeders, and entomologists. The actual crossing, genetic analysis, and evaluation of agronomic characteristics are the responsibilities of plant scientists. Entomologists contribute by identifying resistance sources, characterizing mechanisms of resistance, and performing laboratory and field assays on resistance sources.

There are at least eight traditional methods of breeding for insect resistance (Panda and Khush 1995). The usual procedure of plant breeders is to identify traits in a resistant cultivar and transfer these traits to another cultivar by crossing them. After an initial cross, the offspring or progeny of the cross are evaluated for the resistant trait. This process is sometimes repeated several times with subsequent generations of progeny, among other methods. In addition to levels of resistance in the resulting progeny, plant breeders also attempt to identify the resistance genes and measure their degree of expression in new cultivars.

A relatively new method in improving breeders' efficiency in selecting plant with desirable gene combinations for resistance is **marker-assisted breeding.** A marker is a "genetic tag" that shows a certain location within a plant's genome that possesses an insect-resistant characteristic. The marker then enables a breeder to transfer a single gene into a new cultivar, or it can be used to test a new plant cultivar for inheritance of many genes at once. The time required to locate a specific gene, compared to using a random search approach, is thus reduced phenomenally, getting resistant varieties to producers quickly and with much less of an investment in development. This approach has translated into a great cost savings for seed producers and, subsequently, for producers. Additional technologies that improve plant-breeding efficiency in developing insect-resistant plants include grafting, tissue culture, and micropropagation (see Suslow et al. 2002 for detail on these procedures).

## BIOTECHNOLOGY AND RESISTANT-VARIETY DEVELOPMENT

Today, much excitement is being generated over rapid advances in biotechnology and its possible use in solving many of our most difficult insect pest problems. One major use of biotechnology in insect pest management is in the development of improved insect-resistant cultivars.

With its newfound popularity in the media, the term *biotechnology* has had many different meanings and is represented in different ways. The Office of Technology Assessment (U.S. Congress) defines biotechnology as "any technique that uses living organisms, or substances from those organisms, to make or modify a product, to improve plants or animals, or to develop microorganisms for specific uses." In this sense, biotechnology is not a new technology since it includes methodology for brewing, food fermentation, conventional animal vaccine production, and many other traditional endeavors.

Usually, however, biotechnology is perceived as a new technology that encompasses: (1) genetic engineering for producing vaccines and improving plants and animals; (2) monoclonal antibodies for diagnosis of cell proteins; and (3) new cell and tissue culture techniques for rapid propagation of living cells. Particularly, genetic engineering has created great excitement in agriculture because it can allow for the manipulation and control of genes in heretofore unimaginable ways. However, all three aspects of modern biotechnology are very important to agriculture and can be used to aid in the development of new plant cultivars resistant to insect pests.

### Basics of Genetic Engineering

An understanding of how plants are "engineered" begins with elementary information about molecular biology and proceeds with manipulations of genetic material to arrive at a transformed plant.

**DNA: The blueprint of life.** All living organisms can be subdivided into several functional components beginning with organ systems, organs, tissues, cells, and, ultimately, cell contents. Of particular interest here is the nucleus of the cell, which contains a slightly acid substance, deoxyribonucleic acid, or DNA. This substance controls every biochemical process within cells and, consequently, the whole organism. Moreover, DNA possesses all information necessary for all phases of an organism's development, and this information is passed on to offspring from parents through a sophisticated process of replication.

The structure of a DNA molecule resembles a helix with two strands. The two strands are attached by the chemical bases adenine (A), thymine (T), cytosine (C), and guanine (G), which occur in pairs. In these pairs, A occurs with T and G occurs with C (Fig. 13.10). A portion of a DNA strand that encodes sufficient information to make a single protein is called a **gene.**

The gene passes instructions to the cell through a complex decoding process. To accomplish this, the DNA double helix separates. New bases then pair with the open DNA strands to form a molecule like DNA, called messenger RNA (ribonucleic acid) or mRNA. Finally, the messenger RNA commands the cell to assemble amino acids in certain ways to make proteins. The proteins give rise to new cells, and, subsequently, to tissues, organs, organ systems, and the whole organism (Fig. 13.10).

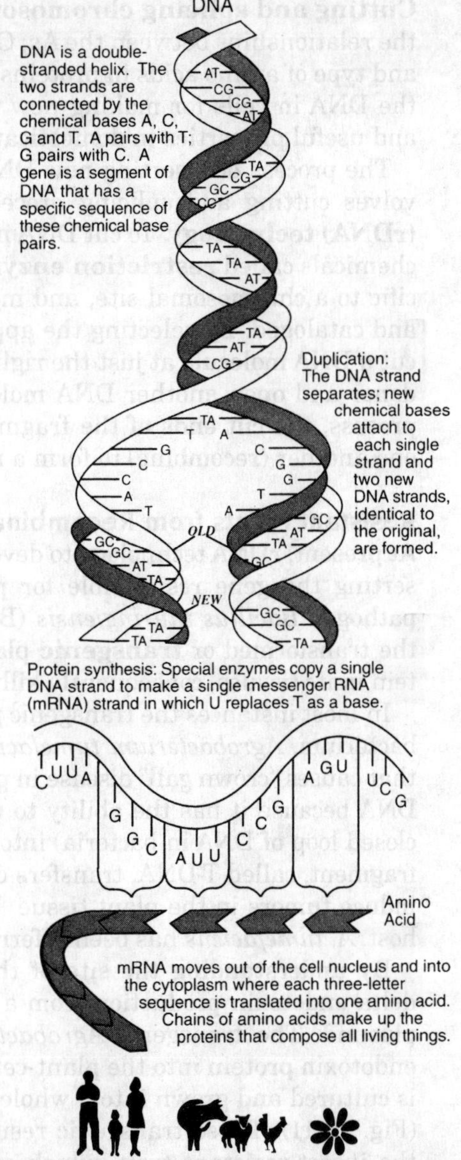

DNA
DNA is a double-stranded helix. The two strands are connected by the chemical bases A, C, G and T. A pairs with T; G pairs with C. A gene is a segment of DNA that has a specific sequence of these chemical base pairs.
Duplication: The DNA strand separates; new chemical bases attach to each single strand and two new DNA strands, identical to the original, are formed.
OLD
NEW
Protein synthesis: Special enzymes copy a single DNA strand to make a single messenger RNA (mRNA) strand in which U replaces T as a base.
Amino Acid
mRNA moves out of the cell nucleus and into the cytoplasm where each three-letter sequence is translated into one amino acid. Chains of amino acids make up the proteins that compose all living things.


**Figure 13.10** A DNA molecule (*top*) showing double-stranded helix and paired bases, A, C, G, and T. Duplication is occurring in the bottom portion of the helix. Protein synthesis (*middle*) occurs when special enzymes copy a single DNA strand and produce messenger RNA. Here U replaces the T base. The messenger RNA produces sequences of amino acids (*bottom*) that build proteins and subsequently whole organisms. (Redrawn from *The New Biology: The Science and Its Applications*, Monsanto Company)

**Cutting and splicing chromosomes.** Molecular biologists now understand the relationships between the As, Cs, Gs, and Ts of genes and the specific order and type of amino acids in proteins. In knowing this, they proceeded to change the DNA in cells for making new proteins, with the goal of conferring novel and useful properties in domesticated plants and animals.

The process of inserting new DNA along the DNA strand (chromosome) involves cutting and splicing procedures referred to as **recombinant DNA (rDNA) technology.** To cut DNA into reproducible pieces at specific locations, chemicals called **restriction enzymes** are used. Restriction enzymes are specific to a chromosomal site, and many of such enzymes have been discovered and cataloged. By selecting the appropriate restriction enzyme, biologists can cut a DNA molecule at just the right location to obtain a desired gene from the donor and open another DNA molecule of the recipient for insertion. In this process, the cut ends of the fragments are chemically "sticky" and attach to one another (recombine) to form a new molecule.

## Resistant Plants from Recombinant DNA Technology

At present, rDNA technology to develop resistant plants has mostly involved inserting the gene responsible for producing delta endotoxin from the insect pathogen *Bacillus thuringiensis* (Bt) into plant genomes. With this approach, the transformed or **transgenic** plant has the power to produce the toxic protein in its tissues, consequently killing feeding insects and protecting the plant.

In most instances the transgenic plants are developed with the use of a vector bacterium, *Agrobacterium tumefaciens.* This is a natural soil-borne bacterium that causes "crown gall" disease in plants. *A. tumefaciens* is an efficient vector of DNA because it has the ability to transmit a fragment of its large plasmid (a closed loop of DNA in bacteria) into the nuclear genome of an infected cell. The fragment, called T-DNA, transfers contained genes, known as "oncogenes," that induce tumors in the plant tissue. With the ability to insert its own DNA in a host, *A. tumefaciens* has been referred to as a natural "genetic engineer."

By understanding the site of the T-DNA, a fragment of DNA controlling delta endotoxin production from a Bt cell is inserted into the *A. tumafaciens* plasmid. The transgenic *Agrobacterium* is then used to transfer the delta endotoxin protein into the plant-cell chromosome. Subsequently, the plant cell is cultured and grown into a whole plant whose cells contain the toxic protein (Fig. 13.11). These transgenic resistant plants then produce seed expressing the insect-resistant trait, which can be commercialized. This procedure has been used to transfer Bt delta endotoxin gene to many plants, including cotton, tobacco, tomato, and potato.

Subsequently, new transgenic cottons were developed and marketed for the 1996 growing season using Bollgard® genetics. More than 2 million acres of Bt cotton were planted in 1996 for management of cotton bollworm, tobacco budworm, and pink bollworm. Although most of the plantings were successful in preventing losses that year, additional pesticide applications were required on several thousand acres in Texas and the Southeast. Seemingly, hot weather and unusually high acreage of corn fostered dense populations of the cotton bollworm, which were able to overwhelm the resistance in some areas. Although this problem occurred on only a small fraction of the Bt cotton acreage, it emphasizes the need for vigilance when implementing new tactics.

Additionally, Bt transgenic potatoes have been developed and marketed using the *A. tumafaciens* method of gene transfer. Transgenic potatoes were

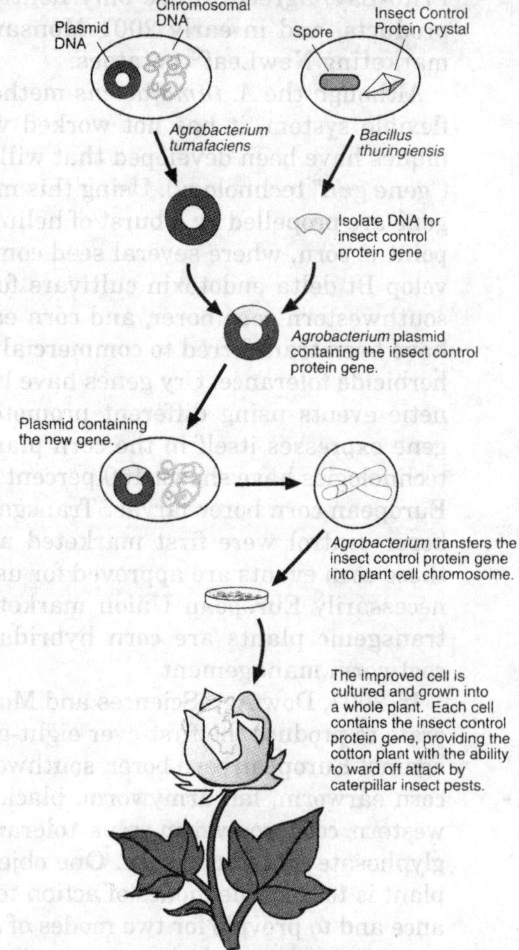

DEVELOPING COTTON RESISTANT TO INSECT PESTS
Chromosomal DNA
Plasmid DNA
Spore
Insect Control Protein Crystal
Agrobacterium tumafaciens
Bacillus thuringiensis
Isolate DNA for insect control protein gene
Agrobacterium plasmid containing the insect control protein gene.
Plasmid containing the new gene.
Agrobacterium transfers the insect control protein gene into plant cell chromosome.
The improved cell is cultured and grown into a whole plant. Each cell contains the insect control protein gene, providing the cotton plant with the ability to ward off attack by caterpillar insect pests.

**Figure 13.11** The process of developing a transgenic plant using *Agrobacterium tumafaciens* as a vector of delta-endotoxin-expressing DNA from *Bacillus thuringiensis.* (Redrawn from *The New Biology: The Science and Its Applications,* Monsanto Company)

developed by Monsanto Company and marketed as NewLeaf® varieties. The first transgenic potatoes were approved by U.S. regulatory agencies in 1995 and conferred resistance to control the Colorado potato beetle. Genetically engineered Russet Burbank potatoes appeared in groceries in Washington state in 1997, accompanied with brochures that explained the new technology to consumers. The following year Monsanto received regulatory approval for a second transgenic potato (NewLeaf® Plus) that combined the Bt resistance for Colorado potato beetles with resistance to the Potato Leaf Roll Virus. Just before transgenic potatoes were available to farmers, a new insecticide was approved that controlled the Colorado potato beetle and several other pests. Most farmers chose the insecticide over the transgenic potato, and the new varieties never exceeded 3 percent of the potato market. Additionally, several large food processors (McDonald's, Burger King, Frito-Lay) agreed to use only nontransgenic potato varieties in their food products, and in early 2001 Monsanto reported that it would discontinue marketing NewLeaf® potatoes.

Although the *A. tumafaciens* method of gene transfer has been a good and flexible system, it has not worked very well with cereals. Even still, techniques have been developed that will allow transformation, such as biolistics ("gene gun" technology). Using this method, tiny particles coded with the new gene are propelled by a burst of helium gas through plant-cell walls. A case in point is corn, where several seed companies have used this technology to develop Bt delta endotoxin cultivars for use against the European corn borer, southwestern corn borer, and corn earworm. Several genes, (see Table 13.1) have been transferred to commercial hybrids with both insect resistance and herbicide tolerance. Cry genes have been transformed by several different genetic events using different promoters that regulate when and where the gene expresses itself in the corn plant. Tests with hybrids using YieldGard® technologies have shown 100 percent mortality of first- and second-generation European corn borer larvae. Transgenic Bt corn hybrids with European corn borer control were first marketed and commercially grown in 1996. All of these corn events are approved for use in U.S. and Japanese markets but not necessarily European Union markets. More recent developments in the Bt transgenic plants are corn hybrids carrying Cry1F and Cry3Bb for corn rootworm management.

In 2007, Dow AgroSciences and Monsanto announced that they would cooperate to produce the first-ever eight-gene stack in corn that confers resistance against European corn borer, southwestern corn borer, western bean cutworm, corn earworm, fall armyworm, black cutworm, northern corn rootworm, and western corn rootworm, plus tolerance to two commonly used herbicides—glyphosate and glufosinate. One objective of this eight-gene stack in a single plant is to combine modes of action to reduce the development of insect resistance and to provide for two modes of action for weed control.

A more recent technique in gene transfer involves the use of engineered enzymes called *zinc finger nucleases* (Bibikova et al. 2003). A zinc finger is a protein that is held together by a zinc ion. Each finger fits into the DNA helix at a specific spot. Arrays of zinc fingers can latch onto specific sequences of DNA to deliver strand-breaking enzymes to a desired location. Once the strand is broken, the nuclease drops replacement DNA carrying the desired character sequence into the break. This approach is said to potentially reduce development time of a new cultivar by 3 to 6 years.

Even though the transfer of Bt genes has been most prevalent in resistance development, genes that confer other modes of resistance are also being investigated. Most notable among these are genes that control the production of a-amylase inhibitor in bruchid weevils, pests of stored grain. By inhibiting a-amylase enzyme, the weevils are prevented from digesting the starch in the grain. Many weevils feeding on the grain starve to death, and developmental time is increased in others. The delay in development keeps insect numbers from increasing to damaging levels.

A gene from common bean seed, responsible for the inhibitor, also has been transferred to garden peas, making the pea seeds resistant to weevil feeding.

**Table 13.1 Field Corn Events for Insect Resistance and Herbicide Tolerance and Market Acceptance.[a]**

| Product Registrant/ Trade Name | Characteristic | Event | Japan Approved | EU Food Approval | EU Processed Feed Approval |
|---|---|---|---|---|---|
| Syngenta *Agrisure* CB *YieldGard Liberty Link* | Cry1Ab: Corn borer protection Glufosinate herbicide tolerance | Bt11 | Yes | Yes | Yes |
| DowAgrosciences, and Pioneer Hi-Bred *Herculex I* | Cry1F: Western bean cutworm, corn borer, black cutworm, and fall armyworm resistance Glufosinate herbicide tolerance | TC1507 | Yes | Yes | Yes |
| Monsanto *YieldGard* | Cry1Ab: Corn borer protection | MON 810 | Yes | Yes | Yes |
| Monsanto *YieldGard Roundup Ready 2* | Cry1Ab: Corn borer protection Glyphosate herbicide tolerance | MON 810+NK603 | Yes | No | Yes |
| *YieldGard Corn Rootworm Protection Roundup Ready 2* | Corn rootworm protection Glyphosate herbicide tolerance | MON 863+NK603 | Yes | No | Yes |
| *YieldGard Corn Rootworm Protection* | Corn rootworm protection | MON 863 | Yes | Yes | Yes |
| Monsanto *YieldGard Plus* | Cry1Ab: Corn borer protection Corn rootworm protection | MON 810+MON 863 | Yes | No | Yes |
| Monsanto *YieldGard Plus* with *Roundup Ready 2* | Cry1Ab: Corn borer protection Corn rootworm protection Glyphosate herbicide tolerance | MON 810+MON 863+NK603 | Yes | No | No |
| *Herculex I Roundup Ready 2* | Cry1F: Western bean cutworm, corn borer, black cutworm, and fall armyworm resistance Glyphosate herbicide tolerance Glufosinate herbicide tolerance | TC1507+NK603 | Yes | No | No |

***(Continued)***

**Table 13.1 Field Corn Events for Insect Resistance and Herbicide Tolerance and Market Acceptance.[a] *(Continued)***

| Product Registrant/ Trade Name | Characteristic | Event | Japan Approved | EU Food Approval | EU Processed Feed Approval |
|---|---|---|---|---|---|
| Syngenta *Agrisure* GT/CB *YieldGard Liberty Link* | Cry1Ab: Corn borer protection Glyphosate herbicide tolerance | SYTGA21+Bt11 | No | No | No |
| Monsanto *YieldGard Roundup Ready* | Cry1Ab: Corn borer resistance Glyphosate herbicide tolerance | MON 810+SYTGA21 | Yes | No | Yes |
| Dow AgroSciences and Pioneer Hi-Bred *Herculex RW* | Cry34/35Ab1: Western corn rootworm, northern corn rootworm, and Mexican corn protection Glufosinate herbicide tolerance | DAS-59122-7 | Yes | No | No |
| Dow AgroSciences and Pioneer Hi-Bred *Herculex XTRA* | Cry1F: Western bean cutworm, corn borer, black cutworm, and fall armyworm resistance Northern corn rootworm, western corn rootworm, and Mexican corn rootworm protection Glufosinate herbicide tolerance | TC1507+DAS-59122-7 | Yes | No | No |
| Dow AgroSciences and Pioneer Hi-Bred *Herculex Rootworm* Monsanto *Roundup Ready 2* | Cry34/35Ab1: Western corn rootworm, northern corn rootworm, and Mexican corn protection Glyphosate herbicide tolerance | DAS-59122-7+NK603 | Yes | No | No |
| Dow AgroSciences and Pioneer Hi-Bred *Herculex XTRA* Monsanto *Roundup Ready 2* | Cry1F: Western bean cutworm, corn borer, black cutworm, and fall armyworm resistance Glufosinate herbicide tolerance | TC1507+DAS-59122-7+ NK603 | Yes | No | No |

**Table 13.1 Field Corn Events for Insect Resistance and Herbicide Tolerance and Market Acceptance.[a] *(Continued)***

| Product Registrant/ Trade Name | Characteristic | Event | Japan Approved | EU Food Approval | EU Processed Feed Approval |
|---|---|---|---|---|---|
| | Cry34/35Ab1: Western corn rootworm, northern corn rootworm, and Mexican corn protection<br>Glyphosate herbicide tolerance | | | | |
| *YieldGard VT™ Rootworm/RR2* | Corn rootworm protection<br>Glyphosate herbicide tolerance | MON 88017 | Yes | No | No |
| *YieldGard VT™ Triple* | Cry1Ab: Corn borer protection<br>Corn rootworm protection<br>Glyphosate herbicide tolerance | MON 810+MON 88017 | Yes | No | No |
| Syngenta *Agrisure RW* | Modified Cry3A: Western, northern, and Mexican corn rootworm protection | MIR604 | No | No | No |
| Syngenta *Agrisure GT/RW* | Modified Cry3A: Western, northern, and Mexican corn rootworm protection<br>Glyphosate herbicide tolerance | MIR604+SYTGA21 | No | No | No |
| Syngenta *Agrisure CB/LL/RW* | Modified Cry3A: Western, northern, and Mexican corn rootworm protection<br>Glufosinate herbicide tolerance | Bt11+MIR604 | No | No | No |

[a]All products have full food and feed approval in the United States. Approval for Japan and the European Union (EU) are designated.

SOURCE: From National Corn Growers Association; current as of November 5, 2007.

The hope is that this gene can be moved to other legumes and to cereals such as corn and rice, thus preventing serious storage losses.

Other transgenic plant developments include plant virus protection. Aphids transmit many viruses to crops, and once viral symptoms appear on the plant, it is too late to effectively control the insects with insecticides. The development of transgenic virus-protected crops helps to reduce the use of broad-spectrum insecticides. The first transgenic virus-protected crop to gain nonregulated status and be sold commercially was yellow crookneck squash.

Domesticated varieties of *Cucurbita pepo,* including acorn squash, yellow crookneck squash, and zucchini, are susceptible to stunt and mottle viral infections that can reduce yields or cause total crop loss. Squash is susceptible to several viruses including watermelon mosaic virus 2 and zucchini yellow mosaic virus. Transgenic squash developed by Asgrow Seed Company uses viral coat protein genes to confer resistance against both of these viruses. This transgenic squash, known as Freedom II®, deregulated in 1994, showed better protection from viruses than other commercial varieties in early field tests in North Carolina. Later Asgrow developed CZW-3, a transgenic pest-protected squash that was deregulated in 1996, with a marker gene for resistance to the antibiotic kanamycin and coat-protein genes for protection against three virus species (National Academy of Sciences 2000). Virus-resistant squash is planted on approximately 5,000 acres in the southeastern United States.

Another crop with transgenic development is papaya, *Carica papaya.* Papaya, especially important in tropical and subtropical regions, is often infected by papaya ringspot virus (PRSV), which is transmitted by aphids. The virus causes severe stunting of the plants, low yields, inhibits year-long crop production, and may result in abandonment of infected plantations. Both control of the aphid and traditional breeding for resistance have been unsuccessful. Papaya was introduced to the Hawaiian island of Oahu in 1940, and infections of PRSV started in 1945. The industry then moved to the island of Hawaii, and the crop remained virus-free until 1992. Gonsalves and colleagues developed transgenic papaya to protect the crop in a small virus-free area on the islands. The transgenic plant was developed by fusing a coat protein gene from a mild strain of PRSV to a kanamycin-resistance marker and then inserting the linked transgenes into the genomes of two cultivars, UH Rainbow and SunUp. The transgenic papaya, which was deregulated in 1996, provided protection against PRSV and allowed the recovery of this industry on Hawaii (National Academy of Sciences 2000), and it currently is planted on about 2,000 acres.

Numerous other transgenic cultivars of fruits and vegetables are being tested for their potential role in pest management. These include insect-resistant broccoli; virus-resistant citrus, plums, raspberries and tomatoes; herbicide-tolerant lettuce, strawberries, and processing tomatoes; bacteria-resistant apples and grapes; and nematode-resistant pineapple.

## Deployment of Engineered Resistant Plant Varieties

The future looks very bright for the development of transgenic resistant plants. We can expect the release of plant varieties with both Bt- and perhaps, proteinase-inhibiting traits in many crops. However, there is considerable concern about the durability of these varieties because of the potential for resistance-breaking insect biotypes. This potential is heightened because of their effectiveness, which produces strong selective pressures on pest insect populations, and the nature of the inheritance. Since resistance in most transgenic plants has been monogenic or oligogenic, it can be overcome more easily than polygenic traits.

For sustainability, the deployment of the engineered resistance factor should involve strategies to avoid the development of resistance-breaking biotypes early on. Some of these strategies include: (1) mixes of resistant and susceptible plants in the plant stand, leaving a refuge for some individuals; (2)

sublethal doses that make insects more vulnerable to other environmental factors; and (3) expression of the resistance factor only in the plant part needing protection, usually reproductive structures. In all of these strategies, the objective is to reduce total insect mortality, thereby reducing selection pressures.

Clearly, with continued advances in biotechnology, we are on the threshold of a new era, marked by effective new tools for managing pests. However, the most significant challenge ahead is not the development of new transgenic plants but finding ways to deploy resistance factors to help these plants remain effective in the long term.

## Insect Resistance Management

Insects have developed resistance to many control tactics, therefore the risk of insect resistance to transgenic technology is a realistic possibility, and steps must be taken to minimize the risk. To preserve the benefits of the insect protection provided by biotechnology, growers must implement an insect resistance management (IRM) plan on their farms.

A good example of IRM can be found in corn production. Here, corn growers can plant transgenic hybrids that provide control of a spectrum of pests. This biotechnology, known as YieldGard Plus® and developed by Monsanto Company, provides control of European corn borer, southwestern corn borer, and several species of *Diabrotica* rootworm larvae. IRM is a requirement when purchasing any transgenic corn, as mandated by the EPA. A key component of the IRM strategy is a refuge, and Monsanto Company (2007) produces a grower guide that outlines the concept of the refuge.

A refuge is a strip or block planting of corn that does not contain the Bt technology for controlling the target corn pest(s). The primary purpose of a refuge is to maintain a population of insect pests that is not exposed to the Bt proteins found in transgenic corn. The lack of exposure to the Bt proteins allows susceptible insects in the refuge corn to potentially mate with any rare resistant insects that may have survived and emerged from the transgenic Bt corn. Susceptibility to the Bt protein would then be genetically passed to the progeny, helping to preserve the biotechnology.

Growers planting YieldGard Plus can implement one of two different refuge options to meet the IRM requirements. The first option is to plant a *common refuge* that will serve as the refuge for both European corn borers and corn rootworms. The common refuge must be planted with corn hybrids that do not contain the Bt technology for the control of corn borers or corn rootworms. The common refuge requirement for each farm is that no more than 80 percent of corn acres can be planted with YieldGard Plus corn (or no more than 50 percent of the acres in cotton-growing areas). The refuge must represent at least 20 percent of a grower's corn acres (50 percent in cotton-growing areas). The refuge must be within or adjacent to the YieldGard Plus field and can be planted as a block, strips within the field, or as a perimeter around the field. If perimeter or in-field strips are used for the refuge, the strips must be at least six, and preferably twelve, consecutive rows wide.

The common refuge can be treated with a soil-applied, seed-applied, or foliar-applied insecticide to control corn rootworm larvae and other soil insects. The refuge also can be treated with a non-Bt foliar insecticide for control of late-season pests, if pest pressure reaches an economic threshold for injury.

However, if corn rootworm adults are present at the time of the foliar application, the YieldGard Plus field must be treated in a similar manner.

In some corn-growing regions, large populations of European corn borers may pose a significant risk of yield loss with planting a common refuge for YieldGard Plus. In these situations, growers have a second IRM option that allows for planting and managing separate refuges for both corn borers and corn rootworms. The *separate refuge* option provides greater flexibility to manage corn borers in both the corn rootworm and corn borer refuge areas without the need to also spray the YieldGard Plus field.

Additional requirements for planting a refuge are that a refuge must be planted on every farm where YieldGard Plus is planted; the refuge must be planted at the same time as the YieldGard Plus corn; mixing non-Bt seed with YieldGard Plus seed is not permitted; adjacent refuge fields must be owned and managed by the same grower; and refuge fields and YieldGard Plus fields must have similar cropping history.

## Transgenic Plants as Trap Crops

Biotechnology may offer unique opportunities for pest control in perennial tree and vine crops. The concept is to incorporate the Bt insecticidal protein into a trap crop, in this case, apple trees (California Agriculture 2004). The codling moth is a serious pest of apple, pear, and walnut in California but prefers to lay its eggs on apple trees. The apple trees were engineered to express Bt protein, which is toxic to the codling moth larvae. The transgenic apple trees were planted in and adjacent to a 90-acre walnut orchard. Over a 5-year period the codling moth was almost completely eliminated in the walnuts, a level of control that was similar to three separate insecticide applications.

## Benefits and Risks of Transgenic Crops

Although using biotechnology to create insect resistant plants seems desirable, this approach is not without criticism. Much of the criticism emanates from environmental groups, seemingly having its most significant influence in Europe and Japan. In Britain, food from genetically modified crops (GMOs) is derided as "Frankenfood," and Japanese beer makers and food processors in the past have banned purchases of GMO products from the United States and other nations.

In the United States, the response to GMOs has been more moderate, but groups promoting the high-risk factor are having an influence. Some major food processors have begun to refuse GMO crops, causing growers to cut back on their production, particularly with potatoes but also other crops, such as corn.

Critics have raised concerns that GMOs could harm the environment, be dangerous to human health, and are an unnatural way of producing food (a problem of ethics). Opponents argue that when genes from one species are transferred to another, there is a risk that allergens may also be transferred, as well as other health-degrading substances. They also believe environmental risks could be significant, including: (1) unintended cross-pollination and subsequent weediness of GMOs, (2) loss of biological diversity, (3) pest resistance to GMOs (already discussed), (4) increased herbicide use with herbicide-tolerant crops, and (5) adverse impacts on nontarget species, such as butterflies.

The latter was spotlighted in a sensationalized news release on the effects of Bt corn on monarch butterflies. The news release was based on the finding of a study that pollen from Bt corn has toxic effects on butterfly larvae. Recall that monarch larvae feed on milkweed. Because milkweed can grow in and next to corn in the Midwest, it was suggested that Bt-laden pollen could drift onto milkweed, be eaten by the larvae, and adversely impact the monarch populations.

Reports also have attempted to predict the effects of widespread production and food use of these crops. For instance, transgenic Bollgard® cotton was commercially released in 1996 and targeted major caterpillar pests including cotton bollworm, *Helicoverpa zea;* tobacco budworm, *Heliothis virescens;* and pink bollworm, *Pectinophora gossypiella.* After 5 years (1996–2000) of commercial production, Edge and his colleagues (2001) examined the potential economic, environmental, and social benefits of Bt cotton compared with broad-spectrum insecticide use. Additionally, they determined whether the benefits were directly (primary) or indirectly (secondary) related to growing Bt cotton. They found that the direct benefits of Bt cotton include reduced broad-spectrum insecticide use, improved control of target pests, better yield and profitability, lower production costs and farming risk, expanded opportunities to grow cotton, and a brighter economic outlook for the cotton industry. The indirect benefits of Bt cotton are associated with a reduction in broad-spectrum insecticide use and include increased effectiveness of beneficial arthropods as pest conrol agents, reduced risks for farmland wildlife species, less runoff of insecticides, reduced fuel usage, lower levels of air pollution and related waste production, and improved safety for farm workers. They concluded that after 5 years of commercial use on $>2 \times 10^6$ hectares globally, Bt cotton provides an effective method for catepillar control that is safer to humans and the environment than conventional broad-spectrum insecticides.

Similar conclusions were expected for the potential agronomic, economic, and environmental benefits of transgenic YieldGard rootworm corn. The potential benefits are expected to include increased root protection; increased intangible benefits to farmers (safety of not being exposed to insecticides, ease of use and handling, time and labor savings, better pest control than using a soil insecticide or seed treatment); increased economic benefits to farmers ($231 million annually from yield gains [$25 to $75/acre relative to no insecticide control, $4 to $12/acre relative to control with a soil insecticide] and $58 million annually in reduced insecticide risks and time savings); reduced incidence of corn stalk rot pathogens; and increased yield protection (9 to 28 percent increase relative to no insecticide use, 1.5 to 4.5 percent relative to control with a soil insecticide). If transgenic rootworm corn is planted on 10 million acres, the annual impact will be a reduction of 5,344,462 pounds active ingredient (75.2 percent) of insecticide use; increased resource conservation (3.07 to 5.23 million gallons of diesel fuel equivalents conserved that would have been consumed in the manufacture and delivery of insecticides); increased water conservation (5,657,734 gallons of water not used in insecticide application); conservation of aviation fuel (68,845 gallons of aviation fuel not used); reduced farm waste (1,187,035 fewer insecticide containers used); increased planting efficiency; and improved safety to wildlife and other nontarget organisms (Rice 2004).

Certainly, developing and growing such crops could significantly reduce insecticide use and potentially improve nutrient content in food for underdeveloped nations. Yet, risks should be considered. We can question, then, the proper path to follow in future development. A report of the U.S. House Committee on Science (Subcommittee on Basic Research) sheds some light on this subject and may help in making decisions.

The report, titled "Seeds of Opportunity: An Assessment of the Benefits, Safety, and Oversight of Plant Genomics and Agricultural Biotechnology," presents findings and recommendations on how the United States should proceed with applications of biotechnology. It specifically addresses safety concerns of the critics. Highlights of the report include the following:

1. The promise of agricultural biotechnology is immense. Advances in this technology will result in crops with a wide range of desirable traits that will directly benefit farmers, consumers, and the environment and increase global food production and quality.
2. There is no evidence that transferring genes from unrelated organisms to plants poses unique risks. The risks associated with plant varieties developed using agricultural biotechnology are the same as those for similar varieties developed using classical breeding methods. As the new methods are more precise and allow for better characterization of the changes being made, plant developers and food producers are in a better position to assess safety than when using classical breeding methods.
3. The threat posed by pest-resistant crop varieties developed using agricultural biotechnology to the monarch butterfly and other nontarget species has been vastly overblown and is probably insignificant.
4. There is no scientific justification for labeling foods based on the method by which they are produced. Labeling of agricultural biotechnology products would confuse, not inform, consumers and send a misleading message on safety.
5. Federal regulations should focus on the characteristics of the plant, its intended use, and the environment into which it will be introduced, not the method used to produce it. Regulations developed specifically for products of agricultural biotechnology do not reflect the scientific consensus on risk, are overly burdensome, and can stifle scientific research.

The report then goes on to make several important recommendations, including maintaining its current science-based policy on food labeling and regulations based on the product, not how the product was created. It also recommended that the United States should work to ensure that markets for biotechnology products exist and refuse to accept international agreements that limit trade in, or mandate labeling of, a plant or food product based on the method used to develop it.

Presently, the status of GMO crops and products is in flux, and worldwide acceptance is far from being assured. With the great potential of these as pest management tools, we will keep a watchful eye on developments and support scientifically, not politically, based decisions on their use.

## SUCCESSFUL USES OF INSECT-RESISTANT CULTIVARS

In addition to the initial successes with transgenic plants, other outstanding achievements in plant resistance to insects from conventional plant breeding should be mentioned. Major developments have taken place, particularly in grain and forage crops, including wheat, rice, corn, sorghum, sugarcane, beans,

**Figure 13.12** Hessian fly, *Mayetiola destructor,* on a wheat stem. (Courtesy Kansas Agricultural Experiment Station)

barley, and alfalfa. Progress, with varying degrees of success, has also occurred in vegetables, flowers, and other field crops.

Many specific examples of success with plant resistance against insects could be mentioned; in the United States, however, five stand out as having provided significant monetary gains to users. These include wheats resistant to Hessian fly and wheat stem sawfly, corn resistant to the European corn borer, alfalfa resistant to the spotted alfalfa aphid, and barley resistant to the greenbug.

### Resistance to Hessian Fly

The search for wheat resistant to the Hessian fly (Fig. 13.12) began in 1914 in Kansas. An early variety, Kawvale, developed from a soft-wheat variety, Indiana Swamp, was released as resistant in 1928. Subsequently, Kawvale was superseded by higher-quality and more resistant varieties like Pawnee and, later, Ponca. In Kansas, growing resistant varieties has resulted in the virtual disappearance of the insect, except where susceptible varieties are present. Annual savings from growing resistant varieties has been estimated at $238 million, produced from reductions in fly populations of 95 percent.

### Resistance to European Corn Borer

The release of corn varieties resistant to the European corn borer began in the 1940s, and by the 1950s, workers in Iowa and Minnesota had developed more than forty insect-resistant lines and hybrids. Subsequently, many other varieties have been bred and released that have significantly reduced the threat of this introduced pest.

One of the more popular varieties, CI31A, has shown good resistance to the European corn borer in the United States, Romania, Canada, Yugoslavia, Hungary, and Russia. Although many resistant cultivars do not show activity against second-generation borers, one line, B52, offers a degree of resistance. B52 has been successfully transferred to sweet corn, where second-generation injury is particularly important. However, heavy infestations can seriously impede development of the primary ear of B52 and other resistant lines, signifying the need for higher levels of resistance. Two other cultivars, B86 and BS9C5, are also resistant to both generations of borers, and B86 has been successfully transferred to popcorn, which suffers injury from both generations.

Today, many commercial field corn hybrids possess genes for resistance to the European corn borer. It has been estimated that resistant cultivars are grown on about one-third of the corn area of the United States, resulting in annual savings of approximately $150 million. These figures have increased as the acreage of Bt varieties has increased.

### BOX 13.3 Spotted Alfalfa Aphid

**SPECIES:** *Therioaphis maculata* (Buckton) (Hemiptera: Aphididae)

**DISTRIBUTION:** The spotted alfalfa aphid is native to the Middle East and was introduced into the United States (New Mexico) in 1954. The species spread rapidly, and by 1957, it was distributed over most of the United States.

**IMPORTANCE:** The species is a significant alfalfa pest, especially in the southwestern United States. In just 2 years, from 1954 to 1956, the pest caused $81 million in damage to alfalfa. Both adults and nymphs inject toxic salivary secretions and suck juices from leaves and stems, causing disruption of phloem flow. This injury causes leaves to become curled and yellowed. Individuals typically feed on lower leaves, then move up the plant. With large infestations, plants are severely stunted and often die.

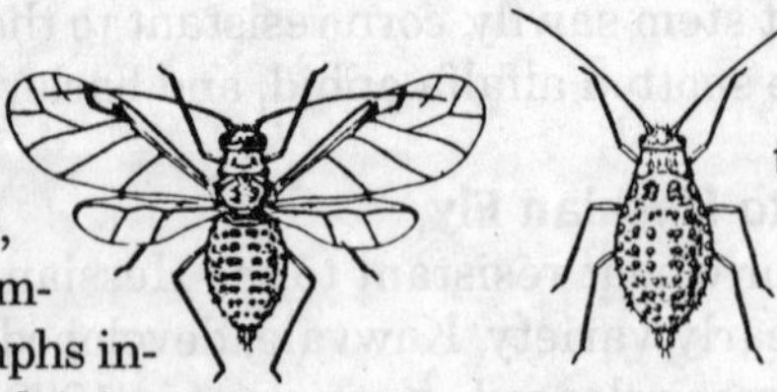

**APPEARANCE:** Individuals are approximately 2 mm long and yellow, with four to six rows of black spots (short spines) on the abdomen. Alate (winged) individuals have clear wings with smoky-gray veins.

**LIFE CYCLE:** Eggs overwinter in the northern part of the species range, whereas in the South, the aphids are parthenogenetic the entire year. Females produce active young at rates of one to eight per day, depending on temperature. Each female can produce up to 140 offspring during her lifespan. Nymphs reach the adult stages in 1 to 4 weeks (four stadia), depending on climatic conditions. In colder regions, winged males and females appear in fall and mate, then females oviposit. There are twenty to thirty generations per year.

### Resistance to Spotted Alfalfa Aphid

The spotted alfalfa aphid (Box 13.3) probably was introduced into the United States about 1954. Shortly after the introduction of this insect, the alfalfa variety Lahontan, along with three of its five parental clones, was discovered to have resistance to the insect in Kansas. Further work allowed the development of the alfalfa variety Cody, which possessed even higher levels of resistance and was well adapted to Kansas growing conditions. Other varieties developed with good resistance and adapted to regional growing conditions include Mopa in California and Zia in New Mexico. Reselection within existing varieties is continuing, and yearly savings from growing resistant cultivars are estimated at $60 million.

### Resistance to Wheat Stem Sawfly

One of the most dramatic instances of success in host-plant resistance has occurred with the wheat stem sawfly. This insect was a very destructive pest of wheat grown in the western United States and Canada until the use of fly-resistant wheat, beginning in 1946. At that time the solid-stemmed resistant variety, Rescue, was developed and released in Canada. Subsequently, the variety has been grown extensively in Canada and areas of the northern United States. Rescue has continued to offer protection against the wheat stem sawfly since its introduction, and biotypes have not been known to develop. However, relatively low-quality agronomic characteristics have prompted recommendations of Rescue's rotation with susceptible varieties in some instances. Nevertheless, savings of $4 million per year have been estimated because of the use of plant resistance.

### Resistance to the Greenbug

Much plant breeding activity has been invested in developing greenbug (Fig. 13.13) resistance in both wheat and barley. More than 7,000 wheat

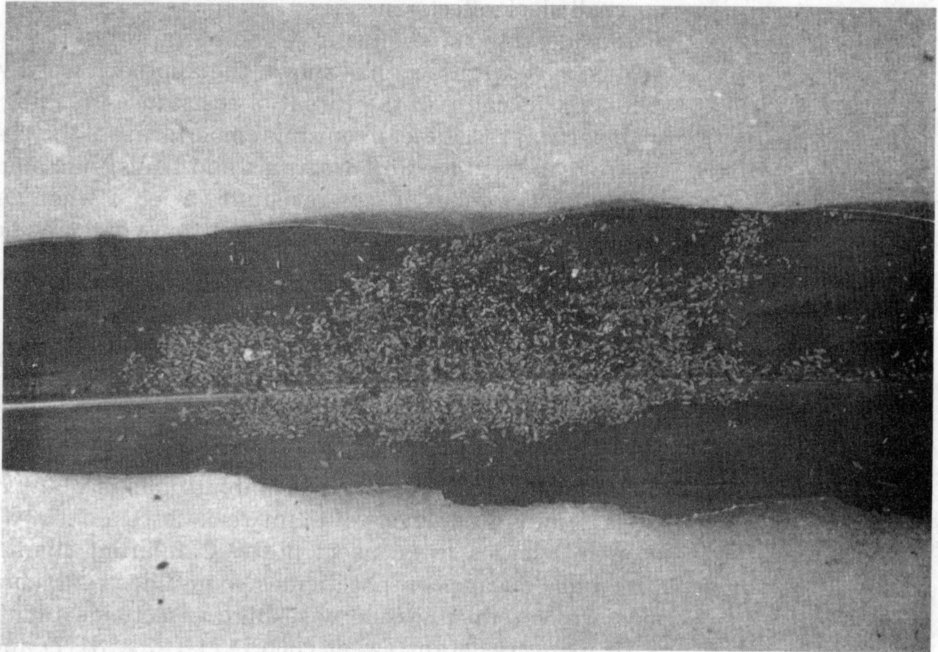

**Figure 13.13** Greenbug, *Schizaphis graminum*, injury to susceptible sorghum. (Photo by Marlin E. Rice)

genotypes have been screened by the United States Department of Agriculture in Kansas and Oklahoma, and eight spring wheat cultivars have shown substantial levels of resistance. Triticale, a hybrid cross of wheat and rye, is also significantly resistant.

However, barleys have shown the highest levels of resistance. These include the American varieties Dicktoo and Will, as well as other varieties developed in China, South Korea, and Japan. Although resistance in barley is controlled by a single dominant gene, biotypes have not been a major problem. The savings from growing greenbug-resistant barley in the United States are estimated at $500,000 annually.

## USE OF PLANT RESISTANCE IN INSECT PEST MANAGEMENT

In insect pest management, plant resistance is used as a preventive measure, and most resistant cultivars have been developed for severe pests where economically significant populations are probable. When varieties providing nonpreference, or antibiosis, are grown, the carrying capacity of the agroecosystem is reduced and the general equilibrium position of the pest population drops. The use of tolerant varieties does not reduce environmental carrying capacities; instead, loss is managed by ensuring yield output. In the first two instances selective pressure is placed on the pest population, and in the third it is not.

### Plant Resistance as the Sole or Primary Tactic

Resistant plants have been used as the primary management tactic where losses are heavy and economics will not allow such curative measures as pesticides. In dealing with most plant diseases, it is the only feasible approach because curative tactics are usually absent and other preventive measures generally lack effectiveness.

Considerable success using resistant varieties as the sole or primary tactic against several insects has shown the approach worthy of continued development. But because of the threat of resistance-breaking biotypes (particularly with vertical resistance), continual monitoring of the pest's status is necessary, as are dynamic breeding programs and transgenic plant development to provide timely supplies of new resistant varieties when needed. The use of horizontal resistance seems not to present these dangers; however, continued pest population assessment is still advisable.

### Plant Resistance Integrated with Other Tactics

A basic principle of insect pest management is that long-lasting solutions to pest problems depend on several harmonious tactics. Plant resistance can serve as an excellent component tactic, because it is usually compatible with pesticides and sometimes with biological agents.

One of the most important combinations is plant resistance and insecticides. Quite often, levels of plant resistance alone are not sufficient to avoid economic losses from insect pests. Combining plant resistance with well-timed low dosages of insecticides sometimes can achieve adequate suppression while reducing otherwise high insecticide inputs. Such was the case with using resistant sweet corn (471-U6 × 81-1) and insecticides against the corn earworm in Georgia. Here, the resistant cultivar plus insecticide resulted in about 93 percent protection, compared with 86 percent for an insecticide-treated susceptible cultivar and 78 percent for the untreated resistant cultivar.

Although plant resistance and natural enemies are not always compatible, integrating them has been effective in certain instances. Theoretical studies have indicated that even small reductions in the growth rate of a pest population feeding on a resistant plant would allow significant increases in the action of a natural enemy. This has been established with resistant varieties of sorghum and barley against greenbugs. Resistant cultivars enhanced the depressive influence of a parasitoid, *Lysiphlebus testaceipes,* on greenbugs severalfold compared with that on susceptible plants.

The reason for complementary and sometimes synergistic effects of plant resistance and natural enemies is not always understood. In some instances secondary plant metabolites may be involved. Some of the chemicals are incorporated into body odors of the pest, and this body odor is detected by natural enemies to locate their hosts. In other instances, natural enemies may be attracted directly by plant volatiles, where subsequently they find their prey or host. Moreover, resistance may prolong the susceptible host stages, giving parasitoids more time to locate and parasitize pests.

It should be mentioned, however, that some characteristics of resistant plants, either morphological or chemical, may be deleterious to natural enemies. This situation could significantly alter the ultimate utility of such cultivars. Much research toward understanding plant-insect-natural enemy relationships, so-called tritrophic interactions, is being conducted. This research should enhance our ability to design new cultivars with greater resistance and more environmental compatibility.

## CONCLUSIONS

Plant resistance has many advantages as a primary tactic in insect pest management strategies. Among the most important are effectiveness, selectivity against the pest, relatively long stability, compatibility with other tactics, and human and environmental safety. In addition, resistant varieties can be adopted into crop production schemes easily and economically, resulting in both short-term and long-term gains.

Although the gains of developing and using resistant plants far outweigh the disadvantages, some of the most important limitations of the approach should be mentioned. Two of these are the time required for development and problems with biotypes.

The time required for the development of new resistant varieties may be as long as 15 to 20 years, as demonstrated by some wheats. Even more time may be required with woody plants. This time is exceptionally long compared with the 6 to 9 years required for insecticide registration. However, the development of plant resistance can be reduced substantially by increasing the rate of funding for research activities. Also, new techniques in selection and breeding are being developed that are expected to shorten this time considerably. Additionally, continued developments in biotechnology are expected to significantly reduce development time.

The problem of biotypes is another worry, particularly with the use of vertical resistance, but this side effect has been much more important with pathogenic microorganisms than with insects. In most instances, entomologists and plant breeders have been able to overcome the hazards of this problem with aggressive, dynamic breeding programs. Genetically engineered plants should make a difference here as well.

Other limitations of plant resistance to insects sometimes include poor agronomic characteristics of resistant cultivars and conflicts of resistant characters between different pest species. In the latter instance, a trait resistant to one insect may be attractive to another. For example, leaf pubescence of soybean leaves is the primary mechanism for resistance to the potato leafhopper, *Empoasca fabae*. However, the hairy surface also stimulates egg laying by the green cloverworm, *Hypena scabra*. In instances of conflicting characters, the relative impact of one pest must be weighed against others and decisions made accordingly. In the example, the destructive potential of the potato leafhopper is much greater than that of the green cloverworm, making resistance to the more important pest preeminent.

Looking toward the near future, we generally concede that plant resistance to insects will continue to rely heavily on identification of sources of resistance and traditional plant breeding. However, with recent successes in biotechnology, we can expect transgenic resistant plants to play an increasingly significant role.

## Further Reading

Ainsworth, G. C. 1981. *Introduction to the History of Plant Pathology.* New York: Cambridge University Press.

Bibikova, M., K. Beumer, J. K. Trautman, and D. Carroll. 2003. Enhancing gene targeting with designed zinc finger nucleases. *Science* 300:764.

California Agriculture. 2004. *Fruits of Biotechnology Struggle to Emerge,* April–June, vol. 58, no. 2. Oakland: University of California, Division of Agriculture and Natural Resources.

Day, P. R. 1972. Crop resistance to pests and pathogens, pp. 257–271. In *Pest Control Strategies for the Future.* Washington, D.C.: National Academy of Sciences.

Edge, J. M., J. H. Benedict, J. P. Carroll, and H. K. Redding. 2001. Bollgard cotton: An assessment of global economic, environmental, and social benefits. *The Journal of Cotton Science* 5:121–136.

Gallun, R. L., and G. S. Khush. 1980. Genetic factors affecting expression and stability of resistance, pp. 63–85. In F. G. Maxwell and P. R. Jennings, eds., *Breeding Plants Resistant to Insects.* New York: Wiley.

Gatehouse, J. A., and A. M. R. Gatehouse. 1998. Genetic engineering of plants for insect resistance, pp. 211–241. In J. E. Rechcigl and N. A. Rechcigl, eds., *Biological and Biotechnological Control of Insect Pests.* Boca Raton, Fla.: Lewis Publishers.

Gianessi, L. P., C. S. Silvers, S. Sankula, and J. E. Carpenter. 2002. Plant biotechnology: Current and potential impact for improving pest management in U.S. agriculture; an analysis of 40 case studies. Washington, D.C.: National Center for Food and Agricultural Policy. Available online at http://www.ncfap.org/40CaseStudies.htm.

Kogan, M. 1994. Plant resistance in pest management, pp. 73–128. In R. L. Metcalf and W. H. Luckmann, eds., *Introduction to Insect Pest Management,* 3rd ed. New York: Wiley.

Letourneau, D. K., and B. E. Burrows, eds. 2002. *Genetically Engineered Organisms.* Boca Raton, Fla.: CRC Press.

Maxwell, F. G., and P. R. Jennings, eds. 1980. *Breeding Plants Resistant to Insects.* New York: Wiley.

Meeusen, R. L., and G. Warren. 1989. Insect control with genetically engineered crops. *Annual Review of Entomology* 34:373–381.

Monsanto Company. 2007. *YieldGard® 2007 TRM Guide,* St. Louis, Mo.

National Academy of Sciences. 1969. *Insect Pest Management and Control, Principles of Plant and Animal Pest Control,* vol. 3. U.S. National Academy of Sciences Publication 1695, pp. 64–99.

National Academy of Sciences, Committee on Genetically Modified Pest-Protected Plants, Board on Agriculture and Natural Resources, National Research Council, P. Adkisson, Chair. 2000. *Genetically Modified Pest-Protected Plants: Science and Regulation,* Washington, D.C.

Norris, D. M., and M. Kogan. 1980. Biochemical and morphological bases of resistance, pp. 23–61. In F. G. Maxwell and P. R. Jennings, eds., *Breeding Plants Resistant to Insects.* New York: Wiley.

Painter, R. H. 1951. *Insect Resistance in Crop Plants.* New York: Macmillan.

Panda, N., and G. S. Khush. 1995. *Host Plant Resistance to Insects.* Wallingford, UK: CAB International.

Rice, M. E. 2004. Transgenic rootworm corn: Assessing potential agronomic, economic, and environmental benefits. *Plant Health Progress*. Available online at: http://www.plantmanagementnetwork.org/pub/php/review/2004/rootworm/

Russell, G. E. 1978. *Plant Breeding for Pest and Disease Resistance.* London: Butterworths.

Shatters, R. G., Jr. 1998. Environmental impact of biotechnology, pp. 281–302. In J. R. Rechcigl and N. A. Rechcigl, eds., *Biological and Biotechnological Control of Insect Pests.* Boca Raton, Fla.: Lewis Publishers.

Smith, C. M. 1989. *Plant Resistance to Insects.* New York: Wiley.

Smith, C. M. 1990. Adaptation of biochemical and genetic techniques to the study of plant resistance to insects. *American Entomologist* 36:141–146.

Smith, C. M. 1998. Plant resistance to insects, pp. 171–208. In J. E. Rechcigl and N. A. Rechcigl, eds., *Biological and Biotechnological Control of Insect Pests.* Boca Raton, Fla.: Lewis Publishers.

Thottappilly, G., L. M. Monti, D. R. Mohan Raj, and A. W. Moore, eds. 1992. *Biotechnology: Enhancing Research on Tropical Crops in Africa.* Ibadan, Nigeria: International Institute of Tropical Agriculture.

Tingey, W. M., and J. C. Steffens. 1991. The environmental control of insects using plant resistance, pp. 131–155. In D. Pimentel, ed., *CRC Handbook of Pest Management in Agriculture,* vol. 1, 2nd ed. Boca Raton, Fla.: CRC Press.

van der Plank, J. E. 1963. *Plant Diseases: Epidemics and Control.* New York: Academic Press.

## Favorite Web Sites

http://www.ipmworld.umn.edu/chapters/teetes.htm
Principles of plant resistance to insects, focusing on specific examples of sorghum resistance to sorghum midge and greenbug.

http://sustainable.tamu.edu/slidesets/ipm/ipm30.html
Examples of plant varieties resistant to insects, with color photos.

http://www.ncfap.org/40CaseStudies.htm
Reviews case studies of transgenic crops and potential impact on pest management in the United States.

http://pewagbiotech.org/
Presents latest news and summaries of developments in biotechnology.

http://www.monsanto.com/monsanto/us_ag/content/stewardship/irm/2007/yieldgard.pdf
Monsanto's 2007 grower guidelines for insect resistance management of their YieldGard® products.

CHAPTER 14

# MANAGEMENT BY MODIFYING INSECT DEVELOPMENT AND BEHAVIOR

SINCE THE EARLY 1960s much activity has been focused on the search for safer and more effective insecticides. The goal has been to discover and develop compounds that are truly selective—that is, to find a means of effectively curing an insect problem without exposing applicators to hazards or causing undue harm to nontarget organisms in the environment.

In their search for new insecticidal compounds, investigators have delved deeply into insect physiology for innovative ways to upset normal life processes. Most conventional insecticides affect the insect's nervous system, which functions similarly to our own and that of other animals. This characteristic gives these chemicals broad powers to poison and, consequently, presents important safety risks. To obtain selectivity, disruptions of processes that are unique to the Hexapoda (or at least to the Arthropoda) seem to offer the greatest potential. Presently, the unique life processes receiving the most attention involve growth and development and behavior.

Significant advances in chemical technology in the last three decades have allowed the discovery, identification, and synthesis of specific chemicals that regulate or mediate growth and development and species behavior. These chemicals, not previously identified, potentially provide new means of reducing an insect population. The mode of action with some of these chemicals is to cause premature death from abnormal molting or metamorphosis. Yet others may be used to repel insects from a source or attract them for our advantage.

Because these chemicals operate on systems that are different from those of warm-blooded animals, they are believed to be safe to use in most situations. Several of these new agents have been developed, and their uses are expanding. Many are still in experimental stages. As the desire for environmentally safe management tactics increases, we can expect even wider application of these rather specialized agents.

Therefore, it is important to understand the nature and possible uses of these chemicals. Such understanding will provide a better grasp of pest management objectives and allow realistic evaluations of future developments.

## DISRUPTING NORMAL GROWTH AND DEVELOPMENT

Of the chemicals discussed in this chapter, those that alter normal growth and development are some of the most promising. These chemicals have been called **insect growth regulators (IGRs)**, a name in keeping with the familiar **plant growth regulators (PGRs)** that influence plant growth and phenology. IGRs are also known as biorationals, or third-generation insecticides, to reflect their environmental safety and advanced development (first-generation insecticides are the stomach poisons, and second-generation insecticides are the contact poisons).

IGRs are like conventional insecticides in many ways. They are used to kill or sterilize insects in a given area and can be applied with insecticide equipment to obtain suppression. Because they act on growth and development, however, IGRs are effective only when immature insects are exposed to them. Therefore, timing of applications becomes even more important and limiting than with conventional insecticides. Moreover, because their mode of action requires some time, pest presence, and sometimes levels of injury, must be tolerated longer than with most conventional insecticides.

### The Basis for IGR Development

IGRs affect insects mainly by disturbing the normal activity of insect endocrine systems. To understand their mode of action, we must rely on an understanding of developmental physiology.

**Functions of the principal growth hormones.** Recall that hormones produced by endocrine glands initiate and regulate molting and metamorphosis. The principal hormones involved in these life processes include brain hormone, ecdysone, and juvenile hormone. Neurosecretory cells in the back of the insect brain (in the location of the pars intercerebralis) secrete brain hormone that links environmental stimuli with other hormone systems. When brain hormone is present in the blood, the prothoracic gland is stimulated to secrete ecdysone, which causes the insect to molt. The body form of the insect after the molt is determined by the concentration of juvenile hormone from the corpora allata in the blood.

Insects with gradual metamorphosis experience a gradual reduction in juvenile hormone concentration, which causes subsequent nymphal stages to appear more and more adultlike. At the time of final molt, juvenile hormone is absent, and the adult emerges. The larvae of insects with complete metamorphosis have groups of cells called imaginal disks with the potential of expressing adult characters. Concentrations of juvenile hormone inhibit the growth of these disks, and the insect retains its larval form in subsequent molts. When a larva reaches full growth, juvenile hormone production is drastically reduced, and its level in the blood declines. This causes the insect to take the pupal form with the next molt. The juvenile hormone level declines still further during the pupal stage, allowing active growth of the imaginal disks and resulting in the adult form (Fig. 14.1). In young adults of many species, the corpora allata increase activity once more, producing levels of juvenile hormone that stimulate ovary development and egg yolk production.

In addition to their metamorphic and reproductive functions, juvenile hormones can also have important influences on diapause, behavior, and communication (through pheromone production). Moreover, in social insects they are known to regulate caste differentiation and have special influences on morphogenesis.

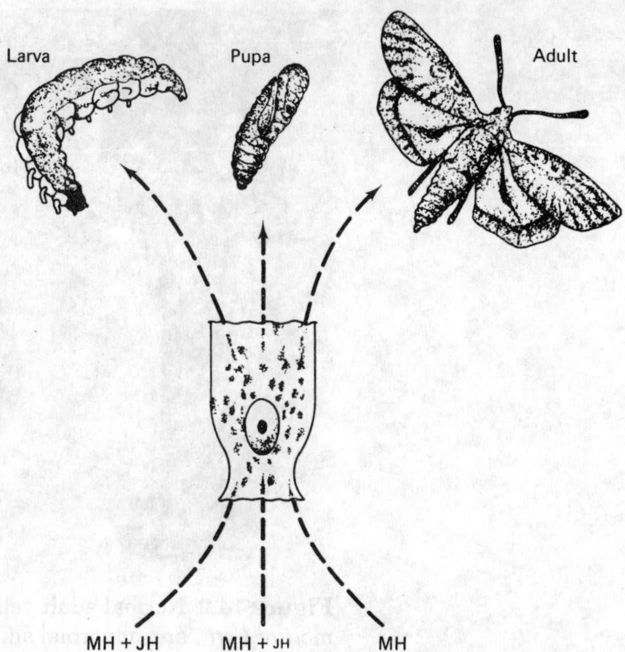


**Figure 14.1** Diagrammatic representation of the role of the principal hormones in complete metamorphosis. MH symbolizes molting hormone and JH symbolizes juvenile hormone. Symbol size represents relative concentrations. (Reprinted with permission of Macmillan Publishing Company from *Insects in Perspective* by Michael D. Atkins. © 1978 by Michael D. Atkins)

**Experimental modification of growth hormones.** Significant modifications in the balance among the principal hormones causes aberrations and malformations of body structures. Several effects of hormone modification have been demonstrated experimentally. In studies where the brain was removed from immature insects, production of brain hormone ceased, and further molting was suspended. When active brains from other individuals were implanted into these immatures, molting resumed. Removal of the corpora allata from young immatures caused cessation of juvenile hormone production, followed by premature pupation to form dwarfed adults. When active corpora allata from young immatures were implanted into full-grown immatures, the titer of juvenile hormone increased, and the insects remained in an immature stage with the next molt. When active corpora allata were implanted into pupae, monstrosities developed, with mixtures of pupal and adult characteristics. More advanced studies with extracts containing juvenile hormone from male cecropia moths, *Hyalophora cecropia,* showed that the material could penetrate the insect cuticle and that topical applications to immatures could produce lethal developmental derangements (see also the yellow mealworm, Fig. 14.2). In instances where insects survived to reach adulthood, they were sterile.

**Hormone mimics.** In other laboratory observations, it was found quite by chance that some plant compounds produce results in insects similar to increasing the concentration of juvenile hormone. While attempting to rear the

**Figure 14.2** Normal adult yellow mealworm, *Tenebrio molitor* (*left*), and abnormal adult treated topically with low doses of a juvenile hormone mimic. The abnormal adult retains a number of pupal structures along with the adult structures. The pupal skin of the abnormal individual was removed artificially because normal emergence would not have occurred. (Courtesy USDA)

bug *Pyrrhocoris apterus* in cages with a paper towel lining, researchers obtained extra-large nymphs that were unable to molt into adults. By process of elimination, they found that the paper used, made from balsam fir, caused this reaction in the insects. Later, it was discovered that the "paper factor" resulted from the presence of a chemical, now called **juvabione,** that is structurally similar to juvenile hormone. With continued research, it has been shown that such compounds, called hormone mimics, are present in a variety of plants, where they may function in part as defensive mechanisms against insect herbivores. For example, many ferns have high concentrations of these chemicals and are fed upon by only a few species of insects that have overcome this defense.

**Synthetic hormones.** With the discovery of hormone functions, particularly those controlling metamorphosis and reproduction, isolation and identification of natural hormones were undertaken in serious detail. Glandular and other extracts from many insect species produced a variety of compounds; however, the discovery of significant juvenile hormone activity in the frass (excrement) of the yellow mealworm, *Tenebrio molitor,* provided a breakthrough. With this discovery, the active compound **farnesol** was identified, and this eventually culminated in the development of the first synthetic hormone useful in pest management.

**Other potential IGRs.** In attempting to upset hormone balances, several possibilities exist. Potentially, concentrations in insect blood could be either increased or decreased to cause disruption. Increasing levels of juvenile hormone analogs or mimics has seemed more easily achieved than attempting to decrease it. However, a problem with agricultural pests is that adding juvenile hormone also keeps the insects in an immature and potentially injurious stage longer than normal. Therefore, if a method of shutting off juvenile hormone production could be found, it would be more useful because it could prevent damage initially and reduce populations in the long run.

With the knowledge that some plants produce compounds that mimic juvenile and molting hormones, researchers tested plant extracts to see if they may also produce antagonists that inhibit juvenile hormone. Subsequently, one such antagonist was discovered in the bedding plant, *Ageratum houstonianum*. Two compounds were identified and named precocene I and precocene II. The name precocene was given to these compounds because they induce precocious metamorphosis and because of their chromene chemical structure. In contact feeding and fumigation studies, these so-called antijuvenile hormones were found to induce premature metamorphosis in insects with gradual metamorphosis but not in those with complete metamorphosis. In many species with complete metamorphosis, however, the adults became sterilized after exposure. The mode of action of these compounds seemingly is lethal activation within the corpora allata, thus destroying the glands.

Researchers have investigated several other materials, prototypes shown to interfere with juvenile hormone synthesis, in an attempt to develop practical materials. These include FMev (fluoromevalonate), piperonyl butoxide, ETB (4-[2-[(*tert*-butylcarbonyl)oxy]butoxy]benzoate), and EMD (ethyl (*E*)-3-methyl-2-dodecanoate). But none has been shown to be sufficiently active for practical purposes. Also, it was demonstrated that some allylic alcohols offer promise as antijuvenile hormone agents against Lepidoptera.

Development of antijuvenile hormones for actual use in pest management remains a question. However, because of the diversity of chemicals that show juvenile hormone disruption, there is still hope that discoveries of practical materials will be forthcoming.

Finally, another potential IGR that has been considered is ecdysone. Like juvenile hormone mimics, ecdysones are found in plants, where they may also play a part in insect/plant relationships. But ecdysones are steroidal compounds with complex molecules, and many tests produced only low-level responses to application. Together, these characteristics tended to discourage development. More recently, however, **azadirachtin**, an ingredient in the seeds of neem trees (a subtropical species), has been found to disrupt molting and has been formulated as a liquid and a dust from ethanol extracts. This IGR, a botanical insecticide that antagonizes ecdyson, is being marketed (see Chapter 11).

## Practical IGRs

In addition to azadirachtin, there are several other practical IGRs available for pest management programs. Indeed, some truly exciting developments have taken place recently that seem to be ushering in a new era of safe and effective insecticides. These new IGRs are ideal pest management tools because they are usually selective for the targeted pests and can be used against insects resistant to conventional insecticides.

Practical IGRs can be divided into two categories: those that disrupt metamorphosis and those that disrupt molting. IGRs that disrupt metamorphosis are juvenile hormone analogs or mimics that kill or sterilize by acting like juvenile hormone. In other words, they force insects to remain in the immature stage

**Table 14.1 Insect Growth Regulators and Their Activity Characteristics.**

| Name | Activity |
|---|---|
| Bistfluron | Chitin synthesis inhibitor |
| Buprofezin | Chitin synthesis inhibitor |
| Chlorfluazorun | Chitin synthesis inhibitor |
| Cyromazine | Chitin synthesis inhibitor |
| Diflubenzuron | Chitin synthesis inhibitor |
| Flucycloxuron | Chitin synthesis inhibitor |
| Flufenoxuron | Chitin synthesis inhibitor |
| Hexaflumuron | Chitin synthesis inhibitor |
| Lufenuron | Chitin synthesis inhibitor |
| Noyaturon | Chitin synthesis inhibitor |
| Novaluron | Chitin synthesis inhibitor |
| Penfluron | Chitin synthesis inhibitor |
| Teflubenzuron | Chitin synthesis inhibitor |
| Triflumuron | Chitin synthesis inhibitor |
| Epofenonane | Juvenile hormone mimic |
| Fenoxycarb | Juvenile hormone mimic |
| Hydroprene | Juvenile hormone mimic |
| Kinoprene | Juvenile hormone mimic |
| Methoprene | Juvenile hormone mimic |
| Pyriproxyfen | Juvenile hormone mimic |
| Triprene | Juvenile hormone mimic |
| Juvenile hormone I | Juvenile hormone anolog |
| Juvenile hormone II | Juvenile hormone anolog |
| Juvenile hormone III | Juvenile hormone anolog |
| Chromafenozide | Molting hormone agonist |
| Halofenozide | Molting hormone agonist |
| Methoxyfenozide | Molting hormone agonist |
| Tebufenozide | Molting hormone agonist |
| α-ecdysone | Molting hormone anolog |
| Ecdysterone | Molting hormone anolog |
| Diofenoian | Molting inhibitor |

SOURCE: From H. Tunaz, and N. Uygun, 2004, Insect growth regulators for insect pest control, *Turkish Journal of Agriculture and Forestry* 28:377–387.

abnormally long. IGRs that disrupt molting are chemicals that either inhibit chitin synthesis or accelerate the onset of molting, thereby killing the insect.

IGRs are practical in insect pest management where instantaneous control is not needed. Additionally, they are most appropriate for insects that have short life cycles and in instances where low densities can be tolerated. For a complete listing, see Table 14.1.

**Methoprene.** Methoprene and its S enantiomer are IGRs with good activity against many Diptera, Siphonaptera, and Coleoptera and some Lepidoptera and Hemiptera. It is sold under the trade name Altosid® for suppression of mosquito larvae and horn flies, *Haematobia irritans.* In the latter instance, Altosid is a food additive for cattle, through which it passes with sufficient integrity to affect developing maggots in cow dung. It has been used with cattle in feedlots. Other uses for methoprene include Coleoptera and Lepidoptera in stored tobacco (Kabat®), sciarid flies in mushrooms (Apex®), pests of stored products (Dianex®), and leaf-mining flies (Agromyzidae) in vegetable and flower crops

(Minex® IGR). Methoprene is also used to enhance silk production in silkworms (*Bombyx mori*).

Methoprene (Altosid®, others)

isopropyl (2*E*-4*E*)-11-methoxy-3,7,11-trimethyl-2,4,-dodecadienoate

**Hydroprene.** Hydroprene and its S enantiomer IGRs are sold under the trade names Gentrol® and Mator®. This IGR is related to methoprene and used indoors against cockroaches. In this instance, exposure of nymphs to the material causes them to become sterile adults. Outward symptoms of sterile individuals are twisted and deformed wings. Hydroprene may offer a new solution for pest species that have become resistant to conventional insecticides.

Hydroprene (Gentrol®, Mator®)

(*E,E*)-ethyl (2*E*,4*E*)-3,7,11-trimethyl-2,4-dodecadienoate

**Kinoprene.** Kinoprene and its S enantiomer are strong, highly selective juvenile hormone analogs effective against some Hemiptera. Sold under the trade name Enstar II®, it is effective against aphids, whiteflies, scales, and mealybugs. Because it lacks environmental stability, it has been used most successfully with ornamental plants and vegetable seed crops in greenhouses and shade houses.

Kinoprene (Enstar II®)

2-propynyl (*E,E*)-3,7,11-trimethyl-2,4-dodecadienoate

**Pyriproxyfen.** This IGR is a juvenile hormone mimic used against public health pests such as flies, beetles, midges, and mosquitoes. For these, it is applied to breeding sites, including swamps and livestock houses. Also, it is active against scales and whiteflies on cotton, citrus, vegetables, and pome fruit. Pyriproxyfen is marketed as Knack®, Admiral®, Adeal®, Atominal®, Epingle®, Juvinal®, Lano®, and Nemesis®.

Pyriproxyfen (Knack®, others)

2-[1-methyl-2-(4-phenoxyphenoxy)ethoxy]pyridine

**Diflubenzuron.** This compound (see Figs. 14.3, 14.4), sold as Dimilin®, Adept®, Micromite®, Agmilin®, and Difuse®, has greater environmental stability than most juvenile hormone analogs and mimics and is registered for use against the gypsy moth, *Lymantria dispar,* in forests, and boll weevil, *Anthonomus grandis,* in cotton. Moreover, registration has been obtained for a variety of other pests on several crops, including cotton, soybeans, citrus, and vegetables. It is also very useful against nuisance and medical pests like flies, gnats, midges, and mosquitoes.

Diflubenzuron (Dimilin®, Adept®, Micromite®, others)

1-(4-chlorophenyl) 3-(2,6-difluorobenzoyl)urea

The mode of action of diflubenzuron is quite different from the juvenile hormone analogs. Diflubenzuron acts on the larvae of most insects by inhibiting chitin synthesis and thus affects the integrity of the insect exoskeleton. Exposure causes improper attachment of the new cuticle during molting and produces a cuticle that lacks some of the layers that normally occur. Most larvae die from ruptures of the new malformed cuticle or from starvation.

**Lufenuron.** Like diflubenzuron, lufenuron is a chitin synthesis inhibitor. This IGR is prescribed by veterinarians for flea control on dogs. It is administered as a tablet once per month at meal time, after which it enters the dog's circulatory system. When a female flea bites a treated dog, lufenuron enters the flea's system and is deposited in her eggs. This prevents some eggs from hatching and prevents any eclosed larvae from becoming adults. Therefore, flea populations do not build up in households, thus eliminating flea problems. Lufenuron is available from veterinarians under the trade name Program®.

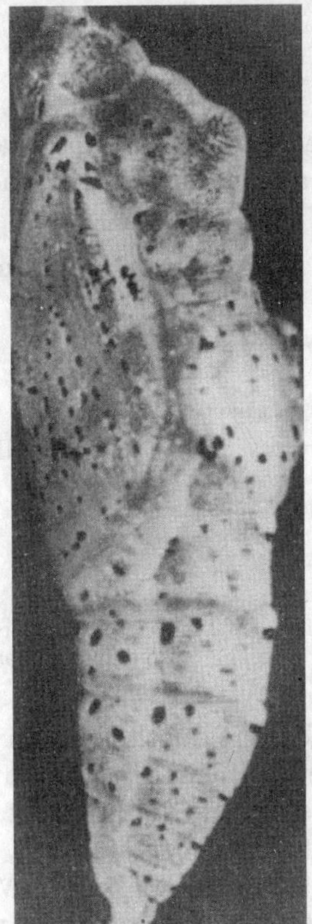
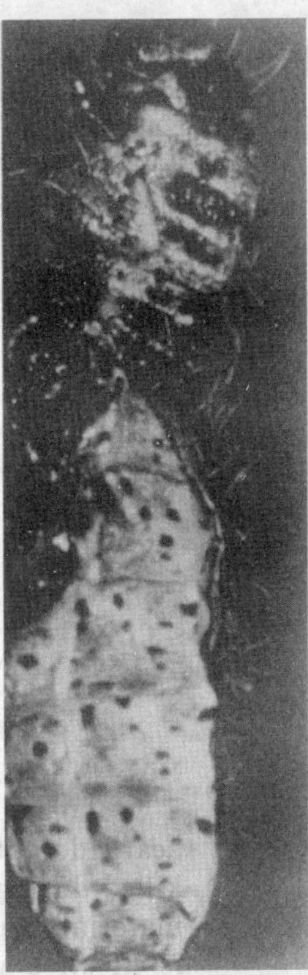

**Figure 14.3** Pupa of a butterfly (Lepidoptera) in normal condition (*left*) and abnormal condition. Abnormal insect was treated with diflubenzuron during the last stage of the larva. In the abnormal pupa, the abdomen has a pupal form, and the head and thorax have larval forms. (Courtesy Philips-Duphar, B.V., Amsterdam, the Netherlands)

**Figure 14.4** Dying larvae of the beet armyworm, *Spodoptera exigua*, that were in contact with diflubenzuron. Chitin production for the insects' cuticles was inhibited. Consequently, the molting process was upset, and the insects could not escape the old skin. The chemical can also upset cuticle development in embryos still in the egg, which causes hatching failure. (Courtesy R. T. Weiland and Uniroyal Chemical)

Lufenuron is also used against Lepidoptera and Coleoptera larvae on cotton, corn, and vegetables, and citrus whitefly and rust mites on citrus.

Lufenuron (Program®, others)

$$CF_3CHFCF_2O\text{—}C_6H_2Cl_2\text{—}NHCONHCO\text{—}C_6H_3F_2$$

*N*-[[[2,5-dichloro-4-(1,1,2,3,3,3-hexafluoropropoxy)phenyl]amino]carbonyl]-2,6-difluorobenzamide

**Buprofezin.** Buprofezin is another chitin synthesis inhibitor that produces a weakened exoskeleton in molting immatures. This is an insecticide and acaricide with both stomach and contact action. Nymphs and larvae are killed during the last immature stadium, and adults lay sterile eggs. The material, marketed as Buprolex®, Butyl®, Maestro®, Profezon®, and Viappla®, is effective against certain Hemiptera in rice, potatoes, cotton, vegetables, and citrus.

Buprofezin (Buprolex®, Butyl®, others)

S, N, $=NC(CH_3)_3$, N, O, $CH(CH_3)_2$

2-[(1,1-dimethylethyl)imino]tetrahydro-3-(1-methylethyl)-5-phenyl-4*H*-1,3,5-thiadiazin-4-one

**Hexaflumuron.** This is one of the most useful IGRs in urban pest control. It is another chitin synthesis inhibitor, and is used primarily for termite control around houses and other buildings. Termite colonies are eliminated with this IGR using a Sentricon® System (Fig. 14.5). The system is made up of a plastic tube with side openings, which is placed in the soil around the structure to be protected. At first, a monitoring material is placed in the tube to detect and monitor termite activity. When termites are found at the station, the monitoring material is replaced with a bait containing hexaflumuron. Termites feed on the hexaflumuron bait and die within 6 to 8 weeks, during molting. For effectiveness, the system depends on the termite behavior of trail marking. When termites find food, they return to the colony and leave a trail marked with pheromone, which leads other termites to the food source. With the Sentricon® System, termites feeding at the bait station are eventually killed, but before they die they lead others from the colony to the

**Figure 14.5** Use of the Sentricon® System by a pest control professional. Here, monitoring material is being placed in the buried plastic tube. Such monitoring stations are placed near houses to detect termite activity. If termites are detected, the monitoring material is replaced with a hexaflumuron bait. (Courtesy S. H. Hutchins and Dow AgroSciences)

bait. Eventually the whole colony is eliminated, which may require 1 to 7 months (Fig. 14.6).

Hexaflumuron is also effective against certain Lepidoptera, Coleoptera, Hemiptera, and Diptera on top fruit, cotton, and potatoes. Where registered, the principal trade names are Consult®, Trueno®, and Recruit®.

Hexaflumuron (Sentricon®, Consult®, others)

F, F (2,6-difluorophenyl) — CONHCONH — (3,5-dichlorophenyl: Cl, Cl) — $OCF_2CHF_2$

*N*-[[[3,5-dichloro-4-(1,1,2,2-tetrafluoroethoxy)phenyl]amino]carbonyl]-2,6-difluorobenzamide

**Novaluron.** This IGR is another chitin synthesis inhibitor that acts by ingestion but also has some contact activity. It is sold under the trade names Rimon®, Pedestal®, and Oscar®. It is used for codling moth in pears, whiteflies on tomato, and Lepidoptera and grasshoppers on mustard greens, among other uses.

Novaluron (Rimon®, others)

*N*-[[[3-chloro-4-[1,1,2-trifluoro-2-(trifluoromethoxy)ethoxy]phenyl]amino]carbonyl]-2,6-difluorobenzamide

**Figure 14.6** Operation of the Sentricon® System. Monitoring stations are placed at several stations near the foundation of a house (*top left*). When termites find and feed on the monitoring material, they are collected and transferred to the hexaflumuron bait. The bait and termites are placed in the bait stations (*top right*). The termites return to the colony before dying and recruit other termites to the bait stations (*bottom left*). This process continues repeatedly until the entire termite colony is eliminated (*bottom right*). Once the colony is eliminated, the stations are maintained in the monitoring mode to detect any new termite invasions.

**Tebufenozide.** This material represents a new class of IGRs known as ecdysone agonists. Tebufenozide acts by binding to the ecdysone receptor protein, which accelerates the molting process and causes insect death. In other words, it imitates the action of ecdysone. Once a larva consumes a lethal dose, it ceases feeding and produces a new but malformed cuticle beneath the old cuticle. The larva, unable to shed its old cuticle, dies of dehydration and starvation. Tebufenozide is effective against lepidopteran pests in rice, fruit, row crops, nut crops, vegetables, vine crops, and forests. The IGR is marketed as Confirm®, Mimic®, and Fimic® and seems to have little direct effect on insect predators and parasitoids.

Tebufenozide (Confirm®, Mimic®, Fimic®)

$CH_3CH_2$ — [benzene ring] — $CONHNCO$ — [benzene ring with two $CH_3$]; $C(CH_3)_3$

3,5-dimethylbenzoic acid 1-(1,1-dimethylethyl)-2-(4-ethylbenzoyl)hydrazide

**Methoxyfenozide.** This IGR, marketed as Intrepid®, acts similarly to tebufenozide but is a plant systemic, entering through plant roots. Lepidopteran pests, including codling moth, oriental fruit moth, and European corn borer, are affected. Current crops are cotton and pome fruits.

Methoxyfenozide (Intrepid®)

$H_3C$, $H_3C$ — [benzene ring] — $C(=O)$—N—NH—$C(=O)$ — [benzene ring with $H_3C$ and O—$CH_3$]; $H_3C$—C—$CH_3$, $CH_3$

3-methoxy-2-methylbenzoic acid 2-(3,5-dimethylbenzoyl)-2-(1,1-dimethylethyl)hydrazide

**Halofenozide.** Halofenozide is another plant systemic with activity for cutworms, sod webworms, armyworms, and white grubs. It does not have the stomach and contact poison characteristics of tebufenozide and methoxyfenozide. Feeding stops soon after the material is ingested. Halofenozide is marketed as Mach-2® for use on turf.

### Compatibility of IGRs with Other Tactics

It is not yet possible to tell if the IGRs will be compatible with other pest management tactics against insects, although some evidence suggests they will. Certainly, the juvenile hormone analogs satisfy many of the important selectivity requirements and would not be expected to strongly impact arthropod enemies. However, some have broad activity against insects and may reduce natural enemy populations, as would conventional insecticides.

In either case, much development remains before these compounds can be widely integrated with other tactics. Lack of environmental stability is the most serious weakness for some juvenile hormone analogs in agriculture. Also, slower activity may also preclude their use in certain crops where blemish-free produce is a priority, although antifeeding activity in some is a plus.

The great advantage of IGRs is safety. They all display low toxicity to warm-blooded animals, which makes them safe to apply and, it would seem, environmentally compatible.

## MODIFYING BEHAVIOR PATTERNS

So far in this discussion, we have dealt with means of disrupting physiological processes within the insect's body, specifically, endocrine-related processes. Other advanced pest management tactics attempt to mediate processes outside the body to change insect behavior patterns.

As we have noted in previous chapters, insects are involved in many interactions with individuals of their own species and others in the environment. Such interactions involve complex and predictable behavior patterns. For example, mating involves complex intraspecies (within a species) behavior, and host finding and natural-enemy avoidance include intricate interspecies behaviors.

In an elementary sense, insect behavior patterns that result in many interactions can be related to degrees of attraction and repellency; for example, males are attracted to females, and individuals are repelled by a noxious plant. Associated behaviors where shorter distances are involved include stimulation and deterrence. These are sometimes more appropriate terms to use when we speak of attraction or repellency.

Attraction, repellency, stimulation, and deterrence are often mediated by chemicals from the exocrine systems of the interacting individuals. As mentioned earlier, these mediating chemicals are called semiochemicals. They can be further subdivided into pheromones, involved in intraspecific communication, and allelochemics, involved in interspecific communication.

In insect pest management, some of the newest tactics involve manipulating insect behavior through the application of semiochemicals in the insect's environment. Such applications have been used to disrupt insect perception and compel individuals to behave according to our desires.

### Tactics Involving Insect Attraction

The idea of using attractants in pest control is not a new one. As far back as the early 1900s, food lures or baits treated with insecticide were used to suppress a variety of agricultural pests. Such poisonous baits serve to attract insects, entice them to feed, and kill them outright. This approach is still used widely in households and other situations where exposure to broad applications of an insecticide presents unacceptable risks.

From a tactical standpoint, attractants can be aimed at luring insects to a source where they are killed, as with poison baits, or simply used to draw them away from a valued commodity to one that is expendable, such as a trap crop.

However, the attract-and-kill approach has been the most common. Another notion is to attract insects with a chemical so compelling as to divert them from performing a vital activity, specifically, mating. Theoretically, a successful application in this way would reduce numbers by lowering reproductive rates.

Although the tactic of attraction is not new, some of the agents involved are. Therefore, in addition to traditional food lures, potent attractants like pheromones, totally unknown in the past, are increasingly being applied. Moreover, new formulations and methods of dispensing attractants have spurred a renewed interest in these compounds.

**Use of pheromones in attraction.** Although observations of individual insects attracting other insects of the same species were reported hundreds of years ago, it was not until 1959 that A. Butenandt and colleagues first reported the isolation and identification of a pheromone. This pheromone (*trans, cis*-10,12-hexadecadien-l-ol), isolated from silkworms (Box 14.1), was a sex attractant involved in successful matings of the species. Since this first report,

## BOX 14.1 SILKWORM

**SPECIES:** *Bombyx mori* (Linnaeus) (Lepidoptera: Bombycidae)

**DISTRIBUTION:** This species, the only member of the Bombycidae, is native to Asia. The insect has been reared artificially for centuries in silk production. After years of rearing, it is now mostly a domestic species and probably would not survive in nature. Many different varieties of silkworms have been developed for silk production, and rearing occurs in many countries, including the United States.

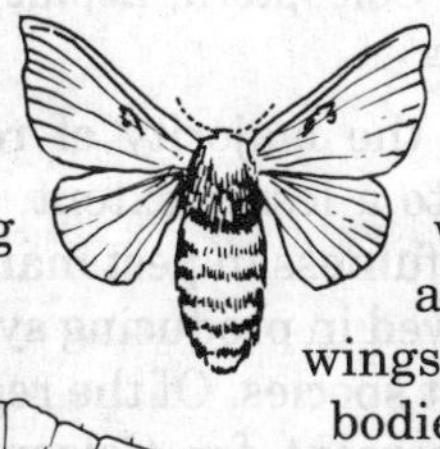

**IMPORTANCE:** The process of obtaining silk produced by silkworms, called sericulture, has been practiced for more than thirty-five centuries. For more than 2,000 years only the Chinese knew the origin of silk (discovered by Lotzu, Empress of Kwang-Ti, about 2697 B.C.). Anyone who tried to reveal the secret of silk production outside of China was quickly executed. It was not until A.D. 555 that two monks, sent as spies to China, discovered the secrets and returned some silkworm eggs to Europe in a hollow staff. Today, sericulture is practiced extensively in Japan, China, Spain, France, and Italy and has a commercial value from \$200 to \$500 million annually (around 100 billion larvae are cultured each year). Silk is formed when full-grown larvae spin cocoons in preparation for pupation. In this process, approximately 914 meters of silk are produced by a single larva. For commercial purposes, pupae are killed prior to emergence because eclosion will break the valuable silk fibers. Approximately 3,000 cocoons are required to make one pound of commercial silk.

**APPEARANCE:** Adult moths are creamy white and have several faint brown lines across the forewings. They have a wingspan of about 50 mm and have heavy, hairy bodies. Larvae are naked but possess a short anal horn. Pupae are large and elliptical. Eggs are yellowish white and semispherical.

**LIFE CYCLE:** Pupae allowed to survive to maintain the colony emerge as moths that do not feed and that live only 2 to 3 days. In this period a single moth will lay 300 to 500 eggs. Colonies are maintained at 25°C, and larvae molt at 6, 12, 18, and 26 days after eggs hatch. During larval growth, mulberry leaves are supplied for food, with each larva consuming about 90 grams of foliage. Approximately 3 days are required to produce the silk cocoon.

hundreds of pheromones have been identified in many organisms, ranging from algae to primates. Such advancements have been possible with the many significant developments in analytical chemistry.

Although sex pheromones are the most widely discussed, insects possess several other types. A categorization of insect pheromones is as follows:

1. **Sex pheromones.** These substances are often produced by females to attract males for mating, but they may also be produced by males to attract females. They seem to be the most highly developed in Lepidoptera and are frequently produced by eversible glands at the tip of the abdomen. The release of sex pheromones is a complex physiological process, often associated with sexual maturity and environmental stimuli such as photoperiod and light intensity. Female sex pheromones are usually received by sensory sensillae on male antennae, and males search upwind, following the odor corridor of the female.
2. **Alarm pheromones.** These pheromones are common in social insects such as ants and bees. They elicit attack or retreat behavior. For instance, the remains of a sting apparatus of a honey bee left in a victim's body releases an alarm pheromone that attracts other bees and stimulates them to sting.
3. **Trail-marking pheromones.** These chemicals are produced by foraging ants and termites to indicate sources of requisites to other members of the colony. As mentioned, these are important in the efficacy of the Sentricon® System.
4. **Aggregation pheromones.** These pheromones are prominent in some species of beetles and cause insects to aggregate or congregate at food sites, reproductive habitats, hibernation sites, and the like. They are particularly well known in bark beetles, *Ips* species and *Dendroctonus* species, where they are involved in tree attacks.
5. **Epideictic pheromones.** Epideictic pheromones, also called spacing pheromones, elicit dispersal away from potentially crowded food sources, thereby reducing numbers. Hence, they are one of the few pheromones that serve to repel, rather than attract. They are produced by bark beetles, as well as other Coleoptera, Lepidoptera, Diptera, Hemiptera, Orthoptera, and Hymenoptera.

Of these types, the majority of research to date has been done with sex pheromones and, to a lesser extent, aggregation pheromones, both because of their potential usefulness in pest management. Today, more than forty business concerns are involved in producing synthetic pheromones (parapheromones) for more than 250 pest species. Of the registered products, about 80 percent are for Lepidoptera, 10 percent for Coleoptera, and the remaining 10 percent for Diptera, Blattodea (cockroaches), and Hymenoptera (see Ridgway et al. 1990 for a listing).

These products have been used in three basic ways: (1) in sampling and detection, (2) to attract and kill, and (3) to disrupt mating. By far most of the products (about 90 percent) are used in sampling and detection.

**Pheromones in sampling and detection.** The use of sex pheromones as attractants in traps is one of the oldest practical applications of semiochemicals in pest management. Presently, either sex or aggregation pheromones may be employed to monitor insect activity and obtain detection, phenology, and relative density information.

Pheromone traps (Fig. 14.7) are used frequently to gain information about pests for making tactical decisions. The first insect caught can serve as the beginning point for the accumulation of degree days, or catches over a period of time are useful in predicting population peaks or egg hatching times. Such

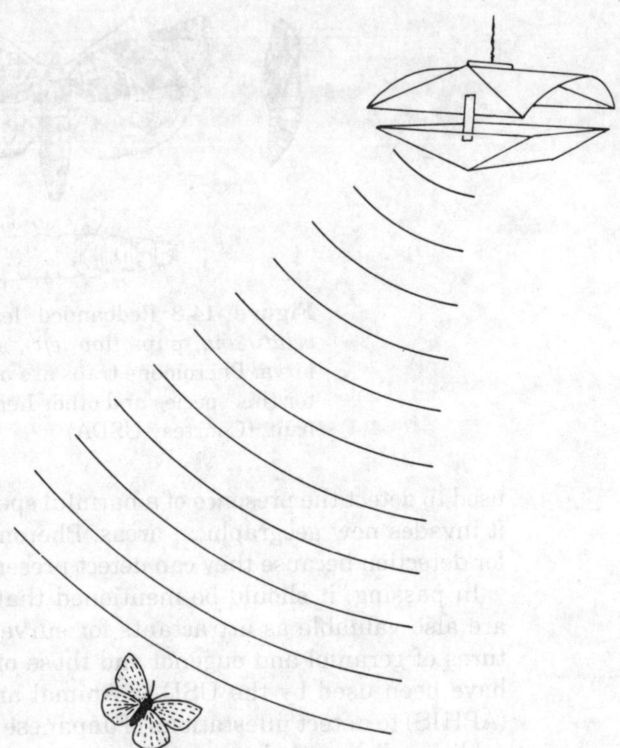

**Figure 14.7** A so-called wing trap used to monitor Lepidoptera and other insects. Some models have reusable plastic tops and disposable, sticky-coated bottoms. Insects are lured inside by pheromone bait containers and stick to the bottom of the trap. Counts are made at regular intervals.

predictions are useful in deciding if insecticides are necessary and, if so, when they should be applied.

Some of the most extensive uses of pheromone traps for making pest management decisions have occurred in apple orchards. Strategically placed pheromone traps, continually monitored, have reduced reliance on pesticide sprays and increased grower profits, as compared with calendar-spray schedules. For example, in a New York pest management program, growers sampled five major lepidopterous pests with pheromone traps and applied pesticides according to trap data, natural enemies, and weather information. Compared with other growers, those following the trapping regimen spent $60 less per hectare, without decrease in quality or quantity of fruit, and pesticide inputs (insecticides and acaricides) were reduced by 50 percent. In the Netherlands such programs in apple orchards allowed the total number of seasonal sprays to be reduced by two or three.

Pheromone traps are used regularly to monitor codling moth (*Cydia pomonella*), redbanded leafroller (*Argyrotaenia velutinana*) (Fig. 14.8), and other leafrollers in deciduous fruit; boll weevil, pink bollworm (*Pectinophora gossypiella*), and corn earworm (*Helicoverpa zea*) in cotton; black cutworm (*Agrotis ipsilon*) in corn; and California red scale (*Aonidiella aurantii*) in citrus. In addition to these examples of pest monitoring for decision making, pheromone traps are

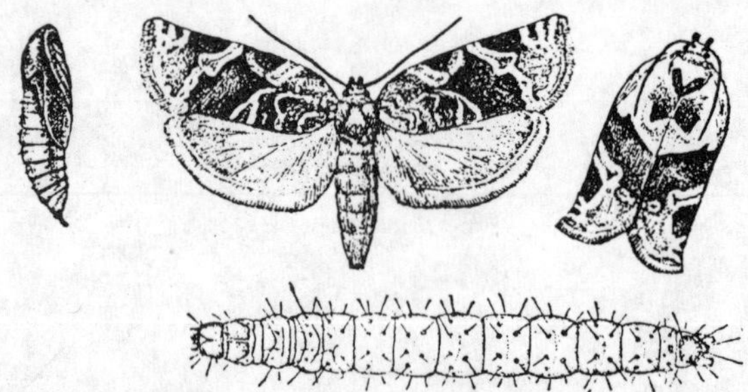

**Figure 14.8** Redbanded leafroller, *Argyrotaenia velutinana,* pupa (*top left*), adults (*top right*), and larva. Pheromone traps are used regularly to monitor this species and other Lepidoptera in deciduous fruit. (Courtesy USDA)

used to detect the presence of a harmful species, for example, the gypsy moth, as it invades new geographical areas. Pheromone traps are particularly valuable for detection because they can detect presence when pest numbers are very low.

In passing, it should be mentioned that chemicals other than pheromones are also valuable as attractants for survey and detection. For instance, mixtures of geraniol and eugenol and those of phenethyl propionate and eugenol have been used by the USDA's Animal and Plant Health Inspection Service (APHIS) to detect infestations of Japanese beetle, *Popillia japonica* (Fig. 14.9). Additionally, methyl eugenol has been used to detect oriental fruit fly, *Dacus dorsalis,* and mango fruit fly, *Dacus zonatus,* and trimedlure has been used to

**Figure 14.9** Japanese beetle, *Popillia japonica,* trap baited with phenethyl propionate and eugenol. (Courtesy USDA)

monitor the activity of Mediterranean fruit fly, *Ceratitis capitata*. Several other chemicals also are used to form an early warning network for action against many exotic and potentially harmful species.

Pheromone traps and traps with other attractants provide inexpensive, effective monitoring for many pests. However, the interpretation of trap catch is sometimes difficult, particularly when attempting to estimate relative density. This is because numbers caught are not always directly proportional to actual population size. Some factors to consider in utilizing pheromone traps and interpreting their catch include: (1) attractiveness of the pheromone, (2) pheromone concentration and release rate, (3) trap design (including color), (4) trap placement, (5) trap durability, and (6) area of trap influence.

**Pheromones used in attract-and-kill programs.** With this approach, insects are attracted to a source and killed by one means or another. Techniques in attract-and-kill programs range from entanglement in sticky materials to outright killing with insecticides or pathogenic microorganisms. The conventional approach to attract-and-kill has been to use traps lined with sticky material, and others have allowed slow-release formulations of small particles that emit both a pheromone and an insecticide.

In theory, when sex pheromones are used, pheromone sources affect half the population (for insects with a 1:1 sex ratio). If a large portion of one sex is attracted and killed, those individuals are eliminated, mating success is reduced, and numbers in the next generation fall. In species whose members mate more than once, lures that attract females have the greatest effect on subsequent generations. When males are attracted and killed, a very high percentage must be removed to have an effect. For instance, if each male mates with ten females, 90 percent of the males would need to be removed at high population densities to cause a significant reduction in the next generation.

When pheromone traps are used in attract-and-kill programs, the approach is often referred to as mass trapping. Traps used in mass trapping are often the same as those used in sampling and detection programs. They may or may not have an active killing agent. Operationally, the distinction between mass trapping and sampling is usually in the number of traps used. For mass trapping, trap densities as great as 100 per hectare may be utilized, compared with 5 to 10 per hectare in sampling.

Mass trapping, or trap out, has been used in tree fruit, field, and forest crops, as well as with stored product and household pests. For example, tests with the boll weevil showing promise involved pheromone traps placed in strips of early planted cotton to aggregate the beetles. After aggregation was accomplished, the strips were treated with an insecticide. Trapping has also shown promise for use against some stored grain pests, like the khapra beetle, *Trogoderma granarium,* and with dermestid beetles in stored products. In the latter instance when combined with spores of the protozoan *Mattesia trogodermae,* mortalities of 100 percent have been observed. To date, one of the largest mass-trapping programs has been conducted with a very destructive forest pest, the spruce bark beetle (*Ips typographus*), in Norway and Sweden.

Another approach to attract-and-kill is the application of slow-release particles. With this technique, small particles are applied over an area by airplane. Each particle is a dispenser for both pheromone and insecticide, which are released gradually. Insects are attracted and exposed to a contact insecticide when they attempt to mate with the particles. Such formulations are sometimes referred to as *attracticides*.

**Mating disruption by air permeation.** A common method proposed for the use of pheromones has been called the confusion, or decoy, method. This approach attempts to permeate the air with sex pheromone. Theoretically, insects entering the area cannot locate mates emitting natural pheromone because the synthetic pheromone permeates the whole environment. This would seemingly cause a reduction of reproductive rates and achieve crop protection without the use of insecticides. This basic idea was one of the earliest suggestions for the use of pheromones in pest management.

The mechanism of the confusion approach is not always completely understood. Some possibilities include: (1) camouflage or covering up the natural pheromone scent of females, (2) misdirection of males to bogus scents from multiple point sources of synthetic pheromone, and (3) adaptation/habituation by desensitizing male antennal receptors through constant pheromone-like exposure.

The first preliminary field test demonstrating the potential of this approach was conducted in 1967 with the cabbage looper, *Trichoplusia ni*. In this test, pheromone concentrations were shown to thwart males from being lured to female moths. Following this success, many studies were conducted to apply the approach but with little success. These included work with fruit, vegetable, field, forest, and stored product pests. In these studies, pheromone dispensing seemed the greatest obstacle to success.

Subsequently, controlled-release dispensers were developed, and these have paved the way for successes in pest suppression. An early example of such a dispenser was the Hercon® flake, produced by Hercon Environmental Company (Fig. 14.10). Hercon flakes were multilayered plastic laminates, about 3 $mm^2$ (1/8 $in.^2$), that contained pheromone in the inner layer, or reservoir. The

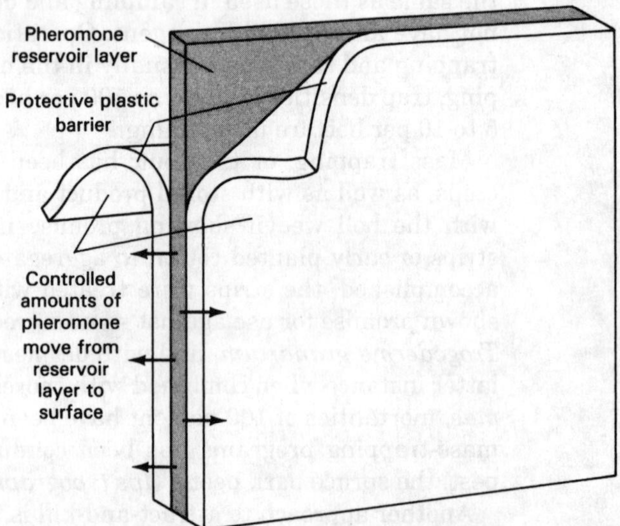


**Figure 14.10** Hercon® controlled-release dispenser (3 $mm^2$) containing pheromone. Pheromone is in each dispenser or flake, which provides protection from degradation. (Redrawn from Hercon brochure)

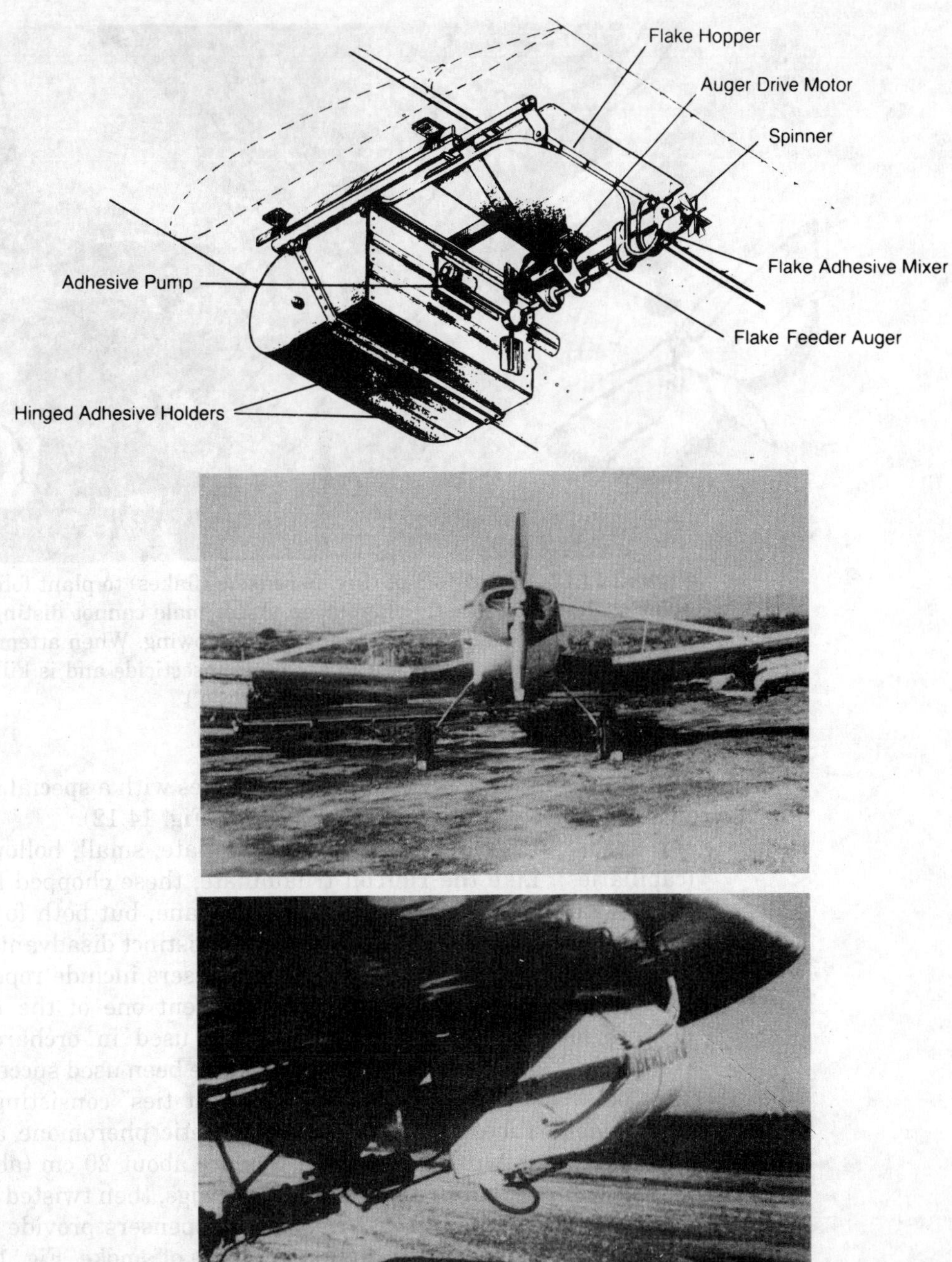


**Figure 14.11** Hercon aerial applicator for applying Hercon flakes. (Courtesy Hercon Environmental Co.)

outer layers of the flake served as a protective barrier but allowed the mixture to diffuse into the air. As the material was dissipated into the air, a replacement quantity automatically moved outward from the reservoir to maintain the desired surface concentration and to give long-lasting effectiveness. The flakes were applied somewhat like a conventional insecticide with the use of a special device, the Hercon dispenser pod, attached to an airplane (Fig. 14.11).

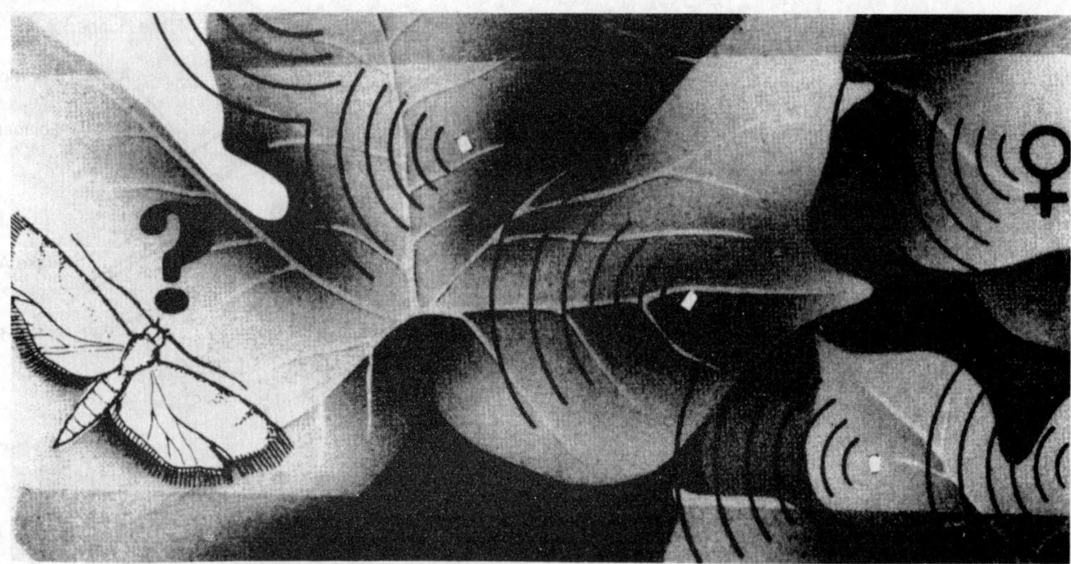

**Figure 14.12** Application of tiny dispensers (flakes) to plant foliage causes multipie-point sources of synthetic pheromone. Adult male cannot distinguish the dispensers from the actual female, causing false-trail following. When attempting to mate with a dispenser, the male is exposed to a contact insecticide and is killed or reproductively impaired. (Courtesy Hercon Environmental Co.)

This unit automatically combined the flakes with a special sticker that caused them to adhere to foliage as they landed (Fig. 14.12).

Dispensers used today include trilaminate, small, hollow, polymeric fibers (capillaries). Like the Hercon trilaminate, these chopped hollow fibers, combined with a sticker, are applied by airplane, but both formulations require specialized equipment for application—a distinct disadvantage.

Other types of controlled-release dispensers include ropes (Fig. 14.13), microcapsules, and flowables. Ropes represent one of the most effective dispensers for mating disruption and are used in orchards, tomatoes, and vineyards (Fig. 14.14). Moreover, they have been used successfully against the pink bollworm in cotton. Ropes are "twist ties" consisting of hollow, sealed polyethylene tubes that contain a synthetic pheromone and are reinforced with aluminum wire. These twist ties are about 20 cm (about 8 inches) long and are wrapped around plant stems or twigs, then twisted in a way similar to using a bread-wrapper twist tie. Such dispensers provide a strong synthetic pheromone plume (something like a plume of smoke, Fig. 14.15), which drifts away from the source for as many as 60 to 90 days.

Yet another development in pheromone dispensers is the Metered Semiochemical Timed Release System (MSTRS), which uses an aerosol canister. Similar to MSTRS are *puffers.* Puffers look like brown rectangular boxes with nozzles that shoot pheromone from the top. A device inside programs when the box will puff the pheromone. Puffers are being used successfully in northern California pear orchards for mating disruption of codling moth mating. In this situation, the puffers are programmed to release codling moth pheromone every 15 minutes, with a canister of the puffer lasting about 200 days.

**Figure 14.13** Rope (twist-tie) dispenser. The rope dispenser consists of a hollow, sealed polyethylene tube, which contains a synthetic pheromone and is reinforced with aluminum wire. (Courtesy Pacific Biocontrol)

Microcapsules and flowables also represent recent formulations developed and can be sprayed on foliage in liquid form with conventional equipment. Because they can be mixed with other chemicals, such as foliar fertilizers and insecticides, their application can be very cost effective. However, multiple applications may be required compared to a single rope application (three applications compared to one for the pink bollworm).

Yet other dispensing devices continue to be developed. One of those is a paraffin dispenser that is glued to trees for a controlled release of coddling moth and Oriental fruit moth pheromone. A novel approach by a British group, Exosect, uses the pest itself in an attempt to confuse male moths. With their unique system, a paperboard dispenser containing specific sex pheromone attracts newly emerged males. Upon entering the dispenser, the males are coated with electrostatically charged particles (powder) containing the pheromone. The treated males then leave the dispenser and subsequently attract other males, who mistake them for females. In mating attempts, the untreated males also become contaminated with the powder and further attract males in failed mating activity. The approach has a compounding effect on the population, being flying-point

**Figure 14.14** Workers applying twist-tie ropes (Isomate–M®) to peach trees in an oriental fruit moth management program. (Courtesy Pacific Biocontrol)

sources of female attractant, causing mating confusion and reducing population fertility.

All types of dispensers have been used to permeate atmospheres for mating disruption of several insect species. Often, mating disruption is most successful when population levels are low and the synthetic pheromone is applied over a broad area. In orchard situations, some manufacturers recommend using their products in well-managed situations without a history of high pest pressure. Otherwise, an insecticide spray is recommended first, to initially reduce insect population density, followed by applications of the synthetic pheromone to hold the population down. This approach extends the effectiveness of the insecticide and can significantly reduce the number of insecticide applications in a cost-effective manner.

The use of synthetic pheromones to disrupt mating has been successful in management programs for the codling moth, oriental fruit moth (*Grapholita molesta*), grape berry moth, tomato pinworm (*Kieferia lycopersicella*), pink bollworm, and several other insect pests. One of the most important programs has been that of the pink bollworm on cotton in the Imperial Valley of California, where moderate to heavy infestations occur. A study in 1985 showed that treatments of the synthetic pheromone gossyplure, using ropes, reduced the number of insecticide applications from eleven to six while lowering cotton damage by more than half compared to fields treated only with insecticides.

Although the mating-disruption method has received much attention for more than 40 years, progress toward implementation has been slow. Part of the reason for this has been the lack of formulations that are effective, inexpensive, and simple to apply. There have been several breakthroughs in formulation, however, including ropes and sprayables, which should improve adoption considerably. Even so, mating disruption is a specialized tactic that

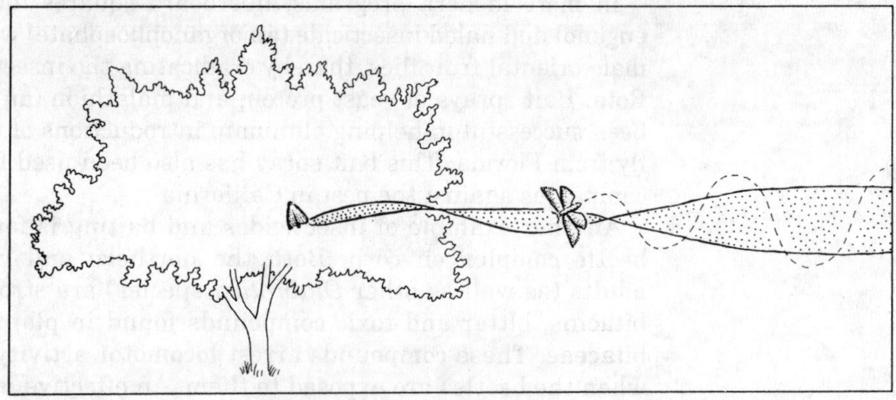

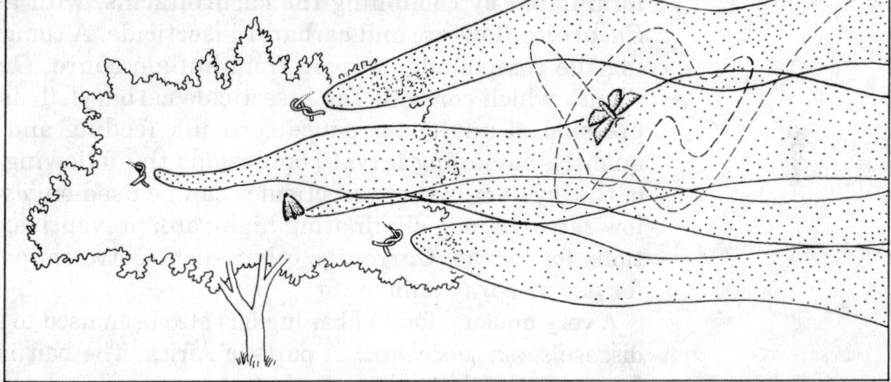

**Figure 14.15** Schematic showing the principle of camouflage/habituation. In the top figure, the male detects the pheromone plume of a female and is attracted to mate with her. In the bottom figure, rope dispensers have been applied, which produce multiple plumes, thus camouflaging the original scent of the female. The male becomes desensitized and cannot find the female. (Redrawn from Pacific Biocontrol brochure)

has application only with certain pest types and in certain crops (see Knight and Weissling 1999 for details). Conversely, the scope of synthetic pheromones for pest monitoring and decision making is very broad, and we should thus expect a continued expansion of use in these areas.

**Use of traditional baits.** Traditional baits or food lures may also be considered for attract-and-kill programs. Baits have been used for decades, and as mentioned in Chapter 11, some insecticides are sold commercially in bait formulations.

Some of the earliest food lures included wheat bran. The bran was moistened with water, and stomach poisons or botanicals were added as the killing agent. Other bait carriers have included hardwood sawdust, fresh horse manure, ground corncobs, and seed hulls. Additives like molasses, chopped oranges or other fruit, amyl acetate, sugar, and honey were often included with these basic ingredients for attraction. With species that are attracted to fat—for example, some ants (Hymenoptera: Formicidae)—ground fatty meat or plant products high in vegetable oil have been added to the insecticide carrier. Liquid bait sprays with sugar and molasses added to the insecticide solution have been commonly used with many species of flies.

In more modern programs, fiberboard squares impregnated with methyl eugenol and naled insecticide (an organophosphate) were used to lure and kill male oriental fruit flies, thereby eradicating the insect on the Pacific island of Rota. Bait sprays of yeast protein and malathion (an organophosphate) have been successful in helping eliminate introductions of the Mediterranean fruit fly from Florida. This bait spray has also been used frequently in continuing campaigns against the pest in California.

Another example of insecticides and baiting is found with the rootworm beetle complex on corn. Both the northern and western corn rootworm adults (as well as other *Diabrotica* species) are strongly affected by cucurbitacins, bitter and toxic compounds found in plants of the family Cucurbitaceae. These compounds arrest locomotor activity and stimulate feeding when the beetles are exposed to them. An effective insecticide bait has been formulated by combining the cucurbitacins, with attractants derived from *Cucurbita* flowers, and carbaryl insecticide. A commercial product containing the compounds, derived from buffalo gourd, *Cucurbita foetidissima,* is Slam®, which contains the insecticide carbaryl. This bait is applied for suppression of adults to reduce corn-silk feeding and egg laying and, consequently, damaging larval populations the following season. The advantage of this system is that insecticides can be used only when necessary and at a low rate, thereby eliminating high-rate, preventive, soil-insecticide applications for larvae. Currently, other such rootworm baits are being developed for use as spray adjuvants.

A very unusual form of baiting has also been used to all but eliminate the fatal disease *sleeping sickness* in parts of Africa. The bait is an artificial cow that is treated with a kairomone (food attractant) and impregnated with an insecticide. Tsetse flies, which transmit the disease-causing trypanosome, are attracted to the artificial cow, feed on it, and receive a lethal dose of insecticide. The artificial cows were introduced in Zimbabwe in the mid-1980s and cases of the disease in the area are now nearly nonexistent. The artificial cows continue to be maintained as a barrier to stop the tsetse from reinvading areas already cleared of flies.

In addition to these baits used with insect pests, many new uses of well-known plant products offer potential in pest management programs. As mentioned earlier, chemicals isolated from flowers and other plant parts are highly attractive to some insects. Several of these chemicals have been registered as biopesticides and are being used in attract-and-kill programs (see Chapter 12).

## Insect Repellents

**Repellents**, strictly speaking, are chemicals that cause insects to orient their movements away from a source. Allied materials that do not cause movement away but do prevent feeding or oviposition by insects are called **deterrents**. Both repellents and deterrents have been important in specialized areas of pest management. In concept, repellents are conventionally viewed as chemicals applied to surfaces to drive insects away. Deterrents, on the other hand, are usually thought of as natural plant constituents that are most important in host plant resistance. It should be realized, however, that many synthetic chemicals, including some insecticides, can have deterrent or irritant properties (for example, some pyrethroids) for certain insects and other arthropods. Actual repellents will be emphasized in this discussion.

Repellents are usually volatile chemicals that express their activity in the vapor phase. A strong repellent will be sensed by insects a few centimeters distant, causing them to fly or crawl away. Less-active repellents may allow insects to alight or touch the surface before being repelled.

Insect repellents based on such natural substances as wood smoke, oils, pitches, tars, various earths, and camel urine were used by humans hundreds of years ago to ward off insects and other arthropods. In the early 1900s, repellency of several essential oils, including oil of citronella, was discovered, and these were widely used as mosquito repellents. Before World War II, there were only four principal repellents, but significant advances were made after the beginning of the war to protect military personnel, particularly in the tropics. Subsequently, the early repellents were replaced by more effective synthetic compounds.

The advantages of using repellents are twofold. They often have low toxicity, and therefore they can be used safely on humans, plants, and domestic animals. Additionally, effective repellents protect the desired source, and because insects are not killed directly, undesirable side effects like resistance to the chemical are not as likely. The limitations of repellents include the need to completely cover all susceptible surfaces with repeated applications and the possibility of increasing infestations on nearby untreated surfaces.

**Traditional repellents.** Presently this group of repellents is the most widely used. Repellants include chemicals applied mostly for protection of humans and human possessions. Insect repellents to protect the body are primarily aimed at repelling mosquitoes, biting flies, fleas, ticks, and chiggers (larval trombiculid mites). These repellents are produced in several formulations, including aerosols, creams, lotions, grease sticks, powders, suntan oils, and clothes-impregnating laundry emulsions. Most repellents last only a few hours because they evaporate, are absorbed by the skin, are abraded by clothing, or are diluted by perspiration. Some of the most common repellents today are **DEET** (*N,N*-diethyl-*m*-toluamide, Delphene, OFF !®) for biting flies and mosquitoes and **benzyl benzoate** for ticks and chiggers. DEET, in particular, has been an important personal repellent, but issues have been raised over some uses. Research with rats has shown that frequent and prolonged exposure to topical applications, especially in conjunction with other pesticides, causes brain cell death and behavioral changes. Therefore, DEET is to be used with caution, particularly for children at risk to brain defects.

Repellents for human possessions include chemicals to protect fabrics and wood products. Protection of woolens against clothes moths (Lepidoptera: Tineidae) and carpet beetles (Coleoptera: Dermestidae) has been a particularly important problem that can be handled effectively with repellents. Colorless dyestuffs like Mitin FF® (a sulfonate) can be applied to wool during the dyeing operation, and these can effectively protect the fabric over its lifetime.

Several chemicals are also used as repellents or feeding deterrents against wood-damaging insects. For instance, creosote is a repellent to many insects and is commonly used to protect such products as telephone poles and railroad ties. Synthetic chemicals like pentachlorophenol are also used widely as surface treatments or pressure impregnants of telephone poles and foundation timbers to deter termite feeding.

The use of traditional repellents with crop and domestic animal protection has been limited. One reason is the high cost of repeated treatments and

DEET (Delphene, OFF!®)

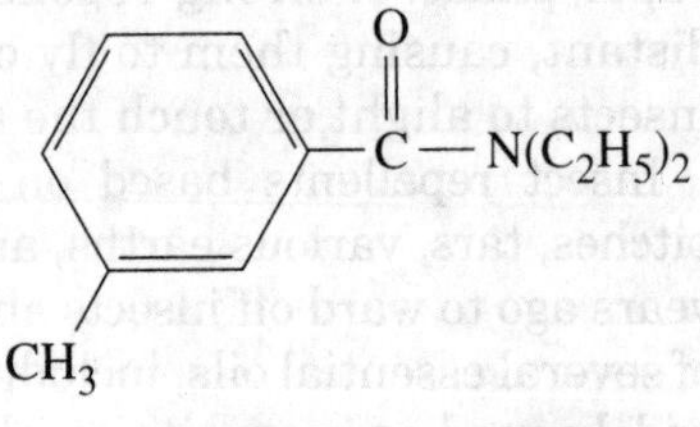

$C_6H_4(CH_3)-C(=O)-N(C_2H_5)_2$

*N,N*-diethyl-*m*-toluamide

Benzyl Benzoate

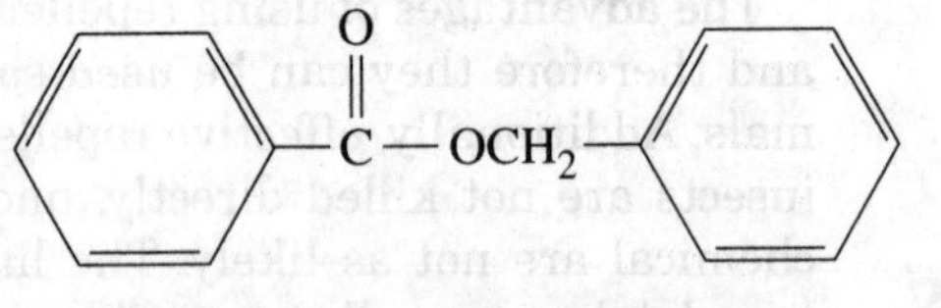

$C_6H_5-C(=O)-OCH_2-C_6H_5$

benzyl benzoate

another is the necessity of complete coverage. However, one exception is pine oil from wood pulp residue. Pine oil serves as a repellent and/or deterrent for bark beetles (Coleoptera: Scolytidae) when applied to individual trees.

**Plant allomones as repellents.** You may recall that allomones are natural substances from plants and animals that produce a response in the receiving species favorable to the emitting species. Most allomones that serve as repellents are usually considered from the aspect of host plant resistance; in other words, some resistant plants emit repellents naturally, and this may be the primary mode of nonpreference. But the use of pine oil, mentioned earlier, also could be considered an application of a plant allomone for protection; however, it is applied exogenously (outwardly).

The exogenous application of plant allomones for repellency and deterrence has not been widely practiced. Although many of these chemicals have been identified as showing activity, they suffer from the same limitations as traditional repellents. It is possible that future research with these plant compounds will allow them to be used more widely, especially where economics permit and the desire for chemical-free produce is paramount, that is, organic farming. With the growing availability of natural plant products registered as biochemical pesticides (see Chapter 12), we might expect to see increased interest in these materials and, subsequently, more widespread use.

Another possibility may be the development of synthetic deterrents with antifeeding activity. A material developed in this regard is *pymetrozine,* which affects the behavior of certain Hemiptera, causing them to stop feeding and die. This antifeedant is used for aphid and whitefly management in vegetables, ornamentals, cotton, field crops, deciduous fruit, and citrus. Current trade names include Chess®, Relay®, and Fulfill®.

**Epideictic pheromones as repellents.** Another group of semiochemicals that may become future repellents are epideictic pheromones. Because these substances reduce population density of a species in an area, it stands to

Pymetrozine (Chess®, others)

$CH_3$ N=CH N N N O H N

(*E*)-4,5-dihydro-6-methyl-4-[(3-pyridinylmethylene)amino]-1,2,4-triazin-3(2*H*)-one.

reason that when applied to a surface, there is potential for repellency. To date, epideictic pheromones have been used in this way against only two insect groups, *Dendroctonus* bark beetles and *Rhagoletis* fruit flies.

With bark beetles, various pheromones have been dispensed to limit their landing on trees. Here, 3,2-MCH dispensed near freshly felled Douglas fir reduced attacks as much as 96 percent and lowered subsequent larval population as much as 91 percent. Other tests with granular and liquid formulations of the pheromone also showed effective results against other bark beetles on felled Sitka spruce trees. Similar encouraging results were obtained in Switzerland, where oviposition-deterring pheromones were sprayed on cherry trees to reduce fruit damage. Although promising, much work with these materials remains before they can be used in practical pest management.

### Integration of Behavior Modification with Other Tactics

In general, chemicals that attract and repel are well suited to be combined with other pest management tactics. As mentioned, one of the most effective combinations is an attractant integrated with a killing agent. This agent most often is an insecticide, but some success with pathogenic-microorganism combinations indicates that a potential also exists here.

Another potentially important combination is special attractant kairomones to increase populations of natural enemies in crops for enhanced biological control. For example, green lacewings, *Chrysoperla carnea,* can be attracted by applying artificial honeydew (an aphid exudate) to plants. This promotes egg laying in the desired location and potentially increases predation of plant pests. Parasitoids may also be managed in future years by the addition of kairomones to attract them to target areas, where a stronger than usual depressive influence may be gained. Although this approach is largely tentative, significant research is being conducted to achieve practical implementation.

Plant resistance and use of semiochemicals is yet another potentially viable combination for insect pest management. By understanding the role of semiochemicals in conferring plant resistance, we can become more efficient in selecting genotypes with higher levels of resistance than have been obtained in the past. In addition, synthetic allomones similar to the plant's natural products may be employed to artificially enhance a resistant plant's overall effect on a pest population.

## CONCLUSIONS

Using chemicals to modify insect development and behavior represents a unique approach to the management of insect populations. These chemicals offer several advantages, of which the foremost is environmental and human safety. For this reason, much research is being directed toward their improvement. Many of these compounds are still under development, and only time will tell whether or not they will be widely adopted. To succeed, they will have to measure up to the standards of other established management procedures. In other words, they must be effective in suppressing insect populations or preventing colonization, be compatible with other management tactics, and be comparable in cost with conventional insecticides.

### Further Reading

Beckage, N. E. 1998. Insect growth regulators, pp. 123–137. In J. E. Rechcigl and N. A. Rechcigl, eds., *Biological and Biotechnological Control of Insect Pests.* Boca Raton, Fla.: Lewis Publishers.

Berryman, A. A. 1986. *Forest Insects—Principles and Practice of Population Management.* New York: Plenum Press, pp. 182–188.

Bowers, W. S. 1982. Endocrine strategies for insect control. *Entomologia Experimentalis et Applicata* 31:3–14.

Cardé, R. T., and A. K. Minks. 1995. Control of moth pests by mating disruption: Successes and constraints. *Annual Review of Entomology* 40:559–585.

Cook, S. M., Z. R. Khan, and J. A. Pickett. 2007. The use of push-pull strategies in integrated pest management. *Annual Review of Entomology* 52:375–400.

Debboun, M., S. P. Frances, and D, Stickman, eds. 2007. *Insect Repellents: Principles, Methods, and Uses.* Boca Raton, Fla.: CRC Press.

Howse, P. E., I. D. R. Stevens, and O. T. Jones. 1998. *Insect Pheromones and Their Use in Pest Management.* New York: Chapman & Hall.

Jacobson, M. 1986. The neem tree: Natural resistance par excellence, pp. 220–232. In M. B. Green and P. A. Hedin, eds., *Natural Resistance of Plants to Pests—Roles of Allelochemicals,* ACS Symposium Series 296. Washington, D.C.: American Chemical Society.

Knight, A. L., and T. J. Weissling. 1999. Behavior-modifying chemicals in management of arthropos pests, pp. 521–545. In J. R. Ruberson, ed., *Handbook of Pest Management.* New York: Marcel Dekker.

Kubo, I., and J. Klocke. 1986. Insect ecdysis inhibitors, pp. 206–219. In M. B. Green and P. A. Hedin, eds., *Natural Resistance of Plants to Pests—Roles of Allelochemicals,* ACS Symposium Series 296. Washington, D.C.: American Chemical Society.

Kydydonieus, A. F., and M. Beroza, eds. 1982. *Insect Suppression with Controlled Release Pheromone Systems,* vol. 2. Boca Raton, Fla.: CRC Press.

Metcalf, R. L., and R. A. Metcalf. 1994. Attractants, repellents, and genetic control in pest management, pp. 315–354. In R. L. Metcalf and W. H. Luckmann, eds., *Introduction to Insect Pest Management,* 3rd ed. New York: Wiley.

National Academy of Sciences. 1969. *Insect Pest Management and Control. Principles of Plant and Animal Pest Control,* vol. 3. U.S. National Academy of Sciences Publication 1695, pp. 290–323.

Nordlund, D. A., R. L. Jones, and W. J. Lewis. 1981. *Semiochemicals—Their Role in Pest Control.* New York: Wiley.

Quistad, G. B., D. C. Cerf, S. J. Kramer, B. J. Bergot, and D. A. Schooley. 1985. Design of novel insect antijuvenile hormones: Allylic alcohol derivatives. *Journal of Agricultural Food Chemistry* 33:47–50.

Ridgway, R. L., R. M. Silverstein, and M. N Inscoe, eds. 1990. *Behavior-Modifying Chemicals for Insect Management.* New York: Marcel Dekker.

Suckling, D. M., and G. Karg. 1998. Pheromones and other semiochemicals, pp. 63–99. In J. E. Rechcigl and N. A. Rechcigl, eds., *Biological and Biotechnological Control of Insect Pests.* Boca Raton, Fla.: Lewis Publishers.

Tunaz, H., and N. Uygun. 2004. Insect growth regulators for insect pest control. *Turkish Journal of Agriculture and Forestry* 28:377–387.

## Favorite Web Sites

http://www.epa.gov/pesticides/biopesticides/ingredients/index.htm
Contains current information on biochemical pesticides. The EPA considers biopesticides naturally occurring substances that control pests by nontoxic mechanisms. Biochemical pesticides include substances that interfere with growth, such as insect growth regulators, or substances that repel or attract pests, such as pheromones.

http://www.pherotech.com
Web site of PheroTech, Inc., a company that deals in pheromone products, including traps for monitoring pest populations in IPM decision making. Gives pictures of many types of pheromone traps.

http://www.cals.ncsu.edu/course/ent425/text19/semiochem.html
Site discusses semiochemicals and how they are used in IPM.

CHAPTER 15

# STERILE-INSECT TECHNIQUE AND OTHER PEST GENETIC TACTICS

THE USE OF STERILIZED INSECTS and manipulations of pest genetics have developed gradually since about 1916. These tactics are pest-selective, aimed mostly at reducing insect populations by lowering reproductive potentials, and they include some of the most innovative and unusual procedures in insect pest technology. Pest insects are used against members of their own species to reduce population levels, which is a type of **autocidal control** called SIT, the sterile-insect technique.

SIT received major input and support from the work of E. F. Knipling, a USDA entomologist, in the late 1930s. In the 1950s, R. Bushland began it as a strategy of replacing normal matings in a population with infertile ones, in effect inducing sterility. Fundamentally, the sterility principle aims at flooding a population with sterile mates, which seek out and mate with normal females. Such matings result in inviable (unfertilized) eggs, and with continued sterile-male releases, the population declines. As decline occurs, the ratio of sterile to normal males increases until virtually no normal males remain. At this point, the population becomes extinct for lack of progeny. Consequently, as originally devised, the main objective of SIT was eradication, not merely suppression. However, suppression using SIT is a distinct possibility in pest management, as it is analogous to using sex pheromones (Chapter 14) to reduce population natality by preventing successful mating.

SIT has evolved into other ideas that emphasize genetic manipulations of the pest population. Whereas the sterile-insect technique involves competition between sterile and normal males, other approaches include competition between gametes (usually sperm) of mutant and normal insects. Collectively, these approaches have been termed **genetic control.** Genetic control focuses on altering genetic makeup of pests in such a way as to: (1) produce sterility of progeny, (2) reduce fecundity, or (3) reduce survival in otherwise favorable environments.

Although SIT is treated separately from genetic control in this discussion, many authorities group the tactics together as genetic control. The rationale for a combined grouping is that genetic elements—gametes—are involved in both approaches. Distinguishing between them is appropriate, however, because successful use of SIT has been demonstrated, whereas most other genetic methods are still largely theoretical.

## THE STERILE-INSECT TECHNIQUE

The sterile-insect technique developed out of studies of the screwworm fly (Box 15.1), *Cochliomyia hominivorax,* a parasite of cattle primarily in tropical and subtropical regions of the New World. Knipling and coworkers noted that adult females mated only once (were monogamous) to fertilize their eggs. Using this knowledge, these workers postulated that if males could be sterilized without impairing their mating behavior, releases into the wild population would result in infertile matings, and isolated populations could be eliminated.

### SIT Theoretical Background

The theory behind the SIT approach can be illustrated by a number of simple models, such as tables developed by Knipling, to show the behavior of population numbers alone and when subjected to sterile-insect releases. In these models, a hypothetical population, having a 1:1 (female to male) sex ratio and growing five-fold each generation, is convenient for presentation. This exponentially growing

### BOX 15.1 SCREWWORM

**SPECIES:** *Cochliomyia hominivorax* (Coquerel) (Diptera: Calliphoridae)

**DISTRIBUTION**: Before eradication, the screwworm was found throughout Texas, New Mexico, Arizona, Mexico, and Central and South America. Through migration (female flies travel up to 300 miles) and transport by humans, the species has been found as far north as Canada, but the flies cannot successfully overwinter in northern areas and die with the onset of cold weather. Presently, the species is found in Central and South America.

**IMPORTANCE**: The screwworm, commonly referred to as primary screwworm or obligate screwworm, is a serious pest of livestock and wildlife. The species infests warm-blooded animals with open wounds, which occur with dehorning, castration, or fly and tick bites. Adult female flies oviposit on wounds, and developing larvae (maggots) feed directly on flesh and wound exudate. The odor from infested animals attracts more female flies, which lay additional eggs on the wounds. Moreover, secondary screwworms, such as *Cochliomyia macellaria* (which usually oviposit on carcasses), will oviposit on the primary screwworm-infested animals.

Infested animals abstain from feeding and may die if not treated. Before substantial control of primary screwworm was initiated, livestock losses were between $50 million and $100 million per year in the southwestern United States. The sterile-insect technique (SIT) has been used to eradicate the species in the United States and Mexico.

**APPEARANCE**: Adults are metallic blue to bluish-green, with an orange to brown head and three dark stripes on the top of the thorax. Larvae (maggots) are white and about 16 mm long when full grown. Eggs are 1 mm long and white. Puparia are dark brown and about 10 mm long.

**LIFE CYCLE**: Adults are active year-round in warm climates; in cool climates, however, pupae hibernate in soil for about 2 months, until warmer weather. Adults feed on manure liquids and wound exudates. Females oviposit from 250 to 300 eggs, and the maggots feed gregariously on live flesh and fluids of the open wound. After 4 to 9 days (three stadia), larvae drop to the ground and pupate in soil. Adults emerge in 7 to 54 days, depending upon temperature.

**Table 15.1 Trend of a Hypothetical Population with Each Female Producing Five Female Offspring Each Generation (Sex Ratio 1:1).**

| Generation | Number Females per Unit Area | Total Number in Population |
|---|---|---|
| 1 | 1,000,000 | 2,000,000 |
| 2 | 5,000,000 | 10,000,000 |
| 3 | 25,000,000 | 50,000,000 |
| 4 | 125,000,000 | 250,000,000 |
| 5 | 625,000,000 | 1,250,000,000 |

population is shown in Table 15.1. (This simplistic pattern is not prevalent in the environment because of density-dependent factors and randomly operating density-independent factors. It is used here for explanatory purposes only.)

If the same population is subjected to releases of 9 million sterile males each generation, as shown in Table 15.2, numbers are essentially eliminated (eradicated) after the fifth generation. The reason for eradication is that every time a sterile male mates with a female, she is prevented from laying fertile eggs. Therefore, in generation 1, only 10 percent (100,000) of the 1 million females mated with a normal male, and even though these females increased their numbers fivefold, the population still began a downward trend. Continued inundations of the same number of sterile males caused increasing rates of decline because of increases in the sterile-to-normal-male ratio. In other words, the *reproduction penalty* (genetic load of geneticists) placed on the population was so great that it could not be overcome and numbers dropped. As they dropped, the method increased in effectiveness.

A primary principle of the sterile-insect technique is shown from this simple example: The initial number of sterile releases must be great enough to overcome reproduction and start a downward trend in the population. The higher the initial sterile-to-normal-male ratio, the quicker the pest eradication. Had only 4 million sterile males been released each generation in the example, the population would still have grown, the method would have become less effective with time, and the pest would not have been eliminated. Theoretically, the initial ratio of sterile-to-normal males must be greater than the pest's net increase potential from generation to generation to be effective. In other words, if the population is

**Table 15.2 Trend of the Hypothetical Population Subjected to Sterile-Insect Releases.**

| Generation | No. Females without Releases | No. Sterile Males Released | No. Females with Releases | Ratio Sterile to Normal Males | No. Fertile Females |
|---|---|---|---|---|---|
| 1 | 1,000,000 | 9,000,000 | 1,000,000 | 9:1 | 100,000 |
| 2 | 5,000,000 | 9,000,000 | 500,000 | 18:1 | 26,316 |
| 3 | 25,000,000 | 9,000,000 | 131,579 | 68:1 | 1,907 |
| 4 | 125,000,000 | 9,000,000 | 9,535 | 944:1 | 10 |
| 5 | 625,000,000 | 9,000,000 | 50 | 180,000:1 | 0 |

**Table 15.3 Trend of the Hypothetical Population When Sterile Releases Are Compared with an Insecticide Treatment.**

| Generation | No. Females with No Treatment | No. Females with Sterile Releases 9:1 | No. Females with Insecticide—90% Kill |
|---|---|---|---|
| 1 | 1,000,000 | 1,000,000 | 1,000,000 |
| 2 | 5,000,000 | 500,000 | 500,000 |
| 3 | 25,000,000 | 131,579 | 250,000 |
| 4 | 125,000,000 | 9,535 | 125,000 |
| 5 | 625,000,000 | 50 | 62,500 |
| 6 | 3,125,000,000 | 0 | 31,250 |

increasing by a factor of five each generation, the ratio would have to be greater than 5:1 for effectiveness. In practice, however, the ratio may need to be double the increase factor to compensate for environmental and genetic factors.

The value of the sterile-insect technique can be viewed, at least theoretically, by comparing the hypothetical example with management using another tactic, insecticide applications. Let us assume that an insecticide is applied each generation to a pest population, causing a 90 percent reduction in numbers. The results, shown in Table 15.3, indicate that even though numbers are reduced by the insecticide, a residual population still remains after generation five. If applications are suspended at that time, we would assume the pest population would rebound. On the other hand, the population would be completely eliminated with the sterile-insect technique, and the program could be terminated.

An important principle from this example is that tactics such as insecticide applications tend to become less efficient as numbers are reduced, whereas the sterile-insect technique becomes more efficient with small numbers. Conversely, when pest populations are large, insecticides are most efficient. Therefore, an effective tactic is to integrate insecticide applications with sterile-male releases in eradication programs. In this tactic, an insecticide should be applied to cause initial number reduction, followed by successive sterile-male releases. As Table 15.4 shows, pest numbers with this integrated approach are reduced very rapidly, compared with the sterile-insect technique alone.

**Table 15.4 Trend of the Hypothetical Population When Insecticides Are Used, Followed by Sterile-Insect Releases.**

| | | Without Insecticides | | With Initial Insecticide Application—90% Kill | |
|---|---|---|---|---|---|
| Generation | No. Sterile Males Released | No. Females | Ratio Sterile to Normal Males | No. Females | Ratio Sterile to Normal Males |
| 1 | 9,000,000 | 1,000,000 | 9:1 | 100,000 | 90:1 |
| 2 | 9,000,000 | 500,000 | 18:1 | 5,495 | 1,638:1 |
| 3 | 9,000,000 | 131,000 | 68:1 | 17 | 529,412:1 |
| 4 | 9,000,000 | 9,535 | 944:1 | 0 | |
| 5 | 9,000,000 | 50 | 180,000:1 | | |
| 6 | 9,000,000 | 0 | | | |

**Circumstances for application.** Several pest situations are envisaged for use of the sterile-insect technique: (1) against well-established pests when they reach low points in their density cycles (either within a season or between seasons); (2) against newly introduced pests or established pests spreading into new areas; (3) with other tactics like insecticides and cultural procedures, which precede sterile-insect releases; and (4) against isolated populations like those on islands and other such situations. In all situations, an areawide program must be imposed for complete success.

### Sterilizing Insects in a Natural Population

The idea of sterilizing insects in a natural population was advanced decades ago when it was found that certain chemicals, called **chemosterilants,** had potential in this regard. The subsequent discovery that these chemicals presented unacceptable human health and environmental hazards discouraged further advancement of the approach. Even though not practical presently, future developments could change the acceptability of the method; therefore, a brief explanation of this idea is appropriate.

Sterilizing insects in a natural population should not be confused with the sterile-insect release procedure. Although sterility is involved in both instances, the mechanism of suppression of each is quite different. Whereas mating is required for suppression in sterile-insect release programs, population reproductive rates are reduced directly by the sterilization of individuals in the natural population. Here, both males and females of the population are sterilized, which in regard to reproductive capacity is the same as killing them; they will not add individuals to the next generation. Moreover, the sterilized individuals, still being active, can mate with individuals of the population that were not sterilized, further reducing the reproductive potential. This latter phenomenon behaves similarly to the release method and, with this approach, has been called the *bonus effect.*

To achieve this bonus effect, the sterility technique must not eliminate mating behavior and competitiveness of the sterilized insects. Indeed, unless a high level of competitiveness is retained, sterilizing **feral** (wild) individuals generally would have the same effect on reproduction as killing them outright with an insecticide. Conversely, fully competitive sterile insects theoretically would be capable of reducing reproduction in normal individuals of the population in proportion to the sterile-insect numbers. For instance, if 90 percent sterility in the population is achieved, that percentage would be similar to mortality from an insecticide. However, unlike the insecticide, the sterilized individuals would further reduce reproduction in the 10 percent of normal females by 90 percent. A hypothetical example of the rapid population reduction that could be achieved with an effective chemosterilant, along with a comparison with insecticides and sterile-male releases, is shown in Table 15.5.

The method of sterilizing feral insects in their environment has been tested extensively on a worldwide basis. Most research on this approach has focused on the house fly (Box 15.2), *Musca domestica,* and several mosquitoes (Diptera: Culicidae) in isolated areas. In house fly studies, chemosterilants have been combined with highly selective baits to maximize exposure of the target organisms and reduce risk to other species. With pilot studies, using the chemosterilant TEPA in a granular bait (67 percent cornmeal, 15 percent sugar, 15 percent powdered milk, 2.5 percent powdered eggs, and

**Table 15.5 Trend of the Hypothetical Population Comparing Application with an Insecticide, Sterile-Male Release, and Application of a Chemosterilant.**

| Generation | No. Females with Insecticide—90% Kill | No. Females with Sterile Releases 9:1 | No. Females with Chemosterilant—90% Sterility |
|---|---|---|---|
| 1 | 1,000,000 | 1,000,000 | 1,000,000 |
| 2 | 500,000 | 500,000 | 50,000 |
| 3 | 250,000 | 131,579 | 2,500 |
| 4 | 125,000 | 9,535 | 125 |
| 5 | 62,500 | 50 | 6 |
| 6 | 31,250 | 0 | 0 |

0.5 percent TEPA), house flies were virtually eliminated around a dump on Bahia Honda Key in Florida during a 5-week period. Another, even more ambitious, program with the house fly on Grand Turk Island in the West Indies caused large reductions in the fly population; however, the program was considered only moderately successful because final eradication was not achieved.

## BOX 15.2 House Fly

**SPECIES:** *Musca domestica* (Linnaeus) (Diptera: Muscidae)

**DISTRIBUTION:** The house fly is a cosmopolitan species found throughout the world. Populations are particularly prevalent at locations in or near human dwellings, where they are well adapted to survive.

**IMPORTANCE:** House flies are very annoying to both humans and animals. In addition, house flies are known to carry disease organisms. For example, typhoid fever, anthrax, cholera, diarrhea, and dysentery have been associated with mechanical transmission by house flies. They also serve as intermediate hosts for parasitic tapeworms (helminths).

**APPEARANCE:** Adult flies have sponging mouthparts that they use to pick up liquids or solids that they liquify. Larvae (maggots) are tapered exteriorly and are creamy white. Full-grown larvae are about 12 mm long. Puparia are shorter than larvae and more robust. They are dark brown and about 6 mm long when mature. Eggs are pearly white, elongate, and about 1 mm long. Larvae are found in manure, rotting food, and other fermenting vegetable matter.

**LIFE CYCLE:** House flies overwinter in all stages, even in the northern United States, where poultry houses, dairies, and other such structures allow survival. After maturing in spring, female flies oviposit about 100 eggs in any kind of animal excrement or decaying matter. Eggs hatch within 24 hours. Maggots feed in decaying material for 3 to 7 days, move to the margins of the food source, and pupate. Pupae in puparia develop in 3 to 7 days, and adults emerge to begin the next cycle. The entire life cycle requires about 16 days under ideal temperatures but may take considerably longer if cooler temperatures occur. The reproductive potential is tremendous. Average population doubling times are estimated at 5 to 7 days. Indeed, house flies usually reproduce to the limit of available food resources. This suggests that sanitation is one of the best methods of limiting house fly populations.

The advantages of sterilizing individuals in the natural population are very clear. With the bonus effect, much higher levels of effectiveness can be achieved than with direct mortality agents like insecticides. In addition, many of the obstacles of the sterile-insect release technique are eliminated, particularly the expensive and sometimes formidable task of rearing large numbers of insects for release. Yet the problem remains of finding safe, efficient sterilizing agents that can be applied in the environment, even with baits. Until this problem is solved, we should not expect much use of this tactic.

## Methods of Sterilization

The concept of autocidal control is sound in principle, but its application relies greatly on acceptable procedures to sterilize insects. Without these, as seen in the concept of sterilizing natural populations, progress toward application can be agonizingly slow or halted completely. During the development of the approach, both ionizing radiation and chemicals have been found to cause insect sterility, but the former has been used more successfully.

**Ionizing radiation.** The sterilizing effects of X-rays on insects were observed as early as 1916 with adult cigarette beetles, *Lasioderma serricorne,* and were the first form of radiation investigated. These and other investigations with X-ray-caused mutations in pomace flies (Box 15.3), *Drosophila melanogaster,* eventually led to the discovery by R. C. Bushland that pupae of the screwworm, when irradiated near adult emergence, could result in competitive, sterile adults.

### BOX 15.3 FRUIT FLY, POMACE FLY, OR VINEGAR FLY

**SPECIES:** *Drosophila melanogaster* (Meigen) (Diptera: Drosophilidae)

**DISTRIBUTION:** The species is found throughout the world in both urban and rural areas.

**IMPORTANCE:** *Drosophila melanogaster* is the most extensively studied insect and one of the best genetically understood organisms in the world. The species is particularly attractive to geneticists because it has very large chromosomes in the nuclei of larval salivary gland cells, allowing researchers to gain information regarding gene mapping, chromosome structure, and gene expression. The flies are ideal laboratory subjects because they have short life cycles (10 to 14 days) and are easy to rear and because mutations occur often in relatively small populations. Thomas Hunt Morgan, the first geneticist to receive the Nobel Prize, experimented with the species in the early part of the twentieth century, revealing detailed information on the genetics of *Drosophila* embryology.

**APPEARANCE:** Adults are yellow to brown and 3 to 4 mm long. Males have three bands on the top of their abdomen, whereas females have five bands. Larvae (maggots) are white and about 5 mm long when full grown. Eggs are about 0.5 mm long, with two projections at one end that prevent the egg from sinking into the decaying fruit. Puparia are brown.

**LIFE CYCLE:** Females oviposit in decaying fruit or fermenting vegetable matter. After about 2 days, eggs hatch, and the larvae feed for about 4 to 5 days (three stadia). Larvae then move out of the food material and attach themselves to a dry surface before pupating. Adults emerge in about three days. Generation time is only 10 to 14 days.

With the development of the atom bomb after World War II, it became much more efficient to work with manufactured isotopes, primarily those producing gamma rays. Further studies showed little difference between certain X-rays and gamma rays in treating insects, and most subsequent programs have utilized gamma radiation, with cobalt or cesium as the source.

High-energy radiation exerts its genetic effects by chromosome breakage and point mutations of DNA. After irradiation, gametes are produced, but they carry dominant lethal mutations so that the zygotes formed die early in development.

The main objective of irradiation is to sterilize insects without greatly affecting their ability to live, mate, and carry on otherwise normal life functions. In practice, both sexes are irradiated, sterilized, and released, but sterile females have no desired effect on the outcome. Because species vary in the radiation dose required to achieve sterilization, considerable laboratory study must be conducted with various doses on different insect stages to establish the appropriate procedures.

In many instances, the pupal stage is the most appropriate for irradiation. Usually less active than larvae and adults, pupae are easy to handle and can be sterilized with relatively low radiation doses after most adult structures are formed. Doses required for sterility in adults are often about the same as those in older pupae, but adults are usually more difficult to manipulate. Both eggs and larvae are radiosensitive, and the radiation levels required to induce sterility often cause premature death or abnormal adults. In some organisms, higher levels of radiation are required to sterilize females than males. Sterilizing doses also vary greatly among insect species. In particular, the Lepidoptera require high doses, which in many instances causes a great decrease in competitiveness. Overall, doses required for radiosterilization of insects are much greater than those for mammals.

**Chemosterilization.** The ability of some chemicals to sterilize insects as well as some of the drawbacks of this method have already been mentioned. When applied before the onset of meiosis, chemosterilants prevent gamete production. Knowing this, we are not surprised to learn that such chemicals prevent all types of cell reproduction and have received much development in cancer therapy (chemotherapy). Indeed, chemotherapy research probably gave impetus to the suggestions of chemical use for insect sterilization.

Chemosterilants can be divided into four basic groups. These include the alkylating agents, phosphorus amides, triazines, and antimetabolites. Presently, the alkylating agents represent the largest class of chemosterilants, and their effects are similar to those of X-rays and gamma rays. These agents cause multiple dominant lethal mutations or severely injured genetic material in the sperm or egg.

Busulfan (Alkylating Agent)

$$CH_3-\overset{\overset{O}{\|}}{\underset{\underset{O}{\|}}{S}}-O-C_4H_8-O-\overset{\overset{O}{\|}}{\underset{\underset{O}{\|}}{S}}-CH_3$$

1,4-butanediol dimethane-sulfonate

Alkylating agents are unstable in the environment and degrade rapidly. Possible contamination of food and water with even small residues, however, makes crop applications presently unfeasible. Safe applications are possible only under laboratory conditions, where these materials may be used to sterilize insects in sterile-release programs. Such an approach has been attempted with the boll weevil, *Anthonomus grandis,* by incorporating busulfan, an alkylating agent, into the diet of mass-reared beetles and releasing these into test areas. More recently, an insect growth regulator, diflubenzuron, has been used in the adult diet, followed by gamma irradiation to achieve effective male sterilization.

It is unfortunate that chemosterilants are dangerous to humans and wildlife. Most present compounds, particularly the alkylating agents, are strongly carcinogenic or mutagenic and thereby create undue risk if applied over broad areas. Therefore, we would not expect them to be registered for widespread use. But the goal of achieving safe chemosterilants, although formidable, is a worthy one.

## Sterile-Insect Release Programs

Much effort has been directed toward developing the sterile-insect release technique for practical use. The USDA has been particularly responsible for many studies and pilot programs that have advanced the concept, and the agency has been the primary force behind large-scale implementations.

A development from these areawide programs has been the so-called concept of **total population management (TPM).** The TPM approach attempts to use all available means to eradicate a pest over a broad area. This objective is not new, but until the development of the sterile-male technique, it had been considered nearly impossible.

Pest eradication has been realized in only a few instances where sterile-insect releases were made, but successful degrees of suppression have been achieved in several other instances. Some of the most important examples of the technique follow.

**Screwworm eradication and suppression.** The most dramatic success with the sterile-insect technique was achieved with eradication of the screwworm fly from the United States, Mexico, and most of Central America. As a parasite of livestock, this fly lays 250 to 300 eggs in wounds, and developing larvae feed on tissues and enlarge the open wound (Fig. 15.1). Such feeding attracts still other flies to oviposit, and death of the parasitized animal may occur in as few as 10 days. Before the sterile-male release program, livestock losses were estimated at $20 million annually in the southeastern United States and $50 to $100 million in the southwestern states. During a particularly severe outbreak in 1935, 1.2 million cases of infestation and 180,000 livestock deaths were reported in the United States.

Use of the sterile-male technique against this insect was conceived in the early 1950s. After 2 years of research, field trials on the island of Sanibel (off the west coast of Florida) gave encouraging results. This test prompted a larger pilot program in 1954 on the larger, more isolated island of Curaçao, off the coast of Venezuela. Here, 400 sterilized males per square mile per week were released over a period of about 3.5 months (four to five generations). The program resulted in eradication from the island and demonstrated the potential of the technique.

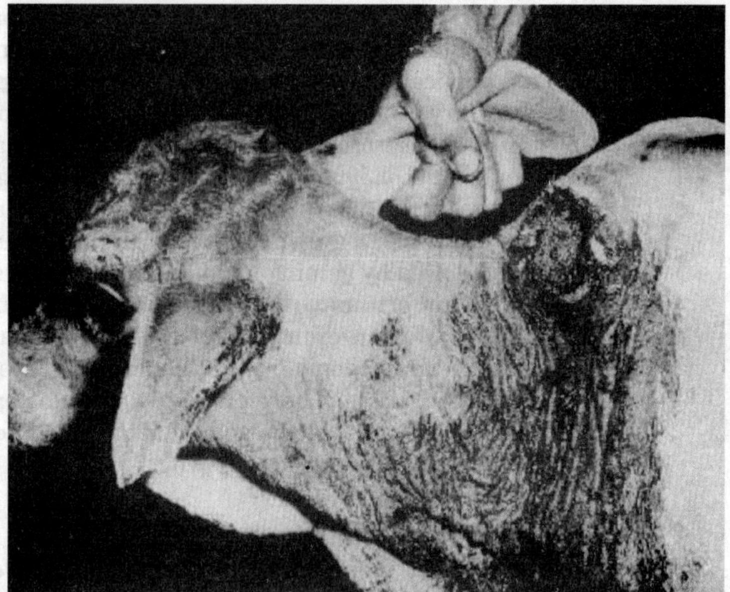

**Figure 15.1** Wound of a calf infested with screwworm, *Cochliomyia hominivorax.* (Courtesy USDA)

An even larger program, financed jointly by the USDA and the livestock industry in the Southeast, was implemented in 1958. Up to 50 million sterile flies of both sexes were produced each week (Figs. 15.2, 15.3, 15.4), and more than 2 billion were released (Fig. 15.5) over an 18-month period. The area involved 85,000 square miles, including Florida and parts of Georgia and Alabama. In this program, more than 40 tons of ground meat were required each week to rear the flies, and twenty aircraft were used to release them.

**Figure 15.2** Hatched screwworm larvae in a petri dish are transferred to a starting tray. Larvae feed on a medium consisting of lean ground meat, bovine plasma, water, and formalin. (Courtesy USDA)

**Figure 15.3** When full-grown, mass-reared screwworm larvae migrate from vats to a trough. Subsequently, they fall through a funnel and grating into a water channel under the flooring. The water transports them to a separator and dryer. (Courtesy USDA)

**Figure 15.4** Accumulations of full-grown screwworm larvae are placed in trays of sawdust where they pupate. Trays are emptied onto a machine that screens out sawdust; pupae and remaining larvae are then placed on a moving belt. Pupae remain on the belt for packaging, and larvae are driven off by lights into a collecting trough at the edge of the belt to be returned to a sawdust tray for pupation. (Courtesy USDA)

**Figure 15.5** An insect container used in aerial release of sterilized screwworms. (Courtesy USDA)

This campaign resulted in complete eradication of screwworm populations from Florida. Subsequent recurrences of the pest from importation of infested animals have been eradicated quickly by treatment of the animals and additional sterile-fly releases. The cost of the Florida program was estimated at $10 million, but it has been credited with saving southeastern livestock producers many times that amount.

With eradication achieved in Florida, attention turned to screwworm infestations in the southern parts of Texas, New Mexico, Arizona, and California. In these instances, flies invaded the United States from Mexico and produced significant infestations. To eliminate the pest and prevent future invasions, the strategy of creating a fly-free barrier zone was advanced. Subsequently, a fly-rearing plant was established in Mission, Texas, capable of producing more than 150 million flies per week. Beginning in February 1962, sterile flies were released along the border between Texas and Mexico. In 1964 the releases were extended to include the Mexican border of Arizona and California (Fig. 15.6). These latter releases created a fly-free zone 2,000 miles long by 300 to 500 miles deep; the zone served as a "barrier" to northward movement into the United States. Using this strategy, fly infestation reports in the United States dropped from more than 50,000 in 1962 to about 150 by 1970.

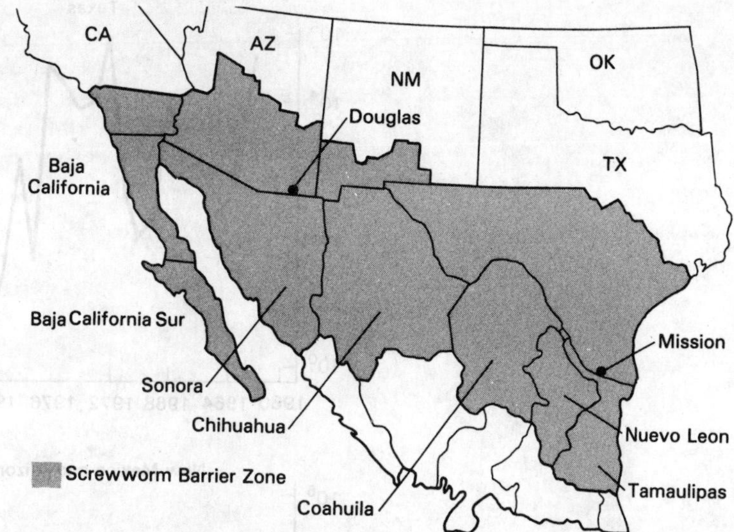


**Figure 15.6** Map showing the original fly-free barrier zone of the Southwestern Screwworm Eradication Program in 1962. Mission, Texas was the original location of the fly-production facility. Both Mission and Douglas, Arizona were packaging and dispersal centers. Northern Mexico frontier states were Baja California, Sonora, Chihuahua, Coahuila, and Tamaulipas. (Redrawn from Krafsur et al., 1987, *Parasitology Today* 3:131–137, with permission)

Unfortunately, the screwworm populations did not remain low. Infestation frequency began a slow increase in 1971 and reached an alarming 95,642 reports by 1972, despite continued fly releases. Release rates were increased as much as fivefold, but infestations remained at an unacceptable level. Later, it was hypothesized that new strains of laboratory flies were inadvertently produced that were less active and competitive for mates than feral flies. Following modified procedures and repeated efforts, the effectiveness of the program was renewed by 1977 (Fig. 15.7).

In 1981 the Mission, Texas, rearing facility was closed and replaced by a more efficient plant in Tuxtla Gutierrez, Mexico, opened in 1976. The Mexican plant is capable of producing up to 500 million flies per week.

Along with increased rearing capabilities, strategies were developed to gradually push the fly-free barrier zone south to the Tehuantepec Isthmus, where a shorter barrier (only 140 miles) was necessary. This objective was established by a joint Mexico–United States agreement in 1972 (Fig. 15.8), and the program was initiated in 1976. Among the new tactics used in this program is the application of poison baits, followed by releases of new-strain competitive flies into the reduced populations.

Case reports since this joint program was initiated indicate that the original barrier zone (southwestern United States and northern Mexico) became effective in 1979, and a plan of eradication south to the Isthmus of Tehuantepec was approved in 1984. In 1986, declining reports of cases from continued releases indicated that Mexico was free of screwworm. Although somewhat behind the original schedule, by mid-October 1994, Belize, El Salvador, and Guatemala were also reported free of the pest. Subsequently, Honduras,

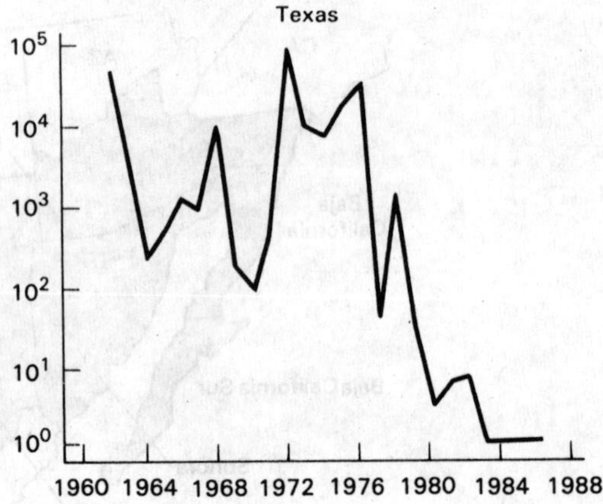

**Figure 15.7** Laboratory-confirmed screwworm reports, with data logarithmically transformed [$\log_{10}$ (cases + 1)]. Reports from Texas (*top*), Arizona and New Mexico (*center*), and northern Mexico. The arrow in the northern Mexico graph indicates an episode of sabotage where unirradiated flies were believed deliberately released. The dotted line shows the trend without this episode. No native screwworms have been detected in the northern Mexico zone since August 10, 1985, but two cases were intercepted in San Luis Potosi in November 1985 in animals transported from southeastern Mexico. (Redrawn from Krafsur et al., 1987, *Parasitology Today* 3:131–137)

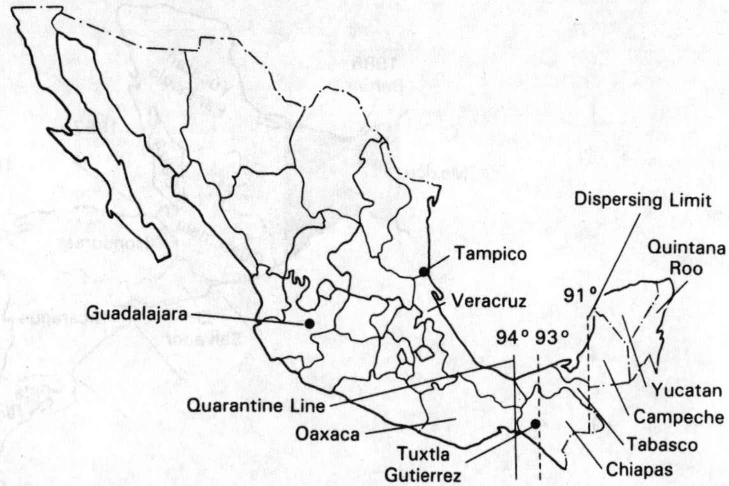


**Figure 15.8** The 1972 Mexico–United States Screwworm Eradication Agreement set 93°W longitude as the fly-free barrier objective to be attained. The eastern limit of sterile fly dispersal was 91°W. Primary packaging and dispersal centers were Tampico, Guadalajara, and Tuxtla Gutierrez. Tuxtla is the only center used presently. (Redrawn from Krafsur et al., 1987, *Parasitology Today* 3:131–137, with permission)

Nicaragua, and Costa Rica were added to this list. Currently, sterile flies are being released in Jamaica and Panama, where a fly-free barrier is being maintained to prevent reinfestation of previously eradicated areas. This barrier occurs in the Darien gap between the Panama Canal and the border with Colombia (Fig. 15.9).

In addition to the North American campaign, SIT has also been used against screwworm infestations in Libya. The pest became established there in 1987, and a campaign led by the Food and Agriculture Organization (FAO) of the United Nations succeeded in eradicating the pest, mostly with the use of SIT, by 1991.

**Tropical fruit fly programs.** On a worldwide basis, tropical fruit flies (Diptera: Tephritidae) are among the most destructive pests of fruits. Some of the most significant are the following: Mediterranean fruit fly (Box 15.4) (*Ceratitis capitata*), olive fruit fly (*Dacus oleae*), Mexican fruit fly (*Anastrepha ludens*), melon fly (*Dacus cucurbitae*), oriental fruit fly (*Dacus dorsalis*), and Queensland fruit fly (*Dacus tryoni*). Since the 1950s, much research has been concentrated on the use of sterile-male releases against these species, and a great deal of the basic technology has been established. Moreover, programs have been implemented against some of these species outside the United States. In these programs, isolated populations of the melon fly and oriental fruit fly have been eradicated, and nonisolated populations of the Queensland fruit fly have been suppressed.

Of the tropical and subtropical fruit flies, introductions of the Mediterranean and Mexican fruit flies have been of greatest concern in the United States. Here, sterile-male releases have been practiced to help prevent introductions and eliminate early infestations of both species. With the particularly

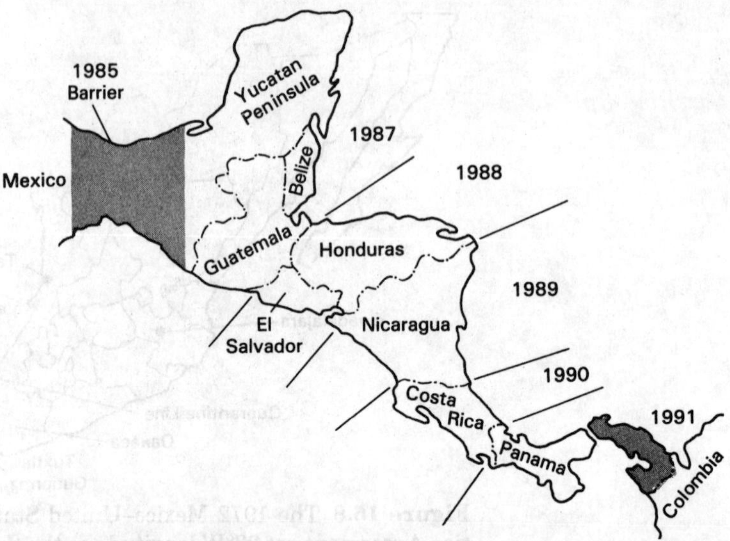


**Figure 15.9** Map of Central America showing original goals of barrier establishment based on sterile-fly production from the plant in southern Mexico. Although the schedule shown was delayed, the ultimate goal was achieved, saving billions of dollars for livestock producers. The Darien barrier (stipled area in eastern Panama) requires only about 50 million flies per week to maintain compared to the present 150 million flies per week. (Redrawn from Krafsur et al., 1987, *Parasitology Today* 3:131–137, with permission)

difficult Mediterranean fruit fly (medfly) problem in California, sterile-male releases have been combined with several other tactics. In recent campaigns, fruit was stripped from trees and fallen fruit picked up. This was followed by weekly aerial spraying of the infested area using bait and ground sprays. Afterward, approximately 100 sterile males for each wild male were released in strategic locations, a reactive use of SIT.

In 1996, the California Department of Food and Agriculture established a new SIT program, the Preventative Release Program, to prevent new infestations from becoming established. In this program, sterile medflies were released each week over a 2,155-square-mile area in southern California. This preventive program is believed to have reduced medfly infestations in the Los Angeles area by about 97 percent. However, there were continued detections of the pest in one location of the release area (1997) and several locations outside the release area (1998). Research continues to make the Preventative Release Program effective.

**Other insects.** In addition to the screwworm and fruit fly programs, sterile-insect releases have been studied for an array of other insects. Some of these include boll weevil, tsetse flies (*Glossina* species), horn fly (Box 15.5) (*Haematobia irritans*), mosquitoes, codling moth (*Cydia pomonella*), and pink bollworm (Box 15.6) (*Pectinophora gossypiella*). Of these, the pink bollworm program stands out as being one of the most successful.

As mentioned earlier, the pink bollworm is a major pest of cotton in the southwestern United States. The pest spread from Texas into New Mexico, Arizona, southern California, and northwest Mexico during the mid-1960s and

## BOX 15.4 MEDITERRANEAN FRUIT FLY

**SPECIES:** *Ceratitis capitata* (Wiedemann) (Diptera: Tephritidae)

**DISTRIBUTION:** The Mediterranean fruit fly is native to central Africa and, as a result of transport by humans, it is now found in practically all the subtropical regions of the world, except North America.

**IMPORTANCE:** The Mediterranean fruit fly, popularly known as the medfly, is among the most destructive pests of more than 100 species of fruit, including orange, peach, grapefruit, plum, apple, and pear. In subtropical areas, larvae (maggots) feed on the pulp of the fruits, causing the development of infection courts for fruit diseases. Also, holes made by female flies during oviposition scar the fruits, which lowers quality. In 1929, the pest was found in Florida and eradicated at a cost of $7 million. In 1956, also in Florida, the species was detected and subsequently eradicated at a cost of $10 million. In 1975, the pest was discovered in southern California and eradicated. In 1980, the pest was again found in southern California, and from 1980 to 1982, an intensive, much publicized program, undertaken to eliminate the pest, resulted in more than $100 million in eradication costs. Eradication efforts against this pest continue.

**APPEARANCE:** Adults, similar in size to house flies, have a yellow and black thorax and a yellow abdomen with two silver crossbands. The wings are transparent, with bands of yellow, brown, and black. Larvae, about 10 mm long when full-grown, are white maggots. Puparia are brown.

**LIFE CYCLE:** Adults or puparia overwinter in cool regions, but in warmer areas, the species is active year-round. Female flies, which may produce up to 600 eggs, oviposit two to ten eggs in holes under the skins of fruits. Eggs hatch in 2 to 20 days, and larvae feed in the fruits for about 1 week. After completion of development, larvae fall to the ground and pupate in soil for approximately 10 days. Generation time is variable (3 weeks to 3 months), depending on climatic conditions.

---

threatened to invade major cotton-growing areas of the San Joaquin valley in California. Beginning in 1967, sterile moths were released into the valley at a rate of about 100 million per year, which produced an estimated sterile-male to invading-male ratio exceeding 1,000:1 in early season. Such a ratio resulted in an extremely low probability of successful matings and is believed to be responsible for preventing the establishment of pink bollworm in the area.

Another lepidoperan pest success has occurred with the codling moth in British Columbia, Canada. After 10 years of operation, this program allowed quality fruit production and a drastic reduction in insecticide use.

### Requirements and Limitations of Sterile-Insect Programs

One of the foremost requirements for success with this approach is to have capabilities for mass-rearing the insect economically. Even with low numbers in the natural population, most programs call for millions of insects to be produced and released weekly over a period of several weeks. This capability sometimes necessitates development of a synthetic diet on which healthy and environmentally competitive insects can be reared. To understand net reproductive rates and low-density points, sampling programs that will yield precise absolute estimates are also required. In addition, the approach is applicable only when releases of sterile insects do not produce major injury or

## BOX 15.5 HORN FLY

**SPECIES:** *Haematobia irritans* (Linnaeus) (Diptera: Muscidae)

**DISTRIBUTION:** The horn fly is native to Europe and was introduced into the United States (New Jersey) about 1887. Currently, the species is distributed throughout all of North America.

**IMPORTANCE:** The horn fly is a common cattle pest throughout its range. The flies receive their common name from their resting behavior, which is typically on the horns of cattle. Several thousand individuals can be found on a single cow, piercing the skin and feeding on the blood. Each fly takes many blood meals both day and night; indeed, it is not uncommon for females to feed up to thirty-eight times per day, with each feeding session lasting up to 4 minutes. Cattle are irritated and annoyed, which causes significant reductions in both weight gain and milk production.

**APPEARANCE:** Adults are 4 mm long and dark gray. Larvae (maggots) are white and about 10 mm long when fully grown. Eggs are oval, about 1.2 mm long, and reddish brown. Puparia are barrel-shaped, about 3.3 mm long, and brown.

**LIFE CYCLE:** Pupae overwinter in soil, and adults emerge in early spring. Females leave the cattle for a short period of time to oviposit one to fourteen eggs in fresh dung. Females can oviposit from 350 to 400 eggs. After 1 day, the eggs hatch, and larvae feed in dung for 4 to 5 days. Pupation occurs in puparia in the soil. After 5 to 7 days, adults emerge. Each generation is about 14 days long, and there are many generations per year.

cause an undue nuisance. For instance, the inundation of insects capable of transmitting arboviruses or plant pathogens in an area probably would be unacceptable.

Experience in suppressing or eradicating populations with the sterile-insect technique is still too limited to determine which species would be highly susceptible to the approach. Even for existing large-scale programs, like those for the screwworm and fruit flies, results are not yet fully understood. A great deal of ecological, production, and operational research will be required before suitability of the sterile-insect technique can be adequately predicted. However, the benefits of research in this area extend to other management tactics and so would not be wasted effort even in instances where the approach proves impractical.

### Other Genetic Tactics

Manipulations of insect genetics for human benefit have been practiced for hundreds of years, since the domestication of the silkworm, *Bombyx mori,* and semidomestication of the honey bee, *Apis mellifera.* Superior strains, with increased production and efficiency, have been obtained through selection and hybridization, and such breeding activities are expected to show even greater progress in the future.

The manipulation of insect genetics for suppression, however, is quite a different idea. This activity, described earlier as part of genetic control, includes any kind of artificial manipulation of insect gene composition to reduce population numbers. It is genetics used in self-destruction of populations as opposed to, for instance, plant protection.

## BOX 15.6 Pink Bollworm

**SPECIES:** *Pectinophora gossypiella* (Saunders) (Lepidoptera: Gelechiidae)

**DISTRIBUTION:** The pink bollworm is native to India, and because of shipments of cottonseed, is now found in North America. The species, moving north from Mexico, was first detected in the United States by 1917 in Texas. Currently, it is distributed in the states of California, Nevada, Arizona, New Mexico, Oklahoma, Missouri, Arkansas, and Louisiana.

**IMPORTANCE:** The species is a serious cotton pest throughout the world, and in some areas, it can cause total crop destruction. Early in the growing season, larvae feed in the squares, attacking developing flower structures. Usually, this damage is not severe. Later in the growing season, however, larvae feed in the bolls on lint, carpel tissues, and seeds, which causes reductions in both yield and quality. Also, larval feeding provides infection courts for pathogens like boll rot.

**APPEARANCE:** Adults have a 12-mm wingspan and are grayish brown, with inconsistent markings. Young larvae are white, with a brown head, whereas older larvae (fourth instars) display a distinctive pink coloration. Full-grown larvae are approximately 13 mm long. Eggs are oval, about 0.5 mm long, and white. Before hatching, they become red. Pupae are brown.

**LIFE CYCLE:** Larvae overwinter in hollowed-out cottonseeds or in plant debris in the field. In early spring, larvae pupate and in about 10 days, adults emerge. Adults are active at night and feed on nectar. Females oviposit up to 200 eggs on the host plant near the bolls. After 5 days, the eggs hatch, and developing larvae undergo three or four molts (10 to 14 days). Larvae then leave the boll or remain in damaged seed to pupate. Generation time is approximately 25 to 30 days, and there may be up to six generations per year.

These genetic tactics for autocidal control are tentative at present; most have not been applied. However, several genetic processes are understood, and their use has been suggested. Fundamentally, the proposal is to alter genetic processes in such a way as to make insects less fecund, less vigorous, or altogether sterile; the effect would be to decrease population fitness. The objective would be suppression or complete eradication of the species in a large area.

The use of genetic principles to improve the fitness of insect natural enemies is another area that recently has engendered a great deal of interest. Much of the emphasis here is in the development of beneficial insects (and other arthropods) that are resistant to pesticides. This aspect of the genetic approach is discussed in more detail in Chapter 17.

Several properties make insects amenable to autocide through genetic manipulations. One of the foremost properties is genetic plasticity. Most species maintain many forms throughout their geographical range, each representing a different genotype; potentially, these are critical resources in selection and breeding programs. Additionally, insects have relatively short life cycles and high reproductive potentials, which enhance breeding programs and genetic experimentation. Moreover, technologies developed for rearing large numbers of some species may lead to practical implementation of programs in the field.

Research to achieve programs of genetic control has been concerned with several specific processes. The most important of these include conditional lethal mutations, inherited sterility, hybrid sterility, cytoplasmic incompatibility, chromosomal rearrangements, and meiotic drive mechanisms.

### Conditional Lethal Mutations

With this process researchers have attempted to breed insects that are less fit than normal for certain kinds of environmental conditions. The approach relies on specific alleles of genes that determine fitness of individuals for factors like temperature tolerance. These alleles result in inherited traits that do not allow survival in all conditions encountered by the insect. For example, susceptibility to cold winter temperatures could be a conditionally lethal trait in southern populations of an insect species. The trait has no effect on individuals of these populations in their normal habitat. However, when individuals bearing the trait are released into the northern part of the species range, they are killed.

Of the conditionally lethal traits, temperature sensitivity has received much of the experimental attention. Theoretically, male insects with a dominant, homozygous, cold-sensitivity trait could be reared and released into the environment. There, they would mate with feral females, which in turn would produce progeny with the trait. The progeny would then be killed by normally low winter temperatures.

Sensitivity to low temperatures is often associated with obligatory dormancy (diapause). In insects with wide geographic ranges, both diapausing and nondiapausing genotypes may be found. Releasing nondiapausing genotypes in populations of obligatory diapausing genotypes could produce populations of progeny that would not diapause and would be killed by winter extremes. Such a tactic has been suggested for the European corn borer, *Ostrinia nubilalis,* in the United States and crickets, *Teleogryllus commodus* and *Teleogryllus oceanicus,* in Australia, but only limited studies have been conducted.

### Inherited Sterility

Inherited sterility, also called delayed sterility, has been suggested as a method to substantially increase sterility ratios in populations, as compared with the conventional sterile-insect release technique. Particularly, it has been recommended for Lepidoptera because these insects usually require enormous dosages of irradiation to achieve high levels of dominant lethality, and such high levels impair competitiveness of the released insects.

In theory, if a 9:1 ratio of sterile-to-fertile males were established by releases in a conventional program, this would result in a 90 percent reduction of fertile crosses. If, on the other hand, the sterility produced by the release was not expressed until the $F_1$ generation, it would be enhanced by 9 percent. This is because a 9:1 original release would create a 9:1 ratio in both males and females of the $F_1$ generation. When $F_1$ individuals cross, the probabilities of all possible outcomes would be as follows:

0.9 sterile male × 0.9 sterile female = 0.81 sterile matings
0.9 sterile male × 0.1 fertile female = 0.09 sterile matings
0.1 fertile male × 0.9 sterile female = 0.09 sterile matings
0.1 fertile male × 0.1 fertile female = 0.01 fertile matings

Therefore, genetic death can be increased from 90 percent to as much as 99 percent by delaying it until the next generation. Such an approach has been recommended for continued investigation.

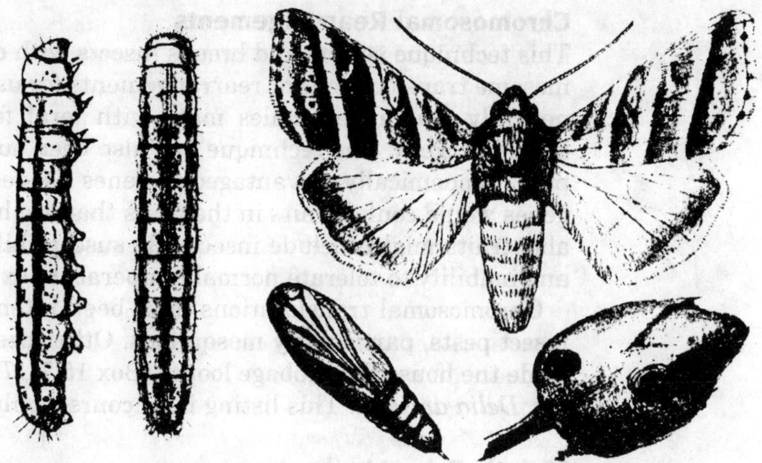

**Figure 15.10** Tobacco budworm, *Heliothis virescens*, larva (side and top aspect *left*), adult (*top center*), pupa, and injured tobacco bud. (Courtesy USDA)

### Hybrid Sterility

Hybrid sterility has been observed in laboratory studies with closely related species. Similar to horse-by-donkey crosses that result in sterile mules, this idea has been suggested as a method of obtaining sterile insects for release.

Particularly notable in this regard have been studies of hybridization between the tobacco budworm (Fig. 15.10), *Heliothis virescens*, and a related species, *Heliothis subflexa*. These species cross quite readily in the laboratory, producing partially sterile $F_1$ offspring. Most important in this phenomenon is that the hybrid males are sterile when they mate with normal *H. virescens* and *H. subflexa* females. Although technical problems exist, this approach has been suggested for use in sterile-insect releases against populations of the tobacco budworm.

### Cytoplasmic Incompatibility

This phenomenon occurs when individuals from different populations are crossed, and reproductive potentials are reduced. The reduction is derived because of incompatibility factors in the cytoplasm causing sterility in individual eggs. Here, sterility results when a sperm enters an egg and stimulates meiosis, but it does not fuse with the egg pronucleus to form a zygote.

The principle of cytoplasmic incompatibility has been studied extensively with a mosquito, *Culex fatigans*. Programs for suppression of the mosquito would require the release of only males because releasing both sexes simply would result in replacing one strain with another. Moreover, because females bite and transmit pathogens, their release would be unacceptable. Unfortunately, sorting out large numbers of male mosquitoes from a population for release presents many technical problems. Research into "sex-killing systems," as accomplished with house flies, Mediterranean fruit flies, and other Diptera, has offered a possible solution to this problem. The sex-killing approach involves chromosomal rearrangements.

### Chromosomal Rearrangements

This technique selects and breeds insects with certain genetic defects or chromosome translocations or rearrangements for use in release programs. The genetically rearranged males mate with feral females and produce partially sterile progeny. The technique has also been suggested as a means to "transport" economically advantageous genes to pest populations; advantageous genes would confer traits in the pests that are helpful to humans. Some desirable traits might include insecticide susceptibility, avoidance of a crop plant, and inability to tolerate normal temperature extremes.

Chromosomal translocations have been intensively studied in a number of insect pests, particularly mosquitoes. Other pests with research emphasis include the house fly; cabbage looper (Box 15.7), *Trichoplusia ni;* and onion maggot, *Delia antiqua.* This listing is, of course, only a partial one.

### Meiotic Drive Mechanisms

*Meiotic drive* refers to the preferential recovery of one of a pair of homologous chromosomes during meiosis. An example of the mechanism can be made with the XY system of sex determination present in many insect species. In this system, zygotes bearing two X chromosomes become females and those with an X and Y chromosome become males. If a Y-linked mutation is induced, only Y sperm would be produced, rather than a 1:1 ratio of X:Y sperm. When mutant males, carrying only Y sperm, mate with normal females, only sons are produced. A distortion of the male-to-female sex ratio would result, favoring males, and the population would decline. As it declines, the drive-mechanism effect increases because there is an accompanying increase of mutant Y chromosomes in the population. Such a

#### BOX 15.7 Cabbage Looper

**SPECIES:** *Trichoplusia ni* (Hübner) (Lepidoptera: Noctuidae)

**DISTRIBUTION:** The cabbage looper is a native North American species, common to all of Canada, Mexico, and the United States.

**IMPORTANCE:** The cabbage looper is a common pest of many plant species, such as cabbage, cauliflower, celery, lettuce, broccoli, tomato, tobacco, potato, bean, and turnip. Larvae defoliate host plants, causing both reduced yield and quality. Under heavy infestations, twenty-five to thirty-five larvae can infest a single cabbage head. Larvae move in a characteristic looping fashion from which the species gets its common name. Adults migrate into Canada, occasionally causing outbreaks.

**APPEARANCE:** Adults have brown mottled forewings and brown hind wings, with a wingspan of 40 mm. A distinctive silver marking is displayed on each forewing. Larvae, about 38 mm long when full grown, are green, with white lines running lengthwise along the body. Eggs are greenish white and dome shaped. Pupae are brown and enclosed in a white silken cocoon.

**LIFE CYCLE:** Pupae overwinter attached to leaf debris. Adults emerge in early spring, and females oviposit 275 to 350 eggs singly on leaf surfaces. Larvae feed about 14 to 28 days, then pupate. After about 10 days, adults emerge. There are three or more generations per year, depending on climatic conditions.

population should eventually become extinct. Although meiotic drive has been induced in laboratory house flies with substerilizing doses of gamma radiation, some researchers have raised serious questions as to its practicality.

### Replacement by Innocuous Forms

Up to this point, the techniques mentioned for genetic control have emphasized establishing genetic loads on insect populations that result in numerical reductions and, in some instances, extinction. A conceivable alternative to this autocidal approach would be replacement of the pest by forms that are not harmful. In this instance, genetically altered replacements could be mass-reared and released with the expressed purpose of lowering pest status through changes in species characteristics.

One of the most appropriate subjects of this potential approach are nonvectoring strains of insects that normally vector pathogenic microorganisms of humans. For example, a strain of the mosquito (Box 15.8), *Anopheles gambiae,*

## BOX 15.8 ANOPHELES MOSQUITOES

**SPECIES:** Numerous *Anopheles* species (Diptera: Culicidae)

**DISTRIBUTION:** *Anopheles* mosquitoes are distributed throughout the tropic regions of the world, and some species occur in temperate regions. There are about 390 species worldwide, with 90 in the western hemisphere and 15 in North America.

**IMPORTANCE:** The genus *Anopheles* vectors malaria, the most significant human parasitic disease of the tropics and subtropics. These pests have been largely responsible for deterring economic development in the tropical areas of the world. Presently, more than 200 million humans are infected with malaria per year. Malaria is currently a significant disease in Africa, Asia, and Central and South America; however, in the past, the disease was distributed throughout the world. The disease is caused by four species of protozoa in the genus *Plasmodium.* The life cycle of malaria is complex, but basically, the mosquito ingests the sexual forms (gametocytes) of the *Plasmodium* when taking a blood meal from a human. Then the malaria forms sporozoites in the mosquito body, which pass to the salivary glands. When the mosquito feeds again, the sporozoites are introduced into human blood. In humans, the malaria attains various forms (merozoites and trophozoites), causing fever and chill cycles. Gametocytes are then produced, which are ingested by the mosquito, and the cycle is repeated. There are several types or species of malaria that vary in severity to humans.

**APPEARANCE:** Adults are frail flies, with long maxillary palps and spotted wings. Males have plumose antennae. Adults rest with their bodies angled to the surface on which they are resting. Larvae (wrigglers) lie parallel to the water surface and lack a breathing tube, which is a common feature in other genera of Culicidae. Pupae (tumblers) lie at the surface of the water.

**LIFE CYCLE:** Where year-round activity does not occur, inseminated female adults overwinter; however, some species overwinter as larvae in standing water in tree holes. Females take a blood meal and oviposit from 150 to 300 eggs singly on still water. Each female can oviposit more than 1,000 eggs. Adult males do not feed on blood, and both sexes maintain themselves on nectar and plant juices. Eggs hatch in 2 to 6 days, and larvae feed on algae for about 14 days. Larvae then pupate in the water, and in about 3 days, adults emerge. Typically, generation time is from 21 to 30 days.

has been discovered that is refractory (immune) to *Plasmodium* species, malaria-causing pathogens. After ingesting infected blood, this mosquito strain is capable of encapsulating the infectious stage of the microorganism, which causes its death. Therefore, this strain is unable to transmit the infection. Production of a fully refractory strain has been achieved by selective breeding, encouraging further research on potential releases in replacement campaigns. If natural populations could be diluted or completely replaced by breeding with the released refractory strain, it is conceivable that malaria incidence could be reduced significantly. However, annoyance from biting mosquitoes would remain. See Beard et al. (1993) for a complete discussion on this topic.

### Use of Molecular Genetic Techniques

The recent expansion of approaches to manipulate the genetics of organisms through biotechnology has led some to speculate that similar approaches could be used to modify the insect genome for human benefit. Some possibilities include developing insecticide susceptibility and lack of winter hardiness in pests, immunity to disease in silkworms and honey bees, and insecticide resistance in natural enemies. Transformation, or the insertion of recombinant DNA into insects, is the technique most actively investigated. Currently, research has demonstrated that foreign DNA can be introduced into the pomace fly, *Drosophila melanogaster*. Here, two antifreeze protein genes isolated from the Atlantic wolffish have been expressed in adult females, which produced the antifreeze protein in their hemolymph and became more cold hardy. This suggests that such genetic changes may be possible with other insects, and many believe that it is only a matter of time before desired traits can be engineered (see Hoy 1994 for a detailed discussion on this topic).

## CONCLUSIONS

Using sterile insects and genetic techniques to achieve autocide or replacement in insect populations is a highly specialized tactic. Objectives of the approach are to cause complete eradication or at least to reduce population levels over a broad area. Because the areawide procedures involved require a combined effort, the tactics are usually beyond the realm of decision making by individual producers. More likely, such programs will be initiated and conducted by federal and state agencies in cooperation with various interest groups. In agriculture, economic benefits of such areawide programs to individuals are substantial, because they represent tax-based subsidies to crop production.

Except where chemosterilants are considered, sterile-insect and genetic tactics would seem to present few personal or environmental hazards. Moreover, if eradication is achieved with the approach, continuing management inputs are reduced significantly, with only surveillance programs and occasional corrective actions necessary. Yet other advantages of the approach may be high returns for expenditures and applicability with low-level infestations.

Unfortunately, many factors limit the practicality of the method. Some of the most significant obstacles to development include the absence of economical mass-rearing technology for certain species and inability to produce fully competitive sterile insects. Many of the newer genetic tactics may allow the latter limitation to be overcome. However, more basic research is needed to make the newer approaches operational, and this research will require substantial expenditures.

## Further Reading


Bartlett, A. C. 1991. Insect sterility, insect genetics, and insect control, pp. 279–287. In D. Pimentel, ed., *CRC Handbook of Pest Management in Agriculture,* vol. 2, 2nd ed. Boca Raton, Fla.: CRC Press.

Beard, C. B., S. L. O'Neill, R. B. Tesh, F. F. Richards, and S. Aksoy. 1993. Modification of arthropod vector competence via symbiotic bacteria. *Parasitology Today* 9:179–183.

Carpenter, J. E., and A. C. Bartlett. 1999. Genetic approaches to managing arthropod pests, pp. 487–519. In J. R. Ruberson, ed., *Handbook of Pest Management.* New York: Marcel Dekker.

Collins, F. H., R. K. Sakai, K. D. Vernick, S. Paskewitz, D. C. Seeley, L. H. Miller, W. E. Collins, C. C. Cambell, and R. W. Gwadz. 1986. Genetic selection of a plasmodium-refractory strain of the malaria vector *Anopheles gambiae. Science* 234:607–609.

Davidson, G. 1974. *Genetic Control of Insect Pests.* New York: Academic Press.

Dyck, V. A., J. Hendrichs, and A. Robinson, eds. 2005. *Sterile Insect Technique: Principles and Practice in Area-Wide Integrated Pest Management.* New York: Springer.

Enkerlin, W., A. Bakri, C. Caceres, J. Cayol, A. Dyck, U. Feldmann, G. Franz, A. Parker, A. Robinson, M. Vreysen, and J. Hendrichs. 2003. Insect pest intervention using the sterile insect technique. Current status on research and on operational programs in the world, pp. 11–24. In 2003 Report, *Recent Trends on Sterile Insect Technique and Area-Wide Integrated Pest Management.* Okinawa, Japan: Research Institute for Subtropics.

Gould, F., and P. Schliekelman. 2004. Population genetics of autocidal control and strain replacement. *Annual Review of Entomology* 49:193–217.

Hendrichs, J. 2000. Use of the sterile insect technique against key insect pests. *Sustainable Development International* 2:75–79.

Hoy, A., and J. J. McKelvey, Jr., eds. 1979. *Genetics in Relation to Insect Management.* Rockefeller Foundation Conference, March 31 to April 5, 1978, Bellagio, Italy. New York: Rockefeller Foundation.

Hoy, M. A. 1994. *Insect Molecular Genetics.* San Diego, Calif.: Academic Press.

Knipling, E. F. 1979. *The Basic Principles of Insect Population Suppression and Management.* United States Department of Agriculture Handbook 512, chapter 10.

Knipling, E. F. 1982. Present status and future trends of the SIT approach to the control of arthropod pests, pp. 3–23. In *Sterile Insect Technique and Radiation in Insect Control.* Proceedings International Symposium on the Sterile Insect Technique and Use of Radiation in Genetic Insect Control, June 29 to July 3, 1981. Vienna, Austria: International Atomic Energy Agency.

Krafsur, E. S. 1998. Sterile insect technique for suppressing and eradicating insect population: 55 years and counting. *Journal Agricultural Entomology* 15:303–317.

Krafsur, E. S., C. J. Whitten, and J. E. Novy. 1987. Screwworm eradication in North and Central America. *Parasitology Today* 3:131–137.

Labrecque, G. C., and C. N. Smith, eds. 1968. *Principles of Insect Chemosterilization.* New York: Appleton-Century-Crofts.

Lloyd, E. P. 1989. The boll weevil: recent research developments and progress towards eradication in the USA, pp. 1–27. In G. E. Russell, ed., *Management and Control of Invertebrate Crop Pests.* Andover, Hampshire, U.K.: Intercept.

Lorimer, N. 1981. Genetic means for controlling agricultural insects, pp. 299–305. In D. Pimentel, ed., *CRC Handbook of Pest Management in Agriculture,* vol. 2. Boca Raton, Fla.: CRC Press.

Metcalf, R. L., and R. A. Metcalf. 1994. Attractants, repellents, and genetic control in pest management, pp. 337–349. In R. L. Metcalf, and W. H. Luckmann, eds., *Introduction to Insect Pest Management*, 3rd ed. New York: Wiley.

National Academy of Sciences. 1969. *Insect Pest Management and Control Principles of Plant and Animal Pest Control,* vol. 3. U.S. National Academy of Sciences Publication 1695, chapters 9 and 15.

Pal, R., and M. J. Whitten, eds. 1974. *The Use of Genetics in Insect Control.* New York: American Elsevier.

Robinson, A. S. 1998. Genetic control of insect pests, pp. 142–169. In J. E. Rechcigl and N. A. Rechcigl, eds., *Biological and Biotechnological Control of Insect Pests.* Boca Raton, Fla.: Lewis Publishers.

Ware, G. W. 1983. *Pesticides—Theory and Application.* San Francisco: Freeman, pp. 225–228.

Whitten, C. J. 1982. The sterile insect technique in the control of the screwworm, pp. 79–84. In *Sterile Insect Technique and Radiation in Insect Control.* Proceedings International Symposium on the Sterile Insect Technique and Use of Radiation in Genetic Insect Control, June 29 to July 3, 1981. Vienna, Austria: International Atomic Energy Agency.

## Favorite Web Sites

http://www.cdfa.ca.gov/phpps/pdep/prpinfo/index.htm
Table of contents provides links to important aspects of the Mediterrean fruit fly and the exclusion program in southern California.

http://www.ars.usda.gov/is/timeline/worm.htm
USDA site giving background on the sterile-male technique and its use with the screwworm. Gives an image of the screwworm larva.

CHAPTER 16

# THE PRACTICE OF INSECT PEST MANAGEMENT

TO SOLVE ANY PROBLEM, we must first analyze the facts, then combine our knowledge and skills to develop an overall plan for implementation. Pest problems are solved in the same way. Up to this point you have learned about insect biology and pest management theory, and you have analyzed the various tactics that can be used to subdue pests and prevent losses. The next step is to consider how this information can be put together into a practical plan for implementation. This is the practice of pest management.

Recall that there are four fundamental strategies to consider in dealing with insect pests: do nothing, reduce numbers, reduce susceptibility of the host, or use combinations of the last two. Implementation of these strategies depends on several primary activities involving pest identification, pest population assessment, and economic evaluation. Except where no action is taken, the pest management strategies assume that economic damage is imminent and that combinations of tactics need to be employed to reduce the threat.

In developing a plan to put into practice, we should focus on the goals of reducing pest status, conserving environmental quality, accepting tolerable pest densities, and improving net profits from crop production. To achieve these goals we need efficient sampling programs, valid decision guidelines, and a number of effective tactics that can be integrated into an overall plan of attack. Integrating several tactics will be one of our most challenging problems.

## CONCEPTS OF INTEGRATION

Integration of several compatible tactics has been a central theme of pest management since its inception and is one of its primary tenets. Indeed, even the name *pest management* is widely referred to as integrated pest management (IPM) to emphasize this point.

The importance of integrating several tactics lies in the desire for sustainability or durability of a management program. Although the use of a single management tactic may be successful in the short term, such programs often fail in time. This is because pest populations can adjust to a tactic, particularly when that tactic significantly reduces a pest's ability to reproduce and survive. Often a change in the frequency of a single gene in a population will allow the tactic to be overcome. Conversely, a pest has much greater difficulty overcoming the destructive influences of several tactics, usually requiring the frequency increase of several new genes in the population. In other words, the pest can be kept off guard when exposed to several destructive factors instead of just one and has less capability to recover.

In addition to sustainability, pest management with several integrated tactics often results in better environmental conservation. This is particularly true when pesticides are an important component in the program. The addition of other tactics to the program reduces the frequency of pesticide application and, consequently, potentially harmful residues in the environment.

In discussions of pest management, integration may also refer to managing the impact of many different types of pests. At this level, all types of pests, including insects, weeds, and plant diseases, may be considered. Where the combined actions of different pests interact, the term *integrated pest management,* or IPM, is again used. However, most commonly, different types of pests are managed separately. Nevertheless, integrating management across different pests is one of the long-term goals of pest management. In our discussions, we will continue to discuss integration in the context of several tactics used against a single pest or closely related pest complex.

### Basis for Integration

In order to integrate several tactics into the overall pest management program, a set of principles is needed as a guide. Without principles, integration may proceed along some serendipitous route, which is usually ineffective and generally unacceptable.

The first approach to integration is to consider the EIL and the variables that determine it. Recall that the EIL is given as:

$$EIL = C / VIDK$$

where
$C$ = cost of the management tactic
$V$ = market value of the produce
$I$ = injury per insect
$D$ = damage per unit of injury
$K$ = proportional reduction in pest injury or damage (see Chapter 7 for more detail).

As represented here, the EIL presents a framework for organizing tactics and is the truly unifying principle of pest management. Our objective in combining tactics is to obtain the highest EIL possible for the pest population using our current technology, without reducing producer profitability. In most instances, our tactics will influence the $K$ variable, but we should not overlook possible inclusion of tactics that change the remaining variables. For instance, $D$ can be reduced by growing healthier plants or planting tolerant varieties, thus increasing the EIL. Additionally, environmental costs could (and should) be included in $C$, which also increases the EIL. Combined tactics to increase the EIL to its highest level possible usually will mean less intervention in the agroecosystem and, therefore, greater sustainability of the pest management program.

The next step is to select a combination of tactics and determine how they will be used. What is the logical basis for doing this? When considering this principle of integration, it is useful to consider other disciplines in which integration of tactics is a crucial part of problem solving.

One such discipline, which is roughly analogous to pest management, is human medicine. Medicine uses a broad range of tactics in managing diseases, and some of its most spectacular successes involve combining both preventive and curative procedures. In fact, the time-proven method of integrating tactics from preventive medicine and therapeutics is the cornerstone of modern medical practice. For example, to reduce the risk of getting cancer, specific diets and limited exposure to carcinogens are recommended. But, if these preventives fail, therapy is applied in the form of chemotherapy or surgery.

Management of agricultural pests is not unlike management of many human diseases. Consequently, we can use this powerful dualistic concept of prevention and cure to produce programs in which the integration of tactics is effective and practical. The major subdivisions of the insect pest management program, therefore, are preventive practice and therapeutic practice. Our tactics are organized under these subdivisions to develop an overall strategy.

### Preventive Practice

The old adage that "an ounce of prevention is worth a pound of cure" is certainly true in insect pest management. Our goal of reducing pest status in a way that is environmentally compatible is quite often satisfied by preventing a problem, rather than curing an existing one.

Preventive practice recently has been called **preventive pest management** (preventive PM) and has been a major thrust in research. Preventive PM is, or at least should be, the ultimate form of IPM; it is our first line of defense against a pest population.

Preventive PM implies that action should be taken against a pest before injury occurs. Therefore, the tactics are often employed without knowledge of pest presence or status of the pest population. An example is selecting and planting seeds of insect-resistant plants. In this example, seeds of resistant plants are chosen and plans made in advance of the growing season, often long before a pest's economic importance in a particular growing season is known. Prevention can be achieved by focusing on either the pest or the host. If the pest population is the focus of our tactics, we aim to lower its average level of fluctuation. If the host, usually a plant population, is the focus of our tactics, we aim to raise the level at which economic damage occurs.

**Lowering the pest's general equilibrium position.** Very often our available tactics focus on the pest. The ecological objective here is to lower the pest's general equilibrium position (average density) and, subsequently, the average level of crop damage (Fig. 16.1) to a position below the economic damage level (level where costs equal benefits).

The general equilibrium position can be lowered either by prohibiting pest establishment or by quickly limiting pest population growth once pest establishment occurs. Insect pest management tactics for this purpose include many biological controls, crop rotations, sanitation and tillage, planting dates to prevent or reduce colonization, trap cropping, plant spatial arrangements (for example, row widths and adjacent crops), and cultivar selection (antibiosis, antixenosis).

Most of these tactics are compatible with each other for integrating into the overall preventive PM program. Seemingly, the least compatible tactic in the

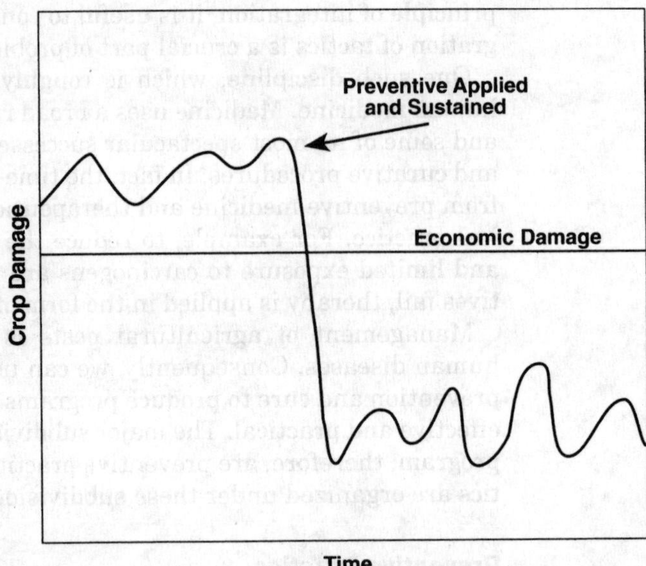


**Figure 16.1** Preventive pest management where the ecological objective is to lower the average level of fluctuation of a pest population, with subsequent reduction in the average level of crop damage.

context of all other tactics is the use of insects as biological control agents. Although a prevailing viewpoint is that these biological controls are compatible, it would not seem so. This is because tactics used to suppress the pest can suppress insect natural-enemy populations as well. At the very least, reduction of the pest population, the enemy's food source, can reduce the size and possible effectiveness of the natural enemy. In general, if tactics work in a density-independent manner, they tend to be compatible for inclusion in the pest management program. Conversely, if effectiveness of the tactic is promoted by high pest density, as with many insect natural enemies and disease-causing organisms, then the tactic becomes less compatible when used in conjunction with other density-reducing tactics. (See Quisenberry and Schotzko 1994 for a discussion of compatibility of host-plant resistance with other tactics.)

This observation is not to imply that insects should not be used in biological control but rather that these agents are some of the most difficult to integrate into the management program. Therefore, special emphasis is needed for this integration. Prospects for natural-enemy compatibility have improved in recent years with more efficient rearing procedures and methods of augmentation. Consequently, there will be a greater potential for augmentation, thereby making the natural enemies less density dependent for effectiveness. Moreover, genetic improvements through conventional selection and bioengineering to produce natural enemies resistant to insecticides and other tactics will enhance compatibility.

**Raising the level at which economic damage occurs.** The other area of prevention focuses on the crop itself. In this instance, attempts are made to manage losses from the pest rather than manage the pest density. Instead of reducing damage by reducing numbers of pests and total injury, this approach seeks to reduce damage by reducing the amount of damage per unit of injury.

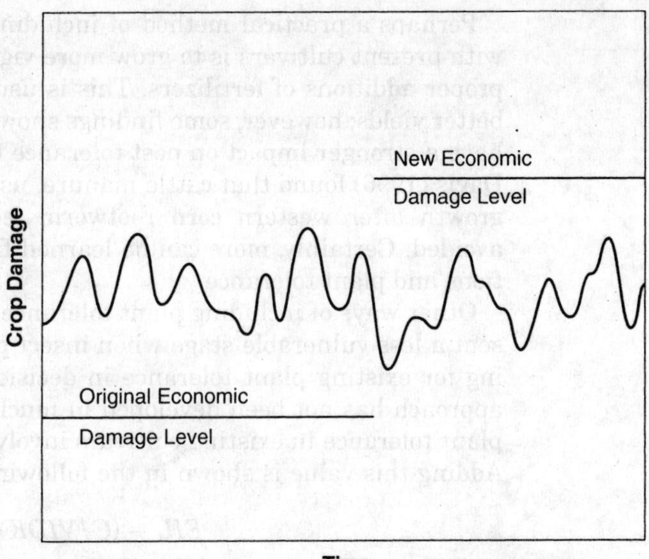


**Figure 16.2** Preventive pest management where the ecological objective is to modify crop susceptibility to pest injury, resulting in an increase of the level at which economic damage occurs.

The latter objective is completely dependent on the plant and is not affected by the insect pest.

The ecological objective of this approach is to raise the level at which economic damage occurs (Fig. 16.2). Using this strategy, the manager looks for ways to make the host less vulnerable or more tolerant to an existing pest population.

Tolerance to pest injury is a tactic often ignored by plant breeders and entomologists. Primarily, this is because cultivar tolerance (the ability to grow and reproduce in spite of injury) is not an obvious way to deal with a pest problem. Outwardly, the pest is still present and feeding, seemingly still causing damage, although it is not. Another drawback is that tolerance mechanisms are not well understood and are controlled by several genes, making plant breeding for tolerance difficult.

Yet, plant tolerance may be the most sustainable tactic that can be used. According to J. C. Reese and coworkers (1994), the advantages of tolerance are twofold. Tolerance is usually compatible with insect natural enemies and disease-causing organisms. It allows higher pest densities without significant yield loss and, therefore, is conducive to factors acting in a density-dependent manner. The second advantage is that no mortality pressure is placed on the pest population, producing no pressure to develop resistance to the tactic. Indeed, because of the latter, *tolerance may be the premier pest management tactic*.

To integrate plant tolerance into IPM programs, the most obvious approach would be to choose and plant tolerant varieties. However, there is a paucity of varieties with a high degree of tolerance for most cropping systems (Velusamy and Heinrichs 1986) because of the reasons already mentioned. This situation may change with the development of new transgenic plants that can tolerate insects and disease.

Perhaps a practical method of including plant tolerance in IPM programs with present cultivars is to grow more vigorous plants through irrigation and proper additions of fertilizers. This is usually done as a matter of course for better yields; however, some findings show that certain types of fertilizers may have a stronger impact on pest tolerance than others. For instance, Allee and Davis (1996) found that cattle manure, used as fertilizer, stimulates corn-root growth after western corn rootworm feeding, allowing some losses to be avoided. Certainly, more can be learned from research into fertilizer applications and plant tolerance.

Other ways of including plant tolerance are changing planting dates to present a less vulnerable stage when insect population peaks occur and accounting for existing plant tolerance in decision making. In particular, the latter approach has not been developed in much detail. A method of accounting for plant tolerance in existing cultivars involves including a tolerance parameter. Adding this value is shown in the following modification of the EIL equation:

$$EIL = (C/VIDK) + E_0$$

where variables in the equation have the same meaning as mentioned previously and $E_0$ = tolerance parameter. In the equation, the **tolerance parameter** is the maximum number of insects or insect equivalents where no detectable yield loss occurs. EILs that include the tolerance parameter, explicitly accounting for plant tolerance, are called **tolerance EILs** (Fig. 16.3). By defining the tolerance parameter and adding it in the EIL equation, the EIL would be expected to be raised and have greater accuracy. As an example, such EILs calculated for the green cloverworm on soybean were raised as much as 54 percent in situations with high market values and low insecticide costs. Moreover, the analysis to determine yield loss and injury in this instance showed a higher degree of accuracy than with conventional methods. With greater EIL values and more accurate accounting of potential losses, management actions are required less frequently, saving money through unnecessary applications and causing unnecessary pesticide residues in the environment.

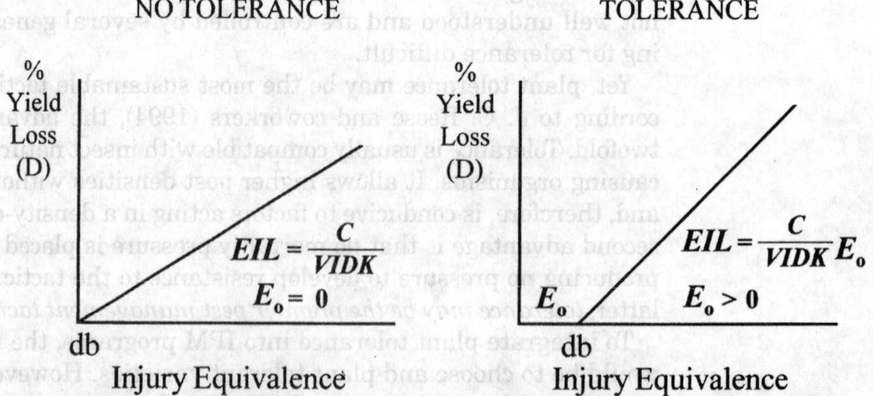


**Figure 16.3** Relationship between yield loss and insect injury in plants with and without tolerance. The tolerance parameter, $E_0$, shown along the $x$-axis is a measure of a plant's ability to maintain normal yield in the presence of insect injury. The $E_0$ in plants without tolerance is 0. The db is the damage boundary where yield loss is first detected.

**Pesticides not recommended as a preventive tactic.** In seeking preventive tactics, pesticides are not recommended for preventive pest management. They have been used widely for prevention, and many producers still use them in this way today. However, as we will discuss more thoroughly in Chapter 17, the use of pesticides in regularly scheduled applications frequently results in pesticide resistance and other forms of ecological backlash. Therefore, this approach is not considered a sustainable pest management practice.

**Plant and animal quarantine as a preventive tactic.** In addition to recommended preventive tactics, an important component not previously discussed in detail is plant and animal quarantine. Quarantines are legal instruments to prevent or slow the spread of pests from infested into uninfested areas by restricting movement of infested stock.

The first legislation on quarantine in the United States was established with the Insect Pest Act in 1905, which provided the authority to regulate the entry and interstate movement of articles that might spread injurious insects. In 1912, the Plant Quarantine Act was passed, authorizing the Secretary of Agriculture to enforce regulations to protect agriculture in the United States from the introduction of exotic insects and plant diseases. This act and additional legislation gives federal agents authority to examine goods as they cross international boundaries into the United States or as they traverse regions of the country that are under pest quarantine. Another regulatory law was the Federal Plant Pest Act, enacted in 1957. This law regulates the movement of plant pests along man-made pathways such as sea-lanes and highways. It gives authority for inspectors to board ships, trucks, aircraft, and other such vehicles and treat cargo that may be infested with pests and disease organisms.

The objective of quarantine is containment of the pest population. In the United States, quarantines are administered by the Animal and Plant Health Inspection Service (APHIS), which regulates both domestic quarantines (in force within the United States and its territories) and foreign quarantines, directed against pests associated with agricultural commodities of foreign countries. Quarantines are enforced by inspectors at state borders and at eighty-five foreign ports of entry (Fig. 16.4). Quarantined goods without import or export certification may be seized by federal agents. This approach to prevention is part of a broader practice called **regulatory control.** Examples of insects under quarantine are the Asian longhorned beetle and the glassy-winged sharpshooter (see Boxes 16.1, 16.2).

### Therapeutic Practice

Although prevention should form the basis and first line of defense in the ideal insect pest management program, prevention alone is often not adequate. Therapy is also necessary in most programs.

Therapy seeks to cure an acute or chronic crop disorder. It differs from prevention in that a pest population is present, and injury is occurring when management decisions are made. In other words, therapy is applied after pest sampling is conducted and an assessment of pest status indicates that economic damage is imminent.

Although therapy is considered distinct from prevention, it includes a preventive aspect. Its primary goal is to prevent future losses; otherwise, the activity would not be economically feasible.

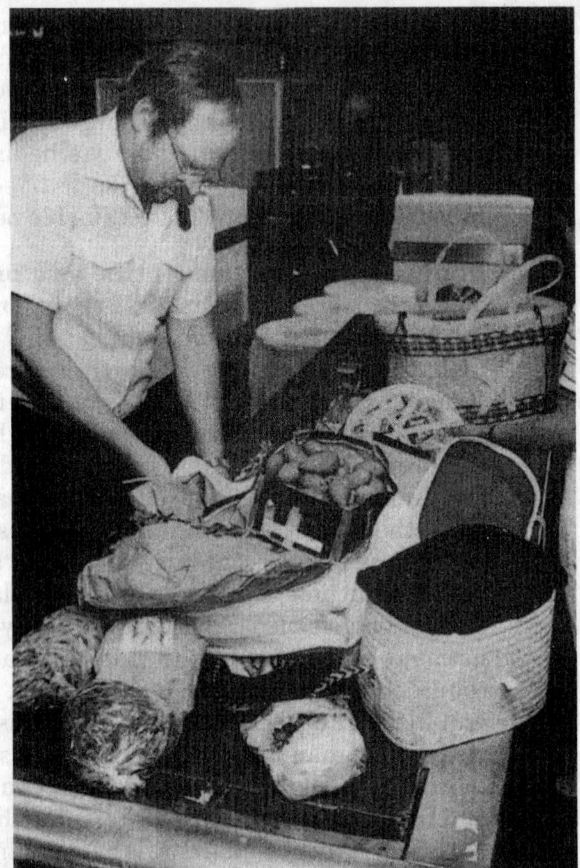

**Figure 16.4** Travelers often bring back quarantined agricultural products that are seized at international ports of entry by inspectors. Many of these items harbor exotic insects and disease organisms. (Courtesy USDA)

Often, therapy constitutes the first phase in the design of insect pest management because therapeutic programs can be formulated quickly to alleviate ongoing problems. From a developmental standpoint, therapy may be the final phase of insect pest management for many occasional pests. With these pests, the general equilibrium position is much lower than the EIL, being suppressed naturally. There may be no need for preventives because nature supplies them. All that may be needed is an infrequent "correction" to population density to achieve the management goal. Therefore, the judicious use of pesticides may be the sole requirement because such use is effective and usually will not cause unanticipated pest problems. Examples of situations where only a therapeutic program is needed include spider mite and grasshopper outbreaks associated with drought in the upper Midwest.

**Use of therapeutics in insect pest management.** The ecological objective of therapy is to interrupt ongoing pest population growth and injury. The goal of therapy can be achieved either by dampening pest population peaks or by truncating growth of injury with altered crop exposure. Dampening pest density peaks is usually accomplished by proper application of a pesticide to kill pests, and reducing crop exposure often is achieved by early harvest. An example of

## BOX 16.1 Asian Longhorned Beetle, Starry Sky Beetle

**SPECIES:** *Anoplophora glabripennis* (Motschulsky) (Coleoptera: Cerambycidae)

**DISTRIBUTION:** This species is native to China and Korea. It was introduced into the United States between 1985 and 1998 and spread as immatures in untreated lumber and shipping crates. Localized, quarantined populations exist in New York and Illinois.

**APPEARANCE:** Individuals are 0.75 to 1.25 inches long, with a jet-black body and about twenty irregular white spots on each wing cover. The long antennae are 1.5 to 2.5 times the body length and have distinctive black and white bands on each segment. The feet have a bluish tinge. Larvae are elongate, cylindrical, and pale yellow, attaining a maximum length of about almost 2 inches, and undergo three molts.

**IMPORTANCE:** This beetle has been discovered attacking trees in the United States. Tunneling by beetle larvae girdles tree stems and branches. Repeated attacks lead to dieback of the tree crown and, eventually, cause death of the tree. The pest probably traveled to the United States inside solid-wood packing material from China. The beetle has been intercepted at ports and found in warehouses throughout the United States. This insect is a serious pest in China, where it kills hardwood trees in roadside plantings, shelterbelts, and plantations. In the United States, the beetle prefers maple species (*Acer* spp.), including boxelder, Norway, red, silver, and sugar maples. Other known hosts are alders, birches, elms, horse-chestnut, poplars, and willows. A complete list of host trees in the United States has not been determined. Currently, the only effective means to eliminate the Asian longhorned beetle is to remove infested trees and destroy them by chipping or burning. To prevent further spread of the insect, quarantines are established to avoid transporting infested trees and branches from the area. Early detection of infestations and rapid treatment response are crucial to successful eradication of the beetle.

**LIFE CYCLE:** This species has one generation per year. Adult beetles are usually present from July to October but can be found later in the fall, if temperatures are warm. Adults usually stay on the trees from which they emerged or they may disperse short distances to a new host to feed and reproduce. Each female is capable of laying up to 160 eggs. The eggs hatch in 10 to 15 days and the larvae tunnel under the bark and into the wood where they eventually pupate. The adults emerge from pupation sites by boring a tunnel in the wood and creating a round exit hole in the tree.

---

the latter approach is found with dense populations of alfalfa weevil, *Hypera postica,* in alfalfa stands with advanced growth. In this instance, early cutting terminates further foliage loss and often is economically feasible (see Table 6.3).

For key insect pests, therapeutics can be used in conjunction with a preventive pest management program. Preventives are applied to reduce the general equilibrium density of the pest. However, if or when the preventives fail, therapy serves as a "correction" in the system (Fig. 16.5). In this pest management system, population sampling and consultation of economic thresholds are the primary activities to determine the need for the correction.

Several tactics are available for use in pest management therapy. Some of these include: (1) selective pesticides; (2) fast-acting, nonpersistent biological controls such as microbial insecticides; (3) early harvest; and (4) mechanical removal of pests (for example, hand picking and pruning infested branches). Of these tactics, pesticides are by far the most significant in pest management therapy.

## BOX 16.2 GLASSYWINGED SHARPSHOOTER

**SPECIES:** *Homalodisca coagulata* (Say) (Hemiptera: Cicadellidae)

**DISTRIBUTION:** This species occurs in northern Mexico and the southern United States, including Alabama, Arkansas, Florida, Georgia, Louisiana, Mississippi, North and South Carolina, and Texas. More recently, it spread to southern and central California.

**IMPORTANCE:** The glassywinged sharpshooter feeds on a wide range of ornamental and crop plants but prefers woody or perennial plants. On most plants, it feeds by sucking sap from stems and, in so doing, produces small droplets of watery excrement. The droplets can be a nuisance in urban areas, and when dry, the excrement gives plants a white-washed look. The pest is of particular importance in California, transmitting Pierce's disease, which kills grapevines in wine, table, and raisin grape crops.

**APPEARANCE:** This insect is large for a sharpshooter, being almost 0.5 inch long. It is dark brown to black, and the underside is light colored. The upper regions of the head and back are stippled with whitish or yellowish spots. The wings are semitransparent and have reddish veins. The watery excrement usually collects on both sides of the insect, appearing as large white spots.

**LIFE CYCLE:** Glassywinged sharpshooters overwinter as adults and begin laying eggs in late February. Egg laying continues through May. Eggs are laid in groups of ten to twelve eggs. These groups appear as small, greenish blisters on plant leaves. After egg hatch, the blisters look like brown scars. Nymphs are small and white, and undergo three molts before reaching maturity. The first generation peaks in late May, with adults producing a second generation. It is these adults that overwinter.

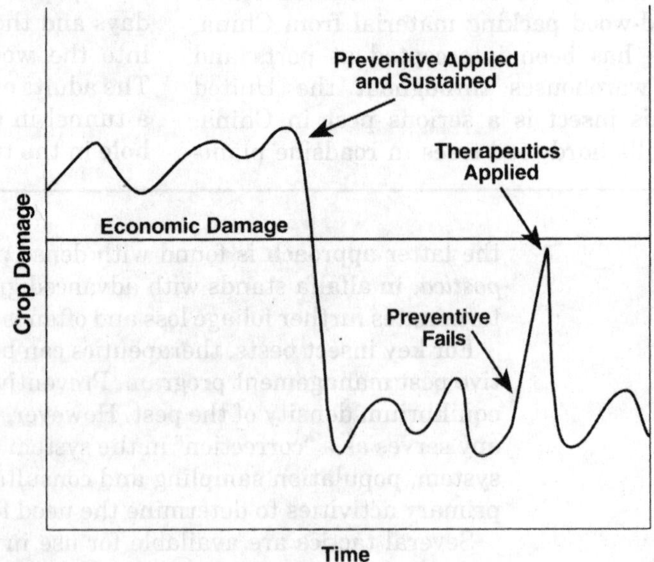


**Figure 16.5** Use of a pest management program integrating both preventive and therapeutic tactics. Therapy is applied only when preventives fail.

## DEVELOPMENT OF AN INTEGRATED MANAGEMENT PROGRAM

When insect pest management is perceived as encompassing both preventive and therapeutic practices, developmental objectives for new programs become sharper, and a much better scheme for combining several tactics emerges; in other words, principles for combining tactics result. The principles suggest that compatible tactics, based on an understanding of pest ecology and pest/plant relationships, be combined as a first line of defense in the management program. Further, a second line of defense is needed, involving therapeutic tactics, should the preventives fail.

To develop an integrated program having both preventive and therapeutic elements, a number of steps can be envisioned:

1. Identify potential preventive and therapeutic tactics. The therapeutic program may need to be developed first for early relief from the problem.
2. Test tactics individually.
3. Formulate mathematical expressions or, more simply, conceptual plans of potential systems.
4. Conduct field trials with the systems to determine costs, compatibility of tactics, and system effectiveness.
5. Deploy successful programs that offer on-farm flexibility (alternatives), such that the system can be tailored to suit individual production practices.

In developing new programs, a special focus on prevention is necessary. Notably, the more severe the pest, the greater the need will be for effective preventive procedures. For occasional (sporadic) pests, a therapeutic program is all that may be required.

### Selection of Tactics

Design of the insect pest management program calls for selection of a set of tactics to achieve prevention on the one hand and therapy on the other. In making selections, considering problems with resistance to tactics and natural enemy destruction (discussed in Chapter 17) becomes paramount. In these considerations, we would favor therapeutic tactics with the narrowest spectrum or those with potential to apply selectively. The major objective in this approach is to protect natural enemies and conserve environmental quality. In a similar vein, to accomplish our objectives, we also would choose the tactics that are relatively nonpersistent.

Moreover, because several tactics are to be used, they must be compatible with one another; that is, conflicts between tactics of prevention and therapeutics need to be avoided or minimized. It would do little good to apply a tactic only to have its potential benefits reduced or nullified by another tactic.

Finally, the least expensive alternatives to achieve an objective should be chosen. Prices of tactics may vary widely, and efforts to reduce costs are an important aspect of successful pest management in a frugal agricultural production environment.

### The Management Plan and Crop Values

Thus far, we have addressed development of the insect pest management plan in the context of all available tactics and severity of the pest attack. Another important consideration is the value of the commodity being protected. The ratio of management costs to commodity prices (market values) can be used as a realistic index for devising a plan. This ratio forms a primary constraint on choice, integration, and employment of useful tactics. Furthermore, because costs of tactics probably are less variable between crops than are commodity prices, it is the latter that mostly limits our alternatives.

For discussion, it is meaningful to locate crops from low to high along a commodity-value continuum. The continuum represents potential value per area, where value has little relationship to importance in terms of human consumption.

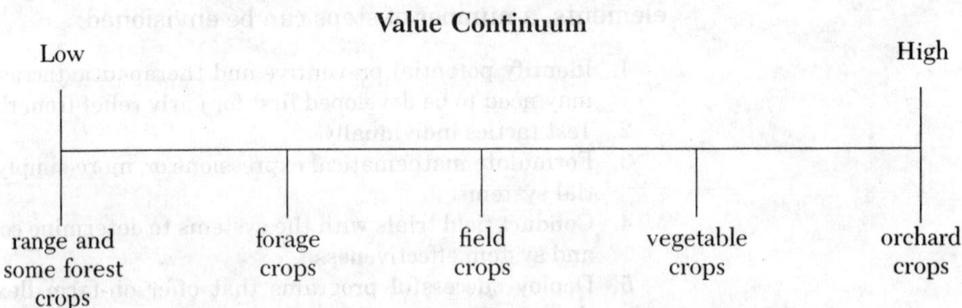


Along the continuum, low-value crops are represented by forests and pastures; moderate value by forage crops, field crops, and vegetables; and high value by orchards. Low-value crops require large land areas for production and have lower economic values per area than other types. Substantial injury of these crops usually can be tolerated before economic damage occurs because each unit injured has a relatively low price. Therefore, expensive management practices, especially frequent use of therapeutic tactics, usually are not practical. Rather, preventive tactics like sanitation, harvest practices, and controlled burning are the more likely tactics to be integrated into pest management programs.

Moderate-value crops have greater value per land area; therefore, injury is less tolerable than with low-value crops. Higher values tend to create lower economic-injury levels, and as a result, pest populations exceed these levels more often. Whereas preventive tactics are useful and important, there is a greater need for and use of therapeutic tactics.

High-value crops present some of the most difficult pest management challenges of all because high prices and quality considerations usually place economic-injury levels just above or even below a pest's general equilibrium position. Consequently, the need for pest population corrections is frequent and continuing. For these crops, preventive practices are useful, but the need for therapeutics often becomes prevalent. Therefore, we might anticipate changes in reliance on prevention and therapeutics as value per land area increases and economic-injury levels fall. However, with advances in pest technology, such as the more economical and effective uses of insect natural enemies and host-plant resistance, we can predict the greater use of nonpesticidal preventives in pest management of high-value crops.

### An Integrated Management Program for the Bean Leaf Beetle

An example of an integrated strategy involving both preventive pest management and therapy can be seen in a case history of the bean leaf beetle on soybean.

The bean leaf beetle is a widespread pest of soybean in the United States and has caused considerable concern in the Midwest since the early 1980s.

The bean leaf beetle begins its activity in early spring when adults emerge from overwintering quarters in woodlots, unmanaged field borders, and other locations. These adults immigrate to wild legumes or alfalfa, feed, and lay a few eggs. Upon soybean emergence and first cutting of alfalfa, beetles move into the soybean and feed on leaves (see Fig. 16.6).

Some of the most obvious injury occurs in the soybean seedling stage when these overwintered beetles chew holes in leaves of seedlings. However, research has shown that soybean plants are very tolerant of this early-season injury, and management is almost never necessary at this time. This

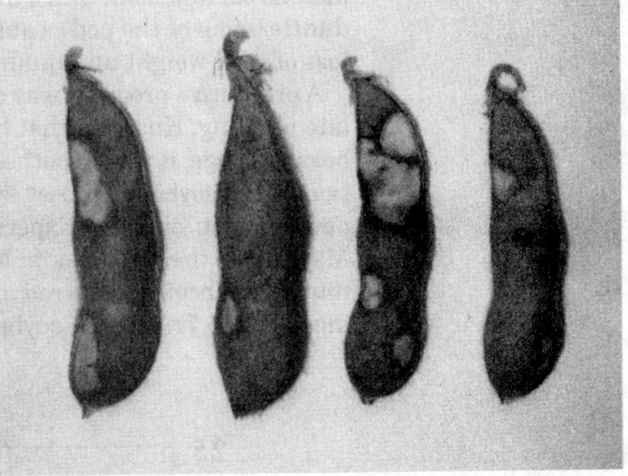

**Figure 16.6** An adult bean leaf beetle, *Ceratoma trifurcata*. This insect defoliates soybean but also feeds on pod surfaces. The result of pod feeding is discolored and shriveled seeds and pod loss. (Photos Marlin E. Rice)

finding illustrates the importance of defining pest status through the EIL because some apparent problems are not problems at all, and unnecessary action can be avoided.

While feeding, overwintered adults lay their main complement of eggs in the soil. These eggs constitute the first generation, with larvae and pupae developing in the soil. First generation adults emerge in July, feeding on foliage and laying eggs to produce the second generation. This foliage feeding almost never causes economic damage.

Second generation adults emerge in late August and early September. At this time the soybean plants are in the pod-filling stages, and the beetles feed both on leaves and pods (Fig. 16.6).

Pod feeding usually is the most significant injury caused by the bean leaf beetle. Adults feed on the pod surface, consuming tissue down to the endocarp, which directly encloses the seed. Seeds beneath the lesion become shrunken, discolored, and sometimes moldy. Additionally, beetles can feed on the pod peduncle (stem of the pod), causing breakage and complete pod loss. The result is loss of seed weight and quality.

A preventive program was developed for the bean leaf beetle using the tactic of late planting. Knowing that beetles move from alfalfa to soybean when the soybeans emerge, it was hypothesized that a delay in planting could reduce colonization in the soybeans, upset synchrony of the second beetle generation with pod development, or both. Experiments comparing seed planting the first week in May versus the last week in May in central Iowa found that consistently greater numbers of beetles occurred in the early soybean plantings than in the late plantings (Fig. 16.7) and that soybean pod injury was lower in the late plantings.

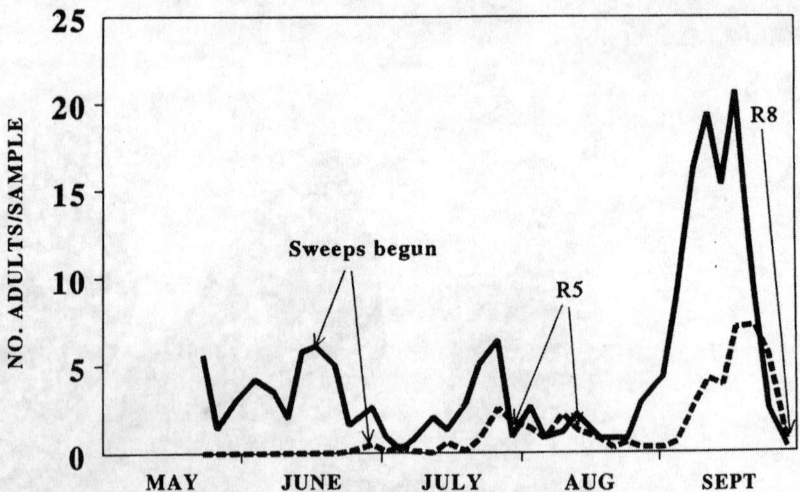


**Figure 16.7** Graph showing the effect of late planting of soybeans on bean leaf beetle densities. Adults were counted first by direct observation, followed by sweep-netting as the plants grew. The solid line represents beetles in soybeans planted the first week of May versus soybeans planted the last week of May in central Iowa. Stage R5 is the pod-fill stage (seeds growing in size) and R8 is maturity. June, July, and September peaks represent overwintered, first-, and second-generation beetles, respectively.

**Table 16.1 Bean Leaf Beetle Economic Thresholds for Soybean in Number/Foot of Row.[a, b]**

| Market Value | Management Costs ($/acre) | | | | | |
|---|---|---|---|---|---|---|
| ($/bushel) | 7.00 | 8.00 | 9.00 | 10.00 | 11.00 | 12.00 |
| 5.00 | 5.5 | 6.3 | 7.1 | 7.9 | 8.7 | 9.5 |
| 6.00 | 4.6 | 5.2 | 5.9 | 6.5 | 7.2 | 7.8 |
| 7.00 | 3.9 | 4.4 | 5.0 | 5.6 | 6.1 | 6.7 |
| 8.00 | 3.5 | 4.0 | 4.5 | 5.0 | 5.5 | 6.0 |
| 9.00 | 3.1 | 3.5 | 4.0 | 4.4 | 4.9 | 5.3 |
| 10.00 | 2.8 | 3.2 | 3.6 | 4.0 | 4.3 | 4.7 |

[a]Economic thresholds set at 75 percent of the EIL.
[b]Based on a soybean row spacing of 30 inches.

A therapeutic program was developed for the pest by studying the feeding of bean leaf beetle adults in field cages and quantifying their injury and resulting pod damage. Using this information, economic thresholds were developed that provided appropriate guidelines for a given soybean market value and specific management cost (see Table 16.1). These thresholds were combined with an effective sampling plan and minimal-rate insecticide recommendations to form the therapeutic program.

Consequently, the recommended management strategy for the bean leaf beetle on soybean is to plant soybean as late as possible in the recommended planting period and to locate them adjacent to a nonhost crop, such as corn, or as far away from alfalfa as is possible. However, the commercial application of systemic seed treatments (neonicotinoids) in recent years has allowed soybean producers to plant earlier and avoid or reduce beetle injury. When soybean plants reach the pod-development stage, sampling is recommended on a weekly basis to detect any failures of the late-planting tactic. Subsequently, estimates of population density are compared to tabular economic thresholds, and an insecticide is applied if necessary. Insecticides allowing greatest personal and environmental safety are selected and lowest effective rates are used. Savings from use of this IPM program in 1998 were estimated at $490,000 and 6,000 pounds of insecticide from unnecessary spraying in Iowa alone.

As you can see, both preventive and therapeutic tactics make up the bean leaf beetle management program. Prevention is the foundation of the program, and pest scouting is used to assess the effectiveness of the preventive tactic. Moreover, a backup program is also available in the form of insecticides, should the preventive tactics fail. With such a program, the risk of economic loss is nil.

## AREAWIDE PEST TECHNOLOGY

For the most part, we have thus far discussed insect pest management as an approach that is practiced by private individuals or businesses for use in a limited area, such as a single farm or orchard. However, as R. L. Rabb (1978) has pointed out, implementation of management from an areawide perspective may be much more effective in solving insect problems than addressing the problem on an individual basis.

The greatest problem with individual initiatives is pest migration. If an individual grower mounts a successful pest management program, but the grower's neighbors do not, the program may eventually fail because of insect colonization from the surrounding areas. Therefore, it has been proposed that pest programs be practiced at a community, regional, or even interregional level.

Areawide pest technology includes coordinated programs that involve federal, state, commodity, and local organizations, as well as producers. In many instances they are publicly funded. E. F. Knipling and G. G. Rohwer outlined some of the basic characteristics of such programs: (1) implementation over large geographical areas; (2) organizational, rather than individual, coordination; (3) pest eradication, if practical (otherwise pest population reduction and maintenance at low densities); and (4) mandatory involvement of growers in the geographic area.

Because of the cost of areawide pest programs, they are most feasible for key pests in certain agroecosystems. Recall that a key pest is a severe or perennial pest that dominates the cultural activities in a production system. If key pests are present at all, most systems would only have one or a few of them. Therefore, even when areawide programs are successful, there is a continuing need for farm-level pest management programs to deal with the remaining pests in the overall pest complex.

Another focus of areawide programs has been on introduced or exotic pests. These pests may not have achieved key pest status when the areawide program is initiated. Rather, program implementation is based on the potential of these species to become key pests. In these instances, areawide programs are employed to eradicate the pests before they can expand their ranges. Because state and federal agencies and legal restrictions are involved with these pests, the practice has been termed *regulatory control,* of which quarantines are a part.

There have been many successful areawide pest programs, most spearheaded by the United States Department of Agriculture. Some of the noteworthy successes include the screwworm and Mediterranean and Mexican fruit flies, as discussed in Chapter 15, as well as the cattle ticks, *Boophilus annulatus* and *Boophilus microplus,* and the khapra beetle, *Trogoderma granarium.* In addition, classical biological control of introduced pests, such as the cottony-cushion scale, can be considered a type of areawide pest technology, except in this instance eradication is not the objective; rather, it is a low-input sustainable system.

### The Boll Weevil Eradication Program

An areawide program having shown considerable success is the Boll Weevil Eradication Program. This program can serve as an excellent example of areawide pest technology concepts.

The boll weevil is a key pest of cotton that does most of its damage when squares begin to form. At this time, females gouge out small cavities with their snouts and lay eggs in the feeding cavities (usually one per square) (Fig. 16.8). The combined injury causes bracts to open or "flare" and squares to shed. Feeding and egg laying continue in bolls (the fruit) as the plant matures, but bolls do not fall from the plant. Larvae develop inside the square or boll (Fig. 16.8) and pupate in about 7 to 12 days. Pupae remain in the

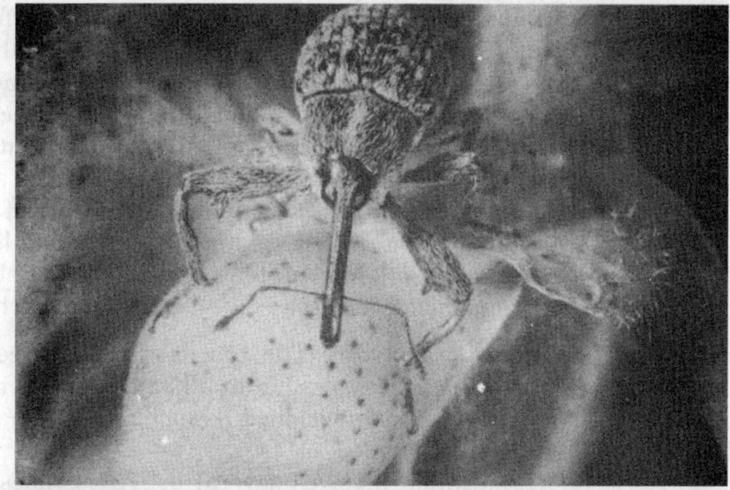

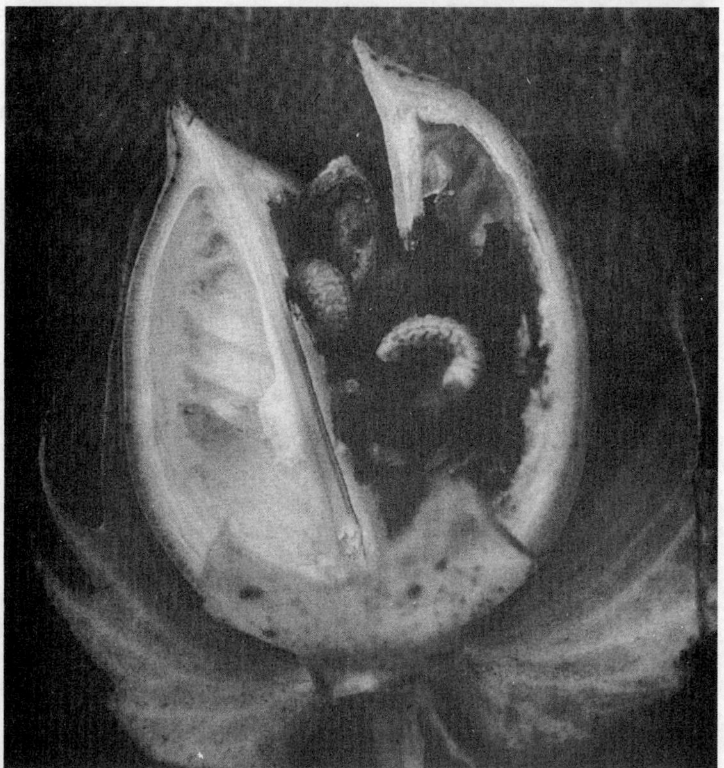

**Figure 16.8** Adult boll weevil, *Anthonomus grandis,* feeding on cotton square (flower bud) and damage to cotton boll from boll weevil larvae. (Photos: boll weevil, Texas A&M University; larvae, Marlin E. Rice)

square or boll for 3 to 6 days, after which the adult weevil chews its way out. From two to seven generations occur annually, depending on latitude, and the fall generation migrates to overwintering sites under ground litter in cotton fields, at edges of woods, in fence rows, and in other protected places. Winter is spent in dormancy, and most adults become active and begin feeding again in June.

**Pilot project.** The boll weevil eradication effort began in 1971 with a large pilot experiment in southern Mississippi, Alabama, and Louisiana to assess the technical and operational feasibility of eradication. The experiment, conducted through 1973, included in-season insecticides applied by growers, reproduction and diapause control of adults in late summer and fall, pheromone trapping in spring, application of insecticides at the bud stage in spring, and release of sexually sterile males. At the end of the test period, boll weevil reproduction was suppressed below detectable levels in 203 of 236 fields. The experiment was interpreted as showing that elimination of the boll weevil as an economic pest in the United States was technologically feasible.

**Current program.** The goal of the Boll Weevil Eradication Program is to eradicate this key pest from all cotton-producing areas of the United States. The program is envisioned as moving in manageable increments throughout the Cotton Belt.

Eradication involves several activities. These include pheromone trapping (Fig. 16.9) to delimit populations judicious use of the insecticide malathion, and grower-implemented cultural controls such as stalk destruction and uniform planting dates. More specifically, the insecticide program calls for fall applications to reduce overwintering adult populations, spring applications to

**Figure 16.9** Boll Weevil Eradication Program trap with the synthetic pheromone attractant, grandlure. Such traps are checked each week from early spring to killing frost to monitor pest populations. Insecticides are applied when active weevil populations are found. (Photo Marlin E. Rice)

prevent feeding and reproduction, and in-season applications based on field inspection and trap catches. The net result of these activities has been a progressive reduction in pest populations as well as a constant reduction in the amount of insecticide used in an area.

The Boll Weevil Eradication Program has been conducted along two major fronts, one in the Southeast, initiated in 1977, and one in the Southwest, initiated in 1985. The Southeast Program began on 15,000 acres of cotton in North Carolina and Virginia and progressed in increments through the remainder of North Carolina and South Carolina (300,000 acres) by 1983. At that time, the weevil was declared eradicated from this region. In 1987, eradication activities were begun on more than 500,000 acres in southern Georgia, Florida, and Alabama, and in 1991, another 580,000 acres were added to the program. Currently, eradication programs are active in Arkansas, Missouri, Mississippi, Louisiana, Texas, Tennessee, and Oklahoma in the United States and in northern Mexico.

In the Southwest, the program was designed to eradicate the boll weevil in approximately 233,000 acres of cotton in western Arizona, southern California, and northwestern New Mexico. Eradication in the western region has now been completed.

As of 2007, the Boll Weevil Eradication Program has eliminated this important pest on over 14.3 million acres of cotton. The program is active on yet another 2.1 million cotton acres (Fig. 16.10).

Such an eradication program was only possible by cooperative efforts of producers and federal, state, and local agencies. Also critical to this effort was legislation, which allowed producer referenda on mandated control activities and financial support. Without this legislation and majority acceptance by cotton producers, an areawide program on this scale would not have been feasible. Additionally, well-organized foundations for collecting funds and federal cost sharing have been an indispensable part of the program's success.

### The Pink Bollworm Eradication Program

An areawide Pink Bollworm Eradication Program also has been established that roughly parallels the approach of the Boll Weevil Eradication Program.

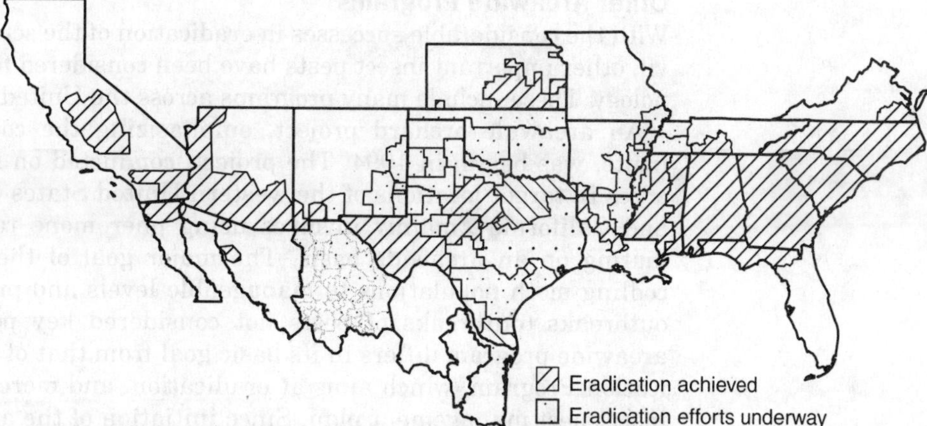


**Figure 16.10** Areawide boll weevil program status in the United States as of 2007. Programs are underway in the mid-South and the Texas/Oklahoma region to eliminate this pest. (Redrawn from a map provided by the USDA/PPQ)

The pink bollworm program was devised by the National Cotton Council's Pink Bollworm Action Committee. The objective of the program is to eradicate this very important lepidopterous pest from infested areas of the United States, that is, the southwestern portion of the Cotton Belt.

Like the boll weevil program, the Pink Bollworm Eradication Program coordinates the efforts of cotton producers and federal, state, and local entities in the United States and Mexico. The design of the program involves three implementation phases, relating to region. Phase I began in 2001/2002 and involved the El Paso/Trans Pecos region of west Texas, south-central New Mexico, and northern Chihuahua, Mexico. Phase II began in 2006, involving cotton-growing areas in southeastern and central Arizona and in southeastern California. Phase II continued and moved into the Phase III region in 2007. Phase III includes Yuma, Arizona, southern California, and the Mexicali Valley of northwest Mexico. The entire Phase III area will be addressed in 2008.

The operational elements of the Pink Bollworm Eradication Program include mapping, detection, and control. In the mapping element, cotton field locations are identified and field acreages and cotton varieties are recorded. For detection, pink bollworm delta traps, baited with the pink bollworm pheromone gossyplure, are placed around each field. The traps yield population density information useful in control decisions and in assessing success of control tactics.

With the information from mapping and detection, multifaceted control tactics are initiated. The control tactics include the following: (1) cultural activities such as uniform planting and harvesting dates for a location and sanitation, (2) mating disruption using gossyplure sprays and rope dispensers, (3) planting of Bt (*Bacillus thuringiensis*) transgenic cotton varieties following EPA refuge requirements, (4) releasing sterile pink bollworm moths, and (5) application of the insecticide chlorpyrifos if the previous four tactics fail to prevent economic loss.

The Pink Bollworm Eradication Program continues and exemplifies a well-balanced integration of diverse tactics to achieve its goal. If successful, the program is expected to provide significant economic gains for cotton producers while conserving environmental quality through reduced pesticide use.

### Other Areawide Programs

With the considerable successes in eradication of the screwworm and boll weevil, other important insect pests have been considered for areawide pest technology. These include many programs across the United States.

An areawide orchard project, emphasizing the codling moth on pome fruits, was begun in 1994. The project, conducted on more than 3,000 crop acres in target locations of the western United States (Washington, Oregon, and California), mainly involves using pheromone ropes to disrupt adult mating on an areawide basis. The major goal of the project is to reduce codling moth populations to manageable levels and prevent secondary pest outbreaks (outbreaks of pests not considered key pests). Therefore, this areawide program differs in its basic goal from that of the Boll Weevil Eradication Program, which aims at eradication, and more closely resembles an insect pest management plan. Since initiation of the areawide orchard project, seventeen sites were added to the original five, gaining more producer participation. Currently, the areawide use of mating disruption covers 100,000 acres of orchards in Washington, California, and Oregon. Moreover,

## BOX 16.3 Russian Wheat Aphid (See Color Plate 6)

**SPECIES:** *Diuraphis noxia* (Mordvilko) (Hemiptera: Aphididae)

**DISTRIBUTION:** The Russian wheat aphid is indigenous to the Mediterranean region of the Old World, including southern Russia. It was introduced into the Republic of South Africa in 1978 and has since become established in central Mexico and northern regions. It is distributed widely across the western half of the United States and southwestern Canada.

**IMPORTANCE:** The Russian wheat aphid was first detected in Texas South Plains in 1986. Since that time, it has spread across the eastern half of the United States and to some southern Canadian provinces. Preferred hosts of the pest are wheat, barley, and triticale, but the insect also may be found in rye and oats. Several native grasses are also suitable as summer hosts of the pest. Russian wheat aphids damage plants by injecting a toxin during feeding that prevents production of chlorophyll. Symptoms of feeding include leaf rolling, with white, yellow, or purple longitudinal stripes along the rolled leaf. The result of this injury is deformed, partially sterile grain heads, and reduced grain weight. The insect is also capable of transmitting several viruses, including brome mosaic, barley stripe mosaic, and a new picorna-like virus. Yearly losses since introduction of this pest have varied from $5.3 million to $70 million.

**APPEARANCE:** Adults are small, elongated, spindle-shaped, and lime-green. Often, they are covered with a powdery coating of wax. There are two forms, a winged form and a wingless form. The winged form is much darker than the wingless form, sometimes appearing black. The species is easily distinguished from other aphids by extremely short antennae, the absence of prominent cornicles (dorsal tubes near the end of the abdomen), and a projection or tail above the cauda (which gives a "double tail" look). Nymphs resemble adults but are smaller. Nymphs destined to be winged adults produce wing pads, which are noticeable just before maturity.

**LIFE CYCLE:** Winged females are produced when the pest population is crowded or the host plant is under stress or passes on to an advanced stage of development. Winged females disperse to uninfested plants, with the aid of wind, and begin feeding. They give birth to active neonates (viviparous birth), which mature into wingless females. These wingless females develop to maturity in about a week, depending on temperature, and then give birth to other neonates. Each female can produce up to four nymphs a day. There are multiple generations each growing season.

---

the amount of damaged apples has dropped to less than 1 percent, compared to 1 to 2 percent when insecticides were used.

Other areawide insect-pest projects sponsored by the USDA and cooperating states include corn rootworms in the Midwest (1995–2003), stored grain insects in the Midwest (1996–2003), tephritid fruit flies in Hawaii (1999–2008), fire ants in the southern United States (2001–present), Russian wheat aphid (Box 16.3) and greenbug in the Great Plains (2001–2008), and the tarnished plant bug in Louisiana and Mississippi (2001–2007).

## SITE-SPECIFIC FARMING AND PEST MANAGEMENT PRACTICE

Site-specific crop management, also known as precision farming, recently has received a great deal of attention. This approach offers the potential for increasing production efficiency and, at the same time, conserving environmental

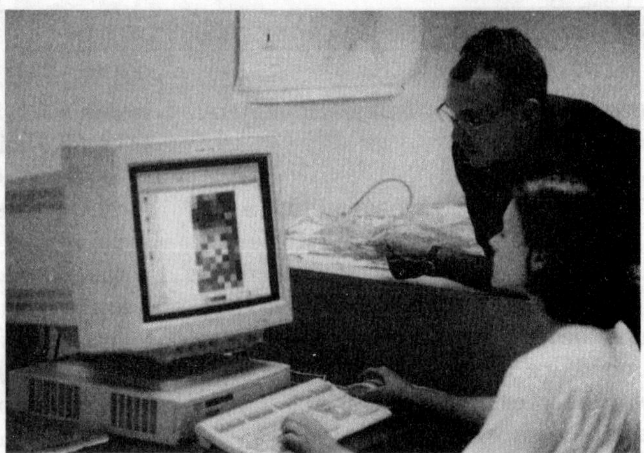

**Figure 16.11** Geographical Information Systems (GIS) can be used to help analyze in-field variability of insect populations and environmental factors causing this variability.

quality by reducing chemical inputs. The idea of site-specific farming involves the measurement and analysis of within-field yield variability. Geographical information systems (GIS), global positioning systems (GPS), electronic sensors, and computer systems are used to produce maps that detail variable crop yields across a field. Factors that potentially influence the variable yields are then measured and correlated with yield differences to determine cause-and-effect relationships. When causal factors are discovered, they are addressed on a site-specific basis, rather than on a whole-field basis, to maximize yields and minimize inputs, usually of chemicals.

Some of the key factors measured as possible yield determinants are fertility and other soil characteristics, weeds, plant diseases, plant-parasitic nematodes, and insects. Addressing these on a site-specific basis requires "smart" field equipment that applies management tactics such as fertilizers and herbicides on an as-needed and correct-amount basis.

Site-specific technology is in its infancy, and therefore its potential for use in insect management is still largely unknown. Sampling populations is at the heart of this application. For insect pests, sampling will be necessary on a within-field basis, followed by mapping (Figs. 16.11, 16.12) to show economically important patches of pests. After diagnosis, variable-rate applications of pesticides or microbials could be applied or, perhaps, variable plantings of resistant cultivars could be done for prevention of spatially persistent problems. To date, very little work exists comparing whole-field pest management with site-specific pest management. However, findings with the Colorado potato beetle in potatoes indicate that site-specific management has potential in reducing insecticide inputs for pests, while allowing within-field refuges for insects' natural enemies and pesticide-susceptible pests (Weisz et al. 1996). We can look forward to hearing more about site-specific farming as it relates to insect pest management in the not-too-distant future.

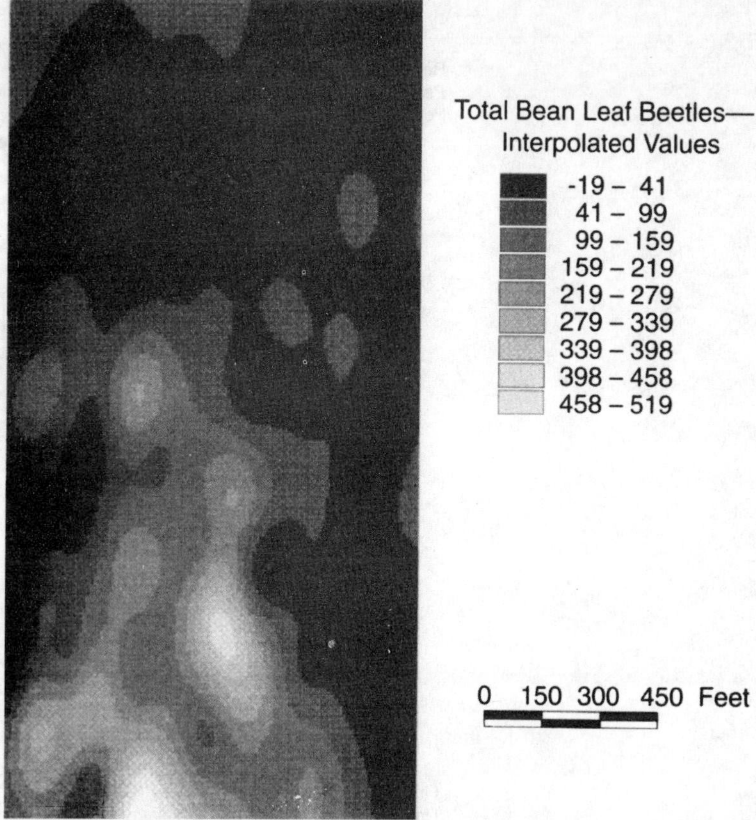


**Figure 16.12** A map of in-field bean leaf beetle densities in soybean using GIS analysis. A strong clustering of the beetles in the lower part of the figure (light area) indicates that action is needed to avoid losses. By treating only this area with insecticides, yield losses could be avoided. In addition to reducing costs, natural enemies and insecticide-susceptible beetles would be conserved in other parts of the field.

## CONCLUSIONS

Programs of pest technology can have different overall goals and be practiced in different ways (see Fig. 16.13). Initially, priorities force us to focus on our own enterprise and its operations, as well as our personal property. At this level, we have the direct responsibility of dealing with pest problems in an economically and environmentally sound way. Our best approach is to use multiple tactics in averting potential pest losses. For insect pests that cause perennial losses, the most sustainable practice is to develop nonpesticidal preventives as our first line of defense in reducing average population densities and/or loss per pest. The preventive program is less risky if it is backstopped by careful diagnosis and therapeutic procedures to avoid losses if the preventives fail. For occasional pests, which are prevented from causing losses most years by nature, a program of sampling, diagnosis, and therapeutic action may be all that is required. In practicing such approaches, we deal with pests on a local level and administer the programs privately. This overall strategy is applicable for the majority of our insect pests.

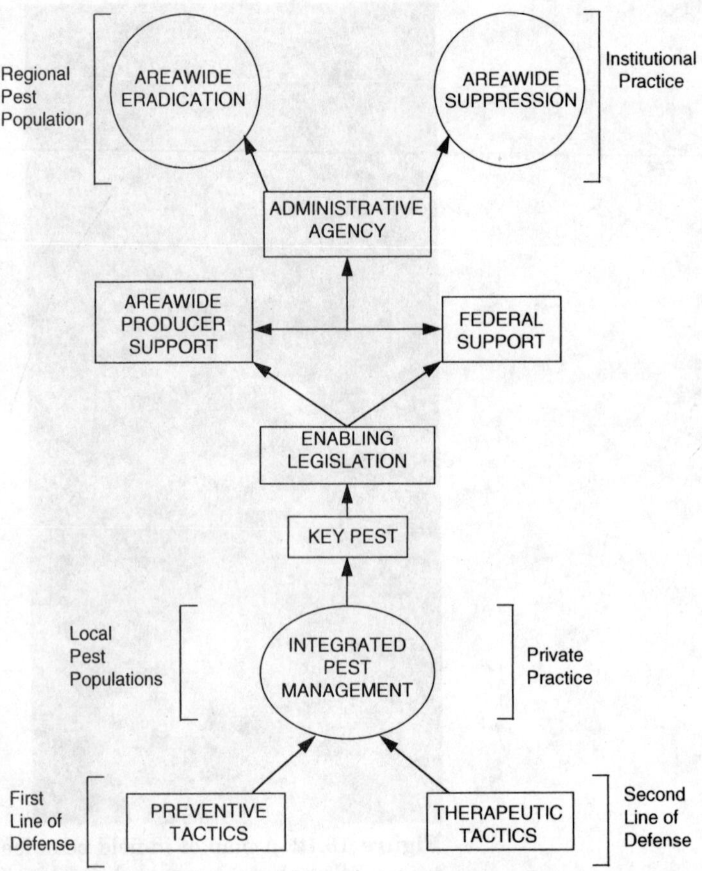


**Figure 16.13** Factors involved and relationships between integrated pest management and areawide pest technology.

In certain production systems, a geographically broader program implementation is more desirable. Some key pests are so severe or potential losses from newly introduced species so great that a regional approach gives an economic advantage. An areawide approach is preferable for these pests because established pests emigrate from unmanaged areas and introduced pests have the potential to spread into uncolonized areas. Additionally, some tactics (for example, the sterile-insect technique) are most applicable when practiced across broad expanses.

Introductions of exotic pests are addressed by public agencies such as the Animal and Plant Health Inspection Service of the United States Department of Agriculture (APHIS/USDA) and state departments of agriculture. These agencies use a variety of means to prevent introductions and limit or eradicate introduced exotic species. In these instances, producers have little involvement in program administration.

For selected key pests, where an areawide approach is economically feasible, goals may vary from complete eradication of populations to reducing average densities. For areawide programs to occur, many steps are necessary. Most areawide programs begin with a federally funded pilot project, implemented over several hundred to several thousand acres. If successful, the procedure is considered for a geographically broader area. Before implementation can occur, however, enabling legislation is required to allow grower referenda. If a referendum outcome is positive, federal support is combined with producer support to conduct the program, which is usually administered by a foundation or executive committee. The administering agency oversees collection of funds, producer compliance, and program direction.

It should be reiterated that even when areawide programs are successful, private insect pest management programs play a vital role in overall pest strategy. This is because these pest management programs are used to manage pests other than the targeted areawide pest in a production system. Moreover, after the goal of areawide suppression is achieved, privately practiced management programs are required to maintain local populations at acceptable densities.

## Further Reading

All, J. N. 1989. Importance of designating prevention and suppression control strategies for insect pest management programs in conservation tillage, pp. 1–3. In *Proceedings, 1989 Southern Conservation Tillage Conference.* Tallahassee, Fla.

All, J. N., and M. F. Treacy. 2006. *Use and Management of Insecticides Acaracides, and Transgenic Crops.* Lanham, Md.: Entomological Society of America, 156 pp.

Allee, L. L., and P. M. Davis. 1996. Effect of manure on maize tolerance to western corn 89:1608–1620.

Aronoff, S. 1993. *Geographic Information Systems: A Management Perspective.* Ottawa, Canada: WDL Pub.

Calkins, C. O. 1997. Areawide IPM as a tool for the future, pp. 154–158. In S. Lynch, C. Greene, and C. Kramer-Leblanc, eds., *Proceedings Third National IPM Symposium / Workshop.* Washington, D.C.: Economic Research Service, U.S. Department of Agriculture.

Frisbie, R. E., H. T. Reynolds, P. L. Adkisson, and R. F. Smith. 1994. Cotton insect pest management, pp. 421–468. In R. L. Metcalf and W. H. Luckmann, eds., *Introduction to Insect Pest Management,* 3rd ed. New York: Wiley.

Funderburk, J. E., L. G. Higley, and G. D. Buntin. 1993. Concepts and directions in arthropod pest management. *Advances in Agronomy* 51:125–172.

Hardee, D. D., and F. A. Harris. 2003. Eradicating the boll weevil (Coleoptera: Curculionidae). *American Entomologist* 49:82–97.

Klassen, W. 1989. Eradication of introduced arthropod pests: Theory and historical practice. *Entomological Society of America Miscellaneous Publications,* vol. 73.

Liebhold, T., T. Work, D. G. McCullough, and J. F. Cavey. 2006. Airlines baggage as a pathway for alien insect species invading the United States. *American Entomologist* 52:48–54.

National Cotton Council of America. 1994. *Boll Weevil Eradication: A National Strategy for Success.* National Cotton Council of America.

Pedigo, L. P. 1992. Integrating preventive and therapeutic tactics in soybean insect management, pp. 10–19. In L. G. Copping, M. B. Green, and R. T. Reese, eds., *Pest Management in Soybean.* London: Elsevier.

Perkins, J. H. 1982. *Insects, Experts, and the Insecticide Crisis.* New York: Plenum.

Quisenberry, S. S., and D. J. Schotzko. 1994. Integration of plant resistance with pest management methods in crop production systems. *Journal Agricultural Entomology* 11:279–290.

Rabb, R. L. 1972. Principles and concepts of pest management, pp. 4–7. In *Implementing Practical Pest Management Strategies.* West Lafayette, Ind.: Purdue University Press.

Rabb, R. L. 1978. A sharp focus on insect populations and pest management from a wide-area view. *Bulletin Entomological Society America* 24:55–61.

Reese, J. C., J. R. Schwenke, P. S. Lamont, and D. D. Zehr. 1994. Importance and quantification of plant tolerance in crop pest management programs for aphids: Greenbug resistance in sorghum. *Journal Agricultural Entomology* 11:255–270.

Velusamy, R., and E. A. Heinrichs. 1986. Tolerance in crop plants to insect pests. *Insect Science and Its Application* 7:689–696.

Weisz, R., S. Fleischer, and Z. Smilowitz. 1996. Site-specific integrated pest management for high-value crops: Impact on potato pest management. *Journal Economic Entomology* 89:501–509.

## Favorite Web Sites

http://www.reeusda.gov/ipm/
Site of the Cooperative State Research, Education, and Extension Service on IPM. Gives links to federal and state IPM programs, basic concepts, and success stories.

http://www.ipm.iastate.edu/ipm/nipmn/
Site of the National Integrated Pest Management Network. Presents the latest information on pest management by state and region of the United States through appropriate links.

http://www.aphis.usda.gov/plant_health/plant_pest_info/cotton_pests/index.shtml
Site of the USDA/Plant Protection and Quarantine that gives links to details, maps, and current status of the Boll Weevil Eradication Program and the Pink Bollworm Eradication Program.

CHAPTER 17

# MANAGING ECOLOGICAL BACKLASH

THE OBJECTIVE OF PEST MANAGEMENT practice is to lower pest status by reducing pest numbers and/or damage potential. That is well and good, but this objective is not always achieved, and if achieved, the effect may be all too brief. Some common reasons for failure are selecting inappropriate or incompatible tactics and improper application of tactics. Even without these problems, failure to achieve lasting, economical suppression may occur because of ecological backlash to a tactic or combination of tactics.

**Ecological backlash** involves the counterresponses of pest populations or other biotic factors in the environment that diminish the effectiveness of management tactics. Many of these responses result from high mortality placed on species, a burden not unlike that of nature. Others arise from disruption of ecological processes or changes in resource levels of the biotic community, causing readjustment in community structure. Such readjustments are counterproductive to the purposes of agriculture and medicine.

An important feature of ecological backlash is that it is usually delayed. Therefore, we may apply one or more management tactics with seemingly good results, encouraging us to continue with the same management program. However, the program gradually—or perhaps abruptly—loses effectiveness and may even become totally useless.

Ecological backlash has occurred with shocking frequency in the history of pest technology. Yet many growers and other agricultural specialists seem surprised and are usually caught off guard when it occurs. Agriculturists with knowledge of the phenomenon and contingency plans to deal with it may avoid significant economic penalties if it occurs. Moreover, understanding this potential problem can aid in the design of management programs that are durable, because such problems are not inevitable.

An understanding of ecological backlash rests on a knowledge of the major sources of the phenomenon. These sources can be called the "three Rs" of pest management awareness. They are population resistance, resurgence, and replacement. All three are ecologically based population phenomena.

## RESISTANCE OF POPULATIONS TO PEST MANAGEMENT TACTICS

Because of its pervasiveness, **resistance** of populations to pest management tactics is the most important of the three Rs. It is a genetic, evolutionary phenomenon most frequently associated with the rise of pesticide use. However, the phenomenon has been demonstrated with almost every pest management tactic and, theoretically, it is possible with any factor that causes high mortality and/or reduced natality in populations.

Resistance to a pest management tactic occurs when the susceptibility of a population to a tactic changes. For example, resistance may occur after exposure of several generations to an insecticide applied at a given rate that renders the insecticide ineffective in subsequent applications at that dose. Formally, resistance is the ability of certain individuals to tolerate or avoid factors that would prove lethal or reproductively degrading to the majority of individuals in a normal population and pass this ability on to offspring. To put it another way, the Insect Resistance Action Committee (see Web site at the end of chapter) defines resistance as *a heritable change in the sensitivity of a pest population that is reflected in the repeated failure of a product to achieve the expected level of control when used according to the label recommendation for that pest species.* Of course, use of the terms "product" and "label" show the orientation of this definition toward registered pesticides, including plant-incorporated protectants. But indeed, resistance has been shown to be broader than the pesticide use alone.

### Principles of Resistance

Resistance in a population can be explained by the same principles that explain evolution by natural selection. Briefly, every biological population possesses genotypes with differential abilities to survive and reproduce in the environment. A given set of environmental conditions favors genotypes with traits that are best adapted to the situation. Although several genotypes usually are present in a population at any one time, the best-adapted genotype assumes numerical dominance. As environmental factors change, other genotypes may be favored that subsequently displace the dominant ones (Fig. 17.1). Natural selection, sometimes called survival of the fittest or Darwinian selection, is a selection of the fittest genotypes for a given set of environmental conditions. This selection has led to physiological and morphological changes in biological species and has allowed their persistence for long (geological) periods of time, despite unusual and sometimes violent changes in the environment.

Insects are some of the most adaptable animals on earth. Their adaptability has allowed them to occupy nearly every terrestrial habitat and has resulted in an unequaled diversity of species. Indeed, proof of this adaptability is seen in the species of today, which have existed for millennia and survived all sorts of natural disasters.

Just as nature selects for well-adapted genotypes, so do we when we apply pest management tactics. We do this unwittingly by imposing heavy mortality or reproductive antagonisms on the pest population. Consequently, the population responds as it would to natural adversities: It adapts to overcome the adversity to survive and reproduce. This is something most insect populations do very well.

In other words, resistance is preadaptive. It is inherited from parents and never acquired through habituation during the lifetime of an individual (that is, it is not postadaptive). For instance, with rare exception, it is not possible to produce resistance within a single generation by exposing insects to sublethal doses of an insecticide.

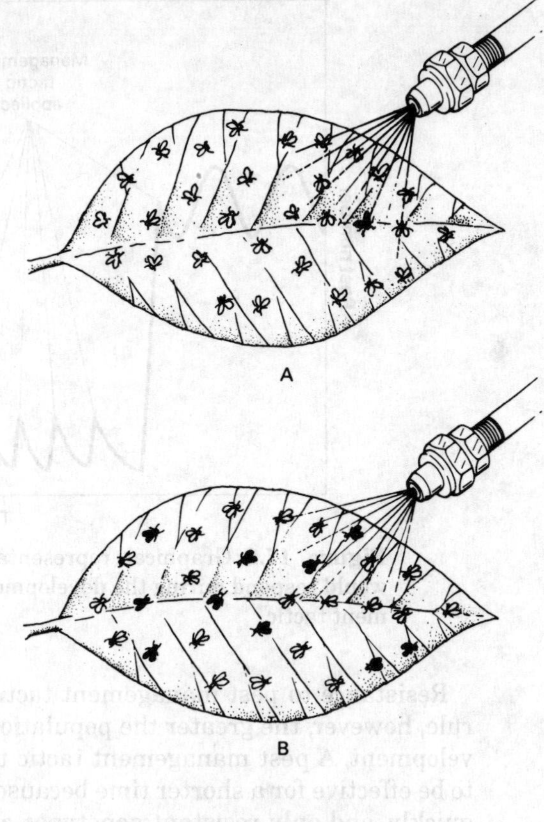


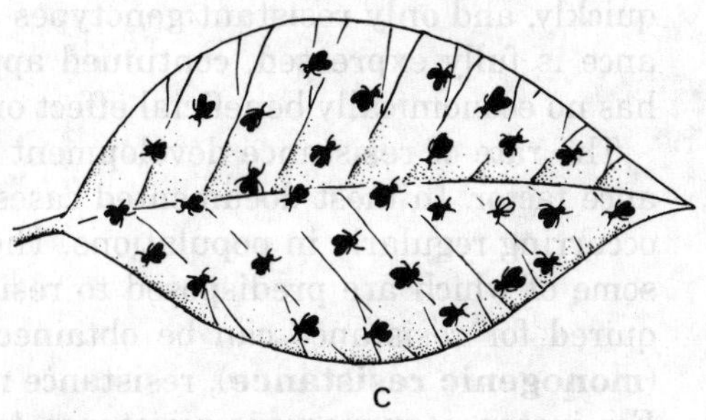

**Figure 17.1** Schematic drawing of development of resistance in a pest population. A. Some individuals in a pest population are genetically adapted to survive applications of a pest management tactic. B. Some of the survivors' offspring inherit resistance traits and will survive other applications. C. If the same tactic is continued, the pest population consists mostly of resistant individuals. (Redrawn from University of California Integrated Pest Management Publication 3303, *Integrated Pest Management for Citrus*)

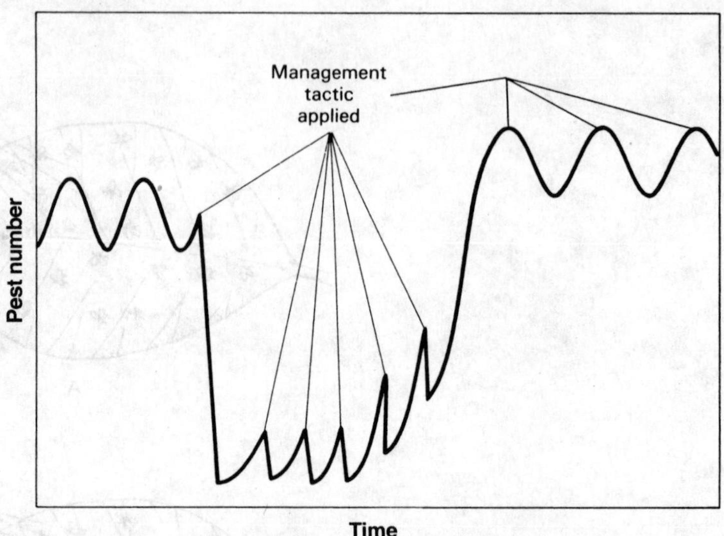


**Figure 17.2** Graphical representation of insect numbers as they would respond during the development of resistance to a pest management tactic.

Resistance to pest management tactics is difficult to predict. As a general rule, however, the greater the population adversity, the greater the rate of development. A pest management tactic that causes greatest mortalities tends to be effective for a shorter time because susceptible genotypes are eliminated quickly, and only resistant genotypes are left in the population. Once resistance is fully expressed, continued application of a pest management tactic has no economically beneficial effect on a population (see Fig. 17.2).

The rate of resistance development also depends on genetics of the resistance factor. In most documented cases, resistance originates with mutations occurring regularly in populations. These mutations result in new genotypes, some of which are predisposed to resist adverse factors. If the character required for resistance can be obtained through expression of a single gene (**monogenic resistance**), resistance may occur after only a few generations. For instance, monogenic resistance to some insecticides has occurred after only a few years of use (house flies, *Musca domestica,* resistant to DDT). However, if many genes are required (**polygenic resistance**), development may be much slower.

## Resistance to Conventional Insecticides

Resistance to insecticides is the preeminent form of resistance in insects. The prevalence of this phenomenon reflects the widespread effectiveness of these compounds in causing pest mortality.

**Magnitude of the insecticide resistance problem.** The United Nations Environment Program stated in 1979 that pesticide resistance ranked as one of the four major environmental problems in the world. Furthermore, the problem has become more acute since that time.

The first instance of insecticide resistance was noted in 1908 in Washington with the San Jose scale, *Quadraspidiotus perniciosus,* and lime-sulfur sprays.

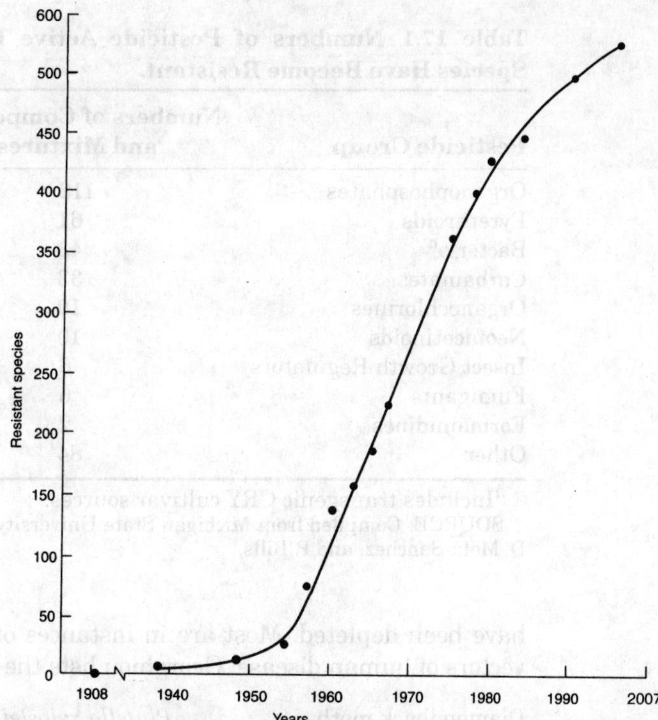


**Figure 17.3** Historical development of insecticide resistance worldwide. (Redrawn, with updates, from Georghiou, 1986, *Pesticide Resistance: Strategies and Tactics for Management,* Courtesy National Academy of Sciences)

This case was followed by infrequently reported incidents of resistance until the development of DDT insecticide.

In 1946, house flies were discovered resistant to DDT in Sweden. Through the early 1950s, the golden period of insecticides, resistance was still rare, and most insect populations were fully susceptible to insecticides. Since then, however, the number of instances has grown exponentially. In the early 1990s, approximately 500 arthropod species were proven resistant to one or more insecticides of the major groups, including DDT, cyclodienes, organophosphates, carbamates, and pyrethroids. This represents a 13 percent increase since 1984. A more recent tally (June 2007) brings the total to 553 (Fig. 17.3) species and 382 active ingredients (Table 17.1).

Of the resistant insect species, about 56 percent are of agricultural importance, 37 percent are of medical or veterinary importance, and 5 percent are natural enemies. Of the insect groups, Diptera have evolved the greatest number of resistant species (35 percent), which mainly reflects the intense insecticidal pressure on mosquitoes. Indeed, insecticide resistance is the major problem in control of arthropod-borne diseases, particularly malaria, in the world today. Resistant Diptera species are followed by Lepidoptera at 16 percent, Acari (mostly mites) at 14 percent, Coleoptera at 14 percent, and Hemiptera at 11 percent.

According to G. P. Georghiou (1990), there are a number of critical cases of resistance in which nearly all of the affordable, previously effective insecticides

**Table 17.1 Numbers of Pesticide Active Ingredients to Which Arthropod Species Have Become Resistant.**

| Pesticide Group | Numbers of Compounds and Mixtures | Percent of Total (List of 382) |
|---|---|---|
| Organophosphates | 118 | 30.9 |
| Pyrethroids | 61 | 16.0 |
| Bacteria[a] | 43 | 11.3 |
| Carbamates | 33 | 8.6 |
| Organochlorines | 19 | 5.0 |
| Neonicotinoids | 10 | 2.6 |
| Insect Growth Regulators | 6 | 1.6 |
| Fumigants | 6 | 1.6 |
| Formamidines | 2 | 0.5 |
| Other | 84 | 22.0 |

[a]Includes transgenic CRY cultivar sources.
SOURCE: Compiled from Michigan State University Resistance Data Base, 2007, M. Whalon, D. Mota-Sanchez, and P. Bills.

have been depleted. Most are in instances of high-intensity cropping and with vectors of human disease. Georghiou lists the following:

| | |
|---|---|
| Diamondback moth | *Plutella xylostella* |
| Whitefly | *Bemisia tabaci* |
| Green peach aphid | *Myzus persicae* |
| Leafminer | *Liriomyza trifolii* |
| Budworms | *Helicoverpa virescens, Helicoverpa armigera* |
| Colorado potato beetle | *Leptinotarsa decemlineata* |
| Twospotted spider mite | *Tetranychus urticae* |
| European red mite | *Panonychus ulmi* |
| Malaria mosquitoes | *Anopheles* species |
| House fly | *Musca domestica* |
| German cockroach | *Blattella germanica* |
| Black fly | *Simulium damnosum* |

Additionally, many resistant species resist compounds in more than one insecticide group, which shows an additional dimension of the problem. Moreover, pesticide resistance is not limited to the arthropods; there are at least 200 species of plant pathogens (fungi), over 273 species of weeds, and several species of nematodes and rodents also resistant to one or more pesticides formerly effective in their control.

**Mechanisms of resistance to insecticides.** Throughout their evolutionary history, insects have had to deal with a plethora of naturally occurring environmental toxicants. To survive in the face of these natural hazards, species have a variety of mechanisms to make the materials innocuous to them. Just as these mechanisms have reduced the toxicity of natural substances, they have also reduced the toxicity of synthetic chemicals. The result is what we call insecticide resistance. As C. F. Wilkinson (1983) stated, insects were "forewarned and forearmed to meet the challenges presented them" by modern synthetic organic insecticides. Insect species deal with toxic

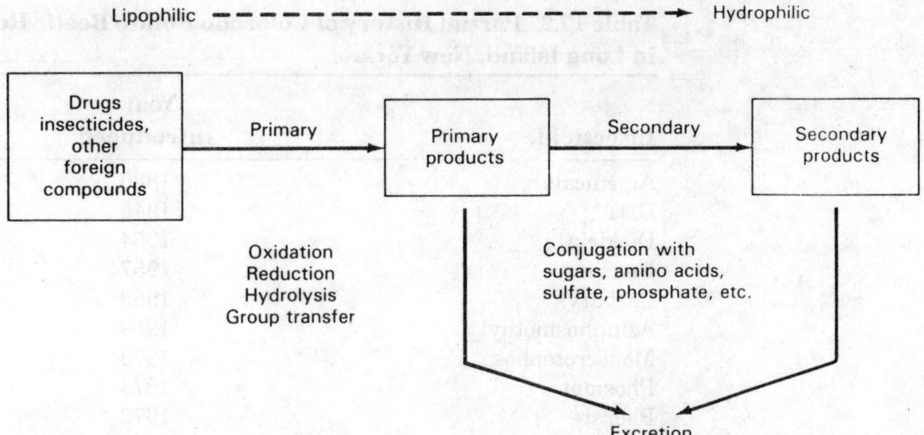


**Figure 17.4** Metabolism of lipophilic foreign compounds, including insecticides, in insects. (Reprinted with permission of Plenum Press from *Pest Resistance to Pesticides*, Georghiou and Saito, eds., © 1986)

chemicals through three major mechanisms: biochemical resistance, physiological resistance, and behavioral resistance. All three forms may occur simultaneously to produce the fittest animals.

**Biochemical resistance is probably the most common type of resistance in insects.** With this form, an insecticide is usually attacked by one or more enzymes that detoxify it before it can reach its site of action. This metabolic process is often accomplished in two stages with high concentrations of **mixed-function oxidases** and other enzymes that produce primary products. Although these primary products may be excreted directly, most often they undergo a secondary metabolism, forming water-soluble conjugates (covalent addition of sugar, amino acids, sulfates, phosphates, and other materials) that in turn are excreted (Fig. 17.4).

**Physiological resistance** is any form of resistance that reduces toxicity through changes in basic physiology. With this mechanism, the chemical is not broken down into a less toxic form by the insect. Rather, it is accommodated by one or more physiological functions. One of the most important of these involves alterations at the site of insecticide activity. For instance, knockdown resistance in house flies to DDT and pyrethroids is believed to be caused by reductions in numbers of target-site receptors, making nerve sheaths less sensitive to the toxicants (a quantitative factor based on numbers of receptors). In other species, cholinesterases may be altered so that they are not affected by organophosphates and carbamates. Physiological resistance may also include decreased penetration of the insecticide through the body wall. This resistance usually is conferred through modifications in composition or structure of the cuticle (for example, additional waxy layers). Other physiological mechanisms include increased rates of excretion of the insecticide and sequestering the chemical. In the latter instance, lipophilic insecticides like DDT may be stored in body fat and thereby are prevented from reaching their site of action.

**Behavioral resistance** involves changes in behavior by which insects avoid insecticides. One of the best-known examples of behavioral resistance occurs with mosquitoes that vector malaria-causing plasmodia, namely *Anopheles*

**Table 17.2 Partial History of Colorado Potato Beetle Resistance to Insecticides in Long Island, New York.**

| Insecticide | Year Introduced | Year First Detected |
|---|---|---|
| Arsenicals | 1880 | 1940s |
| DDT | 1945 | 1952 |
| Dieldrin | 1954 | 1957 |
| Endrin | 1957 | 1958 |
| Carbaryl | 1959 | 1963 |
| Azinphosmethyl | 1959 | 1964 |
| Monocrotophos | 1973 | 1973 |
| Phosmet | 1973 | 1973 |
| Phorate | 1973 | 1974 |
| Disulfoton | 1973 | 1974 |
| Carbofuran | 1974 | 1976 |
| Oxamyl | 1978 | 1978 |
| Fenvalerate | 1979 | 1981 |
| Permethrin | 1979 | 1981 |
| Fenvalerate + piperonyl butoxide | 1982 | 1983 |

SOURCE: Data of A. J. Forgash, New Jersey Agricultural Experiment Station, Cook College, Rutgers University. Updated from Lafarge, 1985, with permission.

*gambiae* and other mosquitoes in Africa. In these species, an endophilic strain, inhabiting human structures, was susceptible to sprays of DDT applied to walls indoors. An exophilic strain not inhabiting buildings became dominant because its behavior allowed it to avoid exposure to the insecticide. Among other pests, behavioral resistance has been shown in the tobacco budworm, where resistant larvae slow their movements in the presence of pyrethroids and thus receive less exposure to otherwise lethal doses.

**Cross-resistance.** When insecticide resistance occurs in a species, pest managers are faced with the difficult problem of how to deal with it. Historically, they have turned to insecticides in other classes to which the species is still susceptible. An example of this strategy and its outcome can be seen with the Colorado potato beetle, *Leptinotarsa decemlineata,* on potatoes in Long Island, New York, where growers have used a single-factor insecticide strategy since the late 1800s. One insecticide after another produced resistance in the population (Table 17.2) beginning with the arsenicals in the 1940s, followed by the chlorinated hydrocarbons, organophosphates, and carbamates through the 1970s, and continuing in resistance to the pyrethroids in the 1980s. Insecticide resistance led growers to alternate to the microbial insecticide, *Bacillus thuringiensis tenebrionis,* the botanical, rotenone, plus piperonyl butoxide, and the inorganic, cryolite. Both rotenone and cryolite predate modern synthetic insecticides and are not effective against older larvae and adults. Additionally, some populations of the pest have shown resistance to such new insecticides as abamectin and imidicloprid, and the potential of resistance to *Bacillus thuringeinsis* sprays is believed to exist. Geographically, by 1994 there were no labeled insecticides effective against the Colorado potato beetle on over 44 percent of the potato acreage in the United States. This makes the pest one of the most insecticide-resistant insects in North America.

Changing to an insecticide in another class when resistance occurs is frequently necessitated by cross-resistance. **Cross-resistance** implies that an insect with resistance to one insecticide is able to resist other insecticides. Therefore, insects that have become resistant to DDT also have cross-resistance to methoxychlor, and those resistant to dieldrin have cross-resistance to heptachlor and toxaphene.

This *within-class* cross-resistance has led to categories of resistance according to insecticide class. Some of the major classes mentioned in the literature include: (1) DDT resistance, (2) dieldrin or cyclodiene resistance, (3) organophosphate resistance, (4) carbamate resistance, and (5) pyrethroid resistance.

Although most instances of cross-resistance have been observed within each of these classes, it may also exist *between classes*. For instance, house flies with knockdown resistance to DDT have cross-resistance to all DDT analogs, pyrethrins, and pyrethroids. Likewise, flies with resistance to organophosphates may have cross-resistance to carbamates. Cross-resistance to some of the most recently developed insecticides, the avermectins, also has been discovered and produced experimentally in the Colorado potato beetle.

Cross-resistance usually occurs because of similar toxicity modes of the chemicals involved and the inheritance of a resistance gene for that mode. DDT and pyrethroid, for example, are both axonic poisons (they upset impulse transmission along the nerve axon), and in house flies, cross-resistance to axonic poisons is inherited from the so-called *kdr* gene. Similarly, house flies, which are also cross-resistant to organophosphates and carbamates through altered cholinesterases, receive this trait through inheritance of the *AChE-R* gene. Other identified genes that function in resistance and cross-resistance include *pen* (for decreased-penetration resistance) and *dld-r* (for dieldrin and cyclodiene resistance). All these genes except *AChE-R* are incompletely or fully recessive, but *AChE-R* and others are codominant.

**Dangers and costs of the resistance phenomenon.** The problem of insecticide resistance is acute. The rate of growth in resistant arthropod species has increased more than 13 percent since 1984. Moreover, most major pests have become resistant to at least one insecticide group, and many show multiple resistance (resistance to more than one group). Indeed, an increasingly larger proportion of new records of resistance involves species that are already resistant to at least one insecticide group (Table 17.3).

Therefore, we face a dilemma: the more often we use insecticides, the quicker they will become useless. They "wear out." At least in some instances when an insecticide begins to lose effectiveness, a response is to increase dosage and/or frequency of application. These increases enhance the rate of resistance, and when the compound is no longer effective, it is discarded for another insecticide, probably from another chemical group. The repetition of this cycle and the ever-increasing need for more poisons has been called the **pesticide treadmill** because it requires much activity with little forward motion.

Although switching to other groups has been a somewhat effective tactic in the past, such a policy cannot be sustained in the long run. One reason is that cross-resistance produces a shortened effective life of each succeeding insecticide. For example, DDT resistance provides a foundation for development of pyrethroid resistance, and the pyrethroid may be effective for a shorter time than would otherwise have been expected.

**Table 17.3 Critical Instances of Cross Resistance in the United States.**

| Pest | OP | C | P | Other |
|---|---|---|---|---|
| Twospotted spider mite (*Tetranychus urticae*) | X | | X | X |
| Colorado potato beetle (*Leptinotarsa decemlineata*) | X | X | X | X |
| Southern house mosquito (*Culex quinquefasciatus*) | X | X | X | X |
| Little house fly (*Fannia canicularis*) | X | | X | X |
| Sweetpotato whitefly (*Bemisia tabaci*) | X | | X | |
| Silverleaf whitefly (*Bemisia argentifolii*) | X | | | X |
| Greenhouse whitefly (*Trialeurodes vaporariorum*) | X | X | X | |
| Cotton aphid (*Aphis gossypii*) | X | X | X | |
| Pear psylla (*Cacopsylla pyricola*) | X | X | X | |
| Tobacco budworm (*Heliothis virescens*) | X | X | X | |
| Soybean looper (*Pseudoplusia includens*) | X | X | X | |
| Beet armyworm (*Spodoptera exigua*) | X | X | X | |
| Fall armyworm (*Spodoptera frugiperda*) | X | X | X | |
| Diamondback moth (*Plutella xylostella*) | X | X | X | X |
| German cockroach (*Blattella germanica*) | X | X | X | X |
| Cat flea (*Ctenocephalides felis*) | X | X | X | X |

*OP* = organophosphates, *C* = carbamates, *P* = pyrethroids, *Other* = miscellaneous categories, including microbials

SOURCE: *Biologically Based Technologies for Pest Control.* 1995. U.S. Congress, Office of Technology Assessment.

Useful insecticides are a valuable resource, and monetary loss can be attributed to their demise. Chemical companies may spend tens of millions of dollars in developing a new compound, only to see diminishing resistance returns, perhaps to the point of not recouping original investments. In addition, growers may spend millions of dollars for extra chemicals and additional treatments when resistance occurs.

## Resistance to Other Pest Management Tactics

Although resistance to insecticides is the form most often mentioned, resistance certainly is not restricted to conventional insecticides. Indeed, insects, having survived so well for so long, eventually may overcome any pest management tactic we employ. Moreover, the principles of selection to explain insecticide resistance apply equally well to these other forms. As was true with insecticides, the more widespread and destructive the management tactic, the more likely these other forms of resistance will occur.

**Resistance to insect growth regulators.** Although originally believed refractory to resistance problems, several insect growth regulators (IGRs) recently have been added to the long list of insecticides to which insects are resistant. At first, resistance to juvenile hormones and their mimics seemed unlikely because insects would be required to overcome the effects of molecules similar to those already present in their bodies. However, we can understand resistance more readily when we recognize that insects regulate their own hormones by metabolism and that exogenous molecules usually are sensed as foreign materials subject to deactivation processes.

Cross-resistance in various insecticide-resistant strains has been a major factor in IGR resistance. Indeed, at least ten insect species, representing the Diptera, Coleoptera, Hemiptera, and Lepidoptera, show cross-resistance to IGRs. Included in the list of common IGR compounds are methoprene, hydroprene, kinoprene, pyriproxyfen, and diflubenzuron. Although resistance mechanisms are still under investigation, it seems as if most IGR resistance results from reduced penetration and increased metabolism of the compounds.

**Resistance to microbial insecticides.** Insect resistance to pathogenic microorganisms is yet another concern. Over thirty instances of such resistance have been reported. Resistance to toxins of the most widely used microbial insecticide, *Bacillus thuringiensis* (Bt), has been observed in experiments with the Indianmeal moth (Box 17.1, Fig. 17.5) (*Plodia interpunctella*), almond moth (*Cadra cautella*), tobacco budworm, diamondback moth (*Plutella xylostella*), Colorado potato beetle, house fly, and other insects. Lower levels of resistance have been reported in the mosquitoes *Aedes aegypti* and *Culex quinquefasciatus*. Although significant resistance in field populations has been reported only for the diamondback moth, experiments with the remaining species indicate significant potential for future development with more widespread use of Bt.

One of the most widely studied species for Bt resistance is the Indianmeal moth. This insect, a major pest of stored grain, has been shown to increase resistance nearly 100 times over fifteen generations when larvae were fed diets containing the pathogen. House flies exposed to beta-exotoxin-producing Bt, causing 50 to 90 percent mortality each generation, developed a thirteenfold resistance level in fifty generations.

Increased levels of resistance have also been observed in Lepidoptera and Hymenoptera populations that suffered viral epidemics. Pests with increased resistance to viruses include the spruce budworm, *Choristoneura fumiferana,* and sawflies, among others. From a beneficial standpoint, silkworms, *Bombyx mori,* are also capable of virus resistance, and strains have been purposefully selected for this trait.

## BOX 17.1 INDIANMEAL MOTH

**SPECIES:** *Plodia interpunctella* (Hübner) (Lepidoptera: Crambidae)

**DISTRIBUTION:** The Indianmeal moth is a cosmopolitan insect. It is probably found in all 50 states because of human transport.

**IMPORTANCE:** This insect is an important grain-infesting moth feeding on whole grains and cereal food products. Larvae damage a wide variety of human foods and feeds, including flour, cornmeal, chocolate, dried fruits, nuts, pastas, crackers, dry dog and cat food, birdseed, powdered milk, and so on. In addition to the feeding damage, the larvae web food particles together with silken threads contaminate the food items with their fecal droppings. When infestations are large, food items may be heavily matted with webbing. Infestations may occur in homes where food or feed items are usually discarded. Infestations in commercial businesses, such as warehouses, restaurants, pet stores, feed mills, and seed companies, may result in considerable loss of the marketable product.

**APPERANCE:** Larvae are about 17 mm when full grown and have a dark brown head. The body is a dirty-white color that may be tinged with green, yellow, brown, or pink, depending on the source of food. The adult moth is about 10 mm long when the wings are folded over the back. The base of the forewing is a pale gray, with the distal half of the wing being reddish brown or copper color. The gray and reddish brown areas are separated by a thin gray band, and another three broken gray bands occur toward the wing tip. The adults are attracted to lights and may fly in houses during the evening.

**LIFE CYCLE:** The female moth lays from 60 to 300 eggs on or near food items. Eggs hatch in 2 to 14 days, with the larvae dispersing shortly thereafter. Larvae will feed inside a tunnel-shaped case of silken webbing and excrement. They often leave the webbing and wander from the food source, sometimes crawling up walls and across ceilings, prior to pupating. The complete life cycle may take from 27 to 305 days, depending on temperature, and under favorable conditions there can be seven to eight generations per year.

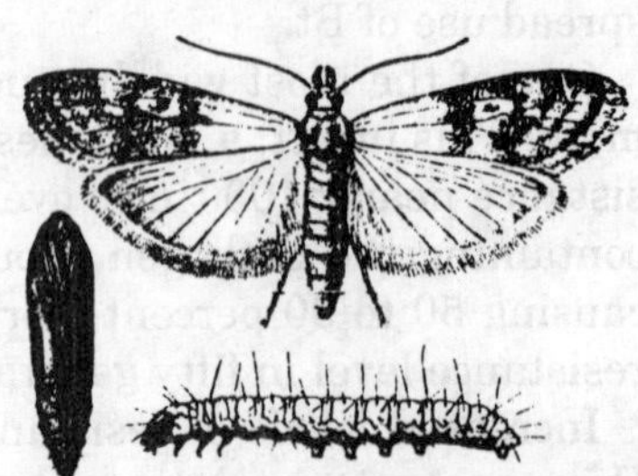

**Figure 17.5** Indianmeal moth, *Plodia interpunctella*, adult *(top center)*, pupa *(left)*, and larva. Resistance to applications of the microbial pesticide *Bacillus thuringiensis* has been observed with this stored-grain pest. (Courtesy USDA)

**Resistance to parasites.** Insect resistance to internal parasites is possible but has not been commonly reported. An instance of resistance to an introduced ichneumonid parasitoid, *Mesoleius tenthredinis,* has been recorded with larch sawfly, *Pristiphora erichsonii,* populations in central and eastern North America. In laboratory studies, parasite resistance has been induced in house flies to a pteromalid parasitoid, *Nasonia vitripennis,* and in an anopheline mosquito to a nematode parasite, *Diximermis peterseni.* Moreover, studies have shown that some mosquitoes and Coleoptera have the ability to encapsulate the nematode, *Steinernema feltiae,* thus allowing them to survive infections. Therefore, development of resistance of insects to parasitic nematodes is possible should they become more widely used as biological control agents.

**Virulence to resistant plants.** The development of virulent biotypes that overcome the defenses of normally resistant plants was discussed previously. The occurrence of this form of resistance (resistance to resistant plants) has been of concern to plant breeders and growers for many years, although it seems most probable with parthenogenetic insect species. To date, most insect biotypes have developed among aphids, although one of the best-known examples of the phenomenon occurs in the Hessian fly, *Mayetiola destructor,* a dipteran.

Additionally, there is much concern over the widespread deployment of transgenic plants that produce Bt endotoxin for pest management. The concern is that Bt resistance has already been documented for foliar applications and that continuous expression of the toxin in plants will cause even stronger selection pressure for resistance. Based on the history of insecticide resistance, such concerns are well founded. Current thought suggests that several tactics should be used to prevent resistance as the transgenic plants are introduced. These tactics include: (1) developing plants that express the Bt genes only in the structure needing protection (such as cotton bolls); (2) using promoters that express the Bt genes only at particular times in the season (like reducing numbers of insecticide applications); (3) using seed mixtures containing Bt transgenic plants and susceptible plants; and (4) planting Bt resistant and susceptible plants at different times and/or locations (see example of the European corn borer in Chapter 13). These tactics all aim at reducing selection pressures in an effort to avoid, or at least delay, resistance. In practice, there are both advantages and disadvantages to each of these approaches.

Another approach to avoid or delay insects of overcoming resistance in transgenic cultivars uses gene pyramiding. Here, research has shown that plants transformed with two Bt-toxin genes substantially delayed development of virulent pests of corn and cotton cultivars compared to those with single-toxin genes. Mathematical models simulating plantings showed that those with the pyramided genes delayed resistance longer than planting a mixture of two single-toxin plants in the field or sequentially using two single-toxin plants in crop rotation. We may expect to see further work on this approach in future years, which could allow more sustainable systems with transgenic plants.

**Resistance to crop rotations.** This very unusual form of resistance exemplifies the persistent capability of insects. As mentioned earlier, a genotype of the northern corn rootworm, *Diabrotica barberi,* has been selected that shows resistance to a 2-year rotation of corn and soybeans (or small grain) in the northcentral United States. The resistant genotype displays a so-called **extended diapause,** a state of dormancy in the egg stage that

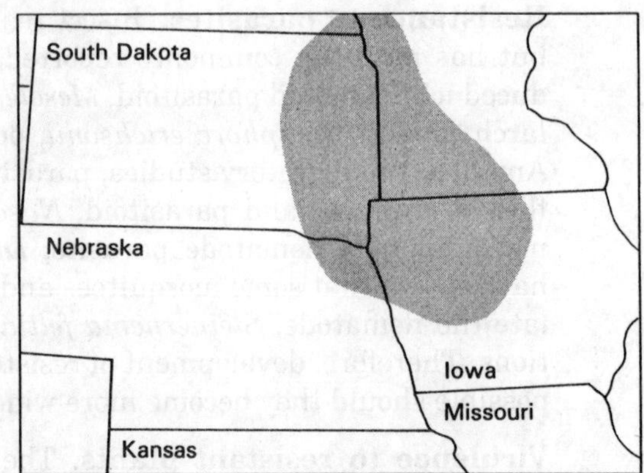


**Figure 17.6** Estimated distribution of a subpopulation of the northern corn rootworm, *Diabrotica barberi,* showing area of highest concentration of extended diapause (2-year life cycle). This species normally has a 1-year life cycle.

lasts for two winters rather than one. The presence of the extended-diapause genotype was reported in 1965, but problematic resistance was not a factor until about 1985, when the incidence of damage to corn grown in rotation became widespread. The adaptive feature of this trait is that eggs laid in cornfields one season could survive in a nonhost habitat the alternate season and subsequently hatch in the presence of the proper host the following season. Studies conducted in the problem region (Fig. 17.6) showed that only 10 percent of the rootworm eggs from continuously grown cornfields underwent extended diapause, versus nearly 50 percent in cornfields from a rotation scheme.

Yet another problem of resistance has begun to occur with a related pest, the western corn rootworm, *Diabrotica virgifera virgifera*. The problem was first noted in Illinois and has since been detected in other parts of the Corn Belt, where the species damaged corn the year following soybean planting. It seems as though a strain of western corn rootworms has been selected with the behavior of ovipositing in soybeans. Therefore, eggs in the soil with soybeans subsequently hatched the following season, when corn has been planted. This problem adds a disturbing dimension to rootworm management in the Midwest and tends to discount the use of corn/soybean rotation as a tactic.

This recurring resistance problem means that previously effective 2-year rotation schemes are becoming less practical for rootworm management. Where problems exist, growers who rotate are asked to consider: (1) planting corn in a 3-year rotation cycle (1 year in corn and 2 years in a nonhost crop), although this is often impractical for economic reasons, and therefore rarely done; or (2) applying a preventive soil insecticide or planting a transgenic corn rootworm hybrid if the adult population reached an economic threshold of five per plant the last year in corn. Two-year rotations used against this pest in many regions of the Corn Belt usually are still effective.

**Resistance to sterile-male releases.** Certain outbreaks of the screwworm, *Cochliomyia hominivorax,* in the southwestern United States and Mexico during the early 1970s were claimed by some workers to have been caused by replacement types not reproductively compatible with released sterile insects. R. H. Richardson and colleagues (1982) reported that at least nine screwworm types existed (five in the United States), with type intermating being low or nonexistent. According to their hypothesis, after the screwworm eradication program was begun in the Southwest, the dominant types were gradually suppressed and replaced by incompatible types. The remaining pest population in certain areas then became a mixture of types, with previously rare types being the more common. Subsequent releases of sterile males then became less successful, with performance being dependent on an adequate match of laboratory flies with the dominant type (or types) in a given area. Thus, mismatches were said to explain program failures in local areas. Richardson and his coworkers suggested that future recurrences of these outbreaks could be reduced by prevention of reinfestations in the United States and by facilitation of rapid, appropriate responses (with properly matched releases) when outbreaks occur.

Although resistance to sterile-insect releases is believed possible, the likelihood of screwworm outbreaks seems to be contradicted by the successful eradication of the pest from the United States, Mexico, and Central America.

**Resistance to pheromones.** Insect resistance to pheromones or pheromone mimics has not yet been recorded. However, these compounds have not been used against insects on a widespread or intensive scale. Many experts agree that, if used widely and successfully for insect suppression, resistance is a distinct possibility.

## Management of Resistance

Because of its pervasiveness, insecticide resistance has received the most attention with regard to management approaches, and we emphasize it here. However, it should be noted that many principles of insecticide management used to slow resistance development apply generally to other types of resistance.

**Conditions that promote resistance.** The development of resistance management is undergirded by an understanding of the causes of resistance. The basic cause is artificial selection of individuals who can survive the management tactic (for example, insecticide exposure) and successfully reproduce to pass on their resistant genes to the next generation. Resistance can be expected to occur against any tactic that imposes a significant population burden; the greatest question is how quickly it will develop. Therefore, resistance management is mostly a matter of anticipating and slowing the rate of development.

Factors that influence the rate of insecticide-resistance development have been discussed by several authors and summarized by T. C. Sparks and B. D. Hammock (1983). A partial list of the factors involved in rapid development is as follows:

Operational Factors

1. The insect has a prolonged exposure to a single insecticide, or the insecticide is used in a slow-release form.
2. Every generation of the insect is selected.
3. Insecticide selection pressure is high (mortality is great).

4. No functional refuges exist; that is, coverage by the insecticide is effectively complete so that no part of the population remains unselected.
5. A large geographical area is covered; all populations in a given area are likely to have been treated.
6. Selection occurs prior to mating.
7. The insecticide is closely related to one used earlier.
8. A low population threshold (economic threshold) is recommended for application of the insecticide.
9. The insecticide is inherently irritating and/or repellent.

Biological Factors

1. No (or little) migration occurs between populations.
2. The species has a monophagous habit.
3. The species has a relatively short generation time.
4. Numerous offspring per generation are produced.
5. The species is highly mobile, increasing the probability of exposure.

Of course, not all of these factors are required to produce resistance, but the greater the number present in a situation, the greater the chance of occurrence and the more rapidly resistance will develop.

**Slowing the development of resistance.** The prevention of resistance to any effective pest management tactic is practically impossible in many situations. However, its rate of development can be slowed by considering the operational factors that enhance it and modifying the pest management program accordingly.

The most basic resistance management routine is the use of combined tactics to achieve suppression. By integrating ecological tactics, natural enemy suppression, and resistant plants with chemical insecticides, undue reliance is not placed on any one tactic. Multiple tactics place diverse pressures on the pest population, making it more difficult for the species to overcome the effects of any one tactic. Moreover, if resistance develops to one tactic in the integrated scheme (such as insecticide resistance), its effects will be lessened because other tactics will still contribute to suppression. Therefore, tactic diversification is the key to insect pest management.

Another possible approach to slow resistance is to employ passive tactics in the management program, when possible. Passive tactics place no known burden on pest populations. These tactics include such measures as irrigation and fertilization to produce thrifty plants, which can better tolerate insect injury (see Chapter 10). The tactics also involve developing plant varieties capable of acceptable yields despite pest injury (that is, growing tolerant varieties) and accounting for existing levels of tolerance when making decisions (see tolerance EILs in Chapter 16). When passive tactics are used, selective pressures are lessened and resistance development is less likely. Indeed, passive tactics provide the most enduring management solutions. Unfortunately, these tactics are not always practical, particularly when high-value, blemish-free produce is desired.

Perhaps a more practical approach to slowing resistance, particularly with insecticides, is to modify use patterns. Several methods have been suggested and employed that disrupt or significantly delay resistance in pest populations. According to G. P. Georghiou (1983, 1990), these methods, which he calls "chemical strategies of resistance management," can be summarized as follows:

### Management by Moderation

Objective: reduce selection pressure and conserve susceptible genes in the population.

Advantage: helps conserve environmental quality and natural enemies.

Limitation: less practical with high-value crops and medical pests.

1. Use low dosages, sparing a proportion of susceptible genotypes.
2. Use less-frequent applications.
3. Use chemicals of brief environmental persistence.
4. Avoid slow-release formulations.
5. Apply selection against adults after reproduction, allowing susceptible genes to be passed on.
6. Make local, rather than areawide, applications. The use of trap crops fits here.
7. Leave some generations or population segments untreated.
8. Preserve refuges (untreated pockets for population segments). This relates to point 7 and is sometimes accomplished by treating only "hot spots" in a field or directing nozzles toward only part of a plant canopy. Site-specific farming fits here.
9. Utilize higher economic thresholds for insecticide application. This relates to point 2 because it should result in fewer applications.

### Management by Saturation

Objective: saturate insect defense mechanisms by doses that can overcome resistance.

Advantage: applicable to high-value crops like blemish-free apples and vegetables and medical pests.

Limitations: possible adverse impact on environment.

1. Render the resistance gene functionally recessive by applying higher dosages on target. This tactic is based on computer simulations and laboratory tests that have shown rapid resistance development when the resistance gene (R gene) is dominant. Resistance develops slowly when the resistance gene is recessive: Because there are few R genes in the population initially, resistance develops quickly with a high proportion of heterozygotes (hybrids with the R gene and the susceptible gene). Applying very high dosages of insecticides kills both susceptible individuals (ss) and heterozygotes (Rs), keeping the frequency of the R gene low. This tactic is not believed practical if some selection already has begun and there is a large proportion of heterozygotes already present in the population.
2. Suppress detoxification mechanisms with the use of synergists. Synergists function by inhibiting specific detoxification enzymes. For example, the common synergist, piperonyl butoxide, is an oxidase inhibitor commonly added to insecticides to enhance action with low to moderate levels of resistance. Although synergists may eliminate the selective advantage of individuals with detoxifying enzymes, some insects like house flies have been known to develop resistance to insecticide-plus-synergist mixtures.

### Management by Multiple Attack

Objective: reduce selection pressure by imposing several independently acting forces, such that pressure from any one is below the level to achieve resistance. This is similar to using several different management tactics as discussed earlier; only here, several different insecticides are involved.

Advantage: potentially useful against pests of high-value crops and medical pests.

Limitations: imposes environmental risks, destruction of natural enemies, and risk of "super" resistance (resistance to several compounds at once).

1. Use insecticide mixtures. The idea of using mixtures to slow resistance assumes that the mechanisms of resistance are different for each member chemical and, furthermore, that all the mechanisms required to resist the mixture are not present in any one individual. In other words, insects developing resistance to one compound would succumb to another in the mixture. Success with this approach is enhanced when the insecticides mixed are synergistic and have similar decay rates and short environmental stability. Also, mixtures should be applied early, before resistance occurs to any one of the components. Although not many definitive studies have been conducted with mixtures, it has been shown that when chemicals with contrasting modes of action of detoxification pathways are used, some delay in the onset of resistance occurs. An example in which a mixture delayed resistance in the laboratory involved a combination of temephos (an organophosphate), propoxur (a carbamate), and permethrin (a pyrethroid) against the mosquito *Culex quinquefasciatus.*
2. Apply insecticides in a mosaic pattern across an area. This is somewhat like a mixture, but in a spatial sense. Planned applications of unrelated compounds are applied in different parts of the pest population range to avoid selection of the whole population for the same resistance mechanism. With this approach, insects not killed by an insecticide treatment in one sector are killed by a different insecticide when they migrate to another sector. This approach is relatively new and has been proposed for use with mosquitoes, where opposite walls of a house would be treated with different chemicals, and with horn flies, *Haematobia irritans,* where different chemicals would be applied strategically to pasture-sized areas. Versions of this approach can be seen in the tactics suggested to prevent Bt resistance in transgenic plants (see section on "Virulence to resistant plants", earlier in this chapter).
3. Apply insecticides of dissimilar modes of action in a rotation scheme. The idea of rotations of dissimilar insecticides to combat resistance is based on the assumption that individuals resistant to a chemical have substantially lower survival potential to natural environmental factors than susceptible individuals. Consequently, the frequency of resistant individuals would be expected to decline between the times the "resistant" chemical is used. In other words, the proportion of resistant individuals would increase during the season with a chemical but would return to its original low level by the time that chemical is used again. Insecticide rotations are presently receiving much attention as a resistance-management tool. A recent successful example is using insect growth regulators with different modes of action. Here, buprofezin, a chitin synthesis inhibitor, was rotated with pyriproxyfen, a juvenile hormone mimic, to avoid resistance of the sweetpotato whitefly.

## PEST POPULATION RESURGENCE AND REPLACEMENT

Resurgence and replacement are ecological backlash phenomena observed in a number of agricultural systems and other situations. Although evidence has shown that they are most often associated with conventional insecticides, the possibility of their occurrence exists with any management tactic. This is particularly

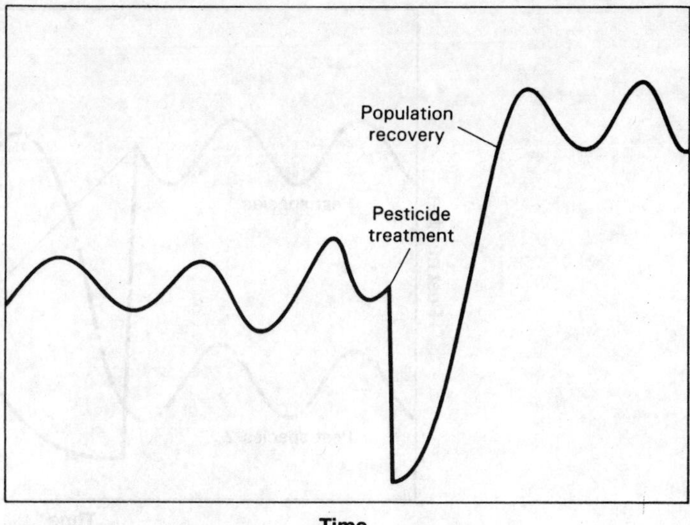


**Figure 17.7** Diagrammatic representation of insect numbers as they would respond during pest resurgence. Note that numbers are higher after population recovery than before treatment.

true if the tactic is directly favorable to the physiology of the insect (for example, enhanced nutrition) or has an adverse effect on important natural enemies.

## Dynamics of Resurgence and Replacement

Resurgence and replacement problems have been discussed by various authors and have been a particularly important consideration in pest management. **Resurgence** can be defined as a situation in which a population, after having been suppressed, rebounds to numbers greater than before suppression occurred (Fig. 17.7).

**Replacement,** also frequently referred to as a secondary pest outbreak, occurs when a major pest is suppressed and continues to be suppressed by a tactic but is replaced by another pest, previously with minor status (Fig. 17.8). In this instance, the primary pest is strongly affected by the tactic, but the secondary pest is not.

These so-called pest upsets and the bases for them were reviewed in detail by W. E. Ripper (1956) in his review of effects of pesticides on the balance of arthropod populations. Although dated, Ripper's review summarizes the important principles of pest upsets that continue to be valid today. He extends three possible causes that should be examined when attempting to explain pest upsets:

1. Reduction of natural enemies by pesticides, along with the pest,
2. Direct favorable influences of pesticides on physiology and behavior of arthropods, and
3. Removal of competitive species.

## Upsets from Reduction of Natural Enemies

Presently, most experimental evidence supports the first cause listed. In many instances, reductions in numbers of arthropod natural enemies after insecticide applications are believed the basis for the upset.

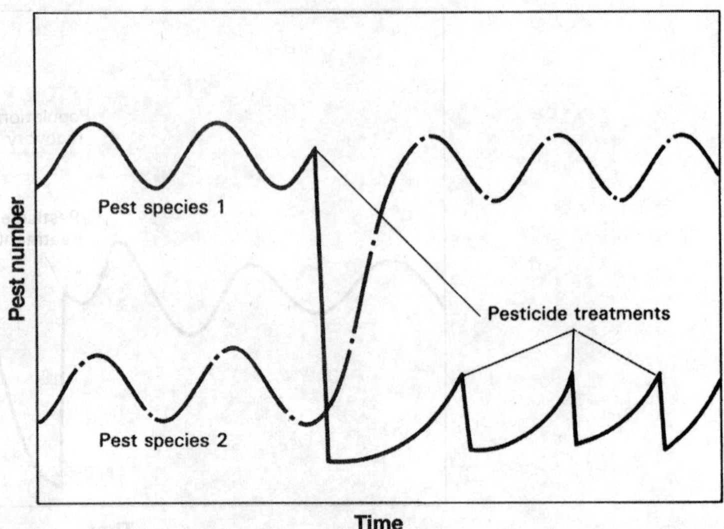


**Figure 17.8** Diagrammatic representation of insect numbers of two pest species as they would respond during pest replacement. Species 2 was a subeconomic pest before replacing species 1.

**Susceptibility of arthropod natural enemies to insecticides.** Insecticides may reduce natural enemies either by direct or indirect means. Mortality from direct exposure to an insecticide is perhaps the most obvious means of natural-enemy reduction. As a group, predators and parasites have been shown to differ in their susceptibility. Records indicate that predators, particularly lady beetles (Coleoptera: Coccinellidae), may have greater tolerance to insecticides than parasitoids, and in some instances predators are no more susceptible than pests. Conversely, parasitoids have been found more susceptible than their hosts. From toxicological studies, it appears that insecticides detoxified oxidatively (several of the organophosphates, including parathion) are much more toxic to natural enemies than to their phytophagous prey (hosts) because the former lack the detoxifying enzymes of the latter. On the other hand, some insecticide groups like the pyrethroids are often less toxic to natural enemies than to pests. B. A. Croft (1990) presents a comprehensive list of natural enemies and selectivity of insecticides for them over their hosts or prey (targeted pest species).

Natural enemies can be suppressed by insecticides in a number of indirect ways. The more important of these for predators are: (1) starvation because of host removal, (2) death from secondary poisoning after feeding on contaminated prey, and (3) repellency of the insecticide. Of these, the first is thought to have the most impact.

In general, parasitoids are suppressed in some of the same indirect ways as predators. For internal parasitoids, insecticide effects are determined by host response. If the host is killed before development is completed, death by starvation is assured. If the host survives treatment, the insecticide effect may be mediated by the host's physiology. The host may detoxify the poison or possibly convert it to an even more toxic metabolite.

**Paradigm for resurgence.** A likely paradigm for pest resurgence could be described as follows. The pest population is fluctuating at an intolerable level for a grower. However, that level is still lower than its potential because of natural enemies. The grower sprays the crop and reduces both pest and natural-enemy populations at the same time. As the effects of spraying wane, the pest population grows, but because the natural enemies must wait for an adequate food supply, their populations lag in regrowth. Another factor for lag in natural enemy growth is that their intrinsic rate of increase is usually lower than their hosts (pest) (Stark et al. 2004). Unencumbered by natural enemies, pest population growth accelerates and exceeds previous levels (Fig. 17.9). If left untreated, the natural enemies would finally return the pest to its original level; however, additional sprays are usually applied before this happens. Such continued activity results in a more or less permanent displacement of the natural-enemy populations.

**Examples of resurgence from natural-enemy reduction.** Most evidence for resurgence from natural-enemy destruction has been based on studies that show negative correlations of pest numbers with enemy numbers. In other words, a reduced natural-enemy-to-pest ratio is offered as proof of the cause of pest upset (this method of proof has been criticized by some researchers as not really demonstrating cause and effect).

Accordingly, resurgence was studied in a springtail, the lucerne flea, *Sminthurus viridis,* in Australia after applications of DDT in alfalfa plots. Pest reduction was observed for 12 days, followed by rapid growth of the population. On the other hand, populations of predaceous mites were reduced significantly for 23 to 45 days after treatment. This predator reduction was used to explain the resurgence of the springtail population in treated plots to levels five times greater than in untreated plots.

In another study with the cyclamen mite, *Steneotarsonemus pallidus,* on strawberries in California, plots were kept free of a predator mite, *Typhlodromus reticulatus,* with treatments of a selective insecticide. Because the insecticide had little or no effect on the cyclamen mite, numbers in treated plots increased fifteen to thirty-five times, but there was no significant increase in untreated plots. In greenhouse studies, hand-picking predator mites from plants gave similar results as removal by the insecticide, indicating that pest resurgence was not caused by any stimulatory effect of the compound.

**Paradigm for pest replacement.** We now describe a possible paradigm for pest replacement. An agroecosystem contains a number of arthropod species that feed on the crop. Many of these species are present at low population levels and are held in check by the action of arthropod natural enemies. These so-called **secondary pests** go largely unnoticed by the grower. The grower applies insecticide to suppress the most destructive pest in the crop, the **key pest,** and continues applications to prevent further damage. In addition to killing the key pest, the pesticide is destructive of secondary-pest natural enemies but causes little mortality in secondary-pest populations. Lacking the natural-enemy check, secondary pests quickly multiply and attain key pest status (Fig. 17.10). In other words, one key pest is exchanged for one or more different pests, with no net gain from the insecticide program.

**Examples of replacement from natural-enemy reduction.** There are many examples of pest replacement in a variety of crops. One of the most infamous occurred in the mid-1960s with attempts to eradicate invading populations of the

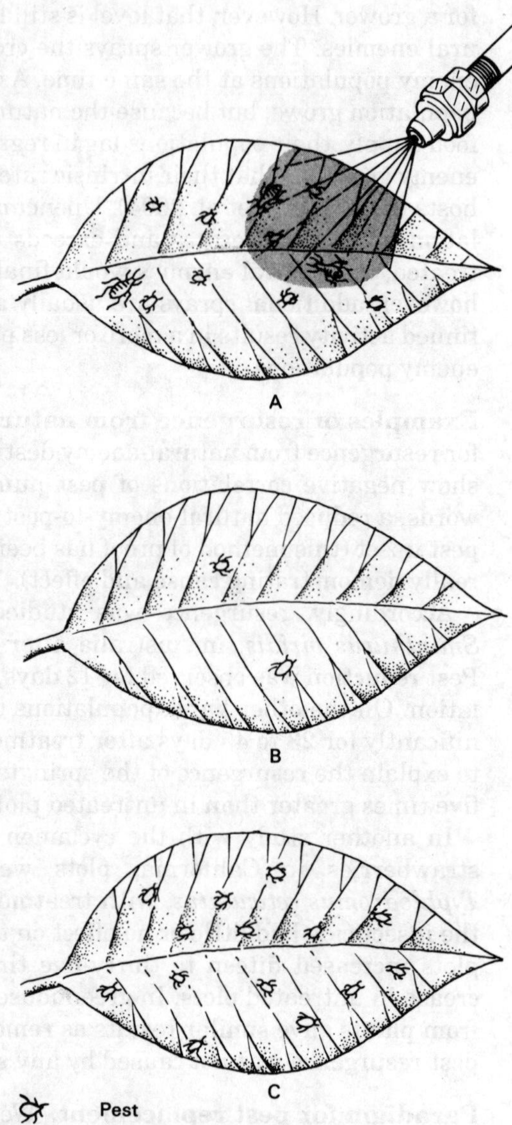


**Figure 17.9** A schematic drawing of development of pest resurgence caused by destruction of natural enemies. A. Both pests and natural enemies are killed by the pesticide application. B. Plant pests recolonize first. C. Plant pests, unrestrained by the natural enemy, increase to a level greater than in A.

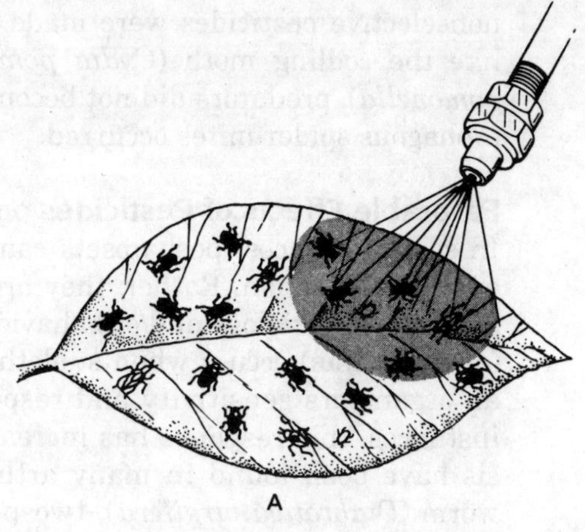


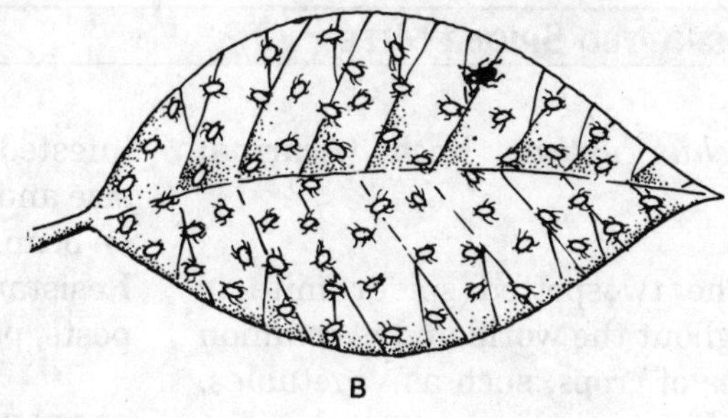


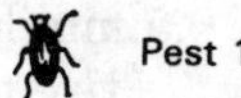




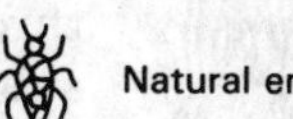

**Figure 17.10** A schematic drawing of development of pest replacement caused by destruction of natural enemies. A. Pesticide is applied to suppress pest 1 but also kills natural enemies that feed on pest 2. B. Released from regulation by natural enemies, pest 2 becomes a major problem. (Redrawn from University of California Integrated Pest Management Publication 3303, *Integrated Pest Management for Citrus*)

pink bollworm, *Pectinophora gossypiella,* in California cotton (Imperial Valley) with carbaryl sprays. In this program, destruction of natural enemies resulted in severe and costly outbreaks of the cotton leafperforator, *Bucculatrix thurberiella,* and, to a lesser extent, certain spider mites.

Another common replacement problem has occurred in Michigan apple orchards, where spider mites (*Panonychus ulmi, Tetranychus urticae,* and *Aculus schlechtendali*) are under natural control by a complex of predators, including three predaceous mites (*Amblyseius fallacis, Agistemus fleshneri, Zetzellia mali*) and a lady beetle, *Stethorus punctum.* When frequent sprays of nonselective pesticides were made during mid-to-late season against insects like the codling moth (*Cydia pomonella*) or the apple maggot (*Rhagoletis pomonella*), predators did not become established, and replacement with phytophagous spider mites occurred.

### Favorable Effects of Pesticides on Arthropod Physiology and Behavior

In some instances, pest upsets cannot be explained on the basis of natural-enemy destruction. Rather, they are caused by direct favorable influences on arthropod physiology and/or behavior. This phenomenon, called **hormoligosis,** (or **hormosis**) occurs when sublethal quantities of a stressful agent increase an organism's sensitivity and response to environmental factors. In several instances, this response has increased reproduction. Examples of hormoligosis have been found in many arthropods, including the western corn rootworm (*Diabrotica virgifera*), twospotted spider mite (Box 17.2) (*Tetranychus*

## BOX 17.2 TWOSPOTTED SPIDER MITE

**SPECIES:** *Tetranychus urticae* Koch (Acarina: Tetranychidae)

**DISTRIBUTION:** The twospotted spider mite is distributed throughout the world and is common on nearly all types of crops, such as vegetables, legumes, cotton, fruits, and ornamentals. The pest can be especially numerous in greenhouses.

**IMPORTANCE:** The twospotted spider mite is a problem on greenhouse crops and, under long-term hot and dry environmental conditions, can cause significant damage to vegetable, field, and orchard crops. The advent of synthetic organic pesticides, such as DDT, helped elevate the status of this species by killing naturally occurring predators, which played a large role in keeping mite populations under control in certain areas. The pest is always a problem in eastern and midwestern field crops when droughts occur.

The species damages plants by feeding on tissues on the undersides of the leaves. Chlorophyll from host tissue is removed, causing a decrease in photosynthesis and increased water loss. Under light infestations, leaves become covered with pale white specks, which eventually turn necrotic. Under heavy infestations, leaves become covered with silk webbing produced by the mites. In addition, the leaves turn bronze and then drop from the plant. Fruit produced from the plants or trees infested by these mites may be reduced in both size and quality. The pest is especially destructive on bean, cotton, strawberry, curcubits, and apples. Resistance to insecticides is common with these pests, particularly in greenhouse situations.

**APPEARANCE:** Adult mites, which are only 0.3 to 0.5 mm long, are greenish yellow with a distinctive pair of dark red spots on the back. These spots, from which the species gets its name, are actually internal food materials seen through the integument. This pest also has stages that include a six-legged larva, an eight-legged protonymph, and an eight-legged deutonymph. All these stages appear somewhat similar to that of the adult stage. The pale yellow eggs are attached to the undersides of leaves.

**LIFE CYCLE:** Adults overwinter as bright orange females under ground litter or in tree bark. After becoming active in early spring, females oviposit, and eggs hatch in 3 to 10 days. The mites pass through larval, protonymph, and deutonymph stages before becoming adults. A typical life cycle takes from 7 to 21 days, and adult females may live up to 9 weeks. There are numerous generations per year, with population peaks occurring in August and September.

*urticae*), granary weevil (*Sitophilus granarius*), and elongate hemlock scale (*Fiorinia externa*).

Resurgence from sublethal sprays also has been observed experimentally with *Helicoverpa* species on cotton and the brown planthopper, *Nilaparvata lugens,* on rice. Resurgence of *Helicoverpa* species after applications of monocrotophos (Azodrin®) and aldicarb (Temik®) was caused by increased ovipositional preference for treated plants. Researchers hypothesized that the insecticides may have changed growth rate, fruiting rate, and physiology of the plants, making them more attractive. Resurgence in the brown planthopper following methyl parathion and decamethrin treatments was also believed to be caused by increased plant attractiveness, but reduction in length of nymphal stadia and longer adult ovipositional period also seemed to contribute to the phenomenon.

As evidence has accumulated, it seems that hormoligosis may be more important than was previously believed. In some instances of resurgence, hormoligosis may play as great a role, if not more so, than the destruction of natural enemies. Particularly crucial to understanding the problem is more research into the effect of pesticides on the growth and phenology characteristics of plants and how these characteristics change survival and reproductive habits of arthropod pests.

### Upsets from Removal of Competitors

If two or more species are competing in an area for the same requisite and one species is dominant, some workers believe that removal of the dominant will allow replacement by subordinate species, regardless of changes in natural-enemy populations. Although suggested as an alternative cause when replacement occurs, experimental evidence to support this hypothesis is not very impressive. However, competition is a very difficult phenomenon to evaluate experimentally.

### Managing Resurgence and Replacement

As with insecticide resistance, the basic principle in dealing with resurgence and replacement is to combine several management tactics and thereby reduce the need for insecticide application. In addition, the manner of application can be modified to take these problems into account. The fundamental objective of these modifications is to avoid hormoligosis and the destruction of natural enemies in the agroecosystem. Another approach is to make inoculative releases of natural enemies after insecticide treatments.

**Avoiding hormoligosis.** At present, less is known and, therefore, less has been written about avoiding hormoligosis. From a practical standpoint, a crucial step would involve eliminating sublethal doses where the problem has occurred or is expected. Sublethal doses can be minimized by applying insecticides on target and attempting to avoid drift. In other words, following the rules of good pesticide practice, including proper chemical and formulation selection, can significantly lessen the problem. Perhaps a more realistic solution to hormoligosis is to apply sprays at a time when the pest is most active and exposed to maximum residues. Other tactics are yet to be developed.

**Avoiding natural-enemy destruction.** Much has been written about reducing destruction of natural enemies to avoid pest upsets. Most of the approaches suggested are based on insecticide selectivity. The many levels of selectivity range from the ability of a poison to select between insects and warm-blooded animals to those that are selective between species of arthropods. It is the latter, more specific type of selectivity that is important in conserving natural enemies in ecosystems.

Insecticide selectivity can be achieved in one of two ways. One is to use selective chemicals that have a stronger depressive influence on the pest population than on its natural enemies. This form of selectivity has been termed *physiological selectivity.* Another way is to use otherwise nonselective chemicals in a selective manner. This method has been called *ecological selectivity.*

**Physiological selectivity.** Although much has been accomplished in medicine with the selective use of chemicals in therapy (for example, chemotherapy to cure cancer), progress has been slow with physiological selectivity in insect pest management. To date, few insecticides can achieve the desired result of killing pests without also killing their natural enemies. Some of the more important examples of physiological selectivity are found with nonconventional materials like microbial insecticides, avermectins, naturalytes, and some insect growth regulators. The list of existing conventional insecticides becomes very small indeed, with such selective materials as the aphicide pirimicarb (Pirimor®), cryolite (Kryocide®, a stomach poison), trichlorfon (Dylox®), phosmet (Imidan®), several neonicotinoids, and certain acaricides being noteworthy exceptions. Additionally, pyrethroids may offer a degree of selectivity with certain pest/natural-enemy complexes.

The reasons for the lack of development of selective insecticides are well known. The most important of these involves economics. Because these compounds are developed by private industry, profits become an overriding consideration. The present cost of development of a new pesticide is $30 to $40 million, and up to $100 million or more may be needed for production facilities. These costs prevent most companies from developing and marketing a candidate compound unless potential sales are at least $40 to $50 million per year. Consequently, broad-spectrum compounds (nonselective among arthropods) are most desirable from an industry perspective because these have market potentials necessary to cover costs of research, development, and manufacture.

However, because of changing government policies and more stringent registration requirements, new insecticides with novel modes of action and acceptable degrees of selectivity are currently in the offing. Therefore, we can anticipate a much greater selection and availability of these safe and selective pest management tools in the near future.

**Ecological selectivity.** Specific selectivity will often need to be devised through using broad-spectrum insecticides. In other words, we may need to use these materials as a dagger, rather than as a sickle. Several ecological approaches to achieve selectivity have been developed through the years. The major ones can be summarized as follows:

1. Monitor populations and follow recommended economic thresholds. Very often, objective determinations of treatment need will allow reductions in amount of insecticide applied, conserve natural enemies, and still provide optimal economic returns.
2. Treat with lowest rates possible. In some instances, high mortalities are not required to produce acceptable suppression. Actually, the pest needs only to be

reduced to a level where surviving individuals will not cause any appreciable loss. This may call for mortality levels of 75 percent, rather than 95 percent. Leaving a residual pest population usually results in natural-enemy conservation and provides hosts for population recovery. The greatest risks with low rates are weather factors such as rainfall that may reduce low-dosage treatment to unacceptable levels and sublethal doses that could cause hormoligosis. Another risk of low doses is killing all homozygous susceptible individuals and some heterozygous individuals, thus hastening population resistance. Most often, however, the potential benefits of reduced doses outweigh the risks.

3. Avoid treating broad areas. If trap crops can be treated or parts of a field sprayed for acceptable suppression, untreated areas can serve as a refuge for natural enemies. Subsequently, these untreated reservoirs supply natural enemies to repopulate treated areas. This approach works best with relatively immobile pests like scales and highly mobile natural enemies like braconid parasitoids.
4. Time treatments to avoid heaviest natural-enemy destruction. Sometimes treatments can be timed to avoid natural-enemy activity periods and peaks. In the previous example with mite resurgence on apples, the most advantageous spray period was early season. In cotton, however, a delayed scheme is deemed best, with late application resulting in better natural-enemy conservation and fewer pest upsets. Understanding important natural enemies and their phenology is crucial in determining proper timing.
5. Use the most selective insecticide formulation. In some instances, changing to another insecticide formulation can aid in achieving a degree of selectivity. For instance, granules applied to corn for management of second-generation European corn borer, *Ostrinia nubilalis,* concentrate in the whorl where larvae are most likely to be found. Therefore, because this formulation concentrates the insecticide at the target site, it exerts minimal effects on natural enemies on other parts of the plant. Baits are another selective formulation. These give selectivity by concentrating a lethal dose in certain areas and utilize pest activity to optimize contact.

## Inoculative Releases of Natural Enemies

Another management approach that has been used to help prevent pest upsets is to repopulate an area with inoculative releases of natural enemies after insecticide application. This tactic involves applying an insecticide when pest populations cannot be restrained by natural enemies, realizing that the natural-enemy population will be destroyed. After insecticide residues have subsided and pest populations are recovering, releases of insectary-reared natural enemies are made, which affect natural control once more. This approach is not unlike providing untreated refuges for recolonization, but it requires more overt management activity and is potentially more effective.

The idea of inoculative releases in biological control is not a new one. As discussed previously, natural-enemy augmentation has been practiced for many years, made most feasible by successful rearing programs that provide an inexpensive supply of common predators and parasitoids. Many of these natural enemies are available from commercial insectaries.

The newest, most unusual modification of the inoculative-release tactic is to release insecticide-resistant natural enemies. This tactic greatly improves the potential for integration of biological control with insecticides. It is based on the finding that some natural-enemy populations can develop insecticide resistance with repeated exposure to the toxicant. Although the list of known resistant natural enemies is not yet long, it includes diverse groups, the most notable of

**Figure 17.11** Adult lady beetle, *Coleomegilla maculata,* feeding on an aphid. Resistance to insecticides has been found in this predator. (Photo Marlin E. Rice)

which are the phytoseiid mites, *Metaseiulus* species (= *Typhlodromus*) and *Amblyseius.* Resistance has also been found in lady beetles (Fig. 17.11) (*Coleomegilla maculata*), braconid and aphelinid parasitoids, a common predator, the green lacewing (*Crysoperla carnea*), and a few other species.

Predatory mites in the family Phytoseiidae are a group where selection of resistance has been adequate for use in practical programs. Organophosphate-resistant genotypes of *Metaseiulus occidentalis* and *Amblyseius fallacis* have been used in conjunction with certain organophosphates, like azinphosmethyl (Guthion®) and phosmet (Imidan®), for management of key pests (for example, codling moth, plum curculio, *Conotrachelus nenuphar,* apple maggot, *Ragoletis pomonella,* and others) in apples, pears, plums, peaches, grapes, and hops. With these programs, substantial reductions of acaricides have been possible and the potential for pest replacements minimized. In California almonds, certain genotypes of *M. occidentalis* have developed resistance to organophosphates in the field and to carbamates and pyrethroids in the laboratory. These resistant genotypes have been released against mite pests on the crop; this has achieved biological control in the presence of sprays directed toward the navel orangeworm, *Amyelois transitella.* Also, genotypes of the wasp parasitoid *Aphytis melinus* have been selected in the laboratory that are resistant to the organophosphates methidathion, chlorpyrifos, and dimethoate, and the carbamate formetanate. This parasitoid is being produced commercially for use in citrus to prevent secondary outbreaks (replacements) of the California red scale, *Aonidiella aurantii.*

Also under consideration is genetic improvement of natural enemies by translocating resistance genes in them. In theory, if a pesticide resistance gene can be found and cloned, it could be used in several species of natural enemies to make them resistant to a group or several groups of insecticides. Although this approach has not been demonstrated, it may be technologically feasible because genetically engineered fruit flies, *Drosophila melanogaster,* have been produced (see Chapter 15). Future genetic improvement to produce insecticide-resistant natural enemies could greatly further the integration of biological control tactics into insect pest management programs.

## OTHER FORMS OF ECOLOGICAL BACKLASH

Although the three Rs are the most obvious forms of ecological backlash, many other biological factors can respond to changes imposed by pest management tactics. Two other such factors are enhanced microbial degradation, which involves unusually rapid degradation of soil insecticides, and upsets in community balance.

### Enhanced Microbial Degradation

The problem of **enhanced microbial degradation** applies to insecticides and herbicides placed on or in soil. Degradation of these compounds is a natural process affected by numerous chemical, physical, and biological factors that interact to modulate pesticide persistence and, therefore, efficacy against intended pests.

Various microorganisms, especially bacteria, are significant components of the degradation process. These microorganisms secrete enzymes that biochemically alter pesticides placed in contact with the soil. The altered compounds are inactive as pesticides and may serve as nutrients for growth of microbial populations. This degradation process is usually beneficial because it prevents the long-term accumulation of pesticides in the environment. With normal microbial degradation, pesticides applied one season will not affect pests or cropping plans the following season. Indeed, relatively rapid degradation after killing action is attained has been a desirable characteristic in development of modern, biodegradable pesticides.

Problems arise, however, when the degradation rate is so rapid that the pesticide does not achieve its primary purpose—to kill the targeted pest population. In some instances, repeated application of the same chemical to the same soil encourages buildup of specific pesticide-degrading microorganisms to the point where succeeding applications have a progressively shorter active life. Continuing applications of the same chemical eventually can lead to the complete loss of pesticide efficacy. Fundamentally, this is the description of enhanced microbial degradation, and the soils where this phenomenon occurs are called **problem soils,** or aggressive soils.

Problem soils have been identified in the midwestern United States and elsewhere where continuous cropping and preventive pesticide treatments are common. Most evidence for enhanced microbial degradation of soil insecticides to date has accumulated on the carbamate insecticide carbofuran (Furadan®). This insecticide was applied in granular form at planting time as a preventive against the corn-rootworm complex, *Diabrotica* species, in fields continuously cropped with corn. The first failures of this previously effective insecticide began in the late 1970s, and the problem became worse with time.

Laboratory and field studies on enhanced microbial degradation of other rootworm insecticides have indicated that the problem also can occur with organophosphates. Failure of isofenphos (Amaze®) to suppress the corn-rootworm complex in Iowa prompted in-depth studies into the cause. These studies showed that enhanced microbial degradation could occur quickly in some soils; subsequently, this insecticide was withdrawn from the market. Other organophosphates shown susceptible to enhanced microbial degradation are diazinon, ethoprophos, mephospholan, fonofos, and coumaphos.

Enhanced microbial degradation is a significant development that could become as important as insecticide resistance in some situations. Although objective guidelines for managing the problem are not developed at this time, tentative measures are similar to those for slowing the rate of insecticide resistance. Most recommendations call for rotating insecticides from one year to the next with those in different classes. Of course, the rotation scheme may not be possible when problem soils already have developed. Here, switching to another class of compounds on a continuing basis may be the only option available. But there are no assurances that either procedure will provide a long-term solution.

The incorporation of microbial inhibitors into pesticide formulations and the use of slow-release formulations are two additional tactics employed to inhibit or delay enhanced degradation. Unfortunately, susceptibility of several available carbamate and organophosphate insecticides to enhanced microbial degradation seems to be encouraging the development of more persistent, and thus environmentally unacceptable, replacements. Probably the best solution for problem soils is to base applications on economic thresholds and rotate crops, thus reducing the frequency of insecticide application at a location.

### Upsets in Community Balance

Basically, community balance exists when the elements in a biotic community (interacting species in an area) have reached a stable condition with regard to one another; that is, proportions of biomass (amount of living matter) for individual species have reached a status quo. **Community upsets** invariably occur when we change the status quo by applying management tactics. We already have discussed an example of this with the pest replacement problem.

Many times, changes in community structure may go unnoticed, particularly if there is no immediate deleterious effect on the management program. However, unplanned changes created by management tactics sometimes can produce belated problems. Such problems are exemplified by the successful eradication of the screwworm fly in Florida and Texas. Elimination of the screwworm was believed to have accounted for significant increases in white-tailed deer populations, which were partly regulated by the parasite. The white-tailed deer is also a good host for *Boophilus microplus,* a tick parasite of cattle. Some specialists believe that white-tailed deer increase due to screwworm eradication caused subsequent tick outbreaks in areas where the tick had previously been controlled.

Dealing with community upsets is very difficult, because the full consequences of community modification can rarely be predicted. Simply following good pest management practice would seem to be the most logical course to take.

## CONCLUSIONS

The results of pest management tactics are varied and sometimes insidious. The main purpose in applying them is to reduce the status of pests to tolerable levels; even when recommendations are followed, however, results may be unexpected. These unexpected, often significant events—ecological backlash—can result from tactic-induced changes in the innate properties of the pest population and/or ecological modifications in the biotic community.

The ramifications of inept management and failure to recognize and deal with ecological backlash are costly, both from continued management and environmental quality perspectives. In general, using tactics in moderation, integrating several tactics, emphasizing selectivity, and accepting the presence of low pest numbers will reduce significant backlash. These are the fundamental principles of insect pest management. Sustainable dissolution of pest problems requires their adoption.

## Further Reading

Brattsten, L. B., C. W. Holyoke, Jr., J. R. Leeper, and K. F. Raffa. 1986. Insecticide resistance: Challenge to pest management and basic research. *Science* 231:1255–1260.

Byford, R. L., J. A. Lockwood, and T. C. Sparks. 1987. A novel resistance management strategy for horn flies (Diptera: Muscidae). *Journal Economic Entomology* 80: 291–296.

Chelliah, S., and E. A. Heinrichs. 1980. Factors affecting insecticide-induced resurgence of the brown planthopper, *Nilaparvata lugens*, on rice. *Environmental Entomology* 9:773–777.

Croft, B. A. 1990. *Arthropod Biological Control Agents and Pesticides.* New York: Wiley.

Croft, B. A., and A. W. A. Brown. 1975. Responses of arthropod natural enemies to insecticides. *Annual Review Entomology* 20:285–335.

Croft, B. A., and K. Strickler. 1983. Natural enemy resistance to pesticides: Documentation, characterization, theory, and application, pp. 669–702. In G. P. Georghiou and T. Saito, eds., *Pest Resistance to Pesticides.* New York: Plenum Press.

Denholm, I., and M. W. Rowland. 1992. Tactics for managing pesticide resistance in arthropods: Theory and practice. *Annual Review of Entomology* 37:91–112.

Desneux, N., A. Decourtye, and J. Delpuech. 2007. The sublethal effects of pesticides on beneficial arthropods. *Annual Review of Entomology* 52:81–106.

Felsot, A., J. V. Maddox, and W. Bruce. 1981. Enhanced microbial degradation of carbofuran in soils with histories of Furadan® use. *Bulletin Environmental Contamination and Toxicology* 26:781–788.

Georghiou, G. P. 1983. Management of resistance in arthropods, pp. 769–792. In G. P. Georghiou and T. Saito, eds., *Pest Resistance to Pesticides.* New York: Plenum Press.

Georghiou, G. P. 1986. The magnitude of the resistance problem, pp. 14–43. In U.S. National Academy of Sciences, *Pesticide Resistance: Strategies and Tactics for Management.* Washington, D.C.: National Academy Press.

Georghiou, G. P. 1990. Overview of insecticide resistance, pp. 18–41. In M. B. Green, H. M. LeBaron, and W. K. Moberg, eds., *Managing Resistance to Agrochemicals.* Washington, D.C.: American Chemical Society.

Grafton-Cardwell, E. E., and M. A. Hoy. 1986. Genetic improvement of common green lacewing, *Chrysoperla carnea* (Neuroptera: Chrysopidae): Selection for carbaryl resistance. *Environmental Entomology* 15:1130–1136.

James, E. G., and T. S. Price. 2002. Imidacloprid boosts TSSM egg production. *Agrichemical and Environmental News,* Washington State University Cooperative Extension Service, Issue 189.

Johnson, M. W., and B. E. Tabashnik. 1999. Enhanced biological control through pesticide selectivity, pp. 297–354. In T. S. Bellows and T. W. Fisher, eds., *Handbook of Biological Control.* San Diego: Academic Press.

Kaufman, D. D., and D. F. Edwards. 1983. Pesticide-microbe interactive effects on persistence of pesticides in soil. In J. Miyamoto and P. C. Kearney, eds., *Pesticide Chemistry: Human Welfare and the Environment* 3:177–182.

Krysan, J. L., D. E. Foster, T. F. Branson, K. R. Ostlie, and W. S. Cranshaw. 1986. Two years before the hatch: Rootworms adapt to crop rotation. *Bulletin Entomological Society America* 32:250–253.

Lafarge, A. D. 1985. The persistence of resistance. *Agrichemical Age* 29:10–24.

McClure, M. S. 1977. Resurgence of the scale, *Fiorinia externa* (Hemiptera: Diaspididae) on hemlock following insecticide application. *Environmental Entomology* 6:482–483.

Newsom, L. D., R. F. Smith, and W. H. Whitcomb. 1976. Selective pesticides and selective use of pesticides, pp. 565–591. In C. B. Huffaker, and P. S. Messenger, eds., *Theory and Practice of Biological Control.* New York: Academic Press.

Ostlie, K. R. 1987. Extended diapause. *Crops and Soils Magazine,* 23–26, June–July.

Pedigo, L. P. 1985. Integrated pest management, pp. 22–31. In *McGraw-Hill Yearbook of Science and Technology.* New York: McGraw-Hill.

Pfeiffer, D. G., 2000. Selective insecticides, pp. 131–144. In J. E. Rechcigl and N. A. Rechcigl, eds., *Insect Pest Management.* Boca Raton, Fla: Lewis Publishers.

Plapp, F. W., JR. 1986. Genetics and biochemistry of insecticide resistance in arthropods: Prospects for the future, pp. 74–86. In U.S. National Academy of Sciences, *Pesticide Resistance: Strategies and Tactics for Management.* Washington, D.C.: National Academy Press.

Plapp, F. W., JR. 1991. The nature, modes of action, and toxicity of insecticides, pp. 447–459. In D. Pimentel, ed., *CRC Handbook of Pest Management in Agriculture,* vol. 2, 2nd ed. Boca Raton, Fla.: CRC Press.

Racke, K. D., and J. R. Coats. 1987. Enhanced degradation of isofenphos by soil microorganisms. *Journal Agriculture and Food Chemistry* 35:94–99.

Racke, K. D., and J. R. Coats, eds. 1990. *Enhanced Biodegradation of Pesticides in the Environment.* Washington, D.C.: American Chemical Society.

Richardson, R. H., J. R. Ellison, and W. W. Averhoff. 1982. Autocidal control of screwworms in North America. *Science* 215:361–370.

Ripper, W. E. 1956. Effect of pesticides on balance of arthropod populations. *Annual Review Entomology* 1:403–436.

Roush, R. T. 1991. Management of pesticide resistance, pp. 721–740. In D. Pimentel, ed., *CRC Handbook of Pest Management in Agriculture,* vol. 2, 2nd ed. Boca Raton, Fla.: CRC Press.

Sparks, T. C., and B. D. Hammock. 1983. Insect growth regulators: Resistance and the future, pp. 615–668. In G. P. Georghiou and T. Saito, eds., *Pest Resistance to Pesticides.* New York: Plenum Press.

Sparks, T. C., S. S. Quisenberry, J. A. Lockwood, R. L. Byford, and R. T. Roush. 1985. Insecticide resistance in the horn fly, *Haematobia irritans. Journal Agricultural Entomology* 2:217–233.

Staal, G. B. 1975. Insect growth regulators with juvenile hormone activity. *Annual Review Entomology* 20:417–460.

Stark, J. D., J. E. Banks, and R. Vargas. 2004. How risky is risk assessment: The role and life history strategies play in susceptibility of species to stress. *Proceedings of the National Academy of Sciences of the United States of America* 101:732–736.

Tabashnik, B. E., and B. A. Croft. 1982. Managing pesticide resistance in crop-arthropod complexes: Interactions between biological and operational factors. *Environmental Entomology* 11:1137–1144.

Watson, T. F. 1975. Practical considerations in use of selective insecticides against major crop pests, pp. 47–65. In J. C. Street, ed., *Pesticide Selectivity.* New York: Marcel Dekker.

Whitten, M. J.1, and M. A. Hoy. 1999. Genetic improvement and other genetic considerations for improving the efficacy and success rate of biological control, pp. 271–296. In T. S. Bellows and T. W. Fisher, eds., *Handbook of Biological Control.* San Diego: Academic Press.

Wilkinson, C. F. 1983. Role of mixed-function oxidases in insecticide resistance, pp. 175–205. In G. P. Georghiou and T. Saito, eds., *Pest Resistance to Pesticides.* New York: Plenum Press.

## Favorite Web Sites

http://www.irac-online.org/
The Insecticide Resistance Action Committee (IRAC) Web site provides information on the growing threat of insect and mite resistance around the world.

http://www.irac-online.org/Crop_Protection/Database.asp.
Official site of resistant pest management at Michigan State University. Provides a database of resistant pests, including mites, spiders, and insects, that have had one or more documented cases of resistance. The database can be searched by species, pesticides, regions, or citations. Great for up-to-date information.

CHAPTER 18

# INSECT PEST MANAGEMENT CASE HISTORIES

PEST MANAGEMENT, OFTEN CALLED integrated pest management or IPM, is one of the most widely practiced approaches to pest technology in the United States. A survey conducted by A. Vandeman and coworkers (1994) indicated that IPM is being applied on more than half of U.S. acreage of fruits and nuts, vegetables, and major field crops (corn, soybean, and fall potatoes). A goal is to increase this acreage to 75 percent. In addition, use of pest management has been shown to reduce pesticide use, cost of production, and risk, and to increase yield and profitability (Norton and Mullen 1994).

The purpose of this chapter is to give you an overview of the types of pest management programs being practiced. You will note that the case histories presented show pest management as an information-based technology that uses multiple tactics to suppress pest populations, not eradicate them. Furthermore, you will notice that in nearly all instances, elements of both prevention and therapy are combined to form the overall program. Finally, you will note contrasts in emphasis on prevention or therapy, according to commodity value per land area.

## INSECT PEST MANAGEMENT IN A LOW-VALUE PRODUCTION SYSTEM

An example of insect pest management in a low-value production system can be seen with the spruce bark beetle, *Ips typographus,* on Norway spruce in Europe. This insect is identified as the most destructive bark beetle of coniferous forests on the continent. Management approaches for this pest have been analyzed by A. A. Berryman (1986).

**Pest life cycle and biology.** Spruce bark beetles overwinter as adults in forest litter or in the bark of spruce trees. Activity begins in mid-to-late May, when males attack spruce trees or logs. These colonizing males release aggregating pheromones that attract other males and females, which subsequently attack trees or logs. Females bore into trees, where they construct individual egg galleries in phloem tissue and lay eggs in notches cut into walls of the gallery. If the host site is crowded, some adults may leave, and one or two more broods are produced at other sites.

Following egg hatch, larvae feed on the phloem tissue and pupate at the tip of tunnels formed during their feeding. Adults emerge, feed for a short time beneath the bark, and bore out of their tunnel either to continue another generation or to overwinter (Fig. 18.1). In northern Europe, only one generation occurs each year, whereas two or three may develop in southern Europe.

**Injury and interactions with the tree.** As with many bark beetles, spruce bark beetles transmit several fungi pathogenic to the tree as they bore beneath the bark. One of these fungi, *Ceratocystis polonica,* is particularly destructive and causes considerable tree mortality, especially in unhealthy trees.

**Figure 18.1** *Ips typographus* life cycle. A. Adults overwinter in litter and under bark. B. Adults emerge and attack trees, logs, and windthrown trees in early spring. C. Parent beetles may reemerge in summer and attack new trees or logs. D. Offspring emerge in fall and enter overwintering habitats. (Reprinted with permission of Plenum Press from *Forest Insects* by A. A. Berryman. © 1986)

Healthy trees can resist damage to a degree. In these trees, resin flow from severed ducts isolates and kills some beetles and repels others. Wound response to the fungus involves development of dead resin-impregnated tissue that lacks nutrients necessary for fungal growth and development. These responses in healthy trees are usually sufficient to prevent significant damage. If beetle attack is intense, however, this resistance can be overwhelmed, and the tree will succumb. Trees stressed by other factors have little or no resistance to beetle attack.

**Beetle outbreaks.** Outbreaks of the spruce bark beetle have occurred irregularly through the years and in various regions. In many instances, outbreaks are caused by unusual storms that result in windthrown trees. These trees, when knocked down by high winds, often continue to survive for a time (because some roots remain in the soil) but do so in a weakened state. Such windthrown trees become excellent sources for beetle reproduction and survival. Other factors implicated in outbreaks include poor management practices and trees growing on unfavorable soils or on steep north- and east-facing mountain slopes.

**Pest management program for spruce bark beetle.** The suggested pest management program for this pest proceeds through a number of steps that Berryman (1986) calls diagnosis, prognosis, prescription, and conducting management operations (implementation). **Diagnosis** involves recognizing the presence and extent of an infestation and determining the causal agent. **Prognosis** includes anticipating problems before they occur and predicting their outcome, if unattended. **Prescription** is recommending a strategy or course of action. Conducting the actual program is self-explanatory.

Spruce bark beetle outbreaks are diagnosed on the basis of symptoms expressed by injured trees. These symptoms include dead and discolored (orange) trees, which drop their needles a few weeks to several months after infestation. The damaged trees are usually found in clumps of several to many individuals. Diagnosis involves inspections that may be conducted in intensive programs by ground personnel or extensive programs by airplane surveys. The intensive program is desirable for accessible, high-value stands, whereas extensive programs are used with inaccessible or low-value stands.

Based on these surveys, prognoses of risk to bark beetle outbreaks are made. Subsequently, individual stands may be grouped into treatment or no-treatment categories on the basis of this risk.

In the prescription phase, recommendations are made that attempt to prevent eruptive outbreaks in individual stands or groups of stands. These recommendations are based on stand prognosis and depend on the course of an outbreak in an area. They include: (1) thinning the stand before outbreaks occur; (2) clear-cutting (removing all trees in one cut) or shelterwood harvesting (two or more cuttings at different times) if the stand is old or stagnating, (slash from logging operations must be handled by burning or other means to eliminate reproductive habitats for the pest); (3) selective logging of spruce and encouragement of resistant species like Scots pine, where growing conditions for spruce are poor; and (4) sanitation-salvage cuts to remove damaged trees. These silvicultural practices are preventive approaches aimed at avoiding eruptive outbreaks. If preventive measures are impractical because of terrain or threat of the current outbreak, use of pheromone traps is suggested. These traps (pipe traps) contain pheromone mixtures impregnated in plastic

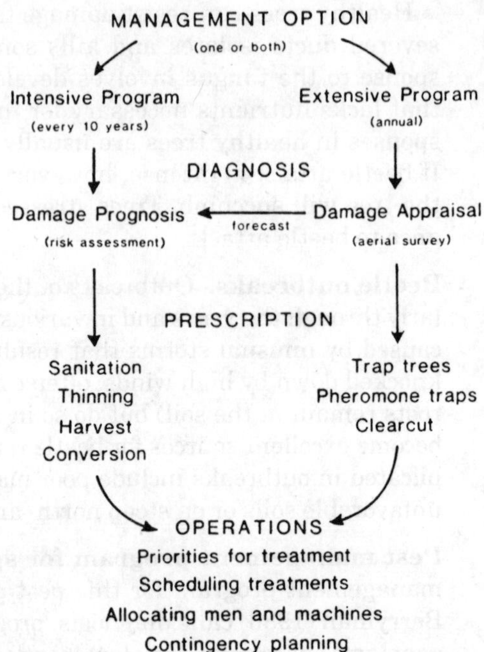


**Figure 18.2** Pest management plan suggested for *Ips typographus* in spruce forests. (Reprinted with permission of Plenum Press from *Forest Insects* by A. A. Berryman. © 1986)

strips or in plastic bags that efficiently capture large numbers of flying beetles. They usually are established in open areas to reduce attacks on adjacent trees.

Finally, scheduling of pest management programs is done by forest managers, who allocate personnel, equipment, and material to accomplish the planned activities. In scheduling and allocation, contingency plans are also in place to accommodate unexpected windthrows or other disturbances that produce new outbreak potentials (Fig. 18.2).

## INSECT PEST MANAGEMENT IN MODERATE-VALUE PRODUCTION SYSTEMS

Moderate-value crops include some of the most important with regard to feeding and clothing the world's population. Although there are many examples of insect pest management in this category, we discuss two, which involve cotton in Texas and corn in the northcentral United States.

### Insect Pest Management in Cotton

Insects are one of the major limiting factors of cotton production in the United States. Yield losses of more than 8 percent per annum can be attributed to these pests. Cotton is attacked by insects from planting until harvest, with nearly 100 species recorded as pests. All parts of the plant may be attacked; however, injury to reproductive structures, including squares (flower buds), blooms, and bolls, accounts for the greatest damage (estimated at 80 percent).

The key insect pests of cotton in the West include the pink bollworm, *Pectinophora gossypiella,* and plant bugs, *Lygus* species (Fig. 18.3), whereas the

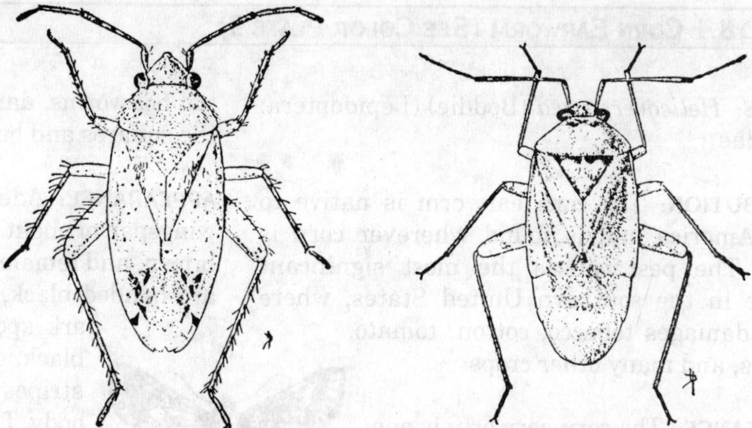

**Figure 18.3** Two important species of the family Miridae (Hemiptera), attacking cotton. Cotton fleahopper, *Pseudatomoscelis seriatus* (*left*), and *Lygus* species. (Courtesy USDA)

boll weevil, *Anthonomus grandis grandis,* and certain plant bugs (Hemiptera: Miridae) dominate in Texas and the Midsouth. As discussed in Chapter 16, the boll weevil has nearly been eliminated in many parts of the Southeast and West.

In most areas, repeated insecticide applications have caused replacement of key pests with other species, particularly the corn earworm (also known as the cotton bollworm) (Box 18.1), *Helicoverpa zea,* and tobacco budworm, *Heliothis virescens,* which have assumed major pest status in most areas.

To understand the development and current status of insect pest management in cotton, a brief survey of important historical events is helpful.

**Historical background of cotton-insect control.** The historical development of pest control in cotton has been shaped to a great degree by the boll weevil. This insect invaded the United States from Mexico in 1892 and spread rapidly through Texas to the Atlantic coast (Fig. 18.4). Invasion by this pest resulted in yield reductions of up to one-third to one-half in some areas. Because much of the South's economic welfare at the time depended on cotton, bankruptcies of growers, merchants, and bankers followed in the weevil's aftermath.

Early pest control programs relied mainly on preventive practices for reducing damage, including early planting of rapidly fruiting cotton varieties and stalk destruction after harvest to eliminate overwintering sites. Before World War II, these procedures were supplemented by low-key applications of inorganic insecticides (mostly calcium arsenate), but these insecticides had limited effectiveness.

Following the discovery of the insecticidal properties of DDT and its development for agriculture, use of insecticides became widespread. Growers began to apply vast quantities of DDT, HCH, aldrin, dieldrin, endrin, and toxaphene on a regularly scheduled and frequent basis. With the great successes of this approach, shifts were made from short-season cotton varieties to long-season, higher-yielding varieties. These long-season varieties called for even more insecticide treatments, and for about 5 to 7 years, yields with this system were unparalleled. By the mid-1950s, boll weevils developed resistance to many

## BOX 18.1 CORN EARWORM (SEE COLOR PLATE 3)

**SPECIES:** *Helicoverpa zea* (Boddie) (Lepidoptera: Noctuidae)

**DISTRIBUTION:** The corn earworm is native to North America and is found wherever corn is grown. The pest causes the most significant damage in the southern United States, where it also damages tobacco, cotton, tomato, legumes, and many other crops.

**IMPORTANCE:** The corn earworm is one of the most serious sweet corn and field corn pests in the United States; it causes an estimated $75 million to $140 million damage annually. The larvae feed on corn silks, which interferes with pollination and results in the formation of barren ears or "nubbins." Developing kernels are also fed on, providing entry wounds for secondary pests like sap beetles and molds. Heavy growths of mold on damaged ears result in unsafe feed for livestock. Damaged sections of sweet corn ears must be cut out, which reduces market value and quality. In field corn, larvae burrow into germ tissue under kernels, causing the kernels to drop from the ear. Besides field and sweet corn, larvae are also significant tomato and cotton pests. Known to growers as the tomato fruitworm, corn earworm larvae burrow into the tomato fruits, causing fruit drop. The larvae, also known as cotton bollworms, damage cotton by burrowing into the squares and bolls, thus reducing yield.

**APPEARANCE:** Adult moths are large, with a wingspan of about 38 mm. Male moths are yellow-brown, and females are orange-brown. The wings are banded black, and each forewing displays a dark spot. Larvae vary from green to black, with alternating light and dark stripes running lengthwise along the body. The larvae, having yellow heads and black legs, may be 50 mm long. Eggs are yellow, hemispherical, and are oviposited singly. Pupae are about 20 mm long and turn from green to brown during development.

**LIFE CYCLE:** Pupae overwinter in soil south of 40 degrees North latitude. Moths begin migration to the northern United States and Canada in early spring from the southern overwintering areas. Female moths oviposit from 350 to 3,000 eggs singly on host plants. Eggs hatch in 2 to 8 days, and the larvae initially feed on their empty eggshells. After five molts (about 14 to 30 days), they cease feeding, drop to the ground, and pupate in the soil. After 14 to 21 days, adults emerge and feed on nectar. There are from two to seven generations per year, depending on the climate and on the type of host plant.

organochlorine compounds in Louisiana, and this resistance spread quickly through the South and Southwest.

At this time, growers turned to other insecticide groups, especially organophosphates and carbamates, for solutions. Compounds from these groups included methyl parathion, azinphosmethyl, ethyl parathion, malathion, EPN, and carbaryl, and resistant weevil populations were suppressed effectively once more. Although low rates of these compounds would control the boll weevil, they were not effective for cotton bollworms (also called corn earworms) (Fig. 18.5) and tobacco budworms. These species had only been an occasional problem before widespread insecticide use, having been kept mostly in check by insect natural enemies and effectively suppressed by the organochlorines. When growers switched to organophosphates and carbamates and there was no adequate natural-enemy check, bollworm and budworm populations erupted in a classic example of the replacement phenomenon.

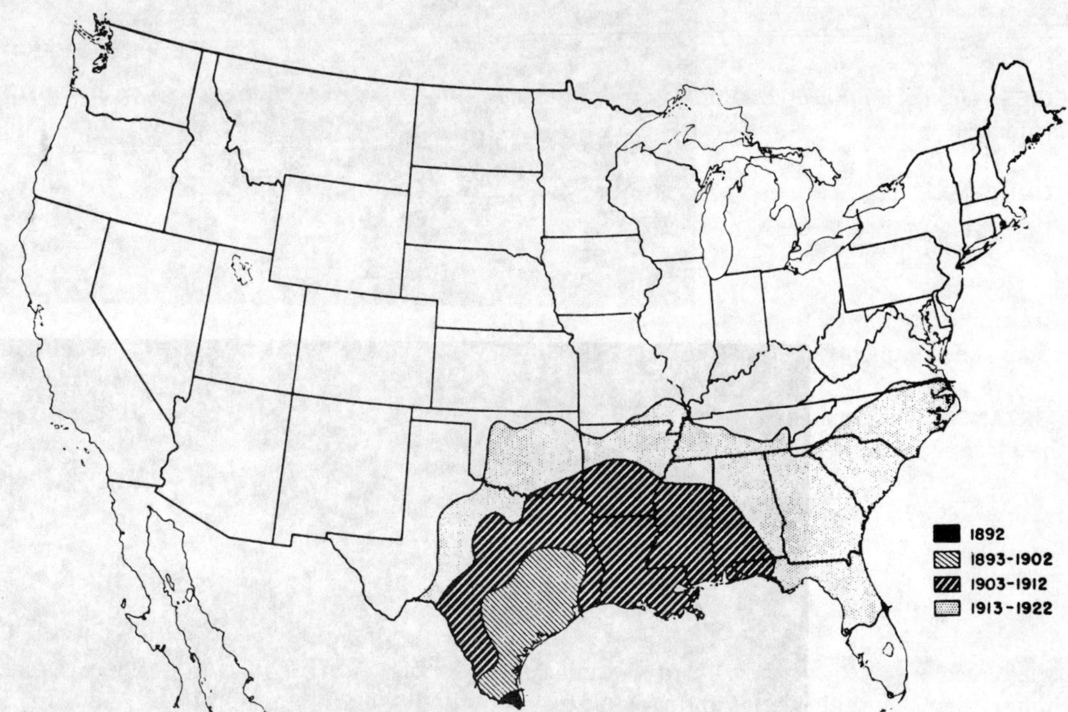


**Figure 18.4** Invasion and spread of boll weevil, *Anthonomus grandis grandis,* populations in the United States from 1892 through 1922. (Courtesy USDA)

To protect the long-season cotton, growers used higher doses of organophosphates mixed with chlorinated hydrocarbons. Although boll weevil and bollworm/budworm suppression was achieved for a time, by the early 1960s resistance of bollworms and budworms to the mixtures began to develop in Texas. Other cotton-producing areas were affected by the early 1970s. This cross-resistance problem was addressed with increased insecticide rates and more frequent applications. The more intense insecticide program was costly and resulted in reduced profits and a search for alternative practices.

**Insect pest management in Texas cotton.** Declining profits from programs of heavy insecticide use contributed greatly to changes in cotton-pest technology. Although high yields could be maintained in long-season cotton, greater costs of production reduced profits significantly.

An example of current strategies can be made with pest management recommendations for cotton production in southern Texas. The foundation of the program begins with preventive tactics aimed at boll weevils. The basic tactic in prevention is a return to early planting of short-season cotton cultivars, moderate fertilizer use, and well-timed irrigation. In addition, plant thinning is delayed or not implemented, which suppresses vegetative growth and stimulates early fruiting. These practices shorten the production season and the period of time cotton is vulnerable to insect attack.

Early harvesting and stalk destruction are the other major preventive tactics of the program. Early harvest is facilitated by defoliants in some operations, after which stalks are cut and shredded. These late-season tactics prevent

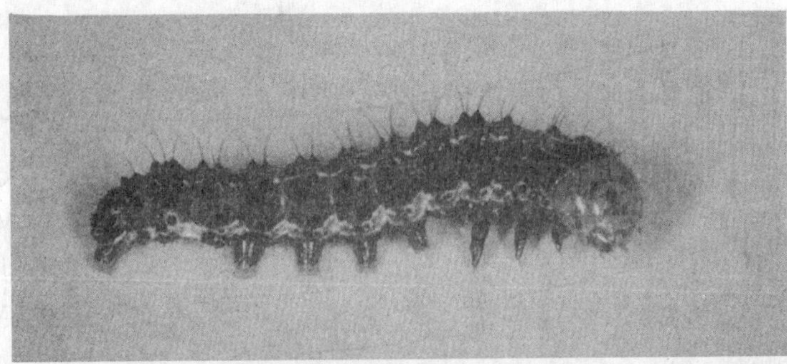

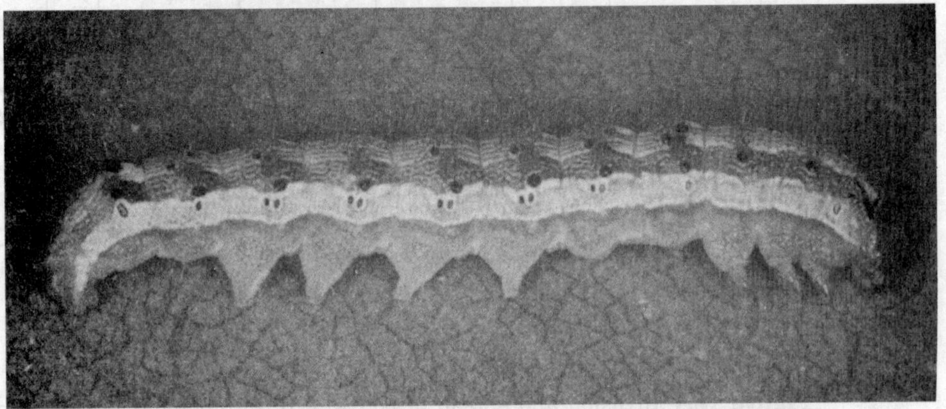

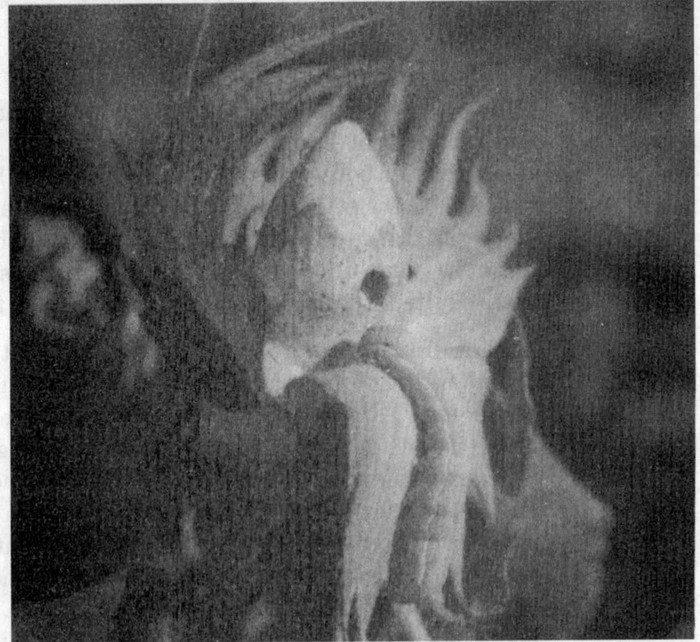

**Figure 18.5** The cotton bollworm, *Helicoverpa zea* (*top,* early instar; *middle,* late instar), and the beet armyworm, *Spodoptera exigua.* Both species can be serious pests of cotton squares (shown) and bolls. (Photos: bollworms, Marlin E. Rice; armyworm, Texas A&M University)

further weevil reproduction and weaken or starve weevils going into hibernation. Consequently, hibernating populations suffer higher-than-normal mortalities.

Such preventive tactics are supplemented by pest surveillance (scouting) and therapeutic treatments of organophosphates against the boll weevil and cotton fleahopper (Fig. 18.3), *Pseudatomoscelis seriatus.* Additionally, pyrethroids (among other insecticides) are applied when cotton bollworm and cotton budworm outbreaks occur. In making applications, natural enemy conservation is practiced by closely following economic thresholds, using lowest possible rates, and, when possible, treating only "hot spots" in a field.

The cotton pest management programs discussed here have been effective and more profitable than traditional full-season management programs. In one cost/benefit study, profits from growing short-season cotton and practicing insect pest management were from 56 to 130 percent greater than those in a high-input, full-season program.

Some changes to this management scheme may occur with progress on weevil eradication through the U.S. Boll Weevil Eradication Program. Recall from Chapter 16 that this program has eliminated the boll weevil from millions of cotton acres, particularly in the southeastern United States. Since the beginning of the eradication effort in Texas, the amount of insecticides used has declined steadily, giving opportunities to adjust programs against other insect pests in the production systems. Additionally, another important pest group, the cotton bollworm/tobacco budworm complex, can be managed by growing new transgenic cotton varieties. Even with these new developments, however, continued threats from remaining pests, such as the pink bollworm and plant bugs, remain, which will require vigilance in dealing with outbreaks and continued dedication to effective prevention.

### Insect Pest Management in Corn

Corn is the major cash grain crop grown in the United States; it is produced for silage, grain, seed, ethanol, high fructose corn syrup, and fresh market produce. Nearly 19 percent is exported to foreign markets. Of these products, corn grown for feed grain (field corn) occupies the majority of the total acreage (usually around 78 million acres but expected to reach 90 million acres in 2007 because of ethanol demand), and nearly 60 percent of this is grown in the northcentral United States (Illinois, Iowa, Indiana, Minnesota, Nebraska, Ohio, and Wisconsin).

A large complex of insects attack corn, with more than thirty species or species groups causing economic damage. Some of the most significant problems occur in field corn when it is grown continuously on the same land. In the northcentral states (Corn Belt), this practice gives rise to problems from a key beetle complex comprising the northern corn rootworm, *Diabrotica barberi,* and the western corn rootworm, *Diabrotica virgifera virgifera.* Another key pest is the European corn borer *Ostrinia nubilalis,* which causes damage to field corn grown continuously or in rotation with other crops. Together, these species account for most insecticide use in field corn in the Corn Belt, and they contribute to this crop's top ranking in the United States for insecticide use.

**Life history and injury from key pests.** Both northern and western rootworm species have similar life histories and behavior patterns (Fig. 18.6). In these species, eggs from the previous season hatch in May and June, and larvae move in the soil until a corn root is encountered. Northern and western corn

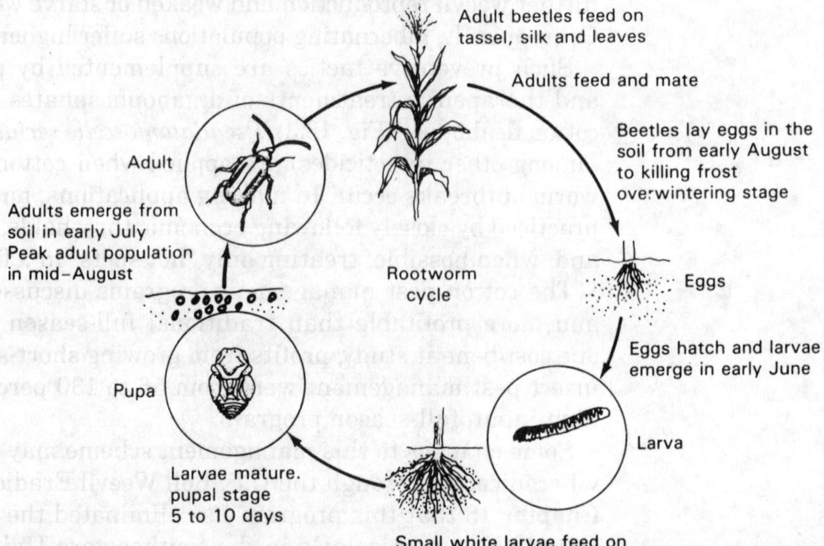


**Figure 18.6** Life cycle of the western corn rootworm, *Diabrotica virgifera*. (Courtesy J. Tollefson and Iowa State University)

**Figure 18.7** Uninjured (*left*) and corn rootworm (*Diabrotica* spp.) injured corn roots. (Photo Marlin E. Rice)

rootworms are monophagous; that is, they feed mostly on one species, corn. However, western corn rootworms are also known to survive on foxtail and a limited number of other grasses. Corn is injured by larvae as they chew off and tunnel inside roots, including main roots and brace roots touching the soil (Fig. 18.7). Root pruning is believed to reduce the flow of water and nutrients to the plant, which disrupts plant physiology and causes yield loss. An additional, sometimes more serious loss occurs when root pruning allows lodging during rain and wind storms. Lodged plants lose yield because of modified architecture (goose necking), and further losses occur from harvest difficulties (Fig. 18.8).

Upon reaching full growth in July, larvae pupate in earthen cells. Adults emerge in late July and August, crawl to the soil surface, and feed on corn pollen and silks or pollen of other plants. If adult densities are high, corn silk clipping can reduce kernel set and grain yield by interfering with pollination. After mating, females lay clusters of eggs in the soil from late July to September. These eggs hibernate until the following spring, when the next corn crop is planted.

The European corn borer (Box 18.2), a species introduced to the United States and first found near Boston, Massachusetts, in 1917, is probably the

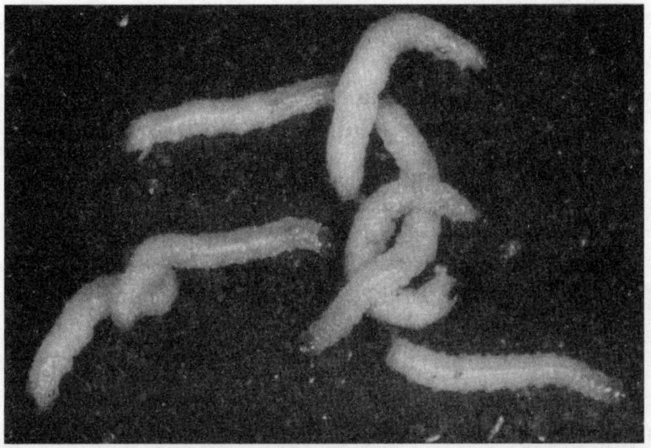

**Figure 18.8** Corn rootworm larvae, *Diabrotica* spp. (*upper left*), injury to corn roots (*right*), and lodged corn caused by excessive feeding on the roots (*lower left*). (Photos by Marlin E. Rice)

## BOX 18.2 European Corn Borer (See Color Plate 3)

**SPECIES:** *Ostrinia nubilalis* (Hübner) (Lepidoptera: Crambidae)

**DISTRIBUTION:** The European corn borer entered the United States at Boston harbor prior to 1917 and has successfully moved across the continent to the Rocky Mountains. The species has evolved into ecotypes based on the number of generations per year. Southern populations may have four generations per year, eastern and lower midwestern populations have two generations, and northern populations typically only have one generation per year. The European corn borer is also distributed over most of Europe and parts of Asia.

**IMPORTANCE:** The European corn borer is one of the most serious corn pests in the United States. In addition to feeding on corn, the species is capable of surviving in many herbaceous plants with stems large enough for the larvae to enter. Extensive larval boring in plant stalks or stems results in premature breakage or lodging that leads to complications in harvesting and ultimately reduced yield. Later in the season, larvae tunnel into ear shanks, causing ears to drop. Several control strategies have been investigated to limit losses to this pest. Host-plant resistance and cultural manipulation have proven effective on many occasions. The use of insecticides, when timed on the basis of egg-mass and larval scouting, is a major tactic for the management of the second generation.

**APPEARANCE:** Adult female moths are pale yellowish brown, with irregular darker bands running in wavy lines across the wings. Male moths are distinctly darker, with olive-brown wings. Eggs are present in masses of five to fifty. They are initially white but turn yellow with continued development. Young larvae are yellow, with several rows of brown spots. Full-grown larvae are about 25 mm long, gray to light brown or pink, and faintly spotted on the dorsal surface.

**LIFE CYCLE:** Full-grown larvae overwinter inside their tunnels in stubble, stalks, ears, or weeds. Prior to pupation, larvae cut escape holes at the base of the plants to allow the moth to exit following pupation. Moths emerge in spring and mate within 24 hours, usually at dusk. Three days after mating, adults begin to oviposit egg masses on the lower surfaces of host leaves. Eggs hatch in 3 to 12 days, and larvae feed on the surface of the leaves initially before entering the whorl, where they develop. The borers have five or six larval stadia and become fully grown in 25 to 35 days. Larvae of second and third generations readily bore in the plant or feed on ears. Diapause in fully-grown larvae is induced by temperature and photoperiod.

---

most important pest across all corn-growing regions. Different genetic strains, or so-called ecotypes, produce various numbers of generations, with only one generation in Canada and northern Minnesota and four in Georgia and other southern states. Two or three generations are the most common in the Corn Belt (Fig. 18.9).

The life cycle of the two-generation ecotype can be described as beginning with full-grown larvae, which hibernate through the winter mostly in corn stalks and cobs (Fig. 18.10). These mature larvae pupate in May, and adults emerge in late May and June. Adults mate in grassy areas near cornfields and fly back into the cornfield for egg laying. Eggs are laid in masses on the undersides of leaves. After egg hatch, surviving larvae feed in the leaf whorl for a few days and subsequently burrow into the stalk. Once in the stalk, they tunnel in conductive tissues and cause physiological stress that ultimately reduces yields. First-generation larvae

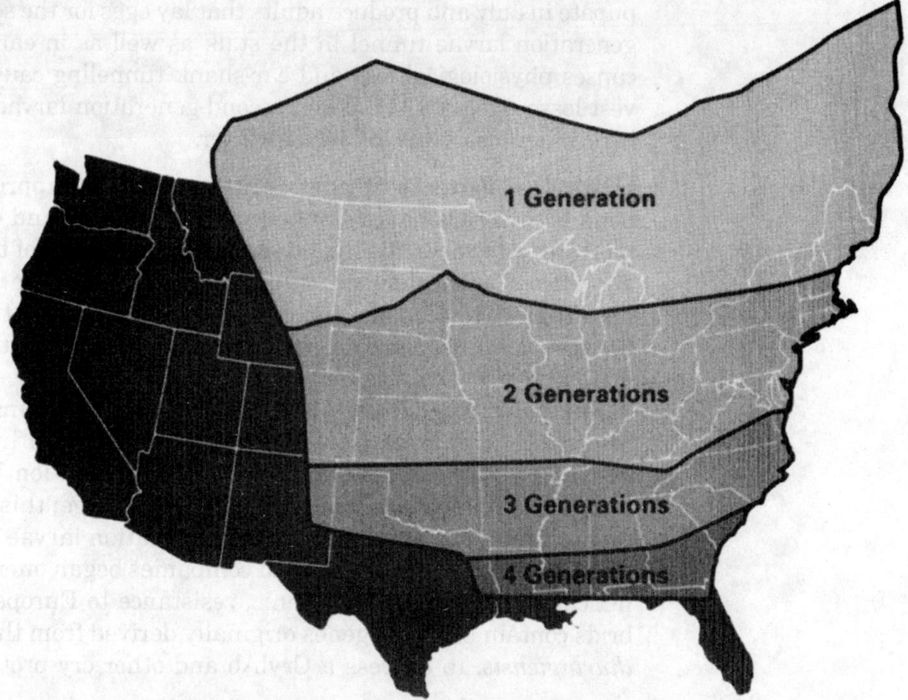


**Figure 18.9** Approximate distribution of generations of the European corn borer, *Ostrinia nubilalis,* in the United States and Canada. (After Mason et al., 1996, *European Corn Borer Ecology and Management,* North Central Regional Extension Publication No. 327)

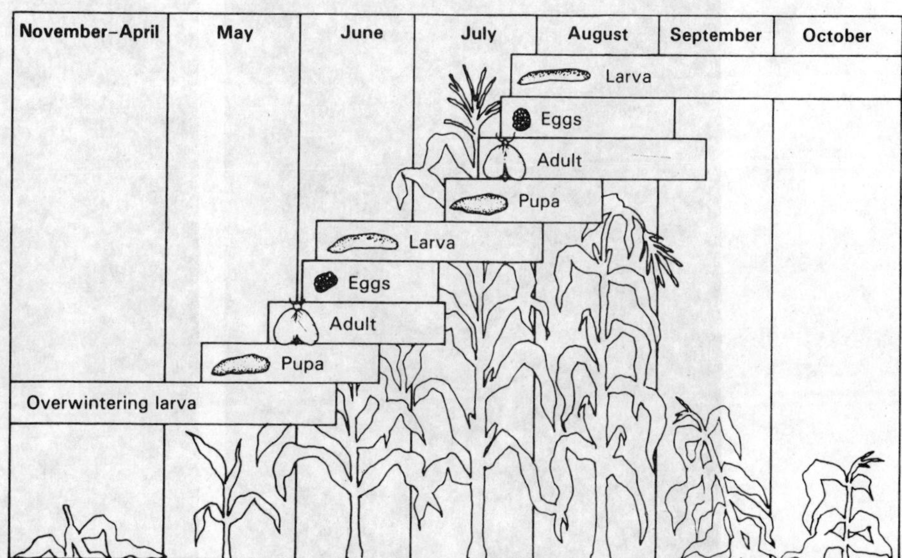


**Figure 18.10** Typical life history of the European corn borer in the northcentral United States. (After Showers et al., 1989, North Central Regional Publication 327, revised 1996)

pupate in July and produce adults that lay eggs for the second generation. Second-generation larvae tunnel in the stalk as well as in ear shanks. Stalk tunneling causes physiological loss and ear-shank tunneling causes ears to drop and harvest losses (Fig. 18.11). When second-generation larvae become full grown, they hibernate to pass the following winter.

**Historical aspects of pest control.** Cultural approaches to suppress European corn borers have included stalk shredding and deep plowing to increase mortality of overwintering larvae. However, many of these practices have been abandoned with reduced tillage and the need for soil conservation. Early planting to avoid heaviest losses from second-generation corn borers and early harvest to reduce yield losses from dropped ears and lodged plants remain as attractive preventive practices.

Resistant hybrids are also practical for use against the corn borer. Today, most seed companies have hybrids available with intermediate "traditional" resistance to leaf-feeding by small, first-generation larvae. The resistance is lost before the hybrids reach the pretassel stage; this makes them susceptible to late first-generation and second-generation larvae.

Beginning in 1996, many seed companies began offering corn hybrids with genetically modified, or transgenic, resistance to European corn borer. These hybrids contain modified genes originally derived from the soil bacterium, *Bacillus thuringiensis,* to express a CrylAb and other cry proteins that are toxic to the

**Figure 18.11** Leaf and whorl injury to corn from first-generation (*left*) European corn borer larvae, *Ostrinia nubilalis,* and cornstalk breakage caused by second-generation larvae. (Photos by Marlin E. Rice)

larvae when they feed upon the corn plant. These proteins are expressed in the Bt corn hybrid, season-long, and kill both first- and second-generation larvae.

Through the years, insecticides were a mainstay in dealing with both corn rootworms and the European corn borer. Insecticides were used as a preventive measure against rootworms (applied regularly at planting time) in continuous corn and were often applied as a therapeutic treatment for corn borers. In the latter instance, prediction of phenology and proper timing of treatments was critical, because larvae are vulnerable for only a short time before they enter the plant. The development of degree-day models, scouting procedures, and accurate economic-injury levels contributed significantly to effective therapeutics for corn borer problems.

Historically, insecticide programs did not result in major pest upsets with corn in the northcentral United States, as has occurred with cotton in the South. A likely reason is that rootworms have few known natural enemies, and it does not seem that the natural enemy complex strongly regulates corn borer populations. But rootworms have become resistant to a number of insecticides in each of the major groups, and this nagging problem is accompanied by the growing threat of enhanced microbial degradation and insect resistance to rotation in some areas. Such events argue for the adoption of a durable insect pest management program.

**Insect pest management program for corn in the northcentral United States.** The key element of management design for corn rootworms and the European corn borer is prevention. The prevention program begins in most regions with a plan of crop rotation to eliminate losses from the corn-rootworm complex. This plan would include alternate year rotations with soybeans or other nonhost crops like oats or involve a 3-year rotation scheme. The 3-year plan would be necessary in areas of extended rootworm diapause.

In the preventive program, a hybrid like A619 × A632 or A619 × H99, with intermediate resistance to European corn borer, would be chosen and planted as early as possible. This preventive program could be rounded out by early harvest and grain drying, if on-farm facilities are available, or by selling the grain at a moisture-discounted price, if they are not. Alternatively, transgenic Bt hybrids with high levels of resistance to the European corn borer may be planted, with 20 percent of the acreage being reserved for susceptible hybrids. The latter serves as a refuge in attempts to prevent borer resistance to the Bt hybrids, but insecticide treatments may be used on this acreage.

During many years, such a preventive program would eliminate the need for insecticides in corn completely. However, occasional therapeutic treatments would be required for second-generation corn borers in susceptible hybrids and for pests like black cutworms, *Agrotis ipsilon,* when outbreaks occur. The therapeutic program relies on scouting and early detection of these problems and is particularly refined for second-generation corn borers.

Conventional guidelines with second-generation corn borers call for insecticide treatment when 50 percent of the plants in pre- to post-tasseling stages of development have an egg mass on them. A more advanced procedure is to use egg scouting information with a detailed cost/benefit analysis to arrive at treatment decisions (see North Central Regional Publication 327, revised 1996, for procedures). If the benefit gained from the insecticide exceeds the cost of suppression, then applying the tactic is cost effective. If therapeutics are required, aerial applications with a granular insecticide can be made to minimize drift and reduce deleterious effects on natural enemies and honey bees.

This example of insect pest management in corn shows very well how several tactics and pests can be integrated into a farm-level management system. Moreover, it shows that both prevention and therapy are crucial in achieving program objectives.

## INSECT PEST MANAGEMENT IN HIGH-VALUE PRODUCTION SYSTEMS

Insect pest management for high-value crops is both tedious and complex. These crops usually require significant economic inputs for establishment and continued production, enhancing risks of insect losses and leaving little margin for error in decision making. High values of produce, accompanied by consumer demand for cosmetically "perfect" or blemish-free products, raise the stakes for pest management and create a system where therapeutic corrections are prevalent. Some of the best examples of pest management in this type of production system is that for potatoes and apples.

### Insect Pest Management in Potatoes

Many insect pests infest potatoes grown in the garden and commercially. Losses to these can be severe, both from a yield standpoint and in appearance. Some of the more important species are green peach aphids (*Myzus persicae*), flea beetles (several species, Coleoptera: Chysomelidae), potato leafhoppers (*Empoasca fabae*), potato aphids (*Macrosiphum euphorbiae*), potato tuberworms (*Phthorimaea operculella*), and Colorado potato beetles (*Leptinotarsa decemlineata*).

The Colorado potato beetle (see Box 5.4, Fig. 13.9) is one of the best known of all North American insects. This pest has been particularly important in commercial potato production and is an overriding consideration in many insect management programs, and thus is emphasized here.

**Pest biology.** Both adult and larval Colorado potato beetles feed on potato foliage and, if not properly managed, can completely defoliate the plant. At certain plant growth stages, this defoliation can cause a complete loss in tuber production.

Adults overwinter in the soil. They walk or fly to their hosts and, after 5 to 10 days, females start laying clusters of twenty to sixty eggs on foliage. Females may live 55 to 120 days, and in the laboratory total fecundity can exceed 4,000 eggs. All eggs in an egg mass hatch about the same time, and new larvae begin feeding immediately. The larvae develop through four stadia, which last from 8 to 28 days, depending on temperature. Larvae of the third and fourth stages consume 15 and 77 percent, respectively, of total leaf consumption. At the end of the fourth stage, larvae crawl from the plant and burrow into the soil, where they pupate. The new adults emerge about 1 to 3 weeks later. Depending on several environmental and host-plant factors, adults may mate and start another generation or they may fly from the field and enter diapause.

In more northern latitudes, there is usually a partial second generation in which summer adults feed only briefly and then migrate to sheltered locations, where they burrow into the ground and overwinter. Some summer adults will feed voraciously, mate, and lay eggs. The primary trigger that determines whether individual beetles mate or go into reproductive dormancy is day length at the time of adult emergence. By mid summer, a potato field in Minnesota, for example, can have all stages of Colorado potato beetles present.

The native host for this insect is a relative of potato, buffalo bur (*Solanum rostratum*), and Colorado potato beetles were found feeding on this plant near the Iowa–Nebraska border in 1819. The beetle was first recognized as a pest of potato in 1859 in eastern Nebraska. Once Colorado potato beetles adapted to feeding on potato, the beetles migrated eastward throughout the Great Plains and Ohio River Valley, expanding their range at a rate of 85 miles per year. They reached the Atlantic Coast of the United States in 1874. The insect became established in Western Europe during World War I and has since moved into Eastern Europe and western China and Iran. The first large-scale use of insecticides in an agricultural crop was for control of Colorado potato beetle, and early successes prompted widespread use of insecticides in other crops.

**Early control efforts.** Nineteenth-century entomologists C. V. Riley and Benjamin Walsh recommended in the 1860s and 1870s a variety of tactics to manage the Colorado potato beetle, including a horse-drawn beetle collector, potato varieties with "immunity," conservation of natural enemies, early-planted trap-crop potatoes, crop isolation, early-maturing varieties, and crop rotation. But in 1871, Riley tested Paris green, a copper-based arsenical, and found it effective against the Colorado potato beetle. Its use became widespread, and in 1875 Riley reported that "the American farmer by means of intelligence and a little Paris green is pretty much master of the [Colorado potato beetle]."

In the early 1900s calcium arsenate was used to control the insect before DDT was introduced in the mid 1940s. Insecticides reduced the Colorado potato beetle to minor pest status until it developed resistance to DDT in 1952. Additional insecticides were developed during the next three decades to control the insect but it quickly developed resistance to all of them, except rotenone, by 1986. In addition to the resistance to synthetic insecticides, the insect has the capacity to develop resistance to the *Bacillus thuringiensis tenebrionis* delta-endotoxin. Widespread development of insecticide resistance has necessitated a more thorough understanding of the insect's ecology.

**Insect pest management in the northern United States.** There are a number of insecticides available for suppression of Colorado potato beetle populations, but the development of insecticide resistance is still a concern for commercial potato producers. If the application rate of an insecticide must be increased to achieve adequate control, then resistance to that insecticide is likely developing in the population.

The baseline in dealing with this important pest is ecological management. Crop rotation is first and foremost as an effective method of reducing the risk of loss. Also, locating a new potato field some distance from the previous year's crop greatly reduces the damage potential because of delayed colonization from old fields. Delayed colonization results because adults cannot fly unless temperatures reach 70°F. Early spring daytime temperatures in the northern United States can be considerably below 70°F, thereby making beetles walk from overwintering sites to find a potato field. A potato field 1,500 feet from a previous year's field can delay infestation by about 7 days. Straw-mulched fields also help to reduce populations, possibly by increasing predation rates within the field or interfering with beetle movement into the field from overwintering habitats.

As important as ecological management is, most commercial potato production is difficult without using insecticides. Systemic insecticides have been applied to the soil or to seed at planting, particularly if there has been a history of damage each season. Foliar-applied insecticides are appropriate if insect pressure is not too severe. Such applications necessitate field scouting to assess the need for treatment, as adult beetles may invade the field from only one side, requiring only local spraying.

Insecticides are most effective on young larvae; eggs and pupae are not susceptible to insecticides. Spraying young larvae helps prevent damage and allows resistant adults present in the field to mate with susceptible ones. This tactic keeps selection pressure for insecticide resistance to a minimum. If insecticides are sprayed against adults ready to overwinter, the only survivors will be resistant individuals. When resistant adults mate, their progeny are resistant, consequently fixing the resistance gene in that population. Once insecticide resistance becomes established in a population, reversion to susceptibility occurs only after many generations of nonexposure to the insecticide, or may never revert to preexposure levels. Moreover, multiple applications of foliar insecticides to control summer adults and larvae of Colorado potato beetle can trigger green peach aphid outbreaks because insecticides disrupt the predators and parasitoids that help hold these populations in check.

Managing insecticide resistance in the Colorado potato beetle is the essence of a good pest management program for this pest. Resistance management requires that the same insecticide or other insecticides with a similar mode of action not be used in successive years at the same location. In states such as Minnesota, potato producers are encouraged not to use the same insecticide to control the overwintering generation and summer adults or any larvae they produce (second generation) during a single growing season. Repeated exposure of the pest to the same or similar insecticide has created resistance in as little as four to ten generations. Many newer insecticides have a unique mode of action. This allows producers the opportunity to delay the development of resistance by rotating insecticides of different chemical classes.

In the recent past, very effective potato varieties that resisted infestations were developed for use against the Colorado potato beetle and were introduced to commercial potato production. These varieties were genetically modified to express a bacterial gene from *Bacillus thuringiensis* var. *tenebrionis* that produces a cry protein in the plant that is toxic to the pest. The varieties were marketed as NewLeaf® potatoes and were available for commercial use only. NewLeaf potatoes produced the protein toxin in all green tissues, and larval mortality was essentially 100 percent. One major advantage of using genetically engineered potato varieties is that beneficial insects were not harmed, as is the case with many conventional insecticides. Beneficial insects that could be conserved in fields then could be used to help control other pests such as aphids.

As effective as NewLeaf technology was, however, its developer discontinued NewLeaf varieties beginning with the 2001 growing season. This decision was partly based on seemingly unfounded fears of genetically engineered crops by certain public sectors. Fearing loss of customers, several large potato processors, including McDonald's, Burger King, Frito-Lay, and Procter & Gamble, agreed to use only potatoes without genetic-engineering technology. This development and the labeling of new, effective insecticides for potatoes contributed to the demise of these potato varieties.

As can be envisioned, the management of insects in commercial potato production is complex and challenging to say the least. Yield of tubers is an overriding consideration, but appearance of the produce to the consumer cannot be overlooked. Several tactics are available for inclusion in management programs, and integrating them in a thoughtful manner is paramount. In particular, managing Colorado potato beetle is demanding because of the ability of the pest to develop resistance to insecticides. Incorporating an effective resistance management scheme into pest management programs becomes the hallmark of a durable strategy.

### Insect Pest Management in Apples

Insect pests have traditionally received great attention from apple producers because these pests have had a significant impact on apple crops. More than 500 species feed on trees and fruit on a worldwide basis and cause millions of dollars in losses. In the United States, forty-three species have been designated as economic pests, but fewer than ten are considered serious pests.

Most apple pest management focuses on integrating chemical pesticides with natural-enemy control. In particular, the major emphasis has been to suppress key pests without causing replacements from secondary pests. Preventive practices are implemented with these programs to some extent, but they would not be expected to eliminate the need for therapeutics as the growing season progresses.

For brevity, the management examples to be mentioned here are taken from production areas of the eastern and midwestern United States. Although key pests may vary in other regions, pest types, relationships, and management approaches are similar.

**Key pests and injury.** There are four key insect pests of apples in the midwestern and eastern United States: codling moth (*Cydia pomonella*), plum curculio (*Conotrachelus nenuphar*), apple maggot (*Rhagoletis pomonella*), and redbanded leafroller (*Argyrotaenia velutinana*) (Fig. 18.12). In addition to these insects, two diseases, apple scab and powdery mildew, are considered key to management operations in commercial orchards.

All of these pest problems involve direct injury of fruit, and combined pest injury cannot exceed 1 percent of the harvest. This low economic-injury level reflects the high market value of the produce and gives a perspective of how little injury can be tolerated economically. Because of the importance of these key pests, most pest management programs are directed primarily toward them.

Codling moth larvae injure fruit when they tunnel and feed on seeds. There are usually two or more generations per season. Plum curculios make crescentlike punctures on fruit surfaces by laying eggs, and emerging adults later make additional holes as they feed. The apple maggot may produce the heaviest infestations of all, with larvae feeding, often undetected, inside the fruit. Both apple maggot and plum curculio produce one generation per year. Redbanded leafroller larvae feed on leaves and the surface of fruit and produce two to four generations per season.

**Secondary pests and injury.** Because of serious problems with insecticides and pest replacements, secondary pests are a major consideration in designing pest management programs for apple production. These secondary pests

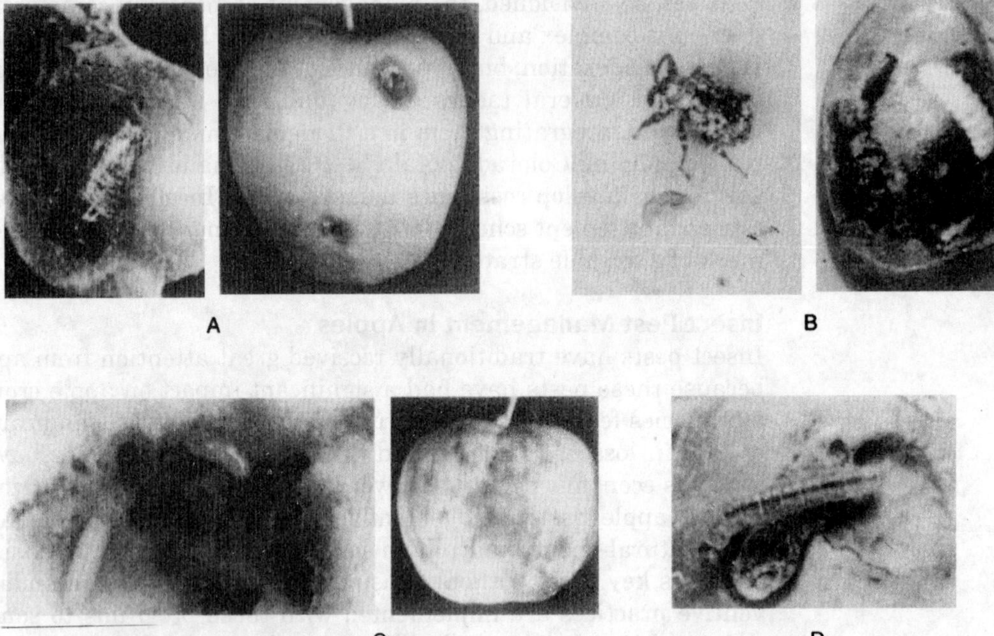


**Figure 18.12** Key insect pests of apples in the midwestern and eastern United States. A. Codling moth, *Cydia pomonella* (*left*), and injury. B. Plum curculio, *Conotrachelus nenuphar,* on cherry (*left*) and larva with injury. C. Apple maggot, *Rhagoletis pomonella,* and injury (*left*) and blotched and streaked apple infested with larvae. D. Redbanded leafroller, *Argyrotaenia velutinana,* larva and injury to apple. (Courtesy USDA)

include various species of scales, aphids, leafhoppers, and mites. Of these, phytophagous mites especially have been a problem.

The major mite pests of concern are the European red mite (*Panonychus ulmi*), twospotted spider mite (*Tetranychus urticae*), and the apple rust mite (*Aculus schlechtendali*). These species injure foliage by sucking the contents of individual cells, eventually producing a yellow-green (spider mites) or silvery (rust mites) look to leaves. A bronze appearance of foliage may occur with very heavy infestations. A result of this indirect injury can be reduced fruit size, poor fruit color, and excessive fruit drop in the year of initial infestation. Reduced blossoms and fruit set and poor tree vigor may occur in following years.

Studies of unmanaged orchards have shown that European red mite and twospotted spider mite populations are strongly regulated by predatory mites and insects. One of the major predators is *Amblyseius fallacis* (Fig. 18.13), a pear-shaped mite that can virtually eliminate phytophagous mites from apple trees. Very often the phytophagous mites have developed resistance to many of the broad-spectrum insecticides used against key pests, but predators may succumb after applications. As mentioned earlier, phytophagous mite populations allowed to reproduce without pressure from their natural enemies can quickly become major pests.

**Insect pest management in commercial apple orchards.** Because natural control of the key pests is unacceptable, much of the insect pest management program in apples is focused on proper therapeutics. Preventive measures are frequently (but not exclusively) directed toward secondary

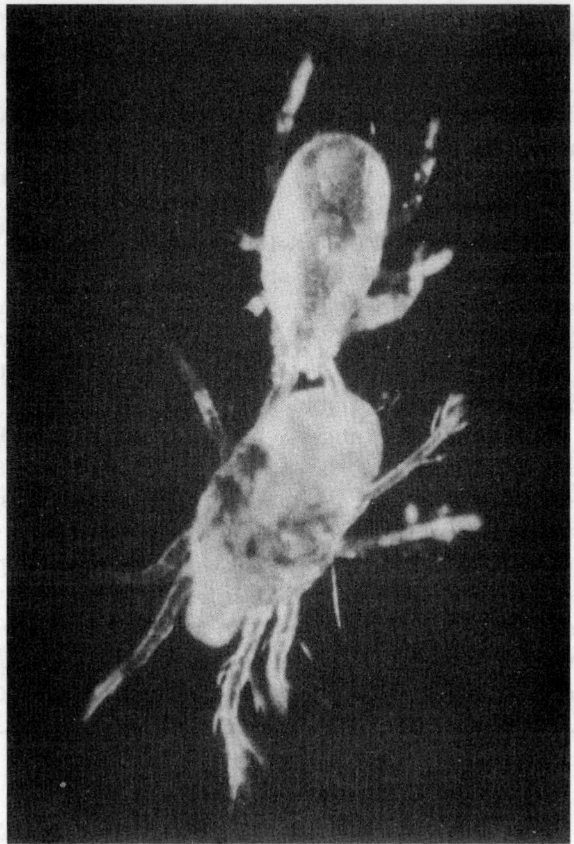

**Figure 18.13** A predator mite, *Amblyseius fallacis* (*top*), about to feed on a plant-feeding mite, the twospotted spider mite, *Tetranychus urticae.* (After Michigan State University Extension Bulletin E-825)

mite pests and reducing the threat of sporadic outbreaks from occasional pests. According to R. J. Prokopy and B. A. Croft (1994), these measures include: (1) application of petroleum oil against overwintering European red mite eggs and overwintering San Jose scale before bloom; (2) application of organophosphate insecticides against plant bugs before bloom and against European apple sawfly and plum curculio at petal fall and again about 2 weeks thereafter; (3) removal of principal unattended host trees (especially apple) within 100 meters of the orchard perimeter to prevent immigration of codling moth females; (4) ringing perimeter orchard trees with odor-baited visual traps to intercept immigrating apple maggot flies; and (5) removal of dropped fruit at harvest to prevent within-orchard buildup of codling moth and apple maggot.

Preventive measures may also be achieved through ground-cover management. In particular, where *Amblyseius fallacis* is an important regulator of European red mite, maintenance of ground cover with some broadleaf weeds is beneficial because the broadleaves serve as hosts for twospotted spider mites on which *Amblyseius fallacis* can feed. After populations of these predators increase in the ground cover, they invade the tree habitat and subdue populations

## BOX 18.3 San Jose Scale

**SPECIES:** *Quadraspidiotus perniciosus* (Comstock) (Hemiptera: Diaspididae)

**DISTRIBUTION:** The San Jose scale is native to China but was introduced from Japan into San Jose, California, about 1880. Currently, the species is distributed throughout southern Canada and the United States.

**IMPORTANCE:** The species is a serious tree fruit pest throughout its range. Both adults and nymphs suck sap from the wood and leaves of trees, reducing vigor and subsequently crop yield. Developing fruits are also attacked, leaving gray, mottled blemishes that reduce quality. Under high infestations, trees may be killed. The scales also attack ornamental trees and shrubs. Humans have aided in transport and dispersal of this pest.

**APPEARANCE:** Adult females are round, about 2 mm in diameter, and covered with waxy scales secreted by the body. Adult males are oval and about 1 mm long. Both males and females are brown to black and have a raised nipple at the top of the scale cover. Nymphal scales are light but turn dark with age. Young nymphs that have not produced scales are commonly called "crawlers"; they are yellow and resemble mites.

**LIFE CYCLE:** Partially grown nymphs (sooty-black stage) overwinter on bark. In spring, nymphs resume feeding and are fully grown during fruit blossom. Two-winged males appear and mate with scale-covered females, which produce active offspring. The crawlers begin feeding on hosts and soon after molt into the recognizable scale form. There are from two to six overlapping generations per year.

---

of the European red mite. Consequently, management recommendations emphasize proper timing of mowing, herbicide applications, and cultivation to reduce deleterious effects on predator-and-prey interactions.

As with other pest management programs, the basis for therapeutics is surveillance of pest populations. Therefore, weekly scouting and inspections are conducted in an orchard-by-orchard program to alert producers of impending problems and to serve as a foundation for decision making. According to J. R. Leeper and J. P. Tette (1981), scouting should be started in the preseason by searching for the San Jose scale (Box 18.3), *Quadraspidiotus perniciosus,* where it occurs, and for eggs of the European red mite. Population densities of these pests above economic thresholds may require oil applications during the prebloom period. After trees break dormancy, scouts check for developing aphid populations and, where redbanded leafroller has been a problem, establish pheromone traps to continuously monitor populations.

Beginning with the tight-cluster to pink stage, inspections are made for young European red mites on interior leaves of the tree, along with predator mites. Other pests that might be anticipated at this time include the tarnished plant bug, *Lygus lineolaris,* and the green fruitworm, *Lithophane antennata.* Most orchards would require an insecticide spray for aphids and mites during this stage.

During the bloom stage, codling moth and oblique-banded leafroller, *Choristoneura rosaceana,* traps are placed in the orchard for surveillance. Checks are also continued on mites, aphids, leafhoppers, and other pests, and petal-fall management strategies are developed from projections of pest status.

In the petal-fall period, most orchards would receive a selective insecticide spray to reduce populations of insect pests that have built up by this time.

This spray may be delayed for up to 10 days, depending on the status of mites and other pests, to coincide with activity of the plum curculio.

During the post-petal-fall period, aphids, mites, and apple maggots may require corrective action, as well as other species as indicated by pheromone traps. If pesticides are used effectively in the earlier part of the season and discriminately later on, most arthropod problems should be kept under control without mite upsets. This is because early-season applications have little effect on *Amblyseius fallacis* because the species inhabits ground vegetation during the heaviest spray periods.

Checks of phytophagous mites are made through the season, however, and a close watch is kept on pest-to-predator ratios. As spider mites increase, pest status is determined from sample counts of pest and predator. Subsequent action is taken based on pest status and specific decision categories. Additionally, computer software designed to analyze data inputs for pest status and to conduct cost/benefit analyses of spider mite situations has been developed by several states.

This example of insect pest management is important for showing the significance of therapeutics in dealing with high-value crops. Previously, scheduled sprays to offer protection have been most common because of pest omnipresence. However, by following the spray-as-needed program advocated by pest management, production costs can be reduced and environmental quality maintained.

## Insect Pest Management in Almonds

One of the best examples of insect pest management in a high-value crop is that in almonds. The United States is the world's leading producer of almonds, with most of the production in the San Joaquin and Sacramento Valleys of California. Because the value of almonds is high, insect pests can cause significant losses, even at relatively low densities.

**Key pests and injury.** The key insect pests of California almonds include the peach twig borer (*Anarsia lineatella*), navel orangeworm (*Amyelois transitella*), San Jose scale (*Quandraspidiotus perniciosus*), web-spinning mites, and ants.

The peach twig borer is a lepidopterous pest that was imported into the United States about 1860. In addition to almonds, it is an important pest of peaches, apricots, cherries, and plums. It causes injury when larvae feed on fruit and tunnel into twigs. Nuts are damaged by peach twig borers in March and can contribute to later navel orangeworm infestations. The species overwinters as larvae and up to four generations develop each year.

The navel orangeworm is also a lepidopterous pest, which damages almond when larvae feed directly on the fruit or nutmeats. First instars bore into the nutmeat and can consume most of the nut as the larvae grow. There may be more than one larva feeding in a nut. When full grown, a larva weaves a cocoon in the nut or between the hulls and shells, and becomes a pupa. Larvae overwinter in mummy nuts (unharvested nuts on the tree or ground). There are three to four generations produced each year.

The San Jose scale is a pest that increases in density over a period of time, perhaps more than 1 year. Nymphs of this scale (sooty-black stage) overwinter and begin active feeding from May through mid-November. This feeding causes branches to die, which decreases nut production, and may even kill trees.

Web-spinning mites in almonds include the Pacific spider mite (*Tetranychus pacificus*), the twospotted spider mite (*Tetranychus urticae*), and the strawberry spider mite (*Tetranychus turkesani*). These mites affect tree vigor by sucking out cell contents of leaves, causing necrosis and, consequently, loss of photosynthetic efficiency. Consistent decline of tree health may cause yield loss with time. These mites are often held in check by predaceous mites such as the western orchard mite (*Galandromus occidentalis*). Upsets of these predators, however, can release the web-spinning mites, which causes losses.

Two ant species, the pavement ant (*Tetramerium caespitum*) and the southern fire ant (*Solenopsis xyloni*), can cause losses by feeding on almond kernels that are drying on the ground during harvest. Particularly, they may be a problem in production systems with drip and microsprinkler irrigation.

### Insect Pest Management in Commerical Orchards

Management of almond insects relies on several tactics, including, sanitation, monitoring (usually by pest control advisors), biological controls, and treatment with insecticides when necessary. The management program is based on a seasonal timeline for initiation of tactics, which includes: (1) *a dormancy period,* (2) *a bloom/postbloom period,* (3) *an in-season period,* (4) *a harvest period,* and (5) *a postharvest period.*

*Dormancy* occurs during the winter months, and management activities in this period are preventive, focusing on the navel orangeworm and sanitation. Mummy nuts may be found on trees at this time that contain overwintering larvae. Mummy nuts are removed to less than two per tree before February 1 and destroyed on the ground naturally from wet weather or by flail mowing. Another preventive activity during this period is to spray a dormant oil for San Jose scale nymphs (low to moderate densities) and eggs of secondary pests, such as European red mites and brown almond mites. A broad-spectrum insecticide may be added to the dormant oil, if necessary, to additionally suppress populations of peach twig borer larvae.

The *bloom/postbloom* period occurs from February through May. Many pest management activities at this time emphasize fungicides for disease prevention. Insect management activities mostly involve monitoring of major pests and accumulating degree days for prediction of hull-split and sprays needed at that time.

The following period, *in-season,* relies on continual monitoring of major insect pests and natural enemies with traps. Based on this monitoring, peach twig borer populations exceeding the economic threshold may be treated in May or hull split, as indicated by degree-day accumulations. Environmentally safe insecticides (for example, microbials, growth regulators, spinosad) may be applied at this time. A hull-split spray may also be necessary for large populations of navel orangeworm. Avoiding water stress at this time helps to suppress web-spinning mites, and predatory mites may be released along with summer oil sprays to prevent losses from these pests. Ants are also monitored at this time by looking for ant mounds and applying baits, if necessary.

The next activity period, *harvest,* uses mostly ecologically based management tactics. Producers are advised to harvest early to avoid a third or fourth generation of egg laying by navel orangeworm and to reduce damage from hull rot. Nuts are picked up promptly to prevent exposure to ants and other pests. Nuts are also sampled at harvest by cracking them and recording damage from peach twig borer, navel orangeworm, ants, and other insects. This information is used to determine likely problems in the following season.

Finally, during the *postharvest* period, analyses are made of potential insect problems and disease pressure, and management plans for the next year are crafted. The insect focus at this time is on San Jose scale, particularly looking for yellow leaves or dead leaves stuck to spurs (growths with buds). This monitoring helps determine the need for spraying dormant oils, sometimes with other insecticides, during the dormant period.

## CONCLUSIONS

All insect pest management programs have elements of both prevention and therapy. Preventive practice is based on a past history of pest problems and relies on a detailed understanding of insect life history and ecology. Preventive tactics are often less expensive to employ and become routine in producing crops.

Therapeutic tactics are integrated into the pest management programs when preventive practices fail to maintain a pest within tolerable economic bounds. Most often, therapeutics involve killing insects outright with conventional broad-spectrum insecticides. Less frequently, selective materials like microbial insecticides and insect growth regulators are practical alternatives. As crop values increase, economic-injury levels fall; consequently, there is a greater need for therapeutic tactics.

The key to durable insect pest management is to understand pest life cycles and ecology, then to integrate several nonchemical preventive tactics with minimal therapeutics. Integration may occur at different levels, but several tactics applied to one pest is the most basic. As more pests are added to the system and greater areas and time periods are encompassed, integration becomes more difficult. Difficulty arises from incompatible tactics and countereffects in the system. Much greater understanding of the entire agroecosystem and total crop management will be required to resolve these conflicts in the future and to allow development of truly integrated pest management systems.

### Further Reading

Benbrook, C. M. 1996. *Pest Management at the Crossroads.* Yonkers, N.Y.: Consumers Union, pp. 125–141, 175–195.

Berryman, A. A. 1986. *Forest Insects—Principles and Practice of Population Management.* New York: Plenum Press, pp. 145–245.

Casagrande, R. A. 1987. The Colorado potato beetle: 125 years of mismanagement. *Bulletin of the Entomological Society of America* 33:142–150.

Croft, B. A., and L. A. Hull. 1983. The orchard as an ecosystem, pp. 19–42. In B. A. Croft and S. C. Hoyt, eds., *Integrated Management of Insect Pests of Pome and Stone Fruits.* New York: Wiley.

Croft, B. A., and W. M. Bode. 1983. Tactics for deciduous fruit IPM, pp. 219–270. In B. A. Croft and S. C. Hoyt, eds., *Integrated Management of Insect Pests of Pome and Stone Fruits.* New York: Wiley.

Frisbie, R. E., H. T. Reynolds, P. L. Adkisson, and R. F. Smith. 1994. Cotton insect pest management, pp. 421–468. In R. L. Metcalf and W. H. Luckmann, eds., *Introduction to Insect Pest Management,* 3rd ed. New York: Wiley.

Frisbie, R. E., and J. K. Walker. 1981. Pest management systems for cotton insects, pp. 187–202. In D. Pimentel, ed., *CRC Handbook of Pest Management in Agriculture,* vol. 3. Boca Raton, Fla: CRC Press.

Frisbie, R. E., and P. L. Adkisson, eds. 1985. *Integrated Pest Management on Major Agricultural Systems.* College Station, Tex.: Texas Agricultural Experiment Station MP–1616.

Gray, M. E., and W. H. Luckmann. 1994. Integrating the cropping system for corn insect pest management, pp. 507–541. In R. L. Metcalf and W. H. Luckmann, eds., *Introduction to Insect Pest Management,* 3rd ed. New York: Wiley.

Hare, J. D. 1990. Ecology and management of the Colorado potato beetle. *Annual Review of Entomology* 35:81–100.

Leeper, J. R., and J. P. Tette. 1981. Pest management systems for apple insects, pp. 243–255. In D. Pimentel, ed., *CRC Handbook of Pest Management in Agriculture,* vol. 3. Boca Raton, Fla.: CRC Press.

Mason, C. E., M. E. Rice, D. D. Calvin, J. W. Van Duyn, W. B. Showers, W. D. Hutchison, J. F. Witkowski, R. A. Higgins, D. W. Onstad, and G. P. Dively. 1996. European corn borer development and management. In *North Central Regional Publication 327.* Ames: Iowa State University.

Norton, G. W., and J. Mullen. 1994. Economic Evaluation of Integrated pest management Programs. Virginia Cooperative Extension Service Publication 448-120.

Pfadt, R. E. 1985. Insect pests of cotton, pp. 339–370. In R. E. Pfadt, ed., *Fundamentals of Applied Entomology,* 4th ed. New York: Macmillan.

Prokopy, R. J., and B. A. Croft. 1994. Apple insect pest management, pp. 543–585. In R. L. Metcalf and W. H. Luckmann, eds., *Introduction to Insect Pest Management,* 3rd ed. New York: Wiley.

Rabb, R. L., G. K. Defoliart, and G. G. Kennedy. 1984. An ecological approach to managing insect populations, pp. 691–728. In C. B. Huffaker and R. L. Rabb, eds., *Ecological Entomology.* New York: Wiley.

Showers, W. B., J. F. Witkowski, C. E. Mason, D. D. Calvin, R. A. Higgins, and G. P. Dively. 1989. European corn borer development and management. In *North Central Regional Publication 327.* Ames, Iowa: Iowa State University.

Tanigoshi, L. K., S. C. Hoyt, and B. A. Croft. 1983. Basic biology and management components for insect IPM, pp. 153–202. In B. A. Croft and S. C. Hoyt, eds., *Integrated Management of Insect Pests of Pome and Stone Fruits.* New York: Wiley.

Vandeman, A., J. Fernandez-Cornejo, S. Jans, and B. H. Lin. 1994. *Adoption of Integrated Pest Management in U.S. Agriculture.* U.S. Economic Research Service Agriculture Information Bulletin 707.

Wintersteen, W. K., and L. G. Higley. 1993. Advancing IPM systems in corn and soybeans, pp. 9–32. In A. R. Leslie and G. W. Cuperus, eds., *Successful Implementation of Integrated Pest Management for Agricultural Crops.* Boca Raton, Fla.: Lewis Publishers.

## Favorite Web Sites

http://www.ceris.purdue.edu/napis/pests/index.html

Site allows user to search for information on insects and diseases. Selections are made from an alphabetized list, and photos, fact sheets, and links to other sites are given.

http://www.ipm.iastate.edu/ipm/nipmn/

Site of the National Integrated Pest Management Network. Presents the latest information on pest management by state and region of the United States through appropriate links.

http://ipmworld.umn.edu/aphidalert/CPB~DWR.htm

Site gives in-depth treatment of Colorado potato beetle biology and management.

http://anrcatalog.ucdavis.edu/pdf/21619.pdf

Site presents a guide to environmentally responsible pest management practices for almond production in California.

# APPENDIX 1

# Key to the Orders of Hexapoda

The following key is slightly modified from that of Steyskal, Murphy, and Hoover (1986, Agricultural Research Service, Miscellaneous Publication 1443) and is useful in identifying most insect specimens. For uncommon and aberrant forms that do not fit the couplets, a more comprehensive key such as that found in *Borror and DeLong's Introduction to the Study of Insects,* 7th ed. by C. A. Triplehorn and N. F. Johnson (2005, Belmont, Ca: Thomson Brooks/Cole) should be used. In certain instances suborder and superfamily names are listed, which indicates that some specialists believe that the group should have ordinal status. The new order Mantophasmatodea, from Africa, is not presented in this key.

1. Wings present and well-developed ........ 2
   Wings absent or unsuitable for flight (wingless adults and immature stages)...... 32
2. Forewings (on mesothorax) wholly or partly horny, leathery, or otherwise strongly differing from wholly membranous hind wings; hind wings sometimes lacking ..... 3
   Forewings wholly membranous, at least at base ........ 11
3. Forewings (wing covers, elytra) uniformly horny, without apparent veins; hind wings, if present, folded both lengthwise and crosswise, hidden under forewings when at rest; mouth with mandibles ........ 4
   Forewings (hemelytra or tegmina) with veins; hind wings not folded crosswise ..... 5
4. End of abdomen with heavy, forcepslike cerci; wings short, leaving most of abdomen exposed; hind wings very delicate, almost circular, radially folded ... DERMAPTERA
   Without such cerci; wings usually covering most of abdomen, or if forewings short, then hind wings elongate, but sometimes absent ........ COLEOPTERA
5. Mouthparts fitted for sucking, forming jointed beak ........ HEMIPTERA
   Mouthparts fitted for chewing, with mandibles moving sideways........ 6
6. Hind wings not folded, similar to forewings; both wings with thickened, very short basal part separated from rest of wing by suture, so that most of wing can easily be broken off; social insects living in colonies ........ ISOPTERA
   Hind wings folded fanwise, broader than forewings; wings without breaking suture ........ 7
7. Minute insects, usually less than 6 mm long; forewings small, clublike; antennae short, with few segments; parasites of other insects........ STREPSIPTERA males
   Usually large or moderately large insects; forewings usually flat and long; antennae usually lengthened and slender, many-segmented ........ 8
8. Hindfemora enlarged, modified for jumping ........ ORTHOPTERA
   Hindfemora not enlarged, similar to other legs ........ 9

9. Cerci short, unsegmented; body usually elongate and slender (sticklike) .......... PHASMATODEA
Cerci long or short but segmented; body usually not sticklike .......... 10

10. Shape oval; all legs similar, adapted to walking.......... BLATTODEA
Shape elongate; fore legs raptorial (with spines and modified for grasping prey).......... MANTODEA

11. With only two well-developed wings, forepair functional, hindpair not winglike and sometimes small and clublike .......... 12
With four wings, hindpair sometimes small but flat or straplike, not clublike .......... 14

12. Mouthparts forming a sucking or lapping proboscis, rarely rudimentary or virtually lacking; hind wings replaced by clublike halteres; abdomen without tail filaments.......... DIPTERA
Mouthparts not functional; hind wings not formed into clublike halteres; abdomen with tail filaments .......... 13

13. Without halteres; antennae inconspicuous, with small scape and pedicel and bristlelike flagellum; forewings with numerous crossveins (few mayflies).......... EPHEMEROPTERA
With hind wings reduced to halterlike structures; antennae evident, not bristlelike; venation of forewings apparently reduced to one forked vein (male scale insects) .......... HEMIPTERA

14. Wings long, narrow, almost veinless, with long marginal fringes; tarsi one- or two-segmented, with swollen tip; mouthparts conical, fitted for piercing and sucking plant tissues (minute insects).......... THYSANOPTERA
Wings broader; if fringed, fringe not longer than width of wing; veins usually conspicuous and at least one crossvein present; tarsi with more than two segments and tip not swollen .......... 15

15. Wings, legs, and body at least in part with elongate, flattened scales and often also with hairs; wings hyaline (glasslike) under color pattern formed by scales; mouthparts consisting of tongue (rarely rudimentary) formed of helically coiled tube; mandibles (fitted for chewing) present only in a few families of small moths with wingspread not over 12 mm.......... LEPIDOPTERA
Wings, legs, and body not covered with scales, although a few scales sometimes present; color pattern of wing involving wing membrane and/or hair.......... 16

16. Hind wings with broad anal area, plaited when wings folded, usually larger than forewings; antennae prominent .......... 17
Hind wings without plaited anal area, not larger than forewings; antennae often inconspicuous, bristlelike .......... 19

17. Tarsi three-segmented; cerci well developed, usually long and many-segmented .......... PLECOPTERA
Tarsi five-segmented; cerci not prominent .......... 18

18. Wings with several subcostal crossveins, surface without hairs or scales .......... NEUROPTERA, suborder MEGALOPTERA
Wings without subcostal crossveins, surface with hairs or scales .... TRICHOPTERA

19. Antennae short, bristlelike; wings with numerous crossveins forming overall network; mouthparts with mandibles close to eyes .......... 20
Antennae larger or wings with few crossveins or mouthparts at end of beak .......... 21

20. Hind wings much smaller than forewings; abdomen with long tail filaments .......... EPHEMEROPTERA
Hind wings very similar to forewings; abdomen without long tail filaments .......... ODONATA

21. Head extended into beak with mandibles at end; hind wings not folded; wings usually with color pattern and numerous crossveins; male genitalia usually swollen, turned forward, and with strong pair of forceps .......... MECOPTERA
Head not extended into beak; male genitalia without conspicuous forceps .......... 22

22. Mouthparts (sometimes lacking) consisting of proboscis without chewing mandibles; cerci lacking; wings with few crossveins . . . . . . . . . . . . 23
Mouthparts including mandibles fitted for chewing . . . . . . . . . . . . 24
23. Wings covered with scales forming color pattern; antennae many-segmented; mouthparts (when present) consisting of helically coiled haustellum (tongue) . . . . . . . . . . . . LEPIDOPTERA
Wings not covered with scales; antennae with few segments; mouthparts consisting of segmented piercing beak . . . . . . . . . . . . HEMIPTERA
24. Body and wings covered with whitish powder; wings bordered anteriorly by very narrow cell without row of crossveins; insects less than 5 mm long . . . . . . NEUROPTERA, suborder PLANIPENNIA (Coniopterygidae)
Body and wings not covered with whitish powder; otherwise differing. . . . . . . . . . . . 25
25. Tarsi five-segmented . . . . . . . . . . . . 26
Tarsi with four or fewer segments . . . . . . . . . . . . 29
26. Prothorax fused with mesothorax; hind wings smaller than forewings, latter with no more than twenty cells; abdomen often constricted at base . . . . . HYMENOPTERA
Prothorax more or less free, sometimes long; forewings and hind wings approximately equal in size, with more than twenty cells . . . . . . . . . . . . 27
27. Prothorax much longer than head, cylindrical; forelegs similar to others, not enlarged . . . . . . . . . . . . NEUROPTERA, suborder RAPHIDIODEA
Prothorax not longer than head; if longer, then forelegs enlarged for grasping prey . . . . . . . . . . . . 28
28. Costal cell with many crossveins . . . . . . . . NEUROPTERA, suborder PLANIPENNIA
Costal cell without series of crossveins. . . . . . . . . . . . MECOPTERA
29. Wings normally equal in size, hind wings occasionally larger; tarsi three- or four-segmented . . . . . . . . . . . . 30
Hind wings smaller than forewings; tarsi two- or three-segmented . . . . . . . . . . . . 31
30. Tarsi apparently four-segmented; forebasitarsi unswollen; wings dehiscent (see also couplet 7) . . . . . . . . . . . . ISOPTERA
Tarsi three-segmented, forebasitarsi swollen. . . . . . . . . . . . EMBIOPTERA
31. Cerci absent; wings remaining attached; antennae slender, with thirteen or more segments. . . . . . . . . . . . PSOCOPTERA
Cerci evident, although short, ending in bristle; wings shed eventually; antennae with nine beadlike segments; seldom-encountered small insects. . . . . . ZORAPTERA
32. Body with more or less distinct head, thorax, and abdomen; with jointed legs and ability to move about . . . . . . . . . . . . 33
Without distinctly separate body parts, or without legs, or not able to move about . . . . . . . . . . . . 75
33. Parasites of warm-blooded animals . . . . . . . . . . . . 34
Not parasites of warm-blooded animals . . . . . . . . . . . . 38
34. Body strongly flattened sideways; mouth a sharp, downturned beak (jumping insects) . . . . . . . . . . . . SIPHONAPTERA
Body flattened dorsoventrally or maggots of more or less cylindrical form . . . . . . . . 35
35. Mouthparts with mandibles for chewing, directed forward; generally oval insects with more or less triangular head; parasites of birds and mammals . . . . . . . . . . . . PHTHIRAPTERA
Mouthparts in form of beak for piercing and sucking . . . . . . . . . . . . 36
36. Antennae inserted in pits, not visible from above (also maggot-shaped larvae without antennae) . . . . . . . . . . . . DIPTERA (a few families)
Antennae present, although short, not in pits . . . . . . . . . . . . 37
37. Beak not jointed; tarsi forming hook for grasping hairs of host; parasites remaining on host . . . . . . . . . . . . PHTHIRATERA
Beak jointed; tarsi not hooked; parasites not remaining on host (bedbugs and related insects) . . . . . . . . . . . . HEMIPTERA

38. Aquatic, usually breathing by gills; larval and some pupal forms . . . . . . . . . . . . . . . . 39
Terrestrial, breathing by spiracles or rarely without breathing organs . . . . . . . . . . . 47
39. Mouth forming a strong, pointed, downcurved beak. . . . . . . . Immature HEMIPTERA
Mouth with mandibles. . . . . . . . . . . . . . . . . . . . . . . . . . . . . . . . . . . . . . . . . . . . 40
40. Mandibles extending straight forward, united with maxillae to form piercing jaws . . . . . . . . . . . . . . . . . . . . . . . . . . . . . . . Some larval NEUROPTERA
Mandibles moving sideways, forming biting jaws . . . . . . . . . . . . . . . . . . . . . . . . . 41
41. Living in case formed of sand, pebbles, leaves, twigs, and so on; usually with external tracheae serving as gills. . . . . . . . . . . . . Larvae of some TRICHOPTERA
Not living in case . . . . . . . . . . . . . . . . . . . . . . . . . . . . . . . . . . . . . . . . . . . . . 42
42. Abdomen with lateral organs serving as gills (a few larval Trichoptera and Coleoptera key out here also) . . . . . . . . . . . . . . . . . . . . . . . . . . . . . . . . . . . 43
Abdomen without external gills (some larval Trichoptera will key out here also) . . . 44
43. Abdomen with two or three long tail filaments. . . . . . . Naiads of EPHEMEROPTERA
Abdomen with short end processes (larvae of some Trichoptera will run to this point) . . . . . . . . . . . . . . . . Larvae of NEUROPTERA, suborder MEGALOPTERA
44. Lower lip (labium) folded backward, extensible, and furnished with pair of jawlike hooks . . . . . . . . . . . . . . . . . . . . . . . . . . . . . . . . . . . . . . . Naiads of ODONATA
Labium not so constructed . . . . . . . . . . . . . . . . . . . . . . . . . . . . . . . . . . . . . . . . 45
45. Abdomen with nonjointed false legs (prolegs) in pairs on several segments. . . . . . . . . . . . . . . . . . . . . . . . . . . . . Few larvae of LEPIDOPTERA
Abdomen without prolegs . . . . . . . . . . . . . . . . . . . . . . . . . . . . . . . . . . . . . . . . 46
46. Thorax in three loosely united divisions; antennae and tail filaments long and slender . . . . . . . . . . . . . . . . . . . . . . . . . . . . . . . . . . . Naiads of PLECOPTERA
Thoracic divisions without constrictions; antennae and tail filaments short (larvae of some aquatic Diptera and Trichoptera also run to this point) . . . . . . . . . . . . . . . . . . . . . . . . . . . . . . . . . . Larvae of COLEOPTERA
47. Mouthparts retracted into head and hardly, or not at all, visible; underside of abdomen with appendages; very delicate, small, or minute insects, sometimes without antennae . . . . . . . . . . . . . . . . . . . . . . . . . . . . . . . . . . . . . . . . . . 48
External mouthparts conspicuous; antennae always present; underside of abdomen rarely with appendages . . . . . . . . . . . . . . . . . . . . . . . . . . . . . . . . . . . . . . . 50
48. Head pear shaped, without antennae; abdomen without long cerci, pincers, jumping apparatus, or basal ventral sucker . . . . . . . . . . . . . . . . . . . . . . . PROTURA
Head usually not pear shaped, antennae conspicuous; abdomen with long cerci, pincers, or basal ventral sucker . . . . . . . . . . . . . . . . . . . . . . . . . . . . . . . . . . 49
49. Abdomen consisting of six or fewer segments, with forked sucker at base below and usually with conspicuous jumping apparatus near end, but without conspicuous long cerci or pincers . . . . . . . . . . . . . . . . . . . . . . . . . COLLEMBOLA
Abdomen with more than eight evident segments and ending in long, many-jointed cerci or strong pincers; eyes and ocelli lacking . . . . . . . . . . . . . DIPLURA
50. Mouthparts with mandibles fitted for chewing . . . . . . . . . . . . . . . . . . . . . . . . . . 51
Mouthparts in form of proboscis fitted for sucking . . . . . . . . . . . . . . . . . . . . . . . . 72
51. Body usually covered with scales; abdomen with three prominent tail filaments and at least two pairs of ventral appendages (styles) . . . . . . . . . . . . . . . . . . . . . . . . . . . . THYSANURA and MICROCORYPHIA
Body never covered with scales; abdomen never with three tail filaments nor ventral styles. . . . . . . . . . . . . . . . . . . . . . . . . . . . . . . . . . . . . . . . . . . . . 52
52. Abdomen bearing pairs of false legs (prolegs) beneath, not jointed, and differing from true legs on thorax, which is not distinctly separated from abdomen; body caterpillarlike; larval forms . . . . . . . . . . . . . . . . . . . . . . . . . . . . . . . . . . . . . 53
Underside of abdomen without legs or prolegs . . . . . . . . . . . . . . . . . . . . . . . . . . 55
53. Prolegs five pairs or less, none on first, second, or seventh segments; prolegs tipped with many tiny hooklets and rarely present on 2nd and 7th segments . . . . . . . . . . . . . . . . . . . . . . . . . . . . . . . . . . Larvae of most LEPIDOPTERA

Prolegs six to ten pairs, not tipped with tiny hooks; one pair of prolegs on second segment .................................................. 54
54. Head with single ocellus on each side .............. Larvae of some HYMENOPTERA
Head with several ocelli on each side .................... Larvae of MECOPTERA
55. Antennae long and distinct .................................................. 56
Antennae short; larval forms .................................................. 69
56. Abdomen ending in strong pincerlike forceps; prothorax free ....... DERMAPTERA
Abdomen not ending in forceps .................................................. 57
57. Abdomen strongly constricted at base; prothorax fused with mesothorax .................................................. HYMENOPTERA
Abdomen not strongly constricted at base, broadly joined to thorax .............. 58
58. Head produced into beak with mandibles at end .................... MECOPTERA
Head not produced into beak .................................................. 59
59. Very small insects with soft body; tarsi two- or three-segmented ................ 60
Usually much larger insects; tarsi usually with more than five segments, or body hard and cerci absent .................................................. 61
60. Cerci absent .................................................. PSOCOPTERA
Cerci of single segment, prominent .................................................. ZORAPTERA
61. Hind legs fitted for jumping, femora enlarged; wing pads of immatures inverted, hindpads overlapping forepads .................................................. ORTHOPTERA
Hind legs not enlarged for jumping; wing pads, if present, in normal position ..... 62
62. Prothorax much longer than mesothorax; front legs modified for grasping prey (raptorial) .................................................. MANTODEA
Prothorax not greatly lengthened .................................................. 63
63. Without cerci; body often hard shelled; antennae usually with eleven segments .................................................. COLEOPTERA
Cerci present; antennae usually with more than fifteen segments .............. 64
64. Cerci with more than three segments .................................................. 65
Cerci with one to three segments .................................................. 67
65. Body flattened, oval; head turned down and backward .............. BLATTODEA
Body elongate; head nearly horizontal .................................................. 66
66. Cerci long; ovipositor evident, hardened; tarsi five-segmented .................................................. GRYLLOBLATTODEA
Cerci short; ovipositor lacking; tarsi four-segmented .................. ISOPTERA
67. Tarsi five-segmented (three-segmented only in *Timema,* in Pacific Coast States, most antennal segments several times as long as wide); body usually slender and long .................................................. PHASMATODEA
Tarsi two- or three-segmented; antennal segments beadlike .................... 68
68. Front tarsi with first segment swollen, containing silk-spinning gland for producing web in which insects live; cerci conspicuous .................... EMBIOPTERA
Front tarsi not so, not producing silk; cerci inconspicuous .............. ISOPTERA
69. Body cylindrical, caterpillarlike .................................................. 70
Body more or less depressed, not caterpillarlike .................................. 71
70. Head with six ocelli on each side; antennae inserted in membranous area at base of mandibles .............................. Some larvae of LEPIDOPTERA
Head with more than six ocelli on each side; third pair of legs distinctly larger than first pair .......................... MECOPTERA (larvae of Boreidae)
71. Mandibles united with maxillae to form sucking jaws ...................... Larvae of NEUROPTERA, suborder PLANIPENNIA
Mandibles nearly always separate from maxillae..........................................
Larvae of some COLEOPTERA; NEUROPTERA, suborder RAPHIDIODEA; STREPSIPTERA; DIPTERA
72. Body densely clothed with scales and hairs; proboscis, if present, coiled under head .................................................. LEPIDOPTERA
Body bare, with scattered hairs or waxy coating .................................. 73

73. Last tarsal segment bladderlike, without claws; mouth a triangular unsegmented beak; very small insects .......................... THYSANOPTERA
Tarsi not swollen at tip, with distinct claws .......................... 74
74. Prothorax small, hidden when viewed from above.......................... DIPTERA
Prothorax evident when viewed from above.......................... HEMIPTERA
75. Legless grubs or maggots, moving about by squirming ......... larvae of DIPTERA
.......... HYMENOPTERA; LEPIDOPTERA; COLEOPTERA; SIPHONAPTERA;
........... STREPSIPTERA (in body of wasps or bees with flattened head exposed)
Forms legless or with a single claw .......................... 76
76. Small forms with little resemblance to most insects, with filamentous mouth-parts inserted in plant tissue; usually covered with waxy scale, powder, or cottony tufts .......................... HEMIPTERA
Body unable to move or only able to bend from side to side, enclosed in tight skin, sometimes wholly covering body or sometimes with appendages free, but rarely movable; sometimes enclosed in cocoon (pupae) .......................... 77
77. Legs, wings, and so on, more or less free from body; biting mouthparts visible..... 78
Skin enclosing body holding appendages tightly against body; mouthparts evident as proboscis, without mandibles.......................... 80
78. Prothorax small, fused with mesothorax; sometimes enclosed in thin cocoon .......................... Pupae of HYMENOPTERA
Prothorax larger and not fused with mesothorax.......................... 79
79. Wing cases with few or no veins .......................... Pupae of COLEOPTERA
Wing cases with several branched veins .......................... Pupae of NEUROPTERA
80. Proboscis usually long, rarely absent; wing cases four; often in cocoon .......................... Pupae of LEPIDOPTERA
Proboscis usually short; wing cases two; rarely in cocoon, but often tightly enclosed in hardened last larval skin.......................... Pupae of DIPTERA

APPENDIX 2

# List of Some Insects and Related Species Alphabetized by Common Name

## A

| | | |
|---|---|---|
| abbreviated wireworm | *Hypolithus abbreviatus* (Say) | COLEOPTERA: Elateridae |
| acacia psyllid | *Psylla uncatoides* (Ferris & Klyver) | HEMIPTERA: Psyllidae |
| achemon sphinx | *Eumorpha achemon* (Drury) | LEPIDOPTERA: Sphingidae |
| acuminate scale | *Kilifia acuminata* (Signoret) | HEMIPTERA: Coccidae |
| acute-angled fungus beetle | *Gryptophagus acutangulus* (Gyllenhal) | COLEOPTERA: Cryptophagidae |
| aerial yellowjacket | *Dolichovespula arenaria* (Fabricius) | HYMENOPTERA: Vespidae |
| African mole cricket | *Gryllotalpa africana* Palisot de Beauvois | ORTHOPTERA: Gryllotalpidae |
| ailanthus webworm | *Atteva punctella* (Cramer) | LEPIDOPTERA: Yponomeutidae |
| alder bark beetle | *Alniphagus aspericollis* (LeConte) | COLEOPTERA: Scolytidae |
| alder flea beetle | *Altica ambiens* LeConte | COLEOPTERA: Chrysomelidae |
| alder spittlebug | *Clastoptera obtusa* (Say) | HEMIPTERA: Cercopidae |
| alfalfa blotch leafminer | *Agromyza frontella* (Rondani) | DIPTERA: Agromyzidae |
| alfalfa caterpillar | *Colias eurytheme* Boisduval | LEPIDOPTERA: Pieridae |
| alfalfa gall midge | *Asphondylia websteri* Felt | DIPTERA: Cecidomyiidae |
| alfalfa leafcutting bee | *Megachile rotundata* (Fabricius) | HYMENOPTERA: Megachilidae |
| alfalfa leaftier | *Dichomeris ianthes* (Meyrick) | LEPIDOPTERA: Gelechiidae |
| alfalfa looper | *Autographa californica* (Speyer) | LEPIDOPTERA: Noctuidae |
| alfalfa plant bug | *Adelphocoris lineolatus* (Goeze) | HEMIPTERA: Miridae |
| alfalfa seed chalcid | *Bruchophagus roddi* (Gussakovsky) | HYMENOPTERA: Eurytomidae |
| alfalfa snout beetle | *Otiorhynchus ligustici* (Linnaeus) | COLEOPTERA: Curculionidae |
| alfalfa webworm | *Loxostege commixtalis* (Walker) | LEPIDOPTERA: Crambidae |
| alfalfa weevil | *Hypera postica* (Gyllenhal) | COLEOPTERA: Curculionidae |
| alkali bee | *Nomia melanderi* Cockerell | HYMENOPTERA: Halictidae |
| Allegheny mound ant | *Formica exsectoides* Forel | HYMENOPTERA: Formicidae |
| Allegheny spruce beetle | *Dendroctonus punctatus* LeConte | COLEOPTERA: Scolytidae |
| almond moth | *Cadra cautella* (Walker) | LEPIDOPTERA: Crambidae |
| aloe mite | *Eriophyes aloinis* Keifer | ACARI: Eriophyidae |
| American aspen beetle | *Gonioctena americana* (Schaeffer) | COLEOPTERA: Chrysomelidae |
| American black flour beetle | *Tribolium audax* Halstead | COLEOPTERA: Tenebrionidae |
| American cockroach | *Periplaneta americana* (Linnaeus) | BLATTODEA: Blattidae |
| American dagger moth | *Acronicta americana* (Harris) | LEPIDOPTERA: Noctuidae |
| American dog tick | *Dermacentor variabilis* (Say) | ACARI: Ixodidae |
| American grasshopper | *Schistocerca americana* (Drury) | ORTHOPTERA: Acrididae |
| American hornet moth | *Sesia tibialis* (Harris) | LEPIDOPTERA: Sesiidae |

| | | |
|---|---|---|
| American house dust mite | *Dermatophagoides farinae* Hughes | ACARI: Epidermoptidae |
| American plum borer | *Euzophera semifuneralis* (Walker) | LEPIDOPTERA: Crambidae |
| American spider beetle | *Mezium americanum* (Laporte) | COLEOPTERA: Ptinidae |
| Angora goat biting louse | *Bovicola crassipes* (Rudow) | PHTHIRAPTERA: Trichodectidae |
| Angoumois grain moth | *Sitotroga cerealella* (Olivier) | LEPIDOPTERA: Gelechiidae |
| angularwinged katydid | *Microcentrum retinerve* (Burmeister) | ORTHOPTERA: Tettigoniidae |
| angulate leafhopper | *Acinopterus angulatus* Lawson | HEMIPTERA: Cicadellidae |
| apple aphid | *Aphis pomi* DeGeer | HEMIPTERA: Aphididae |
| apple bark borer | *Synanthedon pyri* (Harris) | LEPIDOPTERA: Sesiidae |
| apple barkminer | *Marmara elotella* (Busck) | LEPIDOPTERA: Gracillariidae |
| apple blotch leafminer | *Phyllonorycter crataegella* (Clemens) | LEPIDOPTERA: Gracillariidae |
| apple curculio | *Anthonomus quadrigibbus* (Say) | COLEOPTERA: Curculionidae |
| apple flea weevil | *Rhynchaenus pallicornis* (Say) | COLEOPTERA: Curculionidae |
| apple fruit moth | *Argyresthia conjugella* Zeller | LEPIDOPTERA: Argyresthiidae |
| apple fruitminer | *Marmara pomonella* Busck | LEPIDOPTERA: Gracillariidae |
| apple grain aphid | *Rhopalosiphum fitchii* (Sanderson) | HEMIPTERA: Aphididae |
| apple leafhopper | *Empoasca maligna* (Walsh) | HEMIPTERA: Cicadellidae |
| apple maggot | *Rhagoletis pomonella* (Walsh) | DIPTERA: Tephritidae |
| apple mealybug | *Phenacoccus aceris* (Signoret) | HEMIPTERA: Pseudococcidae |
| apple red bug | *Lygidea mendax* Reuter | HEMIPTERA: Miridae |
| apple rust mite | *Aculus schlechtendali* (Nalepa) | ACARI: Eriophyidae |
| apple seed chalcid | *Torymus varians* (Walker) | HYMENOPTERA: Torymidae |
| apple sucker | *Psylla mali* (Schmidberger) | HEMIPTERA: Psyllidae |
| apple twig beetle | *Hypothenemus obscurus* (Fabricius) | COLEOPTERA: Scolytidae |
| apple twig borer | *Amphicerus bicaudatus* (Say) | COLEOPTERA: Bostrichidae |
| apple-and-thorn skeletonizer | *Choreutis pariana* (Clerck) | LEPIDOPTERA: Choreutidae |
| appleleaf skeletonizer | *Psorosina hammondi* (Riley) | LEPIDOPTERA: Crambidae |
| appleleaf trumpet miner | *Tischeria malifoliella* Clemens | LEPIDOPTERA: Tischeriidae |
| araucaria aphid | *Neophyllaphis araucariae* Takahashi | HEMIPTERA: Aphididae |
| arborvitae leafminer | *Argyresthia thuiella* (Packard) | LEPIDOPTERA: Argyresthiidae |
| arborvitae weevil | *Phyllobius intrusus* Kono | COLEOPTERA: Curculionidae |
| Argentine ant | *Iridomyrmex humilis* (Mayr) | HYMENOPTERA: Formicidae |
| argus tortoise beetle | *Chelymorpha cassidea* (Fabricius) | COLEOPTERA: Chrysomelidae |
| army cutworm | *Euxoa auxiliaris* (Grote) | LEPIDOPTERA: Noctuidae |
| armyworm | *Pseudaletia unipuncta* (Haworth) | LEPIDOPTERA: Noctuidae |
| artichoke plume moth | *Platyptilia carduidactyla* (Riley) | LEPIDOPTERA: Pterophoridae |
| ash plant bug | *Tropidosteptes amoenus* Reuter | HEMIPTERA: Miridae |
| ashgray blister beetle | *Epicauta fabricii* (LeConte) | COLEOPTERA: Meloidae |
| Asiatic garden beetle | *Maladera castanea* (Arrow) | COLEOPTERA: Scarabaeidae |
| Asiatic oak weevil | *Cyrtepistomus castaneus* (Roelofs) | COLEOPTERA: Curculionidae |
| Asiatic rice borer | *Chilo suppressalis* (Walker) | LEPIDOPTERA: Crambidae |
| Asiatic rose scale | *Aulacaspis rosarum* Borchsenius | HEMIPTERA: Diaspididae |
| asparagus beetle | *Crioceris asparagi* (Linnaeus) | COLEOPTERA: Chrysomelidae |
| asparagus miner | *Ophiomyia simplex* (Loew) | DIPTERA: Agromyzidae |
| asparagus spider mite | *Schizotetranychus asparagi* (Oudemans) | ACARI: Tetranychidae |
| aspen blotchminer | *Phyllonorycter tremuloidiella* (Braun) | LEPIDOPTERA: Gracillariidae |
| aspen leaf beetle | *Chrysomela crotchi* Brown | COLEOPTERA: Chrysomelidae |
| aster leafhopper | *Macrosteles fascifrons* (Stål) | HEMIPTERA: Cicadellidae |
| aster leafminer | *Calycomyza humeralis* (Roser) | DIPTERA: Agromyzidae |
| Australian cockroach | *Periplaneta australasiae* (Fabricius) | BLATTODEA: Blattidae |
| Australian fern weevil | *Syagrius fulvitarsis* Pascoe | COLEOPTERA: Curculionidae |
| Australian mantid | *Tenodera australasiae* (Leach) | MANTODEA: Mantidae |
| Australian rat flea | *Xenopsylla vexabilis* Jordan | SIPHONAPTERA: Pulicidae |
| Australian spider beetle | *Ptinus ocellus* Brown | COLEOPTERA: Ptinidae |
| Australian pine borer | *Chrysobothris tranquebarica* (Gmelin) | COLEOPTERA: Buprestidae |
| avocado brown mite | *Oligonychus punicae* (Hirst) | ACARI: Tetranychidae |
| avocado red mite | *Oligonychus yothersi* (McGregor) | ACARI: Tetranychidae |
| avocado whitefly | *Trialeurodes floridensis* (Quaintance) | HEMIPTERA: Aleyrodidae |
| azalea bark scale | *Eriococcus azaleae* Comstock | HEMIPTERA: Eriococcidae |
| azalea lace bug | *Stephanitis pyrioides* (Scott) | HEMIPTERA: Tingidae |

| | | |
|---|---|---|
| azalea leafminer | *Caloptilia azaleella* (Brants) | LEPIDOPTERA: Gracillariidae |
| azalea plant bug | *Rhinocapsus vanduzeei* Uhler | HEMIPTERA: Miridae |
| azalea whitefly | *Pealius azaleae* (Baker & Moles) | HEMIPTERA: Aleyrodidae |

## B

| | | |
|---|---|---|
| bagworm | *Thyridopteryx ephemeraeformis* (Haworth) | LEPIDOPTERA: Psychidae |
| Bahaman swallowtail | *Papilio andraemon bonhotei* Sharpe | LEPIDOPTERA: Papilionidae |
| baldcypress coneworm | *Dioryctria pygmaella* Ragonot | LEPIDOPTERA: Crambidae |
| baldfaced hornet | *Dolichovespula maculata* (Linnaeus) | HYMENOPTERA: Vespidae |
| balsam fir sawfly | *Neodiprion abietis* (Harris) | HYMENOPTERA: Diprionidae |
| balsam fir sawyer | *Monochamus marmorator* Kirby | COLEOPTERA: Cerambycidae |
| balsam gall midge | *Paradiplosis tumifex* Gagné | DIPTERA: Cecidomyiidae |
| balsam shootboring sawfly | *Pleroneura brunneicornis* Rohwer | HYMENOPTERA: Xyelidae |
| balsam twig aphid | *Mindarus abietinus* Koch | HEMIPTERA: Aphididae |
| balsam woolly adelgid | *Adelges piceae* (Ratzeburg) | HEMIPTERA: Adelgidae |
| bamboo borer | *Chlorophorus annularis* (Fabricius) | COLEOPTERA: Cerambycidae |
| bamboo mealybug | *Chaetococcus bambusae* (Maskell) | HEMIPTERA: Pseudococcidae |
| bamboo powderpost beetle | *Dinoderus minutus* (Fabricius) | COLEOPTERA: Bostrichidae |
| bamboo spider mite | *Schizotetranychus celarius* (Banks) | ACARI: Tetranychidae |
| banana aphid | *Pentalonia nigronervosa* Coquerel | HEMIPTERA: Aphididae |
| banana root borer | *Cosmopolites sordidus* (Germar) | COLEOPTERA: Curculionidae |
| banana skipper | *Pelopidas thrax* (Linnaeus) | LEPIDOPTERA: Hesperiidae |
| banded alder borer | *Rosalia funebris* Motschulsky | COLEOPTERA: Cerambycidae |
| banded ash clearwing | *Podosesia aureocincta* Purrington & Nielsen | LEPIDOPTERA: Sesiidae |
| banded cucumber beetle | *Diabrotica balteata* LeConte | COLEOPTERA: Chrysomelidae |
| banded greenhouse thrips | *Hercinothrips femoralis* (O. M. Reuter) | THYSANOPTERA: Thripidae |
| banded hickory borer | *Knulliana cincta* (Drury) | COLEOPTERA: Cerambycidae |
| banded sunflower moth | *Cochylis hospes* Walsingham | LEPIDOPTERA: Cochylidae |
| banded woollybear | *Pyrrharctia isabella* (J. E. Smith) | LEPIDOPTERA: Arctiidae |
| bandedwinged whitefly | *Trialeurodes abutilonea* (Haldeman) | HEMIPTERA: Aleyrodidae |
| Banks grass mite | *Oligonychus pratensis* (Banks) | ACARI: Tetranychidae |
| banyan aphid | *Thoracaphis fici* (Takahashi) | HEMIPTERA: Aphididae |
| Barber brown lacewing | *Sympherobius barberi* (Banks) | NEUROPTERA: Hemerobiidae |
| barberpole caterpillar | *Mimoschinia rufofascialis* (Stephens) | LEPIDOPTERA: Crambidae |
| barley jointworm | *Tetramesa hordei* (Harris) | HYMENOPTERA: Eurytomidae |
| barnacle scale | *Ceroplastes cirripediformis* Comstock | HEMIPTERA: Coccidae |
| basswood lace bug | *Gargaphia tiliae* (Walsh) | HEMIPTERA: Tingidae |
| basswood leafminer | *Baliosus ruber* (Weber) | COLEOPTERA: Chrysomelidae |
| basswood leafroller | *Pantographa limata* Grote & Robinson | LEPIDOPTERA: Crambidae |
| beachgrass scale | *Eriococcus carolinae* Williams | HEMIPTERA: Eriococcidae |
| bean aphid | *Aphis fabae* Scopoli | HEMIPTERA: Aphididae |
| bean butterfly | *Lampides boeticus* (Linnaeus) | LEPIDOPTERA: Lycaenidae |
| bean capsid | *Pycnoderes quadrimaculatus* Guérin-Méneville | HEMIPTERA: Miridae |
| bean fly | *Ophiomyia phaseoli* (Tryon) | DIPTERA: Agromyzidae |
| bean leaf beetle | *Cerotoma trifurcata* (Forster) | COLEOPTERA: Chrysomelidae |
| bean leafroller | *Urbanus proteus* (Linnaeus) | LEPIDOPTERA: Hesperiidae |
| bean leafskeletonizer | *Autoplusia egena* (Guenée) | LEPIDOPTERA: Noctuidae |
| bean pod borer | *Maruca testulalis* (Geyer) | LEPIDOPTERA: Crambidae |
| bean stalk weevil | *Sternechus paludatus* (Casey) | COLEOPTERA: Curculionidae |
| bean thrips | *Caliothrips fasciatus* (Pergande) | THYSANOPTERA: Thripidae |
| bean weevil | *Acanthoscelides obtectus* (Say) | COLEOPTERA: Bruchidae |
| Beardsley leafhopper | *Balclutha beardsleyi* Namba | HEMIPTERA: Cicadellidae |
| bed bug | *Cimex lectularius* Linnaeus | HEMIPTERA: Cimicidae |
| beech blight aphid | *Fagiphagus imbricator* (Fitch) | HEMIPTERA: Aphididae |
| beech scale | *Cryptococcus fagisuga* Lindinger | HEMIPTERA: Eriococcidae |
| beet armyworm | *Spodoptera exigua* (Hübner) | LEPIDOPTERA: Noctuidae |
| beet leaf beetle | *Erynephala puncticollis* (Say) | COLEOPTERA: Chrysomelidae |
| beet leafhopper | *Circulifer tenellus* (Baker) | HEMIPTERA: Cicadellidae |
| beet leafminer | *Pegomya hyoscyami* (Panzer) | DIPTERA: Anthomyiidae |

| | | |
|---|---|---|
| beet webworm | *Loxostege sticticalis* (Linnaeus) | LEPIDOPTERA: Crambidae |
| bella moth | *Utetheisa bella* (Linnaeus) | LEPIDOPTERA: Arctiidae |
| Bermudagrass mite | *Eriophyes cynodoniensis* Say | ACARI: Eriophyidae |
| bertha armyworm | *Mamestra configurata* Walker | LEPIDOPTERA: Noctuidae |
| bidens borer | *Epiblema otiosana* (Clemens) | LEPIDOPTERA: Tortricidae |
| bigheaded ant | *Pheidole megacephala* (Fabricius) | HYMENOPTERA: Formicidae |
| bigheaded grasshopper | *Aulocara elliotti* (Thomas) | ORTHOPTERA: Acrididae |
| birch bark beetle | *Dryocoetes betulae* Hopkins | COLEOPTERA: Scolytidae |
| birch casebearer | *Coleophora serratella* (Linnaeus) | LEPIDOPTERA: Coleophoridae |
| birch leafminer | *Fenusa pusilla* (Lepeletier) | HYMENOPTERA: Tenthredinidae |
| birch sawfly | *Arge pectoralis* (Leach) | HYMENOPTERA: Argidae |
| birch skeletonizer | *Bucculatrix canadensisella* Chambers | LEPIDOPTERA: Lyonetiidae |
| birch tubemaker | *Acrobasis betulella* Hulst | LEPIDOPTERA: Crambidae |
| bird tick | *Haemaphysalis chordeilis* (Packard) | ACARI: Ixodidae |
| black army cutworm | *Actebia fennica* (Tauscher) | LEPIDOPTERA: Noctuidae |
| black blister beetle | *Epicauta pennsylvanica* (DeGeer) | COLEOPTERA: Meloidae |
| black blow fly | *Phormia regina* (Meigen) | DIPTERA: Calliphoridae |
| black carpenter ant | *Camponotus pennsylvanicus* (DeGeer) | HYMENOPTERA: Formicidae |
| black carpet beetle | *Attagenus megatoma* (Fabricius) | COLEOPTERA: Dermestidae |
| black cherry aphid | *Myzus cerasi* (Fabricius) | HEMIPTERA: Aphididae |
| black cherry fruit fly | *Rhagoletis fausta* (Osten Sacken) | DIPTERA: Tephritidae |
| black citrus aphid | *Toxoptera aurantii* (Fonscolombe) | HEMIPTERA: Aphididae |
| black cockroach wasp | *Dolichurus stantoni* (Ashmead) | HYMENOPTERA: Ampulicidae |
| black cutworm | *Agrotis ipsilon* (Hufnagel) | LEPIDOPTERA: Noctuidae |
| black dung beetle | *Copris incertus prociduus* (Say) | COLEOPTERA: Scarabaeidae |
| black earwig | *Chelisoches morio* (Fabricius) | DERMAPTERA: Chelisochidae |
| black elm bark weevil | *Magdalis barbita* (Say) | COLEOPTERA: Curculionidae |
| black flower thrips | *Haplothrips gowdeyi* (Franklin) | THYSANOPTERA: Phlaeothripidae |
| black fungus beetle | *Alphitobius laevigatus* (Fabricius) | COLEOPTERA: Tenebrionidae |
| black grain stem sawfly | *Trachelus tabidus* (Fabricius) | HYMENOPTERA: Cephidae |
| black horse fly | *Tabanus atratus* Fabricius | DIPTERA: Tabanidae |
| black hunter thrips | *Leptothrips mali* (Fitch) | THYSANOPTERA: Phlaeothripidae |
| black imported fire ant | *Solenopsis richteri* Forel | HYMENOPTERA: Formicidae |
| black ladybird beetle | *Rhizobius ventralis* (Erichson) | COLEOPTERA: Coccinellidae |
| black larder beetle | *Dermestes ater* DeGeer | COLEOPTERA: Dermestidae |
| black peach aphid | *Brachycaudus persicae* (Passerini) | HEMIPTERA: Aphididae |
| black pecan aphid | *Melanocallis caryaefoliae* (Davis) | HEMIPTERA: Aphididae |
| black pineleaf scale | *Nuculaspis californica* (Coleman) | HEMIPTERA: Diaspididae |
| black potter wasp | *Delta pyriformis philippinensis* (Bequaert) | HYMENOPTERA: Vespidae |
| black scale | *Saissetia oleae* (Olivier) | HEMIPTERA: Coccidae |
| black soldier fly | *Hermetia illucens* (Linnaeus) | DIPTERA: Stratiomyidae |
| black stink bug | *Coptosoma xanthogramma* (White) | HEMIPTERA: Plataspidae |
| black swallowtail | *Papilio polyxenes asterius* Stoll | LEPIDOPTERA: Papilionidae |
| black thread scale | *Ischnaspis longirostris* (Signoret) | HEMIPTERA: Diaspididae |
| black turfgrass ataenius | *Ataenius spretulus* (Haldeman) | COLEOPTERA: Scarabaeidae |
| black turpentine beetle | *Dendroctonus terebrans* (Olivier) | COLEOPTERA: Scolytidae |
| black twig borer | *Xylosandrus compactus* (Eichhoff) | COLEOPTERA: Scolytidae |
| black vine weevil | *Otiorhynchus sulcatus* (Fabricius) | COLEOPTERA: Curculionidae |
| black walnut curculio | *Conotrachelus retentus* (Say) | COLEOPTERA: Curculionidae |
| black widow spider | *Latrodectus mactans* (Fabricius) | ARANEAE: Theridiidae |
| black witch | *Ascalapha odorata* (Linnaeus) | LEPIDOPTERA: Noctuidae |
| blackbellied clerid | *Enoclerus lecontei* (Wolcott) | COLEOPTERA: Cleridae |
| blackberry skeletonizer | *Schreckensteinia festaliella* (Hübner) | LEPIDOPTERA: Heliodinidae |
| Blackburn butterfly | *Vaga blackburni* (Tuely) | LEPIDOPTERA: Lycaenidae |
| Blackburn damsel bug | *Nabis blackburni* (White) | HEMIPTERA: Nabidae |
| Blackburn dragonfly | *Nesogonia blackburni* (McLachlan) | ODONATA: Libellulidae |
| blackfaced leafhopper | *Graminella nigrifrons* (Forbes) | HEMIPTERA: Cicadellidae |
| blackheaded ash sawfly | *Tethida barba* (Say) | HYMENOPTERA: Tenthredinidae |
| blackheaded fireworm | *Rhopobota unipunctana* (Haworth) | LEPIDOPTERA: Tortricidae |
| blackheaded pine sawfly | *Neodiprion excitans* Rohwer | HYMENOPTERA: Diprionidae |

| | | |
|---|---|---|
| blackhorned pine borer | *Callidium antennatun hesperum* Casey | COLEOPTERA: Cerambycidae |
| blackhorned tree cricket | *Oecanthus nigricornis* Walker | ORTHOPTERA: Gryllidae |
| blackjacket | *Vespula consobrina* (Saussure) | HYMENOPTERA: Vespidae |
| blacklegged tick | *Ixodes scapularis* Say | ACARI: Ixodidae |
| blacklegged tortoise beetle | *Jonthonota nigripes* (Olivier) | COLEOPTERA: Chrysomelidae |
| blackmargined aphid | *Monellia caryella* (Fitch) | HEMIPTERA: Aphididae |
| blister coneworm | *Dioryctria clarioralis* (Walker) | LEPIDOPTERA: Crambidae |
| bloodsucking conenose | *Triatoma sanguisuga* (LeConte) | HEMIPTERA: Reduviidae |
| blue alfalfa aphid | *Acyrthosiphon kondi* Shinji | HEMIPTERA: Aphididae |
| blue cactus borer | *Melitara dentata* (Grote) | LEPIDOPTERA: Crambidae |
| blue horntail | *Sirex cyaneus* Fabricius | HYMENOPTERA: Siricidae |
| blue soldier fly | *Neoexaireta spinigera* (Wiedemann) | DIPTERA: Stratiomyidae |
| blueberry bud mite | *Acalitus vaccinii* (Keifer) | ACARI: Eriophyidae |
| blueberry case beetle | *Neochlamisus cribripennis* (LeConte) | COLEOPTERA: Chrysomelidae |
| blueberry flea beetle | *Altica sylvia* Malloch | COLEOPTERA: Chrysomelidae |
| blueberry maggot | *Rhagoletis mendax* (Curran) | DIPTERA: Tephritidae |
| blueberry thrips | *Frankliniella vaccinii* Morgan | THYSANOPTERA: Thripidae |
| blueberry tip midge | *Contarinia vaccinii* (Felt) | DIPTERA: Cecidomyiidae |
| bluegrass billbug | *Sphenophorus parvulus* Gyllenhal | COLEOPTERA: Curculionidae |
| bluegrass webworm | *Parapediasia teterrella* (Zincken) | LEPIDOPTERA: Crambidae |
| bluntnosed cranberry leafhopper | *Scleroracus vaccinii* (Van Duzee) | HEMIPTERA: Cicadellidae |
| body louse | *Pediculus humanus humanus* Linnaeus | PHTHIRAPTERA: Pediculidae |
| Boisduval scale | *Diaspis boisduvalli* Signoret | HEMIPTERA: Diaspididae |
| boll weevil | *Anthonomus grandis grandis* Boheman | COLEOPTERA: Curculionidae |
| bollworm | *Heliothis zea* (Boddie) | LEPIDOPTERA: Noctuidae |
| booklouse | *Liposcelis corrodens* Heymons | PSOCOPTERA: Liposcelidae |
| boxelder aphid | *Periphyllus negundinis* (Thomas) | HEMIPTERA: Aphididae |
| boxelder bug | *Leptocoris trivittatus* (Say) | HEMIPTERA: Rhopalidae |
| boxelder leafroller | *Caloptilia negundella* (Chambers) | LEPIDOPTERA: Gracillariidae |
| boxelder psyllid | *Psylla negundinis* Mally | HEMIPTERA: Psyllidae |
| boxelder twig borer | *Proteoteras willingana* (Kearfott) | LEPIDOPTERA: Tortricidae |
| boxwood leafminer | *Monarthropalpus buxi* (Laboulbène) | DIPTERA: Cecidomyiidae |
| boxwood psyllid | *Psylla buxi* (Linnaeus) | HEMIPTERA: Psyllidae |
| bramble leafhopper | *Ribautiana tenerrima* (Herrich Schäffer) | HEMIPTERA: Cicadellidae |
| Brasilian leafhopper | *Protalebrella brasiliensis* (Baker) | HEMIPTERA: Cicadellidae |
| bristly cutworm | *Lacinipolia renigera* (Stephens) | LEPIDOPTERA: Noctuidae |
| bristly roseslug | *Cladius difformis* (Panzer) | HYMENOPTERA: Tenthredinidae |
| broad mite | *Polyphagotarsonemus latus* (Banks) | ACARI: Tarsonemidae |
| broadbean weevil | *Bruchus rufimanus* Boheman | COLEOPTERA: Bruchidae |
| broadhorned flour beetle | *Gnathocerus cornutus* (Fabricius) | COLEOPTERA: Tenebrionidae |
| broadnecked root borer | *Prionus laticollis* (Drury) | COLEOPTERA: Cerambycidae |
| broadnosed grain weevil | *Caulophilus oryzae* (Gyllenhal) | COLEOPTERA: Curculionidae |
| broadwinged katydid | *Microcentrum rhombifolium* (Saussure) | ORTHOPTERA: Tettigoniidae |
| bromegrass seed midge | *Contarinia bromicola* (Marikovskij & Agafonova) | DIPTERA: Cecidomyiidae |
| bronze appletree weevil | *Magdalis aenescens* LeConte | COLEOPTERA: Curculionidae |
| bronze birch borer | *Agrilus anxius* Gory | COLEOPTERA: Buprestidae |
| bronze leaf beetle | *Diachus auratus* (Fabricius) | COLEOPTERA: Chrysomelidae |
| bronze poplar borer | *Agrilus liragus* Barter & Brown | COLEOPTERA: Buprestidae |
| bronzed cutworm | *Nephelodes minians* (Guenée) | LEPIDOPTERA: Noctuidae |
| brown chicken louse | *Goniodes dissimilis* Denny | PHTHIRAPTERA: Philopteridae |
| brown citrus aphid | *Toxoptera citricida* (Kirkaldy) | HEMIPTERA: Aphididae |
| brown cockroach | *Periplaneta brunnea* Burmeister | BLATTODEA: Blattidae |
| brown cotton leafworm | *Acontia dacia* Druce | LEPIDOPTERA: Noctuidae |
| brown dog tick | *Rhipicephalus sanguineus* (Latreille) | ACARI: Ixodidae |
| brown dung beetle | *Onthophagus gazella* Fabricius | COLEOPTERA: Scarabaeidae |
| brown flour mite | *Gohieria fusca* (Oudemans) | ACARI: Glycyphagidae |
| brown house moth | *Hofmannophila pseudospretella* (Stainton) | LEPIDOPTERA: Oecophoridae |
| brown mite | *Bryobia rubrioculus* (Scheuten) | ACARI: Tetranychidae |
| brown pineapple scale | *Melanaspis bromeliae* (Leonardi) | HEMIPTERA: Diaspididae |
| brown recluse spider | *Loxosceles reclusa* Gertsch & Mulaik | ARANEAE: Loxoscelidae |

| | | |
|---|---|---|
| brown saltmarsh mosquito | *Aedes cantator* (Coquillett) | DIPTERA: Culicidae |
| brown soft scale | *Coccus hesperidum* Linnaeus | HEMIPTERA: Coccidae |
| brown spider beetle | *Ptinus clavipes* Panzer | COLEOPTERA: Ptinidae |
| brown stink bug | *Euschistus servus* (Say) | HEMIPTERA: Pentatomidae |
| brown wheat mite | *Petrobia latens* (Müller) | ACARI: Tetranychidae |
| brown widow spider | *Latrodectus geometricus* (Fabricius) | ARANEAE: Theridiidae |
| brownbanded cockroach | *Supella longipalpa* (Fabricius) | BLATTODEA: Blattellidae |
| brownheaded ash sawfly | *Tomostethus multicinctus* (Rohwer) | HYMENOPTERA: Tenthredinidae |
| brownheaded jack pine sawfly | *Neodiprion dubiosus* Schedl | HYMENOPTERA: Diprionidae |
| brownlegged grain mite | *Aleuroglyphus ovatus* (Troupeau) | ACARI: Acaridae |
| browntail moth | *Euproctis chrysorrhoea* (Linnaeus) | LEPIDOPTERA: Lymantriidae |
| Bruce spanworm | *Operophtera bruceata* (Hulst) | LEPIDOPTERA: Geometridae |
| buck moth | *Hemileuca maia* (Drury) | LEPIDOPTERA: Saturniidae |
| buckthorn aphid | *Aphis nasturtii* Kaltenbach | HEMIPTERA: Aphididae |
| buffalo treehopper | *Stictocephala bisonia* Kopp & Yonke | HEMIPTERA: Membracidae |
| buffalograss webworm | *Surattha indentella* Kearfott | LEPIDOPTERA: Crambidae |
| bulb mite | *Rhizoglyphus echinopus* (Fumouze & Robin) | ACARI: Acaridae |
| bulb scale mite | *Steneotarsonemus laticeps* (Halbert) | ACARI: Tarsonemidae |
| bumble flower beetle | *Euphoria inda* (Linnaeus) | COLEOPTERA: Scarabaeidae |
| bumelia fruit fly | *Pseudodacus pallens* (Coquillett) | DIPTERA: Tephritidae |
| burdock borer | *Papaipema cataphracta* (Grote) | LEPIDOPTERA: Noctuidae |
| Burmeister mantid | *Orthodera burmeisteri* Wood-Mason | MANTODEA: Mantidae |
| butternut curculio | *Conotrachelus juglandis* LeConte | COLEOPTERA: Curculionidae |

## C

| | | |
|---|---|---|
| cabbage aphid | *Brevicoryne brassicae* (Linnaeus) | HEMIPTERA: Aphididae |
| cabbage curculio | *Ceutorhynchus rapae* Gyllenhal | COLEOPTERA: Curculionidae |
| cabbage looper | *Trichoplusia ni* (Hübner) | LEPIDOPTERA: Noctuidae |
| cabbage maggot | *Delia radicum* (Linnaeus) | DIPTERA: Anthomyiidae |
| cabbage seedpod weevil | *Ceutorhynchus assimilis* (Paykull) | COLEOPTERA: Curculionidae |
| cabbage seedstalk curculio | *Ceutorhynchus quadridens* (Panzer) | COLEOPTERA: Curculionidae |
| cabbage webworm | *Hellula rogatalis* (Hulst) | LEPIDOPTERA: Crambidae |
| cactus moth | *Cactoblastis cactorum* (Berg) | LEPIDOPTERA: Crambidae |
| cactus scale | *Diaspis echinocacti* (Bouché) | HEMIPTERA: Diaspididae |
| cadelle | *Tenebroides mauritanicus* (Linnaeus) | COLEOPTERA: Trogositidae |
| Caledonia seed bug | *Nysius caledoniae* Distant | HEMIPTERA: Lygaeidae |
| calico scale | *Eulecanium cerasorum* (Cockerell) | HEMIPTERA: Coccidae |
| California fivespined ips | *Ips paraconfusus* Lanier | COLEOPTERA: Scolytidae |
| California flatheaded borer | *Melanophila californica* Van Dyke | COLEOPTERA: Buprestidae |
| California harvester ant | *Pogonomyrmex californicus* (Buckley) | HYMENOPTERA: Formicidae |
| California oakworm | *Phryganidia californica* Packard | LEPIDOPTERA: Dioptidae |
| California pear sawfly | *Pristiphora abbreviata* (Hartig) | HYMENOPTERA: Tenthredinidae |
| California prionus | *Prionus californicus* Motschulsky | COLEOPTERA: Cerambycidae |
| California red scale | *Aonidiella aurantii* (Maskell) | HEMIPTERA: Diaspididae |
| California saltmarsh mosquito | *Aedes squamiger* (Coquillett) | DIPTERA: Culicidae |
| California tortoiseshell | *Nymphalis californica* (Boisduval) | LEPIDOPTERA: Nymphalidae |
| camellia scale | *Lepidosaphes camelliae* Hoke | HEMIPTERA: Diaspididae |
| camphor scale | *Pseudaonidia duplex* (Cockerell) | HEMIPTERA: Diaspididae |
| camphor thrips | *Liothrips floridensis* (Watson) | THYSANOPTERA: Phlaeothripidae |
| caragana aphid | *Acyrthosiphon caraganae* (Cholodkovsky) | HEMIPTERA: Aphididae |
| caragana blister beetle | *Epicauta subglabra* (Fall) | COLEOPTERA: Meloidae |
| caragana plant bug | *Lopidea dakota* Knight | HEMIPTERA: Miridae |
| Caribbean black scale | *Saissetia neglecta* De Lotto | HEMIPTERA: Coccidae |
| Caribbean pod borer | *Fundella pellucens* Zeller | LEPIDOPTERA: Crambidae |
| carmine spider mite | *Tetranychus cinnabarinus* (Boisduval) | ACARI: Tetranychidae |
| carnation maggot | *Delia brunnescens* (Zetterstedt) | DIPTERA: Anthomyiidae |
| carnation tip maggot | *Delia echinata* (Séguy) | DIPTERA: Anthomyiidae |
| Carolina conifer aphid | *Cinara atlantica* (Wilson) | HEMIPTERA: Aphididae |
| Carolina grasshopper | *Dissosteira carolina* (Linnaeus) | ORTHOPTERA: Acrididae |

| | | |
|---|---|---|
| Carolina mantid | *Stagmomantis carolina* (Johannson) | MANTODEA: Mantidae |
| carpenter bee | *Xylocopa virginica* (Linnaeus) | HYMENOPTERA: Xylocopidae |
| carpenterworm | *Prionoxystus robiniae* (Peck) | LEPIDOPTERA: Cossidae |
| carpet beetle | *Anthrenus scrophulariae* (Linnaeus) | COLEOPTERA: Dermestidae |
| carpet moth | *Trichophaga tapetzella* (Linnaeus) | LEPIDOPTERA: Tineidae |
| carrot beetle | *Bothynus gibbosus* (DeGeer) | COLEOPTERA: Scarabaeidae |
| carrot rust fly | *Psila rosae* (Fabricius) | DIPTERA: Psilidae |
| carrot weevil | *Listronotus oregonensis* (LeConte) | COLEOPTERA: Curculionidae |
| casemaking clothes moth | *Tinea pellionella* Linnaeus | LEPIDOPTERA: Tineidae |
| cat flea | *Ctenocephalides felis* (Bouché) | SIPHONAPTERA: Pulicidae |
| cat follicle mite | *Demodex cati* Mégnin | ACARI: Demodicidae |
| cat louse | *Felicola subrostratus* (Burmeister) | PHTHIRAPTERA: Trichodectidae |
| catalpa midge | *Contarinia catalpae* (Comstock) | DIPTERA: Cecidomyiidae |
| catalpa sphinx | *Ceratomia catalpae* (Boisduval) | LEPIDOPTERA: Sphingidae |
| cattle biting louse | *Bovicola bovis* (Linnaeus) | PHTHIRAPTERA: Trichodectidae |
| cattle follicle mite | *Demodex bovis* Stiles | ACARI: Demodicidae |
| cattle itch mite | *Sarcoptes bovis* Robin | ACARI: Sarcoptidae |
| cattle tail louse | *Haematopinus quadripertusus* Fahrenholz | PHTHIRAPTERA: Haematopinidae |
| cattle tick | *Boophilus annulatus* (Say) | ACARI: Ixodidae |
| Cayenne tick | *Amblyomma cajennense* (Fabricius) | ACARI: Ixodidae |
| ceanothus silk moth | *Hyalophora euryalus* (Boisduval) | LEPIDOPTERA: Saturniidae |
| cecropia moth | *Hyalophora cecropia* (Linnaeus) | LEPIDOPTERA: Saturniidae |
| cedartree borer | *Semanotus ligneus* (Fabricius) | COLEOPTERA: Cerambycidae |
| celery aphid | *Brachycolus heraclei* Takahashi | HEMIPTERA: Aphididae |
| celery leaftier | *Udea rubigalis* (Guenée) | LEPIDOPTERA: Crambidae |
| celery looper | *Syngrapha falcitera* (Kirby) | LEPIDOPTERA: Noctuidae |
| cereal leaf beetle | *Oulema melanopus* (Linnaeus) | COLEOPTERA: Chrysomelidae |
| chaff scale | *Parlatoria pergandii* Comstock | HEMIPTERA: Diaspididae |
| chainspotted geometer | *Cingilia catenaria* (Drury) | LEPIDOPTERA: Geometridae |
| changa | *Scapteriscus vicinus* Scudder | ORTHOPTERA: Gryllotalpidae |
| charcoal beetle | *Melanophila consputa* LeConte | COLEOPTERA: Buprestidae |
| cheese mite | *Tyrolichus casei* Oudemans | ACARI: Acaridae |
| cheese skipper | *Piophila casei* (Linnaeus) | DIPTERA: Piophilidae |
| cherry casebearer | *Coleophora pruniella* Clemens | LEPIDOPTERA: Coleophoridae |
| cherry fruit fly | *Rhagoletis cingulata* (Loew) | DIPTERA: Tephritidae |
| cherry fruit sawfly | *Hoplocampa cookei* (Clarke) | HYMENOPTERA: Tenthredinidae |
| cherry fruitworm | *Grapholita packardi* Zeller | LEPIDOPTERA: Tortricidae |
| cherry leaf beetle | *Pyrrhalta cavicollis* (LeConte) | COLEOPTERA: Chrysomelidae |
| cherry maggot | *Rhagoletis cingulata* (Loew) | DIPTERA: Tephritidae |
| chestnut timberworm | *Melittomma sericeum* (Harris) | COLEOPTERA: Lymexylonidae |
| chicken body louse | *Menacanthus stramineus* (Nitzsch) | PHTHIRAPTERA: Menoponidae |
| chicken dung fly | *Fannia pusio* (Wiedemann) | DIPTERA: Muscidae |
| chicken head louse | *Cuclotogaster heterographus* (Nitzsch) | PHTHIRAPTERA: Philopteridae |
| chicken mite | *Dermanyssus gallinae* (DeGeer) | ACARI: Dermanyssidae |
| chigoe | *Tunga penetrans* Linnaeus | SIPHONAPTERA: Tungidae |
| chinch bug | *Blissus leucopterus leucopterus* (Say) | HEMIPTERA: Lygaeidae |
| Chinese dryinid | *Pseudogonatopus hospes* Perkins | HYMENOPTERA: Dryinidae |
| Chinese mantid | *Tenodera aridifolia sinensis* Saussure | MANTODEA: Mantidae |
| Chinese obscure scale | *Parlatoreopsis chinensis* (Marlatt) | HEMIPTERA: Diaspididae |
| Chinese rose beetle | *Adoretus sinicus* Burmeister | COLEOPTERA: Scarabaeidae |
| Christmas berry webworm | *Cryptoblabes gnidiella* (Millière) | LEPIDOPTERA: Crambidae |
| chrysanthemum aphid | *Macrosiphoniella sanborni* (Gillette) | HEMIPTERA: Aphididae |
| chrysanthemum flower borer | *Lorita abornana* Busck | LEPIDOPTERA: Cochylidae |
| chrysanthemum gall midge | *Rhopalomyia chrysanthemi* (Ahlberg) | DIPTERA: Cecidomyiidae |
| chrysanthemum lace bug | *Corythucha marmorata* (Uhler) | HEMIPTERA: Tingidae |
| chrysanthemum leafminer | *Phytomyza syngenesiae* (Hardy) | DIPTERA: Agromyzidae |
| chrysanthemum thrips | *Thrips nigropilosus* Uzel | THYSANOPTERA: Thripidae |
| cicada killer | *Sphecius speciosus* (Drury) | HYMENOPTERA: Sphecidae |
| cigar casebearer | *Coleophora serratella* (Linnaeus) | LEPIDOPTERA: Coleophoridae |
| cigarette beetle | *Lasioderma serricorne* (Fabricius) | COLEOPTERA: Anobiidae |

| | | |
|---|---|---|
| cinereous cockroach | *Nauphoeta cinerea* (Olivier) | BLATTODEA: Blaberidae |
| cinnabar moth | *Tyria jacobaeae* (Linnaeus) | LEPIDOPTERA: Arctiidae |
| citricola scale | *Coccus pseudomagnoliarum* (Kuwana) | HEMIPTERA: Coccidae |
| citrophilus mealybug | *Pseudococcus calceolariae* (Maskell) | HEMIPTERA: Pseudococcidae |
| citrus blackfly | *Aleurocanthus woglumi* Ashby | HEMIPTERA: Aleyrodidae |
| citrus bud mite | *Eriophyes sheldoni* Ewing | ACARI: Eriophyidae |
| citrus flat mite | *Brevipalpus lewisi* McGregor | ACARI: Tenuipalpidae |
| citrus mealybug | *Planococcus citri* (Risso) | HEMIPTERA: Pseudococcidae |
| citrus red mite | *Panonychus citri* (McGregor) | ACARI: Tetryanychidae |
| citrus root weevil | *Pachnaeus litus* (Germar) | COLEOPTERA: Curculionidae |
| citrus rust mite | *Phyllocoptruta oleivora* (Ashmead) | ACARI: Eriophyidae |
| citrus snow scale | *Unaspis citri* (Comstock) | HEMIPTERA: Diaspididae |
| citrus swallowtail | *Papilio xuthus* Linnaeus | LEPIDOPTERA: Papilionidae |
| citrus thrips | *Scirtothrips citri* (Moulton) | THYSANOPTERA: Thripidae |
| citrus whitefly | *Dialeurodes citri* (Ashmead) | HEMIPTERA: Aleyrodidae |
| claybacked cutworm | *Agrotis gladiaria* Morrison | LEPIDOPTERA: Noctuidae |
| claycolored billbug | *Sphenophorus aequalis aequalis* Gyllenhal | COLEOPTERA: Curculionidae |
| claycolored leaf beetle | *Anomoea laticlavia* (Forster) | COLEOPTERA: Chrysomelidae |
| Clear Lake gnat | *Chaoborus astictopus* Dyar & Shannon | DIPTERA: Chaoboridae |
| clearwinged grasshopper | *Camnula pellucida* (Scudder) | ORTHOPTERA: Acrididae |
| clematis blister beetle | *Epicauta cinerea* (Forster) | COLEOPTERA: Meloidae |
| clidemia leafroller | *Blepharomastix ebulealis* (Guenée) | LEPIDOPTERA: Crambidae |
| clidemia thrips | *Liothrips urichi* Karny | THYSANOPTERA: Phlaeothripidae |
| clouded plant bug | *Neurocolpus nubilus* (Say) | HEMIPTERA: Miridae |
| clouded sulfur | *Colias philodice* Godart | LEPIDOPTERA: Pieridae |
| cloudywinged whitefly | *Dialeurodes citrifolii* (Morgan) | HEMIPTERA: Aleyrodidae |
| clover aphid | *Nearctaphis bakeri* (Cowen) | HEMIPTERA: Aphididae |
| clover aphid parasite | *Aphelinus lapisligni* Howard | HYMENOPTERA: Encyrtidae |
| clover cutworm | *Scotogramma trifolii* (Hufnagel) | LEPIDOPTERA: Noctuidae |
| clover hayworm | *Hypsopygia costalis* (Fabricius) | LEPIDOPTERA: Crambidae |
| clover head caterpillar | *Grapholita interstinctana* (Clemens) | LEPIDOPTERA: Tortricidae |
| clover head weevil | *Hypera meles* (Fabricius) | COLEOPTERA: Curculionidae |
| clover leaf midge | *Dasineura trifolii* (Loew) | DIPTERA: Cecidomyiidae |
| clover leaf weevil | *Hypera punctata* (Fabricius) | COLEOPTERA: Curculionidae |
| clover leafhopper | *Aceratagallia sanguinolenta* (Provancher) | HEMIPTERA: Cicadellidae |
| clover looper | *Caenurgina crassiuscula* (Haworth) | LEPIDOPTERA: Noctuidae |
| clover mite | *Bryobia praetiosa* Koch | ACARI: Tetranychidae |
| clover root borer | *Hylastinus obscurus* (Marsham) | COLEOPTERA: Scolytidae |
| clover root curculio | *Sitona hispidulus* (Fabricius) | COLEOPTERA: Curculionidae |
| clover seed chalcid | *Bruchophagus platypterus* (Walker) | HYMENOPTERA: Eurytomidae |
| clover seed midge | *Dasineura leguminicola* (Lintner) | DIPTERA: Cecidomyiidae |
| clover seed weevil | *Tychius stephensi* Schoenherr | COLEOPTERA: Curculionidae |
| clover stem borer | *Languria mozardi* Latreille | COLEOPTERA: Languriidae |
| cluster fly | *Pollenia nudis* (Fabricius) | DIPTERA: Calliphoridae |
| cochineal insect | *Dactylopius coccus* Costa | HEMIPTERA: Dactylopiidae |
| cocklebur weevil | *Rhodobaenus tredecimpunctatus* (Illiger) | COLEOPTERA: Curculionidae |
| coconut leafminer | *Agonoxena argaula* Meyrick | LEPIDOPTERA: Agonoxenidae |
| coconut leafroller | *Hedylepta blackburni* (Butler) | LEPIDOPTERA: Crambidae |
| coconut mealybug | *Nipaecoccus nipae* (Maskell) | HEMIPTERA: Pseudococcidae |
| coconut scale | *Aspidiotus destructor* Signoret | HEMIPTERA: Diaspididae |
| codling moth | *Cydia pomonella* (Linnaeus) | LEPIDOPTERA: Tortricidae |
| coffee bean weevil | *Araecerus fasciculatus* (DeGeer) | COLEOPTERA: Anthribidae |
| Colorado potato beetle | *Leptinotarsa decemlineata* (Say) | COLEOPTERA: Chrysomelidae |
| Columbia Basin wireworm | *Limonius subauratus* LeConte | COLEOPTERA: Elateridae |
| Columbian timber beetle | *Corthylus columbianus* Hopkins | COLEOPTERA: Scolytidae |
| columbine borer | *Papaipema purpurifascia* (Grote & Robinson) | LEPIDOPTERA: Noctuidae |
| Comanche lacewing | *Chrysopa comanche* Banks | NEUROPTERA: Chrysopidae |
| common Australian lady beetle | *Coelophora inaequalis* (Fabricius) | COLEOPTERA: Coccinellidae |
| common cattle grub | *Hypodenna lineatum* (Villers) | DIPTERA: Oestridae |
| common damsel bug | *Reduviolus americoferus* (Carayon) | HEMIPTERA: Nabidae |
| common green darner | *Anax junius* (Drury) | ODONATA: Aeschnidae |

| | | |
|---|---|---|
| common green lacewing | *Chrysoperla carnea* Stephens | NEUROPTERA: Chrysopidae |
| common malaria mosquito | *Ahopheles quadrimaculatus* Say | DIPTERA: Culicidae |
| composite thrips | *Microcephalothrips abdominalis* (D. L. Crawford) | THYSANOPTERA: Thripidae |
| Comstock mealybug | *Pseudococcus comstocki* (Kuwana) | HEMIPTERA: Pseudococcidae |
| conchuela | *Chlorochroa ligata* (Say) | HEMIPTERA: Pentatomidae |
| confused flour beetle | *Tribolium confusum* Jacquelin du Val | COLEOPTERA: Tenebrionidae |
| conifer spider mite | *Oligonychus coniferarum* (McGregor) | ACARI: Tetranychidae |
| convergent ladybird beetle | *Hippodamia convergens* Guérin-Méneville | COLEOPTERA: Coccinellidae |
| Cooley spruce gall adelgid | *Adelges cooleyi* (Gillette) | HEMIPTERA: Adelgidae |
| corn delphacid | *Peregrinus maidis* (Ashmead) | HEMIPTERA: Delphacidae |
| corn earworm | *Helicoverpa zea* (Boddie) | LEPIDOPTERA: Noctuidae |
| corn flea beetle | *Chaetocnema pulicaria* Melsheimer | COLEOPTERA: Chrysomelidae |
| corn leaf aphid | *Rhopalosiphum maidis* (Fitch) | HEMIPTERA: Aphididae |
| corn root aphid | *Anuraphis maidiradicis* (Forbes) | HEMIPTERA: Aphididae |
| corn root webworm | *Crambus caliginosellus* Clemens | LEPIDOPTERA: Crambidae |
| corn sap beetle | *Carpophilus dimidiatus* (Fabricius) | COLEOPTERA: Nitidulidae |
| corn silk beetle | *Calomicrus brunneus* (Crotch) | COLEOPTERA: Chrysomelidae |
| cornfield ant | *Lasius alienus* (Foerster) | HYMENOPTERA: Formicidae |
| cosmopolitan grain psocid | *Lachesilla pedicularia* (Linnaeus) | PSOCOPTERA: Lachesillidae |
| cotton aphid | *Aphis gossypii* Glover | HEMIPTERA: Aphididae |
| cotton blister mite | *Acalitus gossypii* (Banks) | ACARI: Eriophyidae |
| cotton fleahopper | *Pseudatomoscelis seriatus* (Reuter) | HEMIPTERA: Miridae |
| cotton lace bug | *Corythucha gossypii* (Fabricius) | HEMIPTERA: Tingidae |
| cotton leafminer | *Stigmella gossypii* (Forbes & Leonard) | LEPIDOPTERA: Nepticulidae |
| cotton leafperforator | *Bucculatrix thurberiella* Busck | LEPIDOPTERA: Lyonetiidae |
| cotton leafworm | *Alabama argillacea* (Hübner) | LEPIDOPTERA: Noctuidae |
| cotton square borer | *Strymon melinus* Hübner | LEPIDOPTERA: Lycaenidae |
| cotton stainer | *Dysdercus suturellus* (Herrich-Schäffer) | HEMIPTERA: Pyrrhocoridae |
| cotton stem moth | *Platyedra subcinerea* (Haworth) | LEPIDOPTERA: Gelechiidae |
| cottonwood borer | *Plectrodera scalator* (Fabricius) | COLEOPTERA: Cerambycidae |
| cottonwood dagger moth | *Acronicta lepusculina* Guenée | LEPIDOPTERA: Noctuidae |
| cottonwood leaf beetle | *Chrysomela scripta* Fabricius | COLEOPTERA: Chrysomelidae |
| cottonwood twig borer | *Gypsonoma haimbachiana* (Kearfott) | LEPIDOPTERA: Tortricidae |
| cottony maple scale | *Pulvinaria innumerabilis* (Rathvon) | HEMIPTERA: Coccidae |
| cottony peach scale | *Pulvinaria amygdali* Cockerell | HEMIPTERA: Coccidae |
| cottony cushion scale | *Icerya purchasi* (Maskell) | HEMIPTERA: Margarodidae |
| coulee cricket | *Peranabrus scabricollis* (Thomas) | ORTHOPTERA: Tettigoniidae |
| cowpea aphid | *Aphis craccivora* Koch | HEMIPTERA: Aphididae |
| cowpea curculio | *Chalcodermus aeneus* Boheman | COLEOPTERA: Curculionidae |
| cowpea weevil | *Callosobruchus maculatus* (Fabricius) | COLEOPTERA: Bruchidae |
| crab louse | *Pthirus pubis* (Linnaeus) | PHTHIRAPTERA: Pediculidae |
| crabhole mosquito | *Deinocerites cancer* Theobald | DIPTERA: Culicidae |
| cranberry fruitworm | *Acrobasis vaccinii* Riley | LEPIDOPTERA: Crambidae |
| cranberry girdler | *Chrysoteuchia topiaria* (Zeller) | LEPIDOPTERA: Crambidae |
| cranberry rootworm | *Rhabdopterus picipes* (Olivier) | COLEOPTERA: Chrysomelidae |
| cranberry weevil | *Anthonomus musculus* Say | COLEOPTERA: Curculionidae |
| crapemyrtle aphid | *Tinocallis kahawaluokalani* (Kirkaldy) | HEMIPTERA: Aphididae |
| crazy ant | *Paratrechina longicornis* (Latreille) | HYMENOPTERA: Formicidae |
| crescentmarked lily aphid | *Neomyzus circumfexus* (Buckton) | HEMIPTERA: Aphididae |
| cribrate weevil | *Otiorhynchus cribricollis* Gyllenhal | COLEOPTERA: Curculionidae |
| crinkled flannel moth | *Lagoa crispata* (Packard) | LEPIDOPTERA: Megalopygida |
| cross-striped cabbageworm | *Evergestis rimosalis* (Guenée) | LEPIDOPTERA: Crambidae |
| croton caterpillar | *Achaea janata* (Linnaeus) | LEPIDOPTERA: Noctuidae |
| croton mussel scale | *Lepidosaphes tokionis* (Kuwana) | HEMIPTERA: Diaspididae |
| Cuban cockroach | *Panchlora nivea* (Linnaeus) | BLATTODEA: Blaberidae |
| Cuban laurel thrips | *Gynaikothrips ficorum* (Marchal) | THYSANOPTERA: Phlaeothripidae |
| cucurbit longicorn | *Apomecyna saltator* (Fabricius) | COLEOPTERA: Cerambycidae |
| cucurbit midge | *Prodiplosis citrulli* (Felt) | DIPTERA: Cecidomyiidae |
| curled rose sawfly | *Allantus cinctus* (Linnaeus) | HYMENOPTERA: Tenthredinidae |
| currant aphid | *Cryptomyzus ribis* (Linnaeus) | HEMIPTERA: Aphididae |
| currant borer | *Synanthedon tipuliformis* (Clerck) | LEPIDOPTERA: Sesiidae |

| | | |
|---|---|---|
| currant bud mite | *Cecidophyopsis ribis* (Westwood) | ACARI: Eriophyidae |
| currant fruit fly | *Epochra canadensis* (Loew) | DIPTERA: Tephritidae |
| currant fruit weevil | *Pseudanthonomus validus* Dietz | COLEOPTERA: Curculionidae |
| currant spanworm | *Itame ribearia* (Fitch) | LEPIDOPTERA: Geometridae |
| currant stem girdler | *Janus integer* (Norton) | HYMENOPTERA: Cephidae |
| cyclamen mite | *Steneotarsonemus pallidus* (Banks) | ACARI: Tarsonemidae |
| cynthia moth | *Samia cynthia* (Drury) | LEPIDOPTERA: Saturniidae |

## D

| | | |
|---|---|---|
| dandelion gall wasp | *Phanacis taraxaci* (Ashmead) | HYMENOPTERA: Cynipidae |
| dark mealworm | *Tenebrio obscurus* Fabricius | COLEOPTERA: Tenebrionidae |
| darksided cutworm | *Euxoa messoria* (Harris) | LEPIDOPTERA: Noctuidae |
| datebug | *Asarcopus palmarum* Horvath | HEMIPTERA: Issidae |
| deodar weevil | *Pissodes nemorensis* Germar | COLEOPTERA: Curculionidae |
| depluming mite | *Knemidokoptes gallinae* (Railliet) | ACARI: Sarcoptidae |
| depressed flour beetle | *Palorus subdepressus* (Wollaston) | COLEOPTERA: Tenebrionidae |
| desert corn flea beetle | *Chaetocnema ectypa* Horn | COLEOPTERA: Chrysomelidae |
| desert spider mite | *Tetranychus desertorum* Banks | ACARI: Tetranychidae |
| devastating grasshopper | *Melanoplus devastator* Scudder | ORTHOPTERA: Acrididae |
| diamondback moth | *Plutella xylostella* (Linnaeus) | LEPIDOPTERA: Plutellidae |
| diamondbacked spittlebug | *Lepyronia quadrangularis* (Say) | HEMIPTERA: Cercopidae |
| dictyospermum scale | *Chrysomphalus dictyospermi* (Morgan) | HEMIPTERA: Diaspididae |
| differential grasshopper | *Melanoplus differentialis* (Thomas) | ORTHOPTERA: Acrididae |
| dingy cutworm | *Feltia ducens* Walker | LEPIDOPTERA: Noctuidae |
| dobsonfly | *Corydalus cornutus* (Linnaeus) | NEUROPTERA: Corydalidae |
| dock sawfly | *Ametastegia glabrata* (Fallén) | HYMENOPTERA: Tenthredinidae |
| dodder gall weevil | *Smicronyx sculpticollis* Casey | COLEOPTERA: Curculionidae |
| dog biting louse | *Trichodectes canis* (DeGeer) | PHTHIRAPTERA: Trichodectidae |
| dog flea | *Ctenocephalides canis* (Curtis) | SIPHONAPTERA: Pulicidae |
| dog follicle mite | *Demodex canis* Leydig | ACARI: Demodicidae |
| dog sucking louse | *Linognathus setosus* (Olfers) | PHTHIRAPTERA: Linognathidae |
| dogwood borer | *Synanthedon scitula* (Harris) | LEPIDOPTERA: Sesiidae |
| dogwood clubgall midge | *Resseliella clavula* (Beutenmüller) | DIPTERA: Cecidomyiidae |
| dogwood scale | *Chionaspis corni* Cooley | HEMIPTERA: Diaspididae |
| dogwood spittlebug | *Clastoptera proteus* Fitch | HEMIPTERA: Cercopidae |
| dogwood twig borer | *Oberea tripunctata* (Swederus) | COLEOPTERA: Cerambycidae |
| Douglas-fir beetle | *Dendroctonus pseudotsugae* Hopkins | COLEOPTERA: Scolytidae |
| Douglas-fir cone moth | *Barbara colfaxiana* (Kearfott) | LEPIDOPTERA: Tortricidae |
| Douglas-fir engraver | *Scolytus unispinosus* LeConte | COLEOPTERA: Scolytidae |
| Douglas-fir pitch moth | *Synanthedon novaroensis* (Edwards) | LEPIDOPTERA: Sesiidae |
| Douglas-fir tussock moth | *Orgyia pseudotsugata* (McDunnough) | LEPIDOPTERA: Lymantriidae |
| Douglas-fir twig weevil | *Cylindrocopturus furnissi* Buchanan | COLEOPTERA: Curculionidae |
| driedfruit beetle | *Carpophilus hemipterus* (Linnaeus) | COLEOPTERA: Nitidulidae |
| driedfruit mite | *Carpoglyphus lactis* (Linnaeus) | ACARI: Carpoglyphidae |
| driedfruit moth | *Vitula edmandsae serratilineella* Ragonot | LEPIDOPTERA: Crambidae |
| drone fly | *Eristalis tenax* (Linnaus) | DIPTERA: Syrphidae |
| drugstore beetle | *Stegobium paniceum* (Linnaeus) | COLEOPTERA: Anobiidae |
| dryberry mite | *Phyllocoptes gracilis* (Nalepa) | ACARI: Eriophyidae |
| dryland wireworm | *Ctenicera glauca* (Germar) | COLEOPTERA: Elateridae |
| dusky birch sawfly | *Croesus latitarsus* Norton | HYMENOPTERA: Tenthredinidae |
| dusky sap beetle | *Carpophilus lugubris* Murray | COLEOPTERA: Nitidulidae |
| dusky stink bug | *Euschistus tristigmus* (Say) | HEMIPTERA: Pentatomidae |

## E

| | | |
|---|---|---|
| ear tick | *Otobius megnini* (Dugès) | ACARI: Argasidae |
| eastern blackheaded budworm | *Acleris variana* (Fernald) | LEPIDOPTERA: Tortricidae |
| eastern field wireworm | *Limonius agonus* (Say) | COLEOPTERA: Elateridae |
| eastern Hercules beetle | *Dynastes tityus* (Linnaeus) | COLEOPTERA: Scarabaeidae |

| | | |
|---|---|---|
| eastern larch beetle | *Dendroctonus simplex* LeConte | COLEOPTERA: Scolytidae |
| eastern lubber grasshopper | *Romalea guttata* (Latreille) | ORTHOPTERA: Acrididae |
| eastern pine seedworm | *Laspeyresia toreuta* (Grote) | LEPIDOPTERA: Tortricidae |
| eastern pine shoot borer | *Eucosma gloriola* Heinrich | LEPIDOPTERA: Tortricidae |
| eastern raspberry fruitworm | *Byturus rubi* Barber | COLEOPTERA: Byturidae |
| eastern spruce gall adelgid | *Adelges abietis* (Linnaeus) | HEMIPTERA: Adelgidae |
| eastern subterranean termite | *Reticulitermes flavipes* (Kollar) | ISOPTERA: Rhinotermitidae |
| eastern tent caterpillar | *Malacosoma americanum* (Fabricius) | LEPIDOPTERA: Lasiocampidae |
| eastern yellowjacket | *Vespula maculifrons* (Buysson) | HYMENOPTERA: Vespidae |
| eggplant flea beetle | *Epitrix fuscula* Crotch | COLEOPTERA: Chrysomelidae |
| eggplant lace bug | *Gargaphia solani* Heidemann | HEMIPTERA: Tingidae |
| eggplant leafminer | *Tildenia inconspicuella* (Murtfeldt) | LEPIDOPTERA: Gelechiidae |
| Egyptian alfalfa weevil | *Hypera brunneipennis* (Boheman) | COLEOPTERA: Curculionidae |
| eightspotted forester | *Alypia octomaculata* (Fabricius) | LEPIDOPTERA: Noctuidae |
| El Segundo blue | *Euphilotes battoides allyni* (Shields) | LEPIDOPTERA: Lycaenidae |
| elder shoot borer | *Achatodes zeae* (Harris) | LEPIDOPTERA: Noctuidae |
| elm borer | *Saperda tridentata* Olivier | COLEOPTERA: Cerambycidae |
| elm calligrapha | *Calligrapha scalaris* (LeConte) | COLEOPTERA: Chrysomelidae |
| elm casebearer | *Coleophora ulmifoliella* McDunnough | LEPIDOPTERA: Coleophoridae |
| elm cockscombgall aphid | *Colopha ulmicola* (Fitch) | HEMIPTERA: Aphididae |
| elm flea beetle | *Altica carinata* Germar | COLEOPTERA: Chrysomelidae |
| elm lace bug | *Corythucha ulmi* Osborn & Drake | HEMIPTERA: Tingidae |
| elm leaf aphid | *Tinocallis ulmifolii* (Monell) | HEMIPTERA: Aphididae |
| elm leaf beetle | *Pyrrhalta luteola* (Müller) | COLEOPTERA: Chrysomelidae |
| elm leafminer | *Fenusa ulmi* Sundevall | HYMENOPTERA: Tenthredinidae |
| elm sawfly | *Cimbex americana* Leach | HYMENOPTERA: Cimbicidae |
| elm scurfy scale | *Chionaspis americana* Johnson | HEMIPTERA: Diaspididae |
| elm spanworm | *Ennomos subsignarius* (Hübner) | LEPIDOPTERA: Geometridae |
| elm sphinx | *Ceratomia amyntor* (Geyer) | LEPIDOPTERA: Sphingidae |
| elongate flea beetle | *Systena elongata* (Fabricius) | COLEOPTERA: Chrysomelidae |
| elongate hemlock scale | *Fiorinia externa* (Ferris) | HEMIPTERA: Diaspididae |
| emerald cockroach wasp | *Ampulex compressa* (Fabricius) | HYMENOPTERA: Ampulicidae |
| Engelmann spruce weevil | *Pissodes strobi* (Peck) | COLEOPTERA: Curculionidae |
| English grain aphid | *Macrosiphum avenae* (Fabricius) | HEMIPTERA: Aphididae |
| erigeron root aphid | *Aphis middletonii* (Thomas) | HEMIPTERA: Aphididae |
| ermine moth | *Yponomeuta padella* (Linnaeus) | LEPIDOPTERA: Yponomeutidae |
| eugenia caterpillar | *Phlegetonia delatrix* Guenée | LEPIDOPTERA: Noctuidae |
| euonymus scale | *Unaspis euonymi* (Comstock) | HEMIPTERA: Diaspididae |
| eupatorium gall fly | *Procecidochares utilis* Stone | DIPTERA: Tephritidae |
| Eurasian pine adelgid | *Pineus pini* (Macquart) | HEMIPTERA: Adelgidae |
| European alder leafminer | *Fenusa dohrnii* (Tischbein) | HYMENOPTERA: Tenthredinidae |
| European apple sawfly | *Hoplocampa testudinea* (Klug) | HYMENOPTERA: Tenthredinidae |
| European chafer | *Rhizotrogus majalis* (Razoumowsky) | COLEOPTERA: Scarabaeidae |
| European chicken flea | *Ceratophyllus gallinae* (Schrank) | SIPHONAPTERA: Ceratophyllidae |
| European corn borer | *Ostrinia nubilalis* (Hübner) | LEPIDOPTERA: Crambidae |
| European crane fly | *Tipula paludosa* Meigen | DIPTERA: Tipulidae |
| European earwig | *Forficula auricularia* Linnaeus | DERMAPTERA: Forficulidae |
| European elm scale | *Gossyparia spuria* (Modeer) | HEMIPTERA: Eriococcidae |
| European fruit lecanium | *Parthenolecanium corni* (Bouché) | HEMIPTERA: Coccidae |
| European fruit scale | *Quadraspidiotus ostreaeformis* (Curtis) | HEMIPTERA: Diaspididae |
| European grain moth | *Nemapogon granella* (Linnaeus) | LEPIDOPTERA: Tineidae |
| European honeysuckle leafroller | *Ypsolophus dentella* (Fabricius) | LEPIDOPTERA: Plutellidae |
| European hornet | *Vespa crabro germana* Christ | HYMENOPTERA: Vespidae |
| European house dust mite | *Dermatophagoides pteronyssinus* (Trouessart) | ACARI: Epidermoptidae |
| European mantid | *Mantis religiosa* Linnaeus | MANTODEA: Mantidae |
| European mouse flea | *Leptopsylla segnis* (Schönherr) | SIPHONAPTERA: Leptopsyllidae |
| European peach scale | *Parthenolecanium persicae* (Fabricius) | HEMIPTERA: Coccidae |
| European pine sawfly | *Neodiprion sertifer* (Geoffroy) | HYMENOPTERA: Diprionidae |
| European pine shoot moth | *Rhyacionia buoliana* (Denis & Schiffermüller) | LEPIDOPTERA: Tortricidae |
| European red mite | *Panonychus ulmi* (Koch) | ACARI: Tetranychidae |

| | | |
|---|---|---|
| European spruce beetle | *Dendroctonus micans* (Kugelann) | COLEOPTERA: Scolytidae |
| European spruce sawfly | *Gilpinia hercyniae* (Hartig) | HYMENOPTERA: Diprionidae |
| European wheat stem sawfly | *Cephus pygmaeus* (Linnaeus) | HYMENOPTERA: Cephidae |
| eyed click beetle | *Alaus oculatus* (Linnaeus) | COLEOPTERA: Elateridae |
| eyespotted bud moth | *Spilonota ocellana* (Denis & Schiffermüller) | LEPIDOPTERA: Tortricidae |

## F

| | | |
|---|---|---|
| face fly | *Musca autumnalis* DeGeer | DIPTERA: Muscidae |
| fall armyworm | *Spodoptera frugiperda* (J. E. Smith) | LEPIDOPTERA: Noctuidae |
| fall cankerworm | *Alsophila pometaria* (Harris) | LEPIDOPTERA: Geometridae |
| fall webworm | *Hyphantria cunea* (Drury) | LEPIDOPTERA: Arctiidae |
| false celery leaftier | *Udea profundalis* (Packard) | LEPIDOPTERA: Crambidae |
| false chinch bug | *Nysius raphanus* Howard | HEMIPTERA: Lygaeidae |
| false German cockroach | *Blattella lituricollis* (Walker) | BLATTODEA: Blattellidae |
| false hemlock looper | *Nepytia canosaria* (Walker) | LEPIDOPTERA: Geometridae |
| false potato beetle | *Leptinotarsa juncta* (Germar) | COLEOPTERA: Chrysomelidae |
| false stable fly | *Muscina stabulans* (Fallén) | DIPTERA: Muscidae |
| feather mite | *Megninia cubitalis* (Mégnin) | ACARI: Analgidae |
| fern aphid | *Idiopterus nephrelepidis* Davis | HEMIPTERA: Aphididae |
| fern caterpillar | *Callopistria* spp. | LEPIDOPTERA: Noctuidae |
| fern scale | *Pinnaspis aspidistrae* (Signoret) | HEMIPTERA: Diaspididae |
| field crickets | *Gryllus* spp. | ORTHOPTERA: Gryllidae |
| fiery hunter | *Calosoma calidum* (Fabricius) | COLEOPTERA: Carabidae |
| fiery skipper | *Hylephila phyleus* (Drury) | LEPIDOPTERA: Hesperiidae |
| fig mite | *Eriophyes ficus* Cotte | ACARI: Eriophyidae |
| fig scale | *Lepidosaphes conchiformis* (Gmelin) | HEMIPTERA: Diaspididae |
| fig wasp | *Blastophaga psenes* (Linnaeus) | HYMENOPTERA: Agaonidae |
| Fijian ginger weevil | *Elytroteinus subtruncatus* (Fairmaire) | COLEOPTERA: Curculionidae |
| filament bearer | *Nematocampa filamentaria* Guenée | LEPIDOPTERA: Geometridae |
| filbert aphid | *Myzocallis coryli* (Goetze) | HEMIPTERA: Aphididae |
| filbert bud mite | *Phytocoptella avellanae* (Nalepa) | ACARI: Nalepellidae |
| filbert weevil | *Curculio uniformis* (LeConte) | COLEOPTERA: Curculionidae |
| filbertworm | *Melissopus latiferreanus* (Walsingham) | LEPIDOPTERA: Tortricidae |
| fir cone looper | *Eupithecia spermaphaga* (Dyar) | LEPIDOPTERA: Geometridae |
| fir engraver | *Scolytus ventralis* LeConte | COLEOPTERA: Scolytidae |
| fir seed moth | *Laspeyresia bracteatana* (Fernald) | LEPIDOPTERA: Tortricidae |
| fire ant | *Solenopsis geminata* (Fabricius) | HYMENOPTERA: Formicidae |
| firebrat | *Thermobia domestica* (Packard) | THYSANURA: Lepismatidae |
| firtree borer | *Semanotus litigiosus* (Casey) | COLEOPTERA: Cerambycidae |
| flat grain beetle | *Cryptolestes pusillus* (Schönherr) | COLEOPTERA: Cucujidae |
| flatheaded appletree borer | *Chrysobothris femorata* (Olivier) | COLEOPTERA: Buprestidae |
| flatheaded cone borer | *Chrysophana placida conicola* Van Dyke | COLEOPTERA: Buprestidae |
| flatheaded fir borer | *Melanophila drummondi* (Kirby) | COLEOPTERA: Buprestidae |
| flax bollworm | *Heliothis ononis* (Denis & Schiffermüller) | LEPIDOPTERA: Noctuidae |
| Fletcher scale | *Parthenolecanium fletcheri* (Cockerell) | HEMIPTERA: Coccidae |
| floodwater mosquito | *Aedes sticticus* (Meigen) | DIPTERA: Culicidae |
| Florida carpenter ant | *Camponotus abdominalis* (Fabricius) | HYMENOPTERA: Formicidae |
| Florida fern caterpillar | *Callopistria floridensis* (Guenée) | LEPIDOPTERA: Noctuidae |
| Florida harvester ant | *Pogonomyrmex badius* (Latreille) | HYMENOPTERA: Formicidae |
| Florida red scale | *Chrysomphalus aonidum* (Linnaeus) | HEMIPTERA: Diaspididae |
| Florida wax scale | *Ceroplastes floridensis* Comstock | HEMIPTERA: Coccidae |
| flower thrips | *Frankliniella tritici* (Fitch) | THYSANOPTERA: Thripidae |
| fluff louse | *Goniocotes gallinae* (DeGeer) | PHTHIRAPTERA: Philopteridae |
| follicle mite | *Demodex folliculorum* (Simon) | ACARI: Demodicidae |
| forage looper | *Caenurgina erechtea* (Cramer) | LEPIDOPTERA: Noctuidae |
| Forbes scale | *Quadraspidiotus forbesi* (Johnson) | HEMIPTERA: Diaspididae |
| foreign grain beetle | *Ahasverus advena* (Waltl) | COLEOPTERA: Cucujidae |
| forest day mosquito | *Aedes albopictus* (Skuse) | DIPTERA: Culicidae |
| forest tent caterpillar | *Malacosoma disstria* Hübner | LEPIDOPTERA: Lasiocampidae |

| | | |
|---|---|---|
| forest tree termite | *Neotennes connexus* Snyder | ISOPTERA: Kalotermitidae |
| forktailed bush katydid | *Scudderia furcata* Brunner von Wattenwyl | ORTHOPTERA: Tettigoniidae |
| Formosan subterranean termite | *Coptotermes formosanus* Shiraki | ISOPTERA: Rhinotermitidae |
| fourlined plant bug | *Poecilocapsus lineatus* (Fabricius) | HEMIPTERA: Miridae |
| fourspotted spider mite | *Tetranychus canadensis* (McGregor) | ACARI: Tetranychidae |
| fourspotted tree cricket | *Oecanthus quadrtpunctatus* Beutenmüller | ORTHOPTERA: Gryllidae |
| fowl tick | *Argas persicus* (Oken) | ACARI: Argasidae |
| foxglove aphid | *Acyrthosiphon solani* (Kaltenbach) | HEMIPTERA: Aphididae |
| frigate bird fly | *Olfersia spinifera* (Leach) | DIPTERA: Hippoboscidae |
| fringed orchid aphid | *Cerataphis orchidearum* (Westwood) | HEMIPTERA: Aphididae |
| frit fly | *Oscinella frit* (Linnaeus) | DIPTERA: Chloropidae |
| fruittree leafroller | *Archips argyrospila* (Walker) | LEPIDOPTERA: Tortricidae |
| Fuller rose beetle | *Pantomorus cenvinus* (Boheman) | COLEOPTERA: Curculionidae |
| furniture beetle | *Anobium punctatum* (DeGeer) | COLEOPTERA: Anobiidae |
| furniture carpet beetle | *Anthrenus flavipes* LeConte | COLEOPTERA: Dermestidae |

## G

| | | |
|---|---|---|
| gallmaking maple borer | *Xylotrechus aceris* Fisher | COLEOPTERA: Cerambycidae |
| garden fleahopper | *Halticus bractatus* (Say) | HEMIPTERA: Miridae |
| garden millipede | *Oxidus gracilis* Koch | POLYDESMIDA: Paradoxosomatidae |
| garden springtail | *Bourletiella hortensis* (Fitch) | COLLEMBOLA: Sminthuridae |
| garden symphylan | *Scutigerella immaculata* (Newport) | SYMPHYLA: Scutigerellidae |
| garden webworm | *Achyra rantalis* (Guenée) | LEPIDOPTERA: Crambidae |
| gardenia bud mite | *Colomerus gardeniella* (Keifer) | ACARI: Eriophyidae |
| genista caterpillar | *Uresiphita reversalis* (Guenée) | LEPIDOPTERA: Crambidae |
| German cockroach | *Blattella germanica* (Linnaeus) | BLATTODEA: Blattellidae |
| giant African snail | *Achatina fulica* Bowdich | STYLOMMATOPHORA: Achatinidae |
| giant bark aphid | *Longistigma caryae* (Harris) | HEMIPTERA: Aphididae |
| giant Hawaiian dragonfly | *Anax strenuus* Hagen | ODONATA: Aeschnidae |
| giant stag beetle | *Lucanus elaphus* Fabricius | COLEOPTERA: Lucanidae |
| giant water bug | *Lethocerus americanus* (Leidy) | HEMIPTERA: Belostomatidae |
| Giffard whitefly | *Bemisia giffardi* (Kotinsky) | HEMIPTERA: Aleyrodidae |
| ginger maggot | *Eumerus figurans* Walker | DIPTERA: Syrphidae |
| gladiolus thrips | *Thrips simplex* (Morison) | THYSANOPTERA: Thripidae |
| glassy cutworm | *Crymodes devastator* (Brace) | LEPIDOPTERA: Noctuidae |
| globose scale | *Sphaerolecanium prunastri* (Fonscolombe) | HEMIPTERA: Coccidae |
| globular spider beetle | *Trigonogenius globulum* Solier | COLEOPTERA: Ptinidae |
| gloomy scale | *Melanaspis tenebricosa* (Comstock) | HEMIPTERA: Diaspididae |
| Glover scale | *Lepidosaphes gloveri* (Packard) | HEMIPTERA: Diaspididae |
| goat biting louse | *Bovicola caprae* Gurlt | PHTHIRAPTERA: Trichodectidae |
| goat follicle mite | *Demodex caprae* Railliet | ACARI: Demodicidae |
| goat sucking louse | *Linognathus stenopsis* (Burmeister) | PHTHIRAPTERA: Linognathidae |
| golden buprestid | *Buprestis aurulenta* Linnaeus | COLEOPTERA: Buprestidae |
| golden cricket wasp | *Liris aurulenta* (Fabricius) | HYMENOPTERA: Sphecidae |
| golden oak scale | *Asterolecanium variolosum* (Ratzeburg) | HEMIPTERA: Asterolecaniidae |
| golden paper wasp | *Polistes fuscatus aurifer* Saussure | HYMENOPTERA: Vespidae |
| golden spider beetle | *Niptus hololeucus* (Faldermann) | COLEOPTERA: Ptinidae |
| golden tortoise beetle | *Metriona bicolor* (Fabricius) | COLEOPTERA: Chrysomelidae |
| goldeneye lacewing | *Chrysoperla oculata* Say | NEUROPTERA: Chrysopidae |
| goldenglow aphid | *Dactynotus rudbeckiae* (Fitch) | HEMIPTERA: Aphididae |
| goose body louse | *Trinoton anserinum* (Fabricius) | PHTHIRAPTERA: Menoponidae |
| gooseberry fruitworm | *Zophodia convolutella* (Hübner) | LEPIDOPTERA: Crambidae |
| gooseberry witchbroom aphid | *Kakimia houghtonensis* (Troop) | HEMIPTERA: Aphididae |
| gophertortoise tick | *Amblyomma tuberculatum* Marx | ACARI: Ixodidae |
| gorse seed weevil | *Apion ulicis* (Forster) | COLEOPTERA: Curculionidae |
| grain mite | *Acarus siro* Linnaeus | ACARI: Acaridae |
| grain rust mite | *Abacarus hystrix* (Nalepa) | ACARI: Eriophyidae |
| grain thrips | *Limothrips cerealium* (Haliday) | THYSANOPTERA: Thripidae |
| granary weevil | *Sitophilus granarius* (Linnaeus) | COLEOPTERA: Curculionidae |

| | | |
|---|---|---|
| granulate cutworm | *Feltia subterranea* (Fabricius) | LEPIDOPTERA: Noctuidae |
| grape berry moth | *Endopiza viteana* Clemens | LEPIDOPTERA: Tortricidae |
| grape blossom midge | *Contarinia johnsoni* Felt | DIPTERA: Cecidomyiidae |
| grape cane gallmaker | *Ampeloglypter sesostris* (LeConte) | COLEOPTERA: Curculionidae |
| grape colaspis | *Colaspsis brunnea* (Fabricius) | COLEOPTERA: Chrysomelidae |
| grape curculio | *Craponius inaequalis* (Say) | COLEOPTERA: Curculionidae |
| grape erineum mite | *Colomerus vitis* (Pagenstecher) | ACARI: Eriophyidae |
| grape flea beetle | *Altica chalybea* Illiger | COLEOPTERA: Chrysomelidae |
| grape leaffolder | *Destnia funeralis* (Hübner) | LEPIDOPTERA: Crambidae |
| grape mealybug | *Pseudococcus maritimus* (Ehrhorn) | HEMIPTERA: Pseudococcidae |
| grape phylloxera | *Daktalosphaira vitifoliae* (Fitch) | HEMIPTERA: Phylloxeridae |
| grape plume moth | *Pterophorus periscelidactylus* Fitch | LEPIDOPTERA: Pterophoridae |
| grape root borer | *Vitacea polistiformis* (Harris) | LEPIDOPTERA: Sesiidae |
| grape rootworm | *Fidia viticida* Walsh | COLEOPTERA: Chrysomelidae |
| grape sawfly | *Erythraspides vitis* (Harris) | HYMENOPTERA: Tenthredinidae |
| grape scale | *Diaspidiotus uvae* (Comstock) | HEMIPTERA: Diaspididae |
| grape seed chalcid | *Eroxysoma vitis* (Saunders) | HYMENOPTERA: Eurytomidae |
| grape trunk borer | *Clytoleptus albofasciatus* (Laporte & Gory) | COLEOPTERA: Cerambycidae |
| grape whitefly | *Trialeurodes vittata* (Quaintance) | HEMIPTERA: Aleyrodidae |
| grapeleaf skeletonizer | *Harrisina americana* (Guérin) | LEPIDOPTERA: Zygaenidae |
| grapevine aphid | *Aphis illinoisensis* Shimer | HEMIPTERA: Aphididae |
| grapevine looper | *Eulythis diversilineata* (Hübner) | LEPIDOPTERA: Geometridae |
| grass fleahopper | *Halticus chrysolepis* Kirkaldy | HEMIPTERA: Miridae |
| grass mite | *Siteroptes graminum* (Reuter) | ACARI: Siteroptidae |
| grass sawfly | *Pachynematus extensicornis* (Norton) | HYMENOPTERA: Tenthredinidae |
| grass scolytid | *Hypothenemus pubescens* Hopkins | COLEOPTERA: Scolytidae |
| grass sharpshooter | *Draeculacephala minerva* Ball | HEMIPTERA: Cicadellidae |
| grass thrips | *Anaphothrips obscurus* (Müller) | THYSANOPTERA: Thripidae |
| grass webworm | *Herpetogramma licarsisalis* (Walker) | LEPIDOPTERA: Crambidae |
| grasshopper bee fly | *Systoechus vulgaris* Loew | DIPTERA: Bombyliidae |
| grasshopper maggots | *Blaesoxipha* spp. | DIPTERA: Sarcophagidae |
| gray lawn leafhopper | *Exitianus exitiosus* (Uhler) | HEMIPTERA: Cicadellidae |
| gray pineapple mealybug | *Dysmicoccus neobrevipes* Beardsley | HEMIPTERA: Pseudococcidae |
| gray sugarcane mealybug | *Dysmicoccus boninsis* (Kuwana) | HEMIPTERA: Pseudococcidae |
| gray willow leaf beetle | *Pyrrhalta decora decora* (Say) | COLEOPTERA: Chrysomelidae |
| graybanded leafroller | *Argyrotaenia mariana* (Fernald) | LEPIDOPTERA: Tortricidae |
| great ash sphinx | *Sphinx chersis* (Hübner) | LEPIDOPTERA: Sphingidae |
| Great Basin wireworm | *Ctenicera pruinina* (Horn) | COLEOPTERA: Elateridae |
| greater wax moth | *Galleria mellonella* (Linnaeus) | LEPIDOPTERA: Crambidae |
| greedy scale | *Hemiberlesia rapax* (Comstock) | HEMIPTERA: Diaspididae |
| green budworm | *Hedya nubiferana* (Haworth) | LEPIDOPTERA: Tortricidae |
| green cloverworm | *Plathypena scabra* (Fabricius) | LEPIDOPTERA: Noctuidae |
| green fruitworm | *Lithophane antennata* (Walker) | LEPIDOPTERA: Noctuidae |
| green garden looper | *Chrysodeixis chalcites* (Esper) | LEPIDOPTERA: Noctuidae |
| green June beetle | *Cotinis nitida* (Linnaeus) | COLEOPTERA: Scarabaeidae |
| green peach aphid | *Myzus persicae* (Sulzer) | HEMIPTERA: Aphididae |
| green rose chafer | *Dichelonyx backi* (Kirby) | COLEOPTERA: Scarabaeidae |
| green scale | *Coccus viridis* (Green) | HEMIPTERA: Coccidae |
| green shield scale | *Pulvinaria psidii* Maskell | HEMIPTERA: Coccidae |
| green sphinx | *Tinostoma smaragditis* (Meyrick) | LEPIDOPTERA: Sphingidae |
| green spruce aphid | *Cinara fornacula* Hottes | HEMIPTERA: Aphididae |
| green stink bug | *Acrosternum hilare* (Say) | HEMIPTERA: Pentatomidae |
| greenbug | *Schizaphis graminum* (Rondani) | HEMIPTERA: Aphididae |
| greenheaded spruce sawfly | *Pikonema dimmockii* (Cresson) | HYMENOPTERA: Tenthredinidae |
| greenhouse leaftier | *Udea rubigalis* (Guenée) | LEPIDOPTERA: Crambidae |
| greenhouse orthezii | *Orthezia insignis* Browne | HEMIPTERA: Ortheziidae |
| greenhouse stone cricket | *Tachycines asynamorus* Adelung | ORTHOPTERA: Gryllacrididae |
| greenhouse thrips | *Heliothrips haemorrhoidalis* (Bouché) | THYSANOPTERA: Thripidae |
| greenhouse whitefly | *Trialeurodes vaporariorum* (Westwood) | HEMIPTERA: Aleyrodidae |
| greenstriped grasshopper | *Chortophaga viridifasciata* (DeGeer) | ORTHOPTERA: Acrididae |

| | | |
|---|---|---|
| greenstriped mapleworm | *Dryocampa rubicunda* (Fabricius) | LEPIDOPTERA: Saturniidae |
| gregarious oak leafminer | *Cameraria cincinnatiella* (Chambers) | LEPIDOPTERA: Gracillariidae |
| ground mealybug | *Rhizoecus falcifer* Kunckel d'Herculais | HEMIPTERA: Pseudococcidae |
| Guinea ant | *Tetramorium guineense* (Fabricius) | HYMENOPTERA: Formicidae |
| Guinea feather louse | *Goniodes numidae* Mjöberg | PHTHIRAPTERA: Philopteridae |
| Gulf Coast tick | *Amblyomma maculatum* Koch | ACARI: Ixodidae |
| Gulf wireworm | *Conoderus amplicollis* (Gyllenhal) | COLEOPTERA: Elateridae |
| gypsy moth | *Lymantria dispar* (Linnaeus) | LEPIDOPTERA: Lymantriidae |

## H

| | | |
|---|---|---|
| hackberry engraver | *Scolytus muticus* Say | COLEOPTERA: Scolytidae |
| hackberry lace bug | *Corythucha celtidis* Osborn & Drake | HEMIPTERA: Tingidae |
| hackberry nipplegall maker | *Pachypsylla celtidismamma* (Riley) | HEMIPTERA: Psyllidae |
| hag moth | *Phobetron pithecium* (J. E. Smith) | LEPIDOPTERA: Limacodidae |
| hairy chinch bug | *Blissus leucopterus hirtus* Montandon | HEMIPTERA: Lygaeidae |
| hairy fungus beetle | *Typhaea stercorea* (Linnaeus) | COLEOPTERA: Mycetophagidae |
| hairy maggot blow fly | *Chrysomya rufifacies* (Macquart) | DIPTERA: Calliphoridae |
| hairy rove beetle | *Creophilus maxillosus* (Linnaeus) | COLEOPTERA: Staphylinidae |
| hairy spider beetle | *Ptinis villiger* (Reitter) | COLEOPTERA: Ptinidae |
| Hall scale | *Nilotaspis halli* (Green) | HEMIPTERA: Diaspididae |
| hard maple budminer | *Obrussa ochrefasciella* (Chambers) | LEPIDOPTERA: Nepticulidae |
| harlequin bug | *Murgantia histrionica* (Hahn) | HEMIPTERA: Pentatomidae |
| harlequin cockroach | *Neostylopyga rhombifolia* (Stoll) | BLATTODEA: Blattidae |
| hau leafminer | *Philodoria hauicola* (Swezey) | LEPIDOPTERA: Gracillariidae |
| Hawaiian antlion | *Eidoleon wilsoni* (McLachlan) | NEUROPTERA: Myrmeleontidae |
| Hawaiian beet webworm | *Spoladea recurvalis* (Fabricius) | LEPIDOPTERA: Crambidae |
| Hawaiian bud moth | *Heliothis hawaiiensis* (Quaintance & Brues) | LEPIDOPTERA: Noctuidae |
| Hawaiian carpenter ant | *Camponotus variegatus* (F. Smith) | HYMENOPTERA: Formicidae |
| Hawaiian flower thrips | *Thrips hawaiiensis* (Morgan) | THYSANOPTERA: Thripidae |
| Hawaiian grass thrips | *Anaphothrips swezeyi* Moulton | THYSANOPTERA: Thripidae |
| Hawaiian pelagic water strider | *Halobates hawaiiensis* Usinger | HEMIPTERA: Gerridae |
| Hawaiian sphinx | *Hyles calida* (Butler) | LEPIDOPTERA: Sphingidae |
| hawthorn lace bug | *Corythucha cydoniae* (Fitch) | HEMIPTERA: Tingidae |
| hazelnut weevil | *Curculio neocorylus* Gibson | COLEOPTERA: Curculionidae |
| head louse | *Pediculus humanus capitis* DeGeer | PHTHIRAPTERA: Pediculidae |
| heath spittlebug | *Clastoptera saintcyri* Provancher | HEMIPTERA: Cercopidae |
| hellgrammite | *Corydalus cornutus* (Linnaeus) | NEUROPTERA: Corydalidae |
| hemispherical scale | *Saissetia coffeae* (Walker) | HEMIPTERA: Coccidae |
| hemlock borer | *Melanophila fulvoguttata* (Harris) | COLEOPTERA: Buprestidae |
| hemlock looper | *Lambdina fiscellaria fiscellaria* (Guenée) | LEPIDOPTERA: Geometridae |
| hemlock sawfly | *Neodiprion tsugae* Middleton | HYMENOPTERA: Diprionidae |
| hemlock scale | *Abgrallaspis ithacae* (Ferris) | HEMIPTERA: Diaspididae |
| Hessian fly | *Mayetiola destructor* (Say) | DIPTERA: Cecidomyiidae |
| hibiscus leafminer | *Philodoria hibiscella* (Swezey) | LEPIDOPTERA: Gracillariidae |
| hibiscus mealybug | *Nipaecoccus vastator* (Maskell) | HEMIPTERA: Pseudococcidae |
| hibiscus whitefly | *Pealius hibisci* (Kotinsky) | HEMIPTERA: Aleyrodidae |
| hickory bark beetle | *Scolytus quadrispinosus* Say | COLEOPTERA: Scolytidae |
| hickory horned devil | *Citheronia regalis* (Fabricius) | LEPIDOPTERA: Saturniidae |
| hickory leafroller | *Argyrotaenia juglandana* (Fernald) | LEPIDOPTERA: Tortricidae |
| hickory plant bug | *Lygocoris caryae* (Knight) | HEMIPTERA: Miridae |
| hickory shuckworm | *Laspeyresia caryana* (Fitch) | LEPIDOPTERA: Tortricidae |
| hickory tussock moth | *Lophocampa caryae* (Harris) | LEPIDOPTERA: Arctiidae |
| hide beetle | *Dermestes maculatus* DeGeer | COLEOPTERA: Dermestidae |
| High Plains grasshopper | *Dissosteira longipennis* (Thomas) | ORTHOPTERA: Acrididae |
| hog follicle mite | *Demodex phylloides* Csokor | ACARI: Demodicidae |
| hog louse | *Haematopinus suis* (Linnaeus) | PHTHIRAPTERA: Haematopinidae |
| holly leafminer | *Phytomyza ilicis* Curtis | DIPTERA: Agromyzidae |
| holly scale | *Dynaspidiotus britannicus* (Newstead) | HEMIPTERA: Diaspididae |
| hollyhock plant bug | *Brooksetta althaeae* (Hussey) | HEMIPTERA: Miridae |

| | | |
|---|---|---|
| hollyhock weevil | *Apion longirostre* Olivier | COLEOPTERA: Curculionidae |
| honey bee | *Apis mellifera* Linnaeus | HYMENOPTERA: Apidae |
| honey bee mite | *Acarapis woodi* (Rennie) | ACARI: Tarsonemidae |
| honeylocust plant bug | *Diaphnocoris chlorionis* (Say) | HEMIPTERA: Miridae |
| honeysuckle leafminer | *Swezeyula lonicerae* Zimmerman & Bradley | LEPIDOPTERA: Elachistidae |
| honeysuckle sawfly | *Zaraea inflata* Norton | HYMENOPTERA: Cimbicidae |
| hop aphid | *Phorodon humuli* (Schrank) | HEMIPTERA: Aphididae |
| hop flea beetle | *Psylliodes punctulata* Melsheimer | COLEOPTERA: Chrysomelidae |
| hop looper | *Hypena humuli* (Harris) | LEPIDOPTERA: Noctuidae |
| hop plant bug | *Taedia hawleyi* (Knight) | HEMIPTERA: Miridae |
| hop vine borer | *Hydraecia immanis* Guenée | LEPIDOPTERA: Noctuidae |
| horn fly | *Haematobia irritans* (Linnaeus) | DIPTERA: Muscidae |
| horned passalus | *Odontotaenius disjunctus* (Illiger) | COLEOPTERA: Passalidae |
| horned squash bug | *Anasa armigera* (Say) | HEMIPTERA: Coreidae |
| hornet moth | *Sesia apiformis* (Clerck) | LEPIDOPTERA: Sesiidae |
| hornets | *Vespidae* spp. | HYMENOPTERA: Vespidae |
| horse biting louse | *Bovicola equi* (Denny) | PHTHIRAPTERA: Trichodectidae |
| horse bot fly | *Gasterophilus intestinalis* (DeGeer) | DIPTERA: Gasterophilidae |
| horse follicle mite | *Demodex equi* Railliet | ACARI: Demodicidae |
| horse sucking louse | *Haematopinus asini* (Linnaeus) | PHTHIRAPTERA: Haematopinidae |
| horseradish flea beetle | *Phyllotreta armoraciae* (Koch) | COLEOPTERA: Chrysomelidae |
| house centipede | *Scutigera coleoptrata* (Linnaeus) | SCUTIGEROMORPHA: Scutigeridae |
| house cricket | *Acheta domesticus* (Linnaeus) | ORTHOPTERA: Gryllidae |
| house fly | *Musca domestica* Linnaeus | DIPTERA: Muscidae |
| house mite | *Glycyphagus domesticus* (DeGeer) | ACARI: Glycyphagidae |
| house mouse mite | *Lyponyssoides sanguineus* (Hirst) | ACARI: Macronyssidae |
| household casebearer | *Phereoeca uterella* Walsingham | LEPIDOPTERA: Tineidae |
| human flea | *Pulex irritans* Linnaeus | SIPHONAPTERA: Pulicidae |
| hunting billbug | *Sphenophorus venatus vestitus* Chittenden | COLEOPTERA: Curculionidae |
| hyaline grass bug | *Liorhyssus hyalinus* (Fabricius) | HEMIPTERA: Rhopalidae |

## I

| | | |
|---|---|---|
| ilima leafminer | *Philodoria marginestrigata* (Walsingham) | LEPIDOPTERA: Gracillariidae |
| ilima moth | *Amyna natalis* (Walker) | LEPIDOPTERA: Noctuidae |
| imbricated snout beetle | *Epicaerus imbricatus* (Say) | COLEOPTERA: Curculionidae |
| immigrant acacia weevil | *Orthorhinus klugi* Boheman | COLEOPTERA: Curculionidae |
| imperial moth | *Eacles imperialis* (Drury) | LEPIDOPTERA: Saturniidae |
| imported cabbageworm | *Artogeia rapae* (Linnaeus) | LEPIDOPTERA: Pieridae |
| imported crucifer weevil | *Baris lepidii* Germar | COLEOPTERA: Curculionidae |
| imported currantworm | *Nematus ribesii* (Scopoli) | HYMENOPTERA: Tenthredinidae |
| imported longhorned weevil | *Calomycterus setarius* Roelofs | COLEOPTERA: Curculionidae |
| imported willow leaf beetle | *Plagiodera versicolora* (Laicharting) | COLEOPTERA: Chrysomelidae |
| incense-cedar wasp | *Syntexis libocedrii* Rohwer | HYMENOPTERA: Anaxyelidae |
| Indianmeal moth | *Plodia interpunctella* (Hübner) | LEPIDOPTERA: Crambidae |
| inornate scale | *Aonidiella inornata* McKenzie | HEMIPTERA: Diaspididae |
| introduced pine sawfly | *Diprion similis* (Hartig) | HYMENOPTERA: Diprionidae |
| io moth | *Automeris io* (Fabricius) | LEPIDOPTERA: Saturniidae |
| iris borer | *Macronoctua onusta* Grote | LEPIDOPTERA: Noctuidae |
| iris thrips | *Frankliniella iridis* (Watson) | THYSANOPTERA: Thripidae |
| iris weevil | *Mononychus vulpeculus* (Fabricius) | COLEOPTERA: Curculionidae |
| Italian pear scale | *Epidiaspis leperii* (Signoret) | HEMIPTERA: Diaspididae |
| itch mite | *Sarcoptes scabiei* (DeGeer) | ACARI: Sarcoptidae |
| ivy aphid | *Aphis hederae* Kaltenbach | HEMIPTERA: Aphididae |

## J

| | | |
|---|---|---|
| jack pine budworm | *Choristoneura pinus* Freeman | LEPIDOPTERA: Tortricidae |
| jack pine sawfly | *Neodiprion pratti banksianae* Rohwer | HYMENOPTERA: Diprionidae |
| jack pine tip beetle | *Conophthorus banksianae* McPherson | COLEOPTERA: Scolytidae |

| | | |
|---|---|---|
| Japanese beetle | *Popillia japonica* Newman | COLEOPTERA: Scarabaeidae |
| Japanese broadwinged katydid | *Holochlora japonica* Brunner von Wattenwyl | ORTHOPTERA: Tettigoniidae |
| Japanese grasshopper | *Oxya japonica* (Thunberg) | ORTHOPTERA: Acrididae |
| Jeffrey pine beetle | *Dendroctonus jeffreyi* Hopkins | COLEOPTERA: Scolytidae |
| Jerusalem cricket | *Stenopelmatus fuscus* Haldeman | ORTHOPTERA: Stenopelmatidae |
| juniper midge | *Contarinia juniperina* Felt | DIPTERA: Cecidomyiidae |
| juniper scale | *Carulaspis juniperi* (Bouché) | HEMIPTERA: Diaspididae |
| juniper tip midge | *Oligotrophus betheli* Felt | DIPTERA: Cecidomyiidae |
| juniper webworm | *Dichomeris marginella* (Fabricius) | LEPIDOPTERA: Gelechiidae |

## K

| | | |
|---|---|---|
| Kamehameha butterfly | *Vanessa tameamea* Eschscholtz | LEPIDOPTERA: Nymphalidae |
| keyhole wasp | *Pachodynerus nasidens* (Latreille) | HYMENOPTERA: Vespidae |
| khapra beetle | *Trogoderma granarium* Everts | COLEOPTERA: Dermestidae |
| kiawe bean weevil | *Algarobius bottimeri* Kingsolver | COLEOPTERA: Bruchidae |
| kiawe flower moth | *Ithome concolorella* (Chambers) | LEPIDOPTERA: Cosmopterigidae |
| kiawe roundheaded borer | *Placosternus cyclene crinicornis* (Chevrolat) | COLEOPTERA: Cerambycidae |
| kiawe scolytid | *Hypothenemus birmanus* (Eichhoff) | COLEOPTERA: Scolytidae |
| Kirkaldy whitefly | *Dialeurodes kirkaldyi* (Kotinsky) | HEMIPTERA: Aleyrodidae |
| Klamathweed beetle | *Chrysolina quadrigemina* (Suffrian) | COLEOPTERA: Chrysomelidae |
| koa bug | *Coleotichus blackburniae* White | HEMIPTERA: Pentatomidae |
| koa haole seed weevil | *Araecerus levipennis* Jordan | COLEOPTERA: Anthribidae |
| koa moth | *Scotorythra paludicola* (Butler) | LEPIDOPTERA: Geometridae |
| koa seedworm | *Cryptophlebia illepida* (Butler) | LEPIDOPTERA: Tortricidae |
| kou leafworm | *Ethmia nigroapicella* (Saalmüller) | LEPIDOPTERA: Oecophoridae |

## L

| | | |
|---|---|---|
| Lange metalmark | *Apodemia mormo langei* Comstock | LEPIDOPTERA: Riodinidae |
| lantana cerambycid | *Plagiohammus spinipennis* Thomson | COLEOPTERA: Cerambycidae |
| lantana defoliator caterpillar | *Hypena strigata* (Fabricius) | LEPIDOPTERA: Noctuidae |
| lantana gall fly | *Eutreta xanthochaeta* Aldrich | DIPTERA: Tephritidae |
| lantana hispid | *Uroplata girardi* Pic | COLEOPTERA: Chrysomelidae |
| lantana lace bug | *Teleonemia scrupulosa* Stål | HEMIPTERA: Tingidae |
| lantana leaf beetle | *Octotoma scabripennis* Guérin-Méneville | COLEOPTERA: Chrysomelidae |
| lantana leafminer | *Cremastobombycia lantanella* (Schrank) | LEPIDOPTERA: Gracillariidae |
| lantana leaftier | *Salbia haemorrhoidalis* Guenée | LEPIDOPTERA: Crambidae |
| lantana plume moth | *Lantanophaga pusillodactyla* (Walker) | LEPIDOPTERA: Pterophoridae |
| lantana seed fly | *Ophiomyia lantanae* (Froggatt) | DIPTERA: Agromyzidae |
| lantana stick caterpillar | *Neogalea esula* (Druce) | LEPIDOPTERA: Noctuidae |
| lappet moth | *Phyllodesma americana* (Harris) | LEPIDOPTERA: Lasiocampidae |
| larch aphid | *Cinara laricis* (Walker) | HEMIPTERA: Aphididae |
| larch casebearer | *Coleophora laricella* (Hübner) | LEPIDOPTERA: Coleophoridae |
| larch sawfly | *Pristiphora erichsonii* (Hartig) | HYMENOPTERA: Tenthredinidae |
| larder beetle | *Dermestes lardarius* Linnaeus | COLEOPTERA: Dermestidae |
| large aspen tortrix | *Choristoneura conflictana* (Walker) | LEPIDOPTERA: Tortricidae |
| large bigeyed bug | *Geocoris bullatus* (Say) | HEMIPTERA: Lygaeidae |
| large brown spider | *Heteropoda venatoria* (Linnaeus) | ARANEAE: Sparassidae |
| large chestnut weevil | *Curculio caryatrypes* (Boheman) | COLEOPTERA: Curculionidae |
| large chicken louse | *Goniodes gigas* (Taschenberg) | PHTHIRAPTERA: Philopteridae |
| large cottony scale | *Pulvinaria mammeae* Maskell | HEMIPTERA: Coccidae |
| large duck louse | *Trinoton querquedulae* (Linnaeus) | PHTHIRAPTERA: Menoponidae |
| large kissing bug | *Triatoma rubrofasciata* (DeGeer) | HEMIPTERA: Reduviidae |
| large milkweed bug | *Oncopeltus fasciatus* (Dallas) | HEMIPTERA: Lygaeidae |
| large turkey louse | *Chelopistes meleagridis* (Linnaeus) | PHTHIRAPTERA: Philopteridae |
| larger black flour beetle | *Cynaeus angustus* (LeConte) | COLEOPTERA: Tenebrionidae |
| larger canna leafroller | *Calpodes ethlius* (Stoll) | LEPIDOPTERA: Hesperiidae |
| larger elm leaf beetle | *Monocesta coryli* (Say) | COLEOPTERA: Chrysomelidae |
| larger grain borer | *Prostephanus truncatus* (Horn) | COLEOPTERA: Bostrichidae |

| | | |
|---|---|---|
| larger Hawaiian cutworm | *Agrotis crinigera* (Butler) | LEPIDOPTERA: Noctuidae |
| larger lantana butterfly | *Strymon echion* (Linnaeus) | LEPIDOPTERA: Lycaenidae |
| larger pale trogiil | *Trogium pulsatorium* (Linnaeus) | PSOCOPTERA: Trogiidae |
| larger shothole borer | *Scolytus mali* (Bechstein) | COLEOPTERA: Scolytidae |
| larger yellow ant | *Acanthomyops interjectus* (Mayr) | HYMENOPTERA: Formicidae |
| latania scale | *Hemiberlesia lataniae* (Signoret) | HEMIPTERA: Diaspididae |
| latrine fly | *Fannia scalaris* (Fabricius) | DIPTERA: Muscidae |
| lawn armyworm | *Spodoptera mauritia* (Boisduval) | LEPIDOPTERA: Noctuidae |
| lawn leafhopper | *Recilia hospes* (Kirkaldy) | HEMIPTERA: Cicadellidae |
| leadcable borer | *Scobicia declivis* (LeConte) | COLEOPTERA: Bostrichidae |
| leaf crumpler | *Acrobasis indigenella* (Zeller) | LEPIDOPTERA: Crambidae |
| leaffooted bug | *Leptoglossus phyllopus* (Linnaeus) | HEMIPTERA: Coreidae |
| leaffooted pine seed bug | *Leptoglossus corculus* (Say) | HEMIPTERA: Coreidae |
| leafhopper assassin bug | *Zelus renardii* Kolenati | HEMIPTERA: Reduviidae |
| leek moth | *Acrolepiopsis assectella* (Zeller) | LEPIDOPTERA: Acrolepiidae |
| leopard moth | *Zeuzera pyrina* (Linnaeus) | LEPIDOPTERA: Cossidae |
| lespedeza webworm | *Tetralopha scortealis* (Lederer) | LEPIDOPTERA: Crambidae |
| lesser appleworm | *Grapholita prunivora* (Walsh) | LEPIDOPTERA: Tortricidae |
| lesser bud moth | *Recunvaria nanella* (Denis & Schiffermüller) | LEPIDOPTERA: Gelechiidae |
| lesser bulb fly | *Eumerus tuberculatus* Rondani | DIPTERA: Syrphidae |
| lesser canna leafroller | *Geshna cannalis* (Quaintance) | LEPIDOPTERA: Crambidae |
| lesser clover leaf weevil | *Hypera nigrirostris* (Fabricius) | COLEOPTERA: Curculionidae |
| lesser cornstalk borer | *Elasmopalpus lignosellus* (Zeller) | LEPIDOPTERA: Crambidae |
| lesser ensign wasp | *Szepligetella sericia* (Cameron) | HYMENOPTERA: Evaniidae |
| lesser follicle mite | *Demodex brevis* Bulanova | ACARI: Demodicidae |
| lesser grain borer | *Rhyzopertha dominica* (Fabricius) | COLEOPTERA: Bostrichidae |
| lesser lawn leafhopper | *Graminella sonorus* (Ball) | HEMIPTERA: Cicadellidae |
| lesser mealworm | *Alphitobius diaperinus* (Panzer) | COLEOPTERA: Tenebrionidae |
| lesser orchid weevil | *Orchidophilus peregrinator* Buchanan | COLEOPTERA: Curculionidae |
| lesser peachtree borer | *Synanthedon pictipes* (Grote & Robinson) | LEPIDOPTERA: Sesiidae |
| lesser wax moth | *Achroia grisella* (Fabricius) | LEPIDOPTERA: Crambidae |
| lettuce root aphid | *Pemphigus bursarius* (Linnaeus) | HEMIPTERA: Aphididae |
| light brown apple moth | *Austrotortrix postvittana* (Walker) | LEPIDOPTERA: Tortricidae |
| lilac borer | *Podosesia syringae* (Harris) | LEPIDOPTERA: Sesiidae |
| lilac leafminer | *Caloptilia syringella* (Fabricius) | LEPIDOPTERA: Gracillariidae |
| lily bulb thrips | *Liothrips vaneeckei* Priesner | THYSANOPTERA: Phlaeothripidae |
| lily weevil | *Agasphaerops nigra* Horn | COLEOPTERA: Curculionidae |
| limabean pod borer | *Etiella zinckenella* (Treitschke) | LEPIDOPTERA: Crambidae |
| limabean vine borer | *Monoptilota pergratialis* (Hulst) | LEPIDOPTERA: Crambidae |
| linden borer | *Saperda vestita* Say | COLEOPTERA: Cerambycidae |
| linden looper | *Erannis tiliaria* (Harris) | LEPIDOPTERA: Geometridae |
| lined click beetle | *Agriotes lineatus* (Linnaeus) | COLEOPTERA: Elateridae |
| lined spittlebug | *Neophilaenus lineatus* (Linnaeus) | HEMIPTERA: Cercopidae |
| lined stalk borer | *Oligia fractilinea* (Grote) | LEPIDOPTERA: Noctuidae |
| lion beetle | *Ulochaetes leoninus* LeConte | COLEOPTERA: Cerambycidae |
| litchi fruit moth | *Cryptophlebia ombrodelta* (Lower) | LEPIDOPTERA: Tortricidae |
| litchi mite | *Eriophyes litchii* Keifer | ACARI: Eriophyidae |
| little black ant | *Monomorium minimum* (Buckley) | HYMENOPTERA: Formicidae |
| little carpenterworm | *Prionoxystus macmurtrei* (Guérin) | LEPIDOPTERA: Cossidae |
| little fire ant | *Ochetomyrmex auropunctatus* (Roger) | HYMENOPTERA: Formicidae |
| little green leafhopper | *Balclutha hospes* (Kirkaldy) | HEMIPTERA: Cicadellidae |
| little house fly | *Fannia canicularis* (Linnaeus) | DIPTERA: Muscidae |
| little yellow ant | *Plagiolepis alluaudi* Emery | HYMENOPTERA: Formicidae |
| loblolly pine sawfly | *Neodiprion taedae linearis* Ross | HYMENOPTERA: Diprionidae |
| locust borer | *Megacyllene robiniae* (Forster) | COLEOPTERA: Cerambycidae |
| locust leafminer | *Odontota dorsalis* (Thunberg) | COLEOPTERA: Chrysomelidae |
| locust leafroller | *Nephopterix subcaesiella* (Clemens) | LEPIDOPTERA: Crambidae |
| locust twig borer | *Ecdytolopha insiticiana* Zeller | LEPIDOPTERA: Tortricidae |
| lodgepole cone beetle | *Conophthorus contortae* Hopkins | COLEOPTERA: Scolytidae |

| | | |
|---|---|---|
| lodgepole needleminer | *Coleotechnites milleri* (Busck) | LEPIDOPTERA: Gelechiidae |
| lodgepole pine beetle | *Dendroctonus murrayanae* Hopkins | COLEOPTERA: Scolytidae |
| lodgepole sawfly | *Neodiprion burkei* Middleton | HYMENOPTERA: Diprionidae |
| lodgepole terminal weevil | *Pissodes tenninalis* Hopping | COLEOPTERA: Curculionidae |
| lone star tick | *Amblyomma americanum* (Linnaeus) | ACARI: Ixodidae |
| long brown scale | *Coccus longulus* (Douglas) | HEMIPTERA: Coccidae |
| longheaded flour beetle | *Latheticus oryzae* Waterhouse | COLEOPTERA: Tenebrionidae |
| longleaf pine seedworm | *Laspeyresia ingens* Heinrich | LEPIDOPTERA: Tortricidae |
| longlegged ant | *Anoplolepis longipes* (Jerdon) | HYMENOPTERA: Formicidae |
| longnosed cattle louse | *Linognathus vituli* (Linnaeus) | PHTHIRAPTERA: Linognathidae |
| longtailed fruit fly parasite | *Opius longicaudatus* Ashmead | HYMENOPTERA: Braconidae |
| longtailed mealybug | *Pseudococcus longispinus* (Targioni-Tozzetti) | HEMIPTERA: Pseudococcidae |
| lotis blue | *Lycaeides argyrognomon lotis* (Lintner) | LEPIDOPTERA: Lycaenidae |
| lowland tree termite | *Incisitermes immigrans* (Snyder) | ISOPTERA: Kalotermitidae |
| lubber grasshopper | *Brachystola magna* (Girard) | ORTHOPTERA: Acrididae |
| luna moth | *Actias luna* (Linnaeus) | LEPIDOPTERA: Saturniidae |

## M

| | | |
|---|---|---|
| Macao paper wasp | *Polistes macaensis* (Fabricius) | HYMENOPTERA: Vespidae |
| Madeira cockroach | *Leucophaea maderae* (Fabricius) | ORTHOPTERA: Blaberidae |
| magnolia scale | *Neolecanium cornuparvum* (Thro) | HEMIPTERA: Coccidae |
| maize billbug | *Sphenophorus maidis* Chittenden | COLEOPTERA: Curculionidae |
| maize weevil | *Sitophilus zeamais* Motschulsky | COLEOPTERA: Curculionidae |
| mango bark beetle | *Hypocryphalus mangiferae* (Stebbing) | COLEOPTERA: Scolytidae |
| mango bud mite | *Eriophyes mangiferae* (Sayed) | ACARI: Eriophyidae |
| mango flower beetle | *Protaetia fusca* (Herbst) | COLEOPTERA: Scarabaeidae |
| mango shoot caterpillar | *Bombotelia jocosatrix* Guenée | LEPIDOPTERA: Noctuidae |
| mango spider mite | *Oligonychus mangiferus* (Rahman & Sapra) | ACARI: Tetranychidae |
| mango weevil | *Cryptorhynchus mangiferae* (Fabricius) | COLEOPTERA: Curculionidae |
| maple bladdergall mite | *Vasates quadripedes* Shimer | ACARI: Eriophyidae |
| maple callus borer | *Synanthedon acerni* (Clemens) | LEPIDOPTERA: Sesiidae |
| maple leafcutter | *Paraclemensia acerifoliella* (Fitch) | LEPIDOPTERA: Incurvariidae |
| maple petiole borer | *Caulocampus acericaulis* (MacGillivray) | HYMENOPTERA: Tenthredinidae |
| maple trumpet skeletonizer | *Epinotia aceriella* (Clemens) | LEPIDOPTERA: Tortricidae |
| margined blister beetle | *Epicauta pestifera* Werner | COLEOPTERA: Meloidae |
| masked hunter | *Reduvius personatus* (Linnaeus) | HEMIPTERA: Reduviidae |
| mauna loa bean beetle | *Araeocorynus cumingi* Jekel | COLEOPTERA: Anthribidae |
| McDaniel spider mite | *Tetranychus mcdanieli* McGregor | ACARI: Tetranychidae |
| meadow plant bug | *Leptopterna dolabrata* (Linnaeus) | HEMIPTERA: Miridae |
| meadow spittlebug | *Philaenus spumarius* (Linnaeus) | HEMIPTERA: Cercopidae |
| meal moth | *Pyralis farinalis* Linnaeus | LEPIDOPTERA: Crambidae |
| mealy plum aphid | *Hyalopterus pruni* (Geoffroy) | HEMIPTERA: Aphididae |
| mealybug destroyer | *Cryptolaemus montrouzieri* Mulsant | COLEOPTERA: Coccinellidae |
| Mediterranean flour moth | *Anagasta kuehniella* (Zeller) | LEPIDOPTERA: Crambidae |
| Mediterranean fruit fly | *Ceratitis capitata* (Wiedemann) | DIPTERA: Tephritidae |
| melastoma borer | *Selca brunella* (Hampson) | LEPIDOPTERA: Noctuidae |
| melon aphid | *Aphis gossypii* Glover | HEMIPTERA: Aphididae |
| melon fly | *Dacus cucurbitae* Coquillett | DIPTERA: Tephritidae |
| melonworm | *Diaphania hyalinata* (Linnaeus) | LEPIDOPTERA: Crambidae |
| merchant grain beetle | *Oryzaephilus mercator* (Fauvel) | COLEOPTERA: Cucujidae |
| Mexican bean beetle | *Epilachna varivestis* Mulsant | COLEOPTERA: Coccinellidae |
| Mexican bean weevil | *Zabrotes subfasciatus* (Boheman) | COLEOPTERA: Bruchidae |
| Mexican black scale | *Saissetia miranda* (Cockerell & Parrott) | HEMIPTERA: Coccidae |
| Mexican corn rootworm | *Diabrotica virgifera zeae* Krysan & Smith | COLEOPTERA: Chrysomelidae |
| Mexican fruit fly | *Anastrepha ludens* (Loew) | DIPTERA: Tephritidae |
| Mexican leafroller | *Amorbia emigratella* Busck | LEPIDOPTERA: Tortricidae |
| Mexican mealybug | *Phenacoccus gossypii* Townsend & Cockerell | HEMIPTERA: Pseudococcidae |
| Mexican pine beetle | *Dendroctonus approximatus* Dietz | COLEOPTERA: Scolytidae |

| | | |
|---|---|---|
| migratory grasshopper | *Melanoplus sanguinipes* (Fabricius) | ORTHOPTERA: Acrididae |
| mimosa webworm | *Homadaula anisocentra* Meyrick | LEPIDOPTERA: Plutellidae |
| mining scale | *Howardia biclavis* (Comstock) | HEMIPTERA: Diaspididae |
| mint aphid | *Ovatus crataegarius* (Walker) | HEMIPTERA: Aphididae |
| minute egg parasite | *Trichogramma minutum* Riley | HYMENOPTERA: Trichogrammatidae |
| minute pirate bug | *Orius tristicolor* (White) | HEMIPTERA: Anthocoridae |
| mission blue | *Icaricia icarioides missionensis* (Hovanitz) | LEPIDOPTERA: Lycaenidae |
| mold mite | *Tyrophagus putrescentiae* (Schrank) | ACARI: Acaridae |
| monarch butterfly | *Danaus plexippus* (Linnaeus) | LEPIDOPTERA: Danaidae |
| monkeypod moth | *Polydesma umbricola* Boisduval | LEPIDOPTERA: Noctuidae |
| monkeypod roundheaded borer | *Xystrocera globosa* (Olivier) | COLEOPTERA: Cerambycidae |
| Monterey pine cone beetle | *Conophthorus radiatae* Hopkins | COLEOPTERA: Scolytidae |
| Monterey pine resin midge | *Cecidomyia resinicoloides* Williams | DIPTERA: Cecidomyiidae |
| Monterey pine weevil | *Pissodes radiatae* Hopkins | COLEOPTERA: Curculionidae |
| Mormon cricket | *Anabrus simplex* Haldeman | ORTHOPTERA: Tettigoniidae |
| morningglory leafminer | *Bedellia somnulentella* (Zeller) | LEPIDOPTERA: Lyonetiidae |
| Morrill lace bug | *Corythucha morrilli* Osborn & Drake | HEMIPTERA: Tingidae |
| mossyrose gall wasp | *Diplolepis rosae* (Linnaeus) | HYMENOPTERA: Cynipidae |
| mottled tortoise beetle | *Deloyala guttata* (Olivier) | COLEOPTERA: Chrysomelidae |
| mountain leafhopper | *Colladonus montanus* (Van Duzee) | HEMIPTERA: Cicadellidae |
| mountain pine beetle | *Dendroctonus ponderosae* Hopkins | COLEOPTERA: Scolytidae |
| mountain pine coneworm | *Dioryctria yatesi* Mutuura & Munroe | LEPIDOPTERA: Crambidae |
| mountain-ash sawfly | *Pristiphora geniculata* (Hartig) | HYMENOPTERA: Tenthredinidae |
| mourningcloak butterfly | *Nymphalis antiopa* (Linnaeus) | LEPIDOPTERA: Nymphalidae |
| mulberry whitefly | *Tetraleurodes mori* (Quaintance) | HEMIPTERA: Aleyrodidae |
| mullein thrips | *Haplothrips verbasci* (Osborn) | THYSANOPTERA: Phlaeothripidae |

## N

| | | |
|---|---|---|
| Nantucket pine tip moth | *Rhyacionia frustrana* (Comstock) | LEPIDOPTERA: Tortricidae |
| narcissus bulb fly | *Merodon equestris* (Fabricius) | DIPTERA: Syrphidae |
| narrownecked grain beetle | *Anthicus floralis* (Linnaeus) | COLEOPTERA: Anthicidae |
| narrowwinged mantid | *Tenodera augustipennis* Saussure | MANTODEA: Mantidae |
| native elm bark beetle | *Hylurgopinus opaculus* (LeConte) | COLEOPTERA: Scolytidae |
| native holly leafminer | *Phytomyza ilicicola* Loew | DIPTERA: Agromyzidae |
| navel orangeworm | *Amyelois transitella* (Walker) | LEPIDOPTERA: Crambidae |
| negro bug | *Corimelaena pulicaria* (Germar) | HEMIPTERA: Thyreocoridae |
| Nevada sage grasshopper | *Melanoplus rugglesi* Gurney | ORTHOPTERA: Acrididae |
| New Guinea sugarcane weevil | *Rhabdoscelus obscurus* (Boisduval) | COLEOPTERA: Curculionidae |
| new house borer | *Arhopalus productus* (LeConte) | COLEOPTERA: Cerambycidae |
| New York weevil | *Ithycerus noveboracensis* (Forster) | COLEOPTERA: Curculionidae |
| nigra scale | *Parasaissetia nigra* (Nietner) | HEMIPTERA: Coccidae |
| northeastern sawyer | *Monochamus notatus* (Drury) | COLEOPTERA: Cerambycidae |
| northern cattle grub | *Hypoderma bovis* (Linnaeus) | DIPTERA: Oestridae |
| northern corn rootworm | *Diabrotica barberi* Smith & Lawrence | COLEOPTERA: Chrysomelidae |
| northern fowl mite | *Ornithonyssus sylviarum* (Canestrini & Fanzago) | ACARI: Macronyssidae |
| northern house mosquito | *Culex pipiens* Linnaeus | DIPTERA: Culicidae |
| northern masked chafer | *Cyclocephala borealis* Arrow | COLEOPTERA: Scarabaeidae |
| northern mole cricket | *Neocurtilla hexadactyla* (Perty) | ORTHOPTERA: Gryllotalpidae |
| northern pine weevil | *Pissodes approximatus* Hopkins | COLEOPTERA: Curculionidae |
| northern pitch twig moth | *Petrova albicapitana* (Busck) | LEPIDOPTERA: Tortricidae |
| northern rat flea | *Nosopsyllus fasciatus* (Bosc) | SIPHONAPTERA: Ceratophyllidae |
| northwest coast mosquito | *Aedes aboriginis* Dyar | DIPTERA: Culicidae |
| Norway maple aphid | *Periphyllus lyropictus* (Kessler) | HEMIPTERA: Aphididae |
| nose bot fly | *Gasterophilus haemorrhoidalis* (Linnaeus) | DIPTERA: Gasterophilidae |
| nutgrass armyworm | *Spodoptera exempta* (Walker) | LEPIDOPTERA: Noctuidae |
| nutgrass billbug | *Sphenophorus cariosus* (Olivier) | COLEOPTERA: Curculionidae |
| nutgrass borer moth | *Bactra venosoma* (Zeller) | LEPIDOPTERA: Tortricidae |
| nutgrass weevil | *Athesapeuta cyperi* Marshall | COLEOPTERA: Curculionidae |
| Nuttall blister beetle | *Lytta nuttalli* (Say) | COLEOPTERA: Meloidae |

# O

| | | |
|---|---|---|
| oak clearwing moth | *Paranthrene asilipennis* (Boisduval) | LEPIDOPTERA: Sesiidae |
| oak lace bug | *Corythucha arcuata* (Say) | HEMIPTERA: Tingidae |
| oak leafroller | *Archips semiferana* (Walker) | LEPIDOPTERA: Tortricidae |
| oak leaftier | *Croesia semipurpurana* (Kearfott) | LEPIDOPTERA: Tortricidae |
| oak lecanium | *Parthenolecanium quercifex* (Fitch) | HEMIPTERA: Coccidae |
| oak sapling borer | *Goes tesselatus* (Haldeman) | COLEOPTERA: Cerambycidae |
| oak skeletonizer | *Bucculatrix ainsliella* Murtfeldt | LEPIDOPTERA: Lyonetiidae |
| oak timberworm | *Arrhenodes minutus* (Drury) | COLEOPTERA: Brentidae |
| oak webworm | *Archips fervidana* (Clemens) | LEPIDOPTERA: Tortricidae |
| oatbird-cherry aphid | *Rhopalosiphum padi* (Linnaeus) | HEMIPTERA: Aphididae |
| obliquebanded leafroller | *Choristoneura rosaceana* (Harris) | LEPIDOPTERA: Tortricidae |
| obscure mealybug | *Pseudococcus obscurus* Essig | HEMIPTERA: Pseudococcidae |
| obscure root weevil | *Sciopithes obscurus* Horn | COLEOPTERA: Curculionidae |
| obscure scale | *Melanaspis obscura* (Comstock) | HEMIPTERA: Diaspididae |
| oceanic burrower bug | *Geotomus pygmaeus* (Dallas) | HEMIPTERA: Cydnidae |
| oceanic embiid | *Oligotoma oceanica* Ross | EMBIOPTERA: Oligotomidae |
| oceanic field cricket | *Teleogryllus oceanicus* (Le Guillou) | ORTHOPTERA: Gryllidae |
| odd beetle | *Thylodrias contractus* Motschulsky | COLEOPTERA: Dermestidae |
| odorous house ant | *Tapinoma sessile* (Say) | HYMENOPTERA: Formicidae |
| old house borer | *Hylotrupes bajulus* (Linnaeus) | COLEOPTERA: Cerambycidae |
| oleander aphid | *Aphis nerii* Fonscolombe | HEMIPTERA: Aphididae |
| oleander hawk moth | *Daphnis nerii* (Linnaeus) | LEPIDOPTERA: Sphingidae |
| oleander pit scale | *Asterolecanium pustulans* (Cockerell) | HEMIPTERA: Asterolecaniidae |
| oleander scale | *Aspidiotus nerii* Bouché | HEMIPTERA: Diaspididae |
| olive fruit fly | *Dacus oleae* (Gmelin) | DIPTERA: Tephritidae |
| olive scale | *Parlatoria oleae* (Colvée) | HEMIPTERA: Diaspididae |
| omnivorous leaftier | *Cnephasia longana* (Haworth) | LEPIDOPTERA: Tortricidae |
| omnivorous looper | *Sabulodes caberata* Guenée | LEPIDOPTERA: Geometridae |
| onespotted stink bug | *Euschistus variolarius* (Palisot de Beauvois) | HEMIPTERA: Pentatomidae |
| onion aphid | *Keotoxoptera formosana* (Takahashi) | HEMIPTERA: Aphididae |
| onion bulb fly | *Kumerus strigatus* (Fallén) | DIPTERA: Syrphidae |
| onion maggot | *Delia antiqua* (Meigen) | DIPTERA: Anthomyiidae |
| onion plant bug | *Labopidicola allii* (Knight) | HEMIPTERA: Miridae |
| onion thrips | *Thrips tabaci* Lindeman | THYSANOPTERA: Thripidae |
| orange spiny whitefly | *Aleurocanthus spiniferus* (Quaintance) | HEMIPTERA: Aleyrodidae |
| orange tortrix | *Argyrotaenia citrana* (Fernald) | LEPIDOPTERA: Tortricidae |
| orangedog | *Papilio cresphontes* Cramer | LEPIDOPTERA: Papilionidae |
| orangehumped mapleworm | *Symmerista leucitys* Franclemont | LEPIDOPTERA: Notodontidae |
| orangestriped oakworm | *Anisota senatoria* (J. E. Smith) | LEPIDOPTERA: Saturniidae |
| orangetailed potter wasp | *Delta latreillei petiolaris* (Schulz) | HYMENOPTERA: Vespidae |
| orchid aphid | *Macrosiphum luteum* (Buckton) | HEMIPTERA: Aphididae |
| orchidfly | *Eurytoma orchidearum* (Westwood) | HYMENOPTERA: Eurytomidae |
| Oregon fir sawyer | *Monochamus oregonensis* (LeConte) | COLEOPTERA: Cerambycidae |
| Oregon wireworm | *Melanotus longulus oregonensis* (LeConte) | COLEOPTERA: Elateridae |
| oriental beetle | *Anomala orientalis* Waterhouse | COLEOPTERA: Scarabaeidae |
| oriental cockroach | *Blatta orientalis* Linnaeus | BLATTODEA: Blattidae |
| oriental fruit fly | *Dacus dorsalis* Hendel | DIPTERA: Tephritidae |
| oriental fruit moth | *Grapholita molesta* (Busck) | LEPIDOPTERA: Tortricidae |
| oriental house fly | *Musca domestica vicina* Macquart | DIPTERA: Muscidae |
| oriental moth | *Cnidocampa flavescens* (Walker) | LEPIDOPTERA: Limacodidae |
| oriental rat flea | *Xenopsylla cheopis* (Rothschild) | SIPHONAPTERA: Pulicidae |
| oriental stink bug | *Plautia stali* Scott | HEMIPTERA: Pentatomidae |
| ornate aphid | *Myzus ornatus* Laing | HEMIPTERA: Aphididae |
| orthezia lady beetle | *Hyperaspis jocosa* (Mulsant) | COLEOPTERA: Coccinellidae |
| oval guineapig louse | *Gyropus ovalis* Burmeister | PHTHIRAPTERA: Gyropidae |
| oxalis spider mite | *Tetranychina harti* (Ewing) | ACARI: Tetranychidae |
| oxalis whitefly | *Aleyrodes shizuokensis* Kuwana | HEMIPTERA: Aleyrodidae |
| oystershell scale | *Lepidosaphes ulmi* (Linnaeus) | HEMIPTERA: Diaspididae |

## P

| | | |
|---|---|---|
| Pacific beetle cockroach | *Diploptera punctata* (Eschscholtz) | BLATTODEA: Blaberidae |
| Pacific Coast tick | *Dermacentor occidentalis* Marx | ACARI: Ixodidae |
| Pacific Coast wireworm | *Limonius canus* LeConte | COLEOPTERA: Elateridae |
| Pacific cockroach | *Euthyrrhapha pacifica* (Coquebert) | BLATTODEA: Polyphagidae |
| Pacific dampwood termite | *Zootermopsis angusticollis* (Hagen) | ISOPTERA: Hodotermitidae |
| Pacific flatheaded borer | *Chrysobothris mali* Horn | COLEOPTERA: Buprestidae |
| Pacific kissing bug | *Oncocephalus pacificus* (Kirkaldy) | HEMIPTERA: Reduviidae |
| Pacific pelagic water strider | *Halobates sericeus* Eschscholtz | HEMIPTERA: Gerridae |
| Pacific spider mite | *Tetranychus pacificus* McGregor | ACARI: Tetranychidae |
| Pacific tent caterpillar | *Malacosoma constrictum* (H. Edwards) | LEPIDOPTERA: Lasiocampidae |
| Pacific willow leaf beetle | *Pyrrhalta decora carbo* (LeConte) | COLEOPTERA: Chrysomelidae |
| Packard grasshopper | *Melanoplus packardii* Scudder | ORTHOPTERA: Acrididae |
| painted beauty | *Vanessa virginiensis* (Drury) | LEPIDOPTERA: Nymphalidae |
| painted hickory borer | *Megacyllene caryae* (Gahan) | COLEOPTERA: Cerambycidae |
| painted lady | *Vanessa cardui* (Linnaeus) | LEPIDOPTERA: Nymphalidae |
| painted leafhopper | *Endria inimica* (Say) | HEMIPTERA: Cicadellidae |
| painted maple aphid | *Drepanaphis acerifoliae* (Thomas) | HEMIPTERA: Aphididae |
| pale damsel bug | *Tropiconabis capsiformis* (Germar) | HEMIPTERA: Nabidae |
| pale juniper webworm | *Aethes rutilana* (Hübner) | LEPIDOPTERA: Cochylidae |
| pale leafcutting bee | *Megachile concinna* Smith | HYMENOPTERA: Megachilidae |
| pale legume bug | *Lygus elisus* Van Duzee | HEMIPTERA: Miridae |
| pale tussock moth | *Halysidota tessellaris* (J. E. Smith) | LEPIDOPTERA: Arctiidae |
| pale western cutworm | *Agrotis orthogonia* Morrison | LEPIDOPTERA: Noctuidae |
| pales weevil | *Hylobius pales* (Herbst) | COLEOPTERA: Curculionidae |
| palesided cutworm | *Agrotis malefida* Guenée | LEPIDOPTERA: Noctuidae |
| palestriped flea beetle | *Systena blanda* Melsheimer | COLEOPTERA: Chrysomelidae |
| palm leafskeletonizer | *Homaledra sabalella* (Chambers) | LEPIDOPTERA: Coleophoridae |
| palm mealybug | *Palmicultor palmarum* (Ehrhorn) | HEMIPTERA: Pseudococcidae |
| palmerworm | *Dichomeris ligulella* Hübner | LEPIDOPTERA: Gelechiidae |
| pandanus mealybug | *Laminicoccus pandani* (Cockerell) | HEMIPTERA: Pseudococcidae |
| pandora moth | *Coloradia pandora* Blake | LEPIDOPTERA: Saturniidae |
| papaya fruit fly | *Toxotrypana curvicauda* Gerstaecker | DIPTERA: Tephritidae |
| paper wasps | *Polistes* spp. | HYMENOPTERA: Vespidae |
| parasitic grain wasp | *Cephalonomia waterstoni* Gahan | HYMENOPTERA: Bethylidae |
| parlatoria date scale | *Parlatoria blanchardi* (Targioni-Tozzetti) | HEMIPTERA: Diaspididae |
| parsleyworm | *Papilio polyxenes asterius* Stoll | LEPIDOPTERA: Papilionidae |
| parsnip webworm | *Depressaria pastinacella* (Duponchel) | LEPIDOPTERA: Oecophoridae |
| pavement ant | *Tetramorium caespitum* (Linnaeus) | HYMENOPTERA: Formicidae |
| pea aphid | *Acyrthosiphon pisum* (Harris) | HEMIPTERA: Aphididae |
| pea leaf weevil | *Sitona lineatus* (Linnaeus) | COLEOPTERA: Curculionidae |
| pea leafminer | *Liriomyza huidobrensis* (Blanchard) | DIPTERA: Agromyzidae |
| pea moth | *Cydia nigricana* (Fabricius) | LEPIDOPTERA: Tortricidae |
| pea weevil | *Bruchus pisorum* (Linnaeus) | COLEOPTERA: Bruchidae |
| peach bark beetle | *Phloeotribus liminaris* (Harris) | COLEOPTERA: Scolytidae |
| peach silver mite | *Aculus cornutus* (Banks) | ACARI: Eriophyidae |
| peach twig borer | *Anarsia lineatella* Zeller | LEPIDOPTERA: Gelechiidae |
| peachtree borer | *Synanthedon exitiosa* (Say) | LEPIDOPTERA: Sesiidae |
| pear midge | *Contarinia pyrivora* (Riley) | DIPTERA: Cecidomyiidae |
| pear plant bug | *Lygocoris communis* (Knight) | HEMIPTERA: Miridae |
| pear psylla | *Psylla pyricola* Foerster | HEMIPTERA: Psyllidae |
| pear rust mite | *Epitrimerus pyri* (Nalepa) | ACARI: Eriophyidae |
| pear sawfly | *Caliroa cerasi* (Linnaeus) | HYMENOPTERA: Tenthredinidae |
| pear thrips | *Taeniothrips inconsequens* (Uzel) | THYSANOPTERA: Thripidae |
| pearleaf blister mite | *Phytoptus pyri* Pagenstecher | ACARI: Eriophyidae |
| pecan bud moth | *Gretchena bolliana* (Slingerland) | LEPIDOPTERA: Tortricidae |
| pecan carpenterworm | *Cossula magnifica* (Strecker) | LEPIDOPTERA: Cossidae |
| pecan cigar casebearer | *Coleophora laticornella* Clemens | LEPIDOPTERA: Coleophoridae |

| | | |
|---|---|---|
| pecan leaf casebearer | *Acrobasis juglandis* (LeBaron) | LEPIDOPTERA: Crambidae |
| pecan leaf phylloxera | *Phylloxera notabilis* Pergande | HEMIPTERA: Phylloxeridae |
| pecan leaf scorch mite | *Eotetranychus hicoriae* (McGregor) | ACARI: Tetranychidae |
| pecan leafroll mite | *Eriophyes caryae* Keifer | ACARI: Eriophyidae |
| pecan nut casebearer | *Acrobasis nuxvorella* Neunzig | LEPIDOPTERA: Crambidae |
| pecan phylloxera | *Phylloxera devastatrix* Pergande | HEMIPTERA: Phylloxeridae |
| pecan serpentine leafminer | *Stigmella juglandifoliella* (Clemens) | LEPIDOPTERA: Nepticulidae |
| pecan spittlebug | *Clastoptera achatina* Germar | HEMIPTERA: Cercopidae |
| pecan weevil | *Curculio caryae* (Horn) | COLEOPTERA: Curculionidae |
| pepper maggot | *Zonosemata electa* (Say) | DIPTERA: Tephritidae |
| pepper weevil | *Anthonomus eugenii* Cano | COLEOPTERA: Curculionidae |
| pepper-and-salt moth | *Biston betularia cognataria* (Guenée) | LEPIDOPTERA: Geometridae |
| peppergrass beetle | *Galeruca browni* Blake | COLEOPTERA: Chrysomelidae |
| periodical cicada | *Magicicada septendecim* (Linnaeus) | HEMIPTERA: Cicadidae |
| persimmon borer | *Sannina uroceriformis* Walker | LEPIDOPTERA: Sesiidae |
| persimmon psylla | *Trioza diospyri* (Ashmead) | HEMIPTERA: Psyllidae |
| phantom hemlock looper | *Nepytia phantasmaria* (Strecker) | LEPIDOPTERA: Geometridae |
| Pharaoh ant | *Monomorium pharaonis* (Linnaeus) | HYMENOPTERA: Formicidae |
| Philippine katydid | *Phaneroptera furcifera* Stål | ORTHOPTERA: Tettigoniidae |
| phlox plant bug | *Lopidea davisi* Knight | HEMIPTERA: Miridae |
| pickleworm | *Diaphania nitidalis* (Stoll) | LEPIDOPTERA: Crambidae |
| pigeon fly | *Pseudolynchia canariensis* (Macquart) | DIPTERA: Hippoboscidae |
| pigeon tremex | *Tremex columba* (Linnaeus) | HYMENOPTERA: Siricidae |
| pine bark adelgid | *Pineus strobi* (Hartig) | HEMIPTERA: Phylloxeridae |
| pine bud mite | *Trisetacus pini* (Nalepa) | ACARI: Nalepellidae |
| pine butterfly | *Neophasia menapia* (Felder & Felder) | LEPIDOPTERA: Pieridae |
| pine candle moth | *Exoteleia nepheos* Freeman | LEPIDOPTERA: Gelechiidae |
| pine chafer | *Anomala oblivia* Horn | COLEOPTERA: Scarabaeidae |
| pine colaspis | *Colaspis pini* Barber | COLEOPTERA: Chrysomelidae |
| pine conelet looper | *Nepytia semiclusaria* (Walker) | LEPIDOPTERA: Geometridae |
| pine engraver | *Ips pini* (Say) | COLEOPTERA: Scolytidae |
| pine false webworm | *Acantholyda erythrocephala* (Linnaeus) | HYMENOPTERA: Pamphiliidae |
| pine gall weevil | *Podapion gallicola* Riley | COLEOPTERA: Curculionidae |
| pine leaf adelgid | *Pineus pinifoliae* (Fitch) | HEMIPTERA: Adelgidae |
| pine needle scale | *Chionaspis pinifoliae* (Fitch) | HEMIPTERA: Diaspididae |
| pine needle sheathminer | *Zelleria haimbachi* Busck | LEPIDOPTERA: Yponomeutidae |
| pine needleminer | *Exoteleia pinifoliella* (Chambers) | LEPIDOPTERA: Gelechiidae |
| pine root collar weevil | *Hylobius radicis* Buchanan | COLEOPTERA: Curculionidae |
| pine root tip weevil | *Hylobius rhizophagus* Millers, Benjamin & Warner | COLEOPTERA: Curculionidae |
| pine rosette mite | *Trisetacus gemmavitians* Styer | ACARI: Nalepellidae |
| pine spittlebug | *Aphrophora parallela* (Say) | HEMIPTERA: Cercopidae |
| pine tortoise scale | *Toumeyella parvicornis* (Cockerell) | HEMIPTERA: Coccidae |
| pine tube moth | *Argyrotaenia pinatubana* (Kearfott) | LEPIDOPTERA: Tortricidae |
| pine tussock moth | *Dasychira pinicola* (Dyar) | LEPIDOPTERA: Lymantriidae |
| pine webworm | *Tetralopha robustella* Zeller | LEPIDOPTERA: Crambidae |
| pineapple false spider mite | *Dolichotetranychus floridanus* (Banks) | ACARI: Tenuipalpidae |
| pineapple mealybug | *Dysmicoccus brevipes* (Cockerell) | HEMIPTERA: Pseudococcidae |
| pineapple scale | *Diaspis bromeliae* (Kerner) | HEMIPTERA: Diaspididae |
| pineapple tarsonemid | *Steneotarsonemus ananas* (Tryon) | ACARI: Tarsonemidae |
| pineapple weevil | *Metamasius ritchiei* Marshall | COLEOPTERA: Curculionidae |
| pink bollworm | *Pectinophora gossypiella* (Saunders) | LEPIDOPTERA: Gelechiidae |
| pink scavenger caterpillar | *Pyroderces rileyi* (Walsingham) | LEPIDOPTERA: Cosmopterigidae |
| pink sugarcane mealybug | *Saccharicoccus sacchari* (Cockerell) | HEMIPTERA: Pseudococcidae |
| pinkstriped oakworm | *Anisota virginiensis* (Drury) | LEPIDOPTERA: Saturniidae |
| pinkwinged grasshopper | *Atractomorpha sinensis* Bolivar | ORTHOPTERA: Pyrgomorphidae |
| pinon cone beetle | *Conophthorus edulis* Hopkins | COLEOPTERA: Scolytidae |
| pipevine swallowtail | *Battus philenor* (Linnaeus) | LEPIDOPTERA: Papilionidae |
| pistol casebearer | *Coleophora malivorella* Riley | LEPIDOPTERA: Coleophoridae |
| pitch mass borer | *Synanthedon pini* (Kellicott) | LEPIDOPTERA: Sesiidae |

| | | |
|---|---|---|
| pitch pine tip moth | *Rhyacionia rigidana* (Fernald) | LEPIDOPTERA: Tortricidae |
| pitch twig moth | *Petrova comstockiana* (Fernald) | LEPIDOPTERA: Tortricidae |
| pitch-eating weevil | *Pachylobius picivorus* (Germar) | COLEOPTERA: Curculionidae |
| pitcherplant mosquito | *Wyeomyia smithii* (Coquillett) | DIPTERA: Culicidae |
| plains false wireworm | *Eleodes opacus* (Say) | COLEOPTERA: Tenebrionidae |
| plaster beetle | *Cartodere constricta* (Gyllenhal) | COLEOPTERA: Lathridiidae |
| plum curculio | *Conotrachelus nenuphar* (Herbst) | COLEOPTERA: Curculionidae |
| plum gouger | *Coccotorus scutellaris* (LeConte) | COLEOPTERA: Curculionidae |
| plum leafhopper | *Macropsis trimaculata* (Fitch) | HEMIPTERA: Cicadellidae |
| plum rust mite | *Aculus fockeui* (Nalepa & Trouessart) | ACARI: Eriophyidae |
| plum webspinning sawfly | *Neurotoma inconspicua* (Norton) | HYMENOPTERA: Pamphiliidae |
| plumeria borer | *Lagocheirus undatus* (Voet) | COLEOPTERA: Cerambycidae |
| plumeria whitefly | *Paraleyrodes perseae* (Quaintance) | HEMIPTERA: Aleyrodidae |
| poinciana looper | *Pericyma cruegeri* (Butler) | LEPIDOPTERA: Noctuidae |
| polyphemus moth | *Antheraea polyphemus* (Cramer) | LEPIDOPTERA: Saturniidae |
| ponderosa pine bark borer | *Canonura princeps* (Walker) | COLEOPTERA: Cerambycidae |
| ponderosa pine cone beetle | *Conophthonus ponderosae* Hopkins | COLEOPTERA: Scolytidae |
| poplar borer | *Saperda calcarata* Say | COLEOPTERA: Cerambycidae |
| poplar leaffolding sawfly | *Phyllocolpa bozemani* (Cooley) | HYMENOPTERA: Tenthredinidae |
| poplar petiolegall aphid | *Pemphigus populitransversus* Riley | HEMIPTERA: Aphididae |
| poplar tentmaker | *Ichthyura inclusa* (Hübner) | LEPIDOPTERA: Notodontidae |
| poplar twig gall aphid | *Pemphigus populiramulorum* Riley | HEMIPTERA: Aphididae |
| poplar vagabond aphid | *Mordvilkoja vagabunda* (Walsh) | HEMIPTERA: Aphididae |
| poplar-and-willow borer | *Cryptorhynchus lapathi* (Linnaeus) | COLEOPTERA: Curculionidae |
| portulaca leafmining weevil | *Hypurus bertrandi* (Perris) | COLEOPTERA: Curculionidae |
| potato aphid | *Macrosiphum euphorbiae* (Thomas) | HEMIPTERA: Aphididae |
| potato flea beetle | *Epitrix cucumeris* (Harris) | COLEOPTERA: Chrysomelidae |
| potato leafhopper | *Empoasca fabae* (Harris) | HEMIPTERA: Cicadellidae |
| potato psyllid | *Paratrioza cockerelli* (Sulc) | HEMIPTERA: Psyllidae |
| potato scab gnat | *Pnyxia scabiei* (Hopkins) | DIPTERA: Sciaridae |
| potato stalk borer | *Trichobaris trinotata* (Say) | COLEOPTERA: Curculionidae |
| potato stem borer | *Hydraecia micacea* (Esper) | LEPIDOPTERA: Noctuidae |
| potato tuberworm | *Phthorimaea operculella* (Zeller) | LEPIDOPTERA: Gelechiidae |
| poultry bug | *Haematosiphon inodorus* (Dugès) | HEMIPTERA: Cimicidae |
| poultry house moth | *Niditinea fuscipunctella* (Haworth) | LEPIDOPTERA: Tineidae |
| powderpost bostrichid | *Amphicerus cornutus* (Pallas) | COLEOPTERA: Bostrichidae |
| prairie flea beetle | *Altica canadensis* Gentner | COLEOPTERA: Chrysomelidae |
| prairie grain wireworm | *Ctenicera aeripennis destructor* (Brown) | COLEOPTERA: Elateridae |
| privet aphid | *Myzus ligustri* (Mosley) | HEMIPTERA: Aphididae |
| privet leafminer | *Caloptilia cuculipennella* (Hübner) | LEPIDOPTERA: Gracillariidae |
| privet mite | *Brevipalpus obovatus* Donnadieu | ACARI: Tenuipalpidae |
| privet thrips | *Dendrothrips ornatus* (Jablonowski) | THYSANOPTERA: Thripidae |
| promethea moth | *Callosamia promethea* (Drury) | LEPIDOPTERA: Saturniidae |
| pruinose bean weevil | *Stator pruininus* (Horn) | COLEOPTERA: Bruchidae |
| prune leafhopper | *Edwardsiana prunicola* (Edwards) | HEMIPTERA: Cicadellidae |
| Puget Sound wireworm | *Ctenicera aeripennis aeripennis* (Kirby) | COLEOPTERA: Elateridae |
| puncturevine seed weevil | *Microlarinus lareynii* (Jacquelin du Val) | COLEOPTERA: Curculionidae |
| puncturevine stem weevil | *Microlarinus lypriformis* (Wollaston) | COLEOPTERA: Curculionidae |
| purple scale | *Lepidosaphes beckii* (Newman) | HEMIPTERA: Diaspididae |
| purplebacked cabbageworm | *Evergestis pallidata* (Hufnagel) | LEPIDOPTERA: Crambidae |
| purplespotted lily aphid | *Macrosiphum lilii* (Monell) | HEMIPTERA: Aphididae |
| puss caterpillar | *Megalopyge opercularis* (J. E. Smith) | LEPIDOPTERA: Megalopygidae |
| Putnam scale | *Diaspidiotus ancylus* (Putnam) | HEMIPTERA: Diaspididae |
| pyramid ant | *Conomyrma insana* (Buckley) | HYMENOPTERA: Formicidae |
| pyriform scale | *Protopulvinaria pyriformis* (Cockerell) | HEMIPTERA: Coccidae |

## Q

| | | |
|---|---|---|
| quince curculio | *Conotrachelus crataegi* Walsh | COLEOPTERA: Curculionidae |
| quince treehopper | *Glossonotus crataegi* (Fitch) | HEMIPTERA: Membracidae |

## R

| | | |
|---|---|---|
| rabbit louse | *Haemodipsus ventricosus* (Denny) | PHTHIRAPTERA: Hoplopleuridae |
| rabbit tick | *Haemaphysalis leporispalustris* (Packard) | ACARI: Ixodidae |
| ragweed borer | *Epiblema strenuana* (Walker) | LEPIDOPTERA: Tortricidae |
| ragweed plant bug | *Chlamydatus associatus* (Uhler) | HEMIPTERA: Miridae |
| raisin moth | *Cadra figulilella* (Gregson) | LEPIDOPTERA: Crambidae |
| range caterpillar | *Hemileuca oliviae* Cockerell | LEPIDOPTERA: Saturniidae |
| range crane fly | *Tipula simplex* Doane | DIPTERA: Tipulidae |
| rapid plant bug | *Adelphocoris rapidus* (Say) | HEMIPTERA: Miridae |
| raspberry bud moth | *Lampronia rubiella* (Bjerkander) | LEPIDOPTERA: Incurvariidae |
| raspberry cane borer | *Oberea bimaculata* (Olivier) | COLEOPTERA: Cerambycidae |
| raspberry cane maggot | *Pegomya rubivora* (Coquillett) | DIPTERA: Anthomyiidae |
| raspberry crown borer | *Pennisetia marginata* (Harris) | LEPIDOPTERA: Sesiidae |
| raspberry leafroller | *Olethreutes permundana* (Clemens) | LEPIDOPTERA: Tortricidae |
| raspberry sawfly | *Monophadnoides geniculatus* (Hartig) | HYMENOPTERA: Tenthredinidae |
| red admiral | *Vanessa atalanta rubria* (Fruhstorfer) | LEPIDOPTERA: Nymphalidae |
| red and black flat mite | *Brevipalpus phoenicis* (Geijskes) | ACARI: Tenuipalpidae |
| red assassin bug | *Haematoloecha rubescens* Distant | HEMIPTERA: Reduviidae |
| red carpenter ant | *Camponotus ferrugineus* (Fabricius) | HYMENOPTERA: Formicidae |
| red clover seed weevil | *Tychius stephensi* (Schönherr) | COLEOPTERA: Curculionidae |
| red date scale | *Phoenicococcus marlatti* Cockerell | HEMIPTERA: Phoenicococcidae |
| red elm bark weevil | *Magdalis armicollis* (Say) | COLEOPTERA: Curculionidae |
| red flour beetle | *Tribolium castaneum* (Herbst) | COLEOPTERA: Tenebrionidae |
| red grasshopper mite | *Eutrombidium trigonum* (Hermann) | ACARI: Trombidiidae |
| red harvester ant | *Pogonomyrmex barbatus* (F. Smith) | HYMENOPTERA: Formicidae |
| red imported fire ant | *Solenopsis invicta* Buren | HYMENOPTERA: Formicidae |
| red milkweed beetle | *Tetraopes tetrophthalmus* (Forster) | COLEOPTERA: Cerambycidae |
| red oak borer | *Enaphalodes rufulus* (Haldeman) | COLEOPTERA: Cerambycidae |
| red orchid scale | *Furcaspis biformis* (Cockerell) | HEMIPTERA: Diaspididae |
| red pine cone beetle | *Conophthorus resinosae* Hopkins | COLEOPTERA: Scolytidae |
| red pine sawfly | *Neodiprion nanulus nanulus* Schedl | HYMENOPTERA: Diprionidae |
| red pine scale | *Matsucoccus resinosae* Bean & Godwin | HEMIPTERA: Margarodidae |
| red turnip beetle | *Entomoscelis americana* Brown | COLEOPTERA: Chrysomelidae |
| red turpentine beetle | *Dendroctonus valens* LeConte | COLEOPTERA: Scolytidae |
| red wax scale | *Ceroplastes rubens* Maskell | HEMIPTERA: Coccidae |
| redbacked cutworm | *Euxoa ochrogaster* (Guenée) | LEPIDOPTERA: Noctuidae |
| redbanded leafroller | *Argyrotaenia velutinana* (Walker) | LEPIDOPTERA: Tortricidae |
| redbanded thrips | *Selenothrips rubrocinctus* (Giard) | THYSANOPTERA: Thripidae |
| redberry mite | *Acalitus essigi* (Hassan) | ACARI: Eriophyidae |
| redblack oedemerid | *Eobia bicolor* (Fairmaire) | COLEOPTERA: Oedemeridae |
| redbud leaffolder | *Fascista cercerisella* (Chambers) | LEPIDOPTERA: Gelechiidae |
| redheaded ash borer | *Neoclytus acuminatus* (Fabricius) | COLEOPTERA: Cerambycidae |
| redheaded jack pine sawfly | *Neodiprion rugifrons* Middleton | HYMENOPTERA: Diprionidae |
| redheaded pine sawfly | *Neodiprion lecontei* (Fitch) | HYMENOPTERA: Diprionidae |
| redhumped caterpillar | *Schizura concinna* (J. E. Smith) | LEPIDOPTERA: Notodontidae |
| redlegged flea beetle | *Derocrepis erythropus* (Melsheimer) | COLEOPTERA: Chrysomelidae |
| redlegged grasshopper | *Melanoplus femurrubrum* (DeGeer) | ORTHOPTERA: Acrididae |
| redlegged ham beetle | *Necrobia rufipes* (DeGeer) | COLEOPTERA: Cleridae |
| redmargined assassin bug | *Scadra rufidens* Stål | HEMIPTERA: Reduviidae |
| rednecked cane borer | *Agrilus ruficollis* (Fabricius) | COLEOPTERA: Buprestidae |
| rednecked peanutworm | *Stegasta bosqueella* (Chambers) | LEPIDOPTERA: Gelechiidae |
| redshouldered ham beetle | *Necrobia ruficollis* (Fabricius) | COLEOPTERA: Cleridae |
| redshouldered stink bug | *Thyanta accerra* McAtee | HEMIPTERA: Pentatomidae |
| redtailed spider wasp | *Tachypompilus analis* (Fabricius) | HYMENOPTERA: Pompilidae |
| redtailed tachina | *Winthemia quadripustulata* (Fabricius) | DIPTERA: Tachinidae |
| regal moth | *Citheronia regalis* (Fabricius) | LEPIDOPTERA: Saturniidae |
| relapsing fever tick | *Ornithodoros turicata* (Dugès) | ACARI: Argasidae |
| resplendent shield bearer | *Coptodisca splendoriferella* (Clemens) | LEPIDOPTERA: Heliozelidae |
| reticulate mite | *Lorryia reticulata* (Oudemans) | ACARI: Tydeidae |

| | | |
|---|---|---|
| reticulatewinged trogiid | *Lepinotus reticulatus* Enderlein | PSOCOPTERA: Trogiidae |
| rhinoceros beetle | *Xylorystes jamaicensis* (Drury) | COLEOPTERA: Scarabaeidae |
| Rhodesgrass mealybug | *Antonina graminis* (Maskell) | HEMIPTERA: Pseudococcidae |
| rhododendron borer | *Synanthedon rhododendri* Beutenmüller | LEPIDOPTERA: Sesiidae |
| rhododendron lace bug | *Stephanitis rhododendri* Horvath | HEMIPTERA: Tingidae |
| rhododendron whitefly | *Dialeurodes chittendeni* Laing | HEMIPTERA: Aleyrodidae |
| rhubarb curculio | *Lixus concavus* Say | COLEOPTERA: Curculionidae |
| rice delphacid | *Sogatodes orizicola* (Muir) | HEMIPTERA: Delphacidae |
| rice leaffolder | *Lerodea eufala* (Edwards) | LEPIDOPTERA: Hesperiidae |
| rice leafhopper | *Nephotettix nigropictus* (Stål) | HEMIPTERA: Cicadellidae |
| rice root aphid | *Rhopalosiphum rufiabdominalis* (Sasaki) | HEMIPTERA: Aphididae |
| rice stalk borer | *Chilo plejadellus* Zincken | LEPIDOPTERA: Crambidae |
| rice stink bug | *Oebalus pugnax* (Fabricius) | HEMIPTERA: Pentatomidae |
| rice water weevil | *Lissorhoptrus oryzophilus* Kuschel | COLEOPTERA: Curculionidae |
| rice weevil | *Sitophilus oryzae* (Linnaeus) | COLEOPTERA: Curculionidae |
| ridgewinged fungus beetle | *Thes bergrothi* (Reitter) | COLEOPTERA: Lathridiidae |
| ringlegged earwig | *Euborellia annulipes* (Lucas) | DERMAPTERA: Labiduridae |
| robust leafhopper | *Stragania robusta* (Uhler) | HEMIPTERA: Cicadellidae |
| Rocky Mountain grasshopper | *Melanoplus spretus* (Walsh) | ORTHOPTERA: Acrididae |
| Rocky Mountain wood tick | *Dermacentor andersoni* Stiles | ACARI: Ixodidae |
| rose aphid | *Macrosiphum rosae* (Linnaeus) | HEMIPTERA: Aphididae |
| rose chafer | *Macrodactylus subspinosus* (Fabricius) | COLEOPTERA: Scarabaeidae |
| rose curculio | *Merhynchites bicolor* (Fabricius) | COLEOPTERA: Attelabidae |
| rose leaf beetle | *Nodonota puncticollis* (Say) | COLEOPTERA: Chrysomelidae |
| rose leafhopper | *Edwardsiana rosae* (Linnaeus) | HEMIPTERA: Cicadellidae |
| rose midge | *Dasineura rhodophaga* (Coquillett) | DIPTERA: Cecidomyiidae |
| rose scale | *Aulacaspis rosae* (Bouché) | HEMIPTERA: Diaspididae |
| rose stem girdler | *Agrilus aurichalceus* Redtenbacher | COLEOPTERA: Buprestidae |
| roseroot gall wasp | *Diplolepis radicum* (Osten Sacken) | HYMENOPTERA: Cynipidae |
| roseslug | *Endelomyia aethiops* (Fabricius) | HYMENOPTERA: Tenthredinidae |
| rosy apple aphid | *Dysaphis plantaginea* (Passerini) | HEMIPTERA: Aphididae |
| rotund tick | *Ixodes kingi* Bishopp | ACARI: Ixodidae |
| rough stink bug | *Brochymena quadripustulata* (Fabricius) | HEMIPTERA: Pentatomidae |
| roughskinned cutworm | *Proxenus mindara* Barnes & McDunnough | LEPIDOPTERA: Noctuidae |
| roundheaded appletree borer | *Saperda candida* Fabricius | COLEOPTERA: Cerambycidae |
| roundheaded cone borer | *Paratimia conicola* Fisher | COLEOPTERA: Cerambycidae |
| roundheaded fir borer | *Tetropium abietis* Fall | COLEOPTERA: Cerambycidae |
| roundheaded pine beetle | *Dendroctonus adjunctus* Blandford | COLEOPTERA: Scolytidae |
| Russian wheat aphid | *Diuraphis noxiu* (Mordvilko) | HEMIPTERA: Aphididae |
| rustic borer | *Xylotrechus colonus* (Fabricius) | COLEOPTERA: Cerambycidae |
| rusty banded aphid | *Dysaphis apiifolia* (Theobald) | HEMIPTERA: Aphididae |
| rusty grain beetle | *Cryptolestes ferrugineus* (Stephens) | COLEOPTERA: Cucujidae |
| rusty plum aphid | *Hysteroneura setariae* (Thomas) | HEMIPTERA: Aphididae |
| rusty tussock moth | *Orgyia antiqua* (Linnaeus) | LEPIDOPTERA: Lymantriidae |

## S

| | | |
|---|---|---|
| saddleback caterpillar | *Sibine stimulea* (Clemens) | LEPIDOPTERA: Limacodidae |
| saddled leafhopper | *Colladonus clitellarius* (Say) | HEMIPTERA: Cicadellidae |
| saddled prominent | *Heterocampa guttivitta* (Walker) | LEPIDOPTERA: Notodontidae |
| sagebrush defoliator | *Aroga websteri* Clarke | LEPIDOPTERA: Gelechiidae |
| saltmarsh caterpillar | *Estigmene acrea* (Drury) | LEPIDOPTERA: Arctiidae |
| saltmarsh mosquito | *Aedes sollicitans* (Walker) | DIPTERA: Culicidae |
| San Bruno elfin | *Incisalia fotis bayensis* (R. M. Brown) | LEPIDOPTERA: Lycaenidae |
| San Jose scale | *Quadraspidiotus perniciosus* (Comstock) | HEMIPTERA: Diaspididae |
| sand wireworm | *Horistonotus uhlerii* Horn | COLEOPTERA: Elateridae |
| sandcherry weevil | *Coccotorus hirsutus* Bruner | COLEOPTERA: Curculionidae |
| sapwood timberworm | *Hylecoetus lugubris* Say | COLEOPTERA: Lymexylonidae |
| Saratoga spittlebug | *Aphrophora saratogensis* (Fitch) | HEMIPTERA: Cercopidae |
| saskatoon borer | *Saperda bipunctata* R. Hopping | COLEOPTERA: Cerambycidae |

| | | |
|---|---|---|
| satin moth | *Leucoma salicis* (Linnaeus) | LEPIDOPTERA: Lymantriidae |
| Saunders embiid | *Oligotoma saundersii* (Westwood) | EMBIOPTERA: Oligotomidae |
| sawtoothed grain beetle | *Oryzaephilus surinamensis* (Linnaeus) | COLEOPTERA: Cucujidae |
| Say blister beetle | *Lytta sayi* LeConte | COLEOPTERA: Meloidae |
| Say stink bug | *Chlorochroa sayi* Stål | HEMIPTERA: Pentatomidae |
| scab mite | *Psoroptes equi* (Raspail) | ACARI: Psoroptidae |
| scaly grain mite | *Suidasia nesbitti* Hughes | ACARI: Acaridae |
| scalyleg mite | *Knemidokoptes mutans* (Robin & Lanquetin) | ACARI: Sarcoptidae |
| scarlet oak sawfly | *Caliroa quercuscoccineae* (Dyar) | HYMENOPTERA: Tenthredinidae |
| Schaus swallowtail | *Papilio aristodemus ponceanus* Schaus | LEPIDOPTERA: Papilionidae |
| Schoene spider mite | *Tetranychus schoenei* McGregor | ACARI: Tetranychidae |
| screwworm | *Cochliomyia hominivorax* (Coquerel) | DIPTERA: Calliphoridae |
| sculptured pine borer | *Chalcophora angulicollis* (LeConte) | COLEOPTERA: Buprestidae |
| scurfy scale | *Chionaspis furfura* (Fitch) | HEMIPTERA: Diaspididae |
| secondary screwworm | *Cochliomyia macellaria* (Fabricius) | DIPTERA: Calliphoridae |
| seed bugs | *Nysius* spp. | HEMIPTERA: Lygaeidae |
| seedcorn beetle | *Stenolophus lecontei* (Chaudoir) | COLEOPTERA: Carabidae |
| seedcorn maggot | *Delia platura* (Meigen) | DIPTERA: Anthomyiidae |
| sequoia pitch moth | *Synanthedon sequoiae* (H. Edwards) | LEPIDOPTERA: Sesiidae |
| shaft louse | *Menopon gallinae* (Linnaeus) | PHTHIRAPTERA: Menoponidae |
| shallot aphid | *Myzus ascalonicus* Doncaster | HEMIPTERA: Aphididae |
| sheep biting louse | *Bovicola ovis* (Schrank) | PHTHIRAPTERA: Trichodectidae |
| sheep bot fly | *Oestrus ovis* Linnaeus | DIPTERA: Oestridae |
| sheep follicle mite | *Demodex ovis* Railliet | ACARI: Demodicidae |
| sheep ked | *Melophagus ovinus* (Linnaeus) | DIPTERA: Hippoboscidae |
| sheep scab mite | *Psoroptes ovis* (Hering) | ACARI: Psoroptidae |
| shieldbacked pine seed bug | *Tetyra bipunctata* (Herrich-Schäffer) | HEMIPTERA: Pentatomidae |
| shortleaf pine cone borer | *Eucosma cocana* Kearfott | LEPIDOPTERA: Tortricidae |
| shortnosed cattle louse | *Haematopinus eurysternus* (Nitzsch) | PHTHIRAPTERA: Haematopinidae |
| shothole borer | *Scolytus rugulosus* (Müller) | COLEOPTERA: Scolytidae |
| sigmoid fungus beetle | *Cryptophagus varus* Woodroffe & Coombs | COLEOPTERA: Cryptophagidae |
| silkworm | *Bombyx mori* (Linnaeus) | LEPIDOPTERA: Bombycidae |
| silky ant | *Formica fusca* Linnaeus | HYMENOPTERA: Formicidae |
| silky cane weevil | *Metamasius hemipterus sericeus* (Olivier) | COLEOPTERA: Curculionidae |
| silverfish | *Lepisma saccharina* Linnaeus | THYSANURA: Lepismatidae |
| silverspotted skipper | *Epargyreus clarus* (Cramer) | LEPIDOPTERA: Hesperiidae |
| silverspotted tiger moth | *Lophocampa argentata* (Packard) | LEPIDOPTERA: Arctiidae |
| sinuate lady beetle | *Hippodamia sinuata* Mulsant | COLEOPTERA: Coccinellidae |
| sinuate peartree borer | *Agrilus sinuatus* (Olivier) | COLEOPTERA: Buprestidae |
| Sitka spruce weevil | *Pissodes strobi* (Peck) | COLEOPTERA: Curculionidae |
| sixspotted mite | *Eotetranychus sexmaculatus* (Riley) | ACARI: Tetryanychidae |
| sixspotted thrips | *Scolothrips sexmaculatus* (Pergande) | THYSANOPTERA: Thripidae |
| slash pine flower thrips | *Gnophothrips fuscus* (Morgan) | THYSANOPTERA: Phlaeothripidae |
| slash pine sawfly | *Neodiprion merkeli* Ross | HYMENOPTERA: Diprionidae |
| slash pine seedworm | *Laspeyresia anaranjada* Miller | LEPIDOPTERA: Tortricidae |
| slender duck louse | *Anaticola crassicornis* (Scopoli) | PHTHIRAPTERA: Philopteridae |
| slender goose louse | *Anaticola anseris* (Linnaeus) | PHTHIRAPTERA: Philopteridae |
| slender guinea louse | *Lipeurus numidae* (Denny) | PHTHIRAPTERA: Philopteridae |
| slender guineapig louse | *Gliricola porcelli* (Schrank) | PHTHIRAPTERA: Gyropidae |
| slender pigeon louse | *Columbicola columbae* (Linnaeus) | PHTHIRAPTERA: Philopteridae |
| slender seedcorn beetle | *Clivina impressifrons* LeConte | COLEOPTERA: Carabidae |
| slender turkey louse | *Oxylipeurus polytrapezius* (Burmeister) | PHTHIRAPTERA: Philopteridae |
| slenderhorned flour beetle | *Gnathocerus maxillosus* (Fabricius) | COLEOPTERA: Tenebrionidae |
| small chestnut weevil | *Curculio sayi* (Gyllenhal) | COLEOPTERA: Curculionidae |
| small milkweed bug | *Lygaeus kalmii* Stål | HEMIPTERA: Lygaeidae |
| small pigeon louse | *Campanulotes bidentanis compar* (Burmeister) | PHTHIRAPTERA: Philopteridae |
| small southern pine engraver | *Ips avulsus* (Eichhoff) | COLEOPTERA: Scolytidae |
| smaller European elm bark beetle | *Scolytus multistriatus* (Marsham) | COLEOPTERA: Scolytidae |
| smaller Hawaiian cutworm | *Agrotis dislocata* (Walker) | LEPIDOPTERA: Noctuidae |
| smaller lantana butterfly | *Strymon bazochii gundlachianus* (Bates) | LEPIDOPTERA: Lycaenidae |

| | | |
|---|---|---|
| smaller yellow ant | *Acanthomyops claviger* (Roger) | HYMENOPTERA: Formicidae |
| smalleyed flour beetle | *Palorus ratzeburgi* (Wissmann) | COLEOPTERA: Tenebrionidae |
| smartweed borer | *Ostrinia obumbratalis* (Lederer) | LEPIDOPTERA: Crambidae |
| smeared dagger moth | *Acronicta oblinita* (J. E. Smith) | LEPIDOPTERA: Noctuidae |
| Smith blue | *Euphilotes enoptes smithi* (Mattoni) | LEPIDOPTERA: Lycaenidae |
| smokybrown cockroach | *Periplaneta fuliginosa* (Serville) | BLATTODEA: Blattidae |
| smut beetle | *Phalacrus politus* Melsheimer | COLEOPTERA: Phalacridae |
| snowball aphid | *Neoceruraphis viburnicola* (Gillette) | HEMIPTERA: Aphididae |
| snowy tree cricket | *Oecanthus fultoni* Walker | ORTHOPTERA: Gryllidae |
| solanaceous treehopper | *Antianthe expansa* (Germar) | HEMIPTERA: Membracidae |
| solitary oak leafminer | *Cameraria hamadryadella* (Clemens) | LEPIDOPTERA: Gracillariidae |
| sonchus fly | *Ensina sonchi* (Linnaeus) | DIPTERA: Tephritidae |
| Sonoran tent caterpillar | *Malacosoma tigris* (Dyar) | LEPIDOPTERA: Lasiocampidae |
| sorghum midge | *Contarinia sorghicola* (Coquillett) | DIPTERA: Cecidomyiidae |
| sorghum webworm | *Celama sorghiella* (Riley) | LEPIDOPTERA: Noctuidae |
| sourbush seed fly | *Acinia picturata* (Snow) | DIPTERA: Tephritidae |
| South African emex weevil | *Apion antiquum* Gyllenhal | COLEOPTERA: Curculionidae |
| south coastal coneworm | *Diorystria ebeli* Mutuura & Munroe | LEPIDOPTERA: Crambidae |
| southern armyworm | *Spodoptera eridania* (Cramer) | LEPIDOPTERA: Noctuidae |
| southern beet webworm | *Herpetogramma bipunctalis* (Fabricius) | LEPIDOPTERA: Crambidae |
| southern buffalo gnat | *Cnephia pecuarum* (Riley) | DIPTERA: Simuliidae |
| southern cabbageworm | *Pontia protodice* (Boisduval & LeConte) | LEPIDOPTERA: Pieridae |
| southern cattle tick | *Boophilus microplus* (Canestrini) | ACARI: Ixodidae |
| southern chinch bug | *Blissus insularis* Barber | HEMIPTERA: Lygaeidae |
| southern corn billbug | *Sphenophorus callosus* (Olivier) | COLEOPTERA: Curculionidae |
| southern corn rootworm | *Diabrotica undecimpunctata howardi* Barber | COLEOPTERA: Chrysomelidae |
| southern cornstalk borer | *Diatraea crambidoides* (Grote) | LEPIDOPTERA: Crambidae |
| southern fire ant | *Solenopsis xyloni* McCook | HYMENOPTERA: Formicidae |
| southern garden leafhopper | *Empoasca solana* DeLong | HEMIPTERA: Cicadellidae |
| southern green stink bug | *Nezara viridula* (Linnaeus) | HEMIPTERA: Pentatomidae |
| southern house mosquito | *Culex quinquefasciatus* Say | DIPTERA: Culicidae |
| southern lyctus beetle | *Lyctus planicollis* LeConte | COLEOPTERA: Lyctidae |
| southern masked chafer | *Cyclocephala immaculata* (Olivier) | COLEOPTERA: Scarabaeidae |
| southern mole cricket | *Scapteriscus acletus* Rehn & Hebard | ORTHOPTERA: Gryllotalpidae |
| southern pine beetle | *Dendroctonus frontalis* Zimmermann | COLEOPTERA: Scolytidae |
| southern pine coneworm | *Dioryctria amatella* (Hulst) | LEPIDOPTERA: Crambidae |
| southern pine root weevil | *Hylobius aliradicis* Warner | COLEOPTERA: Curculionidae |
| southern pine sawyer | *Monochamus titillator* (Fabricius) | COLEOPTERA: Cerambycidae |
| southern potato wireworm | *Conoderus falli* Lane | COLEOPTERA: Elateridae |
| southern red mite | *Oligonychus ilicis* (McGregor) | ACARI: Tetranychidae |
| southwestern corn borer | *Diatraea grandiosella* (Dyar) | LEPIDOPTERA: Crambidae |
| southwestern Hercules beetle | *Dynastes granti* Horn | COLEOPTERA: Scarabaeidae |
| southwestern pine tip moth | *Rhyacionia neomexicana* (Dyar) | LEPIDOPTERA: Tortricidae |
| southwestern squash vine borer | *Melittia calabaza* Duckworth & Eichlin | LEPIDOPTERA: Sesiidae |
| southwestern tent caterpillar | *Malacosoma incurvum* (H. Edwards) | LEPIDOPTERA: Lasiocampidae |
| sow thistle aphid | *Nasonovia lactucae* (Linnaeus) | HEMIPTERA: Aphididae |
| soybean aphid | *Aphis glycines* Matsumura | HEMIPTERA: Aphididae |
| soybean leafminer | *Odontota horni* Smith | COLEOPTERA: Chrysomelidae |
| soybean looper | *Pseudoplusia includens* (Walker) | LEPIDOPTERA: Noctuidae |
| soybean thrips | *Sericothrips variabilis* (Beach) | THYSANOPTERA: Thripidae |
| Spanishfly | *Lytta vesicatoria* (Linnaeus) | COLEOPTERA: Meloidae |
| spicebush swallowtail | *Papilio troilus* Linnaeus | LEPIDOPTERA: Papilionidae |
| spider mite destroyer | *Stethorus picipes* Casey | COLEOPTERA: Coccinellidae |
| spinach flea beetle | *Disonycha xanthomelas* (Dalman) | COLEOPTERA: Chrysomelidae |
| spinach leafminer | *Pegomya hyoscyami* (Panzer) | DIPTERA: Anthomyiidae |
| spined assassin bug | *Sinea diadema* (Fabricius) | HEMIPTERA: Reduviidae |
| spined rat louse | *Polyplax spinulosa* (Burmeister) | PHTHIRAPTERA: Hoplopleuridae |
| spined soldier bug | *Podisus maculiventris* (Say) | HEMIPTERA: Pentatomidae |
| spined stilt bug | *Jalysus spinosus* (Say) | HEMIPTERA: Berytidae |
| spiny assassin bug | *Polididus armatissimus* Stål | HEMIPTERA: Reduviidae |

| | | |
|---|---|---|
| spiny oakworm | *Anisota stigma* (Fabricius) | LEPIDOPTERA: Saturniidae |
| spirea aphid | *Aphis citricola* Van der Goot | HEMIPTERA: Aphididae |
| spotted alfalfa aphid | *Therioaphis maculata* (Buckton) | HEMIPTERA: Aphididae |
| spotted asparagus beetle | *Crioceris duodecimpunctata* (Linnaeus) | COLEOPTERA: Chrysomelidae |
| spotted beet webworm | *Hymenia perspectalis* (Hübner) | LEPIDOPTERA: Crambidae |
| spotted blister beetle | *Epicauta maculata* (Say) | COLEOPTERA: Meloidae |
| spotted cucumber beetle | *Diabrotica undecimpunctata howardi* Barber | COLEOPTERA: Chrysomelidae |
| spotted cutworm | *Amathes c-nigrum* (Linnaeus) | LEPIDOPTERA: Noctuidae |
| spotted hairy fungus beetle | *Mycetophagus quadriguttatus* Müller | COLEOPTERA: Mycetophagidae |
| spotted Mediterranean cockroach | *Ectobius pallidus* (Olivier) | BLATTODEA: Blattellidae |
| spotted pine sawyer | *Monochamus maculosus* Haldeman | COLEOPTERA; Cerambycidae |
| spotted tentiform leafminer | *Phyllonorycter blancardella* (Fabricius) | LEPIDOPTERA: Gracillariidae |
| spotted tussock moth | *Lophocampa maculata* Harris | LEPIDOPTERA: Arctiidae |
| spottedwinged antlion | *Dendroleon obsoletus* (Say) | NEUROPTERA: Myrmeleontidae |
| spring cankerworm | *Paleacrita vernata* (Peck) | LEPIDOPTERA: Geometridae |
| spruce aphid | *Elatobium abietinum* (Walker) | HEMIPTERA: Aphididae |
| spruce beetle | *Dendroctonus rufipennis* (Kirby) | COLEOPTERA: Scolytidae |
| spruce bud midge | *Rhabdophaga swainei* Felt | DIPTERA: Cecidomyiidae |
| spruce bud moth | *Zeiraphera canadensis* Mutuura & Freeman | LEPIDOPTERA: Tortricidae |
| spruce bud scale | *Physokermes piceae* (Schrank) | HEMIPTERA: Coccidae |
| spruce budworm | *Choristoneura fumiferana* (Clemens) | LEPIDOPTERA: Tortricidae |
| spruce coneworm | *Dioryctria reniculelloides* Mutuura & Munroe | LEPIDOPTERA: Crambidae |
| spruce mealybug | *Puto sandini* Washburn | HEMIPTERA: Pseudococcidae |
| spruce needleminer | *Endothenia albolineana* (Kearfott) | LEPIDOPTERA: Tortricidae |
| spruce seed moth | *Laspeyresia youngana* (Kearfott) | LEPIDOPTERA: Tortricidae |
| spruce spider mite | *Oligonychus ununguis* (Jacobi) | ACARI: Tetranychidae |
| squarenecked grain beetle | *Cathartus quadricollis* (Guérin-Méneville) | COLEOPTERA: Cucujidae |
| squarenosed fungus beetle | *Lathridius minutus* (Linnaeus) | COLEOPTERA: Lathridiidae |
| squash beetle | *Epilachna borealis* (Fabricius) | COLEOPTERA: Coccinellidae |
| squash bug | *Anasa tristis* (DeGeer) | HEMIPTERA: Coreidae |
| squash vine borer | *Melittia cucurbitae* (Harris) | LEPIDOPTERA: Sesiidae |
| stable fly | *Stomoxys calcitrans* (Linnaeus) | DIPTERA: Muscidae |
| stalk borer | *Papaipema nebris* (Guenée) | LEPIDOPTERA: Noctuidae |
| star jasmine thrips | *Thrips orientalis* (Bagnall) | THYSANOPTERA: Thripidae |
| steelblue ladybird beetle | *Orcus chalybeus* (Boisduval) | COLEOPTERA: Coccinellidae |
| Stevens leafhopper | *Empoasca stevensi* Young | HEMIPTERA: Cicadellidae |
| sticktight flea | *Echidnophaga gallinacea* (Westwood) | SIPHONAPTERA: Pulicidae |
| stinging rose caterpillar | *Parasa indetermina* (Boisduval) | LEPIDOPTERA: Limacodidae |
| stink beetle | *Nomius pygmaeus* (Dejean) | COLEOPTERA: Carabidae |
| stored nut moth | *Aphomia gularis* (Zeller) | LEPIDOPTERA: Crambidae |
| straw itch mite | *Pyemotes tritici* (Lagrèze-Fossat & Montané) | ACARI: Pyemotidae |
| strawberry aphid | *Chaetosiphon fragaefolii* (Cockerell) | HEMIPTERA: Aphididae |
| strawberry bud weevil | *Anthonomus signatus* Say | COLEOPTERA: Curculionidae |
| strawberry crown borer | *Tyloderma fragariae* (Riley) | COLEOPTERA: Curculionidae |
| strawberry crown moth | *Synanthedon bibionipennis* (Boisduval) | LEPIDOPTERA: Sesiidae |
| strawberry crownminer | *Aristotelia fragariae* Busck | LEPIDOPTERA: Gelechiidae |
| strawberry leafroller | *Ancylis comptana* (Froelich) | LEPIDOPTERA: Tortricidae |
| strawberry root aphid | *Aphis forbesi* Weed | HEMIPTERA: Aphididae |
| strawberry root weevil | *Otiorhynchus ovatus* (Linnaeus) | COLEOPTERA: Curculionidae |
| strawberry rootworm | *Paria fragariae* Wilcox | COLEOPTERA: Chrysomelidae |
| strawberry sap beetle | *Stelidota geminata* (Say) | COLEOPTERA: Nitidulidae |
| strawberry spider mite | *Tetranychus turkestani* Ugarov & Nikolski | ACARI: Tetranychidae |
| strawberry whitefly | *Trialeurodes packardi* (Morrill) | HEMIPTERA: Aleyrodidae |
| striped alder sawfly | *Hemichroa crocea* (Geoffroy) | HYMENOPTERA: Tenthredinidae |
| striped ambrosia beetle | *Trypodendron lineatum* (Olivier) | COLEOPTERA: Scolytidae |
| striped blister beetle | *Epicauta vittata* (Fabricius) | COLEOPTERA: Meloidae |
| striped cucumber beetle | *Acalymma vittatum* (Fabricius) | COLEOPTERA: Chrysomelidae |
| striped cutworm | *Euxoa tessellata* (Harris) | LEPIDOPTERA: Noctuidae |
| striped earwig | *Labidura riparia* Pallas | DERMAPTERA: Labiduridae |
| striped flea beetle | *Phyllotreta striolata* (Fabricius) | COLEOPTERA: Chrysomelidae |

| | | |
|---|---|---|
| striped garden caterpillar | *Lacanobia legitima* (Grote) | LEPIDOPTERA: Noctuidae |
| striped horse fly | *Tabanus lineola* Fabricius | DIPTERA: Tabanidae |
| striped mealybug | *Ferrisia virgata* (Cockerell) | HEMIPTERA: Pseudococcidae |
| subtropical pine tip moth | *Rhyacionia subtropica* Miller | LEPIDOPTERA: Tortricidae |
| suckfly | *Cyrtopeltis notatus* (Distant) | HEMIPTERA: Miridae |
| sugar maple borer | *Glycobius speciosus* (Say) | COLEOPTERA: Cerambycidae |
| sugar pine cone beetle | *Conophthorus lambertianae* Hopkins | COLEOPTERA: Scolytidae |
| sugarbeet crown borer | *Hulstia undulatella* (Clemens) | LEPIDOPTERA: Crambidae |
| sugarbeet root aphid | *Pemphigus populivenae* Fitch | HEMIPTERA: Aphididae |
| sugarbeet root maggot | *Tetanops myopaeformis* (Röder) | DIPTERA: Otitidae |
| sugarbeet wireworm | *Limonius californicus* (Mannerheim) | COLEOPTERA: Elateridae |
| sugarcane aphid | *Melanaphis sacchari* (Zehntner) | HEMIPTERA: Aphididae |
| sugarcane beetle | *Euetheola humilis rugiceps* (LeConte) | COLEOPTERA: Scarabaeidae |
| sugarcane borer | *Diatraea saccharalis* (Fabricius) | LEPIDOPTERA: Crambidae |
| sugarcane bud moth | *Neodecadarachis flavistriata* (Walsingham) | LEPIDOPTERA: Tineidae |
| sugarcane delphacid | *Perkinsiella saccharicida* Kirkaldy | HEMIPTERA: Delphacidae |
| sugarcane leaf mite | *Oligonychus indicus* (Hirst) | ACARI: Tetranychidae |
| sugarcane leafroller | *Hedylepta accepta* (Butler) | LEPIDOPTERA: Crambidae |
| sugarcane stalk mite | *Steneotarsonemus bancrofti* (Michael) | ACARI: Tarsonemidae |
| sugarcane thrips | *Baliothrips minutus* (van Deventer) | THYSANOPTERA: Thripidae |
| sunflower beetle | *Zygogramma exclamationis* (Fabricius) | COLEOPTERA: Chrysomelidae |
| sunflower bud moth | *Suleima helianthana* (Riley) | LEPIDOPTERA: Tortricidae |
| sunflower maggot | *Strauzia longipennis* (Wiedemann) | DIPTERA: Tephritidae |
| sunflower moth | *Homoeosoma electellum* (Hulst) | LEPIDOPTERA: Crambidae |
| sunflower seed midge | *Neolasioptera murtfeldtiana* (Felt) | DIPTERA: Cecidomyiidae |
| sunflower spittlebug | *Clastoptera xanthocephala* Germar | HEMIPTERA: Cercopidae |
| superb plant bug | *Adelphocoris superbus* (Uhler) | HEMIPTERA: Miridae |
| Surinam cockroach | *Pycnoscelus surinamensis* (Linnaeus) | BLATTODEA: Blaberidae |
| Swaine jack pine sawfly | *Neodiprion swainei* Middleton | HYMENOPTERA: Diprionidae |
| swallow bug | *Oeciacus vicarius* Horvath | HEMIPTERA: Cimicidae |
| sweetclover aphid | *Therioaphis riehmi* (Börner) | HEMIPTERA: Aphididae |
| sweetclover root borer | *Walshia miscecolorella* (Chambers) | LEPIDOPTERA: Cosmopterigidae |
| sweetclover weevil | *Sitona cylindricollis* Fåhraeus | COLEOPTERA: Curculionidae |
| sweetfern leaf casebearer | *Acrobasis comptoniella* (Hulst) | LEPIDOPTERA: Crambidae |
| sweetpotato flea beetle | *Chaetocnema confinis* Crotch | COLEOPTERA: Chrysomelidae |
| sweetpotato hornworm | *Agrius cingulatus* (Fabricius) | LEPIDOPTERA: Sphingidae |
| sweetpotato leaf beetle | *Typophorus nigritus viridicyaneus* (Crotch) | COLEOPTERA: Chrysomelidae |
| sweetpotato leafminer | *Bedellia orchilella* Walsingham | LEPIDOPTERA: Lyonetiidae |
| sweetpotato leafroller | *Pilocrocis tripunctata* (Fabricius) | LEPIDOPTERA: Crambidae |
| sweetpotato vine borer | *Omphisa anastomosalis* (Guenée) | LEPIDOPTERA: Crambidae |
| sweetpotato weevil | *Cylas formicarius elegantulus* (Summers) | COLEOPTERA: Curculionidae |
| sweetpotato whitefly | *Bemisia tabaci* (Gennadius) | HEMIPTERA: Aleyrodidae |
| sycamore lace bug | *Corythucha ciliata* (Say) | HEMIPTERA: Tingidae |
| sycamore tussock moth | *Halysidota harrisii* Walsh | LEPIDOPTERA: Arctiidae |

## T

| | | |
|---|---|---|
| Tahitian coconut weevil | *Diocalandra taitensis* (Guérin-Menéville) | COLEOPTERA: Curculionidae |
| tamarind weevil | *Sitophilus linearis* (Herbst) | COLEOPTERA: Curculionidae |
| tamarix leafhopper | *Opsius stactogalus* Fieber | HEMIPTERA: Cicadellidae |
| tarnished plant bug | *Lygus lineolaris* (Palisot de Beauvois) | HEMIPTERA: Miridae |
| tea scale | *Fiorinia theae* Green | HEMIPTERA: Diaspididae |
| tenlined June beetle | *Polyphylla decemlineata* (Say) | COLEOPTERA: Scarabaeidae |
| tenspotted ladybird beetle | *Coelophora pupillata* (Swartz) | COLEOPTERA: Coccinellidae |
| terrapin scale | *Mesolecanium nigrofasciatum* (Pergande) | HEMIPTERA: Coccidae |
| tessellated scale | *Eucalymnatus tessellatus* (Signoret) | HEMIPTERA: Coccidae |
| Texas citrus mite | *Eutetranychus banksi* (McGregor) | ACARI: Tetranychidae |
| Texas leafcutting ant | *Atta texana* (Buckley) | HYMENOPTERA: Formicidae |
| thief ant | *Solenopsis molesta* (Say) | HYMENOPTERA: Formicidae |
| thirteenspotted ladybird beetle | *Hippodamia tredecimpunctata tibialis* (Say) | COLEOPTERA: Coccinellidae |

| | | |
|---|---|---|
| thistle aphid | *Brachycaudus cardui* (Linnaeus) | HEMIPTERA: Aphididae |
| thread bug | *Empicoris rubromaculatus* (Blackburn) | HEMIPTERA: Reduviidae |
| threebanded leafhopper | *Erythroneura tricincta* Fitch | HEMIPTERA: Cicadellidae |
| threecornered alfalfa hopper | *Spissistilus festinus* (Say) | HEMIPTERA: Membracidae |
| threelined leafroller | *Pandemis limitata* (Robinson) | LEPIDOPTERA: Tortricidae |
| threelined potato beetle | *Lema trilineata* (Olivier) | COLEOPTERA: Chrysomelidae |
| threespotted flea beetle | *Disonycha triangularis* (Say) | COLEOPTERA: Chrysomelidae |
| threestriped blister beetle | *Epicauta lemniscata* (Fabricius) | COLEOPTERA: Meloidae |
| threestriped ladybird beetle | *Brumoides suturalis* (Fabricius) | COLEOPTERA: Coccinellidae |
| throat bot fly | *Gasterophilus nasalis* (Linnaeus) | DIPTERA: Gasterophilidae |
| thurberia weevil | *Anthonomus grandis thurberiae* Pierce | COLEOPTERA: Curculionidae |
| tiger swallowtail | *Papilio glaucus* Linnaeus | LEPIDOPTERA: Papilionidae |
| tilehorned prionus | *Prionus imbricornis* (Linnaeus) | COLEOPTERA: Cerambycidae |
| tipdwarf mite | *Calepiterimerus thujae* (Garman) | ACARI: Eriophyidae |
| toad bug | *Gelastocoris oculatus* (Fabricius) | HEMIPTERA: Gelastocoridae |
| tobacco budworm | *Helicoverpa virescens* (Fabricius) | LEPIDOPTERA: Noctuidae |
| tobacco flea beetle | *Epitrix hirtipennis* (Melsheimer) | COLEOPTERA: Chrysomelidae |
| tobacco hornworm | *Manduca sexta* (Linnaeus) | LEPIDOPTERA: Sphingidae |
| tobacco moth | *Ephestia elutella* (Hübner) | LEPIDOPTERA: Crambidae |
| tobacco stalk borer | *Trichobaris mucorea* (LeConte) | COLEOPTERA: Curculionidae |
| tobacco thrips | *Frankliniella fusca* (Hinds) | THYSANOPTERA: Thripidae |
| tobacco wireworm | *Conoderus vespertinus* (Fabricius) | COLEOPTERA: Elateridae |
| tomato bug | *Cyrtopeltis modestus* (Distant) | HEMIPTERA: Miridae |
| tomato fruitworm | *Helicoverpa zea* (Boddie) | LEPIDOPTERA: Noctuidae |
| tomato hornworm | *Manduca quinquemaculata* (Haworth) | LEPIDOPTERA: Sphingidae |
| tomato pinworm | *Keiferia lycopersicella* (Walsingham) | LEPIDOPTERA: Gelechiidae |
| tomato psyllid | *Paratrioza cockerelli* (Sulc) | HEMIPTERA: Psyllidae |
| tomato russet mite | *Aculops lycopersici* (Massee) | ACARI: Eriophyidae |
| toothed flea beetle | *Chaetocnema denticulata* (Illiger) | COLEOPTERA: Chrysomelidae |
| torsalo | *Dermatobia hominis* (Linnaeus, Jr.) | DIPTERA: Cuterebridae |
| transparentwinged plant bug | *Hyalopeplus pellucidus* (Stål) | HEMIPTERA: Miridae |
| transverse lady beetle | *Coccinella transversoguttata richardsoni* Brown | COLEOPTERA: Coccinellidae |
| trefoil seed chalcid | *Bruchophagus kolobovae* Fedoseeva | HYMENOPTERA: Eurytomidae |
| tropical fowl mite | *Ornithonyssus bursa* (Berlese) | ACARI: Macronyssidae |
| tropical horse tick | *Anocentor nitens* (Neumann) | ACARI: Ixodidae |
| tropical rat louse | *Hoplopleura pacifica* Ewing | PHTHIRAPTERA: Hoplopleuridae |
| tropical rat mite | *Ornithonyssus bacoti* (Hirst) | ACARI: Macronyssidae |
| tropical sod webworm | *Herpetogramma phaeopteralis* Guenée | LEPIDOPTERA: Crambidae |
| tuber flea beetle | *Epitrix tuberis* Gentner | COLEOPTERA: Chrysomelidae |
| tule beetle | *Agonum maculicolle* Dejean | COLEOPTERA: Carabidae |
| tulip bulb aphid | *Dysaphis tulipae* (Fonscolombe) | HEMIPTERA: Aphididae |
| tuliptree aphid | *Macrosiphum liriodendri* (Monell) | HEMIPTERA: Aphididae |
| tuliptree scale | *Toumeyella liriodendri* (Gmelin) | HEMIPTERA: Coccidae |
| tumid spider mite | *Tetranychus tumidus* Banks | ACARI: Tetranychidae |
| tupelo leafminer | *Antispila nysaefoliella* Clemens | LEPIDOPTERA: Heliozelidae |
| turkey chigger | *Neoschoengastia americana* (Hirst) | ACARI: Trombiculidae |
| turkey gnat | *Simulium meridionale* Riley | DIPTERA: Simuliidae |
| turnip aphid | *Hyadaphis erysimi* (Kaltenbach) | HEMIPTERA: Aphididae |
| turnip maggot | *Delia fioralis* (Fallén) | DIPTERA: Anthomyiidae |
| turpentine borer | *Buprestis apricans* Herbst | COLEOPTERA: Buprestidae |
| twicestabbed ladybird beetle | *Chilocorus stigma* (Say) | COLEOPTERA: Coccinellidae |
| twig girdler | *Oncideres cingulata* (Say) | COLEOPTERA: Cerambycidae |
| twig pruner | *Elaphidionoides villosus* (Fabricius) | COLEOPTERA: Cerambycidae |
| twobanded fungus beetle | *Alphitophagus bifasciatus* (Say) | COLEOPTERA: Tenebrionidae |
| twobanded Japanese weevil | *Callirhopalus bifasciatus* (Roelofs) | COLEOPTERA: Curculionidae |
| twolined chestnut borer | *Agrilus bilineatus* (Weber) | COLEOPTERA: Buprestidae |
| twolined spittlebug | *Prosapia bicincta* (Say) | HEMIPTERA: Cercopidae |
| twomarked treehopper | *Enchenopa binotata* (Say) | HEMIPTERA: Membracidae |
| twospotted ladybird beetle | *Adalia bipunctata* (Linnaeus) | COLEOPTERA: Coccinellidae |
| twospotted spider mite | *Tetranychus urticae* Koch | ACARI: Tetranychidae |

| | | |
|---|---|---|
| twospotted stink bug | *Perillus bioculatus* (Fabricius) | HEMIPTERA: Pentatomidae |
| twostriped grasshopper | *Melanoplus bivittatus* (Say) | ORTHOPTERA: Acrididae |
| twostriped walkingstick | *Anisomorpha buprestoides* (Stoll) | PHASMATODEA: Phasmatidae |

## U

| | | |
|---|---|---|
| uglynest caterpillar | *Archips cerasivorana* (Fitch) | LEPIDOPTERA: Tortricidae |
| unicorn caterpillar | *Schizura unicornis* (J. E. Smith) | LEPIDOPTERA: Notodontidae |

## V

| | | |
|---|---|---|
| vagabond crambus | *Agriphila vulgivagella* (Clemens) | LEPIPOPTERA: Crambidae |
| vagrant grasshopper | *Schistocerca nitens nitens* (Thunberg) | ORTHOPTERA: Acrididae |
| Van Duzee treehopper | *Vanduzea segmentata* (Fowler) | HEMIPTERA: Membracidae |
| vanda thrips | *Dichromothrips corbetti* (Priesner) | THYSANOPTERA: Thripidae |
| variable oakleaf caterpillar | *Heterocampa manteo* (Doubleday) | LEPIDOPTERA: Notodontidae |
| varied carpet beetle | *Anthrenus verbasci* (Linnaeus) | COLEOPTERA: Dermestidae |
| variegated cutworm | *Peridroma saucia* (Hübner) | LEPIDOPTERA: Noctuidae |
| vedalia | *Rodolia cardinalis* (Mulsant) | COLEOPTERA: Coccinellidae |
| vegetable leafminer | *Liriomyza sativae* Blanchard | DIPTERA: Agromyzidae |
| vegetable weevil | *Listroderes costirostris obliquus* Klug | COLEOPTERA: Curculionidae |
| velvetbean caterpillar | *Anticarsia gemmatalis* (Hübner) | LEPIDOPTERA: Noctuidae |
| verbena bud moth | *Endothenia hebesana* (Walker) | LEPIDOPTERA: Tortricidae |
| vespiform thrips | *Franklinothrips vespiformis* (Crawford) | THYSANOPTERA: Aeolothripidae |
| vetch bruchid | *Bruchus brachialis* Fåhraeus | COLEOPTERA: Bruchidae |
| vexans mosquito | *Aedes vexans* (Meigen) | DIPTERA: Culicidae |
| viburnum aphid | *Aphis viburniphila* Patch | HEMIPTERA: Aphididae |
| viceroy | *Basilarchia archippus* (Cramer) | LEPIDOPTERA: Nymphalidae |
| violet aphid | *Micromyzus violae* (Pergande) | HEMIPTERA: Aphididae |
| violet sawfly | *Ametastegia pallipes* (Spinola) | HYMENOPTERA: Tenthredinidae |
| Virginia pine sawfly | *Neodiprion pratti pratti* (Dyar) | HYMENOPTERA: Diprionidae |
| Virginiacreeper leafhopper | *Erythroneura ziczac* Walsh | HEMIPTERA: Cicadellidae |
| Virginiacreeper sphinx | *Darapsa myron* (Cramer) | LEPIDOPTERA: Sphingidae |

## W

| | | |
|---|---|---|
| w-marked cutworm | *Spaelotis clandestina* (Harris) | LEPIDOPTERA: Noctuidae |
| walkingstick | *Diapheromera femorata* (Say) | ORTHOPTERA: Phasmatidae |
| walnut aphid | *Chromaphis juglandicola* (Kaltenbach) | HEMIPTERA: Aphididae |
| walnut blister mite | *Eriophyes erinea* (Nalepa) | ACARI: Eriophyidae |
| walnut caterpillar | *Datana integerrima* Grote & Robinson | LEPIDOPTERA: Notodontidae |
| walnut husk fly | *Rhagoletis completa* Cresson | DIPTERA: Tephritidae |
| walnut scale | *Quadraspidiotus juglansregiae* (Comstock) | HEMIPTERA: Diaspididae |
| walnut shoot moth | *Acrobasis demotella* Grote | LEPIDOPTERA: Crambidae |
| walnut sphinx | *Cressonia juglandis* (J. E. Smith) | LEPIDOPTERA: Sphingidae |
| wardrobe beetle | *Attagenus fasciatus* (Thunberg) | COLEOPTERA: Dermestidae |
| warehouse beetle | *Trogoderma variabile* Ballion | COLEOPTERA: Dermestidae |
| warty grain mite | *Aeroglyphus robustus* (Banks) | ACARI: Glycyphagidae |
| watercress leaf beetle | *Phaedon viridus* (Melsheimer) | COLEOPTERA: Chrysomelidae |
| watercress sharpshooter | *Draeculacephala mollipes* (Say) | HEMIPTERA: Cicadellidae |
| waterlily aphid | *Rhopalosiphum nymphaeae* (Linnaeus) | HEMIPTERA: Aphididae |
| waterlily leaf beetle | *Pyrrhalta nymphaeae* (Linnaeus) | COLEOPTERA: Chrysomelidae |
| waterlily leafcutter | *Synclita obliteralis* (Walker) | LEPIDOPTERA: Crambidae |
| webbing clothes moth | *Tineola bisselliella* (Hummel) | LEPIDOPTERA: Tineidae |
| webbing coneworm | *Dionystria disclusa* Heinrich | LEPIDOPTERA: Crambidae |
| West Indian cane weevil | *Metamasius hemipterus hemipterus* (Linnaeus) | COLEOPTERA: Curculionidae |
| West Indian flatid | *Melormensis antillarium* (Kirkaldy) | HEMIPTERA: Flatidae |
| West Indian fruit fly | *Anastrepha obliqua* (Macquart) | DIPTERA: Tephritidae |

| | | |
|---|---|---|
| West Indian sweetpotato weevil | *Euscepes postfasciatus* (Fairmaire) | COLEOPTERA: Curculionidae |
| western balsam bark beetle | *Dryocoetes confusus* Swaine | COLEOPTERA: Scolytidae |
| western bean cutworm | *Loxagrotis albicosta* Smith | LEPIDOPTERA: Noctuidae |
| western bigeyed bug | *Geocoris pallens* Stål | HEMIPTERA: Lygaeidae |
| western black flea beetle | *Phyllotreta pusilla* Horn | COLEOPTERA: Chrysomelidae |
| western blackheaded budworm | *Acleris gloverana* (Walsingham) | LEPIDOPTERA: Tortricidae |
| western bloodsucking conenose | *Triatoma protracta* (Uhler) | HEMIPTERA: Reduviidae |
| western boxelder bug | *Leptocoris rubrolineatus* Barber | HEMIPTERA: Rhopalidae |
| western brown stink bug | *Euschistus impictiventris* Stål | HEMIPTERA: Pentatomidae |
| western cedar bark beetle | *Phloeosinus punctatus* LeConte | COLEOPTERA: Scolytidae |
| western cedar borer | *Trachykele blondeli* Marseul | COLEOPTERA: Buprestidae |
| western cherry fruit fly | *Rhagoletis indifferens* Curran | DIPTERA: Tephritidae |
| western chicken flea | *Ceratophyllus niger* Fox | SIPHONAPTERA: Ceratophyllidae |
| western chinch bug | *Blissus occiduus* Barber | HEMIPTERA: Lygaeidae |
| western corn rootworm | *Diabrotica virgifera virgifera* LeConte | COLEOPTERA: Chrysomelidae |
| western damsel bug | *Reduviolus alternatus* (Parshley) | HEMIPTERA: Nabidae |
| western drywood termite | *Incisitermes minor* (Hagen) | ISOPTERA: Kalotermitidae |
| western field wireworm | *Limonius infuscatus* Motschulsky | COLEOPTERA: Elateridae |
| western flower thrips | *Frankliniella occidentalis* (Pergande) | THYSANOPTERA: Thripidae |
| western grape rootworm | *Bromius obscurus* (Linnaeus) | COLEOPTERA: Chrysomelidae |
| western grapeleaf skeletonizer | *Harrisina brillians* Barnes & McDunnough | LEPIDOPTERA: Zygaenidae |
| western harvester ant | *Pogonomyrmex occidentalis* (Cresson) | HYMENOPTERA: Formicidae |
| western hemlock looper | *Lambdina fiscellaria lugubrosa* (Hulst) | LEPIDOPTERA: Geometridae |
| western lawn moth | *Tehama bonifatella* (Hulst) | LEPIDOPTERA: Crambidae |
| western lily aphid | *Macrosiphum scoliopi* Essig | HEMIPTERA: Aphididae |
| western lygus bug | *Lygus hesperis* Knight | HEMIPTERA: Miridae |
| western oak looper | *Lambdina fiscellaria somniaria* (Hulst) | LEPIDOPTERA: Geometridae |
| western pine beetle | *Dendroctonus brevicomis* LeConte | COLEOPTERA: Scolytidae |
| western pine shoot borer | *Eucosma sonomana* Kearfott | LEPIDOPTERA: Tortricidae |
| western plant bug | *Rhinacloa forticornis* Reuter | HEMIPTERA: Miridae |
| western poplar clearwing | *Paranthrene robiniae* (H. Edwards) | LEPIDOPTERA: Sesiidae |
| western potato flea beetle | *Epitrix subcrinita* LeConte | COLEOPTERA: Chrysomelidae |
| western potato leafhopper | *Empoasca abrupta* DeLong | HEMIPTERA: Cicadellidae |
| western predatory mite | *Galandromus occidentalis* (Nesbitt) | ACARI: Phytoseiidae |
| western raspberry fruitworm | *Bytunus bakeri* Barber | COLEOPTERA: Byturidae |
| western spotted cucumber beetle | *Diabrotica undecimpunctata undecimpunctata* Mannerheim | COLEOPTERA: Chrysomelidae |
| western spruce budworm | *Choristoneura occidentalis* Freeman | LEPIDOPTERA: Tortricidae |
| western striped cucumber beetle | *Acalymma trivittatum* (Mannerheim) | COLEOPTERA: Chrysomelidae |
| western striped flea beetle | *Phyllotreta ramosa* (Crotch) | COLEOPTERA: Chrysomelidae |
| western subterranean termite | *Reticulitermes hesperus* Banks | ISOPTERA: Rhinotermitidae |
| western tent caterpillar | *Malacosoma californicum* (Packard) | LEPIDOPTERA: Lasiocampidae |
| western thatching ant | *Formica obscuripes* Forel | HYMENOPTERA: Formicidae |
| western treehole mosquito | *Aedes sierrensis* (Ludlow) | DIPTERA: Culicidae |
| western tussock moth | *Orgyia vetusta* (Boisduval) | LEPIDOPTERA: Lymantriidae |
| western w-marked cutworm | *Spaelotis havilae* (Grote) | LEPIDOPTERA: Noctuidae |
| western wheat aphid | *Brachycolus tritici* (Gillette) | HEMIPTERA: Aphididae |
| western yellowjacket | *Vespula pensylvanica* (Saussure) | HYMENOPTERA: Vespidae |
| western yellowstriped armyworm | *Spodoptera praefica* (Grote) | LEPIDOPTERA: Noctuidae |
| wharf borer | *Nacerdes melanura* (Linnaeus) | COLEOPTERA: Oedemeridae |
| wheat curl mite | *Eriophyes tulipae* Keifer | ACARI: Eriophyidae |
| wheat head armyworm | *Faronta diffusa* (Walker) | LEPIDOPTERA: Noctuidae |
| wheat jointworm | *Tetramesa tritici* (Fitch) | HYMENOPTERA: Eurytomidae |
| wheat midge | *Sitodiplosis mosellana* (Géhin) | DIPTERA: Cecidomyiidae |
| wheat stem maggot | *Meromyza americana* Fitch | DIPTERA: Chloropidae |
| wheat stem sawfly | *Cephus cinctus* Norton | HYMENOPTERA: Cephidae |
| wheat strawworm | *Tetramesa grandis* (Riley) | HYMENOPTERA: Eurytomidae |
| wheat wireworm | *Agriotes mancus* (Say) | COLEOPTERA: Elateridae |
| wheel bug | *Arilus cristatus* (Linnaeus) | HEMIPTERA: Reduviidae |
| white apple leafhopper | *Typhlocyba pomaria* McAtee | HEMIPTERA: Cicadellidae |

| | | |
|---|---|---|
| white cutworm | *Euxoa scandens* (Riley) | LEPIDOPTERA: Noctuidae |
| white fir needleminer | *Epinotia meritana* Heinrich | LEPIDOPTERA: Tortricidae |
| white oak borer | *Goes tigrinus* (DeGeer) | COLEOPTERA: Cerambycidae |
| white peach scale | *Pseudaulacaspis pentagona* (Targioni-Tozzetti) | HEMIPTERA: Diaspididae |
| white pine aphid | *Cinara strobi* (Fitch) | HEMIPTERA: Aphididae |
| white pine cone beetle | *Conophthorus coniperda* (Schwarz) | COLEOPTERA: Scolytidae |
| white pine cone borer | *Eucosma tocullionana* Heinrich | LEPIDOPTERA: Tortricidae |
| white pine sawfly | *Neodiprion pinetum* (Norton) | HYMENOPTERA: Diprionidae |
| white pine weevil | *Pissodes strobi* (Peck) | COLEOPTERA: Curculionidae |
| whitebanded elm leafhopper | *Scaphoideus luteolus* (Van Duzee) | HEMIPTERA: Cicadellidae |
| whitecrossed seed bug | *Neacoryphus bicrucis* (Say) | HEMIPTERA: Lygaeidae |
| whitefringed beetles | *Graphognathus* spp. | COLEOPTERA: Curculionidae |
| whitelined sphinx | *Hyles lineata* (Fabricius) | LEPIDOPTERA: Sphingidae |
| whitemargined cockroach | *Melanozosteria soror* (Brunner) | BLATTODEA: Blattidae |
| whitemarked fleahopper | *Spanagonicus albofasciatus* (Reuter) | HEMIPTERA: Miridae |
| whitemarked spider beetle | *Ptinus fur* (Linnaeus) | COLEOPTERA: Ptinidae |
| whitemarked treehopper | *Tricentrus albomaculatus* Distant | HEMIPTERA: Membracidae |
| whitemarked tussock moth | *Orgyia leucostigma* (J. E. Smith) | LEPIDOPTERA: Lymantriidae |
| whiteshouldered house moth | *Endrosis sarcitrella* (Linnaeus) | LEPIDOPTERA: Oecophoridae |
| whitespotted sawyer | *Monochamus scutellatus* (Say) | COLEOPTERA: Cerambycidae |
| willow beaked-gall midge | *Mayetiola rigidae* (Osten Sacken) | DIPTERA: Cecidomyiidae |
| willow flea weevil | *Rhynchaenus rufipes* (LeConte) | COLEOPTERA: Curculionidae |
| willow redgall sawfly | *Pontania promixa* (Lepeletier) | HYMENOPTERA: Tenthredinidae |
| willow sawfly | *Nematus ventralis* Say | HYMENOPTERA: Tenthredinidae |
| willow shoot sawfly | *Janus abbreviatus* (Say) | HYMENOPTERA: Cephidae |
| Wilson sphinx | *Hyles wilsoni* (Rothschild) | LEPIDOPTERA: Sphingidae |
| wing louse | *Lipeurus caponis* (Linnaeus) | PHTHIRAPTERA: Philopteridae |
| winter grain mite | *Penthaleus major* (Dugès) | ACARI: Eupodidae |
| winter moth | *Operophtera brumata* (Linnaeus) | LEPIDOPTERA: Geometridae |
| winter tick | *Dermacentor albipictus* (Packard) | ACARI: Ixodidae |
| wood cockroaches | *Parcoblatta* spp. | BLATTODEA: Blattellidae |
| woodrose bug | *Graptostethus manillensis* (Stål) | HEMIPTERA: Lygaeidae |
| woods weevil | *Nemocestes incomptus* Horn | COLEOPTERA: Curculionidae |
| woolly alder aphid | *Paraprociphilus tessellatus* (Fitch) | HEMIPTERA: Aphididae |
| woolly apple aphid | *Eriosoma lanigerum* (Hausmann) | HEMIPTERA: Aphididae |
| woolly elm aphid | *Eriosoma americanum* (Riley) | HEMIPTERA: Aphididae |
| woolly pear aphid | *Eriosoma pyricola* Baker & Davidson | HEMIPTERA: Aphididae |
| woolly whitefly | *Aleurothrixus floccosus* (Maskell) | HEMIPTERA: Aleyrodidae |

## Y

| | | |
|---|---|---|
| yellow and black potter wasp | *Delta campaniformis campaniformis* (Fabricius) | HYMENOPTERA: Vespidae |
| yellow clover aphid | *Therioaphis trifolii* (Monell) | HEMIPTERA: Aphididae |
| yellow garden spider | *Argiope aurantia* Lucas | ARANEAE: Araneidae |
| yellow mealworm | *Tenebrio molitor* Linnaeus | COLEOPTERA: Tenebrionidae |
| yellow rose aphid | *Acyrthosiphon porosum* (Sanderson) | HEMIPTERA: Aphididae |
| yellow scale | *Aonidiella citrina* (Coquillett) | HEMIPTERA: Diaspididae |
| yellow spider mite | *Eotetranychus carpini borealis* (Ewing) | ACARI: Tetranychidae |
| yellow sugarcane aphid | *Sipha flava* (Forbes) | HEMIPTERA: Aphididae |
| yellow woollybear | *Spilosoma virginica* (Fabricius) | LEPIDOPTERA: Arctiidae |
| yellowfaced leafhopper | *Scaphytopius loricatus* (Van Duzee) | HEMIPTERA: Cicadellidae |
| yellowfever mosquito | *Aedes aegypti* (Linnaeus) | DIPTERA: Culicidae |
| yellowheaded cutworm | *Apamea amputatrix* (Fitch) | LEPIDOPTERA: Noctuidae |
| yellowheaded fireworm | *Acleris minuta* (Robinson) | LEPIDOPTERA: Tortricidae |
| yellowheaded leafhopper | *Carneocephala flaviceps* (Riley) | HEMIPTERA: Cicadellidae |
| yellowheaded spruce sawfly | *Pikonema alaskensis* (Rohwer) | HYMENOPTERA: Tenthredinidae |
| yellowjackets | *Dolichovespula* spp. and *Vespula* spp. | HYMENOPTERA: Vespidae |
| yellowmargined leaf beetle | *Microtheca ochroloma* Stål | COLEOPTERA: Chrysomelidae |
| yellownecked caterpillar | *Datana ministra* (Drury) | LEPIDOPTERA: Notodontidae |

| | | |
|---|---|---|
| yellowshouldered ladybird beetle | *Scymnodes lividigaster* (Mulsant) | COLEOPTERA: Coccinellidae |
| yellowstriped armyworm | *Spodoptera ornithogalli* (Guenée) | LEPIDOPTERA: Noctuidae |
| Yosemite bark weevil | *Pissodes schwarzi* Hopkins | COLEOPTERA: Curculionidae |
| yucca moth | *Tegeticula yuccasella* (Riley) | LEPIDOPTERA: Incurvariidae |
| yucca plant bug | *Halticotoma valida* Townsend | HEMIPTERA: Miridae |
| Yuma spider mite | *Eotetranychus yumensis* (McGregor) | ACARI: Tetranychidae |

## Z

| | | |
|---|---|---|
| zebra caterpillar | *Melanchra picta* (Harris) | LEPIDOPTERA: Noctuidae |
| Zimmerman pine moth | *Dioryctria zimmermani* (Grote) | LEPIDOPTERA: Crambidae |

APPENDIX 3

# World Wide Web Sites of Entomological Resources

BY JOHN K. VANDYK

This appendix consists of a wide range of Internet resources gleaned from The Entomology Index of Internet Resources (www.ent.iastate.edu/list/), an online database of entomological resources from Iowa State University.

The Entomology Index began in 1994 as a collaborative effort of John VanDyk (Iowa State University) and Lou Bjostad (Colorado State University) to create a comprehensive central directory of Internet resources of use to entomologists. VanDyk continues to maintain the site, which has grown to over 1,500 resources.

Anyone can submit their online entomological resource to the Index, which stands as an example to the power of collaborative subject-specific databases.

The links given here were all correct as of October 19, 2007.

## APICULTURE (BEEKEEPING) AND SOCIAL INSECTS

**Africanized Honey Bees.** Africanized honey bee information, especially in relation to California. Includes bibliography with over 900 references (http://bees.ucr.edu/index.html).

**All about Bees.** (http://www.beekeeping.com/info/sommaires/index_us.htm).

**American Association of Professional Apiculturists.** (http://entomology.ucdavis.edu/aapa/).

**American Bee Journal.** The journal has the honor of being the oldest English language beekeeping publication in the world (http://www.dadant.com/journal/).

**American Beekeeping Federation.** (http://www.abfnet.org/).

**Ant Colony Developers Association.** For those with an interest in ant colonies (http://www.antcolony.org/).

**Apiculture Newsletter.** From the Department of Entomology, University of California–Davis (http://entomology.ucdavis.edu/faculty/mussen/news.cfm).

**APIS: Apicultural Information and Issues.** Archive of individual issues of the *Apicultural Information and Issues* monthly newsletter, published by the Florida Cooperative Extension Service (http://apis.ifas.ufl.edu/).

**Apiservices—Virtual Beekeeping Gallery.** Information, products, and services for beekeeping, bees, and honey. Trilingual site: in English, French, and Spanish (http://www.beekeeping.com/).

**ARS Insect Locations.** Includes stock centers (http://www.ars-grin.gov/nigrp/ars_insects.html).

**Bee Alert.** Using bees to assess environmental hazards. Online video cameras, electronic hive monitoring. Kids' section (http://beekeeper.dbs.umt.edu/bees/).

**Bee Briefs.** Short articles on topics of interest to beekeepers. From the Department of Entomology, University of California–Davis (http://entomology.ucdavis.edu/faculty/mussen/beebriefs/index.cfm).

**Bee Genera of the World.** Synonymic listing, including type species and subgenera; based on Michener (2000), with updates (http://cache.ucr.edu/~heraty/beepage.html).

**Bee Improvement.** From Bee Improvement and Bee Breeders' Association (BIBBA), a U.K.-based organization dedicated to conservation and improvement of native honey bees (http://www.bibba.co.uk/).

**Bee Research Lab.** Beltsville, MD. Research projects and instruction for submission of samples to the diagnostic service. From USDA/ARS (http://www.ars.usda.gov/main/site_main.htm?modecode=12-75-05-00).

**Bee Tidings Newsletter.** Image-rich newsletter for Midwestern beekeepers produced by the University of Nebraska, edited by Marion Ellis (http://entomology.unl.edu/beekpg/).

**BEE-L.** For discussion of bee research and biology. Subjects include sociobiology, behavior, ecology, genetics, taxonomy, physiology, pollination, and many others (http://www.honeybeeworld.com/bee-l/).

**Beehoo.** World beekeeping directory. More than 1,000 Web sites about beekeeping, bees, apitherapy, honey recipes, queen breeding (http://www.beehoo.com/).

**Beekeeper's Reference.** Regional beekeeping groups, multimedia, books (http://hive-mind.com/bee/).

**Beekeeping in Top-Bar Hives.** An alternative method of beekeeping. Plans for hives, helpful photographs (http://www2.gsu.edu/~biojdsx/main.htm).

**Beekeeping: The Beekeeper's Home Page.** Information about the honey bee and honey production: beginner tips and photographs. Canadian and North American content (http://ourworld.compuserve.com/homepages/Beekeeping/).

**Benefit of Manuka Honey.** An informational site that describes the benefits of using manuka honey for medicinal purposes (http://www.benefitofmanukahoney.com).

**BOMBUS-L.** Mailing list dedicated to discussion of bumblebees. To subscribe, send a message to LISTSERV @ umdd.umd.edu with SUB BOMBUS-L in the body of your message. Maintained by David Inouye (http://www.beesource.com/bee-l/bombusl.htm).

**Bug Tutorials.** Authorized for pesticide applicator training CEUs in Arizona, Florida, Vermont, and West Virginia. Excellent for training in many categories. Requires Windows only (http://pests.ifas.ufl.edu/software/det_bugs.htm).

**Buzzwords and New Zealand Beekeeping.** Beekeeping news, bees, statistics, honey descriptions (http://www.beekeeping.co.nz/).

**Canadian Association of Professional Apiculturists.** Beekeeping and pollination (http://www.capabees.ca/).

**Carl Hayden Bee Research Center GEARS.** Research on bees and pollination. From USDA/ARS (http://www.ars.usda.gov/Main/docs.htm?docid=12371).

**Danish Beekeepers' Association.** (http://www.biavl.dk/).

**E. H. Thorne (Beehives) Ltd.** Wragby, Lincoln, U.K. Beekeeping events in the United Kingdom and courses for beginning beekeepers (http://www.thorne.co.uk/).

**Espacio Apicola.** Argentine beekeepers' magazine (http://www.apicultura.com.ar/).

**Featured Creatures.** A continuously growing Florida Web site providing detailed information on description, distribution, damage, and management (if necessary) of various Florida organisms (insects, spiders, mites, snails, nematodes, and so on) as well as scientific references, links to recommended pesticides, and more. Over 350 publications with thousands of color photographs and line drawings (http://creatures.ifas.ufl.edu/).

**Gruppo di Studio per Gli Insetti Sociali UniversitÃƒÂ di Firenze.** Study Group for Social Wasps of the Department of Animal Biology of the University of Florence, Italy. Discovering social wasps. Biology and behavior of social wasps (mainly *Polistes*). In Italian and English. (http://www.dbag.unifi.it/wasps/).

**Honey Bees and Pollination.** Bee information site for those in the Virginia/Mid-Atlantic region of the United States (http://www.virginiafruit.ento.vt.edu/VAFS-bees.html).

**Honeybee.** Students identify honey bees' responsibilities inside the hive and importance of the honey bee to humans, and discover how honey bees communicate. For pre-K to grade 4 (http://www.ent.iastate.edu/zoo/lessonplans/honeybee.html).

**Insects in Motion.** QuickTime video clips of various insects showing feeding behavior and life stage development, including a bee stinging (http://everest.ento.vt.edu/~carroll/insect_video_home.html).

**International Bee Research Association.** IBRA aims to increase awareness of the vital role of bees in the environment and encourages the use of bees as wealth creators (http://www.ibra.org.uk/).

**International Union for the Study of Social Insects.** (http://www.iussi.org/).

**Internet Apiculture and Beekeeping Archive.** Articles from the Usenet newsgroup sci.agriculture.beekeeping, the logs from the listserv bee-l, FAQ files, and pointers to other beekeeping and apicultural resources, on and off the Internet (http://www.ibiblio.org/bees/home.html).

**Japanese Ants Image Database.** Taxonomy of Japanese ants (http://ant.edb.miyakyo-u.ac.jp/E/index.html).

**John's Beekeeping Notebook.** Observation of beehives, cell-plug queen rearing, experiences of a Peace Corps beekeeper in Fiji, top-bar beekeeping (http://outdoorplace.org/beekeeping/).

**Korea Beekeeping Association.** (http://www.korapis.or.kr/).

**Maine Beekeeping.** Articles, reports, and tips on beekeeping; calendar of events; recommended reading (http://www.mainebee.com/).

**Mexican Bee Database.** In FoxPro format (http://www.inhs.uiuc.edu/cbd/PCAM/readme_PCAM.html).

**National Beekeepers Association of New Zealand.** (http://www.nba.org.nz/).

**NECTAR.** Netherlands Expertise Centre for Tropical Agricultural Resources. Nongovernmental, nonprofit association of beekeeping experts in the Netherlands (http://www.xs4all.nl/~jtemp/nectar_index.html).

**Ohio State University Bee Breeding Program.** Bee breeding and instrumental insemination of honey bees, including procedures, equipment, and training (http://www174.pair.com/birdland/Breeding/).

**Raising Bees.** From the Small Farm Resource (http://www.farminfo.org/bees/bees.htm).

**SIWeb.** Social Insects Web. Systematics, phylogeny, distribution, images, literature, and conservation of social insects. From the Department of Entomology, American Museum of Natural History (http://antbase.org/).

**Small Hive Beetle.** *Aethina tumida*, a new pest of honey bees established in the southeastern United States (http://www.doacs.state.fl.us/pi/enpp/ento/aethinanew.html).

**Solitary Bees: An Addition to Honey Bees.** Use of *Osmia* (orchard bees) for pollination. Includes a list of suppliers (http://www.pollinatorparadise.com/Solitary_Bees/SOLITARY.HTM).

**Stridulation Sounds of Black Fire Ants.** *Solenopsis richteri* in different situations (http://home.olemiss.edu/~hickling/).

**SwisTrack: A Tracking Tool for Multiunit Biological and Artificial Systems.** SwisTrack is an open-source tracking tool for tracking multiple, markerless objects. SwisTrack has been successfully used for tracking cockroaches and miniature robots, and can easily be extended due to its open, object-oriented architecture (http://swistrack.sourceforge.net).

**The Amazing Beecam.** Video feed from a beehive (http://gears.tucson.ars.ag.gov/beecam/).

**The Bumblebee Pages.** Life cycle, FAQs, behavior (http://www.bumblebee.org/).

**The Hive and the Honeybee.** Phillips' Beekeeping Collection online apiculture library (http://bees.library.cornell.edu/).

**The Pollination Home Page.** Portal to pollination information and images (http://pollinator.com/).

**Torre Bueno's Glossary for Social Insects.** Entomological glossary (http://antbase.org/databases/glossary_files/glossary_A.htm).

**Varroa Mites and How to Catch Them.** Treatment, drone method (http://www.xs4all.nl/~jtemp/dronemethod.html).

**Woyke Honey Bee Page.** Describes genetic expression of body color in different species of honey bees. From Warsaw Agricultural University, Poland (http://jerzy_woyke.users.sggw.pl/).

## BIBLIOGRAPHIES, BY TAXONOMIC GROUP

### Blattodea (Cockroaches)

**Australian Faunal Directory.** Checklists and bibliographies for Australian insects (http://www.environment.gov.au/biodiversity/abrs/online-resources/fauna/index.html#insects).

**The International Database on Insect Disinfestation and Sterilization.** Provides information on disinfestation and sterilization of arthropod species, with dosage levels and conditions, efficacy, and references. For researchers, phytosanitary regulators, plant protection services, and SIT facility and food irradiation operations personnel (http://www-ididas.iaea.org/).

### Coleoptera (Beetles)

**A Bibliography of the Cucujidae (sens. lat).** A worldwide bibliography to the flat bark beetles (Coleoptera: Cucujidae [*sens. str.*, Silvanidae, Laemophloeidae, Passandridae]), with an emphasis on the systematic literature (http://fsca.entomology.museum/Coleoptera/Mike/cucujidbib.htm).

**Australian Faunal Directory.** Checklists and bibliographies for Australian insects (http://www.environment.gov.au/biodiversity/abrs/online-resources/fauna/index.html#insects).

**Calodema Web Site of Dr. T. J. Hawkeswood.** Provides free PDF files of the biological research of T. J. Hawkeswood and his coworkers (http://www.calodema.com).

**Coleoptera Bibliography.** (http://www.coleoptera.org/p156.htm).

**Coleoptera of Great Smoky Mountain National Park.** (http://www2.lsuagcenter.com/Inst/Research/Departments/arthropodmuseum/smokybeetles.htm).

**Coleoptera Taxonomic Literature Database.** References and papers on higher classification of beetles (http://134.60.85.50:591/ColeRefDB/ColReferencesN_su.html).

**Colorado Potato Beetle.** *Leptinotarsa decemlineata* (Say), an important worldwide pest of potatoes. Life history, a bibliography of about 500 references, and links to related sites (http://www.umit.maine.edu/~andrei_alyokhin/CPBWeb/frpage.htm).

**Colorado Potato Beetle Bibliography.** About 500 references devoted to the Colorado potato beetle, *Leptinotarsa decemlineata* (Say) (http://www.umit.maine.edu/~andrei_alyokhin/CPBWeb/bibintr.htm).

**Elateridae of Boreal North America.** Click beetles (http://nathist.sdstate.edu/SMIRCOL/elna.htm).

**Longhorn Beetles (Cerambycidae) of the West Palearctic Region.** Taxonomy, biology, and macro photography of longhorn beetles of the West Palearctic region (http://www.uochb.cas.cz/~natur/cerambyx/index.htm).

**Nicrophorus Central.** For all those whose research involves nicrophorine species (burying beetles) (Coleoptera: Silphidae, Nicrophorinae). Taxonomy and photo gallery (http://collections2.eeb.uconn.edu/nicroweb/nicrophorus.htm).

**The International Database on Insect Disinfestation and Sterilization.** Provides information on disinfestation and sterilization of arthropod species, with dosage levels and conditions, efficacy, and references. For researchers, phytosanitary regulators, plant protection services, and SIT facility and food irradiation operations personnel (http://www-ididas.iaea.org/).

**Trictenotomidae of the World.** Includes full bibliography, distribution, and other interesting information (http://www.coleoptera.org/tricte~1.htm).

**Water Beetle World.** Newsletter for water beetle workers worldwide. Over 7,000 citations on water beetles (http://www.zo.utexas.edu/faculty/sjasper/beetles/index.htm).

**Worldwide Bibliography of Aquatic and Semiaquatic Dryopoidea.** Aquatic and semiaquatic Dryopoidea (families Dryopidae, Elmidae, Eulichadidae, Limnichidae, Psephenidae, and Ptilodactylidae) (http://www.calacademy.org/research/entomology/Entomology_Resources/Coleoptera/dryopoidea/index.html).

## Collembola (Springtails)

**Australian Faunal Directory.** Checklists and bibliographies for Australian insects (http://www.environment.gov.au/biodiversity/abrs/online-resources/fauna/index.html#insects).

**Synanthropic Collembola: Springtails in Association with Man.** (http://www.collembola.org/publicat/sidney.htm).

## Dermaptera (Earwigs)

**Australian Faunal Directory.** Checklists and bibliographies for Australian insects (http://www.environment.gov.au/biodiversity/abrs/online-resources/fauna/index.html#insects).

## Diptera (Flies)

**Asilidae Publications.** Publications about Diptera: Asilidae by F. Geller-Grimm (http://www.geller-grimm.de/publi.htm).

**Australian Faunal Directory.** Checklists and bibliographies for Australian insects (http://www.environment.gov.au/biodiversity/abrs/online-resources/fauna/index.html#insects).

**Bibliography Update 1977–1995 for the Asilidae.** PDF files of R. J. Lavigne, 1999. Includes short translations from Japanese and Russian. University of Wyoming Agricultural Experiment Station Science Monograph SM55 (http://ces.uwyo.edu/PUBS/SM-55.pdf).

**BioSystematic Database of World Diptera.** Fly names and information about those names and the taxa to which they apply. Taxonomic bibliography. From National Museum of Natural History, Smithsonian Institution (http://www.sel.barc.usda.gov/Diptera/biosys.htm).

**Calodema Web Site of Dr. T. J. Hawkeswood.** Provides free PDF files of the biological research of T. J. Hawkeswood and his coworkers (http://www.calodema.com).

**Chironomid Home Page.** All about Chironomidae (Diptera): bibliographies, listserver, newsletter, and so on (http://insects.ummz.lsa.umich.edu/~ethanbr/chiro/).

**PHEREC Publications.** Publications of the John A. Mulrennan, Sr., Public Health Entomology Research and Education Center, on mosquito control, stable flies, sandflies, nonchemical mosquito control, and pesticide environmental impact research (http://www.pherec.org/EntGuides/).

**Syrphidae of Lower Saxony.** Species list, red data list, literature. In German (http://www.schwebfliegen.de/).

**Tachnid Bibliography.** 1980–present, gleaned from *The Tachnid Times* (http://www.nadsdiptera.org/Tach/Bib/biblio.htm).

**Tephritid Worker Database.** A directory of tephritid fruit fly researchers and plant protection officers in the world, providing expertise on species and speciality field, and recent publications, bibliographic references, news, directories, meetings, and projects (http://www.tephritid.org/).

**The International Database on Insect Disinfestation and Sterilization.** Provides information on disinfestation and sterilization of arthropod species, with dosage levels and conditions, efficacy, and references. For researchers, phytosanitary regulators, plant protection services, and SIT facility and food irradiation operations personnel (http://www-ididas.iaea.org/).

## Embioptera (Web-Spinners)

**Australian Faunal Directory.** Checklists and bibliographies for Australian insects (http://www.environment.gov.au/biodiversity/abrs/online-resources/fauna/index.html#insects).

## Ephemeroptera (Mayflies)

**Australian Faunal Directory.** Checklists and bibliographies for Australian insects (http://www.environment.gov.au/biodiversity/abrs/online-resources/fauna/index.html#insects).

**Ephemeroptera Galactica.** Central site for information on mayflies (Ephemeroptera) of interest to researchers and aquatic biologists (http://www.famu.org/mayfly/).

## Grylloblattodea (Rock Crawlers)

**Grylloblattodea.** Tree of Life (http://tolweb.org/tree?group=Grylloblattidae&contgroup=Neoptera).

## Hemiptera (True Bugs, Aphids, Scale Insects, Leafhoppers, Cicadas)

**Australian Faunal Directory.** Checklists and bibliographies for Australian insects (http://www.environment.gov.au/biodiversity/abrs/online-resources/fauna/index.html#insects).

**Bibliography of Tobacco and Silverleaf Whitefly.** Searchable and downloadable (http://www.ars.usda.gov/Services/docs.htm?docid=10916).

**Calodema Web Site of Dr. T. J. Hawkeswood.** Provides free PDF files of the biological research of T. J. Hawkeswood and his coworkers (http://www.calodema.com).

**Podisus Online.** Bibliography of the predatory bugs of the genus *Podisus* (Heteroptera: Pentatomidae) (http://users.ugent.be/~padclerc/).

**ScaleNet.** Comprehensive information on the scale insects of the world, including queriable information on their classification, nomenclatural history, distribution, hosts, and literature (http://www.sel.barc.usda.gov/scalenet/scalenet.htm).

**The International Database on Insect Disinfestation and Sterilization.** Provides information on disinfestation and sterilization of arthropod species, with dosage levels and conditions, efficacy, and references. For researchers, phytosanitary regulators, plant protection services, and SIT facility and food irradiation operations personnel (http://www-ididas.iaea.org/).

**TYMBAL.** Web site on the Auchenorrhyncha: cicadas, spittlebugs, leafhoppers, treehoppers, and planthoppers (http://www.agric.nsw.gov.au/Hort/ascu/tymbal/tymbal.htm).

## Hymenoptera (Wasps, Ants, Bees)

**Africanized Honey Bees.** Africanized honey bee information, especially in relation to California. Includes bibliography with over 900 references (http://bees.ucr.edu/index.html).

**Australian Faunal Directory.** Checklists and bibliographies for Australian insects (http://www.environment.gov.au/biodiversity/abrs/online-resources/fauna/index.html#insects).

**Bees of the World.** Over 2,200 research references on all aspects of bees (http://geocities.com/BeesInd/).

**Biological Control: A Guide to Natural Enemies in North America.** Tutorial on the concept and practice of biological control, with photographs and descriptions of biological control (or biocontrol) agents of insect, disease, and weed pests in North America (http://www.nysaes.cornell.edu/ent/biocontrol/).

**Calodema Web Site of Dr. T. J. Hawkeswood.** Provides free PDF files of the biological research of T. J. Hawkeswood and his coworkers (http://www.calodema.com).

**Chrysis.net.** Image galleries, articles, research, hosts on cuckoo wasps (Hymenoptera: Chrysididae). (http://www.chrysis.net/index_en.php).

**FORMIS.** A composite of several ant literature databases. It contains citations for a large fraction of the world's ant literature (about 32,000 references). FORMIS contains all known ant taxonomic literature (through 1996). It also contains comprehensive bibliographies of leaf-cutting ants, fire ants, and Russian wood ants (http://www.ars.usda.gov/research/docs.htm?docid=10003).

## Isoptera (Termites)

**Australian Faunal Directory.** Checklists and bibliographies for Australian insects (http://www.environment.gov.au/biodiversity/abrs/online-resources/fauna/index.html#insects).

## Lepidoptera (Butterflies, Moths)

**Almond Moth Bibliography.** Bibliography of *Ephestia cautella* (Walker) (http://www.oardc.ohio-state.edu/cocoa/almond.htm).

**Australian Faunal Directory.** Checklists and bibliographies for Australian insects (http://www.environment.gov.au/biodiversity/abrs/online-resources/fauna/index.html#insects).

**Bibliography of Pink Bollworm.** Searchable and downloadable (http://www.ars.usda.gov/Services/docs.htm?docid=10935).

**Calodema Web Site of Dr. T. J. Hawkeswood.** Provides free PDF files of the biological research of T. J. Hawkeswood and his coworkers (http://www.calodema.com).

**CATE Sphingidae.** Project aimed to develop the methodology and sociology for transferring the enterprise of taxonomy onto the Internet using hawkmoths (Lepidoptera: Sphingidae) and aroid lilies (Alismatales: Araceae) as exemplar taxa (http://www.cate-sphingidae.org/).

**Lepidopterology.com.** Butterfly and moth news, glossary, directory, electronic museum, and almanac (http://www.lepidopterology.com/).

**The International Database on Insect Disinfestation and Sterilization.** Provides information on disinfestation and sterilization of arthropod species, with dosage levels and conditions, efficacy, and references. For researchers, phytosanitary regulators, plant protection services, and SIT facility and food irradiation operations personnel (http://www-ididas.iaea.org/).

## Mantodea (Mantids)

**Australian Faunal Directory.** Checklists and bibliographies for Australian insects (http://www.environment.gov.au/biodiversity/abrs/online-resources/fauna/index.html#insects).

**Bibliography of Mantids and Phasmids in North America.** (http://www.herper.com/insects/mpbiblio.html).

## Mantophasmatodea (Gladiators, Heelwalkers)

**Mantophasmatodea.** Tree of Life (http://tolweb.org/tree?group=Mantophasmatodea&contgroup=Neoptera).

## Mecoptera (Scorpionflies and Hangingflies)

**Australian Faunal Directory.** Checklists and bibliographies for Australian insects (http://www.environment.gov.au/biodiversity/abrs/online-resources/fauna/index.html#insects).

## Neuroptera (Lacewings, Antlions, Dobsonflies)

**Australian Faunal Directory.** Checklists and bibliographies for Australian insects (http://www.environment.gov.au/biodiversity/abrs/online-resources/fauna/index.html#insects).

**Calodema Web Site of Dr. T. J. Hawkeswood.** Provides free PDF files of the biological research of T. J. Hawkeswood and his coworkers (http://www.calodema.com).

**Neuropterida.** A working bibliography of the literature on extant and fossil Neuroptera, Megaloptera, and Raphidioptera (Insecta: Neuropterida) of the world (http://entowww.tamu.edu/research/neuropterida/neur_bibliography/bibhome.html).

## Odonata (Dragonflies, Damselflies)

**Australian Faunal Directory.** Checklists and bibliographies for Australian insects (http://www.environment.gov.au/biodiversity/abrs/online-resources/fauna/index.html#insects).

**Calodema Web Site of Dr. T. J. Hawkeswood.** Provides free PDF files of the biological research of T. J. Hawkeswood and his coworkers (http://www.calodema.com).

**Odonata Bibliography.** Based on Roy J. Beckemeyer's odonatological library (http://www.windsofkansas.com/odbib.html).

**Odonatological Bibliography.** About 1,300 references, mainly from 1990 to 1996 (http://mitglied.lycos.de/GBechly/index.htm).

## Orthoptera (Grasshoppers, Crickets)

**Australian Faunal Directory.** Checklists and bibliographies for Australian insects (http://www.environment.gov.au/biodiversity/abrs/online-resources/fauna/index.html#insects).

**Orthoptera of the Netherlands.** Book, *De sprinkhanen en krekels van Nederland (Orthoptera)* by Roy Kleukers, Erik van Nieukerken, Baudewijn Odé, Luc Willemse, and Walter van Wingerden. Accompanying CD is titled "De zingende sprinkhanen en krekels van de Benelux" [The singing Orthoptera of the Benelux-countries] by Baudewijn Odé. Book has 416 pages, 14 color plates, bound, with dust-jacket. ISBN 90-5011-100-9. Publication: 15 May 1997. Price: HFL 82,50 (excl. P&P). Order in writing from: KNNV Uitgeverij, Oudegracht 237, 3511 NK Utrecht, The Netherlands. Fax: +31-30-236 89 07 (http://www.pensoft.net/treesubj/h130.stm).

**The International Database on Insect Disinfestation and Sterilization.** Provides information on disinfestation and sterilization of arthropod species, with dosage levels and conditions, efficacy, and references. For researchers, phytosanitary regulators, plant protection services, and SIT facility and food irradiation operations personnel (http://www-ididas.iaea.org/)

## Phasmatodea (Stick Insects)

**Australian Faunal Directory.** Checklists and bibliographies for Australian insects (http://www.environment.gov.au/biodiversity/abrs/online-resources/fauna/index.html#insects).

**Bibliography of Mantids and Phasmids in North America.** (http://www.herper.com/insects/mpbiblio.html).

**Phasmida Species File Online.** Taxonomic database of the world's stick and leaf insects, known as walkingsticks and walking leaves in the United States. Full synonymic and taxonomic information for over 3,400 valid species and over 4,100 taxonomic names (all ranks, valid and not valid). There are over 23,000 citations to references. There are also images for many species (http://phasmida.orthoptera.org/).

## Phthiraptera (Lice)

**Australian Faunal Directory.** Checklists and bibliographies for Australian insects (http://www.environment.gov.au/biodiversity/abrs/online-resources/fauna/index.html#insects).

## Plecoptera (Stoneflies)

**Australian Faunal Directory.** Checklists and bibliographies for Australian insects (http://www.environment.gov.au/biodiversity/abrs/online-resources/fauna/index.html#insects).

## Protura (Proturans)

**Australian Faunal Directory.** Checklists and bibliographies for Australian insects (http://www.environment.gov.au/biodiversity/abrs/online-resources/fauna/index.html#insects).

## Psocoptera (Psocids)

**Australian Faunal Directory.** Checklists and bibliographies for Australian insects (http://www.environment.gov.au/biodiversity/abrs/online-resources/fauna/index.html#insects).

## Siphonaptera (Fleas)

**Australian Faunal Directory.** Checklists and bibliographies for Australian insects (http://www.environment.gov.au/biodiversity/abrs/online-resources/fauna/index.html#insects).

## Strepsiptera (Twisted-Wing Parasites)

**Australian Faunal Directory.** Checklists and bibliographies for Australian insects (http://www.environment.gov.au/biodiversity/abrs/online-resources/fauna/index.html#insects).

## Thysanoptera (Thrips)

**Australian Faunal Directory.** Checklists and bibliographies for Australian insects (http://www.environment.gov.au/biodiversity/abrs/online-resources/fauna/index.html#insects).

**The International Database on Insect Disinfestation and Sterilization.** Provides information on disinfestation and sterilization of arthropod species, with dosage levels and conditions, efficacy, and references. For researchers, phytosanitary regulators, plant protection services, and SIT facility and food irradiation operations personnel (http://www-ididas.iaea.org/).

## Trichoptera (Caddisflies)

**Australian Faunal Directory.** Checklists and bibliographies for Australian insects (http://www.environment.gov.au/biodiversity/abrs/online-resources/fauna/index.html#insects).

## Zoraptera (Angel Insects)

**Australian Faunal Directory.** Checklists and bibliographies for Australian insects (http://www.environment.gov.au/biodiversity/abrs/online-resources/fauna/index.html#insects).

**Catalog of Zoraptera.** A catalog and bibliography of the Zoraptera (http://www.famu.org/zoraptera/).

**Zoraptera.** Tree of Life (http://tolweb.org/tree?group=Zoraptera&contgroup=Neoptera).

## General

**AGRICOLA.** National Agricultural Library's (NAL) Web Gateway to AGRICOLA (AGRICultural OnLine Access) (http://agricola.nal.usda.gov/).

**Blattaria, Mantodea, Orthoptera, and Dermaptera of the Czech and Slovak Republics.** Nearly 500 papers published to the end of 1998 (http://web.archive.org/web/20030710093320/http://www.mujweb.cz/www/petr_kocarek/Bibliography.htm).

**Ecological Database of the World's Insect Pathogens.** Information on fungi, viruses, protozoa, mollicutes, and bacteria (other than *Bacillus thuringiensis*) that are infectious in insects, mites, and related arthropods (http://cricket.inhs.uiuc.edu/edwipweb/edwipabout.htm).

**Entomopathogenic Nematode Bibliography Database.** Citations for *Steinernema* and *Heterorhabditis* spp. nematodes and their symbiotic bacteria. Citations come from journals, periodicals, chapters, series, books, thesis, patents, and important abstracts from various international symposia (http://www.oardc.ohio-state.edu/nematodes/insect_parasitic_nematode_public.htm).

**Hawaiian Terrestrial Arthropod Bibliography.** Covers terrestrial and freshwater Hawaiian arthropods, and also marine insects (http://hbs.bishopmuseum.org/hibib/arthbib.html).

**Insect Bibliography Server.** Citations on bot flies, Chironomidae, Cuterebridae, *Dermatobia*, FORMIS 2001 (ants), face fly, fire ant—2001, Gasterophilidae, horn fly, hypoderma, ITS, insect genetics, insect nematodes, leaf-cutting ants, NCB-ESA, 1996, Oestridae, parasitoids, reprint list, restriction enzymes, screwworm, and stable fly (http://entobib.unl.edu/).

**Jean-Henri Fabre.** E-museum and bibliography (http://www.e-fabre.com/).

**Plant Protection Database.** Formerly CABPESTCD. Over 30 years of research from over seventy-five countries. Provides an ideal source for professionals and students requiring an international perspective on crop protection, pesticide research, and the agricultural sciences (http://www.cabi.org/AbstractDatabases.asp?PID=108).

**Popular Classics in Entomology.** Unabashedly idiosyncratic collection of some of the best books about the natural history of insects and the personal experiences of the entomologists who work on them. These books emphasize the literary and human side of entomology. They are not highly technical and would be good reading for backyard collectors as well as those with professional interests in entomology (http://www.colostate.edu/Depts/Entomology/readings.html).

**The BUGS Project.** New Zealand terrestrial invertebrate literature 1775–1993 (http://www.landcareresearch.co.nz/research/biodiversity/invertebratesprog/nzac/BUGS_project.asp).

**Viral Diseases of Insects in the Literature.** Includes almost all literature published up to 1984, including the 733 references contained in the review articles by Hughes (1957) and Martignoni and Langston (1960) (http://insectweb.inhs.uiuc.edu/Pathogens/VIDIL/index.html).

**World Literature of Arthropods Associated with Soybeans.** (http://insectweb.inhs.uiuc.edu/Soy/Siric/Home.htm).

## BIOLOGICAL CONTROL

**ANBP Newsletter.** Publication of the Association of Natural Biocontrol Producers (http://www.anbp.org/ANBPnewsletter.htm).

**Animal and Plant Health Inspection Service.** Responsible for protecting and promoting U.S. agricultural health (http://www.aphis.usda.gov/).

**Areawide Suppression of Fire Ants.** Biological control techniques used by the USDA to control red imported fire ants (http://fireant.ifas.ufl.edu/).

**ARS National Invertebrate Genetic Resources Program.** List of stock centers and other databases relevant to invertebrate germplasm (http://www.ars-grin.gov/nigrp/).

**Arthropod Biological Control NCR-125.** Biological control working group in the northcentral region of the United States (http://www.ncera125.ent.msu.edu/).

**Association of Natural Biocontrol Producers.** Professional association representing the biological pest management industry. Augmentative biological control utilizes beneficial insects, mites, and nematodes to manage agricultural, horticultural, and plant pests (http://www.anbp.org/).

**Biocontrol Web Site.** USDA/ARS laboratory searches for alternative methods to reduce crop losses caused by insect pests. Research on the introduction and conservation of pests' natural enemies and use of trap crops (http://www.ars.usda.gov/main/site_main.htm?modecode=12-75-21-00).

**Biological Control: A Guide to Natural Enemies in North America.** Tutorial on-the concept and practice of biological control with photographs and descriptions of biological control (or biocontrol) agents of insect, disease, and weed pests in North America (http://www.nysaes.cornell.edu/ent/biocontrol/).

**Biology of Cameraria ohridella.** Video, by Ulrich Zunke and Gerhard Doobe, chronicles the feeding behavior and morphology of the horse-chestnut leaf miner in Hamburg, Germany (http://www.mactode.com/Pages/Entopix_insects.html).

**Bug Tutorials.** Authorized for pesticide applicator training CEUs in Arizona, Florida, Vermont, and West Virginia. Excellent for training in many categories. Requires Windows only (http://pests.ifas.ufl.edu/software/det_bugs.htm).

**Coccinellidae of the Indian Subcontinent.** Images of common species of lady beetles (Coleoptera: Coccinellidae) found in the agroecosystems of the Indian region, and their natural enemies. An annotated checklist of the Indian fauna and an illustrated key to common *Chilocorus* species of India are also available (http://www.angelfire.com/bug2/j_poorani/index.html).

**Colorado Potato Beetle Bibliography.** About 500 references devoted to the Colorado potato beetle, *Leptinotarsa decemlineata* (Say) (http://www.umit.maine.edu/~andrei_alyokhin/CPBWeb/bibintr.htm).

**Cooperative Agricultural Pest Survey.** Wyoming program maintaining databases on biocontrol for weeds and insects and surveys for cereal leaf beetle, gypsy moths, and karnal bunt (http://w3.uwyo.edu/~caps/caps.html).

**Cooperative Agricultural Pest Survey.** Combined effort by U.S. federal and state agricultural organizations to conduct surveillance, detection, and monitoring of agricultural crop pests and biological control agents (http://www.ceris.purdue.edu/napis/docs/caps.html).

**CSIRO European Laboratory.** Provides the capacity for research in entomology, biological control, chemical control, pest ecology, pathology, and related disciplines. The major impetus for this development has been biological control (http://www.csiro-europe.org/).

**Ecological Database of the World's Insect Pathogens.** Information on fungi, viruses, protozoa, mollicutes, and bacteria (other than *Bacillus thuringiensis*) that are infectious in insects, mites, and related arthropods (http://cricket.inhs.uiuc.edu/edwipweb/edwipabout.htm).

**Entomopathogenic Nematode Taxonomy.** How to identify entomopathogenic nematodes (http://kbn.ifas.ufl.edu/kbnstein.htm).

**Entopix Vol. 1.** Over 1,000 images and 7 video clips of plant parasitic arthropods and their natural enemies (http://www.mactode.com/Pages/Entopix_insects.html).

**European Biological Control Laboratory.** USDA/ARS laboratory to develop biological control technologies, which can be used to suppress invading weeds and insect pests. This is done through explorations to find natural enemies (insects, mites, and pathogens) (http://www.ars-ebcl.org/).

**Evaniidae Research.** The Evaniidae (Hymenoptera) are a medium-sized family of wasps without stings whose solitary larvae "parasitize" the egg cases of cockroaches (http://world.std.com/~mhuben/evaniidae.html).

**Featured Creatures.** A continuously growing Florida Web site providing detailed information on description, distribution, damage, and management (if necessary) of various Florida organisms (insects, spiders, mites, snails, nematodes, and so on) as well as scientific references, links to recommended pesticides, and more. Over 350 publications with thousands of color photographs and line drawings (http://creatures.ifas.ufl.edu/).

**Florida Integrated Pest Management.** IPM information and technology, with an emphasis on biological control, as they apply to Florida (http://ipm.ifas.ufl.edu/).

**Florida IPM.** For discussion of IPM and biological control as it applies to Florida pest and beneficial organisms (http://ipm.ifas.ufl.edu/resources/listserv/index.shtml).

**FRASS.** Insect-rearing newsletter (http://users.ugent.be/~padclerc/AMRQC/frass.htm).

**Global Insecticide Directory.** Includes biological control agents (http://www.agranova.co.uk/gid.asp).

**Insect Field Collection Photos.** Arizona Crop Information photographs 2000–2003 (http://cals.arizona.edu/crops/images/insectidaz/).

**Institute of Zoology of National Ukrainian Academy of Sciences.** Department of Taxonomy of Entomophagous Insects and Biocontrol, Kiev, Ukraine (http://www.icfcst.kiev.ua/siz/depart/taxonomy/dep-taxonomy.htm).

**International Database on Insect Disinfestation and Sterilization.** Mass rearing facilities of sterile pest insects, ticks, and mites. Production size, radiation process, quality control parameters, dosimetry, program objective, transboundary shipment, field release data, and the facility's full address. International Atomic Energy Agency, Joint FAO/IAEA Division, Insect Pest Control Section (http://www-ididas.iaea.org/IDIDAS/).

**McClay Ecoscience.** Research and consulting services in biological control, invasive species, and insect/plant ecology (http://www.mcclay-ecoscience.com).

**Midwest Biological Control News.** Archives of this publication available from 1994 to 2000. No longer in publication (http://www.entomology.wisc.edu/mbcn/mbcn.html).

**Mole Cricket Knowledgebase.** Florida's pest management success story. Mole cricket knowledge base on all ten species found in the United States, including Hawaii, Puerto Rico, and the U.S. Virgin Islands. Distribution, description, life cycle, damage, and biological controls. Identification key for these species that makes heavy use of graphics and photographs. Large lists of references in every area. HTML format for Macintosh or PC (http://molecrickets.ifas.ufl.edu/).

**National Agriculture Pest Information System PestTracker.** Database for the Cooperative Agricultural Pest Survey (CAPS) (http://www.ceris.purdue.edu/napis/).

**Nematodes as Biological Control Agents of Insects.** (http://nematode.unl.edu/wormepns.htm).

**Pheromones and Other Semiochemicals in Integrated Production Working Group.** International Organization for Biological and Integrated Control of Noxious Animals and Plants, West Palearctic Regional Section. Announcements and proceedings of working group meetings (http://phero.net/iobc/).

**Plant and Insect Parasitic Nematodes.** Database from University of Nebraska—Lincoln (http://nematode.unl.edu/).

**Podisus Online.** Bibliography of the predatory bugs of the genus *Podisus* (Heteroptera: Pentatomidae) (http://users.ugent.be/~padclerc/).

**Queensland Museum.** Australia. 500,000 specimens. Special collections of rain forest insects and parasitic wasps (http://www.qm.qld.gov.au/organisation/sections/insects/).

**ROBO.** Releases of Beneficial Organisms in the United States and Territories documents importations and releases of beneficial invertebrates, both biological control agents and pollinators (http://www.ars-grin.gov/nigrp/robo.html).

**Society for *In Vitro* Biology.** The invertebrate section of the SIVB is devoted to research *in vitro* using invertebrate cells to study problems concerning growth, development, toxicology, parasites and pathogens, molecular analyses, mass rearing for bioproduction of biological products, and insect control (http://www.sivb.org/).

**Suppliers of Beneficial Organisms in North America.** Publication of California/EPA's Department of Pesticide Regulation. Lists 143 commercial suppliers of 130 beneficial organisms used for biological control (http://www.cdpr.ca.gov/docs/pestmgt/ipminov/ben_supp/contents.htm).

**Tephritid Worker Database.** A directory of tephritid fruit fly researchers and plant protection officers in the world, providing expertise on species and speciality field, and recent publications, bibliographic references, news, directories, meetings, and projects (http://www.tephritid.org/).

**The Fly in Your Eye.** A small book (published in 1989) about the war deliberately started in Australia between dung beetles and bush flies. Based on C.S.I.R.O. research. Hard facts, technical drawings, cartoons, jokes. Much used in Australian schools (http://www.viacorp.com/flybook/fulltext.html/).

**The International Database on Insect Disinfestation and Sterilization.** Provides information on disinfestation and sterilization of arthropod species, with dosage levels and conditions, efficacy, and references. For researchers, phytosanitary regulators, plant protection services, and SIT facility and food irradiation operations personnel (http://www-ididas.iaea.org/).

**USDA Cooperative State Research, Education, and Extension Service Pest Management Emphasis Areas.** Summary of pest management programs sponsored by CSREES (http://www.csrees.usda.gov/nea/pest/pest.cfm).

**USDA/APHIS Plant Protection and Quarantine.** Safeguards agriculture and natural resources from the risks associated with the entry, establishment, or spread of animal and plant pests and noxious weeds (http://www.aphis.usda.gov/plant_health/index.shtml).

**Veg-Edge.** Vegetable IPM resource for the upper midwestern United States. From the University of Minnesota (http://www.vegedge.umn.edu/).

**World Educational Films.** Entomological videos documenting life histories and parasitic hymenoptera. Some clips available online (http://www.worldeducationalfilms.com/films.htm).

## CD-ROM, BY TAXONOMIC GROUP

### Blattodea (Cockroaches)

**Myths and Science of Cricket Chirps.** Hissing cockroach and cricket chirping (http://www.cricketscience.com/).

### Coleoptera (Beetles)

**Calodema Web Site of Dr. T. J. Hawkeswood.** Provides free PDF files of the biological research of T. J. Hawkeswood and his coworkers (http://www.calodema.com).

**Carabini of Yugoslavia and Adjacent Areas.** The first in the series is the "Catalogue of the Fauna of Yugoslavia: Carabini," representing an interactive, multimedia guide through 140 taxa of Coleoptera from the tribe Carabini registered in Yugoslavia and the surrounding countries (http://www.ecolibribionet.co.yu/Izdanja/Karabini.htm).

**Forest Insects and Their Damage.** Images from the Southern Forest Insect Work Conference Slide Series (The Southern Extension and Research Activity-Information Exchange Group-12 [SERA-IEG-12]). Images may be used freely for nonprofit educational purposes when accompanied by an appropriate copyright notice (http://www.bugwood.org/forestcd/).

**Forest Pests of North America.** Integrated pest management Photo CD series. 300 images in Kodak Photo CD format that can be used to support and assist the implementation of forest IPM in North America (http://www.bugwood.org/ipmcd/).

**Hforest.** Hypermedia Forest Insect and Disease Knowledge Base and Diagnosis (http://www.pfc.forestry.ca/diseases/hforest/About/howto_e.html).

## Diptera (Flies)

**Calodema Web Site of Dr. T. J. Hawkeswood.** Provides free PDF files of the biological research of T. J. Hawkeswood and his coworkers (http://www.calodema.com).

**CulicID.** Series of computerized interactive identification and information tools for Australasian mosquitoes (http://www.sph.uq.edu.au/Divisions/acithn/buycd.htm).

**Photographic Atlas and Identification Key to the Robber Flies of Germany (Diptera: Asilidae).** Includes more than 1,900 photographs of 81 robber fly species (Diptera: Asilidae). The identification key for determination is applicable to all species of Asilidae found in the following countries: Belgium, Denmark, Germany, the United Kingdom, the Netherlands, and Sweden (http://www.geller-grimm.de/key/).

## Hemiptera (True Bugs, Aphids, Scale Insects, Leafhoppers, Cicadas)

**Calodema Web Site of Dr. T. J. Hawkeswood.** Provides free PDF files of the biological research of T. J. Hawkeswood and his coworkers (http://www.calodema.com).

**Hforest.** Hypermedia Forest Insect and Disease Knowledge Base and Diagnosis (http://www.pfc.forestry.ca/diseases/hforest/About/howto_e.html).

## Hymenoptera (Wasps, Ants, Bees)

**All about Bees.** (http://www.beekeeping.com/info/sommaires/index_us.htm).

**Calodema Web Site of Dr. T. J. Hawkeswood.** Provides free PDF files of the biological research of T. J. Hawkeswood and his coworkers (http://www.calodema.com).

**World Chalcidoidea Database on CD-ROM.** Complete world catalog of all 26,000 names used in Chalcidoidea, more than 95,000 host records, over 110,000 distribution records, more than 300 color images of adults and immature stages of most families, and an immense amount of information on their use in biocontrol, biology, physiology, and so on. Requires Windows only (http://www.ent.iastate.edu/List/info/chalcidoideacd.html).

## Lepidoptera (Butterflies, Moths)

**Biology of Cameraria ohridella.** Video, by Ulrich Zunke and Gerhard Doobe, chronicles the feeding behavior and morphology of the horse-chestnut leaf miner in Hamburg, Germany (http://www.mactode.com/Pages/Entopix_insects.html).

**British Butterflies.** Covers sixty-one species from Britain and Ireland. Over 1 hour of motion video (http://www.birdguides.com/html/catalog/CDROMs/bfcd.htm).

**British Butterfly Prints by Humphreys.** Scans of forty-two original antique prints from Henry Noel Humphrey's British Butterflies. The prints were over 155 years old and had original hand coloring. They feature the butterflies on the wildflowers and plants they inhabit (http://www.finerareprints.com/cds/vol_cd_humphreys.htm).

**Butterflies of North America CD-ROM: A Natural History and Field Guide.** By Dr. James A. Scott. Pictures of 679 species, maps, life cycles, behavior, host plants, identification tools, and glossary. Requires Windows only (http://www.hoptechno.com/buttrfly.htm).

**Butterfly Tutorials.** Caterpillars and adults of cloudless sulpher, giant swallowtail, Gulf fritillary, longtailed skipper, zebra longwing, black swallowtail, monarch, viceroy, European cabbage butterfly, and Florida atala (http://pests.ifas.ufl.edu/software/det_bfly.htm).

**Calodema Web Site of Dr. T. J. Hawkeswood.** Provides free PDF files of the biological research of T. J. Hawkeswood and his coworkers (http://www.calodema.com).

**Canegrub.** Training package for the Australian sugar industry (http://www.ctpm.uq.edu.au/software/canegrub/).

**Colour Atlas of the Siberian Lepidoptera.** Photographs by Eduard Berlov (http://www.geocities.com/rainforest/jungle/5695/butt.html).

**Forest Insects and Their Damage.** Images from the Southern Forest Insect Work Conference Slide Series (The Southern Extension and Research Activity-Information Exchange Group-12 [SERA-IEG-12]). Images may be used freely for nonprofit educational purposes when accompanied by an appropriate copyright notice (http://www.bugwood.org/forestcd/).

**Forest Pests of North America.** Integrated pest management Photo CD series. 300 images in Kodak Photo CD format that can be used to support and assist the implementation of forest IPM in North America (http://www.bugwood.org/ipmcd/).

**Hforest.** Hypermedia Forest Insect and Disease Knowledge Base and Diagnosis (http://www.pfc.forestry.ca/diseases/hforest/About/howto_e.html).

**Key to Butterflies of the Lake Baikal Region.** Almost 2,000 high-resolution, full-color digital images of all 197 Baikalian species and many of its caterpillars and pupae. Includes pictured key to butterfly families and pictured keys to all species. Detailed information of every species (in Russian). HTML format for Macintosh or PC. First version 31 December 2001. Price: $30.00 U.S. (incl. postage). For more information, contact olegberlov@narod.ru (http://babochki.narod.ru/).

**Lepibase.** Butterflies of Europe. Author is Antti Roine from Finland. Includes high-resolution photographs of almost all European butterflies and many caterpillars as well as distribution maps (http://www.netti.fi/~avanto/lepibase.html).

**Schmetterlingsfotos auf CD.** European Butterflies. Two Photo CDs. CD 1: Nymphalidae, Papilionidae, Pieridae. CD 2: Lycaenidae, Nemeobiidae, Danaidae, Satyridae (http://www.geocities.com/Paris/Cafe/1508/photocds.htm).

## Neuroptera (Lacewings, Antlions, Dobsonflies)

**Calodema Web Site of Dr. T. J. Hawkeswood.** Provides free PDF files of the biological research of T. J. Hawkeswood and his coworkers (http://www.calodema.com).

## Odonata (Dragonflies, Damselflies)

**Calodema Web Site of Dr. T. J. Hawkeswood.** Provides free PDF files of the biological research of T. J. Hawkeswood and his coworkers (http://www.calodema.com).

## Orthoptera (Grasshoppers, Crickets)

**An Illustrated Catalog of Orthoptera.** Complete taxonomic and synonymic catalog of the superfamily Tettigonioidea (katydids, bush crickets, and their allies) of the world. Includes all taxa, from family down to subspecies, of Tettigonioidea described until the end of 1998. Requires Macintosh or PC (http://net-28187.mcb.harvard.edu/CD).

**Katydids of Costa Rica, Vol. 1.** Systematics and bioacoustics of the cone-head katydids by Piotr Naskrecki (http://140.247.119.145/book/book.htm).

**Mole Cricket Knowledgebase.** Florida's pest management success story. Mole cricket knowledge base on all ten species found in the United States, including Hawaii, Puerto Rico, and the U.S. Virgin Islands. Distribution, description, life cycle, damage, and biological controls. Identification key for these species that makes heavy use of graphics and photographs. Large lists of references in every area. HTML format for Macintosh or PC (http://molecrickets.ifas.ufl.edu/).

**Myths and Science of Cricket Chirps.** Hissing cockroach and cricket chirping (http://www.cricketscience.com/).

**Orthoptera of the Netherlands.** Book *De sprinkhanen en krekels van Nederland (Orthoptera)* by Roy Kleukers, Erik van Nieukerken, Baudewijn Odé, Luc Willemse, and Walter van Wingerden. Accompanying CD is titled "De zingende sprinkhanen en krekels van de Benelux" [The singing Orthoptera of the Benelux-countries] by Baudewijn Odé. Book has 416 pages, 14 color plates, bound, with dust-jacket. ISBN 90-5011-100-9. Publication: 15 May 1997. Price: HFL 82,50 (excl. P&P). Order in writing from: KNNV Uitgeverij, Oudegracht 237, 3511 NK Utrecht, The Netherlands. Fax: +31-30-236 89 07 (http://www.pensoft.net/treesubj/h130.stm).

**Songs of Crickets and Katydids of the Mid-Atlantic States.** Audio CD (http://members.aol.com/_ht_a/whershberg/page/).

## General

**Annotated Invertebrate Clip Art.** 780 colorful graphics in 17 categories, including Arthropoda, Insecta, and insect anatomy and behavior (http://www.educationalimages.com/it020006.htm).

**Antique Prints of Insects by Charles Orbigny.** From original antique prints from the first edition of *Dictionnaire D'Histoire Naturelle* by Charles Orbigny, published in 1849 (http://www.finerareprints.com/cds/vol_cd_hn_insects.htm).

**Backyard Bugs.** Natural history of the monarch butterfly, dragonfly, praying mantis, stick insects, luna moth, viceroy butterfly, antlion, caddisflies, whirligig beetle, giant water bug, hickory horned devil caterpillar, millipedes, centipedes, tarantulas, honey bee, mosquitoes, cockroaches, and flies via macro photography videos. Included activities are designed for 5- to 10-year-olds. $79.95 (http://www.totallybuggin.com/site/1586384/page/814871).

**Belgian Spiders.** Descriptions available for the species from Belgium, the Netherlands, Luxembourg, France, and Germany (http://users.skynet.be/spinnen/CD-Rom/CD-Romsale.htm).

**BioEd.** Multimedia tool to help teach taxonomy and systematics to undergraduates (http://www.ctpm.uq.edu.au/software/bioed/).

**Bug Tutorials.** Authorized for pesticide applicator training CEUs in Arizona, Florida, Vermont, and West Virginia. Excellent for training in many categories. Requires Windows only (http://pests.ifas.ufl.edu/software/det_bugs.htm).

**BugMatch Citrus.** Allows you to identify pests, pathogens, and deficiency diseases in citrus. Color images, video, sound, and text are combined to give you a comprehensive coverage of some of the factors affecting citrus production as well as a greater understanding of pest management in citrus. This product was produced in 1995. From the University of Queensland, Australia (http://www.ctpm.uq.edu.au/software/bugmatch/bugmatch_citrus.htm).

**BugMatch Cotton.** For cotton production in Australia. Pest keys, photographs and video, IPM strategies (http://www.ctpm.uq.edu.au/software/bugmatch/).

**BugMatch Grapes.** For grape production in Australia and New Zealand. Keys for diagnosing diseases and pests, and fact sheets (http://www.ctpm.uq.edu.au/software/bugmatch/).

**Entomology Abstracts.** Global coverage on the geographic distribution, nomenclature, new species, and more of insects and insectlike species (http://www.nisc.com/factsheets/qent.asp).

**Entopix Vol. 1.** Over 1,000 images and 7 video clips of plant parasitic arthropods and their natural enemies (http://www.mactode.com/Pages/Entopix_insects.html).

**Florida Vegetable Pest Photographic Gallery CD-ROMs.** Three CD-ROMs in HTML format contain over 260 high-quality images of vegetable pests and their damage. Pests are common to Florida. However, many are common throughout the United States and the world. Images are provided in three different sizes and resolutions: print quality, display for large audiences, and Web-optimized (http://pests.ifas.ufl.edu/software/det_veggies.htm).

**Global Insecticide Directory.** Includes biological control agents (http://www.agranova.co.uk/gid.asp).

**Harmful Effects and Emergency Response to Pesticide Poisoning, Heat Stress, and Heat Stroke.** Tutorial based on Chapter 10 of the "Applying Pesticide Correctly" manual. Requires Windows only (http://pests.ifas.ufl.edu/software/det_core6.htm).

**Insect Jigsaw Puzzles.** (http://www.bugpeople.org/curriculum/jigsaw/50jigsawpuzzles.htm).

**Nomina Insecta Nearctica.** Checklist of the insects of North America. Abbreviated online version of the publication of the same title published by Entomological Information Services in 1996 and 1997 in four volumes and a CD-ROM (http://www.nearctica.com/nomina/main.htm).

**Ozpest.** Reference system for urban insects and their control. Keys to urban insect pests, timber damage, and stored product damage. From the University of Queensland, Australia (http://www.ctpm.uq.edu.au/software/ozpest/).

**Pecan Pest Management.** Pecan Pest Management. Available from the Texas Pecan Growers Association. Send check for $30.00 plus $3.00 shipping (Texas residents add $4.95 [8.25 percent sales tax]) to: Olde Pecan Book Store, P.O. Drawer CC, College Station, TX 77841 (http://insects.tamu.edu/extension/bulletins/l-5362.html).

**Pesticide Labeling Tutorial.** Based on Chapter 2 of the "Applying Pesticides Correctly" manual. Requires Windows only (http://pests.ifas.ufl.edu/software/det_core2.htm).

**Pests in and around the Home.** Knowledge base of pests of humans, pets, structures, lawns, and landscapes. CD-ROM contains information on pest biology, life cycle, identification keys, distribution, damage, and management. Hundreds of color photographs. HTML version for Macintosh or PC. Version 2.0 (http://pests.ifas.ufl.edu/software/det_pests.htm).

**Plant Protection Database.** Formerly CABPESTCD. Over 30 years of research from over seventy-five countries. Provides an ideal source for professionals and students requiring an international perspective on crop protection, pesticide research, and the agricultural sciences (http://www.cabi.org/AbstractDatabases.asp?PID=108).

**RiceIPM.** Interactive training to learn about integrated pest management in tropical rice. From a collaboration between Ministries of Agriculture in Thailand, Malaysia, and Vietnam; the International Rice Research Institute in the Philippines; and the University of Queensland, Australia (http://www.ctpm.uq.edu.au/software/riceipm/).

**School IPM.** A CD-ROM capture of the National School IPM site (http://schoolipm.ifas.ufl.edu/cdrom.html).

**Science Snoops—The Monarch Case.** Students play the role of new researchers as they work to solve each life science mystery case as part of a team of real and virtual scientists. Virtual coworkers are portrayed by actors who interact with the student to give information, assign work, and give feedback on students' performance (http://www.totallybuggin.com/site/1586384/product/KM-MC-802).

**Survey of the Invertebrates.** Part of the extensive Survey of the Animal Kingdom series. Requires Macintosh or PC. Discount for school or library. Over 600 color photographs (http://www.educationalimages.com/it020001.htm).

**The BUGS Project.** New Zealand terrestrial invertebrate literature 1775–1993 (http://www.landcareresearch.co.nz/research/biodiversity/invertebratesprog/nzac/BUGS_project.asp).

**The UC Guide to Solving Garden and Landscape Problems.** Forty fruit and vegetable crops and eighty ornamental plants, each with its own pest species list. Management strategies for more than 600 common pests with 2,800 separate information screens and 4,800 color photographs (http://www.ipm.ucdavis.edu/IPMPROJECT/ADS/cd_solvinggarden.html).

## DATABASES, BY TAXONOMIC GROUP

### Blattodea (Cockroaches)

**Blattodea Species File Online.** A comprehensive, up-to-date taxonomic database of world cockroaches (http://blattodea.speciesfile.org/HomePage.aspx).

**The International Database on Insect Disinfestation and Sterilization.** Provides information on disinfestation and sterilization of arthropod species, with dosage levels and conditions, efficacy, and references. For researchers, phytosanitary regulators, plant protection services, and SIT facility and food irradiation operations personnel (http://www-ididas.iaea.org/).

### Coleoptera (Beetles)

**The International Database on Insect Disinfestation and Sterilization.** Provides information on disinfestation and sterilization of arthropod species, with dosage levels and conditions, efficacy, and references. For researchers, phytosanitary regulators, plant protection services, and SIT facility and food irradiation operations personnel (http://www-ididas.iaea.org/).

### Dermaptera (Earwigs)

**Earwig Research Centre.** Web site contains the most complete collection of checklists, photographs, and drawings and an almost complete literature database on the Dermaptera, the earwigs. It is designed to help identification of specimens in any region of the world. By Fabian Haas, ICIPE Nairobi (http://www.earwigs-online.de).

### Diptera (Flies)

**AnoBase.** A database containing genomic/biological information on anopheline mosquitoes, with an emphasis on *Anopheles gambiae*, the world's most important malaria vector (http://www.anobase.org/).

**Asilidae Predator-Prey Database.** A database containing more than 11,000 prey records of robber flies taken from the scientific literature (http://www.geller-grimm.de/catalog/lavigne.htm).

**Catalogue of the Craneflies of the World (Diptera, Tipuloidea: Pediciidae, Limoniidae, Cylindrotomidae, Tipulidae).** CCW covers all 17,334 genus-group and species-group taxa of the families Pediciidae, Limoniidae, Cylindrotomidae, and Tipulidae (Insecta, Diptera, Tipuloidea). Its author is Pjotr Oosterbroek, honorary staff member of the Zoological Museum Amsterdam, section Entomology (http://ip30.eti.uva.nl/ccw/).

**Flybase.** Database of the *Drosophila* genome, Indiana University mirror (http://flybase.bio.indiana.edu/).

**Flybrain: Drosophila Nervous System Database.** Atlas and database of the *Drosophila* nervous system (http://flybrain.neurobio.arizona.edu/).

**Mosquito and Moth Cell Lines.** Cell culture availability database (http://www.biotech.ist.unige.it/cldb/spestr.html).

**Stiletto Flies.** Describes fly family Therevidae, including ongoing research and participants in the research (http://www.inhs.uiuc.edu/cee/therevid/).

**The International Database on Insect Disinfestation and Sterilization.** Provides information on disinfestation and sterilization of arthropod species, with dosage levels and conditions, efficacy, and references. For researchers, phytosanitary regulators, plant protection services, and SIT facility and food irradiation operations personnel (http://www-ididas.iaea.org/).

### Hemiptera (True Bugs, Aphids, Scale Insects, Leafhoppers, Cicadas)

**The International Database on Insect Disinfestation and Sterilization.** Provides information on disinfestation and sterilization of arthropod species, with dosage levels and conditions, efficacy, and references. For researchers, phytosanitary regulators, plant protection services, and SIT facility and food irradiation operations personnel (http://www-ididas.iaea.org/).

### Hymenoptera (Wasps, Ants, Bees)

**Antbase.** Provides access to information on all the ant species of the world (http://antbase.org).

**BeeBase.** Honey bee genome database (http://racerx00.tamu.edu/bee_resources.html).

**ECatSym: Electronic World Catalog of Symphyta.** Supported by GBIF (Global Biodiversity Information Facility), Copenhagen, the EcatSym project is working on the first Web-based database for the approximately 10,000 valid sawfly taxa worldwide (http://www.zalf.de/home_zalf/institute/dei/php_e/ecatsym/ecatsym.php).

**The International Database on Insect Disinfestation and Sterilization.** Provides information on disinfestation and sterilization of arthropod species, with dosage levels and conditions, efficacy, and references. For researchers, phytosanitary regulators, plant protection services, and SIT facility and food irradiation operations personnel (http://www-ididas.iaea.org/).

### Lepidoptera (Butterflies, Moths)

**CATE Sphingidae.** Project aimed to develop the methodology and sociology for transferring the enterprise of taxonomy onto the Internet using hawkmoths

(Lepidoptera: Sphingidae) and aroid lilies (Alismatales: Araceae) as exemplar taxa (http://www.cate-sphingidae.org/).

**HOSTS.** Database of host plants for rearing Lepidoptera (http://www.nhm.ac.uk/research-curation/projects/hostplants/).

**Lepidoptera Electronic Resources.** Butterfly and moth resources organized alphabetically, geographically, and topically (http://www.chebucto.ns.ca/Environment/NHR/lepidoptera.html).

**Mosquito and Moth Cell Lines.** Cell culture availability database (http://www.biotech.ist.unige.it/cldb/spestr.html).

**Pherolist.** Database of sex pheromones of Lepidoptera and related attractants by families/subfamilies/tribes, genus, species, or chemical component (http://www.pherolist.slu.se/pherolist.php).

**The International Database on Insect Disinfestation and Sterilization.** Provides information on disinfestation and sterilization of arthropod species, with dosage levels and conditions, efficacy, and references. For researchers, phytosanitary regulators, plant protection services, and SIT facility and food irradiation operations personnel (http://www.ididas.iaea.org/).

## Orthoptera (Grasshoppers, Crickets)

**Orthoptera Species File Online.** Taxonomic database of the Orthoptera of the world. Includes full synonymic and taxonomic data as well as type data for all species of Tettigonioidea (katydids, bush crickets), Grylloidea (crickets), Tridactyloidea (sand crickets), and Tetrigoidea (pygmy grasshoppers) (http://140.247.119.145/Orthoptera/).

**Orthoptera Species File Online.** Taxonomic database of the world's grasshoppers, locusts, katydids, and crickets, both living and fossil. Full synonymic and taxonomic information for 23,700 valid species, 39,999 taxonomic names, 145,100 citations to 11,850 references, 44,000 images, 184 sound recordings, 37,980 specimens, and keys to 2,100 taxa (http://osf2.orthoptera.org/).

**The International Database on Insect Disinfestation and Sterilization.** Provides information on disinfestation and sterilization of arthropod species, with dosage levels and conditions, efficacy, and references. For researchers, phytosanitary regulators, plant protection services, and SIT facility and food irradiation operations personnel (http://www-ididas.iaea.org/).

## Phasmatodea (Stick Insects)

**Phasmida Species File.** A comprehensive, regularly updated taxonomic database of worldwide Phasmida (stick and leaf insects) (http://phasmida.orthoptera.org).

**Phasmida Species File Online.** Taxonomic database of the world's stick and leaf insects, known as walking sticks and walking leaves in the United States. Full synonymic and taxonomic information for over 3,400 valid species and over 4,100 taxonomic names (all ranks, valid and not valid). There are over 23,000 citations to references. There are also images for many species (http://phasmida.orthoptera.org/).

## Thysanoptera (Thrips)

**The International Database on Insect Disinfestation and Sterilization.** Provides information on disinfestation and sterilization of arthropod species, with dosage levels and conditions, efficacy, and references. For researchers, phytosanitary regulators, plant protection services, and SIT facility and food irradiation operations personnel (http://www-ididas.iaea.org/).

## General

**AgNIC.** Agriculture Network Information Center. Searchable metadata directory of quality agricultural databases, data sets, and information systems—many related to pest management (http://www.agnic.org/).

**ARS Insect Locations.** Includes stock centers (http://www.ars-grin.gov/nigrp/ars_insects.html).

**ARS National Invertebrate Genetic Resources Program.** List of stock centers and other databases relevant to invertebrate germplasm (http://www.ars-grin.gov/nigrp/).

**Ecological Database of the World's Insect Pathogens.** Information on fungi, viruses, protozoa, mollicutes, and bacteria (other than *Bacillus thuringiensis*) that are infectious in insects, mites, and related arthropods (http://cricket.inhs.uiuc.edu/edwipweb/edwipabout.htm).

**Featured Creatures.** A continuously growing Florida Web site providing detailed information on description, distribution, damage, and management (if necessary) of various Florida organisms (insects, spiders, mites, snails, nematodes, and so on) as well as scientific references, links to recommended pesticides and more. Over 350 publications with thousands of color photographs and line drawings (http://creatures.ifas.ufl.edu/).

**Glossary: Entomology and Crop protection.** Words and technical terms related to entomology and crop protection (http://bijlmakers.com/glossary/glossary.htm).

**Integrated Taxonomic Information System.** A U.S. government database of current names and classification (http://www.itis.gov/).

**International Database on Insect Disinfestation and Sterilization.** Mass rearing facilities of sterile pest insects, ticks, and mites. Production size, radiation process, quality control parameters, dosimetry, program objective, transboundary shipment, field release data, and the facility's full address. International Atomic Energy Agency, Joint FAO/IAEA Division, Insect Pest Control Section (http://www-ididas.iaea.org/IDIDAS/).

**Invertebrate Fossil Type Specimen Catalog.** Taxonomic (genus, species, and subspecies names) and publication (author, date, citation) information. Locality searches provide information on location and age (http://www.ucmp.berkeley.edu/science/invertebrate_coll.php).

**Malaria Database.** From WHO/TDR (http://www.wehi.edu.au/MalDB-www/who.html).

**Microbial Germplasm Database.** Listed by host insect order (http://mgd.nacse.org/cgi-bin/mgd).

**National Agriculture Pest Information System PestTracker.** Database for the Cooperative Agricultural Pest Survey (CAPS) (http://www.ceris.purdue.edu/napis/).

**NC State AgNIC Systematic Entomology.** A collaborative effort to collect, organize, and provide access to the best academic, scholarly, research, and practical resources available on the identification, classification, nomenclature, and evolution of insects and related arthropods. For both the public and the researchers (http://www.lib.ncsu.edu/agnic/sys_entomology/).

**Online Insect Database.** Brief descriptions of insect orders. From the University of Delaware (http://ag.udel.edu/enwc/).

**Parasitic Organisms Codon Usage Tables.** All about *Brugia*, *Onchocerca*, *Plasmodium*, *Trypanosoma*, and *Leishmania* (http://www.ebi.ac.uk/parasites/cutg.html).

**Pherobase.** Database of insect pheromones and semiochemicals (http://www.pherobase.com/).

**PHYTO Resources.** Database on insect and disease of ornamental plants of Quebec, Canada (http://www.phyto.qc.ca/).

**ROBO.** Releases of Beneficial Organisms in the United States and Territories documents importations and releases of beneficial invertebrates, both biological control agents and pollinators (http://www.ars-grin.gov/nigrp/robo.html).

## DIRECTORIES, BY TAXONOMIC GROUP

### Coleoptera (Beetles)

**Scarab Workers: World Directory.** International directory of scarab workers (alive and deceased), with full contact information, biographies, interests, images. Organized alphabetically, taxonomically, and geographically (http://www-museum.unl.edu/research/entomology/workers/index2.htm).

**Tiger Beetle World.** Information on tiger beetles (Coleoptera: Cicindelidae) of North America: biology, ecology, identification, techniques for study, research library, resources, cicindelophile directory (http://members.aol.com/YESedu/home.html).

**World Directory of Workers on Tenebrionidae (Coleoptera).** Authors who worked and are working on Tenebrionidae and allied families (http://www.coleoptera.org/p151.htm).

### Diptera (Flies)

**Directory of Ceratopogonid Workers.** Over 100 workers worldwide (http://campus.belmont.edu/cienews/director.html).

**Directory of Chironomid Workers.** (http://insects.ummz.lsa.umich.edu/~ethanbr/chiro/Directory/direct_new.html).

**Directory of North American Dipterists.** Over 250 North American fly workers listed, including addresses, phone numbers, fax numbers, e-mail addresses, and interests. Updated regularly (http://www.nadsdiptera.org/Directory/Directhome.htm).

**Historical World Directory of Diptera Systematists.** Names and dates of birth and death for over 3,000 persons who have described new taxa in Diptera since Linnaeus, with links to Web pages or portraits (http://hbs.bishopmuseum.org/dipterists/).

**Tephritid Worker Database.** A directory of tephritid fruit fly researchers and plant protection officers in the world, providing expertise on species and speciality field, and recent publications, bibliographic references, news, directories, meetings, and projects (http://www.tephritid.org/).

**World Dipterists.** Directory of world dipterists (http://www.sel.barc.usda.gov/Diptera/workers.htm).

### Ephemeroptera (Mayflies)

**Directory of Ephemeroptera Workers.** From Mayfly Central, Purdue University (http://www.entm.purdue.edu/entomology/mayfly/mayfly.html).

**Mayfly Researchers and Enthusiasts Directory.** (http://www.entm.purdue.edu/entomology/mayfly/dir.html).

### Hemiptera (Aphids, Scale Insects, Leafhoppers, Cicadas)

**Leafhoppers.** Classification and biology of leafhoppers and their relatives. Directory of leafhopper specialists. Key to major groups (http://www.inhs.uiuc.edu/~dietrich/Leafhome.html).

### Hymenoptera (Wasps, Ants, Bees)

**Non-native Ants: Invasive and Exotic Species.** Biology and geographic distributions of invasive ants, as well as other exotic species that may become invasive (http://www.leaflitter.org/exotic.htm).

**Symphytos.** Web forum for sawfly and horntail (Hymenoptera: Symphyta) workers (http://www.kean.edu/~symphytos/).

### Lepidoptera (Butterflies, Moths)

**Butterfly Art.** Butterfly art designs: artistically arranged specimens of butterflies, beetles, and arachnids set either mounted in frames or in acrylic domes (http://www.newman-art-designs.com.au).

**Lepidopterology.com.** Butterfly and moth news, glossary, directory, electronic museum, and almanac (http://www.lepidopterology.com/).

### Neuroptera (Lacewings, Antlions, Dobsonflies)

**Neuropterists' Directory.** Individuals and companies interested in neuropterid insects (orders Neuroptera, Megaloptera, and Raphidioptera) (http://entowww.tamu.edu/research/neuropterida/ndi-home.html).

### General

**Acarologists of the World: Online Directory.** With registration form for acarologists and lists of acarologists of the world (http://www.nhm.ac.uk/hosted_sites/acarology/directory.html).

**Addresses of European Dipterists.** (http://www.geller-grimm.de/address/europe.htm).

**Best of the Bugs.** Listing of outstanding insect and nematode Web sites. From the Department of Entomology and Nematology, University of Florida (http://pests.ifas.ufl.edu/bestbugs/).

**Collection Managers Online.** Directory of collection managers in all areas of natural history (http://www.msb.unm.edu/herbarium/collmgrs_online.html).

**Photo Album of Entomologists.** Photographs of over 100 entomologists from around the world (http://www.nhm.ku.edu/ksem/features/picttour.htm).

**World Taxonomist Database.** Currently 4,402 taxonomists/specialists are registered in the World Taxonomist Database. You can search for a person, an institute, a specific country, a taxonomic group, or for a combination of these criteria (http://www.eti.uva.nl/tools/wtd.php).

## HOST PLANT RESISTANCE

**Kearney Agricultural Center Citrus Entomology Laboratory.** Provides degree-day and resistance management information to citrus growers (http://citrusent.uckac.edu/).

**Phytoparasitica.** Israel Journal of Plant Protection Sciences (http://www.phytoparasitica.org/).

**Resistance Pest Management Newsletter.** From the Center for Integrated Plant Systems (CIPS) (http://whalonlab.msu.edu/rpmnews/).

**Tephritid Worker Database.** A directory of tephritid fruit fly researchers and plant protection officers in the world, providing expertise on species and speciality field, and recent publications, bibliographic references, news, directories, meetings, and projects (http://www.tephritid.org/).

## IMAGE GALLERIES, BY TAXONOMIC GROUP

### Blattodea (Cockroaches)

**Allpet Roaches.** Cockroaches of the world. Rearing information. Photographs of many species (http://www.angelfire.com/oh2/Roaches/).

**BugGuide.Net.** Community site with user-submitted images for identification and information on insect groups (http://www.bugguide.net/).

**Cockroach Home Page.** By Joseph Kunkel. Includes FAQs on cockroaches (http://www.bio.umass.edu/biology/kunkel/cockroach.html).

**Cockroach Images CD-ROM.** CD-ROM containing sixty different images of cockroaches commonly encountered or found in culture. Images are provided in three different sizes and resolutions: print quality, display for large audiences, and Web-optimized (http://pests.ifas.ufl.edu/software/det_roaches.htm).

**German Cockroach Mutants.** From the Genetic Stock Center for the German cockroach at Virginia Tech (http://everest.ento.vt.edu/~watson/pages/Photoalbum.html).

**MicroAngela.** Scanning electron microscope images of insects (http://www5.pbrc.hawaii.edu/microangela/).

### Coleoptera (Beetles)

**Aleocharine Staphylinid Image Database.** The photographs, scanning electron micrographs, half-tone drawings, and line drawings in this database are being accumulated as potential illustrations for a guide to the aleocharine genera of North America and Mexico (http://web.nhm.ku.edu/ashe/aleo/).

**An Illustrated Atlas of the Laemophloeidae Genera of the World.** Illustrations of all the described laemophloeid genera (Coleoptera: Laemophloeidae), with notes on their distribution and number of species (http://fsca.entomology.museum/Coleoptera/Mike/LaemophloeidaeLink.html).

**Beetle Science.** Virtual labs, virtual beetles, videos, carbon dust illustrations, and interactive science features (http://www.cornell.edu/explorecornell/?scene=Beetle%2520Science).

**BugGuide.Net.** Community site with user-submitted images for identification and information on insect groups (http://www.bugguide.net/).

**Calodema Web Site of Dr. T. J. Hawkeswood.** Provides free PDF files of the biological research of T. J. Hawkeswood and his coworkers (http://www.calodema.com).

**Cerambycidae of Florida.** Photographic atlas of the longhorn beetles (Coleoptera: Cerambycidae) of Florida, with an AutoMontage habitus photo of each species and a Florida distribution map (http://www.fsca-dpi.org/Coleoptera/Mike/FloridaCerambycids/openingpage.htm).

**Cerambycids.com.** Images of nearctic and neotropical Cerambycidae (http://www.cerambycids.com/).

**Chilean Insects.** Universidad Catolica, Chile. South American insect photographs with habitats, links to Chilean entomologists, and checklists of Buprestidae and Tenebrionidae (http://www.entomologia.cl/).

**Coccinellidae of the Indian Subcontinent.** Images of common species of lady beetles (Coleoptera: Coccinellidae) found in the agroecosystems of the Indian region, and their natural enemies. An annotated checklist of the Indian fauna and an illustrated key to common *Chilocorus* species of India are also available (http://www.angelfire.com/bug2/j_poorani/index.html).

**Common North American Arthropods.** Extensive catalog of copyrighted high-resolution arthropod photographs (http://www.cirrusimage.com/index.htm).

**Forest Insects and Their Damage.** Images from the Southern Forest Insect Work Conference Slide Series [The Southern Extension and Research Activity-Information Exchange Group-12 (SERA-IEG-12)]. Images may be used freely for nonprofit educational purposes when accompanied by an appropriate copyright notice (http://www.bugwood.org/forestcd/).

**Forest Pests of North America.** Integrated pest management Photo CD series. 300 images in Kodak Photo CD format that can be used to support and assist the implementation of forest IPM in North America (http://www.bugwood.org/ipmcd/).

**Goliathus.** Goliath beetles, some of the largest beetles on earth (http://www.naturalworlds.org/goliathus/index.htm).

**Goliathus.com: Beetle Breeding Project.** Beetle breeding site and image gallery of Cetonidae, Lucanidae, Dynastidae. Over 600 close-up photographs of 120+ species of live insects (http://www.goliathus.com/).

**Goliathus.cz: Online Entomological Experiences and Insect Museum.** Online insect museum of Coleoptera and Lepidoptera, including bilateral gynandromorphs (http://www.goliathus.cz/en/).

**Insect Field Collection Photos.** Arizona Crop Information photographs 2000–2003 (http://cals.arizona.edu/crops/images/insectidaz/).

**Lady Beetle Pictures.** Images of ladybird beetles (http://www.ent.iastate.edu/imagegal/coleoptera/lady/defaulttn.html).

**Longhorn Beetles (Cerambycidae) of the West Palearctic Region.** Taxonomy, biology, and macro photography of longhorn beetles of the West Palearctic region (http://www.uochb.cas.cz/~natur/cerambyx/index.htm).

**Nicrophorus Central.** For all those whose research involves nicrophorine species (burying beetles) (Coleoptera: Silphidae: Nicrophorinae). Taxonomy and photo gallery (http://collections2.eeb.uconn.edu/nicroweb/nicrophorus.htm).

**Picture Gallery of Carabid Beetles.** Pictures and videos of beetles (Coleoptera: Carabidae, Buprestidae, Cerambycidae) from Europe (http://volny.cz/midge/carabus/carabus.htm).

**Quaternary Entomology Laboratory.** Fossil insects. North Dakota State University (http://www.ndsu.nodak.edu/instruct/schwert/qel/qel.htm).

**Stag Beetles of Taiwan.** Images, information, and links about stag beetles (Coleoptera: Lucanidae) and various other beetles of Taiwan (http://www.geocities.com/RainForest/1803/).

**Stored Product Insect Images Database.** High-quality images of *Sitophilus*, *Tribolium*, and other stored product pests from the Biological Research Unit of the Grain Marketing and Production Research Center in Manhattan, Kansas (http://bru.gmprc.ksu.edu/db/insect/index.asp).

**Stored Product Pest Images CD-ROM.** CD-ROM containing eighty-one different images of stored product pests commonly encountered. Images are provided in three different sizes and resolutions: print quality, display for large audiences, and Web-optimized (http://pests.ifas.ufl.edu/software/det_stored.htm).

**Thais Entomology.** Two exhibits: Beautiful World of Butterflies and Shapes and Colors from the World of Beetles (http://www.thais.it/entomologia/default_uk.htm).

**The Darkling Beetles of Florida and Eastern United States.** Contains species checklists, identification keys, species profiles and images, distributions in Florida and eastern United States, and literature (http://entnemdept.ifas.ufl.edu/teneb/).

## Collembola (Springtails)

**BugGuide.Net.** Community site with user-submitted images for identification and information on insect groups (http://www.bugguide.net/).

**Collembola Image Gallery.** Images of springtails (http://www.ent.iastate.edu/imagegal/collembola/defaulttn.html).

**The Collembola.** Brief description and images of body types (http://www.biosci.missouri.edu/carrel/photos/Arthropods/Collembola.htm).

## Dermaptera (Earwigs)

**BugGuide.Net.** Community site with user-submitted images for identification and information on insect groups (http://www.bugguide.net/).

## Diptera (Flies)

**BugGuide.Net.** Community site with user-submitted images for identification and information on insect groups (http://www.bugguide.net/).

**Calodema Web Site of Dr. T. J. Hawkeswood.** Provides free PDF files of the biological research of T. J. Hawkeswood and his coworkers (http://www.calodema.com).

**Common North American Arthropods.** Extensive catalog of copyrighted high-resolution arthropod photographs (http://www.cirrusimage.com/index.htm).

**Electron Microscopy of Drosophila Polytene Chromosomes.** (http://www.helsinki.fi/~saura/EM/index.html).

**Fly Images.** From the Young Diptera site (http://www.sel.barc.usda.gov/Diptera/safari.htm).

**FlyView and FlyMove.** A *Drosophila* image database and movies of developmental processes (http://pbio07.uni-muenster.de/).

**Harlan's Big Bugtography Web Page.** Photographs and observations about insects found in Iowa. Nontechnical (http://users.crosspaths.net/~bugs/bugtography/homepage.html).

**Images of Mosquito Vectors of Arboviruses.** Images of *Aedes*, *Culex*, and *Anopheles*. From the CDC (http://www.cdc.gov/ncidod/dvbid/arbor/mosqpics.htm).

**Insect Field Collection Photos.** Arizona Crop Information photographs 2000–2003 (http://cals.arizona.edu/crops/images/insectidaz/).

**MicroAngela.** Scanning electron microscope images of insects (http://www5.pbrc.hawaii.edu/microangela/).

**Mosquito Images.** Mosquitoes of the midwestern United States (http://www.ent.iastate.edu/imagegal/diptera/culicidae/).

**Syrphidae Europe.** Hoverflies (Syrphidae) of Europe: photographs, range maps, keys. In English and Dutch (http://www.syrphidae.com/).

## Ephemeroptera (Mayflies)

**BugGuide.Net.** Community site with user-submitted images for identification and information on insect groups (http://www.bugguide.net/).

## Hemiptera (True Bugs Aphids, Scale Insects, Leafhoppers, Cicadas)

**BugGuide.Net.** Community site with user-submitted images for identification and information on insect groups (http://www.bugguide.net/).

**Calodema Web Site of Dr. T. J. Hawkeswood.** Provides free PDF files of the biological research of T. J. Hawkeswood and his coworkers (http://www.calodema.com).

**Cicada Mania.** Sightings, videos, sounds, and news (http://www.cicadamania.net/).

**Cicadas of Michigan.** Sounds, photographs, and a key (http://insects.ummz.lsa.umich.edu/fauna/michigan_cicadas/Michigan/Index.html).

**Cirrus Digital Imaging—True Bugs—Insect Order Hemiptera Photographs.** High-resolution images of common true bugs, photographed live, in the wild. Identified with common and scientific names (http://www.cirrusimage.com/bugs.htm).

**Common North American Arthropods.** Extensive catalog of copyrighted high-resolution arthropod photographs (http://www.cirrusimage.com/index.htm).

**Insect Field Collection Photos.** Arizona Crop Information photographs 2000–2003 (http://cals.arizona.edu/crops/images/insectidaz/).

**Kids' Cicada Hunt.** Follow two young brothers as they hunt for cicada nymphs, watch cicadas shed their skins, discover cicada predators, and try to solve cicada mysteries (http://saltthesandbox.org/cicada_hunt/).

**MicroAngela.** Scanning electron microscope images of insects (http://www5.pbrc.hawaii.edu/microangela/).

**Periodical Cicada.** Information on 1998 emergences of 17- and 13-year cicadas, with photographs, sound samples, and brood distribution maps (http://insects.ummz.lsa.umich.edu/fauna/michigan_cicadas/).

## Hymenoptera (Wasps, Ants, Bees)

**BugGuide.Net.** Community site with user-submitted images for identification and information on insect groups (http://www.bugguide.net/).

**Calodema Web Site of Dr. T. J. Hawkeswood.** Provides free PDF files of the biological research of T. J. Hawkeswood and his coworkers (http://www.calodema.com).

**Chrysis.net.** Image galleries, articles, research, and hosts of cuckoo wasps (Hymenoptera: Chrysididae) (http://www.chrysis.net/index_en.php).

**Common North American Arthropods.** Extensive catalog of copyrighted high-resolution arthropod photographs (http://www.cirrusimage.com/index.htm).

**European Pine Sawfly Images.** (http://zoologie.forst.tu-muenchen.de/PHERODIP/DIPRIONIDAE/NEODIPRIONSERTIFER/neodiprion.sertifer.html).

**HANABACHI Japanese Bee Image Database.** Each record is composed of family, genus and species (or subspecies), Japanese name, distribution data, type locality, type depository, source of the original description, synonymies, remarks, visiting or associated flowers, references, various images, and distribution map (http://konchudb.agr.agr.kyushu-u.ac.jp/hanabachi/).

**Harlan's Big Bugtography Web Page.** Photographs and observations about insects found in Iowa. Nontechnical (http://users.crosspaths.net/~bugs/bugtography/homepage.html).

**Insect Field Collection Photos.** Arizona Crop Information photographs 2000–2003 (http://cals.arizona.edu/crops/images/insectidaz/).

**Japanese Ants Image Database.** Taxonomy of Japanese ants (http://ant.edb.miyakyo-u.ac.jp/E/index.html).

**Myrmecology.** The scientific study of ants. Habitat, biology, image gallery, and recommended reading (http://www.myrmecology.info/portal/news.php).

**Symphytos.** Web forum for sawfly and horntail (Hymenoptera: Symphyta) workers (http://www.kean.edu/~symphytos/).

**The Ants of Cachoeira Nature Reserve.** By Jochen H. Bihn. Information on ecology and taxonomy of ants from Cachoeira Nature Reserve, Brazil (http://www.ants-cachoeira.de).

## Isoptera (Termites)

**BugGuide.Net.** Community site with user-submitted images for identification, and information on insect groups (http://www.bugguide.net/).

**MicroAngela.** Scanning electron microscope images of insects (http://www5.pbrc.hawaii.edu/microangela/).

**Termites of North America and the World.** Wealth of urban entomology information. From the University of Toronto (http://www.utoronto.ca/forest/termite/termite.htm).

## Lepidoptera (Butterflies, Moths)

**Atlas of Caterpillars from Europe and Asia Minor.** Photographs of caterpillars of butterflies and moths from Europe and Asia Minor (http://www.raupenatlas.de/english.html).

**Biology of Cameraria ohridella.** Video, by Ulrich Zunke and Gerhard Doobe, chronicles the feeding behavior and morphology of the horse-chestnut leaf miner in Hamburg, Germany (http://www.mactode.com/Pages/Entopix_insects.html).

**Borboletas GaÚchas.** Fotos de borboletas (Lepidoptera: Papilionoidea) de Rio Grande Do Sul, Brasil (http://www.geocities.com/RainForest/8617/).

**BugGuide.Net.** Community site with user-submitted images for identification and information on insect groups (http://www.bugguide.net/).

**Butterflies and Moths.** Biology and ecology of butterflies and moths, with basic information about differences between the two, hearing, and communication (http://www.butterflies-moths.com/).

**Butterflies of North America.** Distributional data, thumbnails, and checklists (http://www.butterfliesandmoths.org/).

**Butterfly Info in the Netherlands and Surrounding Areas.** Brief description and plates of each Dutch butterfly species, as well as a literature listing and butterfly-related poems (http://www.vlinderstichting.nl/index.asp?CatID=14).

**Butterfly Pictures.** By Philip Greenspun (http://photo.net/philg/photo/butterflies/index.html).

**Calodema Web Site of Dr. T. J. Hawkeswood.** Provides free PDF files of the biological research of T. J. Hawkeswood and his coworkers (http://www.calodema.com).

**CATE Sphingidae.** Project aimed to develop the methodology and sociology for transferring the enterprise of taxonomy onto the Internet using hawkmoths (Lepidoptera: Sphingidae) and aroid lilies (Alismatales: Araceae) as exemplar taxa (http://www.cate-sphingidae.org/).

**Caterpillars of La Selva.** Images of caterpillars in a tropical wet forest in Costa Rica (http://www.tulane.edu/~ldyer/lsacat/index.htm).

**Colour Atlas of the Siberian Lepidoptera.** Photographs by Eduard Berlov (http://www.geocities.com/rainforest/jungle/5695/butt.html).

**Common North American Arthropods.** Extensive catalog of copyrighted high-resolution arthropod photographs (http://www.cirrusimage.com/index.htm).

**Forest Insects and Their Damage.** Images from the Southern Forest Insect Work Conference Slide Series (The Southern Extension and Research Activity-Information Exchange Group-12 [SERA-IEG-12]). Images may be used freely for nonprofit educational purposes when accompanied by an appropriate copyright notice (http://www.bugwood.org/forestcd/).

**Forest Pests of North America.** Integrated pest management Photo CD series. 300 images in Kodak Photo CD format that can be used to support and assist the implementation of forest IPM in North America (http://www.bugwood.org/ipmcd/).

**Goliathus.cz: Online Entomological Experiences and Insect Museum.** Online insect museum of Coleoptera and Lepidoptera, including bilateral gynandromorphs (http://www.goliathus.cz/en/).

**Images of New Zealand Lepidoptera-Type Specimens.** Combined multipage checklist and image collection. Provides images of type specimens for the vast majority of New Zealand Lepidoptera species (http://www.landcareresearch.co.nz/research/biosystematics/invertebrates/lepidoptera/).

**Images of the Furman University Collection of North American Lepidoptera.** (http://facweb.furman.edu/~snyderjohn/furmanleps/index.htm).

**Indian Butterflies.** Basic photographic guide to the birds and butterflies found in India (http://www.nerdybirders.com).

**Key to Butterflies of the Lake Baikal Region.** Almost 2,000 high-resolution full-color digital images of all 197 Baikalian species and many of its caterpillars and pupae. Includes pictured key to butterfly families and pictured keys to all species. Detailed information of every species (in Russian). HTML format for Macintosh or PC. First version 31 December 2001. Price: $30.00 U.S. (incl. postage). For more information, contact olegberlov@narod.ru (http://babochki.narod.ru/).

**Lepidopterology.com.** Butterfly and moth news, glossary, directory, electronic museum, and almanac (http://www.lepidopterology.com/).

**Lepibase.** Butterflies of Europe. Author is Antti Roine from Finland. Includes high-resolution photographs of almost all European butterflies and many caterpillars, as well as distribution maps (http://www.netti.fi/~avanto/lepibase.html).

**Lepidoptera Larvae (Caterpillars) of Australia.** Images, biology, behavior, and life history of over 1,300 species, both butterflies and moths (http://www.usyd.edu.au/museums/larvae/).

**Lepidoptera.Net.** Butterflies of Georgia and butterflies and moths from around the world (http://www.lepidoptera.net/).

**Lynn Scott's Lepidoptera Images.** Photographs of over 300 moth species from Ottawa, Canada (http://www.theperfecthouse.biz/dls/mothhome.html).

**Moth Photographers Group.** Hosted by Mississippi Entomological Museum. Plates for both living moth photographs and spread specimens for more than 5,500 species found in North America. Collaborative effort among those who photograph moths with more than 200 contributors (http://mothphotographersgroup.msstate.edu/Plates.shtml).

**Moths in a Connecticut Yard.** Photographs and comments on the various moths appearing in John Himmelman's yard in Connecticut (http://booksandnature.homestead.com/NewCTMoth.html).

**Moths of North Dakota.** Identification guide, photographs of adults, and larvae (http://www.ndsu.nodak.edu/ndsu/ndmoths/).

**Olethreutinae (Lepidoptera: Torticidae) of the Plum Brook NASA Station.** Ohio, U.S.A. Photographs of the adult of fifty-eight species along with male genitalia (if needed) for identification. (http://hymfiles.biosci.ohio-state.edu/plumbrook/plumbrok.html).

**Schmetterlingsfotos auf CD.** European Butterflies. Two Photo CDs. CD 1: Nymphalida, Papilionidae, Pieridae. CD 2: Lycaenidae, Nemeobiidae, Danaidae, Satyridae (http://www.geocities.com/Paris/Cafe/1508/photocds.htm).

**Thais Entomology.** Two exhibits: Beautiful World of Butterflies and Shapes and Colors from the World of Beetles (http://www.thais.it/entomologia/default_uk.htm).

**Tortricid.net.** Photo database of tortricid moths (Lepidoptera: Tortricidae) (http://www.tortricidae.com/).

**UKmoths.** Photographic database of moths in the United Kingdom displayed in their natural resting position. About 1,500 of the 2,400 U.K. moths are featured (http://www.ukmoths.force9.co.uk/).

**Web Images of North American Moth Species.** Links to images of over 3,000 species (http://facweb.furman.edu/~snyderjohn/leplist/).

**Wisconsin Butterflies.** Field guide with photographs and maps for Wisconsin butterfly species (http://wisconsinbutterflies.org/butterflies/).

**Woolly Bears: Arctiidae Caterpillars.** Tiger moth identification and rearing (http://www3.islandtelecom.com/~oehlkew/indexarc.htm).

## Mantodea (Mantids)

**BugGuide.Net.** Community site with user-submitted images for identification and information on insect groups (http://www.bugguide.net/).

**Common North American Arthropods.** Extensive catalog of copyrighted high-resolution arthropod photographs (http://www.cirrusimage.com/index.htm).

## Mantophasmatodea (Gladiators, Heelwalkers)

**Fossil Mantophasmatodea.** (http://www.fossilmuseum.net/Fossil_Galleries/Insect_Galleries_by_Order/Mantophasmatodea/Mantophasmatodea.htm).

## Mecoptera (Scorpionflies and Hangingflies)

**BugGuide.Net.** Community site with user-submitted images for identification and information on insect groups (http://www.bugguide.net/).

## Neuroptera (Lacewings, Antlions, Dobsonflies)

**BugGuide.Net.** Community site with user-submitted images for identification and information on insect groups (http://www.bugguide.net/).

**Calodema Web Site of Dr. T. J. Hawkeswood.** Provides free PDF files of the biological research of T. J. Hawkeswood and his coworkers (http://www.calodema.com).

**Insect Field Collection Photos.** Arizona Crop Information photographs 2000–2003 (http://cals.arizona.edu/crops/images/insectidaz/).

## Odonata (Dragonflies, Damselflies)

**BugGuide.Net.** Community site with user-submitted images for identification and information on insect groups (http://www.bugguide.net/).

**Calodema Web Site of Dr. T. J. Hawkeswood.** Provides free PDF files of the biological research of T. J. Hawkeswood and his coworkers (http://www.calodema.com).

**Checklist of Odonata Found in Oregon.** Checklists, county record maps, photographs, links to Odonata sites, Odonata-related projects in Oregon, notes on commercial nymph harvesting for fishing bait, notes on dragonfly migrations (http://www.ent.orst.edu/ore_dfly/).

**Common North American Arthropods.** Extensive catalog of copyrighted high-resolution arthropod photographs (http://www.cirrusimage.com/index.htm).

**Digital Dragonflies.** Digital images of dragonflies with instructions on how to make your own (http://www.dragonflies.org/).

**Dragonflies and Damselflies of India.** Photographic guide to dragonflies and damselflies found in India (http://www.nerdybirders.com/html/dragonflies/dragonflies.html).

**Fossil Dragonflies.** Photographs of fossil dragonflies (http://www.bernstein.naturkundemuseum-bw.de/odonata/gallfoss.htm).

**Iowa Odonates.** Photographs, maps, and checklists of dragonflies and damselflies recorded in Iowa (http://www.iowaodes.com/).

**Odonata Central.** Searchable county-level database of the dragonflies and damselflies of the southcentral United States, including maps and photographs. Texas checklist (http://www.odonatacentral.org/).

**Ray Bruun Photography.** Pictures of Odonata (dragonflies and damselflies) (http://www.bruunphotography.com).

**Swedish Dragonflies.** Dragonfly photography, folklore, list of Swedish dragonflies, key, rearing information (http://home9.swipnet.se/~w-90582/dragonfly/dragonfly.html).

**Wisconsin Dragonflies and Damselflies.** (http://wisconsinbutterflies.org/dragonflies/).

## Orthoptera (Grasshoppers, Crickets)

**BugGuide.Net.** Community site with user-submitted images for identification, and information on insect groups (http://www.bugguide.net/).

**Common North American Arthropods.** Extensive catalog of copyrighted high-resolution arthropod photographs (http://www.cirrusimage.com/index.htm).

**Orthoptera of the Northern Great Plains.** Anatomy, key, photographs (http://www.ndsu.nodak.edu/entomology/hopper/orthoptera_home.htm).

## Phasmatodea (Stick Insects)

**BugGuide.Net.** Community site with user-submitted images for identification and information on insect groups (http://www.bugguide.net/).

## Phthiraptera (Lice)

**MicroAngela.** Scanning electron microscope images of insects (http://www5.pbrc.hawaii.edu/microangela/).

## Plecoptera (Stoneflies)

**American Stoneflies.** Photographs of plecopterans (http://www.mc.edu/campus/users/stark/american.html).

**BugGuide.Net.** Community site with user-submitted images for identification and information on insect groups (http://www.bugguide.net/).

## Psocoptera (Psocids)

**BugGuide.Net.** Community site with user-submitted images for identification and information on insect groups (http://www.bugguide.net/).

**Insect Field Collection Photos.** Arizona Crop Information photographs 2000–2003 (http://cals.arizona.edu/crops/images/insectidaz/).

## Siphonaptera (Fleas)

**Fleas.** (Siphonaptera) Taxonomy, morphology, distribution, host-association and ecology, scanning electron micrographs. Hosted by Zoological Institute, St. Petersburg, Russia (http://www.zin.ru/Animalia/Siphonaptera/index.htm).

**MicroAngela.** Scanning electron microscope images of insects (http://www5.pbrc.hawaii.edu/microangela/).

## Thysanoptera (Thrips)

**Eocene Insect Fossil Images.** Collected in northwest Colorado. They date at about 48 million years ago (http://www.coloradomtn.edu/campus_rfl/staff_rfl/kohls/eocene.shtml).

## Trichoptera (Caddisflies)

**BugGuide.Net.** Community site with user-submitted images for identification and information on insect groups (http://www.bugguide.net/).

## General

**Agriculture Western Australia Insect Reference Collection Database.** 250,000 specimens (http://agspsrv34.agric.wa.gov.au/ento/icdb/icdb1.htm).

**Animal Diversity Web: Insecta.** Large image gallery of insects as well as a few sounds (http://animaldiversity.ummz.umich.edu/site/accounts/information/Insecta.html).

**Arthropa.** Insects and other terrestrial arthropods. In French (http://arthropa.free.fr/).

**BIODIDAC.** Some digital resources for biology teaching. Diagrams and images of insects (http://biodidac.bio.uottawa.ca/thumbnails/catquery.htm?Kingdom=Animalia&phylum=Insecta).

**Bug Pictures.** High-quality pictures of bugs, insects, and spiders (http://www.bugpics.com/?id=44040192&ix=0).

**Bugs in Cyberspace.** Pictures and information for various North American backyard bugs and guides for keeping them in captivity (http://www.bugsincyberspace.com/backyard_bugs.html).

**Bugscope.** Students control a scanning electron microscope to view insects over the Web. From the University of Illinois (http://bugscope.beckman.uiuc.edu/).

**California's Endangered Insects.** Exhibit at the Essig Museum of Entomology (http://essig.berkeley.edu/endins/endins.htm).

**Cirrus Digital Imaging.** High-resolution digital photographs of butterflies, moths, and other insects (http://www.cirrusimage.com/).

**DPDx.** Identification and diagnosis of parasites and parasitic diseases of public health concern. Includes an image library and a review of recommended diagnostic procedures. From the CDC (http://www.dpd.cdc.gov/dpdx/).

**eNature.com.** Species accounts with color photographs of insects and spiders from the National Audubon Society Field Guide Series. Users can e-mail photographs and species accounts to friends and set up life lists (http://www.enature.com/).

**Entomology Image Gallery.** From Iowa State University (http://www.ent.iastate.edu/imagegallery/).

**Entomopathogenic Nematode Taxonomy.** How to identify entomopathogenic nematodes (http://kbn.ifas.ufl.edu/kbnstein.htm).

**Entopix Vol. 1.** Over 1,000 images and 7 video clips of plant parasitic arthropods and their natural enemies (http://www.mactode.com/Pages/Entopix_insects.html).

**Exploring California Insects.** Web-based field guide to more than 700 insects with over 2,000 photographs. Downloadable curriculum (http://www.bugpeople.org/).

**Florida Insect Photo Gallery.** Florida state collection of arthropods (http://www.fsca-dpi.org/FloridaInsectGallery/A%20Gallery%20of%20Florida%20Insects.htm).

**Florida Vegetable Pest Photographic Gallery CD-ROMs.** Three CD-ROMs in HTML format contain over 260 high-quality images of vegetable pests and their damage. Pests are common to Florida. However, many are common throughout the United States and the world. Images are provided in three different sizes and resolutions: print quality, display for large audiences, and Web-optimized (http://pests.ifas.ufl.edu/software/det_veggies.htm).

**Forest Pests of North America.** Image database from the Bugwood Network (http://www.forestpests.org/).

**Forestry Images: The Source for Forest Health, Natural Resources, and Silviculture Images.** Forestry Images provides users with access to more than 62,000 quality, reviewed images on topics and subjects related to forests and ornamentals, and is operated by the Bugwood Network, University of Georiga's taxonomic, relational database system. All images are available for educational use at no cost as long as appropriate credit is given. Insects are a significant portion of the images available, although Forestry Images also includes images on other forest pests and damage agents, trees and plants encountered in forests, and silvicultural practices and urban forestry. The overall objective of Forestry Images is to provide an accessible and easily used archive of high-quality images related to forest health and silviculture, with particular emphasis on educational applications. Forestry Images is a global resource (http://www.forestryimages.org/).

**Images of Goldenrod Visitors.** Images of insect vistors to goldenrod (*Solidago*) and neighboring flowers (http://pollinator.com/goldenrod.htm).

**Insect Drawings.** Drawn as a Works Progress Administration project during the Great Depression. From the University of Illinois at Champaign–Urbana (http://www.life.uiuc.edu/entomology/illustrations/).

**Insect Illustration.** Scientific illustration of insects and other invertebrates (http://www.scientificillustrator.com/insects.html).

**Insect Images.** From the Clemson University–USDA joint project (http://entweb.clemson.edu/cuentres/cesheets/).

**Insect Images: The Source for Entomology Images.** A fully relational, taxonomically based Web information system operated by the University of Georgia Bugwood Network that makes over 24,000 quality images on a broad array of entomological subjects, subdisciplines, program types, damage symptoms, and identification characteristics on more than 4,700 subjects. Available for download or other no-cost educational uses. The overall objective is to provide an accessible and easily used archive of high-quality images related to insects and entomology, with particular emphasis on educational applications. Images are available from more than eighty countries and from across North America and are used in all levels of educational programs and uses (http://www.insectimages.org/).

**Insect Macro Photography.** Providing an intimate exposure to the behavioral wonders of insects (http://www.insects.org/entophiles/index.html).

**Insect Macro Photography by M. Plonsky.** Very large collection of insect and spider images with lots of extreme close-ups (http://www.pbase.com/mplonsky/insects).

**Insect Photo Gallery.** Virtual collection of various insects (http://www.insect-photo.com).

**Insect Photography Search.** Search thousands of entomological images and species, all at one site. Browsing through the archive is free, and there are no access charges, registration requirements, or usage limits. It is a great source for reference or entertainment (http://www.fotosearch.com/photos-images/insect.html).

**Insect Photos from Southwest Michigan.** By Mark Cassino (http://www.markcassino.com/galleries/photogalleries.htm).

**Insects of Southeast Iowa.** Photographic images of butterflies and other insects commonly found in southeast Iowa (http://www.iowapix.com/main_page/Insects/insect_gallery.htm).

**Macro Photography.** Photography of many European arthropods. Close-ups of heads and details for certain species. In French (http://lioroux.free.fr/).

**Most Wanted Bugs.** Insect microscopy by Dennis Kunkel. Includes rap sheets (http://education.denniskunkel.com/MostWantedBugs.php).

**Nanoworld Image Gallery.** Scanning electron micrographs of insects (http://www.xtalent.com.au/gallery/index.php?cat=3).

**Pan Photo.** Images of insects of British Columbia, Canada (http://www.pansphoto.com/).

**Photo Album of Entomologists.** Photographs of over 100 entomologists from around the world (http://www.nhm.ku.edu/ksem/features/picttour.htm).

**Primary Type Specimen Database.** Insect primary types in the collection of the Museum of Comparative Zoology at Harvard University. Data on more than 28,000 types, representing 29 orders, 565 families, and 7,578 genera of insects. It also includes high-quality images of types—virtually all types of Cerambycidae, Buprestidae, Nymphalidae, Lycaenidae, Pieridae, Tettigoniidae, and a large portion of Formicidae (http://mcz-28168.oeb.harvard.edu/mcztypedb.htm).

**Public Health Image Library.** Includes images, illustrations, and animations of insects of medical importance (http://phil.cdc.gov/Phil/).

**South African Insect Photography.** Noncommercial site with many photographs and articles on South African insect life (http://www.insecta.co.za).

**Spiders in Australia.** Pictures and information about spiders found in Queensland, Australia (http://www.xs4all.nl/~ednieuw/australian/Spidaus.html).

**Spiders in Northwestern Europe.** Pictures of over 100 species and information about spiders (http://www.xs4all.nl/~ednieuw/Spiders/spidhome.htm).

**Thailand's Amazing Insects.** Over 1,800 photographs of insects from Northern Thailand; audio files of insect sounds; video clips of insects in their natural surroundings; articles on insects in Thailand and Thai culture; the results of a survey on Thai people's attitudes to insects; insects in Thai proverbs; insects on Thai postage stamps (http://www.thaibugs.com/).

**The Virtual Insectary.** Pictures of insects with links to their food plants and habitats (http://www.virtualinsectary.com/).

**UIUC Entomology Insect Illustrations.** Around twenty illustrations of various insects (http://www.life.uiuc.edu/entomology/illustrations.html).

**Uniramous Arthropods Exhibit.** At the University of California–Berkeley Museum of Paleontology (http://www.ucmp.berkeley.edu/arthropoda/uniramia/uniramia.html).

**Veterinary Parasitology Images Gallery.** Images of arthropods (Arthropoda: Insecta and Acari) of medical veterinary importance. Built along Marcelo de Campos Pereira's professional career as veterinary parasitologist and amateur photographer (http://www.icb.usp.br/~marcelcp/#INDEX and http://icb.usp.br/~marcelcp/Default.htm).

**Wildlife and Macro Photography—Images of Insects.** Boris Krylov's photographs of insects and other wildlife (http://www.macro-photo.org/photo-gallery-macro.html).

## INSECT PATHOLOGY, MICROBIAL CONTROL

**Biological Control: A Guide to Natural Enemies in North America.** Tutorial on the concept and practice of biological control, with photographs and descriptions of biological control (or biocontrol) agents of insect, disease, and weed pests in North America (http://www.nysaes.cornell.edu/ent/biocontrol/).

**Ecological Database of the World's Insect Pathogens.** Information on fungi, viruses, protozoa, mollicutes, and bacteria (other than *Bacillus thuringiensis*) that are infectious in insects, mites, and related arthropods (http://cricket.inhs.uiuc.edu/edwipweb/edwipabout.htm).

**Global Insecticide Directory.** Includes biological control agents (http://www.agranova.co.uk/gid.asp).

**Gypsy Moth in North America.** USDA Forest Service (http://www.fs.fed.us/ne/morgantown/4557/gmoth/).

**Microbial Germplasm Database.** Listed by host insect order (http://mgd.nacse.org/cgi-bin/mgd).

**Nematodes as Biological Control Agents of Insects.** (http://nematode.unl.edu/wormepns.htm).

**Society for Invertebrate Pathology.** (http://www.sipweb.org/).

**Tephritid Worker Database.** A directory of tephritid fruit fly researchers and plant protection officers in the world, providing expertise on species and speciality field, and recent publications, bibliographic references, news, directories, meetings, and projects (http://www.tephritid.org/).

**Varroa Mites and How to Catch Them.** Treatment, drone method (http://www.xs4all.nl/~jtemp/dronemethod.html).

**Viral Diseases of Insects in the Literature.** Includes almost all literature published up to 1984, including the 733 references contained in the review articles by Hughes (1957) and Martignoni and Langston (1960) (http://insectweb.inhs.uiuc.edu/Pathogens/VIDIL/index.html).

## INTEGRATED PEST MANAGEMENT, BY TAXONOMIC GROUP

### Coleoptera (Beetles)

**Asian Longhorned Beetle.** Broad base of information about the Asian longhorned beetle (*Anoplophora glabripennis*), a devastating exotic pest threatening hardwood trees in North America (http://www.uvm.edu/albeetle/).

**Corn Rootworm Home Page.** Central site for western and northern corn rootworm. From Iowa State University (http://www.ent.iastate.edu/pest/rootworm/).

**Emerald Ash Borer.** Photographs and history of impact of emerald ash borer (*Agrilus planipennis*) on ash trees. By David Roberts, Michigan State University (http://www.msue.msu.edu/reg_se/roberts/ash/index.html).

**SBexpert.** Decision support system aids forest-land managers in the reduction of risk and hazard from spruce beetle outbreaks in Alaska. Also included are an electronic textbook and a literature search program. Requires Windows only (http://www.fsl.orst.edu/usfs/sbexpert/).

**Small Hive Beetle.** *Aethina tumida*, a new pest of honey bees established in the southeastern United States (http://www.doacs.state.fl.us/pi/enpp/ento/aethinanew.html).

**Southern Pine Beetle.** Clearinghouse for information on *Dendroctonus frontalis*. From Virginia Tech (http://whizlab.isis.vt.edu/servlet/sf/spbicc/).

**The Fly in Your Eye.** A small book (published in 1989) about the war deliberately started in Australia between dung beetles and bush flies. Based on C.S.I.R.O. research. Hard facts, technical drawings, cartoons, jokes. Much used in Australian schools (http://www.viacorp.com/flybook/fulltext.html/).

### Diptera (Flies)

**Integrated Fly Management Presentation.** Presentation emphasizing insect light traps but covering all areas need to control filth breeding flies without the use of pesticides. Requires Flash (http://www.actroninc.com/flash/ifmflash.htm).

**Nzi Trap for Biting Flies.** Practical tsetse and biting fly control with environment-friendly trapping technology that does not rely on the use of insecticides (http://www.nzitrap.com/).

**The Fly in Your Eye.** A small book (published in 1989) about the war deliberately started in Australia between dung beetles and bush flies. Based on C.S.I.R.O. research. Hard facts, technical drawings, cartoons, jokes. Much used in Australian schools (http://www.viacorp.com/flybook/fulltext.html/).

**The Hessian Fly.** (http://www.oznet.ksu.edu/hessianfly/).

## Hemiptera (True Bugs, Aphids, Scale Insects, Leafhoppers, Cicadas)

**Bed Bugs: Biology and Management FAQ.** Biology, ecology, and public health significance of bed bugs. Identification resource and integrated pest management strategies for homeowners, tenants, and landlords. From the Harvard School of Public Health (http://www.hsph.harvard.edu/bedbugs/).

**Xylella fastidiosa.** Plant diseases caused by the insect-transmitted bacterium *Xylella fastidiosa*. Included are descriptions of the diseases by geography and plant host, vectors and transmission, news, contacts, and references. From the University of California–Berkeley (http://www.CNR.Berkeley.EDU/xylella/).

## Hymenoptera (Wasps, Ants, Bees)

**Areawide Suppression of Fire Ants.** Biological control techniques used by the USDA to control red imported fire ants (http://fireant.ifas.ufl.edu/).

**Pheromones of Diprionids.** Pine sawfly pheromones for sustainable management of European forests (http://zoologie.forst.tu-muenchen.de/PHERODIP/pherodip.html).

**Varroa Mites and How to Catch Them.** Treatment, drone method (http://www.xs4all.nl/~jtemp/dronemethod.html).

## Isoptera (Termites)

**Dr. Don's Termite Pages.** Basic information for the consumer and student. What termites do, why they do it, and how to avoid problems. Comments on biology and ecology as well as hints for choosing pest control companies (http://www.labyrinth.net.au/~dewart/).

## Lepidoptera (Butterflies, Moths)

**Canegrub.** Training package for the Australian sugar industry (http://www.ctpm.uq.edu.au/software/canegrub/).

**Chestnut Leaf Miner.** Web site for *Cameraria ohridella*, a new pest in Central Europe. From the Department of Applied Entomology, Ludwig-Maximilians-Universität, Munich (http://www.cameraria.de/index.php).

**European Corn Borer Home Page.** Central site for European Corn Borer. From Iowa State University (http://www.ent.iastate.edu/pest/cornborer/).

**Gypsy Moth in North America.** USDA Forest Service (http://www.fs.fed.us/ne/morgantown/4557/gmoth/).

## Orthoptera (Grasshoppers, Crickets)

**Australian Plague Locust Commission.** Role of the Australian Plague Locust Commission (APLC); how we forecast, monitor, and control locusts in Australia; research and extension activities; as well as issues of the monthly *APLC Locust Bulletin* (http://www.daff.gov.au/animal-plant-health/locusts).

**Grasshoppers of Wyoming and the West.** Dr. Robert Pfadt's Field Guide to Common Western Grasshoppers, with photography and detailed life history information on over fifty grasshopper species. Management, distribution, and population dynamics (http://www.sdvc.uwyo.edu/grasshopper/).

**Grasshoppers: Their Biology, Identification and Management.** Comprehensive grasshopper management site containing APHIS' Grasshopper Integrated Pest Management User Handbook, Field Guide to Common Western Grasshoppers, information on new grasshopper control techniques, and new research findings (http://www.sidney.ars.usda.gov/grasshopper/).

**Mole Cricket Knowledgebase.** Florida's pest management success story. Mole cricket knowledge base on all ten species found in the United States, including Hawaii, Puerto Rico, and the U.S. Virgin Islands. Distribution, description, life cycle, damage, and biological controls. Identification key for these species that makes heavy use of graphics and photographs. Large lists of references in every area. HTML format for Macintosh or PC (http://molecrickets.ifas.ufl.edu/).

## Phthiraptera (Lice)

**Head Lice Information and Management.** From the Harvard School of Public Health (http://www.hsph.harvard.edu/headlice.html).

**Head Lice Information Sheet.** Authoritative and evidence-based information site on head lice founded upon daily ongoing research. From the James Cook University, Australia (http://www.jcu.edu.au/school/phtm/PHTM/hlice/hlinfo1.htm).

**Head Lice Resources You Can Trust.** Current, research-based information on head lice and their management. Reproduction-ready fact sheets and information on the video *Removing Head Lice Safely*. From the University of Nebraska (http://lancaster.unl.edu/enviro/pest/Lice.htm).

## General

**ACORN.** Alternative Control Outreach Research Network. Network of master gardeners, extension educators, and university researchers interested in reducing pesticide use in home gardens (http://www.agriculture.purdue.edu/acorn/).

**Armed Forces Pest Management Board.** Resources and policy for pest management for the U.S. Department of Defense (http://www.afpmb.org/).

**Association of Applied IPM Ecologists.** Not-for-profit association of IPM practitioners. Provides information about pest management while encouraging environmentally compatible approaches and an awareness of IPM (http://www.aaie.com/).

**BugMatch Cotton.** For cotton production in Australia. Pest keys, photographs and video, and IPM strategies (http://www.ctpm.uq.edu.au/software/bugmatch/).

**BugMatch Grapes.** For grape production in Australia and New Zealand. Keys for diagnosing diseases and pests, and fact sheets (http://www.ctpm.uq.edu.au/software/bugmatch/).

**Canadian Food Inspection Agency.** For plant pest surveillance and invasive species (http://www.inspection.gc.ca/english/plaveg/pestrava/pestravae.shtml).

**Center for Urban and Industrial Pest Management.** From Purdue University (http://www.entm.purdue.edu/Entomology/urban/home.html).

**China Society of Plant Protection.** (http://www.ipmchina.net/cspp/english.asp).

**Consortium for International Crop Protection.** Purpose is to assist developing nations reduce food crop losses caused by pests while also safeguarding the environment (http://www.ipmnet.org/).

**Diagnosis for Crop Protection.** Training tool for teaching diagnostic skills, aimed primarily at tertiary students and crop consultants. Presents various problem scenarios with graphic displays, video, and sound. For New Zealand. Requires Windows only (http://www.diagnosis.co.nz/).

**ESCAPE.** Exotic Species Curriculum for Agricultural Problem-Solving Education. A series of modules about exotic insect species that have been introduced to America for precollege students and undergraduates (http://www.unk.edu/acad/biology/hoback/escape/introduction.html).

**Featured Creatures.** A continuously growing Florida Web site providing detailed information on description, distribution, damage, and management (if necessary) of various Florida organisms (insects, spiders, mites, snails, nematodes, and so on) as well as scientific references, links to recommended pesticides, and more. Over 350 publications with thousands of color photographs and line drawings (http://creatures.ifas.ufl.edu/).

**FETCH21.** Forest Entomology Challenge for the 21st Century. Provides information on forest insects (http://www.forestry.ubc.ca/fetch21/fetch21/FETCH21.html).

**Florida Insect Management Guide.** Insect management recommendations developed by entomologists of the University of Florida (http://edis.ifas.ufl.edu/TOPIC_GUIDE_Insect_Management_Guide).

**Florida Integrated Pest Management.** IPM information and technology, with an emphasis on biological control as they apply to Florida (http://ipm.ifas.ufl.edu/).

**Forest and Shade Tree Insects of Florida.** For identifying, understanding, and managing the insects affecting forest and shade trees in Florida (http://eny3541.ifas.ufl.edu/).

**Forest Pests of North America.** Image database from the Bugwood Network (http://www.forestpests.org/).

**Global Crop Pest Identification and Information Services in IPM.** To provide information about pest identification, biology, ecology, and IPM control tactics via the Internet to extensionists of Third World countries. Site seems abandoned (http://www.nysaes.cornell.edu/ent/hortcrops/).

**High Plains Integrated Pest Management Guide.** For Montana, western Nebraska, Wyoming, and Colorado (http://highplainsipm.org/).

**Horticulture and Home Pest Newsletter.** Published by Iowa State University (http://www.ipm.iastate.edu/ipm/hortnews/).

**Insect Traps and Sampling.** Information on various insect traps and sampling techniques (http://ufinsect.ifas.ufl.edu/).

**Integrated Crop Management Newsletter.** Published by Iowa State University (http://www.ipm.iastate.edu/ipm/icm/).

**Integrated Pest Management Resource Centre.** Databases on insects for sale and contacts in the pest management world; pages on ecotoxicology, jobs, courses, journals' current contents; and an acronym library. Based in the United Kingdom (http://www.pestmanagement.co.uk/).

**Invasive and Exotic Insects.** Extensive list of insects with associated images as well as information pertaining to identification and management resources (http://www.invasive.org/insects.cfm).

**Iowa State University IPM.** (http://www.ipm.iastate.edu/ipm/).

**IPM for Commercial Nursery/Floral Crops.** Jointly sponsored by Texas A&M, Texas Association of Nurserymen, and Texas Pest Management Association (http://hortipm.tamu.edu/).

**IPM Institute of North America.** Independent nonprofit organization formed in 1998 to foster recognition and rewards in the marketplace for goods and service providers who practice IPM (http://www.ipminstitute.org/).

**IPM Manual for Home and Garden Pests in British Columbia.** (http://www.env.gov.bc.ca/epd/ipm/docs/envirowe/default.htm).

**IPM Online Homestudy Courses.** From the University of Connecticut. Noncredit (http://www.hort.uconn.edu/ipm/homecourse/coursinfo.htm).

**Kentucky IPM.** (http://www.uky.edu/Ag/IPM/ipm.htm).

**Laboratory for Pest Control Application Technology.** Multidisciplinary five-department, twenty-member team focused on developing a better understanding of toxin delivery processes, optimizing pest control systems, and reducing human and environmental exposure (http://www.oardc.ohio-state.edu/lpcat/).

**LivingWithBugs.** Least-toxic solutions to insect and mite problems for homeowners, small business, and the medical community (http://www.livingwithbugs.com/).

**Michigan State University IPM.** (http://ipm.msu.edu/).

**Ministry of Agriculture, Food and Fisheries.** Government of British Columbia, Canada (http://www.agf.gov.bc.ca/cropprot/ipm2.htm).

**Montana State University IPM.** (http://ipm.montana.edu/index.html).

**Mountain West IPM.** IPM for Colorado and Wyoming (http://wsprod.colostate.edu/cwis79/ipminfo/).

**New Jersey IPM.** From Rutgers University (http://pestmanagement.rutgers.edu/).

**New York State IPM.** (http://www.nysipm.cornell.edu/).

**North Central IPM Center.** (http://www.ncipmc.org/).

**North Central Region IPM Center: Minnesota Project.** Pest alerts, invasive species, and pesticide resources (http://pestmanagementcenter-mn.coafes.umn.edu/).

**Northeast IPM Center.** (http://www.neipmc.org/).

**Ohio IPM.** (http://ipm.osu.edu/).

**Pennsylvania IPM.** (http://paipm.cas.psu.edu/).

**Pest Management at the Crossroads.** The book summarizes the origins of current policies and problems related to pest management. Maps out a transition to make biointensive IPM—driven by market forces—the predominant strategy by the year 2020 (http://www.pmac.net/).

**Pests in and around the Home.** Knowledge base of pests of humans, pets, structures, lawns, and landscapes. CD-ROM contains information on pest biology, life cycle, identification keys, distribution, damage, and management. Hundreds of color photographs. HTML version for Macintosh or PC. Version 2.0 (http://pests.ifas.ufl.edu/software/det_pests.htm).

**Public Health Pest Control manual.** Available in HTML and PDF formats. From USDA (http://vector.ifas.ufl.edu/).

**Purdue University Field Crops IPM.** Includes corn, soybeans, alfalfa, and small grains in Indiana (http://www.entm.purdue.edu/entomology/ext/fieldcropsipm/index.html).

**Radcliffe's IPM World Textbook.** Electronic textbook from the University of Minnesota, edited by Ted Radcliffe and Bill Hutchison (http://ipmworld.umn.edu/).

**Regional IPM Resource Manual.** The SCC IPM Web site and the Regional IPM Resource Manual (CD-ROM) provide public agencies and communities, including entomologists with access to information, tools, and networks that maximize opportunities for pesticide reduction in noncrop production agriculture and structural IPM programs. The information will help entomologists to understand how to design, implement, and administer pest management programs in urban, industrial, and nonproduction agriculture environments. An excellent resource for practicing entomologists (http://ipm.sccgov.org/portal/site/ipm/).

**RiceIPM.** Interactive training to learn about integrated pest management in tropical rice. From a collaboration between Ministries of Agriculture in Thailand, Malaysia, and Vietnam; the International Rice Research Institute in the Philippines; and the University of Queensland, Australia (http://www.ctpm.uq.edu.au/software/riceipm/).

**Scaffolds Fruit Journal.** Weekly update on pest management and crop development from Cornell University, New York (http://www.nysaes.cornell.edu/ent/scafolds/).

**School IPM.** Web site on integrated pest management in schools (http://schoolipm.ifas.ufl.edu/).

**Schoolbugs-L.** Updates on additions to the School IPM Web site and communication with others interested in this area (http://schoolipm.ifas.ufl.edu/listsrvr.htm).

**South Carolina IPM.** From Clemson University (http://www.clemson.edu/scg/ipm/).

**Southern IPM Center.** (http://www.sripmc.org/).

**Texas IPM.** (http://txipmnet.tamu.edu/).

**Tree Fruit Pest Management.** From Washington State University (http://www.ncw.wsu.edu/treefruit/pestman.htm).

**University of California IPM.** (http://www.ipm.ucdavis.edu/).

**University of Connecticut IPM.** (http://www.hort.uconn.edu/ipm/).

**University of Delaware IPM.** (http://www.udel.edu/IPM/).

**University of Illinois IPM.** (http://www.ipm.uiuc.edu/).

**University of Maine IPM.** (http://www.umext.maine.edu/topics/pest.htm).

**University of Maryland IPM.** (http://www.mdipm.umd.edu/).

**University of Massachussetts IPM.** (http://www.umass.edu/umext/ipm/).

**University of New Hampshire IPM.** (http://extension.unh.edu/Agric/AGPMP.htm).

**University of Vermont IPM.** (http://pss.uvm.edu/ipm/).

**Urban Entomology.** Pest and pest management information for homeowners and small business in Oregon from Oregon State University (http://www.ent.orst.edu/urban/home.html).

**USDA Cooperative State Research, Education, and Extension Service Pest Management Emphasis Areas.** Summary of pest management programs sponsored by CSREES (http://www.csrees.usda.gov/nea/pest/pest.cfm).

**USDA National Site for Regional IPM Centers.** Includes access to CABI Compendia for land-grant college faculty, staff, and graduate students (http://www.ipmcenters.org/).

**Veg-Edge.** Vegetable IPM resource for the upper Midwestern United States. From the University of Minnesota (http://www.vegedge.umn.edu/).

**Vegetable Insects and Their Management.** From Purdue University (http://www.entm.purdue.edu/entomology/vegisite/).

**Vegetable IPM.** For the home vegetable gardener. Locate vegetable insect pests by name, by photograph, or by feeding mechanism (mouthparts) (http://vegipm.tamu.edu/).

**Virginia Tech IPM.** (http://www.ext.vt.edu/schoolipm/index.shtml).

**Washington State University IPM.** (http://ipm.wsu.edu/).

**West Virginia IPM.** (http://www.wvu.edu/~agexten/ipm/).

**Western IPM Center.** (http://www.wrpmc.ucdavis.edu/).

**Wisconsin IPM.** (http://ipcm.wisc.edu/).

**WoodyBug.** Knowledge base on woody ornamental pests (insects and mites) of the southeastern United States (http://woodypest.ifas.ufl.edu/).

## KEYS, BY TAXONOMIC GROUP

### Coleoptera (Beetles)

**Aquatic Insects of Michigan.** Checklists and identification resources (http://insects.ummz.lsa.umich.edu/~ethanbr/aim/).

**Carabini of Yugoslavia and Adjacent Areas.** The first in the series is the "Catalogue of the Fauna of Yugoslavia: Carabini," representing an interactive, multimedia guide through 140 taxa of Coleoptera from the tribe Carabini registered in Yugoslavia and the surrounding countries (http://www.ecolibribionet.co.yu/Izdanja/Karabini.htm).

**Click Beetles of Hawaii.** (http://nathist.sdstate.edu/SMIRCOL/Hawaii/hawaii.htm).

**Coccinellidae of the Indian Subcontinent.** Images of common species of lady beetles (Coleoptera: Coccinellidae) found in the agroecosystems of the Indian region, and their natural enemies. An annotated checklist of the Indian fauna and an illustrated key to common *Chilocorus* species of India are also available (http://www.angelfire.com/bug2/j_poorani/index.html).

**Cucujidae (sens. str.) of the World.** Taxonomy, distribution, morphology, and biology of the flat bark beetles (Coleoptera: Cucujidae). Includes color photographs of representative species, a list of species, and a key to the genera of the world (http://www.fsca-dpi.org/Coleoptera/Mike/cucujidae1.htm).

**Elateridae of Boreal North America.** Click beetles (http://nathist.sdstate.edu/SMIRCOL/elna.htm).

**Flea Beetles of North Dakota.** (http://www.ndsu.edu/ndsu/gefauske/Fleabeetles/alticini_home.htm).

**Key to Genera of Chlamydopsinae.** (Coleoptera: Histeridae). Requires free Lucid Player (http://www2.lsuagcenter.com/Inst/Research/Departments/arthropodmuseum/chlamydop%20intro.htm).

**Nicrophorus Central.** For all those whose research involves nicrophorine species (burying beetles) (Coleoptera: Silphidae, Nicrophorinae). Taxonomy and photo gallery (http://collections2.eeb.uconn.edu/nicroweb/nicrophorus.htm).

**The Darkling Beetles of Florida and Eastern United States.** Contains species checklists, identification keys, species profiles and images, distributions in Florida and eastern United States, and literature (http://entnemdept.ifas.ufl.edu/teneb/).

**Tiger Beetle World.** Information on tiger beetles (Coleoptera: Cicindelidae) of North America: biology, ecology, identification, techniques for study, research library, resources, cicindelophile directory (http://members.aol.com/YESedu/home.html).

**Tiger Beetles of Connecticut.** Taxonomy, ecology, and conservation status of Connecticut's fourteen *Cicindela* species. Keys to adults and larvae (http://collections2.eeb.uconn.edu/collections/insects/CTBnew/ctb.htm).

**Tiger Beetles of North Dakota.** Biology, morphology, checklist, and key (http://www.ndsu.nodak.edu/ndsu/beauzay/tigerbeetles/).

## Collembola (Springtails)

**Aquatic Insects of Michigan.** Checklists and identification resources (http://insects.ummz.lsa.umich.edu/~ethanbr/aim/).

## Diptera (Flies)

**Aquatic Insects of Michigan.** Checklists and identification resources (http://insects.ummz.lsa.umich.edu/~ethanbr/aim/).

**Asilidae Keys.** Keys for subfamilies and so on (http://www.geller-grimm.de/key.htm).

**Computerized Disease Vector Identification Keys.** Including Key to the Mosquito Genera of North America North of Mexico. From Walter Reed Biosystematics Unit (http://wrbu.si.edu/VecIDService.html).

**CulicID.** Series of computerized interactive identification and information tools for Australasian mosquitoes (http://www.sph.uq.edu.au/Divisions/acithn/buycd.htm).

**Fruit Flies of New South Wales.** (http://www.agric.nsw.gov.au/Hort/ascu/fruitfly/fflyinde.htm).

**Identification Guide to Common Mosquitoes of Florida.** Larval and adult keys (http://fmel.ifas.ufl.edu/Key/).

**Key to Conopidae (Diptera) of the Netherlands.** (http://home.hccnet.nl/mp.van.veen/conopidae/index.html).

**Syrphidae Europe.** Hoverflies (Syrphidae) of Europe: photographs, range maps, keys. In English and Dutch (http://www.syrphidae.com/).

**The Laphriini Pages.** All about the North American members of this tribe of robber flies (Diptera: Asilidae: Laphriini). Many of the members in this tribe are large bumblebee mimics. Includes species-level descriptions and online keys (http://users.usachoice.net/~swb/Laphriini/Laphriini.htm).

**Walter Reed Biosystematics Unit.** Dedicated to mosquito systematics. Systematic catalog of Culicidae, identification resources, and mosquito and tick publications (http://wrbu.si.edu/).

## Ephemeroptera (Mayflies)

**Aquatic Insects of Michigan.** Checklists and identification resources (http://insects.ummz.lsa.umich.edu/~ethanbr/aim/).

**Ephemeroptera Germanica.** All about German mayflies. In German and English (http://www.ephemeroptera.de/Eph_int/about/about.html).

**Ephemeroptera of Michigan.** Key to adults and nymphs (http://insects.ummz.lsa.umich.edu/~ethanbr/aim/Keys/Ephemeroptera/id_eom.html).

## Hemiptera (True Bugs, Aphids, Scale Insects, Leafhoppers, Cicadas)

**Aquatic Insects of Michigan.** Checklists and identification resources (http://insects.ummz.lsa.umich.edu/~ethanbr/aim/).

**Checklists and Keys to Australian Planthoppers.** (Hemiptera: Auchenorrhynca). From New South Wales Agricultural Scientific Collections Unit (http://www.agric.nsw.gov.au/Hort/ascu/keys.htm).

**Cicadas of Michigan.** Sounds, photographs, and a key (http://insects.ummz.lsa.umich.edu/fauna/michigan_cicadas/Michigan/Index.html).

**Illustrated Key to the Economically Important Leafhoppers of Australia.** (Hemiptera: Cicadellidae) (http://www.agric.nsw.gov.au/Hort/ascu/cicadell/ecokey0.htm).

**Leafhoppers.** Classification and biology of leafhoppers and their relatives; directory of leafhopper specialists; and key to major groups (http://www.inhs.uiuc.edu/~dietrich/Leafhome.html).

## Hymenoptera (Wasps, Ants, Bees)

**Aquatic Insects of Michigan.** Checklists and identification resources (http://insects.ummz.lsa.umich.edu/~ethanbr/aim/).

**Automatic Bee Identification System.** (http://www.insects-online.de/projfram.htm).

**Key to Families and Genera of Japanese Bees.** (http://konchudb.agr.agr.kyushu-u.ac.jp/identify/).

**Key to Genera of Ecitoninae.** (Hymenoptera: Formicidae) (http://www.armyants.org/armyants/indexfiles/keys.html).

**Yellowjackets of the Northwestern United States.** Biology, systematics, literature, and keys for yellow jackets (http://www.evergreen.edu/ants/TESCBiota/kingdom/animalia/phylum/arthropoda/class/insecta/order/hymenoptera/family/Vespidae/Kweskin97/main.htm).

## Isoptera (Termites)

**Key to Brazilian Termite Genera.** In Portuguese (http://www.unb.br/ib/zoo/docente/constant/cupins/chave/index.html).

## Lepidoptera (Butterflies, Moths)

**Aquatic Insects of Michigan.** Checklists and identification resources (http://insects.ummz.lsa.umich.edu/~ethanbr/aim/).

**Catacola.** Taxonomy, rearing guide, keys, and egg gathering techniques for underwing moths (http://www.silkmoths.bizland.com/catocala.html).

**CATE Sphingidae.** Project aimed to develop the methodology and sociology for transferring the enterprise of taxonomy onto the Internet using hawkmoths (Lepidoptera: Sphingidae) and aroid lilies (Alismatales: Araceae) as exemplar taxa (http://www.cate-sphingidae.org/).

**Key to Butterflies of the Lake Baikal Region.** Almost 2,000 high-resolution full-color digital images of all 197 Baikalian species and many of its caterpillars and pupae. Includes pictured key to butterfly families and pictured keys to all species. Detailed information of every species (in Russian). HTML format for Macintosh or PC. First version 31 December 2001. Price: $30.00 U.S. (incl. postage). For more information, contact olegberlov@narod.ru (http://babochki.narod.ru/).

**Keys for Butterflies from Venezuela.** Keys for *Morpho* (Lepidoptera: Nymphalidae), *Xylophanes* (Lepidoptera: Sphingidae), *Diaphania* (Lepidoptera: Crambidae) (http://www.miza-fpolar.info.ve/claves/index.php?LNG=1).

**Moths of North Dakota.** Identification guide and photographs of adults and larvae (http://www.ndsu.nodak.edu/ndsu/ndmoths/).

**UKmoths.** Photographic database of moths in the United Kingdom displayed in their natural resting position. About 1,500 of the 2,400 U.K. moths are featured (http://www.ukmoths.force9.co.uk/).

## Mantodea (Mantids)

**Key to Praying Mantids of the World.** Key to the level of genera (http://www.earthlife.net/insects/mant-key.html).

## Mantophasmatodea (Gladiators, Heelwalkers)

**Mantophasmatodea Systematics.** In German and English (http://www.mantophasmatodea.de/).

## Mecoptera (Scorpionflies and Hangingflies)

**Scorpionflies, Hangingflies, and Other Mecoptera.** A summary of the Mecoptera, including a key to North American adults (http://www.emporia.edu/ksn/v48n1-may2002/index.htm).

## Neuroptera (Lacewings, Antlions, Dobsonflies)

**Aquatic Insects of Michigan.** Checklists and identification resources (http://insects.ummz.lsa.umich.edu/~ethanbr/aim/).

## Odonata (Dragonflies, Damselflies)

**Aquatic Insects of Michigan.** Checklists and identification resources (http://insects.ummz.lsa.umich.edu/~ethanbr/aim/).

**Swedish Dragonflies.** Dragonfly photography, folklore, list of Swedish dragonflies, key, rearing information (http://home9.swipnet.se/~w-90582/dragonfly/dragonfly.html).

## Orthoptera (Grasshoppers, Crickets)

**Aquatic Insects of Michigan.** Checklists and identification resources (http://insects.ummz.lsa.umich.edu/~ethanbr/aim/).

**Electronic Key for Common Adult Grasshoppers of the Western United States.** Identifies and finds information for fifty-eight of the most common adult grasshoppers of the western United States which pose the greatest environmental and economic threat (http://www.sidney.ars.usda.gov/grasshopper/Support/lucid.htm).

**Grasshoppers of Florida.** Color guide to grasshoppers in the southeastern United States, including maps. In PDF format (http://entnemdept.ufl.edu/ghopper/ghopper.html).

**Grasshoppers of New Mexico.** (Orthoptera: Acrididae and Romaleidae) (http://www.sdvc.uwyo.edu/grasshopper/ghnmtoc.htm).

**Grasshoppers: Their Biology, Identification, and Management.** Comprehensive grasshopper management site containing APHIS' Grasshopper Integrated Pest Management User Handbook, Field Guide to Common Western Grasshoppers, information on new grasshopper control techniques, and new research findings (http://www.sidney.ars.usda.gov/grasshopper/).

**Guide to Orthoptera of the Netherlands.** In Dutch (http://home.hccnet.nl/mp.van.veen/sprinkhanen/index_pocket.html).

**Guide to the Grasshoppers of Wisconsin.** Keys, photographs, and species distributions (http://www.dnr.state.wi.us/org/es/science/publications/ss1008_2005.htm).

**Katydids of Costa Rica Vol. 1.** Systematics and bioacoustics of the cone-head katydids by Piotr Naskrecki (http://140.247.119.145/book/book.htm).

**Key to Families and Subfamilies of Crickets.** (http://buzz.ifas.ufl.edu/k340k1.htm).

**Key to Wyoming Grasshoppers: Acrididae and Tetrigidae.** Online presentation of a revision made in 1998 by Timothy J. McNary to a key developed by Dr. Robert E. Pfadt, Professor Emeritus, University of Wyoming (http://www.sdvc.uwyo.edu/grasshopper/kwgtoc.htm).

**Mole Cricket Knowledgebase.** Florida's pest management success story. Mole cricket knowledge base on all ten species found in the United States, including

Hawaii, Puerto Rico, and the U.S. Virgin Islands. Distribution, description, life cycle, damage, and biological controls. Identification key for these species that makes heavy use of graphics and photographs. Large lists of references in every area. HTML format for Macintosh or PC (http://molecrickets.ifas.ufl.edu/).

**Orthoptera of the Northern Great Plains.** Anatomy, key, and photographs (http://www.ndsu.nodak.edu/entomology/hopper/orthoptera_home.htm).

**Orthoptera Species File Online.** Taxonomic database of the world's grasshoppers, locusts, katydids, and crickets, both living and fossil. Full synonymic and taxonomic information for 23,700 valid species, 39,999 taxonomic names, 145,100 citations to 11,850 references, 44,000 images, 184 sound recordings, 37,980 specimens, and keys to 2,100 taxa (http://osf2.orthoptera.org/).

### Phasmida (Stick Insects)

**A Guide to the Stick Insects of Australia.** Includes key and distribution map (http://home.swiftdsl.com.au/~pmiller/stick_insects/).

### Plecoptera (Stoneflies)

**Aquatic Insects of Michigan.** Checklists and identification resources (http://insects.ummz.lsa.umich.edu/~ethanbr/aim/).

### Thysanoptera (Thrips)

**ThripsNet.** Checklists, keys, biology, and collection. From the Martin Luther University of Halle, Wittenberg, Germany (http://www.thripsnet.com/).

### Trichoptera (Caddisflies)

**Aquatic Insects of Michigan.** Checklists and identification resources (http://insects.ummz.lsa.umich.edu/~ethanbr/aim/).

### General

**BugMatch Cotton.** Pest keys, photographs and video, and IPM strategies. For cotton production in Australia (http://www.ctpm.uq.edu.au/software/bugmatch/).

**BugMatch Grapes.** Keys for diagnosing diseases and pests, and fact sheets. For grape production in Australia and New Zealand (http://www.ctpm.uq.edu.au/software/bugmatch/).

**Keys to Insect Pests and Beneficials in Tropical Rice.** From the University of Queensland, Australia. Requires free Lucid Player (http://www.ctpm.uq.edu.au/software/riceipm/keys/).

**Ozpest.** Reference system for urban insects and their control. Keys to urban insect pests, timber damage, and stored product damage. From the University of Queensland, Australia (http://www.ctpm.uq.edu.au/software/ozpest/).

## LESSON PLANS

**Alphabetical Insects.** Students find the names of insects that begin with each letter of the alphabet. For elementary to middle school (http://www.ent.iastate.edu/zoo/lessonplans/alphabetical.html).

**Arthropod Proverbs.** A collection of pithy sayings mentioning arthropods; organized taxonomically (http://entnemdept.ifas.ufl.edu/proverbs.htm).

**Crickets in the Classroom.** Instructional package of information, activities, worksheets, drawing, and links (http://www.telusplanet.net/public/ecade/CricketsintheClassroom/cricketsintheclassroom.html).

**Elementary Urban IPM Curriculum.** Exploring urban integrated pest management. For kindergarten to grade 6 (http://www.pested.msu.edu/CommunitySchoolIpm/curriculum.htm).

**Georgia WOWBugs Project.** *Melittobia digitata*, commonly found in nature but never before in the precollege classroom. Educational materials developed to teach fundamental biological concepts to grades 5 to 12 (http://www.wowbugs.com/).

**Honey Bee.** Students identify honey bee's responsibilities inside the hive and importance of the honey bee to humans, and discover how honey bees communicate. For pre-K to grade 4 (http://www.ent.iastate.edu/zoo/lessonplans/honeybee.html).

**Life Cycles.** Students will identify the life stages of four different insects and will care for and observe the life cycle of one species. For pre-K to grade 4 (http://www.ent.iastate.edu/zoo/lessonplans/lifecycles.html).

**School IPM Lesson Plans.** K–12 lesson plans that teach integrated pest management concepts (http://www.ipm.iastate.edu/ipm/schoolipm/lessonplans).

**Social Behavior of Polistine Wasps.** Wasps were marked and videotaped. Students use observation skills to study animal behavior (http://www.ruf.rice.edu/~evolve/Waspweb/wasphome.html).

**Teacher/Parent Resource Materials.** Including Wee Beasties—entomology newsletter for teachers and lesson plans (http://www.uky.edu/Ag/Entomology/ythfacts/resourc/resourc.htm).

**The Manduca Project.** Student explorations of the tobacco hornworm, *Manduca sexta* (http://insected.arizona.edu/manduca/default.html).

**Using Live Insects in Elementary Classrooms.** Lesson plans using insects to teach a variety of subjects: math, language arts, music, art, drama, science, and health (http://insected.arizona.edu/uli.htm).

## MEDICAL AND VETERINARY ENTOMOLOGY

**ACAROLOGY.** Mailing list dedicated to discussion of the Acari (mites and ticks) (http://www.nhm.ac.uk/hosted_sites/acarology/acarolist.html).

**American Board of Forensic Entomology.** Interaction of entomology with the law (http://research.missouri.edu/entomology/).

**American Mosquito Control Association.** (http://www.mosquito.org/).

**AnoBase.** A database containing genomic/biological information on anopheline mosquitoes, with an emphasis on *Anopheles gambiae*, the world's most important malaria vector (http://www.anobase.org/).

**Australian Spider and Insect Bites.** Symptoms and first aid for funnel-web and redback spiders and paralysis ticks (http://www.usyd.edu.au/anaes/venom/spiders.html).

**Bed Bugs: Biology and Management FAQ.** Biology, ecology, and public health significance of bed bugs. Identification resource and integrated pest management strategies for homeowners, tenants, and landlords. From the Harvard School of Public Health (http://www.hsph.harvard.edu/bedbugs/).

**Borrelia burgdorferi Molecular Genetics Server.** (http://www.pasteur.fr/recherche/borrelia/Welcome.html).

**Bristol University Entomology Group.** Information about myiasis-causing flies, mainly *Luccilia sericata*, and parasitic mites (http://www.bio.bris.ac.uk/research/insects/).

**Bug Tutorials.** Authorized for pesticide applicator training CEUs in Arizona, Florida, Vermont, and West Virginia. Excellent for training in many categories. Requires Windows only (http://pests.ifas.ufl.edu/software/det_bugs.htm).

**CDC WONDER.** CDC reports, guidelines, and public health data (http://wonder.cdc.gov/).

**Centers for Disease Control and Prevention.** Division of Vector-Borne Diseases (http://www.cdc.gov/ncidod/dvbid/).

**Chironomid Home Page.** All about Chironomidae (Diptera): bibliographies, listserver, newsletter, and so on (http://insects.ummz.lsa.umich.edu/~ethanbr/chiro/).

**Chironomus Newsletter.** Covers research on Diptera: Chironomidae (http://insects.ummz.lsa.umich.edu/~ethanbr/chiro/newsletter.html).

**Computerized Disease Vector Identification Keys.** Including Key to the Mosquito Genera of North America North of Mexico. From Walter Reed Biosystematics Unit (http://wrbu.si.edu/VecIDService.html).

**CulicID.** Series of computerized interactive identification and information tools for Australasian mosquitoes (http://www.sph.uq.edu.au/Divisions/acithn/buycd.htm).

**Delusional Parasitosis.** Devoted to the mistaken belief that one is being infested by parasites, such as mites, lice, fleas, spiders, worms, bacteria, or other organisms. From the University of California–Davis (http://delusion.ucdavis.edu/).

**Dengue and Dengue Hemorrhagic Fever.** About surveillance and outbreaks. From the World Health Organization (http://www.who.int/topics/dengue/en/).

**Department of Vector Ecology and Environment.** Nagasaki University. Vector ecology in Southeast Asia. (http://www.tm.nagasaki-u.ac.jp/medical/index_eng.html).

**DPDx.** Identification and diagnosis of parasites and parasitic dieseases of public health concern. Includes an image library and a review of recommended diagnostic procedures. From the CDC (http://www.dpd.cdc.gov/dpdx/).

**Emerging Infectious Diseases.** Published by the CDC (http://www.cdc.gov/ncidod/eid/).

**Entomotropica.** Journal of the Venezuelan Entomological Society (publishes original papers on tropical entomology) (http://www.entomotropica.org/).

**Featured Creatures.** A continuously growing Florida Web site providing detailed information on description, distribution, damage, and management (if necessary) of various Florida organisms (insects, spiders, mites, snails, nematodes, and so on) as well as scientific references, links to recommended pesticides, and more. Over 350 publications with thousands of color photographs and line drawings (http://creatures.ifas.ufl.edu/).

**Flea News.** From Iowa State University (http://www.ent.iastate.edu/fleanews/aboutfleanews.html).

**Fleas of South Africa.** Compiled by Joyce Segerman (http://www.ru.ac.za/academic/departments/zooento/Martin/siphonaptera.html).

**Fleas, Ticks and Your Pet FAQ.** From the rec.pets.* Usenet newsgroup (http://www.faqs.org/faqs/pets/fleas-ticks/).

**Florida Medical Entomology Laboratory.** Research and education on arthropods of medical importance and on ecologically sound control strategies (http://fmel.ifas.ufl.edu/).

**Florida Mosquito Control Association.** (http://www.floridamosquito.org/).

**Forensic Entomology.** Mailing list for forensic entomologists and their students. Restricted list: you must e-mail the list owner with information about your forensic entomology background before you can be included on the list (http://tech.groups.yahoo.com/group/Forensic_Entomology/).

**Forensic Entomology Lecture by Martin Hall.** Martin Hall describing the usefulness of entomology in forensic investigation. Requires either Windows Media Player or QuickTime (http://www.nhm.ac.uk/nature-online/science-of-natural-history/forensic-sleuth/webcast-forensicentomology/forensic-entomology.html).

**Head Lice Information and Management.** From the Harvard School of Public Health (http://www.hsph.harvard.edu/headlice.html).

**Head Lice Information Sheet.** Authoritative and evidence-based information site on head lice founded upon daily ongoing research. From James Cook University, Australia (http://www.jcu.edu.au/school/phtm/PHTM/hlice/hlinfo1.htm).

**Head Lice Resources You Can Trust.** Current research-based information on head lice and their management. Reproduction-ready fact sheets and information on the video *Removing Head Lice Safely*. From the University of Nebraska (http://lancaster.unl.edu/enviro/pest/Lice.htm).

**Identification Guide to Common Mosquitoes of Florida.** Larval and adult keys (http://fmel.ifas.ufl.edu/Key/).

**Images of Mosquito Vectors of Arboviruses.** Images of *Aedes*, *Culex*, and *Anopheles*. From the CDC (http://www.cdc.gov/ncidod/dvbid/arbor/mosqpics.htm).

**Insect Bibliography Server.** Citations on bot flies, Chironomidae, Cuterebridae, dermatobia, FORMIS 2001 (ants), face fly, fire ant—2001, Gasterophilidae, horn fly, hypoderma, ITS, insect genetics, insect nematodes, leaf-cutting ants, NCB-ESA, 1996, Oestridae, parasitoids, reprint list, restriction enzymes, screwworm, and stable fly (http://entobib.unl.edu/).

**Insects, Disease, and History.** Devoted to understanding the impact arthropods and diseases have had on historical events (http://scarab.msu.montana.edu/historybug/).

**Integrated Fly Management Presentation.** Presentation emphasizing insect light traps but covering all areas need to control filth breeding flies without the use of pesticides. Requires Flash (http://www.actroninc.com/flash/ifmflash.htm).

**Integrated Management of Arthropod Pests of Livestock and Poultry.** Regional research project S-274, Veterinary Entomology, livestock and poultry pests (http://www.oznet.ksu.edu/pr_LP-pests/welcome.htm).

**International Biotherapy Society.** About use of maggots, leeches, and bees in modern medicine (http://biotherapy.md.huji.ac.il/).

**International Database on Insect Disinfestation and Sterilization.** Mass rearing facilities of sterile pest insects, ticks, and mites. Production size, radiation process, quality control parameters, dosimetry, program objective, transboundary shipment, field release data, and the facility's full address. From the International Atomic Energy Agency, Joint FAO/IAEA Division, Insect Pest Control Section (http://www-ididas.iaea.org/IDIDAS/).

**Japan Society of Medical Entomology and Zoology.** (http://www.jsmez.gr.jp/eng/index.html).

**John W. Hock Company.** Information on insect sampling equipment (http://www.johnwhockco.com/).

**Louisiana Mosquito Control Association.** (http://www.lmca.us/).

**Lyme Disease in the United States and Canada.** Information on Lyme disease in individual states and Canadian provinces (http://www.geocities.com/HotSprings/Oasis/6455/uscanada-links.html).

**Lyme Disease Risk Assessments.** Done by the U.S. Army (http://members.utech.net/users/10766/lyme.htm).

**Maggot Debridement Therapy.** Information on how to cleanse nonhealing wounds with maggots (http://medent.usyd.edu.au/projects/maggott.htm).

**MALARIA.** Mailing list for discussion of malaria. To subscribe, send the message "subscribe MALARIA Jane Doe" to majordomo@wehi.edu.au (replace Jane Doe with your real name) (http://www.wehi.edu.au/MalDB-www/discuss/listserv.html).

**Malaria Database.** From WHO/TDR (http://www.wehi.edu.au/MalDB-www/who.html).

**Medical Entomology.** From U.S. Air Force (http://www.afpmb.org/military_entomology/usafento/af.htm).

**Michigan Mosquito Control Association.** (http://www.mimosq.org/).

**Mid-Atlantic Mosquito Control Association.** (http://www.mamca.org/).

**Mosquito and Moth Cell Lines.** Database on cell culture availability (http://www.biotech.ist.unige.it/cldb/spestr.html).

**Mosquito and Vector Control Association of California.** (http://www.mvcac.org/).

**Mosquito Control Association of Australia.** (http://www.mcaa.org.au/).

**Mosquito Images.** Images of mosquitoes of the midwestern United States (http://www.ent.iastate.edu/imagegal/diptera/culicidae/).

**Mosquito Movies.** Movies showing hatching, emerging, and biting (http://www-rci.rutgers.edu/~insects/mosvid.htm).

**Mosquito-L.** For the discussion of topics related to mosquitoes (Diptera: Culicidae). Possible topics include, but are not limited to, behavior, bionomics, ecology, control, general biology, legislative/regulatory issues, mosquito-borne diseases, physiology, pathogens, sampling methods, and systematics. To subscribe, send an e-mail message to MOSQUITO-L-REQUEST@IASTATE.EDU with the word SUBSCRIBE in the body of the message. The subject field will be ignored. If your e-mail program automatically adds your signature to the end of your message, put the word END on a separate line after the word SUBSCRIBE. Archives are available (http://www.ent.iastate.edu/mailinglist/mosquito-l/).

**New Jersey Mosquito Control Association.** (http://www.rci.rutgers.edu/~insects/njmca.htm).

**Nzi Trap for Biting Flies.** Practical tsetse and biting fly control with environment-friendly trapping technology that does not rely on the use of insecticides (http://www.nzitrap.com/).

**On Maggots and Murders: Forensic Entomology.** Short PDF explains how maggots and other animals that feed on the remains of others give clues to crime scene investigators (http://www.nhm.ac.uk/nature-online/life/insects-spiders/fathom-maggot/assets/22feat_maggots_and_murders.pdf).

**PAG-L.** The Pacific Ant Group was established in September 2003 to facilitate communication toward preventing establishment of red imported fire ant and other invasive ants in Pacific Islands (https://listserv.hawaii.edu/archives/pag-l.html).

**PARASITE-GENOME.** Parasite genome databases and genome research resources mailing list and Web site. List is to ask for probes, offer resources to share, request technical advice, warn of problems encountered, post news and results, and advertise positions available (http://www.ebi.ac.uk/parasites/parasite-genome.html).

**Parasitic Organisms Codon Usage Tables.** All about *Brugia*, *Onchocerca*, *Plasmodium*, *Trypanosoma*, and *Leishmania* (http://www.ebi.ac.uk/parasites/cutg.html).

**Pennsylvania Vector Control Association.** (http://www.pavectorcontrol.org/).

**Pests in and around the Home.** Knowledge base of pests of humans, pets, structures, lawns, and landscapes. CD-ROM contains information on pest biology, life cycle, identification keys, distribution, damage, and management. Hundreds of color photographs. HTML version for Macintosh or PC. Version 2.0 (http://pests.ifas.ufl.edu/software/det_pests.htm).

**PHEREC Publications.** Publications of the John A. Mulrennan, Sr., Public Health Entomology Research and Education Center, on mosquito control, stable flies, sandflies, nonchemical mosquito control, and pesticide environmental impact research (http://www.pherec.org/EntGuides/).

**Phlebotomine Sandfly and Leishmania Research at the Liverpool School of Tropical Medicine.** Introduction to the research on the phlebotomine sandfly vector of the *Leishmania* parasite. Overview of current research, including insect genomics, biology, biochemistry, and field research (http://pcwww.liv.ac.uk/leishmania/index.htm).

**ProMed.** PROgram for Monitoring Emerging Diseases. Reporting of incidents or outbreaks, infectious disease problems of emerging interest, and discussions on how to improve surveillance and response capabilities (http://www.promedmail.org/).

**Public Health Image Library.** Includes images, illustrations, and animations of insects of medical importance (http://phil.cdc.gov/Phil/).

**Public Health Pest Control Manual.** Available in HTML and PDF versions. From USDA (http://vector.ifas.ufl.edu/).

**Quantitative Parasitology.** Free software designed to improve education and research in parasite ecology and epidemiology. Designed to deal with the notoriously left-biased frequency distributions of parasites. Requires Macintosh or PC (http://www.behav.org/QP/%21%21%21start.htm).

**SIMULIIDAE.** An offshoot of the British Simuliid Group—a very informal gathering of scientists of any discipline and from many countries—who have an interest in the Simuliidae. Members include entomologists, parasitologists, and medics, with interests in ecology, bionomics, taxonomy, cytotaxonomy, disease transmission, freshwater biology, and so on. The aim is to assemble as diverse a group as possible in order to get a wide interchange of ideas and information (http://www.jiscmail.ac.uk/lists/SIMULIIDAE.html).

**Society for Vector Ecology.** (http://www.sove.org/).

**Spiders and Other Arachnids.** Spiders, especially venom and spider bites, and misdiagnosis and identification of brown recluse spiders. Debunking of spider-related urband legends (http://spiders.ucr.edu/).

**Synanthropic Collembola: Springtails in Association with Man.** (http://www.collembola.org/publicat/sidney.htm).

**Texas Mosquito Control Association.** (http://www.texasmosquito.org/).

**The International Database on Insect Disinfestation and Sterilization.** Provides information on disinfestation and sterilization of arthropod species, with dosage levels and conditions, efficacy, and references. For researchers, phytosanitary regulators, plant protection services, and SIT facility and food irradiation operations personnel (http://www-ididas.iaea.org/).

**Tick and Lyme Disease Information for Europe.** From European Union Concerted Action on Lyme Borreliosis (http://meduni09.edis.at/eucalb/cms/index.php?lang=en).

**Tick Research Laboratory.** Nonprofit adjunct of the Department of Biological Sciences, University of Rhode Island (http://riaes.cels.uri.edu/resources/ticklab/).

**Tropical Diseases Research Networking Page.** Links to tropical disease resources, including vector-borne diseases (http://www.who.int/tdr/kh/res_link.html).

**University of Sydney.** Department of Medical Entomology. Research and contacts within the department and numerous fact sheets, mostly with an Australian focus (http://medent.usyd.edu.au/).

**Vector-Borne Infectious Diseases.** From Tulane University Department of Tropical Medicine (http://www.tropmed.tulane.edu/vbid/).

**Veterinary Entomology.** Biological information on various insect parasites of animals, and to some extent humans, including tsetse flies, tabanids, screwworms, fleas, and cockroaches (http://www.roberth.u-net.com/).

**Veterinary Parasitology Images Gallery.** Images of arthropods (Arthropoda: Insecta and Acari) of medical veterinary importance. Built along Marcelo de Campos Pereira's professional career as veterinary parasitologist and amateur photographer (http://www.icb.usp.br/~marcelcp/#INDEX and http://icb.usp.br/~marcelcp/Default.htm).

**Virginia Mosquito Control Association.** (http://www.mosquito-va.org/).

**Walter Reed Biosystematics Unit.** Dedicated to mosquito systematics. Systematic catalog of Culicidae, identification resources, and mosquito and tick publications (http://wrbu.si.edu/).

## ONLINE COURSES

**II Curso de Iniciación a la Entomología.** Introductory course about invertebrates and insects. Twenty-one (21) pages and 38 figures. In Spanish (http://scriptusnaturae.8m.com/II_ento/indice.htm).

**III Curso de Introducción a la Entomología.** Introductory course of Entomology. Ed. in 1998 by Asociación Naturalista Altoaragonesa Onso. 16 pages and 43 figures. In Spanish. Noncredit (http://scriptusnaturae.8m.com/III_ento/indice.htm).

**Distance Master's Degree in Entomology.** University of Nebraska–Lincoln (UNL) offers a full curriculum distance M.S. degree in entomology (http://entomology.unl.edu/educatn/distancems.htm).

**Distance Master's Degree in Entomology.** University of Florida (Gainesville) offers M.S. degree (nonthesis) programs in entomology and pest management for place-bound students, and certificate programs/courses for undergraduate and graduate students interested in taking courses but not enrolling in degree programs (http://entnemdept.ifas.ufl.edu/dept_disted.htm#degree).

**Entomology and Acarology.** The University of Queensland, through its Entomology program, now offers instruction in the sciences of entomology and acarology at both the undergraduate and postgraduate levels for Australian and overseas students (http://www.sols.uq.edu.au/index.html?id=18076).

**Forest Entomology and Pathology.** Forest entomology and forest insect pests of Canada. From Sir Sandford Fleming College (http://gaia.flemingc.on.ca/~pbell/eandp.htm).

**Insect Classification.** Three-credit distance education course offered via CD-ROM. From the University of Florida (http://insectclass.ifas.ufl.edu/).

**Insects and Human Society.** Three-credit Web-based course for nonentomology majors. From Virginia Tech (http://www.ento.vt.edu/ihs/distance/).

**Insects and Society.** Two-credit Web-based course covering a variety of interactions between insects and humans. Primarily for nonentomology majors. From Iowa State University (http://www.ent.iastate.edu/dept/courses/ent211/).

**Introduction to Insects.** One-credit, 6-week Web-based introduction to insect diversity, biology, and management. From Iowa State University (http://www.ent.iastate.edu/dept/courses/ent201/).

**IPM Online Homestudy Courses.** From the University of Connecticut. Noncredit (http://www.hort.uconn.edu/ipm/homecourse/coursinfo.htm).

**Management of Insect Pests.** Two-credit course. Introduction to insects and their lifestyles. Theory and application of pest management practices. Examples drawn primarily from field crops. From Iowa State University (http://www.ent.iastate.edu/dept/courses/).

**Principles of Entomology.** Three-credit beginning course for entomology majors or anyone desiring a thorough introduction to the world of insects. Students must meet for lab at the University of Florida/IFAS Research and Education Center (http://www.cals.ufl.edu/distance/showCourse.aspx?ID=56).

**The Snodgrass Tapes.** Three audio lectures by Robert E. Snodgrass given in 1960. Transcripts included (http://www.life.umd.edu/entm/shultzlab/snodgrass/).

## PESTICIDES

**American Association of Pesticide Safety Educators.** Association business and related issues (http://aapse.ext.vt.edu/index.html).

**California Department of Pesticide Regulation.** Includes information about pesticides registered in California (http://www.cdpr.ca.gov/).

**Centre for Entomological Research and Insecticide Technology.** CERIT, University of New South Wales, Sydney, Australia (http://www.cerit.unsw.edu.au/).

**Clemson University Pesticide Information Program.** Pesticide safety information for South Carolina pesticide applicators. Links to pesticide labels, MSDSs, and pesticide fact sheets (http://entweb.clemson.edu/pesticid/).

**EPA Office of Pesticide Programs.** U.S. Environmental Protection Agency (http://www.epa.gov/pesticides/).

**EXTOXNET.** The Extension Toxicology Network, including pesticide information profiles (http://extoxnet.orst.edu/).

**Florida Department of Agriculture and Consumer Services Bureau of Entomology and Pest Control, and Pesticides.** Regulatory information concerning structural pest control licensing (http://www.flaes.org/).

**Global Insecticide Directory.** Includes biological control agents (http://www.agranova.co.uk/gid.asp).

**Hawaii Pesticide Information Retrieval System.** Index to agricultural-use pesticide labels licensed for sale in Hawaii by the Hawaii State Department of Agriculture (http://state.ceris.purdue.edu/doc/hi/statehi.html).

**High Plains Integrated Pest Management Guide.** For Montana, western Nebraska, Wyoming, and Colorado (http://highplainsipm.org/).

**Laguna Pesticide Use Manager.** Label information, field/crop history, inventory, accounting (http://www.pacificaresearch.com/Pesticide.html).

**National Pesticide Information Center.** U.S. national resource for pesticide questions, emergency treatment, cleanup, and disposal (http://npic.orst.edu/).

**New York Pesticide Management Education Program.** From Cornell University (http://pmep.cce.cornell.edu/).

**North Carolina Pesticide Safety Education Program.** (http://ipm.ncsu.edu/pesticidesafety/).

**North Central Region IPM Center: Minnesota Project.** Pest alerts, invasive species, and pesticide resources (http://pestmanagementcenter-mn.coafes.umn.edu/).

**Oklahoma Pesticide Safety Education.** (http://pested.okstate.edu/).

**PAN Pesticide Database.** The most comprehensive online collection of pesticide data in the world, providing detailed information (at no cost to the user) for about 5,400 pesticide active ingredients, breakdown products, and related chemicals. It also contains information on more than 100,000 formulated pesticide products (current and historic registrations) from the U.S. Environmental Protection Agency. Where available, the database provides information on toxicity, regulatory status, aquatic ecotoxicity, and general identification information, including an extensive list of synonyms (http://www.pesticideinfo.org/).

**Pesticide Action Network.** Nonprofit citizen-based NGO that advocates adoption of ecologically sound practices in place of pesticide use (http://www.panna.org/).

**Pesticide Broadcast.** North Carolina Cooperative Extension Service pesticide newsletter containing articles on pesticide registration, use, and safety (http://ipm.ncsu.edu/current_ipm/broadcast.html).

**Pesticide Education.** From Pennsylvania State University (http://www.pested.psu.edu/).

**Pesticide Education Articles.** From Iowa State's *Integrated Crop Management Newsletter* (http://www.ipm.iastate.edu/ipm/icm/indices/pesticideeducation.html).

**Pesticide Education Resources.** From the University of Nebraska (http://pested.unl.edu/).

**Pesticide Labeling Tutorial.** Based on Chapter 2 of the "Applying Pesticides Correctly" manual. Requires Windows only (http://pests.ifas.ufl.edu/software/det_core2.htm).

**Pesticide Reregistration Eligibility Decisions.** From the U.S. Environmental Protection Agency (http://www.epa.gov/pesticides/reregistration/).

**PHEREC Publications.** Publications of the John A. Mulrennan, Sr., Public Health Entomology Research and Education Center, on mosquito control, stable flies, sandflies, nonchemical mosquito control, and pesticide environmental impact research (http://www.pherec.org/EntGuides/).

**Regulatory Information on Pesticide Products in Canada.** From the Canadian Centre for Occupational Health and Safety (http://www.ccohs.ca/).

**South Carolina Department of Pesticide Regulation.** (http://dpr.clemson.edu/).

**University of Arizona Pesticide Information and Training Office.** (http://ag.arizona.edu/pito/).

**Virginia Tech Pesticide Programs.** For pesticide safety educators (http://www.vtpp.ext.vt.edu/).

**Washington State Pesticide Page.** Current pesticide information, certification, and training (http://pep.wsu.edu/).

## JOURNALS AND NEWSLETTERS, BY TAXONOMIC GROUP

### Coleoptera (Beetles)

**Buprestis.** Newsletter about jewel beetles (Coleoptera: Buprestidae), edited by Chuck Bellamy (http://www.fond4beetles.com/Buprestidae/newsletter.htm).

**Calodema Web Site of Dr. T. J. Hawkeswood.** Provides free PDF files of the biological research of T. J. Hawkeswood and his coworkers (http://www.calodema.com).

**Water Beetle World.** Newsletter for water beetle workers worldwide. Over 7,000 citations on water beetles (http://www.zo.utexas.edu/faculty/sjasper/beetles/index.htm).

## Diptera (Flies)

**Calodema Web Site of Dr. T. J. Hawkeswood.** Provides free PDF files of the biological research of T. J. Hawkeswood and his coworkers (http://www.calodema.com).

**Ceratopogonid Information Exchange Newsletter.** For those interested in Diptera: Ceratopogonidae (http://campus.belmont.edu/cienews/cie.html).

**Chironomus Newsletter.** Covers research on Diptera: Chironomidae (http://insects.ummz.lsa.umich.edu/~ethanbr/chiro/newsletter.html).

**Dipterological Research.** An international journal of dipterological research with particular reference to the Palearctic fauna (http://www.dipterologic.com/).

**Fly Times.** Newsletter of the North American Dipterists' Society, edited by Jeffrey M. Cumming and Art Borkent (http://www.nadsdiptera.org/News/FlyTimes/Flyhome.htm).

**Studia dipterologica.** Journal of taxonomy, systematics, ecology, and faunistics of Diptera (http://www.studia-dipt.de/).

**Tachinid Times.** Annual newsletter for persons interested in research on parasitic flies (Diptera) of the family Tachinidae. This newsletter acts as a forum for informal communication about current projects, recent research findings, field trips, and similar types of information relating to the Tachinidae (http://www.nadsdiptera.org/Tach/TTimes/TThome.htm).

**Volucella.** International journal on taxonomy, ecology, and faunistics of Palearctic Syrphidae (Diptera) (http://www.naturkundemuseum-bw.de/stuttgart/volucella/).

## Hemiptera (True Bugs, Aphids, Scale Insects, Leafhoppers, Cicadas)

**Calodema Web Site of Dr. T. J. Hawkeswood.** Provides free PDF files of the biological research of T. J. Hawkeswood and his coworkers (http://www.calodema.com).

**Massachusetts Cicadas.** Web site of an amateur of the Massachusetts annual cicada enthusiast (http://www.mechaworx.com/Cicada/masscic1.asp).

## Hymenoptera (Wasps, Ants, Bees)

**American Bee Journal.** The journal has the honor of being the oldest English language beekeeping publication in the world (http://www.dadant.com/journal/).

**Apiculture Newsletter.** From the Department of Entomology, University of California–Davis (http://entomology.ucdavis.edu/faculty/mussen/news.cfm).

**APIS: Apicultural Information and Issues.** Archive of individual issues of the *Apicultural Information and Issues* monthly newsletter, published by the Florida Cooperative Extension Service (http://apis.ifas.ufl.edu/).

**Bee Briefs.** Short articles on topics of interest to beekeepers. From the University of California–Davis (http://entomology.ucdavis.edu/faculty/mussen/beebriefs/index.cfm).

**Bee Tidings Newsletter.** Image-rich newsletter for Midwestern beekeepers produced by the University of Nebraska, edited by Marion Ellis (http://entomology.unl.edu/beekpg/).

**Buzzwords and New Zealand Beekeeping.** Beekeeping news, bees, statistics, and honey descriptions (http://www.beekeeping.co.nz/).

**Calodema Web Site of Dr. T. J. Hawkeswood.** Provides free PDF files of the biological research of T. J. Hawkeswood and his coworkers (http://www.calodema.com).

**Chalcid Forum.** Published by the USDA Systematic Entomology Laboratory (http://www.sel.barc.usda.gov/hym/chalcforum.html).

**Espacio Apicola.** Argentine beekeepers' magazine (http://www.apicultura.com.ar/).

**Ichnews.** For those interested in Ichneumonoidea (http://hymfiles.biosci.ohio-state.edu/newsletters/ichnews/).

**Proctos.** Newsletter for workers on Proctotrupoidea *s. lat.* (http://hymfiles.biosci.ohio-state.edu/newsletters/proctos/homepg.html).

**Sphecos.** Newsletter for aculeate wasp researchers (now defunct) (http://www.sel.barc.usda.gov/selhome/sphecos/sph30ttl.htm).

## Lepidoptera (Butterflies, Moths)

**Calodema Web Site of Dr. T. J. Hawkeswood.** Provides free PDF files of the biological research of T. J. Hawkeswood and his coworkers (http://www.calodema.com).

**The Journal of Research on the Lepidoptera.** Available in PDF format (http://www.doylegroup.harvard.edu/~carlo/JRL/contents.html).

## Neuroptera (Lacewings, Antlions, Dobsonflies)

**Calodema Web Site of Dr. T. J. Hawkeswood.** Provides free PDF files of the biological research of T. J. Hawkeswood and his coworkers (http://www.calodema.com).

**Journal of Neuropterology.** Ceased publication in 2000 (http://www.ucm.es/info/zoo/JofN.htm).

**Neuropterists Newsletter.** (http://www.calacademy.org/research/entomology/Entomology_Resources/mecoptera/Neuroptera/newsletters/index.html).

## Odonata (Dragonflies, Damselflies)

**Calodema Web Site of Dr. T. J. Hawkeswood.** Provides free PDF files of the biological research of T. J. Hawkeswood and his coworkers (http://www.calodema.com).

**Ode News.** Occasional newsletter about dragonflies and damselflies on Cape Cod (http://www.odenews.net/).

**PETALURA.** Electronic journal of the Specialist Group for Systematic and Phylogenetic Odonatology (http://www.bechly.de/sgspo.htm).

## Plecoptera (Stoneflies)

**Illiesia—International Journal of Stonefly Research.** Completely free of charge international online journal dealing with all aspects of stonefly (Plecoptera) research (http://www2.pms-lj.si/illiesia/).

## Siphonaptera (Fleas)

**Flea News.** From Iowa State University (http://www.ent.iastate.edu/fleanews/aboutfleanews.html).

## General

**Acarology Bulletin.** Newsletter of the Systematic and Applied Acarology Society (http://www.nhm.ac.uk/hosted_sites/acarology/saas/ab.html).

**ANBP Newsletter.** Publication of the Association of Natural Biocontrol Producers (http://www.anbp.org/ANBPnewsletter.htm).

**Aracnet.** Electronic Bulletin of Entomology (in Spanish) (http://entomologia.rediris.es/aracnet/).

**Cultural Entomology Digest.** Beliefs and symbolism of insects within all facets of the humanities (http://www.insects.org/ced1/ced_index.html).

**Cyberbugs Minibeast e-Magazine.** From the Young Entomologists' Society (http://members.aol.com/yesbugs/pubmenu.html).

**Dugesiana.** A journal devoted to the study of insects and other arthropods published by the Universidad de Guadalajara is available free on the Web (in PDF format). Articles on original research are refereed by specialists around the world. Contributions are accepted (http://www.cucba.udg.mx/new/publicaciones/page_dugesiana/dugesiana.htm).

**Emerging Infectious Diseases.** Published by the CDC (http://www.cdc.gov/ncidod/eid/).

**Entomological Society of Ontario Newsletter.** (http://www.entsocont.com/newsletters.htm).

**Entomology Notes.** From the Michigan Entomological Society. Appropriate for classroom use (http://insects.ummz.lsa.umich.edu/MES/notes/noteslist.html).

**Entomotropica.** Journal of the Venezuelan Entomological Society. Publishes original papers on tropical entomology (http://www.entomotropica.org/).

**European Journal of Entomology.** International journal covering the whole field of general, experimental, systematic, and applied entomology (http://www.eje.cz/).

**Florida Entomologist.** International journal for the Americas published by the Florida Entomological Society (http://palmm.fcla.edu/fe/).

**Florida Entomology and Nematology Newsletter.** Monthly newsletter of the University of Florida's Entomology and Nematology Department. Moderated to prevent spam mail. Only two messages a month sent: (1) to request news and (2) to announce when newsletter is posted to Web site (http://entnews.ifas.ufl.edu/).

**Florida Pest Alert.** Regulatory and control information (http://pestalert.ifas.ufl.edu/).

**Folia Entomologica Mexicana.** Abstracts, information for authors, and subscription prices of the scientific journal published by Sociedad Mexicana de Entomologia, A.C. Publishes refereed articles on original research, synthesis or essays, scientific notes, and book revisions on entomology, acarology, and arachnology in America; comparative papers with fauna from other parts of the world (http://www.ecologia.edu.mx/folentmex/).

**FRASS.** Insect rearing newsletter (http://users.ugent.be/~padclerc/AMRQC/frass.htm).

**Horticulture and Home Pest Newsletter.** Published by Iowa State University (http://www.ipm.iastate.edu/ipm/hortnews/).

**Insect Systematics and Evolution.** Published for the Scandinavian Society of Entomology (http://www.zmuc.dk/EntoWeb/InSysEvol/).

**Insecta Mundi.** Publication of the Center for Systematic Entomology (http://centerforsystematicentomology.org/InsectaMundi/InsectaMundiOn-Line.htm).

**Insectarium Virtual.** Electronic magazine in Spanish (http://www.insectariumvirtual.com/blog-iv/).

**Integrated Crop Management Newsletter.** Published by Iowa State University (http://www.ipm.iastate.edu/ipm/icm/).

**Kansas School Naturalist.** Appropriate for both young and older entomologists (http://www.emporia.edu/ksn/).

**Kentucky Pest News Newsletter.** Information on crop and livestock insect, disease, and weed pests. Published weekly during peak season, otherwise biweekly (http://www.uky.edu/Ag/kpn/kpnhome.htm).

**Midwest Biological Control News.** Archives of this publication available from 1994 to 2000. No longer in publication (http://www.entomology.wisc.edu/mbcn/mbcn.html).

**Neotropical Entomology.** Newsletter published by the Entomological Society of Brasil since April 1976 (http://www.scielo.br/scielo.php?script=sci_issues&pid=1519-566X&lng=en&nrm=iso).

**North Carolina Pest News.** North Carolina Cooperative Extension Service newsletter published weekly on the Web from April to September (http://ipm.ncsu.edu/current_ipm/pest_news.html).

**PCT Online.** Pest Control Technology and Service Technician magazines. Includes MSDS label information (http://www.pctonline.com/).

**PEST CABWeb.** Entomology and crop protection journals (http://pest.cabweb.org/).

**Pest Control Magazine.** Monthly publication reaches pest control operators, researchers, entomologists, and others who are interested in learning more about the biology, behavior, business, and control of structural pests (http://www.pestcontrolmag.com/ME2/Default.asp).

**Pesticide Broadcast.** North Carolina Cooperative Extension Service pesticide newsletter containing articles on pesticide registration, use, and safety (http://ipm.ncsu.edu/current_ipm/broadcast.html).

**Phytoparasitica.** Israel Journal of Plant Protection Sciences (http://www.phytoparasitica.org/).

**Resistance Pest Management Newsletter.** From the Center for Integrated Plant Systems (CIPS) (http://whalonlab.msu.edu/rpmnews/).

**Rostrum.** Newsletter of the Entomological Society of Southern Africa (http://search.sabinet.co.za/essa/rostrum.html).

**Scaffolds Fruit Journal.** Weekly update on pest management and crop development from Cornell University, New York (http://www.nysaes.cornell.edu/ent/scafolds/).

**Suppliers of Beneficial Organisms in North America.** Publication of California/EPA's Department of Pesticide Regulation. It lists 143 commercial suppliers of 130 beneficial organisms used for biological control (http://www.cdpr.ca.gov/docs/pestmgt/ipminov/ben_supp/contents.htm).

**Systematic and Applied Acarology Special Publications.** Rapid publication for papers on mites and ticks (http://www.nhm.ac.uk/hosted_sites/acarology/saas/saasp.html).

**Teacher/Parent Resource Materials.** Including Wee Beasties—entomology newsletter for teachers and lesson plans (http://www.uky.edu/Ag/Entomology/ythfacts/resourc/resourc.htm).

**The Kansas School Naturalist.** (http://www.emporia.edu/ksn/index2.htm).

**ZOOTAXA.** Rapid international journal for systematic entomologists and zoologists (http://www.mapress.com/zootaxa/).

## SOFTWARE, BY TAXONOMIC GROUP

### Blattodea (Cockroaches)

**3-D Insects.** Virtual models of insects using the virtual reality modeling language (http://www.ento.vt.edu/~sharov/3d/3dinsect.html).

**MaHiRoSi.** An open-source roach simulator written in FreeBASIC. Just three species and a bunch of sprites right now, but hoping to grow (http://paulvern.newsit.es).

## Coleoptera (Beetles)

**3-D Insects.** Virtual models of insects using the virtual reality modeling language (http://www.ento.vt.edu/~sharov/3d/3dinsect.html).

**SBexpert.** Decision support system aids forest-land managers in the reduction of risk and hazard from spruce beetle outbreaks in Alaska. Also included are an electronic textbook and a literature search program. Requires Windows only (http://www.fsl.orst.edu/usfs/sbexpert/).

## Diptera (Flies)

**3-D Insects.** Virtual models of insects using the virtual reality modeling language (http://www.ento.vt.edu/~sharov/3d/3dinsect.html).

## Hymenoptera (Wasps, Ants, Bees)

**3-D Insects.** Virtual models of insects using the virtual reality modeling language (http://www.ento.vt.edu/~sharov/3d/3dinsect.html).

## Isoptera (Termites)

**3-D Insects.** Virtual models of insects using the virtual reality modeling language (http://www.ento.vt.edu/~sharov/3d/3dinsect.html).

## Lepidoptera (Butterflies, Moths)

**Gypsy Moth Life Systems Models.** Modeling tree growth and mortality, population dynamics, and moth/foliage/natural enemies (http://www.fs.fed.us/ne/morgantown/4557/gypsymth/).

## Mantodea (Mantids)

**3-D Insects.** Virtual models of insects using the virtual reality modeling language (http://www.ento.vt.edu/~sharov/3d/3dinsect.html).

## Odonata (Dragonflies, Damselflies)

**ClubTail.** Dragonfly database software. Requires Windows only (http://www.onmymountain.com/store-products/Software-Nature-Related-Software-10001-CLUBTAIL-1.0,-the-Dragonfly-Database-for-the-Serious-Odonatologist_16661.html).

## Orthoptera (Grasshoppers, Crickets)

**3-D Insects.** Virtual models of insects using the virtual reality modeling language (http://www.ento.vt.edu/~sharov/3d/3dinsect.html).

**CARMA.** Predicts the proportion of available forage that will be consumed by grasshoppers and estimates the economic returns of various treatment options (http://www.sidney.ars.usda.gov/grasshopper/Support/Carma.htm).

**Hopper.** Expert system for selecting appropriate treatments for grasshopper infestations and computer models for economic analyses of those treatments (http://www.sidney.ars.usda.gov/grasshopper/Support/Hopper.htm).

## Siphonaptera (Fleas)

**3-D Insects.** Virtual models of insects using the virtual reality modeling language (http://www.ento.vt.edu/~sharov/3d/3dinsect.html).

## General

**BioLink.** Integrated software for the collection, maintenance, analysis, application, and dissemination of taxonomic, biodiversity, and environmental information. Requires Windows only (http://www.biolink.csiro.au/).

**Biota.** Biodiversity database manager. Requires Macintosh or PC (http://viceroy.eeb.uconn.edu/biota).

**Bug Tutorials.** Authorized for pesticide applicator training CEUs in Arizona, Florida, Vermont, and West Virginia. Excellent for training in many categories. Requires Windows only (http://pests.ifas.ufl.edu/software/det_bugs.htm).

**DELIA.** Integrated DELTA database management software. Requires Windows only (http://www.naturebase.net/content/view/2401/482/).

**DELTA.** Descriptive language for taxonomy (http://delta-intkey.com/).

**Diagnosis for Crop Protection.** Training tool for teaching diagnostic skills, aimed primarily at tertiary students and crop consultants. Presents various problem scenarios with graphic displays, video, and sound. For New Zealand. Requires Windows only (http://www.diagnosis.co.nz/).

**DYMEX.** Modeling package for deterministic population models. Developed by C.S.I.R.O. in Australia (http://www.hearne.co.uk/products/dymex/).

**Faunist.** For faunistic analysis, including distribution, seasonal activity period, trend, and flower visit information. Maintains a database of records (http://home.hccnet.nl/mp.van.veen/index_e.html).

**Insectile Font.** Insect font illustrations and an alphabet made up of bug parts. Requires Macintosh or PC (http://www.p22.com/products/insectile.html).

**Invertebrate Collection Database Screensaver.** Twenty-three high-resolution specimen images. Requires Windows only (http://agspsrv34.agric.wa.gov.au/ento/icdb/icdb1.htm).

**Laguna Pesticide Use Manager.** Maintains label information, field/crop history, inventory, accounting (http://www.pacificaresearch.com/Pesticide.html).

**Ldp Line.** Program devoted to calculate probit analyses according to Finney (1972), and used to illustrate relationship between stimulus and response in biological and toxicological studies (http://embakr.tripod.com/ldpline/ldpline.htm).

**Lucidcentral.** Identification software for publishing interactive keys (http://www.lucidcentral.org/).

**LVB.** Phylogeny program that uses parsimony to reconstruct phylogeny from a nucleotide alignment, using a simulated annealing heuristic search. This is suitable for moderately large data matrices. Requires Unix, Linux, Mac OS X, or Windows (http://www.rubic.reading.ac.uk/lvb/).

**Lysandra.** Software for creating keys and databases on groups of organisms. Distributed on a CD-ROM. Downloadable demo. Requires Windows only (http://www.lysandrasoft.com/).

**Mandala.** FileMaker Pro database for data acquisition and management of systematics and biodiversity studies (http://www.inhs.uiuc.edu/research/mandala/).

**Mantis.** Freeware relational taxonomy and collection database designed for entomologists. Requires FileMaker Pro for Macintosh or PC (http://140.247.119.145/Mantis/).

**Noldus Information Technology.** Software for behavioral research, motion, and image analysis (http://www.noldus.com/).

**PAUP.** Tools for inferring and interpreting phylogenetic trees (http://paup.csit.fsu.edu/index.html).

**Pesticide Labeling Tutorial.** Based on Chapter 2 of the "Applying Pesticides Correctly" manual. Requires Windows only (http://pests.ifas.ufl.edu/software/det_core2.htm).

**PestThreatSaver.** Screen saver of exotic pests of Western Australia (http://agspsrv34.agric.wa.gov.au/ento/pestthreatsaver.htm).

**Platypus.** Database package for taxonomists. Requires Windows only (http://www.environment.gov.au/biodiversity/abrs/online-resources/software/platypus/index.html).

**Quantitative Parasitology.** Free software designed to improve education and research in parasite ecology and epidemiology. Designed to deal with the notoriously left-biased frequency distributions of parasites. Requires Macintosh or PC (http://www.behav.org/QP/%21%21%21start.htm).

**SwisTrack: A Tracking Tool for Multiunit Biological and Artificial Systems.** SwisTrack is an open-source tracking tool for tracking multiple, markerless objects. SwisTrack has been successfully used for tracking cockroaches and miniature robots, and can easily be extended due to its open, object-oriented architecture (http://swistrack.sourceforge.net).

**TAXIS.** Taxonomic Information System is a database management system for recording specimens, taxa, localities, characters, and images. Some of the features include interactive identification, reports on taxa and collection, and printing of labels. Requires Windows only (http://www.bio-tools.net/products/taxis/index.htm).

**Virtual Creatures.** Virtual tarantulas. Feed them virtual crickets. Requires Windows only (http://www.virtualcreatures.com/).

**WORLDMAP.** Quantitative measures for mapping biodiversity value, rarity, and endemism, combined with accountable methods for assessing priority areas for biodiversity conservation, including gap analysis. Web site shows examples from different spatial scales and a free downloadable demo (http://www.nhm.ac.uk/research-curation/projects/worldmap/).

## SOUNDS, BY TAXONOMIC GROUP

### Blattodea (Cockroaches)

**Borror Laboratory of Bioacoustics.** From Ohio State University (http://blb.biosci.ohio-state.edu/).

**Myths and Science of Cricket Chirps.** Hissing cockroach and cricket chirping (http://www.cricketscience.com/).

### Coleoptera (Beetles)

**Borror Laboratory of Bioacoustics.** From Ohio State University (http://blb.biosci.ohio-state.edu/).

**Insect Sounds from the Forests of Northern Thailand.** Sounds of crickets, cicadas, and even a sampling of "the blessed silence when the cicadas have stopped" (http://www.thaibugs.com/sounds.htm).

**Reference Library of Digitized Insect Sounds.** Many diverse insect sounds, including sounds of insect pests of stored products (http://www.ars.usda.gov/sp2UserFiles/person/3559/soundlibrary.html).

### Diptera (Flies)

**Borror Laboratory of Bioacoustics.** From Ohio State University (http://blb.biosci.ohio-state.edu/).

**Reference Library of Digitized Insect Sounds.** Many diverse insect sounds, including sounds of insect pests of stored products (http://www.ars.usda.gov/sp2UserFiles/person/3559/soundlibrary.html).

### Hemiptera (True Bugs, Aphids, Scale Insects, Leafhoppers, Cicadas)

**Borror Laboratory of Bioacoustics.** From Ohio State University (http://blb.biosci.ohio-state.edu/).

**Cicada Mania.** Sightings, videos, sounds, and news (http://www.cicadamania.net/).

**Cicadas of Michigan.** Sounds, photographs, and a key (http://insects.ummz.lsa.umich.edu/fauna/michigan_cicadas/Michigan/Index.html).

**Insect Sounds from the Forests of Northern Thailand.** Sounds of crickets, cicadas, and even a sampling of "the blessed silence when the cicadas have stopped" (http://www.thaibugs.com/sounds.htm).

**Phantastic Songs of the SE Asian Cicadas.** Mainly from Thailand and Malaysia (http://www2.arnes.si/~ljprirodm3/asian_cicadas.html).

**Singing Insects of North America.** Enables users to identify crickets, katydids, and cicadas from America north of Mexico (http://buzz.ifas.ufl.edu/).

**Songs of Cicadas.** From Slovenia, Croatia, and Macedonia (http://www2.arnes.si/~ljprirodm3/cikade.html).

### Hymenoptera (Wasps, Ants, Bees)

**Borror Laboratory of Bioacoustics.** From Ohio State University (http://blb.biosci.ohio-state.edu/).

**Insect Sounds from the Forests of Northern Thailand.** Sounds of crickets, cicadas, and even a sampling of "the blessed silence when the cicadas have stopped" (http://www.thaibugs.com/sounds.htm).

**Reference Library of Digitized Insect Sounds.** Many diverse insect sounds, including sounds of insect pests of stored products (http://www.ars.usda.gov/sp2UserFiles/person/3559/soundlibrary.html).

**Stridulation Sounds of Black Fire Ants.** *Solenopsis richteri* in different situations (http://home.olemiss.edu/~hickling/).

### Isoptera (Termites)

**Borror Laboratory of Bioacoustics.** From Ohio State University (http://blb.biosci.ohio-state.edu/).

**Reference Library of Digitized Insect Sounds.** Many diverse insect sounds, including sounds of insect pests of stored products (http://www.ars.usda.gov/sp2UserFiles/person/3559/soundlibrary.html).

### Lepidoptera (Butterflies, Moths)

**Reference Library of Digitized Insect Sounds.** Many diverse insect sounds, including sounds of insect pests of stored products (http://www.ars.usda.gov/sp2UserFiles/person/3559/soundlibrary.html).

### Odonata (Dragonflies, Damselflies)

**Borror Laboratory of Bioacoustics.** From Ohio State University (http://blb.biosci.ohio-state.edu/).

### Orthoptera (Grasshoppers, Crickets)

**Borror Laboratory of Bioacoustics.** From Ohio State University (http://blb.biosci.ohio-state.edu/).

**Insect Sound World.** Sounds of crickets and katydids from Japan (http://mushinone.cool.ne.jp/English/ENGindex.htm).

**Insect Sounds from the Forests of Northern Thailand.** Sounds of crickets, cicadas, and even a sampling of "the blessed silence when the cicadas have stopped" (http://www.thaibugs.com/sounds.htm).

**Myths and Science of Cricket Chirps.** Hissing cockroach and cricket chirping (http://www.cricketscience.com/).

**Orthoptera Species File Online.** Taxonomic database of the world's grasshoppers, locusts, katydids, and crickets, both living and fossil. Full synonymic and taxonomic information for 23,700 valid species, 39,999 taxonomic names, 145,100 citations to 11,850 references, 44,000 images, 184 sound recordings, 37,980 specimens, and keys to 2,100 taxa (http://osf2.orthoptera.org/).

**Reference Library of Digitized Insect Sounds.** Many diverse insect sounds, including sounds of insect pests of stored products (http://www.ars.usda.gov/sp2UserFiles/person/3559/soundlibrary.html).

**Singing Insects of North America.** Enables users to identify crickets, katydids, and cicadas from America north of Mexico (http://buzz.ifas.ufl.edu/).

**Songs of Crickets and Katydids of the Mid-Atlantic States.** Audio CD (http://members.aol.com/_ht_a/whershberg/page/).

### General

**Animal Diversity Web: Insecta.** Large image gallery of insects as well as a few sounds (http://animaldiversity.ummz.umich.edu/site/accounts/information/Insecta.html).

**Insect Sounds.** Available as .wav files. By Doug Von Gausig (http://www.naturesongs.com/insects.html).

**Thailand's Amazing Insects.** Over 1,800 photographs of insects from Northern Thailand; audio files of insect sounds; video clips of insects in their natural surroundings; articles on insects in Thailand and Thai culture; the results of a survey on Thai people's attitudes to insects; insects in Thai proverbs; insects on Thai postage stamps (http://www.thaibugs.com/).

**The Snodgrass Tapes.** Three audio lectures by Robert E. Snodgrass given in 1960. Transcripts included (http://www.life.umd.edu/entm/shultzlab/snodgrass/).

## VIDEO, BY TAXONOMIC GROUP

### Blattodea (Cockroaches)

**3-D Insects.** Virtual models of insects using the virtual reality modeling language (http://www.ento.vt.edu/~sharov/3d/3dinsect.html).

**Insects in Motion.** QuickTime video clips of various insects showing feeding behavior and life stage development, including a bee stinging (http://everest.ento.vt.edu/~carroll/insect_video_home.html).

### Coleoptera (Beetles)

**3-D Insects.** Virtual models of insects using the virtual reality modeling language (http://www.ento.vt.edu/~sharov/3d/3dinsect.html).

**Beetle Science.** Virtual labs, virtual beetles, videos, carbon dust illustrations, and interactive science features (http://www.cornell.edu/explorecornell/?scene=Beetle%2520Science).

**Picture Gallery of Carabid Beetles.** Pictures and videos of beetles (Coleoptera: Carabidae, Buprestidae, Cerambycidae) from Europe (http://volny.cz/midge/carabus/carabus.htm).

**World Educational Films.** Entomological videos documenting life histories and parasitic hymenoptera. Some clips available online (http://www.worldeducationalfilms.com/films.htm).

### Diptera (Flies)

**3-D Insects.** Virtual models of insects using the virtual reality modeling language (http://www.ento.vt.edu/~sharov/3d/3dinsect.html).

**FlyView and FlyMove.** A *Drosophila* image database and movies of developmental processes (http://pbio07.uni-muenster.de/).

**Insects in Motion.** QuickTime video clips of various insects showing feeding behavior and life stage development, including a bee stinging (http://everest.ento.vt.edu/~carroll/insect_video_home.html).

**Mosquito Movies.** Movies showing hatching, emerging, biting (http://www-rci.rutgers.edu/~insects/mosvid.htm).

### Hemiptera (True Bugs, Aphids, Scale Insects, Leafhoppers, Cicadas)

**Cicada Mania.** Sightings, videos, sounds, and news (http://www.cicadamania.net/).

### Hymenoptera (Wasps, Ants, Bees)

**3-D Insects.** Virtual models of insects using the virtual reality modeling language (http://www.ento.vt.edu/~sharov/3d/3dinsect.html).

**Bee Alert.** Using bees to assess environmental hazards. Online video cameras, electronic hive monitoring. Kids' section (http://beekeeper.dbs.umt.edu/bees/).

**Insects in Motion.** QuickTime video clips of various insects showing feeding behavior and life stage development, including a bee stinging (http://everest.ento.vt.edu/~carroll/insect_video_home.html).

**Social Behavior of Polistine Wasps.** Wasps were marked and videotaped. Students use observation skills to study animal behavior (http://www.ruf.rice.edu/~evolve/Waspweb/wasphome.html).

**The Amazing Beecam.** Video feed from a beehive (http://gears.tucson.ars.ag.gov/beecam/).

**WAXWeb.** The discovery of television among the bees. Not an entomological film, but with beekeeping theme (http://www.iath.virginia.edu/wax/).

**World Educational Films.** Entomological videos documenting life histories and parasitic hymenoptera. Some clips available online (http://www.worldeducationalfilms.com/films.htm).

### Isoptera (Termites)

**3-D Insects.** Virtual models of insects using the virtual reality modeling language (http://www.ento.vt.edu/~sharov/3d/3dinsect.html).

### Lepidoptera (Butterflies, Moths)

**Insects in Motion.** QuickTime video clips of various insects showing feeding behavior and life stage development, including a bee stinging (http://everest.ento.vt.edu/~carroll/insect_video_home.html).

### Mantodea (Mantids)

**3-D Insects.** Virtual models of insects using the virtual reality modeling language (http://www.ento.vt.edu/~sharov/3d/3dinsect.html).

**Insects in Motion.** QuickTime video clips of various insects showing feeding behavior and life stage development, including a bee stinging (http://everest.ento.vt.edu/~carroll/insect_video_home.html).

## Neuroptera (Lacewings, Antlions, Dobsonflies)

**Antlion Pit: A Doodlebug Anthology.** A collection of resources related to the antlion. Videos of feeding behavior and metamorphosis. Antlions in culture (http://www.antlionpit.com/).

## Odonata (Dragonflies, Damselflies)

**Insects in Motion.** QuickTime video clips of various insects showing feeding behavior and life stage development, including a bee stinging (http://everest.ento.vt.edu/~carroll/insect_video_home.html).

## Orthoptera (Grasshoppers, Crickets)

**3-D Insects.** Virtual models of insects using the virtual reality modeling language (http://www.ento.vt.edu/~sharov/3d/3dinsect.html).

**How Grasshoppers Jump.** By W. J. Heitler (http://www.st-andrews.ac.uk/~wjh/jumping/).

## Phthiraptera (Lice)

**Head Lice Resources You Can Trust.** Current, research-based information on head lice and their management. Reproduction-ready fact sheets and information on the video *Removing Head Lice Safely*. From the University of Nebraska (http://lancaster.unl.edu/enviro/pest/Lice.htm).

## Siphonaptera (Fleas)

**3-D Insects.** Virtual models of insects using the virtual reality modeling language (http://www.ento.vt.edu/~sharov/3d/3dinsect.html).

## General

**Entomology Education Videos.** Includes QuickTime clip (http://www.slvideopublishing.com/subject_entomology.cfm).

**Entomology Image Gallery.** From Iowa State University (http://www.ent.iastate.edu/imagegallery/).

**Forensic Entomology Lecture by Martin Hall.** Martin Hall describing the usefulness of entomology in forensic investigation. Requires either Windows Media Player or QuickTime (http://www.nhm.ac.uk/nature-online/science-of-natural-history/forensic-sleuth/webcast-forensicentomology/forensic-entomology.html).

**Thailand's Amazing Insects.** Over 1,800 photographs of insects from Northern Thailand; audio files of insect sounds; video clips of insects in their natural surroundings; articles on insects in Thailand and Thai culture; the results of a survey on Thai people's attitudes to insects; insects in Thai proverbs; insects on Thai postage stamps (http://www.thaibugs.com/).

# GLOSSARY

**Abdomen.** The most posterior of the three main body divisions of insects.

**Absolute population estimate.** An estimate of population size expressed as an absolute number per ground-surface area (for example, number per hectare) or per unit of volume (for example, number per liter of soil).

**Acaricide.** A pesticide that kills mites and ticks.

**Acceptable daily intake (ADI).** The level of a chemical residue to which daily exposure over a lifetime is not thought to cause appreciable risk.

**Accessory gland.** Any secondary gland of a glandular system (Figures 2.44, 2.45).

**Acetylcholine.** A synaptic neurotransmitter, $CH_3COOCH_2CH_2N(CH_3)_3{}^{1}$.

**Acetylcholinesterase.** An enzyme that catalyzes the breakdown of acetylcholine.

**Action potential.** A brief electrical depolarization that propagates along an axon or muscle fiber (Figure 2.43).

**Active ingredient (AI).** The component of a pesticide formulation responsible for the toxic effect.

**Acute poisoning.** Illness or death from a single dose of a toxicant.

**Adjuvant.** Any ingredient that improves the properties of a pesticide formulation.

**Aedeagus.** The copulatory organ of most male insects (Figure 2.22).

**Aerosol.** Pesticide formulation in which propulsion from a pressurized container suspends pesticide particles or droplets in the air.

**Aesthetic pest.** A pest whose mere presence is objectionable for psychological reasons.

**Aestivation.** Summer dormancy.

**African sleeping sickness.** A disease caused by infection with *Trypanosoma* species protozoans; transmitted by the bite of the tsetse fly.

**Age distribution.** The proportions of individuals in different age groups of a population at any given time (Figure 5.3).

**Aggregation pheromone.** An intraspecific chemical messenger that causes insect congregation at food sites, reproductive habitats, or hibernation sites.

**Agrobacterium tumefaciens.** A soilborne bacterium that can naturally transfer genetic information into plant cells, causing crown gall disease. This ability has been harnessed to transform (introduce desired sections of DNA into) many plant species.

**Agroecosystem.** An ecosystem largely created and maintained to satisfy a human want or need (Figure 5.7).

**Air sac.** A pouchlike expansion of a trachea (Figure 2.28).

**Alarm pheromone.** An intraspecific chemical messenger produced to elicit aggressive or defensive behavior.

**Alary muscle.** A wing-shaped muscle that connects the heart to lateral portions of the terga (Figure 2.27).

**Alate.** Winged.

**Algorithm.** A precisely defined sequence of rules stating how to produce specific outputs from specific inputs through a given number of steps (Figure 8.16).

**Alimentary canal.** The tube of the digestive system, extending from the mouth to the anus and including the foregut, midgut, and hindgut.

**Aliphatic.** Organic chemical compound in which the carbon atoms are linked in open chains rather than rings.

**Allele.** One of two or more alternative forms of a gene at corresponding sites on homologous chromosomes.

**Allelochemic.** A substance involved in communication between members of different species.

**Allethrin.** The first pyrethroid insecticide synthesized.

**Allomone.** An interspecific chemical messenger that benefits the releaser but not the receiver (for example, repellents, toxicants).

**Amino acid.** Any organic compound containing an amino group ($NH_2$) and a carboxylic acid group (COOH); building blocks of proteins.

**Amylase.** An enzyme that catalyzes the splitting of starch into maltose units.

**Anamorphosis.** The adding of segments to the body as development progresses.

**Antenna (pl. antennae).** One of a pair of segmented sensory organs located on the head above the mouthparts (Figures 2.5, 2.15).

**Anthropocentrism.** Regarding humans or human objectives as the principal focus of a system or situation.

**Anthropomorphism.** The attribution of human characteristics to nonhumans.

**Antibiosis.** Plant characteristics that affect insects in a negative manner (for example, increase mortality, reduce fecundity); involves activities of both host plant and insect.

**Anticoagulant.** Any substance that inhibits, delays, or suppresses blood clotting.

**Antixenosis (nonpreference).** Plant characteristics that drive insects away from a particular host; involves activities of both host and insect.

**Aorta.** The anterior portion of the dorsal blood vessel (Figure 2.27).

**Aphicide.** An insecticide that kills aphids.

**Apical.** Located at the end or tip.

**Apiculture.** The management of honey bees.

**Apodeme.** Invagination of the body wall that strengthens the exoskeleton and provides areas for muscle attachment.

**Apolysis.** The separation of the old cuticle from the epidermis; an early step in the molting cycle.

**Apparent resistance (ecological resistance, pseudoresistance).** Temporary host-plant resistance characteristics that rely heavily on environmental conditions; includes host evasion, induced resistance, and host escape.

**Approved common name.** Generic name of a pesticide (for example, carbaryl) or English name of an insect as approved by the Entomological Society of America (for example, European corn borer).

**Apterous.** Without wings.

**Apterygota.** The subclass of Hexapoda that comprises the primitively wingless insects.

**Arachnida.** The arthropod class that includes the spiders, scorpions, ticks, and mites (Figure 1.8); characterized by having a cephalothorax and an abdomen, and six pairs of appendages including the chelicerae, pedipalps, and four pairs of walking legs.

**Arbovirus.** A virus transmitted by mosquitoes or other blood-sucking arthropods.

**Areawide pest technology.** Pest technology implemented across broad areas (hundreds of acres to entire regions) to eradicate or suppress pest populations; administered by outside agencies.

**Arolium (pl. arolia).** An adhesive pad between the claws of the tarsus (Figure 2.18).

**Arthropod.** An invertebrate animal with jointed appendages; a member of the phylum Arthropoda.

**Atropine.** An anticholinergic alkaloid drug used as an antidote for organophosphate and carbamate insecticide poisoning.

**Augmentation.** Any biological control practice designed to increase the number or effectiveness of existing natural enemies.

**Autocidal control.** Use of induced sterility for reducing pest population size.

**Axon.** The part of a neuron that carries impulses away from the cell body (Figure 2.42).

**Axonic poison.** A chemical that interrupts the normal transmission of impulses along the axon of a neuron (Figure 11.5).

**Bait (B).** A pesticide formulation that combines an edible or attractive substance with a pesticide.

**Basement membrane.** The noncellular membrane underlying the epidermal cells of the integument (Figure 2.3).

**Behavioral resistance.** The ability of an insect population to change its behavior in order to avoid insecticides or other injurious factors.

**Benzyl benzoate.** A repellent for use against ticks and chiggers.

**Berlese funnel.** A sampling device that uses heat to drive insects from a sample of soil, vegetation, or litter (Figure 6.8).

**Binomial.** Having two names, as in a scientific name, composed of a genus name and a trivial name or specific epithet (for example, *Ostrinia nubilalis*).

**Binomial nomenclature.** The system of classification of organisms, as advanced by Carolus Linnaeus.

**Biochemical resistance.** The ability of an insect population to develop enzymes that detoxify an insecticide or other injurious material before it can reach its site of action.

**Bioeconomics.** The relationship between pest numbers and economic losses.

**Biolistics.** The insertion of DNA into plants by coating the DNA on tiny metal beads and propelling them at high speed into tissues; transformed plants can be regenerated from these single transgenic cells.

**Biological control (biocontrol).** The employment of any biological agent for control of a pest.

**Biomagnification.** The increase in concentration of a substance (usually a pesticide) in animal tissue as related to the animal's higher position in the food chain (Figure 11.4).

**Biomass.** The total dry weight or volume of all living organisms in a given area.

**Biotechnology.** Any technique that uses living organisms, or substances from those organisms, to make or modify a product, to improve plants or animals, or to develop microorganisms for specific uses.

**Biotic potential.** A measure of the innate ability of a population to survive and reproduce.

**Biotype (race).** A population of a pest species that differs from other populations of the species in its ability to attack a particular cultivar.

**Blastoderm.** The one-cell thick layer of cells that surrounds the yolk of an egg early in its embryological development (Figure 4.4).

**Blastula.** An early embryonic form that usually consists of a hollow sphere composed of a single layer of cells.

**Book lung.** The respiratory cavity of some spiders that contains a series of leaflike folds.

**Botanical insecticide.** An insecticide produced from a plant or plant product (for example, pyrethrum).

**Boundary layer.** With respect to insects, the layer of air that extends from ground level upward through increasing wind speeds to a height where wind speed and flight speed are equal.

**Brain hormone (AH, activation hormone; or PTTH, prothoracicotropic hormone).** A hormone secreted by neurosecretory cells of the insect brain; activates prothoracic glands (Figure 4.11).

**Brood.** Several cohorts of offspring produced by a parent or parent population at different times or in different places.

**Bubonic plague.** See **plague**.

**Bursicon.** A hormone secreted by certain cells of the nervous system that plays a role in the control of cuticular hardening and darkening.

**Calyx.** The bulbous junction of the pedicels of the ovarioles.

**Campaniform.** Domelike; campaniform sensilla (Figure 2.39).

**Cantharidin.** An irritating substance produced by blister beetles; used medicinally in keratolytic preparations.

**Capitate.** Head-shaped or with an apical knob (Figure 2.15).

**Carbamate.** An organic compound derived from carbamic acid, $NH_2COOH$; includes many important synthetic insecticides (for example, carbaryl).

**Carbohydrate.** A compound of carbon, hydrogen, and oxygen that follows the general formula $(CH_2O)_n$.

**Carcinogen.** A material that causes cancer.

**Cardiac valve.** The valve at the junction of the foregut and the midgut.

**Carrying capacity.** The maximum population density a given environment will support for a sustained period.

**Caste.** The type or form of an individual in a colony of social insects (for example, worker).

**Catfacing.** Injury resulting from the feeding of piercing-sucking insects on developing fruit; a result of uneven growth and deformation (Figure 7.8).

**Cell body.** The swollen portion of a neuron that contains the nucleus and is responsible for metabolic and maintenance functions of the neuron (Figure 2.42).

**Cellulase.** An enzyme that catalyzes the breakdown of cellulose.

**Cephalothorax.** The united head and thorax of the arachnids and crustaceans.

**Cercus (pl. cerci).** One of a pair of sensory appendages at the posterior of the abdomen (Figure 2.20).

**Cervix.** The membranous neck region of an insect.

**Chemical name.** Name given to a pesticide based on the chemical structure (for example, 1-naphthyl-*N*-methylcarbamate); naming follows rules of the International Union of Pure and Applied Chemistry.

**Chemigation.** Use of irrigation systems for dispensing pesticides (Figure 10.12).

**Chemoreceptor.** A sense organ involved in the perception of tastes and/or odors (Figure 2.38).

**Chemosterilant.** A chemical that induces reproductive sterility.

**Chilopoda.** The arthropod class that includes the centipedes (Figure 1.7); characterized by mandibulate mouthparts and one pair of appendages per body segment, including a pair of venomous claws on the first segment.

**Chitin.** A polymer of the nitrogenous polysaccharide *n*-acetyl-D-glucosamine that occurs in the exoskeleton of arthropods.

**Chlorinated hydrocarbon.** An organic compound containing chlorine, hydrogen, and occasionally oxygen and sulfur; includes the first widely used synthetic organic insecticides (for example, DDT).

**Cholesterol.** A fatlike steroid alcohol, $C_{27}H_{45}OH$, which is an important precursor in the formation of other steroids like ecdysone.

**Cholinesterase.** Shortened term for acetylcholinesterase.

**Chordotonal sensillum.** A mechanoreceptor that consists of a bundle of bipolar nerve cells stretched between two surfaces of the integument.

**Chorion.** The outer covering of an insect egg (Figure 4.3).

**Chromosome.** One of the small rod-shaped elements in the nucleus of a cell, which contains genetic information for a cell; a collection of genes.

**Chronic poisoning.** Illness or death from long-time exposure to low levels of a toxicant.

**Chrysalis.** The pupal stage in lepidopterans that do not construct cocoons.

**Class.** A category of the classification scheme that includes one or more orders (for example, Hexapoda).

**Classification.** The ordering of organisms into a hierarchy of categories.

**Clavate.** Club-shaped or enlarged at the tip; clavate antennae (Figure 2.15).

**Coarctate.** Compacted and oval in form, concealed by a hardened, thickened, cylindrical case; coarctate pupa.

**Cocoon.** Case of silk or silk-bound debris in which a pupa develops.

**Codlemone.** The synthetic sex pheromone of the codling moth.

**Coleoptera.** The order of insects comprising the beetles (Figure 3.27).

**Collembola.** The order of insects comprising the springtails (Figure 3.2).

**Collophore.** Extendable tubelike structure found on the ventral side of the first abdominal segment of collembolans (Figure 3.2).

**Comb.** The hexagonal wax structure made by honey bees in which the young are reared and food is stored.

**Community.** The interacting "web" of different populations in a given area (Figure 5.5).

**Compound eye.** A visual sensory organ composed of many individual elements, each of which is represented externally by a hexagonal facet (Figure 2.16).

**Conditional lethal.** An allele that results in mortality of carriers under certain environmental conditions (for example, cold temperatures).

**Conservation.** A biological control practice that includes any activity designed to protect and maintain existing populations of natural enemies.

**Contact poison.** An insecticide that enters the body when the insect walks or crawls over a treated surface; insecticide that is absorbed through the body wall.

**Copulation.** The joining of male and female genital structures (Figure 4.14); mating.

**Cornea.** The outer surface or lens of an ommatidium (Figure 2.37).

**Cornicle.** One of a pair of dorsal tubular structures on the posterior part of the abdomen of aphids.

**Corpus allatum (pl. corpora allata).** One of a pair of neuroendocrine glands just posterior to the brain (Figure 2.41).

**Corpus cardiacum (pl. corpora cardiaca).** One of a pair of neuroendocrine glands adjacent to the corpora allata and posterior to the brain (Figure 2.41).

**Coxa (pl. coxae).** The basal segment of the leg (Figure 2.18).

**Cranium.** The hardened capsule of the head that bears the mouthparts, antennae, and eyes.

**Crawler.** Newly emerged, active immature form of a scale insect.

**Cremaster.** A spine at the tip of the abdomen of a chrysalis.

**Crochets.** Hooks or claws at the tips of prolegs of Lepidoptera larvae.

**Crop.** The expanded portion of the foregut just posterior to the esophagus (Figure 2.23).

**Crop rotation.** Alternating, by growing season, the type of crop planted at a location.

**Cross-resistance.** The phenomenon of insect resistance to one type of insecticide, providing resistance to other insecticides with similar modes of action.

**Crustacea.** The anthropod class that includes the crayfish, lobsters, pillbugs, crabs, and shrimp (Figures 1.3, 1.4); characterized by having two pairs of antennae, one pair of mandibles, two pairs of maxillae, and gills for respiration.

**Crystalline cone.** A transparent body that lies beneath the cornea in an ommatidium (Figure 2.37).

**Cultivar.** A particular cultivated variety of a plant.

**Cultural management (ecological management, cultural control).** Purposeful manipulation of a cropping environment to reduce rates of pest increase and damage.

**Cuneus.** A triangular section of a hemelytron, set off by a suture.

**Cuticle.** The noncellular, multilayered portion of the integument that overlies the epidermis.

**Cuticulin.** The lipoprotein layer of the epicuticle.

**Cyclodiene.** A cyclic chlorinated hydrocarbon insecticide (for example, chlordane).

**Cytoplasmic incompatibility.** When factors in the egg cytoplasm prevent fusion of the sperm and egg nuclei.

**Cytoplasmic resistance.** Host-plant resistance conferred by mutable substances in cell cytoplasm.

**Damage.** A measurable loss of host utility, most often including yield quantity, quality, or aesthetic appeal (Figure 7.1).

**Damage boundary.** The level of injury where damage can be measured (Figure 7.2).

**DDT (dichlorodiphenyltrichloroethane).** A synthetic chlorinated hydrocarbon insecticide.

**DEET (*N,N*-diethyl-*m*-toluamide).** Commonly used biting fly and mosquito repellent.

**Degree day.** An accumulation of heat units above some threshold temperature for a 24-hour period.

**Delayed voltinism.** Life cycle requiring more than 1 year for completion.

**Delusory parasitosis.** Psychological infliction in which sufferers imagine themselves to be infested with parasites.

**Deme.** A subpopulation of interbreeding individuals.

**Density.** The number of individuals per unit of measure (for example, larvae per square meter).

**Density-dependent factor.** A population-regulating factor that changes proportionally in intensity with changes in population density (for example, intraspecific competition).

**Density-independent factor.** A population-regulating factor that causes mortality and is unrelated to a population's density (for example, weather factors).

**Deodorant.** A material added to a pesticide formulation for masking unpleasant odors (for example, cedar oil, pine oil).

**Depolarization.** The reduction of the electrical potential of a neuron's membrane.

**Dermaptera.** The order of insects comprising the earwigs (Figure 3.15).

**Deterrent.** A chemical that prevents feeding or oviposition by insects.

**Deutocerebrum.** The medial pair of lobes of the brain that innervate the antennae (Figure 2.41).

**Developmental threshold.** The minimum temperature (cardinal temperature) required for development to proceed.

**Diagnosis.** The recognition of the presence and extent of a pest infestation and determination of the causal agent responsible.

**Diapause.** A physiological state of arrested metabolism, growth, and development that occurs at one stage in the life cycle.

**Diaphragm.** A thin membrane that partitions areas of the body cavity (Figure 2.27).

**Diffraction.** The splitting of white light into its component spectral colors by a series of fine grooves or ridges.

**Diflubenzuron.** A chitin synthesis inhibitor for insect control.

**Diluent.** A substance used as a carrier for concentrated pesticide; can be liquid or solid (for example, refined oil, organic flour).

**DIMBOA (2,4-dihydroxy-7-methoxyl-1,4-benzoxazine-3-one).** A compound in corn that confers resistance against first-generation European corn borer.

**Dinitrophenol.** A synthetic organic insecticide characterized by nitro groups ($NO_2$) attached to a phenol ($C_6H_5OH$) ring (for example, dinoseb).

**Dioecious.** Having male and female elements on separate individuals of the same species.

**Diploid.** Double; possessing the full complement of maternal and paternal chromosomes.

**Diplopoda.** The arthropod class that includes the millipedes (Figure 1.6); characterized by cylindrical bodies and two pairs of appendages on each abdominal segment.

**Diplura.** One of the apterygote orders of insects (Figure 3.3).

**Diptera.** The order of insects comprising the true flies (Figures 3.43, 3.44).

**Direct loss (direct injury).** Devaluation of a marketable commodity owing to the presence of insects or insect damage; injury to yield-forming organs.

**Dispersal.** The process of scattering in various directions.

**Dispersion.** The spatial arrangement of individuals (Figure 5.1).

**Distal.** Farthest from the point of attachment or origin.

**DNA.** Deoxyribonucleic acid, a compound of deoxyribose, phosphoric acid, and nitrogen bases. A DNA molecule consists of two strands in the shape of a double helix. A gene is a portion of a DNA molecule.

**Dockage.** Value depreciation of a product owing to the presence of insects or other foreign matter.

**Dormancy.** A seasonally recurring period in the life cycle when growth, development, and reproduction are suppressed.

**Dormant oil.** A petroleum oil applied to trees only when foliage is not present.

**Dorsal.** Pertaining to the back or upper side of the body.

**Dorsal vessel.** Primary organ of hemocoel circulation; extends from the abdomen to the head and consists of a posterior heart and anterior aorta (Figure 2.27).

**Dorsoventral.** From the upper to lower surface; dorsoventral muscle (Figures 2.35, 2.36).

**Drone.** A male bee.

**Dust (D).** A dry pesticide formulation prepared by milling the pesticidal compound into a fine powder that is then diluted with a dry material like organic flour.

**Ecdysis.** The process of shedding the old cuticle.

**Ecdysone (molting hormone, prothoracic hormone).** A hormone produced in the prothoracic gland; initiates growth and molting activities of epidermal cells (Figure 4.11).

**Eclosion.** The act of an insect leaving the egg or emergence of the adult insect at the terminal molt.

**Eclosion hormone.** A substance secreted by certain cells in the nervous system that initiates activities associated with ecdysis.

**Ecological backlash.** The counterresponses of pest populations or other biotic factors in the environment that devalue the effectiveness of insect management tactics.

**Ecological management (cultural management, cultural control).** Purposeful manipulation of the cropping environment to reduce rates of pest increase and damage.

**Economic damage.** The amount of pest-induced injury that justifies the cost of applying pest control measures (Figure 7.2).

**Economic-injury level (EIL).** The lowest number of insects that will cause economic damage (Figures 7.3, 7.5, 7.6).

**Economic poison.** Legal classification for a substance used for controlling, preventing, destroying, repelling, or mitigating any pest.

**Economic threshold (ET, action threshold).** The pest density at which management action should be taken to prevent an increasing pest population from reaching the economic-injury level (Figure 7.5).

**Ecosystem.** An assemblage of interacting plant and animal communities and nonliving environmental components (Figure 5.6).

**Ecotype.** A specific strain of a species adapted to a particular set of environmental conditions.

**Ectoderm.** The outermost of the three primitive germ layers of the developing embryo; gives rise to the body wall, foregut, hindgut, nervous system, respiratory system, and many glands (Figure 4.6).

**Ectognathous.** Having mouthparts that are exposed, not enclosed within a cavity.

**Effective environment.** The elements in an ecosystem that have a direct influence on reproduction, survival, and movements of a subject species.

**Ejaculatory duct.** The terminus of the male sperm duct (Figure 2.45).

**Elytron (pl. elytra).** A thickened, hardened forewing (Figure 2.19).

**Embioptera.** The order of insects comprising the webspinners (Figure 3.17).

**Emigration.** Movement out of an area.

**Emulsifiable concentrate (EC, E).** A concentrated pesticide formulation containing organic solvent and detergentlike emulsifier to facilitate emulsification with water.

**Emulsion.** A suspension of microscopic droplets of one liquid in another.

**Encephalitis** (pl. **encephalitides**). An inflammation of the brain or spinal cord; one type due to infection by mosquito-transmitted viruses.

**Endocuticle.** The innermost layer of the cuticle (Figure 2.3).

**Endoderm.** The innermost of the three primitive germ layers of the developing embryo; gives rise to the midgut.

**Endopeptidase.** An enzyme that catalyzes the splitting of a peptide chain.

**Endopterygota.** The division of Pterygota comprising insects with wings that develop internally.

**Energid.** In the fertilized egg, the daughter nuclei that have migrated to the egg periphery and become compartmentalized (Figure 4.4).

**Enhanced microbial degradation.** Unusually rapid breakdown of soil pesticides by microorganisms.

**Entognathous.** Having mouthparts housed within a cavity.

**Entomophobia.** An irrational, pronounced fear of insects.

**Envenomation.** The poisonous effects caused by the bites, stings, or secretions of insects and other arthropods.

**Environmental fate.** How a substance (usually a pesticide) behaves in the environment, and what becomes of it and its degradation products in the environment.

**Environmental resistance.** The physical and biological restraints that prevent a species from realizing its biotic potential.

**Enzyme.** A protein that regulates the rate of chemical reactions.

**Ephemeroptera.** The order of insects comprising the mayflies (Figure 3.6).

**Epicuticle.** The outermost layers of the cuticle (Figure 2.3).

**Epideictic pheromone (spacing pheromone).** An intraspecific chemical messenger that elicits dispersal from potentially crowded food sources.

**Epidermis.** The cellular layer of the integument that underlies and secretes the cuticle.

**Epipharynx (labrum-epipharynx).** A fleshy mouthpart structure on the inner surface of the labrum or clypeus.

**Epiproct.** The dorsal part of the terminal abdominal segment that often appears as a process situated above the anus (Figure 2.20).

**Epithelial cell.** A cell that covers a body structure or lines a body cavity.

**Eradication.** Complete and total elimination of a group of organisms from an area.

**Esophagus.** The narrow portion of the foregut, between the pharynx and crop (Figure 2.23).

**Ester.** An acid derivative in which the hydrogen of the acid has been replaced by a hydrocarbon group.

**Eversible.** Capable of being turned outward.

**Exocuticle.** The middle layer of the cuticle (Figure 2.3).

**Exopeptidase.** An enzyme that catalyzes the splitting of terminal peptides from a peptide chain.

**Exopterygota.** The division of Pterygota comprising insects with wings that develop externally.

**Exoskeleton.** A skeleton or supporting structure on the outside of a body.

**Exotic.** Introduced from another country or continent.

**Exuviae** (sing. **exuvium**). The cast-off cuticles of an arthropod.

**Facet.** The hexagonal external surface of an individual compound eye unit.

**Fallow.** Land left unseeded and usually clean cultivated during a growing season.

**Family.** A classification category that includes a number of genera sharing one or a number of characteristics and ending in the letters "idae" (for example, Crambidae).

**Farnesol.** An alcohol, $C_{15}H_{25}OH$, with juvenile hormone activity in insects.

**Fat body.** A meshwork of loosely organized cells in the body that serves as a storage depot and functions in intermediate metabolism.

**Fecundity.** The rate at which females produce ova.

**Femur (pl. femora).** The third segment of the leg, between the coxa and the tibia (Figure 2.18).

**Feral.** Existing in the wild; wild.

**Fertility.** The rate at which fertilized eggs (zygotes) are produced.

**Fidelity.** The accuracy with which a population estimate follows actual numbers in a population.

**Filariasis.** Infection with parasitic nematodes, often transmitted by mosquitoes.

**Filiform.** Threadlike; filiform antennae (Figure 2.15).

**Flagellum (pl. flagella).** The antennal segments beyond the scape and pedicel.

**Flaxseed stage.** The puparium of the Hessian fly.

**Flowable (F, L).** A pesticide formulation in which the active ingredient is wet-milled with a clay diluent and water; has a puddinglike consistency and can be mixed with water for spraying.

**Food Quality Protection Act (FQPA).** An act signed into law in 1996 that amends both the Federal Insecticide, Fungicide, and Rodenticide Act and the Federal Food, Drug, and Cosmetic Act. Its purpose was to improve food safety in the United States.

**Forensic entomology.** The employment of entomological information in assessing time, location, or cause of a human death or a criminal act.

**Formamidine.** A synthetic organic amine insecticide; a recent type of insecticide that appears to mimic some insect neurotransmitters (for example, chlordimeform).

**Formulation (pesticide formulation, insecticide formulation).** A mixture of active and inert ingredients for killing pests.

**Frass.** Plant fragments, usually of wood boring insects, mixed with insect excrement.

**Freezing tolerance.** Survival despite freezing of body fluids.

**Frontal ganglion.** The median ganglion of the stomodaeal nervous system; lies above the esophagus.

**Fumigant.** A volatile insecticide that enters an insect via the tracheal system.

**Fungicide.** A substance that kills fungi.

**Furcula.** A forked abdominal structure used in springing or jumping of Collembola (Figure 3.2).

**Gain threshold.** The beginning point of economic damage, expressed in amount of harvestable produce; when cost of suppressing insect injury equals money to be gained from avoiding the damage (Figure 7.2).

**Gall.** An abnormal growth of some plant tissue, often induced by the presence or activities of insects or mites.

**Ganglion (pl. ganglia).** A bundle of nervous tissue (Figure 2.40).

**Gastric caecum (pl. gastric caeca).** A saclike outpocketing of the anterior portion of the midgut (Figure 2.23).

**Gastrula.** An embryo in the process of gastrulation, when a single-layered blastula turns into a three-layered embryo consisting of ectoderm, mesoderm, and endoderm (Figure 4.6).

**Gene.** A piece of a DNA molecule; a portion of a chromosome that contains the hereditary information for the production of a protein.

**Gene-for-gene relationship.** The concept of a resistant (or susceptible) allele at a gene locus in a particular plant cultivar corresponding to a susceptible (or virulent) allele at the same locus in the pest; a one-to-one relationship between host and pest genotypes.

**Gene splicing.** Adding genes to a DNA molecule.

**General equilibrium position (GEP).** A population's long-term average density.

**Generation.** A cohort of offspring from a parent population moving through a life cycle together.

**Genetic control.** Altering the genetic makeup of organisms to inhibit their reproduction and survival.

**Genetic engineering (gene splicing, recombinant DNA technology).** The technique of removing, modifying, or adding genes to a DNA molecule.

**Genetic transformation.** A change in the genetic structure of an organism via genetic engineering.

**Genetically modified organism (GMO).** An organism that has been modified by the application of recombinant DNA technology.

**Geniculate.** Elbowed or abruptly bent, as with antennae.

**Genus** (pl. **genera**). An assemblage of evolutionarily related species sharing a number of characteristics.

**Geographical Information System (GIS).** A computer-based system for gathering, storing, analyzing, and displaying spatial information, for example, insect densities across a field.

**Geomagnetic receptor.** A sense organ involved in the perception of magnetic fields.

**Germ band (germ disk).** A thickened area of the blastoderm that becomes the embryo (Figure 4.5).

**Gill.** A respiratory organ in the aquatic immature stage of many insects that obtains dissolved oxygen from water (Figure 2.25).

**Global Positioning System (GPS).** A technology that uses receivers to read radio signals transmitted from aerial satellites to fix a position on the earth's surface. Characteristics of these fixed positions can serve as inputs in a GIS for decision making.

**Graft.** A plant bud, shoot, or scion that is inserted in a slit or groove in the stem or stock of another plant, where it continues to grow.

**Granular (G).** A type of pesticide formulation prepared by applying liquid insecticide to coarse particles of porous material like clay, corncobs, or walnut shells.

**Growth form (population growth form).** The shape of a population density curve (Figure 5.4).

**Gustation.** The sense of taste.

**Haltere** (pl. **halteres**). A metathoracic, knoblike guidance organ in Diptera (Figure 2.19).

**Hamuli.** Small hooks used to couple together fore- and hindwings in Hymenoptera.

**Haplodiploidy (facultative parthenogenesis).** Reproduction via a combination of sexual and asexual modes resulting in both haploid offspring (from unfertilized eggs) and diploid offspring (from fertilized eggs).

**Haploid.** Having only a single (maternal) complement of chromosomes.

**Hazard.** The danger that injury will occur with the use of a particular pesticide; depends on both toxicity and exposure.

**Head.** The most anterior of the three main body divisions of insects; bears the mouthparts and antennae and houses the brain.

**Heart.** The posterior portion of the dorsal blood vessel (Figure 2.27).

**Hellgrammite.** A larva of a dobsonfly.

**Hemelytron** (pl. **hemelytra**). The forewing of hemipterans, the basal portion of which is thickened and the apical portion membranous.

**Hemiptera.** The order of insects comprising the true bugs, cicadas, leafhoppers, aphids, scale insects, and others (Figure 3.25).

**Hemocoel.** The body cavity in which the blood, or hemolymph, flows.

**Hemocyte.** A cell of the hemolymph.

**Hemolymph.** The blood or lymphlike fluid of invertebrates, including insects.

**Herbicide.** A substance that kills weeds.

**Heterocyclic.** Containing more than one type of atom joined in a ring.

**Heteroptera.** The suborder of insects comprising the true bugs (Figure 3.24).

**Hibernation.** Winter dormancy.

**Histogenesis.** The formation of new body tissues.

**Histolysis.** A breakdown of body tissues.

**Homogeneous.** Uniform in structure or composition.

**Honeydew.** A sugary liquid exuded through the anus of many Hemiptera.

**Hopperburn.** A leaf necrosis resulting from the feeding of leafhoppers.

**Horizontal resistance.** Host-plant resistance to a broad range of pest genotypes (Figure 13.8).

**Hormoligosis.** Reproductive stimulation by sublethal doses of pesticides.

**Host.** An organism in or on which a parasite lives; a plant on which an insect feeds.

**Host escape.** A type of apparent host-plant resistance in which an unexplained lack of infestation on a susceptible plant occurs in a population of otherwise-infested plants.

**Host evasion.** A type of apparent host-plant resistance in which the plant passes through a susceptible stage quickly or at a time such that its exposure to potentially injurious pests is reduced.

**Host-plant resistance.** The relative amount of heritable qualities possessed by a plant that reduces the degree of damage done to the plant by a pest or pests.

**Hexapoda.** The class of arthropods comprising the insects; characterized by three body regions (head, thorax, and abdomen), three pairs of thoracic legs, and, usually, one pair of antennae.

**Hydroprene.** A juvenile hormone analog.

**Hygroreceptor.** A sense organ involved in the perception of moisture.

**Hymenoptera.** The order of insects comprising bees, wasps, ants, and sawflies (Figures 3.46–3.53).

**Hypermetamorphosis.** A type of development in which there are two or more distinct forms of larvae (Figure 4.22).

**Hyperparasite.** A parasite whose host is another parasite.

**Hypognathus.** Head positioned vertically and mouthparts pointed downward (Figure 2.6).

**Hypopharynx.** A median, tonguelike mouthpart structure anterior to the labium; associated with the ducts from the salivary glands (Figure 2.7).

**Imaginal disk.** Groups of cells set aside in the embryo and maintained through larval development, as sites from which adult structures arise (Figure 4.21).

**Immigration.** Movement into an area.

**Immunity.** Total and complete resistance to a pest or pests.

**Imperfectly density-dependent factor.** An agent that is usually density dependent but whose action sometimes fails; includes parasites, predators, and pathogens.

**Indigenous.** Native to an area.

**Indirect loss (indirect injury).** Decline in quality or quantity of a marketable commodity (for example, seeds) owing to the feeding of insects on nonmarketable portions (sometimes roots, stems, leaves) of the plant; damage resulting from injury to nonyield-forming organs.

**Induced resistance.** A type of apparent host-plant resistance in which a particular plant condition or environmental state (for example, nutrient level, soil moisture) makes a plant more resistant to pests than under other circumstances.

**Inert ingredient.** Auxiliary ingredient in a pesticide formulation that has no direct effect on pests.

**Injury.** The effects of pest activities on host physiology that are usually deleterious (Figure 7.1).

**Inoculative release.** Release of natural enemies that are expected to colonize and spread throughout an area naturally.

**Inorganic.** Lacking carbon atoms.

**Insect equivalent.** The amount of injury that could be produced by an individual pest through its complete life cycle.

**Insect growth regulator (IGR, biorational, third-generation insecticide).** Substance effective in upsetting or modifying normal insect growth processes.

**Insecticide.** A substance, usually a chemical, that kills insects.

**Insemination.** Introduction of sperm into the female reproductive tract.

**Instar.** The insect between molts (for example, sixth instar).

**Integrated control.** Control of pests that emphasizes selective use of insecticides so as to conserve natural enemies in the agroecosystem.

**Integrated Pest Management (IPM).** A comprehensive pest technology that uses combined means to reduce the status of pests to tolerable levels while maintaining a quality environment.

**Integument.** The outer covering of an insect body that includes the epidermis and the cuticle (Figure 2.3).

**Interference coloration.** Coloration resulting from the reflection of light from a series of superimposed surfaces separated by distances comparable with the wavelengths of light (Figure 2.4).

**Interneuron.** Neuron that serves as a connection between sensory and motor neurons (Figure 2.42).

**Interspecific.** Between two or more species.

**Intestine.** The anterior portion of the hindgut (Figure 2.23).

**Intima.** The cuticular lining of the tracheae (Figure 2.29), foregut, and hindgut.

**Intraspecific.** Within a species.

**Intraspecific competition.** Situation in which individuals within a population compete for a limited resource.

**Introduction (importation).** A biological control practice involving identification of the natural enemies that regulate a pest in its original location and introduction of these into the pest's new location.

**Inundative release.** Widespread distribution of natural enemies for the purpose of suppressing a pest population, with little or no impact expected from the progeny of the released individuals.

**Invertase.** An enzyme that catalyzes the breakdown of sucrose into fructose and glucose units.

**Isolines.** Varieties of a crop plant that differ in only a single gene.

**Isoptera.** The order of insects comprising the termites (Figure 3.16).

**Johnston's organ.** A chordotonal organ located in the pedicel of the antenna.

**Juvabione.** A juvenile hormone analog.

**Juvenile.** The immature stages in primitively wingless insects with no metamorphosis (Figure 4.16).

**Juvenile hormone (JH).** Hormone produced by the corpora allata; serves several functions including the suppression of adult characteristics.

**Kairomone.** An interspecific messenger substance that benefits the receiver but not the releaser (for example, attractants, excitants).

**Keratolytic.** Pertaining to or promoting the loosening or separation of the outer layer of the skin.

**Key.** A tabular arrangement of species, genera, orders, or other classification categories according to characters and traits that serve to identify them (Table 3.1).

**Key pest.** A perennial, severe pest that causes serious and difficult crop production problems; a pest that dominates cultural activities.

**Kinoprene.** A juvenile hormone analog.

***K* strategist.** A species characterized by low reproductive rate and high survival rate (Figure 5.11).

**Label.** The printed information on, or attached to, a pesticide container; federal regulations stipulate the exact information to be included on it (Figures 11.11, 11.12).

**Labium.** The lower lip; the fused pair of mouthpart structures just posterior to the maxillae (Figures 2.5, 2.7).

**Labrum.** The upper lip, lying below the clypeus and in front of the mandibles (Figures 2.5, 2.7).

**Lamella (pl. lamellae).** One in a series of layers or plates.

**Lamellate.** Platelike; lamellate antennae (Figure 2.15).

**Larva** (pl. **larvae**). The immature stage between egg and pupa in insects with complete metamorphosis (Figures 4.19, 4.20).

**Lateral.** Pertaining to the side of the body.

**$LC_{50}$.** The concentration of a toxicant in some medium (air, water, soil) that will kill 50 percent of the test organisms exposed; generally expressed as mg or $cm^3$ per animal or as parts per million or billion in the medium.

**$LD_{50}$.** The dose of a toxicant that will kill 50 percent of the test organisms to which it is administered; generally expressed as milligrams of toxicant per kilogram of body weight (Figure 11.7).

**Lepidoptera.** The order of insects comprising the butterflies, skippers, and moths (Figures 3.33–3.41).

**Life cycle.** The chain or sequence of events that occurs during a lifetime of an individual organism (Figure 4.1).

**Life system.** A subdivision of an ecosystem; the subject species and its effective environment (Figure 5.6).

**Life table.** A tabulation that accounts for age-specific deaths in a population.

**Lipase.** Any enzyme that catalyzes the splitting of fats into glycerol and fatty acids.

**Longitudinal.** Lengthwise in the body or an appendage (for example, dorsal longitudinal muscle) (Figures 2.35, 2.36).

**Malaise trap.** A tentlike sampling device that intercepts and captures flying insects (Figure 6.19).

**Malaria.** An infectious, fever-inducing disease caused by protozoans of the genus *Plasmodium* and transmitted by *Anopheles* mosquitoes.

**Malpighian tubules.** Excretory organs that arise near the anterior end of the hindgut and extend into the body cavity (Figure 2.23).

**Maltase.** An enzyme that catalyzes the breakdown of the disaccharide maltose into glucose molecules.

**Mandible.** The jaw; one of a pair of mouthpart structures located immediately posterior to the labrum (Figures 2.5, 2.7).

**Marker (genetic).** A distinguishing feature that can be used to identify a particular gene location on a chromosome. Markers may be morphological (e.g., a physical trait such as growth habit , leaf form, color, flower shape), biochemical (e.g., isozymes, enzymes, or other proteins), or molecular (DNA fragment size or specific DNA sequence).

**Marker assisted breeding.** Use of DNA markers to increase the efficiency of selection in a population.

**Market value.** The amount of money that a seller can expect to obtain for a commodity.

**Maxilla** (pl. **maxillae**). One of the paired mouthpart structures just posterior to the mandibles (Figures 2.5, 2.7).

**Mean.** The arithmetic average of a group of samples.

**Mechanoreceptor.** A sense organ involved in the perception of tactile stimuli and/or vibration (Figure 2.39).

**Mecoptera.** The order of insects comprising the scorpionflies (Figure 3.30).

**Median caudal filament.** In Thysanura, a threadlike process at the posterior end of the abdomen that lies between the cerci (Figure 3.4).

**Medical pest.** A pest that causes a pathological condition or transmits pathogenic organisms to humans or animals.

**Meiosis.** The two successive divisions of a cell's nucleus in which the chromosome number is reduced by one-half.

**Meiotic drive.** The unequal recovery of homologous chromosomes during meiosis.

**Melanin.** A pigment that produces black, amber, and dark brown colors in the cuticle.

**Mesenteron.** The midgut; primary site of digestion and absorption (Figure 2.24).

**Mesoderm.** One of the three primitive germ layers in the developing embryo; gives rise to muscles, fat body, heart, blood cells, and reproductive organs (Figure 4.6).

**Mesothorax.** The second of the three segments of the thorax (Figure 2.17).

**Metamorphosis.** A change in body form during development.

**Metathorax.** The third, or posterior, of the three segments of the thorax.

**Methoprene.** A juvenile hormone analog.

**Microbial insecticide.** A biological preparation of viruses or microorganisms or their products, applied and used in ways similar to conventional chemical insecticides.

**Microcoryphia.** The order of insects comprising the jumping bristletails (Figure 3.5).

**Microencapsulation.** A pesticide formulation technique in which pesticide-containing microscopic spheres or capsules are created to permit pesticide release at a slow, steady, and effective rate.

**Microorganism (microbe).** An organism that can be seen only with the aid of a microscope.

**Micropropagation.** The mass production of clonal copies of a donor or parent plant by tissue culture techniques. The initial steps in micropropagation take place in synthetic solid or liquid growth media. Some processes can be conducted on a mass scale in culture tanks.

**Micropyle.** A pore in the egg through which sperm may enter (Figure 4.3).

**Migration.** Movements over great distances, during which locomotion is not inhibited by food, mates, or oviposition sites; an adaptation to periodically transport insects beyond the boundaries of their old reproductive sites and into new ones.

**Milky disease.** A bacterial disease of Japanese beetle caused by *Bacillus popilliae* and *Bacillus lentimorbis* (Figure 12.1).

**Mitochondrion (pl. mitochondria).** An organelle that is the principal site of energy synthesis in the cell.

**Mixed-function oxidase (MFO).** An enzyme that metabolizes many foreign substances.

**Mode of action.** Means by which a toxin affects the anatomy, physiology, or biochemistry of an organism.

**Molt.** To shed the old cuticle.

**Molting fluid.** A mixture of enzymes produced by the epidermal cells capable of digesting the endocuticle.

**Moniliform.** (See Figure 2.15). Beadlike antennal type.

**Monogamy.** Mating with a single individual.

**Monogenic resistance.** Resistance resulting through expression of a single gene.

**Monophagus.** Feeding on one type of food.

**Mortality.** Death rate or numbers dying per unit of time.

**Multivoltine.** Having more than one generation per year (Figure 4.25).

**Mutagen.** A material that induces genetic changes.

**Myofibril.** A contractile element of a muscle fiber.

**Naiad.** The immature stage in insects with incomplete metamorphosis; typically feed and develop in water (Figure 4.18).

**Narcotic.** A chemical that causes reversible depression of the nervous system.

**Nasute (pl. nasuti).** In some termites, a type of soldier that has a snout through which a defensive fluid may be ejected.

**Natality.** Rate of birth; often measured as total number of eggs or eggs per female per unit of time (Figure 5.2).

**Natural control.** The suppression of pest populations by naturally occurring biological and other environmental agents.

**Natural enemies.** Living organisms found in nature that kill insects outright, weaken them, or reduce their reproductive potential.

**Negative binomial model.** A mathematical model used to describe a clumped, or contagious, dispersion of organisms.

**Nematicide.** A substance that kills nematodes.

**Neoptera.** The infraclass of Pterygota comprising insects capable of wing flexion.

**Neuron.** A nerve cell; the basic unit of nerve impulse transmission.

**Neuroptera.** The order of insects comprising the lacewings, antlions, and dobsonflies (Figure 3.26).

**Neurotoxin.** A substance that is poisonous or destructive to nerve tissues.

**Neurotransmitter.** A substance released from an axon terminal that either excites or inhibits a target cell.

**Nicotine.** A botanical alkaloid insecticide derived from the tobacco plant.

**Nit.** The egg of a louse, often attached to a feather or hair.

**Nomenclature.** The naming of organisms.

**Nonpreference (antixenosis).** Plant characteristics that lead insects away from a particular host; involves activities of both host plant and insect.

**Notum (pl. nota).** A dorsal surface of a thoracic segment.

**Nymph.** The immature stage in insects with gradual metamorphosis (Figure 4.17).

**Occasional pest.** A pest with a general equilibrium position substantially below the economic-injury level; highest pest population fluctuations occasionally and sporadically exceed the economic-injury level (Figure 8.11).

**Ocellus (pl. ocelli).** The small, simple eye of an insect or other arthropod.

**Odonata.** The order of insects comprising the dragonflies and damselflies (Figure 3.7).

**Olfaction.** The sense of smell.

**Oligogenic resistance (major-gene resistance).** Host-plant resistance conferred by one or a few genes.

**Oligophagus.** Feeding on a few types of food.

**Ommatidium** (pl. **ommatidia**). An individual sensory unit of a compound eye.

**Omnivorous.** Feeding on a variety of foods.

**Ootheca** (pl. **oothecae**). A protective pod produced by some female insects that contains several to many eggs.

**Opisthognathous.** Head positioned obliquely and mouthparts pointed backward (Figure 2.6).

**Order.** A category of classification; one of the principal divisions of the class Hexapoda, based largely on wing structure and usually ending in the letters "ptera" (for example, Lepidoptera).

**Organic.** Possessing carbon atoms.

**Organophosphate.** An organic compound containing phosphorus; includes many important insecticides (for example, malathion).

**Organosulfur.** A sulfur-containing organic compound used as an acaricide (for example, tetradifon).

**Organotin.** A tin-containing organic compound used as an acaricide (for example, cyhexatin) and/or fungicide.

**Orthoptera.** The order of insects comprising the grasshoppers, crickets, walkingsticks, mantids, and cockroaches (Figures 3.8–3.14).

**Ostium** (pl. **ostia**). A slitlike opening in the heart (Figure 2.27).

**Ovariole.** A tubular division of an ovary (Figure 2.44).

**Ovary.** The egg-producing organ (Figure 2.44).

**Oviduct.** The tube leading away from the ovary through which eggs pass (Figure 2.44).

**Oviparous.** Egg-laying.

**Oviposition.** The act of egg laying.

**Ovipositor.** The egg-laying structure of female insects (Figure 2.21).

**Ovoviviparous.** Producing active young immediately following the hatching of the eggs within the female.

**Paedogenesis.** Reproduction by the juvenile form.

**Paleoptera.** The infraclass of Pterygota comprising insects incapable of wing flexion.

**Palpus (pl. palpi).** A sensory organ associated with mouthpart structures (Figures 2.5, 2.7).

**Paraproct.** One of a pair of lobes adjacent to the anus (Figure 2.20).

**Parasite.** An organism that lives in or on the body of another organism (the host) during some portion of its life cycle.

**Parasitoid.** An arthropod that parasitizes and kills an arthropod host; is parasitic in its immature stages but is free-living as an adult (Figure 9.5).

**Parthenogenesis.** Reproduction by the development of an egg that has not been fertilized with sperm.

**Pasture spelling.** Leaving a pasture idle for a period of time by removing livestock from it.

**Pathogen.** A disease-causing organism; usually a microorganism.

**Pectinate.** With processes like the teeth of a comb; pectinate antennae (Figure 2.15).

**Pedicel.** The second segment of the antenna; the base of the ovariole.

**Penis (pl. penes).** The copulatory organ of the male.

**Perennial pest.** A pest with a general equilibrium position below but close to the economic-injury level; highest pest population fluctuations regularly reach the economic-injury level (Figure 8.12).

**Perfectly density-dependent factor.** A factor that never fails to control the increase in a population's numbers (i.e., intraspecific competition).

**Peritrophic membrane.** A chitinous membrane secreted by cells lining the midgut; forms a semipermeable envelope around the food (Figure 2.24).

**Pest.** A species that interferes with human activities, property, or health or is objectionable.

**Pest control.** The application of technology, in the context of biological knowledge, to achieve satisfactory reduction of pest numbers or effects.

**Pesticide.** Any substance for controlling, preventing, destroying, disabling, or repelling a pest.

**Pest management (integrated pest management, IPM).** A comprehensive approach to dealing with pests that strives to reduce pest status to tolerable levels by using methods that are effective, economically sound, and ecologically compatible; most often involving multiple tactics; usually a practice of individuals and private enterprises.

**Pest management strategy.** An overall plan for eliminating or alleviating a pest problem.

**Pest management tactic.** A method of implementing a pest management strategy.

**Pest status.** The ranking of a pest relative to the economics of dealing with the species (Figure 1.30).

**Pest technology.** Approaches to eliminate or reduce pest numbers and/or to prevent losses from undesirable species.

**Petiole.** The waistlike constriction in apocritan hymenopterans.

**Phagocyte.** A blood cell capable of degrading tissues and foreign matter.

**Pharynx.** The anterior portion of the foregut, between the mouth and the esophagus (Figure 2.23).

**Phasic.** Characterized by short-duration stimulation; phasic receptors.

**Phenology.** The periodicity of biological phenomena.

**Phenotype.** The sum total of the observable or measurable characteristics of an organism without reference to its genetic nature.

**Phenyl.** An organic radical, $C_6H_5$, derived from benzene via the removal of a hydrogen atom.

**Pheromone.** A substance secreted by one animal that causes a specific reaction upon reception by another animal of the same species (for example, sex pheromone).

**Photoperiod.** Day length.

**Photoreceptor.** A sense organ involved in the perception of light.

**Phthiraptera.** The order of insects comprising the chewing lice and sucking lice (Figure 3.21, 3.22 respectively).

**Phylum (pl. phyla).** One of the major divisions of the animal kingdom.

**Physiological resistance.** The ability of an insect strain to tolerate insecticides or other factors through physiological change (for example, via modification of cuticle structure).

**Phytoalexins.** Phenolic compounds produced by plants when they become diseased or injured by insects; enable plants, once damaged, to resist further attack by insects.

**Phytophagous.** Feeding on plants.

**Phytotoxic.** Injures or kills plants.

**Piroplasmosis.** Infection with tick-transmitted piroplasma protozoans, *Babesia* species, that attack the red blood cells of cattle, dogs, and other animals.

**Pitfall trap.** A sampling device designed to capture insects moving over the ground (Figure 6.25).

**Placoid.** Platelike; placoid sensilla.

**Plague (bubonic plague, black death).** An infectious disease caused by the bacterium *Yersinia pestis* and spread via the oriental rat flea *Xenopsylla cheopis.*

**Plasma.** The noncellular fluid portion of the hemolymph.

**Plasmid.** A piece of circular DNA found outside the chromosome in bacteria; principal tool for inserting new genetic information into microorganisms or plants.

**Plasmodium** (pl. **plasmodia**). The protozoans that may cause malaria.

**Plastron.** A film of air over the body of certain aquatic insects that serves as a physical gill for respiration.

**Plecoptera.** The order of insects comprising the stoneflies (Figure 3.18).

**Pleuron (pl. pleura).** A lateral plate of a thoracic segment.

**Plumose.** Featherlike; plumose antenna (Figure 2.15).

**Poikilothermic.** Having body temperatures that fluctuate with environmental temperatures; "cold-blooded."

**Poisson model.** A mathematical model that describes a random dispersion of organisms.

**Polychloroterpene.** A chlorinated hydrocarbon insecticide prepared from camphene, a pine tree derivative (for example, toxaphene).

**Polyembryony.** More than one embryo formed in a single egg.

**Polygenic resistance (minor-gene resistance).** Host-plant resistance conferred by many genes.

**Polyphagous.** Feeding on many types of food.

**Population.** A group of individuals of the same species within given space and time constraints.

**Population dynamics.** The interplay between populations and the environmental forces that influence them.

**Population index.** An estimate of population size based on insect effects or products.

**Population intensity estimate.** A type of absolute estimate of population size expressed as number per unit habitat (for example, number per leaf).

**Pore canal.** A small channel involved in the transport of materials from the epidermis to the body surface (Figure 2.3).

**Postdiapause quiescence.** A transitional period of dormancy, following diapause, that is not sensitive to token stimuli.

**Potential natality.** The reproductive rate of individuals in an optimal environment.

**Precision.** A measure of the degree of error in making estimates of a population's size.

**Precision farming.** See **site-specific farming.**

**Precocene.** An antijuvenile hormone isolated from the plant *Ageratum houstonianum.*

**Predator.** An organism that attacks and feeds on other animals (usually smaller or less powerful than itself) and consumes more than one animal in its lifetime.

**Prepupa.** The last larval stage.

**Prescription.** Recommendation of a strategy or course of action.

**Preventive practice (preventive pest management).** A pest management approach that attempts to prevent problems before they become economically important; in pest management, it involves use of other-than-traditional pesticides (Figures 16.1, 16.2).

**Prey.** An animal caught for food.

**Primary metabolite.** A metabolic product necessary for growth and reproduction of a plant.

**Proboscis.** The extended beaklike mouthparts of sucking insects.

**Proctodaeum.** The hindgut; plays important roles in elimination and water and salt resorption.

**Prognathous.** Head positioned horizontally and mouthparts pointed forward (Figure 2.6).

**Prognosis.** Anticipation of pest problems before they occur and prediction of the outcome of these if left unattended.

**Proleg.** A fleshy, unsegmented abdominal walking appendage of some insect larvae.

**Pronotum (pl. pronota).** The dorsal surface of the prothorax.

**Proprioreceptor.** A mechanoreceptor that provides information about an insect's position in its environment and/or the position of the insect's body parts.

**Protein.** A chain of amino acids that makes up cell structure and controls cell function.

**Proteolysis.** The splitting of proteins into smaller components by breaking peptide bonds.

**Prothoracic gland.** An endocrine organ in the prothorax of immature insects that secretes ecdysone (Figure 4.11).

**Prothorax.** The first, or anterior, of the three segments of the thorax.

**Protocerebrum.** The anterior pair of lobes of the brain that innervate the compound eyes and ocelli (Figure 2.41).

**Protura.** The order of insects comprising the telsontails or proturans (Figure 3.1).

**Proventriculus.** The most posterior portion of the foregut (Figure 2.23).

**Proximal.** Near the point of attachment or origin.

**Psocoptera.** The order of insects comprising the booklice and barklice (Figure 3.20).

**Pterygota.** The subclass of Hexapoda comprising the winged and secondarily wingless insects.

**Pulvillus (pl. pulvilli).** A pad at the base of each tarsal claw (Figure 2.18).

**Pupa (pl. pupae).** The stage between the larva and adult in insects with complete metamorphosis (Figure 4.19).

**Puparium.** A protective enclosure for the pupa of some Diptera formed from the last larval cuticle.

**Pyloric valve.** The valve at the junction of the midgut and hindgut.

**Pyrethroid (synthetic pyrethroid).** An organic synthetic insecticide with a structure based on that of pyrethrum (for example, allethrin).

**Pyrethrum.** A botanical insecticide derived from *Chrysanthemum* flowers.

**Quarantine.** A legal instrument to prevent or slow the spread of pests from infested into uninfested areas by restricting movement of infested stock.

**Quiescence (torpor).** Inactivity induced by unfavorable environmental conditions; activity resumes immediately upon the return of favorable conditions.

**Quinone.** A benzene derivative in which two hydrogen atoms are replaced by two oxygen atoms. Quinones are utilized in the hardening and darkening of the insect cuticle.

**Range.** The difference between the smallest number and the largest number in a sample.

**Receptor fibril.** The portion of a neuron that transmits impulses toward the cell body.

**Recombinant DNA technology (genetic engineering, gene splicing).** Technique of isolating DNA molecules and inserting them into the DNA of a cell.

**Rectum.** The posterior portion of the hindgut (Figure 2.23).

**Reflex.** An involuntary movement or other response elicited by a stimulus applied peripherally, conducted through the central nervous system, and conducted back peripherally.

**Regression.** The amount of change in one variable associated with a unit change in another variable (Figure 7.4).

**Regulation (population regulation).** The natural phenomenon of population numbers fluctuating within certain bounds for long periods of time, thus keeping the population in existence.

**Relapsing fever.** A disease caused by the *Borrelia* species of bacteria and transmitted by body lice.

**Relative population estimate.** An estimate of population size based on the kind of sampling technique used (for example, number per trap).

**Relative variation (RV).** Percentage standard error of a mean; used as a measure of sampling precision.

**Repellent.** A chemical that causes insects to orient their movements away from a source.

**Replacement (pest replacement, secondary pest outbreak).** When a major pest is suppressed and continues to be suppressed by a tactic but is replaced in importance by another pest, previously with minor status (Figures 17.8, 17.10).

**Resistance.** The ability of a strain to tolerate or avoid factors that would prove lethal or reproductively degrading to the majority of strains in a normal population (Figures 17.1, 17.2).

**Restricted-use pesticide.** A pesticide that may be applied only by applicators certified by the state in which they work.

**Restriction enzymes (restriction endonucleases).** Enzymes used to "cut" a gene from a piece of DNA.

**Resurgence (pest resurgence).** A situation in which a population, after having been suppressed, rebounds to numbers higher than before suppression occurred (Figures 17.7, 17.9).

**Retinula cells.** One of a group of cells comprising the receptor unit of an ommatidium (Figure 2.37).

**Rhabdom.** A rodlike structure formed by the inner surfaces of adjacent retinula cells in an ommatidium (Figure 2.37).

**Rickettsia.** Obligate intracellular parasitic bacterialike microorganisms that infect arthropods and also may be pathogenic in vertebrates.

**Risk.** The probability that a pesticide will cause harm.

**RNA.** Ribonucleic acid, a messenger molecule that contains instructions for protein synthesis.

**Rotenone.** A botanical insecticide extracted from the roots of the *Derris* and *Lonchocarpus* species of plants.

**Royal jelly.** A glandular secretion of worker bees on which larvae are fed.

***r* strategist.** A species characterized by high reproductive rate and low survival rate (Figure 5.10).

**Ryania.** A botanical alkaloid insecticide extracted from the stem and roots of the shrub *Ryania speciosa.*

**Sabadilla.** A botanical insecticide extracted from the seeds of *Schoenocaulon officinale.*

**Saprophagus.** Feeding on dead and decaying animals or plants.

**Sarcoplasm.** The cytoplasm of striated muscle fibers.

**Scabies.** A contagious skin disorder caused by burrowing of the itch mite *Sarcoptes scabiei.*

**Scape.** The basal segment of the antenna.

**Scavenger.** An animal that feeds on dead and decaying animals or plants or on animal wastes.

**Sclerite.** A hardened integumentary plate surrounded by membranous areas or sutures.

**Sclerotin.** A protein that contributes to the hardness of the cuticle.

**Sclerotization.** The hardening of the body wall.

**Scutellum.** A sclerite of the thoracic notum; appears as a triangular plate between the base of the wings in Hemiptera.

**Seasonal cycle.** The sequence of insect life cycles of a species that occurs over a 1-year period (Figure 4.1).

**Secondary metabolite.** A plant-produced chemical that is nonessential in primary metabolism but may play a role in defense against herbivory.

**Secondary pest.** Pest species that are usually present at low levels and are held in check by the action of natural enemies; can assume full-pest status when natural enemies are destroyed by a pest management tactic (for example, insecticide application).

**Segment.** A subdivision of the body or of an appendage with its own musculature.

**Semen.** Fluid discharged at ejaculation by the male, consisting of spermatozoa and secretions of glands associated with the genital tract.

**Seminal vesicle.** A saclike storage structure that holds seminal fluid of the male prior to discharge.

**Semiochemical.** Any chemical involved in communications among organisms (Figure 13.3).

**Sensillum** (pl. **sensilla**). A simple sense organ.

**Septum** (pl. **septa**). A wall or partition.

**Sericulture.** The propagation of silkworms for silk production.

**Serrate.** (See Figure 2.15). Sawlike antennal type.

**Seta** (pl. **setae**). A bristle or hairlike process.

**Severe pest.** A pest with a general equilibrium position above the economic-injury level, which makes the pest a constant problem (Figure 8.13).

**Sex pheromone.** An intraspecific chemical messenger emitted by one sex that attracts the opposite sex.

**Sex ratio.** Within a population, the proportion of males compared with that of females.

**Signal word.** A word required on every pesticide label to denote the relative toxicity of the material (for example, "danger–poison" for highly toxic compounds).

**Silviculture.** The care and cultivation of forest trees.

**Sinus.** A partitioned chamber of the body cavity.

**Siphonaptera.** The order of insects comprising the fleas (Figure 3.45).

**Site-specific farming (precision farming, precision agriculture, site-specific crop management).** A farm management technology that identifies spatial production variability and suggests a means of managing this variability to increase yields and profits. Most often, site-specific farming uses GPS and GIS with variable rate applicators to implement precise inputs at specific locations.

**Soluble powder (SP).** A pesticide formulation consisting of a finely ground solid material that dissolves in water or some other liquid.

**Solution (S).** A concentrated liquid pesticide formulation that may be used directly or which could require diluting.

**Species.** A group of organisms whose members are similar in structure and physiology, that actually or potentially interbreed and produce fertile offspring and that are reproductively isolated from all other such groups.

**Spermatheca** (pl. **spermathecae**). A saclike structure in which sperm from the male is received and stored in the female (Figure 2.44).

**Spermatophore.** A packet containing sperm produced by some male insects.

**Spiracle.** An external opening of the tracheal system (Figures 2.28, 2.29).

**Stadium.** The time between molts.

**Stage.** The insect's developmental status with characteristic body form (for example, egg stage).

**Standard deviation.** The square root of the arithmetic averages of the squares of the deviations from the mean; a measure of variation in a sample.

**Standard error.** The measure of standard deviations of a population of means.

**Stem mother.** A wingless, parthenogenetic, viviparous female aphid that arises from an overwintering aphid egg.

**Sternum** (pl. **sterna**). The ventral surface of a thoracic segment.

**Sticker.** An ingredient added to a pesticide formulation to improve its adherence to a surface (for example, casein).

**Stomach poison.** An insecticide that is fatal only after being eaten, entering the insect body through the gut (for example, boric acid).

**Stomodaeal nervous system.** The main component of the visceral nervous system; controls activities of the anterior gut and dorsal vessel.

**Stomodaeum.** The foregut, beginning with the mouth and ending with the proventriculus.

**Strepsiptera.** The order of insects comprising the twisted-winged parasites (Figure 3.29).

**Stridulation.** Sound production accomplished by rubbing body parts together.

**Strip harvesting.** Harvesting of different areas of a single crop at different times (Figure 10.17).

**Stylet.** One of the piercing structures in piercing-sucking mouthparts.

**Subeconomic pest (noneconomic pest).** A pest with a general equilibrium position far below the economic-injury level; highest pest populations do not reach the economic-injury level (Figure 8.10).

**Subesophageal ganglion.** A swelling of nerve tissue at the anterior end of the ventral nerve cord, below the esophagus (Figures 2.40, 2.41).

**Summer oil.** A highly refined petroleum oil that can be applied to trees in full foliage without having phytotoxic effects.

**Supercooling.** Resistance to freezing by lowering the temperature at which freezing of body fluids begins.

**Supraesophageal ganglion.** A nerve mass that lies in the head above the esophagus; the insect brain.

**Surfactant.** An agent used to improve emulsifying, wetting, and spreading properties of a pesticide formulation.

**Surveillance (scouting).** The watch kept on a pest for detection of the species' presence and determination of population density, dispersion, and dynamics.

**Survivorship curve.** A graphic representation of deaths over time for a given population (Figure 5.13).

**Suture.** A seam or line indicating the division of distinct parts of the body wall.

**Swarm.** A concerted departure or association of a large group of insects from a nest or breeding habitat.

**Symbiont.** One species living in symbiosis with another.

**Symbiosis.** A close association between two or more different species that benefits one or both of the species.

**Synapse.** The junction between two neurons or between a neuron and another cell (Figure 2.42).

**Synaptic poison.** A chemical that interrupts the normal transmission of impulses across synapses (Figure 11.6).

**Synergism.** The action of two or more factors together achieving a greater effect than the sum of the individual effects of the factors.

**Synergist.** A nontoxic chemical used to increase the effectiveness of an insecticide (for example, piperonyl butoxide).

**System.** An interrelated set of elements that can be identified and around which a boundary can be drawn (Figure 8.15).

**Systematics.** The study of the diversity and classification of organisms.

**Systemic insecticide.** An insecticide that is taken up and translocated within plants and animals (for example, aldicarb).

**Systems technology.** The assembly, organization, and employment of an identifiable set of interrelated elements that constitute a system.

**Taenidium** (pl. **taenidia**). A spiral or circular thickening of the inner surface of the tracheae.

**Tagmosis.** The differentiation of the body into distinct functional regions, or tagmata (Figure 2.1).

**Tarsomeres.** The subdivisions of the tarsus (Figure 2.18).

**Tarsus (pl. tarsi).** The terminal leg segment, immediately distal to the tibia; consists of one or more subdivisions or tarsomeres (Figure 2.18).

**Taxonomy.** The theory and principles of classifying organisms.

**Technical-grade material.** The relatively pure form of a pesticidal compound that comprises the active ingredient in the final pesticide mixture.

**Tegmen (pl. tegmina).** A leathery or parchmentlike forewing (for example, Orthoptera).

**Telson.** The terminal body segment in some primitive arthropods.

**Tenaculum.** The clasplike structure associated with the furcula of Collembola.

**Tentorium.** The endoskeleton of the head, formed from invaginations of the body wall (Figure 2.5).

**Teratogen.** A material that causes birth defects.

**Tergum (pl. terga).** The dorsal surface of a segment.

**Termiticide.** An insecticide that kills termites.

**Terpene.** A naturally occurring compound in which the basic structural component is isoprene, $CH_2C(CH_3)CH_2CH_2$.

**Testis (pl. testes).** The male sex organ that produces sperm (Figure 2.45).

**Therapeutic practice.** A pest management approach that attempts to cure existing problems (Figure 16.5).

**Thermal constant.** The number of degree days required for an event (for example, pupation) to occur.

**Thiocyanate.** An organic insecticide that contains the SCN group; has a creosotelike odor; interferes with cellular respiration and metabolism.

**Thorax.** The body region between the head and abdomen that bears the legs and wings (Figure 1.2).

**Thysanoptera.** The order of insects comprising the thrips (Figure 3.23).

**Thysanura.** The order of insects comprising the silverfish (Figure 3.4).

**Tibia (pl. tibiae).** The fourth segment of the leg, between the femur and tarsus (Figure 2.18).

**Tillage.** The cultivation of land.

**Token stimuli.** Specific environmental cues that induce diapause or elicit a specific behavioral response only at certain times.

**Tolerance.** The ability of a host to withstand injury by pests; involves a plant response only.

**Tolerance EIL.** An economic-injury level that explicitly recognizes the tolerance of plants to pest injury by including a tolerance value in the EIL equation.

**Tolerance parameter.** A value of maximum injury (insect density) that can be tolerated without any measurable loss in plant yield or quality. The parameter is designated as $E_0$ and is added in the conventional EIL equation.

**Tonic.** Characterized by continuous or long-term stimulation; tonic receptors.

**Toxicity.** The inherent poisonous potency of a material.

**Toxicology.** The science or study of poisons.

**Trachea (pl. tracheae).** A tube of the respiratory system that begins with the spiracle and ends with the tracheoles (Figures 2.28, 2.29).

**Tracheoles.** The minute terminal branches of the tracheal system (Figure 2.29).

**Trade name (proprietary name, brand name).** Name given to a pesticide by the manufacturer or formulator (for example, Sevin®).

**Trail-marking pheromone.** An intraspecific chemical messenger produced by foraging ants and termites to indicate sources of requisites to other colony members.

**Transgene, Transgenic organism.** An organism containing genetic material from other species introduced via the process of transformation.

**Trap crop.** A small area of a crop used to divert pests from a larger area of the same or another crop (Figure 10.15).

**Trench fever.** A disease caused by infection with the rickettsia *Rochalimaea quintana* and transmitted via the feces or crushed bodies of the louse *Pediculus humanus humanus.*

**Trichoptera.** The order of insects comprising the caddisflies (Figures 3.31, 3.32).

**Tritocerebrum.** The posterior pair of lobes of the brain that connect to the visceral nervous system (Figure 2.41).

**Triungulin.** An active, first instar of insects that undergoes hypermetamorphosis.

**Trivial movement.** Displacement of insects within, or close to, the breeding habitat; nonmigratory movement.

**Trochanter.** The second segment of the leg, between the coxa and femur (Figure 2.18).

**Trypanosome.** A protozoan parasite of the blood and lymph.

**Tympanum** (pl. **tympana).** An auditory membrane.

**Typhus fever.** A human disease caused by the bacteriumlike microorganism *Rickettsia prowazekii* and transmitted by the body louse *Pediculus humanus humanus.*

**Ultra low volume (ULV) concentrate.** Pesticide sprays applied at 0.6 to 4.7 liters per hectare or sprays applied in an undiluted, low-volume formulation.

**Univoltine.** Having one generation per year (Figure 4.24).

**Urate cell.** A cell of the fat body that stores uric acid.

**Uric acid.** The primary nitrogenous waste in insects, $C_5H_4N_4O_3$.

**Vagina.** The terminus of the female reproductive tract that opens to the outside.

**Vas deferens** (pl. **vasa deferentia).** The sperm duct that leads away from a testis (Figure 2.45).

**Vas efferens** (pl. **vasa efferentia).** A short duct connecting a sperm tube in the testis to the vas deferens (Figure 2.45).

**Vector.** An organism capable of transmitting pathogens from one host to another.

**Vein.** A thickened line in the insect wing.

**Ventral.** Pertaining to the underside of the body.

**Vertical resistance.** Host-plant resistance to one or a narrow group of pest genotypes (Figure 13.8).

**Virulent gene.** A gene that allows a pest species to overcome resistance factors in a plant cultivar.

**Visceral.** Pertaining to or affecting the internal organs of the body.

**Vitamin.** An organic substance necessary in trace amounts for normal metabolic functioning.

**Viviparous.** Giving birth to active young, which undergo growth and development inside the mother.

**Voltinity.** The number of generations that occur in one year.

**Wettable powder (WP, W).** A pesticide formulation of active ingredient mixed with inert dust and a surfactant that mixes readily with water and forms a short-term suspension.

**Wing pads.** Externally visible beginnings of wings observed on insect nymphs.

**Wireworm.** The larva of a click beetle (family Elateridae).

**Yellow fever.** An infectious, mosquito-transmitted viral disease.

**Zoophagous.** Feeding on animals.

**Zoraptera.** An order of Pterygota that consists of rare, small, and termitelike insects (Figure 3.19).

**Zygote.** A fertilized egg.

# INDEX